QA
179
B266
798

D0906783

Selected Titles in This Series

(See the AMS catalog for earlier titles)

American Mathematical Society

Colloquium Publications

Volume 44

BELL LIBRARY
TEXAS A&M UNIVERSITY
CORPUS CHRISTI

The Book of Involutions

Max-Albert Knus

Alexander Merkurjev

Markus Rost

Jean-Pierre Tignol

With a preface by J. Tits

Editorial Board

Joan S. Birman
Susan J. Friedlander, Chair
Stephen Lichtenbaum

1991 *Mathematics Subject Classification.* Primary 11E39, 11E57, 11E72;
Secondary 11E88, 16K20, 16W10, 17A75, 17C40, 20G15.

This monograph yields a comprehensive exposition of the theory of central simple algebras with involution, in relation with linear algebraic groups. It aims to provide the algebra-theoretic foundations for much of the recent work on linear algebraic groups over arbitrary fields. Involutions are viewed as twisted forms of similarity classes of hermitian or bilinear forms, leading to new developments on the model of the algebraic theory of quadratic forms. Besides classical groups, phenomena related to triality are also discussed, as well as groups of type F_4 or G_2 arising from exceptional Jordan or composition algebras. Several results and notions appear here for the first time, notably the discriminant algebra of an algebra with unitary involution and the algebra-theoretic counterpart to linear groups of type D_4.

For research mathematicians and graduate students working in central simple algebras, algebraic groups, nonabelian Galois cohomology or Jordan algebras.

Cover design by Herbert Rost.

Library of Congress Cataloging-in-Publication Data

The book of involutions / Max-Albert Knus ... [et al.].
p.   cm. — (Colloquium publications / American Mathematical Society, ISSN 0065-9258 ; v. 44)
Includes bibliographical references and index.
ISBN 0-8218-0904-0 (alk. paper)
1. Linear algebraic groups. 2. Hermitian forms. 3. Homology theory. 4. Galois theory.
I. Knus, Max-Albert. II. Series: Colloquium publications (American Mathematical Society) ; v. 44.

QA179.B66   1998
512′.5–dc21

98-22202
CIP

Copying and reprinting. Individual readers of this publication, and nonprofit libraries acting for them, are permitted to make fair use of the material, such as to copy a chapter for use in teaching or research. Permission is granted to quote brief passages from this publication in reviews, provided the customary acknowledgment of the source is given.

Republication, systematic copying, or multiple reproduction of any material in this publication (including abstracts) is permitted only under license from the American Mathematical Society. Requests for such permission should be addressed to the Assistant to the Publisher, American Mathematical Society, P. O. Box 6248, Providence, Rhode Island 02940-6248. Requests can also be made by e-mail to reprint-permission@ams.org.

© 1998 by the American Mathematical Society. All rights reserved.
The American Mathematical Society retains all rights
except those granted to the United States Government.
Printed in the United States of America.

♾ The paper used in this book is acid-free and falls within the guidelines
established to ensure permanence and durability.
Visit the AMS home page at URL: http://www.ams.org/

10 9 8 7 6 5 4 3 2 1   03 02 01 00 99 98

Contents

Préface

Quatre des meilleurs algébristes d'aujourd'hui (j'aimerais dire, comme jadis, «géomètres», au sens noble mais hélas désuet du terme) nous donnent ce beau *Livre des Involutions*, qu'ils me demandent de préfacer.

Quel est le propos de l'ouvrage et à quels lecteurs s'adresse-t-il? Bien sûr il y est souvent question d'involutions, mais celles-ci sont loin d'être omniprésentes et le titre est plus l'expression d'un état d'âme que l'affirmation d'un thème central. En fait, les questions envisagées sont multiples, relevant toutes de domaines importants des mathématiques contemporaines; sans vouloir être exhaustif (ceci n'est pas une introduction), on peut citer :

- les formes quadratiques et les algèbres de Clifford,
- les algèbres associatives centrales simples (ici les involutions, et notamment celles de seconde espèce, se taillent une place de choix!) mais aussi les algèbres alternatives et les algèbres de Jordan,
- les algèbres de Hopf,
- les groupes algébriques, principalement semi-simples,
- la cohomologie galoisienne.

Pour ce qui est du public concerné, la lecture ou la consultation du livre sera profitable à un large éventail de mathématiciens. Le non-initié y trouvera une introduction claire aux concepts fondamentaux des domaines en question; exposés le plus souvent en fonction d'applications concrètes, ces notions de base sont présentées de façon vivante et dépouillée, sans généralités gratuites (les auteurs ne sont pas adeptes de grandes théories abstraites). Le lecteur déjà informé, ou croyant l'être, pourra réapprendre (ou découvrir) quelques beaux théorèmes jadis «bien connus» mais un peu oubliés dans la littérature récente, ou au contraire, voir des résultats qui lui sont en principe familiers exposés sous un jour nouveau et éclairant (je pense par exemple à l'introduction des algèbres trialitaires au dernier chapitre). Enfin, les spécialistes et les chercheurs auront à leur disposition une référence précieuse, parfois unique, pour des développements récents, souvents dûs aux auteurs eux-mêmes, et dont certains sont exposés ici pour la première fois (c'est par exemple le cas pour plusieurs résultats sur les invariants cohomologiques, donnés à la fin du chapitre 7).

Malgré la grande variété des thèmes considérés et les individualités très marquées des quatre auteurs, ce Livre des Involutions a une unité remarquable. Le ciment un peu fragile des involutions n'est certes pas seul à l'expliquer. Il y a aussi, bien sûr, les interconnections multiples entre les sujets traités; mais plus déterminante encore est l'importance primordiale accordée à des structures fortes, se prêtant par exemple à des théorèmes de classification substantiels. Ce n'est pas un hasard si les algèbres centrales simples de petites dimensions (trois et quatre), les groupes exceptionnels de type G_2 et F_4 (on regrette un peu que Sa Majesté E_8

fasse ici figure de parent pauvre), les algèbres de composition, ..., reçoivent autant d'attention.

On l'a compris, ce Livre est tout à la fois un livre de lecture passionnant et un ouvrage de référence d'une extrême richesse. Je suis reconnaissant aux auteurs de l'honneur qu'ils m'ont fait en me demandant de le préfacer, et plus encore de m'avoir permis de le découvrir et d'apprendre à m'en servir.

Jacques Tits

Introduction

For us an involution is an anti-automorphism of order two of an algebra. The most elementary example is the transpose for matrix algebras. A more complicated example of an algebra over $\mathbb{Q}$ admitting an involution is the multiplication algebra of a Riemann surface (see the notes at the end of Chapter I for more details). The central problem here, to give necessary and sufficient conditions on a division algebra over $\mathbb{Q}$ to be a multiplication algebra, was completely solved by Albert (1934/35). To achieve this, Albert developed a theory of central simple algebras with involution, based on the theory of simple algebras initiated a few years earlier by Brauer, Noether, and also Albert and Hasse, and gave a complete classification over $\mathbb{Q}$. This is the historical origin of our subject, however our motivation has a different source. The basic objects are still central simple algebras, i.e., "forms" of matrix algebras. As observed by Weil (1960), central simple algebras with involution occur in relation to classical algebraic simple adjoint groups: connected components of automorphism groups of central simple algebras with involution are such groups (with the exception of a quaternion algebra with an orthogonal involution, where the connected component of the automorphism group is a torus), and, in their turn, such groups are connected components of automorphism groups of central simple algebras with involution.

Even if this is mainly a book on algebras, the correspondence between algebras and groups is a constant leitmotiv. Properties of the algebras are reflected in properties of the groups and of related structures, such as Dynkin diagrams, and vice versa. For example we associate certain algebras to algebras with involution in a functorial way, such as the Clifford algebra (for orthogonal involutions) or the λ-powers and the discriminant algebra (for unitary involutions). These algebras are exactly the "Tits algebras," defined by Tits (1971) in terms of irreducible representations of the groups. Another example is algebraic triality, which is historically related with groups of type D_4 (É. Cartan) and whose "algebra" counterpart is, so far as we know, systematically approached here for the first time.

In the first chapter we recall basic properties of central simple algebras and involutions. As a rule for the whole book, without however going to the utmost limit, we try to allow base fields of characteristic 2 as well as those of other characteristic. Involutions are divided up into orthogonal, symplectic and unitary types. A central idea of this chapter is to interpret involutions in terms of hermitian forms over skew fields. Quadratic pairs, introduced at the end of the chapter, give a corresponding interpretation for quadratic forms in characteristic 2.

In Chapter II we define several invariants of involutions; the index is defined for every type of involution. For quadratic pairs additional invariants are the discriminant, the (even) Clifford algebra and the Clifford module; for unitary involutions we introduce the discriminant algebra. The definition of the discriminant algebra is prepared for by the construction of the λ-powers of a central simple algebra. The last part of this chapter is devoted to trace forms on algebras, which represent an important tool for recent results discussed in later parts of the book. Our method of definition is based on scalar extension: after specifying the definitions "rationally" (i.e., over an arbitrary base field), the main properties are proven by working over a splitting field. This is in contrast to Galois descent, where constructions over a separable closure are shown to be invariant under the Galois group and therefore are defined over the base field. A main source of inspiration for Chapters I and II is the paper [291] of Tits on "Formes quadratiques, groupes orthogonaux et algèbres de Clifford."

In Chapter III we investigate the automorphism groups of central simple algebras with involutions. Inner automorphisms are induced by elements which we call similitudes. These automorphism groups are twisted forms of the classical projective orthogonal, symplectic and unitary groups. After proving results which hold for all types of involutions, we focus on orthogonal and unitary involutions, where additional information can be derived from the invariants defined in Chapter II. The next two chapters are devoted to algebras of low degree. There exist certain isomorphisms among classical groups, known as exceptional isomorphisms. From the algebra point of view, this is explained in the first part of Chapter IV by properties of the Clifford algebra of orthogonal involutions on algebras of degree 3, 4, 5 and 6. In the second part we focus on tensor products of two quaternion algebras, which we call biquaternion algebras. These algebras have many interesting properties, which could be the subject of a monograph of its own. This idea was at the origin of our project.

Algebras with unitary involutions are also of interest for odd degrees, the lowest case being degree 3. From the group point of view algebras with unitary involutions of degree 3 are of type A_2. Chapter V gives a new presentation of results of Albert and a complete classification of these algebras. In preparation for this, we recall general results on étale and Galois algebras.

The aim of Chapter VI is to give the classification of semisimple algebraic groups over arbitrary fields. We use the functorial approach to algebraic groups, although we quote without proof some basic results on algebraic groups over algebraically closed fields. In the central section we describe in detail Weil's correspondence [310] between central simple algebras with involution and classical groups. Exceptional isomorphisms are reviewed again in terms of this correspondence. In the last section we define Tits algebras of semisimple groups and give explicit constructions of them in classical cases.

The theme of Chapter VII is Galois cohomology. We introduce the formalism and describe many examples. Previous results are reinterpreted in this setting and cohomological invariants are discussed. Most of the techniques developed here are also needed for the following chapters.

The last three chapters are dedicated to the exceptional groups of type G_2, F_4 and to D_4, which, in view of triality, is also exceptional. In the Weil correspondence, octonion algebras play the algebra role for G_2 and exceptional simple Jordan algebras the algebra role for F_4.

Octonion algebras are an important class of composition algebras and Chapter VIII gives an extensive discussion of composition algebras. Of special interest from the group point of view are "symmetric" compositions. In dimension 8 these are of two types, corresponding to algebraic groups of type A_2 or type G_2. Triality is defined through the Clifford algebra of symmetric 8-dimensional compositions. As a step towards exceptional simple Jordan algebras, we introduce twisted compositions, which are defined over cubic étale algebras. This generalizes a construction of Springer. The corresponding group of automorphisms in the split case is the semidirect product $\mathrm{Spin}_8 \rtimes S_3$.

In Chapter IX we describe different constructions of exceptional simple Jordan algebras, due to Freudenthal, Springer and Tits (the algebra side) and give interpretations from the algebraic group side. The Springer construction arises from twisted compositions, defined in Chapter VIII, and basic ingredients of Tits constructions are algebras of degree 3 with unitary involutions, studied in Chapter III. We conclude this chapter by defining cohomological invariants for exceptional simple Jordan algebras.

The last chapter deals with trialitarian actions on simple adjoint groups of type D_4. To complete Weil's program for outer forms of D_4 (a case not treated by Weil), we introduce a new notion, which we call a trialitarian algebra. The underlying structure is a central simple algebra with an orthogonal involution, of degree 8 over a cubic étale algebra. The trialitarian condition relates the algebra to its Clifford algebra. Trialitarian algebras also occur in the construction of Lie algebras of type D_4. Some indications in this direction are given in the last section.

Exercises and notes can be found at the end of each chapter. Omitted proofs sometimes occur as exercises. Moreover we included as exercises some results we like, but which we did not wish to develop fully. In the notes we wanted to give complements and to look at some results from a historical perspective. We have tried our best to be useful; we cannot, however, give strong guarantees of completeness or even fairness.

This book is the achievement of a joint (and very exciting) effort of four very different people. We are aware that the result is still quite heterogeneous; however, we flatter ourselves that the differences in style may be viewed as a positive feature.

Our work started out as an attempt to understand Tits' definition of the Clifford algebra of a generalized quadratic form, and ended up including many other topics to which Tits made fundamental contributions, such as linear algebraic groups, exceptional algebras, triality, ... Not only was Jacques Tits a constant source of inspiration through his work, but he also had a direct personal influence, notably through his threat — early in the inception of our project — to speak evil of our work if it did not include the characteristic 2 case. Finally he also agreed to bestow his blessings on our book sous forme de préface. For all that we thank him wholeheartedly.

This book could not have been written without the help and the encouragement of many friends. They are too numerous to be listed here individually, but we hope they will recognize themselves and find here our warmest thanks. Richard Elman deserves a special mention for his comment that the most useful book is not the one to which nothing can be added, but the one which is published. This no-nonsense statement helped us set limits to our endeavor. We were fortunate to get useful advice on various points of the exposition from Ottmar Loos, Antonio Paques, Parimala, Michel Racine, David Saltman, Jean-Pierre Serre and Sridharan. We thank all of them for lending helping hands at the right time. A number of people were nice enough to read and comment on drafts of parts of this book: Eva Bayer-Fluckiger, Vladimir Chernousov, Ingrid Dejaiffe, Alberto Elduque, Darrell Haile, Luc Haine, Pat Morandi, Holger Petersson, Ahmed Serhir, Tony Springer,

Paul Swets and Oliver Villa. We know all of them had better things to do, and we are grateful. Skip Garibaldi and Adrian Wadsworth actually summoned enough grim self-discipline to read a draft of the whole book, detecting many shortcomings, making shrewd comments on the organization of the book and polishing our broken English. Each deserves a medal. However, our capacity for making mistakes certainly exceeds our friends' sagacity. We shall gratefully welcome any comment or correction.

Tignol had the privilege to give a series of lectures on "Central simple algebras, involutions and quadratic forms" in April 1993 at the National Taiwan University. He wants to thank Ming-chang Kang and the National Research Council of China for this opportunity to test high doses of involutions on a very patient audience, and Eng-Tjioe Tan for making his stay in Taiwan a most pleasant experience. The lecture notes from this crash course served as a blueprint for the first chapters of this book.

Our project immensely benefited by reciprocal visits among the authors. We should like to mention with particular gratitude Merkurjev's stay in Louvain-la-Neuve in 1993, with support from the Fonds de Développement Scientifique and the Institut de Mathématique Pure et Appliquée of the Université catholique de Louvain, Rost's stays in Zürich at the Mathematics Research Institute (FIM) in winter 1995–96 and at the Mathematics Department of the ETH Zürich for the winter semester of 1996–97, and Tignol's stay at the Mathematics Department of the ETH Zürich for the winter semester of 1995–96, both with support from the Eidgenössische Technische Hochschule. Moreover, Merkurjev gratefully acknowledges support from the Alexander von Humboldt foundation and the hospitality of the Bielefeld university for the year 1995–96, and Tignol is grateful to the National Fund for Scientific Research of Belgium and to the European Commission for partial support under the "crédit aux chercheurs" and the TMR programmes respectively (contract ERB-FMRX-CT97-0107).

The four authors enthusiastically thank Herbert Rost (Markus' father) for the design of the cover page, in particular for his wonderful and colorful rendition of the Dynkin diagram D_4. They also give special praise to Sergei Gelfand, Director of Acquisitions of the American Mathematical Society, for his helpfulness and patience in taking care of all their wishes for the publication.

Conventions and Notations

Maps. The image of an element x under a map f is generally denoted $f(x)$; the notation x^f is also used however, notably for homomorphisms of left modules. In that case, we also use the right-hand rule for mapping composition; for the image of $x \in X$ under the composite map $X \xrightarrow{f} Y \xrightarrow{g} Z$ we set either $g \circ f(x)$ or x^{fg} and the composite is thus either $g \circ f$ or fg.

As a general rule, module homomorphisms are written on the opposite side of the scalars. (Right modules are usually preferred.) Thus, if M is a module over a ring R, it is also a module (on the opposite side) over $\mathrm{End}_R(M)$, and the R-module structure defines a natural homomorphism:

$$R \to \mathrm{End}_{\mathrm{End}_R(M)}(M).$$

Note therefore that if $S \subset \mathrm{End}_R(M)$ is a subring, and if we endow M with its natural S-module structure, then $\mathrm{End}_S(M)$ is the *opposite* of the centralizer of S in $\mathrm{End}_R(M)$:

$$\mathrm{End}_S(M) = \left(C_{\mathrm{End}_R(M)}S\right)^{\mathrm{op}}.$$

Of course, if R is commutative, every right R-module M_R may also be regarded as a left R-module $_RM$, and every endomorphism of M_R also is an endomorphism of $_RM$. Note however that with the convention above, the canonical map $\mathrm{End}_R(M_R) \to \mathrm{End}_R(_RM)$ is an anti-isomorphism.

The characteristic polynomial and its coefficients. Let F denote an arbitrary field. The characteristic polynomial of a matrix $m \in M_n(F)$ (or an endomorphism m of an n-dimensional F-vector space) is denoted

(0.1) $$P_m(X) = X^n - s_1(m)X^{n-1} + s_2(m)X^{n-2} - \cdots + (-1)^n s_n(m).$$

The trace and determinant of m are denoted $\mathrm{tr}(m)$ and $\det(m)$:

$$\mathrm{tr}(m) = s_1(m), \qquad \det(m) = s_n(m).$$

We recall the following relations between coefficients of the characteristic polynomial:

(0.2) Proposition. *For $m, m' \in M_n(F)$, we have $s_1(m)^2 - s_1(m^2) = 2s_2(m)$ and*

$$s_1(m)s_1(m') - s_1(mm') = s_2(m + m') - s_2(m) - s_2(m').$$

Proof: It suffices to prove these relations for generic matrices $m = (x_{ij})_{1 \le i, j \le n}$, $m' = (x'_{ij})_{1 \le i, j \le n}$ whose entries are indeterminates over $\mathbb{Z}$; the general case follows by specialization. If $\lambda_1, \ldots, \lambda_n$ are the eigenvalues of the generic matrix m (in an algebraic closure of $\mathbb{Q}(x_{ij} \mid 1 \le i, j \le n)$), we have $s_1(m) = \sum_{1 \le i \le n} \lambda_i$ and $s_2(m) = \sum_{1 \le i < j \le n} \lambda_i \lambda_j$, hence

$$s_1(m)^2 - 2s_2(m) = \sum_{1 \le i \le n} \lambda_i^2 = s_1(m^2),$$

proving the first relation. The second relation follows by linearization, since 2 is not a zero-divisor in $\mathbb{Z}[x_{ij}, x'_{ij} \mid 1 \le i, j \le n]$. $\qquad\square$

If L is an associative and commutative F-algebra of dimension n and $\ell \in L$, the characteristic polynomial of multiplication by ℓ, viewed as an F-endomorphism of L, is called the *generic polynomial of ℓ* and is denoted

$$P_{L,\ell}(X) = X^n - s_1(\ell)X^{n-1} + s_2(\ell)X^{n-2} - \cdots + (-1)^n s_n(\ell).$$

The trace and norm of ℓ are denoted $T_{L/F}(\ell)$ and $N_{L/F}(\ell)$ (or simply $T(\ell)$, $N(\ell)$):

$$T_{L/F}(\ell) = s_1(\ell), \qquad N_{L/F}(\ell) = s_n(\ell).$$

We also denote

(0.3) $$S_{L/F}(\ell) = S(\ell) = s_2(\ell).$$

The characteristic polynomial is also used to define a generic polynomial for central simple algebras, called the *reduced characteristic polynomial*: see (1.6). Generalizations to certain nonassociative algebras are given in §32.

Bilinear forms. A bilinear form $b\colon V \times V \to F$ on a finite dimensional vector space V over an arbitrary field F is called *symmetric* if $b(x, y) = b(y, x)$ for all $x, y \in V$, *skew-symmetric* if $b(x, y) = -b(y, x)$ for all $x, y \in V$ and *alternating* if $b(x, x) = 0$ for all $x \in V$. Thus, the notions of skew-symmetric and alternating (resp. symmetric) form coincide if char $F \ne 2$ (resp. char $F = 2$). Alternating forms are skew-symmetric in every characteristic.

If b is a symmetric or alternating bilinear form on a (finite dimensional) vector space V, the induced map

$$\hat{b}\colon V \to V^* = \operatorname{Hom}_F(V, F)$$

is defined by $\hat{b}(x)(y) = b(x, y)$ for $x, y \in V$. The bilinear form b is *nonsingular* (or *regular*, or *nondegenerate*) if $\hat{b}$ is bijective. (It suffices to require that $\hat{b}$ be injective, i.e., that the only vector $x \in V$ such that $b(x, y) = 0$ for all $y \in V$ is $x = 0$, since we are dealing with finite dimensional vector spaces over fields.) Alternately, b is nonsingular if and only if the determinant of its Gram matrix with respect to an arbitrary basis of V is nonzero:

$$\det\big(b(e_i, e_j)\big)_{1 \le i, j \le n} \ne 0.$$

In that case, the square class of this determinant is called the *determinant* of b:

$$\det b = \det\big(b(e_i, e_j)\big)_{1 \le i, j \le n} \cdot F^{\times 2} \in F^\times / F^{\times 2}.$$

The *discriminant* of b is the signed determinant:

$$\operatorname{disc} b = (-1)^{n(n-1)/2} \det b \in F^\times / F^{\times 2} \quad \text{where } n = \dim V.$$

For $\alpha_1, \ldots, \alpha_n \in F$, the bilinear form $\langle \alpha_1, \ldots, \alpha_n \rangle$ on F^n is defined by

$$\langle \alpha_1, \ldots, \alpha_n \rangle\big((x_1, \ldots, x_n), (y_1, \ldots, y_n)\big) = \alpha_1 x_1 y_1 + \cdots + \alpha_n x_n y_n.$$

We also define the *n-fold Pfister bilinear form* $\langle\!\langle \alpha_1, \ldots, \alpha_n \rangle\!\rangle$ by

$$\langle\!\langle \alpha_1, \ldots, \alpha_n \rangle\!\rangle = \langle 1, -\alpha_1 \rangle \otimes \cdots \otimes \langle 1, -\alpha_n \rangle.$$

If $b\colon V \times V \to F$ is a symmetric bilinear form, we denote by $q_b\colon V \to F$ the associated quadratic map, defined by

$$q_b(x) = b(x, x) \qquad \text{for } x \in V.$$

Quadratic forms. If $q\colon V \to F$ is a quadratic map on a finite dimensional vector space over an arbitrary field F, the associated symmetric bilinear form b_q is called the *polar* of q; it is defined by

$$b_q(x, y) = q(x + y) - q(x) - q(y) \qquad \text{for } x,\, y \in V,$$

hence $b_q(x, x) = 2q(x)$ for all $x \in V$. Thus, the quadratic map q_{b_q} associated to b_q is $q_{b_q} = 2q$. Similarly, for every symmetric bilinear form b on V, we have $b_{q_b} = 2b$.

Let $V^\perp = \{\, x \in V \mid b_q(x, y) = 0 \text{ for } y \in V \,\}$. The quadratic map q is called *nonsingular* (or *regular*, or *nondegenerate*) if either $V^\perp = \{0\}$ or $\dim V^\perp = 1$ and $q(V^\perp) \neq \{0\}$. The latter case occurs only if $\operatorname{char} F = 2$ and V is odd-dimensional. Equivalently, a quadratic form of dimension n is nonsingular if and only if it is equivalent over an algebraic closure to $\sum_{i=1}^{n/2} x_{2i-1} x_{2i}$ (if n is even) or to $x_0^2 + \sum_{i=1}^{(n-1)/2} x_{2i-1} x_{2i}$ (if n is odd).

The *determinant* and the *discriminant* of a nonsingular quadratic form q of dimension n over a field F are defined as follows: let M be a matrix representing q in the sense that

$$q(X) = X \cdot M \cdot X^t$$

where $X = (x_1, \ldots, x_n)$ and ^{t} denotes the transpose of matrices; the condition that q is nonsingular implies that $M + M^t$ is invertible if n is even or $\operatorname{char} F \neq 2$, and has rank $n - 1$ if n is odd and $\operatorname{char} F = 2$. The matrix M is uniquely determined by q up to the addition of a matrix of the form $N - N^t$; therefore, $M + M^t$ is uniquely determined by q.

If $\operatorname{char} F \neq 2$ we set

$$\det q = \det\!\left(\tfrac{1}{2}(M + M^t)\right) \cdot F^{\times 2} \in F^\times / F^{\times 2}$$

and

$$\operatorname{disc} q = (-1)^{n(n-1)/2} \det q \in F^\times / F^{\times 2}.$$

Thus, the determinant (resp. the discriminant) of a quadratic form is the determinant (resp. the discriminant) of its polar form divided by 2^n.

If $\operatorname{char} F = 2$ and n is odd we set

(0.4) $$\det q = \operatorname{disc} q = q(y) \cdot F^{\times 2} \in F^\times / F^{\times 2}$$

where $y \in F^n$ is a nonzero vector such that $(M + M^t) \cdot y = 0$. Such a vector y is uniquely determined up to a scalar factor, since $M + M^t$ has rank $n - 1$, hence the definition above does not depend on the choice of y.

If $\operatorname{char} F = 2$ and n is even we set

$$\det q = s_2\!\left((M + M^t)^{-1} M\right) + \wp(F) \in F / \wp(F)$$

and

$$\operatorname{disc} q = \tfrac{m(m-1)}{2} + \det q \in F / \wp(F)$$

where $m = n/2$ and $\wp(F) = \{x + x^2 \mid x \in F\}$. (More generally, for fields of characteristic $p \neq 0$, $\wp$ is defined as $\wp(x) = x + x^p$, $x \in F$.) The following lemma shows that the definition of $\det q$ does not depend on the choice of M:

(0.5) Lemma. *Suppose* char $F = 2$. *Let* $M, N \in M_n(F)$ *and* $W = M + M^t$. *If* W *is invertible, then*

$$s_2\bigl(W^{-1}(M + N + N^t)\bigr) = s_2(W^{-1}M) + s_1(W^{-1}N) + \bigl(s_1(W^{-1}N)\bigr)^2.$$

Proof: The second relation in (0.2) yields

$$s_2\bigl(W^{-1}M + W^{-1}(N + N^t)\bigr) =$$
$$s_2(W^{-1}M) + s_2\bigl(W^{-1}(N + N^t)\bigr) + s_1(W^{-1}M)s_1\bigl(W^{-1}(N + N^t)\bigr)$$
$$+ s_1\bigl(W^{-1}MW^{-1}(N + N^t)\bigr).$$

In order to prove the lemma, we show below:

(0.6) $$s_2\bigl(W^{-1}(N + N^t)\bigr) = \bigl(s_1(W^{-1}N)\bigr)^2$$
(0.7) $$s_1(W^{-1}M)s_1\bigl(W^{-1}(N + N^t)\bigr) = 0$$
(0.8) $$s_1\bigl(W^{-1}MW^{-1}(N + N^t)\bigr) = s_1(W^{-1}N).$$

Since a matrix and its transpose have the same characteristic polynomial, the traces of $W^{-1}N$ and $(W^{-1}N)^t = N^tW^{-1}$ are the same, hence

$$s_1(W^{-1}N^t) = s_1(N^tW^{-1}) = s_1(W^{-1}N).$$

Therefore, $s_1\bigl(W^{-1}(N + N^t)\bigr) = 0$, and (0.7) follows.

Similarly, we have

$$s_1(W^{-1}MW^{-1}N^t) = s_1(NW^{-1}M^tW^{-1}) = s_1(W^{-1}M^tW^{-1}N),$$

hence the left side of (0.8) is

$$s_1(W^{-1}MW^{-1}N) + s_1(W^{-1}M^tW^{-1}N) = s_1\bigl(W^{-1}(M + M^t)W^{-1}N\bigr).$$

Since $M + M^t = W$, (0.8) follows.

The second relation in (0.2) shows that the left side of (0.6) is

$$s_2(W^{-1}N) + s_2(W^{-1}N^t) + s_1(W^{-1}N)s_1(W^{-1}N^t) + s_1(W^{-1}NW^{-1}N^t).$$

Since $W^{-1}N$ and $W^{-1}(W^{-1}N)^tW \ (= W^{-1}N^t)$ have the same characteristic polynomial, we have $s_i(W^{-1}N) = s_i(W^{-1}N^t)$ for $i = 1, 2$, hence the first two terms cancel and the third is equal to $s_1(W^{-1}N)^2$. In order to prove (0.6), it therefore suffices to show

$$s_1(W^{-1}NW^{-1}N^t) = 0.$$

Since $W = M + M^t$, we have $W^{-1} = W^{-1}MW^{-1} + W^{-1}M^tW^{-1}$, hence

$$s_1(W^{-1}NW^{-1}N^t) = s_1(W^{-1}MW^{-1}NW^{-1}N^t) + s_1(W^{-1}M^tW^{-1}NW^{-1}N^t),$$

and (0.6) follows if we show that the two terms on the right side are equal. Since $W^t = W$ we have $(W^{-1}MW^{-1}NW^{-1}N^t)^t = NW^{-1}N^tW^{-1}M^tW^{-1}$, hence

$$s_1(W^{-1}MW^{-1}NW^{-1}N^t) = s_1\bigl((NW^{-1}N^t)(W^{-1}M^tW^{-1})\bigr)$$
$$= s_1(W^{-1}M^tW^{-1}NW^{-1}N^t).$$

$\square$

Quadratic forms are called *equivalent* if they can be transformed into each other by invertible linear changes of variables. The various quadratic forms representing a quadratic map with respect to various bases are thus equivalent. It is easily verified that the determinant $\det q$ (hence also the discriminant $\operatorname{disc} q$) is an invariant of the equivalence class of the quadratic form q; the determinant and the discriminant are therefore also defined for quadratic maps. The discriminant of a quadratic form (or map) of even dimension in characteristic 2 is also known as the *pseudodiscriminant* or the *Arf invariant* of the form. See §8.D for the relation between the discriminant and the even Clifford algebra.

Let $\alpha_1, \ldots, \alpha_n \in F$. If $\operatorname{char} F \neq 2$ we denote by $\langle \alpha_1, \ldots, \alpha_n \rangle$ the diagonal quadratic form

$$\langle \alpha_1, \ldots, \alpha_n \rangle = \alpha_1 x_1^2 + \cdots + \alpha_n x_n^2$$

which is the quadratic form associated to the bilinear form $\langle \alpha_1, \ldots, \alpha_n \rangle$. We also define the *n-fold Pfister quadratic form* $\langle\langle \alpha_1, \ldots, \alpha_n \rangle\rangle$ by

$$\langle\langle \alpha_1, \ldots, \alpha_n \rangle\rangle = \langle 1, -\alpha_1 \rangle \otimes \cdots \otimes \langle 1, -\alpha_n \rangle$$

where $\otimes = \otimes_F$ is the tensor product over F. If $\operatorname{char} F = 2$, the quadratic forms $[\alpha_1, \alpha_2]$ and $[\alpha_1]$ are defined by

$$[\alpha_1, \alpha_2] = \alpha_1 X_1^2 + X_1 X_2 + \alpha_2 X_2^2 \qquad \text{and} \qquad [\alpha_1] = \alpha_1 X^2,$$

and the *n-fold Pfister quadratic form* $\langle\langle \alpha_1, \ldots, \alpha_n]]$ by

$$\langle\langle \alpha_1, \ldots, \alpha_n]] = \langle\langle \alpha_1, \ldots, \alpha_{n-1} \rangle\rangle \otimes [1, \alpha_n].$$

(See Baeza [28, p. 5] or Knus [157, p. 50] for the definition of the tensor product of a bilinear form and a quadratic form.) For instance,

$$\langle\langle \alpha_1, \alpha_2]] = (x_1^2 + x_1 x_2 + \alpha_2 x_2^2) + \alpha_1 (x_3^2 + x_3 x_4 + \alpha_2 x_4^2).$$

Involutions and Hermitian Forms

Our perspective in this work is that involutions on central simple algebras are twisted forms of symmetric or alternating bilinear forms up to a scalar factor. To motivate this point of view, we consider the basic, classical situation of linear algebra.

Let V be a finite dimensional vector space over a field F of arbitrary characteristic. A bilinear form $b \colon V \times V \to F$ is called *nonsingular* if the induced map

$$\hat{b} \colon V \to V^* = \operatorname{Hom}_F(V, F)$$

defined by

$$\hat{b}(x)(y) = b(x, y) \quad \text{for } x, y \in V$$

is an isomorphism of vector spaces. For any $f \in \operatorname{End}_F(V)$ we may then define $\sigma_b(f) \in \operatorname{End}_F(V)$ by

$$\sigma_b(f) = \hat{b}^{-1} \circ f^t \circ \hat{b}$$

where $f^t \in \operatorname{End}_F(V^*)$ is the *transpose* of f, defined by mapping $\varphi \in V^*$ to $\varphi \circ f$. Alternately, $\sigma_b(f)$ may be defined by the following property:

$$(*) \qquad b\big(x, f(y)\big) = b\big(\sigma_b(f)(x), y\big) \quad \text{for } x, y \in V.$$

The map $\sigma_b \colon \operatorname{End}_F(V) \to \operatorname{End}_F(V)$ is then an anti-automorphism of $\operatorname{End}_F(V)$ which is known as the *adjoint anti-automorphism* with respect to the nonsingular bilinear form b. The map σ_b clearly is F-linear.

The basic result which motivates our approach and which will be generalized in (4.2) is the following:

Theorem. *The map which associates to each nonsingular bilinear form b on V its adjoint anti-automorphism σ_b induces a one-to-one correspondence between equivalence classes of nonsingular bilinear forms on V modulo multiplication by a factor in $F^\times$ and linear anti-automorphisms of $\operatorname{End}_F(V)$. Under this correspondence, F-linear involutions on $\operatorname{End}_F(V)$ (i.e., anti-automorphisms of period 2) correspond to nonsingular bilinear forms which are either symmetric or skew-symmetric.*

Proof: From relation $(*)$ it follows that for $\alpha \in F^\times$ the adjoint anti-automorphism $\sigma_{\alpha b}$ with respect to the multiple αb of b is the same as the adjoint anti-automorphism σ_b. Therefore, the map $b \mapsto \sigma_b$ induces a well-defined map from the set of nonsingular bilinear forms on V up to a scalar factor to the set of F-linear anti-automorphisms of $\operatorname{End}(V)$.

To show that this map is one-to-one, note that if b, b' are nonsingular bilinear forms on V, then the map $v = \hat{b}^{-1} \circ \hat{b}' \in \operatorname{GL}(V)$ satisfies

$$b'(x, y) = b\big(v(x), y\big) \quad \text{for } x, y \in V.$$

From this relation, it follows that the adjoint anti-automorphisms σ_b, $\sigma_{b'}$ are related by

$$\sigma_b(f) = v \circ \sigma_{b'}(f) \circ v^{-1} \quad \text{for } f \in \text{End}_F(V),$$

or equivalently

$$\sigma_b = \text{Int}(v) \circ \sigma_{b'},$$

where $\text{Int}(v)$ denotes the *inner automorphism* of $\text{End}_F(V)$ induced by v:

$$\text{Int}(v)(f) = v \circ f \circ v^{-1} \quad \text{for } f \in \text{End}_F(V).$$

Therefore, if $\sigma_b = \sigma_{b'}$, then $v \in F^\times$ and b, b' are scalar multiples of each other.

Moreover, if b is a fixed nonsingular bilinear form on V with adjoint anti-automorphism σ_b, then for any linear anti-automorphism σ' of $\text{End}_F(V)$, the composite $\sigma_b \circ {\sigma'}^{-1}$ is an F-linear automorphism of $\text{End}_F(V)$. Since these automorphisms are inner, by the Skolem-Noether theorem (see (1.4) below), there exists $u \in \text{GL}(V)$ such that $\sigma_b \circ {\sigma'}^{-1} = \text{Int}(u)$. Then σ' is the adjoint anti-automorphism with respect to the bilinear form b' defined by

$$b'(x,y) = b\big(u(x),y\big).$$

Thus, the first part of the theorem is proved.

Observe also that if b is a nonsingular bilinear form on V with adjoint anti-automorphism σ_b, then the bilinear form b' defined by

$$b'(x,y) = b(y,x) \quad \text{for } x,\, y \in V$$

has adjoint anti-automorphism $\sigma_{b'} = \sigma_b^{-1}$. Therefore, $\sigma_b^2 = \text{Id}$ if and only if b and b' are scalar multiples of each other; since the scalar factor ε such that $b' = \varepsilon b$ clearly satisfies $\varepsilon^2 = 1$, this condition holds if and only if b is symmetric or skew-symmetric.

This shows that F-linear involutions correspond to symmetric or skew-symmetric bilinear forms under the bijection above. $\square$

The involution σ_b associated to a nonsingular symmetric or skew-symmetric bilinear form b under the correspondence of the theorem is called the *adjoint involution* with respect to b. Our aim in this first chapter is to give an analogous interpretation of involutions on arbitrary central simple algebras in terms of hermitian forms on vector spaces over skew fields. We first review basic notions concerning central simple algebras. The first section also discusses Severi-Brauer varieties, for use in §3. In §2 we present the basic definitions concerning involutions on central simple algebras. We distinguish three types of involutions, according to the type of pairing they are adjoint to over an algebraic closure: involutions which are adjoint to symmetric (resp. alternating) bilinear forms are called *orthogonal* (resp. *symplectic*); those which are adjoint to hermitian forms are called *unitary*. Involutions of the first two types leave the center invariant; they are called *involutions of the first kind*. Unitary involutions are also called *involutions of the second kind*; they restrict to a nontrivial automorphism of the center. Necessary and sufficient conditions for the existence of an involution on a central simple algebra are given in §3.

The theorem above, relating bilinear forms on a vector space to involutions on the endomorphism algebra, is generalized in §4, where hermitian forms over simple algebras are investigated. Relations between an analogue of the Scharlau transfer for hermitian forms and extensions of involutions are also discussed in this section.

When F has characteristic 2, it is important to distinguish between bilinear and quadratic forms. Every quadratic form defines (by polarization) an alternating form, but not conversely since a given alternating form is the polar of various quadratic forms. The quadratic pairs introduced in the final section may be regarded as twisted analogues of quadratic forms up to a scalar factor in the same way that involutions may be thought of as twisted analogues of nonsingular symmetric or skew-symmetric bilinear forms. If the characteristic is different from 2, every orthogonal involution determines a unique quadratic pair since a quadratic form is uniquely determined by its polar bilinear form. By contrast, in characteristic 2 the involution associated to a quadratic pair is symplectic since the polar of a quadratic form is alternating, and the quadratic pair is not uniquely determined by its associated involution. Quadratic pairs play a central rôle in the definition of twisted forms of orthogonal groups in Chapter VI.

§1. Central Simple Algebras

Unless otherwise mentioned, all the algebras we consider in this work are finite-dimensional with 1. For any algebra A over a field F and any field extension K/F, we write A_K for the K-algebra obtained from A by extending scalars to K:

$$A_K = A \otimes_F K.$$

We also define the *opposite algebra* A^{op} by

$$A^{\mathrm{op}} = \{\, a^{\mathrm{op}} \mid a \in A \,\},$$

with the operations defined as follows:

$$a^{\mathrm{op}} + b^{\mathrm{op}} = (a+b)^{\mathrm{op}}, \quad a^{\mathrm{op}}b^{\mathrm{op}} = (ba)^{\mathrm{op}}, \quad \alpha \cdot a^{\mathrm{op}} = (\alpha \cdot a)^{\mathrm{op}}$$

for a, $b \in A$ and $\alpha \in F$.

A *central simple algebra* over a field F is a (finite dimensional) algebra $A \neq \{0\}$ with center F ($= F \cdot 1$) which has no two-sided ideals except $\{0\}$ and A. An algebra $A \neq \{0\}$ is a *division algebra* (or a *skew field*) if every non-zero element in A is invertible.

1.A. Fundamental theorems. For the convenience of further reference, we summarize without proofs some basic results from the theory of central simple algebras. The structure of these algebras is determined by the following well-known theorem of Wedderburn:

(1.1) Theorem (Wedderburn). *For an algebra A over a field F, the following conditions are equivalent:*

(1) *A is central simple.*

(2) *The canonical map $A \otimes_F A^{\mathrm{op}} \to \mathrm{End}_F(A)$ which associates to $a \otimes b^{\mathrm{op}}$ the linear map $x \mapsto axb$ is an isomorphism.*

(3) *There is a field K containing F such that A_K is isomorphic to a matrix algebra over K, i.e., $A_K \simeq M_n(K)$ for some n.*

(4) *If Ω is an algebraically closed field containing F,*

$$A_\Omega \simeq M_n(\Omega) \quad \text{for some } n.$$

(5) *There is a finite dimensional central division algebra D over F and an integer r such that $A \simeq M_r(D)$.*

Moreover, if these conditions hold, all the simple left (or right) A-modules are isomorphic, and the division algebra D is uniquely determined up to an algebra isomorphism as $D = \mathrm{End}_A(M)$ for any simple left A-module M.

References: See for instance Scharlau [247, Chapter 8] or Draxl [84, §3].   □

The fields K for which condition (3) holds are called *splitting fields* of A. Accordingly, the algebra A is called *split* if it is isomorphic to a matrix algebra $M_n(F)$ (or to $\mathrm{End}_F(V)$ for some vector space V over F).

Since the dimension of an algebra does not change under an extension of scalars, it follows from the above theorem that the dimension of every central simple algebra is a square: $\dim_F A = n^2$ if $A_K \simeq M_n(K)$ for some extension K/F. The integer n is called the *degree* of A and is denoted by $\deg A$. The degree of the division algebra D in condition (5) is called the *index* of A (or sometimes the *Schur index* of A) and denoted by $\mathrm{ind}\, A$. Alternately, the index of A can be defined by the relation

$$\deg A \, \mathrm{ind}\, A = \dim_F M$$

where M is any simple left module over A. This relation readily follows from the fact that if $A \simeq M_r(D)$, then D^r is a simple left module over A.

We rephrase the implication (1) $\Rightarrow$ (5) in Wedderburn's theorem:

(1.2) Corollary. *Every central simple F-algebra A has the form*

$$A \simeq \mathrm{End}_D(V)$$

for some (finite dimensional) central division F-algebra D and some finite-dimensional right vector space V over D. The F-algebra D is uniquely determined by A up to isomorphism, V is a simple left A-module and $\deg A = \deg D \dim_D V$.

In view of the uniqueness (up to isomorphism) of the division algebra D (or, equivalently, of the simple left A-module M), we may formulate the following definition:

(1.3) Definition. Finite dimensional central simple algebras A, B over a field F are called *Brauer-equivalent* if the F-algebras of endomorphisms of any simple left A-module M and any simple left B-module N are isomorphic:

$$\mathrm{End}_A(M) \simeq \mathrm{End}_B(N).$$

Equivalently, A and B are Brauer-equivalent if and only if $M_\ell(A) \simeq M_m(B)$ for some integers ℓ, m.

Clearly, every central simple algebra is Brauer-equivalent to one and only one division algebra (up to isomorphism). If A and B are Brauer-equivalent central simple algebras, then $\mathrm{ind}\, A = \mathrm{ind}\, B$; moreover, $A \simeq B$ if and only if $\deg A = \deg B$.

The tensor product endows the set of Brauer equivalence classes of central simple algebras over F with the structure of an abelian group, denoted $\mathrm{Br}(F)$ and called the *Brauer group* of F. The unit element in this group is the class of F which is also the class of all the matrix algebras over F. The inverse of the class of a central simple algebra A is the class of the opposite algebra A^{op}, as part (2) of Wedderburn's theorem shows.

Uniqueness (up to isomorphism) of simple left modules over central simple algebras leads to the following two fundamental results:

(1.4) Theorem (Skolem-Noether). *Let A be a central simple F-algebra and let $B \subset A$ be a simple subalgebra. Every F-algebra homomorphism $\rho \colon B \to A$ extends to an inner automorphism of A: there exists $a \in A^\times$ such that $\rho(b) = aba^{-1}$ for all $b \in B$. In particular, every F-algebra automorphism of A is inner.*

References: Scharlau [247, Theorem 8.4.2], Draxl [84, §7] or Pierce [219, §12.6]. □

The *centralizer $C_A B$* of a subalgebra $B \subset A$ is, by definition, the set of elements in A which commute with every element in B.

(1.5) Theorem (Double centralizer). *Let A be a central simple F-algebra and let $B \subset A$ be a simple subalgebra with center $K \supset F$. The centralizer $C_A B$ is a simple subalgebra of A with center K which satisfies*

$$\dim_F A = \dim_F B \cdot \dim_F C_A B \qquad and \qquad C_A C_A B = B.$$

If $K = F$, then multiplication in A defines a canonical isomorphism $A = B \otimes_F C_A B$.

References: Scharlau [247, Theorem 8.4.5], Draxl [84, §7] or Pierce [219, §12.7]. □

Let Ω denote an algebraic closure of F. Under scalar extension to Ω, every central simple F-algebra A of degree n becomes isomorphic to $M_n(\Omega)$. We may therefore fix an F-algebra embedding $A \hookrightarrow M_n(\Omega)$ and view every element $a \in A$ as a matrix in $M_n(\Omega)$. Its characteristic polynomial has coefficients in F and is independent of the embedding of A in $M_n(\Omega)$ (see Scharlau [247, Ch. 8, §5], Draxl [84, §22], Reiner [229, §9] or Pierce [219, §16.1]); it is called the *reduced characteristic polynomial* of A and is denoted

(1.6) $$\mathrm{Prd}_{A,a}(X) = X^n - s_1(a)X^{n-1} + s_2(a)X^{n-2} - \cdots + (-1)^n s_n(a).$$

The *reduced trace* and *reduced norm* of a are denoted $\mathrm{Trd}_A(a)$ and $\mathrm{Nrd}_A(a)$ (or simply $\mathrm{Trd}(a)$ and $\mathrm{Nrd}(a)$):

$$\mathrm{Trd}_A(a) = s_1(a), \qquad \mathrm{Nrd}_A(a) = s_n(a).$$

We also write

(1.7) $$\mathrm{Srd}_A(a) = s_2(a).$$

(1.8) Proposition. *The bilinear form $T_A \colon A \times A \to F$ defined by*

$$T_A(x, y) = \mathrm{Trd}_A(xy) \qquad for\ x, y \in A$$

is nonsingular.

Proof: The result is easily checked in the split case and follows in the general case by scalar extension to a splitting field. (See Reiner [229, Theorem 9.9]). □

1.B. One-sided ideals in central simple algebras. A fundamental result of the Wedderburn theory of central simple algebras is that all the finitely generated left (resp. right) modules over a central simple F-algebra A decompose into direct sums of simple left (resp. right) modules (see Scharlau [247, p. 283]). Moreover, as already pointed out in (1.1), the simple left (resp. right) modules are all isomorphic. If $A = M_r(D)$ for some integer r and some central division algebra D, then D^r is a simple left A-module (via matrix multiplication, writing the elements of D^r as column vectors). Therefore, every finitely generated left A-module M is isomorphic to a direct sum of copies of D^r:

$$M \simeq (D^r)^s \quad \text{for some integer } s,$$

hence

$$\dim_F M = rs \dim_F D = s \deg A \operatorname{ind} A.$$

More precisely, we may represent the elements in M by $r \times s$-matrices with entries in D:

$$M \simeq M_{r,s}(D)$$

so that the action of $A = M_r(D)$ on M is the matrix multiplication.

(1.9) Definition. The *reduced dimension* of the left A-module M is defined by

$$\operatorname{rdim}_A M = \frac{\dim_F M}{\deg A}.$$

The reduced dimension $\operatorname{rdim}_A M$ will be simply denoted by $\operatorname{rdim} M$ when the algebra A is clear from the context. Observe from the preceding relation that the reduced dimension of a finitely generated left A-module is always a multiple of $\operatorname{ind} A$. Moreover, every left A-module M of reduced dimension $s \operatorname{ind} A$ is isomorphic to $M_{r,s}(D)$, hence the reduced dimension classifies left A-modules up to isomorphism.

The preceding discussion of course applies also to right A-modules; writing the elements of D^r as row vectors, matrix multiplication also endows D^r with a right A-module structure, and D^r is then a simple right A-module. Every right module of reduced dimension $s \operatorname{ind} A$ over $A = M_r(D)$ is isomorphic to $M_{s,r}(D)$.

(1.10) Proposition. *Every left module of finite type M over a central simple F-algebra A has a natural structure of right module over $E = \operatorname{End}_A(M)$, so that M is an A-E-bimodule. If $M \neq \{0\}$, the algebra E is central simple over F and Brauer-equivalent to A; moreover,*

$$\deg E = \operatorname{rdim}_A M, \quad \operatorname{rdim}_E M = \deg A,$$

and

$$A = \operatorname{End}_E(M).$$

Conversely, if A and E are Brauer-equivalent central simple algebras over F, then there is an A-E-bimodule $M \neq \{0\}$ such that $A = \operatorname{End}_E(M)$, $E = \operatorname{End}_A(M)$, $\operatorname{rdim}_A(M) = \deg E$ and $\operatorname{rdim}_E(M) = \deg A$.

Proof: The first statement is clear. (Recall that endomorphisms of left modules are written on the right of the arguments.) Suppose that $A = M_r(D)$ for some integer r and some central division algebra D. Then D^r is a simple left A-module, hence $D \simeq \operatorname{End}_A(D^r)$ and $M \simeq (D^r)^s$ for some s. Therefore,

$$\operatorname{End}_A(M) \simeq M_s\big(\operatorname{End}_A(D^r)\big) \simeq M_s(D).$$

This shows that E is central simple and Brauer-equivalent to A. Moreover, $\deg E = s \deg D = \operatorname{rdim}_A M$, hence

$$\operatorname{rdim}_E M = \frac{rs \dim D}{s \deg D} = r \deg D = \deg A.$$

Since M is an A-E-bimodule, we have a natural embedding $A \hookrightarrow \operatorname{End}_E(M)$. Computing the degree of $\operatorname{End}_E(M)$ as we computed $\deg \operatorname{End}_A(M)$ above, we get

$$\deg \operatorname{End}_E(M) = \deg A,$$

hence this natural embedding is surjective.

For the converse, suppose that A and E are Brauer-equivalent central simple F-algebras. We may assume that

$$A = M_r(D) \quad \text{and} \quad E = M_s(D)$$

for some central division F-algebra D and some integers r and s. Let $M = M_{r,s}(D)$ be the set of $r \times s$-matrices over D. Matrix multiplication endows M with an A-E-bimodule structure, so that we have natural embeddings

(1.11) $$A \hookrightarrow \operatorname{End}_E(M) \quad \text{and} \quad E \hookrightarrow \operatorname{End}_A(M).$$

Since $\dim_F M = rs \dim_F D$, it is readily computed that $\operatorname{rdim}_E M = \deg A$ and $\operatorname{rdim}_A M = \deg E$. The first part of the proposition then yields

$$\deg \operatorname{End}_A(M) = \operatorname{rdim}_A M = \deg E \quad \text{and} \quad \deg \operatorname{End}_E(M) = \operatorname{rdim}_E M = \deg A,$$

hence the natural embeddings (1.11) are surjective. $\square$

Ideals and subspaces. Suppose now that $A = \operatorname{End}_D(V)$ for some central division algebra D over F and some finite dimensional right vector space V over D. We aim to get an explicit description of the one-sided ideals in A in terms of subspaces of V.

Let $U \subset V$ be a subspace. Composing every linear map from V to U with the inclusion $U \hookrightarrow V$, we identify $\operatorname{Hom}_D(V, U)$ with a subspace of $A = \operatorname{End}_D(V)$:

$$\operatorname{Hom}_D(V, U) = \{ f \in \operatorname{End}_D(V) \mid \operatorname{im} f \subset U \}.$$

This space clearly is a right ideal in A, of reduced dimension

$$\operatorname{rdim} \operatorname{Hom}_D(V, U) = \dim_D U \deg D.$$

Similarly, composing every linear map from the quotient space V/U to V with the canonical map $V \to V/U$, we may identify $\operatorname{Hom}_D(V/U, V)$ with a subspace of $A = \operatorname{End}_D(V)$:

$$\operatorname{Hom}_D(V/U, V) = \{ f \in \operatorname{End}_D(V) \mid \ker f \supset U \}.$$

This space is clearly a left ideal in A, of reduced dimension

$$\operatorname{rdim} \operatorname{Hom}_D(V/U, V) = \dim_D(V/U) \deg D.$$

(1.12) Proposition. *The map $U \mapsto \operatorname{Hom}_D(V, U)$ defines a one-to-one correspondence between subspaces of dimension d in V and right ideals of reduced dimension $d \operatorname{ind} A$ in $A = \operatorname{End}_D(V)$. Similarly, the map $U \mapsto \operatorname{Hom}_D(V/U, V)$ defines a one-to-one correspondence between subspaces of dimension d in V and left ideals of reduced dimension $\deg A - d \operatorname{ind} A$ in A. Moreover, there are canonical isomorphisms of F-algebras:*

$$\operatorname{End}_A\big(\operatorname{Hom}_D(V, U)\big) \simeq \operatorname{End}_D(U) \quad \text{and} \quad \operatorname{End}_A\big(\operatorname{Hom}_D(V/U, V)\big) \simeq \operatorname{End}_D(V/U).$$

Proof: The last statement is clear: multiplication on the left defines an F-algebra homomorphism $\operatorname{End}_D(U) \hookrightarrow \operatorname{End}_A\big(\operatorname{Hom}_D(V, U)\big)$ and multiplication on the right defines an F-algebra homomorphism

$$\operatorname{End}_D(V/U) \hookrightarrow \operatorname{End}_A\big(\operatorname{Hom}_D(V/U, V)\big).$$

Since $\operatorname{rdim}\big(\operatorname{Hom}_D(V, U)\big) = \dim_D U \deg D$, we have

$$\deg \operatorname{End}_A\big(\operatorname{Hom}_D(V, U)\big) = \dim_D U \deg D = \deg \operatorname{End}_D(U),$$

so the homomorphism $\mathrm{End}_D(U) \hookrightarrow \mathrm{End}_A\big(\mathrm{Hom}_D(V,U)\big)$ is an isomorphism. Similarly, the homomorphism $\mathrm{End}_D(V/U) \hookrightarrow \mathrm{End}_A\big(\mathrm{Hom}_D(V/U,V)\big)$ is an isomorphism by dimension count.

For the first part, it suffices to show that every right (resp. left) ideal in A has the form $\mathrm{Hom}_D(V,U)$ (resp. $\mathrm{Hom}_D(V/U,V)$) for some subspace $U \subset V$. This is proved for instance in Baer [27, §5.2]. $\qquad\square$

(1.13) Corollary. *For every left (resp. right) ideal $I \subset A$ there exists an idempotent $e \in A$ such that $I = Ae$ (resp. $I = eA$). Multiplication on the right (resp. left) induces a surjective homomorphism of right (resp. left) $\mathrm{End}_A(I)$-modules:*

$$\rho\colon I \to \mathrm{End}_A(I)$$

which yields an isomorphism: $eAe \simeq \mathrm{End}_A(I)$.

Proof: If $I = \mathrm{Hom}_D(V/U,V)$ (resp. $\mathrm{Hom}_D(V,U)$), choose a complementary subspace U' in V, so that $V = U \oplus U'$, and take for e the projection on U' parallel to U (resp. the projection on U parallel to U'). We then have $I = Ae$ (resp. $I = eA$).

For simplicity of notation, we prove the rest only in the case of a *left* ideal I. Then $\mathrm{End}_A(I)$ acts on I *on the right*. For $x \in I$, define $\rho(x) \in \mathrm{End}_A(I)$ by

$$y^{\rho(x)} = yx.$$

For $f \in \mathrm{End}_A(I)$ we have

$$(yx)^f = yx^f = y^{\rho(x^f)},$$

hence

$$\rho(x^f) = \rho(x) \circ f,$$

which means that ρ is a homomorphism of right $\mathrm{End}_A(I)$-modules. In order to see that ρ is onto, pick an idempotent $e \in A$ such that $I = Ae$. For every $y \in I$ we have $y = ye$; it follows that every $f \in \mathrm{End}_A(I)$ is of the form $f = \rho(e^f)$, since for every $y \in I$,

$$y^f = (ye)^f = ye^f = y^{\rho(e^f)}.$$

Therefore, ρ is surjective.

To complete the proof, we show that the restriction of ρ to eAe is an isomorphism $eAe \xrightarrow{\sim} \mathrm{End}_A(I)$. It is readily verified that this restriction is an F-algebra homomorphism. Moreover, for every $x \in I$ one has $\rho(x) = \rho(ex)$ since $y = ye$ for every $y \in I$. Therefore, the restriction of ρ to eAe is also surjective onto $\mathrm{End}_A(I)$. Finally, if $\rho(ex) = 0$, then in particular

$$e^{\rho(ex)} = ex = 0,$$

so ρ is injective on eAe. $\qquad\square$

Annihilators. For every left ideal I in a central simple algebra A over a field F, the *annihilator* I^0 is defined by

$$I^0 = \{\, x \in A \mid Ix = \{0\} \,\}.$$

This set is clearly a right ideal. Similarly, for every right ideal I, the annihilator I^0 is defined by

$$I^0 = \{\, x \in A \mid xI = \{0\} \,\};$$

it is a left ideal in A.

(1.14) Proposition. *For every left or right ideal* $I \subset A$, $\mathrm{rdim}\, I + \mathrm{rdim}\, I^0 = \deg A$ *and* $I^{00} = I$.

Proof: Let $A = \mathrm{End}_D(V)$. For any subspace $U \subset V$ it follows from the definition of the annihilator that

$$\mathrm{Hom}_D(V, U)^0 = \mathrm{Hom}_D(V/U, V) \quad \text{and} \quad \mathrm{Hom}_D(V/U, V)^0 = \mathrm{Hom}_D(V, U).$$

Since every left (resp. right) ideal $I \subset A$ has the form $I = \mathrm{Hom}_D(V/U, V)$ (resp. $I = \mathrm{Hom}_D(V, U)$), the proposition follows. $\square$

Now, let $J \subset A$ be a right ideal of reduced dimension k and let $B \subset A$ be the *idealizer* of J:

$$B = \{\, a \in A \mid aJ \subset J \,\}.$$

This set is a subalgebra of A containing J as a two-sided ideal. It follows from the definition of J^0 that $J^0 b \subset J^0$ for all $b \in B$ and that $J^0 \subset B$. Therefore, (1.13) shows that the map $\rho\colon B \to \mathrm{End}_A(J^0)$ defined by multiplication on the right is surjective. Its kernel is $J^{00} = J$, hence it induces an isomorphism $B/J \xrightarrow{\sim} \mathrm{End}_A(J^0)$.

For every right ideal $I \subset A$ containing J, let

$$\tilde{I} = \rho(I \cap B).$$

(1.15) Proposition. *The map* $I \mapsto \tilde{I}$ *defines a one-to-one correspondence between right ideals of reduced dimension* r *in* A *which contain* J *and right ideals of reduced dimension* $r - k$ *in* $\mathrm{End}_A(J^0)$. *If* $A = \mathrm{End}_D(V)$ *and* $J = \mathrm{Hom}_D(V, U)$ *for some subspace* $U \subset V$ *of dimension* $r/\mathrm{ind}\, A$, *then for* $I = \mathrm{Hom}_D(V, W)$ *with* $W \supset U$, *we have under the natural isomorphism* $\mathrm{End}_A(J^0) = \mathrm{End}_D(V/U)$ *of (1.12) that*

$$\tilde{I} = \mathrm{Hom}_D(V/U, W/U).$$

Proof: In view of (1.12), the second part implies the first, since the map $W \mapsto W/U$ defines a one-to-one correspondence between subspaces of dimension $r/\mathrm{ind}\, A$ in V which contain U and subspaces of dimension $(r - k)/\mathrm{ind}\, A$ in V/U.

Suppose that $A = \mathrm{End}_D(V)$ and $J = \mathrm{Hom}_D(V, U)$, hence $J^0 = \mathrm{Hom}_D(V/U, V)$ and $B = \{\, f \in A \mid f(U) \subset U \,\}$. Every $f \in B$ induces a linear map $\overline{f} \in \mathrm{End}_D(V/U)$, and the homomorphism $\rho\colon B \to \mathrm{End}_A(J^0) = \mathrm{End}_D(V/U)$ maps f to $\overline{f}$ since for $g \in J^0$ we have

$$g^{\rho(f)} = g \circ f = g \circ \overline{f}.$$

For $I = \mathrm{Hom}_D(V, W)$ with $W \supset U$, it follows that

$$\tilde{I} = \{\, \overline{f} \mid f \in I \text{ and } f(U) \subset U \,\} \subset \mathrm{Hom}_D(V/U, W/U).$$

The converse inclusion is clear, since using bases of U, W and V it is easily seen that every linear map $h \in \mathrm{Hom}_D(V/U, W/U)$ is of the form $h = \overline{f}$ for some $f \in \mathrm{Hom}_D(V, W)$ such that $f(U) \subset U$. $\square$

1.C. Severi-Brauer varieties. Let A be a central simple algebra of degree n over a field F and let r be an integer, $1 \leq r \leq n$. Consider the *Grassmannian* $\mathrm{Gr}(rn, A)$ of rn-dimensional subspaces in A. The Plücker embedding identifies $\mathrm{Gr}(rn, A)$ with a closed subvariety of the projective space on the rn-th exterior power of A (see Harris [117, Example 6.6, p. 64]):

$$\mathrm{Gr}(rn, A) \subset \mathbb{P}(\textstyle\bigwedge^{rn} A).$$

The rn-dimensional subspace $U \subset A$ corresponding to a non-zero rn-vector $u_1 \wedge \cdots \wedge u_{rn} \in \bigwedge^{rn} A$ is

$$U = \{\, x \in A \mid u_1 \wedge \cdots \wedge u_{rn} \wedge x = 0 \,\} = u_1 F + \cdots + u_{rn} F.$$

Among the rn-dimensional subspaces in A, the right ideals of reduced dimension r are the subspaces which are preserved under multiplication on the right by the elements of A. Such ideals may fail to exist: for instance, if A is a division algebra, it does not contain any nontrivial ideal; on the other hand, if $A \simeq M_n(F)$, then it contains right ideals of every reduced dimension $r = 0, \ldots, n$. Since every central simple F-algebra becomes isomorphic to a matrix algebra over some scalar extension of F, this situation is best understood from an algebraic geometry viewpoint: it is comparable to the case of varieties defined over some base field F which have no rational point over F but acquire points over suitable extensions of F.

To make this viewpoint precise, consider an arbitrary basis $(e_i)_{1 \leq i \leq n^2}$ of A. The rn-dimensional subspace represented by an rn-vector $u_1 \wedge \cdots \wedge u_{rn} \in \bigwedge^{rn} A$ is a right ideal of reduced dimension r if and only if it is preserved under right multiplication by $e_1, \ldots, e_{n^2}$, i.e.,

$$u_1 e_i \wedge \cdots \wedge u_{rn} e_i \in u_1 \wedge \cdots \wedge u_{rn} F \quad \text{for } i = 1, \ldots, n^2,$$

or, equivalently,

$$u_1 e_i \wedge \cdots \wedge u_{rn} e_i \wedge u_j = 0 \quad \text{for } i = 1, \ldots, n^2 \text{ and } j = 1, \ldots, rn.$$

This condition translates to a set of equations on the coordinates of the rn-vector $u_1 \wedge \cdots \wedge u_{rn}$, hence the right ideals of reduced dimension r in A form a closed subvariety of $\mathrm{Gr}(rn, A)$.

(1.16) Definition. The (*generalized*) *Severi-Brauer variety* $\mathrm{SB}_r(A)$ is the variety of right ideals of reduced dimension r in A. It is a closed subvariety of the Grassmannian:

$$\mathrm{SB}_r(A) \subset \mathrm{Gr}(rn, A).$$

For $r = 1$, we write simply $\mathrm{SB}(A) = \mathrm{SB}_1(A)$. This is the (usual) Severi-Brauer variety of A, first defined by F. Châtelet [63].

(1.17) Proposition. *The Severi-Brauer variety $\mathrm{SB}_r(A)$ has a rational point over an extension K of F if and only if the index $\mathrm{ind}\, A_K$ divides r. In particular, $\mathrm{SB}(A)$ has a rational point over K if and only if K splits A.*

Proof: From the definition, it follows that $\mathrm{SB}_r(A)$ has a rational point over K if and only if A_K contains a right ideal of reduced dimension r. Since the reduced dimension of any finitely generated right A_K-module is a multiple of $\mathrm{ind}\, A_K$, it follows that $\mathrm{ind}\, A_K$ divides r if $\mathrm{SB}_r(A)$ has a rational point over K. Conversely, suppose $r = m\,\mathrm{ind}\, A_K$ for some integer m and let $A_K \simeq M_t(D)$ for some division algebra D and some integer t. The set of matrices in $M_t(D)$ whose $t - m$ last rows are zero is a right ideal of reduced dimension r, hence $\mathrm{SB}_r(A)$ has a rational point over K. $\qquad\square$

The following theorem shows that Severi-Brauer varieties are twisted forms of Grassmannians:

(1.18) Theorem. *For $A = \mathrm{End}_F(V)$, there is a natural isomorphism*

$$\mathrm{SB}_r(A) \simeq \mathrm{Gr}(r, V).$$

In particular, for $r = 1$,

$$\mathrm{SB}(A) \simeq \mathbb{P}(V).$$

Proof: Let $V^* = \mathrm{Hom}_F(V, F)$ be the dual of V. Under the natural isomorphism $A = \mathrm{End}_F(V) \simeq V \otimes_F V^*$, multiplication is given by

$$(v \otimes \phi) \cdot (w \otimes \psi) = (v \otimes \psi)\phi(w).$$

By (1.12), the right ideals of reduced dimension r in A are of the form $\mathrm{Hom}_F(V, U) = U \otimes V^*$ where U is an r-dimensional subspace in V.

We will show that the correspondence $U \leftrightarrow U \otimes V^*$ between r-dimensional subspaces in V and right ideals of reduced dimension r in A induces an isomorphism of varieties $\mathrm{Gr}(r, V) \simeq \mathrm{SB}_r(A)$.

For any vector space W of dimension n, there is a morphism $\mathrm{Gr}(r, V) \to \mathrm{Gr}(rn, V \otimes W)$ which maps an r-dimensional subspace $U \subset V$ to $U \otimes W \subset V \otimes W$. In the particular case where $W = V^*$ we thus get a morphism $\Phi\colon \mathrm{Gr}(r, V) \to \mathrm{SB}_r(A)$ which maps U to $U \otimes V^*$.

In order to show that Φ is an isomorphism, we consider the following affine covering of $\mathrm{Gr}(r, V)$: for each subspace $S \subset V$ of dimension $n - r$, we denote by $\mathcal{U}_S$ the set of complementary subspaces:

$$\mathcal{U}_S = \{\, U \subset V \mid U \oplus S = V \,\}.$$

The set $\mathcal{U}_S$ is an affine open subset of $\mathrm{Gr}(r, V)$; more precisely, if U_0 is a fixed complementary subspace of S, there is an isomorphism:

$$\mathrm{Hom}_F(U_0, S) \xrightarrow{\sim} \mathcal{U}_S$$

which maps $f \in \mathrm{Hom}_F(U_0, S)$ to $U = \{\, x + f(x) \mid x \in U_0 \,\}$ (see Harris [117, p. 65]). Similarly, we may also consider $\mathcal{U}_{S \otimes V^*} \subset \mathrm{Gr}(rn, A)$. The image of the restriction of Φ to $\mathcal{U}_S$ is

$$\{\, U \otimes V^* \subset V \otimes V^* \mid (U \otimes V^*) \oplus (S \otimes V^*) = V \otimes V^* \,\} = \mathcal{U}_{S \otimes V^*} \cap \mathrm{SB}_r(A).$$

Moreover, there is a commutative diagram:

$$
\begin{array}{ccc}
\mathcal{U}_S & \xrightarrow{\;\Phi|_{\mathcal{U}_S}\;} & \mathcal{U}_{S \otimes V^*} \\[2pt]
\simeq \downarrow & & \downarrow \simeq \\[2pt]
\mathrm{Hom}_F(U_0, S) & \xrightarrow{\;\phi\;} & \mathrm{Hom}_F(U_0 \otimes V^*, S \otimes V^*)
\end{array}
$$

where $\phi(f) = f \otimes \mathrm{Id}_{V^*}$. Since ϕ is linear and injective, it is an isomorphism of varieties between $\mathrm{Hom}_F(U_0, S)$ and its image. Therefore, the restriction of Φ to $\mathcal{U}_S$ is an isomorphism $\Phi|_{\mathcal{U}_S}\colon \mathcal{U}_S \xrightarrow{\sim} \mathcal{U}_{S \otimes V^*} \cap \mathrm{SB}_r(A)$. Since the open sets $\mathcal{U}_S$ form a covering of $\mathrm{Gr}(r, V)$, it follows that Φ is an isomorphism. $\square$

Although Severi-Brauer varieties are defined in terms of *right* ideals, they can also be used to derive information on *left* ideals. Indeed, if J is a left ideal in a central simple algebra A, then the set

$$J^{\mathrm{op}} = \{\, j^{\mathrm{op}} \in A^{\mathrm{op}} \mid j \in J \,\}$$

is a right ideal in the opposite algebra A^{op}. Therefore, the variety of left ideals of reduced dimension r in A can be identified with $\mathrm{SB}_r(A^{\mathrm{op}})$. We combine this observation with the annihilator construction (see §1.B) to get the following result:

(1.19) Proposition. *For any central simple algebra A of degree n, there is a canonical isomorphism*

$$\alpha \colon \mathrm{SB}_r(A) \xrightarrow{\sim} \mathrm{SB}_{n-r}(A^{\mathrm{op}})$$

which maps a right ideal $I \subset A$ of reduced dimension r to $(I^0)^{\mathrm{op}}$.

Proof: In order to prove that α is an isomorphism, we may extend scalars to a splitting field of A. We may therefore assume that $A = \mathrm{End}_F(V)$ for some n-dimensional vector space V. Then $A^{\mathrm{op}} = \mathrm{End}_F(V^*)$ under the identification $f^{\mathrm{op}} = f^t$ for $f \in \mathrm{End}_F(V)$. By (1.18), we may further identify

$$\mathrm{SB}_r(A) = \mathrm{Gr}(r, V), \quad \mathrm{SB}_{n-r}(A^{\mathrm{op}}) = \mathrm{Gr}(n - r, V^*).$$

Under these identifications, the map $\alpha \colon \mathrm{Gr}(r, V) \to \mathrm{Gr}(n - r, V^*)$ carries every r-dimensional subspace $U \subset V$ to $U^0 = \{\, \varphi \in V^* \mid \varphi(U) = \{0\} \,\}$.

To show that α is an isomorphism of varieties, we restrict it to the affine open sets $\mathcal{U}_S$ defined in the proof of Theorem (1.18): let S be an $(n - r)$-dimensional subspace in V and

$$\mathcal{U}_S = \{\, U \subset V \mid U \oplus S = V \,\} \subset \mathrm{Gr}(r, V).$$

Let $U_0 \subset V$ be such that $U_0 \oplus S = V$, so that $\mathcal{U}_S \simeq \mathrm{Hom}_F(U_0, S)$. We also have $U_0^0 \oplus S^0 = V^*$, $\mathcal{U}_{S^0} \simeq \mathrm{Hom}_F(U_0^0, S^0)$, and the map α restricts to $\alpha|_{\mathcal{U}_S} \colon \mathcal{U}_S \to \mathcal{U}_{S^0}$. It therefore induces a map α' which makes the following diagram commute:

$$
\begin{array}{ccc}
\mathcal{U}_S & \xrightarrow{\ \alpha|_{\mathcal{U}_S}\ } & \mathcal{U}_{S^0} \\
\simeq \downarrow & & \downarrow \simeq \\
\mathrm{Hom}_F(U_0, S) & \xrightarrow{\ \alpha'\ } & \mathrm{Hom}_F(U_0^0, S^0).
\end{array}
$$

We now proceed to show that α' is an isomorphism of (affine) varieties.

Every linear form in U_0^0 restricts to a linear form on S, and since $V = U_0 \oplus S$ we thus get a natural isomorphism $U_0^0 \simeq S^*$. Similarly, $S^0 \simeq U_0^*$, so $\mathrm{Hom}_F(U_0^0, S^0) \simeq \mathrm{Hom}_F(S^*, U_0^*)$. Under this identification, a direct calculation shows that the map α' carries $f \in \mathrm{Hom}_F(U_0, S)$ to $-f^t \in \mathrm{Hom}_F(S^*, U_0^*) = \mathrm{Hom}_F(U_0^0, S^0)$. It is therefore an isomorphism of varieties. Since the open sets $\mathcal{U}_S$ cover $\mathrm{Gr}(r, V)$, it follows that α is an isomorphism. $\square$

If V is a vector space of dimension n over a field F and $U \subset V$ is a subspace of dimension k, then for $r = k, \ldots, n$ the Grassmannian $\mathrm{Gr}(r - k, V/U)$ embeds into $\mathrm{Gr}(r, V)$ by mapping every subspace $\overline{W} \subset V/U$ to the subspace $W \supset U$ such that $W/U = \overline{W}$. The image of $\mathrm{Gr}(r - k, V/U)$ in $\mathrm{Gr}(r, V)$ is the *sub-Grassmannian* of r-dimensional subspaces in V which contain U (see Harris [117, p. 66]). There is an analogous notion for Severi-Brauer varieties:

(1.20) Proposition. *Let A be a central simple F-algebra and let $J \subset A$ be a right ideal of reduced dimension k (i.e., a rational point of $\mathrm{SB}_k(A)$). The one-to-one correspondence between right ideals of reduced dimension r in A which contain J and right ideals of reduced dimension $r - k$ in $\mathrm{End}_A(J^0)$ set up in (1.15) defines an embedding:*

$$\mathrm{SB}_{r-k}\big(\mathrm{End}_A(J^0)\big) \hookrightarrow \mathrm{SB}_r(A).$$

The image of $\mathrm{SB}_{r-k}\big(\mathrm{End}_A(J^0)\big)$ in $\mathrm{SB}_r(A)$ is the variety of right ideals of reduced dimension r in A which contain J.

Proof: It suffices to prove the proposition over a scalar extension. We may therefore assume that A is split, i.e., that $A = \mathrm{End}_F(V)$. Let then $J = \mathrm{Hom}_F(V, U)$ for some subspace $U \subset V$ of dimension k. We have $J^0 = \mathrm{Hom}_F(V/U, V)$ and (1.12) shows that there is a canonical isomorphism $\mathrm{End}_A(J^0) = \mathrm{End}_F(V/U)$. Theorem (1.18) then yields canonical isomorphisms $\mathrm{SB}_r(A) = \mathrm{Gr}(r, V)$ and $\mathrm{SB}_{r-k}(\mathrm{End}_A(J^0)) = \mathrm{Gr}(r-k, V/U)$. Moreover, from (1.15) it follows that the map $\mathrm{SB}_{r-k}(\mathrm{End}_A(J^0)) \to \mathrm{SB}_r(A)$ corresponds under these identifications to the embedding $\mathrm{Gr}(r-k, V/U) \hookrightarrow \mathrm{Gr}(r, V)$ described above. $\square$

§2. Involutions

An *involution* on a central simple algebra A over a field F is a map $\sigma \colon A \to A$ subject to the following conditions:

    (a) $\sigma(x + y) = \sigma(x) + \sigma(y)$ for x, $y \in A$.
    (b) $\sigma(xy) = \sigma(y)\sigma(x)$ for x, $y \in A$.
    (c) $\sigma^2(x) = x$ for $x \in A$.

Note that the map σ is *not* required to be F-linear. However, it is easily checked that the center F $(= F \cdot 1)$ is preserved under σ. The restriction of σ to F is therefore an automorphism which is either the identity or of order 2. Involutions which leave the center elementwise invariant are called *involutions of the first kind*. Involutions whose restriction to the center is an automorphism of order 2 are called *involutions of the second kind*.

This section presents the basic definitions and properties of central simple algebras with involution. Involutions of the first kind are considered first. As observed in the introduction to this chapter, they are adjoint to nonsingular symmetric or skew-symmetric bilinear forms in the split case. Involutions of the first kind are correspondingly divided into two types: the orthogonal and the symplectic types. We show in (2.6) how to characterize these types by properties of the symmetric elements. Involutions of the second kind, also called unitary, are treated next. Various examples are provided in (2.20)–(2.23).

2.A. Involutions of the first kind. Throughout this subsection, A denotes a central simple algebra over a field F of arbitrary characteristic, and σ is an involution of the first kind on A. Our basic object of study is the couple (A, σ); from this point of view, a homomorphism of algebras with involution $f \colon (A, \sigma) \to (A', \sigma')$ is an F-algebra homomorphism $f \colon A \to A'$ such that $\sigma' \circ f = f \circ \sigma$. Our main tool is the extension of scalars: if L is any field containing F, the involution σ extends to an involution of the first kind $\sigma_L = \sigma \otimes \mathrm{Id}_L$ on $A_L = A \otimes_F L$. In particular, if L is a splitting field of A, we may identify $A_L = \mathrm{End}_L(V)$ for some vector space V over L of dimension $n = \deg A$. As observed in the introduction to this chapter, the involution σ_L is then the adjoint involution σ_b with respect to some nonsingular symmetric or skew-symmetric bilinear form b on V. By means of a basis of V, we may further identify V with L^n, hence also A with $M_n(L)$. For any matrix m, let m^t denote the transpose of m. If $g \in \mathrm{GL}_n(L)$ denotes the Gram matrix of b with respect to the chosen basis, then

$$b(x, y) = x^t \cdot g \cdot y$$

where x, y are considered as column matrices and $g^t = g$ if b is symmetric, $g^t = -g$ if b is skew-symmetric. The involution σ_L is then identified with the involution σ_g

defined by

$$\sigma_g(m) = g^{-1} \cdot m^t \cdot g \quad \text{for } m \in M_n(L).$$

For future reference, we summarize our conclusions:

(2.1) Proposition. *Let (A, σ) be a central simple F-algebra of degree n with involution of the first kind and let L be a splitting field of A. Let V be an L-vector space of dimension n. There is a nonsingular symmetric or skew-symmetric bilinear form b on V and an invertible matrix $g \in \mathrm{GL}_n(L)$ such that $g^t = g$ if b is symmetric and $g^t = -g$ if b is skew-symmetric, and*

$$(A_L, \sigma_L) \simeq \big(\mathrm{End}_L(V), \sigma_b\big) \simeq \big(M_n(L), \sigma_g\big).$$

As a first application, we have the following result:

(2.2) Corollary. *For all $a \in A$, the elements a and $\sigma(a)$ have the same reduced characteristic polynomial. In particular, $\mathrm{Trd}_A\big(\sigma(a)\big) = \mathrm{Trd}_A(a)$ and $\mathrm{Nrd}_A\big(\sigma(a)\big) = \mathrm{Nrd}_A(a)$.*

Proof: For all $m \in M_n(L)$, $g \in \mathrm{GL}_n(L)$, the matrix $g^{-1} \cdot m^t \cdot g$ has the same characteristic polynomial as m. $\qquad\square$

Of course, in (2.1), neither the form b nor the matrix g (nor even the splitting field L) is determined uniquely by the involution σ; some of their properties reflect properties of σ, however. As a first example, we show in (2.5) below that two types of involutions of the first kind can be distinguished which correspond to symmetric and to alternating[1] forms. This distinction is made on the basis of properties of symmetric elements which we define next.

In a central simple F-algebra A with involution of the first kind σ, the sets of symmetric, skew-symmetric, symmetrized and alternating elements in A are defined as follows:

$$\mathrm{Sym}(A, \sigma) = \{\, a \in A \mid \sigma(a) = a \,\},$$
$$\mathrm{Skew}(A, \sigma) = \{\, a \in A \mid \sigma(a) = -a \,\},$$
$$\mathrm{Symd}(A, \sigma) = \{\, a + \sigma(a) \mid a \in A \,\},$$
$$\mathrm{Alt}(A, \sigma) = \{\, a - \sigma(a) \mid a \in A \,\}.$$

If $\mathrm{char}\, F \neq 2$, then $\mathrm{Symd}(A, \sigma) = \mathrm{Sym}(A, \sigma)$, $\mathrm{Alt}(A, \sigma) = \mathrm{Skew}(A, \sigma)$ and $A = \mathrm{Sym}(A, \sigma) \oplus \mathrm{Skew}(A, \sigma)$ since every element $a \in A$ decomposes as $a = \frac{1}{2}\big(a + \sigma(a)\big) + \frac{1}{2}\big(a - \sigma(a)\big)$. If $\mathrm{char}\, F = 2$, then $\mathrm{Symd}(A, \sigma) = \mathrm{Alt}(A, \sigma) \subset \mathrm{Skew}(A, \sigma) = \mathrm{Sym}(A, \sigma)$, and (2.6) below shows that this inclusion is strict.

(2.3) Lemma. *Let $n = \deg A$; then $\dim \mathrm{Sym}(A, \sigma) + \dim \mathrm{Alt}(A, \sigma) = n^2$. Moreover, $\mathrm{Alt}(A, \sigma)$ is the orthogonal space of $\mathrm{Sym}(A, \sigma)$ for the bilinear form T_A on A induced by the reduced trace:*

$$\mathrm{Alt}(A, \sigma) = \{\, a \in A \mid \mathrm{Trd}_A(as) = 0 \text{ for } s \in \mathrm{Sym}(A, \sigma) \,\}.$$

Similarly, $\dim \mathrm{Skew}(A, \sigma) + \dim \mathrm{Symd}(A, \sigma) = n^2$, and $\mathrm{Symd}(A, \sigma)$ is the orthogonal space of $\mathrm{Skew}(A, \sigma)$ for the bilinear form T_A.

---

[1] If $\mathrm{char}\, F \neq 2$, every skew-symmetric bilinear form is alternating; if $\mathrm{char}\, F = 2$, the notions of symmetric and skew-symmetric bilinear forms coincide, but the notion of alternating form is more restrictive.

Proof: The first relation comes from the fact that $\text{Alt}(A,\sigma)$ is the image of the linear endomorphism $\text{Id} - \sigma$ of A, whose kernel is $\text{Sym}(A,\sigma)$. If $a \in \text{Alt}(A,\sigma)$, then $a = x - \sigma(x)$ for some $x \in A$, hence for $s \in \text{Sym}(A,\sigma)$,

$$\text{Trd}_A(as) = \text{Trd}_A(xs) - \text{Trd}_A\big(\sigma(x)s\big) = \text{Trd}_A(xs) - \text{Trd}_A\big(\sigma(sx)\big).$$

Corollary (2.2) shows that the right side vanishes, hence the inclusion

$$\text{Alt}(A,\sigma) \subset \{\, a \in A \mid \text{Trd}_A(as) = 0 \text{ for } s \in \text{Sym}(A,\sigma) \,\}.$$

Dimension count shows that this inclusion is an equality since T_A is nonsingular (see (1.8)).

The statements involving $\text{Symd}(A,\sigma)$ readily follow, either by mimicking the arguments above, or by using the fact that in characteristic different from 2, $\text{Symd}(A,\sigma) = \text{Sym}(A,\sigma)$ and $\text{Alt}(A,\sigma) = \text{Skew}(A,\sigma)$, and, in characteristic 2, $\text{Symd}(A,\sigma) = \text{Alt}(A,\sigma)$ and $\text{Skew}(A,\sigma) = \text{Sym}(A,\sigma)$. $\quad\square$

We next determine the dimensions of $\text{Sym}(A,\sigma)$ and $\text{Skew}(A,\sigma)$ (and therefore also of $\text{Symd}(A,\sigma)$ and $\text{Alt}(A,\sigma)$).

Consider first the split case, assuming that $A = \text{End}_F(V)$ for some vector space V over F. As observed in the introduction to this chapter, every involution of the first kind σ on A is the adjoint involution with respect to a nonsingular symmetric or skew-symmetric bilinear form b on V which is uniquely determined by σ up to a factor in $F^\times$.

(2.4) Lemma. *Let $\sigma = \sigma_b$ be the adjoint involution on $A = \text{End}_F(V)$ with respect to the nonsingular symmetric or skew-symmetric bilinear form b on V, and let $n = \dim_F V$.*

(1) *If b is symmetric, then $\dim_F \text{Sym}(A,\sigma) = n(n+1)/2$.*
(2) *If b is skew-symmetric, then $\dim_F \text{Skew}(A,\sigma) = n(n+1)/2$.*
(3) *If $\text{char } F = 2$, then b is alternating if and only if $\text{tr}(f) = 0$ for all $f \in \text{Sym}(A,\sigma)$. In this case, n is necessarily even.*

Proof: As in (2.1), we use a basis of V to identify (A,σ) with $\big(M_n(F),\sigma_g\big)$, where $g \in \text{GL}_n(F)$ satisfies $g^t = g$ if b is symmetric and $g^t = -g$ if b is skew-symmetric. For $m \in M_n(F)$, the relation $gm = (gm)^t$ is equivalent to $\sigma_g(m) = m$ if $g^t = g$ and to $\sigma_g(m) = -m$ if $g^t = -g$. Therefore,

$$g^{-1} \cdot \text{Sym}\big(M_n(F),t\big) = \begin{cases} \text{Sym}(A,\sigma) & \text{if } b \text{ is symmetric,} \\ \text{Skew}(A,\sigma) & \text{if } b \text{ is skew-symmetric.} \end{cases}$$

The first two parts then follow from the fact that the space $\text{Sym}\big(M_n(F),t\big)$ of $n \times n$ symmetric matrices (with respect to the transpose) has dimension $n(n+1)/2$.

Suppose now that $\text{char } F = 2$. If b is not alternating, then $b(v,v) \neq 0$ for some $v \in V$. Consider the map $f \colon V \to V$ defined by

$$f(x) = vb(v,x)b(v,v)^{-1} \quad \text{for } x \in V.$$

Since b is symmetric we have

$$b\big(f(x),y\big) = b(v,y)b(v,x)b(v,v)^{-1} = b\big(x,f(y)\big) \quad \text{for } x,\,y \in V,$$

hence $\sigma(f) = f$. Since f is an idempotent in A, its trace is the dimension of its image:

$$\text{tr}(f) = \dim \text{im } f = 1.$$

Therefore, if the trace of every symmetric element in A is zero, then b is alternating.

Conversely, suppose b is alternating; it follows that n is even, since every alternating form on a vector space of odd dimension is singular. Let $(e_i)_{1 \le i \le n}$ be a *symplectic basis* of V, in the sense that $b(e_{2i-1}, e_{2i}) = 1$, $b(e_{2i}, e_{2i+1}) = 0$ and $b(e_i, e_j) = 0$ if $|i - j| > 1$. Let $f \in \mathrm{Sym}(A, \sigma)$; for $j = 1, \ldots, n$ let

$$f(e_j) = \sum_{i=1}^{n} e_i a_{ij} \quad \text{for some } a_{ij} \in F,$$

so that $\mathrm{tr}(f) = \sum_{i=1}^{n} a_{ii}$. For $i = 1, \ldots, n/2$ we have

$$b\big(f(e_{2i-1}), e_{2i}\big) = a_{2i-1, 2i-1} \quad \text{and} \quad b\big(e_{2i-1}, f(e_{2i})\big) = a_{2i, 2i};$$

since $\sigma(f) = f$, it follows that $a_{2i-1, 2i-1} = a_{2i, 2i}$ for $i = 1, \ldots, n/2$, hence

$$\mathrm{tr}(f) = 2 \sum_{i=1}^{n/2} a_{2i, 2i} = 0. \qquad \square$$

We now return to the general case where A is an arbitrary central simple F-algebra and σ is an involution of the first kind on A. Let $n = \deg A$ and let L be a splitting field of A. Consider an isomorphism as in (2.1):

$$(A_L, \sigma_L) \simeq \big(\mathrm{End}_L(V), \sigma_b\big).$$

This isomorphism carries $\mathrm{Sym}(A_L, \sigma_L) = \mathrm{Sym}(A, \sigma) \otimes_F L$ to $\mathrm{Sym}\big(\mathrm{End}_L(V), \sigma_b\big)$ and $\mathrm{Skew}(A_L, \sigma_L)$ to $\mathrm{Skew}\big(\mathrm{End}_L(V), \sigma_b\big)$. Since extension of scalars does not change dimensions, (2.4) shows

(a) $\dim_F \mathrm{Sym}(A, \sigma) = n(n + 1)/2$ if b is symmetric;
(b) $\dim_F \mathrm{Skew}(A, \sigma) = n(n + 1)/2$ if b is skew-symmetric.

These two cases coincide if $\mathrm{char}\, F = 2$ but are mutually exclusive if $\mathrm{char}\, F \ne 2$; indeed, in this case $A = \mathrm{Sym}(A, \sigma) \oplus \mathrm{Skew}(A, \sigma)$, hence the dimensions of $\mathrm{Sym}(A, \sigma)$ and $\mathrm{Skew}(A, \sigma)$ add up to n^2.

Since the reduced trace of A corresponds to the trace of endomorphisms under the isomorphism $A_L \simeq \mathrm{End}_L(V)$, we have $\mathrm{Trd}_A(s) = 0$ for all $s \in \mathrm{Sym}(A, \sigma)$ if and only if $\mathrm{tr}(f) = 0$ for all $f \in \mathrm{Sym}\big(\mathrm{End}_L(V), \sigma_b\big)$, and Lemma (2.4) shows that, when $\mathrm{char}\, F = 2$, this condition holds if and only if b is alternating. Therefore, in arbitrary characteristic, the property of b being symmetric or skew-symmetric or alternating depends only on the involution and not on the choice of L nor of b. We may thus set the following definition:

(2.5) Definition. An involution σ of the first kind is said to be *of symplectic type* (or simply *symplectic*) if for any splitting field L and any isomorphism $(A_L, \sigma_L) \simeq \big(\mathrm{End}_L(V), \sigma_b\big)$, the bilinear form b is alternating; otherwise it is called *of orthogonal type* (or simply *orthogonal*). In the latter case, for any splitting field L and any isomorphism $(A_L, \sigma_L) \simeq \big(\mathrm{End}_L(V), \sigma_b\big)$, the bilinear form b is symmetric (and nonalternating).

The preceding discussion yields an alternate characterization of orthogonal and symplectic involutions:

(2.6) Proposition. *Let (A, σ) be a central simple F-algebra of degree n with involution of the first kind.*

(1) *Suppose that* char $F \neq 2$, *hence* $\mathrm{Symd}(A, \sigma) = \mathrm{Sym}(A, \sigma)$ *and* $\mathrm{Alt}(A, \sigma) = \mathrm{Skew}(A, \sigma)$. *If σ is of orthogonal type, then*

$$\dim_F \mathrm{Sym}(A, \sigma) = \frac{n(n+1)}{2} \quad and \quad \dim_F \mathrm{Skew}(A, \sigma) = \frac{n(n-1)}{2}.$$

If σ is of symplectic type, then

$$\dim_F \mathrm{Sym}(A, \sigma) = \frac{n(n-1)}{2} \quad and \quad \dim_F \mathrm{Skew}(A, \sigma) = \frac{n(n+1)}{2}.$$

Moreover, in this case n is necessarily even.
(2) *Suppose that* char $F = 2$, *hence* $\mathrm{Sym}(A, \sigma) = \mathrm{Skew}(A, \sigma)$ *and* $\mathrm{Symd}(A, \sigma) = \mathrm{Alt}(A, \sigma)$; *then*

$$\dim_F \mathrm{Sym}(A, \sigma) = \frac{n(n+1)}{2} \quad and \quad \dim_F \mathrm{Alt}(A, \sigma) = \frac{n(n-1)}{2}.$$

The involution σ is of symplectic type if and only if $\mathrm{Trd}_A\big(\mathrm{Sym}(A, \sigma)\big) = \{0\}$, *which holds if and only if* $1 \in \mathrm{Alt}(A, \sigma)$. *In this case n is necessarily even.*

Proof: The only statement which has not been observed before is that, if char $F = 2$, the reduced trace of every symmetric element is 0 if and only if $1 \in \mathrm{Alt}(A, \sigma)$. This follows from the characterization of alternating elements in (2.3).   □

Given an involution of the first kind on a central simple algebra A, all the other involutions of the first kind on A can be obtained by the following proposition:

(2.7) Proposition. *Let A be a central simple algebra over a field F and let σ be an involution of the first kind on A.*
(1) *For each unit $u \in A^{\times}$ such that $\sigma(u) = \pm u$, the map $\mathrm{Int}(u) \circ \sigma$ is an involution of the first kind on A.*
(2) *Conversely, for every involution σ' of the first kind on A, there exists some $u \in A^{\times}$, uniquely determined up to a factor in $F^{\times}$, such that*

$$\sigma' = \mathrm{Int}(u) \circ \sigma \quad and \quad \sigma(u) = \pm u.$$

We then have

$$\mathrm{Sym}(A, \sigma') = \begin{cases} u \cdot \mathrm{Sym}(A, \sigma) = \mathrm{Sym}(A, \sigma) \cdot u^{-1} & \text{if } \sigma(u) = u \\ u \cdot \mathrm{Skew}(A, \sigma) = \mathrm{Skew}(A, \sigma) \cdot u^{-1} & \text{if } \sigma(u) = -u \end{cases}$$

and

$$\mathrm{Skew}(A, \sigma') = \begin{cases} u \cdot \mathrm{Skew}(A, \sigma) = \mathrm{Skew}(A, \sigma) \cdot u^{-1} & \text{if } \sigma(u) = u \\ u \cdot \mathrm{Sym}(A, \sigma) = \mathrm{Sym}(A, \sigma) \cdot u^{-1} & \text{if } \sigma(u) = -u. \end{cases}$$

If $\sigma(u) = u$, then $\mathrm{Alt}(A, \sigma') = u \cdot \mathrm{Alt}(A, \sigma) = \mathrm{Alt}(A, \sigma) \cdot u^{-1}$.
(3) *Suppose that $\sigma' = \mathrm{Int}(u) \circ \sigma$ where $u \in A^{\times}$ is such that $u = \pm u$. If* char $F \neq 2$, *then σ and σ' are of the same type if and only if $\sigma(u) = u$. If* char $F = 2$, *the involution σ' is symplectic if and only if $u \in \mathrm{Alt}(A, \sigma)$.*

Proof: A computation shows that $\big(\mathrm{Int}(u) \circ \sigma\big)^2 = \mathrm{Int}\big(u\sigma(u)^{-1}\big)$, proving (1).

If σ' is an involution of the first kind on A, then $\sigma' \circ \sigma$ is an automorphism of A which leaves F elementwise invariant. The Skolem-Noether theorem then yields an element $u \in A^{\times}$, uniquely determined up to a factor in $F^{\times}$, such that $\sigma' \circ \sigma = \mathrm{Int}(u)$, hence $\sigma' = \mathrm{Int}(u) \circ \sigma$. It follows that $\sigma'^2 = \mathrm{Int}\big(u\sigma(u)^{-1}\big)$, hence the relation $\sigma'^2 = \mathrm{Id}_A$ yields

$$\sigma(u) = \lambda u \quad \text{for some } \lambda \in F^{\times}.$$

Applying σ to both sides of this relation and substituting λu for $\sigma(u)$ in the resulting equation, we get $u = \lambda^2 u$, hence $\lambda = \pm 1$. If $\sigma(u) = u$, then for all $x \in A$,

$$x - \sigma'(x) = u \cdot \left(u^{-1}x - \sigma(u^{-1}x) \right) = \left(xu - \sigma(xu) \right) \cdot u^{-1}.$$

This proves $\mathrm{Alt}(A, \sigma') = u \cdot \mathrm{Alt}(A, \sigma) = \mathrm{Alt}(A, \sigma) \cdot u^{-1}$. The relations between $\mathrm{Sym}(A, \sigma')$, $\mathrm{Skew}(A, \sigma')$ and $\mathrm{Sym}(A, \sigma)$, $\mathrm{Skew}(A, \sigma)$ follow by straightforward computations, completing the proof of (2).

If $\mathrm{char}\, F \neq 2$, the involutions σ and σ' have the same type if and only if $\mathrm{Sym}(A, \sigma)$ and $\mathrm{Sym}(A, \sigma')$ have the same dimension. Part (2) shows that this condition holds if and only if $\sigma(u) = u$. If $\mathrm{char}\, F = 2$, the involution σ' is symplectic if and only if $\mathrm{Trd}_A(s') = 0$ for all $s' \in \mathrm{Sym}(A, \sigma')$. In view of (2), this condition may be rephrased as

$$\mathrm{Trd}_A(us) = 0 \quad \text{for } s \in \mathrm{Sym}(A, \sigma).$$

Lemma (2.3) shows that this condition holds if and only if $u \in \mathrm{Alt}(A, \sigma)$. $\square$

(2.8) Corollary. *Let A be a central simple F-algebra with an involution σ of the first kind.*

(1) *If $\deg A$ is odd, then A is split and σ is necessarily of orthogonal type. Moreover, the space $\mathrm{Alt}(A, \sigma)$ contains no invertible elements.*

(2) *If $\deg A$ is even, then the index of A is a power of 2 and A has involutions of both types. Whatever the type of σ, the space $\mathrm{Alt}(A, \sigma)$ contains invertible elements and the space $\mathrm{Sym}(A, \sigma)$ contains invertible elements which are not in $\mathrm{Alt}(A, \sigma)$.*

Proof: Define a homomorphism of F-algebras

$$\sigma_* \colon A \otimes_F A \to \mathrm{End}_F(A)$$

by $\sigma_*(a \otimes b)(x) = ax\sigma(b)$ for a, b, $x \in A$. This homomorphism is injective since $A \otimes_F A$ is simple and surjective by dimension count, hence it is an isomorphism. Therefore, $A \otimes_F A$ splits[2], and the exponent of A is 1 or 2. Since the index and the exponent of a central simple algebra have the same prime factors (see Draxl [84, Theorem 11, p. 66]), it follows that the index of A, $\mathrm{ind}\, A$, is a power of 2. In particular, if $\deg A$ is odd, then A is split. In this case, Proposition (2.6) shows that every involution of the first kind has orthogonal type. If $\mathrm{Alt}(A, \sigma)$ contains an invertible element u, then $\mathrm{Int}(u) \circ \sigma$ has symplectic type, by (2.7); this is impossible.

Suppose henceforth that the degree of A is even. If A is split, then it has involutions of both types, since a vector space of even dimension carries nonsingular alternating bilinear forms as well as nonsingular symmetric, nonalternating bilinear forms. Let σ be an involution of the first kind on A. In order to show that $\mathrm{Alt}(A, \sigma)$ contains invertible elements, we consider separately the case where $\mathrm{char}\, F = 2$. If $\mathrm{char}\, F \neq 2$, consider an involution σ' whose type is different from the type of σ. Proposition (2.7) yields an invertible element $u \in A$ such that $\sigma' = \mathrm{Int}(u) \circ \sigma$ and $\sigma(u) = -u$. Note also that 1 is an invertible element which is symmetric but not alternating. If $\mathrm{char}\, F = 2$, consider a symplectic involution σ' and an orthogonal involution σ''. Again, (2.7) yields invertible elements u, $v \in A^\times$ such that $\sigma' = \mathrm{Int}(u) \circ \sigma$ and $\sigma'' = \mathrm{Int}(v) \circ \sigma$, and shows that $u \in \mathrm{Alt}(A, \sigma)$ and $v \in \mathrm{Sym}(A, \sigma) \smallsetminus \mathrm{Alt}(A, \sigma)$.

---

[2]Alternately, the involution σ yields an isomorphism $A \simeq A^{\mathrm{op}}$ by mapping $a \in A$ to $\left(\sigma(a)\right)^{\mathrm{op}} \in A^{\mathrm{op}}$, hence the Brauer class $[A]$ of A satisfies $[A] = [A]^{-1}$.

Assume next that A is not split. The base field F is then infinite, since the Brauer group of a finite field is trivial (see for instance Scharlau [247, Corollary 8.6.3]). Since invertible elements s are characterized by $\mathrm{Nrd}_A(s) \neq 0$ where Nrd_A is the reduced norm in A, the set of invertible alternating elements is a Zariski-open subset of $\mathrm{Alt}(A, \sigma)$. Our discussion above of the split case shows that this open subset is nonempty over an algebraic closure. Since F is infinite, rational points are dense, hence this open set has a rational point. Similarly, the set of invertible symmetric elements is a dense Zariski-open subset in $\mathrm{Sym}(A, \sigma)$, hence it is not contained in the closed subset $\mathrm{Sym}(A, \sigma) \cap \mathrm{Alt}(A, \sigma)$. Therefore, there exist invertible symmetric elements which are not alternating.

If $u \in \mathrm{Alt}(A, \sigma)$ is invertible, then $\mathrm{Int}(u) \circ \sigma$ is an involution of the type opposite to σ if char $F \neq 2$, and is a symplectic involution if char $F = 2$. If char $F = 2$ and $v \in \mathrm{Sym}(A, \sigma)$ is invertible but not alternating, then $\mathrm{Int}(v) \circ \sigma$ is an orthogonal involution.   □

The existence of involutions of both types on central simple algebras of even degree with involution can also be derived from the proof of (3.1) below.

The following proposition highlights a special feature of symplectic involutions:

(2.9) Proposition. *Let A be a central simple F-algebra with involution σ of symplectic type. The reduced characteristic polynomial of every element in $\mathrm{Symd}(A, \sigma)$ is a square. In particular, $\mathrm{Nrd}_A(s)$ is a square in F for all $s \in \mathrm{Symd}(A, \sigma)$.*

Proof: Let K be a Galois extension of F which splits A and let $s \in \mathrm{Symd}(A, \sigma)$. It suffices to show that the reduced characteristic polynomial $\mathrm{Prd}_{A,s}(X) \in F[X]$ is a square in $K[X]$, for then its monic square root is invariant under the action of the Galois group $\mathrm{Gal}(K/F)$, hence it is in $F[X]$. Extending scalars from F to K, we reduce to the case where A is split. We may thus assume that $A = M_n(F)$. Proposition (2.7) then yields $\sigma = \mathrm{Int}(u) \circ t$ for some invertible alternating matrix $u \in A^\times$, hence $\mathrm{Symd}(A, \sigma) = u \cdot \mathrm{Alt}\big(M_n(F), t\big)$. Therefore, there exists a matrix $a \in \mathrm{Alt}\big(M_n(F), t\big)$ such that $s = ua$. The (reduced) characteristic polynomial of s is then

$$\mathrm{Prd}_{A,s}(X) = \det(X \cdot 1 - s) = (\det u)\big(\det(X \cdot u^{-1} - a)\big).$$

Since u and $X \cdot u^{-1} - a$ are alternating, their determinants are the squares of their pfaffian $\mathrm{pf}\, u$, $\mathrm{pf}(X \cdot u^{-1} - a)$ (see for instance E. Artin [24, Theorem 3.27]), hence

$$\mathrm{Prd}_{A,s}(X) = [(\mathrm{pf}\, u)\, \mathrm{pf}(X \cdot u^{-1} - a)]^2.   □$$

Let $\deg A = n = 2m$. In view of the preceding proposition, for every $s \in \mathrm{Symd}(A, \sigma)$ there is a unique monic polynomial, the *pfaffian characteristic polynomial*, $\mathrm{Prp}_{\sigma,s}(X) \in F[X]$ of degree m such that

$$\mathrm{Prd}_{A,s}(X) = \mathrm{Prp}_{\sigma,s}(X)^2.$$

For $s \in \mathrm{Symd}(A, \sigma)$, we define the *pfaffian trace* $\mathrm{Trp}_\sigma(s)$ and the *pfaffian norm* $\mathrm{Nrp}_\sigma(s) \in F$ as coefficients of $\mathrm{Prp}_{\sigma,s}(X)$:

(2.10) $$\mathrm{Prp}_{\sigma,s}(X) = X^m - \mathrm{Trp}_\sigma(s)X^{m-1} + \cdots + (-1)^m \mathrm{Nrp}_\sigma(s).$$

Thus, Trp_σ and Nrp_σ are polynomial maps of degree 1 and m respectively on $\mathrm{Symd}(A, \sigma)$, and we have

(2.11) $$\mathrm{Trd}_A(s) = 2\,\mathrm{Trp}_\sigma(s) \qquad \text{and} \qquad \mathrm{Nrd}_A(s) = \mathrm{Nrp}_\sigma(s)^2$$

for all $s \in \operatorname{Symd}(A, \sigma)$. Moreover, we have $\operatorname{Prd}_{A,1}(X) = (X - 1)^{2m}$, hence $\operatorname{Prp}_{\sigma,1}(X) = (X - 1)^m$ and therefore

$$(2.12) \qquad\qquad \operatorname{Trp}_\sigma(1) = m \qquad \text{and} \qquad \operatorname{Nrp}_\sigma(1) = 1.$$

Since polynomial maps on $\operatorname{Symd}(A, \sigma)$ form a domain, the map $\operatorname{Nrp}_\sigma$ is uniquely determined by (2.11) and (2.12). (Of course, if $\operatorname{char} F \neq 2$, the map $\operatorname{Trp}_\sigma$ is also uniquely determined by (2.11); in all characteristics it is uniquely determined by the property in (2.13) below.) Note that the pfaffian norm $\operatorname{Nrp}_\sigma$ (or simply *pfaffian*) is an analogue of the classical pfaffian. However it is defined on the space $\operatorname{Symd}(A, \sigma)$ whereas pf is defined on alternating matrices (under the transpose involution). Nevertheless, it shares with the pfaffian the fundamental property demonstrated in the following proposition:

(2.13) Proposition. *For all $s \in \operatorname{Symd}(A, \sigma)$ and all $a \in A$,*

$$\operatorname{Trp}_\sigma\big(\sigma(a) + a\big) = \operatorname{Trd}_A(a) \qquad \text{and} \qquad \operatorname{Nrp}_\sigma\big(\sigma(a)sa\big) = \operatorname{Nrd}_A(a)\operatorname{Nrp}_\sigma(s).$$

Proof: We first prove the second equation. If s is not invertible, then $\operatorname{Nrd}_A(s) = \operatorname{Nrd}_A\big(\sigma(a)sa\big) = 0$, hence $\operatorname{Nrp}_\sigma(s) = \operatorname{Nrp}_\sigma\big(\sigma(a)sa\big) = 0$, proving the equation in this particular case. For $s \in \operatorname{Symd}(A, \sigma) \cap A^\times$ fixed, consider both sides of the equality to be proved as polynomial maps on A:

$$f_1 \colon a \mapsto \operatorname{Nrp}_\sigma\big(\sigma(a)sa\big) \quad \text{and} \quad f_2 \colon a \mapsto \operatorname{Nrd}_A(a)\operatorname{Nrp}_\sigma(s).$$

Since $\operatorname{Nrd}_A\big(\sigma(a)sa\big) = \operatorname{Nrd}_A(a)^2 \operatorname{Nrd}_A(s)$, we have $f_1^2 = f_2^2$, hence $(f_1 + f_2)(f_1 - f_2) = 0$. Since polynomial maps on A form a domain, it follows that $f_1 = \pm f_2$. Taking into account the fact that $f_1(1) = \operatorname{Nrp}_\sigma(s) = f_2(1)$, we get $f_1 = f_2$.

The first equation follows from the second. For, let t be an indeterminate over F and consider the element $1 + ta \in A_{F(t)}$. By the equation just proven, we have

$$\operatorname{Nrp}_\sigma\big((1 - t\sigma(a))(1 - ta)\big) = \operatorname{Nrd}_A(1 - ta) = 1 - \operatorname{Trd}_A(a)t + \operatorname{Srd}_A(a)t^2 - \ldots$$

On the other hand, for all $s \in \operatorname{Symd}(A, \sigma)$ we have

$$\operatorname{Nrp}_\sigma(1 - s) = \operatorname{Prp}_{\sigma,s}(1) = 1 - \operatorname{Trp}_\sigma(s) + \cdots + (-1)^m \operatorname{Nrp}_\sigma(s),$$

hence the coefficient of $-t$ in $\operatorname{Nrp}_\sigma\big((1 - t\sigma(a))(1 - ta)\big)$ is $\operatorname{Trp}_\sigma\big(\sigma(a) + a\big)$; the first equality is thus proved. $\qquad\square$

2.B. Involutions of the second kind. In the case of involutions of the second kind on a simple algebra B, the base field F is usually not the center of the algebra, but the subfield of central invariant elements which is of codimension 2 in the center. Under scalar extension to an algebraic closure of F, the algebra B decomposes into a direct product of two simple factors. It is therefore convenient to extend our discussion of involutions of the second kind to semisimple F-algebras of the form $B_1 \times B_2$, where B_1, B_2 are central simple F-algebras.

Throughout this section, we thus denote by B a finite dimensional F-algebra whose center K is a quadratic étale[3] extension of F, and assume that B is either simple (if K is a field) or a direct product of two simple algebras (if $K \simeq F \times F$). We denote by ι the nontrivial automorphism of K/F and by τ an involution of the second kind on B, whose restriction to K is ι. For convenience, we refer to (B, τ) as a central simple F-algebra with involution of the second kind, even though

---

[3]This simply means that K is either a field which is a separable quadratic extension of F, or $K \simeq F \times F$. See §18 for more on étale algebras.

its center is *not* F and the algebra B may not be simple.[4] A homomorphism $f\colon (B,\tau) \to (B',\tau')$ is then an F-algebra homomorphism $f\colon B \to B'$ such that $\tau' \circ f = f \circ \tau$.

If L is any field containing F, the L-algebra $B_L = B \otimes_F L$ has center $K_L = K \otimes_F L$, a quadratic étale extension of L, and carries an involution of the second kind $\tau_L = \tau \otimes \mathrm{Id}_L$. Moreover, (B_L, τ_L) is a central simple L-algebra with involution of the second kind.

As a parallel to the terminology of types used for involutions of the first kind, and because of their relation with unitary groups (see Chapter III), involutions of the second kind are also called of *unitary type* (or simply *unitary*).

We first examine the case where the center K is not a field.

(2.14) Proposition. *If $K \simeq F \times F$, there is a central simple F-algebra E such that*

$$(B,\tau) \simeq (E \times E^{\mathrm{op}}, \varepsilon),$$

where the involution ε is defined by $\varepsilon(x, y^{\mathrm{op}}) = (y, x^{\mathrm{op}})$. This involution is called the exchange *involution.*

Proof: Let $B = B_1 \times B_2$ where B_1, B_2 are central simple F-algebras. Since the restriction of τ to the center $K = F \times F$ interchanges the two factors, it maps $(1,0)$ to $(0,1)$, hence

$$\tau\big(B_1 \times \{0\}\big) = \tau\big((B_1 \times B_2) \cdot (1,0)\big) = (0,1) \cdot (B_1 \times B_2) = \{0\} \times B_2.$$

It follows that B_1 and B_2 are anti-isomorphic. We may then define an F-algebra isomorphism $f\colon B_1^{\mathrm{op}} \xrightarrow{\sim} B_2$ by the relation

$$\tau(x,0) = \big(0, f(x^{\mathrm{op}})\big),$$

and identify $B_1 \times B_2$ with $B_1 \times B_1^{\mathrm{op}}$ by mapping (x_1, x_2) to $\big(x_1, f^{-1}(x_2)\big)$. Under this map, τ is identified with the involution ε. $\square$

In view of this proposition, we may define the *degree* of the central simple F-algebra (B,τ) with involution of the second kind by

$$\deg(B,\tau) = \begin{cases} \deg B & \text{if } K \text{ is a field,} \\ \deg E & \text{if } K \simeq F \times F \text{ and } (B,\tau) \simeq (E \times E^{\mathrm{op}}, \varepsilon). \end{cases}$$

Equivalently, $\deg(B,\tau)$ is defined by the relation $\dim_F B = 2\big(\deg(B,\tau)\big)^2$.

If the center K of B is a field, (2.14) applies to $B_K = B \otimes_F K$, since its center is $K_K = K \otimes_F K \simeq K \times K$. In this case we get a canonical isomorphism:

(2.15) Proposition. *Suppose that the center K of B is a field. There is a canonical isomorphism of K-algebras with involution*

$$\varphi\colon (B_K, \tau_K) \xrightarrow{\sim} (B \times B^{\mathrm{op}}, \varepsilon)$$

which maps $b \otimes \alpha$ to $\big(b\alpha, (\tau(b)\alpha)^{\mathrm{op}}\big)$ for $b \in B$ and $\alpha \in K$.

---

[4]We thus follow Jacobson's convention in [136, p. 208]; it can be justified by showing that (B,τ) is indeed central simple as an algebra-with-involution.

Proof: It is straightforward to check that φ is a homomorphism of central simple K-algebras with involution of the second kind. It thus suffices to prove that φ has an inverse. Let $\alpha \in K \smallsetminus F$. Then the map $\Psi \colon B \times B^{\mathrm{op}} \to B_K$ defined by

$$\Psi(x, y^{\mathrm{op}}) = \left(\frac{\tau(y)\alpha - x\iota(\alpha)}{\alpha - \iota(\alpha)}\right) \otimes 1 + \left(\frac{x - \tau(y)}{\alpha - \iota(\alpha)}\right) \otimes \alpha$$

is the inverse of φ. $\qquad\square$

In a semisimple algebra of the form $B_1 \times B_2$, where B_1, B_2 are central simple F-algebras of the same degree, the reduced characteristic polynomial of an element (b_1, b_2) may be defined as $\left(\mathrm{Prd}_{B_1, b_1}(X), \mathrm{Prd}_{B_2, b_2}(X)\right) \in (F \times F)[X]$, where $\mathrm{Prd}_{B_1, b_1}(X)$ and $\mathrm{Prd}_{B_2, b_2}(X)$ are the reduced characteristic polynomials of b_1 and b_2 respectively (see Reiner [229, p. 121]). Since the reduced characteristic polynomial of an element does not change under scalar extension (see Reiner [229, Theorem (9.27)]), the preceding proposition yields:

(2.16) Corollary. *For every* $b \in B$, *the reduced characteristic polynomials of* b *and* $\tau(b)$ *are related by*

$$\mathrm{Prd}_{B, \tau(b)} = \iota(\mathrm{Prd}_{B, b}) \quad \text{in } K[X].$$

In particular, $\mathrm{Trd}_B\big(\tau(b)\big) = \iota\big(\mathrm{Trd}_B(b)\big)$ *and* $\mathrm{Nrd}_B\big(\tau(b)\big) = \iota\big(\mathrm{Nrd}_B(b)\big)$.

Proof: If $K \simeq F \times F$, the result follows from (2.14); if K is a field, it follows from (2.15). $\qquad\square$

As for involutions of the first kind, we may define the sets of symmetric, skew-symmetric, symmetrized and alternating elements in (B, τ) by

$$\mathrm{Sym}(B, \tau) = \{\, b \in B \mid \tau(b) = b \,\},$$
$$\mathrm{Skew}(B, \tau) = \{\, b \in B \mid \tau(b) = -b \,\},$$
$$\mathrm{Symd}(B, \tau) = \{\, b + \tau(b) \mid b \in B \,\},$$
$$\mathrm{Alt}(B, \tau) = \{\, b - \tau(b) \mid b \in B \,\}.$$

These sets are vector spaces over F. In contrast with the case of involutions of the first kind, there is a straightforward relation between symmetric, skew-symmetric and alternating elements, as the following proposition shows:

(2.17) Proposition. $\mathrm{Symd}(B, \tau) = \mathrm{Sym}(B, \tau)$ *and* $\mathrm{Alt}(B, \tau) = \mathrm{Skew}(B, \tau)$. *For any* $\alpha \in K^\times$ *such that* $\tau(\alpha) = -\alpha$,

$$\mathrm{Skew}(B, \tau) = \alpha \cdot \mathrm{Sym}(B, \tau).$$

If $\deg(B, \tau) = n$, *then*

$$\dim_F \mathrm{Sym}(B, \tau) = \dim_F \mathrm{Skew}(B, \tau) = \dim_F \mathrm{Symd}(B, \tau) = \dim_F \mathrm{Alt}(B, \tau) = n^2.$$

Proof: The relations $\mathrm{Skew}(B, \tau) = \alpha \cdot \mathrm{Sym}(B, \tau)$ and $\mathrm{Symd}(B, \tau) \subset \mathrm{Sym}(B, \tau)$, $\mathrm{Alt}(B, \tau) \subset \mathrm{Skew}(B, \tau)$ are clear. If $\beta \in K$ is such that $\beta + \iota(\beta) = 1$, then every element $s \in \mathrm{Symd}(B, \tau)$ may be written as $s = \beta s + \tau(\beta s)$, hence $\mathrm{Sym}(B, \tau) = \mathrm{Symd}(B, \tau)$. Similarly, every element $s \in \mathrm{Skew}(B, \tau)$ may be written as $s = \beta s - \tau(\beta s)$, hence $\mathrm{Skew}(B, \tau) = \mathrm{Alt}(B, \tau)$. Therefore, the vector spaces $\mathrm{Sym}(B, \tau)$, $\mathrm{Skew}(B, \tau)$, $\mathrm{Symd}(B, \tau)$ and $\mathrm{Alt}(B, \tau)$ have the same dimension. This dimension is $\frac{1}{2}\dim_F B$, since $\mathrm{Alt}(B, \tau)$ is the image of the F-linear endomorphism $\mathrm{Id} - \tau$ of B, whose kernel is $\mathrm{Sym}(B, \tau)$. Since $\dim_F B = 2\dim_K B = 2n^2$, the proof is complete. $\qquad\square$

As for involutions of the first kind, all the involutions of the second kind on B which have the same restriction to K as τ are obtained by composing τ with an inner automorphism, as we now show.

(2.18) Proposition. *Let (B, τ) be a central simple F-algebra with involution of the second kind, and let K denote the center of B.*

(1) *For every unit $u \in B^\times$ such that $\tau(u) = \lambda u$ with $\lambda \in K^\times$, the map $\mathrm{Int}(u) \circ \tau$ is an involution of the second kind on B.*

(2) *Conversely, for every involution τ' on B whose restriction to K is ι, there exists some $u \in B^\times$, uniquely determined up to a factor in $F^\times$, such that*

$$\tau' = \mathrm{Int}(u) \circ \tau \quad and \quad \tau(u) = u.$$

In this case,

$$\mathrm{Sym}(B, \tau') = u \cdot \mathrm{Sym}(B, \tau) = \mathrm{Sym}(B, \tau) \cdot u^{-1}.$$

Proof: Computation shows that $\left(\mathrm{Int}(u) \circ \tau\right)^2 = \mathrm{Int}\left(u\tau(u)^{-1}\right)$, and (1) follows.

If τ' is an involution on B which has the same restriction to K as τ, the composition $\tau' \circ \tau$ is an automorphism which leaves K elementwise invariant. The Skolem-Noether theorem shows that $\tau' \circ \tau = \mathrm{Int}(u_0)$ for some $u_0 \in B^\times$, hence $\tau' = \mathrm{Int}(u_0) \circ \tau$. Since $\tau'^2 = \mathrm{Id}$, we have $u_0 \tau(u_0)^{-1} \in K^\times$. Let $\lambda \in K^\times$ be such that $\tau(u_0) = \lambda u_0$. Applying τ to both sides of this relation, we get $N_{K/F}(\lambda) = 1$. Hilbert's theorem 90 (see (29.3)) yields an element $\mu \in K^\times$ such that $\lambda = \mu\iota(\mu)^{-1}$. Explicitly one can take $\mu = \alpha + \lambda\iota(\alpha)$ for $\alpha \in K$ such that $\alpha + \lambda\iota(\alpha)$ is invertible. The element $u = \mu u_0$ then satisfies the required conditions. $\square$

2.C. Examples.

Endomorphism algebras. Let V be a finite dimensional vector space over a field F. The involutions of the first kind on $\mathrm{End}_F(V)$ have been determined in the introduction to this chapter: every such involution is the adjoint involution with respect to some nonsingular symmetric or skew-symmetric bilinear form on V, uniquely determined up to a scalar factor. Moreover, it is clear from Definition (2.5) that the involution is orthogonal (resp. symplectic) if the corresponding bilinear form is symmetric and nonalternating (resp. alternating).

Involutions of the second kind can be described similarly. Suppose that V is a finite dimensional vector space over a field K which is a separable quadratic extension of some subfield F with nontrivial automorphism ι. A *hermitian form* on V (with respect to ι) is a bi-additive map

$$h \colon V \times V \to K$$

such that

$$h(v\alpha, w\beta) = \iota(\alpha)h(v, w)\beta \quad \text{for } v, \, w \in V \text{ and } \alpha, \, \beta \in K$$

and

$$h(w, v) = \iota\left(h(v, w)\right) \quad \text{for } v, \, w \in V.$$

The form h is called *nonsingular* if the only element $x \in V$ such that $h(x, y) = 0$ for all $y \in V$ is $x = 0$. If this condition holds, an involution σ_h on $\mathrm{End}_K(V)$ may be defined by the following condition:

$$h\left(x, f(y)\right) = h\left(\sigma_h(f)(x), y\right) \quad \text{for } f \in \mathrm{End}_K(V), \, x, \, y \in V.$$

The involution σ_h on $\mathrm{End}_K(V)$ is the *adjoint involution* with respect to the hermitian form h. From the definition of σ_h, it follows that $\sigma_h(\alpha) = \iota(\alpha)$ for all $\alpha \in K$, hence σ_h is of the second kind. As for involutions of the first kind, one can prove that every involution τ of the second kind on $\mathrm{End}_K(V)$ whose restriction to K is ι is the adjoint involution with respect to some nonsingular hermitian form on V, uniquely determined up to a factor in $F^{\times}$.

We omit the details of the proof, since a more general statement will be proved in §4 below (see (4.2)).

Matrix algebras. The preceding discussion can of course be translated to matrix algebras, since the choice of a basis in an n-dimensional vector space V over F yields an isomorphism $\mathrm{End}_F(V) \simeq M_n(F)$. However, matrix algebras are endowed with a canonical involution of the first kind, namely the transpose involution t. Therefore, a complete description of involutions of the first kind on $M_n(F)$ can also be easily derived from (2.7).

(2.19) Proposition. *Every involution of the first kind σ on $M_n(F)$ is of the form*

$$\sigma = \mathrm{Int}(u) \circ t$$

for some $u \in \mathrm{GL}_n(F)$, uniquely determined up to a factor in $F^{\times}$, such that $u^t = \pm u$. Moreover, the involution σ is orthogonal if $u^t = u$ and $u \notin \mathrm{Alt}\big(M_n(F), t\big)$, and it is symplectic if $u \in \mathrm{Alt}\big(M_n(F), t\big)$.

If $M_n(F)$ is identified with $\mathrm{End}_F(F^n)$, the involution $\sigma = \mathrm{Int}(u) \circ t$ is the adjoint involution with respect to the nonsingular form b on F^n defined by

$$b(x, y) = x^t \cdot u^{-1} \cdot y \quad \text{for } x, y \in F^n.$$

Suppose now that A is an arbitrary central simple algebra over a field F and that $^-$ is an involution (of any kind) on A. We define an involution * on $M_n(A)$ by

$$(a_{ij})^*_{1 \le i, j \le n} = (\overline{a_{ij}})^t_{1 \le i, j \le n}.$$

(2.20) Proposition. *The involution * is of the same type as $^-$. Moreover, the involutions σ on $M_n(A)$ such that $\sigma(\alpha) = \overline{\alpha}$ for all $\alpha \in F$ can be described as follows:*

(1) *If $^-$ is of the first kind, then every involution of the first kind on $M_n(A)$ is of the form $\sigma = \mathrm{Int}(u) \circ ^*$ for some $u \in \mathrm{GL}_n(A)$, uniquely determined up to a factor in $F^{\times}$, such that $u^* = \pm u$. If $\mathrm{char}\, F \ne 2$, the involution $\mathrm{Int}(u) \circ ^*$ is of the same type as $^-$ if and only if $u^* = u$. If $\mathrm{char}\, F = 2$, the involution $\mathrm{Int}(u) \circ ^*$ is symplectic if and only if $u \in \mathrm{Alt}\big(M_n(A), ^*\big)$.*

(2) *If $^-$ is of the second kind, then every involution of the second kind σ on $M_n(A)$ such that $\sigma(\alpha) = \overline{\alpha}$ for all $\alpha \in F$ is of the form $\sigma = \mathrm{Int}(u) \circ ^*$ for some $u \in \mathrm{GL}_n(A)$, uniquely determined up to a factor in $F^{\times}$ invariant under $^-$, such that $u^* = u$.*

Proof: From the definition of * it follows that $\alpha^* = \overline{\alpha}$ for all $\alpha \in F$. Therefore, the involutions * and $^-$ are of the same kind.

Suppose that $^-$ is of the first kind. A matrix $(a_{ij})_{1 \le i, j \le n}$ is *-symmetric if and only if its diagonal entries are $^-$-symmetric and $a_{ji} = \overline{a_{ij}}$ for $i > j$, hence

$$\dim \mathrm{Sym}\big(M_n(A), ^*\big) = n \dim \mathrm{Sym}(A, ^-) + \tfrac{n(n-1)}{2} \dim A.$$

If $\deg A = d$ and $\dim \mathrm{Sym}(A, ^-) = d(d + \delta)/2$, where $\delta = \pm 1$, we thus get

$$\dim \mathrm{Sym}\big(M_n(A), ^*\big) = nd(nd + \delta)/2.$$

Therefore, if char $F \neq 2$ the type of * is the same as the type of $^-$.

Since $\mathrm{Trd}_{M_n(A)}\big((a_{ij})_{1 \leq i,j \leq n}\big) = \sum_{i=1}^n \mathrm{Trd}_A(a_{ii})$, we have

$$\mathrm{Trd}_{M_n(A)}\big(\mathrm{Sym}\big(M_n(A), ^*\big)\big) = \{0\} \quad \text{if and only if} \quad \mathrm{Trd}_A\big(\mathrm{Sym}(A, ^-)\big) = \{0\}.$$

Therefore, when char $F = 2$ the involution * is symplectic if and only if $^-$ is symplectic.

We have thus shown that in all cases the involutions * and $^-$ are of the same type (orthogonal, symplectic or unitary). The other assertions follow from (2.7) and (2.18). $\qquad\square$

In §4 below, it is shown how the various involutions on $M_n(A)$ are associated to hermitian forms on A^n under the identification $\mathrm{End}_A(A^n) = M_n(A)$.

Quaternion algebras. A *quaternion algebra* over a field F is a central simple F-algebra of degree 2. If the characteristic of F is different from 2, it can be shown (see Scharlau [247, §8.11]) that every quaternion algebra Q has a basis $(1, i, j, k)$ subject to the relations

$$i^2 \in F^\times, \quad j^2 \in F^\times, \quad ij = k = -ji.$$

Such a basis is called a *quaternion basis*; if $i^2 = a$ and $j^2 = b$, the quaternion algebra Q is denoted

$$Q = (a, b)_F.$$

Conversely, for any $a, b \in F^\times$ the 4-dimensional F-algebra Q with basis $(1, i, j, k)$ where multiplication is defined through the relations $i^2 = a$, $j^2 = b$, $ij = k = -ji$ is central simple and is therefore a quaternion algebra $(a, b)_F$.

If char $F = 2$, every quaternion algebra Q has a basis $(1, u, v, w)$ subject to the relations

$$u^2 + u \in F, \quad v^2 \in F^\times, \quad uv = w = vu + v$$

(see Draxl [84, §11]). Such a basis is called a *quaternion basis in characteristic* 2. If $u^2 + u = a$ and $v^2 = b$, the quaternion algebra Q is denoted

$$Q = [a, b)_F.$$

Conversely, for all $a \in F$, $b \in F^\times$, the relations $u^2 + u = a$, $v^2 = b$ and $vu = uv + v$ give the span of 1, u, v, uv the structure of a quaternion algebra.

Quaternion algebras in characteristic 2 may alternately be defined as algebras generated by two elements r, s subject to

$$r^2 \in F, \quad s^2 \in F, \quad rs + sr = 1.$$

Indeed, if $r^2 = 0$ the algebra thus defined is isomorphic to $M_2(F)$; if $r^2 \neq 0$ it has a quaternion basis $(1, sr, r, sr^2)$. Conversely, every quaternion algebra with quaternion basis $(1, u, v, w)$ as above is generated by $r = v$ and $s = uv^{-1}$ satisfying the required relations. The quaternion algebra Q generated by r, s subject to the relations $r^2 = a \in F$, $s^2 = b \in F$ and $rs + sr = 1$ is denoted

$$Q = ((a, b))_F.$$

Thus, $((a, b))_F \simeq M_2(F)$ if a (or b) $= 0$ and $((a, b))_F \simeq [ab, a)_F$ if $a \neq 0$. The quaternion algebra $((a, b))_F$ is thus the Clifford algebra of the quadratic form $[a, b]$.

For every quaternion algebra Q, an F-linear map $\gamma \colon Q \to Q$ can be defined by

$$\gamma(x) = \mathrm{Trd}_Q(x) - x \quad \text{for } x \in Q$$

where Trd_Q is the reduced trace in Q. Explicitly, for x_0, x_1, x_2, $x_3 \in F$,

$$\gamma(x_0 + x_1 i + x_2 j + x_3 k) = x_0 - x_1 i - x_2 j - x_3 k$$

if char $F \neq 2$ and

$$\gamma(x_0 + x_1 u + x_2 v + x_3 w) = x_0 + x_1(u+1) + x_2 v + x_3 w$$

if char $F = 2$. For the split quaternion algebra $Q = M_2(F)$ (in arbitrary characteristic),

$$\gamma \begin{pmatrix} x_{11} & x_{12} \\ x_{21} & x_{22} \end{pmatrix} = \begin{pmatrix} x_{22} & -x_{12} \\ -x_{21} & x_{11} \end{pmatrix}$$

Direct computations show that γ is an involution, called the *quaternion conjugation* or the *canonical involution*. If char $F \neq 2$, then $\mathrm{Sym}(Q, \gamma) = F$ and $\mathrm{Skew}(Q, \gamma)$ has dimension 3. If char $F = 2$, then $\mathrm{Sym}(Q, \gamma)$ is spanned by 1, v, w which have reduced trace equal to zero. Therefore, the involution γ is symplectic in every characteristic.

(2.21) Proposition. *The canonical involution γ on a quaternion algebra Q is the unique symplectic involution on Q. Every orthogonal involution σ on Q is of the form*

$$\sigma = \mathrm{Int}(u) \circ \gamma$$

where u is an invertible quaternion in $\mathrm{Skew}(Q, \gamma) \smallsetminus F$ which is uniquely determined by σ up to a factor in $F^\times$.

Proof: From (2.7), it follows that every involution of the first kind σ on Q has the form $\sigma = \mathrm{Int}(u) \circ \gamma$ where u is a unit such that $\gamma(u) = \pm u$. Suppose that σ is symplectic. If char $F = 2$, Proposition (2.7) shows that $u \in \mathrm{Alt}(Q, \gamma) = F$, hence $\sigma = \gamma$. Similarly, if char $F \neq 2$, Proposition (2.7) shows that $\gamma(u) = u$, hence $u \in F^\times$ and $\sigma = \gamma$. $\qquad\square$

Unitary involutions on quaternion algebras also have a very particular type, as we proceed to show.

(2.22) Proposition (Albert). *Let K/F be a separable quadratic field extension with nontrivial automorphism ι. Let τ be an involution of the second kind on a quaternion algebra Q over K, whose restriction to K is ι. There exists a unique quaternion F-subalgebra $Q_0 \subset Q$ such that*

$$Q = Q_0 \otimes_F K \quad and \quad \tau = \gamma_0 \otimes \iota$$

where γ_0 is the canonical involution on Q_0. Moreover, the algebra Q_0 is uniquely determined by these conditions.

Proof: Let γ be the canonical involution on Q. Then $\tau \circ \gamma \circ \tau$ is an involution of the first kind and symplectic type on Q, so $\tau \circ \gamma \circ \tau = \gamma$ by (2.21). From this last relation, it follows that $\tau \circ \gamma$ is a ι-semilinear automorphism of order 2 of Q. The F-subalgebra Q_0 of invariant elements then satisfies the required conditions. Since these conditions imply that every element in Q_0 is invariant under $\tau \circ \gamma$, the algebra Q_0 is uniquely determined by τ. (It is also the F-subalgebra of Q generated by τ-skew-symmetric elements of trace zero, see (2.27).) $\qquad\square$

The proof holds without change in the case where $K \simeq F \times F$, provided that quaternion algebras over K are defined as direct products of two quaternion F-algebras.

Symbol algebras. Let n be an arbitrary positive integer and let K be a field containing a primitive n^{th} root of unity ζ. For a, $b \in K^\times$, write $(a,b)_{\zeta,K}$ for the K-algebra generated by two elements x, y subject to the relations

$$x^n = a, \quad y^n = b, \quad yx = \zeta xy.$$

This algebra is central simple of degree n (see Draxl [84, §11]); it is called a *symbol algebra*.[5] Clearly, quaternion algebras are the symbol algebras of degree 2.

If K has an automorphism ι of order 2 which leaves a and b invariant and maps ζ to ζ^{-1}, then this automorphism extends to an involution τ on $(a,b)_{\zeta,K}$ which leaves x and y invariant.

Similarly, any automorphism ι' of order 2 of K which leaves a and ζ invariant and maps b to b^{-1} (if any) extends to an involution σ' on $(a,b)_{\zeta,K}$ which leaves x invariant and maps y to y^{-1}.

Tensor products.

(2.23) Proposition. (1) *Let* (A_1, σ_1), ..., (A_n, σ_n) *be central simple F-algebras with involution of the first kind. Then* $\sigma_1 \otimes \cdots \otimes \sigma_n$ *is an involution of the first kind on* $A_1 \otimes_F \cdots \otimes_F A_n$. *If* char $F \neq 2$, *this involution is symplectic if and only if an odd number of involutions among* σ_1, ..., σ_n *are symplectic. If* char $F = 2$, *it is symplectic if and only if at least one of* σ_1, ..., σ_n *is symplectic.*

(2) *Let K/F be a separable quadratic field extension with nontrivial automorphism ι and let* (B_1, τ_1), ..., (B_n, τ_n) *be central simple F-algebras with involution of the second kind with center K. Then* $\tau_1 \otimes \cdots \otimes \tau_n$ *is an involution of the second kind on* $B_1 \otimes_K \cdots \otimes_K B_n$.

(3) *Let K/F be a separable quadratic field extension with nontrivial automorphism ι and let* (A, σ) *be a central simple F-algebra with involution of the first kind. Then* $(A \otimes_F K, \sigma \otimes \iota)$ *is a central simple F-algebra with involution of the second kind.*

The proof, by induction on n for the first two parts, is straightforward. In case (1), we denote

$$(A_1, \sigma_1) \otimes_F \cdots \otimes_F (A_n, \sigma_n) = (A_1 \otimes_F \cdots \otimes_F A_n, \sigma_1 \otimes \cdots \otimes \sigma_n),$$

and use similar notations in the other two cases.

Tensor products of quaternion algebras thus yield examples of central simple algebras with involution. Merkurjev's theorem [189] shows that every central simple algebra with involution is Brauer-equivalent to a tensor product of quaternion algebras. However, there are examples of division algebras with involution of degree 8 which do not decompose into tensor products of quaternion algebras, and there are examples of involutions σ on tensor products of two quaternion algebras which are not of the form $\sigma_1 \otimes \sigma_2$ (see Amitsur-Rowen-Tignol [20]). A necessary and sufficient decomposability condition for an involution on a tensor product of two quaternion algebras has been given by Knus-Parimala-Sridharan [162]; see (7.3) and (15.12).

2.D. Lie and Jordan structures. Every associative algebra A over an arbitrary field F is endowed with a Lie algebra structure for the bracket $[x, y] = xy - yx$. We denote this Lie algebra by $\mathfrak{L}(A)$. Similarly, if char $F \neq 2$, a Jordan product can be defined on A by $x \cdot y = \frac{1}{2}(xy + yx)$. If A is viewed as a Jordan algebra for the product $\cdot$, we denote it by A^+.

---

[5]Draxl calls it a *power norm residue algebra*.

The relevance of the Lie and Jordan structures for algebras with involution stems from the observation that for every algebra with involution (A, σ) (of any kind), the spaces $\mathrm{Skew}(A, \sigma)$ and $\mathrm{Alt}(A, \sigma)$ are Lie subalgebras of $\mathfrak{L}(A)$, and the space $\mathrm{Sym}(A, \sigma)$ is a Jordan subalgebra of A^+ if char $F \neq 2$. Indeed, for x, $y \in \mathrm{Skew}(A, \sigma)$ we have

$$[x, y] = xy - \sigma(xy) \in \mathrm{Alt}(A, \sigma) \subset \mathrm{Skew}(A, \sigma)$$

hence $\mathrm{Alt}(A, \sigma)$ and $\mathrm{Skew}(A, \sigma)$ are Lie subalgebras of $\mathfrak{L}(A)$. On the other hand, for x, $y \in \mathrm{Sym}(A, \sigma)$,

$$x \cdot y = \tfrac{1}{2}\big(xy + \sigma(xy)\big) \in \mathrm{Sym}(A, \sigma),$$

hence $\mathrm{Sym}(A, \sigma)$ is a Jordan subalgebra of A^+. This Jordan subalgebra is usually denoted by $\mathcal{H}(A, \sigma)$.

The algebra $\mathrm{Skew}(A, \sigma)$ is contained in the Lie algebra

$$\mathfrak{g}(A, \sigma) = \{\, a \in A \mid a + \sigma(a) \in F \,\};$$

indeed, $\mathrm{Skew}(A, \sigma)$ is the kernel of the Lie algebra homomorphism[6]

$$\dot{\mu} \colon \mathfrak{g}(A, \sigma) \to F$$

defined by $\dot{\mu}(a) = a + \sigma(a)$, for $a \in \mathfrak{g}(A, \sigma)$. The map $\dot{\mu}$ is surjective, except when char $F = 2$ and σ is orthogonal, since the condition $1 \in \mathrm{Symd}(A, \sigma)$ characterizes symplectic involutions among involutions of the first kind in characteristic 2, and $\mathrm{Symd}(A, \sigma) = \mathrm{Sym}(A, \sigma)$ if σ is of the second kind. Thus, $\mathfrak{g}(A, \sigma) = \mathrm{Skew}(A, \sigma)$ if σ is orthogonal and char $F = 2$, and $\dim \mathfrak{g}(A, \sigma) = \dim \mathrm{Skew}(A, \sigma) + 1$ in the other cases.

Another important subalgebra of $\mathfrak{L}(A)$ is the kernel A^0 of the reduced trace map:

$$A^0 = \{\, a \in A \mid \mathrm{Trd}_A(a) = 0 \,\}.$$

If σ is symplectic (in arbitrary characteristic) or if it is orthogonal in characteristic different from 2, we have $\mathrm{Skew}(A, \sigma) \subset A^0$; in the other cases, we also consider the intersection

$$\mathrm{Skew}(A, \sigma)^0 = \mathrm{Skew}(A, \sigma) \cap A^0.$$

(2.24) Example. Let E be an arbitrary central simple F-algebra and let ε be the exchange involution on $E \times E^{\mathrm{op}}$. There are canonical isomorphisms of Lie and Jordan algebras

$$\mathfrak{L}(E) \xrightarrow{\sim} \mathrm{Skew}(E \times E^{\mathrm{op}}, \varepsilon), \quad E^+ \xrightarrow{\sim} \mathcal{H}(E \times E^{\mathrm{op}}, \varepsilon)$$

which map $x \in E$ respectively to $(x, -x^{\mathrm{op}})$ and to (x, x^{op}). Indeed, these maps are obviously injective homomorphisms, and they are surjective by dimension count (see (2.17)). We also have

$$\mathfrak{g}(A, \sigma) = \{\, (x, \alpha - x^{\mathrm{op}}) \mid x \in E, \; \alpha \in F \,\} \simeq \mathfrak{L}(E \times F).$$

Jordan algebras of symmetric elements in central simple algebras with involution are investigated in Chapter IX in relation with twisted compositions and the Tits constructions. Similarly, the Lie algebras of skew-symmetric and alternating elements play a crucial rôle in the study of algebraic groups associated to algebras

---

[6]The notation $\dot{\mu}$ is motivated by the observation that this map is the differential of the multiplier map $\mu \colon \mathrm{Sim}(A, \sigma) \to F^\times$ defined in (12.14).

with involution in Chapter VI. In this section, we content ourselves with a few basic observations which will be used in the proofs of some specific results in Chapters III and IV.

It is clear that every isomorphism of algebras with involution $f\colon (A,\sigma) \xrightarrow{\sim} (A',\sigma')$ carries symmetric, skew-symmetric and alternating elements in A to elements of the same type in A' and therefore induces Lie isomorphisms $\mathrm{Skew}(A,\sigma) \xrightarrow{\sim} \mathrm{Skew}(A',\sigma')$, $\mathrm{Alt}(A,\sigma) \xrightarrow{\sim} \mathrm{Alt}(A',\sigma')$ and a Jordan isomorphism $\mathcal{H}(A,\sigma) \xrightarrow{\sim} \mathcal{H}(A',\sigma')$. Conversely, if the degrees of A and A' are large enough, every isomorphism of Lie or Jordan algebras as above is induced by an automorphism of algebras with involution: see Jacobson [140, Chapter X, §4] and Jacobson [136, Theorem 11, p. 210]. However, this property does not hold for algebras of low degrees; the exceptional isomorphisms investigated in Chapter IV indeed yield examples of nonisomorphic algebras with involution which have isomorphic Lie algebras of skew-symmetric elements. Other examples arise from triality, see (42.6).

The main result of this subsection is the following extension property, which is much weaker than those referred to above, but holds under weaker degree restrictions:

(2.25) Proposition. (1) *Let (A,σ) and (A',σ') be central simple F-algebras with involution of the first kind and let L/F be a field extension. Suppose that $\deg A > 2$ and let*

$$f\colon \mathrm{Alt}(A,\sigma) \xrightarrow{\sim} \mathrm{Alt}(A',\sigma')$$

be a Lie isomorphism which has the following property: there is an isomorphism of L-algebras with involution $(A_L,\sigma_L) \xrightarrow{\sim} (A_L',\sigma_L')$ whose restriction to $\mathrm{Alt}(A,\sigma)$ is f. Then f extends uniquely to an isomorphism of F-algebras with involution $(A,\sigma) \xrightarrow{\sim} (A',\sigma')$.

(2) *Let (B,τ) and (B',τ') be central simple F-algebras with involution of the second kind and let L/F be a field extension. Suppose that $\deg(B,\tau) > 2$ and let*

$$f\colon \mathrm{Skew}(B,\tau)^0 \xrightarrow{\sim} \mathrm{Skew}(B',\tau')^0$$

be a Lie isomorphism which has the following property: there is an isomorphism of L-algebras with involution $(B_L,\tau_L) \xrightarrow{\sim} (B_L',\tau_L')$ whose restriction to $\mathrm{Skew}(B,\tau)^0$ is f. Then f extends uniquely to an isomorphism of F-algebras with involution $(B,\tau) \xrightarrow{\sim} (B',\tau')$.

The proof relies on the following crucial lemma:

(2.26) Lemma. (1) *Let (A,σ) be a central simple F-algebra with involution of the first kind. The set $\mathrm{Alt}(A,\sigma)$ generates A as an associative algebra if $\deg A > 2$.*

(2) *Let (B,τ) be a central simple F-algebra with involution of the second kind. The set $\mathrm{Skew}(B,\tau)^0$ generates B as an associative F-algebra if $\deg(B,\tau) > 2$.*

Proof: (1) Let $S \subset A$ be the associative subalgebra of A generated by $\mathrm{Alt}(A,\sigma)$. For every field extension L/F, the subalgebra of A_L generated by $\mathrm{Alt}(A_L,\sigma_L) = \mathrm{Alt}(A,\sigma) \otimes_F L$ is then S_L; therefore, it suffices to prove $S_L = A_L$ for some extension L/F.

Suppose that L is a splitting field of A. By (2.1), we have

$$(A_L,\sigma_L) \simeq \big(\mathrm{End}_L(V),\sigma_b\big)$$

for some vector space V over L and some nonsingular symmetric or skew-symmetric bilinear form b on V.

Suppose first that σ is symplectic, hence b is alternating. Identifying V with L^n by means of a symplectic basis, we get

$$(A_L, \sigma_L) = \big(M_n(L), \sigma_g\big)$$

where g is the $n \times n$ block-diagonal matrix

$$g = \operatorname{diag}\left(\left(\begin{smallmatrix} 0 & 1 \\ -1 & 0 \end{smallmatrix}\right), \ldots, \left(\begin{smallmatrix} 0 & 1 \\ -1 & 0 \end{smallmatrix}\right)\right)$$

and $\sigma_g(m) = g^{-1} \cdot m^t \cdot g$ for all $m \in M_n(L)$. The σ_g-alternating elements in $M_n(L)$ are of the form

$$g^{-1} \cdot x - \sigma_g(g^{-1} \cdot x) = g^{-1} \cdot (x + x^t),$$

where $x \in M_n(L)$. Let $(e_{ij})_{1 \le i,j \le n}$ be the standard basis of $M_n(L)$. For $i = 1$, $\ldots, n/2$ and $j \ne 2i-1, 2i$ we have

$$e_{2i-1,j} = g^{-1} \cdot (e_{2i-1,2i} + e_{2i,2i-1}) \cdot g^{-1} \cdot (e_{2i,j} + e_{j,2i})$$

and

$$e_{2i,j} = g^{-1} \cdot (e_{2i-1,2i} + e_{2i,2i-1}) \cdot g^{-1} \cdot (e_{2i-1,j} + e_{j,2i-1}),$$

hence $e_{2i-1,j}$ and $e_{2i,j}$ are in the subalgebra S_L of $M_n(L)$ generated by σ_g-alternating elements. Since $n \ge 4$ we may find for all $i = 1, \ldots, n/2$ some $j \ne 2i-1, 2i$; the elements $e_{2i-1,j}$, $e_{2i,j}$ and their transposes are then in S_L, hence also the products of these elements, among which one can find $e_{2i-1,2i-1}$, $e_{2i-1,2i}$, $e_{2i,2i-1}$ and $e_{2i,2i}$. Therefore, $S_L = M_n(L)$ and the proof is complete if σ is symplectic.

Suppose next that σ is orthogonal, hence that b is symmetric but not alternating. The vector space V then contains an orthogonal basis $(v_i)_{1 \le i \le n}$. (If char $F = 2$, this follows from a theorem of Albert, see Kaplansky [150, Theorem 20].) Extending L further, if necessary, we may assume that $b(v_i, v_i) = 1$ for all i, hence

$$(A_L, \sigma_L) \simeq \big(M_n(L), t\big).$$

If i, j, $k \in \{1, \ldots, n\}$ are pairwise distinct, then we have $e_{ij} - e_{ji}$, $e_{ik} - e_{ki} \in \operatorname{Alt}\big(M_n(L), t\big)$ and

$$e_{ii} = (e_{ij} - e_{ji})^2 \cdot (e_{ik} - e_{ki})^2, \quad e_{ij} = e_{ii} \cdot (e_{ij} - e_{ji}),$$

hence alternating elements generate $M_n(L)$.

(2) The same argument as in (1) shows that it suffices to prove the proposition over an arbitrary scalar extension. Extending scalars to the center of B if this center is a field, we are reduced to the case where $(B, \tau) \simeq (E \times E^{\mathrm{op}}, \varepsilon)$ for some central simple F-algebra E, by (2.15). Extending scalars further to a splitting field of E, we may assume that E is split. Therefore, it suffices to consider the case of $\big(M_n(F) \times M_n(F)^{\mathrm{op}}, \varepsilon\big)$. Again, let $(e_{ij})_{1 \le i,j \le n}$ be the standard basis of $M_n(F)$. For i, j, $k \in \{1, \ldots, n\}$ pairwise distinct we have in $M_n(F) \times M_n(F)^{\mathrm{op}}$

$$(e_{ij}, -e_{ij}^{\mathrm{op}}) \cdot (e_{jk}, -e_{jk}^{\mathrm{op}}) = (e_{ik}, 0) \quad \text{and} \quad (e_{jk}, -e_{jk}^{\mathrm{op}}) \cdot (e_{ij}, -e_{ij}^{\mathrm{op}}) = (0, e_{ik}^{\mathrm{op}}).$$

In each case, both factors on the left side are skew-symmetric of trace zero, hence $\operatorname{Skew}\big(M_n(F) \times M_n(F)^{\mathrm{op}}, \varepsilon\big)^0$ generates $M_n(F) \times M_n(F)^{\mathrm{op}}$ if $n \ge 3$. $\qquad\square$

(2.27) Remarks. (1) Suppose that A is a quaternion algebra over F and that σ is an involution of the first kind on A. If char $F = 2$ or if σ is orthogonal, the space $\operatorname{Alt}(A, \sigma)$ has dimension 1, hence it generates a commutative subalgebra of A. However, (2.26.1) also holds when $\deg A = 2$, provided char $F \ne 2$ and σ is symplectic, since then $A = F \oplus \operatorname{Alt}(A, \sigma)$.

(2) Suppose B is a quaternion algebra over a quadratic étale F-algebra K and that τ is an involution of the second kind on B leaving F elementwise invariant. Let ι be the nontrivial automorphism of K/F. Proposition (2.22) shows that there is a quaternion F-algebra Q in B such that

$$(B, \tau) = (Q, \gamma) \otimes_F (K, \iota),$$

where γ is the canonical involution on Q. It is easily verified that

$$\mathrm{Skew}(B, \tau)^0 = \mathrm{Skew}(Q, \gamma).$$

Therefore, the subalgebra of B generated by $\mathrm{Skew}(B, \tau)^0$ is Q and not B.

Proof of (2.25): Since the arguments are the same for both parts, we just give the proof of (1). Let $g \colon (A_L, \sigma_L) \xrightarrow{\sim} (A'_L, \sigma'_L)$ be an isomorphism of L-algebras with involution whose restriction to $\mathrm{Alt}(A, \sigma)$ is f. In particular, g maps $\mathrm{Alt}(A, \sigma)$ to $\mathrm{Alt}(A', \sigma')$. Since $\deg A_L = \deg A'_L$ and the degree does not change under scalar extension, A and A' have the same degree, which by hypothesis is at least 3. Lemma (2.26) shows that $\mathrm{Alt}(A, \sigma)$ generates A and $\mathrm{Alt}(A', \sigma')$ generates A', hence g maps A to A' and restricts to an isomorphism of F-algebras with involution $(A, \sigma) \xrightarrow{\sim} (A', \sigma')$. This isomorphism is uniquely determined by f since $\mathrm{Alt}(A, \sigma)$ generates A. $\square$

It is not difficult to give examples to show that (2.25) does not hold for algebras of degree 2. The easiest example is obtained from quaternion algebras Q, Q' of characteristic 2 with canonical involutions γ, γ'. Then $\mathrm{Alt}(Q, \gamma) = \mathfrak{L}(F) = \mathrm{Alt}(Q', \gamma')$ and the identity map $\mathrm{Alt}(Q, \gamma) \xrightarrow{\sim} \mathrm{Alt}(Q', \gamma')$ extends to an isomorphism $(Q_L, \gamma_L) \xrightarrow{\sim} (Q'_L, \gamma'_L)$ if L is an algebraic closure of F. However, Q and Q' may not be isomorphic.

(2.28) Remark. Inspection of the proof of (2.25) shows that the Lie algebra structures on $\mathrm{Alt}(A, \sigma)$ or $\mathrm{Skew}(B, \tau)^0$ are not explicitly used. Therefore, (2.25) also holds for *any* linear map f; indeed, if f extends to an isomorphism of (associative) L-algebras with involution, then it necessarily is an isomorphism of Lie algebras.

§3. Existence of Involutions

The aim of this section is to give a proof of the following Brauer-group characterization of central simple algebras with involution:

(3.1) Theorem. (1) (Albert) *Let A be a central simple algebra over a field F. There is an involution of the first kind on A if and only if $A \otimes_F A$ splits.*

(2) (Albert-Riehm-Scharlau) *Let K/F be a separable quadratic extension of fields and let B be a central simple algebra over K. There is an involution of the second kind on B which leaves F elementwise invariant if and only if the norm[7] $N_{K/F}(B)$ splits.*

In particular, if a central simple algebra has an involution, then every Brauer-equivalent algebra has an involution of the same kind.

We treat each part separately. We follow an approach based on ideas of T. Tamagawa (oral tradition—see Berele-Saltman [38, §2] and Jacobson [146, §5.2]), starting with the case of involutions of the first kind. For involutions of the second kind, our arguments are very close in spirit to those of Deligne and Sullivan [69, Appendix B].

---

[7]See (3.12) below for the definition of the norm (or corestriction) of a central simple algebra.

3.A. Existence of involutions of the first kind. The fact that $A \otimes_F A$ splits when A has an involution of the first kind is easy to see (and was already observed in the proof of Corollary (2.8)).

(3.2) Proposition. *Every F-linear anti-automorphism σ on a central simple algebra A endows A with a right $A \otimes_F A$-module structure defined by*

$$x *_\sigma (a \otimes b) = \sigma(a)xb \quad \text{for } a, b, x \in A.$$

The reduced dimension of A as a right $A \otimes_F A$-module is 1, hence $A \otimes_F A$ is split.

Proof: It is straightforward to check that the multiplication $*_\sigma$ defines a right $A \otimes_F A$-module structure on A. Since $\dim_F A = \deg(A \otimes_F A)$, we have $\mathrm{rdim}\, A = 1$, hence $A \otimes_F A$ is split, since the index of a central simple algebra divides the reduced dimension of every module of finite type. $\qquad\square$

(3.3) Remark. The isomorphism $\sigma_* \colon A \otimes_F A \xrightarrow{\sim} \mathrm{End}_F(A)$ defined in the proof of Corollary (2.8) endows A with a structure of *left $A \otimes_F A$-module*, which is less convenient in view of the discussion below (see (3.8)).

To prove the converse, we need a special element in $A \otimes_F A$, called the *Goldman element* (see Knus-Ojanguren [160, p. 112] or Rowen [237, p. 222]).

The Goldman element. For any central simple algebra A over a field F we consider the F-linear *sandwich map*

$$\mathrm{Sand} \colon A \otimes_F A \to \mathrm{End}_F(A)$$

defined by

$$\mathrm{Sand}(a \otimes b)(x) = axb \quad \text{for } a, b, x \in A.$$

(3.4) Lemma. *The map Sand is an isomorphism of F-vector spaces.*

Proof: Sand is the composite of the isomorphism $A \otimes_F A \simeq A \otimes_F A^{\mathrm{op}}$ which maps $a \otimes b$ to $a \otimes b^{\mathrm{op}}$ and of the canonical F-algebra isomorphism $A \otimes_F A^{\mathrm{op}} \simeq \mathrm{End}_F(A)$ of Wedderburn's theorem (1.1). $\qquad\square$

Consider the reduced trace $\mathrm{Trd}_A \colon A \to F$. Composing this map with the inclusion $F \hookrightarrow A$, we may view Trd_A as an element in $\mathrm{End}_F(A)$.

(3.5) Definition. The *Goldman element* in $A \otimes_F A$ is the unique element $g \in A \otimes_F A$ such that

$$\mathrm{Sand}(g) = \mathrm{Trd}_A.$$

(3.6) Proposition. *The Goldman element $g \in A \otimes_F A$ satisfies the following properties:*

(1) $g^2 = 1$.
(2) $g \cdot (a \otimes b) = (b \otimes a) \cdot g$ for all $a, b \in A$.
(3) If $A = \mathrm{End}_F(V)$, then with respect to the canonical identification $A \otimes_F A = \mathrm{End}_F(V \otimes_F V)$ the element g is defined by

$$g(v_1 \otimes v_2) = v_2 \otimes v_1 \quad \text{for } v_1, v_2 \in V.$$

Proof: We first check (3) by using the canonical isomorphism $\mathrm{End}_F(V) = V \otimes_F V^*$, where $V^* = \mathrm{Hom}_F(V, F)$ is the dual of V. If $(e_i)_{1 \le i \le n}$ is a basis of V and $(e_i^*)_{1 \le i \le n}$ is the dual basis, consider the element

$$g = \sum_{i,j} e_i \otimes e_j^* \otimes e_j \otimes e_i^* \in V \otimes V^* \otimes V \otimes V^* = \mathrm{End}_F(V) \otimes_F \mathrm{End}_F(V).$$

For all $f \in \mathrm{End}_F(V)$, we have

$$\mathrm{Sand}(g)(f) = \sum_{i,j}(e_i \otimes e_j^*) \circ f \circ (e_j \otimes e_i^*) = \sum_{i,j}(e_i \otimes e_j^*) \cdot e_j^*(f(e_j)).$$

Since $\sum_i e_i \otimes e_i^* = \mathrm{Id}_V$ and $\sum_j e_j^*(f(e_j)) = \mathrm{tr}(f)$, the preceding equation shows that

$$\mathrm{Sand}(g)(f) = \mathrm{tr}(f) \quad \text{for } f \in \mathrm{End}_F(V),$$

hence g is the Goldman element in $\mathrm{End}_F(V) \otimes \mathrm{End}_F(V)$. On the other hand, for $v_1, v_2 \in V$ we have

$$
\begin{aligned}
g(v_1 \otimes v_2) &= \sum_{i,j}(e_i \otimes e_j^*)(v_1) \otimes (e_j \otimes e_i^*)(v_2) \\
&= \left(\sum_i e_i \cdot e_i^*(v_2)\right) \otimes \left(\sum_j e_j \cdot e_j^*(v_1)\right) \\
&= v_2 \otimes v_1.
\end{aligned}
$$

This completes the proof of (3).

In view of (3), parts (1) and (2) are easy to check in the split case $A = \mathrm{End}_F(V)$, hence they hold in the general case also. Indeed, for any splitting field L of A the Goldman element g in $A \otimes_F A$ is also the Goldman element in $A_L \otimes_L A_L$ since the sandwich map and the reduced trace map commute with scalar extensions. Since A_L is split we have $g^2 = 1$ in $A_L \otimes_L A_L$, and $g \cdot (a \otimes b) = (b \otimes a) \cdot g$ for all $a, b \in A_L$, hence also for all $a, b \in A$. $\qquad\square$

Consider the left and right ideals in $A \otimes_F A$ generated by $1 - g$:

$$J_\ell = (A \otimes_F A) \cdot (1 - g), \quad J_r = (1 - g) \cdot (A \otimes_F A).$$

Let

$$\lambda^2 A = \mathrm{End}_{A \otimes A}(J_\ell), \quad s^2 A = \mathrm{End}_{A \otimes A}(J_r^0).$$

If $\deg A = 1$, then $A = F$ and $g = 1$, hence $J_\ell = J_r = \{0\}$ and $\lambda^2 A = \{0\}$, $s^2 A = F$. If $\deg A > 1$, Proposition (1.10) shows that the algebras $\lambda^2 A$ and $s^2 A$ are Brauer-equivalent to $A \otimes_F A$.

(3.7) Proposition. *If $\deg A = n > 1$,*

$$\mathrm{rdim}\, J_\ell = \mathrm{rdim}\, J_r = \deg \lambda^2 A = \tfrac{n(n-1)}{2} \quad and \quad \deg s^2 A = \tfrac{n(n+1)}{2}.$$

For any vector space V of dimension $n > 1$, there are canonical isomorphisms

$$\lambda^2 \mathrm{End}_F(V) = \mathrm{End}_F(\textstyle\bigwedge^2 V) \quad and \quad s^2 \mathrm{End}_F(V) = \mathrm{End}_F(S^2 V),$$

where $\bigwedge^2 V$ and $S^2 V$ are the exterior and symmetric squares of V, respectively.

Proof: Since the reduced dimension of a module and the degree of a central simple algebra are invariant under scalar extension, we may assume that A is split. Let $A = \mathrm{End}_F(V)$ and identify $A \otimes_F A$ with $\mathrm{End}_F(V \otimes V)$. Then

$$J_\ell = \mathrm{Hom}_F\big(V \otimes V / \ker(\mathrm{Id} - g), V \otimes V\big) \quad and \quad J_r = \mathrm{Hom}_F\big(V \otimes V, \mathrm{im}(\mathrm{Id} - g)\big),$$

and, by (1.12),

$$\lambda^2 A = \mathrm{End}_F\big(V \otimes V / \ker(\mathrm{Id} - g)\big) \quad and \quad s^2 A = \mathrm{End}_F\big(V \otimes V / \mathrm{im}(\mathrm{Id} - g)\big).$$

Since $g(v_1 \otimes v_2) = v_2 \otimes v_1$ for $v_1, v_2 \in V$, there are canonical isomorphisms

$$\textstyle\bigwedge^2 V \xrightarrow{\sim} V \otimes V / \ker(\mathrm{Id} - g) \quad and \quad S^2 V \xrightarrow{\sim} V \otimes V / \mathrm{im}(\mathrm{Id} - g)$$

which map $v_1 \wedge v_2$ to $v_1 \otimes v_2 + \ker(\mathrm{Id} - g)$ and $v_1 \cdot v_2$ to $v_1 \otimes v_2 + \mathrm{im}(\mathrm{Id} - g)$ respectively, for v_1, $v_2 \in V$. Therefore, $\lambda^2 A = \mathrm{End}_F(\bigwedge^2 V)$, $s^2 A = \mathrm{End}_F(S^2 V)$ and

$$\mathrm{rdim}\, J_r = \mathrm{rdim}\, J_\ell = \dim\big(V \otimes V/\ker(\mathrm{Id} - g)\big) = \dim {\textstyle\bigwedge^2} V. \qquad \square$$

Involutions of the first kind and one-sided ideals. For every F-linear anti-automorphism σ on a central simple algebra A, we define a map

$$\sigma' \colon A \otimes_F A \to A$$

by

$$\sigma'(a \otimes b) = \sigma(a)b \quad \text{for } a, b \in A.$$

This map is a homomorphism of right $A \otimes_F A$-modules, if A is endowed with the right $A \otimes_F A$-module structure of Proposition (3.2). The kernel $\ker \sigma'$ is therefore a right ideal in $A \otimes_F A$ which we write I_σ:

$$I_\sigma = \ker \sigma'.$$

Since σ' is surjective, we have

$$\dim_F I_\sigma = \dim_F(A \otimes_F A) - \dim_F A.$$

On the other hand, $\sigma'(1 \otimes a) = a$ for $a \in A$, hence $I_\sigma \cap (1 \otimes A) = \{0\}$. Therefore,

$$A \otimes_F A = I_\sigma \oplus (1 \otimes A).$$

As above, we denote by g the Goldman element in $A \otimes_F A$ and by J_ℓ and J_r the left and right ideals in $A \otimes_F A$ generated by $1 - g$.

(3.8) Theorem. *The map $\sigma \mapsto I_\sigma$ defines a one-to-one correspondence between the F-linear anti-automorphisms of A and the right ideals I of $A \otimes_F A$ such that $A \otimes_F A = I \oplus (1 \otimes A)$. Under this correspondence, involutions of symplectic type correspond to ideals containing J_ℓ^0 and involutions of orthogonal type to ideals containing J_r but not J_ℓ^0. In the split case $A = \mathrm{End}_F(V)$, the ideal corresponding to the adjoint anti-automorphism σ_b with respect to a nonsingular bilinear form b on V is $\mathrm{Hom}_F\big(V \otimes V, \ker(b \circ g)\big)$ where b is considered as a linear map $b \colon V \otimes V \to F$. (If b is symmetric or skew-symmetric, then $\ker(b \circ g) = \ker b$.)*

Proof: To every right ideal $I \subset A \otimes_F A$ such that $A \otimes_F A = I \oplus (1 \otimes A)$, we associate the map $\sigma_I \colon A \to A$ defined by projection of $A \otimes 1$ onto $1 \otimes A$ parallel to I; for $a \in A$, we define $\sigma_I(a)$ as the unique element in A such that

$$a \otimes 1 - 1 \otimes \sigma_I(a) \in I.$$

This map is clearly F-linear. Moreover, for a, $b \in A$ we have

$$ab \otimes 1 - 1 \otimes \sigma_I(b)\sigma_I(a) = \big(a \otimes 1 - 1 \otimes \sigma_I(a)\big) \cdot b \otimes 1$$
$$+ \big(b \otimes 1 - 1 \otimes \sigma_I(b)\big) \cdot 1 \otimes \sigma_I(a) \in I,$$

hence $\sigma_I(ab) = \sigma_I(b)\sigma_I(a)$, which proves that σ_I is an anti-automorphism.

For every anti-automorphism σ of A, the definition of I_σ shows that

$$a \otimes 1 - 1 \otimes \sigma(a) \in I_\sigma \quad \text{for } a \in A.$$

Therefore, $\sigma_{I_\sigma} = \sigma$. Conversely, suppose that $I \subset A \otimes_F A$ is a right ideal such that $A \otimes_F A = I \oplus (1 \otimes A)$. If $x = \sum y_i \otimes z_i \in \ker \sigma'_I$, then $\sum \sigma_I(y_i)z_i = 0$, hence

$$x = \sum\big(y_i \otimes z_i - 1 \otimes \sigma_I(y_i)z_i\big) = \sum\big(y_i \otimes 1 - 1 \otimes \sigma_I(y_i)\big) \cdot (1 \otimes z_i).$$

This shows that the right ideal $\ker \sigma'_I$ is generated by elements of the form

$$a \otimes 1 - 1 \otimes \sigma_I(a).$$

Since these elements all lie in I, by definition of σ_I, it follows that $\ker \sigma'_I \subset I$. However, these ideals have the same dimension, hence $\ker \sigma'_I = I$ and therefore $I_{\sigma_I} = I$.

We have thus shown that the maps $\sigma \mapsto I_\sigma$ and $I \mapsto \sigma_I$ define inverse bijections between anti-automorphisms of A and right ideals I in $A \otimes_F A$ such that $A \otimes_F A = I \oplus (1 \otimes A)$.

In order to identify the ideals which correspond to involutions, it suffices to consider the split case. Suppose that $A = \operatorname{End}_F(V)$ and that $\sigma = \sigma_b$ is the adjoint anti-automorphism with respect to some nonsingular bilinear form b on V. By definition of σ_b (see equation $(*)$ in the introduction to this chapter),

$$b \circ \big(\sigma(f) \otimes 1 - 1 \otimes f\big) = 0 \quad \text{for } f \in \operatorname{End}_F(V).$$

Since I_σ is generated as a right ideal by the elements $f \otimes 1 - 1 \otimes \sigma(f)$, and since $g \circ (f_1 \otimes f_2) = (f_2 \otimes f_1)$ for $f_1, f_2 \in \operatorname{End}_F(V)$, it follows that

$$I_\sigma \subset \{\, h \in \operatorname{End}_F(V \otimes V) \mid b \circ g \circ h = 0 \,\} = \operatorname{Hom}_F\big(V \otimes V, \ker(b \circ g)\big).$$

Dimension count shows that the inclusion is an equality.

As observed in the proof of Proposition (3.7), $J^0_\ell = \operatorname{Hom}_F\big(V \otimes V, \ker(\operatorname{Id} - g)\big)$, hence the inclusion $J^0_\ell \subset \operatorname{Hom}_F\big(V \otimes V, \ker(b \circ g)\big)$ holds if and only if $\ker(\operatorname{Id} - g) \subset \ker(b \circ g)$. Since $\ker(\operatorname{Id} - g)$ is generated by elements of the form $v \otimes v$, for $v \in V$, this condition holds if and only if b is alternating or, equivalently, σ is symplectic.

On the other hand, $J_r \subset \operatorname{Hom}_F\big(V \otimes V, \ker(b \circ g)\big)$ if and only if

$$b \circ g \circ (\operatorname{Id} - g)(v_1 \otimes v_2) = 0 \quad \text{for } v_1, v_2 \in V.$$

Since the left side is equal to $b(v_2, v_1) - b(v_1, v_2)$, this relation holds if and only if b is symmetric. Therefore, σ is orthogonal if and only if the corresponding ideal contains J_r but not J^0_ℓ. $\qquad\square$

(3.9) Remark. If $\operatorname{char} F \neq 2$, then $J^0_\ell = (1 + g) \cdot (A \otimes_F A)$. Indeed, $1 + g \in J^0_\ell$ since $(1 - g)(1 + g) = 1 - g^2 = 0$; on the other hand, if $x \in J^0_\ell$ then $(1 - g)x = 0$, hence $x = gx = (1 + g)x/2$. Therefore, an involution σ is orthogonal if and only if the corresponding ideal I_σ contains $1 - g$; it is symplectic if and only I_σ contains $1 + g$.

Let $\deg A = n$. The right ideals $I \subset A \otimes_F A$ such that $I \oplus (1 \otimes A) = A \otimes_F A$ then have reduced dimension $n^2 - 1$ and form an affine open subvariety

$$\mathcal{U} \subset \operatorname{SB}_{n^2-1}(A \otimes_F A).$$

(It is the affine open set denoted by $\mathcal{U}_{1 \otimes A}$ in the proof of Theorem (1.18).)

On the other hand, since $\operatorname{rdim} J_r = n(n-1)/2$ by (3.7) and $s^2 A = \operatorname{End}_{A \otimes A}(J^0_r)$ by definition, Proposition (1.20) shows that the right ideals of reduced dimension $n^2 - 1$ in $A \otimes_F A$ which contain J_r form a closed subvariety $S_o \subset \operatorname{SB}_{n^2-1}(A \otimes_F A)$ isomorphic to $\operatorname{SB}_m(s^2 A)$ where

$$m = (n^2 - 1) - \tfrac{1}{2}n(n-1) = \tfrac{1}{2}n(n+1) - 1 = \deg s^2 A - 1.$$

By (1.19), this variety is also isomorphic to $\operatorname{SB}\big((s^2 A)^{\operatorname{op}}\big)$.

Similarly, for $n > 1$ we write $S_s \subset \mathrm{SB}_{n^2-1}(A \otimes_F A)$ for the closed subvariety of right ideals of reduced dimension $n^2 - 1$ which contain J_ℓ^0. This variety is isomorphic to $\mathrm{SB}_{\frac{1}{2}n(n-1)-1}(\lambda^2 A)$ and to $\mathrm{SB}((\lambda^2 A)^{\mathrm{op}})$.

With this notation, Theorem (3.8) can be rephrased as follows:

(3.10) Corollary. *There are natural one-to-one correspondences between involutions of orthogonal type on A and the rational points on the variety $\mathcal{U} \cap S_o$, and, if $\deg A > 1$, between involutions of symplectic type on A and the rational points on the variety $\mathcal{U} \cap S_s$.*

Inspection of the split case shows that the open subvariety $\mathcal{U} \cap S_o \subset S_o$ is nonempty, and that $\mathcal{U} \cap S_s$ is nonempty if and only if $\deg A$ is even.

We may now complete the proof of part (1) of Theorem (3.1). We first observe that if F is finite, then A is split since the Brauer group of a finite field is trivial (see for instance Scharlau [247, Corollary 8.6.3]), hence A has involutions of the first kind. We may thus assume henceforth that the base field F is infinite.

Suppose that $A \otimes_F A$ is split. Then so are $s^2 A$ and $\lambda^2 A$ and the varieties $\mathrm{SB}((s^2 A)^{\mathrm{op}})$ and $\mathrm{SB}((\lambda^2 A)^{\mathrm{op}})$ (when $\deg A > 1$) are projective spaces. It follows that the rational points are dense in S_o and S_s. Therefore, $\mathcal{U} \cap S_o$ has rational points, so A has involutions of orthogonal type. If $\deg A$ is even, then $\mathcal{U} \cap S_s$ also has rational points[8], so A also has involutions of symplectic type. $\qquad\square$

(3.11) Remark. Severi-Brauer varieties and density arguments can be avoided in the proof above by reducing to the case of division algebras: if $A \otimes_F A$ is split, then $D \otimes_F D$ is also split, if D is the division algebra Brauer-equivalent to A. Let $I \subset D \otimes_F D$ be a maximal right ideal containing $1 - g$. Then $\dim I = d^2 - d$, where $d = \dim_F D$, and I intersects $1 \otimes D$ trivially, since it does not contain any invertible element. Therefore, dimension count shows that $D \otimes_F D = I \oplus (1 \otimes D)$. It then follows from (3.8) that D has an (orthogonal) involution of the first kind which we denote by $^-$. An involution * of the first kind is then defined on $M_r(D)$ by letting $^-$ act entrywise on $M_r(D)$ and setting

$$a^* = \bar{a}^t \quad \text{for } a \in M_r(D).$$

This involution is transported to A by the isomorphism $A \simeq M_r(D)$.

3.B. Existence of involutions of the second kind. Before discussing involutions of the second kind, we recall the construction of the norm of a central simple algebra in the particular case of interest in this section.

The norm (or corestriction) of central simple algebras. Let K/F be a finite separable field extension. For every central simple K-algebra A, there is a central simple F-algebra $N_{K/F}(A)$ of degree $(\deg A)^{[K:F]}$, called the *norm* of A, defined so as to induce a homomorphism of Brauer groups

$$N_{K/F} \colon \mathrm{Br}(K) \to \mathrm{Br}(F)$$

which corresponds to the corestriction map in Galois cohomology.

In view of Theorem (3.1), we shall only discuss here the case where K/F is a quadratic extension, referring to Draxl [84, §8] or Rowen [237, §7.2] for a more general treatment along similar lines.

---

[8]If $\deg A = 2$, the variety S_s has only one point, namely J_ℓ^0; this is a reflection of the fact that quaternion algebras have a unique symplectic involution, see (2.21).

The case of quadratic extensions is particularly simple in view of the fact that separable quadratic extensions are Galois. Let K/F be such an extension, and let

$$\mathrm{Gal}(K/F) = \{\mathrm{Id}_K, \iota\}$$

be its Galois group. For any K-algebra A, we define the conjugate algebra

$${}^\iota A = \{\, {}^\iota a \mid a \in A \,\}$$

with the following operations:

$${}^\iota a + {}^\iota b = {}^\iota(a+b) \quad {}^\iota a\, {}^\iota b = {}^\iota(ab) \quad {}^\iota(\alpha a) = \iota(\alpha){}^\iota a$$

for a, $b \in A$ and $\alpha \in K$. The switch map

$$s\colon {}^\iota A \otimes_K A \to {}^\iota A \otimes_K A$$

defined by

$$s({}^\iota a \otimes b) = {}^\iota b \otimes a$$

is ι-semilinear over K and is an F-algebra automorphism.

(3.12) Definition. The *norm* $N_{K/F}(A)$ of the K-algebra A is the F-subalgebra of ${}^\iota A \otimes_K A$ elementwise invariant under the switch map:

$$N_{K/F}(A) = \{\, u \in {}^\iota A \otimes_K A \mid s(u) = u \,\}.$$

Of course, the same construction can be used to define the norm $N_{K/F}(V)$ of any K-vector space V.

(3.13) Proposition. (1) *For any K-algebra A,*

$$N_{K/F}(A)_K = {}^\iota A \otimes_K A \quad and \quad N_{K/F}({}^\iota A) = N_{K/F}(A).$$

(2) *For any K-algebras A, B,*

$$N_{K/F}(A \otimes_K B) = N_{K/F}(A) \otimes_F N_{K/F}(B).$$

(3) *For any finite dimensional K-vector space V,*

$$N_{K/F}\big(\mathrm{End}_K(V)\big) = \mathrm{End}_F\big(N_{K/F}(V)\big).$$

(4) *If A is a central simple K-algebra, the norm $N_{K/F}(A)$ is a central simple F-algebra of degree $\deg N_{K/F}(A) = (\deg A)^2$. Moreover, the norm induces a group homomorphism*

$$N_{K/F}\colon \mathrm{Br}(K) \to \mathrm{Br}(F).$$

(5) *For any central simple F-algebra A,*

$$N_{K/F}(A_K) \simeq A \otimes_F A.$$

Proof: (1) Since $N_{K/F}(A)$ is an F-subalgebra of ${}^\iota A \otimes_K A$, there is a natural map $N_{K/F}(A) \otimes_F K \to {}^\iota A \otimes_K A$ induced by multiplication in ${}^\iota A \otimes_K A$. This map is a homomorphism of K-algebras. It is bijective since if $\alpha \in K \smallsetminus F$ every element $a \in {}^\iota A \otimes_K A$ can be written in a unique way as $a = a_1 + a_2\alpha$ with a_1, a_2 invariant under the switch map s by setting

$$a_1 = \frac{s(a)\alpha - a\iota(\alpha)}{\alpha - \iota(\alpha)} \quad \text{and} \quad a_2 = \frac{a - s(a)}{\alpha - \iota(\alpha)}.$$

In order to prove the second equality, consider the canonical isomorphism of K-algebras ${}^\iota({}^\iota A) = A$ which maps ${}^\iota({}^\iota a)$ to a for $a \in A$. In view of this isomorphism, $N_{K/F}({}^\iota A)$ may be regarded as the set of switch-invariant elements in $A \otimes_K {}^\iota A$. The isomorphism ${}^{\iota}A \otimes_K A \xrightarrow{\sim} A \otimes_K {}^{\iota}A$ which maps ${}^{\iota}a \otimes b$ to $b \otimes {}^{\iota}a$ commutes with the switch map and therefore induces a canonical isomorphism $N_{K/F}(A) = N_{K/F}({}^{\iota}A)$.

(2) This is straightforward (Draxl [84, p. 55] or Scharlau [247, Lemma 8.9.7]). The canonical map $N_{K/F}(A) \otimes_F N_{K/F}(B) \to N_{K/F}(A \otimes_K B)$ corresponds, after scalar extension to K, to the map

$$({}^{\iota}A \otimes_K A) \otimes_K ({}^{\iota}B \otimes_K B) \to {}^{\iota}(A \otimes_K B) \otimes_K (A \otimes_K B)$$

which carries ${}^{\iota}a_1 \otimes a_2 \otimes {}^{\iota}b_1 \otimes b_2$ to ${}^{\iota}(a_1 \otimes b_1) \otimes (a_2 \otimes b_2)$.

(3) There is a natural isomorphism

$${}^{\iota}\operatorname{End}_K(V) = \operatorname{End}_K({}^{\iota}V)$$

which identifies ${}^{\iota}f$ for $f \in \operatorname{End}_K(V)$ with the endomorphism of ${}^{\iota}V$ mapping ${}^{\iota}v$ to ${}^{\iota}(f(v))$. We may therefore identify

$${}^{\iota}\operatorname{End}_K(V) \otimes_K \operatorname{End}_K(V) = \operatorname{End}_K({}^{\iota}V \otimes_K V),$$

and check that the switch map s is then identified with conjugation by s_V where $s_V \colon {}^{\iota}V \otimes_K V \to {}^{\iota}V \otimes_K V$ is the ι-linear map defined through

$$s_V({}^{\iota}v \otimes w) = {}^{\iota}w \otimes v \quad \text{for } v, \, w \in V.$$

The F-algebra $N_{K/F}\big(\operatorname{End}_K(V)\big)$ of fixed elements under s is then identified with the F-algebra of endomorphisms of the F-subspace elementwise invariant under s_V, i.e., to $\operatorname{End}_F\big(N_{K/F}(V)\big)$.

(4) If A is a central simple K-algebra, then ${}^{\iota}A \otimes_K A$ also is central simple over K, hence $N_{K/F}(A)$ is central simple over F, by part (1) and Wedderburn's Theorem (1.1). If A' is Brauer-equivalent to A, then we may find vector spaces V, V' over K such that

$$A \otimes_K \operatorname{End}_K(V) \simeq A' \otimes_K \operatorname{End}_K(V').$$

It then follows from parts (2) and (3) above that

$$N_{K/F}(A) \otimes_F \operatorname{End}_F\big(N_{K/F}(V)\big) \simeq N_{K/F}(A') \otimes_F \operatorname{End}_F\big(N_{K/F}(V')\big),$$

hence $N_{K/F}(A)$ and $N_{K/F}(A')$ are Brauer-equivalent. Thus $N_{K/F}$ induces a map on Brauer groups and part (2) above shows that it is a homomorphism.

To prove (5), we first note that if A is an F-algebra, then ${}^{\iota}(A_K) = A_K$ under the identification ${}^{\iota}(a \otimes \alpha) = a \otimes \iota(\alpha)$. Therefore,

$${}^{\iota}(A_K) \otimes_K A_K \simeq A \otimes_F A \otimes_F K$$

and $N_{K/F}(A)$ can be identified with the F-algebra elementwise invariant under the F-algebra automorphism s' of $A \otimes_F A \otimes_F K$ defined through

$$s'(a_1 \otimes a_2 \otimes \alpha) = a_2 \otimes a_1 \otimes \iota(\alpha).$$

On the other hand, $A \otimes_F A$ can be identified with the algebra of fixed points under the automorphism s'' defined through

$$s''(a_1 \otimes a_2 \otimes \alpha) = a_1 \otimes a_2 \otimes \iota(\alpha).$$

We aim to show that these F-algebras are isomorphic when A is central simple.

Let $g \in A \otimes_F A$ be the Goldman element (see (3.5)). By (3.6), we have

$$g^2 = 1 \quad \text{and} \quad g \cdot (a_1 \otimes a_2) = (a_2 \otimes a_1) \cdot g \quad \text{for all } a_1, \, a_2 \in A,$$

hence for all $x \in A \otimes_F A$, $s'(x \otimes 1) = gxg^{-1} \otimes 1$. In particular

$$s'(g \otimes 1) = g \otimes 1,$$

and moreover

$$s''(y) = (g \otimes 1) \cdot s'(y) \cdot (g \otimes 1)^{-1} \quad \text{for } y \in A \otimes_F A \otimes_F K.$$

Let $\alpha \in K$ be such that $\iota(\alpha) \neq \pm\alpha$ and let

$$u = \alpha + (g \otimes 1)\iota(\alpha) \in A \otimes_F A \otimes_F K.$$

This element is invertible, since $u \cdot (\alpha - (g \otimes 1)\iota(\alpha)) = \alpha^2 - \iota(\alpha)^2 \in K^\times$; moreover,

$$s'(u) = \iota(\alpha) + (g \otimes 1)\alpha = u \cdot (g \otimes 1).$$

Therefore, for all $x \in A \otimes_F A \otimes_F K$,

$$s'(uxu^{-1}) = u \cdot (g \otimes 1) \cdot s'(x) \cdot (g \otimes 1)^{-1} \cdot u^{-1} = u \cdot s''(x) \cdot u^{-1}.$$

This equation shows that conjugation by u induces an isomorphism from the F-algebra of invariant elements under s'' onto the F-algebra of invariant elements under s', hence

$$A \otimes_F A \simeq N_{K/F}(A_K). \qquad \square$$

(3.14) Remark. Property (5) in the proposition above does not hold for arbitrary F-algebras. For instance, one may check as an exercise that $N_{\mathbb{C}/\mathbb{R}}(\mathbb{C}_{\mathbb{C}}) \simeq \mathbb{R} \times \mathbb{R} \times \mathbb{C}$ whereas $\mathbb{C} \otimes_{\mathbb{R}} \mathbb{C} \simeq \mathbb{C} \times \mathbb{C}$. (This simple example is due to M. Ojanguren). The proof of (3.13.5) in [84, p. 55] is flawed; see the correction in Tignol [284] or Rowen [237, Theorem 7.2.26].

Involutions of the second kind and one-sided ideals. We now come back to the proof of Theorem (3.1). As above, let K/F be a separable quadratic extension of fields with nontrivial automorphism ι. Let B be a central simple K-algebra. As in the case of involutions of the first kind, the necessary condition for the existence of an involution of the second kind on B is easy to prove:

(3.15) Proposition. *Suppose that B admits an involution τ of the second kind whose restriction to K is ι. This involution endows B with a right ${}^\iota B \otimes_K B$-module structure defined by*

$$x *_\tau ({}^\iota a \otimes b) = \tau(a)xb \quad \text{for } a, b, x \in B.$$

The multiplication $*_\tau$ induces a right $N_{K/F}(B)$-module structure on $\mathrm{Sym}(B,\tau)$ for which $\mathrm{rdim}\,\mathrm{Sym}(B,\tau) = 1$. Therefore, $N_{K/F}(B)$ is split.

Proof: It is straightforward to check that $*_\tau$ defines on B a right ${}^\iota B \otimes_K B$-module structure. For $a, b, x \in B$ we have

$$\tau\big(x *_\tau ({}^\iota a \otimes b)\big) = \tau(x) *_\tau ({}^\iota b \otimes a).$$

Therefore, if $u \in {}^\iota B \otimes_K B$ is invariant under the switch map, then multiplication by u preserves $\mathrm{Sym}(B,\tau)$. It follows that $*_\tau$ induces a right $N_{K/F}(B)$-module structure on $\mathrm{Sym}(B,\tau)$. Since $\dim_F \mathrm{Sym}(B,\tau) = \deg N_{K/F}(B)$, we have $\mathrm{rdim}\,\mathrm{Sym}(B,\tau) = 1$, hence $N_{K/F}(B)$ is split. $\qquad \square$

(3.16) Remark. Alternately, the involution τ yields a K-algebra isomorphism $\tau_* \colon B \otimes_K {}^\iota B \to \mathrm{End}_K(B)$ defined by $\tau_*(a \otimes {}^\iota b)(x) = ax\tau(b)$. This isomorphism restricts to an F-algebra isomorphism $N_{K/F}(B) \to \mathrm{End}_F\big(\mathrm{Sym}(B,\tau)\big)$ which shows that $N_{K/F}(B)$ is split. However, the space $\mathrm{Sym}(B,\tau)$ is then considered as a *left* $N_{K/F}(B)$-module; this is less convenient for the discussion below.

Let $\tau' \colon N_{K/F}(B) \to \mathrm{Sym}(B, \tau)$ be defined by

$$\tau'(u) = 1 *_\tau u \quad \text{for } u \in N_{K/F}(A).$$

Since $\mathrm{rdim}\,\mathrm{Sym}(B, \tau) = 1$, it is clear that the map τ' is surjective, hence $\ker \tau'$ is a right ideal of dimension $n^4 - n^2$ where $n = \deg B$. We denote this ideal by I_τ:

$$I_\tau = \ker \tau'.$$

Extending scalars to K, we have $N_{K/F}(B)_K = {}^\iota B \otimes_K B$ and the map $\tau'_K \colon {}^\iota B \otimes_K B \to B$ induced by τ' is

$$\tau'_K({}^\iota a \otimes b) = \tau(a)b.$$

Therefore, the ideal $(I_\tau)_K = I_\tau \otimes_F K = \ker \tau'_K$ satisfies $(I_\tau)_K \cap (1 \otimes B) = \{0\}$, hence also

$$ {}^\iota B \otimes_K B = (I_\tau)_K \oplus (1 \otimes B).$$

(3.17) Theorem. *The map $\tau \mapsto I_\tau$ defines a one-to-one correspondence between involutions of the second kind on B leaving F elementwise invariant and right ideals $I \subset N_{K/F}(B)$ such that*

$$ {}^\iota B \otimes_K B = I_K \oplus (1 \otimes B)$$

where $I_K = I \otimes_F K$ is the ideal of ${}^\iota B \otimes_K B$ obtained from I by scalar extension.

Proof: We have already checked that for each involution τ the ideal I_τ satisfies the condition above. Conversely, suppose I is a right ideal such that ${}^\iota B \otimes_K B = I_K \oplus (1 \otimes B)$. For each $b \in B$, there is a unique element $\tau_I(b) \in B$ such that

(3.18) $${}^\iota b \otimes 1 - 1 \otimes \tau_I(b) \in I_K.$$

The map $\tau_I \colon B \to B$ is ι-semilinear and the same arguments as in the proof of Theorem (3.8) show that it is an anti-automorphism on B.

   In order to check that $\tau_I^2(b) = b$ for all $b \in B$, we use the fact that the ideal I_K is preserved under the switch map $s \colon {}^\iota B \otimes_K B \to {}^\iota B \otimes_K B$ since it is extended from an ideal I in $N_{K/F}(B)$. Therefore, applying s to (3.18) we get

$$1 \otimes b - {}^\iota \tau_I(b) \otimes 1 \in I_K,$$

hence $\tau_I^2(b) = b$.

   Arguing as in the proof of Theorem (3.8), we see that the ideal I_{τ_I} associated to the involution τ_I satisfies $(I_{\tau_I})_K = I_K$, and conclude that $I_{\tau_I} = I$, since I (resp. I_{τ_I}) is the subset of invariant elements in I_K (resp. $(I_{\tau_I})_K$) under the switch map.

   On the other hand, for any given involution τ on B we have

$$ {}^\iota b \otimes 1 - 1 \otimes \tau(b) \in (I_\tau)_K \quad \text{for } b \in B,$$

hence $\tau_{I_\tau} = \tau$. $\square$

   Let $\deg B = n$. The right ideals $I \subset N_{K/F}(B)$ such that ${}^\iota B \otimes_K B = I_K \oplus (1 \otimes B)$ then have reduced dimension $n^2 - 1$ and form a dense open subvariety $\mathcal{V}$ in the Severi-Brauer variety $\mathrm{SB}_{n^2-1}\big(N_{K/F}(B)\big)$. The theorem above may thus be reformulated as follows:

(3.19) Corollary. *There is a natural one-to-one correspondence between involutions of the second kind on B which leave F elementwise invariant and rational points on the variety $\mathcal{V} \subset \mathrm{SB}_{n^2-1}\big(N_{K/F}(B)\big)$.*

We may now complete the proof of Theorem (3.1). If F is finite, the algebras B and $N_{K/F}(B)$ are split, and (2.20) shows that B carries unitary involutions. We may thus assume henceforth that F is infinite. If $N_{K/F}(B)$ splits, then the variety $\text{SB}_{n^2-1}\big(N_{K/F}(B)\big)$ is a projective space. The set of rational points is therefore dense in $\text{SB}_{n^2-1}\big(N_{K/F}(B)\big)$ and so it intersects the nonempty open subvariety $\mathcal{V}$ nontrivially. Corollary (3.19) then shows B has unitary involutions whose restriction to K is ι. $\qquad\square$

(3.20) Remark. As in the case of involutions of the first kind, density arguments can be avoided by reducing to division algebras. Suppose that $B \simeq M_r(D)$ for some central division algebra D over K and some integer r. Since the norm map $N_{K/F}$ is defined on the Brauer group of K, the condition that $N_{K/F}(B)$ splits implies that $N_{K/F}(D)$ also splits. Let I be a maximal right ideal in $N_{K/F}(D)$. We have $\dim_F I = \dim_F N_{K/F}(D) - \deg N_{K/F}(D) = (\dim_K D)^2 - \dim_K D$. Moreover, since D is a division algebra, it is clear that $I_K \cap (1 \otimes D) = \{0\}$, hence

$$^\iota D \otimes_K D = I_K \oplus (1 \otimes D),$$

by dimension count. Theorem (3.17) then shows that D has an involution of the second kind $^-$ leaving F elementwise invariant. An involution τ of the same kind can then be defined on $M_r(D)$ by letting $^-$ act entrywise and setting

$$\tau(a) = \overline{a}^t.$$

This involution is transported to A by the isomorphism $A \simeq M_r(D)$.

Part (2) of Theorem (3.1) can easily be extended to cover the case of semisimple F-algebras $E_1 \times E_2$ with E_1, E_2 central simple over F. The norm $N_{(F\times F)/F}$ is defined by

$$N_{(F\times F)/F}(E_1 \times E_2) = E_1 \otimes_F E_2.$$

This definition is consistent with (3.12), and it is easy to check that (3.13) extends to the case where $K = F \times F$.

If $E_1 \times E_2$ has an involution whose restriction to the center $F \times F$ interchanges the factors, then $E_2 \simeq E_1^{\text{op}}$, by (2.14). Therefore, $N_{(F\times F)/F}(E_1 \times E_2)$ splits. Conversely, if $N_{(F\times F)/F}(E_1 \times E_2)$ splits, then $E_2 \simeq E_1^{\text{op}}$ and the exchange involution on $E_1 \times E_1^{\text{op}}$ can be transported to an involution of the second kind on $E_1 \times E_2$.

§4. Hermitian Forms

In this section, we set up a one-to-one correspondence between involutions on central simple algebras and hermitian forms on vector spaces over division algebras, generalizing the theorem in the introduction to this chapter.

According to Theorem (1.2), every central simple algebra A may be viewed as the algebra of endomorphisms of some finite dimensional vector space V over a central division algebra D:

$$A = \text{End}_D(V).$$

Explicitly, we may take for V any simple left A-module and set $D = \text{End}_A(V)$. The module V may then be endowed with a right D-vector space structure.

Since D is Brauer-equivalent to A, Theorem (3.1) shows that A has an involution if and only if D has an involution. Therefore, in this section we shall work from the perspective that central simple algebras with involution are algebras of endomorphisms of vector spaces over division algebras with involution. More generally, we shall substitute an arbitrary central simple algebra E for D and consider endomorphism algebras of right modules over E. In the second part of this section, we discuss extending of involutions from a simple subalgebra $B \subset A$ in relation to an analogue of the Scharlau transfer for hermitian forms.

4.A. Adjoint involutions. Let E be a central simple algebra over a field F and let M be a finitely generated right E-module. Suppose that $\theta \colon E \to E$ is an involution (of any kind) on E. A *hermitian form* on M (with respect to the involution θ on E) is a bi-additive map

$$h \colon M \times M \to E$$

subject to the following conditions:

(1) $h(x\alpha, y\beta) = \theta(\alpha)h(x,y)\beta$ for all x, $y \in M$ and α, $\beta \in E$,
(2) $h(y,x) = \theta\big(h(x,y)\big)$ for all x, $y \in M$.

It clearly follows from (2) that $h(x,x) \in \mathrm{Sym}(E,\theta)$ for all $x \in M$. If (2) is replaced by

(2′) $h(y,x) = -\theta\big(h(x,y)\big)$ for all x, $y \in M$,

the form h is called *skew-hermitian*. In that case $h(x,x) \in \mathrm{Skew}(E,\theta)$ for all $x \in M$. If a skew-hermitian form h satisfies $h(x,x) \in \mathrm{Alt}(E,\theta)$ for all $x \in M$, it is called *alternating* (or *even*). If $\mathrm{char}\, F \neq 2$, every skew-hermitian is alternating since $\mathrm{Skew}(E,\theta) = \mathrm{Alt}(E,\theta)$. If $E = F$ and $\theta = \mathrm{Id}$, hermitian (resp. skew-hermitian, resp. alternating) forms are the symmetric (resp. skew-symmetric, resp. alternating) bilinear forms.

Similar definitions can be set for *left* modules. It is then convenient to replace (1) by

(1′) $h(\alpha x, \beta y) = \alpha h(x,y)\theta(\beta)$ for all x, $y \in M$ and α, $\beta \in E$.

The results concerning hermitian forms on left modules are of course essentially the same as for right modules. We therefore restrict our discussion in this section to right modules.

The hermitian or skew-hermitian form h on the right E-module M is called *nonsingular* if the only element $x \in M$ such that $h(x,y) = 0$ for all $y \in M$ is $x = 0$.

(4.1) Proposition. *For every nonsingular hermitian or skew-hermitian form h on M, there exists a unique involution σ_h on $\mathrm{End}_E(M)$ such that $\sigma_h(\alpha) = \theta(\alpha)$ for all $\alpha \in F$ and*

$$h\big(x, f(y)\big) = h\big(\sigma_h(f)(x), y\big) \quad \text{for } x,\, y \in M.$$

The involution σ_h is called the adjoint involution with respect to h.

Proof: Consider the dual $M^* = \mathrm{Hom}_E(M,E)$. It has a natural structure of left E-module. We define a right module ${}^\theta M^*$ by

$${}^\theta M^* = \{\, {}^\theta\varphi \mid \varphi \in M^* \,\}$$

with the operations

$${}^\theta\varphi + {}^\theta\psi = {}^\theta(\varphi + \psi) \quad \text{and} \quad ({}^\theta\varphi)\alpha = {}^\theta\big(\theta(\alpha)\varphi\big) \quad \text{for } \varphi,\, \psi \in M^* \text{ and } \alpha \in E.$$

The hermitian or skew-hermitian form h induces a homomorphism of right E-modules

$$\hat{h} \colon M \to {}^\theta M^*$$

defined by

$$\hat{h}(x) = {}^\theta\varphi \quad \text{where } \varphi(y) = h(x, y).$$

If h is nonsingular, the map $\hat{h}$ is injective, hence bijective since M and ${}^\theta M^*$ have the same dimension over F. The unique involution σ_h for which the condition of the proposition holds is then given by

$$\sigma_h(f) = \hat{h}^{-1} \circ {}^\theta f^t \circ \hat{h}$$

where ${}^\theta f^t \colon {}^\theta M^* \to {}^\theta M^*$ is the transpose of f, so that

$${}^\theta f^t({}^\theta\varphi) = {}^\theta\big(f^t(\varphi)\big) = {}^\theta(\varphi \circ f) \quad \text{for } \varphi \in M^*. \qquad \square$$

The following theorem is the expected generalization of the result proved in the introduction.

(4.2) Theorem. *Let $A = \operatorname{End}_E(M)$.*
(1) *If θ is of the first kind on E, the map $h \mapsto \sigma_h$ defines a one-to-one correspondence between nonsingular hermitian and skew-hermitian forms on M (with respect to θ) up to a factor in $F^\times$ and involutions of the first kind on A.*

If $\operatorname{char} F \neq 2$, the involutions σ_h on A and θ on E have the same type if h is hermitian and have opposite types if h is skew-hermitian.

If $\operatorname{char} F = 2$, the involution σ_h is symplectic if and only if h is alternating.
(2) *If θ is of the second kind on E, the map $h \mapsto \sigma_h$ defines a one-to-one correspondence between nonsingular hermitian forms on M up to a factor in $F^\times$ invariant under θ and involutions σ of the second kind on A such that $\sigma(\alpha) = \theta(\alpha)$ for all $\alpha \in F$.*

Proof: We first make some observations which do not depend on the kind of θ. If h and h' are nonsingular hermitian or skew-hermitian forms on M, then the map $v = \hat{h}^{-1} \circ \hat{h}' \in A^\times$ is such that

$$h'(x, y) = h\big(v(x), y\big) \quad \text{for } x, y \in M.$$

Therefore, the adjoint involutions σ_h and $\sigma_{h'}$ are related by

$$\sigma_h = \operatorname{Int}(v) \circ \sigma_{h'}.$$

Therefore, if $\sigma_h = \sigma_{h'}$, then $v \in F^\times$ and the forms h, h' differ by a factor in $F^\times$.

If θ is of the second kind and h, h' are both hermitian, the relation $h' = v \cdot h$ implies that $\theta(v) = v$. We have thus shown injectivity of the map $h \mapsto \sigma_h$ on the set of equivalence classes modulo factors in $F^\times$ (invariant under θ) in both cases (1) and (2).

Let D be a central division algebra Brauer-equivalent to E. We may then identify E with $M_s(D)$ for some integer s, hence also M with $M_{r,s}(D)$ and A with $M_r(D)$, as in the proof of (1.10). We may thus assume that

$$A = M_r(D), \quad M = M_{r,s}(D), \quad E = M_s(D).$$

Theorem (3.1) shows that D carries an involution $\overline{}$ such that $\overline{\alpha} = \theta(\alpha)$ for all $\alpha \in F$. We use the same notation * for the maps $A \to A$, $E \to E$ and $M \to M_{s,r}(D)$ defined by

$$(d_{ij})^*_{i,j} = (\overline{d_{ij}})^t_{i,j}.$$

Proposition (2.20) shows that the maps * on A and E are involutions of the same type as $\overline{}$.

Consider now case (1), where $^-$ is of the first kind. According to (2.7), we may find $u \in E^\times$ such that $u^* = \pm u$ and $\theta = \mathrm{Int}(u) \circ {}^*$. Moreover, for any involution of the first kind σ on A we may find some $g \in A^\times$ such that $g^* = \pm g$ and $\sigma = \mathrm{Int}(g) \circ {}^*$. Define then a map $h \colon M \times M \to E$ by

$$h(x, y) = u \cdot x^* \cdot g^{-1} \cdot y \quad \text{for } x, y \in M.$$

This map is clearly bi-additive. Moreover, for $\alpha, \beta \in E$ and $x, y \in M$ we have

$$h(x\alpha, y\beta) = u \cdot \alpha^* \cdot x^* \cdot g^{-1} \cdot y \cdot \beta = \theta(\alpha) \cdot h(x, y) \cdot \beta$$

and

$$h(y, x) = u \cdot \left(u^* \cdot x^* \cdot (g^{-1})^* \cdot y \right)^* \cdot u^{-1} = \delta\theta\big(h(x, y)\big),$$

where $\delta = +1$ if $u^{-1}u^* = g^{-1}g^*$ $(= \pm 1)$ and $\delta = -1$ if $u^{-1}u^* = -g^{-1}g^*$ $(= \mp 1)$. Therefore, h is a hermitian or skew-hermitian form on M. For $a \in A$ and $x, y \in M$,

$$h(x, ay) = u \cdot x^* \cdot (ga^*g^{-1})^* \cdot g^{-1} \cdot y = h\big(\sigma(a)x, y\big),$$

hence σ is the adjoint involution with respect to h. To complete the proof of (1), it remains to relate the type of σ to properties of h.

Suppose first that $\mathrm{char}\, F \neq 2$. Proposition (2.20) shows that the type of θ (resp. of σ) is the same as the type of $^-$ if and only if $u^{-1}u^* = +1$ (resp. $g^{-1}g^* = +1$). Therefore, σ and θ are of the same type if and only if $u^{-1}u^* = g^{-1}g^*$, and this condition holds if and only if h is hermitian.

Suppose now that $\mathrm{char}\, F = 2$. We have to show that $h(x, x) \in \mathrm{Alt}(E, \theta)$ for all $x \in M$ if and only if σ is symplectic. Proposition (2.7) shows that this last condition is equivalent to $g \in \mathrm{Alt}(A, {}^*)$. If $g = a - a^*$ for some $a \in A$, then $g^{-1} = -g^{-1}g(g^{-1})^* = b - b^*$ for $b = g^{-1}a^*(g^{-1})^*$. It follows that for all $x \in M$

$$h(x, x) = u \cdot x^* \cdot b \cdot x - \theta(u \cdot x^* \cdot b \cdot x) \in \mathrm{Alt}(E, \theta).$$

Conversely, if h is alternating, then $x^* \cdot g^{-1} \cdot x \in \mathrm{Alt}(E, {}^*)$ for all $x \in M$, since (2.7) shows that $\mathrm{Alt}(E, {}^*) = u^{-1} \cdot \mathrm{Alt}(E, \theta)$. In particular, taking for x the matrix e_{i1} whose entry with indices $(i, 1)$ is 1 and whose other entries are 0, it follows that the i-th diagonal entry of g^{-1} is in $\mathrm{Alt}(D, {}^-)$. Let $g^{-1} = (g'_{ij})_{1 \le i, j \le r}$ and $g'_{ii} = d_i - \overline{d_i}$ for some $d_i \in D$. Then $g^{-1} = b - b^*$ where the matrix $b = (b_{ij})_{1 \le i, j \le r}$ is defined by

$$b_{ij} = \begin{cases} g'_{ij} & \text{if } i < j, \\ d_i & \text{if } i = j, \\ 0 & \text{if } i > j. \end{cases}$$

Therefore, $g = -gg^{-1}g^* = gb^*g^* - (gb^*g^*)^* \in \mathrm{Alt}(A, {}^*)$, completing the proof of (1).

The proof of (2) is similar, but easier since there is only one type. Proposition (2.18) yields an element $u \in E^*$ such that $u^* = u$ and $\theta = \mathrm{Int}(u) \circ {}^*$, and shows that every involution σ on A such that $\sigma(\alpha) = \overline{\alpha}$ for all $\alpha \in F$ has the form $\sigma = \mathrm{Int}(g) \circ {}^*$ for some $g \in A^\times$ such that $g^* = g$. The same computations as for (1) show that σ is the adjoint involution with respect to the hermitian form h on M defined by

$$h(x, y) = u \cdot x^* \cdot g^{-1} \cdot y \quad \text{for } x, y \in M. \qquad \square$$

The preceding theorem applies notably in the case where E is a division algebra, to yield a correspondence between involutions on a central simple algebra A

and hermitian and skew-hermitian forms on vector spaces over the division algebra Brauer-equivalent to A. However, (1.10) shows that a given central simple
algebra may be represented as $A = \operatorname{End}_E(M)$ for any central simple algebra E
Brauer-equivalent to A (and for a suitable E-module M). Involutions on A then
correspond to hermitian and skew-hermitian forms on M by the preceding theorem. In particular, if A has an involution of the first kind, then a theorem of
Merkurjev [189] shows that we may take for E a tensor product of quaternion
algebras.

4.B. Extension of involutions and transfer. This section analyzes the
possibility of extending an involution from a simple subalgebra. One type of extension is based on an analogue of the Scharlau transfer for hermitian forms which is
discussed next. The general extension result, due to Kneser, is given thereafter.

The transfer. Throughout this subsection, we consider the following situation: Z/F is a finite extension of fields, E is a central simple Z-algebra and T is a
central simple F-algebra contained in E. Let C be the centralizer of T in E. By
the double centralizer theorem (see (1.5)) this algebra is central simple over Z and

$$E = T \otimes_F C.$$

Suppose that θ is an involution on E (of any kind) which preserves T, hence also C.
For simplicity, we also call θ the restriction of θ to T and to C.

(4.3) Definition. An F-linear map $s\colon E \to T$ is called an *involution trace* if it
satisfies the following conditions (see Knus [157, (7.2.4)]):

(1) $s(t_1 x t_2) = t_1 s(x) t_2$ for all $x \in E$ and $t_1, t_2 \in T$;
(2) $s\big(\theta(x)\big) = \theta\big(s(x)\big)$ for all $x \in E$;
(3) if $x \in E$ is such that $s\big(\theta(x)y\big) = 0$ for all $y \in E$, then $x = 0$.

In view of (2), condition (3) may equivalently be phrased as follows: the only
element $y \in E$ such that $s\big(\theta(x)y\big) = 0$ for all $x \in E$ is $y = 0$. It is also equivalent
to the following:

(3′) $\ker s$ does not contain any nontrivial left or right ideal in E.

Indeed, I is a right (resp. left) ideal in $\ker s$ if and only if $s\big(\theta(x)y\big) = 0$ for all
$\theta(x) \in I$ and all $y \in E$ (resp. for all $x \in E$ and $y \in I$).

For instance, if $T = F = Z$, the reduced trace $\operatorname{Trd}_E\colon E \to F$ is an involution
trace. Indeed, condition (2) follows from (2.2) if θ is of the first kind and from (2.16)
if θ is of the second kind, and condition (3) follows from the fact that the bilinear
(reduced) trace form is nonsingular (see (1.8)).

If $E = Z$ and $T = F$, every nonzero linear map $s\colon Z \to F$ which commutes
with θ is an involution trace. Indeed, if $x \in Z$ is such that $s\big(\theta(x)y\big) = 0$ for all
$y \in Z$, then $x = 0$ since $s \neq 0$ and $Z = \{\, \theta(x)y \mid y \in Z \,\}$ if $x \neq 0$.

The next proposition shows that every involution trace $s\colon E \to T$ can be obtained by combining these particular cases.

(4.4) Proposition. *Fix a nonzero linear map $\ell\colon Z \to F$ which commutes with θ.
For every unit $u \in \operatorname{Sym}(C, \theta)$, the map $s\colon E \to T$ defined by*

$$s(t \otimes c) = t \cdot \ell\big(\operatorname{Trd}_C(uc)\big) \quad \text{for } t \in T \text{ and } c \in C$$

*is an involution trace. Every involution trace from E to T is of the form above for
some unit $u \in \operatorname{Sym}(C, \theta)$.*

Proof: Conditions (1) and (2) are clear. Suppose that $x = \sum_i t_i \otimes c_i \in E$ is such that $s(\theta(x)y) = 0$ for all $y \in E$. We may assume that the elements $t_i \in T$ are linearly independent over F. The relation $s(\theta(x) \cdot 1 \otimes c) = 0$ for all $c \in C$ then yields $\ell(\mathrm{Trd}_C(u\theta(c_i)c)) = 0$ for all i and all $c \in C$. Since ℓ is nonzero, it follows that $\mathrm{Trd}_C(u\theta(c_i)c) = 0$ for all i and all $c \in C$, hence $u\theta(c_i) = 0$ for all i since the bilinear reduced trace form is nonsingular. It follows that $\theta(c_i) = 0$ for all i since u is invertible, hence $x = 0$.

Let $s \colon E \to T$ be an arbitrary involution trace. For $t \in T$ and $c \in C$,

$$t \cdot s(1 \otimes c) = s(t \otimes c) = s(1 \otimes c) \cdot t,$$

hence the restriction of s to C takes values in F and $s = \mathrm{Id}_T \otimes s_0$ where $s_0 \colon C \to F$ denotes this restriction. Since ℓ is nonzero and the bilinear reduced trace form is nonsingular, the linear map $C \to \mathrm{Hom}_F(C, F)$ which carries $c \in C$ to the linear map $x \mapsto \ell(\mathrm{Trd}_C(cx))$ is injective, hence also surjective, by dimension count. Therefore, there exists $u \in C$ such that $s_0(x) = \ell(\mathrm{Trd}_C(ux))$ for all $x \in C$. If u is not invertible, then the annihilator of the left ideal generated by u is a nontrivial right ideal in the kernel of s_0, contrary to the hypothesis that s is an involution trace. Finally, observe that for $c \in C$,

$$s_0(\theta(c)) = \ell\left(\theta(\mathrm{Trd}_C(c\theta(u)))\right) = \ell(\mathrm{Trd}_C(\theta(u)c)),$$

hence the condition $s_0(\theta(c)) = s_0(c)$ for all $c \in C$ implies that $\theta(u) = u$. □

(4.5) Corollary. *For every involution trace $s \colon E \to T$, there exists an involution θ_s on C such that*

$$s(\theta(c)x) = s(x\theta_s(c)) \quad \text{for } c \in C, \, x \in E.$$

The involutions θ_s and θ have the same restriction to Z.

Proof: Fix a nonzero linear map $\ell \colon Z \to F$ which commutes with θ. According to (4.4), we have $s = \mathrm{Id}_T \otimes s_0$ where $s_0 \colon C \to F$ is defined by $s_0(c) = \ell(\mathrm{Trd}_C(uc))$ for some symmetric unit $u \in C^\times$. Let $\theta_s = \mathrm{Int}(u) \circ \theta$. For $c, c' \in C$,

$$\mathrm{Trd}_C(u\theta(c)c') = \mathrm{Trd}_C(\theta_s(c)uc') = \mathrm{Trd}_C(uc'\theta_s(c)),$$

hence for all $t \in T$,

$$s(\theta(c) \cdot (t \otimes c')) = t \cdot \ell(\mathrm{Trd}_C(u\theta(c)c')) = s((t \otimes c') \cdot \theta_s(c)).$$

Therefore, the involution θ_s satisfies

$$s(\theta(c)x) = s(x\theta_s(c)) \quad \text{for } c \in C, \, x \in E.$$

The involution θ_s is uniquely determined by this condition, because if $s(x\theta_s(c)) = s(x\theta_s'(c))$ for all $c \in C$ and $x \in E$, then property (3) of involution traces in (4.3) implies that $\theta_s(c) = \theta_s'(c)$ for all $c \in C$.

Since $\theta_s = \mathrm{Int}(u) \circ \theta$, it is clear that $\theta_s(z) = \theta(z)$ for all $z \in Z$. □

Using an involution trace $s \colon E \to T$, we may define a structure of hermitian module over T on every hermitian module over E, as we proceed to show.

Suppose that M is a finitely generated right module over E. Since $T \subset E$, we may also consider M as a right T-module, and

$$\mathrm{End}_E(M) \subset \mathrm{End}_T(M).$$

The centralizer of $\mathrm{End}_E(M)$ in $\mathrm{End}_T(M)$ is easily determined:

(4.6) Lemma. *For c in the centralizer C of T in E, let $r_c \in \mathrm{End}_T(M)$ be the right multiplication by c. The map $c^{\mathrm{op}} \mapsto r_c$ identifies C^{op} with the centralizer of $\mathrm{End}_E(M)$ in $\mathrm{End}_T(M)$.*

Proof: Every element $f \in \mathrm{End}_T(M)$ in the centralizer of $\mathrm{End}_E(M)$ may be viewed as an endomorphism of M for its $\mathrm{End}_E(M)$-module structure. By (1.10), we have $\mathrm{End}_{\mathrm{End}_E(M)}(M) = E$, hence f is right multiplication by some element $c \in E$. Since f is a T-module endomorphism, $c \in C$. $\qquad\square$

Suppose now that $h\colon M \times M \to E$ is a hermitian or skew-hermitian form with respect to θ. If $s\colon E \to T$ is an involution trace, we define

$$s_*(h)\colon M \times M \to T$$

by

$$s_*(h)(x,y) = s\big(h(x,y)\big) \quad \text{for } x,\, y \in M.$$

In view of the properties of s, the form $s_*(h)$ is clearly hermitian over T (with respect to θ) if h is hermitian, and skew-hermitian if h is skew-hermitian. It is also alternating if h is alternating, since the relation $h(x,x) = e - \theta(e)$ implies that $s_*(h)(x,x) = s(e) - \theta\big(s(e)\big)$.

(4.7) Proposition. *If h is nonsingular, then $s_*(h)$ is nonsingular and the adjoint involution $\sigma_{s_*(h)}$ on $\mathrm{End}_T(M)$ extends the adjoint involution σ_h on $\mathrm{End}_E(M)$:*

$$\big(\mathrm{End}_E(M), \sigma_h\big) \subset \big(\mathrm{End}_T(M), \sigma_{s_*(h)}\big).$$

Moreover, with the notation of (4.5) and (4.6),

$$\sigma_{s_*(h)}\big(r_c\big) = r_{\theta_s(c)}$$

for all $c \in C$.

Proof: If $x \in M$ is such that $s_*(h)(x,y) = 0$ for all $y \in M$, then $h(x,M)$ is a right ideal of E contained in $\ker s$, hence $h(x,M) = \{0\}$. This implies that $x = 0$ if h is nonsingular, proving the first statement.

For $f \in \mathrm{End}_E(M)$ and $x,\, y \in M$ we have

$$h\big(x, f(y)\big) = h\big(\sigma_h(f)(x), y\big).$$

Hence, applying s to both sides,

$$s_*(h)\big(x, f(y)\big) = s_*(h)\big(\sigma_h(f)(x), y\big).$$

Therefore, $\sigma_{s_*(h)}(f) = \sigma_h(f)$.

On the other hand, for $x,\, y \in M$ and $c \in C$,

$$s_*(h)(xc, y) = s\big(\theta(c)h(x,y)\big).$$

The defining property of θ_s shows that the right side is also equal to

$$s\big(h(x,y)\theta_s(c)\big) = s_*(h)\big(x, y\theta_s(c)\big),$$

hence $\sigma_{s_*(h)}(r_c) = r_{\theta_s(c)}$. $\qquad\square$

(4.8) Example. Suppose that E is central over F, hence the centralizer C of T in E also is central over F. Let M be a finitely generated right module over E. The algebra $\mathrm{End}_E(M)$ is a central simple F-subalgebra in $\mathrm{End}_T(M)$, and (4.6) shows that its centralizer is isomorphic to C^{op} under the map which carries $c^{\mathrm{op}} \in C^{\mathrm{op}}$

to the endomorphism r_c of right multiplication by c. Hence there is an F-algebra isomorphism

$$\Psi \colon \operatorname{End}_E(M) \otimes_F C^{\mathrm{op}} \xrightarrow{\sim} \operatorname{End}_T(M)$$

which maps $f \otimes c^{\mathrm{op}}$ to $f \circ r_c = r_c \circ f$ for $f \in \operatorname{End}_E(M)$ and $c \in C$.

Pick an invertible element $u \in \operatorname{Sym}(C, \theta)$ and define an involution trace $s \colon E \to T$ by

$$s(t \otimes c) = t \cdot \operatorname{Trd}_C(uc) \quad \text{for } t \in T,\ c \in C.$$

The proof of (4.5) shows that $\theta_s = \operatorname{Int}(u) \circ \theta$. Moreover, (4.7) shows that for every nonsingular hermitian or skew-hermitian form $h \colon M \times M \to E$, the involution $\sigma_{s_*(h)}$ on $\operatorname{End}_T(M)$ corresponds under Ψ to $\sigma_h \otimes \theta_s^{\mathrm{op}}$ where $\theta_s^{\mathrm{op}}(c^{\mathrm{op}}) = \big(\theta_s(c)\big)^{\mathrm{op}}$ for $c \in C$:

$$\Psi \colon \big(\operatorname{End}_E(M) \otimes_F C^{\mathrm{op}},\, \sigma_h \otimes \theta_s^{\mathrm{op}}\big) \xrightarrow{\sim} \big(\operatorname{End}_T(M),\, \sigma_{s_*(h)}\big).$$

As a particular case, we may consider $T = F$, $E = C$ and $M = C$. Then one sees that $\operatorname{End}_E(M) = C$ by identifying $c \in C$ with left multiplication by c, and the isomorphism Ψ is the same as in Wedderburn's theorem (1.1):

$$\Psi \colon C \otimes_F C^{\mathrm{op}} \xrightarrow{\sim} \operatorname{End}_F(C).$$

If $h \colon C \times C \to C$ is defined by $h(x, y) = \theta(x)y$, then $\sigma_h = \theta$ and the result above shows that $\sigma_{\operatorname{Trd}_*(h)}$ corresponds to $\theta \otimes \theta$ under Ψ.

(4.9) Example. Suppose that C is the center Z of E, so that

$$E = T \otimes_F Z.$$

Let N be a finitely generated right module over T and $h \colon N \times N \to T$ be a nonsingular hermitian form with respect to θ. Extending scalars to Z, we get a module $N_Z = N \otimes_F Z$ over E and a nonsingular hermitian form $h_Z \colon N_Z \times N_Z \to E$. Moreover,

$$\operatorname{End}_E(N_Z) = \operatorname{End}_T(N) \otimes_F Z \quad \text{and} \quad \operatorname{End}_T(N_Z) = \operatorname{End}_T(N) \otimes \operatorname{End}_F(Z).$$

Pick a nonzero linear map $\ell \colon Z \to F$ which commutes with θ and let

$$s = \operatorname{Id}_T \otimes \ell \colon E \to T$$

be the induced involution trace on E. We claim that under the identification above,

$$\sigma_{s_*(h_Z)} = \sigma_h \otimes \sigma_k,$$

where $k \colon Z \times Z \to F$ is the hermitian form defined by

$$k(z_1, z_2) = \ell\big(\theta(z_1)z_2\big) \quad \text{for } z_1,\ z_2 \in Z.$$

Indeed, for $x_1,\ x_2 \in N$ and $z_1,\ z_2 \in Z$ we have

$$h_Z(x_1 \otimes z_1, x_2 \otimes z_2) = h(x_1, x_2) \otimes \theta(z_1)z_2,$$

hence

$$s_*(h_Z) = h \otimes k.$$

We now return to the general case, and show that the involutions on $\operatorname{End}_T(M)$ which are adjoint to transfer forms $s_*(h)$ are exactly those which preserve $\operatorname{End}_E(M)$ and induce θ_s on the centralizer.

(4.10) Proposition. *Let σ be an involution on $\mathrm{End}_T(M)$ such that*

$$\sigma\big(\mathrm{End}_E(M)\big) = \mathrm{End}_E(M) \quad and \quad \sigma(r_e) = r_{\theta_s(e)}$$

for all $e \in C_E T$. There exists a nonsingular hermitian or skew-hermitian form $h\colon M \times M \to E$ with respect to θ such that $\sigma = \sigma_{s_*(h)}$.

Proof: Since $\theta_s|_Z = \theta|_Z$ by (4.5), it follows that $\sigma(r_z) = r_{\theta_s(z)} = r_{\theta(z)}$ for all $z \in Z$. Therefore, Theorem (4.2) shows that the restriction of σ to $\mathrm{End}_E(M)$ is the adjoint involution with respect to some nonsingular hermitian or skew-hermitian form $h_0\colon M \times M \to E$. Proposition (2.7) (if $\theta|_F = \mathrm{Id}_F$) or (2.18) (if $\theta|_F \neq \mathrm{Id}_F$) yields an invertible element $u \in \mathrm{End}_T(M)$ such that

$$\sigma = \mathrm{Int}(u) \circ \sigma_{s_*(h_0)}$$

and $\sigma_{s_*(h_0)}(u) = \pm u$. By (4.7), the restriction of $\sigma_{s_*(h_0)}$ to $\mathrm{End}_E(M)$ is σ_{h_0} which is also the restriction of σ to $\mathrm{End}_E(M)$. Therefore, u centralizes $\mathrm{End}_E(M)$. It follows from (4.6) that $u = r_e$ for some $e \in C^\times$. Proposition (4.7) shows that $\sigma_{s_*(h_0)}(r_c) = r_{\theta_s(c)}$ for all $c \in C$, hence

$$\sigma(r_c) = u \circ r_{\theta_s(c)} \circ u^{-1} = r_{e^{-1}\theta_s(c)e}.$$

Since we assume that $\sigma(r_c) = r_{\theta_s(c)}$ for all $c \in C$, it follows that $e \in Z^\times$. Moreover, $\theta(e) = \pm e$ since $\sigma_{s_*(h_0)}(u) = \pm u$. We may then define a nonsingular hermitian or skew-hermitian form $h\colon M \times M \to E$ by

$$h(x,y) = e^{-1} h_0(x,y) \quad \text{for } x,\, y \in M.$$

If $\delta = \theta(e)e^{-1}\ (= \pm 1)$, we also have

$$h(x,y) = \delta h_0(xe^{-1}, y) = \delta h_0\big(r_{e^{-1}}(x), y\big),$$

hence

$$\sigma_{s_*(h)} = \mathrm{Int}(r_e) \circ \sigma_{s_*(h_0)} = \sigma. \qquad \square$$

(4.11) Example. Suppose E is commutative, so that $E = Z = C$ and suppose that $T = F$. Assume further that $\theta = \mathrm{Id}_E$. Let V be a finite dimensional vector space over F and fix some F-algebra embedding

$$i\colon Z \hookrightarrow \mathrm{End}_F(V).$$

We may then consider V as a vector space over Z by defining

$$v \cdot z = i(z)(v) \quad \text{for } v \in V,\, z \in Z.$$

By definition, the centralizer of $i(Z)$ in $\mathrm{End}_F(V)$ is $\mathrm{End}_Z(V)$.

Suppose that σ is an involution on $\mathrm{End}_F(V)$ which leaves $i(Z)$ elementwise invariant and that $s\colon Z \to F$ is a nonzero linear map. By (4.5), we have $\theta_s = \theta = \mathrm{Id}_Z$. On the other hand, since σ preserves $i(Z)$, it also preserves its centralizer $\mathrm{End}_Z(V)$. We may therefore apply (4.10) to conclude that there exists a nonsingular symmetric or skew-symmetric bilinear form $b\colon V \times V \to Z$ such that $\sigma = \sigma_{s_*(b)}$.

By (4.7), the restriction of σ to $\mathrm{End}_Z(V)$ is σ_b. If b is symmetric, skew-symmetric or alternating, then $s_*(b)$ has the same property. If char $F \neq 2$, the bilinear form $s_*(b)$ cannot be simultaneously symmetric and skew-symmetric, or it would be singular. Therefore, b and $s_*(b)$ are of the same type, and it follows that σ has the same type as its restriction to $\mathrm{End}_Z(V)$. If char $F = 2$, it is still true that σ is symplectic if its restriction to $\mathrm{End}_Z(V)$ is symplectic, since $s_*(b)$ is alternating if b is alternating, but the converse is not true without some further hypotheses.

To construct a specific example, consider a field Z which is finite dimensional over its subfield of squares Z^2, and let $F = Z^2$. Pick a nonzero linear map $s\colon Z \to F$ such that $s(1) = 0$. The nonsingular symmetric bilinear form b on $V = Z$ defined by $b(z_1, z_2) = z_1 z_2$ is not alternating, but $s_*(b)$ is alternating since s vanishes on Z^2. Therefore, the involution $\sigma_{s_*(b)}$ on $\mathrm{End}_F(Z)$ is symplectic, but its restriction σ_b to $\mathrm{End}_Z(Z)$ is orthogonal. (Indeed, $\mathrm{End}_Z(Z) = Z$ and $\sigma_b = \mathrm{Id}_Z$.)

These observations on the type of an involution compared with the type of its restriction to a centralizer are generalized in the next proposition.

(4.12) Proposition. *Let A be a central simple F-algebra with an involution σ of the first kind and let $L \subset A$ be a subfield containing F. Suppose that σ leaves L elementwise invariant, so that it restricts to an involution of the first kind τ on the centralizer $C_A L$.*

(1) *If $\mathrm{char}\, F \neq 2$, the involutions σ and τ have the same type.*
(2) *Suppose that $\mathrm{char}\, F = 2$. If τ is symplectic, then σ is symplectic. If L/F is separable, then σ and τ have the same type.*

Proof: We first consider the simpler case where $\mathrm{char}\, F = 2$. If τ is symplectic, then (2.6) shows that the centralizer $C_A L$ contains an element c such that $c + \tau(c) = 1$. Since $\tau(c) = \sigma(c)$, it also follows from (2.6) that σ is symplectic.

If τ is orthogonal, then $\mathrm{Trd}_{C_A L}\big(\mathrm{Sym}(C_A L, \tau)\big) = L$ by (2.6). If L/F is separable, the trace form $T_{L/F}$ is nonzero, hence

$$T_{L/F} \circ \mathrm{Trd}_{C_A L}\big(\mathrm{Sym}(C_A L, \tau)\big) = F.$$

Since $T_{L/F} \circ \mathrm{Trd}_{C_A L}(c) = \mathrm{Trd}_A(c)$ for all $c \in C_A L$ (see Draxl [84, p. 150]), we have

$$T_{L/F} \circ \mathrm{Trd}_{C_A L}\big(\mathrm{Sym}(C_A L, \tau)\big) \subset \mathrm{Trd}_A\big(\mathrm{Sym}(A, \sigma)\big),$$

hence $\mathrm{Trd}_A\big(\mathrm{Sym}(A, \sigma)\big) \neq \{0\}$, and σ is orthogonal. This completes the proof in the case where $\mathrm{char}\, F = 2$.

In arbitrary characteristic, let F' be a splitting field of A in which F is algebraically closed and such that the field extension F'/F is separable (for instance, the function field of the Severi-Brauer variety $\mathrm{SB}(A)$). The composite $L \cdot F'\ (= L \otimes_F F')$ is then a field, and it suffices to prove the proposition after extending scalars to F'. We may thus assume that $A = \mathrm{End}_F(V)$ for some F-vector space V. If $\mathrm{char}\, F \neq 2$ the result then follows from the observations in (4.11). $\square$

(4.13) Corollary. *Let M be a maximal subfield of degree n in a central simple F-algebra A of degree n. Suppose that $\mathrm{char}\, F \neq 2$ or that M/F is separable. Every involution which leaves M elementwise invariant is orthogonal.*

Proof: We have $C_A M = M$ by (1.5). Therefore, if σ is an involution on A which leaves M elementwise invariant, then $\sigma|_{C_A M} = \mathrm{Id}_M$ and (4.12) shows that σ is orthogonal if $\mathrm{char}\, F \neq 2$ or M/F is separable. $\square$

The result does not hold in characteristic 2 when M/F is not separable, as example (4.11) shows.

Extension of involutions. The following theorem is a kind of "Skolem-Noether theorem" for involutions. The first part is due to M. Kneser [155, p. 37]. (For a different proof, see Scharlau [247, §8.10].)

(4.14) Theorem. *Let B be a simple subalgebra of a central simple algebra A over a field F. Suppose that A and B have involutions σ and τ respectively which have the same restriction to F. Then A has an involution σ' whose restriction to B is τ.*

If σ is of the first kind, the types of σ' and τ are related as follows:

(1) *If char $F \neq 2$, then σ' can be arbitrarily chosen of orthogonal or symplectic type, except when the following two conditions hold: τ is of the first kind and the degree of the centralizer $C_A B$ of B in A is odd. In that case, every extension σ' of τ has the same type as τ.*

(2) *Suppose that char $F = 2$. If τ is of symplectic or unitary type, then σ' is symplectic. If τ is of orthogonal type and the center of B is a separable extension of F, then σ' can be arbitrarily chosen of orthogonal or symplectic type, except when the degree of the centralizer $C_A B$ is odd. In that case σ' is orthogonal.*

Proof: In order to show the existence of σ', we first reduce to the case where the centralizer $C_A B$ is a division algebra. Let $Z = B \cap C_A B$ be the center of B, hence also of $C_A B$ by the double centralizer theorem (see (1.5)). Wedderburn's theorem (1.1) yields a decomposition of $C_A B$:

$$C_A B = M \cdot D \simeq M \otimes_Z D$$

where M is a matrix algebra: $M \simeq M_r(Z)$ for some integer r, and D is a division algebra with center Z. Let $B' = B \cdot M \simeq B \otimes_Z M$ be the subalgebra of A generated by B and M. An involution * on $M_r(Z)$ of the same kind as τ can be defined by letting $\tau|_Z$ act entrywise and setting

$$a^* = \tau(a)^t \quad \text{for } a \in M_r(Z).$$

The involution $\tau \otimes \,^*$ on $B \otimes M_r(Z)$ extends τ and is carried to an involution τ' on B' through an isomorphism $B \otimes_Z M_r(Z) \simeq B \cdot M = B'$. It now remains to extend τ' to A. Note that the centralizer of B' is a division algebra, i.e., $C_A B' = D$.

Since σ and τ' have the same restriction to the center F of A, the Skolem-Noether theorem shows that $\sigma \circ \tau'$ is an inner automorphism. Let $\sigma \circ \tau' = \text{Int}(u)$ for some $u \in A^\times$, so that

$$(4.15) \qquad\qquad \sigma \circ \tau'(x)u = ux \quad \text{for } x \in B'.$$

Substituting $\tau'(x)$ for x, we get

$$\sigma(x)u = u\tau'(x) \quad \text{for } x \in B'$$

and, applying σ to both sides,

$$(4.16) \qquad\qquad \sigma \circ \tau'(x)\sigma(u) = \sigma(u)x \quad \text{for } x \in B'.$$

By comparing (4.15) and (4.16), we obtain $u^{-1}\sigma(u) \in C_A B'$. At least one of the elements $a_{+1} = 1 + u^{-1}\sigma(u)$, $a_{-1} = 1 - u^{-1}\sigma(u)$ is nonzero, hence invertible since $C_A B'$ is a division algebra. If a_ε is invertible (where $\varepsilon = \pm 1$), we have

$$\sigma \circ \tau' = \text{Int}(u) \circ \text{Int}(a_\varepsilon) = \text{Int}(ua_\varepsilon)$$

since $a_\varepsilon \in C_A B'$, and $ua_\varepsilon = u + \varepsilon\sigma(u) \in \text{Sym}(A, \sigma) \cup \text{Skew}(A, \sigma)$. Therefore, $\sigma \circ \text{Int}(ua_\varepsilon)$ $(= \text{Int}\big((ua_\varepsilon)^{-1}\big) \circ \sigma)$ is an involution on A whose restriction to B' is τ'. This completes the proof of the existence of an extension σ' of τ to A.

We now discuss the type of σ' (assuming that it is of the first kind, i.e., $\tau|_F = \text{Id}_F$). Suppose first that char $F \neq 2$. Since σ' extends τ, it preserves B, hence also its centralizer $C_A B$. It therefore restricts to an involution on $C_A B$. If τ is of the second kind, we may find some $v \in Z^\times$ such that $\sigma'(v) = -v$. Similarly, if τ is of the first kind, then (2.8) shows that we may find some $v \in (C_A B)^\times$ such that $\sigma'(v) = -v$, except when the degree of $C_A B$ is odd. Assuming we have such a v, the involution $\sigma'' = \mathrm{Int}(v) \circ \sigma'$ also extends τ since $v \in C_A B$, and it is of the type opposite to σ' since $v \in \mathrm{Skew}(A, \sigma')$ (see (2.7)). Therefore, τ has extensions of both types to A.

If the degree of $C_A B$ is odd and σ' leaves Z elementwise invariant, consider the restriction of σ' to the centralizer $C_A Z$. Since B has center Z we have, by (1.5),

$$C_A Z = B \otimes_Z C_A B.$$

The restriction $\sigma'|_{C_A Z}$ preserves both factors, hence it decomposes as

$$\sigma'|_{C_A Z} = \tau \otimes \sigma'|_{C_A B}.$$

Since the degree of $C_A B$ is odd, the involution $\sigma'|_{C_A B}$ is orthogonal by (2.8). Therefore, it follows from (2.23) that the involution $\sigma'|_{C_A Z}$ has the same type as τ. Proposition (4.12) shows that σ' and $\sigma'|_{C_A Z}$ have the same type and completes the proof in the case where char $F \neq 2$.

Suppose next that char $F = 2$. If τ is unitary, then it induces a nontrivial automorphism of order 2 on Z, hence there exists $z \in Z$ such that $z + \tau(z) = 1$. For every extension σ' of τ to A we have that $z + \sigma'(z) = 1$, hence $1 \in \mathrm{Alt}(A, \sigma')$ and σ' is symplectic by (2.6). Similarly, if τ is symplectic then $1 \in \mathrm{Alt}(B, \tau)$ hence also $1 \in \mathrm{Alt}(A, \sigma')$ for every extension σ' of τ. Therefore, every extension of τ is symplectic.

Suppose finally that τ is orthogonal and that Z/F is separable. As above, we have $C_A B = B \otimes_Z C_A B$. If $\deg C_A B$ is even, then (2.8) shows that we may find an involution θ_1 on $C_A B$ of orthogonal type and an involution θ_2 of symplectic type. By (2.23), the involution $\tau \otimes \theta_1$ (resp. $\tau \otimes \theta_2$) is orthogonal (resp. symplectic). It follows from (4.12) that every extension of this involution to A has the same type.

If $\deg C_A B$ is odd, the same arguments as in the case where char $F \neq 2$ show that every extension of τ is orthogonal. $\qquad \square$

If the subalgebra $B \subset A$ is not simple, much less is known on the possibility of extending an involution from B to A. We have however the following general result:

(4.17) Proposition. *Let A be a central simple algebra with involution of the first kind. Every element of A is invariant under some involution.*

Proof: Let $a \in A$ and let σ be an arbitrary orthogonal involution on A. Consider the vector space

$$V = \{\, x \in A \mid \sigma(x) = x, \; x\sigma(a) = ax \,\}.$$

It suffices to show that V contains an invertible element u, for then $\mathrm{Int}(u) \circ \sigma$ is an involution on A which leaves a invariant.

Let L be a splitting field of A. Fix an isomorphism $A_L \simeq M_n(L)$. The existence of $g \in \mathrm{GL}_n(L)$ such that $g^t = g$ and $ga^t = ag$ is shown in Kaplansky [150, Theorem 66] (see also Exercise 16). The involution $\tau = \mathrm{Int}(g) \circ t$ leaves a invariant, and it is orthogonal if char $F \neq 2$. By (2.7), there exists an invertible element $u \in A_L$ such that $\tau = \mathrm{Int}(u) \circ \sigma_L$ and $\sigma_L(u) = u$. Since τ leaves a invariant, we have $u \in V \otimes L$. We have thus shown that $V \otimes L$ contains an element whose reduced norm is nonzero, hence the Zariski-open subset in V consisting of elements whose reduced norm is nonzero is nonempty. If F is infinite, we may use density of the rational points to conclude that V contains an invertible element. If F is finite, we may take $L = F$ in the discussion above. $\qquad\square$

Note that if char $F \neq 2$ the proof yields a more precise result: every element is invariant under some orthogonal involution. Similar arguments apply to involutions of the second kind, as the next proposition shows.

(4.18) Proposition. *Let (B, τ) be a central simple F-algebra with involution of the second kind of degree n and let K be the center of B. For every $b \in B$ whose minimal polynomial over K has degree n and coefficients in F, there exists an involution of the second kind on B which leaves b invariant.*

Proof: Consider the F-vector space

$$W = \{\, x \in B \mid \tau(x) = x, \; x\tau(b) = bx \,\}.$$

As in the proof of (4.17), it suffices to show that W contains an invertible element. By (2.15), we may find a field extension L/F such that $B_L \simeq M_n(L) \times M_n(L)^{\mathrm{op}}$. Fix such an isomorphism. Since the minimal polynomial of b has degree n and coefficients in F, its image in $M_n(L) \times M_n(L)^{\mathrm{op}}$ has the form (m_1, m_2^{op}) where m_1, m_2 are matrices which have the same minimal polynomial of degree n. We may then find a matrix $u \in \mathrm{GL}_n(L)$ such that $um_2u^{-1} = m_1$. If ε is the exchange involution on $M_n(L) \times M_n(L)^{\mathrm{op}}$, the involution $\mathrm{Int}(u, u^{\mathrm{op}}) \circ \varepsilon$ leaves (m_1, m_2^{op}) invariant. This involution has the form $\mathrm{Int}(v) \circ \tau_L$ for some invertible element $v \in W \otimes L$. If F is infinite, we may conclude as in the proof of (4.17) that W also contains an invertible element, completing the proof.

If F is finite and $K \simeq F \times F$, the arguments above apply with $L = F$. The remaining case where F is finite and K is a field is left to the reader. (See Exercise 17.) $\qquad\square$

§5. Quadratic Forms

This section introduces the notion of a quadratic pair which is a twisted analogue of quadratic form in the same way that involutions are twisted analogues of symmetric, skew-symmetric or alternating forms (up to a scalar factor). The full force of this notion is in characteristic 2, since quadratic forms correspond bijectively to symmetric bilinear forms in characteristic different from 2. Nevertheless we place no restrictions on the characteristic of our base field F.

As a preparation for the proof that quadratic pairs on a split algebra $\mathrm{End}_F(V)$ correspond to quadratic forms on the vector space V (see (5.11)), we first show that every nonsingular bilinear form on V determines a standard identification $\mathrm{End}_F(V) = V \otimes_F V$. This identification is of central importance for the definition of the Clifford algebra of a quadratic pair in §8.

5.A. Standard identifications. In this subsection, D denotes a central division algebra over F and θ denotes an involution (of any kind) on D. Let V be a finite dimensional right vector space over D. We define[9] a left vector space $^{\theta}V$ over D by

$$^{\theta}V = \{\, ^{\theta}v \mid v \in V \,\}$$

---

[9]Note that this definition is consistent with those in §3.B and §4.A.

with the operations

$$^\theta v + {}^\theta w = {}^\theta(v + w) \quad \text{and} \quad \alpha \cdot {}^\theta v = {}^\theta\big(v \cdot \theta(\alpha)\big)$$

for v, $w \in V$ and $\alpha \in D$. We may then consider the tensor product $V \otimes_D {}^\theta V$ which is a vector space over F of dimension

$$\dim_F V \otimes_D {}^\theta V = \frac{\dim_F V \otimes_F V}{\dim_F D} = (\dim_F V)^2 \dim_F D.$$

Now, let $h \colon V \times V \to D$ be a nonsingular hermitian or skew-hermitian form on V with respect to θ. There is an F-linear map

$$\varphi_h \colon V \otimes_D {}^\theta V \to \operatorname{End}_D(V)$$

such that

$$\varphi_h(v \otimes {}^\theta w)(x) = v \cdot h(w, x) \quad \text{for } v,\, w,\, x \in V.$$

(5.1) Theorem. *The map φ_h is bijective. Letting σ_h denote the adjoint involution on $\operatorname{End}_D(V)$ with respect to h, we have*

$$\sigma_h\big(\varphi_h(v \otimes {}^\theta w)\big) = \delta \varphi_h(w \otimes {}^\theta v) \quad \text{for } v,\, w \in V,$$

where $\delta = +1$ if h is hermitian and $\delta = -1$ if h is skew-hermitian. Moreover,

$$\operatorname{Trd}_{\operatorname{End}_D(V)}\big(\varphi_h(v \otimes {}^\theta w)\big) = \operatorname{Trd}_D\big(h(w, v)\big) \quad \text{for } v,\, w \in V$$

and, for v_1, v_2, w_1, $w_2 \in V$,

$$\varphi_h(v_1 \otimes {}^\theta w_1) \circ \varphi_h(v_2 \otimes {}^\theta w_2) = \varphi_h\big(v_1 h(w_1, v_2) \otimes {}^\theta w_2\big).$$

Proof: Let $(e_1, \ldots, e_n)$ be a basis of V over D. Since h is nonsingular, for $i \in \{1, \ldots, n\}$ there exists a unique vector $e_i^\sharp \in V$ such that

$$h(e_i^\sharp, e_j) = \begin{cases} 1 & \text{if } i = j, \\ 0 & \text{if } i \neq j, \end{cases}$$

and $(e_1^\sharp, \ldots, e_n^\sharp)$ is a basis of V over D. Every element $x \in V \otimes_D {}^\theta V$ therefore has a unique expression of the form

$$x = \sum_{i,j=1}^n e_i a_{ij} \otimes {}^\theta e_j^\sharp \quad \text{for some } a_{ij} \in D.$$

The map φ_h takes the element x to the endomorphism of V with the matrix $(a_{ij})_{1 \le i, j \le n}$ (with respect to the basis $(e_1, \ldots, e_n)$), hence φ_h is bijective. Moreover, we have

$$\operatorname{Trd}_{\operatorname{End}_D(V)}\big(\varphi_h(x)\big) = \sum_{i=1}^n \operatorname{Trd}_D(a_{ii}) = \sum_{i,j=1}^n \operatorname{Trd}_D\big(h(e_j^\sharp, e_i a_{ij})\big),$$

hence in particular

$$\operatorname{Trd}_{\operatorname{End}_D(V)}\big(\varphi_h(v \otimes {}^\theta w)\big) = \operatorname{Trd}_D\big(h(w, v)\big) \quad \text{for } v,\, w \in V.$$

For v, w, x, $y \in V$ we have

$$h\big(x, \varphi_h(v \otimes {}^\theta w)(y)\big) = h(x, v) h(w, y)$$

and

$$h\big(\varphi_h(w \otimes {}^\theta v)(x), y\big) = h\big(w h(v, x), y\big) = \theta\big(h(v, x)\big) h(w, y),$$

hence $\sigma_h\big(\varphi_h(v \otimes {}^\theta w)\big) = \delta \varphi_h(w \otimes {}^\theta v)$.

Finally, for v_1, w_1, v_2, w_2, $x \in V$,

$$\varphi_h(v_1 \otimes {}^\theta w_1) \circ \varphi_h(v_2 \otimes {}^\theta w_2)(x) = v_1 h(w_1, v_2) h(w_2, x)$$
$$= \varphi_h\big(v_1 h(w_1, v_2) \otimes {}^\theta w_2\big)(x),$$

hence $\varphi_h(v_1 \otimes {}^\theta w_1) \circ \varphi_h(v_2 \otimes {}^\theta w_2) = \varphi_h\big(v_1 h(w_1, v_2) \otimes {}^\theta w_2\big)$. $\square$

Under φ_h, the F-algebra with involution $\big(\mathrm{End}_D(V), \sigma_h\big)$ is thus identified with $V \otimes_D {}^\theta V$ endowed with the product

$$(v_1 \otimes {}^\theta w_1) \circ (v_2 \otimes {}^\theta w_2) = v_1 h(w_1, v_2) \otimes {}^\theta w_2 \quad \text{for } v_1, v_2, w_1, w_2 \in V$$

and the involution σ defined by

$$\sigma(v \otimes {}^\theta w) = \delta w \otimes {}^\theta v \quad \text{for } v, w \in V,$$

where $\delta = +1$ if h is hermitian and $\delta = -1$ if h is skew-hermitian. We shall refer to the map φ_h in the sequel as the *standard identification* of $\big(\mathrm{End}_D(V), \sigma_h\big)$ with $(V \otimes_D {}^\theta V, \sigma)$. Note that the map φ_h depends on the choice of h and not just on the involution σ_h. Indeed, for any $\alpha \in F^\times$ fixed by θ we have $\sigma_{\alpha h} = \sigma_h$ but $\varphi_{\alpha h} = \alpha \varphi_h$.

The standard identification will be used mostly in the split case where $D = F$. If moreover $\theta = \mathrm{Id}_F$, then ${}^\theta V = V$, hence the standard identification associated to a nonsingular symmetric or skew-symmetric bilinear form b on V is

(5.2) $\varphi_b \colon (V \otimes_F V, \sigma) \xrightarrow{\sim} \big(\mathrm{End}_F(V), \sigma_b\big), \quad \varphi_b(v \otimes w)(x) = vb(w, x),$

where $\sigma(v \otimes w) = w \otimes v$ if b is symmetric and $\sigma(v \otimes w) = -w \otimes v$ if b is skew-symmetric.

(5.3) Example. As in (4.2), for all integers r, s let $^* \colon M_{r,s}(D) \to M_{s,r}(D)$ be the map defined by

$$(a_{ij})^*_{i,j} = \big(\theta(a_{ij})\big)^t_{i,j}.$$

Let $V = D^r \; (= M_{r,1}(D))$ and let $h \colon D^r \times D^r \to D$ be the hermitian form defined by

$$h(x, y) = x^* \cdot g \cdot y \quad \text{for } x, y \in D^r,$$

where $g \in M_r(D)$ is invertible and satisfies $g^* = g$.

Identify $M_r(D)$ with $\mathrm{End}_D(D^r)$ by mapping $m \in M_r(D)$ to the endomorphism $x \mapsto m \cdot x$. The standard identification

$$\varphi_h \colon D^r \otimes_D D^r \xrightarrow{\sim} M_r(D)$$

carries $v \otimes w \in D^r \otimes_D D^r$ to the matrix $v \cdot w^* \cdot g$, since for all $x \in D^r$

$$v \cdot w^* \cdot g \cdot x = vh(w, x).$$

5.B. Quadratic pairs. Let A be a central simple algebra of degree n over a field F of arbitrary characteristic.

(5.4) Definition. A *quadratic pair* on A is a couple (σ, f), where σ is an involution of the first kind on A and $f \colon \mathrm{Sym}(A, \sigma) \to F$ is a linear map, subject to the following conditions:

(1) $\dim_F \mathrm{Sym}(A, \sigma) = n(n+1)/2$ and $\mathrm{Trd}_A\big(\mathrm{Skew}(A, \sigma)\big) = \{0\}$.

(2) $f\big(x + \sigma(x)\big) = \mathrm{Trd}_A(x)$ for all $x \in A$.

Note that the equality $x + \sigma(x) = x' + \sigma(x')$ holds for x, $x' \in A$ if and only if $x - x' \in \mathrm{Skew}(A, \sigma)$. Therefore, condition (2) makes sense only if the reduced trace of every skew-symmetric element is zero, as required in (1).

If char $F \neq 2$, the equality $\mathrm{Trd}_A\big(\mathrm{Skew}(A, \sigma)\big) = \{0\}$ holds for every involution of the first kind, by (2.2), hence condition (1) simply means that σ is of orthogonal type. On the other hand, the map f is uniquely determined by (2) since for $s \in \mathrm{Sym}(A, \sigma)$ we have $s = \frac{1}{2}\big(s + \sigma(s)\big)$, hence $f(s) = \frac{1}{2}\mathrm{Trd}_A(s)$.

If char $F = 2$, Proposition (2.6) shows that condition (1) holds if and only if σ is symplectic (which implies that n is even). Condition (2) determines the value of f on the subspace $\mathrm{Symd}(A, \sigma)$ but not on $\mathrm{Sym}(A, \sigma)$. Indeed, in view of (2.13), condition (2) simply means that f is an extension of the linear form $\mathrm{Trp}_\sigma \colon \mathrm{Symd}(A, \sigma) \to F$. Therefore, there exist several quadratic pairs with the same symplectic involution.

(5.5) Example. Let τ be an involution of orthogonal type on A. Every element $a \in A$ such that $a + \tau(a)$ is invertible determines a quadratic pair (σ_a, f_a) as follows: let $g = a + \tau(a) \in A^\times$ and define

$$\sigma_a = \mathrm{Int}(g^{-1}) \circ \tau \quad \text{and} \quad f_a(s) = \mathrm{Trd}_A(g^{-1}as) \quad \text{for } s \in \mathrm{Sym}(A, \sigma_a).$$

From (2.7), it follows that σ_a is orthogonal if char $F \neq 2$ and symplectic if char $F = 2$. In order to check condition (2) of the definition of a quadratic pair, we compute

$$f_a\big(x + \sigma_a(x)\big) = \mathrm{Trd}_A(g^{-1}ax) + \mathrm{Trd}_A\big(g^{-1}ag^{-1}\tau(x)g\big) \quad \text{for } x \in A.$$

Since $ag^{-1}\tau(x)$ and $\tau\big(ag^{-1}\tau(x)\big) = xg^{-1}\tau(a)$ have the same trace, the last term on the right side is also equal to $\mathrm{Trd}_A\big(xg^{-1}\tau(a)\big)$, hence

$$f_a\big(x + \sigma_a(x)\big) = \mathrm{Trd}_A\big(xg^{-1}\big(a + \tau(a)\big)\big) = \mathrm{Trd}_A(x).$$

We will show in (5.8) that every quadratic pair is of the form (σ_a, f_a) for some $a \in A$ such that $a + \tau(a)$ is invertible.

We start with a couple of general results:

(5.6) Proposition. *For every quadratic pair (σ, f) on A,*

$$f(1) = \tfrac{1}{2} \deg A.$$

Proof: If char $F \neq 2$, we have $f(1) = \frac{1}{2}\mathrm{Trd}_A(1) = \frac{1}{2}\deg A$. If char $F = 2$, the proposition follows from (2.12) since $f(1) = \mathrm{Trp}_\sigma(1)$. $\qquad\square$

(5.7) Proposition. *For every quadratic pair (σ, f) on A, there exists an element $\ell \in A$, uniquely determined up to the addition of an element in $\mathrm{Alt}(A, \sigma)$, such that*

$$f(s) = \mathrm{Trd}_A(\ell s) \quad \text{for all } s \in \mathrm{Sym}(A, \sigma).$$

The element ℓ satisfies $\ell + \sigma(\ell) = 1$.

Proof: Since the bilinear reduced trace form on A is nonsingular by (1.8), every linear form $A \to F$ is of the form $x \mapsto \mathrm{Trd}_A(ax)$ for some $a \in A$. Therefore, extending f arbitrarily to a linear form on A, we may find some $\ell \in A$ such that $f(s) = \mathrm{Trd}_A(\ell s)$ for all $s \in \mathrm{Sym}(A, \sigma)$. If ℓ, $\ell' \in A$ both satisfy this relation, then $\mathrm{Trd}_A\big((\ell - \ell')s\big) = 0$ for all $s \in \mathrm{Sym}(A, \sigma)$, hence (2.3) shows that $\ell - \ell' \in \mathrm{Alt}(A, \sigma)$.

Condition (2) of the definition of a quadratic pair yields

$$\mathrm{Trd}_A\big(\ell\big(x + \sigma(x)\big)\big) = \mathrm{Trd}_A(x) \quad \text{for } x \in A.$$

Since $\mathrm{Trd}_A\big(\ell\sigma(x)\big) = \mathrm{Trd}_A\big(x\sigma(\ell)\big)$, it follows that

$$\mathrm{Trd}_A\big((\ell + \sigma(\ell))x\big) = \mathrm{Trd}_A(x) \quad \text{for } x \in A,$$

hence $\ell + \sigma(\ell) = 1$ since the bilinear reduced trace form on A is nonsingular. $\quad\square$

(5.8) Proposition. *Let τ be an orthogonal involution on A. Every quadratic pair on A is of the form (σ_a, f_a) for some $a \in A$ such that $a + \tau(a) \in A^\times$. If $a, b \in A$ are such that $a + \tau(a) \in A^\times$ and $b + \tau(b) \in A^\times$, then $(\sigma_a, f_a) = (\sigma_b, f_b)$ if and only if there exists $\lambda \in F^\times$ and $c \in \mathrm{Alt}(A, \tau)$ such that $a = \lambda b + c$.*

Proof: Let (σ, f) be a quadratic pair on A. By (2.7), there exists an invertible element $g \in A^\times$ such that $\sigma = \mathrm{Int}(g^{-1}) \circ \tau$ and $g \in \mathrm{Sym}(A, \tau)$ if char $F \neq 2$, and $g \in \mathrm{Alt}(A, \tau)$ if char $F = 2$. Moreover, (5.7) yields an element $\ell \in A$ such that $\sigma(\ell) + \ell = 1$ and $f(s) = \mathrm{Trd}_A(\ell s)$ for all $s \in \mathrm{Sym}(A, \sigma)$. Let $a = g\ell \in A$. Since $\tau(g) = g$ and $\sigma(\ell) + \ell = 1$, we have $a + \tau(a) = g$, hence $\sigma_a = \sigma$. Moreover, for $s \in \mathrm{Sym}(A, \sigma)$,

$$f_a(s) = \mathrm{Trd}_A(g^{-1}as) = \mathrm{Trd}_A(\ell s) = f(s),$$

hence $(\sigma, f) = (\sigma_a, f_a)$.

Suppose now that $a, b \in A$ are such that $a + \tau(a)$, $b + \tau(b)$ are each invertible and that $(\sigma_a, f_a) = (\sigma_b, f_b)$. Writing $g = a + \tau(a)$ and $h = b + \tau(b)$, we have $\sigma_a = \mathrm{Int}(g^{-1}) \circ \tau$ and $\sigma_b = \mathrm{Int}(h^{-1}) \circ \tau$, hence the equality $\sigma_a = \sigma_b$ yields $g = \lambda h$ for some $\lambda \in F^\times$. On the other hand, since $f_a = f_b$ we have

$$\mathrm{Trd}_A(g^{-1}as) = \mathrm{Trd}_A(h^{-1}bs) \quad \text{for } s \in \mathrm{Sym}(A, \sigma_a) = \mathrm{Sym}(B, \sigma_b).$$

Since $\mathrm{Sym}(A, \sigma_b) = \mathrm{Sym}(A, \tau)h$ and $g = \lambda h$, it follows that

$$\lambda^{-1}\,\mathrm{Trd}_A(ax) = \mathrm{Trd}_A(bx) \quad \text{for } x \in \mathrm{Sym}(A, \tau),$$

hence $a - \lambda b \in \mathrm{Alt}(A, \tau)$, by (2.3). $\quad\square$

This proposition shows that quadratic pairs on A are in one-to-one correspondence with equivalence classes of elements $a + \mathrm{Alt}(A, \tau) \in A/\mathrm{Alt}(A, \tau)$ such that $a + \tau(a)$ is invertible, modulo multiplication by a factor in $F^\times$.

In the particular case where $A = M_n(F)$ and $\tau = t$ is the transpose involution, the elements in $A/\mathrm{Alt}(A, \tau)$ may be regarded as quadratic forms of dimension n, by identifying $a + \mathrm{Alt}(A, \tau)$ with the quadratic form $q(X) = X \cdot a \cdot X^t$ where $X = (x_1, \ldots, x_n)$. The matrix $a + a^t$ is invertible if and only if the corresponding quadratic form is nonsingular (of even dimension if char $F = 2$). Therefore, quadratic pairs on $M_n(F)$ are in one-to-one correspondence with equivalence classes of nonsingular quadratic forms of dimension n modulo a factor in $F^\times$ (with n even if char $F = 2$).

(5.9) Example. Suppose that $n = 2$ and char $F = 2$. Straightforward computations show that the quadratic pair on $M_2(F)$ associated to the quadratic form $aX^2 + bXY + cY^2$ is (σ, f) with

$$\sigma\begin{pmatrix} a_{11} & a_{12} \\ a_{21} & a_{22} \end{pmatrix} = \begin{pmatrix} a_{22} & a_{12} \\ a_{21} & a_{11} \end{pmatrix}$$

and

$$f\begin{pmatrix} a_{11} & a_{12} \\ a_{21} & a_{11} \end{pmatrix} = a_{11} + ab^{-1}a_{12} + cb^{-1}a_{21}.$$

The observation above explains why quadratic pairs on central simple algebras may be thought of as twisted forms of nonsingular quadratic forms up to a scalar factor. We next give another perspective on this result by relating quadratic forms on a vector space V to quadratic pairs on the endomorphism algebra $\mathrm{End}_F(V)$.

Let V be a vector space of dimension n over F and let $q: V \to F$ be a quadratic form on V. We recall that b_q is the polar symmetric bilinear form of q,

$$b_q(x, y) = q(x + y) - q(x) - q(y) \quad \text{for } x, y \in V,$$

which we assume to be nonsingular. This hypothesis implies that n is even if $\mathrm{char}\, F = 2$, since the bilinear form b_q is alternating in this case. We write simply σ_q for the adjoint involution σ_{b_q} on $\mathrm{End}_F(V)$ and

(5.10) $$\varphi_q: V \otimes_F V \to \mathrm{End}_F(V), \quad \varphi_q(v \otimes w)(x) = v b_q(w, x)$$

for the standard identification φ_{b_q} of (5.2). Under this identification, we have

$$(V \otimes V, \sigma) = \big(\mathrm{End}_F(V), \sigma_q\big),$$

where $\sigma: V \otimes V \to V \otimes V$ is the switch.

(5.11) Proposition. *There is a unique linear map* $f_q: \mathrm{Sym}\big(\mathrm{End}_F(V), \sigma_q\big) \to F$ *such that*

$$f_q \circ \varphi_q(v \otimes v) = q(v) \quad \text{for } x, y \in V.$$

The couple (σ_q, f_q) *is a quadratic pair on* $\mathrm{End}_F(V)$. *Moreover, assuming that* $\dim_F V$ *is even if* $\mathrm{char}\, F = 2$, *every quadratic pair on* $\mathrm{End}_F(V)$ *is of the form* (σ_q, f_q) *for some nonsingular quadratic form* q *on* V *which is uniquely determined up to a factor in* $F^\times$.

Proof: Let $(e_1, \ldots, e_n)$ be a basis of V. The elements $\varphi_q(e_i \otimes e_i)$ for $i = 1, \ldots, n$ and $\varphi_q(e_i \otimes e_j + e_j \otimes e_i)$ for $i, j \in \{1, \ldots, n\}$, $j \neq i$, form a basis of $\mathrm{Sym}\big(\mathrm{End}_F(V), \sigma_q\big) = \varphi_q\big(\mathrm{Sym}(V \otimes V, \sigma)\big)$. Define

$$f_q\big(\varphi_q(e_i \otimes e_i)\big) = q(e_i), \quad f_q\big(\varphi_q(e_i \otimes e_j + e_j \otimes e_i)\big) = b_q(e_i, e_j)$$

and extend by linearity to a map $f_q: \mathrm{Sym}\big(\mathrm{End}_F(V), \sigma_q\big) \to F$. For $v = \sum_{i=1}^n e_i \alpha_i \in V$ we have

$$f_q \circ \varphi_q(v \otimes v) = f_q \circ \varphi_q\big(\textstyle\sum_{1 \leq i \leq n} e_i \otimes e_i \alpha_i^2 + \sum_{1 \leq i < j \leq n}(e_i \otimes e_j + e_j \otimes e_i)\alpha_i \alpha_j\big)$$
$$= \textstyle\sum_{1 \leq i \leq n} q(e_i)\alpha_i^2 + \sum_{1 \leq i < j \leq n} b_q(e_i, e_j)\alpha_i \alpha_j = q(v),$$

hence the map f_q thus defined satisfies the required condition. Uniqueness of f_q is clear, since $\mathrm{Sym}\big(\mathrm{End}_F(V), \sigma_q\big)$ is spanned by elements of the form $\varphi_q(v \otimes v)$.

Since the bilinear form q is symmetric, and alternating if $\mathrm{char}\, F = 2$, the involution σ_q is orthogonal if $\mathrm{char}\, F \neq 2$ and symplectic if $\mathrm{char}\, F = 2$. To check that (σ_q, f_q) is a quadratic pair, it remains to prove that

$$f_q\big(x + \sigma(x)\big) = \mathrm{Trd}_{\mathrm{End}_F(V)}(x) \quad \text{for } x \in \mathrm{End}_F(V).$$

Since both sides are linear, it suffices to check this formula for $x = \varphi_q(v \otimes w)$ with $v, w \in V$. The left side is then

$$f_q \circ \varphi_q(v \otimes w + w \otimes v) = f_q \circ \varphi_q\big((v + w) \otimes (v + w)\big) - f_q \circ \varphi_q(v \otimes v)$$
$$- f_q \circ \varphi_q(w \otimes w)$$
$$= b_q(v, w)$$

and the claim follows, since (5.1) shows that $b_q(v, w) = b_q(w, v) = \mathrm{Trd}\big(\varphi_q(v \otimes w)\big)$.

Suppose now that (σ, f) is an arbitrary quadratic pair on $\mathrm{End}_F(V)$. As shown in the introduction to this chapter (and in (4.2)), the involution σ is the adjoint involution with respect to some nonsingular symmetric bilinear form $b \colon V \times V \to F$ which is uniquely determined up to a factor in $F^\times$. Use the standard identification φ_b (see (5.2)) to define a map $q \colon V \to F$ by

$$q(v) = f \circ \varphi_b(v \otimes v) \quad \text{for } v \in V.$$

From the definition, it is clear that $q(v\alpha) = q(v)\alpha^2$ for $\alpha \in F$. Moreover, for v, $w \in V$ we have

$$
\begin{aligned}
q(v + w) - q(v) - q(w) &= f \circ \varphi_b(v \otimes w) + f \circ \varphi_b(w \otimes v) \\
&= f\big(\varphi_b(v \otimes w) + \sigma\big(\varphi_b(v \otimes w)\big)\big).
\end{aligned}
$$

Since (σ, f) is a quadratic pair, the right side is equal to

$$\mathrm{Trd}_{\mathrm{End}_F(V)}\big(\varphi_b(v \otimes w)\big) = b(w, v) = b(v, w).$$

Therefore, q is a quadratic form with associated polar form b, and it is clear that the corresponding quadratic pair (σ_q, f_q) is (σ, f). Since b is uniquely determined up to a factor in $F^\times$, the same property holds for q. $\qquad\square$

For later use, we give an explicit description of an element $\ell \in \mathrm{End}_F(V)$ satisfying property (5.7) for the quadratic pair (σ_q, f_q). It suffices to consider the case where $\mathrm{char}\, F = 2$, since otherwise we may take $\ell = \frac{1}{2}$.

(5.12) Proposition. *Let (V, q) be a nonsingular quadratic space of even dimension $n = 2m$ over a field F of characteristic 2 and let $(e_1, \dots, e_n)$ be a symplectic basis of V for the bilinear form b_q, i.e., a basis such that*

$$b_q(e_{2i-1}, e_{2i}) = 1, \quad b_q(e_{2i}, e_{2i+1}) = 0 \quad \text{and} \quad b_q(e_i, e_j) = 0 \quad \text{if } |i - j| > 1.$$

Set

$$\ell = \varphi_q\big(\textstyle\sum_{i=1}^m e_{2i-1} \otimes e_{2i-1} q(e_{2i}) + e_{2i} \otimes e_{2i} q(e_{2i-1}) + e_{2i-1} \otimes e_{2i}\big) \in \mathrm{End}_F(V).$$

(1) *The element ℓ satisfies $\mathrm{tr}(\ell s) = f(s)$ for all $s \in \mathrm{Sym}\big(\mathrm{End}_F(V), \sigma_q\big)$.*
(2) *The characteristic polynomial of ℓ equals*

$$\mathrm{Pc}_\ell(X) = \big(X^2 + X + q(e_1)q(e_2)\big) \cdots \big(X^2 + X + q(e_{2m-1})q(e_{2m})\big),$$

hence

$$s_2(\ell) = \big(\textstyle\sum_{i=1}^m q(e_{2i-1})q(e_{2i})\big) + \frac{m(m-1)}{2}.$$

Proof: It suffices to check the equation for s of the form $\varphi_q(v \otimes v)$ with $v \in V$, since these elements span $\mathrm{Sym}\big(\mathrm{End}_F(V), \sigma\big)$. If $v = \sum_{i=1}^n e_i\alpha_i$, we have $b_q(e_{2i-1}, v) = \alpha_{2i}$ and $b_q(e_{2i}, v) = \alpha_{2i-1}$, hence

$$
\begin{aligned}
&\ell\varphi_q(v \otimes v) = \\
&\quad \varphi_q\big(\textstyle\sum_{i=1}^m (e_{2i-1} \otimes v)\alpha_{2i}q(e_{2i}) + (e_{2i} \otimes v)\alpha_{2i-1}q(e_{2i-1}) + (e_{2i-1} \otimes v)\alpha_{2i-1}\big).
\end{aligned}
$$

It follows that

$$\operatorname{tr}\bigl(\ell\varphi_q(v \otimes v)\bigr) = \sum_{i=1}^{m}\bigl(b_q(v, e_{2i-1})\alpha_{2i}q(e_{2i}) + b_q(v, e_{2i})\alpha_{2i-1}q(e_{2i-1})$$

$$+ \, b_q(v, e_{2i-1})\alpha_{2i-1}\bigr)$$

$$= \sum_{i=1}^{n}\alpha_i^2 q(e_i) + \sum_{i=1}^{m}\alpha_{2i-1}\alpha_{2i} = q(v).$$

Since $q(v) = f \circ \varphi_q(v \otimes v)$, the first assertion is proved. Identifying $\operatorname{End}_F(V)$ with $M_n(F)$ by means of the basis $(e_1, \ldots, e_n)$ maps ℓ to the matrix

$$\ell = \begin{pmatrix} \ell_1 & & 0 \\ & \ddots & \\ 0 & & \ell_m \end{pmatrix} \quad \text{where} \quad \ell_i = \begin{pmatrix} 1 & q(e_{2i}) \\ q(e_{2i-1}) & 0 \end{pmatrix}$$

The characteristic polynomial of ℓ is the product of the characteristic polynomials of $\ell_1, \ldots, \ell_m$. This implies the second assertion. $\qquad\square$

(5.13) Example. Suppose that $\operatorname{char} F = 2$ and let $(1, u, v, w)$ be a quaternion basis of a quaternion F-algebra $Q = [a, b)_F$. In every quadratic pair (σ, f) on Q, the involution σ is symplectic. It is therefore the canonical involution γ. The space $\operatorname{Sym}(Q, \gamma)$ is the span of 1, v, w, and $\operatorname{Alt}(Q, \gamma) = F$. Since $1 = u + \gamma(u)$ and $\operatorname{Trd}_Q(u) = 1$, the map f may be any linear form on $\operatorname{Sym}(Q, \gamma)$ such that $f(1) = 1$. An element ℓ corresponding to f as in (5.7) is

$$\ell = u + f(w)b^{-1}v + f(v)b^{-1}w.$$

(For a given f, the element ℓ is uniquely determined up to the addition of an element in F.)

Quadratic pairs on tensor products. Let A_1, A_2 be central simple F-algebras. Given a quadratic pair (σ_1, f_1) on A_1 and an involution σ_2 on A_2, we aim to define a quadratic pair on the tensor product $A_1 \otimes_F A_2$. If $\operatorname{char} F \neq 2$, this amounts to defining an orthogonal involution on $A_1 \otimes_F A_2$, and it suffices to take $\sigma_1 \otimes \sigma_2$, assuming that σ_2 is orthogonal, see (2.23). For the rest of this section, we may thus focus on the case where $\operatorname{char} F = 2$.

(5.14) Lemma. *Let (A_1, σ_1) and (A_2, σ_2) be central simple algebras with involution of the first kind over a field F of characteristic 2.*

(5.15) $\quad \operatorname{Symd}(A_1, \sigma_1) \otimes \operatorname{Symd}(A_2, \sigma_2) =$

$\qquad\qquad \operatorname{Symd}(A_1, \sigma_1) \otimes \operatorname{Sym}(A_2, \sigma_2) \cap \operatorname{Sym}(A_1, \sigma_1) \otimes \operatorname{Symd}(A_2, \sigma_2);$

(5.16) $\quad \operatorname{Symd}(A_1 \otimes A_2, \sigma_1 \otimes \sigma_2) \cap \bigl(\operatorname{Sym}(A_1, \sigma_1) \otimes \operatorname{Sym}(A_2, \sigma_2)\bigr) =$

$\qquad\qquad \operatorname{Symd}(A_1, \sigma_1) \otimes \operatorname{Sym}(A_2, \sigma_2) + \operatorname{Sym}(A_1, \sigma_1) \otimes \operatorname{Symd}(A_2, \sigma_2);$

(5.17) $\quad \operatorname{Sym}(A_1 \otimes A_2, \sigma_1 \otimes \sigma_2) =$

$\qquad\qquad \operatorname{Symd}(A_1 \otimes A_2, \sigma_1 \otimes \sigma_2) + \bigl(\operatorname{Sym}(A_1, \sigma_1) \otimes \operatorname{Sym}(A_2, \sigma_2)\bigr).$

Proof: Equation (5.15) is clear. For $x_1 \in A_1$ and $s_2 \in \mathrm{Sym}(A_2, \sigma_2)$,

$$\bigl(x_1 + \sigma_1(x_1)\bigr) \otimes s_2 = x_1 \otimes s_2 + (\sigma_1 \otimes \sigma_2)(x_1 \otimes s_2),$$

hence $\mathrm{Symd}(A_1, \sigma_1) \otimes \mathrm{Sym}(A_2, \sigma_2) \subset \mathrm{Symd}(A_1 \otimes A_2, \sigma_1 \otimes \sigma_2)$. Similarly,

$$\mathrm{Sym}(A_1, \sigma_1) \otimes \mathrm{Symd}(A_2, \sigma_2) \subset \mathrm{Symd}(A_1 \otimes A_2, \sigma_1 \otimes \sigma_2),$$

hence the left side of (5.16) contains the right side. To prove the reverse inclusion, consider $x \in A_1 \otimes A_2$. If $x + (\sigma_1 \otimes \sigma_2)(x) \in \mathrm{Sym}(A_1, \sigma_1) \otimes \mathrm{Sym}(A_2, \sigma_2)$, then $x + (\sigma_1 \otimes \sigma_2)(x)$ is invariant under $\sigma_1 \otimes \mathrm{Id}_{A_2}$, hence

$$x + (\sigma_1 \otimes \mathrm{Id}_{A_2})(x) + (\mathrm{Id}_{A_1} \otimes \sigma_2)(x) + (\sigma_1 \otimes \sigma_2)(x) = 0.$$

Therefore, the element $u = x + (\sigma_1 \otimes \mathrm{Id}_{A_2})(x)$ is invariant under $\mathrm{Id}_{A_1} \otimes \sigma_2$, hence it lies in $A_1 \otimes \mathrm{Sym}(A_2, \sigma_2)$. Similarly, the element $v = (\sigma_1 \otimes \mathrm{Id}_{A_2})(x) + (\sigma_1 \otimes \sigma_2)(x)$ is in $\mathrm{Sym}(A_1, \sigma_1) \otimes A_2$. On the other hand, it is clear by definition that $u \in \mathrm{Symd}(A_1, \sigma_1) \otimes A_2$ and $v \in A_1 \otimes \mathrm{Symd}(A_2, \sigma_2)$, hence

$$u \in \mathrm{Symd}(A_1, \sigma_1) \otimes \mathrm{Sym}(A_2, \sigma_2) \qquad \text{and} \qquad v \in \mathrm{Sym}(A_1, \sigma_1) \otimes \mathrm{Symd}(A_2, \sigma_2).$$

Since $x + (\sigma_1 \otimes \sigma_2)(x) = u + v$, the proof of (5.16) is complete.

Since the left side of equation (5.17) obviously contains the right side, it suffices to prove that both sides have the same dimension. Let $n_i = \deg A_i$, so that $\dim_F \mathrm{Sym}(A_i, \sigma_i) = \frac{1}{2} n_i (n_i + 1)$ and $\dim_F \mathrm{Symd}(A_i, \sigma_i) = \frac{1}{2} n_i (n_i - 1)$ for $i = 1, 2$. From (5.15), it follows that

$$\dim_F \bigl(\mathrm{Symd}(A_1, \sigma_1) \otimes \mathrm{Sym}(A_2, \sigma_2) + \mathrm{Sym}(A_1, \sigma_1) \otimes \mathrm{Symd}(A_2, \sigma_2)\bigr) =$$

$$\tfrac{1}{4} n_1 n_2 (n_1 - 1)(n_2 + 1) + \tfrac{1}{4} n_1 n_2 (n_1 + 1)(n_2 - 1) - \tfrac{1}{4} n_1 n_2 (n_1 - 1)(n_2 - 1)$$

$$= \tfrac{1}{4} n_1 n_2 (n_1 n_2 + n_1 + n_2 - 3).$$

Therefore, (5.16) yields

$$\dim_F \bigl(\mathrm{Symd}(A_1 \otimes A_2, \sigma_1 \otimes \sigma_2) + \mathrm{Sym}(A_1, \sigma_1) \otimes \mathrm{Sym}(A_2, \sigma_2)\bigr) =$$

$$\tfrac{1}{2} n_1 n_2 (n_1 n_2 - 1) + \tfrac{1}{4} n_1 n_2 (n_1 + 1)(n_2 + 1) - \tfrac{1}{4} n_1 n_2 (n_1 n_2 + n_1 + n_2 - 3)$$

$$= \tfrac{1}{2} n_1 n_2 (n_1 n_2 + 1) = \dim_F \mathrm{Sym}(A_1 \otimes A_2, \sigma_1 \otimes \sigma_2). \qquad \square$$

(5.18) Proposition. *Suppose that* $\mathrm{char}\, F = 2$. *Let* (σ_1, f_1) *be a quadratic pair on a central simple F-algebra A_1 and let (A_2, σ_2) be a central simple F-algebra with involution of the first kind. There is a unique quadratic pair $(\sigma_1 \otimes \sigma_2, f_{1*})$ on $A_1 \otimes_F A_2$ such that*

$$f_{1*}(s_1 \otimes s_2) = f_1(s_1) \mathrm{Trd}_{A_2}(s_2)$$

for $s_1 \in \mathrm{Sym}(A_1, \sigma_1)$ and $s_2 \in \mathrm{Sym}(A_2, \sigma_2)$.

Proof: Since σ_1 is symplectic, (2.23) shows that $\sigma_1 \otimes \sigma_2$ is symplectic. To prove the existence of a quadratic pair $(\sigma_1 \otimes \sigma_2, f_{1*})$ as above, we have to show that the values that f_{1*} is required to take on $\mathrm{Symd}(A_1 \otimes A_2, \sigma_1 \otimes \sigma_2)$ because of the quadratic pair conditions agree with the prescribed values on $\mathrm{Sym}(A_1, \sigma_1) \otimes \mathrm{Sym}(A_2, \sigma_2)$. In view of the description of $\mathrm{Symd}(A_1 \otimes A_2, \sigma_1 \otimes \sigma_2) \cap \bigl(\mathrm{Sym}(A_1, \sigma_1) \otimes \mathrm{Sym}(A_2, \sigma_2)\bigr)$ in the preceding lemma, it suffices to consider the values of f_{1*} on elements of the form $\bigl(x_1 + \sigma_1(x_1)\bigr) \otimes s_2 = x_1 \otimes s_2 + (\sigma_1 \otimes \sigma_2)(x_1 \otimes s_2)$ or $s_1 \otimes \bigl(x_2 + \sigma_2(x_2)\bigr) = s_1 \otimes x_2 + (\sigma_1 \otimes \sigma_2)(s_1 \otimes x_2)$ with $x_i \in A_i$ and $s_i \in \mathrm{Sym}(A_i, \sigma_i)$ for $i = 1, 2$. Since (σ_1, f_1) is a quadratic pair on A_1, we have

$$f_1\bigl(x_1 + \sigma_1(x_1)\bigr) \mathrm{Trd}_{A_2}(s_2) = \mathrm{Trd}_{A_1}(x_1) \mathrm{Trd}_{A_2}(s_2) = \mathrm{Trd}_{A_1 \otimes A_2}(x_1 \otimes s_2),$$

as required. For the second type of element we have

$$f_1(s_1) \operatorname{Trd}_{A_2}\big(x_2 + \sigma_2(x_2)\big) = 0.$$

On the other hand, since σ_1 is symplectic we have $\operatorname{Trd}_{A_1}(s_1) = 0$, hence

$$\operatorname{Trd}_{A_1 \otimes A_2}(s_1 \otimes x_2) = \operatorname{Trd}_{A_1}(s_1) \operatorname{Trd}_{A_2}(x_2) = 0.$$

Therefore,

$$f_1(s_1) \operatorname{Trd}_{A_2}\big(x_2 + \sigma_2(x_2)\big) = \operatorname{Trd}_{A_1 \otimes A_2}(s_1 \otimes x_2)$$

for $s_1 \in \operatorname{Sym}(A_1, \sigma_1)$ and $x_2 \in A_2$, and the existence of the quadratic pair $(\sigma_1 \otimes \sigma_2, f_{1*})$ is proved.

Uniqueness is clear, since the values of the linear map f_{1*} are determined on the set $\operatorname{Symd}(A_1 \otimes A_2, \sigma_1 \otimes \sigma_2)$ and on $\operatorname{Sym}(A_1, \sigma_1) \otimes \operatorname{Sym}(A_2, \sigma_2)$, and (5.17) shows that these subspaces span $\operatorname{Sym}(A_1 \otimes A_2, \sigma_1 \otimes \sigma_2)$. □

(5.19) Example. Let (V_1, q_1) be a nonsingular quadratic space of even dimension and let (V_2, b_2) be a nonsingular symmetric bilinear space over a field F of characteristic 2. Let (σ_1, f_1) be the quadratic pair on $A_1 = \operatorname{End}_F(V_1)$ associated with q_1 (see (5.11)) and let $\sigma_2 = \sigma_{b_2}$ denote the adjoint involution with respect to b_2 on $A_2 = \operatorname{End}_F(V_2)$. We claim that, under the canonical isomorphism $A_1 \otimes A_2 = \operatorname{End}_F(V_1 \otimes V_2)$, the quadratic pair $(\sigma_1 \otimes \sigma_2, f_{1*})$ is associated with the quadratic form $q_1 \otimes b_2$ on $V_1 \otimes V_2$ whose polar form is $b_{q_1} \otimes b_2$ and such that

$$(q_1 \otimes b_2)(v_1 \otimes v_2) = q_1(v_1) b_2(v_2, v_2) \qquad \text{for } v_1 \in V_1 \text{ and } v_2 \in V_2.$$

Indeed, letting φ_1, φ_2 and φ denote the standard identifications $V_1 \otimes V_1 \xrightarrow{\sim} \operatorname{End}_F(V_1)$, $V_2 \otimes V_2 \xrightarrow{\sim} \operatorname{End}_F(V_2)$ and $(V_1 \otimes V_2) \otimes (V_1 \otimes V_2) \xrightarrow{\sim} \operatorname{End}_F(V_1 \otimes V_2)$ associated with the bilinear forms b_{q_1}, b_2 and $b_{q_1} \otimes b_2$, we have

$$\varphi(v_1 \otimes v_2 \otimes v_1 \otimes v_2) = \varphi_1(v_1 \otimes v_1) \otimes \varphi_2(v_2 \otimes v_2)$$

and

$$f_1\big(\varphi_1(v_1 \otimes v_1)\big) \operatorname{Trd}_{A_2}\big(\varphi_2(v_2 \otimes v_2)\big) = q_1(v_1) b_2(v_2, v_2),$$

hence

$$f_{1*}\big(\varphi(v_1 \otimes v_2 \otimes v_1 \otimes v_2)\big) = q_1 \otimes b_2(v_1 \otimes v_2).$$

(5.20) Corollary. *Let* (A_1, σ_1), (A_2, σ_2) *be central simple algebras with symplectic involutions over a field* F *of arbitrary characteristic. There is a unique quadratic pair* $(\sigma_1 \otimes \sigma_2, f_\otimes)$ *on* $A_1 \otimes A_2$ *such that* $f_\otimes(s_1 \otimes s_2) = 0$ *for all* $s_1 \in \operatorname{Skew}(A_1, \sigma_1)$, $s_2 \in \operatorname{Skew}(A_2, \sigma_2)$.

Proof: If $\operatorname{char} F \neq 2$, the linear form $f_\otimes$ which is the restriction of $\frac{1}{2} \operatorname{Trd}_{A_1 \otimes A_2}$ to $\operatorname{Sym}(A_1 \otimes A_2, \sigma_1 \otimes \sigma_2)$ satisfies

$$f_\otimes(s_1 \otimes s_2) = \tfrac{1}{2} \operatorname{Trd}_{A_1}(s_1) \operatorname{Trd}_{A_2}(s_2) = 0$$

for all $s_1 \in \operatorname{Skew}(A_1, \sigma_1)$, $s_2 \in \operatorname{Skew}(A_2, \sigma_2)$. Suppose next that $\operatorname{char} F = 2$. For any linear form f_1 on $\operatorname{Sym}(A_1, \sigma_1)$ we have $f_1(s_1) \operatorname{Trd}_{A_2}(s_2) = 0$ for all $s_1 \in \operatorname{Sym}(A_1, \sigma_1)$, $s_2 \in \operatorname{Sym}(A_2, \sigma_2)$, since σ_2 is symplectic. Therefore, we may set

$$(\sigma_1 \otimes \sigma_2, f_\otimes) = (\sigma_1 \otimes \sigma_2, f_{1*})$$

for any quadratic pair (σ_1, f_1) on A_1. Uniqueness of $f_\otimes$ follows from (5.17). □

(5.21) Definition. The quadratic pair $(\sigma_1 \otimes \sigma_2, f_\otimes)$ of (5.20) is called the *canonical quadratic pair* on $A_1 \otimes A_2$.

EXERCISES

1. Let A be a central simple algebra over a field F and fix $a \in A$. Show that there is a canonical F-algebra isomorphism $\mathrm{End}_A(aA) \simeq \mathrm{End}_A(Aa)$ which takes $f \in \mathrm{End}_A(aA)$ to the endomorphism $\widehat{f} \in \mathrm{End}_A(Aa)$ defined by $(xa)^{\widehat{f}} = xf(a)$ for $x \in A$, and the inverse takes $g \in \mathrm{End}_A(Aa)$ to the endomorphism $\widetilde{g} \in \mathrm{End}_A(aA)$ defined by $\widetilde{g}(ax) = a^g x$ for $x \in A$.

   Show that there is a canonical F-algebra isomorphism $\mathrm{End}_A(Aa)^{\mathrm{op}} \xrightarrow{\sim} \mathrm{End}_{A^{\mathrm{op}}}(a^{\mathrm{op}} A^{\mathrm{op}})$ which, for $f \in \mathrm{End}_A(Aa)$, maps f^{op} to the endomorphism $\bar{f}$ defined by $\bar{f}(m^{\mathrm{op}}) = (m^f)^{\mathrm{op}}$. Therefore, there is a canonical isomorphism $\mathrm{End}_A(Aa)^{\mathrm{op}} \simeq \mathrm{End}_{A^{\mathrm{op}}}(A^{\mathrm{op}} a^{\mathrm{op}})$. Use it to identify $(\lambda^k A)^{\mathrm{op}} = \lambda^k(A^{\mathrm{op}})$, for $k = 1, \ldots, \deg A$.

2. Let Q be a quaternion algebra over a field F of arbitrary characteristic. Show that the conjugation involution is the only linear map $\sigma \colon Q \to Q$ such that $\sigma(1) = 1$ and $\sigma(x)x \in F$ for all $x \in F$.

3. (Rowen-Saltman [238]) Let V be a vector space of dimension n over a field F and let τ be an involution of the first kind on $\mathrm{End}_F(V)$. Prove that τ is orthogonal if and only if there exist n symmetric orthogonal[10] idempotents in $\mathrm{End}_F(V)$. Find a similar characterization of the symplectic involutions on $\mathrm{End}_F(V)$.

4. Let A be a central simple algebra with involution σ of the first kind. Show that σ is orthogonal if and only if it restricts to the identity on a maximal étale (commutative) subalgebra of A.

   Hint: Extend scalars and use the preceding exercise.

5. Show that in a central simple algebra with involution, every left or right ideal is generated by a symmetric element, unless the algebra is split and the involution is symplectic.

6. (Albert) Let b be a symmetric, nonalternating bilinear form on a vector space V over a field of characteristic 2. Show that V contains an orthogonal basis for b.

7. Let $(a_i)_{i=1,\ldots,n^2}$ be an arbitrary basis of a central simple algebra A, and let $(b_i)_{i=1,\ldots,n^2}$ be the dual basis for the bilinear form T_A, which means that $\mathrm{Trd}_A(a_i b_j) = \delta_{ij}$ for $i, j = 1, \ldots, n^2$. Show that the Goldman element of A is $\sum_{i=1}^{n^2} a_i \otimes b_i$.

   Hint: Reduce by scalar extension to the split case and show that it suffices to prove the assertion for the standard basis of $M_n(F)$.

8. Let $(1, i, j, k)$ be a quaternion basis in a quaternion algebra Q of characteristic different from 2. Show that the Goldman element in $Q \otimes Q$ is $g = \frac{1}{2}(1 \otimes 1 + i \otimes i^{-1} + j \otimes j^{-1} + k \otimes k^{-1})$. Let $(1, u, v, w)$ be a quaternion basis in a quaternion algebra Q of characteristic 2. Show that the Goldman element in $Q \otimes Q$ is $g = 1 \otimes 1 + u \otimes 1 + 1 \otimes u + w \otimes v^{-1} + v^{-1} \otimes w$.

9. Let K/F be a quadratic extension of fields of characteristic different from 2, and let $a \in F^\times$, $b \in K^\times$. Prove the "projection formula" for the norm of the

---

[10]Two idempotents e, f are called *orthogonal* if $ef = fe = 0$.

quaternion algebra $(a, b)_K$:

$$N_{K/F}(a, b)_K \sim \left(a, N_{K/F}(b)\right)_F$$

(where $\sim$ denotes Brauer-equivalence). Prove corresponding statements in characteristic 2: if K/F is a separable quadratic extension of fields and $a \in F$, $b \in K^\times$, $c \in K$, $d \in F^\times$,

$$N_{K/F}[a, b)_K \sim [a, N_{K/F}(b))_F \quad \text{and} \quad N_{K/F}[c, d)_K \sim [T_{K/F}(c), d)_F.$$

Hint (when char $F \neq 2$): Let ι be the non-trivial automorphism of K/F and let $(1, i_1, j_1, k_1)$ (resp. $(1, i_2, j_2, k_2)$) denote the usual quaternion basis of $\left(a, \iota(b)\right)_K = {}^\iota(a, b)_K$ (resp. $(a, b)_K$). Let s be the switch map on ${}^\iota(a, b)_K \otimes_K (a, b)_K$. Let $u = \frac{1}{2}\left(1 + a^{-1} i_1 \otimes i_2 + b^{-1} j_1 \otimes j_2 - (ab)^{-1} k_1 \otimes k_2\right) \in \left(a, \iota(b)\right)_K \otimes_K (a, b)_K$. Show that $s(u)u = 1$, and that $s' = \mathrm{Int}(u) \circ s$ is a semi-linear automorphism of order 2 of ${}^\iota(a, b)_K \otimes_K (a, b)_K$ which leaves invariant $i_1 \otimes 1$, $1 \otimes i_2$ and $j_1 \otimes j_2$. Conclude that the F-subalgebra of elements invariant under s' is Brauer-equivalent to $\left(a, N_{K/F}(b)\right)_F$. If $b \notin F$, let $v = 1 + u$; if $b \in F$, pick $c \in K$ such that $c^2 \notin F$ and let $v = c + u\iota(c)$. Show that $u = s'(v)^{-1} v$ and that $\mathrm{Int}(v)$ maps the subalgebra of s-invariant elements onto the subalgebra of s'-invariant elements.

10. (Knus-Parimala-Srinivas [166, Theorem 4.1]) Let A be a central simple algebra over a field F, let V be an F-vector space and let

$$\rho \colon A \otimes_F A \to \mathrm{End}_F(V)$$

be an isomorphism of F-algebras. We consider V as a left A-module via $av = \rho(a \otimes 1)(v)$ and identify $\mathrm{End}_A(A) = A$, $\mathrm{Hom}_A(A, V) = V$ by mapping every homomorphism f to 1^f. Moreover, we identify $\mathrm{End}_A(V) = A^{\mathrm{op}}$ by setting

$$v^{a^{\mathrm{op}}} = \rho(1 \otimes a)(v) \qquad \text{for } v \in V, \, a \in A$$

and $\mathrm{Hom}_A(V, A) = V^* (= \mathrm{Hom}_F(V, F))$ by mapping $h \in \mathrm{Hom}_A(V, A)$ to the linear form

$$v \mapsto \mathrm{Trd}_A(v^h) \qquad \text{for } v \in V.$$

Let

$$B = \mathrm{End}_A(A \oplus V) = \begin{pmatrix} \mathrm{End}_A(A) & \mathrm{Hom}_A(A, V) \\ \mathrm{Hom}_A(V, A) & \mathrm{End}_A(V) \end{pmatrix} = \begin{pmatrix} A & V \\ V^* & A^{\mathrm{op}} \end{pmatrix};$$

this is a central simple F-algebra which is Brauer-equivalent to A, by Proposition (1.10). Let $\gamma = \rho(g) \in \mathrm{End}_F(V)$, where $g \in A \otimes_F A$ is the Goldman element. Show that

$$\begin{pmatrix} a & v \\ f & b^{\mathrm{op}} \end{pmatrix} \mapsto \begin{pmatrix} b & \gamma(v) \\ \gamma^t(f) & a^{\mathrm{op}} \end{pmatrix}$$

is an involution of orthogonal type of B.

11. (Knus-Parimala-Srinivas [166, Theorem 4.2]) Let K/F be a separable quadratic extension with nontrivial automorphism ι, let A be a central simple K-algebra, let V be an F-vector space and let

$$\rho \colon N_{K/F}(A) \to \mathrm{End}_F(V)$$

be an isomorphism of F-algebras. Set $W = V \otimes_F K$ and write

$$\widetilde{\rho} \colon A \otimes_K {}^\iota A \to \mathrm{End}_K(W)$$

for the isomorphism induced from ρ by extension of scalars. We consider W as a left A-module via $av = \widetilde{\rho}(a \otimes 1)(v)$. Let

$$B = \operatorname{End}_A(A \oplus W) = \begin{pmatrix} \operatorname{End}_A(A) & \operatorname{Hom}_A(A,W) \\ \operatorname{Hom}_A(W,A) & \operatorname{End}_A(W) \end{pmatrix} = \begin{pmatrix} A & W \\ W^* & {}^{\iota}A^{\mathrm{op}} \end{pmatrix}$$

with identifications similar to those in Exercise 10. The K-algebra B is Brauer equivalent to A by (1.10). Show that

$$\begin{pmatrix} a & w \\ f & {}^{\iota}b^{\mathrm{op}} \end{pmatrix} \mapsto \begin{pmatrix} b & \iota(w) \\ \iota(f) & {}^{\iota}a^{\mathrm{op}} \end{pmatrix}$$

is an involution of the second kind of B.

12. Let A be a central simple F-algebra with involution σ of the first kind. Recall the F-algebra isomorphism

$$\sigma_* \colon A \otimes_F A \to \operatorname{End}_F A$$

defined in the proof of Corollary (2.8) by

$$\sigma_*(a \otimes b)(x) = ax\sigma(b) \quad \text{for } a, b, x \in A.$$

Show that the image of the Goldman element $g \in A \otimes A$ under this isomorphism is $\delta\sigma$ where $\delta = +1$ if σ is orthogonal and $\delta = -1$ if σ is symplectic. Use this result to define canonical F-algebra isomorphisms

$$s^2 A \simeq \begin{cases} \operatorname{End}_F\big(A/\operatorname{Alt}(A,\sigma)\big) \simeq \operatorname{End}_F\big(\operatorname{Sym}(A,\sigma)\big)^{\mathrm{op}} & \text{if } \delta = +1, \\ \operatorname{End}_F\big(A/\operatorname{Symd}(A,\sigma)\big) \simeq \operatorname{End}_F\big(\operatorname{Skew}(A,\sigma)\big)^{\mathrm{op}} & \text{if } \delta = -1, \end{cases}$$

$$\lambda^2 A \simeq \begin{cases} \operatorname{End}_F\big(\operatorname{Alt}(A,\sigma)\big) & \text{if } \delta = +1, \\ \operatorname{End}_F\big(\operatorname{Symd}(A,\sigma)\big) & \text{if } \delta = -1. \end{cases}$$

13. (Saltman [241, Proposition 5]) Let A, B be central simple algebras of degrees m, n over a field F. For every F-algebra homomorphism $f \colon A \to B$, define a map $f' \colon A^{\mathrm{op}} \otimes_F B \to B$ by $f'(a^{\mathrm{op}} \otimes b) = f(a)b$. Show that f' is a homomorphism of right $A^{\mathrm{op}} \otimes_F B$-modules if B is endowed with the following $A^{\mathrm{op}} \otimes_F B$-module structure:

$$x *_f (a^{\mathrm{op}} \otimes b) = f(a)xb \quad \text{for } a \in A \text{ and } b, x \in B.$$

Show that $\ker f' \subset A^{\mathrm{op}} \otimes_F B$ is a right ideal of reduced dimension $mn - (n/m)$ generated by the elements $a^{\mathrm{op}} \otimes 1 - 1 \otimes f(a)$ for $a \in A$ and that

$$A^{\mathrm{op}} \otimes_F B = (1 \otimes B) \oplus \ker f'.$$

Conversely, show that every right ideal $I \subset A^{\mathrm{op}} \otimes_F B$ of reduced dimension $mn - (n/m)$ such that

$$A^{\mathrm{op}} \otimes_F B = (1 \otimes B) \oplus I$$

defines an F-algebra homomorphism $f \colon A \to B$ such that $I = \ker f$.

Deduce from the results above that there is a natural one-to-one correspondence between the set of F-algebra homomorphisms $A \to B$ and the rational points in an open subset of the Severi-Brauer variety $\operatorname{SB}_d(A^{\mathrm{op}} \otimes_F B)$ where $d = mn - (n/m)$.

14. Let (A, σ) be a central simple algebra with involution of the first kind over a field F of characteristic different from 2. Let $a \in A$ be an element whose minimal polynomial over F is separable. Show that a is symmetric for some symplectic involution on A if and only if its reduced characteristic polynomial is a square.

15. Let (A, σ) be a central simple algebra with involution of orthogonal type. Show that every element in A is the product of two symmetric elements, one of which is invertible.

    Hint: Use (4.17).

16. Let V be a finite dimensional vector space over a field F of arbitrary characteristic and let $a \in \operatorname{End}_F(V)$. Extend the notion of involution trace by using the structure of V as an $F[a]$-module to define a nonsingular symmetric bilinear form $b \colon V \times V \to F$ such that a is invariant under the adjoint involution σ_b.

17. Let K/F be a separable quadratic extension of fields with nontrivial automorphism ι. Let V be a vector space of dimension n over K and let $b \in \operatorname{End}_K(V)$ be an endomorphism whose minimal polynomial has degree n and coefficients in F. Show that V is a free $F[b]$-module of rank 1 and define a nonsingular hermitian form $h \colon V \times V \to K$ such that b is invariant under the adjoint involution σ_h.

    Show that a matrix $m \in M_n(K)$ is symmetric under some involution of the second kind whose restriction to K is ι if and only if all the invariant factors of m have coefficients in F.

18. Let q be a nonsingular quadratic form of dimension n over a field F, with n even if $\operatorname{char} F = 2$, and let $a \in M_n(F)$ be a matrix representing q, in the sense that $q(X) = X \cdot a \cdot X^t$. After identifying $M_n(F)$ with $\operatorname{End}_F(F^n)$ by mapping every matrix $m \in M_n(F)$ to the endomorphism $x \mapsto m \cdot x$, show that the quadratic pair (σ_q, f_q) on $\operatorname{End}_F(F^n)$ associated to the quadratic map $q \colon F^n \to F$ is the same as the quadratic pair (σ_a, f_a) associated to a.

19. Let Q_1, Q_2 be quaternion algebras with canonical involutions γ_1, γ_2 over a field F of arbitrary characteristic. Show that $\operatorname{Alt}(Q_1 \otimes Q_2, \gamma_1 \otimes \gamma_2) = \{ q_1 \otimes 1 - 1 \otimes q_2 \mid \operatorname{Trd}_{Q_1}(q_1) = \operatorname{Trd}_{Q_2}(q_2) \}$. If $\operatorname{char} F = 2$, show that $f(q_1 \otimes 1 + 1 \otimes q_2) = \operatorname{Trd}_{Q_1}(q_1) = \operatorname{Trd}_{Q_2}(q_2)$ for all $q_1 \otimes 1 + 1 \otimes q_2 \in \operatorname{Symd}(Q_1 \otimes Q_2, \gamma_1 \otimes \gamma_2)$ and for all quadratic pairs $(\gamma_1 \otimes \gamma_2, f)$ on $Q_1 \otimes Q_2$.

20. The aim of this exercise is to give a description of the variety of quadratic pairs on a central simple algebra in the spirit of (3.8). Let σ be a symplectic involution on a central simple algebra A over a field F of characteristic 2 and let $\sigma_* \colon A \otimes A \to \operatorname{End}_F(A)$ be the isomorphism of Exercise 12. Let $I_\sigma \subset A \otimes A$ denote the right ideal corresponding to σ by (3.8) and let $J_\ell \subset A \otimes A$ be the left ideal generated by $1 - g$, where g is the Goldman element. Denote by $A^0 \subset A$ the kernel of the reduced trace: $A^0 = \{ a \in A \mid \operatorname{Trd}_A(a) = 0 \}$. Show that

$$\sigma_*(I_\sigma) = \operatorname{Hom}(A, A^0), \qquad \sigma_*(J_\ell) = \operatorname{Hom}\big(A/\operatorname{Sym}(A, \sigma), A\big)$$

and

$$\sigma_*(J_\ell^0) = \operatorname{Hom}\big(A, \operatorname{Sym}(A, \sigma)\big).$$

Let now $I \subset A \otimes A$ be a left ideal containing J_ℓ, so that

$$\sigma_*(I) = \operatorname{Hom}(A/U, A)$$

for some vector space $U \subset \mathrm{Sym}(A, \sigma)$. Show that $\sigma_*\big(I{\cdot}(1{+}g)\big) = \mathrm{Hom}(A/W, A)$, where $W = \{\, a \in A \mid a + \sigma(a) \in U \,\}$, and deduce that $\sigma_*\big([I \cdot (1 + g)]^0\big) = I_\sigma$ if and only if $W = A^0$, if and only if $U \cap \mathrm{Symd}(A, \sigma) = \ker \mathrm{Trp}_\sigma$.

Observe now that the set of rational points in $\mathrm{SB}\big(s^2(A^{\mathrm{op}})\big)$ is in canonical one-to-one correspondence with the set of left ideals $I \subset A \otimes A$ containing J_ℓ^0 and such that $\mathrm{rdim}\, I - \mathrm{rdim}\, J_\ell = 1$. Consider the subset $\mathcal{U}$ of such ideals which satisfy $[I \cdot (1 + g)]^0 \oplus (1 \otimes A) = A \otimes A$. Show that the map which carries every quadratic pair (σ, f) on A to the left ideal $\sigma_*^{-1}\big(\mathrm{Hom}(A/\ker f, A)\big)$ defines a bijection from the set of quadratic pairs on A onto $\mathcal{U}$.

Hint: For $I \in \mathcal{U}$, the right ideal $[I \cdot (1 + g)]^0$ corresponds by (3.8) to some symplectic involution σ. If $U \subset A$ is the subspace such that $\sigma_*(I) = \mathrm{Hom}(A/U, A)$, there is a unique quadratic pair (σ, f) such that $U = \ker f$.

21. Let (V_1, b_1) and (V_2, b_2) be vector spaces with nonsingular alternating forms over an arbitrary field F. Show that there is a unique quadratic form q on $V_1 \otimes V_2$ whose polar form is $b_1 \otimes b_2$ and such that $q(v_1 \otimes v_2) = 0$ for all $v_1 \in V_1$, $v_2 \in V_2$. Show that the canonical quadratic pair $(\sigma_{b_1} \otimes \sigma_{b_2}, f_\otimes)$ on $\mathrm{End}_F(V_1) \otimes \mathrm{End}_F(V_2) = \mathrm{End}_F(V_1 \otimes V_2)$ is associated with the quadratic form q.

22. Let (A, σ) be a central simple algebra with involution of the first kind over an arbitrary field F. Assume σ is symplectic if $\mathrm{char}\, F = 2$. By (2.6), there exists an element $\ell \in A$ such that $\ell + \sigma(\ell) = 1$. Define a quadratic form $q_{(A, \sigma)} \colon A \to F$ by

$$q_{(A, \sigma)}(x) = \mathrm{Trd}_A\big(\sigma(x)\ell x\big) \qquad \text{for } x \in A.$$

Show that this quadratic form does not depend on the choice of ℓ such that $\ell + \sigma(\ell) = 1$. Show that the associated quadratic pair on $\mathrm{End}_F(A)$ corresponds to the canonical quadratic pair $(\sigma \otimes \sigma, f_\otimes)$ on $A \otimes A$ under the isomorphism $\sigma_* \colon A \otimes A \xrightarrow{\sim} \mathrm{End}_F(A)$ such that $\sigma_*(a \otimes b)(x) = ax\sigma(b)$ for a, b, $x \in A$.

NOTES

§1. Additional references for the material in this section include the classical books of Albert [9], Deuring [72] and Reiner [229]. For Severi-Brauer varieties, see Artin's notes [25]. A self-contained exposition of Severi-Brauer varieties can be found in Jacobson's book [146, Chapter 3].

§2. The first systematic investigations of involutions of central simple algebras are due to Albert. His motivation came from the theory of Riemann matrices: on a Riemann surface of genus g, choose a basis $(\omega_\alpha)_{1 \leq \alpha \leq g}$ of the space of holomorphic differentials and a system of closed curves $(\gamma_\beta)_{1 \leq \beta \leq 2g}$ which form a basis of the first homology group, and consider the matrix of periods $P = (\int_{\gamma_\beta} \omega_\alpha)$. This is a complex $g \times 2g$ matrix which satisfies Riemann's *period relations*: there exists a nonsingular alternating matrix $C \in M_{2g}(\mathbb{Q})$ such that $PCP^t = 0$ and $iPC\overline{P}^t$ is positive definite hermitian. The study of correspondences on the Riemann surface leads one to consider the matrices $M \in M_{2g}(\mathbb{Q})$ for which there exists a matrix $K \in M_g(\mathbb{C})$ such that $KP = PM$. Following Weyl's simpler formulation [314], one considers the matrix $W = \begin{pmatrix} P \\ \overline{P} \end{pmatrix} \in M_{2g}(\mathbb{C})$ and the so-called *Riemann matrix*

$R = W^{-1} \begin{pmatrix} -iI_g & 0 \\ 0 & iI_g \end{pmatrix} W \in M_{2g}(\mathbb{R})$. The matrices M such that $KP = PM$ for some $K \in M_g(\mathbb{C})$ are exactly those which commute with R. They form a subalgebra of $M_{2g}(\mathbb{Q})$ known as the *multiplication algebra*. As observed by Rosati [232], this algebra admits the involution $X \mapsto C^{-1}X^tC$ (see Weyl [314]). Albert completely determined the structure of the multiplication algebra in three papers in the *Annals of Mathematics* in 1934–1935. An improved version, [5], see also [9], laid the foundations of the theory of simple algebras with involutions.

Corollary (4.13) was observed independently by several authors: see Tits [291, Proposition 3], Platonov [220, Proposition 5] and Rowen [236, Proposition 5.3].

The original proof of Albert's theorem on quaternion algebras with involution of the second kind (2.22) is given in [9, Theorem 10.21]. This result will be put in a broader perspective in §10: the subalgebra Q_0 is the *discriminant algebra* of (Q, σ) (see (10.30)).

There is an extensive literature on Lie and Jordan structures in rings with involution; we refer the reader to Herstein's monographs [119] and [120]. In particular, Lemma (2.26) can be proved by ring-theoretic arguments which do not involve scalar extension (and therefore hold for more general simple rings): see Herstein [119, Theorem 2.2, p. 28]. In the same spirit, extension of Lie isomorphisms has been investigated for more general rings: see[11] Martindale [178], Rosen [233] and Beidar-Martindale-Mikhalev [37].

§3. Part (1) of Theorem (3.1) is due to Albert [9, Theorem 10.19]. Albert also proved part (2) for crossed products of a special kind: Albert assumes in [9, Theorem 10.16] the existence of a splitting field of the form $L \otimes_F K$ where L is Galois over F. Part (2) was stated in full generality by Riehm [231] and proved by Scharlau [246] (see also [247, §8.9]). In order to see that every central simple algebra which is Brauer-equivalent to an algebra with involution also has an involution, it is not essential to use (3.1): see Albert [9, Theorem 10.12] or Scharlau [247, Corollary 8.8.3]. By combining this result with Exercises 10 and 11, we obtain an alternate proof of Theorem (3.1).

§4. If E and E' are Brauer-equivalent central simple F-algebras, then Morita theory yields an E-E'-bimodule P and an E'-E-bimodule Q such that $P \otimes_{E'} Q \simeq E$ and $Q \otimes_E P \simeq E'$. If M is a right E-module, then there is a natural isomorphism

$$\text{End}_E(M) = \text{End}_{E'}(M \otimes_E P).$$

Therefore, if E (hence also E') has an involution, (4.2) yields one-to-one correspondences between hermitian or skew-hermitian forms on M (up to a central factor), involutions on $\text{End}_E(M) = \text{End}_{E'}(M \otimes_E P)$ and hermitian or skew-hermitian forms on $M \otimes_E P$ (up to a central factor). The correspondence between hermitian forms can be made more precise and explicit; it is part of a *Morita equivalence* between categories of hermitian modules which is discussed in Knus' book [157, §1.9].

The notion of involution trace was introduced by Fröhlich and McEvett [102, §7]. Special cases of the extension theorem (4.14) have been proved by Rowen [236, Corollary 5.5] and by Lam-Leep-Tignol [170, Proposition 5.1]. Kneser's theorem has been generalized by Held and Scharlau [118] to the case where the subalgebra is semisimple. (A particular case of this situation had also been considered by Kneser in [155, p. 37].)

---

[11]We are indebted to W. S. Martindale III for references to the recent literature.

The existence of involutions for which a given element is symmetric or skew-symmetric is discussed in Shapiro [259].

§5. The definition of quadratic pair in (5.4) is new. While involutions on arbitrary central simple algebras have been related to hermitian forms in §4, the relation between quadratic pairs and quadratic forms is described only in the split case. The nonsplit case requires an extension of the notion of quadratic form. For quaternion algebras such an extension was given by Seip-Hornix [250]. Tits [291] defines a (generalized) quadratic form as an element in the factor group $A/\operatorname{Alt}(A, \tau)$ (compare with (5.8)); a more geometric viewpoint which also extends this notion further, was proposed by Bak [29] (see also for instance Hahn-O'Meara [111, 5.1C], Knus [157, Ch. 1, §5] or Scharlau [247, Ch. 7, §3]).

CHAPTER II

Invariants of Involutions

In this chapter, we define various kinds of invariants of central simple algebras with involution (or with quadratic pair) and we investigate their basic properties. The invariants considered here are analogues of the classical invariants of quadratic forms: the Witt index, the discriminant, the Clifford algebra and the signature. How far the analogy can be pushed depends of course on the nature of the involution: an index is defined for arbitrary central simple algebras with involution or quadratic pair, but the discriminant is defined only for orthogonal involutions and quadratic pairs, and the Clifford algebra just for quadratic pairs. The Clifford algebra construction actually splits into two parts: while it is impossible to define a full Clifford algebra for quadratic pairs, the even and the odd parts of the Clifford algebra can be recovered in the form of an algebra and a bimodule. For unitary involutions, the notion of discriminant turns out to lead to a rich structure: we associate in §10 a discriminant algebra (with involution) to every unitary involution on a central simple algebra of even degree. Finally, signatures can be defined for arbitrary involutions through the associated trace forms. These trace forms also have relations with the discriminant or discriminant algebra. They yield higher invariants for algebras with unitary involution of degree 3 in Chapter V and for Jordan algebras in Chapter IX.

The invariants considered in this chapter are produced by various techniques. The index is derived from a representation of the algebra with involution as the endomorphism algebra of some hermitian or skew-hermitian space over a division algebra, while the definitions of discriminant and Clifford algebra are based on the fact that scalar extension reduces the algebra with quadratic pair to the endomorphism algebra of a quadratic space. We show that the discriminant and even Clifford algebra of the corresponding quadratic form can be defined in terms of the adjoint quadratic pair, and that the definitions generalize to yield invariants of arbitrary quadratic pairs. A similar procedure is used to define the discriminant algebra of a central simple algebra of even degree with unitary involution. Throughout most of this chapter, our method of investigation is thus based on scalar extension: after specifying the definitions "rationally" (i.e., over an arbitrary base field), the main properties are proven by extending scalars to a splitting field. This method contrasts with Galois descent, where constructions over a separable closure are shown to be invariant under the action of the absolute Galois group and are therefore defined over the base field.

§6. The Index

According to (4.2), every central simple F-algebra with involution (A, σ) can be represented as $\left(\mathrm{End}_D(V), \sigma_h\right)$ for some division algebra D, some D-vector space V and some nonsingular hermitian form h on V. Since this representation is essentially unique, it is not difficult to check that the Witt index $w(V, h)$ of the hermitian space (V, h), defined as the maximum of the dimensions of totally isotropic subspaces of V, is an invariant of (A, σ). In this section, we give an alternate definition of this invariant which does not depend on a representation of (A, σ) as $(\operatorname{End}_D(V), \sigma_h)$, and we characterize the involutions which can be represented as adjoint involutions with respect to a hyperbolic form. We define a slightly more general notion of index which takes into account the Schur index of the algebra. A (weak) analogue of Springer's theorem on odd degree extensions is discussed in the final subsection.

6.A. Isotropic ideals. Let (A, σ) be a central simple algebra with involution (of any kind) over a field F of arbitrary characteristic.

(6.1) Definition. For every right ideal I in A, the *orthogonal ideal* $I^\perp$ is defined by
$$I^\perp = \{\, x \in A \mid \sigma(x)y = 0 \text{ for } y \in I \,\}.$$

It is clearly a right ideal of A, which may alternately be defined as the annihilator of the left ideal $\sigma(I)$:
$$I^\perp = \sigma(I)^0.$$

(6.2) Proposition. *Suppose the center of A is a field. For every right ideal $I \subset A$, $\operatorname{rdim} I + \operatorname{rdim} I^\perp = \deg A$ and $I^{\perp\perp} = I$. Moreover, if $(A, \sigma) = (\operatorname{End}_D(V), \sigma_h)$ and $I = \operatorname{Hom}_D(V, W)$ for some subspace $W \subset V$, then*
$$I^\perp = \operatorname{Hom}_D(V, W^\perp).$$

Proof: Since $\operatorname{rdim} \sigma(I) = \operatorname{rdim} I$, the first relation follows from the corresponding statement for annihilators (1.14). This first relation implies that $\operatorname{rdim} I^{\perp\perp} = \operatorname{rdim} I$. Since the inclusion $I \subset I^{\perp\perp}$ is obvious, we get $I = I^{\perp\perp}$. Finally, suppose $I = \operatorname{Hom}_D(V, W)$ for some subspace $W \subset V$. For every $f \in \operatorname{End}_D(V)$, $g \in I$ we have
$$g(y) \in W \quad \text{and} \quad h\big(f(x), g(y)\big) = h\big(x, \sigma(f) \circ g(y)\big) \quad \text{for } x, y \in V.$$
Therefore, $\sigma(f) \circ g = 0$ if and only if $f(x) \in W^\perp$, hence
$$I^\perp = \operatorname{Hom}_D(V, W^\perp). \qquad \square$$

A similar result holds if $(A, \sigma) = (E \times E^{\mathrm{op}}, \varepsilon)$, where E is a central simple F-algebra and ε is the exchange involution, although the reduced dimension of a right ideal is not defined in this case. Every right ideal $I \subset A$ has the form $I = I_1 \times I_2^{\mathrm{op}}$ where I_1 (resp. I_2) is a right (resp. left) ideal in E, and
$$(I_1 \times I_2^{\mathrm{op}})^\perp = I_2^0 \times (I_1^0)^{\mathrm{op}}.$$
Therefore, by (1.14),
$$\dim_F I^\perp = \dim_F A - \dim_F I \quad \text{and} \quad I^{\perp\perp} = I$$
for every right ideal $I \subset A$.

In view of the proposition above, the following definitions are natural:

(6.3) Definitions. Let (A, σ) be a central simple algebra with involution over a field F. A right ideal $I \subset A$ is called *isotropic* (with respect to the involution σ) if $I \subset I^\perp$. This inclusion implies $\operatorname{rdim} I \leq \operatorname{rdim} I^\perp$, hence (6.2) shows that $\operatorname{rdim} I \leq \frac{1}{2} \deg A$ for every isotropic right ideal.

The algebra with involution (A, σ)—or the involution σ itself—is called *isotropic* if A contains a nonzero isotropic ideal.

If the center of A is a field, the *index* of the algebra with involution (A, σ) is defined as the set of reduced dimensions of isotropic right ideals:

$$\mathrm{ind}(A, \sigma) = \{\, \mathrm{rdim}\, I \mid I \subset I^\perp \,\}.$$

Since the (Schur) index of A divides the reduced dimension of every right ideal, the index $\mathrm{ind}(A, \sigma)$ is a set of multiples of $\mathrm{ind}\, A$. More precisely, if $(A, \sigma) \simeq (\mathrm{End}_D(V), \sigma_h)$ for some hermitian or skew-hermitian space (V, h) over a division algebra D and $w(V, h)$ denotes the Witt index of (V, h), then $\mathrm{ind}\, A = \deg D$ and

$$\mathrm{ind}(A, \sigma) = \{\, \ell \deg D \mid 0 \le \ell \le w(V, h) \,\}$$

since (6.2) shows that the isotropic ideals of $\mathrm{End}_D(V)$ are of the form $\mathrm{Hom}_D(V, W)$ with W a totally isotropic subspace of V, and $\mathrm{rdim}\, \mathrm{Hom}_D(V, W) = \dim_D W \deg D$. Thus, if $\mathrm{ind}(A, \sigma)$ contains at least two elements, the difference between two consecutive integers in $\mathrm{ind}(A, \sigma)$ is $\mathrm{ind}\, A$. If $\mathrm{ind}(A, \sigma)$ has only one element, then $\mathrm{ind}(A, \sigma) = \{0\}$, which means that (A, σ) is anisotropic; this is the case for instance when A is a division algebra.

We extend the definition of $\mathrm{ind}(A, \sigma)$ to the case where the center of A is $F \times F$; then $(A, \sigma) \simeq (E \times E^{\mathrm{op}}, \varepsilon)$ for some central simple F-algebra E where ε is the exchange involution, and we define $\mathrm{ind}(A, \sigma)$ as the set of multiples of the Schur index of E in the interval $[0, \frac{1}{2} \deg E]$:

$$\mathrm{ind}(A, \sigma) = \left\{\, \ell \, \mathrm{ind}\, E \,\middle|\, 0 \le \ell \le \frac{\deg E}{2 \, \mathrm{ind}\, E} \,\right\}.$$

(Note that $\deg E = \deg(A, \sigma)$ is not necessarily even.)

(6.4) Proposition. *For every field extension L/F,*

$$\mathrm{ind}(A, \sigma) \subset \mathrm{ind}(A_L, \sigma_L).$$

Proof: This is clear if the center of $A_L = A \otimes_F L$ is a field, since scalar extensions preserve the reduced dimension of ideals and since isotropic ideals remain isotropic under scalar extension. If the center of A is not a field, the proposition is also clear. Suppose the center of A is a field K properly containing F and contained in L; then by (2.15),

$$(A_L, \sigma_L) \simeq \big((A \otimes_K L) \times (A \otimes_K L)^{\mathrm{op}}, \varepsilon\big).$$

Since the reduced dimension of every right ideal in A is a multiple of $\mathrm{ind}\, A$ and since $\mathrm{ind}(A \otimes_K L)$ divides $\mathrm{ind}\, A$, the reduced dimension of every isotropic ideal of (A, σ) is a multiple of $\mathrm{ind}(A \otimes_K L)$. Moreover, the reduced dimension of isotropic ideals is bounded by $\frac{1}{2} \deg A$, hence $\mathrm{ind}(A, \sigma) \subset \mathrm{ind}(A_L, \sigma_L)$. $\square$

For central simple algebras with a quadratic pair, we define isotropic ideals by a more restrictive condition.

(6.5) Definition. Let (σ, f) be a quadratic pair on a central simple algebra A over a field F. A right ideal $I \subset A$ is called *isotropic* with respect to the quadratic pair (σ, f) if the following two conditions hold:

(1) $\sigma(x)y = 0$ for all $x, y \in I$.
(2) $f(x) = 0$ for all $x \in I \cap \mathrm{Sym}(A, \sigma)$.

The first condition means that $I \subset I^\perp$, hence isotropic ideals for the quadratic pair (σ, f) are also isotropic for the involution σ. Condition (1) implies that every $x \in I \cap \mathrm{Sym}(A, \sigma)$ satisfies $x^2 = 0$, hence also $\mathrm{Trd}_A(x) = 0$. If char $F \neq 2$, the map f is the restriction of $\frac{1}{2} \mathrm{Trd}_A$ to $\mathrm{Sym}(A, \sigma)$, hence condition (2) follows from (1). Therefore, in this case the isotropic ideals for (σ, f) are exactly the isotropic ideals for σ.

The algebra with quadratic pair (A, σ, f)—or the quadratic pair (σ, f) itself—is called *isotropic* if A contains a nonzero isotropic ideal for the quadratic pair (σ, f).

(6.6) Example. Let (V, q) be an even-dimensional quadratic space over a field F of characteristic 2, and let (σ_q, f_q) be the corresponding quadratic pair on $\mathrm{End}_F(V)$ (see (5.11)). A subspace $W \subset V$ is totally isotropic for q if and only if the right ideal $\mathrm{Hom}_F(V, W) \subset \mathrm{End}_F(V)$ is isotropic for (σ_q, f_q).

Indeed, the standard identification φ_q (see (5.10)) identifies $\mathrm{Hom}_F(V, W)$ with $W \otimes_F V$. The elements in $W \otimes V$ which are invariant under the switch involution (which corresponds to σ_q under φ_q) are spanned by elements of the form $w \otimes w$ with $w \in W$, and (5.11) shows that $f_q \circ \varphi_q(w \otimes w) = q(w)$. Therefore, the subspace W is totally isotropic with respect to q if and only if f_q vanishes on $\mathrm{Hom}_F(V, W) \cap \mathrm{Sym}(\mathrm{End}_F(V), \sigma_q)$. Proposition (6.2) shows that this condition also implies that $\mathrm{Hom}_F(V, W)$ is isotropic with respect to the involution σ_q.

Mimicking (6.3), we define the *index* of a central simple algebra with quadratic pair (A, σ, f) as the set of reduced dimensions of isotropic ideals:

$$\mathrm{ind}(A, \sigma, f) = \{\, \mathrm{rdim}\, I \mid I \text{ is isotropic with respect to } (\sigma, f) \,\}.$$

6.B. Hyperbolic involutions. Let E be a central simple algebra with involution θ (of any kind) and let U be a finitely generated right E-module. As in (4.1), we use the involution θ to endow the dual of U with a structure of right E-module ${}^\theta U^*$. For $\lambda = \pm 1$, define

$$h_\lambda \colon ({}^\theta U^* \oplus U) \times ({}^\theta U^* \oplus U) \to E$$

by

$$h_\lambda({}^\theta \varphi + x, {}^\theta \psi + y) = \varphi(y) + \lambda \theta(\psi(x))$$

for φ, $\psi \in U^*$ and x, $y \in U$. Straightforward computations show that h_1 (resp. h_{-1}) is a nonsingular hermitian (resp. alternating) form on ${}^\theta U^* \oplus U$ with respect to the involution θ on E. The hermitian or alternating module $({}^\theta U^* \oplus U, h_\lambda)$ is denoted $\mathbb{H}_\lambda(U)$. A hermitian or alternating module (M, h) over (E, θ) is called *hyperbolic* if it is isometric to some $\mathbb{H}_\lambda(U)$.

The following proposition characterizes the adjoint involutions with respect to hyperbolic forms:

(6.7) Proposition. *Let (A, σ) be a central simple algebra with involution (of any kind) over a field F of arbitrary characteristic. Suppose the center K of A is a field. The following conditions are equivalent:*

(1) *for every central simple K-algebra E Brauer-equivalent to A and every involution θ on E such that $\theta|_K = \sigma|_K$, every hermitian or skew-hermitian module (M, h) over (E, θ) such that $(A, \sigma) \simeq (\mathrm{End}_E(M), \sigma_h)$ is hyperbolic;*

(2) *there exists a central simple K-algebra E Brauer-equivalent to A, an involution θ on E such that $\theta|_K = \sigma|_K$ and a hyperbolic hermitian or skew-hermitian module (M, h) over (E, θ) such that $(A, \sigma) \simeq (\mathrm{End}_E(M), \sigma_h)$;*

(3) $\frac{1}{2}\deg A \in \operatorname{ind}(A,\sigma)$ *and further, if* char $F = 2$, *the involution* σ *is either symplectic or unitary;*

(4) *there is an idempotent* $e \in A$ *such that* $\sigma(e) = 1 - e$.

Proof: (1) $\Rightarrow$ (2) This is clear.

(2) $\Rightarrow$ (3) In the hyperbolic module $M = {}^{\theta}U^* \oplus U$, the submodule U is totally isotropic, hence the same argument as in (6.2) shows that the right ideal $\operatorname{Hom}_E(M,U)$ is isotropic in $\operatorname{End}_E(M)$. Moreover, since $\operatorname{rdim} U = \frac{1}{2}\operatorname{rdim} M$, we have $\operatorname{rdim}\operatorname{Hom}_E(M,U) = \frac{1}{2}\deg\operatorname{End}_E(M)$. Therefore, (2) implies that $\frac{1}{2}\deg A \in \operatorname{ind}(A,\sigma)$. If char $F = 2$, the hermitian form $h_{+1} = h_{-1}$ is alternating, hence σ is symplectic if the involution θ on E is of the first kind, and is unitary if θ is of the second kind.

(3) $\Rightarrow$ (4) Let $I \subset A$ be an isotropic ideal of reduced dimension $\frac{1}{2}\deg A$. By (1.13), there is an idempotent $f \in A$ such that $I = fA$. Since I is isotropic, we have $\sigma(f)f = 0$. We shall modify f into an idempotent e such that $I = eA$ and $\sigma(e) = 1 - e$.

The first step is to find $u \in A$ such that $\sigma(u) = 1 - u$. If char $F \neq 2$, we may choose $u = 1/2$; if char $F = 2$ and σ is symplectic, the existence of u follows from (2.6); if char $F = 2$ and σ is unitary, we may choose u in the center K of A, since K/F is a separable quadratic extension and the restriction of σ to K is the nontrivial automorphism of K/F.

We next set $e = f - fu\sigma(f)$ and proceed to show that this element satisfies (4). Since $f^2 = f$ and $\sigma(f)f = 0$, it is clear that e is an idempotent and $\sigma(e)e = 0$. Moreover, since $\sigma(u) + u = 1$,

$$e\sigma(e) = f\sigma(f) - fu\sigma(f) - f\sigma(u)\sigma(f) = 0.$$

Therefore, e and $\sigma(e)$ are orthogonal idempotents; it follows that $e + \sigma(e)$ also is an idempotent, and $\big(e + \sigma(e)\big)A = eA \oplus \sigma(e)A$. To complete the proof of (4), observe that $e \in fA$ and $f = ef \in eA$, hence $eA = fA = I$. Since $\operatorname{rdim} I = \frac{1}{2}\deg A$, it follows that $\dim_F eA = \dim_F \sigma(e)A = \frac{1}{2}\dim_F A$, hence $\big(e + \sigma(e)\big)A = A$ and therefore $e + \sigma(e) = 1$.

(4) $\Rightarrow$ (1) Let E be a central simple K-algebra Brauer-equivalent to A and θ be an involution on E such that $\theta|_K = \sigma|_K$. Let also (M, h) be a hermitian or skew-hermitian module over (E, θ) such that $(A, \sigma) \simeq \big(\operatorname{End}_E(M), \sigma_h\big)$. Viewing this isomorphism as an identification, we may find for every idempotent $e \in A$ a pair of complementary submodules $U = \operatorname{im} e$, $W = \ker e$ in M such that e is the projection $M \to U$ parallel to W; then $1 - e$ is the projection $M \to W$ parallel to U and $\sigma(e)$ is the projection $M \to W^{\perp}$ parallel to $U^{\perp}$. Therefore, if $\sigma(e) = 1 - e$ we have $U = U^{\perp}$ and $W = W^{\perp}$. We then define an isomorphism $W \xrightarrow{\sim} {}^{\theta}U^*$ by mapping $w \in W$ to ${}^{\theta}\varphi_w$ where $\varphi_w \in U^*$ is defined by $\varphi_w(u) = h(w, u)$ for $u \in U$. This isomorphism extends to an isometry $M = W \oplus U \xrightarrow{\sim} \mathbb{H}_{\lambda}(U)$ where $\lambda = +1$ if h is hermitian and $\lambda = -1$ if h is skew-hermitian. $\qquad\square$

(6.8) Definition. A central simple algebra with involution (A, σ) over a field F— or the involution σ itself—is called *hyperbolic* if either the center of A is isomorphic to $F \times F$ or the equivalent conditions of (6.7) hold. If the center is $F \times F$, then the idempotent $e = (1,0)$ satisfies $\sigma(e) = 1 - e$; therefore, in all cases (A, σ) is hyperbolic if and only if A contains an idempotent e such that $\sigma(e) = 1 - e$. This condition is also equivalent to the existence of an isotropic ideal I of dimension $\dim_F I = \frac{1}{2}\dim_F A$ if char $F \neq 2$, but if char $F = 2$ the extra assumption that σ is symplectic or unitary is also needed. For instance, if $A = M_2(F)$ (with char $F = 2$) and σ is the transpose involution, then $\frac{1}{2} \deg A \in \mathrm{ind}(A, \sigma)$ since the right ideal $I = \{ \left(\begin{smallmatrix} x & x \\ y & y \end{smallmatrix} \right) \mid x, y \in F \}$ is isotropic, but (A, σ) is not hyperbolic since σ is of orthogonal type.

Note that (A, σ) may be hyperbolic without A being split; indeed we may have $\mathrm{ind}(A, \sigma) = \{0, \frac{1}{2} \deg A\}$, in which case the index of A is $\frac{1}{2} \deg A$.

From any of the equivalent characterizations in (6.7), it is clear that hyperbolic involutions remain hyperbolic over arbitrary scalar extensions. Characterization (4) readily shows that hyperbolic involutions are also preserved by transfer. Explicitly, consider the situation of §4.B: Z/F is a finite extension of fields, E is a central simple Z-algebra and T is a central simple F-algebra contained in E, so that $E = T \otimes_F C$ where C is the centralizer of T in E. Let θ be an involution on E which preserves T and let $s \colon E \to T$ be an involution trace. Recall from (4.7) that for every hermitian or skew-hermitian module (M, h) over (E, θ) there is a transfer $\big(M, s_*(h)\big)$ which is a hermitian or skew-hermitian module over (T, θ).

(6.9) Proposition. *If h is hyperbolic, then $s_*(h)$ is hyperbolic.*

Proof: If h is hyperbolic, (6.7) yields an idempotent $e \in \mathrm{End}_E(M)$ such that $\sigma_h(e) = 1 - e$. By (4.7), the involution $\sigma_{s_*(h)}$ on $\mathrm{End}_T(M)$ extends σ_h, hence e also is an idempotent of $\mathrm{End}_T(M)$ such that $\sigma_{s_*(h)}(e) = 1 - e$. Therefore, $s_*(h)$ is hyperbolic. $\square$

In the same spirit, we have the following transfer-type result:

(6.10) Corollary. *Let (A, σ) be a central simple algebra with involution (of any kind) over a field F and let L/F be a finite extension of fields. Embed $L \hookrightarrow \mathrm{End}_F(L)$ by mapping $x \in L$ to multiplication by x, and let ν be an involution on $\mathrm{End}_F(L)$ leaving the image of L elementwise invariant. If (A_L, σ_L) is hyperbolic, then $\big(A \otimes_F \mathrm{End}_F(L), \sigma \otimes \nu\big)$ is hyperbolic.*

Proof: The embedding $L \hookrightarrow \mathrm{End}_F(L)$ induces an embedding

$$(A_L, \sigma_L) = (A \otimes_F L, \sigma \otimes \mathrm{Id}_L) \hookrightarrow \big(A \otimes_F \mathrm{End}_F(L), \sigma \otimes \nu\big).$$

The same argument as in the proof of (6.9) applies. (Indeed, (6.10) may be regarded as the special case of (6.9) where $C = Z = L$: see 4.11). $\square$

(6.11) Example. Interesting examples of hyperbolic involutions can be obtained as follows: let (A, σ) be a central simple algebra with involution (of any kind) over a field F of characteristic different from 2 and let $u \in \mathrm{Sym}(A, \sigma) \cap A^\times$ be a symmetric unit in A. Define an involution ν_u on $M_2(A)$ by

$$\nu_u \begin{pmatrix} a_{11} & a_{12} \\ a_{21} & a_{22} \end{pmatrix} = \begin{pmatrix} \sigma(a_{11}) & -\sigma(a_{21})u^{-1} \\ -u\sigma(a_{12}) & u\sigma(a_{22})u^{-1} \end{pmatrix}$$

for $a_{11}, a_{12}, a_{21}, a_{22} \in A$, i.e.,

$$\nu_u = \mathrm{Int} \begin{pmatrix} 1 & 0 \\ 0 & -u \end{pmatrix} \circ (\sigma \otimes t)$$

where t is the transpose involution on $M_2(F)$.

Claim. The involution ν_u is hyperbolic if and only if $u = v\sigma(v)$ for some $v \in A$.

Proof: Let D be a division algebra Brauer-equivalent to A and let θ be an involution on D of the same type as σ. We may identify $(A, \sigma) = \bigl(\operatorname{End}_D(V), \sigma_h\bigr)$ for some hermitian space (V, h) over (D, σ), by (4.2). Define a hermitian form h' on V by

$$h'(x, y) = h\bigl(u^{-1}(x), y\bigr) = h\bigl(x, u^{-1}(y)\bigr) \quad \text{for } x, y \in V.$$

Under the natural identification $M_2(A) = \operatorname{End}_D(V \oplus V)$, the involution ν_u is the adjoint involution with respect to $h \perp (-h')$. The form $h \perp (-h')$ is hyperbolic if and only if (V, h) is isometric to (V, h'). Therefore, by Proposition (6.7), ν_u is hyperbolic if and only if (V, h) is isometric to (V, h'). This condition is also equivalent to the existence of $a \in A^\times$ such that $h'(x, y) = h\bigl(a(x), a(y)\bigr)$ for all $x, y \in V$; in view of the definition of h', this relation holds if and only if $u = a^{-1}\sigma(a^{-1})$. □

The fact that ν_u is hyperbolic if $u = v\sigma(v)$ for some $v \in A$ can also be readily proved by observing that the matrix $e = \frac{1}{2}\left(\begin{smallmatrix} 1 & v^{-1} \\ v & 1 \end{smallmatrix}\right)$ is an idempotent such that $\nu_u(e) = 1 - e$.

Hyperbolic quadratic pairs. By mimicking characterization (3) of hyperbolic involutions, we may define hyperbolic quadratic pairs as follows:

(6.12) Definition. A quadratic pair (σ, f) on a central simple algebra A of even degree over a field F of arbitrary characteristic is called *hyperbolic* if $\frac{1}{2} \deg A \in \operatorname{ind}(A, \sigma, f)$ or, in other words, if A contains a right ideal I such that

$$\dim_F I = \tfrac{1}{2} \dim_F A, \quad \sigma(I)I = \{0\}, \quad f\bigl(I \cap \operatorname{Sym}(A, \sigma)\bigr) = \{0\}.$$

If char $F \neq 2$, the map f is determined by σ and $\operatorname{ind}(A, \sigma, f) = \operatorname{ind}(A, \sigma)$, hence (σ, f) is hyperbolic if and only if σ is hyperbolic. If char $F = 2$, the involution σ is symplectic and $\operatorname{ind}(A, \sigma, f) \subset \operatorname{ind}(A, \sigma)$; therefore, σ is hyperbolic if (σ, f) is hyperbolic.

We proceed to show that the quadratic pair associated to a quadratic space is hyperbolic if and only if the quadratic space is hyperbolic.

Recall that a quadratic space over a field F is called *hyperbolic* if it is isometric to a space $\mathbb{H}(U) = (U^* \oplus U, q_U)$ for some vector space U where $U^* = \operatorname{Hom}_F(U, F)$ and

$$q_U(\varphi + u) = \varphi(u)$$

for $\varphi \in U^*$ and $u \in U$. The corresponding symmetric bilinear space is thus the hyperbolic space denoted $\mathbb{H}_1(U)$ above.

(6.13) Proposition. *Let (V, q) be a nonsingular quadratic space of even dimension over an arbitrary field F. The corresponding quadratic pair (σ_q, f_q) on $\operatorname{End}_F(V)$ is hyperbolic if and only if the space (V, q) is hyperbolic.*

Proof: By (6.6), the quadratic pair (σ_q, f_q) is hyperbolic if and only if V contains a totally isotropic subspace U of dimension $\dim U = \frac{1}{2} \dim V$. This condition is equivalent to $(V, q) \simeq \mathbb{H}(U)$: see Scharlau [247, p. 12] (if char $F \neq 2$) and [247, p. 340] (if char $F = 2$). □

Hyperbolic quadratic pairs can also be characterized by the existence of certain idempotents:

(6.14) Proposition. *A quadratic pair* (σ, f) *on a central simple algebra A of even degree over a field F of arbitrary characteristic is hyperbolic if and only if A contains an idempotent e such that*

$$f(s) = \mathrm{Trd}_A(es) \qquad \textit{for all } s \in \mathrm{Sym}(A, \sigma).$$

Proof: If $f(s) = \mathrm{Trd}_A(es)$ for all $s \in \mathrm{Sym}(A, \sigma)$, then $\sigma(e) = 1 - e$, by (5.7), hence the ideal eA is isotropic for σ and has reduced dimension $\frac{1}{2} \deg A$ if e is an idempotent. Moreover, for $s \in eA \cap \mathrm{Sym}(A, \sigma)$ there exists $x \in A$ such that $s = ex = \sigma(x)\sigma(e)$, hence

$$f(s) = \mathrm{Trd}_A\big(e\sigma(x)\sigma(e)\big) = \mathrm{Trd}_A\big(\sigma(e)e\sigma(x)\big) = 0.$$

Therefore the ideal eA is isotropic for (σ, f), and (σ, f) is hyperbolic.

Conversely, suppose that $I \subset A$ is an isotropic right ideal of reduced dimension $\frac{1}{2} \deg A$. By arguing as in (6.7), we get an idempotent $e_0 \in A$ such that $\sigma(e_0) = 1 - e_0$ and $I = e_0 A$. If $\mathrm{char}\, F \neq 2$, we have $e_0 - \frac{1}{2} \in \mathrm{Skew}(A, \sigma) = \mathrm{Alt}(A, \sigma)$, hence, by (2.3), $\mathrm{Trd}_A(e_0 s) = \frac{1}{2} \mathrm{Trd}_A(s) = f(s)$ for all $s \in \mathrm{Sym}(A, \sigma)$. We may thus set $e = e_0$ if $\mathrm{char}\, F \neq 2$, and assume that $\mathrm{char}\, F = 2$ for the rest of the proof.

For all $x \in \mathrm{Skew}(A, \sigma)$, the element $e = e_0 - e_0 x \sigma(e_0)$ also is an idempotent such that $\sigma(e) = 1 - e$. To complete the proof in the case where $\mathrm{char}\, F = 2$, we show that e satisfies the required condition for a suitable choice of x. Consider the linear form φ on $\mathrm{Sym}(A, \sigma)$ defined by

$$\varphi(s) = f(s) - \mathrm{Trd}_A(e_0 s) \qquad \text{for } s \in \mathrm{Sym}(A, \sigma).$$

This form vanishes on $\mathrm{Symd}(A, \sigma)$, since $\sigma(e_0) = 1 - e_0$, and also on $I \cap \mathrm{Sym}(A, \sigma)$ since for all $s \in I \cap \mathrm{Sym}(A, \sigma)$ we have $s^2 = 0$, hence $\mathrm{Trd}_A(s) = 0$. On the other hand, for all $x \in e_0 \mathrm{Skew}(A, \sigma)\sigma(e_0)$, the linear form $\psi_x \in \mathrm{Sym}(A, \sigma)^*$ defined by $\psi_x(s) = \mathrm{Trd}_A(xs)$ also vanishes on $\mathrm{Symd}(A, \sigma)$, because $x \in \mathrm{Skew}(A, \sigma)$, and on $I \cap \mathrm{Sym}(A, \sigma)$ because $\sigma(e_0)e_0 = 0$. If we show that $\varphi = \psi_x$ for some $x \in e_0 \mathrm{Skew}(A, \sigma)\sigma(e_0)$, then we may set $e = e_0 + x$.

We may thus complete the proof by dimension count: if $\deg A = n = 2m$, it can be verified by extending scalars to a splitting field of A that

$$\dim\big(\mathrm{Symd}(A, \sigma) + \big(I \cap \mathrm{Sym}(A, \sigma)\big)\big) = mn,$$

hence the dimension of the space of linear forms on $\mathrm{Sym}(A, \sigma)$ which vanish on $\mathrm{Symd}(A, \sigma)$ and $I \cap \mathrm{Sym}(A, \sigma)$ is m. On the other hand, by (2.3), the kernel of the map which carries $x \in e_0 \mathrm{Skew}(A, \sigma)\sigma(e_0)$ to $\psi_x \in \mathrm{Sym}(A, \sigma)^*$ is the space $\mathrm{Alt}(A, \sigma) \cap e_0 \mathrm{Skew}(A, \sigma)\sigma(e_0)$, and we may compute its dimension over a splitting field:

$$\dim e_0 \mathrm{Skew}(A, \sigma)\sigma(e_0) = \tfrac{1}{2}m(m + 1)$$

and

$$\dim \mathrm{Alt}(A, \sigma) \cap e_0 \mathrm{Skew}(A, \sigma)\sigma(e_0) = \tfrac{1}{2}m(m - 1).$$

Therefore, the space of linear forms on $\mathrm{Sym}(A, \sigma)$ which vanish on the intersection $\mathrm{Symd}(A, \sigma) + \big(I \cap \mathrm{Sym}(A, \sigma)\big)$ is $\{\, \psi_x \mid x \in e_0 \mathrm{Skew}(A, \sigma)\sigma(e_0) \,\}$, hence $\varphi = \psi_x$ for a suitable element $x \in e_0 \mathrm{Skew}(A, \sigma)\sigma(e_0)$. $\qquad\square$

6.C. Odd-degree extensions. Using the fact that the torsion in the Witt group of central simple algebras with involution is 2-primary (Scharlau [245]), we show in this section that involutions which are not hyperbolic do not become hyperbolic when tensored with a central simple algebra with involution of odd degree, nor after an odd-degree scalar extension. This last statement generalizes a weak version of a theorem of Springer; it is due to Bayer-Fluckiger-Lenstra [33].

Since some of the Witt group arguments do not hold in characteristic 2, we assume that the characteristic of the base field F is different from 2 throughout this subsection.

(6.15) Proposition. *Let (A, σ) be a central simple algebra with involution (of any kind) over a field F and let (B, τ) be a central simple algebra of odd degree with involution of the first kind over F. If $(A, \sigma) \otimes_F (B, \tau)$ is hyperbolic, then (A, σ) is hyperbolic.*

Proof: It follows from (2.8) that the algebra B is split. Let $B = \mathrm{End}_F(W)$ for some odd-dimensional F-vector space W and let b be a nonsingular symmetric bilinear form on W such that $\tau = \sigma_b$. Similarly, let $(A, \sigma) = \big(\mathrm{End}_D(V), \sigma_h\big)$ for some hermitian or skew-hermitian space (V, h) over a central division algebra with involution (D, θ). We then have

$$(A, \sigma) \otimes_F (B, \tau) = \big(\mathrm{End}_D(V \otimes W), \sigma_{h \otimes b}\big),$$

and it remains to show that (V, h) is hyperbolic if $(V \otimes W, h \otimes b)$ is hyperbolic.

We mimic the proof of Corollary 2.6.5 in Scharlau [247]. Suppose there exists a non-hyperbolic hermitian or skew-hermitian space (V, h) over (D, θ) which becomes hyperbolic when tensored by a nonsingular symmetric bilinear space (W, b) of odd dimension. Among all such examples, choose one where the dimension of W is minimal. Let $\dim W = n \ (\geq 3)$. We may assume that b has a diagonalization $\langle 1, a_2, \ldots, a_n \rangle$. Since $h \otimes b$ is hyperbolic, we have in the Witt group $W^\lambda(D, \theta)$ where $\lambda = +1$ if h is hermitian and $\lambda = -1$ if h is skew-hermitian,

$$h \otimes \langle a_3, \ldots, a_n \rangle = h \otimes \langle -1, -a_2 \rangle.$$

Since $\langle 1, -a_2 \rangle \otimes \langle -1, -a_2 \rangle$ is hyperbolic, it follows that $\big(\langle 1, -a_2 \rangle \otimes h\big) \otimes \langle a_3, \ldots, a_n \rangle$ is hyperbolic. By minimality of n, it follows that $\langle 1, -a_2 \rangle \otimes h$ is hyperbolic, hence

$$h \simeq h \otimes \langle a_2 \rangle.$$

Similarly, we have $h \simeq h \otimes \langle a_i \rangle$ for all $i = 2, \ldots, n$, hence

$$n \cdot h = h \otimes \langle 1, a_2, \ldots, a_n \rangle.$$

By hypothesis, this form is hyperbolic; therefore, h has odd order in the Witt group $W^\lambda(D, \theta)$, contrary to Scharlau's result [245, Theorem 5.1]. □

(6.16) Corollary. *Let (A, σ) be a central simple algebra with involution (of any kind) over a field F of characteristic different from 2 and let L/F be a field extension of odd degree. If (A_L, σ_L) is hyperbolic, then (A, σ) is hyperbolic.*

Proof: Embed $L \hookrightarrow \mathrm{End}_F(L)$ by mapping $x \in L$ to multiplication by x and let ν be an involution on $\mathrm{End}_F(L)$ leaving the image of L elementwise invariant. (The existence of such an involution ν follows from (4.14); explicitly, one may pick any nonzero F-linear map $\ell \colon L \to F$ and take for ν the adjoint involution with respect to the bilinear form $b(x, y) = \ell(xy)$ on L.) If (A_L, σ_L) is hyperbolic, then (6.10) shows that $(A, \sigma) \otimes_F \left(\operatorname{End}_F(L), \nu \right)$ is hyperbolic, hence (A, σ) is hyperbolic by (6.15). □

(6.17) Corollary. *Let (A, σ) be a central simple algebra with involution (of any kind) over a field F of characteristic different from 2 and let L/F be a field extension of odd degree. Let $u \in \operatorname{Sym}(A, \sigma) \cap A^\times$ be a symmetric unit. If there exists $v \in A_L$ such that $u = v\sigma_L(v)$, then there exists $w \in A$ such that $u = w\sigma(w)$.*

Proof: Consider the involution ν_u on $M_2(A)$ as in (6.11). The preceding corollary shows that $\left(M_2(A), \nu_u \right)$ is hyperbolic if $\left(M_2(A)_L, (\nu_u)_L \right)$ is hyperbolic. Therefore, the corollary follows from (6.11). □

This result has an equivalent formulation in terms of hermitian forms, which is the way it was originally stated by Bayer-Fluckiger and Lenstra [33, Corollary 1.14]:

(6.18) Corollary (Bayer-Fluckiger-Lenstra). *Let h, h' be nonsingular hermitian forms on a vector space V over a division F-algebra D, where $\operatorname{char} F \neq 2$. The forms h, h' are isometric if they are isometric after an odd-degree scalar extension of F.*

Proof: The forms h and h' are isometric if and only if $h \perp -h'$ is hyperbolic, so that the assertion follows from (6.16). □

§7. The Discriminant

The notion of discriminant considered in this section concerns involutions of orthogonal type and quadratic pairs. The idea is to associate to every orthogonal involution σ over a central simple F-algebra a square class $\operatorname{disc} \sigma \in F^\times/F^{\times 2}$, in such a way that for the adjoint involution σ_b with respect to a symmetric bilinear form b, the discriminant $\operatorname{disc} \sigma_b$ is the discriminant of the form b. If $\operatorname{char} F = 2$, we also associate to every quadratic pair (σ, f) an element $\operatorname{disc}(\sigma, f) \in F/\wp(F)$, generalizing the discriminant (Arf invariant) of quadratic forms.

7.A. The discriminant of orthogonal involutions. Let F be a field of arbitrary characteristic. Recall that if b is a nonsingular bilinear form on a vector space V over F, the *determinant* of b is the square class of the determinant of the Gram matrix of b with respect to an arbitrary basis $(e_1, \ldots, e_n)$ of V:

$$\det b = \det\left(b(e_i, e_j) \right)_{1 \leq i, j \leq n} \cdot F^{\times 2} \in F^\times/F^{\times 2}.$$

The *discriminant* of b is the signed determinant:

$$\operatorname{disc} b = (-1)^{n(n-1)/2} \det b \in F^\times/F^{\times 2}$$

where $n = \dim V$.

If $\dim V$ is odd, then for $\alpha \in F^\times$ we have $\operatorname{disc}(\alpha b) = \alpha \operatorname{disc} b$. Therefore, the discriminant is an invariant of the equivalence class of b modulo scalar factors if and only if the dimension is even. Since involutions correspond to such equivalence classes, the discriminant of an orthogonal involution is defined only for central simple algebras of even degree.

The definition of the discriminant of an orthogonal involution is based on the following crucial result:

(7.1) Proposition. *Let (A, σ) be a central simple algebra with orthogonal involution over F. If $\deg A$ is even, then for any $a, b \in \text{Alt}(A, \sigma) \cap A^\times$,*

$$\text{Nrd}_A(a) \equiv \text{Nrd}_A(b) \mod F^{\times 2}.$$

Proof: Fix some $a, b \in \text{Alt}(A, \sigma) \cap A^\times$. The involution $\sigma' = \text{Int}(a) \circ \sigma$ is symplectic by (2.7). The same proposition shows that $ab \in \text{Alt}(A, \sigma')$ if $\text{char } F = 2$ and $ab \in \text{Sym}(A, \sigma')$ if $\text{char } F \neq 2$; therefore, it follows from (2.9) that $\text{Nrd}_A(ab) \in F^{\times 2}$. $\square$

An alternate proof is given in (8.25) below.

This proposition makes it possible to give the following definition:

(7.2) Definition. Let σ be an orthogonal involution on a central simple algebra A of even degree $n = 2m$ over a field F. The *determinant* of σ is the square class of the reduced norm of any alternating unit:

$$\det \sigma = \text{Nrd}_A(a) \cdot F^{\times 2} \in F^\times / F^{\times 2} \quad \text{for } a \in \text{Alt}(A, \sigma) \cap A^\times$$

and the *discriminant* of σ is the signed determinant:

$$\text{disc } \sigma = (-1)^m \det \sigma \in F^\times / F^{\times 2}.$$

The following properties follow from the definition:

(7.3) Proposition. *Let A be a central simple algebra of even degree over a field F of arbitrary characteristic.*

(1) *Suppose σ is an orthogonal involution on A, and let $u \in A^\times$. If $\text{Int}(u) \circ \sigma$ is an orthogonal involution on A, then $\text{disc}\big(\text{Int}(u) \circ \sigma\big) = \text{Nrd}_A(u) \cdot \text{disc } \sigma$.*

(2) *Suppose σ is a symplectic involution on A, and let $u \in A^\times$. If $\text{Int}(u) \circ \sigma$ is an orthogonal involution on A, then $\text{disc}\big(\text{Int}(u) \circ \sigma\big) = \text{Nrd}_A(u)$.*

(3) *If $A = \text{End}_F(V)$ and σ_b is the adjoint involution with respect to some nonsingular symmetric bilinear form b on V, then $\text{disc } \sigma_b = \text{disc } b$.*

(4) *Suppose σ is an orthogonal involution on A. If (B, τ) is a central simple F-algebra with orthogonal involution, then*

$$\text{disc}(\sigma \otimes \tau) = \begin{cases} \text{disc } \sigma & \text{if } \deg B \text{ is odd,} \\ 1 & \text{if } \deg B \text{ is even.} \end{cases}$$

(5) *Suppose σ is a symplectic involution on A. If (B, τ) is a central simple algebra with symplectic involution and $\text{char } F \neq 2$, then $\text{disc}(\sigma \otimes \tau) = 1$. (If $\text{char } F = 2$, (2.23) shows that $\sigma \otimes \tau$ is symplectic.)*

(6) *Suppose σ is an orthogonal involution on A. If σ is hyperbolic, then $\text{disc } \sigma = 1$. (Since hyperbolic involutions in characteristic 2 are symplectic or unitary, the hypotheses imply $\text{char } F \neq 2$.)*

Proof: (1) If $\text{Int}(u) \circ \sigma$ is an orthogonal involution, then $\sigma(u) = u$ and

$$\text{Alt}\big(A, \text{Int}(u) \circ \sigma\big) = u \cdot \text{Alt}(A, \sigma)$$

by (2.7). The property readily follows.

(2) It suffices to show that $u \in \text{Alt}\big(A, \text{Int}(u) \circ \sigma\big)$. This is clear if $\text{char } F \neq 2$, since the condition that σ is symplectic and $\text{Int}(u) \circ \sigma$ is orthogonal implies $\sigma(u) = -u$, by (2.7). If $\text{char } F = 2$ we have $\text{Alt}\big(A, \text{Int}(u) \circ \sigma\big) = u \cdot \text{Alt}(A, \sigma)$ by (2.7) and $1 \in \text{Alt}(A, \sigma)$ by (2.6).

(3) Let $n = 2m = \dim V$ and identify A with $M_n(F)$ by means of a basis e of V. Let also $b_e \in \mathrm{GL}_n(F)$ be the Gram matrix of the bilinear form b with respect to the chosen basis e. The involution σ_b is then given by

$$\sigma_b = \mathrm{Int}(b_e^{-1}) \circ t,$$

where t is the transpose involution. It is easily seen that $\mathrm{disc}\, t = (-1)^m$ (indeed, it suffices to find an alternating matrix of determinant 1), hence (1) yields:

$$\mathrm{disc}\, \sigma_b = (-1)^m \det(b_e^{-1}) \cdot F^{\times 2} = \mathrm{disc}\, b.$$

(4) If $a \in \mathrm{Alt}(A, \sigma) \cap A^\times$, then $a \otimes 1 \in \mathrm{Alt}(A \otimes B, \sigma \otimes \tau) \cap (A \otimes B)^\times$. The property follows from the relation

$$\mathrm{Nrd}_{A \otimes B}(a \otimes 1) = \mathrm{Nrd}_A(a)^{\deg B}.$$

(5) Since τ is symplectic, $\deg B$ is even, by (2.8). The same argument as in (4) applies to yield $a \otimes 1 \in \mathrm{Alt}(A \otimes B, \sigma \otimes \tau)$ satisfying

$$\mathrm{Nrd}_{A \otimes B}(a \otimes 1) \in F^{\times 2}.$$

(6) Let $\deg A = 2m$ and let $e \in A$ be an idempotent such that $e + \sigma(e) = 1$. We have $\mathrm{rdim}(eA) = m$, hence, over a splitting field, e may be represented by a diagonal matrix

$$e = \mathrm{diag}(\underbrace{1, \ldots, 1}_{m}, \underbrace{0, \ldots, 0}_{m}).$$

Since $\sigma(e) = 1 - e$, we have $2e - 1 \in \mathrm{Alt}(A, \sigma)$; on the other hand, over a splitting field,

$$2e - 1 = \mathrm{diag}(\underbrace{1, \ldots, 1}_{m}, \underbrace{-1, \ldots, -1}_{m}),$$

hence $\mathrm{Nrd}_A(2e - 1) = (-1)^m$ and therefore $\mathrm{disc}\, \sigma = 1$. $\qquad\qquad\square$

(7.4) Example. Let Q be a quaternion algebra with canonical involution γ. By Proposition (2.7), every orthogonal involution on Q has the form $\sigma = \mathrm{Int}(s) \circ \gamma$ for some invertible $s \in \mathrm{Skew}(Q, \gamma) \smallsetminus F$. Proposition (7.3) shows that $\mathrm{disc}\, \sigma = -\mathrm{Nrd}_Q(s) \cdot F^{\times 2}$. Therefore, if two orthogonal involutions $\sigma = \mathrm{Int}(s) \circ \gamma$ and $\sigma' = \mathrm{Int}(s') \circ \gamma$ have the same discriminant, then we may assume that s and s' have the same reduced norm, hence also the same reduced characteristic polynomial since $\mathrm{Trd}_Q(s) = 0 = \mathrm{Trd}_Q(s')$. Therefore,

$$s' = xsx^{-1} = \mathrm{Nrd}_Q(x)^{-1} xs\gamma(x)$$

for some $x \in Q^\times$, and it follows that

$$\sigma' = \mathrm{Int}(x) \circ \sigma \circ \mathrm{Int}(x)^{-1}.$$

This show that orthogonal involutions on a quaternion algebra are classified up to conjugation by their discriminant.

Observe also that if $\sigma = \mathrm{Int}(s) \circ \gamma$ has trivial discriminant, then $s^2 \in F^{\times 2}$. Since $s \notin F$, this relation implies that Q splits, hence quaternion division algebras do not carry any orthogonal involution with trivial discriminant.

The next proposition may be seen as an analogue of the formula for the determinant of an orthogonal sum of two bilinear spaces. Let (A, σ) be a central simple F-algebra with orthogonal involution and let e_1, $e_2 \in A$ be symmetric idempotents such that $e_1 + e_2 = 1$. Denote $A_1 = e_1 A e_1$ and $A_2 = e_2 A e_2$. These algebras are central simple and Brauer-equivalent to A (see (1.13)). They are not subalgebras of A, however, since their unit elements e_1 and e_2 are not the unit 1 of A. The involution σ restricts to involutions σ_1 and σ_2 on A_1 and A_2. If $(A, \sigma) = \big(\operatorname{End}_F(V), \sigma_b\big)$ for some vector space V and some nonsingular symmetric, nonalternating, bilinear form b, then e_1 and e_2 are the orthogonal projections on some subspaces V_1, V_2 such that $V = V_1 \overset{\perp}{\oplus} V_2$. The algebras A_1, A_2 may be identified with $\operatorname{End}_F(V_1)$ and $\operatorname{End}_F(V_2)$, and σ_1, σ_2 are the adjoint involutions with respect to the restrictions of b to V_1 and V_2. These restrictions clearly are symmetric, but if char $F = 2$ one of them may be alternating. Therefore, in the general case, extension of scalars to a splitting field of A shows that σ_1 and σ_2 are both orthogonal if char $F \neq 2$, but one of them may be symplectic if char $F = 2$.

(7.5) Proposition. *With the notation above,*

$$\det \sigma = \det \sigma_1 \det \sigma_2$$

where we set $\det \sigma_i = 1$ *if* σ_i *is symplectic.*

Proof: Let $a_i \in \operatorname{Alt}(A_i, \sigma_i)$ for $i = 1$, 2; then $a_1 + a_2 \in \operatorname{Alt}(A, \sigma)$, and scalar extension to a splitting field of A shows that

$$\operatorname{Nrd}_A(a_1 + a_2) = \operatorname{Nrd}_{A_1}(a_1) \operatorname{Nrd}_{A_2}(a_2).$$

This completes the proof, since $\operatorname{Nrd}_{A_i}(a_i) \cdot F^{\times 2} = \det \sigma_i$ if σ_i is orthogonal, and $\operatorname{Nrd}_{A_i}(a_i) \in F^{\times 2}$ if σ_i is symplectic, by (2.9). $\qquad\square$

7.B. The discriminant of quadratic pairs. Let (σ, f) be a quadratic pair on a central simple F-algebra of even degree. If char $F \neq 2$, the involution σ is orthogonal and the map f is the restriction of $\frac{1}{2} \operatorname{Trd}_A$ to $\operatorname{Sym}(A, \sigma)$; we then set

$$\det(\sigma, f) = \det \sigma \in F^\times / F^{\times 2} \quad \text{and} \quad \operatorname{disc}(\sigma, f) = \operatorname{disc} \sigma \in F^\times / F^{\times 2};$$

this is consistent with the property that the discriminant of a quadratic form of even dimension is equal to the discriminant of its polar bilinear form.

For the rest of this subsection, assume char $F = 2$. Recall that we write $\operatorname{Srd}_A \colon A \to F$ for the map which associates to every element in A the coefficient of $X^{\deg A - 2}$ in its reduced characteristic polynomial (see (1.7)). Recall also that $\wp(x) = x^2 + x$ for $x \in F$.

(7.6) Proposition. *Let* $\ell \in A$ *be such that* $f(s) = \operatorname{Trd}_A(\ell s)$ *for all* $s \in \operatorname{Sym}(A, \sigma)$ *(see (5.7)). For all* $x \in A$,

$$\operatorname{Srd}_A\big(\ell + x + \sigma(x)\big) = \operatorname{Srd}_A(\ell) + \wp\big(\operatorname{Trd}_A(x)\big).$$

Proof: It suffices to prove this formula after scalar extension to a splitting field. We may therefore assume $A = M_n(F)$. By (5.8), we may find an element $a \in M_n(F)$ such that $a + a^t \in \operatorname{GL}_n(F)$ and $(\sigma, f) = (\sigma_a, f_a)$. Letting $g = a + a^t$, we then have $f(s) = \operatorname{tr}(g^{-1} a s)$ for all $s \in \operatorname{Sym}(A, \sigma)$. Since (5.7) shows that the element ℓ is uniquely determined up to the addition of an element in $\operatorname{Alt}(A, \sigma)$, it follows that

$$\ell = g^{-1} a + m + \sigma(m) \quad \text{for some } m \in A,$$

hence

$$\ell + x + \sigma(x) = g^{-1}a + (m + x) + \sigma(m + x).$$

Since $\sigma = \sigma_a = \mathrm{Int}(g^{-1}) \circ t$ and $g^t = g$, the right side may be rewritten as

$$g^{-1}a + g^{-1}y + g^{-1}y^t, \quad \text{where } y = g(m + x).$$

As proved in (0.5), we have $s_2(g^{-1}a + g^{-1}y + g^{-1}y^t) = s_2(g^{-1}a) + \wp(\mathrm{tr}(g^{-1}y))$, hence

$$s_2\big(\ell + x + \sigma(x)\big) = s_2(g^{-1}a) + \wp\big(\mathrm{tr}(m + x)\big) = s_2(g^{-1}a) + \wp\big(\mathrm{tr}(m)\big) + \wp\big(\mathrm{tr}(x)\big).$$

In particular, by letting $x = 0$ we obtain $s_2(\ell) = s_2(g^{-1}a) + \wp\big(\mathrm{tr}(m)\big)$, hence the preceding relation yields

$$s_2\big(\ell + x + \wp(x)\big) = s_2(\ell) + \wp\big(\mathrm{tr}(x)\big). \qquad \square$$

(7.7) Definition. Let (σ, f) be a quadratic pair on a central simple algebra A over a field F of characteristic 2. By (5.7), there exists an element $\ell \in A$ such that $f(s) = \mathrm{Trd}_A(\ell s)$ for all $s \in \mathrm{Sym}(A, \sigma)$, and this element is uniquely determined up to the addition of an element in $\mathrm{Alt}(A, \sigma)$. The preceding proposition shows that the element $\mathrm{Srd}_A(\ell) + \wp(F) \in F/\wp(F)$ does not depend on the choice of ℓ; we may therefore set

$$\det(\sigma, f) = \mathrm{Srd}_A(\ell) + \wp(F) \in F/\wp(F)$$

and, letting $\deg A = 2m$,

$$\mathrm{disc}(\sigma, f) = \det(\sigma, f) + \tfrac{m(m-1)}{2} \in F/\wp(F).$$

The following proposition justifies the definitions above:

(7.8) Proposition. *Let (V, q) be a nonsingular quadratic space of even dimension over an arbitrary field F. For the associated quadratic pair (σ_q, f_q) on $\mathrm{End}_F(V)$ defined in (5.11),*

$$\mathrm{disc}(\sigma_q, f_q) = \mathrm{disc}\, q.$$

Proof: If char $F \neq 2$, the proposition follows from (7.3). For the rest of the proof we therefore assume that char $F = 2$. Let $\dim V = n = 2m$ and consider a basis $(e_1, \ldots, e_n)$ of V which is symplectic for the polar form b_q, i.e.,

$$b_q(e_{2i-1}, e_{2i}) = 1, \quad b_q(e_{2i}, e_{2i+1}) = 0 \quad \text{and} \quad b_q(e_i, e_j) = 0 \quad \text{if } |i - j| > 1.$$

As observed in (5.12), an element $\ell \in \mathrm{End}_F(V)$ such that $f_q(s) = \mathrm{Trd}(\ell s)$ for all $s \in \mathrm{Sym}\big(\mathrm{End}_F(V), \sigma_q\big)$ is given by

$$\ell = \varphi_q\Big(\textstyle\sum_{i=1}^m e_{2i-1} \otimes e_{2i-1}q(e_{2i}) + e_{2i} \otimes e_{2i}q(e_{2i-1}) + e_{2i-1} \otimes e_{2i}\Big)$$

where $\varphi_q \colon V \otimes V \xrightarrow{\sim} \mathrm{End}_F(V)$ is the standard identification (5.10) associated to b_q. Furthermore we have, by (5.12), (2),

$$s_2(\ell) = \Big(\textstyle\sum_{i=1}^m q(e_{2i-1})q(e_{2i})\Big) + \tfrac{m(m-1)}{2},$$

and therefore

$$\mathrm{disc}(\sigma_q, f_q) = \textstyle\sum_{i=1}^m q(e_{2i-1})q(e_{2i}) = \mathrm{disc}\, q. \qquad \square$$

(7.9) Proposition. *The discriminant of any hyperbolic quadratic pair is trivial.*

Proof: If char $F \neq 2$, the proposition follows from (7.3). We may thus assume that char $F = 2$. Let (σ, f) be a hyperbolic quadratic pair on a central simple F-algebra A. By (6.14), there is an idempotent e such that $f(s) = \text{Trd}_A(es)$ for all $s \in \text{Sym}(A, \sigma)$; thus

$$\text{disc}(\sigma, f) = \text{Srd}_A(e) + \tfrac{m(m-1)}{2} + \wp(F)$$

where $m = \frac{1}{2}\deg A$. Since e is an idempotent such that $\text{rdim}(eA) = m$, we have $\text{Prd}_{A,e}(X) = (X-1)^m$, hence $\text{Srd}_A(e) = \binom{m}{2}$, and therefore $\text{disc}(\sigma, f) = 0$. $\square$

The discriminant of the tensor product of a quadratic pair with an involution is calculated in the next proposition. We consider only the case where char $F = 2$, since the case of characteristic different from 2 reduces to the tensor product of involutions discussed in (7.3).

(7.10) Proposition. *Suppose* char $F = 2$. *Let* (σ_1, f_1) *be an orthogonal pair on a central simple F-algebra A_1 of degree $n_1 = 2m_1$ and let (A_2, σ_2) be a central simple F-algebra with involution of the first kind, of degree n_2. The determinant and the discriminant of the orthogonal pair $(\sigma_1 \otimes \sigma_2, f_{1*})$ on $A_1 \otimes A_2$ defined in (5.18) are as follows*:

$$\det(\sigma_1 \otimes \sigma_2, f_{1*}) = n_2 \det(\sigma_1, f_1) + m_1 \binom{n_2}{2};$$
$$\text{disc}(\sigma_1 \otimes \sigma_2, f_{1*}) = n_2 \, \text{disc}(\sigma_1, f_1).$$

In particular, if σ_2 is symplectic, the discriminant of the canonical quadratic pair $(\sigma_1 \otimes \sigma_2, f_\otimes)$ on $A_1 \otimes A_2$ is trivial since n_2 is even.

Proof: Let $\ell_1 \in A_1$ be such that $f_1(s_1) = \text{Trd}_{A_1}(\ell_1 s_1)$ for all $s_1 \in \text{Sym}(A_1, \sigma_1)$. We claim that the element $\ell = \ell_1 \otimes 1$ satisfies $f_{1*}(s) = \text{Trd}_{A_1 \otimes A_2}(\ell s)$ for all $s \in \text{Sym}(A_1 \otimes A_2, \sigma_1 \otimes \sigma_2)$. By (5.14), we have

$$\text{Sym}(A_1 \otimes A_2, \sigma_1 \otimes \sigma_2) = \text{Symd}(A_1 \otimes A_2, \sigma_1 \otimes \sigma_2) + \text{Sym}(A_1, \sigma_1) \otimes \text{Sym}(A_2, \sigma_2),$$

hence it suffices to show

(7.11) $\text{Trd}_{A_1 \otimes A_2}\big(\ell\big(x + \sigma_1 \otimes \sigma_2(x)\big)\big) = \text{Trd}_{A_1 \otimes A_2}(x)$ for $x \in A_1 \otimes A_2$

and

(7.12) $\text{Trd}_{A_1 \otimes A_2}(\ell s_1 \otimes s_2) = f_1(s_1)\,\text{Trd}_{A_2}(s_2)$

for $s_1 \in \text{Sym}(A_1, \sigma_1)$ and $s_2 \in \text{Sym}(A_2, \sigma_2)$. Since $\ell \sigma_1 \otimes \sigma_2(x)$ and $x \sigma_1 \otimes \sigma_2(\ell)$ have the same reduced trace, we have

$$\text{Trd}_{A_1 \otimes A_2}\big(\ell\big(x + \sigma_1 \otimes \sigma_2(x)\big)\big) = \text{Trd}_{A_1 \otimes A_2}\big(\big(\ell + \sigma_1 \otimes \sigma_2(\ell)\big)x\big).$$

Now, $\sigma_1 \otimes \sigma_2(\ell) = \sigma_1(\ell_1) \otimes 1$, hence it follows by (5.7) that $\ell + \sigma_1 \otimes \sigma_2(\ell) = 1$, proving (7.11). To prove (7.12), it suffices to observe

$$\text{Trd}_{A_1 \otimes A_2}(\ell s_1 \otimes s_2) = \text{Trd}_{A_1}(\ell_1 s_1)\,\text{Trd}_{A_2}(s_2) = f_1(s_1)\,\text{Trd}_{A_2}(s_2),$$

hence the claim is proved.

The determinant of $(\sigma_1 \otimes \sigma_2, f_{1*})$ is thus represented by $\text{Srd}_{A_1 \otimes A_2}(\ell)$ in $F/\wp(F)$. Since

$$\text{Prd}_{A_1 \otimes A_2, \ell}(X) = \text{Nrd}_{A_1 \otimes A_2}(X - \ell_1 \otimes 1) = \text{Nrd}_{A_1}(X - \ell_1)^{n_2} = \text{Prd}_{A_1, \ell_1}(X)^{n_2},$$

we have $\text{Srd}_{A_1 \otimes A_2}(\ell) = n_2\,\text{Srd}_{A_1}(\ell_1) + \binom{n_2}{2}\text{Trd}_{A_1}(\ell_1)$. The proposition follows, since $\text{Trd}_{A_1}(\ell_1) = f_1(1)$ and (5.6) shows that $f_1(1) = m_1$. $\square$

As for orthogonal involutions (see (7.4)), quadratic pairs on a quaternion algebra are classified by their discriminant:

(7.13) Proposition. *Let (γ, f_1) and (γ, f_2) be quadratic pairs on a quaternion algebra Q over a field F of characteristic 2. If $\mathrm{disc}(\gamma, f_1) = \mathrm{disc}(\gamma, f_2)$, then there exists $x \in Q^\times$ such that $f_2 = f_1 \circ \mathrm{Int}(x)$.*

Proof: For $i = 1$, 2, let $\ell_i \in Q$ be such that $f_i(s) = \mathrm{Trd}_Q(\ell_i s)$ for all $s \in \mathrm{Sym}(Q, \gamma)$. We have $\mathrm{Trd}_Q(\ell_i) = f_i(1) = 1$, by (5.6), and

$$\mathrm{disc}(\gamma, f_i) = \mathrm{Srd}_Q(\ell_i) + \wp(F) = \mathrm{Nrd}_Q(\ell_i) + \wp(F).$$

Therefore, the hypothesis yields

$$\mathrm{Nrd}_Q(\ell_2) = \mathrm{Nrd}_Q(\ell_1) + (\alpha^2 + \alpha) = \mathrm{Nrd}_Q(\ell_1 + \alpha)$$

for some $\alpha \in F$. Since ℓ_1 is determined up to the addition of an element in $\mathrm{Alt}(Q, \gamma) = F$, we may substitute $\ell_1 + \alpha$ for ℓ_1, and assume $\mathrm{Nrd}_Q(\ell_2) = \mathrm{Nrd}_Q(\ell_1)$. The elements ℓ_1, ℓ_2 then have the same reduced characteristic polynomial, hence we may find $x \in Q^\times$ such that $\ell_1 = x \ell_2 x^{-1}$. For $s \in \mathrm{Sym}(Q, \gamma)$, we then have

$$\mathrm{Trd}_Q(\ell_2 s) = \mathrm{Trd}_Q(x^{-1} \ell_1 x s) = \mathrm{Trd}_Q(\ell_1 x s x^{-1}),$$

hence $f_2(s) = f_1(x s x^{-1})$. $\qquad\qquad\square$

Our final result is an analogue of the formula for the discriminant of an orthogonal sum of quadratic spaces. Let (σ, f) be a quadratic pair on a central simple algebra A over a field F of characteristic 2, and let e_1, $e_2 \in A$ be symmetric idempotents such that $e_1 + e_2 = 1$. As in the preceding section, we let $A_1 = e_1 A e_1$, $A_2 = e_2 A e_2$ and restrict σ to symplectic involutions σ_1 and σ_2 on A_1 and A_2. The degrees of A_1 and A_2 are therefore even. We have $\mathrm{Sym}(A_i, \sigma_i) = \mathrm{Sym}(A, \sigma) \cap A_i$ for $i = 1$, 2, hence we may also restrict f to $\mathrm{Sym}(A_i, \sigma_i)$ and get a quadratic pair (σ_i, f_i) on A_i.

(7.14) Proposition. *With the notation above,*

$$\mathrm{disc}(\sigma, f) = \mathrm{disc}(\sigma_1, f_1) + \mathrm{disc}(\sigma_2, f_2).$$

Proof: For $i = 1$, 2, let $\ell_i \in A_i$ be such that $f_i(s) = \mathrm{Trd}_{A_i}(\ell_i s)$ for all $s \in \mathrm{Sym}(A_i, \sigma_i)$. For $s \in \mathrm{Sym}(A, \sigma)$, we have

$$s = e_1 s e_1 + e_1 s e_2 + \sigma(e_1 s e_2) + e_2 s e_2,$$

hence

$$f(s) = f_1(e_1 s e_1) + \mathrm{Trd}_A(e_1 s e_2) + f_2(e_2 s e_2).$$

Since $\mathrm{Trd}_A(e_1 s e_2) = \mathrm{Trd}_A(s e_2 e_1)$ and $e_2 e_1 = 0$, the middle term on the right side vanishes. Therefore,

$$f(s) = \mathrm{Trd}_{A_1}(\ell_1 e_1 s e_1) + \mathrm{Trd}_{A_2}(\ell_2 e_2 s e_2) \qquad \text{for all } s \in \mathrm{Sym}(A, \sigma).$$

Taking into account the fact that $e_i \ell_i e_i = \ell_i$ for $i = 1$, 2, we obtain

$$f(s) = \mathrm{Trd}_{A_1}(\ell_1 s) + \mathrm{Trd}_{A_2}(\ell_2 s) = \mathrm{Trd}_A\big((\ell_1 + \ell_2)s\big) \qquad \text{for all } s \in \mathrm{Sym}(A, \sigma).$$

We may thus compute $\det(\sigma, f)$:

$$\det(\sigma, f) = \mathrm{Srd}_A(\ell_1 + \ell_2) + \wp(F).$$

Scalar extension to a splitting field of A shows that $\mathrm{Prd}_{A,\ell_1+\ell_2} = \mathrm{Prd}_{A_1,\ell_1}\mathrm{Prd}_{A_2,\ell_2}$, hence

$$\mathrm{Srd}_A(\ell_1 + \ell_2) = \mathrm{Srd}_{A_1}(\ell_1) + \mathrm{Srd}_{A_2}(\ell_2) + \mathrm{Trd}_{A_1}(\ell_1)\,\mathrm{Trd}_{A_2}(\ell_2).$$

Since $\mathrm{Trd}_{A_i}(\ell_i) = f_i(\ell_i) = \frac{1}{2}\deg A_i$, by (5.6), the preceding relation yields

$$\det(\sigma, f) = \det(\sigma_1, f_1) + \det(\sigma_2, f_2) + \tfrac{1}{4}\deg A_1 \deg A_2.$$

The formula for $\mathrm{disc}(\sigma, f)$ is then easily checked, using that $\deg A = \deg A_1 + \deg A_2$. $\square$

§8. The Clifford Algebra

Since the Clifford algebra of a quadratic form is not invariant when the quadratic form is multiplied by a scalar, it is not possible to define a corresponding notion for involutions. However, the *even* Clifford algebra is indeed an invariant for quadratic forms up to similarity, and our aim in this section is to generalize its construction to algebras with quadratic pairs. The first definition of the (generalized, even) Clifford algebra of an algebra with orthogonal involution of characteristic different from 2 was given by Jacobson [134], using Galois descent. Our approach is based on Tits' "rational" definition [291] which includes the characteristic 2 case.

Since our main tool will be scalar extension to a splitting field, we first discuss the case of a quadratic space.

8.A. The split case. Let (V, q) be a nonsingular quadratic space over a field F of arbitrary characteristic. The *Clifford algebra* $C(V, q)$ is the factor of the tensor algebra $T(V)$ by the ideal $I(q)$ generated by all the elements of the form $v \otimes v - q(v) \cdot 1$ for $v \in V$. The natural gradation of $T(V)$ (by natural numbers) induces a gradation by $\mathbb{Z}/2\mathbb{Z}$:

$$T(V) = T_0(V) \oplus T_1(V) = T(V \otimes V) \oplus \big(V \otimes T(V \otimes V)\big).$$

Since generators of $I(q)$ are in $T_0(V)$, the $\mathbb{Z}/2\mathbb{Z}$ gradation of $T(V)$ induces a gradation of $C(V, q)$:

$$C(V, q) = C_0(V, q) \oplus C_1(V, q).$$

We have $\dim_F C(V, q) = 2^{\dim V}$ and $\dim_F C_0(V, q) = 2^{(\dim V)-1}$: see Knus [157, Ch. IV, (1.5.2)].

The *even Clifford algebra* $C_0(V, q)$ may also be defined directly as a factor algebra of $T_0(V) = T(V \otimes V)$:

(8.1) Lemma. *In the tensor algebra $T(V \otimes V)$, consider the following two-sided ideals:*

(1) *$I_1(q)$ is the ideal generated by all the elements of the form*

$$v \otimes v - q(v), \quad for\ v \in V.$$

(2) *$I_2(q)$ is the ideal generated by all the elements of the form*

$$u \otimes v \otimes v \otimes w - q(v)u \otimes w, \quad for\ u,\ v,\ w \in V.$$

Then

$$C_0(V, q) = \frac{T(V \otimes V)}{I_1(q) + I_2(q)}.$$

Proof: The inclusion map $T(V \otimes V) \hookrightarrow T(V)$ maps $I_1(q)$ and $I_2(q)$ into $I(q)$; it therefore induces a canonical epimorphism

$$\frac{T(V \otimes V)}{I_1(q) + I_2(q)} \to C_0(V, q).$$

The lemma follows if we show

$$\dim_F \left(\frac{T(V \otimes V)}{I_1(q) + I_2(q)} \right) \leq \dim_F C_0(V, q).$$

This inequality is easily established by using an orthogonal decomposition of V into subspaces of dimension 1 (if char $F \neq 2$) or of dimension 2 (if char $F = 2$ and $\dim V$ is even) or into one subspace of dimension 1 and subspaces of dimension 2 (if char $F = 2$ and $\dim V$ is odd). $\qquad\square$

Structure of even Clifford algebras. We recall the structure theorem for even Clifford algebras:

(8.2) Theorem. *Let (V, q) be a nonsingular quadratic space over a field F of arbitrary characteristic.*

(1) *If $\dim V$ is odd: $\dim V = 2m + 1$, then $C_0(V, q)$ is central simple F-algebra of degree 2^m.*

(2) *If $\dim V$ is even: $\dim V = 2m$, the center of $C_0(V, q)$ is an étale quadratic F-algebra Z. If Z is a field, then $C_0(V, q)$ is a central simple Z-algebra of degree 2^{m-1}; if $Z \simeq F \times F$, then $C_0(V, q)$ is the direct product of two central simple F-algebras of degree 2^{m-1}. Moreover, the center Z can be described as follows:*

(a) *If char $F \neq 2$, $Z \simeq F[X]/(X^2 - \delta)$ where $\delta \in F^\times$ is a representative of the discriminant: $\operatorname{disc} q = \delta \cdot F^{\times 2} \in F^\times / F^{\times 2}$.*

(b) *If char $F = 2$, $Z \simeq F[X]/(X^2 + X + \delta)$ where $\delta \in F$ is a representative of the discriminant: $\operatorname{disc} q = \delta + \wp(F) \in F/\wp(F)$.*

A proof can be found in Knus [157, Ch. IV] or Lam [169, Ch. 5] (for the case where char $F \neq 2$) or Scharlau [247, Ch. 9] (for the cases where char $F \neq 2$ or char $F = 2$ and $\dim V$ even).

For future reference, we recall an explicit description of the Clifford algebra of hyperbolic quadratic spaces, from which a proof of the theorem above can be derived by scalar extension.

Let U be an arbitrary finite dimensional vector space over F and let $\mathbb{H}(U) = (U^* \oplus U, q_U)$ be the hyperbolic quadratic space defined by

$$q_U(\varphi + u) = \varphi(u)$$

for $\varphi \in U^*$ and $u \in U$, as in §6.B.

In order to give an explicit description of the Clifford algebra of $\mathbb{H}(U)$, consider the exterior algebra $\bigwedge U$. Collecting separately the even and odd exterior powers of U, we get a $\mathbb{Z}/2\mathbb{Z}$-gradation

$$\bigwedge U = \bigwedge_0 U \oplus \bigwedge_1 U,$$

where

$$\bigwedge_0 U = \bigoplus_{i \geq 0} \bigwedge^{2i} U \quad \text{and} \quad \bigwedge_1 U = \bigoplus_{i \geq 0} \bigwedge^{2i+1} U.$$

For $u \in U$, let $\ell_u \in \operatorname{End}_F(\bigwedge U)$ denote (exterior) multiplication on the left by u:

$$\ell_u(x_1 \wedge \cdots \wedge x_r) = u \wedge x_1 \wedge \cdots \wedge x_r.$$

For $\varphi \in U^*$, let $d_\varphi \in \operatorname{End}_F(\bigwedge U)$ be the unique derivation of $\bigwedge U$ extending φ which is explicitly defined by

$$d_\varphi(x_1 \wedge \cdots \wedge x_r) = \sum_{i=1}^{r} (-1)^{i+1} x_1 \wedge \cdots \wedge x_{i-1} \wedge x_{i+1} \wedge \cdots \wedge x_r \varphi(x_i).$$

It is clear that ℓ_u and d_φ interchange the subspaces $\bigwedge_0 U$ and $\bigwedge_1 U$ for all $u \in U$, $\varphi \in U^*$.

(8.3) Proposition. *The map which carries $\varphi + u \in U^* \oplus U$ to $d_\varphi + \ell_u \in \operatorname{End}_F(\bigwedge U)$ induces an isomorphism*

$$\Theta \colon C\big(\mathbb{H}(U)\big) \xrightarrow{\sim} \operatorname{End}_F(\textstyle\bigwedge U).$$

The restriction of this isomorphism to the even Clifford algebra is an isomorphism

$$\Theta_0 \colon C_0\big(\mathbb{H}(U)\big) \xrightarrow{\sim} \operatorname{End}_F(\textstyle\bigwedge_0 U) \times \operatorname{End}_F(\textstyle\bigwedge_1 U).$$

Proof: A computation shows that for $\xi \in \bigwedge^r U$, $\eta \in \bigwedge^s U$ and $\varphi \in U^*$,

$$d_\varphi(\xi \wedge \eta) = d_\varphi(\xi) \wedge \eta + (-1)^r \xi \wedge d_\varphi(\eta).$$

By applying this formula twice in the particular case where $r = 1$, we obtain

$$d_\varphi^2(u \wedge \eta) = u \wedge d_\varphi^2(\eta);$$

by induction on s we conclude that $d_\varphi^2 = 0$. Therefore, for $\eta \in \bigwedge^s U$, $u \in U$ and $\varphi \in U^*$,

$$(d_\varphi + \ell_u)^2(\eta) = d_\varphi(u \wedge \eta) + u \wedge d_\varphi(\eta) = \eta \varphi(u).$$

By the universal property of Clifford algebras, it follows that the map $U^* \oplus U \to \operatorname{End}_F(\bigwedge U)$ which carries $\varphi + u$ to $d_\varphi + \ell_u$ induces an algebra homomorphism $\Theta \colon C\big(\mathbb{H}(U)\big) \to \operatorname{End}_F(\bigwedge U)$. The fact that Θ is an isomorphism is established by induction on $\dim U$ (see Knus [157, Ch. IV, (2.1.1)]). (Alternately, assuming the structure theorem for Clifford algebras, injectivity of Θ follows from the fact that $C\big(\mathbb{H}(U)\big)$ is simple, and surjectivity follows by dimension count).

Let Θ_0 be the restriction of Θ to $C_0\big(\mathbb{H}(U)\big)$. Since $d_\varphi + \ell_u$ exchanges $\bigwedge_0 U$ and $\bigwedge_1 U$ for all $\varphi \in U^*$ and $u \in U$, the elements in the image of Θ_0 preserve $\bigwedge_0 U$ and $\bigwedge_1 U$. Therefore, Θ_0 maps $C_0\big(\mathbb{H}(U)\big)$ into $\operatorname{End}_F(\bigwedge_0 U) \times \operatorname{End}_F(\bigwedge_1 U)$. This map is onto by dimension count. $\qquad\square$

The canonical involution. For every quadratic space (V, q), the identity map on V extends to an involution on the tensor algebra $T(V)$ which preserves the ideal $I(q)$. It therefore induces an involution τ on the Clifford algebra $C(V, q)$. This involution is called the *canonical involution* of $C(V, q)$; it is the unique involution which is the identity on (the image of) V. The involution τ clearly restricts to an involution on $C_0(V, q)$ which we denote by τ_0 and call the *canonical involution* of $C_0(V, q)$. The type of this canonical involution is determined as follows:

(8.4) Proposition. (1) *If $\dim V \equiv 2 \mod 4$, then τ_0 is unitary.*
(2) *If $\dim V \equiv 0 \mod 4$, then τ_0 is the identity on the center Z of $C_0(V, q)$. It is orthogonal if $\dim V \equiv 0 \mod 8$ and $\operatorname{char} F \neq 2$, and symplectic if $\dim V \equiv 4 \mod 8$ or $\operatorname{char} F = 2$. (In the case where $Z \simeq F \times F$, this means that τ_0 is of orthogonal or symplectic type on each factor of $C_0(V, q)$.)*
(3) *If $\dim V \equiv 1, 7 \mod 8$, then τ_0 is orthogonal if $\operatorname{char} F \neq 2$ and symplectic if $\operatorname{char} F = 2$.*

(4) *If* $\dim V \equiv 3,\ 5 \mod 8$, *then* τ_0 *is symplectic.*

Proof: Consider first the case where $\dim V$ is even: $\dim V = 2m$. By extending scalars, we may assume that (V, q) is a hyperbolic quadratic space. Let $(V, q) = \mathbb{H}(U)$ for some m-dimensional vector space U, hence $C(V, q) \simeq \operatorname{End}_F(\bigwedge U)$ by (8.3). Under this isomorphism, the canonical involution τ on $C(V, q)$ corresponds to the adjoint involution with respect to some bilinear form on $\bigwedge U$ which we now describe.

Let $\bar{\ } \colon \bigwedge U \to \bigwedge U$ be the involution such that $\bar{u} = u$ for all $u \in U = \bigwedge^1 U$ and let $s \colon \bigwedge U \to F$ be a nonzero linear map which vanishes on $\bigwedge^r U$ for $r < m$. Define a bilinear form $b \colon \bigwedge U \times \bigwedge U \to F$ by

$$b(\xi, \eta) = s(\bar{\xi} \wedge \eta) \quad \text{for } \xi,\, \eta \in \bigwedge U.$$

We have $b(\eta, \xi) = s(\overline{\bar{\xi} \wedge \eta})$ for $\xi,\, \eta \in \bigwedge U$. Since $\bar{\zeta} = (-1)^{m(m-1)/2} \zeta$ for $\zeta \in \bigwedge^m U$, it follows that b is symmetric if $m \equiv 0,\ 1 \mod 4$, and it is skew-symmetric if $m \equiv 2,\ 3 \mod 4$. If $\operatorname{char} F = 2$, then $\bigwedge U$ is commutative, $\bar{\ }$ is the identity on $\bigwedge U$ and $\xi \wedge \xi = 0$ for all $\xi \in \bigwedge U$, hence b is alternating. In all cases, the form b is nonsingular.

For $u \in U$ and $\xi,\, \eta \in \bigwedge U$ we have $\bar{u} = u$, hence

$$b(u \wedge \xi, \eta) = s(\bar{\xi} \wedge u \wedge \eta) = b(\xi, u \wedge \eta).$$

Similarly, for $\varphi \in U^*$, $\xi,\, \eta \in \bigwedge U$, a simple computation (using the fact that d_φ is a derivation on $\bigwedge U$) shows that

$$b\big(d_\varphi(\xi), \eta\big) = b\big(\xi, d_\varphi(\eta)\big).$$

Therefore, the adjoint involution σ_b on $\operatorname{End}_F(\bigwedge U)$ is the identity on all the endomorphisms of the form $d_\varphi + \ell_u$. It follows that σ_b corresponds to the canonical involution τ under the isomorphism Θ of (8.3). In view of the type of b, the involution τ is orthogonal if $m \equiv 0,\ 1 \mod 4$ and $\operatorname{char} F \neq 2$, and it is symplectic in the other cases.

If m is odd, the complementary subspaces $\bigwedge_0 U$ and $\bigwedge_1 U$ are totally isotropic for b. Therefore, letting $e_0 \in \operatorname{End}_F(\bigwedge U)$ (resp. $e_1 \in \operatorname{End}_F(\bigwedge U)$) denote the projection on $\bigwedge_0 U$ (resp. $\bigwedge_1 U$) parallel to $\bigwedge_1 U$ (resp. $\bigwedge_0 U$), we have $\sigma_b(e_0) = e_1$. It follows that σ_b exchanges $\operatorname{End}_F(\bigwedge_0 U)$ and $\operatorname{End}_F(\bigwedge_1 U)$, hence τ_0 is unitary.

If m is even, b restricts to nonsingular bilinear forms b_0 on $\bigwedge_0 U$ and b_1 on $\bigwedge_1 U$, and the restriction of σ_b to $\operatorname{End}_F(\bigwedge_0 U) \times \operatorname{End}_F(\bigwedge_1 U)$ is $\sigma_{b_0} \times \sigma_{b_1}$. Since b_0 and b_1 have the same type as b, the proof is complete in the case where $\dim V$ is even.

If $\dim V$ is odd: $\dim V = 2m + 1$, we may extend scalars to assume (V, q) decomposes as

$$(V, q) \simeq [-1] \perp (V', q')$$

for some nonsingular quadratic space (V', q') of dimension $2m$ which may be assumed hyperbolic. Considering this isometry as an identification, and letting $e \in V$ denote a basis element of the subspace $[-1]$ such that $q(e) = -1$, we get an isomorphism $C(V', q') \xrightarrow{\sim} C_0(V, q)$ by mapping $x \in V'$ to $e \cdot x \in C_0(V, q)$. If $\operatorname{char} F = 2$, the canonical involution τ_0 on $C_0(V, q)$ corresponds to the canonical involution τ' on $C(V', q')$ under this isomorphism. Therefore, τ_0 is symplectic. If $\operatorname{char} F \neq 2$, the canonical involution τ_0 corresponds to $\operatorname{Int}(\zeta) \circ \tau'$ where $\zeta \in C(V', q')$ is the product of the elements in an orthogonal basis of V'. As observed above, τ' is orthogonal if $m \equiv 0,\ 1 \mod 4$ and is symplectic if $m \equiv 2,\ 3 \mod 4$. On the other hand, $\tau'(\zeta) = (-1)^m \zeta$, hence (2.7) shows that τ_0 is orthogonal if $m \equiv 0, 3 \mod 4$ and symplectic if $m \equiv 1, 2 \mod 4$. $\qquad\square$

(8.5) Proposition. *The involutions τ and τ_0 are hyperbolic if the quadratic space (V, q) is isotropic.*

Proof: If (V, q) is isotropic, it contains a hyperbolic plane; we may thus find in V vectors x, y such that $q(x) = q(y) = 0$ and $b_q(x, y) = 1$. Let $e = x \cdot y \in C_0(V, q) \subset C(V, q)$. The conditions on x and y imply $e^2 = e$ and $\tau(e) = \tau_0(e) = 1 - e$, hence τ and τ_0 are hyperbolic, by (6.8). $\qquad\square$

8.B. Definition of the Clifford algebra. Let (σ, f) be a quadratic pair on a central simple algebra A over a field F of arbitrary characteristic. Our goal is to define an algebra $C(A, \sigma, f)$ in such a way that for every nonsingular quadratic space (V, q) (of even dimension if char $F = 2$),

$$C\big(\operatorname{End}_F(V), \sigma_q, f_q\big) \simeq C_0(V, q)$$

where (σ_q, f_q) is the quadratic pair associated to q by (5.11). The idea behind the definition below (in (8.7)) is that $\operatorname{End}_F(V) \simeq V \otimes V$ under the standard identification φ_q of (5.1); since $C_0(V, q)$ is a factor algebra of $T(V \otimes V)$, we define $C\big(\operatorname{End}_F(V), \sigma_q, f_q\big)$ as a factor algebra of $T\big(\operatorname{End}_F(V)\big)$.

Let $\underline{A}$ denote A viewed as an F-vector space. We recall the "sandwich" isomorphism

$$\operatorname{Sand}: \underline{A} \otimes \underline{A} \xrightarrow{\sim} \operatorname{End}_F(\underline{A})$$

such that $\operatorname{Sand}(a \otimes b)(x) = axb$ for a, b, $x \in A$ (see (3.4)). We use this isomorphism to define a map

$$\sigma_2: \underline{A} \otimes \underline{A} \to \underline{A} \otimes \underline{A}$$

as follows: for fixed $u \in \underline{A} \otimes \underline{A}$ the map $\underline{A} \to \underline{A}$ defined by $x \mapsto \operatorname{Sand}(u)\big(\sigma(x)\big)$ is linear and therefore of the form $\operatorname{Sand}\big(\sigma_2(u)\big)$ for a certain $\sigma_2(u) \in \underline{A}$. In other words, the map σ_2 is defined by the condition

$$\operatorname{Sand}\big(\sigma_2(u)\big)(x) = \operatorname{Sand}(u)\big(\sigma(x)\big) \quad \text{for } u \in \underline{A} \otimes \underline{A}, \, x \in A.$$

(8.6) Lemma. *Let (V, b) be a nonsingular symmetric bilinear space and let σ_b be its adjoint involution on $\operatorname{End}_F(V)$. The map σ_2 on $\underline{\operatorname{End}_F(V)} \otimes \underline{\operatorname{End}_F(V)}$ satisfies*

$$\sigma_2\big(\varphi_b(x_1 \otimes x_2) \otimes \varphi_b(x_3 \otimes x_4)\big) = \varphi_b(x_1 \otimes x_3) \otimes \varphi_b(x_2 \otimes x_4)$$

for x_1, x_2, x_3, $x_4 \in V$ where $\varphi_b: V \otimes V \xrightarrow{\sim} \operatorname{End}_F(V)$ is the standard identification of (5.1).

Proof: It suffices to see that, for x_1, x_2, x_3, x_4, v, $w \in V$,

$$\operatorname{Sand}\big(\varphi_b(x_1 \otimes x_3) \otimes \varphi_b(x_2 \otimes x_4)\big)\big(\varphi_b(v \otimes w)\big) =$$
$$\operatorname{Sand}\big(\varphi_b(x_1 \otimes x_2) \otimes \varphi_b(x_3 \otimes x_4)\big)\big(\varphi_b(w \otimes v)\big).$$

This follows from a straightforward computation: the left side equals

$$\varphi_b(x_1 \otimes x_3) \circ \varphi_b(v \otimes w) \circ \varphi_b(x_2 \otimes x_4) = \varphi_b(x_1 \otimes x_4)b(x_3, v)b(w, x_2)$$

whereas the right side equals

$$\varphi_b(x_1 \otimes x_2) \circ \varphi_b(w \otimes v) \circ \varphi_b(x_3 \otimes x_4) = \varphi_b(x_1 \otimes x_4)b(x_2, w)b(v, x_3). \qquad\square$$

Let $\ell \in A$ be such that $f(s) = \mathrm{Trd}_A(\ell s)$ for all $s \in \mathrm{Sym}(A, \sigma)$. The existence of such an element ℓ is proved in (5.7), where it is also proved that ℓ is uniquely determined up to the addition of an element in $\mathrm{Alt}(A, \sigma)$. If $\ell' = \ell + a - \sigma(a)$ for some $a \in A$, then for all $u \in \underline{A} \otimes \underline{A}$ such that $\sigma_2(u) = u$ we have

$$\mathrm{Sand}(u)(\ell') = \mathrm{Sand}(u)(\ell) + \mathrm{Sand}(u)(a) - \mathrm{Sand}(u)\big(\sigma(a)\big).$$

The last term on the right side is equal to $\mathrm{Sand}\big(\sigma_2(u)\big)(a) = \mathrm{Sand}(u)(a)$, hence

$$\mathrm{Sand}(u)(\ell) = \mathrm{Sand}(u)(\ell').$$

Therefore, the following definition does not depend on the choice of ℓ:

(8.7) Definition. The *Clifford algebra* $C(A, \sigma, f)$ is defined as a factor of the tensor algebra $T(\underline{A})$:

$$C(A, \sigma, f) = \frac{T(\underline{A})}{J_1(\sigma, f) + J_2(\sigma, f)}$$

where

(1) $J_1(\sigma, f)$ is the ideal generated by all the elements of the form $s - f(s) \cdot 1$, for $s \in \underline{A}$ such that $\sigma(s) = s$;

(2) $J_2(\sigma, f)$ is the ideal generated by all the elements of the form $u - \mathrm{Sand}(u)(\ell)$, for $u \in \underline{A} \otimes \underline{A}$ such that $\sigma_2(u) = u$ and for $\ell \in A$ as above.

The following proposition shows that the definition above fulfills our aim:

(8.8) Proposition. *Let (V, q) be a nonsingular quadratic space (of even dimension if $\mathrm{char}\, F = 2$) and let (σ_q, f_q) be the associated quadratic pair. The standard identification $\varphi_q \colon V \otimes V \xrightarrow{\sim} \mathrm{End}_F(V)$ of (5.10) induces an identification*

$$\eta_q \colon C_0(V, q) \xrightarrow{\sim} C\big(\mathrm{End}_F(V), \sigma_q, f_q\big).$$

Proof: It suffices to show that the isomorphism of tensor algebras

$$T(\varphi_q) \colon T(V \otimes V) \xrightarrow{\sim} T\big(\mathrm{End}_F(V)\big)$$

maps $I_1(q)$ to $J_1(\sigma_q, f_q)$ and $I_2(q)$ to $J_2(\sigma_q, f_q)$.

The ideal $T(\varphi_q)\big(I_1(q)\big)$ is generated by all the elements of the form

$$\varphi_q(v \otimes v) - q(v) \cdot 1, \quad \text{for } v \in V \quad .$$

Since σ_q corresponds to the switch map on $V \otimes V$, the elements $s \in \mathrm{End}_F(V)$ such that $\sigma_q(s) = s$ are spanned by elements of the form $\varphi_q(v \otimes v)$. Since moreover $q(v) = f_q \circ \varphi_q(v \otimes v)$ by (5.11), it follows that $J_1(\sigma_q, f_q) = T(\varphi_q)\big(I_1(q)\big)$.

Similarly, (8.6) shows that the elements $u \in \underline{\mathrm{End}_F(V)} \otimes \underline{\mathrm{End}_F(V)}$ such that $\sigma_2(u) = u$ are spanned by elements of the form $\overline{\varphi_q(x \otimes y)} \otimes \overline{\varphi_q(y \otimes z)}$ for x, y, $z \in V$. Therefore, in order to show that $J_2(\sigma_q, f_q) = T(\varphi_q)\big(I_2(q)\big)$, it suffices to prove

$$(8.9) \qquad \mathrm{Sand}\big(\varphi_q(x \otimes y) \otimes \varphi_q(y \otimes z)\big)(\ell) = q(y)\varphi_q(x \otimes z) \quad \text{for } x, y, z \in V.$$

If $\mathrm{char}\, F \neq 2$ we may choose $\ell = \frac{1}{2}$, hence

$$\mathrm{Sand}\big(\varphi_q(x \otimes y) \otimes \varphi_q(y \otimes z)\big)(\ell) = \tfrac{1}{2}\varphi_q(x \otimes y) \circ \varphi_q(y \otimes z).$$

The right side can be evaluated by (5.1):

$$\tfrac{1}{2}\varphi_q(x \otimes y) \circ \varphi_q(y \otimes z) = \tfrac{1}{2}b_q(y, y)\varphi_q(x \otimes z) = q(y)\varphi_q(x \otimes z),$$

proving (8.9) when $\mathrm{char}\, F \neq 2$.

Suppose next char $F = 2$, hence $\dim V$ is even. We may of course assume $y \neq 0$ in (8.9). Let $\dim V = n = 2m$ and let $(e_1, \ldots, e_n)$ be a symplectic basis of V such that $e_1 = y$. We thus assume

$$b_q(e_{2i-1}, e_{2i}) = 1, \quad b_q(e_{2i}, e_{2i+1}) = 0 \quad \text{and} \quad b_q(e_i, e_j) = 0 \quad \text{if } |i - j| > 1.$$

As observed in (5.12), we may then choose

$$\ell = \varphi_q(\textstyle\sum_{i=1}^m e_{2i-1} \otimes e_{2i-1} q(e_{2i}) + e_{2i} \otimes e_{2i} q(e_{2i-1}) + e_{2i-1} \otimes e_{2i}).$$

By (5.1) we have

$$\mathrm{Sand}\big(\varphi_q(x \otimes e_1) \otimes \varphi_q(e_1 \otimes z)\big)\big(\varphi_q(e_{2i-1} \otimes e_{2i-1})\big) =$$
$$\varphi_q(x \otimes e_1) \circ \varphi_q(e_{2i-1} \otimes e_{2i-1}) \circ \varphi_q(e_1 \otimes z) = 0$$

for $i = 1, \ldots, m$, and similarly

$$\mathrm{Sand}\big(\varphi_q(x \otimes e_1) \otimes \varphi_q(e_1 \otimes z)\big)\big(\varphi_q(e_{2i-1} \otimes e_{2i})\big) = 0$$

for $i = 1, \ldots, m$. Moreover,

$$\mathrm{Sand}\big(\varphi_q(x \otimes e_1) \otimes \varphi_q(e_1 \otimes z)\big)\big(\varphi_q(e_{2i} \otimes e_{2i})\big) = \begin{cases} \varphi_q(x \otimes z) & \text{for } i = 1, \\ 0 & \text{for } i > 1, \end{cases}$$

hence

$$\mathrm{Sand}\big(\varphi_q(x \otimes e_1) \otimes \varphi_q(e_1 \otimes z)\big)(\ell) = q(e_1)\varphi_q(x \otimes z). \qquad \square$$

If char $F \neq 2$, the quadratic pair (σ, f) is entirely determined by the involution σ, since $f(s) = \frac{1}{2}\mathrm{Trd}_A(s)$ for all $s \in \mathrm{Sym}(A, \sigma)$. We then simply write $C(A, \sigma, f)$ for $C(A, \sigma)$ Since we may choose $\ell = 1/2$, we have $\mathrm{Sand}(u)(\ell) = \frac{1}{2}\mu(u)$ where $\mu \colon \underline{A} \otimes \underline{A} \to \underline{A}$ is the multiplication map: $\mu(x \otimes y) = xy$ for x, $y \in \underline{A}$.

Examples. Clifford algebras of quadratic pairs on nonsplit central simple algebras are not easy to describe explicitly in general. We have the following results however:

(a) For algebras of degree 2, it readily follows from (8.10) below that $C(A, \sigma, f)$ is the étale quadratic F-algebra determined by the discriminant $\mathrm{disc}(\sigma, f)$.

(b) For the tensor product of two quaternion algebras Q_1, Q_2 with canonical involutions γ_1, γ_2 it is shown in (8.19) below that

$$C(Q_1 \otimes Q_2, \gamma_1 \otimes \gamma_2) \simeq Q_1 \times Q_2.$$

More generally, Tao [281] has determined (in characteristic different from 2) up to Brauer-equivalence the components of the Clifford algebra of a tensor product of two central simple algebras with involution: see the notes at the end of this chapter.

(c) Combining (8.31) and (9.12), one sees that the Clifford algebra of a hyperbolic quadratic pair on a central simple algebra A of degree divisible by 4 decomposes into a direct product of two central simple algebras, of which one is split and the other is Brauer-equivalent to A.

Besides the structure theorem in (8.10) below, additional general information on Clifford algebras of quadratic pairs is given in (9.12).

Structure of Clifford algebras. Although the degree of A is arbitrary in the discussion above (when $\operatorname{char} F \neq 2$), the case where $\deg A$ is odd does not yield anything beyond the even Clifford algebras of quadratic spaces, since central simple algebras of odd degree with involutions of the first kind are split (see (2.8)). Therefore, we shall discuss the structure of Clifford algebras only in the case where $\deg A = n = 2m$.

(8.10) Theorem. *Let (σ, f) be a quadratic pair on a central simple algebra A of even degree $n = 2m$ over a field F of arbitrary characteristic. The center of $C(A, \sigma, f)$ is an étale quadratic F-algebra Z. If Z is a field, then $C(A, \sigma, f)$ is a central simple Z-algebra of degree 2^{m-1}; if $Z \simeq F \times F$, then $C(A, \sigma, f)$ is a direct product of two central simple F-algebras of degree 2^{m-1}. Moreover, the center Z is as follows:*

(1) If $\operatorname{char} F \neq 2$, $Z \simeq F[X]/(X^2 - \delta)$ where $\delta \in F^\times$ is a representative of the discriminant: $\operatorname{disc}(\sigma, f) = \operatorname{disc} \sigma = \delta \cdot F^{\times 2} \in F^\times / F^{\times 2}$.

(2) If $\operatorname{char} F = 2$, $Z \simeq F[X]/(X^2 + X + \delta)$ where $\delta \in F$ is a representative of the discriminant: $\operatorname{disc}(\sigma, f) = \delta + \wp(F) \in F/\wp(F)$.

Proof: Let L be a splitting field of A in which F is algebraically closed (for instance the function field of the Severi-Brauer variety of A). There is a nonsingular quadratic space (V, q) over L such that

$$(A_L, \sigma_L, f_L) \simeq \big(\operatorname{End}_L(V), \sigma_q, f_q\big),$$

by (5.11). Moreover, (7.8) shows that $\operatorname{disc}(\sigma_L, f_L) = \operatorname{disc} q$ (in $L^\times / L^{\times 2}$ if $\operatorname{char} F \neq 2$, in $L/\wp(L)$ if $\operatorname{char} F = 2$). If $\delta \in F$ is a representative of $\operatorname{disc}(\sigma, f)$, we then have

$$\operatorname{disc} q = \begin{cases} \delta \cdot L^{\times 2} & \text{if } \operatorname{char} F \neq 2, \\ \delta + \wp(L) & \text{if } \operatorname{char} F = 2. \end{cases}$$

It is clear from the definition that the construction of the Clifford algebra commutes with scalar extension, hence by (8.8)

$$C(A, \sigma, f) \otimes_F L = C(A_L, \sigma_L, f_L) \simeq C_0(V, q).$$

In particular, it follows that the center Z of $C(A, \sigma, f)$ is a quadratic étale F-algebra which under scalar extension to L becomes isomorphic to $L[X]/(X^2 - \delta)$ if $\operatorname{char} F \neq 2$ and to $L[X]/(X^2 + X + \delta)$ if $\operatorname{char} F = 2$. Since F is algebraically closed in L, it follows that $Z \simeq F[X]/(X^2 - \delta)$ if $\operatorname{char} F \neq 2$ and $Z \simeq F[X]/(X^2 + X + \delta)$ if $\operatorname{char} F = 2$. The other statements also follow from the structure theorem for even Clifford algebras of quadratic spaces: see (8.2). $\qquad\square$

Alternate methods of obtaining the description of Z proven above are given in (8.25) and (8.28).

The canonical involution. Let $\underline{\sigma}\colon T(\underline{A}) \to T(\underline{A})$ be the involution induced by σ on the tensor algebra $T(\underline{A})$; thus, for $a_1, \ldots, a_r \in A$,

$$\underline{\sigma}(a_1 \otimes \cdots \otimes a_r) = \sigma(a_r) \otimes \cdots \otimes \sigma(a_1).$$

Direct computations show that the ideals $J_1(\sigma, f)$ and $J_2(\sigma, f)$ are preserved under $\underline{\sigma}$. Therefore, $\underline{\sigma}$ induces an involution on the factor algebra $C(A, \sigma, f)$ which we also denote by $\underline{\sigma}$ and call the *canonical involution* of $C(A, \sigma, f)$.

The following result justifies this definition:

(8.11) Proposition. *Let (V, q) be a nonsingular quadratic space (of even dimension if* char $F = 2$*) and let (σ_q, f_q) be the associated quadratic pair on* $\mathrm{End}_F(V)$*. Under the standard identification*

$$\eta_q \colon C_0(V, q) \xrightarrow{\sim} C\big(\mathrm{End}_F(V), \sigma_q, f_q\big)$$

of (8.8), the canonical involution τ_0 of $C_0(V, q)$ corresponds to the involution $\underline{\sigma}_q$ of $C\big(\mathrm{End}_F(V), \sigma_q, f_q\big)$*.*

Proof: The canonical involution τ of $C(V, q)$ is induced by the involution of $T(V)$ which is the identity on V. Therefore, τ_0 is induced by the involution of $T(V \otimes V)$ which switches the factors in $V \otimes V$. Under the standard identification of (5.1), this involution corresponds to $\underline{\sigma}_q$. □

By extending scalars to a splitting field of A, we may apply the preceding proposition and (8.4) to determine the type of the involution $\underline{\sigma}$ on $C(A, \sigma, f)$. As in (8.10), we only consider the case of even degree.

(8.12) Proposition. *Let (σ, f) be a quadratic pair on a central simple algebra A of even degree $n = 2m$ over a field F. The canonical involution $\underline{\sigma}$ of $C(A, \sigma, f)$ is unitary if m is odd, orthogonal if $m \equiv 0 \mod 4$ and* char $F \neq 2$*, and symplectic if $m \equiv 2 \mod 4$ or* char $F = 2$*. (In the case where the center of $C(A, \sigma, f)$ is isomorphic to $F \times F$, this means that $\underline{\sigma}$ is of orthogonal or symplectic type on each factor of $C(A, \sigma, f)$.)*

8.C. Lie algebra structures. We continue with the same notation as in the preceding section; in particular, (σ, f) is a quadratic pair on a central simple algebra A over a field F of arbitrary characteristic and $C(A, \sigma, f)$ is the corresponding Clifford algebra.

Since $C(A, \sigma, f)$ is defined as a quotient of the tensor algebra $T(\underline{A})$, the canonical map $A \to \underline{A} \to T(\underline{A})$ yields a canonical map

(8.13) $$c \colon A \to C(A, \sigma, f)$$

which is F-linear but not injective (nor an algebra homomorphism), since $c(s) = f(s)$ for all $s \in \mathrm{Sym}(A, \sigma)$. In particular, (5.6) shows that $c(1) = \frac{1}{2} \deg A$. We will show that the subspace $c(A) \subset C(A, \sigma, f)$ is a Lie subalgebra of $\mathfrak{L}\big(C(A, \sigma, f)\big)$, and relate it to the Lie subalgebra $\mathrm{Alt}(A, \sigma) \subset \mathfrak{L}(A)$.

(8.14) Lemma. *The kernel of c is* $\ker c = \ker f \subset \mathrm{Sym}(A, \sigma)$*, and* $\dim c(A) = \frac{n(n-1)}{2} + 1$ *if* $\deg A = n$*. Moreover, for x_1, $x_2 \in A$ we have*

$$\big[c(x_1), c(x_2)\big] = c\big(\big[x_1 - \sigma(x_1), x_2\big]\big)$$

where $[\,,\,]$ are the Lie brackets.

Proof: Since c and f have the same restriction to $\mathrm{Sym}(A, \sigma)$, it is clear that $\ker f \subset \ker c$. Dimension count shows that this inclusion is an equality if we show $\dim c(A) = \frac{n(n-1)}{2} + 1$.

In order to compute the dimension of $c(A)$, we may extend scalars to a splitting field of A. Therefore, it suffices to consider the case where A is split: let $A = \mathrm{End}_F(V)$ and $(\sigma, f) = (\sigma_q, f_q)$ for some nonsingular quadratic space (V, q) (of even dimension if char $F = 2$). Under the standard identifications $\varphi_q \colon V \otimes V \xrightarrow{\sim} A$

of (5.10) and $\eta_q \colon C_0(V, q) \xrightarrow{\sim} C(A, \sigma, f)$ of (8.8), the map $c \colon V \otimes V \to C_0(V, q)$ is given by the multiplication in $C(V, q)$:

$$c(v \otimes w) = v \cdot w \in C_0(V, q) \quad \text{for } v,\, w \in V.$$

Let $(e_1, \ldots, e_n)$ be an arbitrary basis of V. The Poincaré-Birkhoff-Witt theorem (Knus [157, Ch. IV, (1.5.1)]) shows that the elements 1 and $e_i \cdot e_j$ for $i < j$ are linearly independent in $C_0(V, q)$. Since these elements span $c(V \otimes V)$, it follows that $\dim c(V \otimes V) = \frac{n(n-1)}{2} + 1$, completing the proof of the first part.

In order to prove the last relation, we may also assume that A is split. As above, we identify A with $V \otimes V$ by means of φ_q. Since both sides of the relation are bilinear in x_1, x_2, it suffices to prove it for $x_1 = v_1 \otimes w_1$ and $x_2 = v_2 \otimes w_2$ with $v_1, v_2, w_1, w_2 \in V$. Then

$$\begin{aligned}
\big[x_1 - \sigma(x_1), x_2\big] &= (v_1 \otimes w_1 - w_1 \otimes v_1) \circ (v_2 \otimes w_2) \\
&\quad - (v_2 \otimes w_2) \circ (v_1 \otimes w_1 - w_1 \otimes v_1) \\
&= v_1 \otimes w_2 b_q(w_1, v_2) - w_1 \otimes w_2 b_q(v_1, v_2) \\
&\quad - v_2 \otimes w_1 b_q(w_2, v_1) + v_2 \otimes v_1 b_q(w_2, w_1),
\end{aligned}$$

hence

$$\begin{aligned}
c\big(\big[x_1 - \sigma(x_1), x_2\big]\big) &= v_1 \cdot w_2 b_q(w_1, v_2) - w_1 \cdot w_2 b_q(v_1, v_2) \\
&\quad + v_2 \cdot v_1 b_q(w_2, w_1) - v_2 \cdot w_1 b_q(w_2, v_1).
\end{aligned}$$

For u, $v \in V$, we have $u \cdot v + v \cdot u = b_q(u, v)$; therefore, the four terms on the right side of the last equation can be evaluated as follows:

$$\begin{aligned}
v_1 \cdot w_2 b_q(w_1, v_2) &= v_1 \cdot w_1 \cdot v_2 \cdot w_2 + v_1 \cdot v_2 \cdot w_1 \cdot w_2 \\
w_1 \cdot w_2 b_q(v_1, v_2) &= v_1 \cdot v_2 \cdot w_1 \cdot w_2 + v_2 \cdot v_1 \cdot w_1 \cdot w_2 \\
v_2 \cdot v_1 b_q(w_2, w_1) &= v_2 \cdot v_1 \cdot w_1 \cdot w_2 + v_2 \cdot v_1 \cdot w_2 \cdot w_1 \\
v_2 \cdot w_1 b_q(w_2, v_1) &= v_2 \cdot v_1 \cdot w_2 \cdot w_1 + v_2 \cdot w_2 \cdot v_1 \cdot w_1.
\end{aligned}$$

The alternating sum of the right sides is

$$v_1 \cdot w_1 \cdot v_2 \cdot w_2 - v_2 \cdot w_2 \cdot v_1 \cdot w_1 = \big[c(x_1), c(x_2)\big]. \qquad \square$$

The lemma shows that $c(A)$ is stable under the Lie brackets, and is therefore a Lie subalgebra of $\mathfrak{L}\big(C(A, \sigma, f)\big)$. Moreover, it shows that if x, $y \in A$ are such that $c(x) = c(y)$, then $x - y \in \operatorname{Sym}(A, \sigma)$, hence $x - \sigma(x) = y - \sigma(y)$. We may therefore define a map

$$\delta \colon c(A) \to \operatorname{Alt}(A, \sigma)$$

by

$$\delta\big(c(x)\big) = x - \sigma(x) \quad \text{for } x \in A.$$

(8.15) Proposition. *The map δ is a Lie-algebra homomorphism which fits into an exact sequence*

$$0 \to F \hookrightarrow c(A) \xrightarrow{\delta} \operatorname{Alt}(A, \sigma) \to 0.$$

Proof: For x, $y \in A$ we have $\big[c(x), c(y)\big] = c\big(\big[x - \sigma(x), y\big]\big)$ by (8.14), hence

$$\delta\big(\big[c(x), c(y)\big]\big) = \big[x - \sigma(x), y\big] - \sigma\big(\big[x - \sigma(x), y\big]\big) = \big[x - \sigma(x), y - \sigma(y)\big],$$

proving that δ is a Lie-algebra homomorphism. This map is surjective by definition. In order to show $F \subset \ker \delta$, pick an element $a \in A$ such that $\mathrm{Trd}_A(a) = 1$; we then have

$$c(a + \sigma(a)) = f(a + \sigma(a)) = \mathrm{Trd}_A(a) = 1,$$

hence

$$\delta(1) = a + \sigma(a) - \sigma(a + \sigma(a)) = 0.$$

Therefore, $F \subset \ker \delta$, and dimension count shows that this inclusion is an equality. $\square$

We proceed to define on $c(A)$ another Lie-algebra homomorphism, using the canonical involution $\underline{\sigma}$ on $C(A, \sigma, f)$.

(8.16) Lemma. *For all $x \in A$,*

$$\underline{\sigma}(c(x)) = c(\sigma(x)) \quad and \quad c(x) + \underline{\sigma}(c(x)) = \mathrm{Trd}_A(x).$$

In particular, $\mathrm{Id} + \underline{\sigma}$ *maps $c(A)$ onto F. Therefore, $c(A) \subset \mathfrak{g}(C(A, \sigma, f), \underline{\sigma})$.*

Proof: The first equation is clear from the definition of $\underline{\sigma}$. The second equation follows, since $c(x + \sigma(x)) = f(x + \sigma(x))$. $\square$

Let $c(A)_0 = c(A) \cap \mathrm{Skew}(C(A, \sigma, f), \underline{\sigma})$. As an intersection of Lie subalgebras, $c(A)_0$ is a subalgebra of $\mathfrak{L}(C(A, \sigma, f))$.

(8.17) Proposition. *The map $\mathrm{Id} + \underline{\sigma} \colon c(A) \to F$ is a Lie-algebra homomorphism which fits into an exact sequence*

$$0 \to c(A)_0 \hookrightarrow c(A) \xrightarrow{\mathrm{Id} + \underline{\sigma}} F \to 0.$$

In particular, it follows that $\dim c(A)_0 = \frac{n(n-1)}{2}$ if $\deg A = n$.

Proof: The definition of $c(A)_0$ shows that this set is the kernel of $\mathrm{Id} + \underline{\sigma}$. For x, $y \in A$, we have $[c(x), c(y)] = c([x - \sigma(x), y])$ by (8.14). The preceding lemma shows that the image of this under $\mathrm{Id} + \underline{\sigma}$ is equal to

$$\mathrm{Trd}_A([x - \sigma(x), y]) = 0,$$

hence $\mathrm{Id} + \underline{\sigma}$ is a Lie-algebra homomorphism. $\square$

Special features of the case where $\mathrm{char}\, F \neq 2$ are collected in the following proposition:

(8.18) Proposition. *If $\mathrm{char}\, F \neq 2$, there is a direct sum decomposition*

$$c(A) = F \oplus c(A)_0.$$

The restriction of δ to $c(A)_0$ is an isomorphism of Lie algebras

$$\delta \colon c(A)_0 \xrightarrow{\sim} \mathrm{Alt}(A, \sigma) = \mathrm{Skew}(A, \sigma).$$

The inverse isomorphism is $\frac{1}{2}c$, mapping $x \in \mathrm{Skew}(A, \sigma)$ to $\frac{1}{2}c(x)$.

Proof: The hypothesis that $\mathrm{char}\, F \neq 2$ ensures that $F \cap c(A)_0 = \{0\}$, hence $c(A) = F \oplus c(A)_0$. For $a \in \mathrm{Alt}(A, \sigma)$, we have

$$\underline{\sigma}(c(a)) = c(\sigma(a)) = -c(a)$$

by (8.16), hence $c(a) \in c(A)_0$. On the other hand, the definition of δ yields

$$\delta(c(a)) = a - \sigma(a) = 2a.$$

Since $c(A)_0$ and $\mathrm{Alt}(A, \sigma)$ have the same dimension, it follows that δ is bijective and that its inverse is $\frac{1}{2}c$. $\square$

(8.19) Example. Suppose $A = Q_1 \otimes Q_2$ is a tensor product of two quaternion algebras over a field F of arbitrary characteristic, and let $\sigma = \gamma_1 \otimes \gamma_2$ be the tensor product of the canonical involutions on Q_1 and Q_2. Since γ_1 and γ_2 are symplectic, there is a canonical quadratic pair $(\sigma, f_\otimes)$ on $Q_1 \otimes Q_2$: see (5.21). By (7.3) (if char $F \neq 2$) or (7.10) (if char $F = 2$), the discriminant of $(\sigma, f_\otimes)$ is trivial, hence (8.10) shows that $C(A, \sigma, f_\otimes) = C^+ \times C^-$ for some quaternion algebras C^+, C^-. Moreover, the canonical involution $\underline{\sigma}$ is symplectic (see (8.12)), hence it is the quaternion conjugation on C^+ and C^-. We claim that C^+ and C^- are isomorphic to Q_1 and Q_2.

Let

$$(C^+ \times C^-)' = \{ (x_+, x_-) \in C^+ \times C^- \mid \mathrm{Trd}_{C^+}(x_+) = \mathrm{Trd}_{C^-}(x_-) \}$$
$$= \{ \xi \in C^+ \times C^- \mid \mathrm{Trd}_{C^+ \times C^-}(\xi) \in F \}$$

and

$$(Q_1 \times Q_2)' = \{ (x_1, x_2) \in Q_1 \times Q_2 \mid \mathrm{Trd}_{Q_1}(x_1) = \mathrm{Trd}_{Q_2}(x_2) \}.$$

In view of (8.16), we have $c(A) \subset (C^+ \times C^-)'$, hence $c(A) = (C^+ \times C^-)'$ by dimension count. On the other hand, we may define a linear map $\Theta \colon A \to Q_1 \times Q_2$ by

$$\Theta(x_1 \otimes x_2) = \big(\mathrm{Trd}_{Q_2}(x_2)x_1, \mathrm{Trd}_{Q_1}(x_1)x_2 \big) \qquad \text{for } x_1 \in Q_1,\ x_2 \in Q_2.$$

Clearly, $\mathrm{im}\,\Theta \subset (Q_1 \times Q_2)'$; the converse inclusion follows from the following observation: if $(x_1, x_2) \in Q_1 \times Q_2$ and $\mathrm{Trd}_{Q_1}(x_1) = \mathrm{Trd}_{Q_2}(x_2) = \alpha$, we have

$$(x_1, x_2) = \begin{cases} \Theta(\alpha^{-1}x_1 \otimes x_2) & \text{if } \alpha \neq 0, \\ \Theta(x_1 \otimes \ell_2 + \ell_1 \otimes x_2) & \text{if } \alpha = 0, \end{cases}$$

where $\ell_i \in Q_i$ is an element of reduced trace 1 for $i = 1, 2$. A computation shows that Θ vanishes on the kernel of the canonical map $c \colon A \to C(A, \sigma, f_\otimes)$ (see (8.14)), hence it induces a surjective linear map $c(A) \to (Q_1 \times Q_2)'$ which we call again Θ. Since $c(A)$ and $(Q_1 \times Q_2)'$ have the same dimension, this map is bijective:

$$\Theta \colon (C^+ \times C^-)' = c(A) \xrightarrow{\sim} (Q_1 \times Q_2)'.$$

Using (8.14), one can check that this bijection is an isomorphism of Lie algebras. To complete the proof, we show that this isomorphism extends to an isomorphism of (associative) F-algebras $C^+ \times C^- = C(A, \sigma, f_\otimes) \xrightarrow{\sim} Q_1 \times Q_2$. Since $C^+ \times C^-$ is generated by the subspace $(C^+ \times C^-)'$, the same argument as in the proof of (2.25) shows that it suffices to find an isomorphism extending Θ over an extension of F. We may thus assume that Q_1 and Q_2 are split and identify $Q_1 = Q_2 = \mathrm{End}_F(V)$ for some 2-dimensional F-vector space V. Let b be a nonsingular alternating form on V (such a form is uniquely determined up to a scalar factor) and let q be the quadratic form on $V \otimes V$ whose polar bilinear form is $b \otimes b$ and such that $q(v \otimes w) = 0$ for all v, $w \in V$ (see Exercise 21 of Chapter I). The canonical quadratic pair $(\gamma \otimes \gamma, f_\otimes)$ on $A = \mathrm{End}_F(V) \otimes \mathrm{End}_F(V) = \mathrm{End}_F(V \otimes V)$ is then associated with the quadratic form q, hence the standard identification φ_q induces an F-algebra isomorphism

$$\eta_q \colon C_0(V \otimes V, q) \xrightarrow{\sim} C(A, \sigma, f_\otimes)$$

(see (8.8)). By definition of the canonical map c, we have

$$c(A) = \eta_q\big((V \otimes V) \cdot (V \otimes V)\big).$$

On the other hand, the map $i\colon V \otimes V \to M_2\big(\mathrm{End}_F(V)\big)$ defined by

$$i(v \otimes w) = \begin{pmatrix} 0 & \varphi_b(v \otimes w) \\ -\varphi_b(w \otimes v) & 0 \end{pmatrix} \qquad \text{for } v, w \in V$$

induces an F-algebra homomorphism $i_*\colon C(V \otimes V, q) \to M_2\big(\mathrm{End}_F(V)\big)$ by the universal property of Clifford algebras. This homomorphism is injective because $C(V \otimes V, q)$ is simple, hence also surjective by dimension count. Under the isomorphism i_*, the natural gradation of the Clifford algebra corresponds to the checkerboard grading of $M_2\big(\mathrm{End}_F(V)\big)$, hence i_* induces an F-algebra isomorphism

$$i_*\colon C_0(V \otimes V, q) \xrightarrow{\sim} \begin{pmatrix} \mathrm{End}_F(V) & 0 \\ 0 & \mathrm{End}_F(V) \end{pmatrix} \simeq \mathrm{End}_F(V) \times \mathrm{End}_F(V).$$

For $v_1, v_2, w_1, w_2 \in V$, we have $\varphi_q\big((v_1 \otimes w_1) \otimes (v_2 \otimes w_2)\big) = \varphi_b(v_1 \otimes v_2) \otimes \varphi_b(w_1 \otimes w_2)$, hence

$$\begin{aligned}
\Theta\big(\eta_q(v_1 \otimes w_1 \cdot v_2 \otimes w_2)\big) &= \\
&= \big(\mathrm{tr}\big(\varphi_b(w_1 \otimes w_2)\big)\varphi_b(v_1 \otimes v_2), \mathrm{tr}\big(\varphi_b(v_1 \otimes v_2)\big)\varphi_b(w_1 \otimes w_2)\big) \\
&= \big(b(w_2, w_1)\varphi_b(v_1 \otimes v_2), b(v_2, v_1)\varphi_b(w_1 \otimes w_2)\big).
\end{aligned}$$

On the other hand,

$$\begin{aligned}
i_*(v_1 &\otimes w_1 \cdot v_2 \otimes w_2) = \\
&= \begin{pmatrix} -\varphi_b(v_1 \otimes w_1) \circ \varphi_b(w_2 \otimes v_2) & 0 \\ 0 & -\varphi_b(w_1 \otimes v_1) \circ \varphi_b(v_2 \otimes w_2) \end{pmatrix} \\
&= \begin{pmatrix} -b(w_1, w_2)\varphi_b(v_1 \otimes v_2) & 0 \\ 0 & -b(v_1, v_2)\varphi_b(w_1 \otimes w_2) \end{pmatrix}.
\end{aligned}$$

Therefore, i_* and $\Theta \circ \eta_q$ have the same restriction to $(V \otimes V) \cdot (V \otimes V)$, and it follows that the F-algebra isomorphism $i_* \circ \eta_q^{-1}\colon C(A, \sigma, f_\otimes) \xrightarrow{\sim} \mathrm{End}_F(V) \times \mathrm{End}_F(V) = Q_1 \times Q_2$ extends Θ. This completes the proof of the claim.

In conclusion, we have shown:

$$C(Q_1 \otimes Q_2, \gamma_1 \otimes \gamma_2, f_\otimes) \simeq Q_1 \times Q_2.$$

A more general statement is proved in (15.7) below.

8.D. The center of the Clifford algebra. The center of the Clifford algebra $C(A, \sigma, f)$ of a central simple algebra A with a quadratic pair (σ, f) is described in (8.10) as an étale quadratic F-algebra. In this section, we show how elements of the center can be produced explicitly, thus providing another proof of the second part of (8.10).

We set $Z(A, \sigma, f)$ for the center of $C(A, \sigma, f)$. If $\mathrm{char}\, F \neq 2$, the map f is uniquely determined by σ and we use the shorter notation $C(A, \sigma)$ for the Clifford algebra and $Z(A, \sigma)$ for its center.

As may expected from (8.10), our methods in characteristic 2 and characteristic not 2 are completely different. In characteristic different from 2 they rely on an analogue of the pfaffian, viewed as a map from $\mathrm{Skew}(A, \sigma)$ to $Z(A, \sigma)$. The case of characteristic 2 is simpler; it turns out then that $Z(A, \sigma, f)$ is in the image $c(A)$ of A in $C(A, \sigma, f)$ under the canonical map of §8.C.

Characteristic not 2. Our first result yields a standard form for certain skew-symmetric elements in split algebras with orthogonal involution.

(8.20) Lemma. *Let (V, q) be a nonsingular quadratic space of dimension $n = 2m$ over a field F of characteristic different from 2 and let $a \in \operatorname{End}_F(V)$ satisfy $\sigma_q(a) = -a$. Assume moreover that the characteristic polynomial of a splits into pairwise distinct linear factors:*

$$\operatorname{Pc}_a(X) = (X - \lambda_1)(X + \lambda_1) \cdots (X - \lambda_m)(X + \lambda_m)$$

for some $\lambda_1, \ldots, \lambda_m \in F^\times$. There exists an orthogonal basis $(e_1, \ldots, e_n)$ of V such that the matrix representing a with respect to this basis is

$$\begin{pmatrix} \Lambda_1 & & 0 \\ & \ddots & \\ 0 & & \Lambda_m \end{pmatrix} \quad where \quad \Lambda_i = \begin{pmatrix} 0 & \lambda_i \\ \lambda_i & 0 \end{pmatrix}$$

Letting $\alpha_i = q(e_i)$ for $i = 1, \ldots, n$, we have $\alpha_{2i} = -\alpha_{2i-1}$ for $i = 1, \ldots, m$. Moreover, with $\varphi_q \colon V \otimes V \xrightarrow{\sim} \operatorname{End}_F(V)$ the standard identification (5.10), we have

$$a = \sum_{i=1}^{m} \frac{\lambda_i}{2\alpha_{2i}} \varphi_q(e_{2i-1} \otimes e_{2i} - e_{2i} \otimes e_{2i-1}).$$

Proof: For $i = 1, \ldots, m$, let $V_i \subset V$ be the sum of the eigenspaces of a for the eigenvalues λ_i and $-\lambda_i$. The subspace V_i is thus the eigenspace of a^2 for the eigenvalue λ_i^2. We have

$$V = V_1 \oplus \cdots \oplus V_m$$

and the subspaces $V_1, \ldots, V_m$ are pairwise orthogonal since, for $x \in V_i$ and $y \in V_j$,

$$\lambda_i^2 b_q(x, y) = b_q\big(a^2(x), y\big) = b_q\big(x, a^2(y)\big) = \lambda_j^2 b_q(x, y),$$

and $\lambda_i^2 \neq \lambda_j^2$ for $i \neq j$. It follows that the subspaces $V_1, \ldots, V_m$ are nonsingular. For $i = 1, \ldots, m$, pick an anisotropic vector $e_{2i-1} \in V_i$ and let $e_{2i} = \lambda_i^{-1} a(e_{2i-1})$. Since $\sigma_q(a) = -a$, we have

$$b_q(e_{2i-1}, e_{2i}) = \lambda_i^{-1} b_q\big(e_{2i-1}, a(e_{2i-1})\big)$$
$$= -\lambda_i^{-1} b_q\big(a(e_{2i-1}), e_{2i-1}\big) = -b_q(e_{2i}, e_{2i-1}),$$

hence (e_{2i-1}, e_{2i}) is an orthogonal basis of V_i. It follows that $(e_1, \ldots, e_n)$ is an orthogonal basis of V, and the matrix of a with respect to this basis is as stated above.

The equation $a(e_{2i-1}) = \lambda_i e_{2i}$ yields

$$b_q\big(a(e_{2i-1}), a(e_{2i-1})\big) = \lambda_i^2 b_q(e_{2i}, e_{2i}) = 2\lambda_i^2 q(e_{2i}).$$

On the other hand, since $\sigma_q(a) = -a$ and $a^2(e_{2i-1}) = \lambda_i^2 e_{2i-1}$, the left side is also equal to

$$b_q\big(e_{2i-1}, -a^2(e_{2i-1})\big) = -\lambda_i^2 b_q(e_{2i-1}, e_{2i-1}) = -2\lambda_i^2 q(e_{2i-1}),$$

hence $q(e_{2i}) = -q(e_{2i-1})$. Finally, for $i, j = 1, \ldots, m$ we have

$$\varphi_q(e_{2i-1} \otimes e_{2i} - e_{2i} \otimes e_{2i-1})(e_{2j-1}) = \begin{cases} -2q(e_{2i-1})e_{2i} & \text{if } i = j, \\ 0 & \text{if } i \neq j, \end{cases}$$

and

$$\varphi_q(e_{2i-1} \otimes e_{2i} - e_{2i} \otimes e_{2i-1})(e_{2j}) = \begin{cases} 2q(e_{2i})e_{2i-1} & \text{if } i = j, \\ 0 & \text{if } i \neq j. \end{cases}$$

The last equation in the statement of the lemma follows. □

Let (A, σ) be a central simple algebra with orthogonal involution over a field F of characteristic different from 2. We assume throughout this subsection that the degree of A is even and let $\deg A = n = 2m$. Our first observations also require the field F to be infinite. Under this hypothesis, we denote by $S(A, \sigma)$ the set of skew-symmetric units in $A^\times$ whose reduced characteristic polynomials are separable (i.e., have no repeated root in an algebraic closure). This set is Zariski-open in $\text{Skew}(A, \sigma)$, since it is defined by the condition that the discriminant of the reduced characteristic polynomial does not vanish. By scalar extension to a splitting field L such that $(A_L, \sigma_L) \simeq (M_n(L), t)$, we can see that this open set is not empty, since $S(M_n(L), t) \neq \varnothing$.

Over an algebraic closure, the reduced characteristic polynomial of every $a \in S(A, \sigma)$ splits into a product of pairwise distinct linear factors of the form

$$\text{Prd}_a(X) = (X - \lambda_1)(X + \lambda_1) \cdots (X - \lambda_m)(X + \lambda_m)$$

since $\sigma(a) = -a$. Therefore, the subalgebra $F[a] \subset A$ generated by a has dimension n and $F[a^2] = F[a] \cap \text{Sym}(A, \sigma)$ has dimension m. Clearly, $F[a] \cap \text{Skew}(A, \sigma) = a \cdot F[a^2]$.

(8.21) Lemma. *Let $a \in S(A, \sigma)$. Denote $H = F[a] \cap \text{Skew}(A, \sigma) = a \cdot F[a^2]$ and $E = F[a^2]$. The bilinear form $T \colon H \times H \to F$ defined by*

$$T(x, y) = T_{E/F}(xy) \quad \text{for } x, y \in H$$

is nonsingular. Moreover, the elements in the image $c(H)$ of H in $C(A, \sigma)$ under the canonical map $c \colon A \to C(A, \sigma)$ commute.

Proof: It suffices to check the lemma over a scalar extension. We may therefore assume that A and the reduced characteristic polynomial of a are split[12]. We identify $(A, \sigma) = (\text{End}_F(V), \sigma_q)$ for some nonsingular quadratic space (V, q) of dimension n. By (8.20), there is an orthogonal basis $(e_1, \ldots, e_n)$ of V such that, letting $q(e_i) = \alpha_i \in F^\times$,

$$a = \sum_{i=1}^m \frac{\lambda_i}{2\alpha_{2i}} \varphi_q(e_{2i-1} \otimes e_{2i} - e_{2i} \otimes e_{2i-1})$$

(and $\alpha_{2i-1} = -\alpha_{2i}$ for $i = 1, \ldots, m$). For $i = 1, \ldots, m$, let

$$h_i = \frac{1}{2\alpha_{2i}} \varphi_q(e_{2i-1} \otimes e_{2i} - e_{2i} \otimes e_{2i-1}) \in A.$$

By using the matrix representation with respect to the basis $(e_1, \ldots, e_n)$, it is easily seen that every skew-symmetric element in A which commutes with a is a linear combination of $h_1, \ldots, h_m$. In particular, H is contained in the span of $h_1, \ldots, h_m$. Since $\dim H = \dim F[a^2] = m$, it follows that $(h_1, \ldots, h_m)$ is a basis of H.

---

[12]In fact, the algebra A splits as soon as the reduced characteristic polynomial of a splits.

With the same matrix representation, it is easy to check that $h_1^2, \ldots, h_m^2$ are primitive orthogonal idempotents in E which form a basis of E and that $h_i h_j = 0$ for $i \neq j$. Therefore, the bilinear form T satisfies

$$T(\textstyle\sum_{i=1}^m x_i h_i, \, \sum_{j=1}^m y_j h_j) = \sum_{i=1}^m x_i y_i$$

for $x_1, \ldots, x_m, y_1, \ldots, y_m \in F$. It is therefore nonsingular.

Consider now the subspace $c(H)$ of $C(A, \sigma)$ spanned by $c(h_1), \ldots, c(h_m)$. For $i = 1, \ldots, m$ we have

$$c(h_i) = \frac{1}{2\alpha_{2i}}(e_{2i-1} \cdot e_{2i} - e_{2i} \cdot e_{2i-1}) = \frac{1}{\alpha_{2i}} e_{2i-1} \cdot e_{2i}.$$

These elements commute, since $(e_1, \ldots, e_n)$ is an orthogonal basis of V. $\qquad\square$

Since the bilinear form T on H is nonsingular, every linear form on H is of the type $x \mapsto T_{E/F}(hx)$ for some $h \in H$. Therefore, every homogeneous polynomial map $P \colon H \to F$ of degree d has the form

$$P(x) = \textstyle\sum_i T_{E/F}(h_{i1}x) \cdots T_{E/F}(h_{id}x)$$

for some $h_{i1}, \ldots, h_{id} \in H$, and in the d-th symmetric power $S^d H$ the element $\sum_i h_{i1} \cdots h_{id}$ is uniquely determined by P. In particular, there is a uniquely determined element $\nu = \sum_i h_{i1} \cdots h_{im} \in S^m H$ such that

$$N_{E/F}(ax) = \textstyle\sum_i T_{E/F}(h_{i1}x) \cdots T_{E/F}(h_{im}x) \quad \text{for } x \in H,$$

since the map $x \mapsto N_{E/F}(ax)$ is a homogeneous polynomial map of degree m on H.

Since the elements in $c(H)$ commute, the canonical map c induces a well-defined linear map $S^m H \to C(A, \sigma)$. We set $\pi(a)$ for the image under this induced map of the element $\nu \in S^m H$ defined above.

In summary, the element $\pi(a) \in C(A, \sigma)$ is defined as follows:

(8.22) Definition. For $a \in S(A, \sigma)$, we let

$$\pi(a) = \textstyle\sum_i c(h_{i1}) \cdots c(h_{im})$$

where $h_{i1}, \ldots, h_{im} \in H = a \cdot F[a^2]$ satisfy

$$N_{F[a^2]/F}(ax) = \textstyle\sum_i T_{F[a^2]/F}(h_{i1}x) \cdots T_{F[a^2]/F}(h_{im}x)$$

for all $x \in a \cdot F[a^2]$.

(8.23) Lemma. *Let ι be the nontrivial automorphism of the center $Z(A, \sigma)$ of $C(A, \sigma)$. For $a \in S(A, \sigma)$ we have $\pi(a) \in Z(A, \sigma)$, $\iota\big(\pi(a)\big) = -\pi(a)$ and*

$$\pi(a)^2 = (-1)^m \operatorname{Nrd}_A(a).$$

Proof: It suffices to verify the assertions over a scalar extension. We may thus assume that A and the reduced characteristic polynomial of a are split, and use the same notation as in (8.21). In particular, we let $(A, \sigma) = \big(\operatorname{End}_F(V), \sigma_q\big)$ and choose an orthogonal basis $(e_1, \ldots, e_m)$ of V such that

$$a = \textstyle\sum_{i=1}^m \lambda_i h_i \quad \text{where} \quad h_i = \frac{1}{2\alpha_{2i}} \varphi_q(e_{2i-1} \otimes e_{2i} - e_{2i} \otimes e_{2i-1}).$$

As observed in (8.21), the elements $h_1, \ldots, h_m$ form a basis of H. For $x = \sum_{i=1}^m x_i h_i$, we have

$$N_{F[a^2]/F}(ax) = N_{F[a^2]/F}(\textstyle\sum_{i=1}^m \lambda_i x_i h_i^2) = \lambda_1 \cdots \lambda_m x_1 \cdots x_m.$$

On the other hand, $T_{F[a^2]/F}(h_i x) = x_i$, hence $\nu = h_1 \dots h_m$ and

$$\pi(a) = \lambda_1 \cdots \lambda_m c(h_1) \cdots c(h_m).$$

It was also seen in (8.21) that $c(h_i) = \frac{1}{\alpha_{2i}} e_{2i-1} \cdot e_{2i}$ where $\alpha_{2i} = q(e_{2i}) = -q(e_{2i-1})$, hence

$$\pi(a) = \frac{\lambda_1 \cdots \lambda_m}{\prod_{i=1}^{m} \alpha_{2i}} e_1 \cdots e_n.$$

It is then clear that $\pi(a) \in Z(A, \sigma)$ and $\iota(\pi(a)) = -\pi(a)$.

Since $(e_1 \cdots e_n)^2 = (-1)^m e_1^2 \cdots e_n^2 = \prod_{i=1}^{m} \alpha_{2i}^2$ and $\mathrm{Nrd}_A(a) = (-1)^m \lambda_1^2 \cdots \lambda_m^2$, the last equation in the statement of the proposition follows. □

To extend the definition of π to the whole of $\mathrm{Skew}(A, \sigma)$, including also the case where the base field F is finite (of characteristic different from 2), we adjoin indeterminates to F and apply π to a generic skew-symmetric element.

Pick a basis $(a_1, \dots, a_d)$ of $\mathrm{Skew}(A, \sigma)$ where $d = m(n-1) = \dim \mathrm{Skew}(A, \sigma)$, and let $\xi = \sum_{i=1}^{d} a_i x_i$ where $x_1, \dots, x_d$ are indeterminates over F. We have $\xi \in S\big(A_{F(x_1, \dots, x_d)}, \sigma_{F(x_1, \dots, x_d)}\big)$ and

$$\pi(\xi) \in Z(A_{F(x_1, \dots, x_d)}, \sigma_{F(x_1, \dots, x_d)}) = Z(A, \sigma) \otimes_F F(x_1, \dots, x_d).$$

Since $\pi(\xi)^2 = \mathrm{Nrd}(\xi)$ is a polynomial in $x_1, \dots, x_d$, we have in fact

$$\pi(\xi) \in Z(A, \sigma) \otimes F[x_1, \dots, x_d].$$

We may then define $\pi(a)$ for all $a \in \mathrm{Skew}(A, \sigma)$ by specializing the indeterminates. We call the map $\pi \colon \mathrm{Skew}(A, \sigma) \to Z(A, \sigma)$ thus defined the *generalized pfaffian* of (A, σ) in view of Example (8.26) below.

(8.24) Proposition. *The map $\pi \colon \mathrm{Skew}(A, \sigma) \to Z(A, \sigma)$ is a homogeneous polynomial map of degree m. Denoting by ι the nontrivial automorphism of $Z(A, \sigma)$ over F, we have*

$$\iota(\pi(a)) = -\pi(a) \quad and \quad \pi(a)^2 = (-1)^m \mathrm{Nrd}_A(a)$$

for all $a \in \mathrm{Skew}(A, \sigma)$. Moreover, for all $x \in A$, $a \in \mathrm{Skew}(A, \sigma)$,

$$\pi(xa\sigma(x)) = \mathrm{Nrd}_A(x)\pi(a).$$

Proof: For the generic element ξ we have by (8.23)

$$\iota(\pi(\xi)) = -\pi(\xi) \quad \text{and} \quad \pi(\xi)^2 = (-1)^m \mathrm{Nrd}(\xi).$$

The same formulas follow for all $a \in \mathrm{Skew}(A, \sigma)$ by specialization. Since the reduced norm is a homogeneous polynomial map of degree n, the second formula shows that π is a homogeneous polynomial map of degree m. It also shows that an element $a \in \mathrm{Skew}(A, \sigma)$ is invertible if and only if $\pi(a) \neq 0$.

In order to prove the last property, fix some element $a \in \mathrm{Skew}(A, \sigma)$. If a is not invertible, then $\pi(a) = \pi(xa\sigma(x)) = 0$ for all $x \in A$ and the property is clear. Suppose $a \in A^\times$. Since the F-vector space of elements $z \in Z(A, \sigma)$ such that $\iota(z) = -z$ has dimension 1, we have for all $x \in A$

$$\pi(xa\sigma(x)) = P(x)\pi(a) \quad \text{for some } P(x) \in F.$$

The map $P \colon A \to F$ is polynomial and satisfies

$$P(x)^2 = \frac{\pi(xa\sigma(x))^2}{\pi(a)^2} = \frac{\mathrm{Nrd}(xa\sigma(x))}{\mathrm{Nrd}(a)} = \mathrm{Nrd}(x)^2.$$

By adjoining indeterminates to F if necessary, we may assume F is infinite. The algebra of polynomial maps on A is then a domain, hence the preceding equation yields the alternative: $P(x) = \mathrm{Nrd}(x)$ for all x or $P(x) = -\mathrm{Nrd}(x)$ for all x. Since $P(1) = 1$, we have $P(x) = \mathrm{Nrd}(x)$ for all $x \in A$. $\square$

Using the map π, we may give an alternate proof of (7.1) and of part of (8.10):

(8.25) Corollary. *For all a, $b \in \mathrm{Skew}(A, \sigma) \cap A^\times$,*

$$\mathrm{Nrd}_A(a) \equiv \mathrm{Nrd}_A(b) \quad \mathrm{mod}\ F^{\times 2}$$

and $Z(A, \sigma) \simeq F[X]/\big(X^2 - (-1)^m \mathrm{Nrd}_A(a)\big)$.

Proof: Since the F-vector space of elements $z \in Z(A, \sigma)$ such that $\iota(z) = -z$ is 1-dimensional, we have $\pi(a) \equiv \pi(b) \mod F^\times$. By squaring both sides we obtain the first equation. The second equation follows from the fact that $Z(A, \sigma) = F\big[\pi(a)\big]$.
$\square$

(8.26) Example. In the case where A is split, the map π can be described explicitly in terms of the pfaffian. Let (V, q) be a nonsingular quadratic space of dimension $n = 2m$ over F. For $(A, \sigma) = \big(\mathrm{End}_F(V), \sigma_q\big)$, we identify $A = V \otimes V$ as in (5.10) and $C(A, \sigma) = C_0(V, q)$ as in (8.8). Let $(e_1, \ldots, e_n)$ be an orthogonal basis of V. The elements $\frac{1}{2}(e_i \otimes e_j - e_j \otimes e_i)$ for $1 \le i < j \le n$ form a basis of $\mathrm{Skew}(A, \sigma)$. For $a = \sum_{i<j} \frac{a_{ij}}{2}(e_i \otimes e_j - e_j \otimes e_i) \in \mathrm{Skew}(A, \sigma)$, define a skew-symmetric matrix $a' = (a'_{ij}) \in M_n(F)$ by

$$a'_{ij} = \begin{cases} a_{ij} & \text{if } i < j, \\ 0 & \text{if } i = j \\ -a_{ji} & \text{if } i > j. \end{cases}$$

Claim. The element $\pi(a) \in Z(A, \sigma)$ is related to the pfaffian $\mathrm{pf}(a')$ as follows:

$$\pi(a) = \mathrm{pf}(a')e_1 \cdots e_n.$$

Proof: A computation shows that the matrix representing a with respect to the basis $(e_1, \ldots, e_n)$ is $a' \cdot d$ where

$$d = \begin{pmatrix} q(e_1) & & 0 \\ & \ddots & \\ 0 & & q(e_n) \end{pmatrix}$$

Since $\det a' = \mathrm{pf}(a')^2$, it follows that

$$\det a = \mathrm{pf}(a')^2 q(e_1) \cdots q(e_n) = (-1)^m \big(\mathrm{pf}(a')e_1 \cdots e_n\big)^2.$$

Therefore, $\pi(a) = \pm\mathrm{pf}(a')e_1 \cdots e_n$, since both sides are polynomial maps of degree m whose squares are equal. To prove that the equality holds with the $+$ sign, it suffices to evaluate both sides on a particular unit in $\mathrm{Skew}(A, \sigma)$. Adjoin indeterminates $z_1, \ldots, z_m$ to F and consider

$$\zeta = \sum_{i=1}^m \frac{z_i}{2}(e_{2i-1} \otimes e_{2i} - e_{2i} \otimes e_{2i-1}) \in \mathrm{Skew}\big(A_{F(z_1,\ldots,z_m)}, \sigma_{F(z_1,\ldots,z_m)}\big).$$

The same computation as in (8.23) shows that

$$\pi(\zeta) = z_1 \cdots z_m e_1 \cdots e_n.$$

By setting $z_1 = \cdots = z_m = 1$, we get for $a = \sum_{i=1}^{m} \frac{1}{2}(e_{2i-1} \otimes e_{2i} - e_{2i} \otimes e_{2i-1})$:

$$\pi(a) = e_1 \cdots e_n.$$

On the other hand, the corresponding matrix a' is

$$a' = \begin{pmatrix} J & & 0 \\ & \ddots & \\ 0 & & J \end{pmatrix} \quad \text{where} \quad J = \begin{pmatrix} 0 & 1 \\ -1 & 0 \end{pmatrix},$$

hence $\mathrm{pf}(a') = 1$. $\qquad\qquad\qquad\qquad\qquad\qquad\qquad\qquad\qquad\qquad\qquad$ $\square$

Characteristic 2. Let (σ, f) be a quadratic pair on a central simple algebra A over a field F of characteristic 2. We write $Z(A, \sigma, f)$ for the center of the Clifford algebra $C(A, \sigma, f)$ and by ι the nontrivial automorphism of $Z(A, \sigma, f)$ over F.

Consider the set

$$\Lambda = \{\, \ell \in A \mid f(s) = \mathrm{Trd}_A(\ell s) \text{ for } s \in \mathrm{Sym}(A, \sigma) \,\}.$$

By (5.7), this set is nonempty; it is a coset of $\mathrm{Alt}(A, \sigma)$.

The following proposition shows that the canonical map $c \colon A \to C(A, \sigma, f)$, restricted to Λ, plays a rôle analogous to the map π in characteristic different from 2.

(8.27) Proposition. *Let* $\deg A = n = 2m$. *For all* $\ell \in \Lambda$, *we have* $c(\ell) \in Z(A, \sigma, f)$, $\iota\big(c(\ell)\big) = c(\ell) + 1$ *and*

$$c(\ell)^2 + c(\ell) = \mathrm{Srd}_A(\ell) + \tfrac{m(m-1)}{2}.$$

Proof: It suffices to check the equations above after a scalar extension. We may therefore assume that A is split. Moreover, it suffices to consider a particular choice of ℓ; indeed, if ℓ_0, $\ell \in \Lambda$, then $\ell = \ell_0 + x + \sigma(x)$ for some $x \in A$, hence

$$c(\ell) = c(\ell_0) + c\big(x + \sigma(x)\big).$$

Since c and f have the same restriction to $\mathrm{Sym}(A, \sigma)$, the last term on the right side is $f\big(x + \sigma(x)\big) = \mathrm{Trd}_A(x)$, hence

$$c(\ell) = c(\ell_0) + \mathrm{Trd}_A(x).$$

Therefore, we have $c(\ell) \in Z(A, \sigma, f)$ and $\iota\big(c(\ell)\big) = c(\ell) + 1$ if and only if $c(\ell_0)$ satisfies the same conditions. Moreover, (7.6) yields $\mathrm{Srd}_A(\ell) = \mathrm{Srd}_A(\ell_0) + \wp\big(\mathrm{Trd}_A(x)\big)$, hence $\wp\big(c(\ell)\big) = \mathrm{Srd}_A(\ell) + \tfrac{m(m-1)}{2}$ if and only if the same equation holds for ℓ_0.

We may thus assume $A = \mathrm{End}_F(V)$ and $(\sigma, f) = (\sigma_q, f_q)$ for some nonsingular quadratic space (V, q), and consider only the case of

$$\ell_0 = \varphi_q\big(\textstyle\sum_{i=1}^{m} e_{2i-1} \otimes e_{2i-1} q(e_{2i}) + e_{2i} \otimes e_{2i} q(e_{2i-1}) + e_{2i-1} \otimes e_{2i}\big)$$

where $(e_1, \ldots, e_n)$ is a symplectic basis of V for the polar form b_q:

$$b_q(e_{2i-1}, e_{2i}) = 1, \quad b_q(e_{2i}, e_{2i+1}) = 0 \quad \text{and} \quad b_q(e_i, e_j) = 0 \quad \text{if } |i - j| > 1.$$

Using the standard identification $C(A, \sigma, f) = C_0(V, q)$ of (8.8), we then have

$$c(\ell_0) = \textstyle\sum_{i=1}^{m} \big(e_{2i-1}^2 q(e_{2i}) + e_{2i}^2 q(e_{2i-1}) + e_{2i-1} \cdot e_{2i}\big) = \sum_{i=1}^{m} e_{2i-1} \cdot e_{2i}$$

and the required equations follow by computation (compare with (5.12)). $\qquad$ $\square$

(8.28) Corollary. *For any* $\ell \in \Lambda$,

$$Z(A, \sigma, f) \simeq F[X]/\big(X^2 + X + \mathrm{Srd}_A(\ell) + \tfrac{m(m-1)}{2}\big).$$

Proof: The proposition above shows that $Z(A, \sigma, f) = F[c(\ell)]$ and $c(\ell)^2 + c(\ell) + \mathrm{Srd}_A(\ell) + \frac{m(m-1)}{2} = 0$. $\qquad\square$

8.E. The Clifford algebra of a hyperbolic quadratic pair. Let (σ, f) be a hyperbolic quadratic pair on a central simple algebra A over an arbitrary field F and let $\deg A = n = 2m$. By (7.9), the discriminant of (σ, f) is trivial, hence (8.10) shows that the Clifford algebra $C(A, \sigma, f)$ decomposes as a direct product of two central simple algebras of degree 2^{m-1}:

$$C(A, \sigma, f) = C^+(A, \sigma, f) \times C^-(A, \sigma, f).$$

Our aim is to show that one of the factors $C_\pm(A, \sigma, f)$ is split if m is even.

We start with some observations on isotropic ideals in a central simple algebra with an arbitrary quadratic pair: suppose $I \subset A$ is a right ideal of even reduced dimension $\mathrm{rdim}\, I = r = 2s$ with respect to a quadratic pair (σ, f). Consider the image $c(I\sigma(I)) \subset C(A, \sigma, f)$ of $I\sigma(I)$ under the canonical map $c \colon A \to C(A, \sigma, f)$ and let

$$\rho(I) = c(I\sigma(I))^s = \left\{ \textstyle\sum x_1 \cdots x_s \mid x_1, \ldots, x_s \in c(I\sigma(I)) \right\}.$$

(8.29) Lemma. *The elements in $c(I\sigma(I))$ commute. The F-vector space $\rho(I) \subset C(A, \sigma, f)$ is 1-dimensional; it satisfies $\underline{\sigma}(\rho(I)) \cdot \rho(I) = \{0\}$ and*

$$\dim_F \rho(I) \cdot C(A, \sigma, f) = \dim_F C(A, \sigma, f) \cdot \rho(I) = 2^{n-r-1}.$$

Proof: It suffices to check the lemma over a scalar extension. We may therefore assume that A is split and identify $A = \mathrm{End}_F(V)$, $C(A, \sigma, f) = C_0(V, q)$ for some nonsingular quadratic space (V, q) of dimension n, by (8.8). The ideal I then has the form $I = \mathrm{Hom}_F(V, U)$ for some m-dimensional totally isotropic subspace $U \subset V$. Let $(u_1, \ldots, u_r)$ be a basis of U. Under the identification $A = V \otimes V$ described in (5.10), the vector space $I\sigma(I)$ is spanned by the elements $u_i \otimes u_j$ for $i, j = 1, \ldots, r$, hence $c(I\sigma(I))$ is spanned by the elements $u_i \cdot u_j$ in $C_0(V, q)$. Since U is totally isotropic, we have $u_i^2 = 0$ and $u_i \cdot u_j + u_j \cdot u_i = 0$ for all $i, j = 1, \ldots, r$, hence the elements $u_i \cdot u_j$ commute. Moreover, the space $\rho(I)$ is spanned by $u_1 \cdots u_r$. The dimensions of $\rho(I) \cdot C_0(V, q)$ and $C_0(V, q) \cdot \rho(I)$ are then easily computed, and since $\underline{\sigma}(u_1 \cdots u_r) = u_r \cdots u_1$ we have $\underline{\sigma}(x)x = 0$ for all $x \in \rho(I)$. $\qquad\square$

(8.30) Corollary. *Let $\deg A = n = 2m$ and suppose the center $Z = Z(A, \sigma, f)$ of $C(A, \sigma, f)$ is a field. For any even integer r, the relation $r \in \mathrm{ind}(A, \sigma, f)$ implies $2^{m-r-1} \in \mathrm{ind}(C(A, \sigma, f), \underline{\sigma})$.*

Proof: If r is an even integer in $\mathrm{ind}(A, \sigma, f)$, then A contains an isotropic right ideal I of even reduced dimension r. The lemma shows that $\rho(I) \cdot C(A, \sigma, f)$ is an isotropic ideal for the involution $\underline{\sigma}$. Its reduced dimension is

$$\frac{\dim_Z \rho(I) \cdot C(A, \sigma, f)}{\deg C(A, \sigma, f)} = \frac{\frac{1}{2} \dim_F \rho(I) \cdot C(A, \sigma, f)}{2^{m-1}} = 2^{m-r-1}. \qquad\square$$

We next turn to the case of hyperbolic quadratic pairs:

(8.31) Proposition. *Let (σ, f) be a hyperbolic quadratic pair on a central simple algebra A of degree $2m$ over an arbitrary field F. If m is even, then one of the factors $C_\pm(A, \sigma, f)$ of the Clifford algebra $C(A, \sigma, f)$ is split.*

Proof: Let $I \subset A$ be a right ideal of reduced dimension m which is isotropic with respect to (σ, f), and consider the 1-dimensional vector space $\rho(I) \subset C(A, \sigma, f)$ defined above. Multiplication on the left defines an F-algebra homomorphism

$$\lambda \colon C(A, \sigma, f) \to \operatorname{End}_F\big(C(A, \sigma, f) \cdot \rho(I)\big).$$

Dimension count shows that this homomorphism is not injective, hence the kernel is one of the nontrivial ideals $C^+(A, \sigma, f) \times \{0\}$ or $\{0\} \times C^-(A, \sigma, f)$. Assuming for instance $\ker \lambda = C^+(A, \sigma, f) \times \{0\}$, the homomorphism λ factors through an injective F-algebra homomorphism $C^-(A, \sigma, f) \to \operatorname{End}_F\big(C(A, \sigma, f) \cdot \rho(I)\big)$. This homomorphism is surjective by dimension count. $\qquad\square$

§9. The Clifford Bimodule

Although the odd part $C_1(V, q)$ of the Clifford algebra of a quadratic space (V, q) is not invariant under similarities, it turns out that the tensor product $V \otimes C_1(V, q)$ is invariant, and therefore an analogue can be defined for a central simple algebra with quadratic pair (A, σ, f). The aim of this section is to define such an analogue. This construction will be used at the end of this section to obtain fundamental relations between the Clifford algebra $C(A, \sigma, f)$ and the algebra A (see (9.12)); it will also be an indispensable tool in the definition of spin groups in the next chapter.

We first review the basic properties of the vector space $V \otimes C_1(V, q)$ that we want to generalize.

9.A. The split case. Let (V, q) be a quadratic space over a field F (of arbitrary characteristic). Let $C_1(V, q)$ be the odd part of the Clifford algebra $C(V, q)$. Multiplication in $C(V, q)$ endows $C_1(V, q)$ with a $C_0(V, q)$-bimodule structure. Since V is in a natural way a left $\operatorname{End}(V)$-module, the tensor product $V \otimes C_1(V, q)$ is at the same time a left $\operatorname{End}(V)$-module and a $C_0(V, q)$-bimodule: for $f \in \operatorname{End}(V)$, $v \in V$, $c_0 \in C_0(V, q)$ and $c_1 \in C_1(V, q)$ we set

$$f \cdot (v \otimes c_1) = f(v) \otimes c_1, \quad c_0 * (v \otimes c_1) = v \otimes c_0 c_1, \quad (v \otimes c_1) \cdot c_0 = v \otimes c_1 c_0.$$

These various actions clearly commute.

We summarize the basic properties of $V \otimes C_1(V, q)$ in the following proposition:

(9.1) Proposition. *Let $\dim V = n$.*

(1) *The vector space $V \otimes C_1(V, q)$ carries natural structures of left $\operatorname{End}(V)$-module and $C_0(V, q)$-bimodule, and the various actions commute.*

(2) *The standard identification $\operatorname{End}(V) = V \otimes V$ induced by the quadratic form q (see (5.10)) and the embedding $V \hookrightarrow C_1(V, q)$ define a canonical map*

$$b \colon \operatorname{End}(V) \to V \otimes C_1(V, q)$$

which is an injective homomorphism of left $\operatorname{End}(V)$-modules.

(3) $\dim_F\big(V \otimes C_1(V, q)\big) = 2^{n-1}n$.

The proof follows by straightforward verification.

Until the end of this subsection we assume that the dimension of V is even: $\dim V = n = 2m$. This is the main case of interest for generalization to central simple algebras with involution, since central simple algebras of odd degree with involution of the first kind are split (see (2.8)). Since $\dim V$ is even, the center of

$C_0(V, q)$ is an étale quadratic F-algebra which we denote Z. Let ι be the nontrivial automorphism of Z/F. In the Clifford algebra $C(V, q)$ we have

$$v \cdot \zeta = \iota(\zeta) \cdot v \quad \text{for } v \in V, \ \zeta \in Z,$$

hence

(9.2) $\iota(\zeta) * (v \otimes c_1) = (v \otimes c_1) \cdot \zeta \quad \text{for } v \in V, \ c_1 \in C_1(V, q), \ \zeta \in Z.$

In view of this equation, we may consider $V \otimes C_1(V, q)$ as a right module over the Z-algebra ${}^\iota C_0(V, q)^{\mathrm{op}} \otimes_Z C_0(V, q)$: for $v \in V$, $c_1 \in C_1(V, q)$ and $c_0, c_0' \in C_0(V, q)$ we set

$$(v \otimes c_1) \cdot ({}^\iota c_0^{\mathrm{op}} \otimes c_0') = c_0 * (v \otimes c_1) \cdot c_0' = v \otimes c_0 c_1 c_0'.$$

On the other hand $V \otimes C_1(V, q)$ also is a left module over $\mathrm{End}(V)$; since the actions of $\mathrm{End}(V)$ and $C_0(V, Q)$ commute, there is a natural homomorphism of F-algebras:

$$\nu \colon {}^\iota C_0(V, q)^{\mathrm{op}} \otimes_Z C_0(V, q) \to \mathrm{End}_{\mathrm{End}(V)}\big(V \otimes C_1(V, q)\big) = \mathrm{End}_F\big(C_1(V, q)\big).$$

This homomorphism is easily seen to be injective: this is obvious if Z is a field, because then the tensor product on the left is a simple algebra. If $Z \simeq F \times F$, the only nontrivial ideals in the tensor product are generated by elements in Z. However the restriction of ν to Z is injective, since the condition $v \cdot \zeta = 0$ for all $v \in V$ implies $\zeta = 0$. Therefore, ν is injective.

The image of ν is determined as follows: through ν, the center Z of $C_0(V, q)$ acts on $V \otimes C_1(V, q)$ by $\mathrm{End}(V)$-linear homomorphisms; the set $V \otimes C_1(V, q)$ therefore has a structure of left $\mathrm{End}(V) \otimes Z$-module (where the action of Z is through the *right* action of $C_0(V, q)$): for $f \in \mathrm{End}(V)$, $\zeta \in Z$, $v \in V$ and $c_1 \in C_1(V, q)$,

$$(f \otimes \zeta) \cdot (v \otimes c_1) = f(v) \otimes c_1 \zeta.$$

The map ν may then be considered as an isomorphism of Z-algebras:

$$\nu \colon {}^\iota C_0(V, q)^{\mathrm{op}} \otimes_Z C_0(V, q) \xrightarrow{\sim} \mathrm{End}_{\mathrm{End}(V) \otimes Z}\big(V \otimes C_1(V, q)\big) = \mathrm{End}_Z\big(C_1(V, q)\big).$$

Equivalently, ν identifies ${}^\iota C_0(V, q)^{\mathrm{op}} \otimes_Z C_0(V, q)$ with the centralizer of Z $(= \nu(Z))$ in $\mathrm{End}_{\mathrm{End}(V)}\big(V \otimes C_1(V, q)\big)$.

9.B. Definition of the Clifford bimodule. In order to define an analogue of $V \otimes C_1(V, q)$ for a central simple algebra with quadratic pair (A, σ, f), we first define a canonical representation of the symmetric group S_{2n} on $A^{\otimes n}$.

Representation of the symmetric group. As in §8.B, we write $\underline{A}$ for the underlying vector space of the F-algebra A. For any integer $n \geq 2$, we define a generalized sandwich map

$$\mathrm{Sand}_n \colon \underline{A}^{\otimes n} \to \mathrm{Hom}_F(\underline{A}^{\otimes n-1}, \underline{A})$$

by the condition:

$$\mathrm{Sand}_n(a_1 \otimes \cdots \otimes a_n)(b_1 \otimes \cdots \otimes b_{n-1}) = a_1 b_1 a_2 b_2 \cdots b_{n-1} a_n.$$

(Thus, Sand_2 is the map denoted simply Sand in §3.A).

(9.3) Lemma. *For any central simple F-algebra A, the map Sand_n is an isomorphism of vector spaces.*

Proof: Since it suffices to prove Sand_n is an isomorphism after scalar extension, we may assume that $A = M_n(F)$. It suffices to prove injectivity of Sand_n, since $\underline{A}^{\otimes n}$ and $\mathrm{Hom}_F(\underline{A}^{\otimes n-1}, \underline{A})$ have the same dimension. Let e_{ij} $(i, j = 1, \ldots, n)$ be the matrix units of $M_n(F)$. Take any nonzero $\alpha \in \underline{A}^{\otimes n}$ and write

$$\alpha = \sum_{i_1=1}^{n} \sum_{j_1=1}^{n} \cdots \sum_{i_n=1}^{n} \sum_{j_n=1}^{n} c_{i_1 j_1 \ldots i_n j_n} e_{i_1 j_1} \otimes \cdots \otimes e_{i_n j_n}$$

with the $c_{i_1 j_1 \ldots i_n j_n} \in F$. Some coefficient of α, say $c_{p_1 q_1 \ldots p_n q_n}$ is nonzero. Then,

$$\mathrm{Sand}_n(\alpha)(e_{q_1 p_2} \otimes e_{q_2 p_3} \otimes \cdots \otimes e_{q_i p_{i+1}} \otimes \cdots \otimes e_{q_{n-1} p_n}) =$$

$$\sum_{i_1=1}^{n} \sum_{j_n=1}^{n} c_{i_1 q_1 p_2 q_2 \ldots p_n j_n} e_{i_1 j_n}$$

which is not zero, since its $p_1 q_n$-entry is not zero. $\square$

(9.4) Proposition. *Let (A, σ) be a central simple F-algebra with involution of the first kind. If char $F \neq 2$, suppose further that σ is orthogonal. For all $n \geq 1$ there is a canonical representation $\rho_n \colon S_{2n} \to \mathrm{GL}(\underline{A}^{\otimes n})$ of the symmetric group S_{2n} which is described in the split case as follows: for every nonsingular symmetric bilinear space (V, b) and $v_1, \ldots, v_{2n} \in V$,*

$$\rho_n(\pi)\big(\varphi_b(v_1 \otimes v_2) \otimes \cdots \otimes \varphi_b(v_{2n-1} \otimes v_{2n})\big) =$$

$$\varphi_b\big(v_{\pi^{-1}(1)} \otimes v_{\pi^{-1}(2)}\big) \otimes \cdots \otimes \varphi_b\big(v_{\pi^{-1}(2n-1)} \otimes v_{\pi^{-1}(2n)}\big)$$

for all $\pi \in S_{2n}$ where $\varphi_b \colon V \otimes V \xrightarrow{\sim} \mathrm{End}_F(V)$ is the standard identification (5.2).

Proof: We first define the image of the transpositions $\tau(i) = (i, i+1)$ for $i = 1$, $\ldots$, $2n - 1$.

If i is odd, $i = 2\ell - 1$, let

$$\rho_n\big(\tau(i)\big) = \mathrm{Id}_{\underline{A}} \otimes \cdots \otimes \mathrm{Id}_{\underline{A}} \otimes \underline{\sigma} \otimes \mathrm{Id}_{\underline{A}} \otimes \cdots \otimes \mathrm{Id}_{\underline{A}},$$

where $\underline{\sigma}$ lies in ℓ-th position. In the split case, σ corresponds to the twist under the standard identification $A = V \otimes V$; therefore,

$$\rho_n\big(\tau(2\ell - 1)\big)(v_1 \otimes \cdots \otimes v_{2\ell-1} \otimes v_{2\ell} \otimes \cdots \otimes v_{2n}) =$$

$$v_1 \otimes \cdots \otimes v_{2\ell} \otimes v_{2\ell-1} \otimes \cdots \otimes v_{2n}.$$

If i is even, $i = 2\ell$, we define $\rho_n\big(\tau(i)\big)$ by the condition:

$$\mathrm{Sand}_n\big(\rho_n\big(\tau(i)\big)(u)\big)(x) = \mathrm{Sand}_n(u)\big(\mathrm{Id}_{\underline{A}} \otimes \cdots \otimes \mathrm{Id}_{\underline{A}} \otimes \underline{\sigma} \otimes \mathrm{Id}_{\underline{A}} \otimes \cdots \otimes \mathrm{Id}_{\underline{A}}(x)\big)$$

for $u \in \underline{A}^{\times n}$ and $x \in \underline{A}^{\times n-1}$ where $\underline{\sigma}$ lies in ℓ-th position. The same computation as in (8.6) shows that $\rho_n\big(\tau(2\ell)\big)$ satisfies the required condition in the split case.

In order to define $\rho_n(\pi)$ for arbitrary $\pi \in S_{2n}$, we use the fact that $\tau(1), \ldots, \tau(2n-1)$ generate S_{2n}: we fix some factorization

$$\pi = \tau_1 \circ \cdots \circ \tau_s \quad \text{where } \tau_1, \ldots, \tau_s \in \{\tau(1), \ldots, \tau(2n-1)\}$$

and define $\rho_n(\pi) = \rho_n(\tau_1) \circ \cdots \circ \rho_n(\tau_s)$. The map $\rho_n(\pi)$ thus defined meets the requirement in the split case, hence ρ_n is a homomorphism in the split case. By extending scalars to a splitting field, we see that ρ_n also is a homomorphism in the general case. Therefore, the definition of $\rho_n(\pi)$ does not actually depend on the factorization of π. $\square$

The definition. Let (σ, f) be a quadratic pair on a central simple F-algebra A. For all $n \geq 1$, let $\gamma_n = \rho_n((1, 2, \ldots, 2n)^{-1}) \in GL(\underline{A}^{\otimes n})$ where ρ_n is as in (9.4), and let $\gamma = \oplus\gamma_n \colon T(\underline{A}) \to T(\underline{A})$ be the induced linear map. Thus, in the split case $(A, \sigma, f) = \big(\mathrm{End}_F(V), \sigma_q, f_q\big)$, we have, under the standard identification $A = V \otimes V$ of (5.10):

$$\gamma(v_1 \otimes \cdots \otimes v_{2n}) = \gamma_n(v_1 \otimes \cdots \otimes v_{2n}) = v_2 \otimes \cdots \otimes v_{2n} \otimes v_1$$

for $v_1, \ldots, v_{2n} \in V$.

Let also $T_+(\underline{A}) = \oplus_{n \geq 1} \underline{A}^{\otimes n}$. The vector space $T_+(\underline{A})$ carries a natural structure of left and right module over the tensor algebra $T(\underline{A})$. We define a new left module structure $*$ as follows: for $u \in T(\underline{A})$ and $v \in T_+(\underline{A})$ we set

$$u * v = \gamma^{-1}\big(u \otimes \gamma(v)\big).$$

Thus, in the split case $A = V \otimes V$, the product $*$ avoids the first factor:

$$(u_1 \otimes \cdots \otimes u_{2i}) * (v_1 \otimes \cdots \otimes v_{2j}) = v_1 \otimes u_1 \otimes \cdots \otimes u_{2i} \otimes v_2 \otimes \cdots \otimes v_{2j}$$

for $u_1, \ldots, u_{2i}, v_1, \ldots, v_{2j} \in V$. (Compare with the definition of $*$ in §9.A).

(9.5) Definition. The *Clifford bimodule* of (A, σ, f) is defined as

$$B(A, \sigma, f) = \frac{T_+(\underline{A})}{\big[J_1(\sigma, f) * T_+(\underline{A})\big] + \big[T_+(\underline{A}) \cdot J_1(\sigma, f)\big]}$$

where $J_1(\sigma, f)$ is the two-sided ideal of $T(\underline{A})$ which appears in the definition of the Clifford algebra $C(A, \sigma, f)$ (see (8.7)).

The map $a \in A \mapsto a \in T_+(\underline{A})$ induces a canonical F-linear map

(9.6) $b \colon A \to B(A, \sigma, f)$.

(9.7) Theorem. *Let (A, σ, f) be a central simple F-algebra with a quadratic pair.*
(1) The F-vector space $B(A, \sigma, f)$ carries a natural $C(A, \sigma, f)$-bimodule structure where action on the left is through $*$, and a natural left A-module structure.
(2) In the split case $(A, \sigma, f) = \big(\mathrm{End}_F(V), \sigma_q, f_q\big)$, the standard identification

$$\varphi_q \colon V \otimes V \xrightarrow{\sim} \mathrm{End}_F(V)$$

induces a standard identification of Clifford bimodules

$$V \otimes_F C_1(V, q) \xrightarrow{\sim} B(A, \sigma, f).$$

(3) The canonical map $b \colon A \to B(A, \sigma, f)$ is an injective homomorphism of left[13] A-modules.
(4) $\dim_F B(A, \sigma, f) = 2^{(\deg A)-1} \deg A$.

Proof: By extending scalars to split A, it is easy to verify that

$$J_2(\sigma, f) * T_+(\underline{A}) \subseteq T_+(\underline{A}) \cdot J_1(\sigma, f) \quad \text{and} \quad T_+(\underline{A}) \cdot J_2(\sigma, f) \subseteq J_1(\sigma, f) * T_+(\underline{A}).$$

Therefore, the actions of $T(\underline{A})$ on $T_+(\underline{A})$ on the left through $*$ and on the right through the usual product induce a $C(A, \sigma, f)$-bimodule structure on $B(A, \sigma, f)$.

We define on $T_+(\underline{A})$ a left A-module structure by using the multiplication map $\underline{A}^{\otimes 2} \to \underline{A}$ which carries $a \otimes b$ to ab. Explicitly, for $a \in A$ and $u = u_1 \otimes \cdots \otimes u_i \in \underline{A}^{\otimes i}$, we set

$$a \cdot u = au_1 \otimes u_2 \otimes \cdots \otimes u_i.$$

---

[13] Therefore, the image of $a \in A$ under b will be written a^b.

Thus, in the split case $(A, \sigma, f) = (\mathrm{End}_F(V), \sigma_q, f_q)$, we have, under the standard identification $A = V \otimes V$:

$$a \cdot (v_1 \otimes \cdots \otimes v_{2i}) = a(v_1) \otimes v_2 \otimes \cdots \otimes v_{2i}.$$

It is then clear that the left action of A on $T_+(\underline{A})$ commutes with the left and right actions of $T(\underline{A})$. Therefore, the subspace $\left[J_1(\sigma, f) * T_+(\underline{A}) \right] + \left[T_+(\underline{A}) \cdot J_1(\sigma, f) \right]$ is preserved under the action of A, and it follows that $B(A, \sigma, f)$ inherits this action from $T_+(\underline{A})$.

In the split case, the standard identification $\varphi_q^{-1} \colon A \xrightarrow{\sim} V \otimes V$ induces a surjective linear map from $B(A, \sigma, f)$ onto $V \otimes C_1(V, q)$. Using an orthogonal decomposition of (V, q) into 1- or 2-dimensional subspaces, one can show that

$$\dim_F B(A, \sigma, f) \le \dim_F V \dim_F C_1(V, q).$$

Therefore, the induced map is an isomorphism. This proves (1) and (2), and (4) follows by dimension count. Statement (3) is clear in the split case (see (9.1)), and the theorem follows. $\qquad \square$

As was observed in the preceding section, there is no significant loss if we restrict our attention to the case where the degree of A is even, since A is split if its degree is odd. Until the end of this subsection, we assume $\deg A = n = 2m$. According to (8.10), the center Z of $C(A, \sigma, f)$ is then a quadratic étale F-algebra. Let ι be the non-trivial automorphism of Z/F. By extending scalars to split the algebra A, we derive from (9.2):

(9.8) $\qquad\qquad x \cdot \zeta = \iota(\zeta) * x \quad \text{for } x \in B(A, \sigma, f), \ \zeta \in Z.$

Therefore, we may consider $B(A, \sigma, f)$ as a right module over ${}^{\iota}C(A, \sigma, f)^{\mathrm{op}} \otimes_Z C(A, \sigma, f)$: for $c, c' \in C(A, \sigma, f)$ and $x \in B(A, \sigma, f)$, we set

$$x \cdot ({}^{\iota}c^{\mathrm{op}} \otimes c') = c * x \cdot c'.$$

Thus, $B(A, \sigma, f)$ is an A-${}^{\iota}C(A, \sigma, f)^{\mathrm{op}} \otimes_Z C(A, \sigma, f)$-bimodule, and there is a natural homomorphism of F-algebras:

$$\nu \colon {}^{\iota}C(A, \sigma, f)^{\mathrm{op}} \otimes_Z C(A, \sigma, f) \to \mathrm{End}_A B(A, \sigma, f).$$

By comparing with the split case, we see that the map ν is injective, and that its image is the centralizer of $Z \ (= \nu(Z))$ in $\mathrm{End}_A B(A, \sigma, f)$. Endowing $B(A, \sigma, f)$ with a left $A \otimes_F Z$-module structure (where the action of Z is through ν), we may thus view ν as an isomorphism

(9.9) $\qquad \nu \colon {}^{\iota}C(A, \sigma, f)^{\mathrm{op}} \otimes_Z C(A, \sigma, f) \xrightarrow{\sim} \mathrm{End}_{A \otimes Z} B(A, \sigma, f);$

it is defined by $x^{\nu({}^{\iota}c^{\mathrm{op}} \otimes c')} = c * x \cdot c'$ for $c, c' \in C(A, \sigma, f)$ and $x \in B(A, \sigma, f)$.

The canonical involution. We now use the involution σ on A to define an involutorial A-module endomorphism ω of $B(A, \sigma, f)$. As in §8.B, $\underline{\sigma}$ denotes the involution of $C(A, \sigma, f)$ induced by σ, and τ is the involution on $C(V, q)$ which is the identity on V.

(9.10) Proposition. *The A-module $B(A, \sigma, f)$ is endowed with a canonical endomorphism*[14] ω *such that for $c_1, c_2 \in C(A, \sigma, f)$, $x \in B(A, \sigma, f)$ and $a \in A$:*

$$(c_1 * x \cdot c_2)^{\omega} = \underline{\sigma}(c_2) * x^{\omega} \cdot \underline{\sigma}(c_1) \quad \text{and} \quad (a^b)^{\omega} = a^b,$$

---

[14] Since $B(A, \sigma, f)$ is a left A-module, ω will be written to the right of its arguments.

where $b\colon A \to B(A, \sigma, f)$ *is the canonical map. Moreover, in the split case*

$$(A, \sigma, f) = (\operatorname{End}_F(V), \sigma_q, f_q)$$

we have $\omega = \operatorname{Id}_V \otimes \tau$ *under the standard identifications* $A = V \otimes V$, $B(A, \sigma, f) = V \otimes C_1(V, q)$.

Proof: Let $\widetilde{\omega} = \gamma^{-1} \circ \underline{\sigma}\colon T_+(\underline{A}) \to T_+(\underline{A})$ where $\underline{\sigma}$ is the involution on $T(\underline{A})$ induced by σ. Thus, in the split case $A = V \otimes V$:

$$\widetilde{\omega}(v_1 \otimes \cdots \otimes v_{2n}) = v_1 \otimes v_{2n} \otimes v_{2n-1} \otimes \cdots \otimes v_3 \otimes v_2.$$

By extending scalars to a splitting field of A, it is easy to check that for $a \in A$, u_1, $u_2 \in T(\underline{A})$ and $v \in T_+(\underline{A})$,

$$\widetilde{\omega}(u_1 * v \cdot u_2) = \underline{\sigma}(u_2) * \widetilde{\omega}(v) \cdot \underline{\sigma}(u_1), \quad \widetilde{\omega}(a \cdot v) = a \cdot \widetilde{\omega}(v) \quad \text{and} \quad \widetilde{\omega}(\underline{a}) = \underline{a}.$$

It follows from the first equation that

$$\widetilde{\omega}\big(J_1(\sigma, f) * T_+(\underline{A})\big) = T_+(\underline{A}) \cdot \underline{\sigma}\big(J_1(\sigma, f)\big) \subseteq T_+(\underline{A}) \cdot J_1(\sigma, f)$$

and

$$\widetilde{\omega}\big(T_+(\underline{A}) \cdot J_1(\sigma, f)\big) = \underline{\sigma}\big(J_1(\sigma, f)\big) * T_+(\underline{A}) \subseteq J_1(\sigma, f) * T_+(\underline{A}),$$

hence $\widetilde{\omega}$ induces an involutorial F-linear operator ω on $B(A, \sigma, f)$ which satisfies the required conditions. $\qquad\square$

We thus have $\omega \in \operatorname{End}_A B(A, \sigma, f)$. Moreover, it follows from the first property of ω in the proposition above and from (9.8) that for $x \in B(A, \sigma, f)$ and $\zeta \in Z$,

$$(x \cdot \zeta)^\omega = \underline{\sigma}(\zeta) * x^\omega = x^\omega \cdot \big[\iota \circ \underline{\sigma}(\zeta)\big].$$

The restriction of $\underline{\sigma}$ to Z is determined in (8.12): $\underline{\sigma}$ is of the first kind if m is even and of the second kind if m is odd. Therefore, ω is Z-linear if m is odd, hence it belongs to the image of ${}^\iota C(A, \sigma, f)^{\mathrm{op}} \otimes_Z C(A, \sigma, f)$ in $\operatorname{End}_A B(A, \sigma, f)$ under the natural monomorphism ν. By contrast, when m is even, ω is only ι-semilinear. In this case, we define an F-algebra

$$E(A, \sigma, f) = \big[{}^\iota C(A, \sigma, f) \otimes_Z C(A, \sigma, f)\big] \oplus \big[{}^\iota C(A, \sigma, f) \otimes_Z C(A, \sigma, f)\big] \cdot z$$

where multiplication is defined by the following equations:

$$z({}^\iota c \otimes c') = ({}^\iota c' \otimes c)z \quad \text{for } c, c' \in C(A, \sigma, f), \qquad z^2 = 1.$$

We also define a map $\nu'\colon E(A, \sigma, f) \to \operatorname{End}_A B(A, \sigma, f)$ by

$$x^{\nu'({}^\iota c_1 \otimes c_2 + {}^\iota c_3 \otimes c_4 \cdot z)} = \underline{\sigma}(c_1) * x \cdot c_2 + \big(\underline{\sigma}(c_3) * x \cdot c_4\big)^\omega$$

for $x \in B(A, \sigma, f)$ and $c_1, c_2, c_3, c_4 \in C(A, \sigma, f)$. The fact that ν' is a well-defined F-algebra homomorphism follows from the properties of ω in (9.10), and from the hypothesis that $\deg A$ is divisible by 4 which ensures that $\underline{\sigma}$ is an involution of the first kind: see (8.12).

(9.11) Proposition. *If* $\deg A \equiv 0 \mod 4$, *the map* ν' *is an isomorphism of* F-*algebras.*

Proof: Suppose $\nu'(u + vz) = 0$ for some $u, v \in {}^\iota C(A, \sigma, f) \otimes_Z C(A, \sigma, f)$. Then $\nu'(u) = -\nu'(vz)$; but $\nu'(u)$ is Z-linear while $\nu'(vz)$ is ι-semilinear. Therefore, $\nu'(u) = \nu'(vz) = 0$. It then follows that $u = v = 0$ since the natural map ν is injective. This proves injectivity of ν'. Surjectivity follows by dimension count:

since $\dim_F B(A, \sigma, f) = 2^{\deg A - 1} \deg A$, we have $\mathrm{rdim}_A B(A, \sigma, f) = 2^{2m-1}$ hence $\deg \mathrm{End}_A B(A, \sigma, f) = 2^{2m-1}$ by (1.10). On the other hand,

$$\dim_F E(A, \sigma, f) = 2 \dim_F {}^\iota C(A, \sigma, f) \otimes_Z C(A, \sigma, f) = \left[\dim_F C(A, \sigma, f)\right]^2,$$

hence $\dim_F E(A, \sigma, f) = 2^{2(2m-1)}$. $\square$

9.C. The fundamental relations. In this section, A denotes a central simple F-algebra of even degree $n = 2m$ with a quadratic pair (σ, f). The fundamental relations between the Brauer class $[A]$ of A and the Brauer class of the Clifford algebra $C(A, \sigma, f)$ are the following:

(9.12) Theorem. *Let Z be the center of the Clifford algebra $C(A, \sigma, f)$.*
(1) *If $\deg A \equiv 0 \mod 4$ (i.e., if m is even), then*

(9.13) $$\left[C(A, \sigma, f)\right]^2 = 1 \quad in \ \mathrm{Br}(Z).$$

(9.14) $$N_{Z/F}\left[C(A, \sigma, f)\right] = [A] \quad in \ \mathrm{Br}(F).$$

(2) *If $\deg A \equiv 2 \mod 4$ (i.e., if m is odd), then*

(9.15) $$\left[C(A, \sigma, f)\right]^2 = [A_Z] \quad in \ \mathrm{Br}(Z).$$

(9.16) $$N_{Z/F}\left[C(A, \sigma, f)\right] = 1 \quad in \ \mathrm{Br}(F).$$

(If $Z = F \times F$, the norm $N_{Z/F}$ is defined by $N_{F \times F/F}(C_1 \times C_2) = C_1 \otimes_F C_2$: see the end of §3.B).

Proof: Equations (9.13) and (9.16) follow by (3.1) from the fact that the canonical involution $\underline{\sigma}$ on $C(A, \sigma, f)$ is of the first kind when $\deg A \equiv 0 \mod 4$ and of the second kind when $\deg A \equiv 2 \mod 4$.

To prove equations (9.14) and (9.15), recall the natural isomorphism (9.9):

$$\nu \colon {}^\iota C(A, \sigma, f)^{\mathrm{op}} \otimes_Z C(A, \sigma, f) \xrightarrow{\sim} \mathrm{End}_{A \otimes Z} B(A, \sigma, f).$$

By (1.10), it follows that $A_Z = A \otimes_F Z$ is Brauer-equivalent to ${}^\iota C(A, \sigma, f)^{\mathrm{op}} \otimes_Z C(A, \sigma, f)$. If m is odd, the canonical involution $\underline{\sigma}$ is of the second kind; it therefore defines an isomorphism of Z-algebras:

$$ {}^\iota C(A, \sigma, f)^{\mathrm{op}} \simeq C(A, \sigma, f),$$

and (9.15) follows. Note that the arguments above apply also in the case where $Z \simeq F \times F$; then $C(A, \sigma, f) = C^+ \times C^-$ for some central simple F-algebras C^+, C^-, and there is a corresponding decomposition of $B(A, \sigma, f)$ which follows from its Z-module structure:

$$B(A, \sigma) = B_+ \times B_-.$$

Then $\mathrm{End}_{A \otimes Z} B(A, \sigma, f) = \mathrm{End}_A B_+ \times \mathrm{End}_A B_-$ and ${}^\iota C(A, \sigma, f)^{\mathrm{op}} = C^{-\mathrm{op}} \times C^{+\mathrm{op}}$, and the isomorphism ν can be considered as

$$\nu \colon (C^+ \otimes_F C^{-\mathrm{op}}) \times (C^- \otimes_F C^{+\mathrm{op}}) \xrightarrow{\sim} (\mathrm{End}_A B_+) \times (\mathrm{End}_A B_-).$$

Therefore, A is Brauer-equivalent to $C^+ \otimes_F C^{-\mathrm{op}}$ and to $C^- \otimes C^{+\mathrm{op}}$. Since $\underline{\sigma}$ is of the second kind when m is odd, we have in this case $C^{-\mathrm{op}} \simeq C^+$, hence A is Brauer-equivalent to $C^{+\otimes 2}$ and to $C^{-\otimes 2}$, proving (9.15).

Similarly, (9.14) is a consequence of (9.11), as we proceed to show. Let

$$B'(A, \sigma) = B(A, \sigma)^\omega$$

denote the F-subspace of ω-invariant elements in $B(A, \sigma)$. For the rest of this section, we assume $\deg A \equiv 0 \mod 4$. As observed before (9.11), the involution ω is ι-semilinear, hence the multiplication map

$$B'(A, \sigma) \otimes_F Z \to B(A, \sigma)$$

is an isomorphism of Z-modules. Relation (9.14) follows from (1.10) and the following claim:

Claim. The isomorphism ν' of (9.11) restricts to an F-algebra isomorphism

$$N_{Z/F} C(A, \sigma, f) \xrightarrow{\sim} \operatorname{End}_A B'(A, \sigma, f).$$

To prove the claim, observe that the subalgebra

$$\operatorname{End}_A B'(A, \sigma, f) \subset \operatorname{End}_A B(A, \sigma, f)$$

is the centralizer of Z and ω, hence it is also the F-subalgebra of elements which commute with ω in $\operatorname{End}_{A \otimes Z} B(A, \sigma, f)$. The isomorphism ν' identifies this algebra with the algebra of switch-invariant elements in ${}^\iota C(A, \sigma, f) \otimes_Z C(A, \sigma, f)$, i.e., with $N_{Z/F} C(A, \sigma, f)$. $\qquad \square$

§10. The Discriminant Algebra

A notion of discriminant may be defined for hermitian spaces on the same model as for symmetric bilinear spaces. If (V, h) is a hermitian space over a field K, with respect to a nontrivial automorphism ι of K (of order 2), the determinant of the Gram matrix of h with respect to an arbitrary basis $(e_1, \ldots, e_n)$ lies in the subfield $F \subset K$ elementwise fixed under ι and is an invariant of h modulo the norms of K/F. We may therefore define the *determinant* by

$$\det h = \det\big(h(e_i, e_j)\big)_{1 \le i, j \le n} \cdot N(K/F) \in F^\times / N(K/F)$$

where $N(K/F) = N_{K/F}(K^\times)$ is the group of norms of K/F. The *discriminant* is the signed determinant:

$$\operatorname{disc} h = (-1)^{n(n-1)/2} \det\big(h(e_i, e_j)\big)_{1 \le i, j \le n} \in F^\times / N(K/F).$$

If $\delta \in F^\times$ is a representative of $\operatorname{disc} h$, the quaternion algebra $K \oplus Kz$ where multiplication is defined by $zx = \iota(x)z$ for $x \in K$ and $z^2 = \delta$ does not depend on the choice of the representative δ. We denote it by $(K, \operatorname{disc} h)_F$; thus

$$(K, \operatorname{disc} h)_F = \begin{cases} (\alpha, \delta)_F & \text{if } K = F(\sqrt{\alpha}) \ (\operatorname{char} F \neq 2), \\ [\alpha, \delta)_F & \text{if } K = F\big(\wp^{-1}(\alpha)\big) \ (\operatorname{char} F = 2). \end{cases}$$

Our aim in this section is to generalize this construction, associating a central simple algebra $D(B, \tau)$ to every central simple algebra with involution of unitary type (B, τ) of even degree, in such a way that $D\big(\operatorname{End}_K(V), \sigma_h\big)$ is Brauer-equivalent to $(K, \operatorname{disc} h)_F$ (see (10.35)). In view of this relation with the discriminant, the algebra $D(B, \tau)$ is called the *discriminant algebra*, a term suggested by A. Wadsworth. This algebra is endowed with a canonical involution of the first kind.

As preparation for the definition of the discriminant algebra, we introduce in the first four sections various constructions related to exterior powers of vector spaces. For every central simple algebra A over an arbitrary field F, and for every positive integer $k \le \deg A$, we define a central simple F-algebra $\lambda^k A$ which is Brauer-equivalent to $A^{\otimes k}$ and such that in the split case $\lambda^k \operatorname{End}_F(V) = \operatorname{End}_F(\bigwedge^k V)$. In the second section, we show that when the algebra A has even degree $n = 2m$, the algebra $\lambda^m A$ carries a canonical involution γ of the first kind. In the split case $A = \mathrm{End}_F(V)$, the involution γ on $\lambda^m A = \mathrm{End}_F(\bigwedge^m V)$ is the adjoint involution with respect to the exterior product $\wedge\colon \bigwedge^m V \times \bigwedge^m V \to \bigwedge^n V \simeq F$. The third section is more specifically concerned with the case where $\mathrm{char}\, F = 2$: in this case, we extend the canonical involution γ on $\lambda^m A$ into a canonical quadratic pair (γ, f), when $m \geq 2$. Finally, in §10.D, we show how an involution on A induces an involution on $\lambda^k A$ for all $k \leq \deg A$.

10.A. The λ-powers of a central simple algebra. Let A be a central simple algebra of degree n over an arbitrary field F. Just as for the Clifford bimodule, the definition of $\lambda^k A$ uses a canonical representation of the symmetric group, based on Goldman elements.

(10.1) Proposition. *For all $k \geq 1$, there is a canonical homomorphism $g_k\colon S_k \to \left(A^{\otimes k}\right)^{\times}$ from the symmetric group S_k to the group of invertible elements in $A^{\otimes k}$, such that in the split case $A = \mathrm{End}_F(V)$ we have under the identification $A^{\otimes k} = \mathrm{End}_F(V^{\otimes k})$:*

$$g_k(\pi)(v_1 \otimes \cdots \otimes v_k) = v_{\pi^{-1}(1)} \otimes \cdots \otimes v_{\pi^{-1}(k)}$$

for all $\pi \in S_k$ and $v_1, \ldots, v_k \in V$.

Proof: We first define the image of the transpositions $\tau(i) = (i, i+1)$ for $i = 1, \ldots, k-1$, by setting

$$g_k\big(\tau(i)\big) = \underbrace{1 \otimes \cdots \otimes 1}_{i-1} \otimes g \otimes \underbrace{1 \otimes \cdots \otimes 1}_{k-i-1}$$

where $g \in A \otimes A$ is the Goldman element defined in (3.5). From (3.6), it follows that in the split case

$$g_k\big(\tau(i)\big)(v_1 \otimes \cdots \otimes v_k) = v_1 \otimes \cdots \otimes v_{i+1} \otimes v_i \otimes \cdots \otimes v_k,$$

as required.

In order to define $g_k(\pi)$ for arbitrary $\pi \in S_k$, we fix some factorization

$$\pi = \tau_1 \circ \cdots \circ \tau_s \quad \text{where } \tau_1, \ldots, \tau_s \in \{\tau(1), \ldots, \tau(k-1)\}$$

and set $g_k(\pi) = g_k(\tau_1) \cdots g_k(\tau_s)$. Then, in the split case

$$g_k(\pi)(v_1 \otimes \cdots \otimes v_k) = v_{\pi^{-1}(1)} \otimes \cdots \otimes v_{\pi^{-1}(k)},$$

and it follows that g_k is a homomorphism in the split case. It then follows by scalar extension to a splitting field that g_k is also a homomorphism in the general case. Therefore, the definition of g_k does not depend on the factorization of π. □

(10.2) Corollary. *For all $\pi \in S_k$ and $a_1, \ldots, a_k \in A$,*

$$g_k(\pi) \cdot (a_1 \otimes \cdots \otimes a_k) = (a_{\pi^{-1}(1)} \otimes \cdots \otimes a_{\pi^{-1}(k)}) \cdot g_k(\pi).$$

Proof: The equation follows by scalar extension to a splitting field of A from the description of $g_k(\pi)$ in the split case. □

For all $k \geq 2$, define

$$s_k = \sum_{\pi \in S_k} \mathrm{sgn}(\pi) g_k(\pi) \in A^{\otimes k},$$

where $\mathrm{sgn}(\pi) = \pm 1$ is the sign of π.

(10.3) Lemma. *The reduced dimension of the left ideal* $A^{\otimes k}s_k$ *is given by*

$$\mathrm{rdim}(A^{\otimes k}s_k) = \begin{cases} \binom{n}{k} & \text{for } 2 \leq k \leq n = \deg A, \\ 0 & \text{for } k > n = \deg A. \end{cases}$$

If A *is split:* $A = \mathrm{End}_F(V)$, $A^{\otimes k} = \mathrm{End}_F(V^{\otimes k})$, *then there is a natural isomorphism of* $A^{\otimes k}$-*modules:*

$$A^{\otimes k}s_k = \mathrm{Hom}_F(\textstyle\bigwedge^k V, V^{\otimes k}).$$

Proof: Since the reduced dimension does not change under scalar extension, it suffices to prove the second part. Under the correspondence between left ideals in $A^{\otimes k} = \mathrm{End}_F(V^{\otimes k})$ and subspaces of $V^{\otimes k}$ (see §1.B), we have

$$A^{\otimes k}s_k = \mathrm{Hom}_F(V^{\otimes k}/\ker s_k, V^{\otimes k}).$$

From the description of $g_k(\pi)$ in the split case, it follows that $\ker s_k$ contains the subspace of $V^{\otimes k}$ spanned by the products $v_1 \otimes \cdots \otimes v_k$ where $v_i = v_j$ for some indices $i \neq j$. Therefore, there is a natural epimorphism $\bigwedge^k V \to V^{\otimes k}/\ker s_k$.

To prove that this epimorphism is injective, pick a basis $(e_1, \ldots, e_n)$ of V. For the various choices of indices $i_1, \ldots, i_k$ such that $1 \leq i_1 < i_2 < \cdots < i_k \leq n$, the images $s_k(e_{i_1} \otimes \cdots \otimes e_{i_k})$ are linearly independent, since they involve different basis vectors in $V^{\otimes k}$. Therefore,

$$\dim(V^{\otimes k}/\ker s_k) = \dim \mathrm{im}\, s_k \geq \binom{n}{k},$$

and the epimorphism above is an isomorphism.   □

(10.4) Definition. Let A be a central simple algebra of degree n over a field F. For every integer $k = 2, \ldots, n$ we define the k-*th* λ-*power* of A as

$$\lambda^k A = \mathrm{End}_{A^{\otimes k}}(A^{\otimes k}s_k).$$

We extend this definition by setting $\lambda^1 A = A$. Note that for $k = 2$ we recover the definition of $\lambda^2 A$ given in §3.A (see (3.7)).

The following properties follow from the definition, in view of (10.3), (1.10) and (1.12):

(a) $\lambda^k A$ is a central simple F-algebra Brauer-equivalent to $A^{\otimes k}$, of degree

$$\deg \lambda^k A = \binom{n}{k}.$$

(b) There is a natural isomorphism:

$$\lambda^k \mathrm{End}_F(V) = \mathrm{End}_F(\textstyle\bigwedge^k V).$$

(10.5) Corollary. *If* k *divides the index* $\mathrm{ind}\, A$, *then* $\mathrm{ind}\, A^{\otimes k}$ *divides* $(\mathrm{ind}\, A)/k$.

Proof: By replacing A by a Brauer-equivalent algebra, we may assume that A is a division algebra. Let $n = \deg A = \mathrm{ind}\, A$. Arguing by induction on the number of prime factors of k, it suffices to prove the corollary when $k = p$ is a prime number. If K is a splitting field of A of degree n, then K also splits $\lambda^k A$, hence $\mathrm{ind}\, \lambda^p A$ divides n. On the other hand, $\mathrm{ind}\, \lambda^p A$ divides $\deg \lambda^p A = \binom{n}{p}$, and the greatest common divisor of n and $\binom{n}{p}$ is n/p, hence $\mathrm{ind}\, \lambda^p A$ divides n/p. Since $\lambda^p A$ is Brauer-equivalent to $A^{\otimes p}$, we have $\mathrm{ind}\, \lambda^p A = \mathrm{ind}\, A^{\otimes p}$, and the proof is complete.   □

10.B. The canonical involution. Let V be a vector space of even dimension $n = 2m$ over a field F of arbitrary characteristic. Since $\dim \bigwedge^n V = 1$, the composition of the exterior product

$$\wedge \colon \bigwedge^m V \times \bigwedge^m V \to \bigwedge^n V$$

with a vector-space isomorphism $\bigwedge^n V \xrightarrow{\sim} F$ is a bilinear form on $\bigwedge^m V$, which is uniquely determined up to a scalar factor.

(10.6) Lemma. *The bilinear map $\wedge$ is nonsingular. It is symmetric if m is even and skew-symmetric if m is odd. If $\operatorname{char} F = 2$, the map $\wedge$ is alternating for all m. Moreover, the discriminant of every symmetric bilinear form induced from $\wedge$ through any isomorphism $\bigwedge^n V \simeq F$ is trivial.*

Proof: Let $(e_1, \ldots, e_n)$ be a basis of V. For every subset of m indices

$$\mathcal{S} = \{i_1, \ldots, i_m\} \subset \{1, \ldots, n\} \quad \text{with } i_1 < \cdots < i_m$$

we set $e_{\mathcal{S}} = e_{i_1} \wedge \cdots \wedge e_{i_m} \in \bigwedge^m V$. As $\mathcal{S}$ runs over all the subsets of m indices, the elements $e_{\mathcal{S}}$ form a basis of $\bigwedge^m V$.

Since $e_{\mathcal{S}} \wedge e_{\mathcal{T}} = 0$ when $\mathcal{S}$ and $\mathcal{T}$ are not disjoint, we have for $x = \sum x_{\mathcal{S}} e_{\mathcal{S}}$

$$x \wedge e_{\mathcal{T}} = \pm x_{\mathcal{T}'} e_1 \wedge \cdots \wedge e_n,$$

where $\mathcal{T}'$ is the complementary subset of $\mathcal{T}$ in $\{1, \ldots, n\}$. Therefore, if $x \wedge e_{\mathcal{T}} = 0$ for all $\mathcal{T}$, then $x = 0$. This shows that the map $\wedge$ is nonsingular. Moreover, for all subsets $\mathcal{S}$, $\mathcal{T}$ of m indices we have

$$e_{\mathcal{S}} \wedge e_{\mathcal{T}} = (-1)^{m^2} e_{\mathcal{T}} \wedge e_{\mathcal{S}},$$

hence $\wedge$ is symmetric if m is even and skew-symmetric if m is odd. Since $e_{\mathcal{S}} \wedge e_{\mathcal{S}} = 0$ for all $\mathcal{S}$, the form $\wedge$ is alternating if $\operatorname{char} F = 2$.

Suppose m is even and fix an isomorphism $\bigwedge^n V \simeq F$ to obtain from $\wedge$ a symmetric bilinear form b on $\bigwedge^m V$. The space $\bigwedge^m V$ decomposes into an orthogonal direct sum:

$$\bigwedge^m V = \overset{\perp}{\bigoplus_{\mathcal{S} \in R}} E_{\mathcal{S}}$$

where $E_{\mathcal{S}}$ is the subspace spanned by the basis vectors $e_{\mathcal{S}}$, $e_{\mathcal{S}'}$ where $\mathcal{S}'$ is the complement of $\mathcal{S}$ and R is a set of representatives of the equivalence classes of subsets of m indices under the relation $\mathcal{S} \equiv \mathcal{T}$ if and only if $\mathcal{S} = \mathcal{T}$ or $\mathcal{S} = \mathcal{T}'$. (For instance, one can take $R = \{ \mathcal{S} \mid 1 \in \mathcal{S} \}$.)

On basis elements $e_{\mathcal{S}}$, $e_{\mathcal{S}'}$ the matrix of b has the form $\left(\begin{smallmatrix} 0 & \alpha \\ \alpha & 0 \end{smallmatrix} \right)$ for some $\alpha \in F^{\times}$. Therefore, if $d = \dim \bigwedge^m V$ we have $\det b = (-1)^{d/2} \cdot F^{\times 2}$, hence

$$\operatorname{disc} b = 1. \qquad \qquad \square$$

Since $\wedge$ is nonsingular, there is an adjoint involution γ on $\operatorname{End}_F(\bigwedge^m V)$ defined by

$$\gamma(f)(x) \wedge y = x \wedge f(y)$$

for all $f \in \operatorname{End}_F(\bigwedge^m V)$, x, $y \in \bigwedge^m V$. The involution γ is of the first kind and its discriminant is trivial; it is of orthogonal type if m is even and $\operatorname{char} F \neq 2$, and it is of symplectic type if m is odd or $\operatorname{char} F = 2$. We call γ the *canonical involution* on $\operatorname{End}_F(\bigwedge^m V)$.

Until the end of this subsection, A is a central simple F-algebra of degree $n = 2m$. Our purpose is to define a canonical involution on $\lambda^m A$ in such a way as to recover the definition above in the split case.

We first prove a technical result concerning the elements $s_k \in \left(A^{\otimes k}\right)^\times$ defined in the preceding section:

(10.7) Lemma. *Let A be a central simple F-algebra of degree $n = 2m$. Since $s_m \in A^{\otimes m}$, we may consider $s_m \otimes s_m \in A^{\otimes n}$. Then*

$$s_n \in A^{\otimes n}(s_m \otimes s_m).$$

Proof: In the symmetric group S_n, consider the subgroup $S_{m,m} \simeq S_m \times S_m$ consisting of the permutations which preserve $\{1, \dots, m\}$ (and therefore also the set $\{m+1, \dots, n\}$). The split case shows that $g_n\big((\pi_1, \pi_2)\big) = g_m(\pi_1) \otimes g_m(\pi_2)$ for π_1, $\pi_2 \in S_m$. Therefore,

$$s_m \otimes s_m = \textstyle\sum_{\pi \in S_{m,m}} \operatorname{sgn}(\pi) g_n(\pi).$$

Let R be a set of representatives of the left cosets of $S_{m,m}$ in S_n, so that each $\pi \in S_n$ can be written in a unique way as a product $\pi = \rho \circ \pi'$ for some $\rho \in R$ and some $\pi' \in S_{m,m}$. Since g_n is a homomorphism, it follows that

$$\operatorname{sgn}(\pi) g_n(\pi) = \operatorname{sgn}(\rho) g_n(\rho) \cdot \operatorname{sgn}(\pi') g_n(\pi'),$$

hence, summing over $\pi \in S_n$:

$$s_n = \big(\textstyle\sum_{\rho \in R} \operatorname{sgn}(\rho) g_n(\rho)\big) s_m \otimes s_m. \qquad \square$$

Recall from (10.4) that $\lambda^m A = \operatorname{End}_{A^{\otimes m}}(A^{\otimes m} s_m)$. There is therefore a natural isomorphism:

$$\lambda^m A \otimes_F \lambda^m A = \operatorname{End}_{A^{\otimes n}}\big(A^{\otimes n}(s_m \otimes s_m)\big).$$

Since (10.7) shows that $s_n \in A^{\otimes n}(s_m \otimes s_m)$, we may consider

(10.8) $\qquad I = \{\, f \in \operatorname{End}_{A^{\otimes n}}\big(A^{\otimes n}(s_m \otimes s_m)\big) \mid s_n^f = \{0\} \,\}.$

This is a right ideal in $\operatorname{End}_{A^{\otimes n}}\big(A^{\otimes n}(s_m \otimes s_m)\big) = \lambda^m A \otimes_F \lambda^m A$.

(10.9) Lemma. *If $A = \operatorname{End}_F(V)$, then under the natural isomorphisms*

$$\lambda^m A \otimes_F \lambda^m A = \operatorname{End}_F(\textstyle\bigwedge^m V) \otimes_F \operatorname{End}_F(\textstyle\bigwedge^m V) = \operatorname{End}_F(\textstyle\bigwedge^m V \otimes \textstyle\bigwedge^m V)$$

the ideal I defined above is

$$I = \{\, \varphi \in \operatorname{End}_F(\textstyle\bigwedge^m V \otimes \textstyle\bigwedge^m V) \mid \wedge \circ \varphi = 0 \,\}$$

where $\wedge$ is the canonical bilinear form on $\bigwedge^m V$, viewed as a linear map

$$\wedge \colon \textstyle\bigwedge^m V \otimes \textstyle\bigwedge^m V \to \textstyle\bigwedge^n V.$$

Proof: As observed in (10.3), we have $A^{\otimes m} s_m = \operatorname{Hom}_F(\bigwedge^m V, V^{\otimes m})$, hence

$$A^{\otimes n}(s_m \otimes s_m) = \operatorname{Hom}_F(\textstyle\bigwedge^m V \otimes \textstyle\bigwedge^m V, V^{\otimes n}).$$

Moreover, we may view s_n as a map $s_n \colon V^{\otimes n} \to V^{\otimes n}$ which factors through $\bigwedge^n V$: there is a commutative diagram:

$$
\begin{array}{ccc}
V^{\otimes n} & \xrightarrow{\ s_n\ } & V^{\otimes n} \\[2mm]
\Big\downarrow & \nearrow{\scriptstyle s_n'} & \Big\uparrow \\[2mm]
\textstyle\bigwedge^m V \otimes \textstyle\bigwedge^m V & \xrightarrow{\ \wedge\ } & \textstyle\bigwedge^n V
\end{array}
$$

The image of $s_n \in A^{\otimes n}(s_m \otimes s_m)$ in $\operatorname{Hom}_F(\bigwedge^m V \otimes \bigwedge^m V, V^{\otimes n})$ under the identification above is then the induced map s'_n.

By (1.12) every endomorphism f of the $A^{\otimes n}$-module $A^{\otimes n}(s_m \otimes s_m)$ has the form

$$x^f = x \circ \varphi$$

for some uniquely determined $\varphi \in \operatorname{End}_F(\bigwedge^m V \otimes \bigwedge^m V)$. The correspondence $f \leftrightarrow \varphi$ yields the natural isomorphism

$$\lambda^m A \otimes \lambda^m A = \operatorname{End}_{A^{\otimes n}}\left(A^{\otimes n}(s_m \otimes s_m) \right) = \operatorname{End}_F(\bigwedge^m V \otimes \bigwedge^m V).$$

Under this correspondence, the elements $f \in \operatorname{End}_{A^{\otimes n}}\left(A^{\otimes n}(s_m \otimes s_m) \right)$ which vanish on s_n correspond to endomorphisms $\varphi \in \operatorname{End}_F(\bigwedge^m V \otimes \bigwedge^m V)$ such that $s'_n \circ \varphi = 0$. It is clear from the diagram above that $\ker s'_n = \ker \wedge$, hence the conditions $s'_n \circ \varphi = 0$ and $\wedge \circ \varphi = 0$ are equivalent. $\qquad\square$

(10.10) Corollary. *The right ideal $I \subset \lambda^m A \otimes_F \lambda^m A$ defined in (10.8) above satisfies the following conditions*:

(1) $\lambda^m A \otimes_F \lambda^m A = I \oplus (1 \otimes \lambda^m A)$.

(2) *I contains the annihilator $\left[(\lambda^m A \otimes \lambda^m A) \cdot (1 - g) \right]^0$, where g is the Goldman element of $\lambda^m A \otimes_F \lambda^m A$, if m is odd or $\operatorname{char} F = 2$; it contains $1 - g$ but not $\left[(\lambda^m A \otimes \lambda^m A) \cdot (1 - g) \right]^0$ if m is even and $\operatorname{char} F \neq 2$.*

Proof: It suffices to check these properties after scalar extension to a splitting field of A. We may thus assume $A = \operatorname{End}_F(V)$. The description of I in (10.9) then shows that I is the right ideal corresponding to the canonical involution γ on $\operatorname{End}_F(\bigwedge^m V)$ under the correspondence of (3.8). $\qquad\square$

(10.11) Definition. Let A be a central simple F-algebra of degree $n = 2m$. The *canonical involution* γ on $\lambda^m A$ is the involution of the first kind corresponding to the ideal I defined in (10.8) under the correspondence of (3.8).

The following properties follow from the definition by (10.9) and (10.3), and by scalar extension to a splitting field in which F is algebraically closed:

(a) If $A = \operatorname{End}_F(V)$, the canonical involution γ on $\lambda^m A = \operatorname{End}_F(\bigwedge^m V)$ is the adjoint involution with respect to the canonical bilinear map $\wedge \colon \bigwedge^m V \times \bigwedge^m V \to \bigwedge^n V$.

(b) γ is of symplectic type if m is odd or $\operatorname{char} F = 2$; it is of orthogonal type if m is even and $\operatorname{char} F \neq 2$; in this last case we have $\operatorname{disc}(\gamma) = 1$.

In particular, if A has degree 2 (i.e., A is a quaternion algebra), then the canonical involution on $A = \lambda^1 A$ has symplectic type, hence it is the quaternion conjugation.

10.C. The canonical quadratic pair. Let A be a central simple F-algebra of even degree $n = 2m$. As observed in (10.11), the canonical involution γ on $\lambda^m A$ is symplectic for all m if $\operatorname{char} F = 2$. We show in this section that the canonical involution is actually part of a canonical pair (γ, f) on $\lambda^m A$ for all $m \geq 2$ if $\operatorname{char} F = 2$. (If $\operatorname{char} F \neq 2$, a quadratic pair is uniquely determined by its involution; thus $\lambda^m A$ carries a canonical quadratic pair if and only if γ is orthogonal, i.e., if and only if m is even).

We first examine the split case.

(10.12) Proposition. *Assume* char $F = 2$ *and let* V *be an F-vector space of dimension* $n = 2m \geq 4$. *There is a unique quadratic map*

$$q \colon \textstyle\bigwedge^m V \to \textstyle\bigwedge^n V$$

which satisfies the following conditions:

(1) $q(v_1 \wedge \cdots \wedge v_m) = 0$ *for all* $v_1, \ldots, v_m \in V$;
(2) *the polar form* $b_q \colon \bigwedge^m V \times \bigwedge^m V \to \bigwedge^n V$ *is the canonical pairing* $\wedge$. *In particular, the quadratic map q is nonsingular.*

Moreover, the discriminant of q is trivial.

Proof: Uniqueness of q is clear, since decomposable elements $v_1 \wedge \cdots \wedge v_m$ span $\bigwedge^m V$. To prove the existence of q, we use the same notation as in the proof of (10.6): we pick a basis $(e_1, \ldots, e_n)$ of V and get a basis $e_{\mathcal{S}}$ of $\bigwedge^m V$, where $\mathcal{S}$ runs over the subsets of m indices in $\{1, \ldots, n\}$. Fix a partition of these subsets into two classes C, C' such that the complement $\mathcal{S}'$ of every $\mathcal{S} \in C$ lies in C' and conversely. (For instance, one can take $C = \{\, \mathcal{S} \mid 1 \in \mathcal{S} \,\}$, $C' = \{\, \mathcal{S} \mid 1 \notin \mathcal{S} \,\}$.) We may then define a quadratic form q on $\bigwedge^m V$ by

$$q\bigl(\textstyle\sum_{\mathcal{S}} x_{\mathcal{S}} e_{\mathcal{S}}\bigr) = \bigl(\textstyle\sum_{\mathcal{S} \in C} x_{\mathcal{S}} x_{\mathcal{S}'}\bigr) e_1 \wedge \cdots \wedge e_n.$$

The polar form b_q satisfies

$$b_q\bigl(\textstyle\sum_{\mathcal{S}} x_{\mathcal{S}} e_{\mathcal{S}}, \sum_{T} y_{T} e_{T}\bigr) = \bigl(\textstyle\sum_{\mathcal{S} \in C} x_{\mathcal{S}} y_{\mathcal{S}'} + x_{\mathcal{S}'} y_{\mathcal{S}}\bigr) e_1 \wedge \cdots \wedge e_n$$
$$= \bigl(\textstyle\sum_{\mathcal{S}} x_{\mathcal{S}} y_{\mathcal{S}'}\bigr) e_1 \wedge \cdots \wedge e_n.$$

Since the right side is also equal to $\bigl(\sum_{\mathcal{S}} x_{\mathcal{S}} e_{\mathcal{S}}\bigr) \wedge \bigl(\sum_{T} y_{T} e_{T}\bigr)$, the second condition is satisfied.

It remains to prove that q vanishes on decomposable elements. We show that q actually vanishes on all the elements of the type $v \wedge \eta$, where $v \in V$ and $\eta \in \bigwedge^{m-1} V$.

Let $v = \sum_{i=1}^{n} v_i e_i$ and $\eta = \sum_{I} \eta_I e_I$, where I runs over the subsets of $m - 1$ indices in $\{1, \ldots, n\}$, so that

$$v \wedge \eta = \textstyle\sum_{i, I, i \notin I} v_i \eta_I e_{\{i\} \cup I} = \sum_{\mathcal{S}} \bigl(\sum_{i \in \mathcal{S}} v_i \eta_{\mathcal{S} \smallsetminus \{i\}}\bigr) e_{\mathcal{S}}.$$

We thus get

$$q(v \wedge \eta) = \textstyle\sum_{\mathcal{S} \in C} \bigl(\bigl(\sum_{i \in \mathcal{S}} v_i \eta_{\mathcal{S} \smallsetminus \{i\}}\bigr)\bigl(\sum_{j \in \mathcal{S}'} v_j \eta_{\mathcal{S}' \smallsetminus \{j\}}\bigr)\bigr).$$

The right side is a sum of terms of the form $v_i v_j \eta_I \eta_J$ where I, J are subsets of $m - 1$ indices such that $\{i, j\} \cup I \cup J = \{1, \ldots, n\}$. Each of these terms appears twice: $v_i v_j \eta_I \eta_J$ appears in the term corresponding to $\mathcal{S} = \{i\} \cup I$ or $\mathcal{S} = \{j\} \cup J$ (depending on which one of these two sets lies in C) and in the term corresponding to $\mathcal{S} = \{i\} \cup J$ or $\{j\} \cup I$. Therefore, $q(v \wedge \eta) = 0$.

To complete the proof, we compute the discriminant of q. From the definition, it is clear that q decomposes into an orthogonal sum of 2-dimensional subspaces:

$$q = \perp_{\mathcal{S} \in C} q_{\mathcal{S}},$$

where $q_{\mathcal{S}}(x_{\mathcal{S}} e_{\mathcal{S}} + x_{\mathcal{S}'} e_{\mathcal{S}'}) = x_{\mathcal{S}} x_{\mathcal{S}'} e_1 \wedge \cdots \wedge e_n$. It is therefore easily calculated that $\operatorname{disc} q = 0$. $\square$

(10.13) Remark. The quadratic map q may be defined alternately by representing $\bigwedge^m V$ as the quotient space $F\langle V^m \rangle / W$, where $F\langle V^m \rangle$ is the vector space of formal linear combinations of m-tuples of vectors in V, and W is the subspace generated by all the elements of the form

$$(v_1, \ldots, v_m)$$

where $v_1, \ldots, v_m \in V$ are not all distinct, and

$$(v_1, \ldots, v_i\alpha + v_i'\alpha', \ldots, v_m) - (v_1, \ldots, v_i, \ldots, v_m)\alpha - (v_1, \ldots, v_i', \ldots, v_m)\alpha'$$

where $i = 1, \ldots, m$ and $v_1, \ldots, v_i, v_i', \ldots, v_m \in V$, $\alpha, \alpha' \in F$. Since m-tuples of vectors in V form a basis of $F\langle V^m \rangle$, there is a unique quadratic map $\widetilde{q} \colon F\langle V^m \rangle \to \bigwedge^n V$ whose polar form $b_{\widetilde{q}}$ factors through the canonical pairing $\wedge$ and such that $\widetilde{q}(v) = 0$ for all $v \in V^m$. It is easy to show that this map $\widetilde{q}$ factors through the quadratic map q.

By composing q with a vector-space isomorphism $\bigwedge^n V \xrightarrow{\sim} F$, we obtain a quadratic form on $\bigwedge^m V$ which is uniquely determined up to a scalar factor. Therefore, the corresponding quadratic pair (σ_q, f_q) on $\mathrm{End}_F(\bigwedge^m V)$ is unique. In this pair, the involution σ_q is the canonical involution γ, since the polar form b_q is the canonical pairing.

Given an arbitrary central simple F-algebra A of degree $n = 2m$, we will construct on $\lambda^m A$ a quadratic pair (γ, f) which coincides with the pair (σ_q, f_q) in the case where $A = \mathrm{End}_F(V)$. The first step is to distinguish the right ideals in $\lambda^m A$ which correspond in the split case to the subspaces spanned by decomposable elements.

The following construction applies to any central simple algebra A over an arbitrary field F: if $I \subset A$ is a right ideal of reduced dimension k, we define

$$\psi_k(I) = \{\, f \in \lambda^k A = \mathrm{End}_{A^{\otimes k}}(A^{\otimes k}s_k) \mid s_k^f \in I^{\otimes k} \cdot s_k \,\} \subset \lambda^k A.$$

This set clearly is a right ideal in $\lambda^k A$.

(10.14) Lemma. *If $A = \mathrm{End}_F(V)$ and $I = \mathrm{Hom}(V, U)$ for some k-dimensional subspace $U \subset V$, then we may identify $\psi_k(I) = \mathrm{Hom}(\bigwedge^k V, \bigwedge^k U)$. In particular, $\mathrm{rdim}\,\psi_k(I) = 1$.*

Proof: If $I = \mathrm{Hom}(V, U)$, then $I^{\otimes k} = \mathrm{Hom}(V^{\otimes k}, U^{\otimes k})$ and

$$I^{\otimes k} \cdot s_k = \mathrm{Hom}(\textstyle\bigwedge^k V, U^{\otimes k}).$$

Therefore, for $f \in \lambda^k A = \mathrm{End}_F(\bigwedge^k V)$, we have $s_k^f \in I^{\otimes k} \cdot s_k$ if and only if the image of the composite map

$$\textstyle\bigwedge^k V \xrightarrow{f} \bigwedge^k V \xrightarrow{s_k} V^{\otimes k}$$

is contained in $U^{\otimes k}$. Since $s_k^{-1}(U^{\otimes k}) = \bigwedge^k U$, this condition is fulfilled if and only if $\mathrm{im}\, f \subset \bigwedge^k U$. Therefore, we may identify

$$\psi_k\big(\mathrm{Hom}(V, U)\big) = \mathrm{Hom}(\textstyle\bigwedge^k V, \bigwedge^k U).$$

Since $\dim \bigwedge^k U = 1$ if $\dim U = k$, we have $\mathrm{rdim}\,\psi_k(I) = 1$ for all right ideals I of reduced dimension k. $\qquad\square$

In view of the lemma, we have

$$\psi_k \colon \mathrm{SB}_k(A) \to \mathrm{SB}(\lambda^k A);$$

if $A = \operatorname{End}_F(V)$, this map is the Plücker embedding

$$\psi_k \colon \operatorname{Gr}_k(V) \to \mathbb{P}(\textstyle\bigwedge^k A)$$

which maps every k-dimensional subspace $U \subset V$ to the 1-dimensional subspace $\bigwedge^k U \subset \bigwedge^k V$ (see §1.C).

Suppose now that σ is an involution of the first kind on the central simple F-algebra A. To every right ideal $I \subset A$, we may associate the set $I \cdot \sigma(I) \subset A$.

(10.15) Lemma. *Suppose σ is orthogonal or $\operatorname{char} F = 2$. If $\operatorname{rdim} I = 1$, then $I \cdot \sigma(I)$ is a 1-dimensional subspace in $\operatorname{Sym}(A, \sigma)$.*

Proof: It suffices to prove the lemma in the split case. Suppose therefore that $A = \operatorname{End}_F(V)$ and σ is the adjoint involution with respect to some nonsingular bilinear form b on V. Under the standard identification φ_b, we have $A = V \otimes V$ and $I = v \otimes V$ for some nonzero vector $v \in V$, and σ corresponds to the switch map. Therefore, $I \cdot \sigma(I) = v \otimes v \cdot F$, proving the lemma. $\qquad\square$

Under the hypothesis of the lemma, we thus get a map

$$\varphi \colon \operatorname{SB}(A) \to \mathbb{P}\big(\operatorname{Sym}(A, \sigma)\big)$$

which carries every right ideal $I \subset A$ of reduced dimension 1 to $I \cdot \sigma(I)$. If $A = \operatorname{End}_F(V)$, we may identify $\operatorname{SB}(A) = \mathbb{P}(V)$ and $\mathbb{P}\big(\operatorname{Sym}(A, \sigma)\big) = \mathbb{P}(W)$, where $W \subset V \otimes V$ is the subspace of symmetric tensors; the proof above shows that $\varphi \colon \mathbb{P}(V) \to \mathbb{P}(W)$ maps $v \cdot F$ to $v \otimes v \cdot F$.

The relevance of this construction to quadratic pairs appears through the following lemma:

(10.16) Lemma. *Suppose $\operatorname{char} F = 2$ and σ is symplectic. The map $(\sigma, f) \mapsto \ker f$ defines a one-to-one correspondence between quadratic pairs (σ, f) on A and hyperplanes in $\operatorname{Sym}(A, \sigma)$ whose intersection with $\operatorname{Symd}(A, \sigma)$ is $\ker \operatorname{Trp}_\sigma$.*

Proof: For every quadratic pair (σ, f), the map f extends $\operatorname{Trp}_\sigma$ (this is just condition (2) of the definition of a quadratic pair, see (5.4)); therefore

$$\ker f \cap \operatorname{Symd}(A, \sigma) = \ker \operatorname{Trp}_\sigma .$$

If $U \subset \operatorname{Sym}(A, \sigma)$ is a hyperplane whose intersection with $\operatorname{Symd}(A, \sigma)$ is $\ker \operatorname{Trp}_\sigma$, then $\operatorname{Sym}(A, \sigma) = U + \operatorname{Symd}(A, \sigma)$, hence there is only one linear form on $\operatorname{Sym}(A, \sigma)$ with kernel U which extends $\operatorname{Trp}_\sigma$. $\qquad\square$

Suppose now that A is a central simple algebra of degree $n = 2m$ over a field F of characteristic 2. We consider the composite map

$$\operatorname{SB}_m(A) \xrightarrow{\psi_m} \operatorname{SB}(\lambda^m A) \xrightarrow{\varphi} \mathbb{P}\big(\operatorname{Sym}(\lambda^m A, \gamma)\big),$$

where γ is the canonical involution on $\lambda^m A$.

(10.17) Proposition. *If $m \geq 2$, there is a unique hyperplane in $\operatorname{Sym}(\lambda^m A, \gamma)$ which contains the image of $\varphi \circ \psi_m$ and whose intersection with $\operatorname{Symd}(\lambda^m A, \gamma)$ is $\ker \operatorname{Trp}_\sigma$.*

Proof: The proposition can be restated as follows: the subspace of $\operatorname{Sym}(\lambda^m A, \gamma)$ spanned by the image of $\varphi \circ \psi_m$ and $\ker \operatorname{Trp}_\sigma$ is a hyperplane which does not contain $\operatorname{Symd}(\lambda^m A, \gamma)$. Again, it suffices to prove the result in the split case. We may thus assume $A = \operatorname{End}_F(V)$. From the description of ψ_m and φ in this case, it follows that the image of $\varphi \circ \psi_m$ consists of the 1-dimensional spaces in

$\bigwedge^m V \otimes \bigwedge^m V$ spanned by elements of the form $(v_1 \wedge \cdots \wedge v_m) \otimes (v_1 \wedge \cdots \wedge v_m)$, with $v_1, \ldots, v_m \in V$. By (10.16), hyperplanes whose intersection with $\mathrm{Symd}(\lambda^m A, \gamma)$ coincides with $\ker \mathrm{Trp}_\gamma$ correspond to quadratic pairs on $\lambda^m A$ with involution γ, hence to quadratic forms on $\bigwedge^m V$ whose polar is the canonical pairing, up to a scalar factor. Those hyperplanes which contain the image of $\varphi \circ \psi_m$ correspond to nonsingular quadratic forms which vanish on decomposable elements $v_1 \wedge \cdots \wedge v_m$, and Proposition (10.12) shows that there is one and only one such quadratic form up to a scalar factor. $\qquad \square$

(10.18) Definition. Let A be a central simple algebra of degree $n = 2m$ over a field F of characteristic 2. By (10.16), Proposition (10.17) defines a unique quadratic pair (γ, f) on $\lambda^m A$, which we call the *canonical quadratic pair*. The proof of (10.17) shows that in the case where $A = \mathrm{End}_F(V)$ this quadratic pair is associated with the canonical map q on $\bigwedge^m V$ defined in (10.12). Since A may be split by a scalar extension in which F is algebraically closed, and since the discriminant of the canonical map q is trivial, by (10.12), it follows that $\mathrm{disc}(\gamma, f) = 0$.

If $\mathrm{char}\, F \neq 2$, the canonical involution γ on $\lambda^m A$ is orthogonal if and only if m is even. Letting f be the restriction of $\frac{1}{2} \mathrm{Trd}_A$ to $\mathrm{Sym}(\lambda^m A, \gamma)$, we also call (γ, f) the canonical quadratic pair in this case. Its discriminant is trivial, as observed in (10.11).

10.D. Induced involutions on λ-powers. In this section, ι is an automorphism of the base field F such that $\iota^2 = \mathrm{Id}_F$ (possibly $\iota = \mathrm{Id}_F$). Let V be a (finite dimensional) vector space over F. Every hermitian[15] form h on V with respect to ι induces for every integer k a hermitian form $h^{\otimes k}$ on $V^{\otimes k}$ such that

$$h^{\otimes k}(x_1 \otimes \cdots \otimes x_k, y_1 \otimes \cdots \otimes y_k) = h(x_1, y_1) \cdots h(x_k, y_k)$$

for $x_1, \ldots, x_k, y_1, \ldots, y_k \in V$. The corresponding linear map

$$\widehat{h^{\otimes k}} \colon V^{\otimes k} \to {}^\iota (V^{\otimes k})^*$$

(see (4.1)) is $(\hat{h})^{\otimes k}$ under the canonical identification ${}^\iota (V^{\otimes k})^* = ({}^\iota V^*)^{\otimes k}$, hence $h^{\otimes k}$ is nonsingular if h is nonsingular. Moreover, the adjoint involution $\sigma_{h^{\otimes k}}$ on $\mathrm{End}_F(V^{\otimes k}) = \mathrm{End}_F(V)^{\otimes k}$ is the tensor product of k copies of σ_h:

$$\sigma_{h^{\otimes k}} = (\sigma_h)^{\otimes k}.$$

The hermitian form h also induces a hermitian form $h^{\wedge k}$ on $\bigwedge^k V$ such that

$$h^{\wedge k}(x_1 \wedge \cdots \wedge x_k, y_1 \wedge \cdots \wedge y_k) = \det \big(h(x_i, y_j) \big)_{1 \leq i, j \leq k}$$

for $x_1, \ldots, x_k, y_1, \ldots, y_k \in V$. The corresponding linear map

$$\widehat{h^{\wedge k}} \colon \bigwedge^k V \to {}^\iota \big(\bigwedge^k V \big)^*$$

is $\bigwedge^k \hat{h}$ under the canonical isomorphism $\bigwedge^k ({}^\iota V^*) \xrightarrow{\sim} {}^\iota \big(\bigwedge^k V \big)^*$ which maps ${}^\iota \varphi_1 \wedge \cdots \wedge {}^\iota \varphi_k$ to ${}^\iota \psi$ where $\psi \in \big(\bigwedge^k V \big)^*$ is defined by

$$\psi(x_1 \wedge \cdots \wedge x_k) = \det \big(\varphi_i(x_j) \big)_{1 \leq i, j \leq k},$$

for $\varphi_1, \ldots, \varphi_k \in V^*$ and $x_1, \ldots, x_k \in V$. Therefore, $h^{\wedge k}$ is nonsingular if h is nonsingular.

---

[15] By convention, a hermitian form with respect to Id_F is a symmetric bilinear form.

We will describe the adjoint involution $\sigma_{h^{\wedge k}}$ on $\operatorname{End}_F(\bigwedge^k V)$ in a way which generalizes to the λ^k-th power of an arbitrary central simple F-algebra with involution.

We first observe that if $\epsilon\colon V^{\otimes k} \to \bigwedge^k V$ is the canonical epimorphism and $s_k\colon V^{\otimes k} \to V^{\otimes k}$ is the endomorphism considered in §10.A:

$$s_k(v_1 \otimes \cdots \otimes v_k) = \sum_{\pi \in S_k} \operatorname{sgn}(\pi) v_{\pi^{-1}(1)} \otimes \cdots \otimes v_{\pi^{-1}(k)},$$

then for all u, $v \in V^{\otimes k}$ we have

$$\textbf{(10.19)} \qquad h^{\otimes k}\big(s_k(u), v\big) = h^{\wedge k}\big(\epsilon(u), \epsilon(v)\big) = h^{\otimes k}\big(u, s_k(v)\big).$$

In particular, it follows that $\sigma_h^{\otimes k}(s_k) = s_k$.

(10.20) Definition. Let A be a central simple F-algebra with an involution σ such that $\sigma(x) = \iota(x)$ for all $x \in F$. Recall from (10.4) that for $k = 2, \ldots, \deg A$,

$$\lambda^k A = \operatorname{End}_{A^{\otimes k}}(A^{\otimes k} s_k).$$

According to (1.13), every $f \in \lambda^k A$ has the form $f = \rho(u s_k)$ for some $u \in A^{\otimes k}$, i.e., there exists $u \in A^{\otimes k}$ such that

$$x^f = x u s_k \quad \text{for } x \in A^{\otimes k} s_k.$$

We then define $\sigma^{\wedge k}(f) = \rho\big(\sigma^{\otimes k}(u) s_k\big)$, i.e.,

$$x^{\sigma^{\wedge k}(f)} = x \sigma^{\otimes k}(u) s_k \quad \text{for } x \in A^{\otimes k} s_k.$$

To check that the definition of $\sigma^{\wedge k}(f)$ does not depend on the choice of u, observe first that if $f = \rho(u s_k) = \rho(u' s_k)$, then

$$s_k^f = s_k u s_k = s_k u' s_k.$$

By applying $\sigma^{\otimes k}$ to both sides of this equation, and taking into account the fact that s_k is symmetric under $\sigma^{\otimes k}$ (see (10.19)), we obtain

$$s_k \sigma^{\otimes k}(u) s_k = s_k \sigma^{\otimes k}(u') s_k.$$

Since every $x \in A^{\otimes k} s_k$ has the form $x = y s_k$ for some $y \in A^{\otimes k}$, it follows that

$$x \sigma^{\otimes k}(u) s_k = y s_k \sigma^{\otimes k}(u) s_k = y s_k \sigma^{\otimes k}(u') s_k = x \sigma^{\otimes k}(u') s_k.$$

This shows that $\sigma^{\wedge k}(f)$ is well-defined. Since $\sigma(x) = \iota(x)$ for all $x \in F$, it is easily verified that $\sigma^{\wedge k}$ also restricts to ι on F.

For $k = 1$, we have $\bigwedge^1 A = A$ and we set $\sigma^{\wedge 1} = \sigma$.

(10.21) Proposition. *If $A = \operatorname{End}_F(V)$ and $\sigma = \sigma_h$ is the adjoint involution with respect to some nonsingular hermitian form h on V, then under the canonical isomorphism $\lambda^k A = \operatorname{End}_F(\bigwedge^k V)$, the involution $\sigma^{\wedge k}$ is the adjoint involution with respect to the hermitian form $h^{\wedge k}$ on $\bigwedge^k V$.*

Proof: Recall the canonical isomorphism of (10.3):

$$A^{\otimes k} s_k = \operatorname{Hom}_F(\bigwedge^k V, V^{\otimes k}), \quad \text{hence} \quad \lambda^k A = \operatorname{End}_F(\bigwedge^k V).$$

For $f = \rho(u s_k) \in \operatorname{End}_{A^{\otimes k}}(A^{\otimes k} s_k)$, the corresponding endomorphism φ of $\bigwedge^k V$ is defined by

$$\varphi(x_1 \wedge \cdots \wedge x_k) = \epsilon \circ u \circ s_k(x_1 \otimes \cdots \otimes x_k)$$

or

$$\varphi \circ \epsilon = \epsilon \circ u \circ s_k$$

where $\epsilon \colon V^{\otimes k} \to \bigwedge^k V$ is the canonical epimorphism. In order to prove the proposition, it suffices, therefore, to show:

$$h^{\wedge k}\big(\epsilon(x), \epsilon \circ u \circ s_k(y)\big) = h^{\wedge k}\big(\epsilon \circ \sigma^{\otimes k}(u) \circ s_k(x), \epsilon(y)\big)$$

for all x, $y \in V^{\otimes k}$. From (10.19) we have

$$h^{\wedge k}\big(\epsilon(x), \epsilon \circ u \circ s_k(y)\big) = h^{\otimes k}\big(s_k(x), u \circ s_k(y)\big)$$

and

$$h^{\wedge k}\big(\epsilon \circ \sigma^{\otimes k}(u) \circ s_k(x), \epsilon(y)\big) = h^{\otimes k}\big(\sigma^{\otimes k}(u) \circ s_k(x), s_k(y)\big).$$

The proposition then follows from the fact that $\sigma^{\otimes k}$ is the adjoint involution with respect to $h^{\otimes k}$. $\qquad\square$

The next proposition is more specifically concerned with symmetric bilinear forms b. In the case where $\dim V = n = 2m$, we compare the involution $\sigma_b^{\wedge m}$ with the canonical involution γ on $\operatorname{End}_F(\bigwedge^m V)$.

(10.22) Proposition. *Let b be a nonsingular symmetric, nonalternating, bilinear form on an F-vector space V of dimension $n = 2m$. Let $(e_1, \ldots, e_n)$ be an orthogonal basis of V and let $e = e_1 \wedge \cdots \wedge e_n \in \bigwedge^n V$; let also*

$$\delta = (-1)^m \prod_{i=1}^{n} b(e_i, e_i),$$

so that $\operatorname{disc} b = \delta \cdot F^{\times 2}$. There is a map $u \in \operatorname{End}_F(\bigwedge^m V)$ such that

(10.23) $$b^{\wedge m}\big(u(x), y\big)e = x \wedge y = (-1)^m b^{\wedge m}\big(x, u(y)\big)e$$

for all x, $y \in \bigwedge^m V$, and

(10.24) $$u^2 = \delta^{-1} \cdot \operatorname{Id}_{\wedge^m V}.$$

If $\sigma = \sigma_b$ is the adjoint involution with respect to b, then the involution $\sigma^{\wedge m}$ on $\operatorname{End}_F(\bigwedge^m V)$ is related to the canonical involution γ by

$$\sigma^{\wedge m} = \operatorname{Int}(u) \circ \gamma.$$

In particular, the involutions $\sigma^{\wedge m}$ and γ commute.

Moreover, if $\operatorname{char} F = 2$ and $m \geq 2$, the map u is a similitude of the canonical quadratic map $q \colon \bigwedge^m V \to \bigwedge^n V$ of (10.12) with multiplier δ^{-1}, i.e.,

$$q\big(u(x)\big) = \delta^{-1} q(x)$$

for all $x \in \bigwedge^m V$.

Proof: Let $a_i = b(e_i, e_i) \in F^\times$ for $i = 1, \ldots, n$. As in (10.6), we set

$$e_{\mathcal{S}} = e_{i_1} \wedge \cdots \wedge e_{i_m} \in \bigwedge^m V \quad \text{and let} \quad a_{\mathcal{S}} = a_{i_1} \cdots a_{i_m}$$

for $\mathcal{S} = \{i_1, \ldots, i_m\} \subset \{1, \ldots, n\}$ with $i_1 < \cdots < i_m$. If $\mathcal{S} \neq \mathcal{T}$, the matrix $\big(b(e_i, e_j)\big)_{(i,j) \in \mathcal{S} \times \mathcal{T}}$ has at least one row and one column of 0's, namely the row corresponding to any index in $\mathcal{S} \setminus \mathcal{T}$ and the column corresponding to any index in $\mathcal{T} \setminus \mathcal{S}$. Therefore, $b^{\wedge m}(e_{\mathcal{S}}, e_{\mathcal{T}}) = 0$. On the other hand, the matrix $\big(b(e_i, e_j)\big)_{(i,j) \in \mathcal{S} \times \mathcal{S}}$ is diagonal, and $b^{\wedge m}(e_{\mathcal{S}}, e_{\mathcal{S}}) = a_{\mathcal{S}}$. Therefore, as $\mathcal{S}$ runs over all the subsets of m

indices, the elements $e_{\mathcal{S}}$ are anisotropic and form an orthogonal basis of $\bigwedge^m V$ with respect to the bilinear form $b^{\wedge m}$.

On the other hand, if $\mathcal{S}' = \{1, \ldots, n\} \smallsetminus \mathcal{S}$ is the complement of $\mathcal{S}$, we have

$$e_{\mathcal{S}} \wedge e_{\mathcal{S}'} = \varepsilon_{\mathcal{S}} e$$

for some $\varepsilon_{\mathcal{S}} = \pm 1$. Since $\wedge$ is symmetric when m is even and skew-symmetric when m is odd (see (10.6)), it follows that

(10.25) $$\varepsilon_{\mathcal{S}} \varepsilon_{\mathcal{S}'} = (-1)^m.$$

Define u on the basis elements $e_{\mathcal{S}}$ by

(10.26) $$u(e_{\mathcal{S}}) = \varepsilon_{\mathcal{S}} a_{\mathcal{S}'}^{-1} e_{\mathcal{S}'}$$

and extend u to $\bigwedge^m V$ by linearity. We then have

$$b^{\wedge m}\big(u(e_{\mathcal{S}}), e_{\mathcal{S}'}\big)e = \varepsilon_{\mathcal{S}} e = e_{\mathcal{S}} \wedge e_{\mathcal{S}'}$$

and

$$b^{\wedge m}\big(e_{\mathcal{S}}, u(e_{\mathcal{S}'})\big) = \varepsilon_{\mathcal{S}'} e = (-1)^m e_{\mathcal{S}} \wedge e_{\mathcal{S}'}.$$

Moreover, if $\mathcal{T} \neq \mathcal{S}$, then

$$b\big(u(e_{\mathcal{S}}), e_{\mathcal{T}}\big) = 0 = e_{\mathcal{S}} \wedge e_{\mathcal{T}} = b\big(e_{\mathcal{S}}, u(e_{\mathcal{T}})\big).$$

The equations (10.23) thus hold when x, y run over the basis $(e_{\mathcal{S}})$; therefore they hold for all x, $y \in V$ by bilinearity.

For all $\mathcal{S}$ we have $a_{\mathcal{S}} a_{\mathcal{S}'} = \prod_{i=1}^{n} b(e_i, e_i)$, hence (10.24) follows from (10.26) and (10.25).

From (10.23), it follows that for all $f \in \mathrm{End}_F(V)$ and all x, $y \in V$,

$$b^{\wedge m}\big(u(x), f(y)\big)e = x \wedge f(y),$$

hence

$$b^{\wedge m}\big(\sigma^{\wedge m}(f) \circ u(x), y\big)e = \gamma(f)(x) \wedge y.$$

The left side also equals

$$b^{\wedge m}\big(u \circ u^{-1} \circ \sigma^{\wedge m}(f) \circ u(x), y\big)e = \big(u^{-1} \circ \sigma^{\wedge m}(f) \circ u\big)(x) \wedge y,$$

hence

$$u^{-1} \circ \sigma^{\wedge m}(f) \circ u = \gamma(f) \quad \text{for } f \in \mathrm{End}_F(V).$$

Therefore, $\sigma^{\wedge m} = \mathrm{Int}(u) \circ \gamma$.

We next show that $\sigma^{\wedge m}$ and γ commute. By (10.23), we have $\sigma^{\wedge m}(u) = (-1)^m u$, hence $\gamma(u) = (-1)^m u$. Therefore, $\gamma \circ \sigma^{\wedge m} = \mathrm{Int}(u^{-1})$, while $\sigma^{\wedge m} \circ \gamma = \mathrm{Int}(u)$. Since $u^2 \in F^\times$, we have $\mathrm{Int}(u) = \mathrm{Int}(u^{-1})$, and the claim is proved.

Finally, assume char $F = 2$ and $m \geq 2$. The proof of (10.12) shows that the quadratic map q may be defined by partitioning the subsets of m indices in $\{1, \ldots, n\}$ into two classes C, C' such that the complement of every subset in C lies in C' and vice versa, and letting

$$q(x) = \Big(\textstyle\sum_{\mathcal{S} \in C} x_{\mathcal{S}} x_{\mathcal{S}'}\Big)e$$

for $x = \sum_{\mathcal{S}} x_{\mathcal{S}} e_{\mathcal{S}}$. By definition of u, we have $u(x) = \sum_{\mathcal{S}} x_{\mathcal{S}'} a_{\mathcal{S}}^{-1} e_{\mathcal{S}}$, hence

$$q\big(u(x)\big) = \Big(\textstyle\sum_{\mathcal{S} \in C} x_{\mathcal{S}'} a_{\mathcal{S}}^{-1} x_{\mathcal{S}} a_{\mathcal{S}'}^{-1}\Big)e.$$

Since $a_{\mathcal{S}} a_{\mathcal{S}'} = \delta$ for all $\mathcal{S}$, it follows that $q\big(u(x)\big) = \delta^{-1} q(x)$. $\qquad \square$

10.E. Definition of the discriminant algebra. Let (B, τ) be a central simple algebra with involution of the second kind over a field F. We assume that the degree of (B, τ) is even: $\deg(B, \tau) = n = 2m$. The center of B is denoted K; it is a quadratic étale F-algebra with nontrivial automorphism ι. We first consider the case where K is a field, postponing to the end of the section the case where $K \simeq F \times F$. The K-algebra B is thus central simple. The K-algebra $\lambda^m B$ has a canonical involution γ, which is of the first kind, and also has the involution $\tau^{\wedge m}$ induced by τ, which is of the second kind. The definition of the discriminant algebra $D(B, \tau)$ is based on the following crucial result:

(10.27) Lemma. *The involutions γ and $\tau^{\wedge m}$ on $\lambda^m B$ commute. Moreover, if $\operatorname{char} F = 2$ and $m \geq 2$, the canonical quadratic pair (γ, f) on $\lambda^m B$ satisfies*

$$f\big(\tau^{\wedge m}(x)\big) = \iota\big(f(x)\big)$$

for all $x \in \operatorname{Sym}(\lambda^m B, \gamma)$.

Proof: We reduce to the split case by a scalar extension. To construct a field extension L of F such that $K \otimes_F L$ is a field and $B \otimes_F L$ is split, consider the division K-algebra D which is Brauer-equivalent to B. By (3.1), this algebra has an involution of the second kind θ. We may take for L a maximal subfield of $\operatorname{Sym}(D, \theta)$.

We may thus assume $B = \operatorname{End}_K(V)$ for some n-dimensional vector space V over K and $\tau = \sigma_h$ for some nonsingular hermitian form h on V. Consider an orthogonal basis $(e_i)_{1 \leq i \leq n}$ of V and let $V_0 \subset V$ be the F-subspace of V spanned by $e_1, \ldots, e_n$. Since $h(e_i, e_i) \in F^\times$ for $i = 1, \ldots, n$, the restriction h_0 of h to V_0 is a nonsingular symmetric bilinear form which is not alternating. We have $V = V_0 \otimes_F K$, hence

$$B = \operatorname{End}_F(V_0) \otimes_F K.$$

Moreover, since τ is the adjoint involution with respect to h,

$$\tau = \tau_0 \otimes \iota$$

where τ_0 is the adjoint involution with respect to h_0 on $\operatorname{End}_F(V_0)$. Therefore, there is a canonical isomorphism

$$\lambda^m B = \operatorname{End}_F(\textstyle\bigwedge^m V_0) \otimes_F K$$

and

$$\tau^{\wedge m} = \tau_0^{\wedge m} \otimes \iota.$$

On the other hand, the canonical bilinear map

$$\wedge \colon \textstyle\bigwedge^m V \times \bigwedge^m V \to \bigwedge^n V$$

is derived by scalar extension to K from the canonical bilinear map $\wedge$ on $\bigwedge^m V_0$, hence $\gamma = \gamma_0 \otimes \operatorname{Id}_K$ where γ_0 is the canonical involution on $\operatorname{End}_F(\bigwedge^m V_0)$. By Proposition (10.22), $\tau_0^{\wedge m}$ and γ_0 commute, hence $\tau^{\wedge m}$ and γ also commute.

Suppose now that $\operatorname{char} F = 2$ and $m \geq 2$. Let $z \in K \smallsetminus F$. In view of the canonical isomorphism $\lambda^m B = \operatorname{End}_F(\bigwedge^m V_0) \otimes_F K$, every element $x \in \operatorname{Sym}(\lambda^m B, \gamma)$ may be written in the form $x = x_0 \otimes 1 + x_1 \otimes z$ for some x_0, $x_1 \in \operatorname{End}_F(\bigwedge^m V_0)$ symmetric under γ_0. Proposition (10.22) yields an element $u \in \operatorname{End}_F(\bigwedge^m V_0)$ such that $\tau_0^{\wedge m} = \operatorname{Int}(u) \circ \gamma_0$, hence

$$\tau^{\wedge m}(x) = \tau_0^{\wedge m}(x_0) \otimes 1 + \tau_0^{\wedge m}(x_1) \otimes \iota(z) = (u x_0 u^{-1}) \otimes 1 + (u x_1 u^{-1}) \otimes \iota(z).$$

To prove $f\big(\tau^{\wedge m}(x)\big) = \iota\big(f(x)\big)$, it now suffices to show that $f(uyu^{-1}) = f(y)$ for all $y \in \mathrm{Sym}\big(\mathrm{End}_F(\bigwedge^m V_0), \gamma_0\big)$.

Let $q \colon \bigwedge^m V_0 \to F$ be the canonical quadratic form uniquely defined (up to a scalar multiple) by (10.12). Under the associated standard identification, the elements in $\mathrm{Sym}\big(\mathrm{End}_F(\bigwedge^m V_0), \gamma_0\big)$ correspond to symmetric tensors in $\bigwedge^m V_0 \otimes \bigwedge^m V_0$, and we have $f(v \otimes v) = q(v)$ for all $v \in \bigwedge^m V_0$. Since symmetric tensors are spanned by elements of the form $v \otimes v$, it suffices to prove

$$f\big(u \circ (v \otimes v) \circ u^{-1}\big) = f(v \otimes v) \qquad \text{for all } v \in \bigwedge^m V_0.$$

The proof of (10.22) shows that $\gamma_0(u) = u$ and $u^2 = \delta^{-1} \in F^\times$, hence

$$u \circ (v \otimes v) \circ u^{-1} = \delta u \circ (v \otimes v) \circ \gamma_0(u) = \delta u(v) \otimes u(v);$$

therefore, by (10.22),

$$f\big(u \circ (v \otimes v) \circ u^{-1}\big) = \delta q\big(u(v)\big) = q(v) = f(v \otimes v),$$

and the proof is complete. $\square$

The lemma shows that the composite map $\theta = \tau^{\wedge m} \circ \gamma$ is an automorphism of order 2 on the F-algebra B. Note that $\theta(x) = \iota(x)$ for all $x \in K$, since $\tau^{\wedge m}$ is an involution of the second kind while γ is of the first kind.

(10.28) Definition. The *discriminant algebra* $D(B, \tau)$ of (B, τ) is the F-subalgebra of θ-invariant elements in $\lambda^m B$. It is thus a central simple F-algebra of degree

$$\deg D(B, \tau) = \deg \lambda^m B = \binom{n}{m}.$$

The involutions γ and $\tau^{\wedge m}$ restrict to the same involution of the first kind $\underline{\tau}$ on $D(B, \tau)$:

$$\underline{\tau} = \gamma|_{D(B,\tau)} = \tau^{\wedge m}|_{D(B,\tau)}.$$

Moreover, if $\mathrm{char}\, F = 2$ and $m \geq 2$, the canonical quadratic pair (γ, f) on $\lambda^m B$ restricts to a canonical quadratic pair $(\underline{\tau}, f_D)$ on $D(B, \tau)$; indeed, for an element $x \in \mathrm{Sym}\big(D(B, \tau), \underline{\tau}\big)$ we have $\tau^{\wedge m}(x) = x$, hence, by (10.27), $f(x) = \iota\big(f(x)\big)$, and therefore $f(x) \in F$.

The following number-theoretic observation on $\deg D(B, \tau)$ is useful:

(10.29) Lemma. *Let m be an integer, $m \geq 1$. The binomial coefficient $\binom{2m}{m}$ satisfies*

$$\binom{2m}{m} \equiv \begin{cases} 2 \mod 4 & \text{if } m \text{ is a power of } 2; \\ 0 \mod 4 & \text{if } m \text{ is not a power of } 2. \end{cases}$$

Proof: For every integer $m \geq 1$, let $v(m) \in \mathbb{N}$ be the exponent of the highest power of 2 which divides m, i.e., $v(m)$ is the 2-adic valuation of m. The equation $(m + 1)\binom{2(m+1)}{m+1} = 2(2m + 1)\binom{2m}{m}$ yields

$$v\big(\tbinom{2(m+1)}{m+1}\big) = v\big(\tbinom{2m}{m}\big) + 1 - v(m + 1) \quad \text{for } m \geq 1.$$

On the other hand, let $\ell(m) = m_0 + \cdots + m_k$ where the 2-adic expansion of m is $m = m_0 + 2m_1 + 2^2 m_2 + \cdots + 2^k m_k$ with $m_0, \ldots, m_k = 0$ or 1. It is easily seen that the function $\ell(m)$ satisfies the same recurrence relation as $v\big(\binom{2m}{m}\big)$ and $\ell(1) = 1 = v\big(\binom{2}{1}\big)$, hence $\ell(m) = v\big(\binom{2m}{m}\big)$ for all $m \geq 1$. In particular, $v\big(\binom{2m}{m}\big) = 1$ if m is a power of 2, and $v\big(\binom{2m}{m}\big) \geq 2$ otherwise. $\square$

(10.30) Proposition. *Multiplication in $\lambda^m B$ yields a canonical isomorphism*

$$D(B, \tau) \otimes_F K = \lambda^m B$$

such that $\underline{\tau} \otimes \mathrm{Id}_K = \gamma$ and $\underline{\tau} \otimes \iota = \tau^{\wedge m}$. The index $\mathrm{ind}\, D(B, \tau)$ divides 4, and $\mathrm{ind}\, D(B, \tau) = 1$ or 2 if m is a power of 2.

The involution $\underline{\tau}$ is of symplectic type if m is odd or $\mathrm{char}\, F = 2$; it is of orthogonal type if m is even and $\mathrm{char}\, F \neq 2$.

Proof: The first part follows from the definition of $D(B, \tau)$ and its involution $\underline{\tau}$. By (10.5) we have $\mathrm{ind}\, \lambda^m B = 1$ or 2 since $\lambda^m B$ and $B^{\otimes m}$ are Brauer-equivalent, hence $\mathrm{ind}\, D(B, \tau)$ divides $2[K : F] = 4$. However, if m is a power of 2, then $\mathrm{ind}\, D(B, \tau)$ cannot be 4 since the preceding lemma shows that $\deg D(B, \tau) \equiv 2 \mod 4$.

Since $\gamma = \underline{\tau} \otimes \mathrm{Id}_K$, the involutions γ and $\underline{\tau}$ have the same type, hence $\underline{\tau}$ is orthogonal if and only if m is even and $\mathrm{char}\, F \neq 2$. $\qquad \square$

For example, if B is a quaternion algebra, i.e., $n = 2$, then $m = 1$ hence $\lambda^m B = B$ and $\tau^{\wedge m} = \tau$. The algebra $D(B, \tau)$ is the unique quaternion F-subalgebra of B such that $B = D(B, \tau) \otimes_F K$ and $\tau = \gamma_0 \otimes \iota$ where γ_0 is the canonical (conjugation) involution on $D(B, \tau)$: see (2.22).

To conclude this section, we examine the case where $K = F \times F$. We may then assume $B = E \times E^{\mathrm{op}}$ for some central simple F-algebra E of degree $n = 2m$ and $\tau = \varepsilon$ is the exchange involution. Note that there is a canonical isomorphism

$$(\lambda^m E)^{\mathrm{op}} = \mathrm{End}_{E^{\otimes m}}(E^{\otimes m} s_m)^{\mathrm{op}} \xrightarrow{\sim} \mathrm{End}_{(E^{\mathrm{op}})^{\otimes m}}\left((E^{\mathrm{op}})^{\otimes m} s_m^{\mathrm{op}}\right) = \lambda^m(E^{\mathrm{op}})$$

which identifies $f^{\mathrm{op}} \in (\lambda^m E)^{\mathrm{op}}$ with the endomorphism of $(E^{\mathrm{op}})^{\otimes m} s_m^{\mathrm{op}}$ which maps s_m^{op} to $(s_m^f)^{\mathrm{op}}$ (thus, $(s_m^{\mathrm{op}})^{f^{\mathrm{op}}} = (s_m^f)^{\mathrm{op}}$) (see Exercise 1 of Chapter I). Therefore, the notation $\lambda^m E^{\mathrm{op}}$ is not ambiguous. We may then set

$$\lambda^m B = \lambda^m E \times \lambda^m E^{\mathrm{op}}$$

and define the canonical involution γ on $\lambda^m B$ by means of the canonical involution γ_E on $\lambda^m E$:

$$\gamma(x, y^{\mathrm{op}}) = \left(\gamma_E(x), \gamma_E(y)^{\mathrm{op}}\right) \qquad \text{for } x, y \in \lambda^m E.$$

Similarly, if $\mathrm{char}\, F = 2$ and $m \geq 2$, the canonical pair (γ_E, f_E) on $\lambda^m E$ (see (10.18)) induces a canonical quadratic pair (γ, f) on $\lambda^m B$ by

$$f(x, y^{\mathrm{op}}) = \left(f_E(x), f_E(y)\right) \in F \times F \qquad \text{for } x, y \in \mathrm{Sym}(\lambda^m E, \gamma).$$

We also define the involution $\varepsilon^{\wedge m}$ on $\lambda^m B$ as the exchange involution on $\lambda^m E \times \lambda^m E^{\mathrm{op}}$:

$$\varepsilon^{\wedge m}(x, y^{\mathrm{op}}) = (y, x^{\mathrm{op}}) \qquad \text{for } x, y \in \lambda^m E.$$

The involutions $\varepsilon^{\wedge m}$ and γ thus commute, hence their composite $\theta = \varepsilon^{\wedge m} \circ \gamma$ is an F-automorphism of order 2 on $\lambda^m B$. The invariant elements form the F-subalgebra

(10.31) $D(B, \varepsilon) = \left\{\, \left(x, \gamma(x)^{\mathrm{op}}\right) \mid x \in \lambda^m E \,\right\} \simeq \lambda^m E.$

The involutions $\varepsilon^{\wedge m}$ and γ coincide on this subalgebra and induce an involution which we denote $\underline{\varepsilon}$.

The following proposition shows that these definitions are compatible with the notions defined previously in the case where K is a field:

(10.32) Proposition. *Let (B, τ) be a central simple algebra with involution of the second kind over a field F. Suppose the center K of B is a field. The K-algebra isomorphism $\varphi\colon (B_K, \tau_K) \xrightarrow{\sim} (B \times B^{\mathrm{op}}, \varepsilon)$ of (2.14) which maps $x \otimes k$ to $\left(xk, \left(\tau(x)k\right)^{\mathrm{op}}\right)$ induces a K-algebra isomorphism*

$$\lambda^m \varphi\colon (\lambda^m B)_K \xrightarrow{\sim} \lambda^m B \times \lambda^m B^{\mathrm{op}}$$

mapping $x \otimes k$ to $\left(xk, \left(\tau^{\wedge m}(x)k\right)^{\mathrm{op}}\right)$. This isomorphism is compatible with the canonical involution and the canonical quadratic pair (if $\operatorname{char} F = 2$ and $m \geq 2$), and satisfies $\lambda^m \varphi \circ \tau_K^{\wedge m} = \varepsilon^{\wedge m} \circ \lambda^m \varphi$. Therefore, $\lambda^m \varphi$ induces an isomorphism of K-algebras with involution

$$\left(D(B, \tau)_K, \underline{\tau}_K\right) \xrightarrow{\sim} \left(D(B \times B^{\mathrm{op}}, \varepsilon), \underline{\varepsilon}\right)$$

and also, if $\operatorname{char} F = 2$ and $m \geq 2$, an isomorphism of K-algebras with quadratic pair

$$\left(D(B, \tau)_K, \underline{\tau}_K, (f_D)_K\right) \xrightarrow{\sim} \left(D(B \times B^{\mathrm{op}}, \varepsilon), \underline{\varepsilon}, f\right).$$

Proof: The fact that $\lambda^m \varphi$ is compatible with the canonical involution and the canonical pair follows from (10.27); the equation $\lambda^m \varphi \circ \tau_K^{\wedge m} = \varepsilon^{\wedge m} \circ \lambda^m \varphi$ is clear from the definition of $\lambda^m \varphi$. $\qquad\square$

10.F. The Brauer class of the discriminant algebra. An explicit description of the discriminant algebra of a central simple algebra with involution of the second kind is known only in very few cases: quaternion algebras are discussed after (10.30) above, and algebras of degree 4 are considered in §15.D. Some general results on the Brauer class of a discriminant algebra are easily obtained however, as we proceed to show. In particular, we establish the relation between the discriminant algebra and the discriminant of hermitian forms mentioned in the introduction.

Notation is as in the preceding subsection. Thus, let (B, τ) be a central simple algebra with involution of the second kind of even degree $n = 2m$ over an arbitrary field F, and let K be the center of B. For any element $d = \delta \cdot N(K/F) \in F^\times / N(K/F)$, we denote by $(K, d)_F$ (or $(K, \delta)_F$) the quaternion algebra $K \oplus Kz$ where multiplication is defined by $zx = \iota(x)z$ for $x \in K$ and $z^2 = \delta$. Thus,

$$(K, d)_F = \begin{cases} (\alpha, \delta)_F & \text{if } \operatorname{char} F \neq 2 \text{ and } K \simeq F[X]/(X^2 - \alpha), \\ [\alpha, \delta)_F & \text{if } \operatorname{char} F = 2 \text{ and } K \simeq F[X]/(X^2 + X + \alpha). \end{cases}$$

(In particular, $(K, d)_F$ splits if $K \simeq F \times F$). We write $\sim$ for Brauer-equivalence.

(10.33) Proposition. *Suppose $B = B_0 \otimes_F K$ and $\tau = \tau_0 \otimes \iota$ for some central simple F-algebra B_0 with involution τ_0 of the first kind of orthogonal type; then*

$$D(B, \tau) \sim \lambda^m B_0 \otimes_F (K, \operatorname{disc} \tau_0)_F.$$

Proof: We have $\lambda^m B = \lambda^m B_0 \otimes_F K$, $\tau^{\wedge m} = \tau_0^{\wedge m} \otimes \iota$ and $\gamma = \gamma_0 \otimes \operatorname{Id}_K$ where γ_0 is the canonical involution on $\lambda^m B_0$, hence also $\theta = \theta_0 \otimes \iota$ where $\theta_0 = \tau_0^{\wedge m} \circ \gamma_0$. Since θ_0 leaves F elementwise invariant and $\theta_0^2 = \operatorname{Id}$, we have $\theta_0 = \operatorname{Int}(t)$ for some $t \in (\lambda^m B_0)^\times$ such that $t^2 \in F^\times$. After scalar extension to a splitting field L of B_0 in which F is algebraically closed, (10.22) yields $t = u\xi$ for some $\xi \in L^\times$ and some $u \in (\lambda^m B_0 \otimes L)^\times$ such that $u^2 \cdot L^{\times 2} = \operatorname{disc} \tau_0$. Therefore, letting $\delta = t^2 \in F^\times$, we have $\delta \cdot L^{\times 2} = \operatorname{disc} \tau_0$, hence

$$\delta \cdot F^{\times 2} = \operatorname{disc} \tau_0,$$

since F is algebraically closed in L. The proposition then follows from the following general result:

(10.34) Lemma. *Let $A = A_0 \otimes_F K$ be a central simple K-algebra and let $t \in A_0^\times$ be such that $t^2 = \delta \in F^\times$. The F-subalgebra $A' \subset A$ of elements invariant under $\mathrm{Int}(t) \otimes \iota$ is Brauer-equivalent to $A_0 \otimes_F (K, \delta)_F$.*

Proof: Let $(K, \delta)_F = K \oplus Kz$ where $zx = \iota(x)z$ for $x \in K$ and $z^2 = \delta$, and let $A_1 = A_0 \otimes (K, \delta)_F$. The centralizer of $K \subset (K, \delta)_F$, viewed as a subalgebra in A_1, is

$$C_{A_1} K = A_0 \otimes_F K,$$

which may be identified with A. The algebra A' is then identified with the centralizer of K and tz.

Claim. *The subalgebra $M \subset A_1$ generated by K and tz is a split quaternion algebra.*

Since $t \in A_0^\times$, the elements t and z commute, and $tzx = \iota(x)tz$ for $x \in K$. Moreover, $t^2 = \delta = z^2$, hence $(tz)^2 = \delta^2 \in F^{\times 2}$. Therefore, $M \simeq (K, \delta^2)_F$, proving the claim.

Since A' is the centralizer of M in A_1, Theorem (1.5) yields

$$A_1 \simeq A' \otimes_F M.$$

The lemma then follows from the claim. $\square$

The split case $B = \mathrm{End}_K(V)$ is a particular case of (10.33):

(10.35) Corollary. *For every nonsingular hermitian space (V, h) of even dimension over K,*

$$D\big(\mathrm{End}_K(V), \sigma_h\big) \sim (K, \mathrm{disc}\, h)_F$$

where $\mathrm{disc}\, h$ is defined in the introduction to this section.

Proof: Let $V_0 \subset V$ be the F-subspace spanned by an orthogonal K-basis of V. The hermitian form h restricts to a nonsingular symmetric bilinear form h_0 on V_0 and we have

$$\mathrm{End}_K(V) = \mathrm{End}_F(V_0) \otimes_F K, \quad \sigma_h = \sigma_0 \otimes \iota$$

where $\sigma_0 = \sigma_{h_0}$ is the adjoint involution with respect to h_0. By (10.33),

$$D\big(\mathrm{End}_K(V), \sigma_h\big) \sim (K, \mathrm{disc}\, h_0)_F.$$

The corollary follows, since $\mathrm{disc}\, h = \mathrm{disc}\, h_0 \cdot N(K/F) \in F^\times / N(K/F)$. $\square$

(10.36) Corollary. *For all $u \in \mathrm{Sym}(B, \tau) \cap B^\times$,*

$$D\big(B, \mathrm{Int}(u) \circ \tau\big) \sim D(B, \tau) \otimes_F \big(K, \mathrm{Nrd}_B(u)\big)_F.$$

Proof: If $K \simeq F \times F$, each side is split. We may thus assume K is a field. Suppose first $B = \mathrm{End}_K(V)$ for some vector space V, and let h be a nonsingular hermitian form on V such that $\tau = \sigma_h$. The involution $\mathrm{Int}(u) \circ \tau$ is then adjoint to the hermitian form h' defined by

$$h'(x, y) = h\big(x, u^{-1}(y)\big) \quad \text{for } x, y \in V.$$

Since the Gram matrix of h' is the product of the Gram matrix of h by the matrix of u^{-1}, it follows that

$$\operatorname{disc} h' = \operatorname{disc} h \det u^{-1} = \operatorname{disc} h \det u \in F^\times/N(K/F).$$

The corollary then follows from (10.35) by multiplicativity of the quaternion symbol.

The general case is reduced to the split case by a suitable scalar extension. Let $X = R_{K/F}(\operatorname{SB}(B))$ be the Weil transfer (or restriction of scalars) of the Severi-Brauer variety of B (see Scheiderer [248, §4] for a discussion of the Weil transfer) and let $L = F(X)$ be the function field of X. We have

$$B \otimes_F L = B \otimes_K K\big(\operatorname{SB}(B) \times_K \operatorname{SB}({}^\iota B)\big),$$

hence B_L is split. Therefore, the split case considered above shows that the F-algebra

$$A = D(B, \tau) \otimes_F \big(K, \operatorname{Nrd}_B(u)\big)_F \otimes_F D\big(B, \operatorname{Int}(u) \circ \tau\big)^{\mathrm{op}}$$

is split by L. However, the kernel of the scalar extension map $\operatorname{Br}(F) \to \operatorname{Br}(L)$ is the image under the norm map of the kernel of the scalar extension map $\operatorname{Br}(K) \to \operatorname{Br}\big(K(\operatorname{SB}(B))\big)$ (see Merkurjev-Tignol [196, Corollary 2.12]). The latter is known to be generated by the Brauer class of B (see for instance Merkurjev-Tignol [196, Corollary 2.7]), and $N_{K/F}(B)$ splits since B has an involution of the second kind (see (3.15)). Therefore, the map $\operatorname{Br}(F) \to \operatorname{Br}(L)$ is injective, hence A is split. $\quad\square$

§11. Trace Form Invariants

The invariants of involutions defined in this section are symmetric bilinear forms derived from the reduced trace. Let A be a central simple algebra over an arbitrary field F and let σ be an involution of any kind on A. Our basic object of study is the form

$$T_{(A,\sigma)} \colon A \times A \to F$$

defined by

$$T_{(A,\sigma)}(x, y) = \operatorname{Trd}_A\big(\sigma(x)y\big) \quad \text{for } x, y \in A.$$

Since $\sigma\big(\operatorname{Trd}_A(\sigma(y)x)\big) = \operatorname{Trd}_A\big(\sigma(x)y\big)$, by (2.2) and (2.16), the form $T_{(A,\sigma)}$ is symmetric bilinear if σ is of the first kind and hermitian with respect to $\sigma|_F$ if σ is of the second kind. It is nonsingular in each case, since the *bilinear trace* form $T_A(x, y) = \operatorname{Trd}_A(xy)$ is nonsingular, as is easily seen after scalar extension to a splitting field of A.

More generally, for any $u \in \operatorname{Sym}(A, \sigma)$ we set

$$T_{(A,\sigma,u)}(x, y) = \operatorname{Trd}_A\big(\sigma(x)uy\big) \quad \text{for } x, y \in A.$$

The form $T_{(A,\sigma,u)}$ also is symmetric bilinear if σ is of the first kind and hermitian if σ is of the second kind, and it is nonsingular if and only if u is invertible.

How much information on σ can be derived from the form $T_{(A,\sigma)}$ is suggested by the following proposition, which shows that $T_{(A,\sigma)}$ determines $\sigma \otimes \sigma$ if σ is of the first kind. To formulate a more general statement, we denote by $\iota = \sigma|_F$ the restriction of σ to F, by ${}^\iota A = \{\, {}^\iota a \mid a \in A \,\}$ the conjugate algebra of A (see §3.B) and by ${}^\iota\sigma$ the involution on ${}^\iota A$ defined by

$${}^\iota\sigma({}^\iota a) = {}^\iota\big(\sigma(a)\big) \quad \text{for } a \in A.$$

(11.1) Proposition. *Under the isomorphism* $\sigma_* \colon A \otimes {}^\iota A \xrightarrow{\sim} \operatorname{End}_F(A)$ *such that*

$$\sigma_*(a \otimes {}^\iota b)(x) = ax\sigma(b),$$

the involution $\sigma \otimes {}^\iota\sigma$ *corresponds to the adjoint involution with respect to the form* $T_{(A,\sigma)}$. *More generally, for all* $u \in \operatorname{Sym}(A,\sigma) \cap A^\times$, *the involution* $\big(\operatorname{Int}(u^{-1}) \circ \sigma\big) \otimes {}^\iota\sigma$ *corresponds to the adjoint involution with respect to the form* $T_{(A,\sigma,u)}$ *under the isomorphism* σ_*.

Proof: The proposition follows by a straightforward computation: for a, b, x, $y \in A$,

$$T_{(A,\sigma,u)}\big(\sigma_*(a \otimes b)(x), y\big) = \operatorname{Trd}_A\big(b\sigma(x)\sigma(a)uy\big)$$

and

$$T_{(A,\sigma,u)}\big(x, \sigma_*\big(\operatorname{Int}(u^{-1}) \circ \sigma(a) \otimes \sigma(b)\big)(y)\big) = \operatorname{Trd}_A\big(\sigma(x)u\big(u^{-1}\sigma(a)u\big)yb\big).$$

The equality of these expressions proves the proposition. (Note that the first part (i.e., the case where $u = 1$) was already shown in (4.8)). $\square$

On the basis of this proposition, we define below the *signature* of an involution σ as the square root of the signature of $T_{(A,\sigma)}$. We also show how the form $T_{(A,\sigma)}$ can be used to determine the discriminant of σ (if σ is of orthogonal type and char $F \neq 2$) or the Brauer class of the discriminant algebra of (A,σ) (if σ is of the second kind).

11.A. Involutions of the first kind. In this section, σ denotes an involution of the first kind on a central simple algebra A over a field F. We set T_σ^+ and T_σ^- for the restrictions of the bilinear trace form $T_{(A,\sigma)}$ to $\operatorname{Sym}(A,\sigma)$ and $\operatorname{Skew}(A,\sigma)$ respectively; thus

$$T_\sigma^+(x,y) = \operatorname{Trd}_A\big(\sigma(x)y\big) = \operatorname{Trd}_A(xy) \qquad \text{for } x,\, y \in \operatorname{Sym}(A,\sigma)$$

$$T_\sigma^-(x,y) = \operatorname{Trd}_A\big(\sigma(x)y\big) = -\operatorname{Trd}_A(xy) \quad \text{for } x,\, y \in \operatorname{Skew}(A,\sigma).$$

Also let T_A denote the symmetric bilinear trace form on A:

$$T_A(x,y) = \operatorname{Trd}_A(xy) \quad \text{for } x,\, y \in A,$$

so that $T_\sigma^+(x,y) = T_A(x,y)$ for x, $y \in \operatorname{Sym}(A,\sigma)$ and $T_\sigma^-(x,y) = -T_A(x,y)$ for x, $y \in \operatorname{Skew}(A,\sigma)$.

(11.2) Lemma. $\operatorname{Alt}(A,\sigma)$ *is the orthogonal space of* $\operatorname{Sym}(A,\sigma)$ *in* A *for each of the bilinear forms* $T_{(A,\sigma)}$ *and* T_A. *Consequently,*
(1) *if* char $F = 2$, *the form* $T_\sigma^+ = T_\sigma^-$ *is singular;*
(2) *if* char $F \neq 2$, *the forms* T_σ^+ *and* T_σ^- *are nonsingular and there are orthogonal sum decompositions*

$$\big(A, T_{(A,\sigma)}\big) = \big(\operatorname{Sym}(A,\sigma), T_\sigma^+\big) \overset{\perp}{\oplus} \big(\operatorname{Skew}(A,\sigma), T_\sigma^-\big),$$

$$\big(A, T_A\big) = \big(\operatorname{Sym}(A,\sigma), T_\sigma^+\big) \overset{\perp}{\oplus} \big(\operatorname{Skew}(A,\sigma), -T_\sigma^-\big).$$

Proof: For $x \in A$ and $y \in \operatorname{Sym}(A,\sigma)$, we have $\operatorname{Trd}_A\big(\sigma(x)y\big) = \operatorname{Trd}_A\big(\sigma(yx)\big) = \operatorname{Trd}_A(xy)$, hence

$$T_A\big(x - \sigma(x), y\big) = \operatorname{Trd}_A(xy) - \operatorname{Trd}_A\big(\sigma(x)y\big) = 0.$$

This shows $\text{Alt}(A, \sigma) \subset \text{Sym}(A, \sigma)^\perp$ (the orthogonal space for the form T_A); the equality $\text{Alt}(A, \sigma) = \text{Sym}(A, \sigma)^\perp$ follows by dimension count. Since

$$T_{(A,\sigma)}\big(x - \sigma(x), y\big) = -T_A\big(x - \sigma(x), y\big) \quad \text{for } x \in A,\ y \in \text{Sym}(A, \sigma),$$

the same arguments show that $\text{Alt}(A, \sigma)$ is the orthogonal space of $\text{Sym}(A, \sigma)$ for the form $T_{(A,\sigma)}$. $\qquad\qquad\qquad\qquad\qquad\qquad\qquad\qquad\qquad\qquad\qquad\qquad\qquad\square$

(11.3) Examples. (1) Quaternion algebras. Let $Q = (a, b)_F$ be a quaternion algebra with canonical involution γ over a field F of characteristic different from 2. Let Q^0 denote the vector space of pure quaternions, so that $Q^0 = \text{Skew}(Q, \gamma)$. A direct computation shows that the elements i, j, k of the usual quaternion basis are orthogonal for $T_{(Q,\gamma)}$, hence T_γ^+ and T_γ^- have the following diagonalizations:

$$T_\gamma^+ = \langle 2 \rangle \quad \text{and} \quad T_\gamma^- = \langle 2 \rangle \cdot \langle -a, -b, ab \rangle.$$

Now, let $\sigma = \text{Int}(i) \circ \gamma$. Then $\text{Skew}(Q, \sigma) = i \cdot F$, and $\text{Sym}(Q, \sigma)$ has 1, j, k as orthogonal basis. Therefore,

$$T_\sigma^+ = \langle 2 \rangle \cdot \langle 1, b, -ab \rangle \quad \text{and} \quad T_\sigma^- = \langle -2a \rangle.$$

(2) Biquaternion algebras. Let $A = (a_1, b_1)_F \otimes_F (a_2, b_2)_F$ be a tensor product of two quaternion F-algebras and $\sigma = \gamma_1 \otimes \gamma_2$, the tensor product of the canonical involutions. Let $(1, i_1, j_1, k_1)$ and $(1, i_2, j_2, k_2)$ denote the usual quaternion bases of $(a_1, b_1)_F$ and $(a_2, b_2)_F$ respectively. The element 1 and the products $\xi \otimes \eta$ where ξ and η independently range over i_1, j_1, k_1, and i_2, j_2, k_2, respectively, form an orthogonal basis of $\text{Sym}(A, \sigma)$ for the form T_σ^+. Similarly, $i_1 \otimes 1$, $j_1 \otimes 1$, $k_1 \otimes 1$, $1 \otimes i_2$, $1 \otimes j_2$, $1 \otimes k_2$ form an orthogonal basis of $\text{Skew}(A, \sigma)$. Therefore,

$$T_\sigma^+ = \langle 1 \rangle \perp \langle a_1, b_1, -a_1 b_1 \rangle \cdot \langle a_2, b_2, -a_2 b_2 \rangle$$

and

$$T_\sigma^- = \langle -a_1, -b_1, a_1 b_1, -a_2, -b_2, a_2 b_2 \rangle.$$

(Note T_σ^- is *not* an Albert form of A as discussed in §16.A, unless $-1 \in F^{\times 2}$).

As a further example, consider the split orthogonal case in characteristic different from 2. If b is a nonsingular symmetric bilinear form on a vector space V, we consider the forms b^{S2} and $b^{\wedge 2}$ defined on the symmetric square $S^2 V$ and the exterior square $\bigwedge^2 V$ respectively by

$$b^{S2}(x_1 \cdot x_2, y_1 \cdot y_2) = b(x_1, y_1)b(x_2, y_2) + b(x_1, y_2)b(x_2, y_1),$$

$$b^{\wedge 2}(x_1 \wedge x_2, y_1 \wedge y_2) = b(x_1, y_1)b(x_2, y_2) - b(x_1, y_2)b(x_2, y_1)$$

for x_1, x_2, y_1, $y_2 \in V$. (The form $b^{\wedge 2}$ has already been considered in §10.D). Assuming char $F \neq 2$, we embed $S^2 V$ and $\bigwedge^2 V$ in $V \otimes V$ by mapping $x_1 \cdot x_2$ to $\frac{1}{2}(x_1 \otimes x_2 + x_2 \otimes x_1)$ and $x_1 \wedge x_2$ to $\frac{1}{2}(x_1 \otimes x_2 - x_2 \otimes x_1)$ for x_1, $x_2 \in V$.

(11.4) Proposition. *Suppose char $F \neq 2$ and let $(A, \sigma) = \big(\text{End}_F(V), \sigma_b\big)$. The standard identification $\varphi_b \colon V \otimes V \xrightarrow{\sim} A$ of (5.2) induces isometries of bilinear spaces*

$$\big(V \otimes V, b \otimes b\big) \xrightarrow{\sim} \big(A, T_{(A,\sigma)}\big),$$

$$\big(S^2 V, \tfrac{1}{2} b^{S2}\big) \xrightarrow{\sim} \big(\text{Sym}(A, \sigma), T_\sigma^+\big),$$

$$\big(\textstyle\bigwedge^2 V, \tfrac{1}{2} b^{\wedge 2}\big) \xrightarrow{\sim} \big(\text{Skew}(A, \sigma), T_\sigma^-\big).$$

Proof: As observed in (5.1), we have $\sigma\big(\varphi_b(x_1 \otimes x_2)\big) = \varphi_b(x_2 \otimes x_1)$ and

$$\mathrm{Trd}_A\big(\varphi_b(x_1 \otimes x_2)\big) = b(x_2, x_1) = b(x_1, x_2) \quad \text{for } x_1,\, x_2 \in V.$$

Therefore,

$$T_{(A,\sigma)}\big(\varphi_b(x_1 \otimes x_2), \varphi_b(y_1 \otimes y_2)\big) = b(x_1, y_1)b(x_2, y_2) \quad \text{for } x_1,\, x_2,\, y_1,\, y_2 \in V,$$

proving the first isometry. The other isometries follow by similar computations. $\square$

Diagonalizations of b^{S2} and $b^{\wedge 2}$ are easily obtained from a diagonalization of b: if $b = \langle \alpha_1, \ldots, \alpha_n \rangle$, then

$$b^{S2} = n\langle 2 \rangle \perp (\perp_{1 \le i < j \le n} \langle \alpha_i \alpha_j \rangle) \quad \text{and} \quad b^{\wedge 2} = \perp_{1 \le i < j \le n} \langle \alpha_i \alpha_j \rangle.$$

Therefore, $\det b^{S2} = 2^n (\det b)^{n-1}$ and $\det b^{\wedge 2} = (\det b)^{n-1}$.

(11.5) Proposition. *Let (A, σ) be a central simple algebra with involution of orthogonal type over a field F of characteristic different from 2. If $\deg A$ is even, then*

$$\det T_\sigma^+ = \det T_\sigma^- = 2^{\deg A/2} \det \sigma.$$

Proof: By extending scalars to a splitting field L in which F is algebraically closed (so that the induced map $F^\times / F^{\times 2} \to L^\times / L^{\times 2}$ is injective), we reduce to considering the case where A is split. If $(A, \sigma) = \big(\mathrm{End}_F(V), \sigma_b\big)$, then $\det \sigma = \det b$ by (7.3), and the computations above, together with (11.4), show that

$$\det T_\sigma^+ = \det T_\sigma^- = 2^{-n(n-1)/2}(\det b)^{n-1} = 2^{n/2}\det b \quad \text{in } F^\times / F^{\times 2},$$

where $n = \deg A$. $\square$

As a final example, we compute the form $T_{(A,\sigma,u)}$ for a quaternion algebra with orthogonal involution. This example is used in §17.A (see (17.2)). In the following statement, we denote by WF the Witt ring of nonsingular bilinear forms over F.

(11.6) Proposition. *Let Q be a quaternion algebra over a field F of arbitrary characteristic, let σ be an orthogonal involution on Q and $v \in \mathrm{Sym}(Q, \sigma) \cap Q^\times$. For all $s \in Q^\times$ such that $\sigma(s) = s = -\gamma(s)$,*

$$T_{(Q,\sigma,v)} \simeq \langle \mathrm{Trd}_Q(v) \rangle \cdot \langle\!\langle \mathrm{Nrd}_Q(vs), \mathrm{disc}\, \sigma \rangle\!\rangle \qquad \text{if } \mathrm{Trd}_Q(v) \ne 0;$$

$$T_{(Q,\sigma,v)} = \langle\!\langle \mathrm{Nrd}_Q(vs), \mathrm{disc}\, \sigma \rangle\!\rangle = 0 \quad \text{in } WF \qquad \text{if } \mathrm{Trd}_Q(v) = 0.$$

Proof: Let γ be the canonical (symplectic) involution on Q and let $u \in \mathrm{Skew}(Q, \gamma) \smallsetminus F$ be such that $\sigma = \mathrm{Int}(u) \circ \gamma$. The discriminant $\mathrm{disc}\, \sigma$ is therefore represented in $F^\times / F^{\times 2}$ by $-\mathrm{Nrd}_Q(u) = u^2$. Since $\sigma(v) = v$, we have $v = u\gamma(v)u^{-1}$, hence $\mathrm{Trd}_Q(vu) = 0$. A computation shows that 1, u are orthogonal for the form $T_{(Q,\sigma,v)}$. Since further $T_{(Q,\sigma,v)}(1,1) = \mathrm{Trd}_Q(v)$ and $T_{(Q,\sigma,v)}(u,u) = \mathrm{Trd}_Q(v)\,\mathrm{Nrd}_Q(u)$, the subspace spanned by 1, u is totally isotropic if $\mathrm{Trd}_Q(v) = 0$, hence $T_{(Q,\sigma,v)}$ is metabolic in this case. If $\mathrm{Trd}_Q(v) \ne 0$, a direct calculation shows that for all $s \in Q^\times$ as above, $\big(1, u, \gamma(v)s, \gamma(v)su\big)$ is an orthogonal basis of Q which yields the diagonalization

$$T_{(Q,\sigma,v)} \simeq \langle \mathrm{Trd}_Q(v) \rangle \cdot \langle 1, \mathrm{Nrd}_Q(u), -\mathrm{Nrd}_Q(vs), -\mathrm{Nrd}_Q(vsu) \rangle.$$

To complete the proof, we observe that if $\mathrm{Trd}_Q(v) = 0$, then v and s both anticommute with u, hence $vs \in F[u]$ and therefore $\mathrm{Nrd}_Q(vs)$ is a norm from $F[u]$; it follows that

$$\langle\!\langle \mathrm{Nrd}_Q(vs), u^2 \rangle\!\rangle = 0 \qquad \text{in } WF. \qquad \square$$

The signature of involutions of the first kind. Assume now that the base field F has an ordering P, so $\mathrm{char}(F) = 0$. (See Scharlau [247, §2.7] for background information on ordered fields.) To every nonsingular symmetric bilinear form b there is classically associated an integer $\mathrm{sgn}_P\, b$ called the *signature* of b at P (or with respect to P): it is the difference $m_+ - m_-$ where m_+ (resp. m_-) is the number of positive (resp. negative) entries in any diagonalization of b.

Our goal is to define the signature of an involution in such a way that in the split case $A = \mathrm{End}_F(V)$, the signature of the adjoint involution with respect to a symmetric bilinear form b is the absolute value of the signature of b:

$$\mathrm{sgn}_P\, \sigma_b = |\mathrm{sgn}_P\, b|\,.$$

(Note that $\sigma_b = \sigma_{-b}$ and $\mathrm{sgn}_P(-b) = -\mathrm{sgn}_P\, b$, so $\mathrm{sgn}_P\, b$ is *not* an invariant of σ_b).

(11.7) Proposition. *For any involution σ of the first kind on A, the signature of the bilinear form $T_{(A,\sigma)}$ at P is a square in $\mathbb{Z}$. If A is split: $A = \mathrm{End}_F(V)$ and $\sigma = \sigma_b$ is the adjoint involution with respect to some nonsingular bilinear form b on V, then*

$$\mathrm{sgn}_P\, T_{(A,\sigma)} = \begin{cases} (\mathrm{sgn}_P\, b)^2 & \text{if } \sigma \text{ is orthogonal,} \\ 0 & \text{if } \sigma \text{ is symplectic.} \end{cases}$$

Proof: When A is split and σ is orthogonal, (11.4) yields an isometry $T_{(A,\sigma)} \simeq b \otimes b$ from which the formula for $\mathrm{sgn}_P\, T_{(A,\sigma)}$ follows. When A is split and σ is symplectic, we may find an isomorphism $(A, \sigma) \simeq \bigl(\mathrm{End}_F(V), \sigma_b\bigr)$ for some vector space V and some nonsingular skew-symmetric form b on V. The same argument as in (11.4) yields an isometry $(A, T_{(A,\sigma)}) \simeq (V \otimes V, b \otimes b)$. In this case, $b \otimes b$ is hyperbolic. Indeed, if $U \subset V$ is a maximal isotropic subspace for b, then $\dim U = \frac{1}{2}\dim V$ and $U \otimes V$ is an isotropic subspace of $V \otimes V$ of dimension $\frac{1}{2}\dim(V \otimes V)$. Therefore, $\mathrm{sgn}_P\, T_{(A,\sigma)} = 0$.

In the general case, consider a real closure F_P of F for the ordering P. Since the signature at P of a symmetric bilinear form over F does not change under scalar extension to F_P, we may assume $F = F_P$. The Brauer group of F then has order 2, the nontrivial element being represented by the quaternion algebra $Q = (-1, -1)_F$. Since the case where A is split has already been considered, we may assume for the rest of the proof that A is Brauer-equivalent to Q. According to (4.2), we then have

$$(A, \sigma) \simeq \bigl(\mathrm{End}_Q(V), \sigma_h\bigr)$$

for some (right) vector space V over Q, and some nonsingular form h on V, which is hermitian with respect to the canonical involution γ on Q if σ is symplectic, and skew-hermitian with respect to γ if σ is orthogonal.

Let $(e_i)_{1 \le i \le n}$ be an orthogonal basis of V with respect to h, and let

$$h(e_i, e_i) = q_i \in Q^\times \quad \text{for } i = 1, \ldots, n.$$

Thus $q_i \in F$ if σ is symplectic and q_i is a pure quaternion if σ is orthogonal. For $i, j = 1, \ldots, n$, write $E_{ij} \in \mathrm{End}_Q(V)$ for the endomorphism which maps e_j to e_i and maps e_k to 0 if $k \ne j$. Thus E_{ij} corresponds to the matrix unit e_{ij} under the isomorphism $\mathrm{End}_Q(V) \simeq M_n(Q)$ induced by the choice of the basis $(e_i)_{1 \le i \le n}$.

A direct verification shows that for $i, j = 1, \ldots, n$ and $q \in Q$,

$$\sigma(E_{ij}q) = E_{ji}q_j^{-1}\gamma(q)q_i.$$

Therefore, for i, j, k, $\ell = 1, \ldots, n$ and q, $q' \in Q$,

$$T_{(A,\sigma)}(E_{ij}q, E_{k\ell}q') = \begin{cases} 0 & \text{if } i \neq k \text{ or } j \neq \ell, \\ \mathrm{Trd}_Q\big(q_j^{-1}\gamma(q)q_iq'\big) & \text{if } i = k \text{ and } j = \ell. \end{cases}$$

We thus have an orthogonal decomposition of $\mathrm{End}_Q(V)$ with respect to the form $T_{(A,\sigma)}$:

(11.8) $$\mathrm{End}_Q(V) = \perp_{1 \leq i,j \leq n} E_{ij} \cdot Q.$$

Suppose first that σ is orthogonal, so h is skew-hermitian and q_i is a pure quaternion for $i = 1, \ldots, n$. Fix a pair of indices (i, j). If q_iq_j is a pure quaternion, then E_{ij} and $E_{ij}q_i$ span an isotropic subspace of $E_{ij} \cdot Q$, so $E_{ij} \cdot Q$ is hyperbolic. If q_iq_j is not pure, pick a nonzero pure quaternion $h \in Q$ which anticommutes with $q_jq_iq_j^{-1}$. Since $Q = (-1,-1)_F$ and F is real-closed, the square of every nonzero pure quaternion lies in $-F^{\times 2}$. For $i = 1, \ldots, n$, let $q_i^2 = -\alpha_i^2$ for some $\alpha_i \in F^\times$; let also $h^2 = -\beta^2$ with $\beta \in F$. Then $E_{ij}(\alpha_j\beta + q_jh)$ and $E_{ij}q_i(\alpha_j\beta + q_jh)$ span a 2-dimensional isotropic subspace of $E_{ij} \cdot Q$, so again $E_{ij} \cdot Q$ is hyperbolic. We have thus shown that the form $T_{(A,\sigma)}$ is hyperbolic on $\mathrm{End}_Q(V)$ when σ is orthogonal, hence $\mathrm{sgn}_P T_{(A,\sigma)} = 0$ in this case.

If σ is symplectic, then $q_i \in F^\times$ for all $i = 1, \ldots, n$, hence

$$T_{(A,\sigma)}(E_{ij}q, E_{ij}q) = \mathrm{Trd}_Q\big(\gamma(q)q\big)q_j^{-1}q_i = 2\,\mathrm{Nrd}_Q(q)q_j^{-1}q_i$$

for all i, $j = 1, \ldots, n$. From (11.8) it follows that

$$T_{(A,\sigma)} \simeq \langle 2 \rangle \cdot N_Q \cdot \langle q_1, \ldots, q_n \rangle \cdot \langle q_1, \ldots, q_n \rangle$$

where N_Q is the reduced norm form of Q. Since $Q = (-1,-1)_F$, we have $N_Q \simeq 4\langle 1 \rangle$, hence the preceding relation yields

$$\mathrm{sgn}_P T_{(A,\sigma)} = 4\big(\mathrm{sgn}_P\langle q_1, \ldots, q_n \rangle\big)^2. \qquad \square$$

(11.9) Remark. In the last case, the signature of the F-quadratic form $\langle q_1, \ldots, q_n \rangle$ is an invariant of the hermitian form h on V: indeed, the form h induces a quadratic form h_F on V, regarded as an F-vector space, by

$$h_F(x) = h(x, x) \in F,$$

since h is hermitian. Then

$$h_F \simeq 4\langle q_1, \ldots, q_n \rangle,$$

so $\mathrm{sgn}_P h_F = 4\,\mathrm{sgn}_P\langle q_1, \ldots, q_n \rangle$. Let

$$\mathrm{sgn}_P h = \mathrm{sgn}_P\langle q_1, \ldots, q_n \rangle.$$

The last step in the proof of (11.7) thus shows that if $A = \mathrm{End}_Q(V)$ and $\sigma = \sigma_h$ for some hermitian form h on V (with respect to the canonical involution on Q), then

$$\mathrm{sgn}_P T_{(A,\sigma)} = 4(\mathrm{sgn}_P h)^2.$$

(11.10) Definition. The *signature* at P of an involution σ of the first kind on A is defined by

$$\mathrm{sgn}_P \sigma = \sqrt{\mathrm{sgn}_P T_{(A,\sigma)}}.$$

By (11.7), $\operatorname{sgn}_P \sigma$ is an integer. Since $\operatorname{sgn}_P T_{(A,\sigma)} \leq \dim A$ and $\operatorname{sgn}_P T_{(A,\sigma)} \equiv \dim T_{(A,\sigma)} \mod 2$, we have

$$0 \leq \operatorname{sgn}_P \sigma \leq \deg A \quad \text{and} \quad \operatorname{sgn}_P \sigma \equiv \deg A \mod 2.$$

From (11.7), we further derive:

(11.11) Corollary. *Let F_P be a real closure of F for the ordering P.*
(1) *Suppose A is not split by F_P;*
   (a) *if σ is orthogonal, then $\operatorname{sgn}_P \sigma = 0$;*
   (b) *if σ is symplectic and $\sigma \otimes \operatorname{Id}_{F_P} = \sigma_h$ for some hermitian form h over the quaternion division algebra over F_P, then $\operatorname{sgn}_P \sigma = 2\,|\operatorname{sgn}_P h|$.*
(2) *Suppose A is split by F_P;*
   (a) *if σ is orthogonal and $\sigma \otimes \operatorname{Id}_{F_P} = \sigma_b$ for some symmetric bilinear form b over F_P, then $\operatorname{sgn}_P \sigma = |\operatorname{sgn}_P b|$;*
   (b) *if σ is symplectic, then $\operatorname{sgn}_P \sigma = 0$.*

11.B. Involutions of the second kind. In this section we consider the case of central simple algebras with involution of the second kind (B, τ) over an arbitrary field F. Let K be the center of B and ι the nontrivial automorphism of K over F. The form $T_{(B,\tau)}$ is hermitian with respect to ι. Let T_τ be its restriction to the space of symmetric elements. Thus,

$$T_\tau \colon \operatorname{Sym}(B,\tau) \times \operatorname{Sym}(B,\tau) \to F$$

is a symmetric bilinear form defined by

$$T_\tau(x,y) = \operatorname{Trd}_B\big(\tau(x)y\big) = \operatorname{Trd}_B(xy) \quad \text{for } x,\, y \in \operatorname{Sym}(B,\tau).$$

Since multiplication in B yields a canonical isomorphism of K-vector spaces

$$B = \operatorname{Sym}(B,\tau) \otimes_F K,$$

the hermitian form $T_{(B,\tau)}$ can be recaptured from the bilinear form T_τ:

$$T_{(B,\tau)}\big(\textstyle\sum_i x_i \alpha_i, \sum_j y_j \beta_j\big) = \textstyle\sum_{i,j} \iota(\alpha_i) T_\tau(x_i, y_j)\beta_j$$

for $x_i,\, y_j \in \operatorname{Sym}(B,\tau)$ and $\alpha_i,\, \beta_j \in K$. Therefore, the form T_τ is nonsingular. Moreover, there is no loss of information if we focus on the bilinear form T_τ instead of the hermitian form $T_{(B,\tau)}$.

(11.12) Examples. (1) Quaternion algebras. Suppose $\operatorname{char} F \neq 2$ and let $Q_0 = (a,b)_F$ be a quaternion algebra over F, with canonical involution γ_0. Define an involution τ of the second kind on $Q = Q_0 \otimes_F K$ by $\tau = \gamma_0 \otimes \iota$. (According to (2.22), every involution of the second kind on a quaternion K-algebra is of this type for a suitable quaternion F-subalgebra). Let $K \simeq F[X]/(X^2 - \alpha)$ and let $z \in K$ satisfy $z^2 = \alpha$ (and $\iota(z) = -z$). If $(1, i, j, k)$ is the usual quaternion basis of Q_0, the elements 1, iz, jz, kz form an orthogonal basis of $\operatorname{Sym}(Q,\tau)$ with respect to T_τ, hence

$$T_\tau = \langle 2 \rangle \cdot \langle 1, a\alpha, b\alpha, -ab\alpha \rangle.$$

If $\operatorname{char} F = 2$, $Q_0 = [a,b)_F$ and $K = F[X]/(X^2 + X + \alpha)$, let $(1, i, j, k)$ be the usual quaternion basis of Q_0 and let $z \in K$ be an element such that $z^2 + z = \alpha$ and $\iota(z) = z + 1$. A computation shows that the elements $z + i$, $1 + z + i + j$, $1 + z + i + kb^{-1}$ and $1 + z + i + j + kb^{-1}$ form an orthogonal basis of $\operatorname{Sym}(B,\tau)$ for the form T_τ, with respect to which T_τ has the diagonalization

$$T_\tau = \langle 1, 1, 1, 1 \rangle.$$

(2) *Exchange involution.* Suppose $(B, \tau) = (E \times E^{\mathrm{op}}, \varepsilon)$ where ε is the exchange involution:

$$\varepsilon(x, y^{\mathrm{op}}) = (y, x^{\mathrm{op}}) \quad \text{for } x, y \in E.$$

The space of symmetric elements is canonically isomorphic to E:

$$\mathrm{Sym}(B, \tau) = \{\, (x, x^{\mathrm{op}}) \mid x \in E \,\} = E$$

and since $\mathrm{Trd}_B(x, y^{\mathrm{op}}) = \big(\mathrm{Trd}_E(x), \mathrm{Trd}_E(y)\big)$, the form T_τ is canonically isometric to the reduced trace bilinear form on E:

$$T_\tau\big((x, x^{\mathrm{op}}), (y, y^{\mathrm{op}})\big) = \mathrm{Trd}_E(xy) = T_E(x, y) \quad \text{for } x, y \in E.$$

As a further example, we consider the case of split algebras. Let V be a (finite dimensional) K-vector space with a nonsingular hermitian form h. Define a K-vector space ${}^\iota V$ by

$$ {}^\iota V = \{\, {}^\iota v \mid v \in V \,\}$$

with the operations

$$ {}^\iota v + {}^\iota w = {}^\iota(v + w) \quad ({}^\iota v)\alpha = {}^\iota\big(v\iota(\alpha)\big) \quad \text{for } v, w \in V, \ \alpha \in K.$$

(Compare with §3.B and §5.A). The hermitian form h induces on the vector space $V \otimes_K {}^\iota V$ a nonsingular hermitian form $h \otimes {}^\iota h$ defined by

$$(h \otimes {}^\iota h)(v_1 \otimes {}^\iota v_2, w_1 \otimes {}^\iota w_2) = h(v_1, w_1)\iota\big(h(v_2, w_2)\big) \quad \text{for } v_1, v_2, w_1, w_2 \in V.$$

Let $s \colon V \otimes_K {}^\iota V \to V \otimes_K {}^\iota V$ be the switch map

$$s(v_1 \otimes {}^\iota v_2) = v_2 \otimes {}^\iota v_1 \quad \text{for } v_1, v_2 \in V.$$

The *norm* of V is then defined as the F-vector space of s-invariant elements (see (3.12)):

$$N_{K/F}(V) = \{\, x \in V \otimes_K {}^\iota V \mid s(x) = x \,\}.$$

Since $(h \otimes {}^\iota h)(v_2 \otimes {}^\iota v_1, w_2 \otimes {}^\iota w_1) = \iota\big((h \otimes {}^\iota h)(v_1 \otimes {}^\iota v_2, w_1 \otimes {}^\iota w_2)\big)$, it follows that

$$(h \otimes {}^\iota h)\big(s(x), s(y)\big) = \iota\big((h \otimes {}^\iota h)(x, y)\big) \quad \text{for } x, y \in V \otimes_K {}^\iota V.$$

Therefore, the restriction of the form $h \otimes {}^\iota h$ to the F-vector space $N_{K/F}(V)$ is a symmetric bilinear form

$$N_{K/F}(h) \colon N_{K/F}(V) \times N_{K/F}(V) \to F.$$

The following proposition follows by straightforward computation:

(11.13) Proposition. *Let $z \in K \setminus F$ and let $(e_i)_{1 \le i \le n}$ be an orthogonal K-basis of V. For $i, j = 1, \ldots, n$, let $V_i = (e_i \otimes e_i) \cdot F \subset V \otimes_K {}^\iota V$ and let*

$$V_{ij} = (e_i \otimes {}^\iota e_j + e_j \otimes {}^\iota e_i) \cdot F \oplus \big(e_i z \otimes {}^\iota e_j + e_j \iota(z) \otimes {}^\iota e_i\big) \cdot F \subset V \otimes_K {}^\iota V.$$

There is an orthogonal decomposition of $N_{K/F}(V)$ for the bilinear form $N_{K/F}(h)$:

$$N_{K/F}(V) = \Big(\overset{\perp}{\underset{1 \le i \le n}{\bigoplus}} V_i \Big) \overset{\perp}{\oplus} \Big(\overset{\perp}{\underset{1 \le i < j \le n}{\bigoplus}} V_{ij} \Big).$$

Moreover, $V_i \simeq \langle 1 \rangle$ for all i. If $\operatorname{char} F = 2$, then V_{ij} is hyperbolic; if $\operatorname{char} F \ne 2$, then if $K \simeq F[X]/(X^2 - \alpha)$ we have

$$V_{ij} \simeq \big\langle 2h(e_i, e_i)h(e_j, e_j) \big\rangle \cdot \langle 1, -\alpha \rangle.$$

Therefore, letting $\delta_i = h(e_i, e_i)$ for $i = 1, \ldots, n$,

$$N_{K/F}(h) \simeq \begin{cases} n^2\langle 1 \rangle & \text{if char } F = 2, \\ n\langle 1 \rangle \perp \langle 2 \rangle \cdot \langle 1, -\alpha \rangle \cdot \left(\perp_{1 \leq i < j \leq n} \langle \delta_i \delta_j \rangle \right) & \text{if char } F \neq 2. \end{cases}$$

Consider now the algebra $B = \operatorname{End}_K(V)$ with the adjoint involution $\tau = \sigma_h$ with respect to h.

(11.14) Proposition. *The standard identification $\varphi_h \colon V \otimes_K {}^\iota V \xrightarrow{\sim} B$ of (5.1) is an isometry of hermitian spaces*

$$(V \otimes_F {}^\iota V, h \otimes {}^\iota h) \xrightarrow{\sim} (B, T_{(B,\tau)})$$

and induces an isometry of bilinear spaces

$$\big(N_{K/F}(V), N_{K/F}(h)\big) \xrightarrow{\sim} \big(\operatorname{Sym}(B, \tau), T_\tau\big).$$

Proof: For $x = \varphi_h(v_1 \otimes {}^\iota v_2)$ and $y = \varphi_h(w_1 \otimes {}^\iota w_2) \in B$,

$$T_{(B,\tau)}(x, y) = \operatorname{Trd}_B\big(\varphi_h(v_2 \otimes {}^\iota v_1) \circ \varphi_h(w_1 \otimes {}^\iota w_2)\big) = h(v_1, w_1)\iota\big(h(v_2, w_2)\big),$$

hence

$$T_{(B,\tau)}\big(\varphi_h(\xi), \varphi_h(\eta)\big) = (h \otimes {}^\iota h)(\xi, \eta) \quad \text{for } \xi, \eta \in V \otimes {}^\iota V.$$

Therefore, the standard identification is an isometry

$$\varphi_h \colon (V \otimes {}^\iota V, h \otimes {}^\iota h) \xrightarrow{\sim} (B, T_{(B,\tau)}).$$

Since the involution τ corresponds to the switch map s, this isometry restricts to an isometry between the F-subspaces of invariant elements under s on the one hand and under τ on the other. $\qquad \square$

(11.15) Remark. For $u \in \operatorname{Sym}(B, \tau) \cap B^\times$, the form $h_u(x, y) = h\big(u(x), y\big)$ on V is hermitian with respect to the involution $\tau_u = \operatorname{Int}(u^{-1}) \circ \tau$. The same computation as above shows that φ_h is an isometry of hermitian spaces

$$(V \otimes {}^\iota V, h_u \otimes {}^\iota h) \xrightarrow{\sim} (B, T_{(B,\tau,u)})$$

where $(h_u \otimes {}^\iota h)(v_1 \otimes {}^\iota v_2, w_1 \otimes {}^\iota w_2) = h_u(v_1, w_1)\iota\big(h(v_2, w_2)\big)$ for $v_1, v_2, w_1, w_2 \in V$. In particular, since the Gram matrix of h_u with respect to any basis of V is the product of the Gram matrix of h by the matrix of u, it follows that $\det T_{(B,\tau,u)} = (\det u)^{\dim V} \det(h \otimes {}^\iota h)$, hence

$$\det T_{(B,\tau,u)} = (\det u)^{\dim V} \cdot N(K/F) \in F^\times / N(K/F).$$

(11.16) Corollary. *Let (B, τ) be a central simple algebra of degree n with involution of the second kind over F. Let K be the center of B.*

(1) *The determinant of the bilinear form T_τ is given by*

$$\det T_\tau = \begin{cases} 1 \cdot F^{\times 2} & \text{if char } F = 2, \\ (-\alpha)^{n(n-1)/2} \cdot F^{\times 2} & \text{if char } F \neq 2 \text{ and } K \simeq F[X]/(X^2 - \alpha). \end{cases}$$

(2) *For $u \in \operatorname{Sym}(B, \tau) \cap B^\times$, the determinant of the hermitian form $T_{(B,\tau,u)}$ is*

$$\det T_{(B,\tau,u)} = \operatorname{Nrd}_B(u)^{\deg B} \cdot N(K/F) \in F^\times / N(K/F).$$

Proof: (1) As in (11.5), the idea is to extend scalars to a splitting field L of B in which F is algebraically closed, and to conclude by (11.14). The existence of such a splitting field has already been observed in (10.36): we may take for L the function field of the (Weil) transfer of the Severi-Brauer variety of B if K is a field, or the function field of the Severi-Brauer variety of E if $B \simeq E \times E^{\mathrm{op}}$.

(2) For the same splitting field L as above, the extension of scalars map $\mathrm{Br}(F) \to \mathrm{Br}(L)$ is injective, by Merkurjev-Tignol [196, Corollary 2.12] (see the proof of (10.36)). Therefore, the quaternion algebra

$$\big(K, \det T_{(B,\tau,u)} \operatorname{Nrd}_B(u)^{\deg B}\big)_F$$

splits, since Remark (11.15) shows that it splits over L. $\qquad\qquad\square$

The same reduction to the split case may be used to relate the form T_τ to the discriminant algebra $D(B,\tau)$, which is defined when the degree of B is even. In the next proposition, we assume char $F \neq 2$, so that the bilinear form T_τ defines a nonsingular quadratic form

$$Q_\tau \colon \operatorname{Sym}(B,\tau) \to F$$

by

$$Q_\tau(x) = T_\tau(x,x) \quad \text{for } x \in \operatorname{Sym}(B,\tau).$$

(11.17) Proposition. *Let (B,τ) be a central simple algebra with involution of the second kind over a field F of characteristic different from 2, and let K be the center of B, say $K \simeq F[X]/(X^2 - \alpha)$. Assume that the degree of (B,τ) is even: $\deg(B,\tau) = n = 2m$. Then the (full) Clifford algebra of the quadratic space $\big(\operatorname{Sym}(B,\tau), Q_\tau\big)$ and the discriminant algebra $D(B,\tau)$ are related as follows:*

$$C\big(\operatorname{Sym}(B,\tau), Q_\tau\big) \sim D(B,\tau) \otimes_F \big(-\alpha, 2^m(-1)^{m(m-1)/2}\big)_F,$$

where $\sim$ is Brauer-equivalence.

Proof: Suppose first that K is a field. By extending scalars to the function field L of the transfer of the Severi-Brauer variety of B, we reduce to considering the split case. For, L splits B and the scalar extension map $\mathrm{Br}(F) \to \mathrm{Br}(L)$ is injective, as observed in (10.36).

We may thus assume that B is split: let $B = \operatorname{End}_K(V)$ and $\tau = \sigma_h$ for some nonsingular hermitian form h on V. If $(e_i)_{1 \leq i \leq n}$ is an orthogonal basis of V and $h(e_i, e_i) = \delta_i$ for $i = 1, \ldots, n$, then (10.33) yields

(11.18) $\qquad D(B,\tau) \sim \big(\alpha, (-1)^{n(n-1)/2} d\big)_F = \big(\alpha, (-1)^m d\big)_F$

where we have set $d = \delta_1 \ldots \delta_n$. On the other hand, (11.14) and (11.13) yield

$$Q_\tau \simeq n\langle 1 \rangle \perp \langle 2 \rangle \cdot \langle 1, -\alpha \rangle \cdot q,$$

where $q = \perp_{1 \leq i < j \leq n} \langle \delta_i \delta_j \rangle$. From known formulas for the Clifford algebra of a direct sum (see for instance Lam [169, Chapter 5, §2]), it follows that

(11.19) $\qquad C(Q_\tau) \simeq C\big(n\langle 1 \rangle\big) \otimes_F C\big(\langle 2(-1)^m \rangle \cdot \langle 1, -\alpha \rangle \cdot q\big).$

Let IF be the fundamental ideal of even-dimensional forms in the Witt ring WF and let $I^n F = (IF)^n$. Let $d' \in F^\times$ be a representative of $\operatorname{disc}(q)$. Since $n = 2m$, we have

$$d' \equiv (-1)^{m(m-1)/2} d \mod F^{\times 2}.$$

hence

$$q \equiv \begin{cases} \langle 1, -d' \rangle & \mod I^2 F \quad \text{if } m \text{ is even,} \\ \langle d' \rangle & \mod I^2 F \quad \text{if } m \text{ is odd.} \end{cases}$$

Therefore, the form $\langle 2(-1)^m \rangle \cdot \langle 1, -\alpha \rangle \cdot q$ is congruent modulo $I^3 F$ to

$$\begin{cases} \langle 1, -\alpha \rangle \cdot \langle 1, -d' \rangle & \text{if } m \text{ is even,} \\ \langle 1, -\alpha \rangle \cdot \langle -2d' \rangle & \text{if } m \text{ is odd.} \end{cases}$$

Since quadratic forms which are congruent modulo $I^3 F$ have Brauer-equivalent Clifford algebras (see Lam [169, Chapter 5, Cor. 3.4]) it follows that

$$C\left(\langle 2(-1)^m \rangle \cdot \langle 1, -\alpha \rangle \cdot q\right) \sim \begin{cases} (\alpha, d')_F & \text{if } m \text{ is even,} \\ (\alpha, -2d')_F & \text{if } m \text{ is odd.} \end{cases}$$

On the other hand,

$$C\left(n\langle 1 \rangle\right) \sim (-1, -1)_F^{\otimes m(m-1)/2},$$

hence the required equivalence follows from (11.18) and (11.19).

To complete the proof, consider the case where $K \simeq F \times F$. Then, there is a central simple F-algebra E of degree $n = 2m$ such that $(B, \tau) \simeq (E \times E^{\mathrm{op}}, \varepsilon)$ where ε is the exchange involution. As we observed in (11.12), we then have $\left(\mathrm{Sym}(B, \tau), Q_\tau\right) \simeq (E, Q_E)$ where $Q_E(x) = \mathrm{Trd}_E(x^2)$ for $x \in E$. Moreover, $D(B, \tau) \simeq \lambda^m E \sim E^{\otimes m}$. Since $\alpha \in F^{\times 2}$ and $(-1, 2)_F$ is split, the proposition reduces to

$$C(E, Q_E) \sim E^{\otimes m} \otimes_F (-1, -1)_F^{\otimes m(m-1)/2}.$$

This formula has been proved by Saltman (unpublished), Serre [257, Annexe, p. 167], Lewis-Morales [173] and Tignol [285].   □

Algebras of odd degree. When the degree of B is odd, no discriminant of (B, τ) is defined. However, we may use the fact that B is split by a scalar extension of odd degree, together with Springer's theorem on the behavior of quadratic forms under odd-degree extensions, to get some information on the form T_τ. Since the arguments rely on Springer's theorem, we need to assume char $F \neq 2$ in this section. We may therefore argue in terms of quadratic forms instead of symmetric bilinear forms, associating to the bilinear form T_τ the quadratic form $Q_\tau(x) = T_\tau(x, x)$.

(11.20) Lemma. *Suppose* char $F \neq 2$. *Let* L/F *be a field extension of odd degree and let* q *be a quadratic form over* F. *Let* q_L *be the quadratic form over* L *derived from* q *by extending scalars to* L, *and let* $\alpha \in F^\times \setminus F^{\times 2}$. *If* $q_L \simeq \langle 1, -\alpha \rangle \cdot h$ *for some quadratic form* h *over* L, *of determinant* 1, *then there is a quadratic form* t *of determinant* 1 *over* F *such that*

$$q \simeq \langle 1, -\alpha \rangle \cdot t.$$

Proof: Let $K = F(\sqrt{\alpha})$ and $M = L \cdot K = L(\sqrt{\alpha})$. Let q_{an} be an anisotropic form over F which is Witt-equivalent to q. The form $(q_{\mathrm{an}})_M$ is Witt-equivalent to the form $\left(\langle 1, -\alpha \rangle \cdot h\right)_M$, hence it is hyperbolic. Since the field extension M/K has odd degree, Springer's theorem on the behavior of quadratic forms under field extensions of odd degree (see Scharlau [247, Theorem 2.5.3]) shows that $(q_{\mathrm{an}})_K$ is hyperbolic, hence, by Scharlau [247, Remark 2.5.11],

$$q_{\mathrm{an}} = \langle 1, -\alpha \rangle \cdot t_0$$

for some quadratic form t_0 over F. Let $\dim q = 2d$, so that $\dim h = d$, and let w be the Witt index of q, so that

(11.21) $q \simeq w\mathbb{H} \perp \langle 1, -\alpha \rangle \cdot t_0,$

where $\mathbb{H}$ is the hyperbolic plane. We then have $\dim t_0 = d - w$, hence

$$\det q = (-1)^w (-\alpha)^{d-w} \cdot F^{\times 2} \in F^{\times}/F^{\times 2}.$$

On the other hand, the relation $q_L \simeq \langle 1, -\alpha \rangle \cdot h$ yields

$$\det q_L = (-\alpha)^d \cdot L^{\times 2} \in L^{\times}/L^{\times 2}.$$

Therefore, $\alpha^w \in F^{\times}$ becomes a square in L; since the degree of L/F is odd, this implies that $\alpha^w \in F^{\times 2}$, hence w is even. Letting $t_1 = \frac{w}{2}\mathbb{H} \perp t_0$, we then derive from (11.21):

$$q \simeq \langle 1, -\alpha \rangle \cdot t_1.$$

It remains to prove that we may modify t_1 so as to satisfy the determinant condition. Since $\dim t_1 = d$, we have

$$t_1 \equiv \begin{cases} \langle 1, -(-1)^{d(d-1)/2} \det t_1 \rangle & \mod I^2 F \quad \text{if } d \text{ is even,} \\ \langle (-1)^{d(d-1)/2} \det t_1 \rangle & \mod I^2 F \quad \text{if } d \text{ is odd.} \end{cases}$$

We may use these relations to compute the Clifford algebra of $q \simeq \langle 1, -\alpha \rangle \cdot t_1$ (up to Brauer-equivalence): in each case we get the same quaternion algebra:

$$C(q) \sim \big(\alpha, (-1)^{d(d-1)/2} \det t_1\big)_F.$$

On the other hand, since $\det h = 1$ we derive from $q_L \simeq \langle 1, -\alpha \rangle \cdot h$:

$$C(q_L) \sim \big(\alpha, (-1)^{d(d-1)/2}\big)_L.$$

It follows that the quaternion algebra $(\alpha, \det t_1)_F$ is split, since it splits over the extension L/F of odd degree. Therefore, if $\delta \in F^{\times}$ is a representative of $\det t_1 \in F^{\times}/F^{\times 2}$, we have $\delta \in N(K/F)$. Let $\beta \in F^{\times}$ be a represented value of t_1, so that $t_1 \simeq t_2 \perp \langle \beta \rangle$ for some quadratic form t_2 over F, and let $t = t_2 \perp \langle \delta\beta \rangle$. Then

$$\det t = \delta \cdot \det t_1 = 1.$$

On the other hand, since δ is a norm from the extension K/F we have $\langle 1, -\alpha \rangle \cdot \langle \delta\beta \rangle \simeq \langle 1, -\alpha \rangle \cdot \langle \beta \rangle$, hence

$$\langle 1, -\alpha \rangle \cdot t \simeq \langle 1, -\alpha \rangle \cdot t_1 \simeq q. \qquad \square$$

(11.22) Proposition. *Let B be a central simple K-algebra of odd degree $n = 2m - 1$ with an involution τ of the second kind. Then, there is a quadratic form q_τ of dimension $n(n-1)/2$ and determinant 1 over F such that*

$$Q_\tau \simeq n\langle 1 \rangle \perp \langle 2 \rangle \cdot \langle 1, -\alpha \rangle \cdot q_\tau.$$

Proof: Suppose first $K = F \times F$. We may then assume $(B, \tau) = (E \times E^{\mathrm{op}}, \varepsilon)$ where ε is the exchange involution. In that case $Q_\tau \simeq Q_E$ where $Q_E(x) = \mathrm{Trd}_E(x^2)$, as observed in (11.12). Since $\alpha \in F^{\times 2}$, we have to show that this quadratic form is Witt-equivalent to $n\langle 1 \rangle$. By Springer's theorem, it suffices to prove this relation over an odd-degree field extension. Since the degree of E is odd, we may therefore assume E is split: $E = M_n(F)$. In that case, the relation is easy to check. (Observe that the upper-triangular matrices with zero diagonal form a totally isotropic subspace).

For the rest of the proof, we may thus assume K is a field. Let D be a division K-algebra Brauer-equivalent to B and let θ be an involution of the second kind on D. Let L be a field contained in $\operatorname{Sym}(D, \theta)$ and maximal for this property. The field $M = L \cdot K$ is then a maximal subfield of D, since otherwise the centralizer $C_D M$ contains a symmetric element outside M, contradicting the maximality of L. We have $[L : F] = [M : K] = \deg D$, hence the degree of L/F is odd, since D is Brauer-equivalent to the algebra B of odd degree. Moreover, the algebra $B \otimes_F L = B \otimes_K M$ splits, since M is a maximal subfield of D. By (11.14) and (11.13) the quadratic form $(Q_\tau)_L$ obtained from Q_τ by scalar extension to L has the form

(11.23) $$(Q_\tau)_L \simeq n\langle 1 \rangle \perp \langle 2 \rangle \cdot \langle 1, -\alpha \rangle \cdot h$$

where $h = \perp_{1 \leq i < j \leq n} \langle a_i a_j \rangle$ for some $a_1, \ldots, a_n \in L^\times$. Therefore, the Witt index of the form $(Q_\tau)_L \perp n\langle -1 \rangle$ is at least n:

$$w\big((Q_\tau)_L \perp n\langle -1 \rangle\big) \geq n.$$

By Springer's theorem the Witt index of a form does not change under an odd-degree scalar extension. Therefore,

$$w\big(Q_\tau \perp n\langle -1 \rangle\big) \geq n,$$

and it follows that Q_τ contains a subform isometric to $n\langle 1 \rangle$. Let

$$Q_\tau \simeq n\langle 1 \rangle \perp q$$

for some quadratic form q over F. Relation (11.23) shows that

$$(q)_L \simeq \langle 2 \rangle \cdot \langle 1, -\alpha \rangle \cdot h.$$

Since $\det h = 1$, we may apply (11.20) to the quadratic form $\langle 2 \rangle \cdot q$, obtaining a quadratic form q_τ over F, of determinant 1, such that $\langle 2 \rangle \cdot q \simeq \langle 1, -\alpha \rangle \cdot q_\tau$; hence

$$Q_\tau \simeq n\langle 1 \rangle \perp \langle 2 \rangle \cdot \langle 1, -\alpha \rangle \cdot q_\tau. \qquad \square$$

In the case where $n = 3$, we show in Chapter V that the form q_τ classifies the involutions τ on a given central simple algebra B.

The signature of involutions of the second kind. Suppose that P is an ordering of F which does *not* extend to K; this means that $K = F(\sqrt{\alpha})$ for some $\alpha < 0$. If (V, h) is a hermitian space over K (with respect to ι), the signature $\operatorname{sgn}_P h$ may be defined just as in the case of quadratic spaces (see Scharlau [247, Examples 10.1.6]). Indeed, we may view V as an F-vector space and define a quadratic form $h_0 \colon V \to F$ by

$$h_0(x) = h(x, x) \quad \text{for } x \in V,$$

since h is hermitian. If $(e_i)_{1 \leq i \leq n}$ is an orthogonal K-basis of V for h and $z \in F$ is such that $z^2 = \alpha$, then $(e_i, e_i z)_{1 \leq i \leq n}$ is an orthogonal F-basis of V for h_0. Therefore, if $h(e_i, e_i) = \delta_i$, then

$$h_0 = \langle 1, -\alpha \rangle \cdot \langle \delta_1, \ldots, \delta_n \rangle,$$

hence the signature of the F-quadratic form $\langle \delta_1, \ldots, \delta_n \rangle$ is an invariant for h, equal to $\frac{1}{2} \operatorname{sgn}_P h_0$. We let

$$\operatorname{sgn}_P h = \tfrac{1}{2} \operatorname{sgn}_P h_0.$$

Note that if $\alpha > 0$, then $\operatorname{sgn}_P h_0 = 0$. This explains why the signature is meaningful only when $\alpha < 0$.

(11.24) Proposition. *Let (B,τ) be a central simple algebra with involution of the second kind over F, with center K. Then, the signature of the hermitian form $T_{(B,\tau)}$ on B is the square of an integer. Moreover, if B is split: $B = \mathrm{End}_K(V)$ and $\tau = \sigma_h$ for some hermitian form h on V, then*

$$\mathrm{sgn}_P T_{(B,\tau)} = (\mathrm{sgn}_P h)^2.$$

Proof: If B is split, (11.14) yields an isometry

$$(B, T_{(B,\tau)}) \simeq (V \otimes_K {}^tV, h \otimes {}^th)$$

from which the equation $\mathrm{sgn}_P T_{(B,\tau)} = (\mathrm{sgn}_P h)^2$ follows.

In order to prove the first statement, we may extend scalars from F to a real closure F_P since signatures do not change under scalar extension to a real closure. However, $K \otimes F_P$ is algebraically closed since $\alpha < 0$, hence B is split over F_P. Therefore, the split case already considered shows that the signature of $T_{(B,\tau)}$ is a square. $\qquad\square$

(11.25) Definition. For any involution τ of the second kind on B, we set

$$\mathrm{sgn}_P \tau = \sqrt{\mathrm{sgn}_P T_{(B,\tau)}}.$$

The proposition above shows that if F_P is a real closure of F for P and if $\tau \otimes \mathrm{Id}_{F_P} = \sigma_h$ for some hermitian form h over F_P, then

$$\mathrm{sgn}_P \tau = |\mathrm{sgn}_P h|.$$

EXERCISES

1. Let (A,σ) be a central simple algebra with involution of any kind over a field F. Show that for any right ideals I, J in A,

   $$(I + J)^\perp = I^\perp \cap J^\perp \quad \text{and} \quad (I \cap J)^\perp = I^\perp + J^\perp.$$

   Use this observation to prove that all the maximal isotropic right ideals in A have the same reduced dimension.

   Hint: If J is an isotropic ideal and I is an arbitrary right ideal, show that $\mathrm{rdim}\, J - \mathrm{rdim}(I^\perp \cap J) \leq \mathrm{rdim}\, I - \mathrm{rdim}(I \cap J)$. If I also is isotropic and $\mathrm{rdim}\, I \leq \mathrm{rdim}\, J$, use this relation to show $I^\perp \cap J \not\subset I$, and conclude that $I + (I^\perp \cap J)$ is an isotropic ideal which strictly contains I.

2. (Bayer-Fluckiger-Shapiro-Tignol [36]) Let (A,σ) be a central simple algebra with orthogonal involution over a field F of characteristic different from 2. Show that (A,σ) is hyperbolic if and only if

   $$(A,\sigma) \simeq \big(M_2(F) \otimes A_0, \sigma_h \otimes \sigma_0\big)$$

   for some central simple algebra with orthogonal involution (A_0, σ_0), where σ_h is the adjoint involution with respect to some hyperbolic 2-dimensional symmetric bilinear form. Use this result to give examples of central simple algebras with involution (A,σ), (B,τ), (C,ν) such that $(A,\sigma) \otimes (B,\tau) \simeq (A,\sigma) \otimes (C,\nu)$ and $(B,\tau) \not\simeq (C,\nu)$.

   Let (σ, f) be a quadratic pair on a central simple algebra A over a field F of characteristic 2. Show that (A,σ,f) is hyperbolic if and only if

   $$(A,\sigma,f) \simeq \big(M_2(F) \otimes A_0, \gamma_h \otimes \sigma_0, f_{h*}\big)$$

for some central simple algebra with involution of the first kind (A_0, σ_0), where (γ_h, f_{h*}) is the quadratic pair on $M_2(F)$ associated with a hyperbolic 2-dimensional quadratic form.

Hint: If $e \in A$ is an idempotent such that $\sigma(e) = 1 - e$, use (4.17) to find a symmetric element $t \in A^\times$ such that $t\sigma(e)t^{-1} = e$, and show that e, $et\sigma(e)$, $\sigma(e)te$ and $\sigma(e)$ span a subalgebra isomorphic to $M_2(F)$.

3. Let (A, σ) be a central simple F-algebra with involution of orthogonal type and let $K \subset A$ be a subfield containing F. Suppose K consists of symmetric elements, so that the restriction $\sigma' = \sigma|_{C_A K}$ of σ to the centralizer of K in A is an involution of orthogonal type. Prove that $\operatorname{disc} \sigma = N_{K/F}(\operatorname{disc} \sigma')$.

4. Let $Q = (a, b)_F$ be a quaternion algebra over a field F of characteristic different from 2. Show that the set of discriminants of orthogonal involutions on Q is the set of represented values of the quadratic form $\langle a, b, -ab \rangle$.

5. (Tits [291]) Let (A, σ) be a central simple algebra of even degree with involution of the first kind. Assume that σ is orthogonal if $\operatorname{char} F \neq 2$, and that it is symplectic if $\operatorname{char} F = 2$. For any $a \in \operatorname{Alt}(A, \sigma) \cap A^\times$ whose reduced characteristic polynomial is separable, let

$$H = \{ x \in \operatorname{Alt}(A, \sigma) \mid xa = ax \}.$$

Show that $a^{-1}H$ is an étale subalgebra of A of dimension $\deg A/2$. (The space H is called a *Cartan subspace* in Tits [291].)

Hint: See Lemma (8.21).

6. Let (A, σ) be a central simple algebra with orthogonal involution over a field F of characteristic different from 2. For brevity, write C for $C(A, \sigma)$ its Clifford algebra, Z for $Z(A, \sigma)$ the center of C and B for $B(A, \sigma)$ the Clifford bimodule. Endow $A \otimes_F C$ with the C-bimodule structure such that $c_1 \cdot (a \otimes c) \cdot c_2 = a \otimes c_1 c c_2$ for $a \in A$ and c, c_1, $c_2 \in C$.

   (a) Show that there is an isomorphism of C-bimodules $\psi \colon B \otimes_C B \to A \otimes_F C$ which in the split case satisfies

   $$\psi\big((v_1 \otimes c_1) \otimes (v_2 \otimes c_2)\big) = (v_1 \otimes v_2) \otimes c_1 c_2$$

   under the standard identifications $A = V \otimes V$, $B = V \otimes C_1(V, q)$ and $C = C_0(V, q)$.

   (b) Define a hermitian form $H \colon B \times B \to A \otimes_F Z$ by

   $$H(x, y) = \operatorname{Id}_A \otimes (\iota \circ \operatorname{Trd}_C)\big(\psi(x \otimes y^\omega)\big) \quad \text{for } x, y \in B.$$

   Show that the natural isomorphism ν of (9.9) is an isomorphism of algebras with involution

   $$\nu \colon ({}^\iota C^{\operatorname{op}}, {}^\iota \underline{\sigma}^{\operatorname{op}}) \otimes_Z (C, \underline{\sigma}) \xrightarrow{\sim} \big(\operatorname{End}_{A \otimes Z}(B), \sigma_H\big).$$

7. To each permutation $\pi \in S_k$, associate a permutation π^* of $\{0, 1, \ldots, k-1\}$ by composing the following bijections:

$$\{0, \ldots, k-1\} \xrightarrow{+1} \{1, \ldots, k\} \xrightarrow{\pi^{-1}} \{1, \ldots, k\} \longrightarrow \{0, \ldots, k-1\}$$

where the last map carries k to 0 and leaves every i between 1 and $k-1$ invariant. Consider the decomposition of π^* into disjoint cycles (including the cycles of length 1):

$$\pi^* = (0, \alpha_1, \ldots, \alpha_r)(\beta_1, \ldots, \beta_s) \cdots (\gamma_1, \ldots, \gamma_t).$$

Since the map $\mathrm{Sand}_k \colon \underline{A}^{\otimes k} \to \mathrm{Hom}_F(\underline{A}^{\otimes k-1}, \underline{A})$ is bijective (see (9.3)), there is a unique element $x_\pi \in A^{\otimes k}$ such that for $b_1, \ldots, b_{k-1} \in A$:

$$\mathrm{Sand}_k(x_\pi)(b_1 \otimes \cdots \otimes b_{k-1}) =$$
$$b_{\alpha_1} \cdots b_{\alpha_r} \mathrm{Trd}_A(b_{\beta_1} \cdots b_{\beta_s}) \cdots \mathrm{Trd}_A(b_{\gamma_1} \cdots b_{\gamma_t}).$$

Show that $x_\pi = g_k(\pi)$.

8. Show that $s_k^2 = k! \, s_k$.

9. Show by a direct computation that if A is a quaternion algebra, the canonical involution on $A = \lambda^1 A$ is the quaternion conjugation.

10. Let (B, τ) be a central simple F-algebra with unitary involution. Assume that $\deg(B, \tau)$ is divisible by 4 and that char $F \neq 2$, so that the canonical involution $\underline{\tau}$ on $D(B, \tau)$ has orthogonal type. Show that $\mathrm{disc}\,\underline{\tau} = 1$ if $\deg(B, \tau)$ is not a power of 2 and that $\mathrm{disc}\,\underline{\tau} = \alpha \cdot F^{\times 2}$ if $\deg(B, \tau)$ is a power of 2 and $K \simeq F[X]/(X^2 - \alpha)$.

  Hint: Reduce to the split case by scalar extension to some splitting field of B in which F is algebraically closed (for instance the function field of the Weil transfer of the Severi-Brauer variety of B). Let $\deg(B, \tau) = n = 2m$. Using the same notation as in (10.27), define a map $v \in \mathrm{End}_F(\bigwedge^m V_0)$ as follows: consider a partition of the subsets $S \subset \{1, \ldots, n\}$ of cardinality m into two classes C, C' such that the complement of every $S \in C$ lies in C' and vice-versa; then set $v(e_S) = e_S$ if $S \in C$, $v(e_S) = -e_S$ if $S \in C'$. Show that $v \otimes \sqrt{\alpha} \in \mathrm{Skew}(D(B, \tau), \underline{\tau})$ and use this element to compute $\mathrm{disc}\,\underline{\tau}$.

11. Let K/F be a quadratic extension with non-trivial automorphism ι, and let $\alpha \in F^\times$, $\beta \in K^\times$. Assume F contains a primitive $2m$-th root of unity ξ and consider the algebra B of degree $2m$ over K generated by two elements i, j subject to the following conditions:

$$i^{2m} = \alpha \quad j^{2m} = \iota(\beta)/\beta \quad ji = \xi ij.$$

  (a) Show that there is a unitary involution τ on B such that $\tau(i) = i$ and $\tau(j) = j^{-1}$.
  (b) Show that $D(B, \tau) \sim \big(\alpha, N_{K/F}(\beta)\big)_F \otimes_F \big(K, (-1)^m \alpha\big)_F$.

  Hint: Let $X = R_{K/F}\big(\mathrm{SB}(B)\big)$ be the transfer of the Severi-Brauer variety of B. The algebra B splits over $K \otimes_F F(X)$, but the scalar extension map $\mathrm{Br}(F) \to \mathrm{Br}\big(F(X)\big)$ is injective (see Merkurjev-Tignol [196]); so it suffices to prove the claim when B is split.

12. Let V be a vector space of dimension n over a field F. Fix k with $1 \leq k \leq n-1$, and let $\ell = n - k$. The canonical pairing $\wedge \colon \bigwedge^k V \times \bigwedge^\ell V \to \bigwedge^n V$ induces an isomorphism

$$\big(\textstyle\bigwedge^k V\big)^* \to \textstyle\bigwedge^\ell V$$

which is uniquely determined up to a factor in $F^\times$, hence the pairing also induces a canonical isomorphism

$$\psi_{k,\ell} \colon \mathrm{End}_F\big((\textstyle\bigwedge^k V)^*\big) \to \mathrm{End}_F(\textstyle\bigwedge^\ell V).$$

Our aim in this exercise is to define a corresponding isomorphism for non-split algebras.

Let A be a central simple F-algebra of degree n. For $2 \leq k \leq n$, set:

$$s_k = \sum_{\pi \in S_k} \mathrm{sgn}(\pi) g_k(\pi) \in A^{\otimes k}$$

(as in §10.A), and extend this definition by setting $s_1 = 1$. Let $\ell = n - k$, where $1 \leq k \leq n - 1$.

(a) Generalize (10.7) by showing that $s_n \in A^{\otimes n} \cdot (s_k \otimes s_\ell)$.

We may thus consider the right ideal
$$I = \left\{ f \in \mathrm{End}_{A^{\otimes n}} \left(A^{\otimes n}(s_k \otimes s_\ell) \right) \mid A^{\otimes n} s_n^f = \{0\} \right\} \subset \lambda^k A \otimes \lambda^\ell A.$$

(b) Using exercise 13 of Chapter I, show that this right ideal defines a canonical isomorphism
$$\varphi_{k,\ell} \colon \lambda^k A^{\mathrm{op}} \xrightarrow{\sim} \lambda^\ell A.$$

Show that if $A = \mathrm{End}_F V$, then $\varphi_{k,\ell} = \psi_{k,\ell}$ under the canonical identifications $\lambda^k A^{\mathrm{op}} = \mathrm{End}_F\left((\bigwedge^k V^*) \right)$ and $\lambda^\ell A = \mathrm{End}_F(\bigwedge^\ell V)$.

13. (Wadsworth, unpublished) The aim of this exercise is to give examples of central simple algebras with unitary involution whose discriminant algebra has index 4.

Let F_0 be an arbitrary field of characteristic different from 2 and let $K = F_0(x, y, z)$ be the field of rational fractions in three independent indeterminates over F_0. Denote by ι the automorphism of K which leaves $F_0(x, y)$ elementwise invariant and maps z to $-z$, and let $F = F_0(x, y, z^2)$ be the invariant subfield. Consider the quaternion algebras $Q_0 = (x, y)_F$ and $Q = Q_0 \otimes_F K$, and define an involution θ on Q by $\theta = \gamma_0 \otimes \iota$ where γ_0 is the quaternion conjugation on Q_0. Finally, let $B = M_n(Q)$ for an arbitrary odd integer $n > 1$, and endow B with the involution $*$ defined by
$$(a_{ij})^*_{1 \leq i,j \leq n} = \left(\theta(a_{ij}) \right)^t_{1 \leq i,j \leq n}.$$

(a) Show that $D(B, *) \sim D(Q, \theta)^{\otimes n} \sim D(Q, \theta) \sim Q_0$.

Let $c_1, \ldots, c_n \in \mathrm{Sym}(Q, \theta) \cap Q^\times$ and $d = \mathrm{diag}(c_1, \ldots, c_n) \in B$. Define another involution of unitary type on B by $\tau = \mathrm{Int}(d) \circ *$.

(a) Show that
$$D(B, \tau) \sim D(B, *) \otimes_F \left(z^2, \mathrm{Nrd}_B(d) \right)_F$$
$$\sim (x, y)_F \otimes \left(z^2, \mathrm{Nrd}_Q(c_1) \cdots \mathrm{Nrd}_Q(c_n) \right)_F.$$

(b) Show that the algebra $D(B, \tau)$ has index 4 if $c_1 = z^2 + zi$, $c_2 = z^2 + zj$ and $c_3 = \cdots = c_n = 1$.

14. (Yanchevskiĭ [319, Proposition 1.4]) Let σ, σ' be involutions on a central simple algebra A over a field F of characteristic different from 2. Show that if σ and σ' have the same restriction to the center of A and $\mathrm{Sym}(A, \sigma) = \mathrm{Sym}(A, \sigma')$, then $\sigma = \sigma'$.

Hint: If σ and σ' are of the first kind, use (11.2).

15. Let (A, σ) be a central simple algebra with involution of the first kind over a field F of arbitrary characteristic. Show that a nonsingular symmetric bilinear form on $\mathrm{Symd}(A, \sigma)$ may be defined as follows: for $x, y \in \mathrm{Symd}(A, \sigma)$, pick $y' \in A$ such that $y = y' + \sigma(y')$, and let $T(x, y) = \mathrm{Trd}_A(xy')$. Mimic this construction to define a nonsingular symmetric bilinear form on $\mathrm{Alt}(A, \sigma)$.

Notes

§6. On the same model as Severi-Brauer varieties, varieties of isotropic ideals, known as Borel varieties, or homogeneous varieties, or twisted flag varieties, are associated to an algebra with involution. These varieties can also be defined as varieties of parabolic subgroups of a certain type in the associated simply connected group: see Borel-Tits [47]; their function fields are the generic splitting fields investigated by Kersten and Rehmann [153]. In particular, the variety of isotropic ideals of reduced dimension 1 in a central simple algebra with orthogonal involution (A, σ) of characteristic different from 2 may be regarded as a twisted form of a quadric: after scalar extension to a splitting field L of A, it yields the quadric $q = 0$ where q is a quadratic form whose adjoint involution is σ_L. These twisted forms of quadrics are termed *involution varieties* by Tao [280], who studied their K-groups to obtain index reduction formulas for their function fields. Tao's results were generalized to arbitrary Borel varieties by Merkurjev-Panin-Wadsworth [194], [193]. The Brauer group of a Borel variety is determined in Merkurjev-Tignol [196].

The notion of index in (6.3) is inspired by Tits' definition of index for a semisimple linear algebraic group [290, (2.3)]. Hyperbolic involutions are defined in Bayer-Fluckiger-Shapiro-Tignol [36]. Example (6.11) is borrowed from Dejaiffe [68] where a notion of orthogonal sum for algebras with involution is investigated.

§7. The discriminant of an orthogonal involution on a central simple algebra of even degree over a field of characteristic different from 2 first appeared in Jacobson [134] as the center of the (generalized, even) Clifford algebra. The approach in Tits [291] applies also in characteristic 2; it is based on generalized quadratic forms instead of quadratic pairs. For involutions, the more direct definition presented here is due to Knus-Parimala-Sridharan [163]. Earlier work of Knus-Parimala-Sridharan [164] used another definition in terms of generalized pfaffian maps.

A short, direct proof of (7.1) is given in Kersten [152, (3.1)]; the idea is to split the algebra by a scalar extension in which the base field is algebraically closed.

The set of determinants of orthogonal involutions on a central simple algebra A of characteristic different from 2 has been investigated by Parimala-Sridharan-Suresh [205]. It turns out that, except in the case where A is a quaternion algebra (where the set of determinants is easily determined, see Exercise 4), the set of determinants is the group of reduced norms of A modulo squares:

$$\bigcup_{\sigma} \det \sigma = \mathrm{Nrd}(A^{\times}) \cdot F^{\times 2}.$$

§8. The first definition of Clifford algebra for an algebra with orthogonal involution of characteristic different from 2 is due to Jacobson [134]; it was obtained by Galois descent. A variant of Jacobson's construction was proposed by Seip-Hornix [250] for the case of central simple algebras of Schur index 2. Her definition also covers the characteristic 2 case. Our treatment owes much to Tits [291]. In particular, the description of the center of the Clifford algebra in §8.D and the proof of (8.31) closely follow Tits' paper. Other proofs of (8.31) were given by Allen [17, Theorem 3] and Van Drooge (thesis, Utrecht, 1967).

If $\deg A$ is divisible by 8, the canonical involution $\underline{\sigma}$ on $C(A, \sigma, f)$ is part of a canonical quadratic pair $(\underline{\sigma}, \underline{f})$. If A is split and the quadratic pair (σ, f) is hyperbolic, we may define this canonical pair as follows: representing $A = \mathrm{End}_F\big(\mathbb{H}(U)\big)$ we have as in (8.3)

$$C(A, \sigma, f) = C_0\big(\mathbb{H}(U)\big) \simeq \mathrm{End}(\textstyle\bigwedge_0 U) \times \mathrm{End}(\textstyle\bigwedge_1 U) \subset \mathrm{End}(\textstyle\bigwedge U).$$

Let $m = \dim U$. For $\xi \in \bigwedge U$, let $\xi^{[r]}$ be the component of ξ in $\bigwedge^r U$. Fix a nonzero linear form $s\colon \bigwedge U \to F$ which vanishes on $\bigwedge^r U$ for $r < m$ and define a quadratic form $q_\wedge\colon \bigwedge U \to F$ by

$$q_\wedge(\xi) = s\left(\textstyle\sum_{r<m/2} \overline{\xi^{[r]}} \wedge \xi^{[m-r]} + q(\xi^{[m/2]})\right)$$

where $q\colon \bigwedge^{m/2} U \to \bigwedge^m U$ is the canonical quadratic map of (10.12) and $\bar{}$ is the involution on $\bigwedge U$ which is the identity on U (see the proof of (8.4)). For $i = 0$, 1, let q_i be the restriction of $q_\wedge$ to $\bigwedge_i U$. The pair (q_0, q_1) may be viewed as a quadratic form

$$(q_0, q_1)\colon \textstyle\bigwedge_0 U \times \bigwedge_1 U \to F \times F.$$

The canonical quadratic pair on $\operatorname{End}(\bigwedge_0 U) \times \operatorname{End}(\bigwedge_1 U)$ is associated to this quadratic form. In the general case, the canonical quadratic pair on $C(A, \sigma, f)$ can be defined by Galois descent. The canonical involution on the Clifford algebra of a central simple algebra with hyperbolic involution (of characteristic different from 2) has been investigated by Garibaldi [104].

Clifford algebras of tensor products of central simple algebras with involution have been determined by Tao [281]. Let $(A, \sigma) = (A_1, \sigma_1) \otimes_F (A_2, \sigma_2)$ where A_1, A_2 are central simple algebras of even degree over a field F of characteristic different from 2, and σ_1, σ_2 are involutions which are either both orthogonal or both symplectic, so that σ is an orthogonal involution of trivial discriminant, by (7.3). It follows from (8.10) that the Clifford algebra $C(A, \sigma)$ decomposes into a direct product of two central simple F-algebras: $C(A, \sigma) = C^+(A, \sigma) \times C^-(A, \sigma)$. Tao proves in [281, Theorems 4.12, 4.14, 4.16]:

(a) Suppose σ_1, σ_2 are orthogonal and denote by Q the quaternion algebra $Q = (\operatorname{disc} \sigma_1, \operatorname{disc} \sigma_2)_F$.
  (i) If $\deg A_1$ or $\deg A_2$ is divisible by 4, then one of the algebras $C_\pm(A, \sigma)$ is Brauer-equivalent to $A \otimes_F Q$ and the other one to Q.
  (ii) If $\deg A_1 \equiv \deg A_2 \equiv 2 \mod 4$, then one of the algebras $C_\pm(A, \sigma)$ is Brauer-equivalent to $A_1 \otimes_F Q$ and the other one to $A_2 \otimes_F Q$.
(b) Suppose σ_1, σ_2 are symplectic.
  (i) If $\deg A_1$ or $\deg A_2$ is divisible by 4, then one of the algebras $C_\pm(A, \sigma)$ is split and the other one is Brauer-equivalent to A.
  (ii) If $\deg A_1 \equiv \deg A_2 \equiv 2 \mod 4$, then one of the algebras $C_\pm(A, \sigma)$ is Brauer-equivalent to A_1 and the other one to A_2.

§9. In characteristic different from 2, the bimodule $B(A, \sigma)$ is defined by Galois descent in Merkurjev-Tignol [196]. The fundamental relations in (9.12) between a central simple algebra with orthogonal involution and its Clifford algebra have been observed by several authors: (9.14) was first proved by Jacobson [134, Theorem 4] in the case where $Z = F \times F$. In the same special case, proofs of (9.14) and (9.15) have been given by Tits [291, Proposition 7], [292, 6.2]. In the general case, these relations have been established by Tamagawa [278] and by Tao [281]. See (31.11) for a cohomological proof of the fundamental relations in characteristic different from 2 and Exercise 15 of Chapter VII for another cohomological proof valid in arbitrary characteristic. Note that the bimodule $B(A, \sigma)$ carries a canonical hermitian form which may be used to strengthen (9.9) into an isomorphism of algebras with involution: see Exercise 6.

§10. The canonical representation of the symmetric group S_k in the group of invertible elements of $A^{\otimes k}$ was observed by Haile [112, Lemma 1.1] and Saltman [240].

Note that if $k = \text{ind}\, A$, (10.5) shows that $A^{\otimes k}$ is split; therefore the exponent of A divides its index. Indeed, the purpose of Saltman's paper is to give an easy direct proof (also valid for Azumaya algebras) of the fact that the Brauer group is torsion. Another approach to the λ-construction, using Severi-Brauer varieties, is due to Suslin [276].

The canonical quadratic map on $\bigwedge^m V$, where V is a $2m$-dimensional vector space over a field of characteristic 2 (see (10.12)), is due to Papy [204]. It is part of a general construction of reduced p-th powers in exterior algebras of vector spaces over fields of characteristic p.

The discriminant algebra $D(B, \tau)$ also arises from representations of classical algebraic groups of type ${}^2A_n$: see Tits [292]. If the characteristic does not divide $2 \deg B$, its Brauer class can be obtained by reduction modulo 2 of a cohomological invariant $t(B, \tau)$ called the Tits class, see (31.8). This invariant has been investigated by Quéguiner [226], [227]. In [226, Proposition 11], Quéguiner shows that (10.33) can be derived from (11.17) if char $F \neq 2$; she also considers the analogue of (10.33) where the involution τ_0 is symplectic instead of orthogonal, and proves that $D(B, \tau)$ is Brauer-equivalent to $B_0^{\otimes m}$ in this case. (Note that Quéguiner's "determinant class modulo 2" differs from the Brauer class of $D(B, \tau)$ by the class of the quaternion algebra $(K, -1)_F$ if $\deg B \equiv 2 \mod 4$.)

§11. The idea to consider the form $T_{(A,\sigma)}$ as an invariant of the involution σ dates back to Weil [310]. The relation between the determinant of an orthogonal involution σ and the determinant of the bilinear form T_σ^+ (in characteristic different from 2) was observed by Lewis [172] and Quéguiner [226], who also computed the Hasse invariant $s(Q_\sigma)$ of the quadratic form $Q_\sigma(x) = \text{Trd}_A\big(\sigma(x)x\big)$ associated to $T_{(A,\sigma)}$. The result is the following: for an involution σ on a central simple algebra A of degree n,

$$s(Q_\sigma) = \begin{cases} \frac{n}{2}[A] + \big[(-1, \det \sigma)_F\big] & \text{if } n \text{ is even and } \sigma \text{ is orthogonal,} \\ \frac{n}{2}[A] + \frac{n}{2}\big[(-1, -1)_F\big] & \text{if } n \text{ is even and } \sigma \text{ is symplectic,} \\ 0 & \text{if } n \text{ is odd.} \end{cases}$$

In Lewis' paper [172], these relations are obtained by comparing the Hasse invariant of Q_σ and of $Q_A(x) = \text{Trd}_A(x^2)$ through (11.2). Quéguiner [224] also gives the computation of the Hasse invariant of the quadratic forms Q_σ^+ and Q_σ^- which are the restrictions of Q_σ to $\text{Sym}(A, \sigma)$ and $\text{Skew}(A, \sigma)$ respectively. Just as for Q_σ, the result only depends on the parity of n and on the type and discriminant of σ.

The signature of an involution of the first kind was first defined by Lewis-Tignol [174]. The corresponding notion for involutions of the second kind is due to Quéguiner [225].

Besides the classical invariants considered in this chapter, there are also "higher cohomological invariants" defined by Rost by means of simply connected algebraic groups, with values in Galois cohomology groups of degree 3. See §31.B for a general discussion of cohomological invariants. Some special cases are considered in the following chapters: see §16.B for the case of symplectic involutions on central simple algebras of degree 4 and §19.B for the case of unitary involutions on central simple algebras of degree 3. (In the same spirit, see §40 for an H^3-invariant of Albert algebras.) Another particular instance dates back to Jacobson [129]: if A is a central simple F-algebra of index 2 whose degree is divisible by 4, we may represent

$A = \mathrm{End}_Q(V)$ for some vector space V of even dimension over a quaternion F-algebra Q. According to (4.2), every symplectic involution σ on A is adjoint to some hermitian form h on V with respect to the canonical involution of Q. Assume char $F \neq 2$ and let $h = \langle \alpha_1, \ldots, \alpha_n \rangle$ be a diagonalization of h; then $\alpha_1, \ldots, \alpha_n \in F^\times$ and the element $(-1)^{n/2} \alpha_1 \cdots \alpha_n \cdot \mathrm{Nrd}_Q(Q^\times) \in F^\times / \mathrm{Nrd}_Q(Q^\times)$ is an invariant of σ. There is an alternate description of this invariant, which emphasizes the relation with Rost's cohomological approach: we may associate to σ the quadratic form $q_\sigma = \langle 1, -(-1)^{n/2} \alpha_1 \cdots \alpha_n \rangle \otimes n_Q \in I^3 F$ where n_Q is the reduced norm form of Q, or the cup product $\big((-1)^{n/2} \alpha_1 \cdots \alpha_n\big) \cup [Q] \in H^3(F, \mu_2)$, see (31.47).

CHAPTER III

Similitudes

In this chapter, we investigate the automorphism groups of central simple algebras with involution. The inner automorphisms which preserve the involution are induced by elements which we call *similitudes*, and the automorphism group of a central simple algebra with involution is the quotient of the group of similitudes by the multiplicative group of the center. The various groups thus defined are naturally endowed with a structure of linear algebraic group; they may then be seen as twisted forms of orthogonal, symplectic or unitary groups, depending on the type of the involution. This point of view will be developed in Chapter VI. Here, however, we content ourselves with a more elementary viewpoint, considering the groups of rational points of the corresponding algebraic groups.

After a first section which contains general definitions and results valid for all types, we then focus on quadratic pairs and unitary involutions, where additional information can be derived from the algebra invariants defined in Chapter II. In the orthogonal case, we also use the Clifford algebra and the Clifford bimodule to define Clifford groups and spin groups.

§12. General Properties

To motivate our definition of similitude for an algebra with involution, we first consider the split case, where the algebra consists of endomorphisms of bilinear or hermitian spaces.

12.A. The split case. We treat separately the cases of bilinear, hermitian and quadratic spaces, although the basic definitions are the same, to emphasize the special features of these various cases.

Bilinear spaces. Let (V, b) be a nonsingular symmetric or alternating bilinear space over an arbitrary field F. A *similitude* of (V, b) is a linear map $g\colon V \to V$ for which there exists a constant $\alpha \in F^\times$ such that

$$(12.1) \qquad b\big(g(v), g(w)\big) = \alpha b(v, w) \quad \text{for } v,\, w \in V.$$

The factor α is called the *multiplier* of the similitude g. A similitude with multiplier 1 is called an *isometry*. The similitudes of the bilinear space (V, b) form a group denoted $\mathrm{Sim}(V, b)$, and the map

$$\mu\colon \mathrm{Sim}(V, b) \to F^\times$$

which carries every similitude to its multiplier is a group homomorphism. By definition, the kernel of this map is the group of isometries of (V, b), which we write $\mathrm{Iso}(V, b)$. We also define the group $\mathrm{PSim}(V, b)$ of projective similitudes by

$$\mathrm{PSim}(V, b) = \mathrm{Sim}(V, b)/F^\times.$$

Specific notations for the groups $\mathrm{Sim}(V, b)$, $\mathrm{Iso}(V, b)$ and $\mathrm{PSim}(V, b)$ are used according to the type of b. If b is symmetric nonalternating, we set

$$O(V, b) = \mathrm{Iso}(V, b), \quad GO(V, b) = \mathrm{Sim}(V, b) \quad \text{and} \quad PGO(V, b) = \mathrm{PSim}(V, b);$$

if b is alternating, we let

$$\mathrm{Sp}(V, b) = \mathrm{Iso}(V, b), \quad \mathrm{GSp}(V, b) = \mathrm{Sim}(V, b) \quad \text{and} \quad \mathrm{PGSp}(V, b) = \mathrm{PSim}(V, b).$$

Note that condition (12.1), defining a similitude of (V, b) with multiplier α, can be rephrased as follows, using the adjoint involution σ_b:

(12.2) $$\sigma_b(g) \circ g = \alpha \mathrm{Id}_V.$$

By taking the determinant of both sides, we obtain $(\det g)^2 = \alpha^n$ where $n = \dim V$. It follows that the determinant of an isometry is ± 1 and that, if n is even,

$$\det g = \pm \mu(g)^{n/2} \quad \text{for } g \in \mathrm{Sim}(V, b).$$

A first difference between the orthogonal case and the symplectic case shows up in the following result:

(12.3) Proposition. *If b is a nonsingular alternating bilinear form on a vector space V of dimension n (necessarily even), then*

$$\det g = \mu(g)^{n/2} \quad \text{for } g \in \mathrm{GSp}(V, b).$$

Proof: Let $g \in \mathrm{GSp}(V, b)$ and let G, B denote the matrices of g and b respectively with respect to some arbitrary basis of V. The matrix B is alternating and we have

$$G^t B G = \mu(g) B.$$

By taking the pfaffian of both sides, we obtain, by known formulas for pfaffians (see Artin [24, Theorem 3.28]; compare with (2.13)):

$$\det G \, \mathrm{pf} \, B = \mu(g)^{n/2} \, \mathrm{pf} \, B,$$

hence $\det g = \mu(g)^{n/2}$. $\qquad\qquad\qquad\qquad\qquad\qquad\qquad\qquad\qquad\qquad\quad\square$

By contrast, if b is symmetric and $\mathrm{char}\, F \neq 2$, every hyperplane reflection is an isometry with determinant -1 (see (12.13)), hence it satisfies $\det g = -\mu(g)^{n/2}$. We set

$$O^+(V, b) = \{\, g \in O(V, b) \mid \det g = 1 \,\}.$$

Of course, $O^+(V, b) = O(V, b)$ if $\mathrm{char}\, F = 2$.

Similarly, if $\dim V = n$ is even, we set

$$GO^+(V, b) = \{\, g \in GO(V, b) \mid \det g = \mu(g)^{n/2} \,\},$$

and

$$PGO^+(V, b) = GO^+(V, b)/F^\times.$$

The elements in $GO^+(V, b)$, $O^+(V, b)$ are called *proper* similitudes and *proper* isometries respectively.

If $\dim V$ is odd, there is a close relationship between similitudes and isometries, as the next proposition shows:

(12.4) Proposition. *Suppose that (V, b) is a nonsingular symmetric bilinear space of odd dimension over an arbitrary field F; then*

$$GO(V, b) = O^+(V, b) \cdot F^\times \simeq O^+(V, b) \times F^\times \quad \text{and} \quad PGO(V, b) \simeq O^+(V, b).$$

Proof: If g is a similitude of (V, b) with multiplier $\alpha \in F^\times$, then by taking the determinant of both sides of the isometry $\alpha \cdot b \simeq b$ we get $\alpha \in F^{\times 2}$. If $\alpha = \alpha_1^2$, then $\alpha_1^{-1} g$ is an isometry. Moreover, after changing the sign of α_1 if necessary, we may assume that $\det(\alpha_1^{-1} g) = 1$. The factorization $g = (\alpha_1^{-1} g) \cdot \alpha_1$ shows that $\mathrm{GO}(V, b) = \mathrm{O}^+(V, b) \cdot F^\times$, and the other isomorphisms are clear. $\qquad\square$

Hermitian spaces. Suppose (V, h) is a nonsingular hermitian space over a quadratic separable field extension K of F (with respect to the nontrivial automorphism of K/F). A *similitude* of (V, h) is an invertible linear map $g \colon V \to V$ for which there exists a constant $\alpha \in F^\times$, called the *multiplier* of g, such that

(12.5) $$h\big(g(v), g(w)\big) = \alpha h(v, w) \quad \text{for } v, \, w \in V.$$

As in the case of bilinear spaces, we write $\mathrm{Sim}(V, h)$ for the group of similitudes of (V, h); let

$$\mu \colon \mathrm{Sim}(V, h) \to F^\times$$

be the group homomorphism which carries every similitude to its multiplier; write $\mathrm{Iso}(V, h)$ for the kernel of μ, whose elements are called *isometries*, and let

$$\mathrm{PSim}(V, h) = \mathrm{Sim}(V, h)/K^\times.$$

We also use the following more specific notation:

$$\mathrm{U}(V, h) = \mathrm{Iso}(V, h), \quad \mathrm{GU}(V, h) = \mathrm{Sim}(V, h), \quad \mathrm{PGU}(V, h) = \mathrm{PSim}(V, h).$$

Condition (12.5) can be rephrased as

$$\sigma_h(g) \circ g = \alpha \mathrm{Id}_V.$$

By taking the determinant of both sides, we obtain

$$N_{K/F}(\det g) = \mu(g)^n, \quad \text{where } n = \dim V.$$

This relation shows that the determinant of every isometry has norm 1. Set

$$\mathrm{SU}(V, h) = \{\, g \in \mathrm{U}(V, h) \mid \det g = 1 \,\}.$$

Quadratic spaces. Let (V, q) be a nonsingular quadratic space over an arbitrary field F. A *similitude* of (V, q) is an invertible linear map $g \colon V \to V$ for which there exists a constant $\alpha \in F^\times$, called the *multiplier* of g, such that

$$q\big(g(v)\big) = \alpha q(v) \quad \text{for } v \in V.$$

The groups $\mathrm{Sim}(V, q)$, $\mathrm{Iso}(V, q)$, $\mathrm{PSim}(V, q)$ and the group homomorphism

$$\mu \colon \mathrm{Sim}(V, q) \to F^\times$$

are defined as for nonsingular bilinear forms. We also use the notation

$$\mathrm{O}(V, q) = \mathrm{Iso}(V, q), \quad \mathrm{GO}(V, q) = \mathrm{Sim}(V, q), \quad \mathrm{PGO}(V, q) = \mathrm{PSim}(V, q).$$

It is clear from the definitions that every similitude of (V, q) is also a similitude of its polar bilinear space (V, b_q), with the same multiplier, hence

$$\mathrm{GO}(V, q) \subset \mathrm{Sim}(V, b) = \begin{cases} \mathrm{GO}(V, b_q) & \text{if char } F \neq 2, \\ \mathrm{GSp}(V, b_q) & \text{if char } F = 2, \end{cases}$$

and the reverse inclusion also holds if char $F \neq 2$.

For the rest of this section, we assume therefore char $F = 2$. If $\dim V$ is odd, the same arguments as in (12.4) yield:

(12.6) Proposition. *Suppose (V, q) is a nonsingular symmetric quadratic space of odd dimension over a field F of characteristic 2; then*

$$\mathrm{GO}(V, q) = \mathrm{O}(V, q) \cdot F^\times \simeq \mathrm{O}(V, q) \times F^\times \quad and \quad \mathrm{PGO}(V, q) \simeq \mathrm{O}(V, q).$$

We omit the proof, since it is exactly the same as for (12.4), using the determinant of q defined in (0.4).

If $\dim V$ is even, we may again distinguish proper and improper similitudes, as we now show.

By using a basis of V, we may represent the quadratic map q by a quadratic form, which we denote again q. Let M be a matrix such that

$$q(X) = X^t \cdot M \cdot X.$$

Since q is nonsingular, the matrix $W = M + M^t$ is invertible. Let g be a similitude of V with multiplier α, and let G be its matrix with respect to the chosen basis of V. The equation $q(G \cdot X) = \alpha q(X)$ shows that the matrices $G^t M G$ and αM represent the same quadratic form. Therefore, $G^t M G - \alpha M$ is an alternating matrix. Let $R \in M_n(K)$ be such that

$$G^t M G - \alpha M = R - R^t.$$

(12.7) Proposition. *The element $\mathrm{tr}(\alpha^{-1} W^{-1} R) \in K$ depends only on the similitude g, and not on the choice of basis of V nor on the choices of matrices M and R. It equals 0 or 1.*

Proof: With a different choice of basis of V, the matrix G is replaced by $G' = P^{-1} G P$ for some invertible matrix $P \in \mathrm{GL}_n(K)$, and the matrix M is replaced by a matrix $M' = P^t M P + U - U^t$ for some matrix U. Then $W' = M' + M'^t$ is related to W by $W' = P^t W P$. Suppose R, R' are matrices such that

(12.8) $\qquad G^t M G - \alpha M = R - R^t \quad and \quad G'^t M' G' - \alpha M' = R' - R'^t.$

By adding each side to its transpose, we derive from these equations:

(12.9) $\qquad\qquad\qquad G^t W G = \alpha W \quad and \quad G'^t W' G' = \alpha W'.$

In order to prove that $\mathrm{tr}(\alpha^{-1} W^{-1} R)$ depends only on the similitude g, we have to show $\mathrm{tr}(W^{-1} R) = \mathrm{tr}(W'^{-1} R')$. By substituting for M' its expression in terms of M, we derive from (12.8) that $R' - R'^t = R'' - R''^t$, where

(12.10) $\qquad R'' = P^t R P + P^t G^t (P^{-1})^t U P^{-1} G P - \alpha U,$

hence $R' = R'' + S$ for some symmetric matrix $S \in M_n(K)$. Since $W'^{-1} = W'^{-1} M' W'^{-1} + (W'^{-1} M' W'^{-1})^t$, it follows that W'^{-1} is alternating. By (2.3), alternating matrices are orthogonal to symmetric matrices for the trace bilinear form, hence $\mathrm{tr}(g'^{-1} R') = \mathrm{tr}(g'^{-1} R'')$. In view of (12.10) we have

(12.11) $\quad \mathrm{tr}(W'^{-1} R'') = \mathrm{tr}(P^{-1} W^{-1} R P) + \mathrm{tr}\big(P^{-1} W^{-1} G^t (P^{-1})^t U P^{-1} G P\big)$
$$+ \alpha \, \mathrm{tr}\big(P^{-1} W^{-1} (P^{-1})^t U\big).$$

By (12.9), $W^{-1} G^t = \alpha G^{-1} W^{-1}$, hence the second term on the right side of (12.11) equals

$$\alpha \, \mathrm{tr}\big(P^{-1} G^{-1} W^{-1} (P^{-1})^t U P^{-1} G P\big) = \alpha \, \mathrm{tr}\big(W^{-1} (P^{-1})^t U P^{-1}\big).$$

Therefore, the last two terms on the right side of (12.11) cancel, and we get

$$\text{tr}(W'^{-1}R') = \text{tr}(W'^{-1}R'') = \text{tr}(W^{-1}R),$$

proving that $\text{tr}(\alpha^{-1}W^{-1}R)$ depends only on the similitude g.

In order to prove that this element is 0 or 1, we compute $s_2(W^{-1}M)$, the coefficient of X^{n-2} in the characteristic polynomial of $W^{-1}M$ (see (0.1)). By (12.9), we have $G^{-1}W^{-1} = \alpha^{-1}W^{-1}G^t$, hence $G^{-1}W^{-1}MG = \alpha^{-1}W^{-1}G^tMG$, and therefore

$$s_2(W^{-1}M) = s_2(\alpha^{-1}W^{-1}G^tMG).$$

On the other hand, (12.8) also yields $G^tMG = \alpha M + R - R^t$, hence by substituting this in the right side of the preceding equation we get

$$s_2(W^{-1}M) = s_2(W^{-1}M + \alpha^{-1}W^{-1}R - \alpha^{-1}W^{-1}R^t).$$

By (0.5), we may expand the right side to get

$$s_2(W^{-1}M) = s_2(W^{-1}M) + \text{tr}(\alpha^{-1}W^{-1}R) + \text{tr}(\alpha^{-1}W^{-1}R)^2.$$

Therefore, $\text{tr}(\alpha^{-1}W^{-1}R) + \text{tr}(\alpha^{-1}W^{-1}R)^2 = 0$, hence

$$\text{tr}(\alpha^{-1}W^{-1}R) = 0, 1. \qquad \square$$

(12.12) Definition. Let (V, q) be a nonsingular quadratic space of even dimension over a field F of characteristic 2. Keep the same notation as above. In view of the preceding proposition, we set

$$\Delta(g) = \text{tr}(\alpha^{-1}W^{-1}R) \in \{0, 1\} \quad \text{for } g \in \text{GO}(V, q).$$

Straightforward verifications show that Δ is a group homomorphism

$$\Delta \colon \text{GO}(V, q) \to \mathbb{Z}/2\mathbb{Z},$$

called the *Dickson invariant*. We write $\text{GO}^+(V, q)$ for the kernel of this homomorphism. Its elements are called *proper* similitudes, and the similitudes which are mapped to 1 under Δ are called *improper*. We also let

$$\text{O}^+(V, q) = \{\, g \in \text{O}(V, q) \mid \Delta(g) = 0 \,\} \quad \text{and} \quad \text{PGO}^+(V, q) = \text{GO}^+(V, q)/F^\times.$$

(12.13) Example. Let $\dim V = n = 2m$. For any anisotropic vector $v \in V$, the *hyperplane reflection* $\rho_v \colon V \to V$ is defined in arbitrary characteristic by

$$\rho_v(x) = x - vq(v)^{-1}b_q(v, x) \quad \text{for } x \in V.$$

This map is an isometry of (V, q). We claim that it is improper.

This is clear if $\text{char } F \neq 2$, since the matrix of ρ_v with respect to an orthogonal basis whose first vector is v is diagonal with diagonal entries $(-1, 1, \ldots, 1)$, hence $\det \rho_v = -1$.

If $\text{char } F = 2$, we compute $\Delta(\rho_v)$ by means of a symplectic basis $(e_1, \ldots, e_n)$ of (V, b_q) such that $e_1 = v$. With respect to that basis, the quadratic form q is represented by the matrix

$$M = \begin{pmatrix} M_1 & & 0 \\ & \ddots & \\ 0 & & M_m \end{pmatrix} \quad \text{where} \quad M_i = \begin{pmatrix} q(e_{2i-1}) & 0 \\ 1 & q(e_{2i}) \end{pmatrix},$$

and the map ρ_v is represented by

$$
G = \begin{pmatrix} G_1 & & 0 \\ & \ddots & \\ 0 & & G_m \end{pmatrix} \quad \text{where } G_1 = \begin{pmatrix} 1 & q(e_1)^{-1} \\ 0 & 1 \end{pmatrix}, \ G_i = I, \ i \geq 2.
$$

As a matrix R such that $G^t M G + M = R + R^t$ we may take

$$
R = \begin{pmatrix} R_1 & & 0 \\ & \ddots & \\ 0 & & R_m \end{pmatrix} \quad \text{where } R_1 = \begin{pmatrix} 0 & 1 \\ 0 & 0 \end{pmatrix}, \ R_i = I, \ i \geq 2.
$$

It is readily verified that $\mathrm{tr}(W^{-1}R) = 1$, hence $\Delta(\rho_v) = 1$, proving the claim.

12.B. Similitudes of algebras with involution. In view of the characterization of similitudes of bilinear or hermitian spaces by means of the adjoint involution (see (12.2)), the following definition is natural:

(12.14) Definition. Let (A, σ) be a central simple F-algebra with involution. A *similitude* of (A, σ) is an element $g \in A$ such that

$$
\sigma(g)g \in F^\times.
$$

The scalar $\sigma(g)g$ is called the *multiplier* of g and is denoted $\mu(g)$. The set of all similitudes of (A, σ) is a subgroup of $A^\times$ which we call $\mathrm{Sim}(A, \sigma)$, and the map μ is a group homomorphism

$$
\mu \colon \mathrm{Sim}(A, \sigma) \to F^\times.
$$

It is then clear that similitudes of bilinear spaces are similitudes of their endomorphism algebras:

$$
\mathrm{Sim}\big(\mathrm{End}_F(V), \sigma_b\big) = \mathrm{Sim}(V, b)
$$

if (V, b) is a nonsingular symmetric or alternating bilinear space. There is a corresponding result for hermitian spaces.

Similitudes can also be characterized in terms of automorphisms of the algebra with involution. Recall that an automorphism of (A, σ) is an F-algebra automorphism which commutes with σ:

$$
\mathrm{Aut}_F(A, \sigma) = \{\, \theta \in \mathrm{Aut}_F(A) \mid \sigma \circ \theta = \theta \circ \sigma \,\}.
$$

Let K be the center of A, so that $K = F$ if σ is of the first kind and K is a quadratic étale F-algebra if σ is of the second kind. Define $\mathrm{Aut}_K(A, \sigma) = \mathrm{Aut}_F(A, \sigma) \cap \mathrm{Aut}_K(A)$.

(12.15) Theorem. *With the notation above,*

$$
\mathrm{Aut}_K(A, \sigma) = \{\, \mathrm{Int}(g) \mid g \in \mathrm{Sim}(A, \sigma) \,\}.
$$

There is therefore an exact sequence:

$$
1 \to K^\times \to \mathrm{Sim}(A, \sigma) \xrightarrow{\ \mathrm{Int}\ } \mathrm{Aut}_K(A, \sigma) \to 1.
$$

Proof: By the Skolem-Noether theorem, every automorphism of A over K has the form $\mathrm{Int}(g)$ for some $g \in A^\times$. Since

$$
\sigma \circ \mathrm{Int}(g) = \mathrm{Int}\big(\sigma(g)^{-1}\big) \circ \sigma,
$$

the automorphism $\mathrm{Int}(g)$ commutes with σ if and only if $\sigma(g)^{-1} \equiv g \mod K^\times$, i.e., $\sigma(g)g \in K^\times$. Since $\sigma(g)g$ is invariant under σ, the latter condition is also equivalent to $\sigma(g)g \in F^\times$. $\qquad\square$

Let $\mathrm{PSim}(A, \sigma)$ be the group of *projective similitudes*, defined as

$$\mathrm{PSim}(A, \sigma) = \mathrm{Sim}(A, \sigma)/K^\times.$$

In view of the preceding theorem, the map Int defines a natural isomorphism $\mathrm{PSim}(A, \sigma) \xrightarrow{\sim} \mathrm{Aut}_K(A, \sigma)$.

Specific notations for the groups $\mathrm{Sim}(A, \sigma)$ and $\mathrm{PSim}(A, \sigma)$ are used according to the type of σ, reflecting the notations for similitudes of bilinear or hermitian spaces:

$$\mathrm{Sim}(A, \sigma) = \begin{cases} \mathrm{GO}(A, \sigma) & \text{if } \sigma \text{ is of orthogonal type,} \\ \mathrm{GSp}(A, \sigma) & \text{if } \sigma \text{ is of symplectic type,} \\ \mathrm{GU}(A, \sigma) & \text{if } \sigma \text{ is of unitary type,} \end{cases}$$

and

$$\mathrm{PSim}(A, \sigma) = \begin{cases} \mathrm{PGO}(A, \sigma) & \text{if } \sigma \text{ is of orthogonal type,} \\ \mathrm{PGSp}(A, \sigma) & \text{if } \sigma \text{ is of symplectic type,} \\ \mathrm{PGU}(A, \sigma) & \text{if } \sigma \text{ is of unitary type.} \end{cases}$$

Similitudes with multiplier 1 are *isometries*; they make up the group $\mathrm{Iso}(A, \sigma)$:

$$\mathrm{Iso}(A, \sigma) = \{\, g \in A^\times \mid \sigma(g) = g^{-1} \,\}.$$

We also use the following notation:

$$\mathrm{Iso}(A, \sigma) = \begin{cases} \mathrm{O}(A, \sigma) & \text{if } \sigma \text{ is of orthogonal type,} \\ \mathrm{Sp}(A, \sigma) & \text{if } \sigma \text{ is of symplectic type,} \\ \mathrm{U}(A, \sigma) & \text{if } \sigma \text{ is of unitary type.} \end{cases}$$

For quadratic pairs, the corresponding notions are defined as follows:

(12.16) Definition. Let (σ, f) be a quadratic pair on a central simple F-algebra A. An *automorphism* of (A, σ, f) is an F-algebra automorphism θ of A such that

$$\sigma \circ \theta = \theta \circ \sigma \quad \text{and} \quad f \circ \theta = f.$$

A *similitude* of (A, σ, f) is an element $g \in A^\times$ such that $\sigma(g)g \in F^\times$ and $f(gsg^{-1}) = f(s)$ for all $s \in \mathrm{Sym}(A, \sigma)$. Let $\mathrm{GO}(A, \sigma, f)$ be the group of similitudes of (A, σ, f), let

$$\mathrm{PGO}(A, \sigma, f) = \mathrm{GO}(A, \sigma, f)/F^\times$$

and write $\mathrm{Aut}_F(A, \sigma, f)$ for the group of automorphisms of (A, σ, f). The same arguments as in (12.15) yield an exact sequence

$$1 \to F^\times \to \mathrm{GO}(A, \sigma, f) \xrightarrow{\mathrm{Int}} \mathrm{Aut}_F(A, \sigma, f) \to 1,$$

hence also an isomorphism

$$\mathrm{PGO}(A, \sigma, f) \xrightarrow{\sim} \mathrm{Aut}_F(A, \sigma, f).$$

For $g \in \mathrm{GO}(A, \sigma, f)$ we set $\mu(g) = \sigma(g)g \in F^\times$. The element $\mu(g)$ is called the *multiplier* of g and the map

$$\mu \colon \mathrm{GO}(A, \sigma, f) \to F^\times$$

is a group homomorphism. Its kernel is denoted $\mathrm{O}(A, \sigma, f)$.

It is clear from the definition that $\mathrm{GO}(A, \sigma, f) \subset \mathrm{Sim}(A, \sigma)$. If char $F \neq 2$, the map f is the restriction of $\frac{1}{2}\,\mathrm{Trd}_A$ to $\mathrm{Sym}(A, \sigma)$, hence the condition $f(gsg^{-1}) = f(s)$ for all $s \in \mathrm{Sym}(A, \sigma)$ holds for all $g \in \mathrm{GO}(A, \sigma)$. Therefore, we have in this case

$$\mathrm{GO}(A, \sigma, f) = \mathrm{GO}(A, \sigma), \ \mathrm{PGO}(A, \sigma, f) = \mathrm{PGO}(A, \sigma) \text{ and } \mathrm{O}(A, \sigma, f) = \mathrm{O}(A, \sigma).$$

In particular, if (V, q) is a nonsingular quadratic space over F and (σ_q, f_q) is the associated quadratic pair on $\mathrm{End}_F(V)$ (see (5.11)),

$$\mathrm{GO}\big(\mathrm{End}_F(V), \sigma_q, f_q\big) = \mathrm{GO}\big(\mathrm{End}_F(V), \sigma_q\big) = \mathrm{GO}(V, q).$$

There is a corresponding result if char $F = 2$:

(12.17) Example. Let (V, q) be a nonsingular quadratic space of even dimension over a field F of characteristic 2, and let (σ_q, f_q) be the associated quadratic pair on $\mathrm{End}_F(V)$. We claim that

$$\mathrm{GO}\big(\mathrm{End}_F(V), \sigma_q, f_q\big) = \mathrm{GO}(V, q),$$

hence also $\mathrm{PGO}\big(\mathrm{End}_F(V), \sigma_q, f_q\big) = \mathrm{PGO}(V, q)$ and $\mathrm{O}\big(\mathrm{End}_F(V), \sigma_q, f_q\big) = \mathrm{O}(V, q)$. In order to prove these equalities, observe first that the standard identification φ_q of (5.10) associated with the polar of q satisfies the following property: for all $g \in \mathrm{End}_F(V)$, and for all $v, w \in V$,

$$g \circ \varphi_q(v \otimes w) \circ \sigma_q(g) = \varphi_q\big(g(v) \otimes g(w)\big).$$

Therefore, if $g \in \mathrm{GO}\big(\mathrm{End}_F(V, \sigma_q, f_q)\big)$ and $\alpha = \mu(g) \in F^\times$, the condition

$$f_q\big(g \circ \varphi_q(v \otimes v) \circ g^{-1}\big) = f_q \circ \varphi_q(v \otimes v) \quad \text{for } v \in V$$

amounts to

$$q\big(g(v)\big) = \alpha q(v) \quad \text{for } v \in V,$$

which means that g is a similitude of the quadratic space (V, q), with multiplier α. This shows $\mathrm{GO}\big(\mathrm{End}_F(V), \sigma_q, f_q\big) \subset \mathrm{GO}(V, q)$.

For the reverse inclusion, observe that if g is a similitude of (V, q) with multiplier α, then $\sigma_q(g)g = \alpha$ since g also is a similitude of the associated bilinear space (V, b_q). Moreover, the same calculation as above shows that

$$f_q\big(g \circ \varphi_q(v \otimes v) \circ g^{-1}\big) = f_q \circ \varphi_q(v \otimes v) \quad \text{for } v \in V.$$

Since $\mathrm{Sym}\big(\mathrm{End}_F(V), \sigma_q, f_q\big)$ is spanned by elements of the form $\varphi_q(v \otimes v)$, it follows that $f_q(gsg^{-1}) = f_q(s)$ for all $s \in \mathrm{Sym}(A, \sigma)$, hence $g \in \mathrm{GO}\big(\mathrm{End}_F(V), \sigma_q, f_q\big)$. This proves the claim.

We next determine the groups of similitudes for quaternion algebras.

(12.18) Example. Let Q be a quaternion algebra with canonical (symplectic) involution γ over an arbitrary field F. Since $\gamma(q)q \in F$ for all $q \in Q$, we have

$$\mathrm{Sim}(Q, \gamma) = \mathrm{GSp}(Q, \gamma) = Q^\times.$$

Therefore, γ commutes with all the inner automorphisms of Q. (This observation also follows from the fact that γ is the unique symplectic involution of Q: for every automorphism θ, the composite $\theta \circ \gamma \circ \theta^{-1}$ is a symplectic involution, hence $\theta \circ \gamma \circ \theta^{-1} = \gamma$).

Let σ be an orthogonal involution on Q; by (2.7) we have

$$\sigma = \operatorname{Int}(u) \circ \gamma$$

for some invertible quaternion u such that $\gamma(u) = -u$ and $u \notin F$. Since γ commutes with all automorphisms of Q, an inner automorphism $\operatorname{Int}(g)$ commutes with σ if and only if it commutes with $\operatorname{Int}(u)$, i.e., $gu \equiv ug \mod F^\times$. If $\lambda \in F^\times$ is such that $gu = \lambda ug$, then by taking the reduced norm of both sides of this equation we obtain $\lambda^2 = 1$, hence $gu = \pm ug$. The group of similitudes of (Q, σ) therefore consists of the invertible elements which commute or anticommute with u. If char $F = 2$, we thus obtain

$$\operatorname{GO}(Q, \sigma) = F(u)^\times.$$

If char $F \neq 2$, let v be any invertible element which anticommutes with u; then

$$\operatorname{GO}(Q, \sigma) = F(u)^\times \cup \left(F(u)^\times \cdot v \right).$$

Finally, we consider the case of quadratic pairs on Q. We assume that char $F = 2$ since, if the characteristic is different from 2, the similitudes of a quadratic pair (σ, f) are exactly the similitudes of the orthogonal involution σ. Since char $F = 2$, every involution which is part of a quadratic pair is symplectic, hence every quadratic pair on Q has the form (γ, f) for some linear map $f \colon \operatorname{Sym}(Q, \gamma) \to F$. Take any $\ell \in Q$ satisfying

$$f(s) = \operatorname{Trd}_Q(\ell s) \quad \text{for } s \in \operatorname{Sym}(Q, \gamma)$$

(see (5.7)). The element ℓ is uniquely determined by the quadratic pair (γ, f) up to the addition of an element in $\operatorname{Alt}(Q, \gamma) = F$, and it satisfies $\operatorname{Trd}_Q(\ell) = 1$, by (5.6) and (5.7). Therefore, there exists an element $v \in Q^\times$ such that $v^{-1}\ell v = \ell + 1$. We claim that

$$\operatorname{GO}(Q, \gamma, f) = F(\ell)^\times \cup \left(F(\ell)^\times \cdot v \right).$$

Since $\operatorname{GSp}(Q, \gamma) = Q^\times$, an element $g \in Q^\times$ is a similitude of (Q, γ, f) if and only if $f(gsg^{-1}) = f(s)$ for all $s \in \operatorname{Sym}(Q, \gamma)$. By definition of ℓ, this condition can be rephrased as

$$\operatorname{Trd}_Q(\ell gsg^{-1}) = \operatorname{Trd}_Q(\ell s) \quad \text{for } s \in \operatorname{Sym}(Q, \gamma).$$

Since the left-hand expression equals $\operatorname{Trd}_Q(g^{-1}\ell gs)$, this condition is also equivalent to

$$\operatorname{Trd}_Q\left((\ell - g^{-1}\ell g)s \right) = 0 \quad \text{for } s \in \operatorname{Sym}(Q, \gamma);$$

that is, $\ell - g^{-1}\ell g \in F$, since $F = \operatorname{Alt}(Q, \gamma)$ is the orthogonal space of $\operatorname{Sym}(Q, \gamma)$ for the trace bilinear form (see (2.3)). Suppose that this condition holds and let $\lambda = \ell - g^{-1}\ell g \in F$. We proceed to show that $\lambda = 0$ or 1. Since $\operatorname{Trd}_Q(\ell) = 1$ we have $\operatorname{Nrd}_Q(\ell) = \ell^2 + \ell$ and $\operatorname{Nrd}_Q(\ell + \lambda) = \ell^2 + \ell + \lambda^2 + \lambda$. On the other hand, we must have $\operatorname{Nrd}_Q(\ell) = \operatorname{Nrd}_Q(\ell + \lambda)$, since $\ell + \lambda = g^{-1}\ell g$. Therefore, $\lambda^2 + \lambda = 0$ and $\lambda = 0$ or 1. Therefore,

$$\operatorname{GO}(Q, \gamma, f) = \{ g \in Q^\times \mid g^{-1}\ell g = \ell \} \cup \{ g \in Q^\times \mid g^{-1}\ell g = \ell + 1 \},$$

and the claim is proved.

(12.19) Example. Let $A = Q_1 \otimes_F Q_2$ be a tensor product of two quaternion algebras over a field F of characteristic different from 2, and $\sigma = \gamma_1 \otimes \gamma_2$, the tensor product of the canonical involutions. A direct computation shows that the Lie algebra $\mathrm{Alt}(A, \sigma)$ decomposes as a direct sum of the (Lie) algebras of pure quaternions in Q_1 and Q_2:

$$\mathrm{Alt}(A, \sigma) = (Q_1^0 \otimes 1) \oplus (1 \otimes Q_2^0).$$

Since Lie algebras of pure quaternions are simple and since the decomposition of a semisimple Lie algebra into a direct product of simple subalgebras is unique, it follows that every automorphism $\theta \in \mathrm{Aut}_F(A, \sigma)$ preserves the decomposition above, hence also the pair of subalgebras $\{Q_1, Q_2\}$. If $Q_1 \not\simeq Q_2$, then θ must preserve separately Q_1 and Q_2; therefore, it restricts to automorphisms of Q_1 and of Q_2. Let $q_1 \in Q_1^\times$, $q_2 \in Q_2^\times$ be such that

$$\theta|_{Q_1} = \mathrm{Int}(q_1), \quad \theta|_{Q_2} = \mathrm{Int}(q_2).$$

Then $\theta = \mathrm{Int}(q_1 \otimes q_2)$; so

$$\mathrm{GO}(A, \sigma) = \{\, q_1 \otimes q_2 \mid q_1 \in Q_1^\times, q_2 \in Q_2^\times \,\}$$

and the map which carries $(q_1 \cdot F^\times, q_2 \cdot F^\times)$ to $(q_1 \otimes q_2) \cdot F^\times$ induces an isomorphism

$$(Q_1^\times / F^\times) \times (Q_2^\times / F^\times) \xrightarrow{\sim} \mathrm{PGO}(A, \sigma).$$

If $Q_1 \simeq Q_2$, then we may assume for notational convenience that $A = Q \otimes_F Q$ where Q is a quaternion algebra isomorphic to Q_1 and to Q_2. Under the isomorphism $\gamma_* \colon A \to \mathrm{End}_F(Q)$ such that $\gamma_*(q_1 \otimes q_2)(x) = q_1 x \gamma(q_2)$ for q_1, q_2, $x \in Q$, the involution $\sigma = \gamma \otimes \gamma$ corresponds to the adjoint involution with respect to the reduced norm quadratic form n_Q; therefore $\mathrm{GO}(A, \sigma)$ is the group of similitudes of the quadratic space (Q, n_Q):

$$\mathrm{GO}(A, \sigma) \simeq \mathrm{GO}(Q, n_Q).$$

(These results are generalized in §15.B).

Multipliers of similitudes. Let (A, σ) be a central simple algebra with involution of any kind over an arbitrary field F. Let $G(A, \sigma)$ be the group of multipliers of similitudes of (A, σ):

$$G(A, \sigma) = \{\, \mu(g) \mid g \in \mathrm{Sim}(A, \sigma) \,\} \subset F^\times.$$

If θ is an involution of the same kind as σ on a division algebra D Brauer-equivalent to A, we may represent A as the endomorphism algebra of some vector space V over D and σ as the adjoint involution with respect to some nonsingular hermitian or skew-hermitian form h on V:

$$(A, \sigma) = \bigl(\mathrm{End}_D(V), \sigma_h\bigr).$$

As in the split case (where $D = F$), the similitudes of (A, σ) are the similitudes of the hermitian or skew-hermitian space (V, h). It is clear from the definition that a similitude of (V, h) with multiplier $\alpha \in F^\times$ may be regarded as an isometry $(V, \alpha h) \xrightarrow{\sim} (V, h)$. Therefore, multipliers of similitudes of (A, σ) can be characterized in terms of the Witt group $W(D, \theta)$ of hermitian spaces over D with respect to θ (or of the group $W^{-1}(D, \theta)$ of skew-hermitian spaces over D with respect to θ) (see Scharlau [247, p. 239]). For the next proposition, note that the group $W^{\pm 1}(D, \theta)$ is a module over the Witt ring WF.

(12.20) Proposition. *For $(A, \sigma) \simeq \big(\mathrm{End}_D(V), \sigma_h\big)$ as above,*

$$G(A, \sigma) = \{\, \alpha \in F^\times \mid (V, h) \simeq (V, \alpha h) \,\}$$
$$= \{\, \alpha \in F^\times \mid \langle 1, -\alpha \rangle \cdot h = 0 \text{ in } W^{\pm 1}(D, \theta) \,\}.$$

In particular, if A is split and σ is symplectic, then $G(A, \sigma) = F^\times$.

Proof: The first part follows from the description above of similitudes of (A, σ). The last statement then follows from the fact that $W^{-1}(F, \mathrm{Id}_F) = 0$. $\qquad\square$

As a sample of application, one can prove the following analogue of Scharlau's norm principle for algebras with involution by the same argument as in the classical case (see Scharlau [247, Theorem 2.8.6]):

(12.21) Proposition. *For any finite extension L/F,*

$$N_{L/F}\big(G(A_L, \sigma_L)\big) \subset G(A, \sigma).$$

(12.22) Corollary. *If σ is symplectic, then*

$$F^{\times 2} \cdot \mathrm{Nrd}_A(A^\times) \subset G(A, \sigma).$$

If moreover $\deg A \equiv 2 \mod 4$, then this inclusion is an equality, and

$$G(A, \sigma) = F^{\times 2} \cdot \mathrm{Nrd}_A(A^\times) = \mathrm{Nrd}_A(A^\times).$$

Proof: Let D be the division algebra (which is unique up to a F-isomorphism) Brauer-equivalent to A. Then, $\mathrm{Nrd}_D(D^\times) = \mathrm{Nrd}_A(A^\times)$ by Draxl [84, Theorem 1, p. 146], hence it suffices to show that $\mathrm{Nrd}_D(d) \in G(A, \sigma)$ for all $d \in D^\times$ to prove the first part. Let L be a maximal subfield in D containing d. The algebra A_L is then split, hence (12.20) shows:

$$G(A_L, \sigma_L) = L^\times.$$

From (12.21), it follows that

$$N_{L/F}(d) \in G(A, \sigma).$$

This completes the proof of the first part, since $N_{L/F}(d) = \mathrm{Nrd}_D(d)$.

Next, assume $\deg A = n = 2m$, where m is odd. Since the index of A divides its degree and its exponent, we have $\mathrm{ind}\, A = 1$ or 2, hence $D = F$ or D is a quaternion algebra. In each case, $\mathrm{Nrd}_D(D^\times)$ contains $F^{\times 2}$, hence $F^{\times 2} \cdot \mathrm{Nrd}_A(A^\times) = \mathrm{Nrd}_A(A^\times)$. On the other hand, taking the pfaffian norm of each side of the equation $\sigma(g)g = \mu(g)$, for $g \in \mathrm{GSp}(A, \sigma)$, we obtain $\mathrm{Nrd}_A(g) = \mu(g)^m$ by (2.13). Since m is odd, it follows that

$$\mu(g) = \big(\mu(g)^{-(m-1)/2}\big)^2 \mathrm{Nrd}_A(g) \in F^{\times 2} \cdot \mathrm{Nrd}_A(A^\times),$$

hence $G(A, \sigma) \subset F^{\times 2} \cdot \mathrm{Nrd}_A(A^\times)$. $\qquad\square$

12.C. Proper similitudes. Suppose σ is an involution of the first kind on a central simple F-algebra A of even degree $n = 2m$. For every similitude $g \in \mathrm{Sim}(A, \sigma)$ we have

$$\mathrm{Nrd}_A(g) = \pm \mu(g)^m,$$

as can be seen by taking the reduced norm of both sides of the equation $\sigma(g)g = \mu(g)$.

(12.23) Proposition. *If* σ *is a symplectic involution on a central simple F-algebra A of degree $n = 2m$, then*

$$\mathrm{Nrd}_A(g) = \mu(g)^m \quad \text{for all } g \in \mathrm{GSp}(A, \sigma).$$

Proof: If A is split, the formula is a restatement of (12.3). The general case follows by extending scalars to a splitting field of A. $\qquad\square$

By contrast, if σ is orthogonal, we may distinguish two types of similitudes according to the sign of $\mathrm{Nrd}_A(g)\mu(g)^{-m}$:

(12.24) Definition. Let σ be an orthogonal involution on a central simple algebra A of even degree $n = 2m$ over an arbitrary field F. A similitude $g \in \mathrm{GO}(A, \sigma)$ is called *proper* (resp. *improper*) if $\mathrm{Nrd}_A(g) = +\mu(g)^m$ (resp. $\mathrm{Nrd}_A(g) = -\mu(g)^m$). (Thus, if char $F = 2$, every similitude of (A, σ) is proper; however, see (12.32).)

It is clear that proper similitudes form a subgroup of index at most 2 in the group of all similitudes; we write $\mathrm{GO}^+(A, \sigma)$ for this subgroup. The set of improper similitudes is a coset of $\mathrm{GO}^+(A, \sigma)$, which may be empty.[16] We also set:

$$\mathrm{PGO}^+(A, \sigma) = \mathrm{GO}^+(A, \sigma)/F^\times,$$

and

$$\mathrm{O}^+(A, \sigma) = \mathrm{GO}^+(A, \sigma) \cap \mathrm{O}(A, \sigma) = \{\, g \in A \mid \sigma(g)g = \mathrm{Nrd}_A(g) = 1 \,\}.$$

The elements in $\mathrm{O}^+(A, \sigma)$ are the *proper isometries*.

(12.25) Example. Let Q be a quaternion algebra with canonical involution γ over a field F of characteristic different from 2, and let $\sigma = \mathrm{Int}(u) \circ \gamma$ for some invertible pure quaternion u. Let $v \in A$ be an invertible pure quaternion which anticommutes with u. The group $\mathrm{GO}(A, \sigma)$ has been determined in (12.18); straightforward norm computations show that the elements in $F(u)^\times$ are proper similitudes, whereas those in $F(u)^\times \cdot v$ are improper, hence

$$\mathrm{GO}^+(Q, \sigma) = F(u)^\times.$$

However, no element in $F(u)^\times \cdot v$ has norm 1 unless Q is split, so

$$\mathrm{O}^+(Q, \sigma) = \mathrm{O}(Q, \sigma) = \{\, z \in F(u) \mid N_{F(u)/F}(z) = 1 \,\} \quad \text{if } Q \text{ is not split.}$$

(12.26) Example. Let $A = Q_1 \otimes_F Q_2$, a tensor product of two quaternion algebras over a field F of characteristic different from 2, and $\sigma = \gamma_1 \otimes \gamma_2$ where γ_1, γ_2 are the canonical involutions on Q_1 and Q_2.

If $Q_1 \not\simeq Q_2$, then we know from (12.19) that all the similitudes of (A, σ) are of the form $q_1 \otimes q_2$ for some $q_1 \in Q_1^\times$, $q_2 \in Q_2^\times$. We have

$$\mu(q_1 \otimes q_2) = \gamma_1(q_1)q_1 \otimes \gamma_2(q_2)q_2 = \mathrm{Nrd}_{Q_1}(q_1) \cdot \mathrm{Nrd}_{Q_2}(q_2)$$

and

$$\mathrm{Nrd}_A(q_1 \otimes q_2) = \mathrm{Nrd}_{Q_1}(q_1)^{\deg Q_2} \cdot \mathrm{Nrd}_{Q_2}(q_2)^{\deg Q_1} = \mu(q_1 \otimes q_2)^2,$$

so all the similitudes are proper:

$$\mathrm{GO}(A, \sigma) = \mathrm{GO}^+(A, \sigma) \quad \text{and} \quad \mathrm{O}(A, \sigma) = \mathrm{O}^+(A, \sigma).$$

---

[16]From the viewpoint of linear algebraic groups, one would say rather that this coset may have no rational point. It has a rational point over a splitting field of A however, since hyperplane reflections are improper isometries.

On the other hand, if $Q_1 \simeq Q_2$, then the algebra A is split, hence $\mathrm{GO}^+(A, \sigma)$ is a subgroup of index 2 in $\mathrm{GO}(A, \sigma)$.

Proper similitudes of algebras with quadratic pair. A notion of proper similitudes can also be defined for quadratic pairs. We consider only the characteristic 2 case, since if the characteristic is different from 2 the similitudes of a quadratic pair (σ, f) are the similitudes of the orthogonal involution σ.

Thus let (σ, f) be a quadratic pair on a central simple algebra A of even degree $n = 2m$ over a field F of characteristic 2. Let $\ell \in A$ be an element such that

$$f(s) = \mathrm{Trd}_A(\ell s) \quad \text{for all } s \in \mathrm{Sym}(A, \sigma)$$

(see (5.7)). For $g \in \mathrm{GO}(A, \sigma, f)$, we have $f(gsg^{-1}) = f(s)$ for all $s \in \mathrm{Sym}(A, \sigma)$, hence, as in (12.18),

$$\mathrm{Trd}_A\big((g^{-1}\ell g - \ell)s\big) = 0 \quad \text{for } s \in \mathrm{Sym}(A, \sigma).$$

By (2.3), it follows that

$$g^{-1}\ell g - \ell \in \mathrm{Alt}(A, \sigma).$$

Therefore, $f(g^{-1}\ell g - \ell) \in F$ by property (2) of the definition of a quadratic pair.

(12.27) Proposition. *The element $f(g^{-1}\ell g - \ell)$ depends only on the similitude g, and not on the choice of ℓ. Moreover, $f(g^{-1}\ell g - \ell) = 0$ or 1.*

Proof: As observed in (5.7), the element ℓ is uniquely determined by the quadratic pair (σ, f) up to the addition of an element in $\mathrm{Alt}(A, \sigma)$. If $\ell' = \ell + x + \sigma(x)$, then

$$g^{-1}\ell' g - \ell' = (g^{-1}\ell g - \ell) + (g^{-1}xg - x) + \sigma(g^{-1}xg - x),$$

since $\sigma(g) = \mu(g)g^{-1}$. We have

$$f\big(g^{-1}xg - x + \sigma(g^{-1}xg - x)\big) = \mathrm{Trd}_A(g^{-1}xg - x) = 0,$$

hence the preceding equation yields

$$f(g^{-1}\ell' g - \ell') = f(g^{-1}\ell g - \ell),$$

proving that $f(g^{-1}\ell g - \ell)$ does not depend on the choice of ℓ.

We next show that this element is either 0 or 1. By (5.7), we have $\sigma(\ell) = \ell + 1$, hence $\ell^2 + \ell = \sigma(\ell)\ell$. It follows that

$$g^{-1}\ell^2 g - \ell^2 = \mu(g)^{-1}\sigma(g)\sigma(\ell)\ell g - \sigma(\ell)\ell + (g^{-1}\ell g - \ell),$$

hence $g^{-1}\ell^2 g - \ell^2 \in \mathrm{Sym}(A, \sigma)$. We shall show successively:

(12.28) $$f(g^{-1}\ell g - \ell)^2 = f\big((g^{-1}\ell g - \ell)^2\big),$$

(12.29) $$f\big((g^{-1}\ell g - \ell)^2\big) = f(g^{-1}\ell^2 g - \ell^2),$$

(12.30) $$f(g^{-1}\ell^2 g - \ell^2) = f(g^{-1}\ell g - \ell).$$

By combining these equalities, we obtain

$$f(g^{-1}\ell g - \ell)^2 = f(g^{-1}\ell g - \ell),$$

hence $f(g^{-1}\ell g - \ell) = 0$ or 1.

We first show that $f(x)^2 = f(x^2)$ for all $x \in \mathrm{Alt}(A, \sigma)$; equation (12.28) follows, since $g^{-1}\ell g - \ell \in \mathrm{Alt}(A, \sigma)$. Let $x = y + \sigma(y)$ for some $y \in A$. Since $\sigma(\ell) + \ell = 1$, we have

$$\sigma(y)y = \sigma(y)\ell y + \sigma\big(\sigma(y)\ell y\big),$$

hence

$$f\big(\sigma(y)y\big) = \mathrm{Trd}_A\big(\sigma(y)\ell y\big).$$

The right side also equals

$$\mathrm{Trd}_A\big(\ell y\sigma(y)\big) = f\big(y\sigma(y)\big),$$

hence $f\big(\sigma(y)y + y\sigma(y)\big) = 0$. It follows that

$$f(x^2) = f\big(y^2 + \sigma(y^2)\big) = \mathrm{Trd}_A(y^2).$$

On the other hand, (0.2) shows that $\mathrm{Trd}_A(y^2) = \mathrm{Trd}_A(y)^2$; since $f(x) = \mathrm{Trd}_A(y)$, we thus have $f(x)^2 = f(x^2)$.

To prove (12.29), it suffices to show

$$f(g^{-1}\ell g\ell + \ell g^{-1}\ell g) = 0,$$

since $(g^{-1}\ell g - \ell)^2 = (g^{-1}\ell^2 g - \ell^2) + (g^{-1}\ell g\ell + \ell g^{-1}\ell g)$. By the definition of ℓ, we have

$$f(g^{-1}\ell g\ell + \ell g^{-1}\ell g) = \mathrm{Trd}_A\big(\ell(g^{-1}\ell g\ell + \ell g^{-1}\ell g)\big);$$

the right-hand expression vanishes, since $\mathrm{Trd}_A(\ell g^{-1}\ell g\ell) = \mathrm{Trd}_A(\ell^2 g^{-1}\ell g)$.

To complete the proof, we show (12.30): since g is a similitude and $\ell^2 + \ell = \sigma(\ell)\ell \in \mathrm{Sym}(A,\sigma)$, we have $f\big(g^{-1}(\ell^2 + \ell)g\big) = f(\ell^2 + \ell)$, hence

$$f(g^{-1}\ell^2 g + g^{-1}\ell g + \ell^2 + \ell) = 0$$

and therefore

$$f(g^{-1}\ell^2 g + \ell^2) = f(g^{-1}\ell g + \ell). \qquad \square$$

(12.31) Example. Suppose (V, q) is a nonsingular quadratic space of even dimension $n = 2m$ and let (σ_q, f_q) be the associated quadratic pair on $\mathrm{End}_F(V)$, so that

$$\mathrm{GO}\big(\mathrm{End}_F(V), \sigma_q, f_q\big) = \mathrm{GO}(V, q),$$

as observed in (12.17). If $\ell \in \mathrm{End}_F(V)$ is such that $f_q(s) = \mathrm{tr}(\ell s)$ for all $s \in \mathrm{Sym}\big(\mathrm{End}_F(V), \sigma_q\big)$, we claim that for all $g \in \mathrm{GO}(V, q)$ the Dickson invariant $\Delta(g)$, defined in (12.12), satisfies

$$\Delta(g) = f(g^{-1}\ell g - \ell).$$

Since the right-hand expression does not depend on the choice of ℓ, it suffices to prove the claim for a particular ℓ. Pick a basis $(e_1, \ldots, e_n)$ of V which is symplectic for the alternating form b_q, i.e.,

$$b_q(e_{2i-1}, e_{2i}) = 1, \quad b_q(e_{2i}, e_{2i+1}) = 0 \quad \text{and} \quad b_q(e_i, e_j) = 0 \text{ if } |i - j| > 1,$$

and identify every endomorphism of V with its matrix with respect to that basis. An element $\ell \in \mathrm{End}_F(V)$ such that $\mathrm{tr}(\ell s) = f_q(s)$ for all $s \in \mathrm{Sym}\big(\mathrm{End}_F(V), \sigma_q\big)$ is given in (5.12); the corresponding matrix is (see the proof of (5.12))

$$\ell = \begin{pmatrix} \ell_1 & & 0 \\ & \ddots & \\ 0 & & \ell_m \end{pmatrix} \quad \text{where} \quad \ell_i = \begin{pmatrix} & 1 & q(e_{2i}) \\ q(e_{2i-1}) & 0 \end{pmatrix}.$$

On the other hand, for a matrix M representing the quadratic form q, we may choose

$$M = \begin{pmatrix} M_1 & & 0 \\ & \ddots & \\ 0 & & M_m \end{pmatrix} \quad \text{where} \quad M_i = \begin{pmatrix} q(e_{2i-1}) & 0 \\ 1 & q(e_{2i}) \end{pmatrix}.$$

It is readily verified that $M = W \cdot \ell$ where $W = M + M^t$. Therefore, for all invertible $g \in \operatorname{End}_F(V)$,

$$g^{-1} \ell g + \ell = W^{-1}(W g^{-1} W^{-1} M g + M).$$

Since $\sigma_q(g) = W^{-1} g^t W$, we have $W g^{-1} W^{-1} = \mu(g)^{-1} g^t$ if $g \in \operatorname{GO}(V, q)$, hence the preceding equation may be rewritten as

$$g^{-1} \ell g + \ell = \mu(g)^{-1} W^{-1}(g^t M g + \mu(g) M).$$

Let R be a matrix such that $g^t M g + \mu(g) M = R + R^t$, so that

$$\Delta(g) = \operatorname{tr}(\mu(g)^{-1} W^{-1} R).$$

We then have

$$g^{-1} \ell g + \ell = \mu(g)^{-1} W^{-1}(R + R^t) = \mu(g)^{-1} W^{-1} R + \sigma_q(\mu(g)^{-1} W^{-1} R),$$

hence

$$f_q(g^{-1} \ell g + \ell) = \operatorname{tr}(\mu(g)^{-1} W^{-1} R),$$

and the claim is proved.

Note that this result yields an alternate proof of the part of (12.27) saying that $f(g^{-1} \ell g + \ell) = 0$ or 1 for all $g \in \operatorname{GO}(A, \sigma, f)$. One invokes (12.7) after scalar extension to a splitting field.

(12.32) Definition. Let (σ, f) be a quadratic pair on a central simple algebra A of even degree over a field F of characteristic 2. In view of (12.27), we may set

$$\Delta(g) = f(g^{-1} \ell g - \ell) \in \{0, 1\} \quad \text{for } g \in \operatorname{GO}(A, \sigma, f),$$

where $\ell \in A$ is such that $f(s) = \operatorname{Trd}_A(\ell s)$ for all $s \in \operatorname{Sym}(A, \sigma)$. We call Δ the *Dickson invariant*. By (12.31), this definition is compatible with (12.12) when $(A, \sigma, f) = (\operatorname{End}_F(V), \sigma_q, f_q)$.

It is easily verified that the map Δ is a group homomorphism

$$\Delta \colon \operatorname{GO}(A, \sigma, f) \to \mathbb{Z}/2\mathbb{Z}.$$

We set $\operatorname{GO}^+(A, \sigma, f)$ for its kernel; its elements are called *proper similitudes*. We also set $\operatorname{PGO}^+(A, \sigma, f) = \operatorname{GO}^+(A, \sigma, f)/F^\times$.

(12.33) Example. Let Q be a quaternion algebra with canonical involution γ over a field F of characteristic 2, and let (γ, f) be a quadratic pair on Q. Choose $\ell \in Q$ satisfying $f(s) = \operatorname{Trd}_Q(\ell s)$ for all $s \in \operatorname{Sym}(Q, \gamma)$. As observed in (12.18), we have

$$\operatorname{GO}(Q, \gamma, f) = F(\ell)^\times \cup (F(\ell)^\times \cdot v)$$

where $v \in Q^\times$ satisfies $v^{-1} \ell v = \ell + 1$. For $g \in F(\ell)^\times$ we have $g^{-1} \ell g + \ell = 0$, hence $\Delta(g) = 0$. On the other hand, if $g \in F(\ell)^\times \cdot v$, then $g^{-1} \ell g + \ell = 1$, hence $\Delta(g) = 1$, by (5.6). Therefore (compare with (12.25))

$$\operatorname{GO}^+(Q, \gamma, f) = F(\ell)^\times.$$

12.D. Functorial properties. Elaborating on the observation that similitudes of a given bilinear or hermitian space induce automorphisms of its endomorphism algebra with adjoint involution (see (12.15)), we now show how similitudes between two hermitian spaces induce isomorphisms between their endomorphism algebras. In the case of quadratic spaces of odd dimension in characteristic different from 2, the relationship with endomorphism algebras takes the form of an equivalence of categories.

Let D be a division algebra with involution θ over an arbitrary field F. Let K be the center of D, and assume F is the subfield of θ-invariant elements in K (so $F = K$ if θ is of the first kind). Hermitian or skew-hermitian spaces (V, h), (V', h') over D with respect to θ are called *similar* if there exists a D-linear map $g \colon V \to V'$ and a nonzero element $\alpha \in F^\times$ such that

$$h'\big(g(x), g(y)\big) = \alpha h(x, y) \quad \text{for } x, y \in V.$$

The map g is then called a *similitude* with multiplier α.

Assuming (V, h), (V', h') nonsingular, let σ_h, $\sigma_{h'}$ be their adjoint involutions on $\operatorname{End}_D(V)$, $\operatorname{End}_D(V')$ respectively.

(12.34) Proposition. *Every similitude $g \colon (V, h) \to (V', h')$ induces a K-isomorphism of algebras with involution*

$$g_* \colon \big(\operatorname{End}_D(V), \sigma_h\big) \to \big(\operatorname{End}_D(V'), \sigma_{h'}\big)$$

defined by $g_*(f) = g \circ f \circ g^{-1}$ for $f \in \operatorname{End}_D(V)$. Further, every K-isomorphism of algebras with involution $\big(\operatorname{End}_D(V), \sigma_h\big) \to \big(\operatorname{End}_D(V'), \sigma_{h'}\big)$ has the form g_* for some similitude $g \colon (V, h) \to (V', h')$, which is uniquely determined up to a factor in $K^\times$.

Proof: It is straightforward to check that for every similitude g, the map g_* is an isomorphism of algebras with involution. On the other hand, suppose that

$$\Phi \colon \big(\operatorname{End}_D(V), \sigma_h\big) \to \big(\operatorname{End}_D(V'), \sigma_{h'}\big)$$

is a K-isomorphism of algebras with involution. We then have $\dim_D V = \dim_D V'$, and we use Φ to define on V' the structure of a left $\operatorname{End}_D(V) \otimes_K D^{\mathrm{op}}$-module, by

$$(f \otimes d^{\mathrm{op}}) * v' = \Phi(f)(v')d \quad \text{for } f \in \operatorname{End}_D(V), d \in D, v' \in V'.$$

The space V also is a left $\operatorname{End}_D(V) \otimes_K D^{\mathrm{op}}$-module, with the action defined by

$$(f \otimes d^{\mathrm{op}}) * v = f(v)d \quad \text{for } f \in \operatorname{End}_D(V), d \in D, v \in V.$$

Since $\dim_K V = \dim_K V'$, it follows from (1.9) that V and V' are isomorphic as $\operatorname{End}_D(V) \otimes_K D^{\mathrm{op}}$-modules. Hence, there exists a D-linear bijective map

$$g \colon V \to V'$$

such that $f * \big(g(v')\big) = g \circ f(v')$ for all $f \in \operatorname{End}_D(V)$, $v' \in V'$; this means

$$\Phi(f) \circ g = g \circ f \quad \text{for } f \in \operatorname{End}_D(V).$$

It remains to show that g is a similitude, and that it is uniquely determined up to a factor in $K^\times$. To prove the first part, define a hermitian form h_0 on V by

$$h_0(v, w) = h'\big(g(v), g(w)\big) \quad \text{for } v, w \in V.$$

For all $f \in \operatorname{End}_D(V)$, we then have

$$h_0\big(v, f(w)\big) = h'\big(g(v), \Phi(f) \circ g(w)\big).$$

Since Φ is an isomorphism of algebras with involution, $\sigma_{h'}\big(\Phi(f)\big) = \Phi\big(\sigma_h(f)\big)$, hence the right-hand expression may be rewritten as

$$h'\big(\Phi\big(\sigma_h(f)\big) \circ g(v), g(w)\big) = h_0\big(\sigma_h(f)(v), w\big).$$

Therefore, σ_h is the adjoint involution with respect to h_0. By (4.2), it follows that $h_0 = \alpha h$ for some $\alpha \in F^\times$, hence g is a similitude with multiplier α, and $\Phi = g_*$.

If g, $g' \colon (V, h) \to (V', h')$ are similitudes such that $g_* = g'_*$, then $g^{-1}g' \in \mathrm{End}_D(V)$ commutes with every $f \in \mathrm{End}_D(V)$, hence $g \equiv g' \mod K^\times$. $\qquad\square$

(12.35) Corollary. *All hyperbolic involutions of the same type on a central simple algebra are conjugate. Similarly, all hyperbolic quadratic pairs on a central simple algebra are conjugate.*

Proof: Let A be a central simple algebra, which we represent as $\mathrm{End}_D(V)$ for some vector space V over a division algebra D, and let σ, σ' be hyperbolic involutions of the same type on A. These involutions are adjoint to hyperbolic hermitian or skew-hermitian forms h, h' on V, by (6.7). Since all the hyperbolic forms on V are isometric, the preceding proposition shows that $(A, \sigma) \simeq (A, \sigma')$, hence σ and σ' are conjugate.

Consider next two hyperbolic quadratic pairs (σ, f) and (σ', f') on A. The involutions σ and σ' are hyperbolic, hence conjugate, by the first part of the proof. After a suitable conjugation, we may thus assume $\sigma' = \sigma$. By (6.14), there are idempotents e, $e' \in A$ such that $f(s) = \mathrm{Trd}_A(es)$ and $f'(s) = \mathrm{Trd}_A(e's)$ for all $s \in \mathrm{Sym}(A, \sigma)$.

Claim. There exists $x \in A^\times$ such that $\sigma(x)x = 1$ and $e = xe'x^{-1}$.

The claim yields

$$f'(s) = \mathrm{Trd}_A(x^{-1}exs) = \mathrm{Trd}_A(exsx^{-1}) = f(xsx^{-1}) \quad \text{for all } s \in \mathrm{Sym}(A, \sigma),$$

hence x conjugates (σ, f) into (σ', f').

To prove the claim, choose a representation of A:

$$(A, \sigma) = \big(\mathrm{End}_D(V), \sigma_h\big)$$

for some hyperbolic hermitian space (V, h) over a division algebra D. As in the proof of (6.7), we may find a pair of complementary totally isotropic subspaces U, W (resp. U', W') in V such that e is the projection on U parallel to W and e' is the projection on U' parallel to W'. It is easy to find an isometry of V which maps U' to U and W' to W; every such isometry x satisfies $\sigma(x)x = 1$ and $e = xe'x^{-1}$. $\qquad\square$

There is an analogue to (12.34) for quadratic pairs:

(12.36) Proposition. *Let (V, q) and (V', q') be even-dimensional and nonsingular quadratic spaces over a field F. Every similitude $g \colon (V, q) \to (V', q')$ induces an F-isomorphism of algebras with quadratic pair*

$$g_* \colon \big(\mathrm{End}_F(V), \sigma_q, f_q\big) \to \big(\mathrm{End}_F(V'), \sigma_{q'}, f_{q'}\big)$$

defined by

$$g_*(h) = g \circ h \circ g^{-1} \quad \text{for } h \in \mathrm{End}_F(V).$$

Moreover, every F-isomorphism

$$\big(\mathrm{End}_F(V), \sigma_q, f_q\big) \to \big(\mathrm{End}_F(V'), \sigma_{q'}, f_{q'}\big)$$

of algebras with quadratic pair is of the form g_* for some similitude $g\colon (V, q) \to$
(V', q'), which is uniquely determined up to a factor in $F^\times$.

Proof: The same arguments as in the proof of (12.34) apply here. Details are left
to the reader.   □

Proposition (12.34) shows that mapping every hermitian or skew-hermitian
space (V, h) to the algebra $\big(\operatorname{End}_D(V), \sigma_h\big)$ defines a full functor from the category
of nonsingular hermitian or skew-hermitian spaces over D, where the morphisms
are the similitudes, to the category of central simple algebras with involution where
the morphisms are the isomorphisms. In the particular case where the degree is
odd and the characteristic is different from 2, this functor can be used to set up an
equivalence of categories, as we show in (12.41) below.

A particular feature of the categories we consider here (and in the next chapter)
is that all the morphisms are invertible (i.e., isomorphisms). A category which has
this property is called a *groupoid*. Equivalences of groupoids may be described in a
very elementary way, as the next proposition shows. For an arbitrary category X,
let $\operatorname{Isom}(X)$ be the class[17] of isomorphism classes of objects in X. Every functor
$\mathbf{S}\colon X \to Y$ induces a map $\operatorname{Isom}(X) \to \operatorname{Isom}(Y)$ which we also denote by $\mathbf{S}$.

(12.37) Proposition. *Let X, Y be groupoids. A covariant functor $\mathbf{S}\colon X \to Y$*
defines an equivalence of categories if and only if the following conditions hold:

(1) *the induced map $\mathbf{S}\colon \operatorname{Isom}(X) \to \operatorname{Isom}(Y)$ is a bijection*;
(2) *for each $X \in X$, the induced map $\operatorname{Aut}_X(X) \to \operatorname{Aut}_Y\big(\mathbf{S}(X)\big)$ is a bijection.*

Proof: The conditions are clearly necessary. Suppose that the covariant functor
$\mathbf{S}$ satisfies conditions (1) and (2) above. If X, $X' \in X$ and $g\colon \mathbf{S}(X) \to \mathbf{S}(X')$
is a morphism in Y, then $\mathbf{S}(X)$ and $\mathbf{S}(X')$ are in the same isomorphism class
of Y, hence (1) implies that X and X' are isomorphic. Let $f\colon X \to X'$ be an
isomorphism. Then $g \circ \mathbf{S}(f)^{-1} \in \operatorname{Aut}_Y\big(\mathbf{S}(X')\big)$, hence $g \circ \mathbf{S}(f)^{-1} = \mathbf{S}(h)$ for some
$h \in \operatorname{Aut}_X(X')$. It follows that $g = \mathbf{S}(h \circ f)$, showing that the functor $\mathbf{S}$ is full. It
is also faithful: if f, $g\colon X \to X'$ are morphisms in X, then $\mathbf{S}(f) = \mathbf{S}(g)$ implies
$\mathbf{S}(f^{-1} \circ g) = \operatorname{Id}_{\mathbf{S}(X)}$, hence $f = g$ by (2). Since every object in Y is isomorphic
to an object of the form $\mathbf{S}(X)$ with $X \in X$, it follows that $\mathbf{S}$ is an equivalence of
categories (see Mac Lane [176, p. 91]).   □

(12.38) Remarks. (1) The proof above also applies *mutatis mutandis* to contravariant functors, showing that the same conditions as in (12.37) characterize the
contravariant functors which define anti-equivalence of categories.
(2) The bijection $\operatorname{Aut}_X(X) \xrightarrow{\sim} \operatorname{Aut}_Y\big(\mathbf{S}(X)\big)$ induced by an equivalence of categories is a group isomorphism if the same convention is used in X and Y for
mapping composition. It is an anti-isomorphism if opposite conventions are used
in X and Y. By contrast, the bijection induced by an anti-equivalence of categories
is an anti-isomorphism if the same convention is used in X and Y, and it is an
isomorphism if opposite conventions are used.

For the rest of this section F is a field of characteristic different from 2. For
any integer $n \geq 1$, let B'_n denote the groupoid of central simple F-algebras of

---

[17]For all the categories we consider in the sequel, this class is a set.

degree $2n + 1$ with involution of the first kind,[18] where the morphisms are the F-algebra isomorphisms which preserve the involutions. Note that these algebras are necessarily split, and the involution is necessarily of orthogonal type, by (2.8).

For any integer $n \geq 1$, let Q_n be the groupoid of all nonsingular quadratic spaces of dimension n over the field F, where the morphisms are the isometries, and let Q_n^1 be the full subcategory of quadratic spaces with trivial discriminant. For $(V, q) \in Q_n$, let σ_q denote the adjoint involution on $\mathrm{End}_F(V)$ with respect to (the polar of) q. If $(V, q) \in Q_{2n+1}$, then $\big(\mathrm{End}_F(V), \sigma_q\big) \in B_n'$, and we have a functor

$$\mathbf{End} \colon Q_{2n+1} \to B_n'$$

given by mapping (V, q) to $\big(\mathrm{End}_F(V), \sigma_q\big)$, as observed in (12.34).

(12.39) Proposition. *The functor* **End** *defines a bijection between the sets of isomorphism classes*:

$$\mathbf{End} \colon \mathrm{Isom}(Q_{2n+1}^1) \xrightarrow{\sim} \mathrm{Isom}(B_n').$$

Proof: By (2.8), every algebra with involution in B_n' is isomorphic to an algebra with involution of the form $\big(\mathrm{End}_F(V), \sigma_q\big)$ for some quadratic space (V, q) of dimension $2n + 1$. Since the adjoint involution does not change when the quadratic form is multiplied by a scalar, we may substitute (disc q)q for q and thus assume disc $q = 1$. Therefore, the map induced by **End** on isomorphism classes is surjective.

On the other hand, suppose

$$\Phi \colon \big(\mathrm{End}_F(V), \sigma_q\big) \to \big(\mathrm{End}_F(V'), \sigma_{q'}\big)$$

is an isomorphism, for some quadratic spaces (V, q), $(V', q') \in Q_{2n+1}^1$. By (12.34), we may find a similitude $g \colon (V, q) \to (V', q')$ such that $\Phi = g_*$. This similitude may be regarded as an isometry $(V, \alpha q) \xrightarrow{\sim} (V', q')$, where α is the multiplier of g. Since disc $q = $ disc q' and $\dim V = \dim V'$ is odd, we must have $\alpha = 1$, hence g is an isometry $(V, q) \xrightarrow{\sim} (V', q')$. $\qquad\square$

Even though it defines a bijection between the sets of isomorphism classes, the functor **End** is not an equivalence between Q_{2n+1}^1 and B_n': this is because the group of automorphisms of the algebra with involution $\big(\mathrm{End}_F(V), \sigma_q\big)$ is

$$\mathrm{Aut}_F\big(\mathrm{End}_F(V), \sigma_q\big) = \mathrm{PGO}(V, q) = \mathrm{O}^+(V, q)$$

(the second equality follows from (12.4)), whereas the group of automorphisms of (V, q) is $\mathrm{O}(V, q)$. However, we may define some additional structure on quadratic spaces to restrict the automorphism group and thereby obtain an equivalence of categories.

(12.40) Definition. Let (V, q) be a quadratic space of odd dimension and trivial discriminant over a field F of characteristic different from 2. The center Z of the Clifford algebra $C(V, q)$ is then an étale quadratic extension of F isomorphic to $F \times F$. An *orientation* of (V, q) is an element $\zeta \in Z \smallsetminus F$ such that $\zeta^2 = 1$. Thus, each quadratic space (V, q) as above has two possible orientations which differ by a sign. Triples (V, q, ζ) are called *oriented quadratic spaces*.

---

[18]This notation is motivated by the fact that the automorphism group of each object in this groupoid is a classical group of type B_n: see Chapter VI. However this groupoid is only defined for fields of characteristic different from 2.

Every isometry $g \colon (V, q) \to (V', q')$ induces an isomorphism $g_* \colon C(V, q) \xrightarrow{\sim} C(V', q')$ which carries an orientation of (V, q) to an orientation of (V', q'). The isometries $g \colon (V, q) \to (V, q)$ which preserve a given orientation form the group $O^+(V, q)$.

Let B_n be the groupoid of oriented quadratic spaces of dimension $2n + 1$ over F. The objects of B_n are triples (V, q, ζ) where (V, q) is a quadratic space of dimension $2n + 1$ over F with trivial discriminant and ζ is an orientation of (V, q), and the morphisms are the orientation-preserving isometries. For each $(V, q, \zeta) \in B_n$, the map $-\mathrm{Id}_V \colon V \to V$ defines an isomorphism $(V, q, \zeta) \to (V, q, -\zeta)$, hence two oriented quadratic spaces are isomorphic if and only if the quadratic spaces are isometric. In other words, the functor which forgets the orientation defines a bijection

$$\mathrm{Isom}(B_n) \xrightarrow{\sim} \mathrm{Isom}(Q^1_{2n+1}).$$

(12.41) Theorem. *The functor* **End** *which maps every oriented quadratic space* (V, q, ζ) *in* B_n *to the algebra with involution* $\big(\mathrm{End}_F(V), \sigma_q\big) \in B'_n$ *defines an equivalence of categories*:

$$B_n \equiv B'_n.$$

Proof: Since the isomorphism classes of B_n and Q^1_{2n+1} coincide, (12.39) shows that the functor **End** defines a bijection $\mathrm{Isom}(B_n) \xrightarrow{\sim} \mathrm{Isom}(B'_n)$. Moreover, as we observed above, for every oriented quadratic space (V, q, ζ) of dimension $2n + 1$ we have

(12.42) $\mathrm{Aut}\big(\mathrm{End}_F(V), \sigma_q\big) = O^+(V, q) = \mathrm{Aut}(V, q, \zeta).$

Therefore, (12.37) shows that **End** is an equivalence of categories. $\square$

§13. Quadratic Pairs

In this section, (σ, f) is a quadratic pair on a central simple algebra A over an arbitrary field F. If the degree of A is odd, then A is split, char $F \neq 2$, and the group of similitudes of (A, σ, f) reduces to the orthogonal group of an F-vector space (see (12.4)). We therefore assume throughout this section that the degree is even, and we set

$$\deg A = n = 2m.$$

Our goal is to obtain additional information on the group $GO(A, \sigma, f)$ by relating similitudes of (A, σ, f) to the Clifford algebra $C(A, \sigma, f)$ and the Clifford bimodule $B(A, \sigma, f)$. We use this to define a Clifford group $\Gamma(A, \sigma, f)$, which is a twisted analogue of the special Clifford group of a quadratic space, and also define an extended Clifford group $\Omega(A, \sigma, f)$. These constructions are used to prove an analogue of a classical theorem of Dieudonné on the multipliers of similitudes.

13.A. Relation with the Clifford structures. Since the Clifford algebra $C(A, \sigma, f)$ and the Clifford bimodule $B(A, \sigma, f)$ are canonically associated to (A, σ, f), every automorphism in $\mathrm{Aut}_F(A, \sigma, f)$ induces automorphisms of $C(A, \sigma, f)$ and $B(A, \sigma, f)$. Our purpose in this section is to investigate these automorphisms.

The Clifford algebra. Every automorphism $\theta \in \operatorname{Aut}_F(A, \sigma, f)$ induces an automorphism

$$C(\theta) \in \operatorname{Aut}_F\big(C(A, \sigma), \underline{\sigma}\big).$$

Explicitly, $C(\theta)$ can be defined as the unique automorphism of $C(A, \sigma, f)$ such that

$$C(\theta)\big(c(a)\big) = c\big(\theta(a)\big) \quad \text{for } a \in A,$$

where $c \colon A \to C(A, \sigma, f)$ is the canonical map (8.13). We thereby obtain a canonical group homomorphism

$$C \colon \operatorname{Aut}_F(A, \sigma, f) \to \operatorname{Aut}_F\big(C(A, \sigma, f), \underline{\sigma}\big).$$

Slightly abusing notation, we also call C the homomorphism

$$C \colon \operatorname{GO}(A, \sigma, f) \to \operatorname{Aut}_F\big(C(A, \sigma, f), \underline{\sigma}\big)$$

obtained by composing the preceding map with the epimorphism

$$\operatorname{Int} \colon \operatorname{GO}(A, \sigma, f) \to \operatorname{Aut}_F(A, \sigma, f)$$

of (12.15). Thus, for $g \in \operatorname{GO}(A, \sigma, f)$ and $a \in A$,

$$C(g)\big(c(a)\big) = c(gag^{-1}).$$

(13.1) Proposition. *Suppose A is split; let $(A, \sigma, f) = \big(\operatorname{End}_F(V), \sigma_q, f_q\big)$ for some nonsingular quadratic space (V, q). Then, under the standard identifications $\operatorname{GO}(A, \sigma, f) = \operatorname{GO}(V, q)$ (see (12.17)) and $C(A, \sigma, f) = C_0(V, q)$ (see (8.8)), the canonical map $C \colon \operatorname{GO}(V, q) \to \operatorname{Aut}_F\big(C_0(V, q)\big)$ is defined by*

$$C(g)(v_1 \cdots v_{2r}) = \mu(g)^{-r} g(v_1) \cdots g(v_{2r})$$

for $g \in \operatorname{GO}(V, q)$ and $v_1, \ldots, v_{2r} \in V$.

Proof: It suffices to check the formula above on generators $v \cdot w$ of $C_0(V, q)$. For $v, w \in V$, the product $v \cdot w$ in $C(V, q)$ is the image of $v \otimes w$ under the canonical map c: we thus have $v \cdot w = c(v \otimes w)$, hence

$$C(g)(v \cdot w) = c\big(g \circ (v \otimes w) \circ g^{-1}\big).$$

Let $\alpha = \mu(g)$ be the multiplier of g; then $\sigma(g)^{-1} = \alpha^{-1} g$, hence, for $x \in V$,

$$\big(g \circ (v \otimes w) \circ g^{-1}\big)(x) = g(v) b_q\big(w, g^{-1}(x)\big) = g(v) b_q\big(\alpha^{-1} g(w), x\big).$$

Therefore, $\big(g \circ (v \otimes w) \circ g^{-1}\big)(x) = \big(\alpha^{-1} g(v) \otimes g(w)\big)(x)$, which shows

$$g \circ (v \otimes w) \circ g^{-1} = \alpha^{-1} g(v) \otimes g(w),$$

hence $c\big(g \circ (v \otimes w) \circ g^{-1}\big) = \alpha^{-1} g(v) \cdot g(w)$. $\qquad\square$

Note that, for $g \in \operatorname{GO}(A, \sigma, f)$, the automorphism $C(g)$ of $C(A, \sigma, f)$ is F-linear but is not necessarily the identity on the center of $C(A, \sigma, f)$. The behavior of $C(g)$ on the center in fact determines whether g is proper, as the next proposition shows.

(13.2) Proposition. *A similitude $g \in \operatorname{GO}(A, \sigma, f)$ is proper if and only if $C(g)$ restricts to the identity map on the center Z of $C(A, \sigma, f)$.*

Proof: Suppose first that char $F = 2$. Choose $\ell \in A$ satisfying $f(s) = \mathrm{Trd}_A(\ell s)$ for all $s \in \mathrm{Sym}(A, \sigma)$ (see (5.7)). By (8.27), we have $Z = F[c(\ell)]$, hence it suffices to show

$$C(g)\big(c(\ell)\big) = c(\ell) + \Delta(g) \quad \text{for } g \in \mathrm{GO}(A, \sigma, f).$$

For $g \in \mathrm{GO}(A, \sigma, f)$ we have

$$\Delta(g) = f(g^{-1}\ell g - \ell);$$

since g is a similitude, the right side also equals

$$f\big(g(g^{-1}\ell g - \ell)g^{-1}\big) = f(g\ell g^{-1} - \ell).$$

On the other hand, since $g\ell g^{-1} - \ell \in \mathrm{Sym}(A, \sigma)$, we have

$$f(g\ell g^{-1} - \ell) = c(g\ell g^{-1} - \ell),$$

hence

$$\Delta(g) = c(g\ell g^{-1}) - c(\ell) = C(g)\big(c(\ell)\big) - c(\ell),$$

proving the proposition when char $F = 2$.

Suppose now that char $F \neq 2$. It suffices to check the split case; we may thus assume $(A, \sigma, f) = \big(\mathrm{End}_F(V), \sigma_q, f_q\big)$ for some nonsingular quadratic space (V, q), and use the standard identifications and the preceding proposition. Let $(e_1, \ldots, e_{2m})$ be an orthogonal basis of (V, q). Recall that $e_1 \cdots e_{2m} \in Z \smallsetminus F$. For $g \in \mathrm{GO}(A, \sigma, f) = \mathrm{GO}(V, q)$, we have

$$C(g)(e_1 \cdots e_{2m}) = \mu(g)^{-m} g(e_1) \cdots g(e_{2m}).$$

On the other hand, a calculation in the Clifford algebra shows that

$$g(e_1) \cdots g(e_{2m}) = \det(g) e_1 \cdots e_{2m};$$

hence $e_1 \cdots e_{2m}$ is fixed by $C(g)$ if and only if $\det(g) = \mu(g)^m$. This proves the proposition in the case where char $F \neq 2$. $\qquad\square$

In view of this proposition, the Dickson invariant $\Delta \colon \mathrm{GO}(A, \sigma, f) \to \mathbb{Z}/2\mathbb{Z}$ defined in (12.32) may alternately be defined by

(13.3) $$\Delta(g) = \begin{cases} 0 & \text{if } C(g)|_Z = \mathrm{Id}_Z, \\ 1 & \text{if } C(g)|_Z \neq \mathrm{Id}_Z. \end{cases}$$

The image of the canonical map C has been determined by Wonenburger in characteristic different from 2:

(13.4) Proposition. *If* $\deg A > 2$, *the canonical homomorphism*

$$C \colon \mathrm{PGO}(A, \sigma, f) = \mathrm{Aut}_F(A, \sigma, f) \to \mathrm{Aut}_F\big(C(A, \sigma, f), \underline{\sigma}\big)$$

is injective. If char $F \neq 2$, *the image of* C *is the group of those automorphisms which preserve the image* $c(A)$ *of* A *under the canonical map* $c \colon A \to C(A, \sigma, f)$.

Proof: If $\theta \in \mathrm{Aut}_F(A, \sigma)$ lies in the kernel of C, then

$$c\big(\theta(a)\big) = c(a) \quad \text{for } a \in A,$$

since the left side is the image of $c(a)$ under $C(\theta)$. By applying the map δ of (8.15), we obtain

$$\theta\big(a - \sigma(a)\big) = a - \sigma(a) \quad \text{for } a \in A,$$

hence θ is the identity on $\mathrm{Alt}(A, \sigma)$. Since $\deg A > 2$, (2.26) shows that $\mathrm{Alt}(A, \sigma)$ generates A, hence $\theta = \mathrm{Id}_A$, proving the injectivity of C.

It follows from the definition that every automorphism of the form $C(\theta)$ maps $c(A)$ to itself. Conversely, suppose ψ is an automorphism of $C(A, \sigma, f)$ which preserves $c(A)$, and suppose $\operatorname{char} F \neq 2$. The map f is then uniquely determined by σ, so we may denote $C(A, \sigma, f)$ simply by $C(A, \sigma)$. The restriction of ψ to

$$c(A)_0 = c(A) \cap \mathrm{Skew}\big(C(A, \sigma), \underline{\sigma}\big) = \{\, x \in c(A) \mid \mathrm{Trd}(x) = 0 \,\}$$

is a Lie algebra automorphism. By (8.18), the Lie algebra $c(A)_0$ is isomorphic to $\mathrm{Alt}(A, \sigma)$ via δ, with inverse isomorphism $\frac{1}{2}c$; therefore, there is a corresponding Lie automorphism ψ' of $\mathrm{Alt}(A, \sigma)$ such that

$$c\big(\psi'(a)\big) = \psi\big(c(a)\big) \quad \text{for } a \in \mathrm{Alt}(A, \sigma).$$

Let L be a splitting field of A. A theorem[19] of Wonenburger [316, Theorem 4] shows that the automorphism $\psi_L = \psi \otimes \mathrm{Id}_L$ of $C(A, \sigma)_L = C(A_L, \sigma_L)$ is induced by a similitude g of (A_L, σ_L), hence

$$\psi'_L(a) = gag^{-1} \quad \text{for } a \in \mathrm{Alt}(A_L, \sigma_L).$$

Therefore, the automorphism ψ'_L of $\mathrm{Alt}(A_L, \sigma_L)$ extends to an automorphism of (A_L, σ_L). By (2.25), ψ' extends to an automorphism θ of (A, σ), and this automorphism satisfies $C(\theta) = \psi$ since $c\big(\theta(a)\big) = c\big(\psi'(a)\big) = \psi\big(c(a)\big)$ for all $a \in \mathrm{Alt}(A, \sigma)$. $\qquad \square$

If $\deg A = 2$, then $C(A, \sigma) = Z$, so $\mathrm{Aut}_F\big(C(A, \sigma), \underline{\sigma}\big) = \{\mathrm{Id}, \iota\}$ and the canonical homomorphism C maps $\mathrm{PGO}^+(A, \sigma)$ to Id, so C is not injective.

The Clifford bimodule. The bimodule $B(A, \sigma, f)$ is canonically associated to (A, σ, f), just as the Clifford algebra $C(A, \sigma, f)$ is. Therefore, every automorphism $\theta \in \mathrm{Aut}_F(A, \sigma, f)$ induces a bijective linear map

$$B(\theta) \colon B(A, \sigma, f) \to B(A, \sigma, f).$$

This map satisfies

$$B(\theta)(a^b) = \theta(a)^b \quad \text{for } a \in A$$

(where $b \colon A \to B(A, \sigma, f)$ is the canonical map of (9.6)) and

(13.5) $\qquad B(\theta)\big(a \cdot (c_1 * x \cdot c_2)\big) = \theta(a) \cdot \big(C(\theta)(c_1) * B(\theta)(x) \cdot C(\theta)(c_2)\big)$

for $a \in A$, c_1, $c_2 \in C(A, \sigma, f)$ and $x \in B(A, \sigma, f)$. Explicitly, $B(\theta)$ is induced by the map $\underline{\theta} \colon T^+(\underline{A}) \to T^+(\underline{A})$ such that

$$\underline{\theta}(a_1 \otimes \cdots \otimes a_r) = \theta(a_1) \otimes \cdots \otimes \theta(a_r).$$

As in the previous case, we modify the domain of definition of B to be the group $\mathrm{GO}(A, \sigma, f)$, by letting $B(g) = B\big(\mathrm{Int}(g)\big)$ for $g \in \mathrm{GO}(A, \sigma, f)$. We thus obtain a canonical homomorphism

$$B \colon \mathrm{GO}(A, \sigma, f) \to \mathrm{GL}_F\, B(A, \sigma, f).$$

For $g \in \mathrm{GO}(A, \sigma, f)$, we also define a map

(13.6) $\qquad\qquad\qquad \beta(g) \colon B(A, \sigma, f) \to B(A, \sigma, f)$

---

[19] This theorem is proved under the assumption that $\operatorname{char} F \neq 2$. See Exercise 4 for a sketch of proof.

by

$$x^{\beta(g)} = g \cdot B(g^{-1})(x) \quad \text{for } x \in B(A, \sigma, f).$$

The map $\beta(g)$ is a homomorphism of left A-modules, since for $a \in A$ and $x \in B(A, \sigma, f)$,

$$(a \cdot x)^{\beta(g)} = g \cdot (g^{-1}ag) \cdot B(g^{-1})(x) = a \cdot x^{\beta(g)}.$$

Moreover, the following equation is a straightforward consequence of the definitions:

(13.7) $$(1^b)^{\beta(g)} = g^b.$$

Since b is injective, it follows that the map

$$\beta \colon \operatorname{GO}(A, \sigma, f) \to \operatorname{Aut}_A\big(B(A, \sigma, f)\big)$$

is injective. This map also is a homomorphism of groups; to check this, we compute, for g, $h \in \operatorname{GO}(A, \sigma, f)$ and $x \in B(A, \sigma, f)$:

$$x^{\beta(g) \circ \beta(h)} = \big(g \cdot B(g^{-1})(x)\big)^{\beta(h)}.$$

Since $\beta(h)$ is a homomorphism of left A-modules, the right-hand expression equals

$$\begin{aligned}
g \cdot \big(B(g^{-1})(x)\big)^{\beta(h)} &= g \cdot \big(h \cdot B(h^{-1}) \circ B(g^{-1})(x)\big) \\
&= gh \cdot B(h^{-1}g^{-1})(x) \\
&= x^{\beta(gh)},
\end{aligned}$$

proving the claim.

Let Z be the center of $C(A, \sigma, f)$. Recall the right ${}^{\iota}C(A, \sigma, f)^{\mathrm{op}} \otimes_Z C(A, \sigma, f)$-module structure on $B(A, \sigma, f)$, which yields the canonical map

$$\nu \colon {}^{\iota}C(A, \sigma, f) \otimes_Z C(A, \sigma, f) \to \operatorname{End}_{A \otimes Z} B(A, \sigma, f)$$

of (9.9). It follows from (13.5) that the following equation holds in $\operatorname{End}_A B(A, \sigma, f)$:

(13.8) $$\beta(g) \circ \nu({}^{\iota}c_1^{\mathrm{op}} \otimes c_2) = \nu\big({}^{\iota}C(g)(c_1)^{\mathrm{op}} \otimes C(g)(c_2)\big) \circ \beta(g)$$

for all $g \in \operatorname{GO}(A, \sigma, f)$, c_1, $c_2 \in C(A, \sigma, f)$. In particular, it follows by (13.2) that $\beta(g)$ is Z-linear if and only if g is proper.

The following result describes the maps $B(g)$ and $\beta(g)$ in the split case; it follows by the same arguments as in (13.1).

(13.9) Proposition. *Suppose A is split; let*

$$(A, \sigma, f) = \big(\operatorname{End}_F(V), \sigma_q, f_q\big)$$

for some nonsingular quadratic space (V, q). Under the standard identifications $\operatorname{GO}(A, \sigma, f) = \operatorname{GO}(V, q)$ and $B(A, \sigma, f) = V \otimes C_1(V, q)$ (see (9.7)), the maps B and β are given by

$$B(g)(v \otimes w_1 \cdots w_{2r-1}) = \mu(g)^{-r} g(v) \otimes g(w_1) \cdots g(w_{2r-1})$$

and

$$(v \otimes w_1 \cdots w_{2r-1})^{\beta(g)} = \mu(g)^r v \otimes g^{-1}(w_1) \cdots g^{-1}(w_{2r-1})$$

for $g \in \operatorname{GO}(A, \sigma, f)$ and v, w_1, …, $w_{2r-1} \in V$.

13.B. Clifford groups. For a nonsingular quadratic space (V, q) over an arbitrary field F, the *special Clifford group* $\Gamma^+(V, q)$ is defined by

$$\Gamma^+(V, q) = \{\, c \in C_0(V, q)^\times \mid c \cdot V \cdot c^{-1} \subset V \,\}$$

where the product $c \cdot V \cdot c^{-1}$ is computed in the Clifford algebra $C(V, q)$ (see for instance Knus [157, Ch. 4, §6], or Scharlau [247, §9.3] for the case where char $F \neq 2$).

Although there is no analogue of the (full) Clifford algebra for an algebra with quadratic pair, we show in this section that the Clifford bimodule may be used to define an analogue of the special Clifford group. We also show that an extended Clifford group can be defined by substituting the Clifford algebra for the Clifford bimodule. These constructions are used to define spinor norm homomorphisms on the groups $O^+(A, \sigma, f)$ and $\mathrm{PGO}^+(A, \sigma, f)$.

The special Clifford group in the split case. Let (V, q) be a nonsingular quadratic space of even[20] dimension over an arbitrary field F, and let $\Gamma^+(V, q)$ be the special Clifford group defined above. Conjugation by $c \in \Gamma^+(V, q)$ in $C(V, q)$ induces an isometry of (V, q), since

$$q(c \cdot v \cdot c^{-1}) = (c \cdot v \cdot c^{-1})^2 = v^2 = q(v) \quad \text{for } v \in V.$$

We set $\chi(c)$ for this isometry:

$$\chi(c)(v) = c \cdot v \cdot c^{-1} \quad \text{for } v \in V.$$

The map $\chi \colon \Gamma^+(V, q) \to O(V, q)$ is known as the *vector representation* of the special Clifford group. The next proposition shows that its image is in $O^+(V, q)$.

(13.10) Proposition. *The elements in $\Gamma^+(V, q)$ are similitudes of the even Clifford algebra $C_0 = C_0(V, q)$ for the canonical involution τ_0 (see (8.4)). More precisely, $\tau_0(c) \cdot c \in F^\times$ for all $c \in \Gamma^+(V, q)$. The vector representation χ and the canonical homomorphism C of (13.1) fit into the following commutative diagram with exact rows:*

$$
\begin{array}{ccccccccc}
1 & \longrightarrow & F^\times & \longrightarrow & \Gamma^+(V, q) & \overset{\chi}{\longrightarrow} & O^+(V, q) & \longrightarrow & 1 \\
 & & \downarrow & & \downarrow & & \downarrow{\scriptstyle C} & & \\
1 & \longrightarrow & Z^\times & \longrightarrow & \mathrm{Sim}(C_0, \tau_0) & \overset{\mathrm{Int}}{\longrightarrow} & \mathrm{Aut}_Z(C_0, \tau_0) & \longrightarrow & 1
\end{array}
$$

where Z denotes the center of C_0.

Proof: Let τ be the canonical involution on $C(V, q)$, whose restriction to (the image of) V is the identity. For $c \in \Gamma^+(V, q)$ and $v \in V$, we have $c \cdot v \cdot c^{-1} \in V$, hence

$$c \cdot v \cdot c^{-1} = \tau(c \cdot v \cdot c^{-1}) = \tau_0(c)^{-1} \cdot v \cdot \tau_0(c).$$

This shows that the element $\tau_0(c) \cdot c$ centralizes V; since V generates $C(V, q)$, it follows that $\tau_0(c) \cdot c$ is central in $C(V, q)$, hence $\tau_0(c) \cdot c \in F^\times$. This proves $\Gamma^+(V, q) \subset \mathrm{Sim}(C_0, \tau_0)$.

The elements in $\ker \chi$ centralize V, hence the same argument as above shows $\ker \chi = F^\times$.

---

[20]Clifford groups are also defined in the odd-dimensional case, where results similar to those of this section can be established. Since we are interested in the generalization to the nonsplit case, given below, we consider only the even-dimensional case.

Let $c \in \Gamma^+(V, q)$. By (13.1), the automorphism $C(\chi(c))$ of C_0 maps $v_1 \cdots v_{2r}$ to

$$\chi(c)(v_1) \cdots \chi(c)(v_{2r}) = c \cdot (v_1 \cdots v_{2r}) \cdot c^{-1}$$

for $v_1, \ldots, v_{2r} \in V$, hence

$$C(\chi(c)) = \mathrm{Int}(c).$$

This automorphism is the identity on Z, hence (13.2) shows that $\chi(c) \in O^+(V, q)$. Moreover, the last equation proves that the diagram commutes. Therefore, it remains only to prove surjectivity of χ onto $O^+(V, q)$.

To prove that every proper isometry is in the image of χ, observe that for every $v, x \in V$ with $q(v) \neq 0$,

$$v \cdot x \cdot v^{-1} = v^{-1} b_q(v, x) - x = v q(v)^{-1} b_q(v, x) - x,$$

hence the hyperplane reflection $\rho_v \colon V \to V$ satisfies

$$\rho_v(x) = -v \cdot x \cdot v^{-1} \quad \text{for all } x \in V.$$

Therefore, for anisotropic $v_1, v_2 \in V$, we have $v_1 \cdot v_2 \in \Gamma^+(V, q)$ and

$$\chi(v_1 \cdot v_2) = \rho_{v_1} \circ \rho_{v_2}.$$

The *Cartan-Dieudonné theorem* (see Dieudonné [81, pp. 20, 42], or Scharlau [247, p. 15] for the case where char $F \neq 2$) shows that the group $O(V, q)$ is generated by hyperplane reflections, except in the case where F is the field with two elements, $\dim V = 4$ and q is hyperbolic. Since hyperplane reflections are improper isometries (see (12.13)), it follows that every proper isometry has the form

$$\rho_{v_1} \circ \cdots \circ \rho_{v_{2r}} = \chi(v_1 \cdots v_{2r})$$

for some anisotropic vectors $v_1, \ldots, v_{2r} \in V$, in the nonexceptional case.

Direct computations, which we omit, prove that χ is surjective in the exceptional case as well. $\square$

The proof shows that every element in $\Gamma^+(V, q)$ is a product of an even number of anisotropic vectors in V, except when (V, q) is the 4-dimensional hyperbolic space over the field with two elements.

The Clifford group of an algebra with quadratic pair. Let (σ, f) be a quadratic pair on a central simple algebra A of even degree over an arbitrary field F. The Clifford group consists of elements in $C(A, \sigma, f)$ which preserve the image A^b of A in the bimodule $B(A, \sigma, f)$ under the canonical map $b \colon A \to B(A, \sigma, f)$ of (9.7):

(13.11) Definition. The *Clifford group* $\Gamma(A, \sigma, f)$ is defined by

$$\Gamma(A, \sigma, f) = \{\, c \in C(A, \sigma, f)^\times \mid c^{-1} * A^b \cdot c \subset A^b \,\}.$$

Since the $C(A, \sigma, f)$-bimodule actions on $B(A, \sigma, f)$ commute with the left A-module action and since the canonical map b is a homomorphism of left A-modules, the condition defining the Clifford group is equivalent to

$$c^{-1} * 1^b \cdot c \in A^b.$$

For $c \in \Gamma(A, \sigma, f)$, define $\chi(c) \in A$ by the equation

$$c^{-1} * 1^b \cdot c = \chi(c)^b.$$

(The element $\chi(c)$ is uniquely determined by this equation, since the canonical map b is injective: see (9.7)).

(13.12) Proposition. *In the split case* $(A, \sigma, f) = \big(\operatorname{End}_F(V), \sigma_q, f_q\big)$, *the standard identifications* $C(A, \sigma, f) = C_0(V, q)$, $B(A, \sigma, f) = V \otimes C_1(V, q)$ *of* (8.8), (9.7), *induce an identification* $\Gamma(A, \sigma, f) = \Gamma^+(V, q)$, *and the map* χ *defined above is the vector representation.*

Proof: Under the standard identifications, we have $A = V \otimes V$ and $A^b = V \otimes V \subset V \otimes C_1(V, q)$. Moreover, for $c \in C(A, \sigma, f) = C_0(V, q)$ and v, $w \in V$,

(13.13) $$c^{-1} * (v \otimes w)^b \cdot c = v \otimes (c^{-1} \cdot w \cdot c).$$

Therefore, the condition $c^{-1} * A^b \cdot c \subset A^b$ amounts to:

$$v \otimes (c^{-1} \cdot w \cdot c) \in V \otimes V \quad \text{for } v, \, w \in V,$$

or $c^{-1} \cdot V \cdot c \subset V$. This proves the first assertion.

Suppose now that $c^{-1} * 1^b \cdot c = g^b$. Since b is a homomorphism of left A-modules, we then get for all v, $w \in V$:

(13.14) $$c^{-1} * (v \otimes w)^b \cdot c = (v \otimes w) \cdot (c^{-1} * 1^b \cdot c) = \big((v \otimes w) \circ g\big)^b.$$

By evaluating $(v \otimes w) \circ g$ at an arbitrary $x \in V$, we obtain

$$v b_q\big(w, g(x)\big) = v b_q\big(\sigma(g)(w), x\big) = \big(v \otimes \sigma(g)(w)\big)(x),$$

hence $(v \otimes w) \circ g = v \otimes \sigma(g)(w)$. Therefore, (13.14) yields

$$c^{-1} * (v \otimes w)^b \cdot c = \big(v \otimes \sigma(g)(w)\big)^b.$$

In view of (13.13), this shows: $\sigma(g)(w) = c^{-1} \cdot w \cdot c$, so

$$\chi(c) = \sigma(g)^{-1} = g. \qquad \square$$

By extending scalars to a splitting field of A, it follows from the proposition above and (13.10) that $\chi(c) \in \mathrm{O}^+(A, \sigma, f)$ for all $c \in \Gamma(A, \sigma, f)$. The commutative diagram of (13.10) has an analogue for algebras with quadratic pairs:

(13.15) Proposition. *For brevity, set* $C(A) = C(A, \sigma, f)$. *Let* Z *be the center of* $C(A)$ *and let* $\underline{\sigma}$ *denote the canonical involution on* $C(A)$. *For all* $c \in \Gamma(A, \sigma, f)$ *we have* $\underline{\sigma}(c)c \in F^\times$, *hence* $\Gamma(A, \sigma, f) \subset \operatorname{Sim}\big(C(A), \underline{\sigma}\big)$. *The map* χ *and the restriction to* $\mathrm{O}^+(A, \sigma, f)$ *of the canonical map*

$$C \colon \mathrm{GO}(A, \sigma, f) \to \operatorname{Aut}_F\big(C(A), \underline{\sigma}\big)$$

fit into a commutative diagram with exact rows:

$$
\begin{array}{ccccccccc}
1 & \longrightarrow & F^\times & \longrightarrow & \Gamma(A, \sigma, f) & \overset{\chi}{\longrightarrow} & \mathrm{O}^+(A, \sigma, f) & \longrightarrow & 1 \\
& & \big\downarrow & & \big\downarrow & & \big\downarrow{\scriptstyle C} & & \\
1 & \longrightarrow & Z^\times & \longrightarrow & \operatorname{Sim}\big(C(A), \underline{\sigma}\big) & \overset{\mathrm{Int}}{\longrightarrow} & \operatorname{Aut}_Z\big(C(A), \underline{\sigma}\big) & \longrightarrow & 1
\end{array}
$$

Proof: All the statements follow by scalar extension to a splitting field of A and comparison with (13.10), except for the surjectivity of χ onto $\mathrm{O}^+(A, \sigma, f)$.

To prove this last point, recall the isomorphism ν of (9.9) induced by the $C(A)$-bimodule structure on $B(A, \sigma, f)$:

$$\nu \colon {}^\iota C(A)^{\mathrm{op}} \otimes_Z C(A) \xrightarrow{\sim} \operatorname{End}_{A \otimes Z}\big(B(A, \sigma, f)\big).$$

This isomorphism satisfies

$$x^{\nu({}^{\iota}c_1^{\mathrm{op}} \otimes c_2)} = c_1 * x \cdot c_2 \quad \text{for } x \in B(A, \sigma, f), \, c_1, c_2 \in C(A).$$

For $g \in \mathrm{O}^+(A, \sigma, f)$, it follows from (13.2) that $C(g)$ is the identity on Z, hence $\beta(g)$ is an $A \otimes Z$-endomorphism of $B(A, \sigma, f)$, by (13.8). Therefore, there exists a unique element $\xi \in {}^{\iota}C(A)^{\mathrm{op}} \otimes_Z C(A)$ such that $\nu(\xi) = \beta(g)$.

In the split case, (13.12) shows that $\xi = {}^{\iota}(c^{-1})^{\mathrm{op}} \otimes c$, where $c \in \Gamma^+(V, q)$ is such that $\chi(c) = g$. Since the minimal number of terms in a decomposition of an element of a tensor product is invariant under scalar extension, it follows that $\xi = {}^{\iota}c_1^{\mathrm{op}} \otimes c_2$ for some $c_1, c_2 \in C(A)$. Moreover, if s is the switch map on ${}^{\iota}C(A)^{\mathrm{op}} \otimes_Z C(A)$, defined by

$$s({}^{\iota}c^{\mathrm{op}} \otimes c') = {}^{\iota}c'^{\mathrm{op}} \otimes c \quad \text{for } c, c' \in C(A),$$

then $s(\xi)\xi = 1$, since $\xi = {}^{\iota}(c^{-1})^{\mathrm{op}} \otimes c$ over a splitting field. Therefore, the elements $c_1, c_2 \in C(A)^{\times}$ satisfy

$$c_1 c_2 = \lambda \in Z \quad \text{with } N_{Z/F}(\lambda) = 1.$$

By Hilbert's Theorem 90 (29.3), there exists $\lambda_1 \in Z$ such that $\lambda = \lambda_1 \iota(\lambda_1)^{-1}$. (Explicitly, we may take $\lambda_1 = z + \lambda \iota(z)$, where $z \in Z$ is such that $z + \lambda \iota(z)$ is invertible.) Then $\xi = {}^{\iota}(c_3^{-1})^{\mathrm{op}} \otimes c_3$ for $c_3 = \lambda_1^{-1} c_2$, hence

$$x^{\beta(g)} = c_3^{-1} * x \cdot c_3 \quad \text{for } x \in B(A, \sigma, f).$$

In particular, for $x = 1^b$ we obtain by (13.7):

$$c_3^{-1} * 1^b \cdot c_3 = g^b.$$

This shows that $c_3 \in \Gamma(A, \sigma, f)$ and $\chi(c_3) = g$. $\square$

(13.16) Corollary. *Suppose* $\deg A \geq 4$; *then*

$$\Gamma(A, \sigma, f) \cap Z = \{\, z \in Z^{\times} \mid z^2 \in F^{\times} \,\};$$

thus $\Gamma(A, \sigma, f) \cap Z = F^{\times}$ *if* $\mathrm{char}\, F = 2$, *and* $\Gamma(A, \sigma, f) \cap Z = F^{\times} \cup z \cdot F^{\times}$ *if* $\mathrm{char}\, F \neq 2$ *and* $Z = F[z]$ *with* $z^2 \in F^{\times}$.

Proof: Since χ is surjective, it maps $\Gamma(A, \sigma, f) \cap Z$ to the center of $\mathrm{O}^+(A, \sigma, f)$. It follows by scalar extension to a splitting field of A that the center of $\mathrm{O}^+(A, \sigma, f)$ is trivial if $\mathrm{char}\, F = 2$ and is $\{1, -1\}$ if $\mathrm{char}\, F \neq 2$ (see Dieudonné [81, pp. 25, 45]). Therefore, the factor group $\big(\Gamma(A, \sigma, f) \cap Z\big)/F^{\times}$ is trivial if $\mathrm{char}\, F = 2$ and has two elements if $\mathrm{char}\, F \neq 2$, proving the corollary. $\square$

(13.17) Example. Suppose Q is a quaternion F-algebra with canonical involution γ. If $\mathrm{char}\, F \neq 2$, let $\sigma = \mathrm{Int}(u) \circ \gamma$ for some invertible pure quaternion u. Let $F(u)^1$ be the group of elements of norm 1 in $F(u)$:

$$F(u)^1 = \{\, z \in F(u) \mid z\gamma(z) = 1 \,\}.$$

As observed in (12.25), we have $\mathrm{O}^+(Q, \sigma) = F(u)^1$. On the other hand, it follows from the structure theorem for Clifford algebras (8.10) that $C(Q, \sigma)$ is a commutative algebra isomorphic to $F(u)$. The Clifford group $\Gamma(Q, \sigma)$ is the group of invertible elements in $C(Q, \sigma)$. It can be identified with $F(u)^{\times}$ in such a way that the vector representation χ maps $x \in F(u)^{\times}$ to $x\gamma(x)^{-1} \in F(u)^1$. The upper exact sequence of the commutative diagram in (13.15) thus takes the form

$$1 \to F^{\times} \longrightarrow F(u)^{\times} \xrightarrow{1-\gamma} F(u)^1 \to 1.$$

Similar results hold if char $F = 2$. Using the same notation as in (12.18) and (12.33), we have $O^+(Q, \gamma, f) = F(\ell)^1$, $\Gamma(Q, \gamma, f) \simeq F(\ell)^\times$, and the upper exact sequence of the commutative diagram in (13.15) becomes

$$1 \to F^\times \longrightarrow F(\ell)^\times \xrightarrow{1-\gamma} F(\ell)^1 \to 1.$$

The extended Clifford group. In this subsection, we define an intermediate group $\Omega(A, \sigma, f)$ between $\Gamma(A, \sigma, f)$ and $\mathrm{Sim}\big(C(A, \sigma, f), \underline{\sigma}\big)$, which covers the group $\mathrm{PGO}^+(A, \sigma, f)$ in the same way as $\Gamma(A, \sigma, f)$ covers $O^+(A, \sigma, f)$ by the vector representation χ. This construction will enable us to define an analogue of the spinor norm for the group $\mathrm{PGO}^+(A, \sigma)$.

The notation is as above: (σ, f) is a quadratic pair on a central simple algebra A of even degree over an arbitrary field F. Let Z be the center of the Clifford algebra $C(A, \sigma, f)$. Since (13.4) plays a central rôle almost from the start (we need injectivity of the canonical map C for the definition of χ' in (13.19) below), we exclude the case of quaternion algebras from our discussion. We thus assume

$$\deg A = n = 2m \geq 4.$$

We identify $\mathrm{PGO}(A, \sigma, f)$ with $\mathrm{Aut}_F(A, \sigma, f)$ by mapping $g \cdot F^\times$ to $\mathrm{Int}(g)$, for $g \in \mathrm{GO}(A, \sigma, f)$. By (13.2) and (13.4), the canonical map C induces an injective homomorphism $C \colon \mathrm{PGO}^+(A, \sigma, f) \to \mathrm{Aut}_Z\big(C(A, \sigma, f), \underline{\sigma}\big)$. Consider the following diagram:

$$\mathrm{PGO}^+(A, \sigma, f)$$
$$\downarrow C$$
$$\mathrm{Sim}\big(C(A, \sigma, f), \underline{\sigma}\big) \xrightarrow{\quad \mathrm{Int} \quad} \mathrm{Aut}_Z\big(C(A, \sigma, f), \underline{\sigma}\big).$$

(13.18) Definition. The *extended Clifford group* of (A, σ, f) is the inverse image under Int of the image of the canonical map C:

$$\Omega(A, \sigma, f) = \{\, c \in \mathrm{Sim}\big(C(A, \sigma, f), \underline{\sigma}\big) \mid \mathrm{Int}(c) \in C\big(\mathrm{PGO}^+(A, \sigma, f)\big) \,\}.$$

Thus, $\Omega(A, \sigma, f) \subset \mathrm{Sim}\big(C(A, \sigma, f), \underline{\sigma}\big)$, and there is an exact sequence

$$\textbf{(13.19)} \qquad 1 \to Z^\times \to \Omega(A, \sigma, f) \xrightarrow{\chi'} \mathrm{PGO}^+(A, \sigma, f) \to 1$$

where the map χ' is defined by

$$\chi'(c) = g \cdot F^\times \quad \text{if } \mathrm{Int}(c) = C(g), \text{ with } g \in \mathrm{GO}^+(A, \sigma, f).$$

If char $F \neq 2$, the group $\Omega(A, \sigma, f) = \Omega(A, \sigma)$ may alternately be defined by

$$\Omega(A, \sigma) = \{\, c \in C(A, \sigma)^\times \mid c \cdot c(A) \cdot c^{-1} = c(A) \,\},$$

since the Z-automorphisms of $C(A, \sigma)$ which preserve $c(A)$ are exactly those which are of the form $C(g)$ for some $g \in \mathrm{GO}^+(A, \sigma)$, by (13.4). We shall not use this alternate definition, since we want to keep the characteristic arbitrary.

For $c \in \Gamma(A, \sigma, f)$, we have $\mathrm{Int}(c) = C(g)$ for some $g \in O^+(A, \sigma, f)$, by (13.15), hence $\Gamma(A, \sigma, f) \subset \Omega(A, \sigma, f)$. Our first objective in this subsection is to describe $\Gamma(A, \sigma, f)$ as the kernel of a map $\varkappa \colon \Omega(A, \sigma, f) \to Z^\times / F^\times$.

The multiplier map $\mu \colon \mathrm{GO}(A, \sigma, f) \to F^\times$ induces a map

$$\overline{\mu} \colon \mathrm{PGO}^+(A, \sigma, f) \to F^\times / F^{\times 2},$$

since $\mu(\alpha) = \alpha^2$ for all $\alpha \in F^\times$. This map fits into an exact sequence:

$$O^+(A, \sigma, f) \xrightarrow{\pi} \mathrm{PGO}^+(A, \sigma, f) \xrightarrow{\overline{\mu}} F^\times/F^{\times 2}.$$

(13.20) Lemma. *The kernel of the map $\overline{\mu} \circ \chi' \colon \Omega(A, \sigma, f) \to F^\times/F^{\times 2}$ is the subgroup $Z^\times \cdot \Gamma(A, \sigma, f)$. In particular, if F is algebraically closed, then, since $\overline{\mu}$ is trivial, $\Omega(A, \sigma, f) = Z^\times \cdot \Gamma(A, \sigma, f)$.*

Proof: For $c \in \Gamma(A, \sigma, f)$, we have $\chi'(c) = g \cdot F^\times$ for some $g \in O^+(A, \sigma, f)$, hence $\overline{\mu} \circ \chi'(c) = 1$. Since $\ker \chi' = Z^\times$, the inclusion $Z^\times \cdot \Gamma(A, \sigma, f) \subset \ker(\overline{\mu} \circ \chi')$ follows. In order to prove the reverse inclusion, pick $c \in \ker(\overline{\mu} \circ \chi')$; then $\chi'(c) = g \cdot F^\times$ for some $g \in \mathrm{GO}^+(A, \sigma, f)$ such that $\mu(g) \in F^{\times 2}$. Let $\mu(g) = \alpha^2$ for some $\alpha \in F^\times$. Then $\alpha^{-1} g \in O^+(A, \sigma, f)$, so there is an element $\gamma \in \Gamma(A, \sigma, f)$ such that $\chi(\gamma) = \alpha^{-1} g$. We then have

$$\mathrm{Int}(\gamma) = C(\alpha^{-1} g) = C(g) = \mathrm{Int}(c),$$

hence $c = z \cdot \gamma$ for some $z \in Z^\times$. □

We now define a map

$$\varkappa \colon \Omega(A, \sigma, f) \to Z^\times/F^\times$$

as follows: for $\omega \in \Omega(A, \sigma, f)$, we pick $g \in \mathrm{GO}^+(A, \sigma, f)$ such that $\chi'(\omega) = g \cdot F^\times$; then $\mu(g)^{-1} g^2 \in O^+(A, \sigma, f)$, hence there exists $\gamma \in \Gamma(A, \sigma, f)$ such that $\chi(\gamma) = \mu(g)^{-1} g^2$. By (13.15) it follows that $\mathrm{Int}(\gamma) = C(g^2) = \mathrm{Int}(\omega^2)$, hence

$$\omega^2 = z \cdot \gamma \quad \text{for some } z \in Z^\times.$$

We then set

$$\varkappa(\omega) = z \cdot F^\times \in Z^\times/F^\times.$$

To check that $\varkappa$ is well-defined, suppose $g' \in \mathrm{GO}^+(A, \sigma, f)$ also satisfies $\chi'(\omega) = C(g')$. We then have $g \equiv g' \mod F^\times$, hence $\mu(g)^{-1} g^2 = \mu(g')^{-1} g'^2$. On the other hand, the element $\gamma \in \Gamma(A, \sigma, f)$ such that $\chi(\gamma) = \mu(g)^{-1} g^2$ is uniquely determined up to a factor in $F^\times$, by (13.15), hence the element $z \in Z^\times$ is uniquely determined modulo $F^\times$.

Note that, if $\mathrm{char}\, F \neq 2$, the element z is *not* uniquely determined by the condition that $\omega^2 = z \cdot \gamma$ for some $\gamma \in \Gamma(A, \sigma, f)$, since $\Gamma(A, \sigma, f) \cap Z^\times \neq F^\times$ (see (13.16)).

(13.21) Proposition. *The map $\varkappa \colon \Omega(A, \sigma, f) \to Z^\times/F^\times$ is a group homomorphism and $\ker \varkappa = \Gamma(A, \sigma, f)$.*

Proof: It suffices to prove the proposition over an algebraic closure. We may thus assume F is algebraically closed; (13.20) then shows that

$$\Omega(A, \sigma, f) = Z^\times \cdot \Gamma(A, \sigma, f).$$

For $z \in Z^\times$ and $\gamma \in \Gamma(A, \sigma, f)$ we have $\varkappa(z \cdot \gamma) = z^2 \cdot F^\times$, hence $\varkappa$ is a group homomorphism. Moreover, $\ker \varkappa$ consists of the elements $z \cdot \gamma$ such that $z^2 \in F^\times$. In view of (13.16), this condition implies that $z \in \Gamma(A, \sigma, f)$, hence $\ker \varkappa = \Gamma(A, \sigma, f)$. □

Our next objective is to relate $\varkappa(\omega)$ to $\overline{\mu} \circ \chi'(\omega)$, for $\omega \in \Omega(A, \sigma, f)$. We need the following classical result of Dieudonné, which will be generalized in the next section:

(13.22) Lemma (Dieudonné). *Let (V, q) be a nonsingular even-dimensional quadratic space over an arbitrary field F. Let Z be the center of the even Clifford algebra $C_0(V, q)$. For every similitude $g \in \mathrm{GO}(V, q)$, the multiplier $\mu(g)$ is a norm from Z/F.*

Proof: The similitude g may be viewed as an isometry $\langle \mu(g) \rangle \cdot q \simeq q$. Therefore, the quadratic form $q \perp \langle -\mu(g) \rangle \cdot q$ is hyperbolic. The Clifford algebra of any form $\langle 1, -\alpha \rangle \cdot q$ is Brauer-equivalent to the quaternion algebra[21] $(Z, \alpha)_F$ (see for instance [28, (3.22), p. 47]), hence $(Z, \mu(g))_F$ splits, proving that $\mu(g)$ is a norm from Z/F. $\square$

Continuing with the same notation, and assuming $\dim V \geq 4$, consider $\omega \in \Omega(\mathrm{End}_F(V), \sigma_q, f_q) \subset C_0(V, q)$ and $g \in \mathrm{GO}^+(V, q)$ such that $\chi'(\omega) = g \cdot F^\times$. The preceding lemma yields an element $z \in Z^\times$ such that $\mu(g) = N_{Z/F}(z) = z\iota(z)$, where ι is the nontrivial automorphism of Z/F.

(13.23) Lemma. *There exists an element $z_0 \in Z^\times$ such that $\varkappa(\omega) = (zz_0^2)^{-1} \cdot F^\times$ and, in $C(V, q)$,*

$$\omega \cdot v \cdot \omega^{-1} = \iota(z_0)z_0^{-1}z^{-1}g(v) \quad \text{for } v \in V.$$

Proof: For all $v \in V$, we have in $C(V, q)$

$$\left(z^{-1}g(v)\right)^2 = z^{-1}\iota(z)^{-1}g(v)^2 = \mu(g)^{-1}q\big(g(v)\big) = q(v),$$

hence the map $v \mapsto z^{-1}g(v)$ extends to an automorphism of $C(V, q)$, by the universal property of Clifford algebras. By the Skolem-Noether theorem, we may represent this automorphism as $\mathrm{Int}(c)$ for some $c \in C(V, q)^\times$. For $v, \in V$, we then have

$$\mathrm{Int}(c)(v \cdot w) = z^{-1}g(v) \cdot z^{-1}g(w) = \mu(g)^{-1}g(v) \cdot g(w),$$

hence (13.1) shows that the restriction of $\mathrm{Int}(c)$ to $C_0(V, q)$ is $C(g)$. Since g is a proper similitude, it follows from (13.2) that $\mathrm{Int}(c)$ is the identity on Z, hence $c \in C_0(V, q)^\times$. Moreover, $\mathrm{Int}(c)|_{C_0(V,q)} = \mathrm{Int}(\omega)|_{C_0(V,q)}$ since $\chi'(\omega) = g \cdot F^\times$, hence $c = z_0\omega$ for some $z_0 \in Z^\times$. It follows that for all $v \in V$,

$$\omega \cdot v \cdot \omega^{-1} = z_0^{-1}c \cdot v \cdot c^{-1}z_0 = \iota(z_0)z_0^{-1}z^{-1}g(v).$$

Observe next that

$$\mathrm{Int}(c^2)(v) = z^{-2}g^2(v) \quad \text{for } v \in V,$$

since c commutes with z. If $\gamma \in \Gamma^+(V, q)$ satisfies $\chi(\gamma) = \mu(g)^{-1}g^2$, then

$$\gamma \cdot v \cdot \gamma^{-1} = \mu(g)^{-1}g^2(v) \quad \text{for } v \in V,$$

hence $\gamma^{-1}zc^2$ centralizes V. Since V generates $C(V, q)$, it follows that

$$\gamma \equiv zc^2 \equiv zz_0^2\omega \mod F^\times.$$

By definition of $\varkappa$, these congruences yield $\varkappa(\omega) = (zz_0^2)^{-1} \cdot F^\times$. $\square$

The main result of this subsection is the following:

(13.24) Proposition. *The following diagram is commutative with exact rows and columns*:

$$
\begin{array}{ccccccccc}
1 & \longrightarrow & F^\times & \longrightarrow & \Gamma(A,\sigma,f) & \xrightarrow{\ \chi\ } & O^+(A,\sigma,f) & \longrightarrow & 1 \\
& & \downarrow & & \downarrow & & \downarrow{\scriptstyle \pi} & & \\
1 & \longrightarrow & Z^\times & \longrightarrow & \Omega(A,\sigma,f) & \xrightarrow{\ \chi'\ } & \mathrm{PGO}^+(A,\sigma,f) & \longrightarrow & 1 \\
& & & & \downarrow{\scriptstyle \varkappa} & & \downarrow{\scriptstyle \bar\mu} & & \\
& & & & Z^\times/F^\times & \xrightarrow{\ N_{Z/F}\ } & F^\times/F^{\times 2}. & &
\end{array}
$$

Proof: In view of (13.15) and (13.21), it suffices to prove commutativity of the lower square. By extending scalars to a splitting field of A in which F is algebraically closed, we may assume that A is split. Let (V,q) be a nonsingular quadratic space such that $(A,\sigma,f) = \big(\mathrm{End}_F(V), \sigma_q, f_q\big)$.

Fix some $\omega \in \Omega(A,\sigma,f)$ and $g \in \mathrm{GO}^+(V,q)$ such that $\chi'(\omega) = g \cdot F^\times$. Let $z \in Z^\times$ satisfy $\mu(g) = N_{Z/F}(z)$. The preceding lemma yields $z_0 \in Z^\times$ such that $\varkappa(\omega) = (zz_0^2)^{-1} \cdot F^\times$. Then

$$
N_{Z/F}\big(\varkappa(\omega)\big) = N_{Z/F}(zz_0^2)^{-1} \cdot F^{\times 2} = \mu(g) \cdot F^{\times 2}. \qquad \square
$$

To conclude our discussion of the extended Clifford group, we examine more closely the case where $\deg A$ is divisible by 4. In this case, the map $\varkappa$ factors through the multiplier map, and the homomorphism χ' factors through a homomorphism $\chi_0 \colon \Omega(A,\sigma,f) \to \mathrm{GO}^+(A,\sigma,f)$.

We denote by $\underline{\mu} \colon \Omega(A,\sigma,f) \to Z^\times$ the multiplier map, defined by

$$
\underline{\mu}(\omega) = \underline{\sigma}(\omega)\omega \quad \text{for } \omega \in \Omega(A,\sigma,f).
$$

The element $\underline{\mu}(\omega)$ thus defined belongs to $Z^\times$ since $\Omega(A,\sigma,f) \subset \mathrm{Sim}\big(C(A,\sigma,f),\underline{\sigma}\big)$.

(13.25) Proposition. *Suppose* $\deg A \equiv 0 \mod 4$. *For all* $\omega \in \Omega(A,\sigma,f)$,

$$
\varkappa(\omega) = \underline{\mu}(\omega) \cdot F^\times.
$$

The Clifford group can be characterized as

$$
\Gamma(A,\sigma,f) = \{\, \omega \in \Omega(A,\sigma,f) \mid \underline{\mu}(\omega) \in F^\times \,\}.
$$

Proof: The second part follows from the first, since (13.21) shows that $\Gamma(A,\sigma,f) = \ker \varkappa$.

To prove the first part, we may extend scalars to an algebraic closure. For $\omega \in \Omega(A,\sigma,f)$ we may thus assume, in view of (13.20), that there exist $z \in Z^\times$ and $\gamma \in \Gamma(A,\sigma,f)$ such that $\omega = z \cdot \gamma$. We then have $\varkappa(\omega) = z^2 \cdot F^\times$. On the other hand, since $\deg A \equiv 0 \mod 4$ the involution $\underline{\sigma}$ is of the first kind, by (8.12), hence $\underline{\mu}(\omega) = z^2 \underline{\mu}(\gamma)$. Now, (13.15) shows that $\underline{\mu}(\gamma) \in F^\times$, hence

$$
\underline{\mu}(\omega) \cdot F^\times = z^2 \cdot F^\times = \varkappa(\omega). \qquad \square
$$

Another interesting feature of the case where $\deg A \equiv 0 \mod 4$ is that the extended Clifford group has an alternate description similar to the definition of the Clifford group. In the following proposition, we consider the image A^b of A in the Clifford bimodule $B(A,\sigma,f)$ under the canonical map $b \colon A \to B(A,\sigma,f)$.

(13.26) Proposition. *If* $\deg A \equiv 0 \mod 4$, *then*

$$\Omega(A, \sigma, f) = \{\, \omega \in \mathrm{Sim}\big(C(A, \sigma, f), \underline{\sigma}\big) \mid \underline{\sigma}(\omega) * A^b \cdot \omega = A^b \,\}.$$

Proof: Let $\omega \in \mathrm{Sim}\big(C(A, \sigma, f), \underline{\sigma}\big)$. We have to show that $\mathrm{Int}(\omega) = C(g)$ for some $g \in \mathrm{GO}^+(A, \sigma, f)$ if and only if $\underline{\sigma}(\omega) * A^b \cdot \omega = A^b$.

Assume first that $\omega \in \Omega(A, \sigma, f)$, i.e., $\mathrm{Int}(\omega) = C(g)$ for some $g \in \mathrm{GO}^+(A, \sigma, f)$. To prove the latter equality, we may reduce by scalar extension to the split case. Thus, suppose that (V, q) is a nonsingular quadratic space of dimension divisible by 4 and $(A, \sigma, f) = \big(\mathrm{End}_F(V), \sigma_q, f_q\big)$. Under the standard identifications associated to q we have $A^b = V \otimes V \subset V \otimes C_1(V, q)$ (see (9.7)), hence it suffices to show

$$\underline{\sigma}(\omega) \cdot v \cdot \omega \in V \quad \text{for } v \in V.$$

Let ι be, as usual, the nontrivial automorphism of the center Z of $C(A, \sigma, f) = C_0(V, q)$ over F, and let $z \in Z^\times$ be such that $\mu(g) = z\iota(z)$. By (13.23), there is an element $z_0 \in Z^\times$ such that $\varkappa(\omega) = (zz_0^2)^{-1} \cdot F^\times$ and

$$\omega \cdot v \cdot \omega^{-1} = \iota(z_0) z_0^{-1} z^{-1} g(v) \quad \text{for } v \in V.$$

The canonical involution τ of $C(V, q)$ restricts to $\underline{\sigma} = \tau_0$ on $C_0(V, q)$; by (8.4) this involution is the identity on Z. Therefore, by applying τ to each side of the preceding equation, we obtain

$$\underline{\sigma}(\omega)^{-1} \cdot v \cdot \underline{\sigma}(\omega) = g(v)\iota(z_0) z_0^{-1} z^{-1} = z_0 \iota(z_0)^{-1} \iota(z)^{-1} g(v) \quad \text{for } v \in V,$$

hence

$$\underline{\sigma}(\omega) \cdot v \cdot \omega = z_0^{-1} \iota(z_0) \iota(z) g^{-1}(v) \mu(\omega) \quad \text{for } v \in V.$$

By (13.25), we have $\mu(\omega) \cdot F^\times = \varkappa(\omega)$, hence $\mu(\omega) = \alpha(zz_0^2)^{-1}$ for some $\alpha \in F^\times$. By substituting this in the preceding relation, we get for all $v \in V$

$$\underline{\sigma}(\omega) \cdot v \cdot \omega = \alpha z_0^{-1} \iota(z_0) \iota(z) \iota(zz_0^2)^{-1} g^{-1}(v) = \alpha N_{Z/F}(z_0)^{-1} g^{-1}(v).$$

Since the right-hand term lies in V, we have thus shown $\underline{\sigma}(\omega) * A^b \cdot \omega = A^b$.

Suppose conversely that $\omega \in \mathrm{Sim}\big(C(A, \sigma, f), \underline{\sigma}\big)$ satisfies $\underline{\sigma}(\omega) * A^b \cdot \omega = A^b$. Since b is injective, there is a unique element $g \in A$ such that

(13.27) $$\underline{\sigma}(\omega) * 1^b \cdot \omega = g^b.$$

We claim that $g \in \mathrm{GO}^+(A, \sigma, f)$ and that $\mathrm{Int}(\omega) = C(g)$. To prove the claim, we may extend scalars to a splitting field of A; we may thus assume again that $(A, \sigma, f) = \big(\mathrm{End}_F(V), \sigma_q, f_q\big)$ for some nonsingular quadratic space (V, q) of dimension divisible by 4, and use the standard identifications $A = V \otimes V$, $A^b = V \otimes V \subset B(A, \sigma, f) = V \otimes C_1(V, q)$. Since $B(A, \sigma, f)$ is a left A-module, we may multiply each side of (13.27) by $v \otimes w \in V \otimes V = A$; we thus obtain

$$\underline{\sigma}(\omega) * (v \otimes w)^b \cdot \omega = \big((v \otimes w) \circ g\big)^b \quad \text{for } v, w \in V.$$

Since $(v \otimes w) \circ g = v \otimes \sigma(g)(w)$, it follows that, in $C(V, q)$,

$$\underline{\sigma}(\omega) \cdot w \cdot \omega = \sigma(g)(w) \quad \text{for } w \in V.$$

Since ω is a similitude of $C(A, \sigma, f)$, by squaring this equation we obtain

$$N_{Z/F}\big(\mu(\omega)\big) q(w) = q\big(\sigma(g)(w)\big) \quad \text{for } w \in V.$$

This shows that $\sigma(g)$ is a similitude, hence $g \in \mathrm{GO}(V, q)$, and

(13.28) $\mu\big(\sigma(g)\big) = \mu(g) = N_{Z/F}\big(\underline{\mu}(\omega)\big).$

Moreover, for $v, \in V$,

$$\omega^{-1} \cdot (v \cdot w) \cdot \omega = \underline{\mu}(\omega)^{-1}\big(\underline{\sigma}(\omega) \cdot v \cdot \omega\big)\underline{\mu}(\omega)^{-1}\big(\underline{\sigma}(\omega) \cdot w \cdot \omega\big)$$
$$= N_{Z/F}\big(\underline{\mu}(\omega)\big)^{-1}\sigma(g)(v) \cdot \sigma(g)(w).$$

By (13.28) it follows that

$$\omega^{-1} \cdot (v \cdot w) \cdot \omega = \mu(g)^{-1}\sigma(g)(v) \cdot \sigma(g)(w) \quad \text{for } v,\, w \in V,$$

hence, by (13.1),

$$\omega \cdot (v \cdot w) \cdot \omega^{-1} = \mu(g)^{-1}g(v) \cdot g(w) = C(g)(v \cdot w) \quad \text{for } v,\, w \in V.$$

This equality shows that $\mathrm{Int}(\omega)|_{C_0(V,q)} = C(g)$, hence g is proper, by (13.2), and the proof is complete. $\qquad\square$

(13.29) Definition. Suppose $\deg A \equiv 0 \mod 4$. By using the description of the extended Clifford group $\Omega(A, \sigma, f)$ in the proposition above, we may define a homomorphism

$$\chi_0 \colon \Omega(A, \sigma, f) \to \mathrm{GO}^+(A, \sigma, f)$$

mapping $\omega \in \Omega(A, \sigma, f)$ to the element $g \in \mathrm{GO}^+(A, \sigma, f)$ satisfying (13.27). Thus, for $\omega \in \Omega(A, \sigma, f)$ the similitude $\chi_0(\omega)$ is defined by the relation

$$\underline{\sigma}(\omega) * 1^b \cdot \omega = \chi_0(\omega)^b.$$

The proof above shows that

$$\chi'(\omega) = \chi_0(\omega) \cdot F^\times \quad \text{for } \omega \in \Omega(A, \sigma, f);$$

moreover, by (13.28), the following diagram commutes:

$$\begin{array}{ccc}
\Omega(A, \sigma, f) & \xrightarrow{\chi_0} & \mathrm{GO}^+(A, \sigma, f) \\
\underline{\mu}\downarrow & & \downarrow\mu \\
Z^\times & \xrightarrow{N_{Z/F}} & F^\times.
\end{array}$$

Spinor norms. Let (σ, f) be a quadratic pair on a central simple algebra A of even degree over an arbitrary field F.

(13.30) Definition. In view of (13.15), we may define a homomorphism

$$\mathrm{Sn}\colon \mathrm{O}^+(A, \sigma, f) \to F^\times/F^{\times 2}$$

as follows: for $g \in \mathrm{O}^+(A, \sigma, f)$, pick $\gamma \in \Gamma(A, \sigma, f)$ such that $\chi(\gamma) = g$ and let

$$\mathrm{Sn}(g) = \underline{\sigma}(\gamma)\gamma \cdot F^{\times 2} = \underline{\mu}(\gamma) \cdot F^{\times 2}.$$

This square class depends only on g, since γ is uniquely determined up to a factor in $F^\times$. In other words, Sn is the map which makes the following diagram commute:

$$\begin{array}{ccccccccc}
1 & \longrightarrow & F^\times & \longrightarrow & \Gamma(A, \sigma, f) & \xrightarrow{\chi} & \mathrm{O}^+(A, \sigma, f) & \longrightarrow & 1 \\
 & & \downarrow 2 & & \downarrow\underline{\mu} & & \downarrow\mathrm{Sn} & & \\
1 & \longrightarrow & F^{\times 2} & \longrightarrow & F^\times & \longrightarrow & F^\times/F^{\times 2} & \longrightarrow & 1.
\end{array}$$

We also define the group of *spinor norms*:

$$\mathrm{Sn}(A, \sigma, f) = \{\, \underline{\mu}(\gamma) \mid \gamma \in \Gamma(A, \sigma, f) \,\} \subset F^{\times},$$

so $\mathrm{Sn}\big(\mathrm{O}^{+}(A, \sigma, f)\big) = \mathrm{Sn}(A, \sigma, f)/F^{\times 2}$, and the *spin group*:

$$\mathrm{Spin}(A, \sigma, f) = \{\, \gamma \in \Gamma(A, \sigma, f) \mid \underline{\mu}(\gamma) = 1 \,\} \subset \Gamma(A, \sigma, f).$$

In the split case, if $(A, \sigma, f) = \big(\mathrm{End}_F(V), \sigma_q, f_q\big)$ for some nonsingular quadratic space (V, q) of even dimension, the standard identifications associated to q yield

$$\mathrm{Spin}(A, \sigma, f) = \mathrm{Spin}(V, q) = \{\, c \in \Gamma^{+}(V, q) \mid \tau(c) \cdot c = 1 \,\},$$

where τ is the canonical involution of $C(V, q)$ which is the identity on V. From the description of the spinor norm in Scharlau [247, Chap. 9, §3], it follows that the group of spinor norms $\mathrm{Sn}(V, q) = \mathrm{Sn}(A, \sigma, f)$ consists of the products of any even number of represented values of q.

The vector representation χ induces by restriction a homomorphism

$$\mathrm{Spin}(A, \sigma, f) \to \mathrm{O}^{+}(A, \sigma, f)$$

which we also denote χ.

(13.31) Proposition. *The vector representation χ fits into an exact sequence*:

$$1 \to \{\pm 1\} \longrightarrow \mathrm{Spin}(A, \sigma, f) \xrightarrow{\ \chi\ } \mathrm{O}^{+}(A, \sigma, f) \xrightarrow{\ \mathrm{Sn}\ } F^{\times}/F^{\times 2}.$$

Proof: This follows from the exactness of the top sequence in (13.15) and the definition of Sn. $\qquad\square$

Assume now $\deg A = 2m \geq 4$. We may then use the extended Clifford group $\Omega(A, \sigma, f)$ to define an analogue of the spinor norm on the group $\mathrm{PGO}^{+}(A, \sigma, f)$, as we proceed to show. The map S defined below may be obtained as a connecting map in a cohomology sequence, see §31.A. Its target group is the first cohomology group of the absolute Galois group of F with coefficients in the center of the algebraic group $\mathrm{Spin}(A, \sigma, f)$. The approach we follow in this subsection does not use cohomology, but since the structure of the center of $\mathrm{Spin}(A, \sigma, f)$ depends on the parity of m, we divide the construction into two parts, starting with the case where the degree of A is divisible by 4. As above, we let $Z = Z(A, \sigma, f)$ be the center of the Clifford algebra $C(A, \sigma, f)$.

(13.32) Definition. Assume $\deg A \equiv 0 \mod 4$. We define a homomorphism

$$S \colon \mathrm{PGO}^{+}(A, \sigma, f) \to Z^{\times}/Z^{\times 2}$$

as follows: for $g \cdot F^{\times} \in \mathrm{PGO}^{+}(A, \sigma, f)$, pick $\omega \in \Omega(A, \sigma, f)$ such that $\chi'(\omega) = g \cdot F^{\times}$ and set

$$S(g \cdot F^{\times}) = \underline{\sigma}(\omega)\omega \cdot Z^{\times 2} = \underline{\mu}(\omega) \cdot Z^{\times 2}.$$

Since ω is determined by $g \cdot F^{\times}$ up to a factor in $Z^{\times}$ and $\underline{\sigma}$ is of the first kind, by (8.12), the element $\underline{\mu}(\omega)\omega \cdot Z^{\times 2}$ depends only on $g \cdot F^{\times}$. The map S thus makes the following diagram commute:

$$
\begin{array}{ccccccccc}
1 & \longrightarrow & Z^{\times} & \longrightarrow & \Omega(A, \sigma, f) & \xrightarrow{\ \chi'\ } & \mathrm{PGO}^{+}(A, \sigma, f) & \longrightarrow & 1 \\
 & & \big\downarrow{\scriptstyle 2} & & \big\downarrow{\scriptstyle \mu} & & \big\downarrow{\scriptstyle S} & & \\
1 & \longrightarrow & Z^{\times 2} & \longrightarrow & Z^{\times} & \longrightarrow & Z^{\times}/Z^{\times 2} & \longrightarrow & 1.
\end{array}
$$

Besides the formal analogy between the definition of S and that of the spinor norm Sn, there is also an explicit relationship demonstrated in the following proposition:

(13.33) Proposition. *Assume* $\deg A \equiv 0 \mod 4$. *Let*

$$\pi \colon \mathrm{O}^+(A, \sigma, f) \to \mathrm{PGO}^+(A, \sigma, f)$$

be the canonical map. Then, the following diagram is commutative with exact rows:

$$
\begin{array}{ccccc}
\mathrm{O}^+(A, \sigma, f) & \xrightarrow{\ \pi\ } & \mathrm{PGO}^+(A, \sigma, f) & \xrightarrow{\ \overline{\mu}\ } & F^\times/F^{\times 2} \\
{\scriptstyle \mathrm{Sn}}\downarrow & & \downarrow{\scriptstyle S} & & \| \\
F^\times/F^{\times 2} & \longrightarrow & Z^\times/Z^{\times 2} & \xrightarrow{\ N_{Z/F}\ } & F^\times/F^{\times 2}.
\end{array}
$$

Proof: Consider $g \in \mathrm{O}^+(A, \sigma, f)$ and $\gamma \in \Gamma(A, \sigma, f)$ such that $\chi(\gamma) = g$. We then have $\mathrm{Sn}(g) = \underline{\mu}(\gamma) \cdot F^{\times 2}$. On the other hand, we also have $\chi'(\gamma) = g \cdot F^\times$, hence $S(g \cdot F^\times) = \underline{\mu}(\gamma) \cdot Z^{\times 2}$. This proves that the left square is commutative.

Consider next $g \cdot F^\times \in \mathrm{PGO}^+(A, \sigma, f)$ and $\omega \in \Omega(A, \sigma, f)$ such that $\chi'(\omega) = g \cdot F^\times$, so that $S(g \cdot F^\times) = \underline{\mu}(\omega) \cdot Z^{\times 2}$. By (13.25) we have $\underline{\mu}(\omega) \cdot F^\times = \varkappa(\omega)$ and, by (13.24), $N_{Z/F}\big(\varkappa(\omega)\big) = \underline{\mu}(g) \cdot F^{\times 2}$. Therefore, $N_{Z/F}\big(S(g \cdot F^\times)\big) = \underline{\mu}(g) \cdot F^{\times 2}$ and the right square is commutative.

Exactness of the lower sequence is a consequence of Hilbert's Theorem 90 (29.3): if $z \in Z^\times$ is such that $z\iota(z) = x^2$ for some $x \in F^\times$, then $N_{Z/F}(zx^{-1}) = 1$, hence by (29.3) there exists some $y \in Z^\times$ such that $zx^{-1} = \iota(y)y^{-1}$. (Explicitly, we may take $y = t + xz^{-1}\iota(t)$, where $t \in Z$ is such that $t + xz^{-1}\iota(t)$ is invertible.) Then $zy^2 = xN_{Z/F}(y)$, hence $z \cdot Z^{\times 2}$ lies in the image of $F^\times/F^{\times 2}$. $\square$

Note that the spin group may also be defined as a subgroup of the extended Clifford group: for $\deg A \equiv 0 \mod 4$,

$$\mathrm{Spin}(A, \sigma, f) = \{\, \omega \in \Omega(A, \sigma, f) \mid \underline{\mu}(\omega) = 1 \,\}.$$

Indeed, (13.25) shows that the right side is contained in $\Gamma(A, \sigma, f)$.

The restriction of the homomorphism

$$\chi' \colon \Omega(A, \sigma, f) \to \mathrm{PGO}^+(A, \sigma, f)$$

to $\mathrm{Spin}(A, \sigma, f)$, also denoted χ', fits into an exact sequence:

(13.34) Proposition. *Assume* $\deg A \equiv 0 \mod 4$. *The sequence*

$$1 \to \mu_2(Z) \to \mathrm{Spin}(A, \sigma, f) \xrightarrow{\chi'} \mathrm{PGO}^+(A, \sigma, f) \xrightarrow{S} Z^\times/Z^{\times 2}$$

is exact, where $\mu_2(Z) = \{\, z \in Z \mid z^2 = 1 \,\}$.

Proof: Let $g \cdot F^\times$ be in the kernel of S. Then there exists $\omega \in \Omega(A, \sigma, f)$, with $\underline{\sigma}(\omega)\omega = z^2$ for some $z \in Z^\times$, such that $\chi'(\omega) = g \cdot F^\times$. By replacing ω by ωz^{-1}, we get $\omega \in \mathrm{Spin}(A, \sigma, f)$. Exactness at $\mathrm{Spin}(A, \sigma, f)$ follows from the fact that $Z^\times \cap \mathrm{Spin}(A, \sigma, f) = \mu_2(Z)$ in $\Omega(A, \sigma, f)$. $\square$

In the case where $\deg A \equiv 2 \mod 4$, the involution $\underline{\sigma}$ is of the second kind, hence $\underline{\mu}(\omega) \in F^\times$ for all $\omega \in \Omega(A, \sigma, f)$. The rôle played by $\underline{\mu}$ in the case where $\deg A \equiv 0 \mod 4$ is now played by a map which combines $\underline{\mu}$ and $\varkappa$.

Consider the following subgroup U of $F^\times \times Z^\times$:

$$U = \{\, (\alpha, z) \mid \alpha^4 = N_{Z/F}(z) \,\} \subset F^\times \times Z^\times$$

and its subgroup $U_0 = \{\,\big(N_{Z/F}(z), z^4\big) \mid z \in Z^\times\,\}$. Let[22]

$$H^1(F, \boldsymbol{\mu}_{4[Z]}) = U/U_0,$$

and let $[\alpha, z]$ be the image of $(\alpha, z) \in U$ in $H^1(F, \boldsymbol{\mu}_{4[Z]})$.

For $\omega \in \Omega(A, \sigma, f)$, let $k \in Z^\times$ be a representative of $\varkappa(\omega) \in Z^\times/F^\times$. The element $k\iota(k)^{-1}$ is independent of the choice of the representative k and we define

$$\mu_*(\omega) = \big(\underline{\mu}(\omega), k\iota(k)^{-1}\underline{\mu}(\omega)^2\big) \in U.$$

For $z \in Z^\times$, we have $\varkappa(z) = z^2 \cdot F^\times$ and $\underline{\mu}(z) = N_{Z/F}(z)$, hence

$$\mu_*(z) = \big(N_{Z/F}(z), z^4\big) \in U_0.$$

(13.35) Definition. Assume $\deg A \equiv 2 \mod 4$. Define a homomorphism

$$S\colon \mathrm{PGO}^+(A, \sigma, f) \to H^1(F, \boldsymbol{\mu}_{4[Z]})$$

as follows: for $g \cdot F^\times \in \mathrm{PGO}^+(A, \sigma, f)$, pick $\omega \in \Omega(A, \sigma, f)$ such that $\chi'(\omega) = g \cdot F^\times$ and let $S(g \cdot F^\times)$ be the image of $\mu_*(\omega)$ in $H^1(F, \boldsymbol{\mu}_{4[Z]})$. Since ω is determined up to a factor in $Z^\times$ and $\mu_*(Z^\times) \subset U_0$, the definition of $S(g \cdot F^\times)$ does not depend on the choice of ω. In other words, S is the map which makes the following diagram commute:

$$
\begin{array}{ccccccccc}
1 & \longrightarrow & Z^\times & \longrightarrow & \Omega(A, \sigma, f) & \xrightarrow{\ \chi'\ } & \mathrm{PGO}^+(A, \sigma, f) & \longrightarrow & 1 \\
& & \downarrow{\scriptstyle\mu_*} & & \downarrow{\scriptstyle\mu_*} & & \downarrow{\scriptstyle S} & & \\
1 & \longrightarrow & U_0 & \longrightarrow & U & \longrightarrow & H^1(F, \boldsymbol{\mu}_{4[Z]}) & \longrightarrow & 1.
\end{array}
$$

In order to relate the map S to the spinor norm, we define maps i and j which fit into an exact sequence

$$F^\times/F^{\times 2} \xrightarrow{\ i\ } H^1(F, \boldsymbol{\mu}_{4[Z]}) \xrightarrow{\ j\ } F^\times/F^{\times 2}.$$

For $\alpha \cdot F^{\times 2} \in F^\times/F^{\times 2}$, we let $i(\alpha \cdot F^{\times 2}) = [\alpha, \alpha^2]$. For $[\alpha, z] \in H^1(F, \boldsymbol{\mu}_{4[Z]})$, we pick $z_0 \in Z^\times$ such that $\alpha^{-2}z = z_0\iota(z_0)^{-1}$, and let $j[\alpha, z] = N_{Z/F}(z_0) \cdot F^{\times 2} \in F^\times/F^{\times 2}$. If $N_{Z/F}(z_0) = \beta^2$ for some $\beta \in F^\times$, then we may find $z_1 \in Z^\times$ such that $z_0\beta^{-1} = z_1\iota(z_1)^{-1}$. It follows that $\alpha^{-2}z = z_1^2\iota(z_1)^{-2}$, hence

$$(\alpha, z) = \big(\alpha N_{Z/F}(z_1)^{-1}, \alpha^2 N_{Z/F}(z_1)^{-2}\big) \cdot \big(N_{Z/F}(z_1), z_1^4\big),$$

and therefore $[\alpha, z] = i\big(\alpha N_{Z/F}(z_1)^{-1} \cdot F^{\times 2}\big)$. This proves exactness of the sequence above.

(13.36) Proposition. *Assume* $\deg A \equiv 2 \mod 4$. *Let*

$$\pi\colon \mathrm{O}^+(A, \sigma, f) \to \mathrm{PGO}^+(A, \sigma, f)$$

be the canonical map. Then, the following diagram is commutative with exact rows:

$$
\begin{array}{ccccc}
\mathrm{O}^+(A, \sigma, f) & \xrightarrow{\ \pi\ } & \mathrm{PGO}^+(A, \sigma, f) & \xrightarrow{\ \overline{\mu}\ } & F^\times/F^{\times 2} \\
{\scriptstyle\mathrm{Sn}}\downarrow & & \downarrow{\scriptstyle S} & & \| \\
F^\times/F^{\times 2} & \xrightarrow{\ i\ } & H^1(F, \boldsymbol{\mu}_{4[Z]}) & \xrightarrow{\ j\ } & F^\times/F^{\times 2}.
\end{array}
$$

---

[22]It will be seen in Chapter VII (see (30.13)) that this factor group may indeed be regarded as a Galois cohomology group if $\mathrm{char}\, F \neq 2$. This viewpoint is not needed here, however, and this definition should be viewed purely as a convenient notation.

Proof: Let $g \in \mathrm{O}^+(A, \sigma, f)$ and let $\gamma \in \Gamma(A, \sigma, f)$ be such that $\chi(\gamma) = g$. We then have $\mathrm{Sn}(g) = \underline{\mu}(\gamma) \cdot F^{\times 2}$. On the other hand, we also have $\varkappa(\gamma) = 1$, by (13.21), hence

$$\mu_*(\gamma) = \left[\underline{\mu}(\gamma), \underline{\mu}(\gamma)^2\right] = i\left(\underline{\mu}(\gamma) \cdot F^{\times 2}\right).$$

Since $\chi'(\gamma) = g \cdot F^\times$, this proves commutativity of the left square.

Consider next $g \cdot F^\times \in \mathrm{PGO}^+(A, \sigma, f)$ and $\omega \in \Omega(A, \sigma, f)$ such that $\chi'(\omega) = g \cdot F^\times$. We have $j \circ S(g \cdot F^\times) = N_{Z/F}(k) \cdot F^\times$, where $k \in Z^\times$ is a representative of $\varkappa(\omega) \in Z^\times/F^\times$. Proposition (13.24) shows that $N_{Z/F}(k) \cdot F^{\times 2} = \overline{\mu}(g \cdot F^\times)$, hence the right square is commutative. Exactness of the bottom row was proved above.   □

As in the preceding case, the spin group may also be defined as a subgroup of $\Omega(A, \sigma, f)$: we have for $\deg A \equiv 2 \mod 4$,

$$\mathrm{Spin}(A, \sigma, f) = \{\, \omega \in \Omega(A, \sigma, f) \mid \mu_*(\omega) = (1, 1) \,\},$$

since (13.21) shows that the right-hand group lies in $\Gamma(A, \sigma, f)$. Furthermore we have a sequence corresponding to the sequence (13.34):

(13.37) Proposition. *Assume* $\deg A \equiv 2 \mod 4$ *and* $\deg A \geq 4$. *The sequence*

$$1 \to \boldsymbol{\mu}_{4[Z]}(F) \to \mathrm{Spin}(A, \sigma, f) \xrightarrow{\chi'} \mathrm{PGO}^+(A, \sigma, f) \xrightarrow{S} H^1(F, \boldsymbol{\mu}_{4[Z]}),$$

is exact, where $\boldsymbol{\mu}_{4[Z]}(F) = \{\, z \in Z^\times \mid z^4 = 1 \text{ and } \iota(z)z = 1 \,\}$.

Proof: As in the proof of (13.34) the kernel of S is the image of $\mathrm{Spin}(A, \sigma, f)$ under χ'. Furthermore we have by (13.16)

$$Z^\times \cap \mathrm{Spin}(A, \sigma, f) = \{\, z \in Z^\times \mid z^2 \in F^\times \text{ and } \underline{\sigma}(z)z = 1 \,\} = \boldsymbol{\mu}_{4[Z]}(F)$$

in $\Omega(A, \sigma, f)$.   □

13.C. Multipliers of similitudes. This section is devoted to a generalization of Dieudonné's theorem on the multipliers of similitudes (13.22). As in the preceding sections, let (σ, f) be a quadratic pair on a central simple algebra A of even degree over an arbitrary field F, and let $Z = Z(A, \sigma, f)$ be the center of the Clifford algebra $C(A, \sigma, f)$. The nontrivial automorphism of Z/F is denoted by ι.

For $\alpha \in F^\times$, let $(Z, \alpha)_F$ be the quaternion algebra $Z \oplus Zj$ where multiplication is defined by $jz = \iota(z)j$ for $z \in Z$ and $j^2 = \alpha$. In other words,

$$(Z, \alpha)_F = \begin{cases} (\delta, \alpha)_F & \text{if char } F \neq 2 \text{ and } Z \simeq F[X]/(X^2 - \delta); \\ [\delta, \alpha)_F & \text{if char } F = 2 \text{ and } Z \simeq F[X]/(X^2 + X + \delta). \end{cases}$$

(Compare with §10.F).

(13.38) Theorem. *Let* $g \in \mathrm{GO}(A, \sigma, f)$ *be a similitude of* (A, σ, f).
(1) *If* g *is proper, then* $\left(Z, \mu(g)\right)_F$ *splits.*
(2) *If* g *is improper, then* $\left(Z, \mu(g)\right)_F$ *is Brauer-equivalent to* A.

When A splits, the algebra $\left(Z, \mu(g)\right)_F$ splits in each case, so $\mu(g)$ is a norm from Z/F for every similitude g. We thus recover Dieudonné's theorem (13.22).

In the case where g is proper, the theorem follows from (13.24) (or, equivalently, from (13.33) and (13.36)). For the rest of this section, we fix some improper similitude g. According to (13.2), the automorphism $C(g)$ of $C(A, \sigma)$ then restricts to ι on Z, so $C(g)$ induces a Z-algebra isomorphism $C(A, \sigma, f) \xrightarrow{\sim} {}^\iota C(A, \sigma, f)$ by mapping $c \in C(A, \sigma, f)$ to ${}^\iota\big(C(g)(c)\big) \in {}^\iota C(A, \sigma, f)$. When we view $C(A, \sigma, f)$ as a *left* Z-module, we have the canonical isomorphism

$$C(A, \sigma, f)^{\mathrm{op}} \otimes_Z C(A, \sigma, f) = \operatorname{End}_Z C(A, \sigma, f)$$

which identifies $c_1^{\mathrm{op}} \otimes c_2$ with the endomorphism defined by

$$c^{c_1^{\mathrm{op}} \otimes c_2} = c_1 c c_2 \quad \text{for } c, c_1, c_2 \in C(A, \sigma, f).$$

We then have Z-algebra isomorphisms:

$$\operatorname{End}_Z C(A, \sigma, f) = C(A, \sigma, f)^{\mathrm{op}} \otimes_Z C(A, \sigma, f) \simeq {}^\iota C(A, \sigma, f)^{\mathrm{op}} \otimes_Z C(A, \sigma, f).$$

The embedding ν of ${}^\iota C(A, \sigma, f)^{\mathrm{op}} \otimes_Z C(A, \sigma, f)$ into the endomorphism algebra of the Clifford bimodule $B(A, \sigma, f)$ (see (9.9)) yields an embedding

$$\nu_g \colon \operatorname{End}_Z C(A, \sigma, f) \hookrightarrow \operatorname{End}_A B(A, \sigma, f)$$

defined by

$$x^{\nu_g(c_1^{\mathrm{op}} \otimes c_2)} = C(g)(c_1) * x \cdot c_2 \quad \text{for } c_1, c_2 \in C(A, \sigma, f), \ x \in B(A, \sigma, f).$$

Let $\gamma_g \in \operatorname{End}_F C(A, \sigma, f)$ be the endomorphism $C(g^{-1})$, i.e.,

$$c^{\gamma_g} = C(g^{-1})(c) \quad \text{for } c \in C(A, \sigma, f).$$

Since g is improper, γ_g is not Z-linear, but ι-semilinear. Thus $\operatorname{Int}(\gamma_g)$, which maps $f \in \operatorname{End}_Z C(A, \sigma, f)$ to $\gamma_g \circ f \circ \gamma_g^{-1}$, is an automorphism of $\operatorname{End}_Z C(A, \sigma, f)$.

Define an F-algebra E_g as follows:

$$E_g = \operatorname{End}_Z C(A, \sigma, f) \oplus \operatorname{End}_Z C(A, \sigma, f) \cdot y$$

where y is subject to the following relations:

$$yf = (\gamma_g \circ f \circ \gamma_g^{-1})y \quad \text{for } f \in \operatorname{End}_Z C(A, \sigma, f),$$
$$y^2 = \mu(g)\gamma_g^2.$$

The algebra E_g is thus a *generalized cyclic algebra* (see Albert [9, Theorem 11.11], Jacobson [146, §1.4]); the same arguments as for the usual cyclic algebras show E_g is central simple over F.

(13.39) Proposition. *The homomorphism ν_g extends to an isomorphism of F-algebras:*

$$\nu_g \colon E_g \xrightarrow{\sim} \operatorname{End}_A B(A, \sigma, f)$$

by mapping y to the endomorphism $\beta(g)$ of (13.6).

Proof: Since E_g and $\operatorname{End}_A B(A, \sigma, f)$ are central simple F-algebras of the same dimension, it suffices to show that the map defined above is a homomorphism, i.e., that $\beta(g)$ satisfies the same relations as y:

(13.40)
$$\beta(g) \circ \nu_g(f) = \nu_g(\gamma_g \circ f \circ \gamma_g^{-1}) \circ \beta(g) \quad \text{for } f \in \operatorname{End}_Z C(A, \sigma, f),$$
$$\beta(g)^2 = \mu(g)\nu_g(\gamma_g^2).$$

It suffices to check these relations over an extension of the base field. We may thus assume that F is algebraically closed and $(A, \sigma, f) = \big(\operatorname{End}_F(V), \sigma_q, f_q\big)$ for some nonsingular quadratic space (V, q). Choose $\lambda \in F^\times$ satisfying $\lambda^2 = \mu(g)$. Then $q\big(\lambda^{-1}g(v)\big) = q(v)$ for all $v \in V$, hence there is an automorphism of $C(V, q)$

which maps v to $\lambda^{-1}g(v)$ for all $v \in V$. By the Skolem-Noether theorem we may thus find $b \in C(V,q)^\times$ such that

$$b \cdot v \cdot b^{-1} = \lambda^{-1}g(v) \quad \text{for } v \in V.$$

Then $C(g)$ is the restriction of $\text{Int}(b)$ to $C_0(V,q) = C(A, \sigma, f)$, hence $\gamma_g = \text{Int}(b^{-1})$ and

$$\gamma_g^2 = (b^2)^{\text{op}} \otimes b^{-2} \in C(A, \sigma, f)^{\text{op}} \otimes_Z C(A, \sigma, f).$$

On the other hand, for $v, w_1, \ldots, w_{2r-1} \in V$,

$$\bigl(v \otimes (w_1 \cdots w_{2r-1})\bigr)^{\beta(g)} = \mu(g)^r v \otimes \bigl(g^{-1}(w_1) \cdots g^{-1}(w_{2r-1})\bigr)$$
$$= \lambda\bigl(v \otimes b^{-1} \cdot (w_1 \cdots w_{2r-1}) \cdot b\bigr).$$

The equations (13.40) then follow by explicit computation. $\qquad\square$

To complete the proof of (13.38) we have to show that the algebra E_g is Brauer-equivalent to the quaternion algebra $\bigl(Z, \mu(g)\bigr)_F$. As pointed out by A. Wadsworth, this is a consequence of the following proposition:

(13.41) Proposition. *Let S be a central simple F-algebra, let Z be a quadratic Galois field extension of F contained in S, with nontrivial automorphism ι. Let $s \in S$ be such that $\text{Int}(s)|_Z = \iota$. Let $E = C_S(Z)$ and fix $t \in F^\times$. Let T be the F-algebra with presentation*

$$T = E \oplus Ey, \quad \text{where } yey^{-1} = ses^{-1} \text{ for all } e \in E, \text{ and } y^2 = ts^2.$$

Then $M_2(T) \simeq (Z, t) \otimes_F S$.

Proof: Let j be the standard generator of (Z, t) with $jzj^{-1} = \iota(z)$ for all $z \in Z$ and $j^2 = t$. Let $R = (Z, t) \otimes_F S$, and let

$$T' = (1 \otimes E) + (1 \otimes E)y' \subset R, \quad \text{where } y' = j \otimes s.$$

Then T' is isomorphic to T, since y' satisfies the same relations as y. (That is, for any $e \in E$, $y'(1 \otimes e)y'^{-1} = 1 \otimes (ses^{-1})$ and $y'^2 = 1 \otimes ts^2$.) By the double centralizer theorem (see (1.5)) $R \simeq T' \otimes_F Q$ where $Q = C_R(T')$, and Q is a quaternion algebra over F by a dimension count. It suffices to show that Q is split. For, then

$$R \simeq T' \otimes_F Q \simeq T \otimes_F M_2(F) \simeq M_2(T).$$

Consider $Z \otimes_F Z \subset R$; $Z \otimes_F Z$ centralizes $1 \otimes E$ and $\text{Int}(y')$ restricts to $\iota \otimes \iota$ on $Z \otimes_F Z$. Now $Z \otimes_F Z$ has two primitive idempotents e_1 and e_2, since $Z \otimes_F Z \simeq Z \oplus Z$. The automorphisms $\text{Id} \otimes \iota$ and $\iota \otimes \text{Id}$ permute them, so $\iota \otimes \iota$ maps each e_i to itself. Hence e_1 and e_2 lie in Q since they centralize $1 \otimes E$ and also y'. Because $e_1 e_2 = 0$, Q is not a division algebra, so Q is split, as desired. $\qquad\square$

(13.42) Remark. We have assumed in the above proposition that Z is a field. The argument still works, with slight modification, if $Z \simeq F \times F$.

As a consequence of Theorem (13.38), we may compare the group $G(A, \sigma, f)$ of multipliers of similitudes with the subgroup $G^+(A, \sigma, f)$ of multipliers of proper similitudes. Since the index of $\text{GO}^+(A, \sigma, f)$ in $\text{GO}(A, \sigma, f)$ is 1 or 2, it is clear that either $G(A, \sigma, f) = G^+(A, \sigma, f)$ or $G^+(A, \sigma, f)$ is a subgroup of index 2 in $G(A, \sigma, f)$. If A is split, then

$$[\text{GO}(A, \sigma, f) : \text{GO}^+(A, \sigma, f)] = [\text{O}(A, \sigma, f) : \text{O}^+(A, \sigma, f)] = 2$$

and

$$G(A, \sigma, f) = G^+(A, \sigma, f)$$

since hyperplane reflections are improper isometries (see (12.13)). If A is not split, we deduce from (13.38):

(13.43) Corollary. *Suppose that (σ, f) is a quadratic pair on a central simple F-algebra A of even degree. If A is not split, then $O(A, \sigma, f) = O^+(A, \sigma, f)$ and*

$$[GO(A, \sigma, f) : GO^+(A, \sigma, f)] = [G(A, \sigma, f) : G^+(A, \sigma, f)].$$

If $G(A, \sigma, f) \neq G^+(A, \sigma, f)$, then A is split by $Z = Z(A, \sigma, f)$.

Proof: If $g \in O(A, \sigma, f)$ is an improper isometry, then (13.38) shows that A is Brauer-equivalent to $(Z, \mu(g))_F$, which is split since $\mu(g) = 1$. This contradiction shows that $O(A, \sigma, f) = O^+(A, \sigma, f)$.

If $[GO(A, \sigma, f) : GO^+(A, \sigma, f)] \neq [G(A, \sigma, f) : G^+(A, \sigma, f)]$, then necessarily

$$GO(A, \sigma, f) \neq GO^+(A, \sigma, f) \text{ and } G(A, \sigma, f) = G^+(A, \sigma, f).$$

Therefore, A contains an improper similitude g, and $\mu(g) = \mu(g')$ for some proper similitude g'. It follows that $g^{-1}g'$ is an improper isometry, contrary to the equality $O(A, \sigma, f) = O^+(A, \sigma, f)$.

Finally, if μ is the multiplier of an improper similitude, then (13.38) shows that A is Brauer-equivalent to $(Z, \mu)_F$, hence it is split by Z. $\qquad \square$

(13.44) Corollary. *If $\mathrm{disc}(\sigma, f)$ is trivial, then $G(A, \sigma, f) = G^+(A, \sigma, f)$.*

Proof: It suffices to consider the case where A is not split. Then, if $G(A, \sigma, f) \neq G^+(A, \sigma, f)$, the preceding corollary shows that A is split by Z; this is impossible if $\mathrm{disc}(\sigma, f)$ is trivial, for then $Z \simeq F \times F$. $\qquad \square$

§14. Unitary Involutions

In this section, we let (B, τ) be a central simple algebra with involution of the second kind over an arbitrary field F. Let K be the center of B and ι the nontrivial automorphism of K/F.

We will investigate the group $GU(B, \tau)$ of similitudes of (B, τ) and the unitary group $U(B, \tau)$, which is the kernel of the multiplier map μ (see §12.B). The group $GU(B, \tau)$ has different properties depending on the parity of the degree of B. When $\deg B$ is even, we relate this group to the group of similitudes of the discriminant algebra $D(B, \tau)$.

14.A. Odd degree.

(14.1) Proposition. *If $\deg B$ is odd, the group $G(B, \tau)$ of multipliers of similitudes of (B, τ) is the group of norms of K/F:*

$$G(B, \tau) = N_{K/F}(K^\times).$$

Moreover, $GU(B, \tau) = K^\times \cdot U(B, \tau)$.

Proof: The inclusion $N_{K/F}(K^\times) \subset G(B, \tau)$ is clear, since $K^\times \subset GU(B, \tau)$ and $\mu(\alpha) = N_{K/F}(\alpha)$ for $\alpha \in K^\times$. In order to prove the reverse inclusion, let $\deg B = 2m + 1$ and let $g \in GU(B, \tau)$. By applying the reduced norm to the equation $\tau(g)g = \mu(g)$ we obtain

$$N_{K/F}\big(\mathrm{Nrd}_B(g)\big) = \mu(g)^{2m+1}.$$

Therefore,

$$\mu(g) = N_{K/F}\big(\mu(g)^{-m}\operatorname{Nrd}_B(g)\big) \in N_{K/F}(K^\times),$$

hence $G(B,\tau) \subset N_{K/F}(K^\times)$. This proves the first assertion.

The preceding equation shows moreover that $\mu(g)^m \operatorname{Nrd}_B(g)^{-1}g \in \operatorname{U}(B,\tau)$. Therefore, letting $u = \mu(g)^m \operatorname{Nrd}_B(g)^{-1}g$ and $\alpha = \mu(g)^{-m}\operatorname{Nrd}_B(g) \in K^\times$, we get $g = \alpha u$. Thus, $\operatorname{GU}(B,\tau) = K^\times \cdot \operatorname{U}(B,\tau)$. $\qquad\square$

Note that in the decomposition $g = \alpha u$ above, the elements $\alpha \in K^\times$ and $u \in \operatorname{U}(B,\tau)$ are uniquely determined up to a factor in the group K^1 of norm 1 elements, since $K^\times \cap \operatorname{U}(B,\tau) = K^1$.

14.B. Even degree. Suppose now that $\deg B = 2m$ and let $g \in \operatorname{GU}(B,\tau)$. By applying the reduced norm to the equation $\tau(g)g = \mu(g)$, we obtain

$$N_{K/F}\big(\operatorname{Nrd}_B(g)\big) = \mu(g)^{2m},$$

hence $\mu(g)^m \operatorname{Nrd}_B(g)^{-1}$ is in the group of elements of norm 1. By Hilbert's Theorem 90, there is an element $\alpha \in K^\times$, uniquely determined up to a factor in $F^\times$, such that

$$\mu(g)^{-m}\operatorname{Nrd}_B(g) = \alpha\iota(\alpha)^{-1}.$$

We may therefore define a homomorphism $\nu\colon \operatorname{GU}(A,\sigma) \to K^\times/F^\times$ by

(14.2) $\nu(g) = \alpha \cdot F^\times.$

Let $\operatorname{SGU}(B,\tau)$ be the kernel of ν, and let $\operatorname{SU}(B,\tau)$ be the intersection $\operatorname{SGU}(B,\tau)\cap \operatorname{U}(B,\tau)$:

$$\operatorname{SGU}(B,\tau) = \{\, g \in \operatorname{GU}(B,\tau) \mid \operatorname{Nrd}_B(g) = \mu(g)^m \,\}$$
$$\operatorname{SU}(B,\tau) = \{\, u \in \operatorname{GU}(B,\tau) \mid \operatorname{Nrd}_B(u) = \mu(u) = 1 \,\}.$$

We thus have the following diagram, where all the maps are inclusions:

$$
\begin{array}{ccc}
\operatorname{SU}(B,\tau) & \longrightarrow & \operatorname{SGU}(B,\tau) \\
\downarrow & & \downarrow \\
\operatorname{U}(B,\tau) & \longrightarrow & \operatorname{GU}(B,\tau).
\end{array}
$$

Consider for example the case where $K = F \times F$; we may then assume $B = E \times E^{\mathrm{op}}$ for some central simple F-algebra E of degree $2m$, and $\tau = \varepsilon$ is the exchange involution. We then have

$$\operatorname{GU}(B,\tau) = \{\, \big(x, \alpha(x^{-1})^{\mathrm{op}}\big) \mid \alpha \in F^\times,\ x \in E^\times \,\} \simeq E^\times \times F^\times$$

and the maps μ and ν are defined by

$$\mu\big(x, \alpha(x^{-1})^{\mathrm{op}}\big) = \alpha, \quad \nu\big(x, \alpha(x^{-1})^{\mathrm{op}}\big) = \big(\operatorname{Nrd}_E(x), \alpha^m\big) \cdot F^\times.$$

Therefore,

$$\operatorname{SGU}(B,\tau) \simeq \{\, (x, \alpha) \in E^\times \times F^\times \mid \operatorname{Nrd}_E(x) = \alpha^m \,\},$$
$$\operatorname{U}(B,\tau) \simeq E^\times$$

and the group $\operatorname{SU}(B,\tau)$ is isomorphic to the group of elements of reduced norm 1 in E, which we write $\operatorname{SL}(E)$:

$$\operatorname{SU}(B,\tau) \simeq \{\, x \in E^\times \mid \operatorname{Nrd}_E(x) = 1 \,\} = \operatorname{SL}(E).$$

14.C. Relation with the discriminant algebra. Our first results in this direction do not assume the existence of an involution; we formulate them for an arbitrary central simple F-algebra A:

The canonical map λ^k.

(14.3) Proposition. *Let A be any central simple algebra over a field F. For all integers k such that $1 \leq k \leq \deg A$, there is a homogeneous polynomial map of degree k:*

$$\lambda^k \colon A \to \lambda^k A$$

which restricts to a group homomorphism $A^\times \to (\lambda^k A)^\times$. If the algebra A is split, let $A = \operatorname{End}_F(V)$, then under the identification $\lambda^k A = \operatorname{End}_F(\bigwedge^k V)$ the map λ^k is defined by

$$\lambda^k(f) = \textstyle\bigwedge^k f = f \wedge \cdots \wedge f \quad \text{for } f \in \operatorname{End}_F(V).$$

Proof: Let $g_k \colon S_k \to (A^{\otimes k})^\times$ be the homomorphism of (10.1). By (10.2), it is clear that for all $a \in A^\times$ the element $\otimes^k a = a \otimes \cdots \otimes a$ commutes with $g_k(\pi)$ for all $\pi \in S_k$, hence also with $s_k = \sum_{\pi \in S_k} \operatorname{sgn}(\pi) g_k(\pi)$. Multiplication on the right by $\otimes^k a$ is therefore an endomorphism of the left $A^{\otimes k}$-module $A^{\otimes k} s_k$. We denote this endomorphism by $\lambda^k a$; thus $\lambda^k a \in \operatorname{End}_{A^{\otimes k}}(A^{\otimes k} s_k) = \lambda^k A$ is defined by

$$\big((a_1 \otimes \cdots \otimes a_k) \cdot s_k\big)^{\lambda^k a} = (a_1 \otimes \cdots \otimes a_k) \cdot s_k \cdot \otimes^k a = (a_1 a \otimes \cdots \otimes a_k a) \cdot s_k.$$

If $A = \operatorname{End}_F(V)$, there is a natural isomorphism (see (10.3)):

$$A^{\otimes k} s_k = \operatorname{Hom}_F(\textstyle\bigwedge^k V, V^{\otimes k}),$$

under which s_k is identified with the map $s_k' \colon \bigwedge^k V \to V^{\otimes k}$ defined by

$$s_k'(v_1 \wedge \cdots \wedge v_k) = s_k(v_1 \otimes \cdots \otimes v_k) \quad \text{for } v_1, \ldots, v_k \in V.$$

For $f \in \operatorname{End}_F(V)$ we have $s_k'^{\lambda^k f}(v_1 \wedge \cdots \wedge v_k) = s_k\big(\otimes^k f(v_1 \otimes \cdots \otimes v_k)\big)$, hence

$$s_k'^{\lambda^k f}(v_1 \wedge \cdots \wedge v_k) = s_k'\big(f(v_1) \wedge \cdots \wedge f(v_k)\big).$$

Therefore, $s_k'^{\lambda^k f} = s_k' \circ \bigwedge^k f$, which means that $\lambda^k(f) \in \lambda^k \operatorname{End}_F(V)$ is identified with $\bigwedge^k f \in \operatorname{End}_F(\bigwedge^k V)$. It is then clear that λ^k is a homogeneous polynomial map of degree k, and that its restriction to $A^\times$ is a group homomorphism to $(\lambda^k A)^\times$. $\square$

For the following result, we assume $\deg A = 2m$, so that $\lambda^m A$ has a canonical involution γ of the first kind (see (10.11)).

(14.4) Proposition. *If $\deg A = 2m$, then $\gamma(\lambda^m a)\lambda^m a = \operatorname{Nrd}_A(a)$ for all $a \in A$. In particular, if $a \in A^\times$, then $\lambda^m a \in \operatorname{Sim}(\lambda^m A, \gamma)$ and $\mu(\lambda^m a) = \operatorname{Nrd}_A(a)$.*

Proof: It suffices to check the split case. We may thus assume $A = \operatorname{End}_F(V)$, hence $\lambda^m A = \operatorname{End}_F(\bigwedge^m V)$ and the canonical involution γ is the adjoint involution with respect to the canonical bilinear map $\wedge \colon \bigwedge^m V \times \bigwedge^m V \to \bigwedge^{2m} V$. Moreover, $\lambda^m(f) = \bigwedge^m f$ for $f \in \operatorname{End}_F(V)$. The statement that $\lambda^m(f)$ is a similitude for γ therefore follows from the following identities

$$\textstyle\bigwedge^m f(v_1 \wedge \cdots \wedge v_m) \wedge \bigwedge^m f(w_1 \wedge \cdots \wedge w_m)$$
$$= f(v_1) \wedge \cdots \wedge f(v_m) \wedge f(w_1) \wedge \cdots \wedge f(w_m)$$
$$= \det f \cdot v_1 \wedge \cdots \wedge v_m \wedge w_1 \wedge \cdots \wedge w_m$$

for $v_1, \ldots, v_m, w_1, \ldots, w_m \in V$. $\square$

The canonical map D. We now return to the case of central simple algebras with unitary involution (B, τ). We postpone until after Proposition (14.7) the discussion of the case where the center K of B is isomorphic to $F \times F$; we thus assume for now that K is a field.

(14.5) Lemma. *For $k = 1, \ldots, \deg B$, let $\tau^{\wedge k}$ be the involution on $\lambda^k B$ induced by τ (see (10.20)). For all k, the canonical map $\lambda^k \colon B \to \lambda^k B$ satisfies*

$$\tau^{\wedge k} \circ \lambda^k = \lambda^k \circ \tau.$$

Proof: By extending scalars to a splitting field of B, we reduce to considering the split case. We may thus assume $B = \operatorname{End}_K(V)$ and $\tau = \sigma_h$ is the adjoint involution with respect to some nonsingular hermitian form h on V. According to (10.21), the involution $\tau^{\wedge k}$ is the adjoint involution with respect to $h^{\wedge k}$. Therefore, for $f \in \operatorname{End}_K(V)$, the element $\tau^{\wedge k} \circ \lambda^k(f) \in \operatorname{End}_K(\bigwedge^k V)$ is defined by the condition:

$$h^{\wedge k}\big(\tau^{\wedge k} \circ \lambda^k(f)(v_1 \wedge \cdots \wedge v_k), w_1 \wedge \cdots \wedge w_k\big) =$$
$$h^{\wedge k}\big(v_1 \wedge \cdots \wedge v_k, \lambda^k(f)(w_1 \wedge \cdots \wedge w_k)\big)$$

for $v_1, \ldots, v_k, w_1, \ldots, w_k \in V$. Since $\lambda^k(f) = \bigwedge^k f$, the right-hand expression equals

$$\det\big(h\big(v_i, f(w_j)\big)\big)_{1 \le i, j \le k} = \det\big(h\big(\tau(f)(v_i), w_j\big)\big)_{1 \le i, j \le k}$$
$$= h^{\wedge k}\big(\lambda^k\big(\tau(f)\big)(v_1 \wedge \cdots \wedge v_k), w_1 \wedge \cdots \wedge w_k\big).$$

$\square$

Assume now $\deg B = 2m$; we may then define the discriminant algebra $D(B, \tau)$ as the subalgebra of $\lambda^m B$ of elements fixed by $\tau^{\wedge m} \circ \gamma$, see (10.28).

(14.6) Lemma. *For $g \in \operatorname{GU}(B, \tau)$, let $\alpha \in K^\times$ be such that $\nu(g) = \alpha \cdot F^\times$; then $\alpha^{-1} \lambda^m g \in D(B, \tau)$ and $\underline{\tau}(\alpha^{-1} \lambda^m g) \cdot \alpha^{-1} \lambda^m g = N_{K/F}(\alpha)^{-1} \mu(g)^m$. In particular, $\lambda^m g \in \operatorname{Sim}\big(D(B, \tau), \underline{\tau}\big)$ for all $g \in \operatorname{SGU}(B, \tau)$.*

Proof: By (14.4) we have

$$\gamma(\lambda^m g) = \operatorname{Nrd}_B(g) \lambda^m g^{-1} = \operatorname{Nrd}_B(g) \mu(g)^{-m} \lambda^m\big(\tau(g)\big),$$

hence, by (14.5),

$$\tau^{\wedge m} \circ \gamma(\lambda^m g) = \iota\big(\operatorname{Nrd}_B(g)\big) \mu(g)^{-m} \lambda^m g = \alpha^{-1} \iota(\alpha) \lambda^m g.$$

Therefore, $\alpha^{-1} \lambda^m g \in D(B, \tau)$. Since $\underline{\tau}$ is the restriction of γ to $D(B, \tau)$, we have

$$\underline{\tau}(\alpha^{-1} \lambda^m g) \cdot \alpha^{-1} \lambda^m g = \alpha^{-2} \gamma(\lambda^m g) \lambda^m g,$$

and (14.4) completes the proof. $\square$

The lemma shows that the inner automorphism $\operatorname{Int}(\lambda^m g) = \operatorname{Int}(\alpha^{-1} \lambda^m g)$ of $\lambda^m B$ preserves $D(B, \tau)$ and induces an automorphism of $\big(D(B, \tau), \underline{\tau}\big)$. Since this automorphism is also induced by the automorphism $\operatorname{Int}(g)$ of (B, τ), by functoriality of the discriminant algebra construction, we denote it by $D(g)$. Alternately, under the identification $\operatorname{Aut}\big(D(B, \tau), \underline{\tau}\big) = \operatorname{PSim}\big(D(B, \tau), \underline{\tau}\big)$ of (12.15), we may set $D(g) = \alpha^{-1} \lambda^m g \cdot F^\times$, where $\alpha \in K^\times$ is a representative of $\nu(g)$ as above.

The next proposition follows from the definitions and from (14.6):

(14.7) Proposition. *The following diagram commutes*:

$$\begin{array}{ccc} \mathrm{SGU}(B,\tau) & \longrightarrow & \mathrm{GU}(B,\tau) \\ {\scriptstyle \lambda^m}\downarrow & & \downarrow{\scriptstyle D} \\ \mathrm{Sim}\big(D(B,\tau),\underline{\tau}\big) & \xrightarrow{\ \mathrm{Int}\ } & \mathrm{Aut}_F\big(D(B,\tau),\underline{\tau}\big). \end{array}$$

Moreover, for $g \in \mathrm{SGU}(B,\tau)$, the multipliers of g and $\lambda^m g$ are related by

$$\underline{\mu}(\lambda^m g) = \mu(g)^m = \mathrm{Nrd}_B(g).$$

Therefore, λ^m restricts to a group homomorphism $\mathrm{SU}(B,\tau) \to \mathrm{Iso}\big(D(B,\tau),\underline{\tau}\big)$.

We now turn to the case where $K \simeq F \times F$, which was put aside for the preceding discussion. In this case, we may assume $B = E \times E^{\mathrm{op}}$ for some central simple F-algebra E of degree $2m$ and $\tau = \varepsilon$ is the exchange involution. As observed in §10.E and §14.B, we may then identify

$$\big(D(B,\tau),\underline{\tau}\big) = (\lambda^m E, \gamma) \quad \text{and} \quad \mathrm{GU}(B,\tau) = E^\times \times F^\times.$$

The discussion above remains valid without change if we set $D(x,\alpha) = \mathrm{Int}(\lambda^m x)$ for $(x,\alpha) \in E^\times \times F^\times = \mathrm{GU}(B,\tau)$, a definition which is compatible with the definitions above (in the case where K is a field) under scalar extension.

The canonical Lie homomorphism $\dot\lambda^k$. To conclude this section, we derive from the map λ^m a Lie homomorphism from the Lie algebra $\mathrm{Skew}(B,\tau)^0$ of skew-symmetric elements of reduced trace zero to the Lie algebra $\mathrm{Skew}\big(D(B,\tau),\underline{\tau}\big)$. This Lie homomorphism plays a crucial rôle in §15.D (see (15.22)).

As above, we start with an arbitrary central simple F-algebra A. Let t be an indeterminate over F. For $k = 1, \ldots, \deg A$, consider the canonical map

$$\lambda^k \colon A \otimes F(t) \to \lambda^k A \otimes F(t).$$

Since this map is polynomial of degree k and $\lambda^k(1) = 1$, there is a linear map $\dot\lambda^k \colon A \to \lambda^k A$ such that for all $a \in A$,

$$\lambda^k(t + a) = t^k + \dot\lambda^k(a) t^{k-1} + \cdots + \lambda^k(a).$$

(14.8) Proposition. *The map $\dot\lambda^k$ is a Lie-algebra homomorphism*

$$\dot\lambda^k \colon \mathfrak{L}(A) \to \mathfrak{L}(\lambda^k A).$$

If $A = \mathrm{End}_F(V)$, then under the identification $\lambda^k A = \mathrm{End}_F(\bigwedge^k V)$ we have

$$\dot\lambda^k(f)(v_1 \wedge \cdots \wedge v_k) =$$
$$\big(f(v_1) \wedge v_2 \wedge \cdots \wedge v_k\big) + \big(v_1 \wedge f(v_2) \wedge \cdots \wedge v_k\big) + \cdots + \big(v_1 \wedge v_2 \wedge \cdots \wedge f(v_k)\big)$$

for all $f \in \mathrm{End}_F(V)$ and $v_1, \ldots, v_k \in V$.

Proof: The description of $\dot\lambda^k$ in the split case readily follows from that of λ^k in (14.3). To prove that $\dot\lambda^k$ is a Lie homomorphism, we may reduce to the split case by a scalar extension. The property then follows from an explicit computation: for

$f, g \in \mathrm{End}_F(V)$ and $v_1, \ldots, v_k \in V$ we have

$$\dot{\lambda}^k(f) \circ \dot{\lambda}^k(g)(v_1 \wedge \cdots \wedge v_k) = \sum_{1 \leq i < j \leq k} v_1 \wedge \cdots \wedge f(v_i) \wedge \cdots \wedge g(v_j) \wedge \cdots \wedge v_k$$

$$+ \sum_{1 \leq i \leq k} v_1 \wedge \cdots \wedge f \circ g(v_i) \wedge \cdots \wedge v_k$$

$$+ \sum_{1 \leq j < i \leq k} v_1 \wedge \cdots \wedge g(v_j) \wedge \cdots \wedge f(v_i) \wedge \cdots \wedge v_k,$$

hence $\dot{\lambda}^k(f) \circ \dot{\lambda}^k(g) - \dot{\lambda}^k(g) \circ \dot{\lambda}^k(f)$ maps $v_1 \wedge \cdots \wedge v_k$ to

$$\sum_{1 \leq i \leq k} v_1 \wedge \cdots \wedge (f \circ g - g \circ f)(v_i) \wedge \cdots \wedge v_k = \dot{\lambda}^k([f,g])(v_1 \wedge \cdots \wedge v_k).$$

This shows $\big[\dot{\lambda}^k(f), \dot{\lambda}^k(g)\big] = \dot{\lambda}^k([f,g])$. $\qquad\square$

(14.9) Corollary. *Suppose $k \leq \deg A - 1$. If $a \in A$ satisfies $\dot{\lambda}^k a \in F$, then $a \in F$ and $\dot{\lambda}^k a = ka$. In particular, $\ker \dot{\lambda}^k = \{\, a \in F \mid ka = 0 \,\}$.*

Proof: It suffices to consider the split case; we may thus assume that $A = \mathrm{End}_F(V)$ for some vector space V. If $\dot{\lambda}^k a \in F$, then for all $x_1, \ldots, x_k \in V$ we have $x_1 \wedge \dot{\lambda}^k a(x_1 \wedge \cdots \wedge x_k) = 0$, hence $x_1 \wedge a(x_1) \wedge x_2 \wedge \cdots \wedge x_k = 0$. Since $k < \dim V$, this relation shows that $a(x_1) \in x_1 \cdot F$ for all $x_1 \in V$, hence $a \in F$. The other statements are then clear. $\qquad\square$

In the particular case where $k = \frac{1}{2} \deg A$, we have:

(14.10) Proposition. *Suppose $\deg A = 2m$, and let γ be the canonical involution on $\lambda^m A$. For all $a \in A$,*

$$\dot{\lambda}^m a + \gamma(\dot{\lambda}^m a) = \mathrm{Trd}_A(a).$$

Proof: By (14.4), we have

$$\gamma\big(\lambda^m(t+a)\big) \cdot \lambda^m(t+a) = \mathrm{Nrd}(t+a).$$

The proposition follows by comparing the coefficients of t^{2m-1} on each side. $\qquad\square$

We now consider a central simple algebra with unitary involution (B, τ) over F, and assume that the center K of B is a field. Suppose also that the degree of B is even: $\deg B = 2m$. Since (14.5) shows that

$$\tau^{\wedge m} \circ \lambda^m(t+b) = \lambda^m \circ \tau(t+b) \quad \text{for } b \in B,$$

it follows that

$$\tau^{\wedge m} \circ \dot{\lambda}^m = \dot{\lambda}^m \circ \tau.$$

It is now easy to determine under which condition $\dot{\lambda}^m b \in D(B, \tau)$: this holds if and only if $\gamma(\dot{\lambda}^m b) = \tau^{\wedge m}(\dot{\lambda}^m b)$, which means that

$$\mathrm{Trd}_B(b) - \dot{\lambda}^m b = \dot{\lambda}^m \tau(b).$$

By (14.9), this equality holds if and only if $b + \tau(b) \in F$ and $\mathrm{Trd}_B(b) = m\big(b + \tau(b)\big)$. Let

(14.11) $\qquad \mathfrak{s}(B, \tau) = \{\, b \in B \mid b + \tau(b) \in F \text{ and } \mathrm{Trd}_B(b) = m\big(b + \tau(b)\big) \,\},$

so $\mathfrak{s}(B, \tau) = (\dot{\lambda}^m)^{-1}\big(D(B, \tau)\big)$. The algebra $\mathfrak{s}(B, \tau)$ is contained in $\mathfrak{g}(B, \tau) = \{\, b \in B \mid b + \tau(b) \in F \,\}$ (see §2.D). If $\dot{\mu}_B \colon \mathfrak{g}(B, \tau) \to F$ is the map which carries

$b \in \mathfrak{g}(B,\tau)$ to $b + \tau(b)$, we may describe $\mathfrak{s}(B,\tau)$ as the kernel of the F-linear map $\mathrm{Trd}_B - m\dot\mu_B \colon \mathfrak{g}(B,\tau) \to K$. The image of this map lies in $K^0 = \{\, x \in K \mid T_{K/F}(x) = 0 \,\}$, since taking the reduced trace of both sides of the relation $b + \tau(b) = \dot\mu(b)$ yields $T_{K/F}\big(\mathrm{Trd}_B(b)\big) = 2m\dot\mu_B(b)$. On the other hand, this map is not trivial: since Trd_B is surjective, we may find $x \in B$ such that $\mathrm{Trd}_B(x) \notin F$; then $x - \tau(x) \in \mathfrak{g}(B,\tau)$ is not mapped to 0. Therefore,

$$\dim_F \mathfrak{s}(B,\tau) = \dim_F \mathfrak{g}(B,\tau) - 1 = 4m^2.$$

(14.12) Proposition. *The homomorphism $\dot\lambda^m$ restricts to a Lie algebra homomorphism*

$$\dot\lambda^m \colon \mathfrak{s}(B,\tau) \to \mathfrak{g}\big(D(B,\tau),\underline\tau\big).$$

By denoting by $\dot\mu_D \colon \mathfrak{g}\big(D(B,\tau),\underline\tau\big) \to F$ the map which carries $x \in D(B,\tau)$ to $x + \underline\tau(x)$, we have

$$\dot\mu_D(\dot\lambda^m b) = m\dot\mu_B(b) = \mathrm{Trd}_B(b) \qquad for\ b \in \mathfrak{s}(B,\tau).$$

Therefore, $\dot\lambda^m$ restricts to a Lie algebra homomorphism

$$\dot\lambda^m \colon \mathrm{Skew}(B,\tau)^0 \to \mathrm{Skew}\big(D(B,\tau),\underline\tau\big),$$

where $\mathrm{Skew}(B,\tau)^0$ is the Lie algebra of skew-symmetric elements of reduced trace zero in B.

Proof: Since $\mathfrak{s}(B,\tau) = (\dot\lambda^m)^{-1}\big(D(B,\tau)\big)$, it is clear that $\dot\lambda^m$ restricts to a homomorphism from $\mathfrak{s}(B,\tau)$ to $\mathfrak{L}\big(D(B,\tau)\big)$. To prove that its image lies in $\mathfrak{g}\big(D(B,\tau),\underline\tau\big)$, it suffices to prove

$$\dot\lambda^m b + \underline\tau(\dot\lambda^m b) = \mathrm{Trd}_B(b) \qquad for\ b \in \mathfrak{s}(B,\tau).$$

This follows from (14.10), since $\underline\tau$ is the restriction of γ to $D(B,\tau)$. $\qquad\square$

Suppose finally $K \simeq F \times F$; we may then assume $B = E \times E^{\mathrm{op}}$ for some central simple F-algebra E of degree $2m$, and $\tau = \varepsilon$ is the exchange involution. We have

$$\mathfrak{s}(B,\tau) = \{\, (x, \alpha - x^{\mathrm{op}}) \mid x \in E,\ \alpha \in F,\ \text{and } \mathrm{Trd}_E(x) = m\alpha \,\},$$

and $\big(D(B,\tau),\underline\tau\big)$ may be identified with $(\lambda^m E, \gamma)$. The Lie algebra homomorphism $\dot\lambda^m \colon \mathfrak{s}(B,\tau) \to \mathfrak{g}(\lambda^m E, \gamma)$ then maps $(x, \alpha - x^{\mathrm{op}})$ to $\dot\lambda^m x$. Identifying $\mathrm{Skew}(B,\tau)^0$ with the Lie algebra E^0 of elements of reduced trace zero (see (2.24)), we may restrict this homomorphism to a Lie algebra homomorphism:

$$\dot\lambda^m \colon E^0 \to \mathrm{Skew}(\lambda^m E, \gamma).$$

EXERCISES

1. Let Q be a quaternion algebra with canonical involution γ over a field F of arbitrary characteristic. On the algebra $A = Q \otimes_F Q$, consider the canonical quadratic pair $(\gamma \otimes \gamma, f)$ (see (5.21)). Prove that

$$\mathrm{GO}^+(A, \gamma \otimes \gamma, f) = \{\, q_1 \otimes q_2 \mid q_1, q_2 \in Q^\times \,\}$$

   and determine the group of multipliers $\mu\big(\mathrm{GO}^+(A, \gamma \otimes \gamma, f)\big)$.

2. Let (A, σ) be a central simple F-algebra with involution of any kind with center K and let $\alpha \in \mathrm{Aut}_K(A)$. Prove that the following statements are equivalent:

(a) $\alpha \in \mathrm{Aut}_K(A, \sigma)$.

(b) $\alpha\big(\mathrm{Sym}(A, \sigma)\big) = \mathrm{Sym}(A, \sigma)$.

(c) $\alpha\big(\mathrm{Alt}(A, \sigma)\big) = \mathrm{Alt}(A, \sigma)$.

3. Let (A, σ) be a central simple algebra with orthogonal involution and degree a power of 2 over a field F of characteristic different from 2, and let $B \subset A$ be a proper subalgebra with center $F \neq B$. Prove that every similitude $f \in \mathrm{GO}(A, \sigma)$ such that $fBf^{-1} = B$ is proper.

4. (Wonenburger [316]) The aim of this exercise is to give a proof of Wonenburger's theorem on the image of $\mathrm{GO}(V, q)$ in $\mathrm{Aut}\big(C_0(V, q)\big)$, see (13.4).

Let q be a nonsingular quadratic form on an even-dimensional vector space V over a field F of arbitrary characteristic. Using the canonical identification $\varphi_q \colon V \otimes V \xrightarrow{\sim} \mathrm{End}_F(V)$, we identify $c\big(\mathrm{End}_F(V)\big) = V \cdot V \subset C_0(V, q)$ and $\mathrm{Alt}\big(\mathrm{End}_F(V), \sigma_q\big) = \bigwedge^2 V$. An element $x \in V \cdot V$ is called a *regular plane element* if $x = v \cdot w$ for some vectors $v, w \in V$ which span a nonsingular 2-dimensional subspace of V. The first goal is to show that the regular plane elements are preserved by the automorphisms of $C_0(V, q)$ which preserve $V \cdot V$.

Let $\rho \colon \bigwedge^2 V \to \bigwedge^4 V$ be the quadratic map which vanishes on elements of the type $v \wedge w$ and whose polar is the exterior product (compare with (10.12)), and let τ be the canonical involution on $C(V, q)$ which is the identity on V.

(a) Show that $x \in \bigwedge^2 V$ has the form $x = v \wedge w$ for some $v, w \in V$ if and only if $\rho(x) = 0$.

(b) Show that the Lie homomorphism $\delta \colon V \cdot V \to \bigwedge^2 V$ maps $v \cdot w$ to $v \wedge w$.

(c) Let $W = \{\, x + \tau(x) \mid x \in V \cdot V \cdot V \cdot V \,\} \subset C_0(V, q)$. Show that $F \subset W$ and that there is a surjective map $\omega \colon \bigwedge^4 V \to W/F$ such that

$$\omega(v_1 \wedge v_2 \wedge v_3 \wedge v_4) = v_1 \cdot v_2 \cdot v_3 \cdot v_4 + v_4 \cdot v_3 \cdot v_2 \cdot v_1 + F.$$

Show that for all $x \in V \cdot V$,

$$-\omega \circ \rho \circ \delta(x) = \tau(x) \cdot x + F.$$

(d) Assume that $\mathrm{char}\, F \neq 2$. Show that ω is bijective and use the results above to show that if $x \in V \cdot V$ satisfies $\tau(x) \cdot x \in F$, then $\delta(x) = v \wedge w$ for some $v, w \in V$.

(e) Assume that (V, q) is a 6-dimensional hyperbolic space over a field F of characteristic 2, and let $(e_1, \dots, e_6)$ be a symplectic basis of V consisting of isotropic vectors. Show that $x = e_1 \cdot e_2 + e_3 \cdot e_4 + e_5 \cdot e_6 \in V \cdot V$ satisfies $\tau(x) \cdot x = 0$, $\tau(x) + x = 1$, but $\delta(x)$ cannot be written in the form $v \wedge w$ with $v, w \in V$.

(f) For the rest of this exercise, assume that $\mathrm{char}\, F \neq 2$. Show that $x \in V \cdot V$ is a regular plane element if and only if $\tau(x) \cdot x \in F$, $\tau(x) + x \in F$ and $\big(\tau(x) + x\big)^2 \neq 4\tau(x) \cdot x$. Conclude that every automorphism of $C_0(V, q)$ which commutes with τ and preserves $V \cdot V$ maps regular plane elements to regular plane elements.

(g) Show that regular plane elements $x, y \in V \cdot V$ anticommute if and only if $x = u \cdot v$ and $y = u \cdot w$ for some pairwise orthogonal anisotropic vectors u, $v, w \in V$.

(h) Let $(e_1, \dots, e_n)$ be an orthogonal basis of V. Let $\theta \in \mathrm{Aut}\big(C_0(V, q), \tau\big)$ be an automorphism which preserves $V \cdot V$. Show that there is an orthogonal basis $(v'_1, v_2, \dots, v_n)$ of V such that $\theta(e_1 \cdot e_i) = v'_1 \cdot v_i$ for $i = 2, \dots, n$. Let

$\alpha = q(v_1')^{-1}q(e_1)$. Show that the linear transformation of V which maps e_1 to $\alpha v_1'$ and e_i to v_i for $i = 2, \ldots, n$ is a similitude which induces θ.

5. Let D be a central division algebra with involution $\bar{}$ over a field F of characteristic different from 2 and let V be a (finite dimensional) right vector space over D with a nonsingular hermitian form h. Let $v \in V$ be an anisotropic vector and let $d \in D^\times$ be such that

$$h(v,v) = \bar{d}h(v,v)d \ (= h(vd, vd)).$$

Define $\tau_{v,d} \in \mathrm{End}_F(V)$ by

$$\tau_{v,d}(x) = x + v(d-1)h(v,v)^{-1}h(v,x).$$

Prove: $\tau_{v,d}$ is an isometry of (V,h), $\mathrm{Nrd}_{\mathrm{End}(V)}(\tau_{v,d}) = \mathrm{Nrd}_D(d)$, and show that the group of isometries of (V,h) is generated by elements of the form $\tau_{v,d}$.

   Hint: For the last part, see the proof of Witt's theorem in Scharlau [247, Theorem 7.9.5].

6. (Notation as in the preceding exercise.) Suppose $\bar{}$ is of the first kind. Show that if $d \in D^\times$ is such that $\bar{d}sd = s$ for some $s \in D^\times$ such that $\bar{s} = \pm s$, then $\mathrm{Nrd}_D(d) = 1$, except if D is split and $d = -1$.

   Hint: If $d \neq -1$, set $e = \frac{1-d}{1+d}$. Show that $s^{-1}\bar{e}s = -e$ and $d = \frac{1-e}{1+e}$.

7. Use the preceding two exercises to prove the following special case of (13.38) due to Kneser: assuming char $F \neq 2$, if (A, σ) is a central simple F-algebra with orthogonal involution which contains an improper isometry, then A is split.

8. (Dieudonné) Let (V, q) be a 4-dimensional hyperbolic quadratic space over the field F with two elements, and let $(e_1, \ldots, e_4)$ be a basis of V such that $q(e_1 x_1 + \cdots + e_4 x_4) = x_1 x_2 + x_3 x_4$. Consider the map $\tau: V \to V$ such that $\tau(e_1) = e_3$, $\tau(e_2) = e_4$, $\tau(e_3) = e_1$ and $\tau(e_4) = e_2$. Show that τ is a proper isometry of (V, q) which is not a product of hyperplane reflections. Consider the element

$$\gamma = e_2 \cdot (e_1 + e_3) + (e_1 + e_3) \cdot e_4 \in C_0(V, q).$$

Show that $\gamma \in \Gamma^+(V, q)$, $\chi(\gamma) = \tau$, and that γ is not a product of vectors in V.

9. Let (A, σ) be a central simple algebra with involution (of any type) over a field F of characteristic different from 2. Let

$$U = \{\, u \in A \mid \sigma(u)u = 1 \,\}$$

denote the group of isometries of (A, σ) and let

$$U^0 = \{\, u \in U \mid 1 + u \in A^\times \,\}.$$

Let also

$$\mathrm{Skew}(A, \sigma)^0 = \{\, a \in A \mid \sigma(a) = -a \text{ and } 1 + a \in A^\times \,\}.$$

Show that U is generated (as a group) by the set U^0. Show that U^0 (resp. $\mathrm{Skew}(A, \sigma)^0$) is a Zariski-open subset of U (resp. $\mathrm{Skew}(A, \sigma)$) and that the map $a \mapsto \frac{1-a}{1+a}$ defines a bijection from $\mathrm{Skew}(A, \sigma)^0$ onto U^0. (This bijection is known as the *Cayley parametrization* of U.)

10. Let (A, σ, f) be a central simple algebra of degree $2m$ with quadratic pair and let $g \in \mathrm{GO}(A, \sigma, f)$. Show that

$$\mathrm{Sn}\big(\mu(g)^{-1} g^2\big) = \begin{cases} 1 & \text{if } \begin{cases} m \text{ is odd and } g \text{ is improper,} \\ m \text{ is even and } g \text{ is proper,} \end{cases} \\ \mu(g) \cdot F^{\times 2} & \text{if } \begin{cases} m \text{ is odd and } g \text{ is proper,} \\ m \text{ is even and } g \text{ is improper,} \end{cases} \end{cases}$$

and that

$$\mathrm{Sn}\big(\mathrm{O}^+(A, \sigma, f)\big) \supset \begin{cases} \mu\big(\mathrm{GO}^+(A, \sigma, f)\big) \cdot F^{\times 2} & \text{if } m \text{ is odd,} \\ \mu\big(\mathrm{GO}^-(A, \sigma, f)\big) \cdot F^{\times 2} & \text{if } m \text{ is even,} \end{cases}$$

where $\mathrm{GO}^-(A, \sigma, f)$ is the coset of improper similitudes of (A, σ, f).
Hint: Use the arguments of (13.23).

11. Let (A, σ, f) be a central simple algebra of degree $\deg A \equiv 2 \mod 4$ with a quadratic pair. Show that

$$\big\{\, c \in \mathrm{Sim}\big(C(A, \sigma, f), \underline{\sigma}\big) \mid \underline{\sigma}(c) * A^b \cdot c = A^b \,\big\} = \Gamma(A, \sigma, f).$$

Hint: $\underline{\sigma}(c)c \in F^\times$ for all $c \in \mathrm{Sim}(A, \sigma, f)$.

12. Let (B, τ) be a central simple F-algebra with unitary involution. Let ι be the nontrivial automorphism of the center K of B and assume that $\mathrm{char}\, F \neq 2$.

(a) (Merkurjev [191, Proposition 6.1]) Show that

$$\mathrm{Nrd}_B\big(\mathrm{U}(B, \tau)\big) = \{\, z\iota(z)^{-1} \mid z \in \mathrm{Nrd}_B(B^\times) \,\}.$$

In particular, the subgroup $\mathrm{Nrd}_B\big(U(B, \tau)\big) \subset K^\times$ depends only on the Brauer class of B.

Hint: (Suresh [34, Theorem 5.1.3]) For $u \in \mathrm{U}(B, \tau)$, show that there exists $x \in K$ such that $v = x + u\iota(x)$ is invertible. Then $u = v\tau(v)^{-1}$ and $\mathrm{Nrd}(u) = z\iota(z)^{-1}$ with $z = \mathrm{Nrd}(v)$. To prove the reverse inclusion, let $(B, \tau) = \big(\mathrm{End}_D(V), \sigma_h\big)$ for some hermitian space (V, h) over a division algebra D with unitary involution θ. By considering endomorphisms which have a diagonal matrix representation with respect to an orthogonal basis of V, show that $\mathrm{Nrd}_D\big(\mathrm{U}(D, \theta)\big) \subset \mathrm{Nrd}_B\big(\mathrm{U}(B, \tau)\big)$. Finally, for $d \in D^\times$, show by dimension count that $d \cdot \mathrm{Sym}(D, \theta) \cap \big(F + \mathrm{Skew}(D, \theta)\big) \neq \{0\}$, hence $d = (x + s)t^{-1}$ for some $x \in F$, $s \in \mathrm{Skew}(D, \theta)$ and $t \in \mathrm{Sym}(D, \theta)$. For $u = (x + s)(x - s)^{-1}$, show that $u \in \mathrm{U}(D, \theta)$ and $\mathrm{Nrd}_D(u) = \mathrm{Nrd}_D(d)\iota\big(\mathrm{Nrd}_D(d)\big)^{-1}$.

(b) (Suresh [206, Lemma 2.6]) If $\deg(B, \tau)$ is odd, show that

$$\mathrm{Nrd}_B\big(\mathrm{U}(B, \tau)\big) = \mathrm{Nrd}_B(B^\times) \cap K^1.$$

Hint: Let $\deg(B, \tau) = 2r + 1$. If $\mathrm{Nrd}(b) = \iota(y)y^{-1}$, then

$$y = \mathrm{Nrd}(yb^r)N_{K/F}(y)^{-r} \in F^\times \cdot \mathrm{Nrd}(B^\times).$$

13. Let (A, σ) be a central simple algebra with involution (of any type) over a field F and let L/F be a field extension. Suppose $\varphi, \psi \colon L \to A$ are two embeddings such that $\varphi(L), \psi(L) \subset (A, \sigma)_+$. The Skolem-Noether theorem shows that there exists $a \in A^\times$ such that $\varphi = \mathrm{Int}(a) \circ \psi$. Show that $\sigma(a)a \in C_A\psi(L)$ and find a necessary and sufficient condition on this element for the existence of a similitude $g \in \mathrm{Sim}(A, \sigma)$ such that $\varphi = \mathrm{Int}(g) \circ \psi$.

Notes

§12. The Dickson invariant Δ of (12.12) was originally defined by Dickson [74, Theorem 205, p. 206] by means of an explicit formula involving the entries of the matrix. Subsequently, Dieudonné [80] showed that it can also be defined by considering the action of the similitude on the center of the even Clifford algebra (see (13.3)). The presentation given here is new.

A functor $\mathbf{M} \colon B_n' \to B_n$ such that $\mathbf{End} \circ \mathbf{M} \cong \mathrm{Id}_{B_n'}$ and $\mathbf{M} \circ \mathbf{End} \cong \mathrm{Id}_{B_n}$ (thus providing an alternate proof of (12.41)) can be made explicit as follows. Recall the canonical direct sum decomposition of Clifford algebras (see Wonenburger [316, Theorem 1]): if (V, q) is a quadratic space of dimension d,

$$C(V, q) = M_0 \oplus M_1 \oplus \cdots \oplus M_d$$

where $M_0 = F$, $M_1 = V$ and, for $k \geq 2$, the space M_k is the linear span of the elements $v \cdot m - (-1)^k m \cdot v$ with $v \in V$ and $m \in M_{k-1}$. For $k = 1, \ldots, d$ the vector space M_k is also spanned by the products of k vectors in any orthogonal basis of V. In particular, the dimension of M_k is given by the binomial coefficient:

$$\dim_F M_k = \binom{d}{k}.$$

Clearly, $M_k \subset C_0(V, q)$ if and only if k is even; hence

$$C_0(V, q) = \bigoplus_{i \text{ even}} M_i.$$

Suppose d is odd and $\mathrm{disc}\, q = 1$. We then have

$$M_{d-1} = \zeta \cdot V \subset C_0(V, q)$$

for any orientation ζ of (V, q), hence $x^2 \in F$ for all $x \in M_{d-1}$. Therefore, we may define a quadratic map

$$s \colon M_{d-1} \to F$$

by $s(x) = x^2$. The embedding $M_{d-1} \hookrightarrow C_0(V, q)$ induces an F-algebra homomorphism $e \colon C(M_{d-1}, s) \to C_0(V, q)$, which shows that $\mathrm{disc}\, s = 1$. A canonical orientation η on (M_{d-1}, s) can be characterized by the condition $e(\eta) = 1$.

Since the decomposition of $C_0(V, q)$ is canonical, it can be defined for the Clifford algebra of any central simple algebra with orthogonal involution (A, σ), as Jacobson shows in [134, p. 294]. If the degree of the algebra A is odd: $\deg A = d = 2n + 1$, the construction above associates to (A, σ) an oriented quadratic space (M, s, η) of dimension $2n + 1$ (where $M \subset C(A, \sigma)$ and $s(x) = x^2$ for $x \in M$) and defines a functor $\mathbf{M} \colon B_n' \to B_n$. We leave it to the reader to check that $\mathbf{End} \circ \mathbf{M} \cong \mathrm{Id}_{B_n'}$ and $\mathbf{M} \circ \mathbf{End} \cong \mathrm{Id}_{B_n}$.

§13. If $\mathrm{char}\, F = 0$ and $\deg A \geq 10$, Lie algebra techniques can be used to prove that the Lie-automorphism ψ' of $\mathrm{Alt}(A, \sigma)$ defined in (13.4) extends to an automorphism of (A, σ): see Jacobson [140, p. 307].

The extended Clifford group $\Omega(A, \sigma)$ was first considered by Jacobson [134] in characteristic different from 2. (Jacobson uses the term "even Clifford group.") In the split case, this group has been investigated by Wonenburger [316]. The spin groups $\mathrm{Spin}(A, \sigma, f)$ were defined by Tits [291] in arbitrary characteristic.

The original proof of Dieudonné's theorem on multipliers of similitudes (13.22) appears in [79, Théorème 2]. The easy argument presented here is due to Elman-Lam [92, Lemma 4]. The generalization in (13.38) is due to Merkurjev-Tignol [196].

Another proof of (13.38), using Galois cohomology, has been found by Bayer-Fluckiger [32] assuming that char $F \neq 2$. (This assumption is also made in [196].)

From (13.43), it follows that every central simple algebra with orthogonal involution which contains an improper isometry is split. In characteristic different from 2, this statement can be proved directly by elementary arguments; it was first observed by Kneser [155, Lemma 1.b, p. 42]. (See also Exercise 7; the proof in [155] is different.)

§14. The canonical Lie homomorphism $\dot{\lambda}^k \colon \mathfrak{L}(A) \to \mathfrak{L}(\lambda^k A)$ is defined as the differential of the polynomial map λ^k. It is of course possible to define $\dot{\lambda}^k$ independently of λ^k: it suffices to mimic (14.3), substituting in the proof $a \otimes 1 \otimes \cdots \otimes 1 + 1 \otimes a \otimes \cdots \otimes 1 + \cdots + 1 \otimes 1 \otimes \cdots \otimes a$ for $a^{\otimes k}$. The properties of $\dot{\lambda}^k$ may also be proved directly (by mimicking (14.4) and (14.5)), but the proof that $\tau^{\wedge k} \circ \dot{\lambda}^k = \dot{\lambda}^k \circ \tau$ involves rather tedious computations.

CHAPTER IV

Algebras of Degree Four

Among groups of automorphisms of central simple algebras with involution, there are certain isomorphisms, known as *exceptional isomorphisms*, relating algebras of low degree. (The reason why these isomorphisms are indeed exceptional comes from the fact that in some special low rank cases Dynkin diagrams coincide, see §26.B.) Algebras of degree 4 play a central rôle from this viewpoint: their three types of involutions (orthogonal, symplectic, unitary) are involved with three of the exceptional isomorphisms, which relate them to quaternion algebras, 5-dimensional quadratic spaces and orthogonal involutions on algebras of degree 6 respectively. A correspondence, first considered by Albert [2], between tensor products of two quaternion algebras and quadratic forms of dimension 6 arises as a special case of the last isomorphism.

The exceptional isomorphisms provide the motivation for, and can be obtained as a consequence of, equivalences between certain categories of algebras with involution which are investigated in the first section. In the second section, we focus on tensor products of two quaternion algebras, called *biquaternion algebras*, and their Albert quadratic forms. The third section yields a quadratic form description of the reduced Whitehead group of a biquaternion algebra, making use of symplectic involutions. Analogues of the reduced Whitehead group for algebras with involution are also discussed.

§15. Exceptional Isomorphisms

The exceptional isomorphisms between groups of similitudes of central simple algebras with involution in characteristic different from 2 are easily derived from Wonenburger's theorem (13.4), as the following proposition shows:

(15.1) Proposition. *Let (A, σ) be a central simple algebra with orthogonal involution over a field F of characteristic different from 2. If $2 < \deg A \leq 6$, the canonical homomorphism of (13.4):*

$$C \colon \mathrm{PGO}(A, \sigma) \to \mathrm{Aut}_F\big(C(A, \sigma), \underline{\sigma}\big)$$

is an isomorphism.

Proof: Proposition (13.4) shows that if $\deg A > 2$, then C is injective and its image consists of the automorphisms of $\big(C(A, \sigma), \underline{\sigma}\big)$ which preserve the image $c(A)$ of A under the canonical map c. Moreover, (8.18) shows that $c(A) = F \oplus c(A)_0$ where

$$c(A)_0 = c(A) \cap \mathrm{Skew}\big(C(A, \sigma), \underline{\sigma}\big).$$

Therefore, it suffices to show that every automorphism of $\big(C(A, \sigma), \underline{\sigma}\big)$ preserves $c(A)_0$ if $\deg A \leq 6$.

From (8.14) (or (8.18)), it follows that $\dim c(A)_0 = \dim \mathrm{Skew}(A, \sigma)$. Direct computations show that

$$\dim_F \mathrm{Skew}(A, \sigma) = \dim_F \mathrm{Skew}\big(C(A, \sigma), \underline{\sigma}\big)$$

if $\deg A = 3$, 4, 5; thus $c(A)_0 = \mathrm{Skew}\big(C(A, \sigma), \underline{\sigma}\big)$ in these cases, and every automorphism of $\big(C(A, \sigma), \underline{\sigma}\big)$ preserves $c(A)_0$.

If $\deg A = 6$ we get $\dim_F \mathrm{Skew}\big(C(A, \sigma), \underline{\sigma}\big) = 16$ while $\dim_F c(A)_0 = 15$. However, the involution $\underline{\sigma}$ is unitary; if Z is the center of $C(A, \sigma)$, there is a canonical decomposition

$$\mathrm{Skew}\big(C(A, \sigma), \underline{\sigma}\big) = \mathrm{Skew}(Z, \underline{\sigma}) \oplus \mathrm{Skew}\big(C(A, \sigma), \underline{\sigma}\big)^0$$

where

$$\mathrm{Skew}\big(C(A, \sigma), \underline{\sigma}\big)^0 = \{\, u \in \mathrm{Skew}(C(A, \sigma), \underline{\sigma}) \mid \mathrm{Trd}_{C(A, \sigma)}(u) = 0 \,\}.$$

Inspection of the split case shows that $\mathrm{Trd}_{C(A, \sigma)}(x) = 0$ for all $x \in c(A)_0$. Therefore, by dimension count,

$$c(A)_0 = \mathrm{Skew}\big(C(A, \sigma), \underline{\sigma}\big)^0.$$

Since $\mathrm{Skew}\big(C(A, \sigma), \underline{\sigma}\big)^0$ is preserved under every automorphism of $\big(C(A, \sigma), \underline{\sigma}\big)$, the proof is complete. $\qquad\square$

This proposition relates central simple algebras with orthogonal involutions of degree $n = 3$, 4, 5, 6 with their Clifford algebra. We thus get relations between:

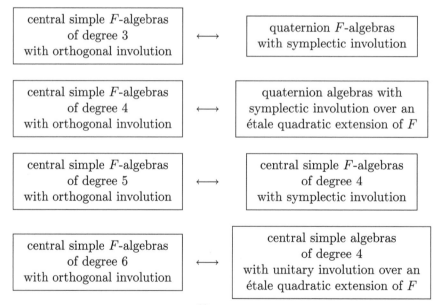

In order to formalize these relations,[23] we introduce various groupoids whose objects are the algebras considered above. The groupoid of central simple F-algebras of degree $2n + 1$ with orthogonal involution has already been considered in §12.D,

---

[23]In the cases $n = 4$ and $n = 6$, the relation also holds if the étale quadratic extension is $F \times F$; central simple algebras of degree d over $F \times F$ should be understood as products $B_1 \times B_2$ of central simple F-algebras of degree d. Similarly, quaternion algebras over $F \times F$ are defined as products $Q_1 \times Q_2$ of quaternion F-algebras.

where it is denoted B'_n. In order to extend the relations above to the case where char $F = 2$, we replace it by the category B_n of oriented quadratic spaces of dimension $2n + 1$: see (12.41). If char $F = 2$, we define an orientation on an odd-dimensional nonsingular quadratic space (V, q) of trivial discriminant as in the case where char $F \neq 2$: an orientation of (V, q) is a central element $\zeta \in C_1(V, q)$ such that $\zeta^2 = 1$. If char $F = 2$, the orientation is unique, hence the category of oriented quadratic spaces is isomorphic to the category of quadratic spaces of trivial discriminant.

We thus consider the following categories, for an arbitrary field F:

- A_1 is the category of quaternion F-algebras, where the morphisms are the F-algebra isomorphisms;
- A_1^2 is the category of quaternion algebras over an étale quadratic extension of F, where the morphisms are the F-algebra isomorphisms;
- A_n, for an arbitrary integer $n \geq 2$, is the category of central simple algebras of degree $n + 1$ over an étale quadratic extension of F with involution of the second kind leaving F elementwise invariant, where the morphisms are the F-algebra isomorphisms which preserve the involutions;
- B_n, for an arbitrary integer $n \geq 1$, is the category of oriented quadratic spaces of dimension $2n + 1$, where the morphisms are the isometries which preserve the orientation (if char $F = 2$, every isometry preserves the orientation, since the orientation is unique);
- C_n, for an arbitrary integer $n \geq 1$, is the category of central simple F-algebras of degree $2n$ with symplectic involution, where the morphisms are the F-algebra isomorphisms which preserve the involutions;
- D_n, for an arbitrary integer $n \geq 2$, is the category of central simple F-algebras of degree $2n$ with quadratic pair, where the morphisms are the F-algebra isomorphisms which preserve the quadratic pairs.

In each case, maps are isomorphisms, hence these categories are groupoids.

Note that there is an isomorphism of groupoids:

$$A_1 = C_1$$

which follows from the fact that each quaternion algebra has a canonical symplectic involution.

In the next sections, we shall successively establish equivalences of groupoids:

$$B_1 \equiv C_1$$
$$D_2 \equiv A_1^2$$
$$B_2 \equiv C_2$$
$$D_3 \equiv A_3.$$

In each case, it is the Clifford algebra construction which provides the functors defining these equivalences from the left-hand side to the right-hand side. Not surprisingly, one will notice deep analogies between the first and the third cases, as well as between the second and the fourth cases.

Our proofs do not rely on (15.1), and indeed provide an alternative proof of that proposition.

15.A. $B_1 \equiv C_1$. In view of the isomorphism $A_1 = C_1$, it is equivalent to prove $A_1 \equiv B_1$.

For every quaternion algebra $Q \in A_1$, the vector space

$$Q^0 = \{\, x \in Q \mid \mathrm{Trd}_Q(x) = 0 \,\}$$

has dimension 3, and the squaring map $s \colon Q^0 \to F$ defined by

$$s(x) = x^2 \quad \text{for } x \in Q^0$$

is a canonical quadratic form of discriminant 1. Moreover, the inclusion $Q^0 \hookrightarrow Q$ induces an orientation π on the quadratic space (Q^0, s), as follows: by the universal property of Clifford algebras, this inclusion induces a homomorphism of F-algebras

$$h_Q \colon C(Q^0, s) \to Q.$$

Since $\dim C(Q^0, s) = 8$, this homomorphism has a nontrivial kernel; in the center of $C(Q^0, s)$, there is a unique element $\pi \in C_1(Q^0, s)$ such that $\pi^2 = 1$ which is mapped to $1 \in Q$. This element π is an orientation on (Q^0, s). Explicitly, if $(1, i, j, k)$ is a quaternion basis of Q in characteristic different from 2, the orientation $\pi \in C(Q^0, s)$ is given by $\pi = i \cdot j \cdot k^{-1}$. If $(1, u, v, w)$ is a quaternion basis of Q in characteristic 2, the orientation is $\pi = \underline{1}$ (the image of $1 \in Q^0$ in $C(Q^0, s)$, *not* the unit of $C(Q^0, s)$).

We define a functor

$$\mathbf{P} \colon A_1 \to B_1$$

by mapping $Q \in A_1$ to the oriented quadratic space $(Q^0, s, \pi) \in B_1$.

A functor $\mathbf{C}$ in the opposite direction is provided by the even Clifford algebra construction: we define

$$\mathbf{C} \colon B_1 \to A_1$$

by mapping every oriented quadratic space $(V, q, \zeta) \in B_1$ to $C_0(V, q) \in A_1$.

(15.2) Theorem. *The functors $\mathbf{P}$ and $\mathbf{C}$ define an equivalence of groupoids*

$$A_1 \equiv B_1.$$

Proof: For any quaternion algebra Q, the homomorphism $h_Q \colon C(Q^0, s) \to Q$ induced by the inclusion $Q^0 \hookrightarrow Q$ restricts to a canonical isomorphism

$$h_Q \colon C_0(Q^0, s) \xrightarrow{\sim} Q.$$

We thus have a natural transformation: $\mathbf{C} \circ \mathbf{P} \cong \mathrm{Id}_{A_1}$.

For $(V, q, \zeta) \in B_1$, we define a bijective linear map $m_\zeta \colon V \to C_0(V, q)^0$ by

$$m_\zeta(v) = v\zeta \quad \text{for } v \in V.$$

Since $s(v\zeta) = (v\zeta)^2 = v^2 = q(v)$, this map is an isometry:

$$m_\zeta \colon (V, q) \xrightarrow{\sim} \big(C_0(V, q)^0, s\big).$$

We claim that m_ζ carries ζ to the canonical orientation π on $\big(C_0(V, q)^0, s\big)$; this map therefore yields a natural transformation $\mathbf{P} \circ \mathbf{C} \cong \mathrm{Id}_{B_1}$ which completes the proof.

To prove the claim, it suffices to consider the case where $\operatorname{char} F \neq 2$, since the orientation is unique if $\operatorname{char} F = 2$. Therefore, for the rest of the proof we assume $\operatorname{char} F \neq 2$. The isometry m_ζ induces an isomorphism

$$\widetilde{m_\zeta} \colon C(V, q) \xrightarrow{\sim} C\big(C_0(V, q)^0, s\big)$$

which maps ζ to π or $-\pi$. Composing this isomorphism with the homomorphism $h_{C_0(V,q)}\colon C\bigl(C_0(V,q)^0, s\bigr) \to C_0(V,q)$ induced by the inclusion $C_0(V,q)^0 \hookrightarrow C_0(V,q)$, we get a homomorphism

$$h_{C_0(V,q)} \circ \widetilde{m_\zeta}\colon C(V,q) \to C_0(V,q)$$

which carries $v \in V$ to $v\zeta \in C_0(V,q)$. Since ζ has the form $\zeta = v_1 \cdot v_2 \cdot v_3$ for a suitable orthogonal basis (v_1, v_2, v_3) of V, we have

$$h_{C_0(V,q)} \circ \widetilde{m_\zeta}(\zeta) = (v_1\zeta)(v_2\zeta)(v_3\zeta) = \zeta^4 = 1.$$

Since the orientation π on $\bigl(C_0(V,q)^0, s\bigr)$ is characterized by the condition

$$h_{C_0(V,q)}(\pi) = 1,$$

it follows that $\widetilde{m_\zeta}(\zeta) = \pi$. $\qquad\square$

(15.3) Corollary. *For every oriented quadratic space (V, q, ζ) of dimension 3, the Clifford algebra construction yields a group isomorphism*

$$O^+(V,q) = \mathrm{Aut}(V, q, \zeta) \xrightarrow{\sim} \mathrm{Aut}_F\bigl(C_0(V,q)\bigr) = \mathrm{PGSp}\bigl(C_0(V,q), \tau\bigr)$$

where τ is the canonical involution on $C_0(V,q)$.

Proof: Since the functor $\mathbf{C}$ defines an equivalence of groupoids, it induces isomorphisms between the automorphism groups of corresponding objects. $\qquad\square$

By combining Theorem (15.2) with (12.41) (in characteristic different from 2), we obtain an equivalence between A_1 and B_1':

(15.4) Corollary. *Suppose $\mathrm{char}\,F \neq 2$. The functor $\mathbf{P}\colon A_1 \to B_1'$, which maps every quaternion algebra Q to the algebra with involution $\bigl(\mathrm{End}_F(Q^0), \sigma_s\bigr)$ where σ_s is the adjoint involution with respect to s, and the functor $\mathbf{C}\colon B_1' \to A_1$, which maps every algebra with involution (A, σ) of degree 3 to the quaternion algebra $C(A, \sigma)$, define an equivalence of groupoids:*

$$A_1 \equiv B_1'.$$

In particular, for every central simple algebra with involution (A, σ) of degree 3, the functor $\mathbf{C}$ induces an isomorphism of groups:

$$O^+(A, \sigma) = \mathrm{PGO}(A, \sigma) = \mathrm{Aut}_F(A, \sigma) \xrightarrow{\sim} \mathrm{Aut}_F\bigl(C(A, \sigma)\bigr) = \mathrm{PGSp}\bigl(C(A, \sigma), \underline{\sigma}\bigr).$$

We thus recover the first case ($\deg A = 3$) of (15.1).

Indices. Let $Q \in A_1$ and $(V, q, \zeta) \in B_1$ correspond to each other under the equivalence $A_1 \equiv B_1$, so that $(V, q) \simeq (Q^0, s)$ and $Q \simeq C_0(V, q)$. Since Q is a quaternion algebra, its (Schur) index may be either 1 or 2; for the canonical involution γ on Q, the index $\mathrm{ind}(Q, \gamma)$ is thus (respectively) either $\{0, 1\}$ or $\{0\}$. On the other hand, since $\dim V = 3$, the Witt index $w(V, q)$ is 1 whenever q is isotropic.

The following correspondence between the various cases is well-known:

(15.5) Proposition. *The indices of Q, (Q, γ) and (V, q) are related as follows:*

$$\mathrm{ind}\,Q = 2 \quad \Longleftrightarrow \quad \mathrm{ind}(Q, \gamma) = \{0\} \quad \Longleftrightarrow \quad w(V, q) = 0;$$
$$\mathrm{ind}\,Q = 1 \quad \Longleftrightarrow \quad \mathrm{ind}(Q, \gamma) = \{0, 1\} \quad \Longleftrightarrow \quad w(V, q) = 1.$$

In other words, Q is a division algebra if and only if q is anisotropic, and Q is split if and only if q is isotropic.

Proof: If Q is a division algebra, then Q^0 does not contain any nonzero nilpotent elements. Therefore, the quadratic form s, hence also q, is anisotropic. On the other hand, q is isotropic if Q is split, since $M_2(F)$ contains nonzero matrices whose square is 0. $\qquad\square$

15.B. $A_1^2 \equiv D_2$. The Clifford algebra construction yields a functor $\mathbf{C}\colon D_2 \to A_1^2$. In order to show that this functor defines an equivalence of groupoids, we first describe a functor $\mathbf{N}\colon A_1^2 \to D_2$ which arises from the norm construction.

Let $Q \in A_1^2$ be a quaternion algebra over some étale quadratic extension K/F. (If $K = F \times F$, then Q should be understood as a direct product of quaternion F-algebras.) Let ι be the nontrivial automorphism of K/F. Recall from (3.12) that $N_{K/F}(Q)$ is the F-subalgebra of ${}^\iota Q \otimes_K Q$ consisting of elements fixed by the switch map

$$s\colon {}^\iota Q \otimes_K Q \to {}^\iota Q \otimes_K Q.$$

The tensor product ${}^\iota \gamma \otimes \gamma$ of the canonical involutions on ${}^\iota Q$ and Q restricts to an involution $N_{K/F}(\gamma)$ of the first kind on $N_{K/F}(Q)$. By (3.13), we have

$$\bigl(N_{K/F}(Q)_K, N_{K/F}(\gamma)_K\bigr) \simeq \bigl({}^\iota Q \otimes_K Q, {}^\iota \gamma \otimes \gamma\bigr),$$

hence $N_{K/F}(\gamma)$ has the same type as ${}^\iota \gamma \otimes \gamma$. Proposition (2.23) thus shows that $N_{K/F}(\gamma)$ is orthogonal if char $F \neq 2$ and symplectic if char $F = 2$. Corollary (5.20) further yields a quadratic pair $({}^\iota \gamma \otimes \gamma, f_\otimes)$ on ${}^\iota Q \otimes Q$, which is uniquely determined by the condition that $f_\otimes$ vanishes on $\mathrm{Skew}({}^\iota Q, {}^\iota \gamma) \otimes_K \mathrm{Skew}(Q, \gamma)$. It is readily seen that $({}^\iota \gamma \otimes \gamma, \iota \circ f_\otimes \circ s)$ is a quadratic pair with the same property, hence $\iota \circ f_\otimes \circ s = f$ and therefore

$$f_\otimes\bigl(s(x)\bigr) = \iota f_\otimes(x) \qquad \text{for all } x \in \mathrm{Sym}({}^\iota Q \otimes Q, {}^\iota \gamma \otimes \gamma).$$

It follows that $f_\otimes(x) \in F$ for all $x \in \mathrm{Sym}\bigl(N_{K/F}(Q), N_{K/F}(\gamma)\bigr)$, hence $({}^\iota \gamma \otimes \gamma, f_\otimes)$ restricts to a quadratic pair on $N_{K/F}(Q)$. We denote this quadratic pair by $\bigl(N_{K/F}(\gamma), f_N\bigr)$. The norm thus defines a functor

$$\mathbf{N}\colon A_1^2 \to D_2$$

which maps $Q \in A_1^2$ to $\mathbf{N}(Q) = \bigl(N_{K/F}(Q), N_{K/F}(\gamma), f_N\bigr)$ where K is the center of Q.

On the other hand, for $(A, \sigma, f) \in D_2$ the Clifford algebra $C(A, \sigma, f)$ is a quaternion algebra over an étale quadratic extension, as the structure theorem (8.10) shows. Therefore, the Clifford algebra construction yields a functor

$$\mathbf{C}\colon D_2 \to A_1^2.$$

The key tool to show that $\mathbf{N}$ and $\mathbf{C}$ define an equivalence of categories is the Lie algebra isomorphism which we define next. For $Q \in A_1^2$, consider the F-linear map $\dot{n}\colon Q \to N_{K/F}(Q)$ defined by

$$\dot{n}(q) = {}^\iota q \otimes 1 + {}^\iota 1 \otimes q \qquad \text{for } q \in Q.$$

This map is easily checked to be a Lie algebra homomorphism; it is in fact the differential of the group homomorphism $n\colon Q^\times \to N_{K/F}(Q)^\times$ which maps $q \in Q^\times$ to ${}^\iota q \otimes q$. We have the nonsingular F-bilinear form $T_{N_{K/F}(Q)}$ on $N_{K/F}(Q)$ and the nonsingular F-bilinear form on Q which is the transfer of T_Q with respect to the trace $T_{K/F}$. Using these, we may form the adjoint linear map

$$\dot{n}^*\colon N_{K/F}(Q) \to Q$$

which is explicitly defined as follows: for $x \in N_{K/F}(Q)$, the element $\dot{n}^*(x) \in Q$ is uniquely determined by the condition

$$T_{K/F}\big(\mathrm{Trd}_Q\big(\dot{n}^*(x)y\big)\big) = \mathrm{Trd}_{N_{K/F}(Q)}\big(x\dot{n}(y)\big) \qquad \text{for all } y \in Q.$$

(15.6) Proposition. *Let $Q' = \{\, x \in Q \mid \mathrm{Trd}_Q(x) \in F \,\}$. The linear map $\dot{n}^*$ factors through the canonical map $c\colon N_{K/F}(Q) \to c\big(\mathbf{N}(Q)\big)$ and induces an isomorphism of Lie algebras*

$$\dot{n}^*\colon c\big(\mathbf{N}(Q)\big) \xrightarrow{\ \sim\ } Q'.$$

This isomorphism is the identity on F.

Proof: Suppose first that $K \simeq F \times F$. We may then assume that $Q = Q_1 \times Q_2$ for some quaternion F-algebras Q_1, Q_2, and $N_{K/F}(Q) = Q_1 \otimes Q_2$. Under this identification, the map $\dot{n}$ is defined by

$$\dot{n}(q_1, q_2) = q_1 \otimes 1 + 1 \otimes q_2 \qquad \text{for } q_1 \in Q_1 \text{ and } q_2 \in Q_2.$$

It is readily verified that

$$\dot{n}^*(q_1 \otimes q_2) = \big(\mathrm{Trd}_{Q_2}(q_2)q_1, \mathrm{Trd}_{Q_1}(q_1)q_2\big) \qquad \text{for } q_1 \in Q_1 \text{ and } q_2 \in Q_2,$$

hence $\dot{n}^*$ is the map Θ of (8.19). From (8.19), it follows that $\dot{n}^*$ factors through c and induces an isomorphism of Lie algebras

$$c\big(\mathbf{N}(Q)\big) = c(Q_1 \otimes Q_2) \to Q' = \{\, (q_1, q_2) \in Q_1 \times Q_2 \mid \mathrm{Trd}_{Q_1}(q_1) = \mathrm{Trd}_{Q_2}(q_2) \,\}.$$

For $i = 1, 2$, let $\ell_i \in Q_i$ be such that $\mathrm{Trd}_{Q_i}(\ell_i) = 1$. Then $\mathrm{Trd}_{Q_1 \otimes Q_2}(\ell_1 \otimes \ell_2) = 1$, hence $f\big(\ell_1 \otimes \ell_2 + \gamma_1(\ell_1) \otimes \gamma_2(\ell_2)\big) = 1$, and therefore $c\big(\ell_1 \otimes \ell_2 + \gamma_1(\ell_1) \otimes \gamma_2(\ell_2)\big) = 1$. On the other hand,

$$\dot{n}^*\big(\ell_1 \otimes \ell_2 + \gamma_1(\ell_1) \otimes \gamma_2(\ell_2)\big) = (\ell_1, \ell_2) + \big(\gamma_1(\ell_1), \gamma_2(\ell_2)\big) = (1, 1),$$

hence $\dot{n}^*$ maps $1 \in c\big(\mathbf{N}(Q)\big)$ to $1 \in Q'$. The map $\dot{n}^*$ thus restricts to the identity on F, completing the proof in the case where $K \simeq F \times F$.

In the general case, it suffices to prove the proposition over an extension of the base field F. Extending scalars to K, we are reduced to the special case considered above, since $K \otimes K \simeq K \times K$. $\qquad\square$

(15.7) Theorem. *The functors $\mathbf{N}$ and $\mathbf{C}$ define an equivalence of groupoids:*

$$A_1^2 \equiv D_2.$$

Moreover, if $Q \in A_1^2$ and $(A, \sigma, f) \in D_2$ correspond to each other under this equivalence, then the center $Z(Q)$ of Q satisfies

$$Z(Q) \simeq \begin{cases} F\big(\sqrt{\mathrm{disc}(A, \sigma, f)}\big) & \text{if } \mathrm{char}\, F \neq 2; \\ F\big(\wp^{-1}\big(\mathrm{disc}(A, \sigma, f)\big)\big) & \text{if } \mathrm{char}\, F = 2. \end{cases}$$

Proof: If the first assertion holds, then the quaternion algebra Q corresponding to $(A, \sigma, f) \in D_2$ is the Clifford algebra $C(A, \sigma, f)$, hence the description of $Z(Q)$ follows from the structure theorem for Clifford algebras (8.10).

In order to prove the first statement, we establish natural transformations $\mathbf{N} \circ \mathbf{C} \cong \mathrm{Id}_{D_2}$ and $\mathbf{C} \circ \mathbf{N} \cong \mathrm{Id}_{A_1^2}$. Thus, for $(A, \sigma, f) \in D_2$ and for $Q \in A_1^2$, we have to describe canonical isomorphisms

$$(A, \sigma, f) \simeq \big(N_{Z(A,\sigma,f)/F}\big(C(A, \sigma, f)\big), N_{Z(A,\sigma,f)/F}(\underline{\sigma}), f_N\big)$$

and

$$Q \simeq C\big(N_{Z(Q)/F}(Q), N_{Z(Q)/F}(\gamma), f_N\big)$$

where $Z(A, \sigma, f)$ is the center of $C(A, \sigma, f)$.

Observe that the fundamental relation (9.14) between an algebra with involution and its Clifford algebra already shows that there is an isomorphism $A \simeq N_{Z(A,\sigma,f)/F}\big(C(A, \sigma, f)\big)$. However, we need a canonical isomorphism which takes the quadratic pairs into account.

Our construction is based on (2.25): we use (15.6) to define isomorphisms of Lie algebras and show that these isomorphisms extend to isomorphisms of associative algebras over an algebraically closed extension, hence also over the base field. Let $(A, \sigma, f) \in D_2$ and let

$$C(A, \sigma, f)' = \{\, x \in C(A, \sigma, f) \mid \mathrm{Trd}_{C(A,\sigma,f)}(x) \in F \,\}.$$

Lemma (8.16) shows that $\mathrm{Trd}_{C(A,\sigma,f)}\big(c(a)\big) = \mathrm{Trd}_A(a)$ for $a \in A$, hence $c(A) \subset C(A, \sigma, f)'$, and dimension count shows that this inclusion is an equality. Proposition (15.6) then yields a Lie algebra isomorphism $\dot{n}^* \colon c\big(\mathbf{N}(C(A, \sigma, f))\big) \to c(A)$ which is the identity on F. By (8.15), it follows that this isomorphism induces a Lie algebra isomorphism

$$\overline{n} \colon \mathrm{Alt}\big(\mathbf{N}(C(A, \sigma, f))\big) \xrightarrow{\;\sim\;} \mathrm{Alt}(A, \sigma).$$

To prove that this isomorphism extends to an isomorphism of algebras with quadratic pairs, it suffices by (2.25) to consider the split case. We may thus assume that A is the endomorphism algebra of a hyperbolic quadratic space $\mathbb{H}(U)$ of dimension 4. Thus

$$A = \mathrm{End}_F\big(\mathbb{H}(U)\big) = \mathrm{End}_F(U^* \oplus U)$$

where U is a 2-dimensional vector space, U^* is its dual, and $(\sigma, f) = (\sigma_{q_U}, f_{q_U})$ is the quadratic pair associated with the hyperbolic quadratic form on $U^* \oplus U$:

$$q_U(\varphi + u) = \varphi(u) \quad \text{for } \varphi \in U^*,\ u \in U.$$

In that case, the Clifford algebra $C(A, \sigma, f)$ can be described as

$$C(A, \sigma, f) = C_0\big(\mathbb{H}(U)\big) = \mathrm{End}_F(\textstyle\bigwedge_0 U) \times \mathrm{End}_F(\textstyle\bigwedge_1 U),$$

where $\bigwedge_0 U$ (resp. $\bigwedge_1 U$) is the 2-dimensional subspace of even- (resp. odd-) degree elements in the exterior algebra of U (see (8.3)):

$$\textstyle\bigwedge_0 U = F \oplus \bigwedge^2 U, \quad \bigwedge_1 U = U.$$

Therefore,

$$N_{Z(A,\sigma,f)/F}\big(C(A, \sigma, f)\big) = \mathrm{End}_F(\textstyle\bigwedge_0 U \otimes \bigwedge_1 U).$$

On the vector space $\bigwedge_0 U \otimes \bigwedge_1 U$, we define a quadratic form q as follows: pick a nonzero element (hence a basis) $e \in \bigwedge^2 U$; for $x, y \in U$, we may then define $q(1 \otimes x + e \otimes y) \in F$ by the equation

$$eq(1 \otimes x + e \otimes y) = x \wedge y.$$

The associated quadratic pair (σ_q, f_q) on $\mathrm{End}_F(\bigwedge_0 U \otimes \bigwedge_1 U)$ is the canonical quadratic pair $\big(N(\underline{\sigma}), f_N\big)$ (see Exercise 21 of Chapter I). A computation shows that the map $g \colon \mathbb{H}(U) \to \bigwedge_0 U \otimes \bigwedge_1 U$ defined by

$$g(\varphi + u) = 1 \otimes x + e \otimes u,$$

where $x \in U$ is such that $x \wedge y = e\varphi(y)$ for all $y \in U$, is a similitude of quadratic spaces

$$g\colon \mathbb{H}(U) \xrightarrow{\sim} (\textstyle\bigwedge_0 U \otimes \bigwedge_1 U, q).$$

By (12.36), this similitude induces an isomorphism of algebras with quadratic pair

$$g_*\colon \big(\mathrm{End}_F\big(\mathbb{H}(U)\big), \sigma_{q_U}, f_{q_U}\big) \xrightarrow{\sim} \big(\mathrm{End}_F(\textstyle\bigwedge_0 U \otimes \bigwedge_1 U), \sigma_q, f_q\big).$$

We leave it to the reader to check that g_*^{-1} extends the Lie algebra homomorphism $\overline{n}$, completing the proof that $\overline{n}$ induces a natural transformation $\mathbf{N} \circ \mathbf{C} \cong \mathrm{Id}_{D_2}$.

We use the same technique to prove that $\mathbf{C} \circ \mathbf{N} \cong \mathrm{Id}_{A_1^2}$. For $Q \in A_1^2$, Proposition (15.6) yields a Lie algebra isomorphism

$$\dot{n}^*\colon c\big(\mathbf{N}(Q)\big) \xrightarrow{\sim} Q'.$$

To prove that $\dot{n}^*$ extends to an isomorphism of F-algebras $C\big(\mathbf{N}(Q)\big) \xrightarrow{\sim} Q$, we may extend scalars, since $\mathbf{N}(Q)$ is generated as an associative algebra by $c\big(\mathbf{N}(Q)\big)$. Extending scalars to $Z(Q)$ if this algebra is a field, we may therefore assume that $Z(Q) \simeq F \times F$. In that case $Q \simeq Q_1 \times Q_2$ for some quaternion F-algebras Q_1, Q_2, hence $N_{Z(Q)/F}(Q) \simeq Q_1 \otimes Q_2$, and $\dot{n}^*$ is the map Θ of (8.19), defined by

$$\Theta\big(c(x_1 \otimes x_2)\big) = \big(\mathrm{Trd}_{Q_2}(x_2)x_1, \mathrm{Trd}_{Q_1}(x_1)x_2\big) \qquad \text{for } x_1 \in Q_1,\ x_2 \in Q_2.$$

Since it was proven in (8.19) that Θ extends to an isomorphism of F-algebras $C\big(\mathbf{N}(Q)\big) = C(Q_1 \otimes Q_2, \gamma_1 \otimes \gamma_2, f_\otimes) \xrightarrow{\sim} Q_1 \times Q_2$, the proof is complete. $\qquad\square$

(15.8) Remark. For $Q \in A_1^2$, the Lie isomorphism $\dot{n}^*\colon c\big(\mathbf{N}(Q)\big) \to Q'$ restricts to an isomorphism $c\big(\mathbf{N}(Q)\big)_0 \xrightarrow{\sim} Q^0$. If char $F \neq 2$, the inverse of this isomorphism is $\frac{1}{2}c \circ \dot{n}$ (see Exercise 1). Similarly, for $(A, \sigma, f) \in D_2$, the inverse of the Lie isomorphism $\overline{n}\colon \mathrm{Alt}\big(\mathbf{N}\big(C(A, \sigma, f)\big)\big) \to \mathrm{Alt}(A, \sigma)$ is $\dot{n} \circ \frac{1}{2}c$ if char $F \neq 2$.

(15.9) Corollary. *For every central simple algebra A of degree 4 with quadratic pair (σ, f), the functor $\mathbf{C}$ induces an isomorphism of groups:*

$$\mathrm{PGO}(A, \sigma, f) = \mathrm{Aut}_F(A, \sigma, f) \xrightarrow{\sim} \mathrm{Aut}_F\big(C(A, \sigma, f), \underline{\sigma}\big) = \mathrm{Aut}_F\big(C(A, \sigma, f)\big)$$

which restricts into an isomorphism of groups:

$$\mathrm{PGO}^+(A, \sigma, f) \xrightarrow{\sim} \mathrm{Aut}_{Z(A, \sigma, f)}\big(C(A, \sigma, f), \underline{\sigma}\big) =$$
$$\mathrm{PGSp}\big(C(A, \sigma, f), \underline{\sigma}\big) = C(A, \sigma, f)^\times / Z(A, \sigma, f)^\times.$$

Proof: The first isomorphism follows from the fact that $\mathbf{C}$ defines an equivalence of groupoids $D_2 \to A_1^2$ (see (12.37)). Under this isomorphism, the proper similitudes correspond to automorphisms of $C(A, \sigma, f)$ which restrict to the identity on the center $Z(A, \sigma, f)$, by (13.2). $\qquad\square$

We thus recover the second case (deg $A = 4$) of (15.1).

Clifford groups. Let $Q \in A_1^2$ and $(A, \sigma, f) \in D_2$. Let Z be the center of Q, and assume that Q and (A, σ, f) correspond to each other under the groupoid equivalence $A_1^2 \equiv D_2$, so that we may identify $Q = C(A, \sigma, f)$ and $(A, \sigma, f) = \big(N_{Z/F}(Q), N_{Z/F}(\gamma), f_N\big)$.

(15.10) Proposition. *The extended Clifford group of (A, σ, f) is $\Omega(A, \sigma, f) = Q^\times$ and the canonical map $\chi_0\colon Q^\times \to \mathrm{GO}^+(A, \sigma, f)$ of (13.29) is given by $\chi_0(q) = {}^\iota q \otimes q \in N_{Z/F}(Q) = A$. For $q \in Q^\times$, the multiplier of $\chi_0(q)$ is $\mu\big(\chi_0(q)\big) = N_{Z/F}\big(\mathrm{Nrd}_Q(q)\big)$.*

The Clifford group of (A, σ, f) *is*

$$\Gamma(A, \sigma, f) = \{\, q \in Q^\times \mid \mathrm{Nrd}_Q(q) \in F^\times \,\},$$

and the vector representation map $\chi \colon \Gamma(A, \sigma, f) \to \mathrm{O}^+(A, \sigma, f)$ *is given by*

$$\chi(q) = \mathrm{Nrd}_Q(q)^{-1}{}^\iota q \otimes q = {}^\iota q \otimes \gamma(q)^{-1}.$$

The spin group is

$$\mathrm{Spin}(A, \sigma, f) = \mathrm{SL}_1(Q) = \{\, q \in Q^\times \mid \mathrm{Nrd}_Q(q) = 1 \,\}.$$

Proof: We identify $\Omega(A, \sigma, f)$ by means of (13.26): the canonical map $b \colon A \to B(A, \sigma, f)$ maps A onto the subspace of invariant elements under the canonical involution ω. Therefore, the condition $\sigma(x) * A^b \cdot x = A^b$ holds for all $x \in Q^\times$.

It suffices to check the description of χ_0 in the split case, where it follows from explicit computations. The Clifford group is characterized in (13.25) by the condition $\mu(q) \in F^\times$, which here amounts to $\mathrm{Nrd}_Q(q) \in F^\times$, and the description of $\mathrm{Spin}(A, \sigma, f)$ follows. □

(15.11) Corollary. *With the same notation as above, the group of multipliers of proper similitudes of* (A, σ, f) *is*

$$G^+(A, \sigma, f) = F^{\times 2} \cdot N_{Z/F}\big(\mathrm{Nrd}_Q(Q^\times)\big)$$

and the group of spinor norms is

$$\mathrm{Sn}(A, \sigma, f) = F^\times \cap \mathrm{Nrd}_Q(Q^\times).$$

Moreover, $G^+(A, \sigma, f) \neq G(A, \sigma, f)$ *if and only if* A *is nonsplit and splits over* Z.

Proof: The description of $G^+(A, \sigma, f)$ follows from (13.24) and the proposition above, since $\chi'(q) = \chi_0(q) \cdot F^\times$ for all $q \in \Omega(A, \sigma, f)$. By definition, the group of spinor norms is $\mathrm{Sn}(A, \sigma, f) = \mu\big(\Gamma(A, \sigma, f)\big)$, and the preceding proposition shows that $\mu\big(\Gamma(A, \sigma, f)\big) = F^\times \cap \mathrm{Nrd}_Q(Q^\times)$.

If $G(A, \sigma, f) \neq G^+(A, \sigma, f)$, then (13.43) shows that A is not split and splits over Z. In order to prove the converse implication, we use the isomorphism $A \simeq N_{Z/F}(Q)$ proved in (15.7) (and also in (9.12), see (9.14)). If A is split by Z, scalar extension to Z shows that ${}^\iota Q \otimes_Z Q$ is split, hence Q is isomorphic to ${}^\iota Q$ as a Z-algebra. It follows that $\mathrm{Aut}_F(Q) \neq \mathrm{Aut}_Z(Q)$, hence (15.9) shows that $\mathrm{PGO}(A, \sigma, f) \neq \mathrm{PGO}^+(A, \sigma, f)$. By (13.43), it follows that $G(A, \sigma, f) \neq G^+(A, \sigma, f)$ if A is not split. □

The case of trivial discriminant. If K is a given étale quadratic extension of F, the equivalence $A_1^2 \equiv D_2$ set up in (15.7) associates quaternion algebras with center K with algebras with quadratic pair (A, σ, f) such that $Z(A, \sigma, f) = K$. In the particular case where $K = F \times F$, we are led to consider the full subgroupoid ${}^1A_1^2$ of A_1^2 whose objects are F-algebras of the form $Q_1 \times Q_2$ where Q_1, Q_2 are quaternion F-algebras, and the full subgroupoid ${}^1D_2$ of D_2 whose objects are central simple F-algebras with quadratic pair of trivial discriminant. Theorem (15.7) specializes to the following statement:

(15.12) Corollary. *The functor* $\mathbf{N} \colon {}^1A_1^2 \to {}^1D_2$ *which maps the object* $Q_1 \times Q_2$ *to* $(Q_1 \otimes Q_2, \gamma_1 \otimes \gamma_2, f_\otimes)$ *(where* γ_1, γ_2 *are the canonical involutions on* Q_1, Q_2 *respectively, and* $(\gamma_1 \otimes \gamma_2, f_\otimes)$ *is the quadratic pair of* (5.21)*) and the Clifford algebra functor* $\mathbf{C} \colon {}^1D_2 \to {}^1A_1^2$ *define an equivalence of groupoids:*

$$^1A_1^2 \equiv {}^1D_2.$$

In particular, every central simple algebra A of degree 4 with quadratic pair (σ, f) of trivial discriminant decomposes as a tensor product of quaternion algebras:

$$(A, \sigma, f) = (Q_1 \otimes Q_2, \gamma_1 \otimes \gamma_2, f_\otimes).$$

Proof: For $(A, \sigma, f) \in {}^1D_2$, we have $\mathbf{C}(A, \sigma, f) = Q_1 \times Q_2$ for some quaternion F-algebras Q_1, Q_2. The isomorphism $(A, \sigma, f) \simeq \mathbf{N} \circ \mathbf{C}(A, \sigma, f)$ yields:

$$(A, \sigma, f) \simeq (Q_1 \otimes Q_2, \gamma_1 \otimes \gamma_2, f_\otimes). \qquad \square$$

Note that the algebras Q_1, Q_2 are uniquely determined by (A, σ, f) up to isomorphism since $\mathbf{C}(A, \sigma, f) = Q_1 \times Q_2$. Actually, they are uniquely determined as subalgebras of A by the relation $(A, \sigma, f) = (Q_1 \otimes Q_2, \gamma_1 \otimes \gamma_2, f_\otimes)$. If char $F \neq 2$, this property follows from the observation that $\mathrm{Skew}(A, \sigma) = \mathrm{Skew}(Q_1, \gamma_1) \otimes 1 + 1 \otimes \mathrm{Skew}(Q_2, \gamma_2)$, since $\mathrm{Skew}(Q_1, \gamma_1) \otimes 1$ and $1 \otimes \mathrm{Skew}(Q_2, \gamma_2)$ are the only simple Lie ideals of $\mathrm{Skew}(A, \sigma)$. See Exercise 2 for the case where char $F = 2$.

The results in (15.9), (15.10) and (15.11) can also be specialized to the case where the discriminant of (σ, f) is trivial. For instance, one has the following description of the group of similitudes and their multipliers:

(15.13) Corollary. *Let $(A, \sigma, f) = (Q_1 \otimes Q_2, \gamma_1 \otimes \gamma_2, f_\otimes) \in {}^1D_2$. The functor $\mathbf{C}$ induces isomorphisms of groups:*

$$\mathrm{PGO}(A, \sigma, f) \xrightarrow{\sim} \mathrm{Aut}_F(Q_1 \times Q_2)$$

and

$$\mathrm{PGO}^+(A, \sigma, f) \xrightarrow{\sim} \mathrm{Aut}_F(Q_1) \times \mathrm{Aut}_F(Q_2) = \mathrm{PGL}(Q_1) \times \mathrm{PGL}(Q_2).$$

Similarly, $\mathrm{Spin}(A, \sigma, f) \simeq \mathrm{SL}_1(Q_1) \times \mathrm{SL}_1(Q_2)$. Moreover,

$$G(A, \sigma, f) = G^+(A, \sigma, f) = \mathrm{Nrd}_{Q_1}(Q_1^\times) \cdot \mathrm{Nrd}_{Q_2}(Q_2^\times)$$

and

$$\mathrm{Sn}(A, \sigma, f) = \mathrm{Nrd}_{Q_1}(Q_1^\times) \cap \mathrm{Nrd}_{Q_2}(Q_2^\times).$$

Indices. Let $Q \in A_1^2$ and $(A, \sigma, f) \in D_2$ correspond to each other under the equivalence $A_1^2 \equiv D_2$. Since $\deg A = 4$, there are four possibilities for $\mathrm{ind}(A, \sigma, f)$:

$$\{0\}, \qquad \{0, 1\}, \qquad \{0, 2\}, \qquad \{0, 1, 2\}.$$

The following proposition describes the corresponding possibilities for the algebra Q. Let K be the center of Q, so $K \simeq F\left(\sqrt{\mathrm{disc}(A, \sigma, f)}\right)$ if char $F \neq 2$ and $K \simeq F\left(\wp^{-1}\left(\mathrm{disc}(A, \sigma, f)\right)\right)$ if char $F = 2$.

(15.14) Proposition. *With the notation above,*

(1) $\mathrm{ind}(A, \sigma, f) = \{0\}$ *if and only if either Q is a division algebra (so K is a field) or $Q \simeq Q_1 \times Q_2$ for some quaternion division F-algebras Q_1, Q_2 (so $K \simeq F \times F$);*
(2) $\mathrm{ind}(A, \sigma, f) = \{0, 1\}$ *if and only if K is a field and $Q \simeq M_2(K)$;*
(3) $\mathrm{ind}(A, \sigma, f) = \{0, 2\}$ *if and only if $Q \simeq M_2(F) \times Q_0$ for some quaternion division F-algebra Q_0;*
(4) $\mathrm{ind}(A, \sigma, f) = \{0, 1, 2\}$ *if and only if $Q \simeq M_2(F) \times M_2(F)$.*

Proof: If $1 \in \mathrm{ind}(A, \sigma, f)$, then A is split and the quadratic pair (σ, f) is isotropic. Thus, $A \simeq \mathrm{End}_F(V)$ for some 4-dimensional F-vector space V, and (σ, f) is the quadratic pair associated with some isotropic quadratic form q on V. Since $\dim V = 4$, the quadratic space (V, q) is hyperbolic if and only if its discriminant is trivial, i.e., $K \simeq F \times F$. Therefore, if $\mathrm{ind}(A, \sigma, f) = \{0, 1\}$, then K is a field;

by (8.5), the canonical involution on $C_0(V, q) \simeq Q$ is hyperbolic, hence Q is split. If $\mathrm{ind}(A, \sigma, f) = \{0, 1, 2\}$, then (V, q) is hyperbolic and $K \simeq F \times F$. By (8.3), it follows that $Q \simeq M_2(F) \times M_2(F)$. Conversely, if $Q \simeq M_2(K)$ (and K is either a field or isomorphic to $F \times F$), then Q contains a nonzero element q which is not invertible. The element $^tq \otimes q \in N_{K/F}(Q) \simeq A$ generates an isotropic right ideal of reduced dimension 1, hence $1 \in \mathrm{ind}(A, \sigma, f)$. This proves (2) and (4).

If $2 \in \mathrm{ind}(A, \sigma, f)$, then (σ, f) is hyperbolic, hence Proposition (8.31) shows that $Q \simeq M_2(F) \times Q_0$ for some quaternion F-algebra Q_0, since $Q \simeq C(A, \sigma, f)$. Conversely, if $Q \simeq M_2(F) \times Q_0$ for some quaternion F-algebra Q_0, then

$$(A, \sigma, f) \simeq \big(M_2(F) \otimes Q_0, \gamma_M \otimes \gamma_0, f_\otimes\big),$$

where γ_M and γ_0 are the canonical (symplectic) involutions on $M_2(F)$ and Q_0 respectively. If $x \in M_2(F)$ is a nonzero singular matrix, then $x \otimes 1$ generates an isotropic right ideal of reduced dimension 2 in A, hence $2 \in \mathrm{ind}(A, \sigma, f)$. This proves (3) and yields an alternate proof of (4).

Since (1), (2), (3) and (4) exhaust all the possibilities for $\mathrm{ind}(A, \sigma, f)$ and for Q, the proof is complete. $\qquad\square$

15.C. $B_2 \equiv C_2$. The arguments to prove the equivalence of B_2 and C_2 are similar to those used in §15.A to prove $B_1 \equiv C_1$.

For any oriented quadratic space $(V, q, \zeta) \in B_2$ (of trivial discriminant), the even Clifford algebra $C_0(V, q)$ is central simple of degree 4, and its canonical involution $\tau\ (= \underline{\sigma}_q)$ is symplectic. We may therefore define a functor

$$\mathbf{C} \colon B_2 \to C_2$$

by

$$\mathbf{C}(V, q, \zeta) = \big(C_0(V, q), \tau\big).$$

On the other hand, let (A, σ) be a central simple F-algebra of degree 4 with symplectic involution. As observed in (2.9), the reduced characteristic polynomial of every symmetrized element is a square; the pfaffian trace Trp_σ is a linear form on V and the pfaffian norm Nrp_σ is a quadratic form on V such that

$$\mathrm{Prd}_{A,a}(X) = \big(X^2 - \mathrm{Trp}_\sigma(a)X + \mathrm{Nrp}_\sigma(a)\big)^2 \quad \text{for } a \in \mathrm{Symd}(A, \sigma).$$

In particular, if $a \in \mathrm{Symd}(A, \sigma)$ is such that $\mathrm{Trp}_\sigma(a) = 0$, then $a^2 = -\mathrm{Nrp}_\sigma(a) \in F$. Let

$$\mathrm{Symd}(A, \sigma)^0 = \big\{\, a \in \mathrm{Symd}(A, \sigma) \mid \mathrm{Trp}_\sigma(a) = 0 \,\big\}$$
$$\big(= \big\{\, a \in \mathrm{Sym}(A, \sigma) \mid \mathrm{Trd}_A(a) = 0 \,\big\} \quad \text{if char } F \neq 2\big).$$

This is a vector space of dimension 5 over F. The map $s_\sigma \colon \mathrm{Symd}(A, \sigma)^0 \to F$ defined by

$$s_\sigma(a) = a^2$$

is a quadratic form on $\mathrm{Symd}(A, \sigma)^0$. Inspection of the split case shows that this form is nonsingular. By the universal property of Clifford algebras, the inclusion $\mathrm{Symd}(A, \sigma) \hookrightarrow A$ induces an F-algebra homomorphism

(15.15) $h_A \colon C\big(\mathrm{Symd}(A, \sigma)^0, s_\sigma\big) \to A,$

which is not injective since $\dim_F C\big(\mathrm{Symd}(A,\sigma)^0, s_\sigma\big) = 2^5$ while $\dim_F A = 2^4$. Therefore, the Clifford algebra $C\big(\mathrm{Symd}(A,\sigma)^0, s_\sigma\big)$ is not simple, hence the discriminant of the quadratic space $\big(\mathrm{Symd}(A,\sigma)^0, s_\sigma\big)$ is trivial. Moreover, there is a unique central element η in $C_1\big(\mathrm{Symd}(A,\sigma)^0, s_\sigma\big)$ such that $\eta^2 = 1$ and $h_A(\eta) = 1$. We may therefore define a functor

$$\mathbf{S} \colon \mathcal{C}_2 \to \mathcal{B}_2$$

by

$$\mathbf{S}(A,\sigma) = \big(\mathrm{Symd}(A,\sigma)^0, s_\sigma, \eta\big).$$

(15.16) Theorem. *The functors $\mathbf{C}$ and $\mathbf{S}$ define an equivalence of groupoids:*

$$\mathcal{B}_2 \equiv \mathcal{C}_2.$$

Proof: For any $(A,\sigma) \in \mathcal{C}_2$, the F-algebra homomorphism h_A of (15.15) restricts to a canonical isomorphism

$$h_A \colon C_0\big(\mathrm{Symd}(A,\sigma)^0, s_\sigma\big) \xrightarrow{\sim} A,$$

and yields a natural transformation $\mathbf{C} \circ \mathbf{S} \cong \mathrm{Id}_{\mathcal{C}_2}$.

For $(V, q, \zeta) \in \mathcal{B}_2$, we define a linear map $m_\zeta \colon V \to C_0(V, q)$ by

$$m_\zeta(v) = v\zeta \quad \text{for } v \in V.$$

Since $s_\tau(v\zeta) = (v\zeta)^2 = v^2 = q(v)$, this map is an isometry:

$$m_\zeta \colon (V, q) \xrightarrow{\sim} \big(\mathrm{Symd}(C_0(V,q), \tau)^0, s_\tau\big).$$

The same argument as in the proof of (15.2) shows that this isometry carries ζ to the canonical orientation η on $\big(\mathrm{Symd}(C_0(V,q), \tau)^0, s_\tau\big)$; therefore, it defines a natural transformation $\mathbf{S} \circ \mathbf{C} \cong \mathrm{Id}_{\mathcal{B}_2}$ which completes the proof. $\square$

(15.17) Corollary. *For every oriented quadratic space (V, q, ζ) of dimension 5, the Clifford algebra construction yields a group isomorphism*

$$\mathbf{O}^+(V,q) = \mathrm{Aut}(V,q,\zeta) \xrightarrow{\sim} \mathrm{Aut}_F\big(C_0(V,q), \tau\big) = \mathrm{PGSp}\big(C_0(V,q), \tau\big).$$

Proof: The functor $\mathbf{C}$ defines an isomorphism between automorphism groups of corresponding objects. $\square$

Suppose now that $(A,\sigma) \in \mathcal{C}_2$ corresponds to $(V, q, \zeta) \in \mathcal{B}_2$ under the equivalence $\mathcal{B}_2 \equiv \mathcal{C}_2$, so that we may identify $(A, \sigma) = \big(C_0(V,q), \tau\big)$ and $(V, q, \zeta) = \big(\mathrm{Symd}(A,\sigma)^0, s_\sigma, \eta\big)$.

(15.18) Proposition. *The special Clifford group of (V, q) is*

$$\Gamma^+(V,q) = \mathrm{GSp}(A, \sigma).$$

Under the identification $V = \mathrm{Symd}(A,\sigma)^0 \subset A$, the vector representation

$$\chi \colon \mathrm{GSp}(A,\sigma) \to \mathbf{O}^+(V,q)$$

is given by $\chi(g)(v) = gvg^{-1}$ for $g \in \mathrm{GSp}(A,\sigma)$ and $v \in V$.

The spin group is $\mathrm{Spin}(V,q) = \mathrm{Sp}(A,\sigma)$ and the group of spinor norms is $\mathrm{Sn}(V,q) = G(A,\sigma)$.

Proof: By definition, $\Gamma^+(V, q)$ is a subgroup of $A^\times$, and it consists of similitudes of (A, σ). We have a commutative diagram with exact rows:

$$
\begin{array}{ccccccccc}
1 & \longrightarrow & F^\times & \longrightarrow & \Gamma^+(V, q) & \xrightarrow{\;\chi\;} & O^+(V, q) & \longrightarrow & 1 \\
 & & \| & & \downarrow & & \downarrow{\scriptstyle C} & & \\
1 & \longrightarrow & F^\times & \longrightarrow & GSp(A, \sigma) & \longrightarrow & PGSp(A, \sigma) & \longrightarrow & 1
\end{array}
$$

(see $(13.10)^{24}$). The corollary above shows that the right-hand vertical map is an isomorphism, hence $\Gamma^+(V, q) = GSp(A, \sigma)$.

For $g \in \Gamma^+(V, q)$, we have $\chi(g)(v) = g \cdot v \cdot g^{-1}$ in $C(V, q)$, by the definition of the vector representation. Under the identification $V = \operatorname{Symd}(A, \sigma)^0$, every vector $v \in V$ is mapped to $v\zeta \in A$, hence the action of $\chi(g)$ is by conjugation by g, since ζ is central in $C(V, q)$. The last assertions are clear. $\qquad\square$

An alternate proof is given in (17.11) below.

If char $F \neq 2$, we may combine (15.16) with the equivalence $B_2 \equiv B_2'$ of (12.41) to get the following relation between groupoids of algebras with involution:

(15.19) Corollary. *Suppose* char $F \neq 2$. *The functor* $\mathbf{S} \colon C_2 \to B_2'$, *which maps every central simple algebra of degree 4 with symplectic involution* (A, σ) *to the algebra of degree 5 with orthogonal involution* $\left(\operatorname{End}_F\left(\operatorname{Symd}(A, \sigma)^0\right), \sigma_{s_\sigma}\right)$ *where* σ_{s_σ} *is the adjoint involution with respect to* s_σ, *and the functor* $\mathbf{C} \colon B_2' \to C_2$, *which maps every central simple algebra of degree 5 with orthogonal involution* (A', σ') *to its Clifford algebra* $\left(C(A', \sigma'), \underline{\sigma}'\right)$, *define an equivalence of groupoids:*

$$C_2 \equiv B_2'.$$

In particular, for every central simple algebra with involution (A', σ') of degree 5, the functor $\mathbf{C}$ induces an isomorphism of groups:

$$O^+(A', \sigma') = PGO(A', \sigma') =$$
$$\operatorname{Aut}_F(A', \sigma') \xrightarrow{\;\sim\;} \operatorname{Aut}_F\left(C(A', \sigma'), \underline{\sigma}'\right)$$
$$= PGSp\left(C(A', \sigma'), \underline{\sigma}'\right).$$

We thus recover the third case (deg $A = 5$) of (15.1).

Indices. Let $(A, \sigma) \in C_2$. In order to relate the index of (A, σ) to the Witt index of the corresponding 5-dimensional quadratic space $\left(\operatorname{Symd}(A, \sigma)^0, s_\sigma\right)$, we establish a one-to-one correspondence between isotropic ideals in (A, σ) and isotropic vectors in $\operatorname{Symd}(A, \sigma)^0$.

(15.20) Proposition. (1) *For every right ideal* $I \subset A$ *of reduced dimension 2, the intersection* $I \cap \operatorname{Symd}(A, \sigma)$ *is a 1-dimensional subspace of* $\operatorname{Symd}(A, \sigma)$ *which is isotropic for the quadratic form* $\operatorname{Nrp}_\sigma$. *This subspace is in* $\operatorname{Symd}(A, \sigma)^0$ *(and therefore isotropic for the form* s_σ*) if* $\sigma(I) \cdot I = \{0\}$.
(2) *For every nonzero vector* $x \in \operatorname{Symd}(A, \sigma)$ *such that* $\operatorname{Nrp}_\sigma(x) = 0$, *the right ideal* xA *has reduced dimension 2. This ideal is isotropic for* σ *if* $x \in \operatorname{Symd}(A, \sigma)^0$.

---

[24]Although (13.10) is stated for even-dimensional quadratic spaces, the arguments used in the proof also apply to odd-dimensional spaces.

Proof: It suffices to prove the proposition over a scalar extension. We may thus assume that A is split; let $(A, \sigma) = \big(\mathrm{End}_F(W), \sigma_b\big)$ for some 4-dimensional vector space W with alternating bilinear form b. The bilinear form b induces the standard identification $\varphi_b \colon W \otimes W = \mathrm{End}_F(W)$ under which

$$\sigma(x \otimes y) = -y \otimes x \quad \text{and} \quad \mathrm{Trd}(x \otimes y) = b(y, x)$$

for x, $y \in W$ (see (5.1)). According to (1.12), every right ideal $I \subset \mathrm{End}_F(W)$ of reduced dimension 2 has the form

$$I = \mathrm{Hom}_F(W, U) = U \otimes W$$

for some 2-dimensional subspace $U \subset W$ uniquely determined by I. If (u_1, u_2) is a basis of U, every element in I has a unique expression of the form $u_1 \otimes v_1 + u_2 \otimes v_2$ for some v_1, $v_2 \in V$. Such an element is symmetrized under σ if and only if $v_1 = u_2 \alpha$ and $v_2 = -u_1 \alpha$ for some $\alpha \in F$. Therefore,

$$I \cap \mathrm{Symd}(A, \sigma) = (u_1 \otimes u_2 - u_2 \otimes u_1) \cdot F,$$

showing that $I \cap \mathrm{Symd}(A, \sigma)$ is 1-dimensional. Since the elements in I are not invertible, it is clear that $\mathrm{Nrp}_\sigma(x) = 0$ for all $x \in I \cap \mathrm{Symd}(A, \sigma)$.

If $\sigma(I) \cdot I = \{0\}$, then $\sigma(u_1 \otimes u_2 - u_2 \otimes u_1) \cdot (u_1 \otimes u_2 - u_2 \otimes u_1) = 0$. We have

$$\sigma(u_1 \otimes u_2 - u_2 \otimes u_1) \cdot (u_1 \otimes u_2 - u_2 \otimes u_1) = (u_1 \otimes u_2 - u_2 \otimes u_1)^2$$
$$= (u_1 \otimes u_2 - u_2 \otimes u_1) b(u_2, u_1)$$

and, by (2.13),

$$\mathrm{Trp}_\sigma(u_1 \otimes u_2 - u_2 \otimes u_1) = \mathrm{Trd}_A(u_1 \otimes u_2) = b(u_2, u_1).$$

Hence the condition $\sigma(I) \cdot I = \{0\}$ implies that $\mathrm{Trp}_\sigma(u_1 \otimes u_2 - u_2 \otimes u_1) = 0$. This completes the proof of (1).

In order to prove (2), we choose a basis of V to identify

$$(A, \sigma) = \big(M_4(F), \mathrm{Int}(u) \circ t\big)$$

for some alternating matrix $u \in \mathrm{GL}_4(F)$. Under this identification, we have

$$\mathrm{Symd}(A, \sigma) = u \cdot \mathrm{Alt}\big(M_4(F), t\big).$$

Since the rank of every alternating matrix is even, it follows that $\mathrm{rdim}(xA) = 0$, 2, or 4 for every $x \in \mathrm{Alt}(A, \sigma)$. If $\mathrm{Nrp}_\sigma(x) = 0$, then x is not invertible, hence $\mathrm{rdim}(xA) < 4$; on the other hand, if $x \neq 0$, then $\mathrm{rdim}(xA) > 0$. Therefore, $\mathrm{rdim}(xA) = 2$ for every nonzero isotropic vector x in $\big(\mathrm{Symd}(A, \sigma), \mathrm{Nrp}_\sigma\big)$. If $x \in \mathrm{Symd}(A, \sigma)^0$, then $x^2 = -\mathrm{Nrp}_\sigma(x)$, hence

$$\sigma(xA) \cdot xA = Ax^2A = \{0\}$$

if x is isotropic. $\qquad\square$

This proposition shows that the maps $I \mapsto I \cap \mathrm{Symd}(A, \sigma)$ and $xF \mapsto xA$ define a one-to-one correspondence between right ideals of reduced dimension 2 in A and 1-dimensional isotropic subspaces in $\big(\mathrm{Symd}(A, \sigma), \mathrm{Nrp}_\sigma\big)$. Moreover, under this bijection isotropic right ideals I for σ correspond to 1-dimensional isotropic subspaces in $\big(\mathrm{Symd}(A, \sigma)^0, s_\sigma\big)$.

If A is split, then σ is adjoint to an alternating bilinear form, hence it is hyperbolic. In particular, if $1 \in \mathrm{ind}(A, \sigma)$, then $\mathrm{ind}(A, \sigma) = \{0, 1, 2\}$. Thus, the only possibilities for the index of (A, σ) are

$$\{0\}, \qquad \{0, 2\} \quad \text{and} \quad \{0, 1, 2\},$$

and the last case occurs if and only if A is split.

(15.21) Proposition. *Let $(V, q, \zeta) \in B_2$ and $(A, \sigma) \in C_2$ correspond to each other under the equivalence $B_2 \equiv C_2$. The index of (A, σ) and the Witt index $w(V, q)$ are related as follows*:

$$\operatorname{ind}(A, \sigma) = \{0\} \quad \Longleftrightarrow \quad w(V, q) = 0;$$
$$\operatorname{ind}(A, \sigma) = \{0, 2\} \quad \Longleftrightarrow \quad w(V, q) = 1;$$
$$\operatorname{ind}(A, \sigma) = \{0, 1, 2\} \quad \Longleftrightarrow \quad w(V, q) = 2.$$

Proof: Proposition (15.20) shows that $2 \in \operatorname{ind}(A, \sigma)$ if and only if $w(V, q) > 0$. Therefore, it suffices to show that A splits if and only if $w(V, q) = 2$. If the latter condition holds, then (V, q) has an orthogonal decomposition $(V, q) \simeq \mathbb{H}(U) \oplus uF$ for some 4-dimensional hyperbolic space $\mathbb{H}(U)$ and some vector $u \in V$ such that $q(u) = 1$, hence $C_0(V, q) \simeq C(\mathbb{H}(U))$. It follows from (8.3) that $C_0(V, q)$, hence also A, is split. Conversely, suppose $(A, \sigma) = (\operatorname{End}_F(W), \sigma_b)$ for some 4-dimensional vector space W with alternating form b. As in (15.20), we identify A with $W \otimes W$ under φ_b. If (e_1, e_2, e_3, e_4) is a symplectic basis of W, the span of $e_1 \otimes e_3 - e_3 \otimes e_1$ and $e_1 \otimes e_4 - e_4 \otimes e_1$ is a totally isotropic subspace of $(\operatorname{Symd}(A, \sigma), s_\sigma)$, hence $w(V, q) = 2$. $\square$

15.D. $A_3 \equiv D_3$. The equivalence between the groupoid A_3 of central simple algebras of degree 4 with involution of the second kind over a quadratic étale extension of F and the groupoid D_3 of central simple F-algebras of degree 6 with quadratic pair is given by the Clifford algebra and the discriminant algebra constructions. Let

$$\mathbf{C} \colon D_3 \to A_3$$

be the functor which maps $(A, \sigma, f) \in D_3$ to its Clifford algebra with canonical involution $(C(A, \sigma, f), \underline{\sigma})$ and let

$$\mathbf{D} \colon A_3 \to D_3$$

be the functor which maps $(B, \tau) \in A_3$ to the discriminant algebra $D(B, \tau)$ with its canonical quadratic pair $(\underline{\tau}, f_D)$.

As in §15.B, the proof that these functors define an equivalence of groupoids is based on a Lie algebra isomorphism which we now describe.

For $(B, \tau) \in A_3$, recall from (14.11) the Lie algebra

$$\mathfrak{s}(B, \tau) = \{\, b \in B \mid b + \tau(b) \in F \text{ and } \operatorname{Trd}_B(b) = 2(b + \tau(b)) \,\}$$

and from (14.8) the Lie algebra homomorphism

$$\dot{\lambda}^2 \colon B \to \lambda^2 B.$$

Endowing B and $\lambda^2 B$ with the nonsingular symmetric bilinear forms T_B and $T_{\lambda^2 B}$, we may consider the adjoint F-linear map

$$(\dot{\lambda}^2)^* \colon \lambda^2 B \to B,$$

which is explicitly defined by the following property: the image $(\dot{\lambda}^2)^*(\xi)$ of $\xi \in \lambda^2 B$ is the unique element of B such that

$$\operatorname{Trd}_B((\dot{\lambda}^2)^*(\xi)y) = \operatorname{Trd}_{\lambda^2 B}(\xi \dot{\lambda}^2 y) \qquad \text{for all } y \in B.$$

(15.22) Proposition. *The map* $(\dot\lambda^2)^*$ *restricts to a linear map*

$$\lambda^*\colon D(B,\tau) \to \mathfrak{s}(B,\tau),$$

which factors through the canonical map $c\colon D(B,\tau) \to C\big(D(B,\tau),\underline\tau,f_D\big)$ *and induces a Lie algebra isomorphism*

$$\lambda^*\colon c\big(D(B,\tau)\big) \xrightarrow{\sim} \mathfrak{s}(B,\tau).$$

This Lie algebra isomorphism extends to an isomorphism of (associative) F-algebras with involution

$$\big(C\big(D(B,\tau),\underline\tau,f_D\big),\underline\tau\big) \xrightarrow{\sim} (B,\tau).$$

Proof: Let γ be the canonical involution on $\lambda^2 B$. For $y \in B$, Proposition (14.10) yields $\gamma(\dot\lambda^2 y) = \mathrm{Trd}_B(y) - \dot\lambda^2 y$. Therefore, for $\xi \in \lambda^2 B$ we have

$$\mathrm{Trd}_{\lambda^2 B}\big(\gamma(\xi)\dot\lambda^2 y\big) = \mathrm{Trd}_B(y)\,\mathrm{Trd}_{\lambda^2 B}(\xi) - \mathrm{Trd}_{\lambda^2 B}(\xi\dot\lambda^2 y).$$

By the definition of $(\dot\lambda^2)^*$, this last equality yields $(\dot\lambda^2)^*\big(\gamma(\xi)\big) = \mathrm{Trd}_{\lambda^2 B}(\xi) - (\dot\lambda^2)^*(\xi)$. Similarly, $(\dot\lambda^2)^*\big(\tau^{\wedge 2}(\xi)\big) = \tau\big((\dot\lambda^2)^*(\xi)\big)$ for $\xi \in \lambda^2 B$. Therefore, if $\xi \in D(B,\tau)$, i.e., $\tau^{\wedge 2}(\xi) = \gamma(\xi)$, then

$$\tau\big((\dot\lambda^2)^*(\xi)\big) + (\dot\lambda^2)^*(\xi) = \mathrm{Trd}_{\lambda^2 B}(\xi) \in F.$$

Since, by the definition of $(\dot\lambda^2)^*$,

$$\mathrm{Trd}_B\big((\dot\lambda^2)^*(\xi)\big) = \mathrm{Trd}_{\lambda^2 B}(\xi\dot\lambda^2 1) = 2\,\mathrm{Trd}_{\lambda^2 B}(\xi),$$

it follows that $(\dot\lambda^2)^*(\xi) \in \mathfrak{s}(B,\tau)$ for $\xi \in D(B,\tau)$, proving the first part.

To prove the rest, we extend scalars to an algebraic closure of F. We may thus assume that $B = \mathrm{End}_F(V) \times \mathrm{End}_F(V^*)$ for some 4-dimensional F-vector space V, and the involution τ is given by

$$\tau(g,h^t) = (h,g^t) \qquad \text{for } g, h \in \mathrm{End}_F(V).$$

We may then identify

$$\big(D(B,\tau),\underline\tau,f_D\big) = \big(\mathrm{End}_F(\textstyle\bigwedge^2 V),\sigma_q,f_q\big)$$

where (σ_q,f_q) is the quadratic pair associated with the canonical quadratic map $q\colon \bigwedge^2 V \to \bigwedge^4 V$ of (10.12). Let $e \in \bigwedge^4 V$ be a nonzero element (hence a basis). We use e to identify $\bigwedge^4 V = F$, hence to view q as a quadratic form on $\bigwedge^2 V$. The standard identification $\varphi_q\colon \bigwedge^2 V \otimes \bigwedge^2 V \xrightarrow{\sim} \mathrm{End}_F(\bigwedge^2 V)$ then yields an identification

$$\eta_q\colon C_0(\textstyle\bigwedge^2 V,q) \xrightarrow{\sim} C\big(D(B,\tau),\underline\tau,f_D\big)$$

which preserves the canonical involutions.

We next define an isomorphism $C_0(\bigwedge^2 V,q) \xrightarrow{\sim} \mathrm{End}_F(V) \times \mathrm{End}_F(V^*) = B$ by restriction of an isomorphism $C(\bigwedge^2 V,q) \xrightarrow{\sim} \mathrm{End}_F(V \oplus V^*)$.

For $\xi \in \bigwedge^2 V$, we define maps $\ell_\xi\colon V \to V^*$ and $r_\xi\colon V^* \to V$ by the following conditions, where $\langle\,,\,\rangle$ is the canonical pairing of a vector space and its dual:

$$\xi \wedge x \wedge y = e \cdot \langle \ell_\xi(x),y\rangle \quad \text{for } x,\, y \in V$$
$$\langle \psi \wedge \varphi,\xi\rangle = \langle \psi, r_\xi(\varphi)\rangle \qquad \text{for } \psi,\, \varphi \in V^*.$$

The map

$$i\colon \bigwedge^2 V \to \operatorname{End}_F(V \oplus V^*) = \begin{pmatrix} \operatorname{End}(V) & \operatorname{Hom}(V^*, V) \\ \operatorname{Hom}(V, V^*) & \operatorname{End}(V^*) \end{pmatrix}$$

which carries $\xi \in \bigwedge^2 V$ to $\left(\begin{smallmatrix} 0 & r_\xi \\ \ell_\xi & 0 \end{smallmatrix}\right)$ induces an isomorphism

(15.23) $i_*\colon C(\bigwedge^2 V, q) \xrightarrow{\sim} \operatorname{End}_F(V \oplus V^*)$

which restricts to an isomorphism of algebras with involution

$$i_*\colon \big(C_0(\textstyle\bigwedge^2 V, q), \tau_0\big) \xrightarrow{\sim} \big(\operatorname{End}_F(V) \times \operatorname{End}_F(V^*), \tau\big).$$

To complete the proof, it now suffices to show that this isomorphism extends λ^*, in the sense that

$$i_*(\xi \cdot \eta) = \lambda^*(\xi \otimes \eta) \qquad \text{for } \xi, \eta \in \textstyle\bigwedge^2 V.$$

In view of the definition of λ^*, this amounts to proving

$$\operatorname{tr}(\xi \otimes \eta \circ \dot\lambda^2 g) = \operatorname{tr}(r_\xi \circ \ell_\eta \circ g) \qquad \text{for all } g \in \operatorname{End}_F(V).$$

Verification of this formula is left to the reader.   □

(15.24) Theorem. *The functors $\mathbf{D}$ and $\mathbf{C}$ define an equivalence of groupoids*

$$\mathsf{A}_3 \equiv \mathsf{D}_3.$$

Moreover, if $(B, \tau) \in \mathsf{A}_3$ and $(A, \sigma, f) \in \mathsf{D}_3$ correspond to each other under this equivalence, the center $Z(B)$ of B satisfies

$$Z(B) \simeq \begin{cases} F\big(\sqrt{\operatorname{disc}(A, \sigma, f)}\big) & \text{if } \operatorname{char} F \neq 2; \\ F\big(\wp^{-1}\big(\operatorname{disc}(A, \sigma, f)\big)\big) & \text{if } \operatorname{char} F = 2. \end{cases}$$

Proof: Once the equivalence of groupoids has been established, then the algebra B of degree 4 corresponding to $(A, \sigma, f) \in \mathsf{D}_3$ is the Clifford algebra $C(A, \sigma, f)$, hence the description of $Z(B)$ follows from the structure theorem for Clifford algebras (8.10).

In order to prove the first part, we show that for $(A, \sigma, f) \in \mathsf{D}_3$ and $(B, \tau) \in \mathsf{A}_3$ there are canonical isomorphisms

$$(A, \sigma, f) \simeq \big(D\big(C(A, \sigma, f), \underline{\sigma}\big), \underline{\sigma}, f_D\big) \quad \text{and} \quad (B, \tau) \simeq \big(C\big(D(B, \tau), \underline{\tau}, f_D\big), \underline{\tau}\big)$$

which yield natural transformations

$$\mathbf{D} \circ \mathbf{C} \cong \operatorname{Id}_{\mathsf{D}_3} \quad \text{and} \quad \mathbf{C} \circ \mathbf{D} \cong \operatorname{Id}_{\mathsf{A}_3}.$$

Proposition (15.22) yields a canonical isomorphism

$$\big(C\big(D(B, \tau), \underline{\tau}, f_D\big), \underline{\tau}\big) \xrightarrow{\sim} (B, \tau).$$

On the other hand, starting with $(A, \sigma, f) \in \mathsf{D}_3$, we may also apply (15.22) to get a Lie algebra isomorphism

$$\lambda^*\colon c\big(D\big(C(A, \sigma, f), \underline{\sigma}\big)\big) \xrightarrow{\sim} \mathfrak{s}\big(C(A, \sigma, f), \underline{\sigma}\big).$$

By (8.16), one may check that $c(A) \subset \mathfrak{s}\big(C(A, \sigma, f), \underline{\sigma}\big)$, and dimension count shows that this inclusion is an equality. Since λ^* extends to an F-algebra isomorphism, it is the identity on F. Therefore, by (8.15) it induces a Lie algebra isomorphism

$$\overline{\lambda}\colon \operatorname{Alt}\big(D\big(C(A, \sigma, f), \underline{\sigma}\big), \underline{\sigma}\big) \xrightarrow{\sim} \operatorname{Alt}(A, \sigma).$$

We aim to show that this isomorphism extends to an isomorphism of (associative) F-algebras with quadratic pair

$$\left(D\big(C(A,\sigma,f),\underline{\sigma}\big),\underline{\sigma},f_D\right) \to (A,\sigma,f).$$

By (2.25), it suffices to prove the property over an algebraic closure of F. We may thus assume that A is the endomorphism algebra of a hyperbolic space, so

$$A = \mathrm{End}_F\big(\mathbb{H}(U)\big) = \mathrm{End}_F(U^* \oplus U)$$

where U is a 3-dimensional vector space, U^* is its dual, and $(\sigma,f) = (\sigma_{q_U}, f_{q_U})$ is the quadratic pair associated with the hyperbolic form q_U. Recall that q_U is defined by

$$q_U(\varphi + u) = \varphi(u) = \langle \varphi, u \rangle \quad \text{for } \varphi \in U^*, \, u \in U.$$

The Clifford algebra of (A, σ, f) may be described as follows:

$$C(A,\sigma,f) = \mathrm{End}_F(\textstyle\bigwedge_0 U) \times \mathrm{End}_F(\textstyle\bigwedge_1 U),$$

where $\bigwedge_0 U = F \oplus \bigwedge^2 U$ and $\bigwedge_1 U = U \oplus \bigwedge^3 U$ (see (8.3)). The involution $\underline{\sigma}$ on $C(A,\sigma)$ is of the second kind; it interchanges $\mathrm{End}_F(\bigwedge_0 U)$ and $\mathrm{End}_F(\bigwedge_1 U)$. Therefore, the discriminant algebra of $\big(C(A,\sigma),\underline{\sigma}\big)$ is

$$D\big(C(A,\sigma),\underline{\sigma}\big) = \mathrm{End}_F\big(\textstyle\bigwedge^2(\bigwedge_1 U)\big),$$

and its quadratic pair $(\underline{\sigma}, f_D)$ is associated with the canonical quadratic map

$$q\colon \textstyle\bigwedge^2(\bigwedge_1 U) \to \bigwedge^4(\bigwedge_1 U) \simeq F.$$

Since $\dim U = 3$, there are canonical isomorphisms

$$\textstyle\bigwedge^2(\bigwedge_1 U) = \bigwedge^2 U \oplus (\bigwedge^3 U \otimes U) \quad \text{and} \quad \bigwedge^4(\bigwedge_1 U) = \bigwedge^3 U \otimes \bigwedge^3 U$$

given by

$$(u_1 + \xi_1) \wedge (u_2 + \xi_2) = u_1 \wedge u_2 + \xi_1 \otimes u_2 - \xi_2 \otimes u_1$$

and

$$(u_1 + \xi_1) \wedge (u_2 + \xi_2) \wedge (u_3 + \xi_3) \wedge (u_4 + \xi_4) =$$
$$\xi_1 \otimes (u_2 \wedge u_3 \wedge u_4) - \xi_2 \otimes (u_1 \wedge u_3 \wedge u_4)$$
$$+ \xi_3 \otimes (u_1 \wedge u_2 \wedge u_4) - \xi_4 \otimes (u_1 \wedge u_2 \wedge u_3)$$

for $u_1, \ldots, u_4 \in U$ and $\xi_1, \ldots, \xi_4 \in \bigwedge^3 U$. Under these identifications, the canonical quadratic map $q\colon \bigwedge^2 U \oplus (\bigwedge^3 U \otimes U) \to \bigwedge^3 U \otimes \bigwedge^3 U$ is given by

$$q(\theta + \xi \otimes u) = \xi \otimes (u \wedge \theta)$$

for $\theta \in \bigwedge^2 U$, $\xi \in \bigwedge^3 U$ and $u \in U$. Picking a nonzero element $\varepsilon \in \bigwedge^3 U$, we identify $\bigwedge^3 U \otimes \bigwedge^3 U$ with F by means of the basis $\varepsilon \otimes \varepsilon$; we may thus regard q as a quadratic form on $\bigwedge^2 U \oplus (\bigwedge^3 U \otimes U)$. The discriminant algebra of $C(A,\sigma,f)$ then has the alternate description

$$\left(D\big(C(A,\sigma,f),\underline{\sigma}\big),\underline{\sigma},f_D\right) = \left(\mathrm{End}_F\big(\textstyle\bigwedge^2 U \oplus (\bigwedge^3 U \otimes U)\big),\sigma_q,f_q\right).$$

In order to define an isomorphism of this algebra with (A,σ,f), it suffices, by (12.36), to define a similitude of quadratic spaces

$$g\colon (U^* \oplus U, q_U) \to \big(\textstyle\bigwedge^2 U \oplus (\bigwedge^3 U \otimes U), q\big).$$

For $\varphi \in U^*$ and $u \in U$, we set

$$g(\varphi + u) = \theta + \varepsilon \otimes u,$$

where $\theta \in \bigwedge^2 U$ is such that $\varepsilon \cdot \langle \varphi, x \rangle = \theta \wedge x$ for all $x \in U$. We then have

$$\varepsilon \otimes \varepsilon \cdot \langle \varphi, u \rangle = \varepsilon \otimes (u \wedge \theta) = q(\theta + \varepsilon \otimes u),$$

hence g is an isometry of quadratic spaces. We claim that the inverse of the induced isomorphism

$$g_* \colon \left(\mathrm{End}_F(U^* \oplus U), \sigma_q \right) \xrightarrow{\sim} \left(\mathrm{End}_F\left(\bigwedge^2 U \oplus (\bigwedge^3 U \otimes U) \right), \sigma_b \right)$$

extends $\overline{\lambda}$. To prove the claim, we use the identifications

$$D\left(C(A, \sigma, f), \underline{\sigma} \right) = \mathrm{End}_F\left(\bigwedge^2 U \oplus (\bigwedge^3 U \otimes U) \right)$$
$$= \left(\bigwedge^2 U \oplus (\bigwedge^3 U \otimes U) \right) \otimes \left(\bigwedge^2 U \oplus (\bigwedge^3 U \otimes U) \right)$$

and

$$A = \mathrm{End}_F(U^* \oplus U) = (U^* \oplus U) \otimes (U^* \oplus U).$$

Since $c(A) = (U^* \oplus U) \cdot (U^* \oplus U) \subset C_0\left(\mathbb{H}(U), q_U \right)$ is spanned by elements of the form $(\varphi + u) \cdot (\psi + v)$ with $\varphi, \psi \in U^*$ and $u, v \in U$, it suffices to show that the corresponding element $(d_\varphi + \ell_u) \circ (d_\psi + \ell_v) \in \mathrm{End}_F(\bigwedge_0 U) \times \mathrm{End}_F(\bigwedge_1 U) = C_0\left(\mathbb{H}(U), q_U \right)$ (under the isomorphism of (8.3)) is $\lambda^*\left(g(\varphi + u) \otimes g(\psi + v) \right)$. This amounts to verifying that

$$\mathrm{tr}\left(h \circ (d_\varphi + \ell_u) \circ (d_\psi + \ell_v) \right) = \mathrm{tr}\left(\dot{\lambda}^2 h \circ g(\varphi + u) \otimes g(\psi + v) \right)$$

for all $h \in \mathrm{End}_F(\bigwedge_0 U) \times \mathrm{End}_F(\bigwedge_1 U)$. Details are left to the reader. $\square$

(15.25) Remark. For $(B, \tau) \in A_3$, the Lie isomorphism $\lambda^* \colon c\left(D(B, \tau) \right) \to \mathfrak{s}(B, \tau)$ restricts to a Lie isomorphism $c\left(D(B, \tau) \right)_0 \xrightarrow{\sim} \mathrm{Skew}(B, \tau)^0$. The inverse of this isomorphism is $\frac{1}{2} c \circ \lambda^2$ if $\mathrm{char}\, F \neq 2$. Similarly, for $(A, \sigma, f) \in D_3$, the inverse of the Lie isomorphism $\overline{\lambda}$ used in the proof of the theorem above is $\dot{\lambda}^2 \circ \frac{1}{2} c$ if $\mathrm{char}\, F \neq 2$.

(15.26) Corollary. *For every central simple algebra A of degree 6 with quadratic pair (σ, f), the functor $\mathbf{C}$ induces an isomorphism of groups*

$$\mathrm{PGO}(A, \sigma, f) = \mathrm{Aut}_F(A, \sigma, f) \xrightarrow{\sim} \mathrm{Aut}_F\left(C(A, \sigma, f), \underline{\sigma} \right)$$

which restricts into an isomorphism of groups:

$$\mathrm{PGO}^+(A, \sigma, f) \xrightarrow{\sim} \mathrm{Aut}_{Z(A, \sigma, f)}\left(C(A, \sigma, f), \underline{\sigma} \right) = \mathrm{PGU}\left(C(A, \sigma, f), \underline{\sigma} \right)$$

where $Z(A, \sigma, f)$ is the center of $C(A, \sigma, f)$.

Proof: The first isomorphism follows from the fact that $\mathbf{C}$ defines an equivalence of groupoids $D_3 \to A_3$ (see (12.37)). Under this isomorphism, the proper similitudes correspond to automorphisms of $C(A, \sigma, f)$ which restrict to the identity on the center $Z(A, \sigma, f)$, by (13.2). $\square$

We thus recover the fourth case ($\deg A = 6$) of (15.1).

Clifford groups. Let $(A, \sigma, f) \in D_3$ and $(B, \tau) \in A_3$. Let K be the center of (B, τ), and assume that (A, σ, f) and (B, τ) correspond to each other under the groupoid equivalence $A_3 \equiv D_3$, so that we may identify $(B, \tau) = \big(C(A, \sigma, f), \underline{\sigma}\big)$ and $(A, \sigma, f) = \big(D(B, \tau), \underline{\tau}, f_D\big)$. Our goal is to relate the Clifford groups of (A, σ, f) to groups of similitudes of (B, τ). We write μ_σ and μ_τ for the multiplier maps for the involutions σ and τ respectively.

(15.27) Proposition. *The extended Clifford group of (A, σ, f) is the group of similitudes of (B, τ), i.e.,*

$$\Omega(A, \sigma, f) = \mathrm{GU}(B, \tau),$$

and the following diagram commutes:

(15.28)
$$
\begin{array}{ccc}
\Omega(A, \sigma, f) & \xrightarrow{\ \chi'\ } & \mathrm{PGO}^+(A, \sigma, f) \\
\| & & \downarrow \\
\mathrm{GU}(B, \tau) & \xrightarrow{\ D\ } & \mathrm{Aut}(A, \sigma, f)
\end{array}
$$

where D is the canonical map of §14.C. The Clifford group of (A, σ, f) is

$$\Gamma(A, \sigma, f) = \mathrm{SGU}(B, \tau) = \{\, g \in \mathrm{GU}(B, \tau) \mid \mathrm{Nrd}_B(g) = \mu_\tau(g)^2 \,\}$$

and the following diagram commutes:

(15.29)
$$
\begin{array}{ccc}
\Gamma(A, \sigma, f) & \xrightarrow{\ \chi\ } & \mathrm{O}^+(A, \sigma, f) \\
\| & & \downarrow \\
\mathrm{SGU}(B, \tau) & \xrightarrow{\ \lambda'\ } & \mathrm{O}(A, \sigma, f)
\end{array}
$$

where $\lambda'(g) = \mu_\tau(g)^{-1} \lambda^2 g \in D(B, \tau) = A$ for $g \in \mathrm{SGU}(B, \tau)$. Moreover,

$$\mathrm{Spin}(A, \sigma, f) = \mathrm{SU}(B, \tau) = \{\, g \in \mathrm{GU}(B, \tau) \mid \mathrm{Nrd}_B(g) = \mu_\tau(g) = 1 \,\}.$$

Proof: Since (15.26) shows that the canonical map

$$C \colon \mathrm{PGO}^+(A, \sigma, f) \to \mathrm{Aut}_K(B, \tau)$$

is surjective, it follows from the definition of the extended Clifford group in (13.18) that $\Omega(A, \sigma, f) = \mathrm{GU}(B, \tau)$. By the definition of χ', the following diagram commutes:

$$
\begin{array}{ccc}
\Omega(A, \sigma, f) & \xrightarrow{\ \chi'\ } & \mathrm{PGO}^+(A, \sigma, f) \\
\| & & \downarrow{\scriptstyle C} \\
\mathrm{GU}(B, \tau) & \xrightarrow{\ \mathrm{Int}\ } & \mathrm{Aut}_K(B, \tau).
\end{array}
$$

The commutativity of (15.28) follows, since the inverse of

$$C \colon \mathrm{PGO}^+(A, \sigma, f) \xrightarrow{\ \sim\ } \mathrm{Aut}_K(B, \tau)$$

is given by the canonical map D; indeed, the groupoid equivalence $A_3 \equiv D_3$ is given by the Clifford and discriminant algebra constructions.

To identify $\Gamma(A, \sigma, f)$, it suffices to prove that the homomorphism

$$\varkappa \colon \Omega(A, \sigma, f) \to K^\times / F^\times$$

whose kernel is $\Gamma(A, \sigma, f)$ (see (13.21)) coincides with the homomorphism

$$\nu \colon \mathrm{GU}(B, \tau) \to K^\times / F^\times$$

whose kernel is $\mathrm{SGU}(B, \tau)$ (see (14.2)). The description of $\mathrm{Spin}(A, \sigma, f)$ also follows, since $\mathrm{Spin}(A, \sigma, f) = \Gamma(A, \sigma, f) \cap \mathrm{U}(B, \tau)$. The following lemma therefore completes the proof:

(15.30) Lemma. *Diagram (15.29) and the following diagram are commutative:*

$$\begin{array}{ccc}
\Omega(A, \sigma, f) & \xrightarrow{\;\varkappa\;} & K^\times/F^\times \\
\| & & \| \\
\mathrm{GU}(B, \tau) & \xrightarrow{\;\nu\;} & K^\times/F^\times.
\end{array}$$

Proof: It suffices to prove commutativity of the diagrams over a scalar extension. We may thus assume that the base field F is algebraically closed.

Let V be a 4-dimensional vector space over F. Pick a nonzero element $e \in \bigwedge^4 V$ to identify $\bigwedge^4 V = F$ and view the canonical quadratic map $q \colon \bigwedge^2 V \to \bigwedge^4 V$ of (10.12) as a quadratic form. Since F is algebraically closed, we have $(A, \sigma, f) \simeq \left(\mathrm{End}_F(\bigwedge^2 V), \sigma_q, f_q\right)$ where (σ_q, f_q) is the quadratic pair associated with q. We fix such an isomorphism and use it to identify until the end of the proof

$$(A, \sigma, f) = \left(\mathrm{End}_F(\textstyle\bigwedge^2 V), \sigma_q, f_q\right).$$

The map $i \colon \bigwedge^2 V \to \mathrm{End}_F(V \oplus V^*)$ defined in (15.23) induces an isomorphism

$$i_* \colon C(\textstyle\bigwedge^2 V, q) \xrightarrow{\sim} \mathrm{End}_F(V \oplus V^*)$$

which identifies the Clifford algebra $B = C(A, \sigma, f) = C_0(\bigwedge^2 V, q)$ with $\mathrm{End}_F(V) \times \mathrm{End}_F(V^*)$. The involution τ is then given by

$$\tau(f_1, f_2^t) = (f_2, f_1^t)$$

for f_1, $f_2 \in \mathrm{End}_F(V)$. Therefore,

$$\mathrm{GU}(B, \tau) = \{\, \left(f, \rho(f^{-1})^t\right) \mid \rho \in F^\times, f \in \mathrm{GL}(V) \,\}.$$

For $g = \left(f, \rho(f^{-1})^t\right) \in \mathrm{GU}(B, \tau)$, we consider

$$f \wedge f \in \mathrm{End}_F(\textstyle\bigwedge^2 V) = A \quad \text{and} \quad \gamma = \left(f^2, \det f (f^{-2})^t\right) \in \mathrm{GU}(B, \tau).$$

A computation shows that

$$f \wedge f \in \mathrm{GO}^+(\textstyle\bigwedge^2 V, q) = \mathrm{GO}^+(A, \sigma, f), \quad \gamma \in \Gamma^+(\textstyle\bigwedge^2 V, q) = \Gamma(A, \sigma, f)$$

and moreover

$$\mathrm{Int}(g) = C(f \wedge f), \quad \chi(\gamma) = \mu(f \wedge f)^{-1}(f \wedge f)^2.$$

Therefore, $\varkappa(g) = z \cdot F^\times$ where $z = \left(1, \rho^2(\det f)^{-1}\right) \in K^\times = F^\times \times F^\times$ is such that $g^2 = z \cdot \gamma$. On the other hand, we have

$$\mu_\tau(g)^{-2} \mathrm{Nrd}_B(g) = \left(\rho^{-2}\det f, \rho^2(\det f)^{-1}\right) = z\iota(z)^{-1},$$

hence $\nu(g) = z \cdot F^\times = \varkappa(g)$.

It remains only to prove the commutativity of (15.29). Since $\nu = \varkappa$, we have $\Gamma(A, \sigma, f) = \mathrm{SGU}(B, \tau)$. Therefore, every element in $\Gamma(A, \sigma, f)$ has the form $\left(f, \rho(f^{-1})^t\right)$ for some $\rho \in F^\times$ and some $f \in \mathrm{GL}(V)$. A computation yields

$$\begin{pmatrix} f & 0 \\ 0 & \rho(f^{-1})^t \end{pmatrix} \cdot i_*(\xi) \cdot \begin{pmatrix} f^{-1} & 0 \\ 0 & \rho^{-1}f^t \end{pmatrix} = \rho^{-1} i_*\left(f \wedge f(\xi)\right) \quad \text{for } \xi \in \textstyle\bigwedge^2 V,$$

hence $\chi\left(f, \rho(f^{-1})^t\right) = \rho^{-1}f \wedge f$ and (15.29) commutes. $\qquad\square$

(15.31) Corollary. *For* $(A, \sigma, f) \in D_3$, *the group of multipliers of proper similitudes and the group of spinor norms of* (A, σ, f) *are given by*

$$G^+(A, \sigma, f) = \{ N_{K/F}(z) \mid z\iota(z)^{-1} = \mu_\tau(g)^{-2} \operatorname{Nrd}_B(g) \text{ for some } g \in \operatorname{GU}(B, \tau) \}$$

and

$$\operatorname{Sn}(A, \sigma, f) = \{ \mu_\tau(g) \mid g \in \operatorname{SGU}(B, \tau) \}.$$

Proof: In view of the description of $\Omega(A, \sigma, f)$ and $\varkappa$ above, it follows from (13.24) that $\overline{\mu}_\sigma(\operatorname{PGO}^+(A, \sigma, f)) = N_{K/F} \circ \nu(\operatorname{GU}(B, \tau))$, proving the first relation. The second relation follows from the description of $\Gamma(A, \sigma, f)$ in (15.27). $\square$

The case of trivial discriminant. If K/F is a given étale quadratic extension, the functors **D** and **C** of (15.24) relate algebras with involution $(B, \tau) \in A_3$ with center K and algebras with involutions $(A, \sigma, f) \in D_3$ whose Clifford algebra has center $Z(A, \sigma, f) \simeq K$. In order to make explicit the special case where $K = F \times F$, let ${}^1A_3$ be the full subgroupoid of A_3 whose objects are algebras of degree 4 over $F \times F$ with involution of the second kind and let ${}^1D_3$ be the full subgroupoid of D_3 whose objects are algebras of degree 6 with quadratic pair of trivial discriminant. Every $(B, \tau) \in {}^1A_3$ is isomorphic to an algebra of the form $(E \times E^{\mathrm{op}}, \varepsilon)$ where E is a central simple F-algebra of degree 4 and ε is the exchange involution, hence ${}^1A_3$ is also equivalent to the groupoid of algebras of the form $(E \times E^{\mathrm{op}}, \varepsilon)$. Since

$$\big(D(E \times E^{\mathrm{op}}, \varepsilon), \underline{\varepsilon}, f_D\big) = (\lambda^2 E, \gamma, f)$$

where (γ, f) is the canonical quadratic pair on $\lambda^2 E$ (see (10.18) if char $F = 2$), the following is a special case of (15.24):

(15.32) Corollary. *The Clifford algebra functor* $\mathbf{C} \colon {}^1D_3 \to {}^1A_3$ *and the functor* $\mathbf{D} \colon {}^1A_3 \to {}^1D_3$, *which maps* $\big(E \times E^{\mathrm{op}}, \varepsilon\big)$ *to* $(\lambda^2 E, \gamma, f)$, *define an equivalence of groupoids*

$$ {}^1A_3 \equiv {}^1D_3.$$

In particular, for every central simple F-*algebra* E *of degree 4,*

$$C(\lambda^2 E, \gamma, f) \simeq (E \times E^{\mathrm{op}}, \varepsilon).$$

Observe that the maps in ${}^1A_3$ are isomorphisms of algebras over F, not over $F \times F$. In particular, $\big(E \times E^{\mathrm{op}}, \varepsilon\big)$ and $\big(E^{\mathrm{op}} \times E, \varepsilon\big)$ are isomorphic in ${}^1A_3$, under the map which interchanges the two factors. Therefore, ${}^1A_3$ is *not* equivalent to the groupoid of central simple F-algebras of degree 4 where the maps are the F-algebra isomorphisms. There is however a correspondence between isomorphism classes which we now describe.

For $(A, \sigma, f) \in {}^1D_3$, the Clifford algebra $C(A, \sigma, f)$ decomposes into a direct product

$$C(A, \sigma, f) = C^+(A, \sigma, f) \times C^-(A, \sigma, f)$$

for some central simple F-algebras $C^+(A, \sigma, f)$, $C^-(A, \sigma, f)$ of degree 4. The fundamental relations (9.15) and (9.16) show that $C^+(A, \sigma, f) \simeq C^-(A, \sigma, f)^{\mathrm{op}}$ and $C^+(A, \sigma, f)^{\otimes 2} \simeq C^-(A, \sigma, f)^{\otimes 2} \sim A$.

If (V, q) is a quadratic space of dimension 6 and trivial discriminant, we also let $C^\pm(V, q)$ denote $C^\pm\big(\operatorname{End}_F(V), \sigma_q, f_q\big)$. The algebras $C^+(V, q)$ and $C^-(V, q)$ are isomorphic central simple F-algebras of degree 4 and exponent 2.

(15.33) Corollary. *Every central simple F-algebra of degree 4 and exponent 2 is of the form $C^\pm(V, q)$ for some quadratic space (V, q) of dimension 6 and trivial discriminant, uniquely determined up to similarity.*

Every central simple F-algebra of degree 4 and exponent 4 is of the form $C^\pm(A, \sigma, f)$ for some $(A, \sigma, f) \in {}^1D_3$ such that $\operatorname{ind} A = 2$, uniquely determined up to isomorphism.

Proof: For every central simple F-algebra E of degree 4, we have $E \times E^{\mathrm{op}} \simeq C^+(\lambda^2 E, \gamma, f) \times C^-(\lambda^2 E, \gamma, f)$ by (15.32), hence

$$E \simeq C^\pm(\lambda^2 E, \gamma, f).$$

Since $\lambda^2 E$ is Brauer-equivalent to $E^{\otimes 2}$, it is split if E has exponent 2 and has index 2 if E has exponent 4, by (10.5). Moreover, if $E \simeq C^\pm(A, \sigma, f)$ for some $(A, \sigma, f) \in {}^1D_3$, then $(E \times E^{\mathrm{op}}, \varepsilon) \simeq \big(C(A, \sigma, f), \underline{\sigma}\big)$ since all involutions on $E \times E^{\mathrm{op}}$ are isomorphic to the exchange involution. Therefore, by (15.32), we have

$$(A, \sigma, f) \simeq \big(D(E \times E^{\mathrm{op}}, \varepsilon), \underline{\varepsilon}, f_D\big) \simeq (\lambda^2 E, \gamma, f).$$

To complete the proof, observe that when $A = \operatorname{End}_F(V)$ we have $(\sigma, f) = (\sigma_q, f_q)$ for some quadratic form q, and the quadratic space (V, q) is determined up to similarity by the algebra with quadratic pair (A, σ, f) by (12.36). $\qquad\square$

Corollaries (15.26) and (15.31), and Proposition (15.27), can also be specialized to the case where the discriminant of (A, σ, f) is trivial. In particular, (15.31) simplifies remarkably:

(15.34) Corollary. *Let $(A, \sigma, f) \in {}^1D_3$ and let $C(A, \sigma, f) \simeq E \times E^{\mathrm{op}}$. The multipliers of similitudes of (A, σ, f) are given by*

$$G(A, \sigma, f) = G^+(A, \sigma, f) = F^{\times 2} \cdot \operatorname{Nrd}_E(E^\times)$$

and the spinor norms of (A, σ, f) by

$$\operatorname{Sn}(A, \sigma, f) = \{\, \rho \in F^\times \mid \rho^2 \in \operatorname{Nrd}_E(E^\times) \,\}.$$

Proof: The equality $G(A, \sigma, f) = G^+(A, \sigma, f)$ follows from the hypothesis that $\operatorname{disc}(A, \sigma, f)$ is trivial by (13.44). Since the canonical involution $\underline{\sigma}$ on $C(A, \sigma, f)$ is the exchange involution, we have under the identification $C(A, \sigma, f) = E \times E^{\mathrm{op}}$ that

$$\operatorname{GU}\big(C(A, \sigma, f), \underline{\sigma}\big) = \{\, \big(x, \rho(x^{-1})^{\mathrm{op}}\big) \mid \rho \in F^\times,\ x \in E^\times \,\},$$

and, for $g = \big(x, \rho(x^{-1})^{\mathrm{op}}\big)$,

$$\underline{\mu}(g)^{-2} \operatorname{Nrd}_{C(A,\sigma,f)}(g) = \big(\rho^{-2} \operatorname{Nrd}_E(x),\ \rho^2 \operatorname{Nrd}_E(x)^{-1}\big) = z\iota(z)^{-1}$$

with $z = \big(\operatorname{Nrd}_E(x), \rho^2\big) \in Z(A, \sigma, f) = F \times F$. Since $N_{Z(A,\sigma,f)/F}(z) = \rho^2 \operatorname{Nrd}_E(x)$, Corollary (15.31) yields the equality $G^+(A, \sigma, f) = F^{\times 2} \cdot \operatorname{Nrd}_E(E^\times)$. Finally, we have by (15.27):

$$\Gamma(A, \sigma, f) = \operatorname{SGU}\big(C(A, \sigma, f), \underline{\sigma}\big)$$
$$= \{\, \big(x, \rho(x^{-1})^{\mathrm{op}}\big) \in \operatorname{GU}\big(C(A, \sigma, f), \underline{\sigma}\big) \mid \rho^2 = \operatorname{Nrd}_E(x) \,\},$$

hence

$$\operatorname{Sn}(A, \sigma, f) = \underline{\mu}\big(\operatorname{SGU}\big(C(A, \sigma, f), \underline{\sigma}\big)\big) = \{\, \rho \in F^\times \mid \rho^2 \in \operatorname{Nrd}_E(E^\times) \,\}. \qquad\square$$

Examples. In this subsection, we explicitly determine the algebra with involution $(B, \tau) \in A_3$ corresponding to $(A, \sigma, f) \in D_3$ when the quadratic pair (σ, f) is isotropic. Since the correspondence is bijective, our computations also yield information on the discriminant algebra of some $(B, \tau) \in A_3$, which will be crucial for relating the indices of (A, σ, f) and (B, τ) in (15.38) and (15.39) below.

(15.35) Example. Let $(A, \sigma, f) = \left(\operatorname{End}_F(V), \sigma_q, f_q \right)$ where (V, q) is a 6-dimensional quadratic space over a field F of characteristic different from 2, and suppose q is isotropic. Suppose that $\operatorname{disc} q = \alpha \cdot F^{\times 2}$, so that the center of $C_0(V, q)$ is isomorphic to $F[X]/(X^2 - \alpha)$. Then multiplying q by a suitable scalar, we may assume that q has a diagonalization of the form

$$q = \langle 1, -1, \alpha, -\beta, -\gamma, \beta\gamma \rangle$$

for some β, $\gamma \in F^\times$. Let $(e_1, \ldots, e_6)$ be an orthogonal basis of V which yields the diagonalization above. In $C_0(V, q)$, the elements $e_1 \cdot e_4$ and $e_1 \cdot e_5$ generate a quaternion algebra $(\beta, \gamma)_F$. The elements $e_1 \cdot e_4 \cdot e_5 \cdot e_6$ and $e_1 \cdot e_4 \cdot e_5 \cdot e_2$ centralize this algebra and generate a split quaternion algebra $\left((\beta\gamma)^2, -\beta\gamma \right)_F$; therefore,

$$C_0(V, q) \simeq M_2\left((\beta, \gamma)_F \right) \otimes F[X]/(X^2 - \alpha)$$

by the double centralizer theorem (see (1.5)), and Proposition (8.5) shows that the canonical involution τ_0 on $C_0(V, q)$ is hyperbolic.

There is a corresponding result in characteristic 2: if the nonsingular 6-dimensional quadratic space (V, q) is isotropic, we may assume (after scaling) that

$$q = [0, 0] \perp [1, \alpha + \beta] \perp \langle \gamma \rangle[1, \beta] = [0, 0] \perp [1, \alpha + \beta] \perp [\gamma, \beta\gamma^{-1}]$$

for some α, $\beta \in F$, $\gamma \in F^\times$. Thus, $\operatorname{disc} q = \alpha + \wp(F)$, hence the center of $C_0(V, q)$ is isomorphic to $F[X]/(X^2 + X + \alpha)$. Let $(e_1, \ldots, e_6)$ be a basis of V which yields the decomposition above. In $C_0(V, q)$, the elements $r = (e_1 + e_2) \cdot e_3$ and $s = (e_1 + e_2) \cdot e_4$ satisfy $r^2 = 1$, $s^2 = \alpha + \beta$ and $rs + sr = 1$, hence they generate a split quaternion algebra $\left((1, \alpha + \beta) \right)_F$ (see §2.C). The elements $(e_1 + e_2) \cdot e_4$ and $(e_1 + e_2) \cdot e_5$ centralize this algebra and generate a quaternion algebra $\left((\gamma, \beta\gamma^{-1}) \right)_F \simeq [\beta, \gamma)_F$. Therefore,

$$C_0(V, q) \simeq M_2\left([\beta, \gamma)_F \right) \otimes F[X]/(X^2 + X + \alpha).$$

As above, Proposition (8.5) shows that the canonical involution τ_0 is hyperbolic.

(15.36) Corollary. *Let $(B, \tau) \in A_3$, with τ hyperbolic.*

(1) *Suppose the center $Z(B)$ is a field, hence B is Brauer-equivalent to a quaternion algebra (so $\operatorname{ind} B = 1$ or 2); then the discriminant algebra $D(B, \tau)$ splits and its canonical quadratic pair $(\underline{\tau}, f_D)$ is associated with an isotropic quadratic form q. The Witt index of q is 1 if $\operatorname{ind} B = 2$; it is 2 if $\operatorname{ind} B = 1$.*

(2) *Suppose $Z(B) \simeq F \times F$, so that $B \simeq E \times E^{\mathrm{op}}$ for some central simple F-algebra E of degree 4. If E is Brauer-equivalent to a quaternion algebra, then $D(B, \tau)$ splits and its canonical quadratic pair $(\underline{\tau}, f_D)$ is associated with an isotropic quadratic form q. The Witt index of q is 1 if $\operatorname{ind} E = 2$; it is 2 if $\operatorname{ind} E = 1$.*

Proof: (1) Since B has an involution of the second kind, Proposition (2.22) shows that the Brauer-equivalent quaternion algebra has a descent to F. We may thus assume that

$$B \simeq \begin{cases} M_2\left((\beta, \gamma)_F \right) \otimes F(\sqrt{\alpha}) & \text{for some } \alpha, \beta, \gamma \in F^\times \text{ if char } F \neq 2, \\ M_2\left([\beta, \gamma)_F \right) \otimes F\left(\wp^{-1}(\alpha) \right) & \text{for some } \alpha, \beta \in F, \gamma \in F^\times \text{ if char } F = 2, \end{cases}$$

hence, by (15.35), $(B, \tau) \simeq \big(C_0(V, q), \tau_0\big)$ where

$$q \simeq \begin{cases} \langle 1, -1, \alpha, -\beta, -\gamma, \beta\gamma \rangle & \text{if char } F \neq 2; \\ [0, 0] \perp [1, \alpha + \beta] \perp \langle \gamma \rangle [1, \beta] & \text{if char } F = 2. \end{cases}$$

If $w(V, q) = 2$, then Corollary (8.30) shows that $1 \in \text{ind}\big(C_0(V, q), \tau_0\big)$, hence B is split. Conversely, if B is split, then we may assume that $\gamma = 1$, and it follows that $w(V, q) = 2$.

(2) The hypothesis yields

$$B \simeq \begin{cases} M_2\big((\beta, \gamma)_F\big) \otimes F[X]/(X^2 - 1) & \text{for some } \beta, \gamma \in F^\times \text{ if char } F \neq 2, \\ M_2\big([\beta, \gamma)_F\big) \otimes F[X]/(X^2 - X) & \text{for some } \beta \in F, \gamma \in F^\times \text{ if char } F = 2, \end{cases}$$

hence $(B, \tau) \simeq \big(C_0(V, q), \tau_0\big)$ where

$$q \simeq \begin{cases} \langle 1, -1, 1, -\beta, -\gamma, \beta\gamma \rangle & \text{if char } F \neq 2; \\ [0, 0] \perp \langle 1, \gamma \rangle [1, \beta] & \text{if char } F = 2. \end{cases}$$

Since q is the orthogonal sum of a hyperbolic plane and the norm form of the quaternion algebra Brauer-equivalent to E, we have $w(V, q) = 1$ if and only if $\text{ind } E = 2$. $\square$

We next consider the case where the algebra A is not split. Since $\deg A = 6$, we must have $\text{ind } A = 2$, by (2.8). We write $Z(A, \sigma, f)$ for the center of the Clifford algebra $C(A, \sigma, f)$.

(15.37) Proposition. *Let* $(A, \sigma, f) \in D_3$ *with* $\text{ind } A = 2$.

(1) *If the quadratic pair* (σ, f) *is isotropic, then* $Z(A, \sigma, f)$ *is a splitting field of* A.

(2) *For each separable quadratic splitting field* Z *of* A, *there is, up to conjugation, a unique quadratic pair* (σ, f) *on* A *such that* $Z(A, \sigma, f) \simeq Z$. *If* $d \in F^\times$ *is such that the quaternion algebra Brauer-equivalent to* A *has the form* $(Z, d)_F$, *then* $C(A, \sigma, f) \simeq M_4(Z)$ *and the canonical involution* $\underline{\sigma}$ *is the adjoint involution with respect to the 4-dimensional hermitian form on* Z *with diagonalization* $\langle 1, -1, 1, -d \rangle$.

Proof: (1) Let $I \subset A$ be a nonzero isotropic right ideal. We have $\text{rdim } I \geq \frac{1}{2} \deg A = 3$, hence $\text{rdim } I = 2$ since $\text{ind } A$ divides the reduced dimension of every right ideal. Let e be an idempotent such that $I = eA$. As in the proof of (6.7), we may assume that $e\sigma(e) = \sigma(e)e = 0$, hence $e + \sigma(e)$ is an idempotent. Let $e_1 = e + \sigma(e)$ and $e_2 = 1 - e_1$; then $e_1 A = eA \oplus \sigma(e)A$, hence $\text{rdim } e_1 A = 4$ and therefore $\text{rdim } e_2 A = 2$. Let $A_i = e_i A e_i$ and let (σ_i, f_i) be the restriction of the quadratic pair (σ, f) to A_i for $i = 1$, 2. By (1.13), we have $\deg A_1 = 4$ and $\deg A_2 = 2$, hence A_2 is a quaternion algebra Brauer-equivalent to A. Moreover, by (7.5) (if char $F \neq 2$) or (7.14) (if char $F = 2$),

$$\text{disc}(\sigma, f) = \begin{cases} \text{disc}(\sigma_1, f_1)\,\text{disc}(\sigma_2, f_2) & \text{if char } F \neq 2, \\ \text{disc}(\sigma_1, f_1) + \text{disc}(\sigma_2, f_2) & \text{if char } F = 2. \end{cases}$$

Since eAe_1 is an isotropic right ideal of reduced dimension 2 in A_1, the quadratic pair (σ_1, f_1) is hyperbolic, and Proposition (7.9) shows that its discriminant is trivial. Therefore, $\text{disc}(\sigma, f) = \text{disc}(\sigma_2, f_2)$, hence $Z(A, \sigma, f) \simeq Z(A_2, \sigma_2, f_2)$. If char $F \neq 2$, it was observed in (7.4) that $Z(A_2, \sigma_2, f_2)$ splits A_2, hence $Z(A, \sigma, f)$ splits A. To see that the same property holds if char $F = 2$, pick $\ell \in A_2$ such that $f_2(s) = \text{Trd}_{A_2}(\ell s)$ for all $s \in \text{Sym}(A_2, \sigma_2)$; then $\text{Trd}_{A_2}(\ell) = 1$ and $\text{Srd}_{A_2}(\ell) =$

$\mathrm{Nrd}_{A_2}(\ell)$ represents $\mathrm{disc}(\sigma_2, f_2)$ in $F/\wp(F)$, so $Z(A_2, \sigma_2, f_2) \simeq F(\ell)$. This completes the proof of (1).

(2) Let Z be a separable quadratic splitting field of A and let $d \in F^\times$ be such that A is Brauer-equivalent to the quaternion algebra $(Z, d)_F$, which we denote simply by Q. We then have $A \simeq M_3(Q)$. To prove the existence of an isotropic quadratic pair (σ, f) on A such that $Z(A, \sigma, f) \simeq Z$, start with a quadratic pair (θ, f_1) on Q such that $Z(Q, \theta, f_1) \simeq Z$, and let $(\sigma, f) = (\theta \otimes \rho, f_{1*})$ on $A = Q \otimes M_3(F)$, where ρ is the adjoint involution with respect to an isotropic 3-dimensional bilinear form. We may choose for instance $\rho = \mathrm{Int}(u) \circ t$ where

$$u = \begin{pmatrix} 0 & 1 & 0 \\ 1 & 0 & 0 \\ 0 & 0 & 1 \end{pmatrix};$$

the involution σ is then explicitly defined by

$$\sigma\big((x_{ij})_{1 \le i,j \le 3}\big) = \begin{pmatrix} \theta(x_{22}) & \theta(x_{12}) & \theta(x_{32}) \\ \theta(x_{21}) & \theta(x_{11}) & \theta(x_{31}) \\ \theta(x_{23}) & \theta(x_{13}) & \theta(x_{33}) \end{pmatrix}$$

and the linear form f by

$$f\begin{pmatrix} x_{11} & s_{12} & x_{13} \\ s_{21} & \theta(x_{11}) & x_{23} \\ \theta(x_{23}) & \theta(x_{13}) & s_{33} \end{pmatrix} = \mathrm{Trd}_Q(x_{11}) + f_1(s_{33}),$$

for $x_{11}, x_{13}, x_{23} \in Q$ and $s_{12}, s_{21}, s_{33} \in \mathrm{Sym}(Q, \theta)$.

It is readily verified that

$$I = \left\{ \begin{pmatrix} x_1 & x_2 & x_3 \\ 0 & 0 & 0 \\ 0 & 0 & 0 \end{pmatrix} \ \middle| \ x_1, x_2, x_3 \in Q \right\}$$

is an isotropic right ideal, and that $\mathrm{disc}(A, \sigma, f) = \mathrm{disc}(Q, \theta, f_1)$, hence $Z(A, \sigma, f) \simeq Z$.

Let $(B, \tau) = \big(C(A, \sigma, f), \underline{\sigma}\big)$. Since $\mathrm{rdim}\, I = 2$, we have $2 \in \mathrm{ind}(A, \sigma, f)$, hence Corollary (8.30) yields $1 \in \mathrm{ind}(B, \tau)$. This relation shows that B is split, hence $B \simeq M_4(Z)$, and τ is the adjoint involution with respect to an isotropic 4-dimensional hermitian form h over Z. Multiplying h by a suitable scalar, we may assume that h has a diagonalization $\langle 1, -1, 1, -a \rangle$ for some $a \in F^\times$. Corollary (10.35) then shows that $D(B, \tau)$ is Brauer-equivalent to the quaternion algebra $(Z, a)_F$. Since $D(B, \tau) \simeq A$, we have $(Z, d)_F \simeq (Z, a)_F$, hence $a \equiv d \mod N(Z/F)$ and therefore

$$h \simeq \langle 1, -1, 1, -d \rangle.$$

The same arguments apply to every isotropic quadratic pair (σ, f) on A such that $Z(A, \sigma, f) \simeq Z$: for every such quadratic pair, we have $\big(C(A, \sigma, f), \underline{\sigma}\big) \simeq \big(M_4(Z), \sigma_h\big)$ where $h \simeq \langle 1, -1, 1, -d \rangle$, hence also

$$(A, \sigma, f) \simeq \big(D\big(M_4(Z), \sigma_h\big), \underline{\sigma_h}, f_D\big).$$

This proves uniqueness of the quadratic pair (σ, f) up to conjugation. $\qquad\square$

Indices. Let $(B, \tau) \in A_3$ and $(A, \sigma, f) \in D_3$ correspond to each other under the equivalence $A_3 \equiv D_3$. Let K be the center of B, which is isomorphic to $F(\sqrt{\text{disc}(\sigma, f)})$ if $\text{char } F \neq 2$ and to $F(\wp^{-1}(\text{disc}(\sigma, f)))$ if $\text{char } F = 2$. Our goal is to relate the indices $\text{ind}(A, \sigma, f)$ and $\text{ind}(B, \tau)$. For clarity, we consider the case where $K \simeq F \times F$ separately.

(15.38) Proposition. *Suppose $K \simeq F \times F$, hence $(B, \tau) \simeq (E \times E^{\text{op}}, \varepsilon)$ for some central simple F-algebra E, where ε is the exchange involution. The only possibilities for $\text{ind}(A, \sigma, f)$ are*

$$\{0\}, \qquad \{0, 1\} \quad and \quad \{0, 1, 2, 3\}.$$

Moreover,

$$\text{ind}(A, \sigma, f) = \{0\} \quad \Longleftrightarrow \quad \text{ind}(B, \tau) = \{0\} \quad \Longleftrightarrow \quad \text{ind } E = 4,$$
$$\text{ind}(A, \sigma, f) = \{0, 1\} \quad \Longleftrightarrow \quad \text{ind}(B, \tau) = \{0, 2\} \quad \Longleftrightarrow \quad \text{ind } E = 2,$$
$$\text{ind}(A, \sigma, f) = \{0, 1, 2, 3\} \quad \Longleftrightarrow \quad \text{ind}(B, \tau) = \{0, 1, 2\} \quad \Longleftrightarrow \quad \text{ind } E = 1.$$

Proof: Since $\deg A = 6$, we have $\text{ind}(A, \sigma, f) \subset \{0, 1, 2, 3\}$. If $3 \in \text{ind}(A, \sigma, f)$, then A splits since $\text{ind } A$ is a power of 2 which divides all the integers in $\text{ind}(A, \sigma, f)$. In that case, we have $(A, \sigma, f) \simeq (\text{End}_F(V), \sigma_q, f_q)$ for some hyperbolic quadratic space (V, q), hence $\text{ind}(A, \sigma, f) = \{0, 1, 2, 3\}$.

Since $K \simeq F \times F$, Proposition (15.37) shows that $\text{ind}(A, \sigma, f) \neq \{0, 2\}$. Therefore, if $2 \in \text{ind}(A, \sigma, f)$, we must also have 1 or $3 \in \text{ind}(A, \sigma, f)$, hence, as above, $(A, \sigma, f) \simeq (\text{End}_F(V), \sigma_q, f_q)$ for some quadratic space (V, q) with $w(V, q) \geq 2$. Since $\text{disc}(\sigma, f) = \text{disc } q$ is trivial, the inequality $w(V, q) \geq 2$ implies q is hyperbolic, hence $\text{ind}(A, \sigma, f) = \{0, 1, 2, 3\}$. Therefore, the only possibilities for $\text{ind}(A, \sigma, f)$ are those listed above.

The relations between $\text{ind}(B, \tau)$ and $\text{ind } E$ readily follow from the definition of $\text{ind}(E \times E^{\text{op}}, \varepsilon)$, and the equivalences $\text{ind}(A, \sigma, f) = \{0, 1\} \iff \text{ind } E = 2$ and $\text{ind}(A, \sigma, f) = \{0, 1, 2, 3\} \iff \text{ind } E = 1$ follow from (15.35) and (15.36). $\square$

(15.39) Proposition. *Suppose K is a field. The only possibilities for $\text{ind}(A, \sigma, f)$ are*

$$\{0\}, \qquad \{0, 1\}, \qquad \{0, 2\} \quad and \quad \{0, 1, 2\}.$$

Moreover,

$$\text{ind}(A, \sigma, f) = \{0\} \iff \text{ind}(B, \tau) = \{0\},$$
$$\text{ind}(A, \sigma, f) = \{0, 1\} \iff \text{ind}(B, \tau) = \{0, 2\},$$
$$\text{ind}(A, \sigma, f) = \{0, 2\} \iff \text{ind}(B, \tau) = \{0, 1\},$$
$$\text{ind}(A, \sigma, f) = \{0, 1, 2\} \iff \text{ind}(B, \tau) = \{0, 1, 2\}.$$

Proof: If $3 \in \text{ind}(A, \sigma, f)$, then (σ, f) is hyperbolic, hence its discriminant is trivial, by (7.9). This contradicts the hypothesis that K is a field. Therefore, we have $\text{ind}(A, \sigma, f) \subset \{0, 1, 2\}$.

To prove the correspondence between $\text{ind}(A, \sigma, f)$ and $\text{ind}(B, \tau)$, it now suffices to show that $1 \in \text{ind}(A, \sigma, f)$ if and only if $2 \in \text{ind}(B, \tau)$ and that $2 \in \text{ind}(A, \sigma, f)$ if and only if $1 \in \text{ind}(B, \tau)$.

If $1 \in \mathrm{ind}(A, \sigma, f)$, then A is split, and Proposition (8.5) shows that $2 \in \mathrm{ind}(B, \tau)$. Conversely, if $2 \in \mathrm{ind}(B, \tau)$, then B is Brauer-equivalent to a quaternion algebra and τ is hyperbolic. It follows from (15.36) that $1 \in \mathrm{ind}(A, \sigma, f)$ in that case. If $2 \in \mathrm{ind}(A, \sigma, f)$, then Corollary (8.30) yields $1 \in \mathrm{ind}(B, \tau)$. Conversely, if $1 \in \mathrm{ind}(B, \tau)$, then B splits and τ is the adjoint involution with respect to some isotropic 4-dimensional hermitian form. By (15.37) (if τ is not hyperbolic) or (15.36) (if τ is hyperbolic), it follows that $2 \in \mathrm{ind}(A, \sigma, f)$. $\qquad\square$

(15.40) Remark. The correspondence between $\mathrm{ind}(A, \sigma, f)$ and $\mathrm{ind}(B, \tau)$ may be summarized in the following relations (which hold when $K \simeq F \times F$ as well as when K is a field):

$$1 \in \mathrm{ind}(A, \sigma, f) \iff 2 \in \mathrm{ind}(B, \tau), \qquad 2 \in \mathrm{ind}(A, \sigma, f) \iff 1 \in \mathrm{ind}(B, \tau),$$

$$3 \in \mathrm{ind}(A, \sigma, f) \iff \mathrm{ind}(A, \sigma, f) = \{0, 1, 2, 3\}$$
$$\iff \big(\mathrm{ind}(B, \tau) = \{0, 1, 2\} \text{ and } K \simeq F \times F\big).$$

§16. Biquaternion Algebras

Algebras which are tensor products of two quaternion algebras are called *biquaternion algebras*. Such algebras are central simple of degree 4 and exponent 2 (or 1). Albert proved the converse:

(16.1) Theorem (Albert [3, p. 369]). *Every central simple algebra of degree 4 and exponent 2 is a biquaternion algebra.*

We present three proofs. The first two proofs rely heavily on the results of §15, whereas the third proof, due to Racine, is more self-contained.

Throughout this section, A is a central simple algebra of degree 4 and exponent 1 or 2 over an arbitrary field F.

First proof (based on $A_3 \equiv D_3$): By (15.33), $A \simeq C^{\pm}(V, q)$ for some 6-dimensional quadratic space (V, q) of trivial discriminant. The result follows from the structure of Clifford algebras of quadratic spaces.

Explicitly, if $\mathrm{char}\, F \neq 2$ we may assume (after a suitable scaling) that q has a diagonalization of the form

$$q = \langle a_1, b_1, -a_1 b_1, -a_2, -b_2, a_2 b_2 \rangle$$

for some a_1, b_1, a_2, $b_2 \in F^{\times}$. Let $(e_1, \ldots, e_6)$ be an orthogonal basis of V which yields that diagonalization. The even Clifford algebra has a decomposition

$$C_0(V, q) = Q_1 \otimes_F Q_2 \otimes_F Z$$

where Q_1 is the F-subalgebra generated by $e_1 \cdot e_2$ and $e_1 \cdot e_3$, Q_2 is the F-subalgebra generated by $e_4 \cdot e_5$ and $e_4 \cdot e_6$, and $Z = F \cdot 1 \oplus F \cdot e_1 \cdot e_2 \cdot e_3 \cdot e_4 \cdot e_5 \cdot e_6$ is the center of $C_0(V, q)$. We have $Z \simeq F \times F$, hence $C^+(V, q) \simeq C^-(V, q) \simeq Q_1 \otimes Q_2$. Moreover, $(1, b_1^{-1} e_2 \cdot e_3, a_1^{-1} e_1 \cdot e_3, e_1 \cdot e_2)$ is a quaternion basis of Q_1 which shows $Q_1 \simeq (a_1, b_1)_F$, and $(1, b_2^{-1} e_5 \cdot e_6, a_2^{-1} e_4 \cdot e_6, e_5 \cdot e_6)$ is a quaternion basis of Q_2 which shows $Q_2 \simeq (a_2, b_2)_F$. Therefore, $C^{\pm}(V, q)$ is a biquaternion algebra.

Similar arguments hold when $\mathrm{char}\, F = 2$. We may then assume

$$q = [1, a_1 b_1 + a_2 b_2] \perp [a_1, b_1] \perp [a_2, b_2]$$

for some a_1, b_1, a_2, $b_2 \in F$. Let $(e_1, \ldots, e_6)$ be a basis of V which yields that decomposition. In $C_0(V, q)$, the elements $e_1 \cdot e_3$ and $e_1 \cdot e_4$ (resp. $e_1 \cdot e_5$ and $e_1 \cdot e_6$)

generate a quaternion F-algebra $Q_1 \simeq ((a_1, b_1))_F$ (resp. $Q_2 \simeq ((a_2, b_2))_F$). We have a decomposition

$$C_0(V, q) = Q_1 \otimes_F Q_2 \otimes_F Z$$

where $Z = F \cdot 1 \oplus F \cdot (e_1 \cdot e_2 + e_3 \cdot e_4 + e_5 \cdot e_6)$ is the center of $C_0(V, q)$. Since $Z \simeq F \times F$, it follows that

$$C^+(V, q) \simeq C^-(V, q) \simeq Q_1 \otimes_F Q_2 \simeq ((a_1, b_1))_F \otimes ((a_2, b_2))_F. \qquad \square$$

Second proof (based on $\mathsf{B}_2 \equiv \mathsf{C}_2$): By (3.1) and (2.8), the algebra A carries an involution σ of symplectic type. In the notation of §15, we have $(A, \sigma) \in \mathsf{C}_2$. The proof of the equivalence $\mathsf{B}_2 \equiv \mathsf{C}_2$ in (15.16) shows that A is isomorphic to the even Clifford algebra of some nonsingular 5-dimensional quadratic form:

$$(A, \sigma) \simeq C_0\big(\mathrm{Symd}(A, \sigma)^0, s_\sigma\big).$$

The result follows from the fact that even Clifford algebras of odd-dimensional quadratic spaces are tensor products of quaternion algebras (Scharlau [247, Theorem 9.2.10]). $\qquad \square$

Third proof (Racine [228]): If A is not a division algebra, the theorem readily follows from Wedderburn's theorem (1.1), which yields a decomposition:

$$A \simeq M_2(F) \otimes_F Q$$

for some quaternion algebra Q. We may thus assume that A is a division algebra.

Our first aim is to find in A a separable quadratic extension K of F. By (3.1), A carries an involution σ. If $\mathrm{char}\, F \neq 2$, we may start with any nonzero element $u \in \mathrm{Skew}(A, \sigma)$; then $u^2 \in \mathrm{Sym}(A, \sigma)$, hence $F(u^2) \subsetneq F(u)$. Since $[F(u) : F] = 4$ or 2, we get $[F(u^2) : F] = 2$ or 1 respectively. We choose $K = F(u^2)$ in the first case and $K = F(u)$ in the second case. In arbitrary characteristic, one may choose a symplectic involution σ on A and take for K any proper extension of F in $\mathrm{Symd}(A, \sigma)$ which is not contained in $\mathrm{Symd}(A, \sigma)^0$, by (2.8).

The theorem then follows from the following proposition, which also holds when A is not a division algebra. Recall that for every étale quadratic F-algebra K with nontrivial automorphism ι and for every $a \in F^\times$, the symbol $(K, a)_F$ stays for the quaternion F-algebra $K \oplus Kz$ where multiplication is defined by $zx = \iota(x)z$ for $x \in K$ and $z^2 = a$.

(16.2) Proposition. *Suppose K is an étale quadratic F-algebra contained in a central simple F-algebra A of degree 4 and exponent 2. There exist an $a \in F^\times$ and a quaternion F-algebra Q such that*

$$A \simeq (K, a)_F \otimes Q.$$

Proof: If K is not a field, then A is not a division algebra, hence Wedderburn's theorem (1.1) yields

$$A \simeq (K, 1)_F \otimes Q$$

for some quaternion F-algebra Q.

If K is a field, the nontrivial automorphism ι extends to an involution τ on A by (4.14). The restriction of τ to the centralizer B of K in A is an involution of the second kind. Since B is a quaternion algebra over K, Proposition (2.22) yields a quaternion F-algebra $Q \subset B$ such that

$$B = Q \otimes_F K.$$

By (1.5), there is a decomposition

$$B = Q \otimes_F C_B Q$$

where $C_B Q$ is the centralizer of Q in B. This centralizer is a quaternion algebra which contains K, hence

$$C_B Q \simeq (K, a)_F$$

for some $a \in F^\times$. We thus get the required decomposition. $\square$

16.A. Albert forms. Let A be a biquaternion algebra over a field F of arbitrary characteristic. The algebra $\lambda^2 A$ is split of degree 6 and carries a canonical quadratic pair (γ, f) of trivial discriminant (see (10.18)). Therefore, there are quadratic spaces (V, q) of dimension 6 and trivial discriminant such that

$$(\lambda^2 A, \gamma, f) \simeq (\operatorname{End}_F(V), \sigma_q, f_q)$$

where (σ_q, f_q) is the quadratic pair associated with q.

(16.3) Proposition. *For a biquaternion algebra A and a 6-dimensional quadratic space (V, q) of discriminant 1, the following conditions are equivalent:*

(1) $(\lambda^2 A, \gamma, f) \simeq (\operatorname{End}_F(V), \sigma_q, f_q)$;
(2) $A \times A \simeq C_0(V, q)$;
(3) $M_2(A) \simeq C(V, q)$.

Moreover, if (V, q) and (V', q') are 6-dimensional quadratic spaces of discriminant 1 which satisfy these conditions for a given biquaternion algebra A, then (V, q) and (V', q') are similar, i.e., $q' \simeq \langle \lambda \rangle \cdot q$ for some $\lambda \in F^\times$.

The quadratic forms which satisfy the conditions of this proposition are called *Albert forms* of the biquaternion algebra A (and the quadratic space (V, q) is called an *Albert quadratic space* of A). As the proposition shows, an Albert form is determined only up to similarity by A. By contrast, it is clear from condition (2) or (3) that any quadratic form of dimension 6 and discriminant 1 is an Albert form for some biquaternion algebra A, uniquely determined up to isomorphism.

Proof: (1) $\Rightarrow$ (2) Condition (1) implies that $C(\lambda^2 A, \gamma, f) \simeq C_0(V, q)$. Since (15.32) shows that $C(\lambda^2 A, \gamma, f) \simeq A \times A^{\operatorname{op}}$ and since $A \simeq A^{\operatorname{op}}$, we get (2).

(2) $\Rightarrow$ (1) Since the canonical involution σ_q on $C_0(V, q) = C(\operatorname{End}_F(V), \sigma_q, f_q)$ is of the second kind, we derive from (2):

$$\left(C(\operatorname{End}_F(V), \sigma_q, f_q), \underline{\sigma_q}\right) \simeq (A \times A^{\operatorname{op}}, \varepsilon),$$

where ε is the exchange involution. By comparing the discriminant algebras of both sides, we obtain

$$\left(D\left(C(\operatorname{End}_F(V), \sigma_q, f_q), \underline{\sigma_q}\right), \underline{\sigma_q}, f_D\right) \simeq (\lambda^2 A, \gamma, f).$$

Corollary (15.32) (or Theorem (15.24)) shows that there is a natural transformation $\mathbf{D} \circ \mathbf{C} \cong \operatorname{Id}_{D_3}$, hence

$$\left(D\left(C(\operatorname{End}_F(V), \sigma_q, f_q), \underline{\sigma_q}\right), \underline{\sigma_q}, f_D\right) \simeq (\operatorname{End}_F(V), \sigma_q, f_q)$$

and we get (1).

(2) $\Leftrightarrow$ (3) This follows from the structure of Clifford algebras of quadratic forms: see for instance Lam [169, Ch. 5, Theorem 2.5] if char $F \neq 2$; similar arguments hold in characteristic 2.

Finally, if (V, q) and (V', q') both satisfy (1), then

$$\left(\mathrm{End}_F(V), \sigma_q, f_q\right) \simeq \left(\mathrm{End}_F(V'), \sigma_{q'}, f_{q'}\right),$$

hence (V, q) and (V', q') are similar, by (12.36). $\square$

(16.4) Example. Suppose char $F \neq 2$. For any a_1, b_1, a_2, $b_2 \in F^\times$, the quadratic form

$$q = \langle a_1, b_1, -a_1 b_1, -a_2, -b_2, a_2 b_2 \rangle$$

is an Albert form of the biquaternion algebra $(a_1, b_1)_F \otimes (a_2, b_2)_F$. This follows from the computation of the Clifford algebra $C(q)$. (See the first proof of (16.1); see also (16.24) below.)

Similarly, if char $F = 2$, then for any a_1, b_1, a_2, $b_2 \in F$, the quadratic form

$$[1, a_1 b_1 + a_2 b_2] \perp [a_1, b_1] \perp [a_2, b_2]$$

is an Albert form of the biquaternion algebra $((a_1, b_1))_F \otimes ((a_2, b_2))_F$, and, for a_1, $a_2 \in F$, b_1, $b_2 \in F^\times$, the quadratic form

$$[1, a_1 + a_2] \perp \langle b_1 \rangle [1, a_1] \perp \langle b_2 \rangle [1, a_2]$$

is an Albert form of $[a_1, b_1)_F \otimes [a_2, b_2)_F$.

Albert's purpose in associating a quadratic form to a biquaternion algebra was to obtain a necessary and sufficient quadratic form theoretic criterion for the biquaternion algebra to be a division algebra.

(16.5) Theorem (Albert [2]). *Let A be a biquaternion algebra and let q be an Albert form of A. The (Schur) index of A, $\mathrm{ind}\, A$, and the Witt index $w(q)$ are related as follows:*

$$\mathrm{ind}\, A = 4 \quad \text{if and only if} \quad w(q) = 0;$$

(in other words, A is a division algebra if and only if q is anisotropic);

$$\mathrm{ind}\, A = 2 \quad \text{if and only if} \quad w(q) = 1;$$

$$\mathrm{ind}\, A = 1 \quad \text{if and only if} \quad w(q) = 3;$$

(in other words, A is split if and only if q is hyperbolic).

Proof: This is a particular case of (15.38). $\square$

Another relation between biquaternion algebras and their Albert forms is the following:

(16.6) Proposition. *The multipliers of similitudes of an Albert form q of a biquaternion algebra A are given by*

$$G(q) = F^{\times 2} \cdot \mathrm{Nrd}_A(A^\times)$$

and the spinor norms by

$$\mathrm{Sn}(q) = \{\, \lambda \in F^\times \mid \lambda^2 \in \mathrm{Nrd}_A(A^\times) \,\}.$$

Proof: This is a direct application of (15.34). $\square$

Even though there is no canonical choice for an Albert quadratic space of a biquaternion algebra A, when an involution of the first kind σ on A is fixed, an Albert form may be defined on the vector space $\operatorname{Symd}(A, \sigma)$ of symmetric elements if σ is symplectic and on the vector space $\operatorname{Skew}(A, \sigma)$ if σ is orthogonal and char $F \neq 2$. Moreover, Albert forms may be used to define an invariant of symplectic involutions, as we now show.

16.B. Albert forms and symplectic involutions. Let σ be a symplectic involution on the biquaternion F-algebra A. Recall from §2.A (see (2.10)) that the reduced characteristic polynomial of every symmetrized element is a square:

$$\operatorname{Prd}_{A,s}(X) = \operatorname{Prp}_{\sigma,s}(X)^2 = \left(X^2 - \operatorname{Trp}_\sigma(s)X + \operatorname{Nrp}_\sigma(s)\right)^2 \qquad \text{for } s \in \operatorname{Symd}(A, \sigma).$$

Since $\deg A = 4$, the polynomial map $\operatorname{Nrp}_\sigma \colon \operatorname{Symd}(A, \sigma) \to F$ has degree 2. We show in (16.8) below that $\left(\operatorname{Symd}(A, \sigma), \operatorname{Nrp}_\sigma\right)$ is an Albert quadratic space of A.

A key tool in this proof is the linear endomorphism $\bar{}$ of $\operatorname{Symd}(A, \sigma)$ defined by

$$\overline{x} = \operatorname{Trp}_\sigma(x) - x \quad \text{for } x \in \operatorname{Symd}(A, \sigma).$$

Since $\operatorname{Prp}_{\sigma,x}(x) = 0$ for all $x \in \operatorname{Symd}(A, \sigma)$, we have

$$\operatorname{Nrp}_\sigma(x) = x\overline{x} = \overline{x}x \qquad \text{for } x \in \operatorname{Symd}(A, \sigma).$$

(16.7) Lemma. *For $x \in \operatorname{Symd}(A, \sigma)$ and $a \in A^\times$,*

$$\overline{ax\sigma(a)} = \operatorname{Nrd}_A(a)\sigma(a)^{-1}\overline{x}a^{-1}.$$

Proof: Since both sides of the equality above are linear in x, it suffices to show that the equality holds for x in some basis of $\operatorname{Symd}(A, \sigma)$. It is readily seen by scalar extension to a splitting field that $\operatorname{Symd}(A, \sigma)$ is spanned by invertible elements. Therefore, it suffices to prove the equality for invertible x. In that case, the property follows by comparing the equalities

$$\operatorname{Nrd}_A(a)\operatorname{Nrp}_\sigma(x) = \operatorname{Nrp}_\sigma\left(ax\sigma(a)\right) = ax\sigma(a) \cdot \overline{ax\sigma(a)}$$

and

$$\operatorname{Nrp}_\sigma(x) = ax\sigma(a) \cdot \sigma(a)^{-1}\overline{x}a^{-1}. \qquad \square$$

(16.8) Proposition. *The quadratic space $\left(\operatorname{Symd}(A, \sigma), \operatorname{Nrp}_\sigma\right)$ is an Albert quadratic space of A.*

Proof: For $x \in \operatorname{Symd}(A, \sigma)$, let $i(x) = \left(\begin{smallmatrix} 0 & \overline{x} \\ x & 0 \end{smallmatrix}\right) \in M_2(A)$. Since $x\overline{x} = \overline{x}x = \operatorname{Nrp}_\sigma(x)$, we have $i(x)^2 = \operatorname{Nrp}_\sigma(x)$, hence the universal property of Clifford algebras shows that i induces an F-algebra homomorphism

$$\textbf{(16.9)} \qquad\qquad i_* \colon C\left(\operatorname{Symd}(A, \sigma), \operatorname{Nrp}_\sigma\right) \to M_2(A).$$

This homomorphism is injective since $C\left(\operatorname{Symd}(A, \sigma), \operatorname{Nrp}_\sigma\right)$ is simple, and it is surjective by dimension count. $\qquad \square$

(16.10) Remark. Proposition (15.20) shows that A contains a right ideal of reduced dimension 2 if and only if the quadratic form $\operatorname{Nrp}_\sigma$ is isotropic. Therefore, A is a division algebra if and only if $\operatorname{Nrp}_\sigma$ is anisotropic; in view of (16.8), this observation yields an alternate proof of a (substantial) part of Albert's theorem (16.5).

The isomorphism i_* of (16.9) may also be used to give an explicit description of the similitudes of the Albert quadratic space $\left(\operatorname{Symd}(A, \sigma), \operatorname{Nrp}_\sigma\right)$, thus yielding an alternative proof of the relation between Clifford groups and symplectic similitudes in (15.18).

(16.11) Proposition. *The proper similitudes of* $\big(\mathrm{Symd}(A,\sigma),\mathrm{Nrp}_\sigma\big)$ *are of the form*

$$x \mapsto \lambda^{-1}ax\sigma(a)$$

where $\lambda \in F^\times$ *and* $a \in A^\times$.

The improper similitudes of $\big(\mathrm{Symd}(A,\sigma),\mathrm{Nrp}_\sigma\big)$ *are of the form*

$$x \mapsto \lambda^{-1}a\overline{x}\sigma(a)$$

where $\lambda \in F^\times$ *and* $a \in A^\times$. *The multiplier of these similitudes is* $\lambda^{-2}\mathrm{Nrd}_A(a)$.

Proof: Since $\mathrm{Nrp}_\sigma(\overline{x}) = \mathrm{Nrp}_\sigma(x)$ for all $x \in \mathrm{Sym}(A,\sigma)$, it follows from (2.13) that the maps above are similitudes with multiplier $\lambda^{-2}\mathrm{Nrd}_A(a)$ for all $\lambda \in F^\times$ and $a \in A^\times$.

Conversely, let $f \in \mathrm{GO}\big(\mathrm{Symd}(A,\sigma),\mathrm{Nrp}_\sigma\big)$ be a similitude and let $\alpha = \mu(f)$ be its multiplier. The universal property of Clifford algebras shows that there is an isomorphism

$$f_* \colon C\big(\mathrm{Symd}(A,\sigma),\mathrm{Nrp}_\sigma\big) \to M_2(A)$$

which maps $x \in \mathrm{Symd}(A,\sigma)$ to $\left(\begin{smallmatrix} 0 & \alpha^{-1}\overline{f(x)} \\ f(x) & 0 \end{smallmatrix}\right)$. By comparing f_* with the isomorphism i_* of (16.9), we get an automorphism $f_* \circ i_*^{-1}$ of $M_2(A)$. Note that the checker-board grading of $M_2(A)$ corresponds to the canonical Clifford algebra grading under both f_* and i_*. Therefore, $f_* \circ i_*^{-1}$ is a graded automorphism, and $f_* \circ i_*^{-1} = \mathrm{Int}(u)$ for some $u \in \mathrm{GL}_2(A)$ of the form

$$u = \begin{pmatrix} v & 0 \\ 0 & w \end{pmatrix} \quad \text{or} \quad \begin{pmatrix} 0 & v \\ w & 0 \end{pmatrix}.$$

Moreover, inspection shows that the automorphism $C(f)$ of $C_0\big(\mathrm{Symd}(A,\sigma),\mathrm{Nrp}_\sigma\big)$ induced by f fits in the commutative diagram

$$
\begin{array}{ccc}
C_0(\mathrm{Nrp}_\sigma) & \xrightarrow{\ C(f)\ } & C_0(\mathrm{Nrp}_\sigma) \\
{\scriptstyle i_*}\downarrow & & \downarrow{\scriptstyle i_*} \\
A \times A & \xrightarrow{\ f_* \circ i_*^{-1}\ } & A \times A
\end{array}
$$

where we view $A \times A$ as $\left(\begin{smallmatrix} A & 0 \\ 0 & A \end{smallmatrix}\right) \subset M_2(A)$. Therefore, in view of (13.2), the similitude f is proper if and only if $f_* \circ i_*^{-1}$ maps each component of $A \times A$ into itself. This means that f is proper if $u = \left(\begin{smallmatrix} v & 0 \\ 0 & w \end{smallmatrix}\right)$ and improper if $u = \left(\begin{smallmatrix} 0 & v \\ w & 0 \end{smallmatrix}\right)$.

In particular, if $f(x) = \lambda^{-1}ax\sigma(a)$ for $x \in \mathrm{Symd}(A,\sigma)$, then

$$f_* \circ i_*^{-1}\big(i(x)\big) = \begin{pmatrix} 0 & \lambda\,\mathrm{Nrd}_A(a)^{-1}\overline{ax\sigma(a)} \\ \lambda^{-1}ax\sigma(a) & 0 \end{pmatrix}$$

and (16.7) shows that the right side is

$$\begin{pmatrix} \lambda\sigma(a)^{-1} & 0 \\ 0 & a \end{pmatrix} \cdot \begin{pmatrix} 0 & \overline{x} \\ x & 0 \end{pmatrix} \cdot \begin{pmatrix} \lambda^{-1}\sigma(a) & 0 \\ 0 & a^{-1} \end{pmatrix} = \mathrm{Int}\begin{pmatrix} \lambda\sigma(a)^{-1} & 0 \\ 0 & a \end{pmatrix}\big(i(x)\big).$$

Since the matrices of the form $i(x)$, $x \in \mathrm{Symd}(A,\sigma)$ generate $M_2(A)$ (as $\mathrm{Symd}(A,\sigma)$ generates $C\big(\mathrm{Symd}(A,\sigma),\mathrm{Nrp}_\sigma\big)$), it follows that

$$f_* \circ i_*^{-1} = \mathrm{Int}\begin{pmatrix} \lambda\sigma(a)^{-1} & 0 \\ 0 & a \end{pmatrix},$$

hence f is proper. Similarly, the same arguments show that the similitudes $x \mapsto \lambda^{-1}a\overline{x}\sigma(a)$ are improper.

Returning to the case where f is an arbitrary similitude of $\big(\mathrm{Symd}(A,\sigma),\mathrm{Nrp}_\sigma\big)$ and $f_* \circ i_*^{-1} = \mathrm{Int}(u)$ with $u = \left(\begin{smallmatrix} v & 0 \\ 0 & w \end{smallmatrix}\right)$ or $\left(\begin{smallmatrix} 0 & v \\ w & 0 \end{smallmatrix}\right)$, we apply $f_* \circ i_*^{-1}$ to $i(x)$ for $x \in \mathrm{Symd}(A,\sigma)$ and get

$$(16.12) \qquad \begin{pmatrix} 0 & \alpha^{-1}\overline{f(x)} \\ f(x) & 0 \end{pmatrix} = u \begin{pmatrix} 0 & \overline{x} \\ x & 0 \end{pmatrix} u^{-1}.$$

Comparing the lower left corners yields that

$$(16.13) \qquad f(x) = \begin{cases} wxv^{-1} & \text{if } f \text{ is proper,} \\ w\overline{x}v^{-1} & \text{if } f \text{ is improper.} \end{cases}$$

Let θ be the involution on $M_2(A)$ defined by

$$\theta \begin{pmatrix} a_{11} & a_{12} \\ a_{21} & a_{22} \end{pmatrix} = \begin{pmatrix} \sigma(a_{22}) & -\sigma(a_{12}) \\ -\sigma(a_{21}) & \sigma(a_{11}) \end{pmatrix}.$$

Applying θ to both sides of (16.12), we get

$$\begin{pmatrix} 0 & \alpha^{-1}\overline{f(x)} \\ f(x) & 0 \end{pmatrix} = \theta(u)^{-1} \begin{pmatrix} 0 & \overline{x} \\ x & 0 \end{pmatrix} \theta(u).$$

Therefore, $\theta(u)u$ commutes with the matrices of the form $i(x)$ for $x \in \mathrm{Symd}(A,\sigma)$. Since these matrices generate $M_2(A)$, it follows that $\theta(u)u \in F^\times$, hence

$$\sigma(w)v = \sigma(v)w \in F^\times.$$

Letting $\sigma(w)v = \lambda$, we have

$$f(x) = \begin{cases} \lambda^{-1}wx\sigma(w) & \text{if } f \text{ is proper,} \\ \lambda^{-1}w\overline{x}\sigma(w) & \text{if } f \text{ is improper.} \end{cases}$$

by (16.13).   □

Since the multipliers of the similitudes $x \mapsto \lambda^{-1}ax\sigma(a)$ and $x \mapsto \lambda^{-1}a\overline{x}\sigma(a)$ are $\lambda^{-2}\mathrm{Nrd}_A(a)$, the multipliers of the Albert form Nrp_σ are

$$G(\mathrm{Nrp}_\sigma) = F^{\times 2} \cdot \mathrm{Nrd}_A(A^\times).$$

We thus get another proof of the first part of (16.6).

(16.14) Example. Suppose Q is a quaternion algebra over F, with canonical involution γ, and $A = M_2(Q)$ with the involution σ defined by $\sigma\big((q_{ij})_{1 \le i,j \le 2}\big) = \big(\gamma(q_{ij})\big)^t_{1 \le i,j \le 2}$; then

$$\mathrm{Symd}(A,\sigma) = \left\{ \begin{pmatrix} \alpha_{11} & a_{12} \\ \gamma(a_{12}) & \alpha_{22} \end{pmatrix} \;\middle|\; \alpha_{11},\, \alpha_{22} \in F,\, a_{12} \in Q \right\}.$$

For $a = \left(\begin{smallmatrix} \alpha_{11} & a_{12} \\ \gamma(a_{12}) & \alpha_{22} \end{smallmatrix}\right) \in \mathrm{Symd}(A,\sigma)$, we have $\mathrm{Trp}_\sigma(a) = \alpha_{11} + \alpha_{22}$, hence $\overline{a} = \left(\begin{smallmatrix} \alpha_{22} & -a_{12} \\ -\gamma(a_{12}) & \alpha_{11} \end{smallmatrix}\right)$ and therefore

$$\mathrm{Nrp}_\sigma(a) = \alpha_{11}\alpha_{22} - \mathrm{Nrd}_Q(a_{12}).$$

This expression is the *Moore determinant* of the hermitian matrix a (see Jacobson [129]). This formula shows that the matrices in $\mathrm{Symd}(A,\sigma)$ whose diagonal entries vanish form a quadratic space isometric to $(Q, -\mathrm{Nrd}_Q)$. On the other hand, the diagonal matrices form a hyperbolic plane $\mathbb{H}$, and

$$\mathrm{Nrp}_\sigma \simeq \mathbb{H} \perp -\mathrm{Nrd}_Q .$$

Also, for the involution θ defined by

$$\theta \begin{pmatrix} a_{11} & a_{12} \\ a_{21} & a_{22} \end{pmatrix} = \begin{pmatrix} \gamma(a_{11}) & -\gamma(a_{21}) \\ -\gamma(a_{12}) & \gamma(a_{22}) \end{pmatrix},$$

we have

$$\mathrm{Symd}(A, \theta) = \left\{ \begin{pmatrix} \alpha_{11} & a_{12} \\ -\gamma(a_{12}) & \alpha_{22} \end{pmatrix} \;\middle|\; \alpha_{11},\, \alpha_{22} \in F,\, a_{12} \in Q \right\}.$$

For $a = \begin{pmatrix} \alpha_{11} & a_{12} \\ -\gamma(a_{12}) & \alpha_{22} \end{pmatrix} \in \mathrm{Symd}(A, \sigma)$, we get

$$\mathrm{Nrp}_\theta(a) = \alpha_{11}\alpha_{22} + \mathrm{Nrd}_Q(a_{12}),$$

hence

$$\mathrm{Nrp}_\theta \simeq \mathbb{H} \perp \mathrm{Nrd}_Q .$$

A more general example is given next.

(16.15) Example. Suppose $A = Q_1 \otimes Q_2$ is a tensor product of quaternion algebras Q_1, Q_2 with canonical (symplectic) involutions γ_1, γ_2. Let v_1 be a unit in $\mathrm{Skew}(Q_1, \gamma_1)$ and $\sigma_1 = \mathrm{Int}(v_1) \circ \gamma_1$. The involution σ_1 on Q_1 is orthogonal, unless $v_1 \in F^\times$, a case which occurs only if $\mathrm{char}\, F = 2$. Therefore, the involution $\sigma = \sigma_1 \otimes \gamma_2$ on A is symplectic in all cases, by (2.23). Our goal is to compute explicitly the quadratic form Nrp_σ.

As a first step, observe that

$$\mathrm{Alt}(A, \gamma_1 \otimes \gamma_2) = \{\, x_1 \otimes 1 - 1 \otimes x_2 \mid \mathrm{Trd}_{Q_1}(x_1) = \mathrm{Trd}_{Q_2}(x_2) \,\},$$

as pointed out in Exercise 19 of Chapter I; therefore,

$$\begin{aligned}
\mathrm{Symd}(A, \sigma) &= (v_1 \otimes 1) \cdot \mathrm{Alt}(A, \gamma_1 \otimes \gamma_2) \\
&= \{\, v_1 x_1 \otimes 1 - v_1 \otimes x_2 \mid \mathrm{Trd}_{Q_1}(x_1) = \mathrm{Trd}_{Q_2}(x_2) \,\}.
\end{aligned}$$

For $x_1 \in Q_1$ and $x_2 \in Q_2$ such that $\mathrm{Trd}_{Q_1}(x_1) = \mathrm{Trd}_{Q_2}(x_2)$, there exist $y_1 \in Q_1$, $y_2 \in Q_2$ such that $x_1 = \mathrm{Trd}_{Q_2}(y_2)y_1$ and $x_2 = \mathrm{Trd}_{Q_1}(y_1)y_2$ (see (8.19)), hence

$$x_1 \otimes 1 - 1 \otimes x_2 = y_1 \otimes \gamma_2(y_2) - \gamma_1(y_1) \otimes y_2$$

and therefore

$$v_1 x_1 \otimes 1 - v_1 \otimes x_2 = v_1 y_1 \otimes \gamma_2(y_2) + \sigma\big(v_1 y_1 \otimes \gamma_2(y_2)\big).$$

By (2.13), it follows that

$$\mathrm{Trp}_\sigma(v_1 x_1 \otimes 1 - v_1 \otimes x_2) = \mathrm{Trd}_A\big(v_1 y_1 \otimes \gamma_2(y_2)\big) = \mathrm{Trd}_{Q_1}(v_1 y_1)\,\mathrm{Trd}_{Q_2}(y_2).$$

Since $\mathrm{Trd}_{Q_2}(y_2)y_1 = x_1$, we get

$$\mathrm{Trp}_\sigma(v_1 x_1 \otimes 1 - v_1 \otimes x_2) = \mathrm{Trd}_{Q_1}(v_1 x_1),$$

hence

$$\overline{v_1 x_1 \otimes 1 - v_1 \otimes x_2} = \gamma_1(v_1 x_1) \otimes 1 + v_1 \otimes x_2$$

and finally

$$\mathrm{Nrp}_\sigma(v_1 x_1 \otimes 1 - v_1 \otimes x_2) = \mathrm{Nrd}_{Q_1}(v_1)\big(\mathrm{Nrd}_{Q_1}(x_1) - \mathrm{Nrd}_{Q_2}(x_2)\big).$$

This shows that the form Nrp_σ on $\mathrm{Symd}(A, \sigma)$ is similar to the quadratic form $q_{\gamma_1 \otimes \gamma_2}$ on $\mathrm{Alt}(A, \gamma_1 \otimes \gamma_2)$ defined by

$$q_{\gamma_1 \otimes \gamma_2}(x_1 \otimes 1 - 1 \otimes x_2) = \mathrm{Nrd}_{Q_1}(x_1) - \mathrm{Nrd}_{Q_2}(x_2)$$

for $x_1 \in Q_1$, $x_2 \in Q_2$ such that $\mathrm{Trd}_{Q_1}(x_1) = \mathrm{Trd}_{Q_2}(x_2)$.

To give a more explicit description of $q_{\gamma_1 \otimes \gamma_2}$, we consider the case where char $F = 2$ separately. Suppose first char $F \neq 2$, and let $Q_1 = (a_1, b_1)_F$, $Q_2 = (a_2, b_2)_F$ with quaternion bases $(1, i_1, j_1, k_1)$ and $(1, i_2, j_2, k_2)$ respectively. Then $(i_1 \otimes 1, j_1 \otimes 1, k_1 \otimes 1, 1 \otimes i_2, 1 \otimes j_2, 1 \otimes k_2)$ is an orthogonal basis of $\mathrm{Alt}(A, \gamma_1 \otimes \gamma_2)$ which yields the following diagonalization of $q_{\gamma_1 \otimes \gamma_2}$:

$$q_{\gamma_1 \otimes \gamma_2} = \langle -a_1, -b_1, a_1 b_1, a_2, b_2, -a_2 b_2 \rangle$$

(compare with (16.4) and (16.24)); therefore,

$$\mathrm{Nrp}_\sigma \simeq \langle \mathrm{Nrd}_{Q_1}(v_1) \rangle \cdot \langle -a_1, -b_1, a_1 b_1, a_2, b_2, -a_2 b_2 \rangle.$$

Suppose next that char $F = 2$, and let $Q_1 = [a_1, b_1)_F$, $Q_2 = [a_2, b_2)_F$ with quaternion bases (in characteristic 2) $(1, u_1, v_1, w_1)$ and $(1, u_2, v_2, w_2)$ respectively. A basis of $\mathrm{Alt}(A, \gamma_1 \otimes \gamma_2)$ is $(1, u_1 \otimes 1 + 1 \otimes u_2, v_1 \otimes 1, w_1 \otimes 1, 1 \otimes v_2, 1 \otimes w_2)$. With respect to this basis, the form $q_{\gamma_1 \otimes \gamma_2}$ has the following expression:

$$q_{\gamma_1 \otimes \gamma_2} = [1, a_1 + a_2] \perp \langle b_1 \rangle \cdot [1, a_1] \perp \langle b_2 \rangle \cdot [1, a_2]$$

(compare with (16.4)); therefore,

$$\mathrm{Nrp}_\sigma \simeq \langle \mathrm{Nrd}_{Q_1}(v_1) \rangle \cdot \big([1, a_1 + a_2] \perp \langle b_1 \rangle \cdot [1, a_1] \perp \langle b_2 \rangle \cdot [1, a_2]\big).$$

The following proposition yields a decomposition of the type considered in the example above for any biquaternion algebra with symplectic involution; it is thus an explicit version of the second proof of (16.1). However, for simplicity we restrict to symplectic involutions which are not hyperbolic.[25] In view of (15.21), this hypothesis means that the space

$$\mathrm{Symd}(A, \sigma)^0 = \{ x \in \mathrm{Symd}(A, \sigma) \mid \mathrm{Trp}_\sigma(x) = 0 \}$$

does not contain any nonzero vector x such that $x^2 = 0$ or, equivalently, $\mathrm{Nrp}_\sigma(x) = 0$. Therefore, all the nonzero elements in $\mathrm{Symd}(A, \sigma)^0$ are invertible.

(16.16) Proposition. *Let (A, σ) be a biquaternion algebra with symplectic involution over an arbitrary field F. Assume that σ is not hyperbolic, and let $V \subset \mathrm{Symd}(A, \sigma)$ be a 3-dimensional subspace such that*

$$F \subset V \not\subset \mathrm{Symd}(A, \sigma)^0.$$

Then there exists a unique quaternion subalgebra $Q_1 \subset A$ containing V. This quaternion algebra is stable under σ, and the restriction $\sigma_1 = \sigma|_{Q_1}$ is orthogonal. Therefore, for $Q_2 = C_A Q_1$ the centralizer of Q_1, we have

$$(A, \sigma) = (Q_1, \sigma_1) \otimes (Q_2, \gamma_2)$$

where γ_2 is the canonical involution on Q_2.

Proof: Choose $v \in V \setminus F$ such that $\mathrm{Trp}_\sigma(v) \neq 0$; then $F[v] = F \oplus vF$ is an étale quadratic F-subalgebra of A, and $^-$ restricts to the nontrivial automorphism of $F[v]$. Pick a nonzero vector $u \in V \cap \mathrm{Symd}(A, \sigma)^0$ which is orthogonal to v for the polar form b_{Nrp_σ}; we then have $\bar{u}v + \bar{v}u = 0$, which means that $uv = \bar{v}u$, since $\mathrm{Trp}_\sigma(u) = 0$. The hypothesis that σ is not hyperbolic ensures that u is

---

[25]See Exercise 2 of Chapter II for the hyperbolic case.

invertible; therefore, u and v generate a quaternion subalgebra $Q_1 = \left(F[v], u^2\right)_F$. This quaternion subalgebra contains V, and is indeed generated by V. Since u and v are symmetric under σ, it is stable under σ, and $\mathrm{Sym}(Q_1, \sigma_1) = V$. Moreover, $\mathrm{Sym}(Q_1, \sigma_1)$ contains the element v such that $\mathrm{Trd}_{Q_1}(v) = v + \bar{v} = \mathrm{Trp}_\sigma(v) \neq 0$, hence σ_1 is orthogonal. The rest follows from (1.5) and (2.23). □

The invariant of symplectic involutions. Let σ be a fixed symplectic involution on a biquaternion algebra A. To every other symplectic involution τ on A, we associate a quadratic form $j_\sigma(\tau)$ over F which classifies symplectic involutions up to conjugation: $j_\sigma(\tau) \simeq j_\sigma(\tau')$ if and only if $\tau' = \mathrm{Int}(a) \circ \tau \circ \mathrm{Int}(a)^{-1}$ for some $a \in A^\times$.

We first compare the Albert forms Nrp_σ and Nrp_τ associated with symplectic involutions σ and τ. Recall from (2.7) that $\tau = \mathrm{Int}(u) \circ \sigma$ for some unit $u \in \mathrm{Symd}(A, \sigma)$. Multiplication on the left by the element u then defines a linear map $\mathrm{Symd}(A, \sigma) \xrightarrow{\sim} \mathrm{Symd}(A, \tau)$.

(16.17) Lemma. *For all $x \in \mathrm{Symd}(A, \sigma)$,*

$$\mathrm{Nrp}_\tau(ux) = \mathrm{Nrp}_\sigma(u) \, \mathrm{Nrp}_\sigma(x).$$

Proof: Both sides of the equation to be established are quadratic forms on the space $\mathrm{Symd}(A, \sigma)$. These quadratic forms differ at most by a factor -1, since squaring both sides yields the equality

$$\mathrm{Nrd}_A(ux) = \mathrm{Nrd}_A(u) \, \mathrm{Nrd}_A(x).$$

On the other hand, for $x = 1$ these quadratic forms take the same nonzero value since from the fact that $\mathrm{Prd}_{A,u} = \mathrm{Prp}_{\sigma,u}^2 = \mathrm{Prp}_{\tau,u}^2$ it follows that $\mathrm{Prp}_{\sigma,u} = \mathrm{Prp}_{\tau,u}$, hence $\mathrm{Nrp}_\sigma(u) = \mathrm{Nrp}_\tau(u)$. Therefore, the quadratic forms are equal. □

Let WF denote the Witt ring of nonsingular bilinear forms over F and write W_qF for the WF-module of even-dimensional nonsingular quadratic forms. For every integer k, the k-th power of the fundamental ideal IF of even-dimensional forms in WF is denoted I^kF; we write I^kW_qF for the product $I^kF \cdot W_qF$. Thus, if $I^kW_qF = I^{k+1}F$ if char $F \neq 2$. From the explicit formulas in (16.4), it is clear that Albert forms are in IW_qF; indeed, if char $F \neq 2$,

$$\langle a_1, b_1, -a_1b_1, -a_2, -b_2, a_2b_2 \rangle = -\langle\!\langle a_1, b_1 \rangle\!\rangle + \langle\!\langle a_2, b_2 \rangle\!\rangle \quad \text{in } WF,$$

and, if char $F = 2$,

$$[1, a_1 + a_2] \perp \langle b_1 \rangle \cdot [1, a_1] \perp \langle b_2 \rangle \cdot [1, a_2] = \langle\!\langle b_1, a_1]\!] + \langle\!\langle b_2, a_2]\!] \quad \text{in } W_qF.$$

(16.18) Proposition. *Let σ, τ be symplectic involutions on a biquaternion F-algebra A and let $\tau = \mathrm{Int}(u) \circ \sigma$ for some unit $u \in \mathrm{Symd}(A, \sigma)$. In the Witt group W_qF,*

$$\langle\!\langle \mathrm{Nrp}_\sigma(u) \rangle\!\rangle \cdot \mathrm{Nrp}_\sigma = \mathrm{Nrp}_\sigma - \mathrm{Nrp}_\tau.$$

There is a 3-fold Pfister form $j_\sigma(\tau) \in I^2W_qF$ and a scalar $\lambda \in F^\times$ such that

$$\langle \lambda \rangle \cdot j_\sigma(\tau) = \langle\!\langle \mathrm{Nrp}_\sigma(u) \rangle\!\rangle \cdot \mathrm{Nrp}_\sigma \quad \text{in } W_qF.$$

The Pfister form $j_\sigma(\tau)$ is uniquely determined by the condition

$$j_\sigma(\tau) \equiv \langle\!\langle \mathrm{Nrp}_\sigma(u) \rangle\!\rangle \cdot \mathrm{Nrp}_\sigma \quad \mod I^3W_qF.$$

Proof: Lemma (16.17) shows that multiplication on the left by u is a similitude:

$$\big(\mathrm{Symd}(A,\sigma),\mathrm{Nrp}_\sigma\big) \to \big(\mathrm{Symd}(A,\tau),\mathrm{Nrp}_\tau\big)$$

with multiplier $\mathrm{Nrp}_\sigma(u)$. Therefore, $\mathrm{Nrp}_\tau \simeq \langle\mathrm{Nrp}_\sigma(u)\rangle \cdot \mathrm{Nrp}_\sigma$, hence

$$\mathrm{Nrp}_\sigma - \mathrm{Nrp}_\tau \simeq \langle\!\langle\mathrm{Nrp}_\sigma(u)\rangle\!\rangle \cdot \mathrm{Nrp}_\sigma .$$

We next show the existence of the 3-fold Pfister form $j_\sigma(\tau)$. Since 1 and u are anisotropic for Nrp_σ, there exist nonsingular 3-dimensional subspaces $U \subset \mathrm{Symd}(A,\sigma)$ containing 1 and u. Choose such a subspace and let q_U be the restriction of Nrp_σ to U. Let q_0 be a 4-dimensional form in IW_qF containing q_U as a subspace: if $\mathrm{char}\, F \neq 2$ and $q_U \simeq \langle a_1, a_2, a_3\rangle$, we set $q_0 = \langle a_1, a_2, a_3, a_1 a_2 a_3\rangle$; if $\mathrm{char}\, F = 2$ and $q_U \simeq [a_1, a_2] \perp [a_3]$, we set $q_0 = [a_1, a_2] \perp [a_3, a_1 a_2 a_3^{-1}]$. Since the quadratic forms Nrp_σ and q_0 have isometric 3-dimensional subspaces, there is a 4-dimensional quadratic form q_1 such that

$$q_0 + q_1 = \mathrm{Nrp}_\sigma \quad \text{in } W_qF.$$

The form q_1 lies in IW_qF, since q_0 and Nrp_σ are in this subgroup. Moreover, since $\mathrm{Nrp}_\sigma(u)$ is represented by q_U, hence also by q_0, we have $\langle\!\langle\mathrm{Nrp}_\sigma(u)\rangle\!\rangle \cdot q_0 = 0$ in W_qF. Therefore, multiplying both sides of the equality above by $\langle\!\langle\mathrm{Nrp}_\sigma(u)\rangle\!\rangle$, we get

$$\langle\!\langle\mathrm{Nrp}_\sigma(u)\rangle\!\rangle \cdot q_1 = \langle\!\langle\mathrm{Nrp}_\sigma(u)\rangle\!\rangle \cdot \mathrm{Nrp}_\sigma \quad \text{in } W_qF.$$

The form on the left is a scalar multiple of a 3-fold Pfister form which may be chosen for $j_\sigma(\tau)$. This Pfister form satisfies $j_\sigma(\tau) \equiv \langle\!\langle\mathrm{Nrp}_\sigma(u)\rangle\!\rangle \cdot q_1 \bmod I^3 W_qF$, hence also

$$j_\sigma(\tau) \equiv \langle\!\langle\mathrm{Nrp}_\sigma(u)\rangle\!\rangle \cdot \mathrm{Nrp}_\sigma \mod I^3 W_qF.$$

It remains only to show that it is uniquely determined by this condition. This follows from the following general observation: if π, π' are k-fold Pfister forms such that

$$\pi \equiv \pi' \mod I^k W_qF,$$

then the difference $\pi - \pi'$ is represented by a quadratic form of dimension $2^{k+1} - 2$ since π and π' both represent 1. On the other hand, $\pi - \pi' \in I^k W_qF$, hence the Hauptsatz of Arason and Pfister (see[26] Lam [169, p. 289] or Scharlau [247, Ch. 4, §5]) shows that $\pi - \pi' = 0$. Therefore, $\pi \simeq \pi'$. $\square$

We next show that the invariant j_σ classifies symplectic involutions up to conjugation:

(16.19) Theorem. *Let σ, τ, τ' be symplectic involutions on a biquaternion algebra A over an arbitrary field F, and let $\tau = \mathrm{Int}(u) \circ \sigma$, $\tau' = \mathrm{Int}(u') \circ \sigma$ for some units u, $u' \in \mathrm{Symd}(A,\sigma)$. The following conditions are equivalent:*

(1) *τ and τ' are conjugate, i.e., there exists $a \in A^\times$ such that $\tau' = \mathrm{Int}(a) \circ \tau \circ \mathrm{Int}(a)^{-1}$;*
(2) $\mathrm{Nrp}_\sigma(u)\,\mathrm{Nrp}_\sigma(u') \in F^{\times 2} \cdot \mathrm{Nrd}(A^\times)$;
(3) $\mathrm{Nrp}_\sigma(u)\,\mathrm{Nrp}_\sigma(u') \in G(\mathrm{Nrp}_\sigma)$;
(4) $j_\sigma(\tau) \simeq j_\sigma(\tau')$;
(5) $\mathrm{Nrp}_\tau \simeq \mathrm{Nrp}_{\tau'}$.

---

[26] The proofs given there are easily adapted to the characteristic 2 case. The main ingredient is the Cassels-Pfister subform theorem, of which a characteristic 2 analogue is given in Pfister [218, Theorem 4.9, Chap. 1].

Proof: (1) $\Rightarrow$ (2) If there exists some $a \in A^\times$ such that $\tau' = \mathrm{Int}(a) \circ \tau \circ \mathrm{Int}(a)^{-1}$, then, since the right-hand side is also equal to $\mathrm{Int}\big(a\tau(a)\big) \circ \tau$, we get

$$\mathrm{Int}(u') = \mathrm{Int}\big(a\tau(a)u\big) = \mathrm{Int}\big(au\sigma(a)\big),$$

hence $u' = \lambda^{-1} a u \sigma(a)$ for some $\lambda \in F^\times$. By (2.13), it follows that

$$\mathrm{Nrp}_\sigma(u') = \lambda^{-2} \mathrm{Nrd}_A(a) \mathrm{Nrp}_\sigma(u),$$

proving (2).

(2) $\Longleftrightarrow$ (3) This readily follows from (16.6), since Nrp_σ is an Albert form of A.

(3) $\Rightarrow$ (1) Suppose $\mathrm{Nrp}_\sigma(u') = \mu \mathrm{Nrp}_\sigma(u)$ for some $\mu \in G(\mathrm{Nrp}_\sigma)$. We may then find a proper similitude $g \in \mathrm{GO}^+(\mathrm{Nrp}_\sigma)$ with multiplier μ. Then

$$\mathrm{Nrp}_\sigma\big(g(u)\big) = \mu \mathrm{Nrp}_\sigma(u) = \mathrm{Nrp}_\sigma(u'),$$

hence there is a proper isometry $h \in \mathrm{O}^+(\mathrm{Nrp}_\sigma)$ such that $h \circ g(u) = u'$. By (16.11), we may find an $a \in A^\times$ and a $\lambda \in F^\times$ such that

$$h \circ g(x) = \lambda^{-1} a x \sigma(a) \qquad \text{for all } x \in \mathrm{Symd}(A, \sigma).$$

In particular, $u' = \lambda^{-1} a u \sigma(a)$, hence $\mathrm{Int}(a) \circ \tau = \tau' \circ \mathrm{Int}(a)$.

(3) $\Longleftrightarrow$ (4) Since $j_\sigma(\tau)$ and $j_\sigma(\tau')$ are the unique 3-fold Pfister forms which are equivalent modulo $I^3 W_q F$ to $\langle\!\langle \mathrm{Nrp}_\sigma(u) \rangle\!\rangle \cdot \mathrm{Nrp}_\sigma$ and $\langle\!\langle \mathrm{Nrp}_\sigma(u') \rangle\!\rangle \cdot \mathrm{Nrp}_\sigma$ respectively, we have $j_\sigma(\tau) \simeq j_\sigma(\tau')$ if and only if $\langle\!\langle \mathrm{Nrp}_\sigma(u) \rangle\!\rangle \cdot \mathrm{Nrp}_\sigma \equiv \langle\!\langle \mathrm{Nrp}_\sigma(u') \rangle\!\rangle \cdot \mathrm{Nrp}_\sigma$ mod $I^3 W_q F$. Using the relation $\langle\!\langle \mathrm{Nrp}_\sigma(u) \rangle\!\rangle - \langle\!\langle \mathrm{Nrp}_\sigma(u') \rangle\!\rangle \equiv \langle\!\langle \mathrm{Nrp}_\sigma(u) \mathrm{Nrp}_\sigma(u') \rangle\!\rangle$ mod $I^2 F$, we may rephrase the latter condition as

$$\langle\!\langle \mathrm{Nrp}_\sigma(u) \mathrm{Nrp}_\sigma(u') \rangle\!\rangle \cdot \mathrm{Nrp}_\sigma \in I^3 W_q F.$$

By the Arason-Pfister Hauptsatz, this relation holds if and only if

$$\langle\!\langle \mathrm{Nrp}_\sigma(u) \mathrm{Nrp}_\sigma(u') \rangle\!\rangle \cdot \mathrm{Nrp}_\sigma = 0,$$

which means that $\mathrm{Nrp}_\sigma(u) \mathrm{Nrp}_\sigma(u') \in G(\mathrm{Nrp}_\sigma)$.

(4) $\Longleftrightarrow$ (5) The relations

$$j_\sigma(\tau) \equiv \mathrm{Nrp}_\sigma - \mathrm{Nrp}_\tau \quad \mathrm{mod}\ I^3 W_q F \quad \text{and} \quad j_\sigma(\tau') \equiv \mathrm{Nrp}_\sigma - \mathrm{Nrp}_{\tau'} \quad \mathrm{mod}\ I^3 W_q F$$

show that $j_\sigma(\tau) \simeq j_\sigma(\tau')$ if and only if $\mathrm{Nrp}_\tau - \mathrm{Nrp}_{\tau'} \in I^3 W_q F$. By the Arason-Pfister Hauptsatz, this relation holds if and only if $\mathrm{Nrp}_\tau - \mathrm{Nrp}_{\tau'} = 0$. $\square$

(16.20) Remark. Theorem (15.16) shows that the conditions in (16.19) are also equivalent to: $\big(\mathrm{Symd}(A, \tau)^0, s_\tau\big) \simeq \big(\mathrm{Symd}(A, \tau')^0, s_{\tau'}\big)$.

(16.21) Example. As in (16.14), consider a quaternion F-algebra Q with canonical involution γ, and $A = M_2(Q)$ with the involution σ defined by $\sigma\big((q_{ij})_{1 \leq i,j \leq 2}\big) = \big(\gamma(q_{ij})\big)^t$. Let $\tau = \mathrm{Int}(u) \circ \sigma$ for some invertible matrix $u \in \mathrm{Symd}(A, \sigma)$. As observed in (16.14), we have $\mathrm{Nrp}_\sigma \simeq \mathbb{H} \perp -\mathrm{Nrd}_Q$; therefore,

$$j_\sigma(\tau) = \langle\!\langle \mathrm{Nrp}_\sigma(u) \rangle\!\rangle \cdot \mathrm{Nrd}_Q.$$

Since $j_\sigma(\tau)$ is an invariant of τ, the image of $\mathrm{Nrp}_\sigma(u)$ in $F^\times / \mathrm{Nrd}_Q(Q^\times)$ also is an invariant of τ up to conjugation. In fact, since $\mathrm{Nrp}_\sigma(u)$ is the Moore determinant of u, this image is the Jacobson determinant of the hermitian form

$$h\big((x_1, x_2), (y_1, y_2)\big) = \big(\gamma(x_1)\ \ \gamma(x_2)\big) \cdot u \cdot \begin{pmatrix} y_1 \\ y_2 \end{pmatrix}$$

on the 2-dimensional Q-vector space Q^2. (See the notes of Chapter II.)

Of course, if Q is split, then Nrd_Q is hyperbolic, hence $j_\sigma(\tau) = 0$ for all symplectic involutions σ, τ. Therefore, all the symplectic involutions are conjugate in this case. (This is clear *a priori*, since all the symplectic involutions on a split algebra are hyperbolic.)

16.C. Albert forms and orthogonal involutions. Let σ be an orthogonal involution on a biquaternion F-algebra A. Mimicking the construction of the Albert form associated to a symplectic involution, in this subsection we define a quadratic form q_σ on the space $\mathrm{Skew}(A, \sigma)$ in such a way that $\big(\mathrm{Skew}(A, \sigma), q_\sigma\big)$ is an Albert quadratic space. By contrast with the symplectic case, the form q_σ is only defined up to a scalar factor, however, and our discussion is restricted to the case where the characteristic is different from 2. We also show how the form q_σ is related to the norm form of the Clifford algebra $C(A, \sigma)$ and to the generalized pfaffian defined in §8.D.

Throughout this subsection, we assume that char $F \neq 2$.

(16.22) Proposition. *There exists a linear endomorphism*

$$p_\sigma \colon \mathrm{Skew}(A, \sigma) \to \mathrm{Skew}(A, \sigma)$$

which satisfies the following two conditions:

(1) $xp_\sigma(x) = p_\sigma(x)x \in F$ *for all* $x \in \mathrm{Skew}(A, \sigma)$;
(2) *an element* $x \in \mathrm{Skew}(A, \sigma)$ *is invertible if and only if* $xp_\sigma(x) \neq 0$.
The endomorphism p_σ *is uniquely determined up to a factor in* $F^\times$. *More precisely, if* $p'_\sigma \colon \mathrm{Skew}(A, \sigma) \to \mathrm{Skew}(A, \sigma)$ *is a linear map such that* $xp'_\sigma(x) \in F$ *for all* $x \in \mathrm{Skew}(A, \sigma)$ *(or* $p'_\sigma(x)x \in F$ *for all* $x \in \mathrm{Skew}(A, \sigma)$*), then*

$$p'_\sigma = \lambda p_\sigma$$

for some $\lambda \in F$.

Proof: By (2.8), the intersection $\mathrm{Skew}(A, \sigma) \cap A^\times$ is nonempty. Let u be a skew-symmetric unit and $\tau = \mathrm{Int}(u) \circ \sigma$. The involution τ is symplectic by (2.7), and we have

$$\mathrm{Sym}(A, \tau) = u \cdot \mathrm{Skew}(A, \sigma) = \mathrm{Skew}(A, \sigma) \cdot u^{-1}.$$

Therefore, for $x \in \mathrm{Skew}(A, \sigma)$ we may consider $\overline{ux} \in \mathrm{Sym}(A, \tau)$ where

$$^- \colon \mathrm{Sym}(A, \tau) \to \mathrm{Sym}(A, \tau)$$

is as in (16.7), and set

$$p_\sigma(x) = \overline{ux}u \in \mathrm{Skew}(A, \sigma) \quad \text{for } x \in \mathrm{Skew}(A, \sigma).$$

We have $p_\sigma(x)x = \overline{ux}ux = \mathrm{Nrp}_\tau(ux) \in F$ and

$$xp_\sigma(x) = u^{-1}(ux\overline{ux})u = u^{-1}\,\mathrm{Nrp}_\tau(ux)u = \mathrm{Nrp}_\tau(ux),$$

hence p_σ satisfies (1). It also satisfies (2), since $\mathrm{Nrp}_\tau(ux)^2 = \mathrm{Nrd}_A(ux)$.

In order to make this subsection independent of §16.B, we give an alternate proof of the existence of p_σ. Consider an arbitrary decomposition of A into a tensor product of quaternion subalgebras:

$$A = Q_1 \otimes_F Q_2$$

and let $\theta = \gamma_1 \otimes \gamma_2$ be the tensor product of the canonical (conjugation) involutions on Q_1 and Q_2. The involution θ is orthogonal since char $F \neq 2$, and

$$\mathrm{Skew}(A, \theta) = (Q_1^0 \otimes 1) \oplus (1 \otimes Q_2^0),$$

where Q_1^0 and Q_2^0 are the spaces of pure quaternions in Q_1 and Q_2 respectively. Define a map $p_\theta\colon \operatorname{Skew}(A,\theta) \to \operatorname{Skew}(A,\theta)$ by

$$p_\theta(x_1 \otimes 1 + 1 \otimes x_2) = x_1 \otimes 1 - 1 \otimes x_2$$

for $x_1 \in Q_1^0$ and $x_2 \in Q_2^0$. For $x = x_1 \otimes 1 + 1 \otimes x_2 \in \operatorname{Skew}(A,\theta)$ we have

$$x p_\theta(x) = p_\theta(x)x = x_1^2 - x_2^2 = -\operatorname{Nrd}_{Q_1}(x_1) + \operatorname{Nrd}_{Q_2}(x_2) \in F,$$

hence (1) holds for p_θ. If $x p_\theta(x) \neq 0$, then x is clearly invertible. Conversely, if x is invertible and $x p_\theta(x) = 0$, then $p_\theta(x) = 0$, hence $x = 0$, a contradiction. Therefore, p_θ also satisfies (2).

If σ is an arbitrary orthogonal involution on A, we have $\sigma = \operatorname{Int}(v) \circ \theta$ for some $v \in \operatorname{Sym}(A,\theta) \cap A^\times$, by (2.7). We may then set

$$p_\sigma(x) = v p_\theta(xv) \quad \text{for } x \in \operatorname{Skew}(A,\sigma)$$

and verify as above that p_σ satisfies the required conditions.

We next prove uniqueness of p_σ up to a scalar factor. The following arguments are based on Wadsworth [299]. For simplicity, we assume that F has more than three elements; the result is easily checked when $F = \mathbb{F}_3$.

Let p_σ be a map satisfying (1) and (2), and let $p'_\sigma\colon \operatorname{Skew}(A,\sigma) \to \operatorname{Skew}(A,\sigma)$ be such that $x p'_\sigma(x) \in F$ for all $x \in \operatorname{Skew}(A,\sigma)$. For $x \in \operatorname{Skew}(A,\sigma)$, we let

$$q_\sigma(x) = x p_\sigma(x) \in F \quad \text{and} \quad q'_\sigma(x) = x p'_\sigma(x) \in F.$$

Let $x \in \operatorname{Skew}(A,\sigma) \cap A^\times$; we have $p_\sigma(x) = q_\sigma(x)x^{-1}$ and $p'_\sigma(x) = q'_\sigma(x)x^{-1}$, hence

$$p'_\sigma(x) = \frac{q'_\sigma(x)}{q_\sigma(x)} p_\sigma(x).$$

Suppose y is another unit in $\operatorname{Skew}(A,\sigma)$, and that it is not a scalar multiple of x. We also have $p_\sigma(y) = q_\sigma(y)y^{-1}$, hence $p_\sigma(y)$ is not a scalar multiple of $p_\sigma(x)$, and $p'_\sigma(y) = \frac{q'_\sigma(y)}{q_\sigma(y)} p_\sigma(y)$. We may find some $\alpha \in F^\times$ such that $x + \alpha y \in A^\times$, since the equation in α

$$q_\sigma(x + \alpha y) = q_\sigma(x) + \alpha q_\sigma(x, y) + \alpha^2 q_\sigma(y) = 0$$

has at most two solutions and F has more than three elements. For this choice of α, we have

$$p'_\sigma(x + \alpha y) = \frac{q'_\sigma(x + \alpha y)}{q_\sigma(x + \alpha y)} p_\sigma(x + \alpha y).$$

By linearity of p_σ and p'_σ, it follows that

$$p'_\sigma(x) + \alpha p'_\sigma(y) = \frac{q'_\sigma(x + \alpha y)}{q_\sigma(x + \alpha y)} p_\sigma(x) + \alpha \frac{q'_\sigma(x + \alpha y)}{q_\sigma(x + \alpha y)} p_\sigma(y).$$

On the other hand, we also have

$$p'_\sigma(x) + \alpha p'_\sigma(y) = \frac{q'_\sigma(x)}{q_\sigma(x)} p_\sigma(x) + \alpha \frac{q'_\sigma(y)}{q_\sigma(y)} p_\sigma(y),$$

hence

$$\frac{q'_\sigma(x)}{q_\sigma(x)} = \frac{q'_\sigma(x + \alpha y)}{q_\sigma(x + \alpha y)} = \frac{q'_\sigma(y)}{q_\sigma(y)}$$

since $p_\sigma(x)$ and $p_\sigma(y)$ are linearly independent.

To conclude, observe that $\operatorname{Skew}(A,\sigma)$ has a basis $(e_i)_{1 \leq i \leq 6}$ consisting of invertible elements: this is clear if σ is the involution θ defined above, and it follows for arbitrary σ since $\mathrm{Skew}(A, \sigma) = v \cdot \mathrm{Skew}(A, \theta)$ if $\sigma = \mathrm{Int}(v) \circ \theta$. Denoting $\lambda = q_\sigma'(e_1) q_\sigma(e_1)^{-1}$, the argument above shows that

$$\lambda = \frac{q_\sigma'(e_i)}{q_\sigma(e_i)} \quad \text{for } i = 1, \ldots, 6,$$

hence $p_\sigma'(e_i) = \lambda p_\sigma(e_i)$ for $i = 1, \ldots, 6$. By linearity of p_σ and p_σ', it follows that $p_\sigma'(x) = \lambda p_\sigma(x)$ for all $x \in \mathrm{Skew}(A, \sigma)$. $\qquad \square$

(16.23) Proposition. *Let p_σ be a non-zero linear endomorphism of $\mathrm{Skew}(A, \sigma)$ such that $x p_\sigma(x) \in F$ for all $x \in \mathrm{Skew}(A, \sigma)$, and let*

$$q_\sigma(x) = x p_\sigma(x) \in F \quad \text{for } x \in \mathrm{Skew}(A, \sigma).$$

The quadratic form q_σ is nonsingular; for $x \in \mathrm{Skew}(A, \sigma)$ we have $q_\sigma(x) = p_\sigma(x) x$, and $q_\sigma(x) \neq 0$ if and only if $x \in A^\times$. Moreover, $\big(\mathrm{Skew}(A, \sigma), q_\sigma\big)$ is an Albert quadratic space of A.

Proof: Proposition (16.22) shows that p_σ is a scalar multiple of an endomorphism satisfying (16.22.1) and (16.22.2); therefore, p_σ also satisfies these conditions. It follows that $q_\sigma(x) = p_\sigma(x) x$ for all $x \in \mathrm{Skew}(A, \sigma)$ and $q_\sigma(x) \neq 0$ if and only if x is invertible.

In order to show that q_σ is nonsingular, we again consider a decomposition of A into a tensor product of two quaternion subalgebras, so

$$A = Q_1 \otimes_F Q_2$$

and set $\theta = \gamma_1 \otimes \gamma_2$, which is the tensor product of the canonical involutions on Q_1 and Q_2. As observed in the proof of (16.22), we have

$$\mathrm{Skew}(A, \sigma) = (Q_1^0 \otimes 1) \oplus (1 \otimes Q_2^0),$$

and we may consider the endomorphism p_θ of $\mathrm{Skew}(A, \sigma)$ defined by

$$p_\theta(x_1 \otimes 1 + 1 \otimes x_2) = x_1 \otimes 1 - 1 \otimes x_2$$

for $x_1 \in Q_1^0$ and $x_2 \in Q_2^0$. Denoting $q_\theta = x p_\theta(x)$ for $x \in \mathrm{Skew}(A, \sigma)$, we then have

$$q_\theta(x_1 \otimes 1 + 1 \otimes x_2) = x_1^2 - x_2^2 = -\mathrm{Nrd}_{Q_1}(x_1) + \mathrm{Nrd}_{Q_2}(x_2),$$

hence q_θ is a nonsingular quadratic form.

If $v \in \mathrm{Sym}(A, \theta) \cap A^\times$ is such that $\sigma = \mathrm{Int}(v) \circ \theta$, (16.22) shows that there exists $\lambda \in F^\times$ such that $p_\sigma(x) = \lambda v p_\theta(xv)$ for all $x \in \mathrm{Skew}(A, \sigma)$. Then $q_\sigma(x) = \lambda q_\theta(xv)$, hence multiplication on the right by v defines a similitude

$$\big(\mathrm{Skew}(A, \sigma), q_\sigma\big) \overset{\sim}{\longrightarrow} \big(\mathrm{Skew}(A, \theta), q_\theta\big)$$

with multiplier λ. Since q_θ is nonsingular, it follows that q_σ is also nonsingular.

To complete the proof, consider the map

$$i \colon \mathrm{Skew}(A, \sigma) \to M_2(A)$$

defined by

$$i(x) = \begin{pmatrix} 0 & p_\sigma(x) \\ x & 0 \end{pmatrix} \quad \text{for } x \in \mathrm{Skew}(A, \sigma).$$

We have $i(x)^2 = q_\sigma(x)$ for all $x \in \mathrm{Skew}(A, \sigma)$, hence the universal property of Clifford algebras shows that i induces an F-algebra homomorphism

$$i_* \colon C\big(\mathrm{Skew}(A, \sigma), q_\sigma\big) \to M_2(A).$$

(Compare with (16.9).) Since q_σ is nonsingular, it follows that the Clifford algebra $C\big(\text{Skew}(A,\sigma), q_\sigma\big)$ is simple, hence i_* is injective. It is also surjective by dimension count, hence (16.3) shows that $\big(\text{Skew}(A,\sigma), q_\sigma\big)$ is an Albert quadratic space of A.

$\square$

(16.24) Example. Let $A = (a_1, b_1)_F \otimes (a_2, b_2)_F$. If $\theta = \gamma_1 \otimes \gamma_2$ is the tensor product of the canonical involutions on the quaternion algebras $(a_1, b_1)_F$ and $(a_2, b_2)_F$, the computations in the proof of (16.23) show that one can take

$$p_\theta(x_1 \otimes 1 + 1 \otimes x_2) = x_1 \otimes 1 - 1 \otimes x_2 \quad \text{and} \quad q_\theta(x_1 \otimes 1 + 1 \otimes x_2) = x_1^2 - x_2^2$$

for $x_i \in (a_i, b_i)_F^0$, $i = 1, 2$. Therefore, q_θ has the following diagonalization:

$$q_\theta = \langle a_1, b_1, -a_1 b_1, -a_2, -b_2, a_2 b_2 \rangle.$$

(Compare with (16.4).)

We now list a few properties of the endomorphism p_σ defined in (16.22).

(16.25) Proposition. *Let p_σ be a non-zero linear endomorphism of $\text{Skew}(A,\sigma)$ such that $x p_\sigma(x) \in F$ for all $x \in \text{Skew}(A,\sigma)$, and let $q_\sigma \colon \text{Skew}(A,\sigma) \to F$ be the (Albert) quadratic map defined by*

$$q_\sigma(x) = x p_\sigma(x) \quad \text{for } x \in \text{Skew}(A,\sigma).$$

(1) *For all $a \in A^\times$ and $x \in \text{Skew}(A,\sigma)$, we have $q_\sigma\big(a x \sigma(a)\big) = \text{Nrd}_A(a) q_\sigma(x)$ and $p_\sigma\big(a x \sigma(a)\big) = \text{Nrd}_A(a) \sigma(a)^{-1} p_\sigma(x) a^{-1}$.*
(2) *There exists some $d_\sigma \in F^\times$ such that*
    (a) *$q_\sigma(x)^2 = d_\sigma \text{Nrd}_A(x)$ for all $x \in \text{Skew}(A,\sigma)$;*
    (b) *$p_\sigma^2 = d_\sigma \cdot \text{Id}_{\text{Skew}(A,\sigma)}$;*
    (c) *$d_\sigma \cdot F^{\times 2} = \text{disc}\,\sigma$;*
    (d) *$[p_\sigma(x), p_\sigma(y)] = d_\sigma[x,y]$ for all $x, y \in \text{Skew}(A,\sigma)$ where $[\,,\,]$ are the Lie brackets (i.e., $[x,y] = xy - yx$).*

Proof: Consider θ, p_θ, q_θ as in (16.24). The relation

$$q_\theta(x)^2 = \text{Nrd}_A(x) \quad \text{for } x \in \text{Skew}(A,\sigma)$$

is easily proved: extending scalars to an algebraic closure, it suffices to show that for any 2×2 matrices m_1, m_2 of trace zero,

$$\det(m_1 \otimes 1 + 1 \otimes m_2) = \big(\det(m_1) - \det(m_2)\big)^2.$$

This follows by a computation which is left to the reader. It is clear that $p_\theta^2 = \text{Id}_{\text{Skew}(A,\theta)}$, and (7.3) shows that $\text{disc}\,\theta = 1$. For $x_1, y_1 \in (a_1, b_1)_F^0$ and $x_2, y_2 \in (a_2, b_2)_F^0$, we have

$$[x_1 \otimes 1 + 1 \otimes x_2, y_1 \otimes 1 + 1 \otimes y_2] = [x_1 \otimes 1, y_1 \otimes 1] + [1 \otimes x_2, 1 \otimes y_2]$$
$$= [x_1 \otimes 1 - 1 \otimes x_2, y_1 \otimes 1 - 1 \otimes y_2],$$

hence $\big[p_\theta(x), p_\theta(y)\big] = [x,y]$ for $x, y \in \text{Skew}(A,\sigma)$. Therefore, the properties in (2) hold with $d_\theta = 1$. The relations

$$q_\theta\big(a x \theta(a)\big) = \text{Nrd}_A(a) q_\theta(x) \quad \text{and} \quad p_\theta\big(a x \theta(a)\big) = \text{Nrd}_A(a) \theta(a)^{-1} p_\theta(x) a^{-1}$$

for all $a \in A^\times$ and $x \in \text{Skew}(A,\sigma)$ follow by the same arguments as in (2.13) and (16.7). The proposition is thus proved for $\sigma = \theta$.

For an arbitrary orthogonal involution σ, there exists $v \in A^\times$ such that $\sigma = \mathrm{Int}(v) \circ \theta$. The proof of (16.22) then yields $\lambda \in F^\times$ such that

$$p_\sigma(x) = \lambda v p_\theta(xv) \quad \text{and} \quad q_\sigma(x) = \lambda q_\theta(xv)$$

for all $x \in \mathrm{Skew}(A, \sigma)$. For $a \in A^\times$ and $x \in \mathrm{Skew}(A, \sigma)$, we then have

$$q_\sigma\big(ax\sigma(a)\big) = \lambda q_\theta\big(axv\theta(a)\big) \quad \text{and} \quad p_\sigma\big(ax\sigma(a)\big) = \lambda v p_\theta\big(axv\theta(a)\big).$$

Since property (1) is already proved for θ, it follows that

$$q_\sigma\big(ax\sigma(a)\big) = \lambda \, \mathrm{Nrd}_A(a) q_\theta(xv) = \mathrm{Nrd}_A(a) q_\sigma(x)$$

and

$$
\begin{aligned}
p_\sigma\big(ax\sigma(a)\big) &= \lambda \, \mathrm{Nrd}_A(a) v\theta(a)^{-1} p_\theta(xv)a^{-1} \\
&= \lambda \, \mathrm{Nrd}_A(a)\sigma(a)^{-1} v p_\theta(xv)a^{-1} \\
&= \mathrm{Nrd}_A(a)\sigma(a)^{-1} p_\sigma(x)a^{-1},
\end{aligned}
$$

proving(1).

To complete the proof, we show that the properties in (2) hold with $d_\sigma = \lambda^2 \, \mathrm{Nrd}_A(v)$.

First, we have for $x \in \mathrm{Skew}(A, \sigma)$ that

$$q_\sigma(x)^2 = \lambda^2 q_\theta(xv)^2 = \lambda^2 \, \mathrm{Nrd}_A(v) \, \mathrm{Nrd}_A(x)$$

and

$$p_\sigma^2(x) = \lambda v p_\theta\big(\lambda v p_\theta(xv)v\big).$$

Since $\theta(v) = v$, we may use property(1) for θ to rewrite the right side as

$$\lambda^2 v \, \mathrm{Nrd}_A(v) v^{-1} p_\theta^2(xv) v^{-1} = \lambda^2 \, \mathrm{Nrd}_A(v)x.$$

Therefore, (a) and (b) hold. Since

$$\mathrm{disc}\,\sigma = \mathrm{Nrd}_A(v) \cdot \mathrm{disc}\,\theta = \mathrm{Nrd}_A(v) \cdot F^{\times 2},$$

by (7.3), we also have (c). Finally, to establish (d), observe that by linearizing the relations

$$q_\sigma(x) = x p_\sigma(x) = p_\sigma(x)x$$

we get

$$b_{q_\sigma}(x, y) = x p_\sigma(y) + y p_\sigma(x) = p_\sigma(x)y + p_\sigma(y)x \quad \text{for } x,\, y \in \mathrm{Skew}(A, \sigma).$$

In particular,

$$b_{q_\sigma}\big(p_\sigma(x), y\big) = p_\sigma(x)p_\sigma(y) + y p_\sigma^2(x) = p_\sigma^2(x)y + p_\sigma(y)p_\sigma(x)$$

for $x,\, y \in \mathrm{Skew}(A, \sigma)$. In view of (b), it follows that

$$p_\sigma(x)p_\sigma(y) + d_\sigma yx = d_\sigma xy + p_\sigma(y)p_\sigma(x),$$

hence

$$p_\sigma(x)p_\sigma(y) - p_\sigma(y)p_\sigma(x) = d_\sigma(xy - yx)$$

for $x,\, y \in \mathrm{Skew}(A, \sigma)$, proving (d).

Alternately, properties (1) and (2) can be established by comparing σ with a symplectic involution instead of θ (see the proof of (16.22)), and using (16.7). Details are left to the reader. $\qquad\square$

With the notation of the preceding proposition, we have for all $x \in \mathrm{Skew}(A, \sigma)$

$$q_\sigma\big(p_\sigma(x)\big) = p_\sigma^2(x)p_\sigma(x) = d_\sigma x p_\sigma(x) = d_\sigma q_\sigma(x),$$

hence p_σ is a similitude of $\big(\mathrm{Skew}(A, \sigma), q_\sigma\big)$ with multiplier d_σ. The group of similitudes of $\big(\mathrm{Skew}(A, \sigma), q_\sigma\big)$ can be described by mimicking (16.11).

(16.26) Proposition. *The proper similitudes of $\big(\mathrm{Skew}(A, \sigma), q_\sigma\big)$ are of the form*

$$x \mapsto \lambda^{-1} a x \sigma(a)$$

where $\lambda \in F^\times$ and $a \in A^\times$.

The improper similitudes of $\big(\mathrm{Skew}(A, \sigma), q_\sigma\big)$ are of the form

$$x \mapsto \lambda^{-1} a p_\sigma(x) \sigma(a)$$

where $\lambda \in F^\times$ and $a \in A^\times$.

The proof is left to the reader.

We now give another point of view on the linear endomorphism p_σ and the Albert form q_σ by relating them to the Clifford algebra $C(A, \sigma)$.

Let $K = F(\sqrt{\mathrm{disc}\,\sigma})$ and let ι be the nontrivial automorphism of K/F. Recall from §15.B that we may identify

$$(A, \sigma) = N_{K/F}(Q, \gamma)$$

for some quaternion K-algebra Q with canonical involution γ. The quaternion algebra Q is canonically isomorphic as an F-algebra to the Clifford algebra $C(A, \sigma)$. Recall also that there is a Lie algebra homomorphism $\dot{n} \colon \mathfrak{L}(Q) \to \mathfrak{L}(A)$ defined by

$$\dot{n}(x) = {}^\iota x \otimes 1 + {}^\iota 1 \otimes x \quad \text{for } x \in Q.$$

This homomorphism restricts to a Lie algebra isomorphism

$$\dot{n} \colon Q^0 \xrightarrow{\sim} \mathrm{Skew}(A, \sigma).$$

(16.27) Proposition. *Let $\alpha \in K^\times$ be such that $\iota(\alpha) = -\alpha$. The linear endomorphism p_σ which makes the following diagram commutative:*

$$
\begin{array}{ccc}
Q^0 & \xrightarrow{\ \dot{n}\ } & \mathrm{Skew}(A, \sigma) \\
{\scriptstyle \alpha \cdot}\downarrow & & \downarrow{\scriptstyle p_\sigma} \\
Q^0 & \xrightarrow{\ \dot{n}\ } & \mathrm{Skew}(A, \sigma)
\end{array}
$$

(where $\alpha \cdot$ is multiplication by α) is such that $x p_\sigma(x) \in F$ for all $x \in \mathrm{Skew}(A, \sigma)$. The corresponding Albert form q_σ satisfies:

$$q_\sigma\big(\dot{n}(x)\big) = \mathrm{tr}_{K/F}(\alpha x^2) \quad \text{for } x \in Q^0.$$

Let $s \colon Q^0 \to K$ be the squaring map defined by $s(x) = x^2$. Then q_σ is the Scharlau transfer of the form $\langle \alpha \rangle \cdot s$ with respect to the linear form $\mathrm{tr}_{K/F} \colon K \to F$:

$$q_\sigma = (\mathrm{tr}_{K/F})_* \big(\langle \alpha \rangle \cdot s\big).$$

Proof: It suffices to prove that

$$\dot{n}(x)\dot{n}(\alpha x) = \mathrm{tr}_{K/F}(\alpha x^2) \quad \text{for } x \in Q^0.$$

This follows from a straightforward computation:

$$\dot{n}(x)\dot{n}(\alpha x) = (^\iota x \otimes 1 + 1 \otimes x)(^\iota(\alpha x) \otimes 1 + 1 \otimes \alpha x)$$
$$= \iota(\alpha x^2) + (\alpha x^2) + {}^\iota(\alpha x) \otimes x + {}^\iota x \otimes \alpha x.$$

Since $\iota(\alpha) = -\alpha$, the last two terms in the last expression cancel. $\qquad\square$

Continuing with the notation of the proposition above and letting $d_\sigma = \alpha^2 \in F^\times$, we obviously have $p_\sigma^2 = d_\sigma \mathrm{Id}_{\mathrm{Skew}(A,\sigma)}$ and $d_\sigma \cdot F^{\times 2} = \mathrm{disc}\,\sigma$, and also

$$\big[p_\sigma(x), p_\sigma(y)\big] = d_\sigma[x, y] \quad \text{for } x, y \in \mathrm{Skew}(A, \sigma)$$

since $\dot{n}$ is an isomorphism of Lie algebras. We may thus recover the properties in (16.25.2).

Conversely, Proposition (16.27) shows that the linear endomorphism p_σ can be used to endow $\mathrm{Skew}(A, \sigma)$ with a structure of K-module, hence to give an explicit description of the Clifford algebra $C(A, \sigma)$.

We may also derive some information on quaternion algebras over quadratic extensions:

(16.28) Corollary. *For a quaternion algebra Q over an étale quadratic extension K/F, the following conditions are equivalent*:

(1) *Q is split by some quadratic extension of F*;
(2) *$Q \simeq (a, b)_K$ for some $a \in F^\times$ and some $b \in K^\times$*;
(3) *$\mathrm{ind}\, N_{K/F}(Q) = 1$ or 2*.

Proof: It suffices to prove the equivalence of (2) and (3) for non-split quaternion algebras Q, since (1) and (2) are clearly equivalent and the three conditions trivially hold if Q is split. We may thus assume that the squaring map $s\colon Q^0 \to K$ is anisotropic. Condition (2) then holds if and only if the transfer $(\mathrm{tr}_{K/F})_*(\langle\alpha\rangle \cdot s)$ is isotropic where $\alpha \in K$ is an arbitrary nonzero element of trace zero. By (16.27) and (16.23), $(\mathrm{tr}_{K/F})_*(\langle\alpha\rangle \cdot s)$ is an Albert form of $N_{K/F}(Q)$; therefore, by Albert's Theorem (16.5), this form is isotropic if and only if condition (3) holds. $\qquad\square$

In the special case where $K = F \times F$, the corollary takes the following form:

(16.29) Corollary (Albert [11]). *For quaternion algebras Q_1, Q_2 over F, the following conditions are equivalent*:

(1) *there is a quadratic extension of F which splits both Q_1 and Q_2*;
(2) *$Q_1 \simeq (a, b_1)_F$ and $Q_2 \simeq (a, b_2)_F$ for some a, b_1, $b_2 \in F^\times$*;
(3) *$\mathrm{ind}(Q_1 \otimes_F Q_2) = 1$ or 2*.

The implication (3) $\Rightarrow$ (1) may be reformulated as follows:

(16.30) Corollary. *Let Q_1, Q_2, Q_3 be quaternion algebras over F. If $Q_1 \otimes Q_2 \otimes Q_3$ is split, then there is a quadratic extension of F which splits Q_1, Q_2, and Q_3.*

Proof: The hypothesis means that $Q_1 \otimes Q_2$ is Brauer-equivalent to Q_3, hence its index is 1 or 2. The preceding corollary yields a quadratic extension of F which splits Q_1 and Q_2, hence also Q_3. $\qquad\square$

Finally, we relate the preceding constructions to the generalized pfaffian defined in §8.D. Let $Z(A, \sigma)$ be the center of the Clifford algebra $C(A, \sigma)$. Let ι be the nontrivial automorphism of $Z(A, \sigma)/F$ and let $Z(A, \sigma)^0$ be the space of elements of trace zero, so

$$Z(a, \sigma)^0 = \{ z \in Z(A, \sigma) \mid \iota(z) = -z \}.$$

By (8.24), the generalized pfaffian is a quadratic map

$$\pi \colon \operatorname{Skew}(A, \sigma) \to Z(A, \sigma)^0.$$

Our goal is to show that this map can be regarded as a canonical Albert form associated to σ.

Let $C(A, \sigma)^0$ be the space of elements of reduced trace zero in $C(A, \sigma)$. Since $C(A, \sigma)$ is a quaternion algebra over $Z(A, \sigma)$, we have $x^2 \in Z(A, \sigma)$ for $x \in C(A, \sigma)^0$. We may then define a map $\varphi \colon C(A, \sigma)^0 \to Z(A, \sigma)^0$ by

$$\varphi(x) = \bigl(x^2 - \iota(x^2)\bigr) \quad \text{for } x \in C(A, \sigma)^0.$$

(16.31) Lemma. *Let* $c \colon A \to C(A, \sigma)$ *be the canonical map. For* $x \in \operatorname{Skew}(A, \sigma)$,

$$\pi(x) = \varphi\bigl(\tfrac{1}{2} c(x)\bigr).$$

Proof: We may extend scalars to an algebraic closure, and assume that

$$(A, \sigma) = \bigl(M_4(F), t\bigr) = \bigl(\operatorname{End}_F(F^4), \sigma_q\bigr)$$

where $q(x_1, x_2, x_3, x_4) = x_1^2 + x_2^2 + x_3^2 + x_4^2$. We use q to identify $\operatorname{End}_F(F^4) = F^4 \otimes F^4$ as in (5.1). Let $(e_i)_{1 \leq i \leq 4}$ be the standard basis of F^4. A basis of $\operatorname{Skew}(A, \sigma)$ is given by

$$h_{ij} = \tfrac{1}{2}(e_i \otimes e_j - e_j \otimes e_i) \quad \text{for } 1 \leq i < j \leq 4,$$

and we have, by (8.26),

$$\pi(\textstyle\sum_{1 \leq i < j \leq 4} x_{ij} h_{ij}) = (x_{12} x_{34} - x_{13} x_{24} + x_{14} x_{23}) e_1 \cdot e_2 \cdot e_3 \cdot e_4$$

for $x_{ij} \in F$. On the other hand,

$$\tfrac{1}{2} c(\textstyle\sum_{1 \leq i < j \leq 4} x_{ij} h_{ij}) = \tfrac{1}{2} \textstyle\sum_{1 \leq i < j \leq 4} x_{ij} e_i \cdot e_j,$$

and a computation shows that

$$\varphi(\textstyle\sum_{1 \leq i < j \leq 4} x_{ij} h_{ij}) = (x_{12} x_{34} - x_{13} x_{24} + x_{14} x_{23}) e_1 \cdot e_2 \cdot e_3 \cdot e_4. \qquad \square$$

Theorem (15.7) yields a canonical isomorphism

$$(A, \sigma) = N_{Z(A, \sigma)/F}\bigl(C(A, \sigma), \underline{\sigma}\bigr);$$

moreover, using the canonical isomorphism as an identification, the map

$$\dot{n} \colon C(A, \sigma)^0 \to \operatorname{Skew}\bigl(N_{Z(A, \sigma)/F}\bigl(C(A, \sigma), \underline{\sigma}\bigr)\bigr)$$

is the inverse of $\tfrac{1}{2} c \colon \operatorname{Skew}(A, \sigma) \to C(A, \sigma)^0$ (see (15.8)). Therefore, the lemma yields:

$$\pi\bigl(\dot{n}(x)\bigr) = \varphi(x) = \bigl(x^2 - \iota(x^2)\bigr) \quad \text{for } x \in C(A, \sigma)^0.$$

Let $\alpha \in Z(A, \sigma)^0$, $\alpha \neq 0$. According to (16.27), the map $q_\sigma \colon \operatorname{Skew}(A, \sigma) \to F$ defined by

$$q_\sigma\bigl(\dot{n}(x)\bigr) = \dot{n}(x) \dot{n}(\alpha x) \quad \text{for } x \in C(A, \sigma)^0$$

is an Albert form of A.

(16.32) Proposition. *For all* $a \in \operatorname{Skew}(A, \sigma)$,

$$q_\sigma(a) = \alpha \pi(a).$$

Proof: It suffices to prove that

$$\dot{n}(x)\dot{n}(\alpha x) = \alpha\big(x^2 - \iota(x^2)\big) \quad \text{for } x \in C(A, \sigma)^0.$$

The left side has been computed in the proof of (16.27):

$$\dot{n}(x)\dot{n}(\alpha x) = \iota(\alpha x^2) + \alpha x^2. \qquad \square$$

§17. Whitehead Groups

For an arbitrary central simple algebra A, we set

$$\mathrm{SL}_1(A) = \{\, a \in A^\times \mid \mathrm{Nrd}_A(a) = 1 \,\}.$$

This group contains the normal subgroup $[A^\times, A^\times]$ generated by commutators $aba^{-1}b^{-1}$. The factor group is denoted

$$\mathrm{SK}_1(A) = \mathrm{SL}_1(A)/[A^\times, A^\times].$$

This group is known in algebraic K-theory as the *reduced Whitehead group* of A.

It is known that $\mathrm{SK}_1(A) = 0$ if A is split (and $A \neq M_2(\mathbb{F}_2)$) or if the index of A is square-free (a result due to Wang [304], see for example Pierce [219, 16.6] or the lecture notes [85]). In the first subsection, we consider the next interesting case where A is a biquaternion algebra. Let F be the center of A, which may be of arbitrary characteristic. Denote by $I^k F$ the k-th power of the fundamental ideal IF of the Witt ring WF of nonsingular bilinear spaces over F, and let $W_q F$ denote the Witt group of nonsingular even-dimensional quadratic spaces over F, which is a module over WF. We write $I^k W_q F$ for $I^k F \cdot W_q F$, so $I^k W_q F = I^{k+1} F$ if char $F \neq 2$. Our objective is to define a canonical injective homomorphism

$$\overline{\alpha} \colon \mathrm{SK}_1(A) \hookrightarrow I^3 W_q F / I^4 W_q F,$$

from which examples where $\mathrm{SK}_1(A) \neq 0$ are easily derived.

In the second subsection, we briefly discuss analogues of the reduced Whitehead group for algebras with involution in characteristic different from 2.

17.A. SK_1 of biquaternion algebras. Although the map $\overline{\alpha}$ that we will define is canonical, it is induced by a map α_σ whose definition depends on the choice of a symplectic involution. Therefore, we start with some general observations on symplectic involutions on biquaternion algebras.

Let A be a biquaternion algebra over a field F of arbitrary characteristic, and let σ be a symplectic involution on A. Recall from §16.B the linear endomorphism $^-$ of $\mathrm{Symd}(A, \sigma)$ defined by $\overline{x} = \mathrm{Trp}_\sigma(x) - x$, and the quadratic form

$$\mathrm{Nrp}_\sigma(x) = x\overline{x} = \overline{x}x \quad \text{for } x \in \mathrm{Symd}(A, \sigma).$$

As in §15.C, we let

$$\mathrm{Symd}(A, \sigma)^0 = \{\, x \in \mathrm{Symd}(A, \sigma) \mid \mathrm{Trp}_\sigma(x) = 0 \,\},$$

and we write $\mathrm{Symd}(A, \sigma)^\times = \mathrm{Symd}(A, \sigma) \cap A^\times$ for simplicity. For $v \in \mathrm{Symd}(A, \sigma)^\times$ and $x \in A$, we have $\sigma(x)vx \in \mathrm{Symd}(A, \sigma)$, because if $v = w + \sigma(w)$, then $\sigma(x)vx = \sigma(x)wx + \sigma\big(\sigma(x)wx\big)$. We may therefore consider the quadratic form $\Phi_v \colon A \to F$ defined by

$$\Phi_v(x) = \mathrm{Trp}_\sigma\big(\sigma(x)vx\big) \quad \text{for } x \in A.$$

(17.1) Proposition. *For each* $v \in \mathrm{Symd}(A, \sigma)^{\times}$, *the quadratic form* Φ_v *is a scalar multiple of a 4-fold Pfister form. This form is hyperbolic if* $\mathrm{Trp}_{\sigma}(v) = 0$. *Moreover, if* σ *is a hyperbolic symplectic involution, then* Φ_v *is hyperbolic for all* $v \in \mathrm{Symd}(A, \sigma)^{\times}$.

Proof: Suppose first that σ is hyperbolic. The algebra A then contains an isotropic right ideal I of reduced dimension 2, i.e., $\dim_F I = 8$. For $x \in I$ we have $\sigma(x)x = 0$, hence for all $v = w + \sigma(w) \in \mathrm{Symd}(A, \sigma)^{\times}$,

$$\Phi_v\big(\sigma(x)\big) = \mathrm{Trp}_{\sigma}\big(xv\sigma(x)\big) = \mathrm{Trd}_A\big(xw\sigma(x)\big) = \mathrm{Trd}_A\big(w\sigma(x)x\big) = 0.$$

Therefore, $\sigma(I)$ is a totally isotropic subspace of A for the form Φ_v, hence this form is hyperbolic.

For the rest of the proof, assume σ is not hyperbolic, and let $V \subset \mathrm{Symd}(A, \sigma)$ be a 3-dimensional subspace containing 1 and v, and not contained in $\mathrm{Symd}(A, \sigma)^0$. By (16.16), there is a decomposition of A into a tensor product of quaternion subalgebras, so that

$$(A, \sigma) = (Q_1, \sigma_1) \otimes_F (Q_2, \gamma_2)$$

where σ_1 is an orthogonal involution, γ_2 is the canonical involution, and $v \in \mathrm{Sym}(Q_1, \sigma_1)$. In view of the computation of the bilinear form $T_{(Q_1, \sigma_1, v)}$ in (11.6), the following lemma completes the proof:

(17.2) Lemma. *Suppose* $(A, \sigma) = (Q_1, \sigma_1) \otimes_F (Q_2, \gamma_2)$, *where* σ_1 *is an orthogonal involution and* γ_2 *is the canonical (symplectic) involution. We have* $\mathrm{Sym}(Q_1, \sigma_1) \subset \mathrm{Symd}(A, \sigma)$ *and, for all* $v \in \mathrm{Sym}(Q_1, \sigma_1)^{\times}$,

$$\Phi_v = T_{(Q_1, \sigma_1, v)} \cdot \mathrm{Nrd}_{Q_2}$$

where Nrd_{Q_2} *is the reduced norm quadratic form on* Q_2.

Proof: Let $\ell_2 \in Q_2$ be such that $\ell_2 + \gamma_2(\ell_2) = 1$. For all $s \in \mathrm{Sym}(Q_1, \sigma_1)$, we have

$$s \otimes 1 = s \otimes \ell_2 + \sigma(s \otimes \ell_2),$$

hence $s \otimes 1 \in \mathrm{Symd}(A, \sigma)$ and

$$\mathrm{Trp}_{\sigma}(s \otimes 1) = \mathrm{Trd}_A(s \otimes \ell_2) = \mathrm{Trd}_{Q_1}(s)\,\mathrm{Trd}_{Q_2}(\ell_2) = \mathrm{Trd}_{Q_1}(s).$$

Let $v \in \mathrm{Sym}(Q_1, \sigma_1)^{\times}$. For $x_1 \in Q_1$ and $x_2 \in Q_2$, we have

$$\sigma(x_1 \otimes x_2)v(x_1 \otimes x_2) = \sigma_1(x_1)vx_1 \otimes \gamma_2(x_2)x_2 = \big(\sigma_1(x_1)vx_1 \otimes 1\big)\,\mathrm{Nrd}_{Q_2}(x_2).$$

Therefore,

$$\Phi_v(x_1 \otimes x_2) = \mathrm{Trp}_{\sigma}\big(\sigma_1(x_1)vx_1 \otimes 1\big)\,\mathrm{Nrd}_{Q_2}(x_2) = \mathrm{Trd}_{Q_1}\big(\sigma_1(x_1)vx_1\big)\,\mathrm{Nrd}_{Q_2}(x_2),$$

hence

$$\Phi_v(x_1 \otimes x_2) = T_{(Q_1, \sigma_1, v)}(x_1, x_1)\,\mathrm{Nrd}_{Q_2}(x_2).$$

To complete the proof, it remains only to show that the polar form b_{Φ_v} of Φ_v is the tensor product $T_{(Q_1, \sigma_1, v)} \otimes b_{\mathrm{Nrd}_{Q_2}}$.

For $x, y \in A$, we have

$$b_{\Phi_v}(x, y) = \mathrm{Trp}_{\sigma}\big(\sigma(x)vy + \sigma(\sigma(x)vy)\big) = \mathrm{Trd}_A\big(\sigma(x)vy\big),$$

hence, for $x = x_1 \otimes x_2$ and $y = y_1 \otimes y_2$,

$$b_{\Phi_v}(x, y) = \mathrm{Trd}_{Q_1}\big(\sigma_1(x_1)vy_1\big)\,\mathrm{Trd}_{Q_2}\big(\gamma_2(x_2)y_2\big).$$

Since $\mathrm{Trd}_{Q_2}\big(\gamma_2(x_2)y_2\big) = b_{\mathrm{Nrd}_{Q_2}}(x_2, y_2)$, the proof is complete. $\qquad \square$

The definition of α_σ uses the following result, which is reminiscent of Hilbert's theorem 90:

(17.3) Lemma. *Suppose σ is not hyperbolic. For every $u \in \operatorname{Symd}(A, \sigma)$ such that $\operatorname{Nrp}_\sigma(u) = 1$, there exists $v \in \operatorname{Symd}(A, \sigma)^\times$ such that $u = v\overline{v}^{-1}$. If $u \neq -1$, the element v is uniquely determined up to multiplication by a factor in $F^\times$.*

Proof: We first prove the existence of v. If $u = -1$, we may take for v any unit in $\operatorname{Symd}(A, \sigma)^0$. If $u \neq -1$, let $v = 1 + u$. We have

$$\overline{v}u = (1 + \overline{u})u = u + \operatorname{Nrp}_\sigma(u) = v,$$

hence v satisfies the required conditions if it is invertible. If v is not invertible, then $v\overline{v} = 0$, since $v\overline{v} = \operatorname{Nrp}_\sigma(v) \in F$. In that case, we also have $v\overline{v}u = 0$, hence $v^2 = 0$ since $\overline{v}u = v$. It follows that $\operatorname{Trp}_\sigma(v) = 0$, and by (15.21) we derive a contradiction with the hypothesis that σ is not hyperbolic. The existence of v is thus proved.

Suppose next that $u \neq -1$ and $v_1, v_2 \in \operatorname{Symd}(A, \sigma)^\times$ are such that

$$u = v_1\overline{v_1}^{-1} = v_2\overline{v_2}^{-1}.$$

Then

$$u + 1 = (v_1 + \overline{v_1})\overline{v_1}^{-1} = (v_2 + \overline{v_2})\overline{v_2}^{-1}.$$

Since $v_i + \overline{v_i} = \operatorname{Trp}_\sigma(v_i) \in F$ for $i = 1, 2$, these equations together with the hypothesis that $u \neq -1$ show that $\operatorname{Trp}_\sigma(v_1) \neq 0$ and $\operatorname{Trp}_\sigma(v_2) \neq 0$. They also yield

$$\operatorname{Trp}_\sigma(v_1)\overline{v_2} = \operatorname{Trp}_\sigma(v_2)\overline{v_1},$$

hence v_1 and v_2 differ by a nonzero factor in F. $\qquad\square$

Now, consider the following subgroup of $F^\times \times A^\times$:

$$\Gamma = \{\, (\lambda, a) \in F^\times \times A^\times \mid \lambda^2 = \operatorname{Nrd}_A(a) \,\}.$$

For $(\lambda, a) \in \Gamma$, we have $-\lambda^{-1}\sigma(a)a \in \operatorname{Symd}(A, \sigma)$, since $1 \in \operatorname{Symd}(A, \sigma)$, and Proposition (2.13) shows that $\operatorname{Nrp}_\sigma(-\lambda^{-1}\sigma(a)a) = 1$. Therefore, if σ is not hyperbolic, the preceding lemma yields $v \in \operatorname{Symd}(A, \sigma)^\times$ such that

(17.4) $$v\overline{v}^{-1} = -\lambda^{-1}\sigma(a)a.$$

If $\lambda^{-1}\sigma(a)a = 1$ (i.e., $a \in \operatorname{GSp}(A, \sigma)$ and $\lambda = \mu(a)$), we have $\overline{v} = -v$, hence $\operatorname{Trp}_\sigma(v) = 0$. Proposition (17.1) then shows that Φ_v is hyperbolic. If $\lambda^{-1}\sigma(a)a \neq 1$, the element v is uniquely determined up to a factor in $F^\times$. Therefore, the quadratic form $\Phi_v \in I^3 W_q F$ is also uniquely determined up to a factor in $F^\times$, and its class in $I^3 W_q F / I^4 W_q F$ is uniquely determined.

We may therefore set the following definition:

(17.5) Definition. Let $\alpha_\sigma \colon \Gamma \to I^3 W_q F / I^4 W_q F$ be defined as follows:

(1) If σ is hyperbolic, let $\alpha_\sigma = 0$.
(2) If σ is not hyperbolic, let

$$\alpha_\sigma(\lambda, a) = \Phi_v + I^4 W_q F$$

where $v \in \operatorname{Symd}(A, \sigma)^\times$ satisfies (17.4).

In particular, the observations above show that $\alpha_\sigma(\mu(g), g) = 0$ for $g \in \operatorname{GSp}(A, \sigma)$.

If L is an extension field of F over which σ is hyperbolic, every quadratic form Φ_v for $v \in \mathrm{Symd}(A, \sigma)^\times$ becomes hyperbolic over L by (17.1), hence the definition above is compatible with scalar extension.

Observe that the group $\mathrm{SL}_1(A)$ embeds in Γ by mapping $a \in \mathrm{SL}_1(A)$ to $(1, a) \in \Gamma$. Our goal is to prove the following theorem:

(17.6) Theorem. *The map α_σ defined above is a homomorphism. Its restriction to $\mathrm{SL}_1(A)$ does not depend on the choice of the symplectic involution σ; letting $\alpha \colon \mathrm{SL}_1(A) \to I^3 W_q F / I^4 W_q F$ denote this restriction, we have $\ker \alpha = [A^\times, A^\times]$, hence α induces an injective homomorphism*

$$\overline{\alpha} \colon \mathrm{SK}_1(A) \hookrightarrow I^3 W_q F / I^4 W_q F.$$

The rest of this subsection consists of the proof, which we break into three parts: we first show that α_σ is a homomorphism, then we investigate the effect of a change of involution, and finally we determine the kernel of α_σ.

α_σ **is a homomorphism.** If σ is hyperbolic, then α_σ is clearly a homomorphism since we set $\alpha_\sigma = 0$. Throughout this part of the proof, we may thus assume that σ is not hyperbolic. Let (λ, a), $(\lambda', a') \in \Gamma$. In order to show that

$$\alpha_\sigma(\lambda, a) + \alpha_\sigma(\lambda', a') = \alpha_\sigma(\lambda\lambda', aa'),$$

we first reduce to the case where a, $a' \in \mathrm{Symd}(A, \sigma)$ and $\lambda = \mathrm{Nrp}_\sigma(a)$, $\lambda' = \mathrm{Nrp}_\sigma(a')$.

(17.7) Lemma. *For $(\lambda, a) \in \Gamma$ and $g \in \mathrm{GSp}(A, \sigma)$,*

$$\alpha_\sigma(\lambda, a) = \alpha_\sigma\big(\mu(g)\lambda, ga\big) = \alpha_\sigma\big(\lambda\mu(g), ag\big).$$

Proof: Let $v \in \mathrm{Symd}(A, \sigma)^\times$ be subject to (17.4); since

$$\lambda^{-1}\mu(g)^{-1}\sigma(ga)ga = \lambda^{-1}\sigma(a)a,$$

the quadratic form Φ_v represents $\alpha_\sigma(\lambda, a)$ as well as $\alpha_\sigma\big(\mu(g)\lambda, ga\big)$, hence the first equation is clear. To prove the second equation, we calculate

$$-\mu(g)^{-1}\lambda^{-1}\sigma(ag)ag = g^{-1}vg \cdot g^{-1}\overline{v}^{\,-1}g.$$

By (16.7), the last factor on the right side is equal to $\overline{g^{-1}vg}^{\,-1}$, hence

$$\alpha_\sigma\big(\lambda\mu(g), ag\big) = \Phi_{g^{-1}vg} + I^4 W_q F \in I^3 W_q F / I^4 W_q F.$$

For $x \in A$, we have $\sigma(x)g^{-1}vgx = \mu(g)^{-1}\sigma(gx)vgx$, hence

$$\Phi_{g^{-1}vg}(x) = \mu(g)^{-1}\Phi_v(gx).$$

This equation shows that the quadratic forms $\Phi_{g^{-1}vg}$ and Φ_v are similar, hence $\alpha_\sigma\big(\lambda\mu(g), ag\big) = \alpha_\sigma(\lambda, a)$. $\qquad\square$

(17.8) Lemma. *For all $(\lambda, a) \in \Gamma$, there exist a similitude $g \in \mathrm{GSp}(A, \sigma)$ and units u, $v \in \mathrm{Symd}(A, \sigma)^\times$ such that*

$$(\lambda, a) = \big(\mu(g), g\big) \cdot \big(\mathrm{Nrp}_\sigma(u), u\big) = \big(\mathrm{Nrp}_\sigma(v), v\big) \cdot \big(\mu(g), g\big).$$

Proof: By (17.3), there exists some $u \in \mathrm{Symd}(A, \sigma)^\times$ such that $u\overline{u}^{\,-1} = \lambda^{-1}\sigma(a)a$. For $g = au^{-1}$ we have

$$\sigma(g)g = u^{-1}\sigma(a)au^{-1} = \lambda \mathrm{Nrp}_\sigma(u)^{-1} \in F^\times,$$

hence $g \in \mathrm{GSp}(A, \sigma)$ and $\mu(g)\mathrm{Nrp}_\sigma(u) = \lambda$. We thus get the first decomposition. In order to get the second, it suffices to choose $v = gug^{-1}$. $\qquad\square$

For (λ, a), $(\lambda', a') \in \Gamma$, we may thus find g, $g' \in \mathrm{GSp}(A, \sigma)$ and u, $v \in \mathrm{Symd}(A, \sigma)^{\times}$ such that

$$(\lambda, a) = \big(\mu(g), g\big) \cdot \big(\mathrm{Nrp}_{\sigma}(u), u\big),$$
$$(\lambda', a') = \big(\mathrm{Nrp}_{\sigma}(v), v\big) \cdot \big(\mu(g'), g'\big),$$

hence also

$$(\lambda \lambda', aa') = \big(\mu(g), g\big) \cdot \big(\mathrm{Nrp}_{\sigma}(u), u\big) \cdot \big(\mathrm{Nrp}_{\sigma}(v), v\big) \cdot \big(\mu(g'), g'\big).$$

From (17.7), it follows that

$$\alpha_{\sigma}(\lambda, a) = \alpha_{\sigma}\big(\mathrm{Nrp}_{\sigma}(u), u\big), \qquad \alpha_{\sigma}(\lambda', a') = \alpha_{\sigma}\big(\mathrm{Nrp}_{\sigma}(v), v\big)$$

and

$$\alpha_{\sigma}(\lambda \lambda', aa') = \alpha_{\sigma}\big(\mathrm{Nrp}_{\sigma}(u)\,\mathrm{Nrp}_{\sigma}(v), uv\big).$$

Therefore, in order to show that

$$\alpha_{\sigma}(\lambda \lambda', aa') = \alpha_{\sigma}(\lambda, a) + \alpha_{\sigma}(\lambda', a'),$$

it suffices to show that

(17.9) $\qquad \alpha_{\sigma}\big(\mathrm{Nrp}_{\sigma}(u)\,\mathrm{Nrp}_{\sigma}(v), uv\big) = \alpha_{\sigma}\big(\mathrm{Nrp}_{\sigma}(u), u\big) + \alpha_{\sigma}\big(\mathrm{Nrp}_{\sigma}(v), v\big).$

We have thus achieved the desired reduction.

If $u \in \mathrm{Symd}(A, \sigma)^0$, then $-\mathrm{Nrp}_{\sigma}(u)^{-1}\sigma(u)u = 1$, hence

$$\alpha_{\sigma}\big(\mathrm{Nrp}_{\sigma}(u), u\big) = \Phi_1 + I^4 W_q F.$$

Therefore, if u and v both lie in $\mathrm{Symd}(A, \sigma)^0$, the right side of (17.9) vanishes. The left side also vanishes, since $uv \in \mathrm{GSp}(A, \sigma)$ and $\mu(uv) = \mathrm{Nrp}_{\sigma}(u)\,\mathrm{Nrp}_{\sigma}(v)$. For the rest of the proof of (17.9), we may thus assume that u and v are not both in $\mathrm{Symd}(A, \sigma)^0$.

Consider a 3-dimensional subspace $V \subset \mathrm{Symd}(A, \sigma)$ which contains 1, u and v, and is therefore not contained in $\mathrm{Symd}(A, \sigma)^0$. By (16.16), there is a decomposition of A into a tensor product of quaternion subalgebras, so

$$(A, \sigma) = (Q_1, \sigma_1) \otimes_F (Q_2, \gamma_2)$$

where σ_1 is an orthogonal involution, γ_2 is the canonical involution, and u, $v \in \mathrm{Sym}(Q_1, \sigma_1)$, hence also $uv \in Q_1$. For $x \in Q_1$, we have $\mathrm{Prd}^2_{Q_1, x} = \mathrm{Prd}_{A, x}$, hence $\mathrm{Prp}_{\sigma, x} = \mathrm{Prd}_{Q_1, x}$ and therefore $\mathrm{Nrp}_{\sigma}(x) = \mathrm{Nrd}_{Q_1}(x)$. To prove (17.9), it now suffices to prove the following lemma:

(17.10) Lemma. *Suppose A decomposes into a tensor product of quaternion algebras stable under the symplectic involution σ, which may be hyperbolic:*

$$(A, \sigma) = (Q_1, \sigma_1) \otimes (Q_2, \gamma_2),$$

where σ_1 is an orthogonal involution and γ_2 is the canonical (symplectic) involution. For all $x \in Q_1^{\times}$,

$$\alpha_{\sigma}\big(\mathrm{Nrd}_{Q_1}(x), x\big) = \langle\!\langle \mathrm{Nrd}_{Q_1}(x), \mathrm{disc}\,\sigma_1 \rangle\!\rangle \cdot \mathrm{Nrd}_{Q_2}.$$

Indeed, assuming the lemma and letting $\theta = \langle\!\langle \mathrm{disc}\,\sigma_1 \rangle\!\rangle \cdot \mathrm{Nrd}_{Q_2}$, the left-hand side of (17.9) is then $\langle\!\langle \mathrm{Nrd}_{Q_1}(uv) \rangle\!\rangle \cdot \theta$, while the right-hand side is $\langle\!\langle \mathrm{Nrd}_{Q_1}(u) \rangle\!\rangle \cdot \theta + \langle\!\langle \mathrm{Nrd}_{Q_1}(v) \rangle\!\rangle \cdot \theta$. The equality follows from the congruence

$$\langle\!\langle \mathrm{Nrd}_{Q_1}(u) \rangle\!\rangle + \langle\!\langle \mathrm{Nrd}_{Q_1}(v) \rangle\!\rangle \equiv \langle\!\langle \mathrm{Nrd}_{Q_1}(u)\,\mathrm{Nrd}_{Q_1}(v) \rangle\!\rangle \mod I^2 F.$$

Proof of (17.10): Suppose first that σ is hyperbolic. We then have to show that the quadratic form $\langle\!\langle \mathrm{Nrd}_{Q_1}(x), \mathrm{disc}\,\sigma_1 \rangle\!\rangle \cdot \mathrm{Nrd}_{Q_2}$ is hyperbolic for all $x \in Q_1$. For $v \in \mathrm{Sym}(Q_1, \sigma_1)^\times$, Proposition (17.1) and Lemma (17.2) show that the quadratic form $T_{(Q_1, \sigma_1, v)} \cdot \mathrm{Nrd}_{Q_2}$ is hyperbolic. In view of (11.6), this means that

$$\langle\!\langle \mathrm{Nrd}_{Q_1}(vs), \mathrm{disc}\,\sigma_1 \rangle\!\rangle \cdot \mathrm{Nrd}_{Q_2} = 0 \qquad \text{in } W_q F$$

for all $v \in \mathrm{Sym}(Q_1, \sigma_1)^\times$ and all $s \in Q_1^\times$ such that $\sigma_1(s) = s = -\gamma_1(s)$, where γ_1 is the canonical involution on Q_1. In particular (for $v = 1$), the quadratic form $\langle\!\langle \mathrm{Nrd}_{Q_1}(s), \mathrm{disc}\,\sigma_1 \rangle\!\rangle \cdot \mathrm{Nrd}_{Q_2}$ is hyperbolic. Adding it to both sides of the equality above, we get

$$\langle\!\langle \mathrm{Nrd}_{Q_1}(v), \mathrm{disc}\,\sigma_1 \rangle\!\rangle \cdot \mathrm{Nrd}_{Q_2} = 0 \qquad \text{for all } v \in \mathrm{Sym}(Q_1, \sigma_1)^\times.$$

To complete the proof in the case where σ is hyperbolic, it now suffices to show that $\mathrm{Sym}(Q_1, \sigma_1)^\times$ generates $Q_1^\times$. For all $x \in Q_1^\times$, the intersection

$$\mathrm{Sym}(Q_1, \sigma_1) \cap x\,\mathrm{Sym}(Q_1, \sigma_1)$$

has dimension at least 2. Since the restriction of Nrd_{Q_1} to $\mathrm{Sym}(Q_1, \sigma_1)$ is a nonsingular quadratic form, this intersection is not totally isotropic for Nrd_{Q_1}, hence it contains an invertible element $s_1 \in \mathrm{Sym}(Q_1, \sigma_1)^\times$. We have $s_1 = xs_2$ for some $s_2 \in \mathrm{Sym}(Q_1, \sigma_1)^\times$, hence $x = s_1 s_2^{-1}$ is in the group generated by $\mathrm{Sym}(Q_1, \sigma_1)^\times$.

For the rest of the proof, assume that σ is not hyperbolic. Let $\sigma_1 = \mathrm{Int}(r) \circ \gamma_1$ for some $r \in \mathrm{Skew}(Q_1, \gamma_1) \smallsetminus F$; thus, $r^2 \in F^\times$ and $\mathrm{disc}\,\sigma_1 = r^2 \cdot F^{\times 2}$.

If $\sigma(x)x = \mathrm{Nrd}_{Q_1}(x)$, then $x \in \mathrm{GSp}(A, \sigma)$ and $\mu(x) = \mathrm{Nrd}_{Q_1}(x)$, hence $\alpha_\sigma(\mathrm{Nrd}_{Q_1}(x), x) = 0$. On the other hand, the condition also implies $\sigma_1(x) = \gamma_1(x)$, hence x commutes with r. Therefore, $x \in F[r]$, and $\langle\!\langle \mathrm{Nrd}_{Q_1}(x), \mathrm{disc}\,\sigma_1 \rangle\!\rangle$ is metabolic. The lemma thus holds in this case.

Assume finally that $\sigma(x)x \neq \mathrm{Nrd}_{Q_1}(x)$, and let

$$w = 1 - \mathrm{Nrd}_{Q_1}(x)^{-1}\sigma(x)x = 1 - \sigma(x)\gamma_1(x)^{-1} \in Q_1,$$

so that $\overline{w}\big(-\mathrm{Nrd}_{Q_1}(x)^{-1}\sigma(x)x\big) = w$. Since σ is not hyperbolic, the same arguments as in the proof of (17.3) show that w is invertible, hence

$$\alpha_\sigma\big(\mathrm{Nrd}_{Q_1}(x), x\big) = \Phi_w + I^4 W_q F.$$

By (17.2), we have $\Phi_w \equiv T_{(Q_1, \sigma_1, w)} \cdot \mathrm{Nrd}_{Q_2} \mod I^4 W_q F$. Moreover, we may compute $T_{(Q_1, \sigma_1, w)}$ by (11.6): since $w\gamma_1(x) = \gamma_1(x) - \sigma_1(x) \in Q_1^\times$ satisfies

$$\sigma_1\big(w\gamma_1(x)\big) = w\gamma_1(x) = -\gamma_1\big(w\gamma_1(x)\big),$$

we may substitute $w\gamma_1(x)$ for s in (11.6) and get

$$T_{(Q_1, \sigma_1, w)} \equiv \langle\!\langle \mathrm{Nrd}_{Q_1}\big(w^2\gamma_1(x)\big), \mathrm{disc}\,\sigma_1 \rangle\!\rangle \equiv \langle\!\langle \mathrm{Nrd}_{Q_1}(x), \mathrm{disc}\,\sigma_1 \rangle\!\rangle \mod I^3 F. \qquad \square$$

Change of involution. We keep the same notation as above, and allow the symplectic involution σ to be hyperbolic. For $x \in \mathrm{Symd}(A, \sigma)^0$, we have $x^2 = -\mathrm{Nrp}_\sigma(x) \in F$. We endow $\mathrm{Symd}(A, \sigma)^0$ with the restriction of Nrp_σ or, equivalently for the next result, with the squaring quadratic form $s_\sigma(x) = x^2$ (see §15.C).

(17.11) Proposition. *Every proper isometry of* $\mathrm{Symd}(A, \sigma)^0$ *has the form* $x \mapsto gxg^{-1}$ *for some* $g \in \mathrm{GSp}(A, \sigma)$. *For every* $g \in \mathrm{GSp}(A, \sigma)$, *one can find two or four anisotropic vectors* $v_1, \ldots, v_r \in \mathrm{Symd}(A, \sigma)^0$ *such that*

$$\big(\mu(g), g\big) = \big(\mathrm{Nrp}_\sigma(v_1), v_1\big) \cdots \big(\mathrm{Nrp}_\sigma(v_r), v_r\big) \quad \text{in } \Gamma.$$

Proof: The proposition readily follows from (15.18). We may however give a short direct argument: for all $v \in \mathrm{Symd}(A, \sigma)^0$ anisotropic, computation shows that the hyperplane reflection ρ_v maps $x \in \mathrm{Symd}(A, \sigma)^0$ to $-vxv^{-1}$. The Cartan-Dieudonné theorem shows that every proper isometry is a product of an even number of hyperplane reflections, and is therefore of the form

$$x \mapsto (v_1 \cdots v_r)x(v_r^{-1} \cdots v_1^{-1})$$

for some anisotropic $v_1, \ldots, v_r \in \mathrm{Symd}(A, \sigma)^0$. Since

$$\sigma(v_1 \cdots v_r) \cdot v_1 \cdots v_r = v_1^2 \cdots v_r^2 \in F^\times,$$

the element $v_1 \cdots v_r$ is in $\mathrm{GSp}(A, \sigma)$, and the first part is proved.

To prove the second part, observe that for $g \in \mathrm{GSp}(A, \sigma)$ and $x \in \mathrm{Symd}(A, \sigma)^0$ we have $gxg^{-1} = \mu(g)^{-1}gx\sigma(g)$, hence by (16.7) and (12.23),

$$\mathrm{Nrp}_\sigma(gxg^{-1}) = \mu(g)^{-2}\,\mathrm{Nrd}_A(g)\,\mathrm{Nrp}_\sigma(x) = \mathrm{Nrp}_\sigma(x).$$

Therefore, the map $x \mapsto gxg^{-1}$ is an isometry of $\mathrm{Symd}(A, \sigma)^0$. If this isometry is improper, then char $F \neq 2$ and $x \mapsto -gxg^{-1}$ is proper, hence of the form $x \mapsto g'xg'^{-1}$ for some $g' \in \mathrm{GSp}(A, \sigma)$. In that case $g^{-1}g'$ anticommutes with every element in $\mathrm{Symd}(A, \sigma)^0$. However, using a decomposition of (A, σ) as in (16.16), it is easily seen that no nonzero element of A anticommutes with $\mathrm{Symd}(A, \sigma)^0$. This contradiction shows that the isometry $x \mapsto gxg^{-1}$ is proper in all cases. By the Cartan-Dieudonné theorem, it is a product of an even number r of hyperplane reflections with $r \leq 5$, hence we may find anisotropic $v_1, \ldots, v_r \in \mathrm{Symd}(A, \sigma)^0$ such that

$$gxg^{-1} = (v_1 \cdots v_r)x(v_r^{-1} \cdots v_1^{-1}) \quad \text{for } x \in \mathrm{Symd}(A, \sigma)^0.$$

Since $\mathrm{Symd}(A, \sigma)^0$ generates A, it follows that $g^{-1}(v_1 \cdots v_r) \in F^\times$. Multiplying v_1 by a suitable factor in $F^\times$, we get $g = v_1 \cdots v_r$; then

$$\mu(g) = v_1^2 \cdots v_r^2 = \mathrm{Nrp}_\sigma(v_1) \cdots \mathrm{Nrp}_\sigma(v_r). \qquad \square$$

Now, let τ be another symplectic involution on A. Recall from (16.18) the 3-fold Pfister form $j_\sigma(\tau)$ uniquely determined by the condition

$$j_\sigma(\tau) \equiv \mathrm{Nrp}_\sigma - \mathrm{Nrp}_\tau \quad \mod I^3 W_q F.$$

Since $j_\sigma(\tau) \equiv -j_\sigma(\tau) \equiv \mathrm{Nrp}_\tau - \mathrm{Nrp}_\sigma \mod I^3 W_q F$, we have $j_\sigma(\tau) \simeq j_\tau(\sigma)$.

(17.12) Proposition. *For all* $(\lambda, a) \in \Gamma$,

$$\alpha_\sigma(\lambda, a) + \alpha_\tau(\lambda, a) = \langle\!\langle \lambda \rangle\!\rangle \cdot j_\sigma(\tau) + I^4 W_q F = \langle\!\langle \lambda \rangle\!\rangle \cdot j_\tau(\sigma) + I^4 W_q F.$$

Proof: If σ and τ are hyperbolic, the proposition is clear since $\alpha_\sigma = \alpha_\tau = 0$ and $j_\sigma(\tau) \simeq j_\sigma(\sigma) = 0$ in $W_q F$ by (16.19), since all the hyperbolic involutions are conjugate. We may thus assume that at least one of σ, τ is not hyperbolic. Let us assume for instance that σ is not hyperbolic, and let $\tau = \mathrm{Int}(u) \circ \sigma$ for some $u \in \mathrm{Symd}(A, \sigma)^\times$.

We consider two cases: suppose first that $\mathrm{Trp}_\sigma(u) \neq 0$. Lemma (17.8) and Proposition (17.11) show that Γ is generated by elements of the form $(\mathrm{Nrp}_\sigma(v), v)$, with $v \in \mathrm{Sym}(A, \sigma)^\times$. Therefore, it suffices to prove

$$\alpha_\sigma(\mathrm{Nrp}_\sigma(v), v) + \alpha_\tau(\mathrm{Nrp}_\sigma(v), v) = \langle\!\langle \mathrm{Nrp}_\sigma(v) \rangle\!\rangle \cdot j_\sigma(\tau) + I^4 W_q F$$

for all $v \in \mathrm{Symd}(A, \sigma)^{\times}$. Let $V \subset \mathrm{Symd}(A, \sigma)$ be a 3-dimensional subspace containing 1, u and v. Since $\mathrm{Trp}_{\sigma}(u) \neq 0$, we have[27] $V \not\subset \mathrm{Symd}(A, \sigma)^0$. By (16.16), there is a decomposition

$$(A, \sigma) = (Q_1, \sigma_1) \otimes_F (Q_2, \gamma_2)$$

where Q_1 is the quaternion subalgebra generated by V and $\sigma_1 = \sigma|_{Q_1}$ is an orthogonal involution. By (17.10), we have

$$\alpha_{\sigma}\big(\mathrm{Nrp}_{\sigma}(v), v\big) = \langle\!\langle \mathrm{Nrp}_{\sigma}(v), \mathrm{disc}\, \sigma_1 \rangle\!\rangle \cdot \mathrm{Nrd}_{Q_2} + I^4 W_q F.$$

Since $\tau = \mathrm{Int}(u) \circ \sigma$ and $u \in Q_1$, the algebra Q_1 is also stable under τ, hence

$$(A, \tau) = (Q_1, \tau_1) \otimes_F (Q_2, \gamma_2)$$

where $\tau_1 = \mathrm{Int}(u) \circ \sigma_1$. The involution τ_1 is orthogonal, since $u \in \mathrm{Sym}(Q_1, \sigma_1)$ and $\mathrm{Trd}_{Q_1}(u) = \mathrm{Trp}_{\sigma}(u) \neq 0$. Therefore, by (17.10) again,

$$\alpha_{\tau}\big(\mathrm{Nrp}_{\sigma}(v), v\big) = \langle\!\langle \mathrm{Nrp}_{\sigma}(v), \mathrm{disc}\, \tau_1 \rangle\!\rangle \cdot \mathrm{Nrd}_{Q_2} + I^4 W_q F.$$

On the other hand, we have $\mathrm{disc}\, \tau_1 = \mathrm{Nrd}_{Q_1}(u)\, \mathrm{disc}\, \sigma_1$ by (7.3), hence

$$\alpha_{\sigma}\big(\mathrm{Nrp}_{\sigma}(v), v\big) + \alpha_{\tau}\big(\mathrm{Nrp}_{\sigma}(v), v\big) = \langle\!\langle \mathrm{Nrp}_{\sigma}(v), \mathrm{Nrd}_{Q_1}(u) \rangle\!\rangle \cdot \mathrm{Nrd}_{Q_2} + I^4 W_q F.$$

This completes the proof in the case where $\mathrm{Trp}_{\sigma}(u) \neq 0$, since the proof of (16.18) shows that

$$\langle\!\langle \mathrm{Nrd}_{Q_1}(u) \rangle\!\rangle \cdot \mathrm{Nrd}_{Q_2} = \langle\!\langle \mathrm{Nrp}_{\sigma}(u) \rangle\!\rangle \cdot \mathrm{Nrd}_{Q_2} = j_{\sigma}(\tau).$$

Consider next the case where $\mathrm{Trp}_{\sigma}(u) = 0$. We then compare α_{σ} and α_{τ} via a third involution ρ, chosen in such a way that the first case applies to σ and ρ on one hand, and to ρ and τ on the other hand. Specifically, let $t \in \mathrm{Symd}(A, \sigma) \setminus \mathrm{Symd}(A, \sigma)^0$ be an invertible element which is not orthogonal to u for the polar form $b_{\mathrm{Nrp}_{\sigma}}$. Let

$$\xi = b_{\mathrm{Nrp}_{\sigma}}(u, t) = \overline{u}t + \overline{t}u \neq 0.$$

Since $\overline{u} = -u$, this relation yields $tutu^{-1} = \mathrm{Nrp}_{\sigma}(t) - \xi tu^{-1}$. Letting $s = ut^{-1}$, we have $s \notin F$ since $\mathrm{Trp}_{\sigma}(t) \neq 0$ while $\mathrm{Trp}_{\sigma}(u) = 0$, and $s^{-2} = \big(\xi s^{-1} - \mathrm{Nrp}_{\sigma}(t)\big) \mathrm{Nrp}_{\sigma}(u)^{-1} \notin F$. Let $\rho = \mathrm{Int}(t) \circ \sigma$, hence $\tau = \mathrm{Int}(s) \circ \rho$ since $u = st$. We have $s \in \mathrm{Symd}(A, \sigma)t^{-1} = \mathrm{Symd}(A, \rho)$, and $\mathrm{Trp}_{\rho}(s) \neq 0$ since $s^2 \notin F$. Therefore, we may apply the first case to compare α_{τ} and α_{ρ}, and also to compare α_{ρ} and α_{σ}, since $t \notin \mathrm{Symd}(A, \sigma)^0$. We thus get

$$\alpha_{\sigma}(\lambda, a) + \alpha_{\rho}(\lambda, a) = \langle\!\langle \lambda \rangle\!\rangle \cdot j_{\sigma}(\rho) + I^4 W_q F,$$

$$\alpha_{\rho}(\lambda, a) + \alpha_{\tau}(\lambda, a) = \langle\!\langle \lambda \rangle\!\rangle \cdot j_{\rho}(\tau) + I^4 W_q F$$

for all $(\lambda, a) \in \Gamma$. The result follows by adding these relations, since $j_{\sigma}(\rho) + j_{\rho}(\tau) \equiv j_{\sigma}(\tau) \mod I^3 W_q F$. $\qquad\square$

(17.13) Corollary. *For all $a \in \mathrm{SL}_1(A)$,*

$$\alpha_{\sigma}(1, a) = \alpha_{\tau}(1, a).$$

Proof: This readily follows from the proposition, since $\langle\!\langle 1 \rangle\!\rangle = 0$ in $W_q F$. $\qquad\square$

---

[27]If char $F \neq 2$, then $V \not\subset \mathrm{Symd}(A, \sigma)^0$ even if $\mathrm{Trp}_{\sigma}(u) = 0$, since $\mathrm{Trp}_{\sigma}(1) = 2 \neq 0$. The arguments in the first case thus yield a complete proof if char $F \neq 2$.

In view of this corollary, we may define a map $\alpha \colon \mathrm{SL}_1(A) \to I^3W_qF/I^4W_qF$ by

$$\alpha(a) = \alpha_\sigma(1, a) \quad \text{for } a \in \mathrm{SL}_1(A),$$

where σ is an arbitrary symplectic involution on A.

(17.14) Example. Let Q be a quaternion F-algebra with canonical involution γ and let $A = M_2(Q)$ with the involution θ defined by

$$\theta \begin{pmatrix} q_{11} & q_{12} \\ q_{21} & q_{22} \end{pmatrix} = \begin{pmatrix} \gamma(q_{11}) & -\gamma(q_{21}) \\ -\gamma(q_{12}) & \gamma(q_{22}) \end{pmatrix}.$$

This involution is symplectic (see (16.14)), and it is hyperbolic since

$$I = \left\{ \begin{pmatrix} q_{11} & q_{12} \\ q_{11} & q_{12} \end{pmatrix} \,\middle|\, q_{11}, q_{12} \in Q \right\}$$

is a right ideal of reduced dimension 2 such that $\theta(I) \cdot I = \{0\}$. Therefore, $\alpha_\theta = 0$ and $\alpha = 0$.

If τ is another symplectic involution on A, Proposition (17.12) yields

$$\alpha_\tau(\lambda, a) = \langle\!\langle \lambda \rangle\!\rangle \cdot j_\theta(\tau)$$

for all $(\lambda, a) \in \Gamma$. More explicitly, if $\tau = \mathrm{Int}(u) \circ \theta$ with $u \in \mathrm{Symd}(A, \theta)^\times$, it follows from (16.14) and (16.18) that $j_\theta(\tau) = \langle\!\langle \mathrm{Nrp}_\theta(u) \rangle\!\rangle \cdot \mathrm{Nrd}_Q$, hence

$$\alpha_\tau(\lambda, a) = \langle\!\langle \lambda, \mathrm{Nrp}_\theta(u) \rangle\!\rangle \cdot \mathrm{Nrd}_Q.$$

Kernel of α. We continue with the notation above. Our objective is to determine the kernel of the homomorphism $\alpha_\sigma \colon \Gamma \to I^3W_qF/I^4W_qF$; the kernel of the induced map $\alpha \colon \mathrm{SL}_1(A) \to I^3W_qF/I^4W_qF$ is then easily identified with $[A^\times, A^\times]$.

(17.15) Lemma. *Suppose the symplectic involution σ is hyperbolic and let U be an arbitrary 2-dimensional subspace in $\mathrm{Symd}(A, \sigma)$. For every $u \in \mathrm{Symd}(A, \sigma)$, there exists an invertible element $x \in A^\times$ such that $\mathrm{Trp}_A\big(\sigma(x)ux\big) = 0$ and $x\sigma(x) \notin U$.*

Note that we do not assume that u is invertible, so the quadratic form $x \mapsto \mathrm{Trp}_A\big(\sigma(x)ux\big)$ may be singular.

Proof: Since σ is hyperbolic, the index of A is 1 or 2, hence $A \simeq M_2(Q)$ for some quaternion F-algebra Q. Consider the involution θ on $M_2(Q)$ defined as in (17.14). Since all the hyperbolic involutions are conjugate by (12.35), we may identify $(A, \sigma) = \big(M_2(Q), \theta\big)$. We have $\mathrm{Nrp}_\theta\big(\begin{smallmatrix} \mathrm{Nrp}_\theta(u) & 0 \\ 0 & 1 \end{smallmatrix}\big) = \mathrm{Nrp}_\theta(u)$, hence Witt's theorem on the extension of isometries (see Scharlau [247, Theorem 7.9.1]) shows that there is an isometry of $\big(\mathrm{Sym}(M_2(Q), \theta), \mathrm{Nrp}_\theta\big)$ which maps u to $\big(\begin{smallmatrix} \mathrm{Nrp}_\theta(u) & 0 \\ 0 & 1 \end{smallmatrix}\big)$. Composing with a suitable hyperplane reflection, we may assume that this isometry is proper. By (16.11), it follows that there exist $a \in A^\times$ and $\lambda \in F^\times$ such that

$$\theta(a)ua = \lambda \begin{pmatrix} \mathrm{Nrp}_\theta(u) & 0 \\ 0 & 1 \end{pmatrix}.$$

Let $b, c \in Q^\times$ be such that $\mathrm{Nrd}_Q(c) = 1$ and $c - 1 \in Q^\times$, and let $x = a\big(\begin{smallmatrix} 1 & 1 \\ b & bc \end{smallmatrix}\big) \in A$. Then

$$\theta(x)ux = \lambda \begin{pmatrix} 1 & -\gamma(b) \\ -1 & \gamma(bc) \end{pmatrix} \begin{pmatrix} \mathrm{Nrp}_\theta(u) & 0 \\ 0 & 1 \end{pmatrix} \begin{pmatrix} 1 & 1 \\ b & bc \end{pmatrix}$$

$$= \lambda \begin{pmatrix} \mathrm{Nrp}_\theta(u) - \mathrm{Nrd}_Q(b) & \mathrm{Nrp}_\theta(u) - \mathrm{Nrd}_Q(b)c \\ -\mathrm{Nrp}_\theta(u) + \mathrm{Nrd}_Q(b)\gamma(c) & -\mathrm{Nrp}_\theta(u) + \mathrm{Nrd}_Q(b) \end{pmatrix}$$

hence $\mathrm{Trp}_\sigma\big(\theta(x)ux\big) = 0$. On the other hand, x is invertible since it is a product of invertible matrices:

$$x = a \begin{pmatrix} 1 & 0 \\ 0 & b \end{pmatrix} \begin{pmatrix} 1 & 0 \\ 1 & 1 \end{pmatrix} \begin{pmatrix} 1 & 0 \\ 0 & c^{-1} \end{pmatrix} \begin{pmatrix} 1 & 1 \\ 0 & 1 \end{pmatrix}.$$

Finally, we have $x\theta(x) = a\big(\begin{smallmatrix} 0 & \gamma(c-1)\gamma(b) \\ b(1-c) & 0 \end{smallmatrix}\big)\theta(a)$, hence $x\theta(x) \in U$ if and only if

$$\begin{pmatrix} 0 & \gamma(c-1)\gamma(b) \\ b(1-c) & 0 \end{pmatrix} \in a^{-1}U\theta(a)^{-1}.$$

Since b is arbitrary in $Q^\times$, it is clear that we can choose b such that this relation does not hold. $\qquad\square$

The following result holds for an arbitrary symplectic involution σ:

(17.16) Lemma. *If x_1, $x_2 \in \mathrm{Symd}(A,\sigma) \smallsetminus F$ satisfy $\mathrm{Trp}_\sigma(x_1) = \mathrm{Trp}_\sigma(x_2)$ and $\mathrm{Nrp}_\sigma(x_1) = \mathrm{Nrp}_\sigma(x_2)$, then there exists some $g \in \mathrm{GSp}(A,\sigma)$ such that $gx_1g^{-1} = x_2$.*

Proof: The hypothesis yields

$$\mathrm{Nrp}_\sigma(\xi + \eta x_1) = \mathrm{Nrp}_\sigma(\xi + \eta x_2) \qquad \text{for } \xi, \eta \in F,$$

hence the 2-dimensional subspace of $\mathrm{Symd}(A,\sigma)$ spanned by 1, x_1 is isometric to the subspace spanned by 1, x_2. By Witt's theorem, the isometry which maps 1 to 1 and x_1 to x_2 extends to an isometry f of $\big(\mathrm{Symd}(A,\sigma), \mathrm{Nrp}_\sigma\big)$, and this isometry may be assumed to be proper. By (16.11), there exist $\lambda \in F^\times$ and $g \in A^\times$ such that $f(x) = \lambda^{-1}gx\sigma(g)$ for all $x \in \mathrm{Symd}(A,\sigma)$. Since $f(1) = 1$, we have $g \in \mathrm{GSp}(A,\sigma)$ and $\lambda = \mu(g)$, hence

$$x_2 = f(x_1) = gx_1g^{-1}. \qquad\square$$

(17.17) Proposition. *For all $(\lambda, a) \in \Gamma$ such that $\alpha_\sigma(\lambda, a) = 0$, there exist $g \in \mathrm{GSp}(A,\sigma)$ and $x, y \in A^\times$ such that*

$$(\lambda, a) = \big(\mu(g), g\big) \cdot (1, xyx^{-1}y^{-1}).$$

Proof: Let $v = 1 - \lambda^{-1}\sigma(a)a \in \mathrm{Symd}(A,\sigma)$. If $v = 0$, then $a \in \mathrm{GSp}(A,\sigma)$ and $\lambda = \mu(a)$, so we may take $g = a$ and $x = y = 1$. For the rest of the proof, we may thus assume that $v \neq 0$.

Claim. *There exists some $y \in A^\times$ such that $\mathrm{Trp}_\sigma\big(\sigma(y)vy\big) = 0$. Moreover, if σ is hyperbolic, we may assume that $y \notin \mathrm{GSp}(A,\sigma)$ and $ay \notin \mathrm{GSp}(A,\sigma)$.*

This readily follows from (17.15) if σ is hyperbolic, since we may find $y \in A^\times$ such that $\mathrm{Trp}_\sigma\big(\sigma(y)vy\big) = 0$ and assume moreover that $y\sigma(y)$ does not lie in the subspace spanned by 1 and $a^{-1}\sigma(a)^{-1}$, hence y, $ay \notin \mathrm{GSp}(A,\sigma)$. If σ is not hyperbolic, then the proof of (17.3) shows that v is invertible and satisfies $v\bar{v}^{-1} = -\lambda^{-1}\sigma(a)a$, hence also $\alpha_\sigma(\lambda, a) = \Phi_v + I^4W_qF$. By hypothesis, $\alpha_\sigma(\lambda, a) = 0$, hence there exists some $y_0 \in A$ such that $\Phi_v(y_0) = 0$. If y_0 is invertible, the claim is proved with $y = y_0$. If y_0 is not invertible, then we have $\mathrm{Nrp}_\sigma\big(\sigma(y_0)vy_0\big) = \mathrm{Nrd}_A(y_0)\,\mathrm{Nrp}_\sigma(v) = 0$ and $\mathrm{Trp}_\sigma\big(\sigma(y_0)vy_0\big) = 0$, hence $\big(\sigma(y_0)vy_0\big)^2 = -\mathrm{Nrp}_\sigma\big(\sigma(y_0)vy_0\big) = 0$. Since σ is anisotropic, this relation implies $\sigma(y_0)vy_0 = 0$ by (15.21), hence also $v^{-1}\sigma(y_0)vy_0 = 0$, showing that the involution $\mathrm{Int}(v^{-1}) \circ \sigma$ is isotropic. Since this involution is symplectic, it is then hyperbolic by (15.21). Therefore, by (17.15) we may find $y_1 \in A^\times$ such that $\mathrm{Trp}_\sigma\big(\mathrm{Int}(v^{-1}) \circ \sigma(y_1)v^{-1}y_1\big) = 0$, i.e.,

$$\mathrm{Trp}_\sigma\big(v^{-1}\sigma(y_1)y_1\big) = 0.$$

For $y = v^{-1}\sigma(y_1) \in A^\times$, we then have

$$\mathrm{Trp}_\sigma\big(\sigma(y)vy\big) = \mathrm{Trp}_\sigma\big(y_1 v^{-1}\sigma(y_1)\big) = 0$$

and the claim is proved.

In view of the definition of v, we derive from $\mathrm{Trp}_\sigma\big(\sigma(y)uy\big) = 0$ that

(17.18) $$\lambda\,\mathrm{Trp}_\sigma\big(\sigma(y)y\big) = \mathrm{Trp}_\sigma\big(\sigma(y)\sigma(a)ay\big).$$

On the other hand, we also have

$$\mathrm{Nrp}_\sigma\big(\lambda\sigma(y)y\big) = \lambda^2\,\mathrm{Nrd}_A(y)$$

and

$$\mathrm{Nrp}_\sigma\big(\sigma(y)\sigma(a)ay\big) = \mathrm{Nrd}_A(a)\,\mathrm{Nrd}_A(y)$$

by (2.13), hence

(17.19) $$\mathrm{Nrp}_\sigma\big(\lambda\sigma(y)y\big) = \mathrm{Nrp}_\sigma\big(\sigma(y)\sigma(a)ay\big).$$

If $\sigma(y)y \notin F$ and $\sigma(ay)ay \notin F$ (which may be assumed if σ is hyperbolic), we may then apply (17.16) to get a similitude $g_0 \in \mathrm{GSp}(A,\sigma)$ such that

$$\sigma(ay)ay = \lambda g_0 \sigma(y)y g_0^{-1}.$$

Multiplying on the left by $\sigma(y)^{-1}\sigma(g_0)$ and on the right by $g_0 y^{-1}$, we derive from the preceding equation

$$\sigma(ayg_0 y^{-1})ayg_0 y^{-1} = \lambda\mu(g_0).$$

Therefore, the element $g_1 = ayg_0 y^{-1}$ is in $\mathrm{GSp}(A,\sigma)$, and $\mu(g_1) = \lambda\mu(g_0)$. We then have

$$a = (g_1 g_0^{-1})(g_0 y g_0^{-1} y^{-1}),$$

which yields the required decomposition with $g = g_1 g_0^{-1}$ and $x = g_0$.

To complete the proof, we examine the cases where σ is not hyperbolic and one of the inclusions $\sigma(y)y \in F$ or $\sigma(ay)ay \in F$ holds.

If $\sigma(y)y = \mu \in F^\times$, we derive from (17.18) and (17.19):

$$\mathrm{Trp}_\sigma\big(\sigma(ay)ay\big) = 2\lambda\mu \qquad \text{and} \qquad \mathrm{Nrp}_\sigma\big(\sigma(ay)ay\big) = \lambda^2\mu^2.$$

Therefore, the element $b = \sigma(ay)ay - \lambda\mu$ satisfies $\mathrm{Trp}_\sigma(b) = \mathrm{Nrp}_\sigma(b) = 0$, hence $b = 0$ since σ is not hyperbolic. We thus have $\sigma(ay)ay = \lambda\mu$ and $\sigma(y)y = \mu$, hence $a \in \mathrm{GSp}(A,\sigma)$ and $\mu(a) = \lambda$. We may take $g = a$ and $x = y = 1$ in this case.

Similarly, if $\sigma(ay)ay = \nu \in F^\times$, then (17.18) and (17.19) yield

$$\mathrm{Trp}_\sigma\big(\sigma(y)y\big) = 2\lambda^{-1}\nu \qquad \text{and} \qquad \mathrm{Nrp}_\sigma\big(\sigma(y)y\big) = \lambda^{-2}\nu^2,$$

and the same argument as above shows that $\sigma(y)y = \lambda^{-1}\nu$. Again, we get that $a \in \mathrm{GSp}(A,\sigma)$ and $\mu(a) = \lambda$, hence we may choose $g = a$ and $x = y = 1$. $\square$

Our next goal is to show that all the elements of the form $\big(\mu(g),g\big)(1, xyx^{-1}y^{-1})$ with $g \in \mathrm{GSp}(A,\sigma)$ and $x,\, y \in A^\times$ are in the kernel of α_σ.

(17.20) Lemma. $\alpha_\sigma\big(\mathrm{Nrd}_A(x), x^2\big) = 0$ for all $x \in A^\times$.

Proof: If σ is hyperbolic, the result is clear since $\alpha_\sigma = 0$. For the rest of the proof, we may thus assume that σ is not hyperbolic. Let $v \in \mathrm{Symd}(A, \sigma)^\times$ be such that $v\bar{v}^{-1} = -\mathrm{Nrd}_A(x)^{-1}\sigma(x)^2 x^2$, so that $\alpha_\sigma\big(\mathrm{Nrd}_A(x), x^2\big) = \Phi_v + I^4 W_q F$. We thus have to show that Φ_v is hyperbolic.

If $\mathrm{Nrd}_A(x)^{-1}\sigma(x)^2 x^2 = 1$, then $\mathrm{Trp}_\sigma(v) = 0$ and the result follows from (17.1). If $\mathrm{Nrd}_A(x)^{-1}\sigma(x)^2 x^2 \neq 1$, the proof of (17.3) shows that we may assume $v = 1 - \mathrm{Nrd}_A(x)^{-1}\sigma(x)^2 x^2$. We then have

$$\Phi_v(x^{-1}) = \mathrm{Trp}_\sigma\big(\sigma(x)^{-1}x^{-1}\big) - \mathrm{Nrd}_A(x)^{-1}\,\mathrm{Trp}_\sigma\big(\sigma(x)x\big).$$

Since $\mathrm{Nrp}_\sigma\big(x\sigma(x)\big) = x\sigma(x)\overline{x\sigma(x)}$, we get by (16.7) that

$$\sigma(x)^{-1}x^{-1} = \mathrm{Nrd}_A(x)^{-1}\overline{x\sigma(x)},$$

hence $\Phi_v(x^{-1}) = 0$. Since every isotropic Pfister form is hyperbolic, it follows that $\Phi_v + I^4 W_q F = 0$. $\qquad\square$

The main result in this part of the proof of Theorem (17.6) is the following:

(17.21) Proposition. *Embedding* $\mathrm{GSp}(A, \sigma)$ *and* $\mathrm{SL}_1(A)$ *in* Γ *by mapping* $g \in \mathrm{GSp}(A, \sigma)$ *to* $\big(\mu(g), g\big)$ *and* $a \in \mathrm{SL}_1(A)$ *to* $(1, a)$, *we have*

$$\ker \alpha_\sigma = \mathrm{GSp}(A, \sigma) \cdot [A^\times, A^\times].$$

Proof: In view of (17.17), it suffices to show that $\alpha_\sigma\big(\mu(g), g\big) = 0$ for all $g \in \mathrm{GSp}(A, \sigma)$ and $\alpha_\sigma(1, xyx^{-1}y^{-1}) = 0$ for all $x, y \in A^\times$.

The first relation is clear, either from the definition of α_σ or from (17.7), since α_σ is a homomorphism. The second relation follows from the equality

$$(1, xyx^{-1}y^{-1}) = \big(\mathrm{Nrd}_A(xy), (xy)^2\big) \cdot \big(\mathrm{Nrd}_A(x)^{-1}, (y^{-1}x^{-1}y)^2\big) \cdot \big(\mathrm{Nrd}_A(y)^{-1}, y^{-2}\big).$$

and the preceding lemma. $\qquad\square$

We proceed to determine the kernel of the induced map

$$\alpha \colon \mathrm{SL}_1(A) \to I^3 W_q F / I^4 W_q F.$$

(17.22) Corollary. $\ker \alpha = [A^\times, A^\times]$. *More precisely, every element in* $\ker \alpha$ *is a product of two commutators.*

Proof: The preceding proposition shows that $[A^\times, A^\times] \subset \ker \alpha$. To derive the converse inclusion from (17.17), it suffices to show that every element $g \in \mathrm{Sp}(A, \sigma)$ is a commutator. By (4.17), there is an involution which leaves g invariant, hence there exists some $x \in A^\times$ such that $x\sigma(g)x^{-1} = g$. Since $\sigma(g) = g^{-1}$, it follows that $x(1 + g)x^{-1}g = 1 + g$. Therefore, if $1 + g$ is invertible, g is a commutator:

$$g = x(1 + g)^{-1}x^{-1}(1 + g).$$

If $g = -1$ and char $F \neq 2$, we have $g = iji^{-1}j^{-1}$, where $(1, i, j, k)$ is a quaternion basis of any quaternion subalgebra of A. The proof is thus complete if A is a division algebra. If $\mathrm{ind}\,A = 2$, we still have to consider the case where $1 + g$ generates a right ideal of reduced dimension 2. Denoting by Q a quaternion division algebra which is Brauer-equivalent to A, we can find a representation $A = M_2(Q)$ such that either

$$g = \begin{pmatrix} -1 & 0 \\ 0 & \alpha \end{pmatrix}$$

for some $\alpha \in Q$ such that $\mathrm{Nrd}_Q(\alpha) = 1$ or

$$g = \begin{pmatrix} -1 & \beta \\ 0 & -1 \end{pmatrix}$$

for some pure quaternion $\beta \neq 0$: see Exercise 14. In the first case, we have $\alpha = \alpha_1 \alpha_2 \alpha_1^{-1} \alpha_2^{-1}$ for some $\alpha_1, \alpha_2 \in Q^\times$ since every quaternion of reduced norm 1 is a commutator (see Exercise 13); hence $g = xyx^{-1}y^{-1}$ where

$$x = \begin{pmatrix} i & 0 \\ 0 & \alpha_1 \end{pmatrix}, y = \begin{pmatrix} j & 0 \\ 0 & \alpha_2 \end{pmatrix} \qquad \text{if char } F \neq 2,$$

and

$$x = \begin{pmatrix} 1 & 0 \\ 0 & \alpha_1 \end{pmatrix}, y = \begin{pmatrix} 1 & 0 \\ 0 & \alpha_2 \end{pmatrix} \qquad \text{if char } F = 2.$$

In the second case, if char $F \neq 2$, pick a quaternion $\gamma \neq 0$ which anticommutes with β. Then g has the following expression as a commutator:

$$g = \begin{pmatrix} \beta & -\beta^2/2 \\ 0 & \beta \end{pmatrix} \begin{pmatrix} \gamma & 0 \\ 0 & \gamma \end{pmatrix} \begin{pmatrix} \beta^{-1} & 1/2 \\ 0 & \beta^{-1} \end{pmatrix} \begin{pmatrix} \gamma^{-1} & 0 \\ 0 & \gamma^{-1} \end{pmatrix}.$$

If char $F = 2$, pick $\gamma \in Q^\times$ such that $1 + \gamma^{-1}\beta \neq 0$. Then

$$g = \begin{pmatrix} 1 & \gamma \\ 0 & 1 \end{pmatrix} \begin{pmatrix} 1+\gamma^{-1}\beta & 0 \\ 0 & 1 \end{pmatrix} \begin{pmatrix} 1 & -\gamma \\ 0 & 1 \end{pmatrix} \begin{pmatrix} (1+\gamma^{-1}\beta)^{-1} & 0 \\ 0 & 1 \end{pmatrix}.$$

We leave the case where A is split as an exercise (see Exercise 15). $\qquad \square$

The proof of Theorem (17.6) is now complete.

(17.23) Example. Let $A = (a, b)_F \otimes (c, d)_F$ be an arbitrary biquaternion algebra over a field F of characteristic different from 2. Suppose F contains a primitive fourth root of unity ζ. We then have $\zeta \in \mathrm{SL}_1(A)$, and we may compute $\alpha(\zeta)$ as follows.

Let $(1, i, j, k)$ be the quaternion basis of $(a, b)_F$, viewed as a subalgebra of A, and let σ be the symplectic involution on A which restricts to the canonical involution on $(c, d)_F$ and such that $\sigma(i) = i$, $\sigma(j) = j$ and $\sigma(k) = -k$. By definition, $\alpha(\zeta) = \alpha_\sigma(1, \zeta) = \Phi_1 + I^5 F$, since $1\bar{1}^{-1} = -\sigma(\zeta)\zeta$. A diagonalization of the form Φ_1 can be derived from (17.2):

$$\Phi_1 \simeq \langle\!\langle -a, -b \rangle\!\rangle \cdot \langle\!\langle c, d \rangle\!\rangle.$$

Since ζ is a square root of -1 in F, we have $-a \equiv a \bmod F^{\times 2}$ and $-b \equiv b \bmod F^{\times 2}$, hence

$$\alpha(\zeta) = \langle\!\langle a, b, c, d \rangle\!\rangle + I^5 F.$$

It is then easy to give an example where $\alpha(\zeta) \neq 0$ (hence $\mathrm{SK}_1(A) \neq 0$): we may start with any field F_0 of characteristic different from 2 containing a primitive fourth root of unity and take for F the field of rational fractions in independent indeterminates a, b, c, d over F_0.

17.B. Algebras with involution. The group SK_1 discussed in the first subsection is based on the linear group SL_1. Analogues of the reduced Whitehead group are defined for other simple algebraic groups (see for instance Platonov-Yanchevskiĭ [223, p. 223]). We give here some basic results for algebras with involution and refer to [223] and to Yanchevskiĭ [319] for further results and detailed references.

We assume throughout that the characteristic of the base field F is different from 2. For any central simple F-algebra A with involution σ we set $\operatorname{Sym}(A, \sigma)^\times$ for the set of symmetric units. The central notion for the definition of analogues of the reduced Whitehead group is the subgroup $\Sigma_\sigma(A)$ of $A^\times$ generated by $\operatorname{Sym}(A, \sigma)^\times$. It turns out that this subgroup is normal and depends only on the type of σ, as the following proposition shows.

(17.24) Proposition. (1) *The subgroup $\Sigma_\sigma(A)$ is normal in $A^\times$.*
(2) *If σ, τ are involutions of the same type (orthogonal, symplectic or unitary) on A, then $\Sigma_\sigma(A) = \Sigma_\tau(A)$. More precisely, any σ-symmetric unit can be written as the product of two τ-symmetric units and conversely.*
(3) *If σ is orthogonal, then $\Sigma_\sigma(A) = A^\times$. More precisely, every unit in A can be written as the product of two σ-symmetric units.*

Proof: (1) This readily follows from the equation

$$asa^{-1} = \big(as\sigma(a)\big)\big(\sigma(a)^{-1}a^{-1}\big)$$

for $a \in A^\times$ and $s \in \operatorname{Sym}(A, \sigma)$.

(2) Since σ and τ have the same type, (2.7) or (2.18) shows that $\tau = \operatorname{Int}(u) \circ \sigma$ and $\operatorname{Sym}(A, \tau) = u \cdot \operatorname{Sym}(A, \sigma)$ for some $u \in \operatorname{Sym}(A, \sigma)^\times$. The last equation shows that every element in $\operatorname{Sym}(A, \tau)$ is a product of two σ-symmetric elements. Interchanging the rôles of σ and τ shows that every element in $\operatorname{Sym}(A, \sigma)$ is a product of two τ-symmetric elements.

(3) By (4.17), every unit $x \in A^\times$ is invariant under some orthogonal involution τ. Let $\tau = \operatorname{Int}(u) \circ \sigma$ for some $u \in \operatorname{Sym}(A, \sigma)^\times$; the equation $\tau(x) = x$ yields $\sigma(xu) = xu$, hence $x = (xu)u^{-1}$ is a decomposition of x into a product of two symmetric units.

If A is a division algebra, we may also prove (3) by the following dimension count argument due to Dieudonné [78, Theorem 3]: since $\dim_F \operatorname{Sym}(A, \sigma) > \frac{1}{2} \deg A$, we have $\operatorname{Sym}(A, \sigma) \cap \big(x \operatorname{Sym}(A, \sigma)\big) \neq \{0\}$ for all $x \in A^\times$. If s_1, $s_2 \in \operatorname{Sym}(A, \sigma)^\times$ are such that $s_1 = xs_2$, then $x = s_1 s_2^{-1}$, a product of two symmetric units. $\square$

The same kind of argument yields the following result for arbitrary involutions:

(17.25) Proposition (Yanchevskiĭ). *Let (A, σ) be a central division algebra with involution over F. Every nonzero element $x \in A$ decomposes as $x = zs$ with $s \in \operatorname{Sym}(A, \sigma)^\times$ and $z \in A^\times$ such that $z\sigma(z) = \sigma(z)z$.*

Proof: We have $A = \operatorname{Sym}(A, \sigma) \oplus \operatorname{Skew}(A, \sigma)$ and $1 \notin \operatorname{Skew}(A, \sigma)$, hence dimension count shows that $\big(x \operatorname{Sym}(A, \sigma)\big) \cap \big(F \cdot 1 \oplus \operatorname{Skew}(A, \sigma)\big) \neq \{0\}$. Therefore, one can find $s_0 \in \operatorname{Sym}(A, \sigma)^\times$, $\lambda \in F$ and $z_0 \in \operatorname{Skew}(A, \sigma)$ such that $xs_0 = \lambda + z_0$. The element $z = \lambda + z_0$ commutes with $\sigma(z) = \lambda - z_0$, and satisfies $x = zs$ for $s = s_0^{-1}$. $\square$

To investigate further the group $\Sigma_\sigma(A)$, we now consider separately the cases where the involution is unitary or symplectic.

Unitary involutions. Let (B, τ) be a central simple algebra with unitary involution over a field F, and let K be the center of B.

(17.26) Proposition (Platonov-Yanchevskiĭ). *Suppose B is a division algebra. Every commutator $xyx^{-1}y^{-1} \in [B^\times, B^\times]$ is a product of five symmetric elements. In particular, $[B^\times, B^\times] \subset \Sigma_\tau(B)$.*

Proof: If $x \in \mathrm{Sym}(B, \tau)^\times$, the formula

$$xyx^{-1}y^{-1} = x\big(yx^{-1}\tau(y)\big)\big(\tau(y)^{-1}y^{-1}\big)$$

shows that $xyx^{-1}y^{-1}$ is a product of three symmetric elements. For the rest of the proof, we may thus assume that $x \notin \mathrm{Sym}(B, \tau)$, hence

$$\dim_F\big(\mathrm{Sym}(B, \tau) + F \cdot x\big) = 1 + \tfrac{1}{2}\dim_F B.$$

Therefore, $\big(\mathrm{Sym}(B, \tau) + F \cdot x\big) \cap \big(\mathrm{Sym}(B, \tau)y\big) \neq \{0\}$, and we may find $s_1 \in \mathrm{Sym}(B, \tau)$, $\lambda \in F$ and $s_2 \in \mathrm{Sym}(B, \tau)^\times$ such that

$$s_1 + \lambda x = s_2 y.$$

If $s_1 = 0$, then $xyx^{-1}y^{-1} = s_2 y s_2^{-1}y^{-1}$, and we are reduced to the case where $x \in \mathrm{Sym}(B, \tau)^\times$. We may thus assume that $s_1 \in B^\times$. A direct computation shows that $xyx^{-1}y^{-1}$ is a product of five symmetric elements:

$$xyx^{-1}y^{-1} = f_1 f_2 f_3 f_4 s_2$$

where

$$f_1 = xs_2^{-1}\tau(x), \quad f_2 = \tau(x)^{-1}s_1 x^{-1},$$
$$f_3 = (1 + \lambda xs_1^{-1})s_1^{-1}\tau(1 + \lambda xs_1^{-1}), \quad f_4 = \tau(1 + \lambda xs_1^{-1})^{-1}(1 + \lambda xs_1^{-1})^{-1}. \qquad \square$$

The group

$$\mathrm{UK}_1(B) = B^\times / \Sigma_\tau(B)$$

is the *unitary Whitehead group* of B. The preceding proposition shows that this group is a quotient of $K_1(B) = B^\times / [B^\times, B^\times]$. We may also consider the group

$$\Sigma'_\tau(B) = \{\, x \in B^\times \mid \mathrm{Nrd}_B(x) \in F^\times \,\},$$

which obviously contains $\Sigma_\tau(B)$. The factor group

$$\mathrm{USK}_1(B) = \Sigma'_\tau(B) / \Sigma_\tau(B)$$

is the *reduced unitary Whitehead group* of B. The following proposition is an analogue of a theorem of Wang:

(17.27) Proposition (Yanchevskiĭ). *If B is a division algebra of prime degree, $\mathrm{USK}_1(B) = 0$.*

Proof: We first consider the case where $\deg B$ is an odd prime p. Let $x \in \Sigma'_\tau(B)$ and $\mathrm{Nrd}_B(x) = \lambda \in F^\times$. Then $\mathrm{Nrd}_B(\lambda^{-1}x^p) = 1$, hence $\lambda^{-1}x^p \in [B^\times, B^\times]$ since $\mathrm{SK}_1(B) = 0$, by a theorem of Wang [304] (see for example Pierce [219, 16.6]). It then follows from (17.26) that $x^p \in \Sigma_\tau(B)$. On the other hand, we have $\mathrm{Nrd}_B\big(\tau(x)^{-1}x\big) = 1$, hence, by the same theorem of Wang, $\tau(x)^{-1}x \in [B^\times, B^\times]$. Therefore, $x^2 = x\tau(x)\big(\tau(x)^{-1}x\big) \in \Sigma_\tau(B)$. Since p is odd we may find $u, v \in \mathbb{Z}$ such that $2u + pv = 1$; then

$$x = (x^2)^u (x^p)^v \in \Sigma_\tau(B)$$

and the proposition is proved in the case where $\deg B$ is odd.

If B is a quaternion algebra, Proposition (15.10) shows that $\Sigma'_\tau(B)$ is the Clifford group of $N_{K/F}(B, \gamma)$, where γ is the canonical involution on B. On the other hand, by (3.16) there is a canonical isomorphism $N_{K/F}(B) = \operatorname{End}_F\big(\operatorname{Sym}(B, \tau)\big)$, hence $\Sigma'_\tau(B)$ is generated by $\operatorname{Sym}(B, \tau)$. To make this argument more explicit, consider the map $i\colon \operatorname{Sym}(B, \tau) \to M_2(B)$ defined by

$$i(x) = \begin{pmatrix} 0 & \gamma(x) \\ x & 0 \end{pmatrix} \qquad \text{for } x \in \operatorname{Sym}(B, \tau).$$

Since $i(x)^2 = \operatorname{Nrd}_B(x)$, this map induces an F-algebra homomorphism

$$i_*\colon C\big(\operatorname{Sym}(B, \tau), \operatorname{Nrd}_B\big) \to M_2(B)$$

which is injective since Clifford algebras of even-dimensional nonsingular quadratic spaces are simple. The image of i_* is the F-subalgebra of invariant elements under the automorphism α defined by

$$\alpha \begin{pmatrix} a_{11} & a_{12} \\ a_{21} & a_{22} \end{pmatrix} = \begin{pmatrix} \gamma \circ \tau(a_{22}) & \gamma \circ \tau(a_{21}) \\ \gamma \circ \tau(a_{12}) & \gamma \circ \tau(a_{11}) \end{pmatrix} \qquad \text{for } a_{ij} \in B,$$

since $\alpha \circ i(x) = i(x)$ for all $x \in \operatorname{Sym}(B, \tau)$. Under i_*, the canonical gradation of $C\big(\operatorname{Sym}(B, \tau), \operatorname{Nrd}_B\big)$ corresponds to the checker-board grading, hence i_* restricts to an isomorphism

$$i_*\colon C_0\big(\operatorname{Sym}(B, \tau), \operatorname{Nrd}_B\big) \xrightarrow{\sim} \left\{ \begin{pmatrix} \gamma \circ \tau(b) & 0 \\ 0 & b \end{pmatrix} \;\middle|\; b \in B \right\} \simeq B.$$

Under this isomorphism, the special Clifford group is mapped to $\Sigma'_\tau(B)$. From the Cartan-Dieudonné theorem, it follows that every element in $\Gamma^+\big(\operatorname{Sym}(B, \tau), \operatorname{Nrd}_B\big)$ is a product of two or four anisotropic vectors, hence for every $b \in \Sigma'_\tau(B)$ there exist $x_1, \ldots, x_r \in \operatorname{Sym}(B, \tau)$ (with $r = 2$ or 4) such that

$$\begin{pmatrix} \gamma \circ \tau(b) & 0 \\ 0 & b \end{pmatrix} = \begin{pmatrix} 0 & \gamma(x_1) \\ x_1 & 0 \end{pmatrix} \cdots \begin{pmatrix} 0 & \gamma(x_r) \\ x_r & 0 \end{pmatrix},$$

hence

$$b = x_1 \gamma(x_2) \qquad \text{or} \qquad b = x_1 \gamma(x_2) x_3 \gamma(x_4).$$

This shows that $\Sigma'_\tau(B) = \Sigma_\tau(B)$, since $\operatorname{Sym}(B, \tau)$ is stable under γ. $\square$

Symplectic involutions. Let (A, σ) be central simple algebra with symplectic involution over F. In view of (2.9), the reduced norm of every element in $\operatorname{Sym}(A, \sigma)$ is a square. Let

$$R(A) = \{\, a \in A^\times \mid \operatorname{Nrd}_A(a) \in F^{\times 2} \,\};$$

we thus have $\Sigma_\sigma \subset R(A)$ and we define, after Yanchevskiĭ [319, p. 437],

$$K_1 \operatorname{Spin}(A) = R(A)/\Sigma_\sigma(A)[A^\times, A^\times].$$

Note that $R(A)$ is in general a proper subgroup of $A^\times$: this is clear if A is a quaternion algebra; examples of degree 4 can be obtained as norms of quaternion algebras by (15.7), since the equality $R(A) = A^\times$ implies that the discriminant of every orthogonal involution on A is trivial.

For every $a \in A^\times$, Proposition (4.17) shows that there is an involution $\operatorname{Int}(g) \circ \sigma$ which leaves a invariant. We then have $\sigma(a) = g^{-1}ag$, hence

$$a^2 = \big(a\sigma(a)\big)(g^{-1}a^{-1}ga) \in \Sigma_\sigma(A)[A^\times, A^\times].$$

This shows that $K_1 \operatorname{Spin}(A)$ is a 2-torsion abelian group.

(17.28) Proposition (Yanchevskiĭ). $K_1 \operatorname{Spin}(A) = 0$ *if* $\deg A \leq 4$.

Proof: Suppose first that A is a quaternion algebra. Let $a \in R(A)$ and $\operatorname{Nrd}_A(a) = \alpha^2$ with $\alpha \in F^\times$; then $\operatorname{Nrd}_A(\alpha^{-1}a) = 1$. Since $\operatorname{SK}_1(A) = 0$ (see Exercise 13), we have $\alpha^{-1}a \in [A^\times, A^\times]$, hence

$$a = \alpha(\alpha^{-1}a) \in \Sigma_\sigma(A)[A^\times, A^\times].$$

Suppose next that $\deg A = 4$. Recall from (16.9) the F-algebra isomorphism

$$i_* \colon C\big(\operatorname{Sym}(A, \sigma), \operatorname{Nrp}_\sigma\big) \xrightarrow{\sim} M_2(A).$$

which maps $x \in \operatorname{Sym}(A, \sigma)$ to $\left(\begin{smallmatrix} 0 & \overline{x} \\ x & 0 \end{smallmatrix}\right) \in M_2(A)$. Under this isomorphism, the gradation of the Clifford algebra corresponds to the checker-board grading of $M_2(A)$, and the canonical involution which is the identity on $\operatorname{Sym}(A, \sigma)$ corresponds to the involution θ on $M_2(A)$ defined by

$$\theta \begin{pmatrix} a_{11} & a_{12} \\ a_{21} & a_{22} \end{pmatrix} = \begin{pmatrix} \sigma(a_{22}) & -\sigma(a_{12}) \\ -\sigma(a_{21}) & \sigma(a_{11}) \end{pmatrix}.$$

Moreover, the special Clifford group $\Gamma^+\big(\operatorname{Sym}(A, \sigma), \operatorname{Nrp}_\sigma\big)$ is mapped to the group

$$\Gamma = \left\{ \begin{pmatrix} \lambda\sigma(a)^{-1} & 0 \\ 0 & a \end{pmatrix} \;\middle|\; \lambda \in F^\times,\, a \in A^\times \text{ and } \operatorname{Nrd}_A(a) = \lambda^2 \right\} \subset \operatorname{GL}_2(A).$$

The map $\Gamma \to A^\times$ which carries

$$\begin{pmatrix} \lambda\sigma(a)^{-1} & 0 \\ 0 & a \end{pmatrix} \in \Gamma$$

to $a \in A^\times$ maps Γ onto $R(A)$. From the Cartan-Dieudonné theorem, it follows that every element in $\Gamma^+\big(\operatorname{Sym}(A, \sigma), \operatorname{Nrp}_\sigma\big)$ is a product of two, four, or six anisotropic vectors in $\operatorname{Sym}(A, \sigma)$, hence for every

$$\begin{pmatrix} \lambda\sigma(a)^{-1} & 0 \\ 0 & a \end{pmatrix} \in \Gamma$$

one can find $x_1, \ldots, x_r \in \operatorname{Sym}(A, \sigma)^\times$, with $r = 2$, 4, or 6, such that

$$\begin{pmatrix} \lambda\sigma(a)^{-1} & 0 \\ 0 & a \end{pmatrix} = \begin{pmatrix} 0 & \overline{x_1} \\ x_1 & 0 \end{pmatrix} \cdots \begin{pmatrix} 0 & \overline{x_r} \\ x_r & 0 \end{pmatrix}.$$

Therefore, every $a \in R(A)$ can be written as $a = x_1\overline{x_2} \ldots x_{r-1}\overline{x_r}$ for some $x_1, \ldots, x_r \in \operatorname{Sym}(A, \sigma)^\times$. $\square$

(17.29) Corollary. *Let* (D, σ) *be a central division F-algebra with involution of degree 4. Any element of the commutator subgroup* $[D^\times, D^\times]$ *is a product of at most*

(1) *two symmetric elements if σ is of orthogonal type,*
(2) *six symmetric elements if σ is of symplectic type,*
(3) *four symmetric elements if σ is of unitary type.*

Proof: The claim follows from (17.24) if σ is orthogonal, from (17.28) if σ is symplectic and from (17.27) if σ is unitary. $\square$

EXERCISES

1. Let Q be a quaternion algebra over an étale quadratic extension K of a field F of arbitrary characteristic. Show that the inverse of the Lie algebra isomorphism $\dot{n}^* \colon c\big(\mathbf{N}(Q)\big) \xrightarrow{\sim} Q'$ of (15.6) maps $q \in Q'$ to $\big({}^\iota q \otimes q_0 + {}^\iota q_0 \otimes q - \mathrm{Trd}_Q(q)\,{}^\iota q_0 \otimes q_0\big)$, where $q_0 \in Q$ is an arbitrary quaternion such that $\mathrm{Trd}_Q(q_0) = 1$.

2. Let Q_1, Q_2 be quaternion algebras over a field F of characteristic 2, with canonical involutions γ_1, γ_2, and let $(A, \sigma, f) = (Q_1 \otimes Q_2, \gamma_1 \otimes \gamma_2, f_\otimes)$. Consider $\mathrm{Symd}(A, \sigma)^0 = \{\, x \in \mathrm{Symd}(A, \sigma) \mid \mathrm{Trp}_\sigma(x) = 0 \,\}$. Suppose V_1, V_2 are 3-dimensional subspaces such that $\mathrm{Symd}(A, \sigma)^0 = V_1 + V_2$ and that the products $v_1 v_2$ with $v_1 \in V_1$ and $v_2 \in V_2$ span the kernel of f. Show that V_1 and V_2 are the spaces Q_1^0 and Q_2^0 of pure quaternions in Q_1 and Q_2. Conclude that Q_1 and Q_2 are uniquely determined as subalgebras of A by the condition $(A, \sigma, f) = (Q_1 \otimes Q_2, \gamma_1 \otimes \gamma_2, f_\otimes)$.

   Hint: Show that if $v_1 = q_{11} + q_{21} \in V_1$ and $v_2 = q_{12} + q_{22} \in V_2$ with $q_{11}, q_{12} \in Q_1^0$ and $q_{21}, q_{22} \in Q_2^0$, then $v_1 v_2 = v_2 v_1$ and $f(v_1 v_2) = [q_{11}, q_{12}] = [q_{21}, q_{22}]$.

3. (Karpenko-Quéguiner [151]) Let (B, τ) be a central simple algebra with unitary involution of degree 4. Let K be the center of B. Show that

$$(B, \tau) = (Q_1, \tau_1) \otimes_K (Q_2, \tau_2)$$

   for some quaternion subalgebras Q_1, Q_2 if and only if the discriminant algebra $D(B, \tau)$ is split.

   Hint: If $D(B, \tau)$ is split, use Theorem (15.24) to represent (B, τ) as the even Clifford algebra of some quadratic space. For the "only if" part, use Propositions (2.22) and (10.33).

4. Suppose $\mathrm{char}\, F \neq 2$. Extensions of the form $F[X, Y]/(X^2 - a, Y^2 - b)$ with a, $b \in F^\times$ are called *biquadratic*. Show that for every central simple F-algebra of degree 6 with orthogonal involution σ of trivial discriminant, there exists an étale biquadratic extension of F over which σ becomes hyperbolic. Deduce that every central simple F-algebra of degree 4 is split by some étale biquadratic extension of F. (This result is due to Albert [9, Theorem 11.9].)

5. Let A be a biquaternion F-algebra. Suppose A is split by an étale extension of the form $K_1 \otimes K_2$, where K_1, K_2 are étale quadratic F-algebras. Show that there exist a_1, $a_2 \in F^\times$ such that

$$A \simeq (K_1, a_1)_F \otimes (K_2, a_2)_F.$$

   Hint: Refine the argument used in the proof of Proposition (16.2).

6. Let σ, τ be distinct symplectic involutions on a biquaternion F-algebra A. Show that $\dim_F\big(\mathrm{Symd}(A, \sigma) \cap \mathrm{Symd}(A, \tau)\big) = 2$. Show also that there is a quaternion algebra $B \subset A$ over some quadratic extension of F such that $\sigma|_B = \tau|_B$ is the conjugation involution, and that the algebra B is uniquely determined by this condition.

7. Let σ, τ, θ be symplectic involutions on a biquaternion F-algebra A. Show that the invariants of these involutions are related by $j_\sigma(\tau) = j_\tau(\sigma)$ and $j_\sigma(\tau) + j_\tau(\theta) + j_\theta(\sigma) \in I^3 W_q F$. Use this result to show that if σ and τ are conjugate, then $j_\sigma(\theta) = j_\tau(\theta)$.

8. Let σ be a symplectic involution on a biquaternion F-algebra A. Let

$$\mathrm{Symd}(A, \sigma)^0 = \{\, x \in \mathrm{Symd}(A, \sigma) \mid \mathrm{Trp}_\sigma(x) = 0 \,\}$$

and let $s_\sigma \colon \operatorname{Symd}(A, \sigma)^0 \to F$ be the squaring map. Show that $\operatorname{ind}(A) \leq 2$ if and only if s_σ is a subform of some (uniquely determined) 3-fold Pfister form π_σ. Suppose these conditions hold; then

(a) show that (A, σ) has a decomposition

$$(A, \sigma) = \big(M_2(F), \sigma_1\big) \otimes_F (Q, \gamma)$$

for some quaternion algebra Q with canonical involution γ and some orthogonal involution σ_1 on $M_2(F)$, and that $\pi_\sigma = \langle\!\langle \operatorname{disc} \sigma_1 \rangle\!\rangle \cdot \operatorname{Nrd}_Q$;

(b) for θ a hyperbolic involution on A, show that $\pi_\sigma = j_\theta(\sigma)$;

(c) show that $G(A, \sigma) = G(\pi_\sigma) = \operatorname{Sn}(s_\sigma)$.

    Hint: The equality $G(A, \sigma) = \operatorname{Sn}(s_\sigma)$ follows from (15.18) and $G(A, \sigma) = G(\pi_\sigma)$ follows from (17.14) and (17.21).

9. Suppose char $F \neq 2$. Let K/F be an étale quadratic extension with nontrivial automorphism ι and let $\delta \in K^\times$ be such that $\iota(\delta) = -\delta$. Let (V, q) be an odd-dimensional quadratic space over K with trivial discriminant and let $\zeta \in C(V, q)$ be an orientation of (V, q). Define an F-linear map $i \colon V \to M_2\big(N_{K/F}(C_0(V, q))\big)$ by

$$i(x) = \begin{pmatrix} 0 & -\delta\big({}^\iota(x \cdot \zeta) \otimes 1 - 1 \otimes (x \cdot \zeta)\big) \\ {}^\iota(x \cdot \zeta) \otimes 1 + 1 \otimes (x \cdot \zeta) & 0 \end{pmatrix}$$

for $x \in V$. Show that the map i induces an F-algebra isomorphism:

$$i_* \colon C\big(V, (\operatorname{tr}_{K/F})_*(\langle \delta \rangle \cdot q)\big) \xrightarrow{\ \sim\ } M_2\big(N_{K/F}(C_0(V, q))\big).$$

Use this result to give a direct proof of the fact that if Q is a quaternion K-algebra and $s \colon Q^0 \to K$ is the squaring map on the space of pure quaternions, then $\big(Q^0, (\operatorname{tr}_{K/F})_*(\langle \delta \rangle \cdot s)\big)$ is an Albert quadratic space of $N_{K/F}(Q)$.

10. Suppose char $F \neq 2$. Let Q_1, Q_2 be quaternion F-algebras with canonical involutions γ_1, γ_2 and let $(A, \theta) = (Q_1, \gamma_1) \otimes_F (Q_2, \gamma_2)$. Define a linear endomorphism p on $\operatorname{Skew}(A, \theta) = (Q_1^0 \otimes 1) \oplus (1 \otimes Q_2^0)$ by $p(x_1 \otimes 1 + 1 \otimes x_2) = x_1 \otimes 1 - 1 \otimes x_2$ and a quadratic form $q \colon \operatorname{Skew}(A, \theta) \to F$ by $q(x) = xp(x)$. Consider another pair of quaternion F-algebras Q_1', Q_2' and $(A', \theta') = (Q_1', \gamma_1') \otimes_F (Q_2', \gamma_2')$, and define p', q' on $\operatorname{Skew}(A', \theta')$ as p, q were defined on $\operatorname{Skew}(A, \theta)$. Show that for every isomorphism $f \colon (A, \theta) \xrightarrow{\ \sim\ } (A', \theta')$ there exists some $\lambda \in F^\times$ such that $f^{-1} \circ p' \circ f = \lambda p$, hence f restricts to a similitude $(\operatorname{Skew}(A, \theta), q) \xrightarrow{\ \sim\ } (\operatorname{Skew}(A', \theta'), q')$ with multiplier λ.

    Hint: Use (16.22).

    This exercise is inspired by Knus-Parimala-Sridharan [164, Theorem 3.4] and Wadsworth [299]. It shows that the forms q and q' are similar without using the fact that they are Albert forms of A and A'.

11. Let σ be a symplectic involution on a biquaternion algebra A. Show that the invariant j_σ of symplectic involutions and the homomorphism $\alpha_\sigma \colon \Gamma \to I^3 W_q F / I^4 W_q F$ are related by

$$\alpha_\sigma\big(\operatorname{Nrp}_\sigma(v), v\big) = j_\sigma\big(\operatorname{Int}(v) \circ \sigma\big) - \langle\!\langle \operatorname{Nrp}_\sigma(v) \rangle\!\rangle \cdot \operatorname{Nrp}_\sigma + I^4 W_q F$$

for all $v \in \operatorname{Symd}(A, \sigma)^\times$.

12. Suppose char $F \neq 2$. Let A be a biquaternion division algebra and let $x \in \operatorname{SL}_1(A)$.

(a) Let σ be an arbitrary symplectic involution and let $L = F(\sqrt{a})$ be a quadratic extension of F in $\operatorname{Sym}(A, \sigma)$ which contains $\sigma(x)x$. Recall from (16.2) that one can find b, c, $d \in F^\times$ such that $A \simeq (a, b)_F \otimes (c, d)_F$.

Show that L contains an element y such that $\sigma(x)x = y\bar{y}^{-1}$ where $\bar{}$ is the nontrivial automorphism of L/F, and that

$$\alpha(x) = \langle\!\langle N_{L/F}(y), b, c, d\rangle\!\rangle + I^5 F.$$

(b) Suppose x is contained in a maximal subfield $E \subset A$ which contains an intermediate quadratic extension $L = F(\sqrt{a})$. Recall from (16.2) that $A \simeq (a,b)_F \otimes (c,d)_F$ for some b, c, $d \in F^\times$. Show that L contains an element y such that $N_{E/L}(x) = y\bar{y}^{-1}$ where $\bar{}$ is the nontrivial automorphism of L/F, and that

$$\alpha(x) = \langle\!\langle N_{L/F}(y), b, c, d\rangle\!\rangle + I^5 F.$$

13. Let Q be a quaternion F-algebra. Show that every element in $\mathrm{SL}_1(Q)$ is a commutator, except if $F = \mathbb{F}_2$ (the field with two elements).

Hint: Argue as in (17.22). If $q \in \mathrm{SL}_1(Q)$ and $1 + q$ is not invertible and nonzero, then show that there is an isomorphism $Q \xrightarrow{\sim} M_2(F)$ which identifies q with $\left(\begin{smallmatrix} -1 & 1 \\ 0 & -1 \end{smallmatrix}\right)$. Then $q = xyx^{-1}y^{-1}$ with $x = \left(\begin{smallmatrix} 1 & 0 \\ 0 & -1 \end{smallmatrix}\right)$ and $y = \left(\begin{smallmatrix} -1/2 & 1 \\ -1 & 0 \end{smallmatrix}\right)$ if char $F \neq 2$, with $x = \left(\begin{smallmatrix} \lambda+1 & 0 \\ 0 & \lambda \end{smallmatrix}\right)$ and $y = \left(\begin{smallmatrix} 1 & \lambda \\ 0 & 1 \end{smallmatrix}\right)$ if char $F = 2$ and $\lambda \neq 0, 1$.

14. Let Q be a quaternion division F-algebra and let (V, h) be a nonsingular hermitian space of dimension 2 over Q with respect to the canonical involution, so that $\big(\mathrm{End}_Q(V), \sigma_h\big)$ is a central simple algebra of degree 4 with symplectic involution. Let $g \in \mathrm{Sp}(V, h)$ be such that $\mathrm{Id}_V + g$ is not invertible, and let $v_1 \in V$ be a nonzero vector such that $g(v_1) = -v_1$.

(a) If v_1 is anisotropic, show that for each vector $v_2 \in V$ which is orthogonal to v_1, there is some $\alpha \in Q$ such that $g(v_2) = v_2\alpha$ and $\mathrm{Nrd}_Q(\alpha) = 1$.

(b) If v_1 is isotropic, show that for each isotropic vector $v_2 \in V$ such that $h(v_1, v_2) = 1$, there is some $\beta \in Q$ such that $g(v_2) = v_1\beta - v_2$ and $\mathrm{Trd}_Q(\beta) = 0$.

15. Let b be a nonsingular alternating bilinear form on a 4-dimensional F-vector space V and let $g \in \mathrm{Sp}(V, b)$ be such that $\mathrm{Id}_V + g$ is not invertible.

(a) Show that $\ker(\mathrm{Id}_V + g) = \mathrm{im}(\mathrm{Id}_V + g)^\perp$.

(b) If the rank of $\mathrm{Id}_V + g$ is 1 or 2, show that $V = V_1 \oplus V_2$ for some 2-dimensional subspaces V_1, V_2 which are preserved under g, hence g can be represented by the matrix

$$g_1 = \begin{pmatrix} -1 & 1 & 0 & 0 \\ 0 & -1 & 0 & 0 \\ 0 & 0 & -1 & 0 \\ 0 & 0 & 0 & -1 \end{pmatrix} \quad \text{or} \quad g_2 = \begin{pmatrix} -1 & 1 & 0 & 0 \\ 0 & -1 & 0 & 0 \\ 0 & 0 & -1 & 1 \\ 0 & 0 & 0 & -1 \end{pmatrix}.$$

Use Exercise 13 to conclude that g is a commutator if $F \neq \mathbb{F}_2$. If char $F = 2$, show that $g_1 = a_1 b_1 a_1^{-1} b_1^{-1}$ and $g_2 = a_2 b_2 a_2^{-1} b_2^{-1}$, where

$$a_1 = \begin{pmatrix} 1 & 0 & 1 & 0 \\ 0 & 1 & 0 & 0 \\ 0 & 0 & 1 & 0 \\ 0 & 0 & 0 & 1 \end{pmatrix}, \quad b_1 = \begin{pmatrix} 1 & 0 & 0 & 0 \\ 0 & 1 & 0 & 0 \\ 0 & 1 & 1 & 0 \\ 0 & 0 & 0 & 1 \end{pmatrix},$$

$$a_2 = \begin{pmatrix} 0 & 0 & 1 & 0 \\ 0 & 0 & 0 & 1 \\ 1 & 0 & 0 & 0 \\ 0 & 1 & 0 & 0 \end{pmatrix}, \quad b_2 = \begin{pmatrix} 1 & 0 & 0 & 0 \\ 0 & 1 & 0 & 0 \\ 0 & 0 & 1 & 1 \\ 0 & 0 & 0 & 1 \end{pmatrix}.$$

(c) If $\mathrm{Id}_V + g$ has rank 3, show that g can be represented by the matrix

$$\begin{pmatrix} -1 & 1 & 0 & 0 \\ 0 & -1 & 1 & 0 \\ 0 & 0 & -1 & 1 \\ 0 & 0 & 0 & -1 \end{pmatrix}$$

(with respect to a suitable basis).

If char $F \neq 2$, let

$$y = \begin{pmatrix} -3 & -8 & 22 & -12 \\ -3 & -9 & 23 & -12 \\ -3 & -9 & 22 & -11 \\ -4 & -5 & 16 & -8 \end{pmatrix}.$$

Show that $(x-1)^3(x+1)$ is the minimum and the characteristic polynomial of both matrices gy and y. Conclude that there is an invertible matrix x such that $gy = xyx^{-1}$, hence g is a commutator. If char $F = 2$, let

$$z = \begin{pmatrix} 0 & 1 & 0 & 0 \\ 1 & 0 & 0 & 1 \\ 1 & 0 & 1 & 1 \\ 1 & 1 & 1 & 0 \end{pmatrix}.$$

Show that $x^4 + x^3 + x^2 + x + 1$ is the minimum and the characteristic polynomial of both matrices gz and z. As in the preceding case, conclude that g is a commutator.

16. Let A be a biquaternion F-algebra with symplectic involution σ. Write simply $V = \mathrm{Symd}(A, \sigma)$, $V^0 = \mathrm{Symd}(A, \sigma)^0$ and $q = \mathrm{Nrp}_\sigma$, and let q^0 be the restriction of q to V^0. Using (15.27), show that there is a canonical isomorphism $\mathrm{SL}_1(A) \simeq \mathrm{Spin}(V, q)$ which maps the subgroup $[A^\times, A^\times]$ to the subgroup

$$\mathrm{Spin}'(V, q) = \mathrm{Spin}(V^0, q^0) \cdot [\Omega, \Gamma^+(V^0, q^0)]$$

where $\Omega = \Omega\big(\mathrm{End}_F(V), \sigma_q, f_q\big)$ is the extended Clifford group. Deduce that the subgroup $\mathrm{Spin}'(V, q) \subset \mathrm{Spin}(V, q)$ is generated by the subgroups $\mathrm{Spin}(U)$ for all the proper nonsingular subspaces $U \subset V$.

   Hint: Use the proof of (17.17).

17. Suppose char $F \neq 2$ and F contains a primitive 4th root of unity ζ. Let $A = (a, b)_F \otimes (c, d)_F$ be a biquaternion division F-algebra.
   (a) Show that if A is cyclic (i.e., if A contains a maximal subfield which is cyclic over F), then $\langle\!\langle a, b, c, d \rangle\!\rangle$ is hyperbolic.
   (b) (Morandi-Sethuraman [198, Proposition 7.3]) Suppose d is an indeterminate over some subfield F_0 containing a, b, c, and $F = F_0(d)$. Show that A is cyclic if and only if $\langle\!\langle a, b, c \rangle\!\rangle$ is hyperbolic.
   Hint: If $\langle\!\langle a, b, c \rangle\!\rangle$ is hyperbolic, the following equation has a nontrivial solution in F_0:

$$a(x_1^2 + cx_2^2) + b(y_1^2 + cy_2^2) - ab(z_1^2 + cz_2^2) = 0.$$

   Let $e = 2\sqrt{c}(ax_1x_2 + by_1y_2 - abz_1z_2) \in F_0(\sqrt{c})$. Show that $F_0(\sqrt{c})(\sqrt{e})$ is cyclic over F_0 and splits $(a, b)_{F(\sqrt{c})}$, hence also A.

NOTES

§15. If char $F \neq 2$, an alternative way to define the canonical isomorphism $\mathbf{N} \circ \mathbf{C}(A, \sigma) \simeq (A, \sigma)$ for $(A, \sigma) \in D_2$ (in (15.7)) is to refine the fundamental relation (9.14) by taking the involutions into account. As pointed out in the notes for Chapter II, one can define a nonsingular hermitian form H' on the left A-submodule $B'(A, \sigma) \subset B(A, \sigma)$ of invariant elements under the canonical involution ω and show that the canonical isomorphism ν' of (9.11) restricts to an isomorphism of algebras with involution:

$$N_{Z/F}\big(C(A, \sigma), \underline{\sigma}\big) \xrightarrow{\sim} \big(\mathrm{End}_A\, B'(A, \sigma), \sigma_{H'}\big)$$

where Z is the center of the Clifford algebra $C(A, \sigma)$. Since $\deg A = 4$, dimension count shows that the canonical map $b \colon A \to B(A, \sigma)$ induces an isomorphism of A-modules $A \xrightarrow{\sim} B'(A, \sigma)$. Moreover, under this isomorphism, $H'(a_1, a_2) = 2a_1\sigma(a_2)$ for $a_1, a_2 \in A$. Therefore, b induces a canonical isomorphism

$$(A, \sigma) \xrightarrow{\sim} \big(\mathrm{End}_A\, B'(A, \sigma), \sigma_{H'}\big).$$

Similarly, in Theorem (15.24) the canonical isomorphism $\mathbf{D} \circ \mathbf{C}(A, \sigma) \simeq (A, \sigma)$ for $(A, \sigma) \in D_3$ can be derived from properties of the bimodule $B(A, \sigma)$. Define a left $C(A, \sigma) \otimes_Z C(A, \sigma)$-module structure on $B(A, \sigma)$ by

$$(c_1 \otimes c_2) \diamond u = c_1 * u \cdot \underline{\sigma}(c_2) \quad \text{for } c_1,\, c_2 \in C(A, \sigma),\, u \in B(A, \sigma).$$

If $g \in C(A, \sigma) \otimes_Z C(A, \sigma)$ is the Goldman element of the Clifford algebra, the map $C(A, \sigma) \otimes_Z C(A, \sigma) \to B(A, \sigma)$ which carries ξ to $\xi \diamond 1^b$ induces an isomorphism of left $C(A, \sigma) \otimes_Z C(A, \sigma)$-modules

$$[C(A, \sigma) \otimes_Z C(A, \sigma)](1 - g) \xrightarrow{\sim} B(A, \sigma).$$

We thus get a canonical isomorphism

$$\lambda^2 C(A, \sigma) \simeq \mathrm{End}_{C(A,\sigma) \otimes_Z C(A,\sigma)} B(A, \sigma).$$

Using (9.9), we may identify the right-hand side with $A \otimes_F Z$ and use this identification to get a canonical isomorphism $D\big(C(A, \sigma), \underline{\sigma}\big) \simeq A$.

In the proofs of Theorems (15.7) and (15.24), it is not really necessary to consider the split cases separately if char $F = 0$; one can instead use results on the extension of Lie algebra isomorphisms in Jacobson [140, Ch. 10, §4] to see directly that the canonical Lie algebra isomorphisms $\dot{n} \circ \frac{1}{2}c$, $\frac{1}{2}c \circ \dot{n}$, $\dot{\lambda}^2 \circ \frac{1}{2}c$ and $\frac{1}{2}c \circ \dot{\lambda}^2$ extend (uniquely) to isomorphisms of the corresponding algebras with involution (see Remarks (15.8) and (15.25)).

The fact that a central simple algebra of degree 4 with orthogonal involution decomposes into a tensor product of stable quaternion subalgebras if and only if the discriminant of the involution is trivial (Corollary (15.12) and Proposition (7.3)) was proved by Knus-Parimala-Sridharan [164, Theorem 5.2], [162]. (The paper [164] deals with Azumaya algebras over rings in which 2 is invertible, whereas [162] focuses on central simple algebras, including the characteristic 2 case).

The description of groups of similitudes of quadratic spaces of dimension 3, 4, 5 and 6 dates back to Van der Waerden [301] and Dieudonné [77, §3]. The case of quadratic spaces over rings was treated by Knus [156, §3] and by Knus-Parimala-Sridharan [164, §6] (see also Knus [157, Ch. 5]). Clifford groups of algebras of degree 4 or 6 with orthogonal involution are determined in Merkurjev-Tignol [196, 1.4.2, 1.4.3].

§16. Albert forms are introduced in Albert [2]. Theorem 3 of that paper yields the criterion for the biquaternion algebra $A = (a, b)_F \otimes (c, d)_F$ to be a division algebra in terms of the associated quadratic form $q = \langle a, b, -ab, -c, -d, cd \rangle$. (See (16.5).) The definition of the form q thus depends on a particular decomposition of the biquaternion algebra A. The fact that quadratic forms associated to different decompositions are similar was first proved by Jacobson [142, Theorem 3.12] using Jordan structures, and later by Knus [156, Proposition 1.14] and Mammone-Shapiro [177] using the algebraic theory of quadratic forms. (See also Knus-Parimala-Sridharan [164, Theorem 4.2].) Other proofs of Albert's Theorem (16.5) were given by Pfister [217, p. 124] and Tamagawa [279] (see also Seligman [252, App. C]). A notion of Albert form in characteristic 2 is discussed in Mammone-Shapiro [177]. Note that the original version of Jacobson's paper [142] does not cover the characteristic 2 case adequately; see the reprinted version in Jacobson's Collected Mathematical Papers [145], where the characteristic is assumed to be different from 2.

From the definition of the quadratic form $q = \langle a, b, -ab, -c, -d, cd \rangle$ associated to $A = (a, b)_F \otimes (c, d)_F$, it is clear that q is isotropic if and only if the quaternion algebras $(a, b)_F$ and $(c, d)_F$ have a common maximal subfield. Thus Corollary (16.29) easily follows from Albert's Theorem (16.5). The proof given by Albert in [11] is more direct and also holds in characteristic 2. Another proof (also valid in characteristic 2) was given by Sah [239]. If the characteristic is 2, the result can be made more precise: if a tensor product of two quaternion division algebras is not a division algebra, then the two quaternion algebras have a common maximal subfield which is a separable extension of the center. This was first shown by Draxl [83]. The proofs given by Knus [158] and by Tits [293] work in all characteristics, and yield the more precise result in characteristic 2.

The fact that $G(q) = F^{\times 2} \cdot \mathrm{Nrd}_A(A^\times)$ for an Albert form q of a biquaternion algebra A is already implicit in Van der Waerden [301, pp. 21–22] and Dieudonné [77, Nos 28, 30, 34]. Knus-Lam-Shapiro-Tignol [159] gives other characterizations of that group. In particular, it is shown that this group is also the set of discriminants of orthogonal involutions on A; see Parimala-Sridharan-Suresh [205] for another proof of that result.

If σ is a symplectic involution on a biquaternion algebra A, the (Albert) quadratic form Nrp_σ defined on the space $\mathrm{Symd}(A, \sigma)$ is the generic norm of $\mathrm{Symd}(A, \sigma)$, viewed as a Jordan algebra, see Jacobson [136]. This is the point of view from which results on Albert forms are derived in Jacobson [142, Ch. 6, §4]. The invariant $j_\sigma(\tau)$ of symplectic involutions is defined in a slightly different fashion in Knus-Lam-Shapiro-Tignol [159, §3]. See (31.46) for the relation between $j_\sigma(\tau)$ and Rost's higher cohomological invariants.

If σ is an orthogonal involution on a biquaternion algebra A, the linear endomorphism p_σ of $\mathrm{Skew}(A, \sigma)$ was first defined by Knus-Parimala-Sridharan [161], [164] (see also Knus [157, Ch. 5]), who called it the *pfaffian adjoint* because of its relation with the pfaffian. Pfaffian adjoints for algebras of degree greater than 4 are considered in Knus-Parimala-Sridharan [163]. If $\deg A = 2m$, the pfaffian adjoint is a polynomial map of degree $m - 1$ from $\mathrm{Skew}(A, \sigma)$ to itself. Knus-Parimala-Sridharan actually treat orthogonal and symplectic involutions simultaneously (and in the context of Azumaya algebras): if σ is a symplectic involution on a biquaternion algebra A, the pfaffian adjoint is the endomorphism $^-$ of $\mathrm{Sym}(A, \sigma)$ defined in §16.B.

Further results on Albert forms can be found in Lam-Leep-Tignol [170], where maximal subfields of a biquaternion algebra are investigated; in particular, necessary and sufficient conditions for the cyclicity of a biquaternion algebra are given in that paper.

§17. Suppose $\operatorname{char} F \neq 2$. As observed in the proof of (17.28), the group $\Gamma = \{ (\lambda, a) \in F^\times \times A^\times \mid \lambda^2 = \operatorname{Nrd}_A(a) \}$, for A a biquaternion F-algebra, can be viewed as the special Clifford group of the quadratic space $\big(\operatorname{Sym}(A, \sigma), \operatorname{Nrp}_\sigma\big)$ for any symplectic involution σ. The map $\alpha_\sigma \colon \Gamma \to I^4 F / I^5 F$ can actually be defined on the full Clifford group $\Gamma\big(\operatorname{Sym}(A, \sigma), \operatorname{Nrp}_\sigma\big)$ by mapping the generators $v \in \operatorname{Sym}(A, \sigma)^\times$ to $\Phi_v + I^5 F$. Showing that this map is well-defined is the main difficulty of this approach.

The homomorphism $\alpha \colon \operatorname{SK}_1(A) \to I^4 F / I^5 F$ (for A a biquaternion algebra) was originally defined by Rost in terms of Galois cohomology, as a map $\alpha' \colon \operatorname{SK}_1(A) \to H^4(F, \mu_2)$. Rost also proved exactness of the following sequence:

$$0 \to \operatorname{SK}_1(A) \xrightarrow{\alpha'} H^4(F, \mu_2) \xrightarrow{h} H^4(F(q), \mu_2)$$

where h is induced by scalar extension to the function field of an Albert form q, see Merkurjev [190, Theorem 4] and Kahn-Rost-Sujatha [148]. The point of view of quadratic forms developed in §17.A is equivalent, in view of the isomorphism $e_4 \colon I^4 F / I^5 F \xrightarrow{\sim} H^4(F, \mu_2)$ proved by Rost (unpublished) and more recently by Voevodsky, which leads to a commutative diagram

$$
\begin{array}{ccccc}
0 \longrightarrow \operatorname{SK}_1(A) & \xrightarrow{\alpha} & I^4 F / I^5 F & \xrightarrow{i} & I^4 F(q) / I^5 F(q) \\
\big\| & & e_4 \big\downarrow & & e_4 \big\downarrow \\
0 \longrightarrow \operatorname{SK}_1(A) & \xrightarrow{\alpha'} & H^4(F, \mu_2) & \xrightarrow{h} & H^4\big(F(q), \mu_2\big).
\end{array}
$$

The fact that the sequences above are zero-sequences readily follows from functoriality of α and α', since $\operatorname{SK}_1(A) = 0$ if A is not a division algebra or, equivalently by (16.5), if q is isotropic. In order to derive exactness of the sequence above from exactness of the sequence below, only "elementary" information on the map e_4 is needed: it is sufficient to use the fact that on Pfister forms e_4 is well-defined (see Elman-Lam [91, 3.2]) and injective (see Arason-Elman-Jacob [23, Theorem 1]). In fact, (17.1) shows that the image of α consists of 4-fold Pfister forms (modulo $I^5 F$) and, on the other hand, the following proposition also shows that every element in $\ker i$ is represented by a single 4-fold Pfister form:

(17.30) Proposition. *If q is an Albert form which represents 1, then*

$$\ker i = \{ \pi + I^5 F \mid \pi \text{ is a 4-fold Pfister form containing } q \}.$$

Proof: By Fitzgerald [98, Corollary 2.3], the kernel of the scalar extension map $WF \to WF(q)$ is an ideal generated by 4-fold and 5-fold Pfister forms. Therefore, every element in $\ker i$ is represented by a sum of 4-fold Pfister forms which become hyperbolic over $F(q)$. By the Cassels-Pfister subform theorem (see Scharlau [247, Theorem 4.5.4]) the Pfister forms which satisfy this condition contain a subform isometric to q. If π_1, π_2 are two such 4-fold Pfister forms, then the Witt index of $\pi_1 \perp -\pi_2$ is at least $\dim q = 6$. By Elman-Lam [91, Theorem 4.5] it follows that there exists a 3-fold Pfister form ϖ and elements $a_1, a_2 \in F^\times$ such that

$$\pi_1 = \varpi \cdot \langle\!\langle a_1 \rangle\!\rangle \quad \text{and} \quad \pi_2 = \varpi \cdot \langle\!\langle a_2 \rangle\!\rangle.$$

Therefore,

$$\pi_1 + \pi_2 \equiv \varpi \cdot \langle\!\langle a_1 a_2 \rangle\!\rangle \quad \mod I^5 F.$$

By induction on the number of terms, it follows that every sum of 4-fold Pfister forms representing an element of $\ker i$ is equivalent modulo $I^5 F$ to a single 4-fold Pfister form. $\square$

The image of the map α_σ can be described in a similar way. Since $\alpha_\sigma = 0$ if σ is hyperbolic or, equivalently by (15.21), if the 5-dimensional form s_σ is isotropic, it follows by functoriality of α_σ that $\mathrm{im}\,\alpha_\sigma$ lies in the kernel of the scalar extension map to $F(s_\sigma)$. One can use the arguments in the proof of Merkurjev [190, Theorem 4] to show that this inclusion is an equality, so that the following sequence is exact:

$$\Gamma_\sigma \xrightarrow{\ \alpha_\sigma\ } I^4 F / I^5 F \longrightarrow I^4 F(s_\sigma) / I^5 F(s_\sigma).$$

Corollary (17.22) shows that every element in $[A^\times, A^\times]$ is a product of two commutators. If A is split, every element in $[A^\times, A^\times]$ can actually be written as a single commutator, as was shown by Thompson [283]. On the other hand, Kursov [168] has found an example of a biquaternion algebra A such that the group $[A^\times, A^\times]$ does not consist of commutators, hence our lower bound for the number of factors is sharp, in general.

The first example of a biquaternion algebra A such that $\mathrm{SK}_1(A) \neq 0$ is due to Platonov [221]. For a slightly different relation between the reduced Whitehead group of division algebras and Galois cohomology of degree 4, see Suslin [277].

Along with the group $\Sigma_\sigma(A)$, the group generated by skew-symmetric units in a central simple algebra with involution (A, σ) is also discussed in Yanchevskiĭ [319]. If σ is orthogonal, it is not known whether this subgroup is normal in $A^\times$. Triviality of the group $\mathrm{USK}_1(B, \tau)$ can be shown not only for division algebras of prime degree, but also for division algebras whose degree is square-free; indeed, the exponent of $\mathrm{USK}_1(B, \tau)$ divides $\deg B / p_1 \ldots p_r$, where $p_1, \ldots, p_r$ are the prime factors of $\deg B$: see Yanchevskiĭ [318]. Examples where the group $K_1 \mathrm{Spin}(A)$ is not trivial are given in Monastyrnyĭ-Yanchevskiĭ [197]. See also Yanchevskiĭ [319] for the relation between $K_1 \mathrm{Spin}$ and decomposability of involutions.

CHAPTER V

Algebras of Degree Three

The main topic of this chapter is central simple algebras of degree 3 with involutions of the second kind and their étale (commutative) subalgebras. To every involution of the second kind on a central simple algebra of degree 3, we attach a 3-fold Pfister form, and we show that this quadratic form classifies involutions up to conjugation. Involutions whose associated Pfister form is hyperbolic form a distinguished conjugacy class; we show that such an involution is present on every central simple algebra of degree 3 with involution of the second kind, and we characterize distinguished involutions in terms of étale subalgebras of symmetric elements.

We start with Galois descent, followed by a general discussion of étale and Galois algebras. These are tools which will also be used in subsequent chapters.

§18. Étale and Galois Algebras

Throughout this section, let F be an arbitrary base field, let F_{sep} be a separable closure of F and let Γ be the *absolute Galois group* of F:

$$\Gamma = \text{Gal}(F_{\text{sep}}/F).$$

The central theme of this section is a correspondence between étale F-algebras and Γ-sets, which is set up in the first subsection. This correspondence is then restricted to Galois algebras and torsors. The final subsection demonstrates the special features of étale algebras of dimension 3.

The key tool for the correspondence between étale F-algebras and Γ-sets is the following Galois descent principle. Let V_0 be a vector space over F. The left action of Γ on $V = V_0 \otimes F_{\text{sep}}$ defined by $\gamma * (u \otimes x) = u \otimes \gamma(x)$ for $u \in V_0$ and $x \in F_{\text{sep}}$ is *semilinear* with respect to Γ, i.e.,

$$\gamma * (vx) = (\gamma * v)\gamma(x)$$

for $v \in V$ and $x \in F_{\text{sep}}$; the action is also *continuous* since for every vector $v \in V$ there is a finite extension M of F in F_{sep} such that $\text{Gal}(F_{\text{sep}}/M)$ acts trivially on v.

The space V_0 can be recovered as the set of fixed elements of V under Γ. More generally:

(18.1) Lemma (Galois descent). *Let V be a vector space over F_{sep}. If Γ acts continuously on V by semilinear automorphisms, then*

$$V^{\Gamma} = \{\, v \in V \mid \gamma * v = v \text{ for all } \gamma \in \Gamma \,\}$$

is an F-vector space and the map $V^{\Gamma} \otimes F_{\text{sep}} \to V$, $v \otimes x \mapsto vx$, is an isomorphism of F_{sep}-vector spaces.

Proof: It is clear that V^Γ is an F-vector space. To prove surjectivity of the canonical map $V^\Gamma \otimes F_{\mathrm{sep}} \to V$, consider an arbitrary vector $v \in V$. Since the action of Γ on V is continuous, we may find a finite Galois extension M of F in F_{sep} such that $\mathrm{Gal}(F_{\mathrm{sep}}/M)$ acts trivially on v. Let $(m_i)_{1 \le i \le n}$ be a basis of M over F and let $(\gamma_i)_{1 \le i \le n}$ be a set of representatives of the left cosets of $\mathrm{Gal}(F_{\mathrm{sep}}/M)$ in Γ, so that the orbit of v in V consists of $\gamma_1 * v, \ldots, \gamma_n * v$, with $\gamma_1 * v = v$, say. Let

$$v_j = \sum_{i=1}^{n} (\gamma_i * v)\gamma_i(m_j) \qquad \text{for } j = 1, \ldots, n.$$

Since for every $\gamma \in \Gamma$ and $i \in \{1, \ldots, n\}$ there exists $\ell \in \{1, \ldots, n\}$ and $\gamma' \in \mathrm{Gal}(F_{\mathrm{sep}}/M)$ such that $\gamma\gamma_i = \gamma_\ell\gamma'$, the action of γ on the right-hand side of the expression above merely permutes the terms of the sum, hence $v_j \in V^\Gamma$ for $j = 1, \ldots, n$. On the other hand, the $(n \times n)$ matrix $(\gamma_i(m_j))_{1 \le i,j \le n}$ with entries in M is invertible, since $\gamma_1, \ldots, \gamma_n$ are linearly independent over M in $\mathrm{End}_F(M)$ (Dedekind's lemma). If $(m'_{ij})_{1 \le i,j \le n}$ is the inverse matrix, we have

$$v = \gamma_1 * v = \sum_{i=1}^{n} v_i m'_{i1},$$

hence v lies in the image of the canonical map $V^\Gamma \otimes F_{\mathrm{sep}} \to V$.

To prove injectivity of the canonical map, it suffices to show that F-linearly independent vectors in V^Γ are mapped to F_{sep}-linearly independent vectors in V. Suppose the contrary; let $v_1, \ldots, v_r \in V^\Gamma$ be F-linearly independent vectors for which there exist nonzero elements $m_1, \ldots, m_r \in F_{\mathrm{sep}}$ such that $\sum_{i=1}^{r} v_i m_i = 0$. We may assume r is minimal, $r > 1$, and $m_1 = 1$. The m_i are not all in F, hence we may assume $m_2 \notin F$. Choose $\gamma \in \Gamma$ satisfying $\gamma(m_2) \ne m_2$. By applying γ to both sides of the linear dependence relation and subtracting, we obtain $\sum_{i=2}^{r} v_i\big(\gamma(m_i) - m_i\big) = 0$, a nontrivial relation with fewer terms. This contradiction proves that the canonical map $V^\Gamma \otimes F_{\mathrm{sep}} \to V$ is injective. $\qquad\square$

(18.2) Remark. Assume that V in Lemma (18.1) admits an F_{sep}-bilinear multiplication $m\colon V \times V \to V$ and that Γ acts on V by (semilinear) algebra automorphisms; then the restriction of m to V^Γ is a multiplication on V^Γ, hence V^Γ is an F-algebra.

Similarly, if $V = V_1 \supset V_2 \supset \cdots \supset V_r$ is a finite *flag* in V, i.e., V_i is a subspace of V_j for $i > j$, and the action of Γ on V preserves $V_2, \ldots, V_r$, then the flag in V descends to a flag $V^\Gamma = V_1^\Gamma \supset V_2^\Gamma \supset \cdots \supset V_r^\Gamma$ in V^Γ.

18.A. Étale algebras. Let Alg_F be the category of unital commutative associative F-algebras with F-algebra homomorphisms as morphisms. For every finite dimensional commutative F-algebra L, let $X(L)$ be the set of F-algebra homomorphisms from L to F_{sep}:

$$X(L) = \mathrm{Hom}_{\mathsf{Alg}_F}(L, F_{\mathrm{sep}}).$$

For any field extension K/F, let L_K be the K-algebra $L \otimes_F K$. If $K \subset F_{\mathrm{sep}}$, then F_{sep} also is a separable closure of K, and every F-algebra homomorphism $L \to F_{\mathrm{sep}}$ extends in a unique way to a K-algebra homomorphism $L_K \to F_{\mathrm{sep}}$; we may therefore identify:

$$X(L_K) = X(L).$$

The following proposition characterizes étale F-algebras:

(18.3) Proposition. *For a finite dimensional commutative F-algebra L, the following conditions are equivalent*:

(1) *for every field extension K/F, the K-algebra L_K is reduced, i.e., L_K does not contain any nonzero nilpotent elements;*

(2) $L \simeq K_1 \times \cdots \times K_r$ *for some finite separable field extensions $K_1, \ldots, K_r$ of F;*

(3) $L_{F_{\mathrm{sep}}} \simeq F_{\mathrm{sep}} \times \cdots \times F_{\mathrm{sep}}$;

(4) *the bilinear form $T \colon L \times L \to F$ induced by the trace:*

$$T(x, y) = T_{L/F}(xy) \quad \text{for } x,\, y \in L$$

is nonsingular;

(5) $\operatorname{card} X(L) = \dim_F L$;

(6) $\operatorname{card} X(L) \geq \dim_F L$.

If the field F is infinite, the conditions above are also equivalent to:

(7) $L \simeq F[X]/(f)$ *for some polynomial $f \in F[X]$ which has no multiple root in an algebraic closure of F.*

References: The equivalences $(1) \Leftrightarrow (2) \Leftrightarrow (3)$ are proven in Bourbaki [49, Théorème 4, p. V.34] and $(3) \Leftrightarrow (5) \Leftrightarrow (6)$ in [49, Corollaire, p. V.29]. The equivalence of (4) with the other conditions is shown in [49, Proposition 1, p. V.47]. (See also Waterhouse [305, §6.2] for the equivalence of (1), (2), (3), and (5)). Finally, to see that (7) characterizes étale algebras over an infinite field, see Bourbaki [49, Proposition 3, p. V.36 and Proposition 1, p. V.47]. $\qquad \square$

If $L \simeq F[X]/(f)$, every F-algebra homomorphism $L \to F_{\mathrm{sep}}$ is uniquely determined by the image of X, which is a root of f in F_{sep}. Therefore, the maps in $X(L)$ are in one-to-one correspondence with the roots of f in F_{sep}.

A finite dimensional commutative F-algebra satisfying the equivalent conditions above is called *étale*. From characterizations (1) or (4), it follows that étale algebras remain étale under scalar extension.

Another characterization of étale algebras is given in Exercise 1.

Étale F-algebras and Γ-sets. If L is an étale F-algebra of dimension n, Proposition (18.3) shows that $X(L)$ consists of exactly n elements. The absolute Galois group $\Gamma = \operatorname{Gal}(F_{\mathrm{sep}}/F)$ acts on this set as follows:

$$^{\gamma}\xi = \gamma \circ \xi \quad \text{for } \gamma \in \Gamma,\, \xi \in X(L).$$

This action is continuous since it factors through a finite quotient $\operatorname{Gal}(M/F)$ of Γ: we may take for M any finite extension of F in F_{sep} which contains $\xi(L)$ for all $\xi \in X(L)$.

The construction of $X(L)$ is functorial, since every F-algebra homomorphism of étale algebras $f \colon L_1 \to L_2$ induces a Γ-equivariant map $X(f) \colon X(L_2) \to X(L_1)$ defined by

$$\xi^{X(f)} = \xi \circ f \quad \text{for } \xi \in X(L_2).$$

Therefore, writing Et_F for the category of étale F-algebras and Sets_Γ for the category of finite sets endowed with a continuous left action of Γ, there is a contravariant functor

$$\mathbf{X} \colon \mathit{Et}_F \to \mathit{Sets}_\Gamma$$

which associates to $L \in \mathit{Et}_F$ the Γ-set $X(L)$.

We now define a functor in the opposite direction. For $X \in \mathit{Sets}_\Gamma$, consider the F_{sep}-algebra $\mathrm{Map}(X, F_{\mathrm{sep}})$ of all functions $X \to F_{\mathrm{sep}}$. For $f \in \mathrm{Map}(X, F_{\mathrm{sep}})$ and $\xi \in X$, it is convenient to write $\langle f, \xi \rangle$ for the image of ξ under f. We define a semilinear action of Γ on $\mathrm{Map}(X, F_{\mathrm{sep}})$: for $\gamma \in \Gamma$ and $f \in \mathrm{Map}(X, F_{\mathrm{sep}})$, the map $^\gamma f$ is defined by

$$\langle {}^\gamma f, \xi \rangle = \gamma \big(\langle f, {}^{\gamma^{-1}}\xi \rangle \big) \quad \text{for } \xi \in X.$$

If γ acts trivially on X and fixes $\langle f, \xi \rangle$ for all $\xi \in X$, then $^\gamma f = f$. Therefore, the action of Γ on $\mathrm{Map}(X, F_{\mathrm{sep}})$ is continuous. Let $\mathrm{Map}(X, F_{\mathrm{sep}})^\Gamma$ be the F-algebra of Γ-invariant maps. This algebra is étale, since by Lemma (18.1)

$$\mathrm{Map}(X, F_{\mathrm{sep}})^\Gamma \otimes_F F_{\mathrm{sep}} \simeq \mathrm{Map}(X, F_{\mathrm{sep}}) \simeq F_{\mathrm{sep}} \times \cdots \times F_{\mathrm{sep}}.$$

Every equivariant map $g \colon X_1 \to X_2$ of Γ-sets induces an F-algebra homomorphism

$$M(g) \colon \mathrm{Map}(X_2, F_{\mathrm{sep}})^\Gamma \to \mathrm{Map}(X_1, F_{\mathrm{sep}})^\Gamma$$

defined by

$$\big\langle M(g)(f), \xi \big\rangle = \langle f, \xi^g \rangle \quad \text{for } f \in \mathrm{Map}(X_2, F_{\mathrm{sep}})^\Gamma, \, \xi \in X_1,$$

hence there is a contravariant functor

$$\mathbf{M} \colon \mathit{Sets}_\Gamma \to \mathit{Et}_F$$

which maps $X \in \mathit{Sets}_\Gamma$ to $\mathrm{Map}(X, F_{\mathrm{sep}})^\Gamma$.

(18.4) Theorem. *The functors* $\mathbf{X}$ *and* $\mathbf{M}$ *define an anti-equivalence of categories*

$$\mathit{Et}_F \equiv \mathit{Sets}_\Gamma.$$

Under this anti-equivalence, the dimension for étale F-algebras corresponds to the cardinality for Γ-sets: if $L \in \mathit{Et}_F$ corresponds to $X \in \mathit{Sets}_\Gamma$, i.e., $X \simeq \mathbf{X}(L)$ and $L \simeq \mathbf{M}(X)$, then

$$\dim_F L = \mathrm{card}\, X.$$

Moreover, the direct product (resp. tensor product) of étale F-algebras corresponds to the disjoint union (resp. direct product) of Γ-sets: for $L_1, \ldots, L_r$ étale F-algebras,

$$X(L_1 \times \cdots \times L_r) = X(L_1) \amalg \cdots \amalg X(L_r)$$

(where $\amalg$ is the disjoint union) and

$$X(L_1 \otimes \cdots \otimes L_r) = X(L_1) \times \cdots \times X(L_r),$$

where Γ acts diagonally on the right side of the last equality.

Proof: For $L \in \mathit{Et}_F$, the canonical F-algebra homomorphism

$$\Phi \colon L \to \mathrm{Map}\big(X(L), F_{\mathrm{sep}}\big)^\Gamma$$

carries $\ell \in L$ to the map e_ℓ defined by

$$\langle e_\ell, \xi \rangle = \xi(\ell) \quad \text{for } \xi \in X(L).$$

Since $\mathrm{card}\, X(L) = \dim_F L$, we have $\dim_{F_{\mathrm{sep}}} \mathrm{Map}\big(X(L), F_{\mathrm{sep}}\big) = \dim_F L$, hence

$$\dim_F \mathrm{Map}\big(X(L), F_{\mathrm{sep}}\big)^\Gamma = \dim_F L$$

by Lemma (18.1). In order to prove that Φ is an isomorphism, it therefore suffices to show that Φ is injective. Suppose $\ell \in L$ is in the kernel of Φ. By the definition of e_ℓ, this means that $\xi(\ell) = 0$ for every F-algebra homomorphism $L \to F_{\text{sep}}$. It follows that the isomorphism $L_{F_{\text{sep}}} \simeq F_{\text{sep}} \times \cdots \times F_{\text{sep}}$ of (18.3) maps $\ell \otimes 1$ to 0, hence $\ell = 0$.

For $X \in \textsf{Sets}_\Gamma$, there is a canonical Γ-equivariant map

$$\Psi \colon X \to X\big(\operatorname{Map}(X, F_{\text{sep}})^\Gamma\big),$$

which associates to $\xi \in X$ the homomorphism e_ξ defined by

$$e_\xi(f) = \langle f, \xi \rangle \quad \text{for } f \in \operatorname{Map}(X, F_{\text{sep}})^\Gamma.$$

Since $X\big(\operatorname{Map}(X, F_{\text{sep}})^\Gamma\big) = X\big(\operatorname{Map}(X, F_{\text{sep}})^\Gamma_{F_{\text{sep}}}\big) = X\big(\operatorname{Map}(X, F_{\text{sep}})\big)$, the map Ψ is easily checked to be bijective. The other equations are clear. $\qquad\square$

Since direct product decompositions of an étale F-algebra L correspond to disjoint union decompositions of $X(L)$, it follows that L is a field if and only if $X(L)$ is indecomposable, which means that Γ acts transitively on $X(L)$. At the other extreme, $L \simeq F \times \cdots \times F$ if and only if Γ acts trivially on $X(L)$.

Traces and norms. Let L be an étale F-algebra of dimension n. Besides the trace $T_{L/F}$ and the norm $N_{L/F}$, we also consider the quadratic map

$$S_{L/F} \colon L \to F$$

which yields the coefficient of X^{n-2} in the generic polynomial (see (0.3)).

(18.5) Proposition. *Let* $X(L) = \{\xi_1, \ldots, \xi_n\}$. *For all* $\ell \in L$,

$$T_{L/F}(\ell) = \sum_{1 \le i \le n} \xi_i(\ell), \quad S_{L/F}(\ell) = \sum_{1 \le i < j \le n} \xi_i(\ell)\xi_j(\ell), \quad N_{L/F}(\ell) = \xi_1(\ell) \cdots \xi_n(\ell).$$

Proof: It suffices to check these formulas after scalar extension to F_{sep}. We may thus assume $L = F \times \cdots \times F$ and $\xi_i(x_1, \ldots, x_n) = x_i$. With respect to the canonical basis of L over F, multiplication by $(x_1, \ldots, x_n)$ is given by the diagonal matrix with entries $x_1, \ldots, x_n$, hence the formulas are clear. $\qquad\square$

When the étale algebra L is fixed, we set T and b_S for the symmetric bilinear forms on L defined by

(18.6) $\quad T(x, y) = T_{L/F}(xy) \quad \text{and} \quad b_S(x, y) = S_{L/F}(x + y) - S_{L/F}(x) - S_{L/F}(y)$

for all $x, y \in L$. From (18.5), it follows that

$$T(x, y) = \sum_{1 \le i \le n} \xi_i(x)\xi_i(y) \quad \text{and} \quad b_S(x, y) = \sum_{1 \le i \ne j \le n} \xi_i(x)\xi_j(y).$$

Therefore,

(18.7) $\qquad T(x, y) + b_S(x, y) = T_{L/F}(x)T_{L/F}(y) \quad \text{for } x, y \in L.$

(This formula also follows readily from the general relations among the coefficients of the characteristic polynomial: see (0.2).) By putting $y = x$ in this equation, we obtain:

$$T_{L/F}(x^2) + 2S_{L/F}(x) = T_{L/F}(x)^2 \quad \text{for } x \in L,$$

hence the quadratic form $T_{L/F}(x^2)$ is singular if $\operatorname{char} F = 2$ and $n \ge 2$. Proposition (18.3) shows however that the bilinear form T is always nonsingular.

Let L^0 be the kernel of the trace map:

$$L^0 = \{\, x \in L \mid T_{L/F}(x) = 0 \,\}$$

and let $S^0 \colon L^0 \to F$ be the restriction of $S_{L/F}$ to L^0. We write b_{S^0} for the polar form of S^0.

(18.8) Proposition. *Suppose L is an étale F-algebra of dimension n.*

(1) *The bilinear form b_S is nonsingular if and only if* char F *does not divide $n - 1$. If* char F *divides $n - 1$, then the radical of b_S is F.*

(2) *The bilinear form b_{S^0} is nonsingular if and only if* char F *does not divide n. If* char F *divides n, then the radical of b_{S^0} is F.*

(3) *If* char $F = 2$, *the quadratic form L/F is nonsingular if and only if $n \not\equiv 1$ mod 4; the quadratic form S^0 is nonsingular if and only if $n \not\equiv 0$ mod 4.*

Proof: (1) It suffices to prove the statements after scalar extension to F_{sep}. We may thus assume that $L = F \times \cdots \times F$, hence

$$b_S\big((x_1, \ldots, x_n), (y_1, \ldots, y_n)\big) = \sum_{1 \le i \ne j \le n} x_i y_j$$

for $x_1, \ldots, x_n, y_1, \ldots, y_n \in F$. The matrix M of b_S with respect to the canonical basis of L satisfies:

$$M + 1 = (1)_{1 \le i, j \le n}.$$

Therefore, $(M + 1)^2 = n(M + 1)$. If char F divides n, it follows that $M + 1$ is nilpotent. If char F does not divide n, the matrix $n^{-1}(M + 1)$ is an idempotent of rank 1. In either case, the characteristic polynomial of $M + 1$ is $X^{n-1}(X - n)$, so that of M is $(X + 1)^{n-1}\big((X + 1) - n\big)$; hence,

$$\det M = (-1)^{n-1}(n - 1).$$

It follows that b_S is nonsingular if and only if char F does not divide $n - 1$.

If char F divides $n - 1$, then the rank of M is $n - 1$, hence the radical of b_S has dimension 1. This radical contains F, since (18.7) shows that for $\alpha \in F$ and $x \in L$,

$$b_S(\alpha, x) = T_{L/F}(\alpha) T_{L/F}(x) - T_{L/F}(\alpha x) = (n - 1)\alpha T_{L/F}(x) = 0.$$

Therefore, the radical of b_S is F.

(2) Equation (18.7) shows that $b_{S^0}(x, y) = -T(x, y)$ for all $x, y \in L^0$ and that $b_S(\alpha, x) = 0 = T(\alpha, x)$ for $\alpha \in F$ and $x \in L^0$.

If char F does not divide n, then $L = F \oplus L^0$; the elements in the radical of b_{S^0} then lie also in the radical of T. Since T is nonsingular, it follows that b_{S^0} must also be nonsingular.

If char F divides n, then F is in the radical of b_{S^0}. On the other hand, the first part of the proposition shows that b_S is nonsingular, hence the radical of b_{S^0} must have dimension 1; this radical is therefore F.

(3) Assume char $F = 2$. From (1), it follows that the quadratic form $S_{L/F}$ is singular if and only if n is odd and $S_{L/F}(1) = 0$. Similarly, it follows from (2) that S^0 is singular if and only if n is even and $S_{L/F}(1) = 0$. Since $S_{L/F}(1) = \frac{1}{2}n(n - 1)$, the equality $S_{L/F}(1) = 0$ holds for n odd if and only if $n \equiv 1$ mod 4; it holds for n even if and only if $n \equiv 0$ mod 4. $\qquad\square$

The separability idempotent. Let L be an étale F-algebra. Recall from (18.4) that we may identify $X(L \otimes_F L) = X(L) \times X(L)$: for ξ, $\eta \in X(L)$, the F-algebra homomorphism $(\xi, \eta) \colon L \otimes_F L \to F_{\mathrm{sep}}$ is defined by

(18.9)
$$(\xi, \eta)(x \otimes y) = \xi(x)\eta(y) \quad \text{for } x,\, y \in L.$$

Theorem (18.4) yields a canonical isomorphism:

$$L \otimes_F L \xrightarrow{\sim} \mathrm{Map}\big(X(L) \times X(L), F_{\mathrm{sep}}\big)^{\Gamma}.$$

The characteristic function on the diagonal of $X(L) \times X(L)$ is invariant under Γ; the corresponding element $e \in L \otimes_F L$ is called the *separability idempotent* of L. This element is indeed an idempotent since every characteristic function is idempotent. By definition, e is determined by the following condition: for all ξ, $\eta \in X(L)$,

$$(\xi, \eta)(e) = \begin{cases} 0 & \text{if } \xi \neq \eta, \\ 1 & \text{if } \xi = \eta. \end{cases}$$

(18.10) Proposition. *Let $\mu \colon L \otimes_F L \to L$ be the multiplication map. The separability idempotent $e \in L \otimes_F L$ is uniquely determined by the following conditions: $\mu(e) = 1$ and $e(x \otimes 1) = e(1 \otimes x)$ for all $x \in L$. The map $\varepsilon \colon L \to e(L \otimes_F L)$ which carries $x \in L$ to $e(x \otimes 1)$ is an F-algebra isomorphism.*

Proof: In view of the canonical isomorphisms

$$L \xrightarrow{\sim} \mathrm{Map}\big(X(L), F_{\mathrm{sep}}\big)^{\Gamma} \quad \text{and} \quad L \otimes_F L \xrightarrow{\sim} \mathrm{Map}\big(X(L) \times X(L), F_{\mathrm{sep}}\big)^{\Gamma},$$

the conditions $\mu(e) = 1$ and $e(x \otimes 1) = e(1 \otimes x)$ for all $x \in L$ are equivalent to

$$\xi\big(\mu(e)\big) = 1 \quad \text{and} \quad (\xi, \eta)\big(e(x \otimes 1)\big) = (\xi, \eta)\big(e(1 \otimes x)\big)$$

for all ξ, $\eta \in X(L)$ and $x \in L$. We have

$$(\xi, \eta)\big(e(x \otimes 1)\big) = (\xi, \eta)(e)\xi(x) \quad \text{and} \quad (\xi, \eta)\big(e(1 \otimes x)\big) = (\xi, \eta)(e)\eta(x).$$

Therefore, the second condition holds if and only if $(\xi, \eta)(e) = 0$ for $\xi \neq \eta$.

On the other hand, $\xi\big(\mu(e)\big) = (\xi, \xi)(e)$, hence the first condition is equivalent to: $(\xi, \xi)(e) = 1$ for all $\xi \in X(L)$. This proves that the separability idempotent is uniquely determined by the conditions of the proposition.

The map ε is injective since $\mu \circ \varepsilon = \mathrm{Id}_L$. It is also surjective since the properties of e imply:

$$e(x \otimes y) = e(xy \otimes 1) = \varepsilon(xy)$$

for all $x,\, y \in L$. $\qquad\qquad\square$

(18.11) Example. Let $L = F[X]/(f)$ for some polynomial

$$f = X^n + a_{n-1}X^{n-1} + \cdots + a_1 X + a_0$$

with no repeated roots in an algebraic closure of F. Let $x = X + (f)$ be the image of X in L and let

$$t_m = \sum_{i=0}^{m-1} x^i \otimes x^{m-1-i} \in L \otimes_F L \quad \text{for } m = 1, \ldots, n.$$

(In particular, $t_1 = 1$.) The hypothesis on f implies that its derivative f' is relatively prime to f, hence $f'(x) \in L$ is invertible.

We claim that the separability idempotent of L is

$$e = (t_n + a_{n-1}t_{n-1} + \cdots + a_1 t_1)\big(f'(x)^{-1} \otimes 1\big).$$

Indeed, we have $\mu(e) = 1$ since

$$\mu(t_n + a_{n-1}t_{n-1} + \cdots + a_1 t_1) = nx^{n-1} + (n-1)a_{n-1}x^{n-1} + \cdots + a_1 = f'(x).$$

Also, $t_m(x \otimes 1 - 1 \otimes x) = x^m \otimes 1 - 1 \otimes x^m$, hence

$$(t_n + a_{n-1}t_{n-1} + \cdots + a_1 t_1)(x \otimes 1 - 1 \otimes x) =$$
$$\big(f(x) - a_0\big) \otimes 1 - 1 \otimes \big(f(x) - a_0\big) = 0.$$

Therefore, $e(x \otimes 1 - 1 \otimes x) = 0$ and, for $m = 1, \ldots, n-1$,

$$e(x^m \otimes 1 - 1 \otimes x^m) = e(x \otimes 1 - 1 \otimes x)t_m = 0.$$

Since $(x^i)_{0 \leq i \leq n-1}$ is a basis of L over F, it follows that $e(\ell \otimes 1 - 1 \otimes \ell) = 0$ for all $\ell \in L$, proving the claim.

An alternate construction of the separability idempotent is given in the following proposition:

(18.12) Proposition. *Let L be an étale F-algebra of dimension $n = \dim_F L$ and let $(u_i)_{1 \leq i \leq n}$ be a basis of L. Suppose $(v_i)_{1 \leq i \leq n}$ is the dual basis for the bilinear form T of (18.6), in the sense that*

$$T(u_i, v_j) = \delta_{ij} \quad \text{(Kronecker delta)} \qquad \text{for } i, j = 1, \ldots, n.$$

The element $e = \sum_{i=1}^n u_i \otimes v_i \in L \otimes L$ is the separability idempotent of L.

Proof: Since $(u_i)_{1 \leq i \leq n}$ and $(v_i)_{1 \leq i \leq n}$ are dual bases, we have for $x \in L$

(18.13) $$x = \sum_{i=1}^n u_i T(v_i, x) = \sum_{i=1}^n v_i T(u_i, x).$$

In particular, $v_j = \sum_i u_i T(v_i, v_j)$ and $u_j = \sum_i v_i T(u_i, u_j)$ for all $j = 1, \ldots, n$, hence

$$e = \sum_{i,j} u_i \otimes u_j T(v_i, v_j) = \sum_{i,j} v_i \otimes v_j T(u_i, u_j).$$

Using this last expression for e, we get for all $x \in L$:

$$e(x \otimes 1) = \sum_{i,j} v_i x \otimes v_j T(u_i, u_j).$$

By (18.13), we have $v_i x = \sum_k u_k T(v_i x, v_k)$, hence

$$e(x \otimes 1) = \sum_{i,j,k} u_k \otimes v_j T(u_i, u_j) T(v_i x, v_k).$$

Since $T(v_i x, v_k) = T_{L/F}(v_i x v_k) = T(v_i, v_k x)$, we have

$$\sum_i T(u_i, u_j) T(v_i x, v_k) = T\big(\sum_i u_i T(v_i, v_k x), u_j\big) = T(v_k x, u_j),$$

hence

$$e(x \otimes 1) = \sum_{j,k} u_k \otimes v_j T(v_k x, u_j).$$

Similarly, by using the expression $e = \sum_{i,j} u_i \otimes u_j T(v_i, v_j)$, we get for all $x \in L$:

$$e(1 \otimes x) = \sum_{i,k} u_i \otimes v_k T(u_k x, v_i).$$

It follows that $e(x \otimes 1) = e(1 \otimes x)$ since for all $\alpha, \beta = 1, \ldots, n$,

$$T(v_\alpha x, u_\beta) = T_{L/F}(v_\alpha x u_\beta) = T(u_\beta x, v_\alpha).$$

By (18.13) we also have for all $x \in L$ and for all $j = 1, \ldots, n$

$$x u_j = \sum_i u_i T(v_i, x u_j) = \sum_i u_i T_{L/F}(x u_j v_i),$$

hence $T_{L/F}(x) = \sum_i T_{L/F}(xu_iv_i)$. It follows that for all $x \in L$

$$T(x,1) = T_{L/F}(x) = T_{L/F}\left(\sum_i xu_iv_i\right) = T\left(x, \sum_i u_iv_i\right),$$

hence $\sum_i u_iv_i = 1$ since the bilinear form T is nonsingular. This proves that $\mu(e) = 1$. We have thus shown that e satisfies the conditions of (18.10). $\qquad\square$

18.B. Galois algebras. In this subsection, we consider étale F-algebras L endowed with an action by a finite group G of F-automorphisms. Such algebras are called G-*algebras* over F. We write L^G for the subalgebra of G-invariant elements:

$$L^G = \{\, x \in L \mid g(x) = x \text{ for all } g \in G \,\}.$$

In view of the anti-equivalence $\mathsf{Et}_F \equiv \mathsf{Sets}_\Gamma$, there is a canonical isomorphism of groups:

$$\mathrm{Aut}_F(L) = \mathrm{Aut}_{\mathsf{Sets}_\Gamma}\big(X(L)\big)$$

which associates to every automorphism α of the étale algebra L the Γ-equivariant permutation of $X(L)$ mapping $\xi \in X(L)$ to $\xi^\alpha = \xi \circ \alpha$. Therefore, an action of G on L amounts to an action of G by Γ-equivariant permutations on $X(L)$.

(18.14) Proposition. *Let L be a G-algebra over F. Then, $L^G = F$ if and only if G acts transitively on $X(L)$.*

Proof: Because of the canonical isomorphism $\Phi\colon L \xrightarrow{\sim} \mathrm{Map}\big(X(L), F_{\mathrm{sep}}\big)^\Gamma$, for $x \in L$ the condition $x \in L^G$ is equivalent to: $\xi \circ g(x) = \xi(x)$ for all $\xi \in X(L)$, $g \in G$. Since $\xi \circ g = \xi^g$, this observation shows that Φ carries L^G onto the set of Γ-invariant maps $X(L) \to F_{\mathrm{sep}}$ which are constant on each G-orbit of $X(L)$. On the other hand, Φ maps F onto the set of Γ-invariant maps which are constant on $X(L)$. Therefore, if G has only one orbit on $X(L)$, then $L^G = F$.

To prove the converse, it suffices to show that if G has at least two orbits, then there is a nonconstant Γ-invariant map $X(L) \to F_{\mathrm{sep}}$ which is constant on each G-orbit of $X(L)$. Since G acts by Γ-equivariant permutations on $X(L)$, the group Γ acts on the G-orbits of $X(L)$. If this action is not transitive, we may find a disjoint union decomposition of Γ-sets $X(L) = X_1 \amalg X_2$ where X_1 and X_2 are preserved by G. The map $f\colon X(L) \to F_{\mathrm{sep}}$ defined by

$$\langle f, \xi \rangle = \begin{cases} 0 & \text{if } \xi \in X_1 \\ 1 & \text{if } \xi \in X_2 \end{cases}$$

is Γ-invariant and constant on each G-orbit of $X(L)$.

For the rest of the proof, we may thus assume that Γ acts transitively on the G-orbits of $X(L)$. Then, fixing an arbitrary element $\xi_0 \in X(L)$, we have

$$X(L) = \{\, {}^\gamma\xi_0^g \mid \gamma \in \Gamma, \, g \in G \,\}.$$

Since G has at least two orbits in $X(L)$, there exists $\rho \in \Gamma$ such that ${}^\rho\xi_0$ does not lie in the G-orbit of ξ_0. Let $a \in F_{\mathrm{sep}}$ satisfy $\rho(a) \neq a$ and $\gamma(a) = a$ for all $\gamma \in \Gamma$ such that ${}^\gamma\xi_0$ belongs to the G-orbit of ξ_0. We may then define a Γ-invariant map $f\colon X(L) \to F_{\mathrm{sep}}$ by

$$\langle f, {}^\gamma\xi_0^g \rangle = \gamma(a) \qquad \text{for all } \gamma \in \Gamma, \, g \in G.$$

The map f is clearly constant on each G-orbit of $X(L)$, but it is not constant since $f({}^\rho\xi_0) \neq f(\xi_0)$. $\qquad\square$

(18.15) Definitions. Let L be a G-algebra over F for which the order of G equals the dimension of L:

$$|G| = \dim_F L.$$

The G-algebra L is said to be *Galois* if $L^G = F$. By the preceding proposition, this condition holds if and only if G acts transitively on $X(L)$. Since $|G| = \operatorname{card} X(L)$, it then follows that the action of G is simply transitive: for all $\xi,\ \eta \in X(L)$ there is a unique $g \in G$ such that $\eta = \xi^g$. In particular, the action of G on L and on $X(L)$ is faithful.

A Γ-set endowed with a simply transitive action of a finite group G is called a G-*torsor* (or a *principal homogeneous set under G*). Thus, a G-algebra L is Galois if and only if $X(L)$ is a G-torsor. (A more general notion of torsor, allowing a nontrivial action of Γ on G, will be considered in §28.D.)

(18.16) Example. Let L be a Galois G-algebra over F. If L is a field, then Galois theory shows $G = \operatorname{Aut}_{Alg_F}(L)$. Therefore, a Galois G-algebra structure on a field L exists if and only if the extension L/F is Galois with Galois group isomorphic to G; the G-algebra structure is then given by an isomorphism $G \xrightarrow{\sim} \operatorname{Gal}(L/F)$.

If L is not a field, then it may be a Galois G-algebra over F for various non-isomorphic groups G. For instance, suppose $L = K \times K$ where K is a quadratic Galois field extension of F with Galois group $\{\operatorname{Id}, \alpha\}$. We may define an action of $\mathbb{Z}/4\mathbb{Z}$ on L by

$$(1 + 4\mathbb{Z})(k_1, k_2) = \big(\alpha(k_2), k_1\big) \quad \text{for } k_1,\ k_2 \in K.$$

This action gives L the structure of a Galois $\mathbb{Z}/4\mathbb{Z}$-algebra over F. On the other hand, L also is a Galois $(\mathbb{Z}/2\mathbb{Z}) \times (\mathbb{Z}/2\mathbb{Z})$-algebra over F for the action:

$$(1 + 2\mathbb{Z}, 0)(k_1, k_2) = (k_2, k_1), \quad (0, 1 + 2\mathbb{Z})(k_1, k_2) = \big(\alpha(k_1), \alpha(k_2)\big)$$

—but *not* for the action

$$(1 + 2\mathbb{Z}, 0)(k_1, k_2) = \big(\alpha(k_1), k_2\big), \quad (0, 1 + 2\mathbb{Z})(k_1, k_2) = \big(k_1, \alpha(k_2)\big),$$

since $L^G = F \times F$.

More generally, if M/F is a Galois extension of fields with Galois group H, the following proposition shows that one can define on $M^r = M \times \cdots \times M$ a structure of Galois G-algebra over F for every group G containing H as a subgroup of index r:

(18.17) Proposition. *Let G be a finite group and $H \subset G$ a subgroup. For every Galois H-algebra M over F there is a Galois G-algebra $\operatorname{Ind}_H^G M$ over F and a homomorphism $\pi \colon \operatorname{Ind}_H^G M \to M$ such that $\pi\big(h(x)\big) = h\big(\pi(x)\big)$ for all $x \in \operatorname{Ind}_H^G M$, $h \in H$. There is an F-algebra isomorphism:*

$$\operatorname{Ind}_H^G M \simeq \underbrace{M \times \cdots \times M}_{[G:H]}.$$

Moreover, the pair $(\operatorname{Ind}_H^G M, \pi)$ is unique in the sense that if L is another Galois G-algebra over F and $\tau \colon L \to M$ is a homomorphism such that $\tau\big(h(x)\big) = h\big(\tau(x)\big)$ for all $x \in L$ and $h \in H$, then there is an isomorphism of G-algebras $m \colon L \to \operatorname{Ind}_H^G M$ such that $\tau = \pi \circ m$.

Proof: Let

$$\operatorname{Ind}_H^G M = \{\, f \in \operatorname{Map}(G, M) \mid \langle f, hg \rangle = h\big(\langle f, g \rangle\big) \text{ for } h \in H, g \in G \,\},$$

which is an F-subalgebra of $\mathrm{Map}(G, M)$. If $g_1, \ldots, g_r \in G$ are representatives of the right cosets of H in G, so that

$$G = Hg_1 \amalg \cdots \amalg Hg_r,$$

then there is an isomorphism of F-algebras: $\mathrm{Ind}_H^G M \to M^r$ which carries every map in $\mathrm{Ind}_H^G M$ to the r-tuple of its values on $g_1, \ldots, g_r$. Therefore,

$$\dim_F \mathrm{Ind}_H^G M = r \dim_F M = |G|.$$

The algebra $\mathrm{Ind}_H^G M$ carries a natural G-algebra structure: for $f \in \mathrm{Ind}_H^G M$ and $g \in G$, the map $g(f)$ is defined by the equation:

$$\langle g(f), g' \rangle = \langle f, g'g \rangle \quad \text{for } g' \in G.$$

From this definition, it follows that $(\mathrm{Ind}_H^G M)^G = F$, proving $\mathrm{Ind}_H^G M$ is a Galois G-algebra over F. There is a homomorphism $\pi \colon \mathrm{Ind}_H^G M \to M$ such that $\pi\big(h(x)\big) = h\big(\pi(x)\big)$ for all $x \in L$, $h \in H$, given by

$$\pi(f) = \langle f, 1 \rangle.$$

If L is a Galois G-algebra and $\tau \colon L \to M$ is a homomorphism as above, we may define an isomorphism $m \colon L \to \mathrm{Ind}_H^G M$ by mapping $\ell \in L$ to the map m_ℓ defined by

$$\langle m_\ell, g \rangle = \tau\big(g(\ell)\big) \quad \text{for } g \in G.$$

Details are left to the reader. $\qquad\square$

It turns out that every Galois G-algebra over F has the form $\mathrm{Ind}_H^G M$ for some Galois field extension M/F with Galois group isomorphic to H:

(18.18) Proposition. *Let L be a Galois G-algebra over F and let $e \in L$ be a primitive idempotent, i.e., an idempotent which does not decompose into a sum of nonzero idempotents. Let $H \subset G$ be the stabilizer subgroup of e and let $M = eL$. The algebra M is a Galois H-algebra and a field, and there is an isomorphism of G-algebras:*

$$L \simeq \mathrm{Ind}_H^G M.$$

Proof: Since e is primitive, the étale algebra M has no idempotent other than 0 and 1, hence (18.3) shows that M is a field. The action of G on L restricts to an action of H on M. Let $e_1, \ldots, e_r$ be the various images of e under the action of G, with $e = e_1$, say. Since each e_i is a different primitive idempotent, the sum of the e_i is an idempotent in $L^G = F$, hence $e_1 + \cdots + e_r = 1$ and therefore

$$L = e_1 L \times \cdots \times e_r L.$$

Choose $g_1, \ldots, g_r \in G$ such that $e_i = g_i(e)$; then $g_i(M) = e_i L$, hence the fields $e_1 L, \ldots, e_r L$ are all isomorphic to M and

$$\dim_F L = r \dim_F M.$$

On the other hand, the coset decomposition $G = g_1 H \amalg \cdots \amalg g_r H$ shows that $|G| = r|H|$, hence

$$\dim_F M = |H|.$$

To complete the proof that M is a Galois H-algebra, we must show that $M^H = F$. Suppose $e\ell \in M^H$, for some $\ell \in L$; then $\sum_{i=1}^r g_i(e\ell) \in L^G = F$, hence

$$e\big(\textstyle\sum_{i=1}^r g_i(e\ell)\big) \in eF.$$

Since $g_1 \in H$ and $eg_i(e) = e_1e_i = 0$ for $i \neq 1$, we have

$$e\left(\sum_{i=1}^{r} g_i(e\ell)\right) = e\ell,$$

hence $e\ell \in eF$ and $M^H = F$.

Multiplication by e defines an F-algebra homomorphism $\tau \colon L \to M$ such that $\tau\big(h(x)\big) = h\big(\pi(x)\big)$ for all $x \in L$, hence (18.17) yields a G-algebra isomorphism $L \simeq \operatorname{Ind}_H^G M$. $\square$

Galois algebras and torsors. Let $G\text{–}Gal_F$ denote the category of Galois G-algebras over F, where the maps are the G-equivariant homomorphisms, and let $G\text{–}Tors_\Gamma$ be the category of Γ-sets with a G-torsor structure (for an action of G on the right commuting with the action of Γ on the left) where the maps are Γ- and G-equivariant functions. As observed in (18.15), we have $X(L) \in G\text{–}Tors_\Gamma$ for all $L \in G\text{–}Gal_F$. This construction defines a contravariant functor

$$\mathbf{X} \colon G\text{–}Gal_F \to G\text{–}Tors_\Gamma.$$

To obtain a functor $\mathbf{M} \colon G\text{–}Tors_\Gamma \to G\text{–}Gal_F$, we define a G-algebra structure on the étale algebra $\operatorname{Map}(X, F_{\mathrm{sep}})^\Gamma$ for $X \in G\text{–}Tors_\Gamma$: for $g \in G$ and $f \in \operatorname{Map}(X, F_{\mathrm{sep}})$, the map $g(f) \colon X \to F_{\mathrm{sep}}$ is defined by

$$\langle g(f), \xi \rangle = \langle f, \xi^g \rangle \quad \text{for } \xi \in X.$$

Since the actions of Γ and G on X commute, it follows that the actions on the algebra $\operatorname{Map}(X, F_{\mathrm{sep}})$ also commute, hence the action of G restricts to an action on $\operatorname{Map}(X, F_{\mathrm{sep}})^\Gamma$. The induced action on $X\big(\operatorname{Map}(X, F_{\mathrm{sep}})^\Gamma\big)$ coincides with the action of G on X under the canonical bijection $\Psi \colon X \xrightarrow{\sim} X\big(\operatorname{Map}(X, F_{\mathrm{sep}})^\Gamma\big)$, hence $\operatorname{Map}(X, F_{\mathrm{sep}})^\Gamma$ is a Galois G-algebra over F for $X \in G\text{–}Tors_\Gamma$. We let $\mathbf{M}(X) = \operatorname{Map}(X, F_{\mathrm{sep}})^\Gamma$, with the G-algebra structure defined above.

(18.19) Theorem. *The functors* $\mathbf{X}$ *and* $\mathbf{M}$ *define an anti-equivalence of categories*:

$$G\text{–}Gal_F \equiv G\text{–}Tors_\Gamma.$$

Proof: For $L \in Et_F$ and $X \in Sets_\Gamma$, canonical isomorphisms are defined in the proof of (18.4):

$$\Phi \colon L \xrightarrow{\sim} \operatorname{Map}\big(X(L), F_{\mathrm{sep}}\big)^\Gamma, \quad \Psi \colon X \xrightarrow{\sim} X\big(\operatorname{Map}(X, F_{\mathrm{sep}})^\Gamma\big).$$

To establish the theorem, it suffices to verify that Φ and Ψ are G-equivariant if $L \in G\text{–}Gal_F$ and $X \in G\text{–}Tors_\Gamma$, which is easy. (For Ψ, this was already observed above). $\square$

The discriminant of an étale algebra. The Galois closure and the discriminant of an étale F-algebra are defined by a construction involving its associated Γ-set. For X a Γ-set of n elements, let $\Sigma(X)$ be the set of all permutations of a list of the elements of X:

$$\Sigma(X) = \big\{ (\xi_1, \ldots, \xi_n) \mid \xi_1, \ldots, \xi_n \in X, \; \xi_i \neq \xi_j \text{ for } i \neq j \big\}.$$

This set carries the diagonal action of Γ:

$$^\gamma(\xi_1, \ldots, \xi_n) = (^\gamma\xi_1, \ldots, {}^\gamma\xi_n) \quad \text{for } \gamma \in \Gamma$$

and also an action of the symmetric group S_n:

$$(\xi_1, \ldots, \xi_n)^\sigma = (\xi_{\sigma(1)}, \ldots, \xi_{\sigma(n)}) \quad \text{for } \sigma \in S_n.$$

Clearly, $\Sigma(X)$ is a torsor under S_n:

$$\Sigma(X) \in S_n\text{-}\mathsf{Tors}_\Gamma,$$

and the projections on the various components define Γ-equivariant maps

$$\pi_i \colon \Sigma(X) \to X$$

for $i = 1, \ldots, n$.

Let $\Delta(X)$ be the set of orbits of $\Sigma(X)$ under the action of the alternating group A_n, with the induced action of Γ:

$$\Delta(X) = \Sigma(X)/A_n \in \mathsf{Sets}_\Gamma.$$

This Γ-set has two elements.

The anti-equivalences $\mathsf{Et}_F \equiv \mathsf{Sets}_\Gamma$ and $S_n\text{-}\mathsf{Gal}_F \equiv S_n\text{-}\mathsf{Tors}_\Gamma$ yield corresponding constructions for étale algebras. If L is an étale algebra of dimension n over F, we set

(18.20) $$\Sigma(L) = \mathrm{Map}\big(\Sigma\big(X(L)\big), F_{\mathrm{sep}}\big)^\Gamma,$$

a Galois S_n-algebra over F with n canonical embeddings $\varepsilon_1, \ldots, \varepsilon_n \colon L \hookrightarrow \Sigma(L)$ defined by the relation

$$\big\langle \varepsilon_i(\ell), (\xi_1, \ldots, \xi_n) \big\rangle = \xi_i(\ell) \quad \text{for } i = 1, \ldots, n,\, \ell \in L,\, (\xi_1, \ldots, \xi_n) \in \Sigma\big(X(L)\big).$$

We also set

$$\Delta(L) = \mathrm{Map}\big(\Delta\big(X(L)\big), F_{\mathrm{sep}}\big)^\Gamma,$$

a quadratic étale algebra over F which may alternately be defined as

$$\Delta(L) = \Sigma(L)^{A_n}.$$

From the definition of $\Delta\big(X(L)\big)$, it follows that an element $\gamma \in \Gamma$ acts trivially on this set if and only if the induced permutation $\xi \mapsto {}^\gamma\xi$ of $X(L)$ is even. Therefore, the kernel of the action of Γ on $\Delta\big(X(L)\big)$ is the subgroup $\Gamma_0 \subset \Gamma$ which acts by even permutations on $X(L)$, and

(18.21) $$\Delta(L) \simeq \begin{cases} F \times F & \text{if } \Gamma_0 = \Gamma, \\ (F_{\mathrm{sep}})^{\Gamma_0} & \text{if } \Gamma_0 \neq \Gamma. \end{cases}$$

The algebra $\Sigma(L)$ is called the *Galois S_n-closure* of the étale algebra L and $\Delta(L)$ is called the *discriminant* of L.

(18.22) Example. Suppose L is a field; it is then a separable extension of degree n of F, by (18.3). We relate $\Sigma(L)$ to the (Galois-theoretic) Galois closure of L.

Number the elements of $X(L)$:

$$X(L) = \{\xi_1, \ldots, \xi_n\}$$

and let M be the subfield of F_{sep} generated by $\xi_1(L), \ldots, \xi_n(L)$:

$$M = \xi_1(L) \cdots \xi_n(L) \subset F_{\mathrm{sep}}.$$

Galois theory shows M is the Galois closure of each of the fields $\xi_1(L), \ldots, \xi_n(L)$. The action of Γ on $X(L)$ factors through an action of the Galois group $\mathrm{Gal}(M/F)$. Letting $H = \mathrm{Gal}(M/F)$, we may therefore identify H with a subgroup of S_n: for $h \in H$ and $i = 1, \ldots, n$ we define $h(i) \in \{1, \ldots, n\}$ by

$$^h\xi_i = h \circ \xi_i = \xi_{h(i)}.$$

We claim that

$$\Sigma(L) \simeq \mathrm{Ind}_H^{S_n} M \quad \text{as } S_n\text{-algebras}.$$

(In particular, $\Sigma(L) \simeq M$ if $H = S_n$). The existence of such an isomorphism follows from (18.17) if we show that there is a homomorphism $\tau \colon \Sigma(L) \to M$ such that $\tau\big(h(f)\big) = h\big(\tau(f)\big)$ for all $f \in \Sigma(L)$, $h \in H$.

For $f \in \Sigma(L) = \mathrm{Map}\big(\Sigma(X(L)), F_{\mathrm{sep}}\big)^\Gamma$, set

$$\tau(f) = \big\langle f, (\xi_1, \ldots, \xi_n) \big\rangle.$$

The right side lies in M since $^\gamma\xi_i = \xi_i$ for all $\gamma \in \mathrm{Gal}(F_{\mathrm{sep}}/M)$. For $h \in H$ and $f \in \Sigma(L)$, we have

$$\tau\big(h(f)\big) = \big\langle h(f), (\xi_1, \ldots, \xi_n) \big\rangle = \big\langle f, (\xi_1, \ldots, \xi_n)^h \big\rangle = \big\langle f, (\xi_{h(1)}, \ldots, \xi_{h(n)}) \big\rangle.$$

On the other hand, since f is invariant under the Γ-action on $\mathrm{Map}\big(\Sigma(X(L)), F_{\mathrm{sep}}\big)$,

$$h\big(\tau(f)\big) = h\big(\langle f, (\xi_1, \ldots, \xi_n) \rangle\big) = \big\langle f, {}^h(\xi_1, \ldots, \xi_n) \big\rangle.$$

Since $^h(\xi_1, \ldots, \xi_n) = ({}^h\xi_1, \ldots, {}^h\xi_n) = (\xi_{h(1)}, \ldots, \xi_{h(n)})$, the claim is proved.

(18.23) Example. Suppose $L = F[X]/(f)$ where f is a polynomial of degree n with no repeated roots in an algebraic closure of F. We give an explicit description of $\Delta(L)$.

Let $x = X + (f)$ be the image of X in L and let x_1, ..., x_n be the roots of f in F_{sep}. An F-algebra homomorphism $L \to F_{\mathrm{sep}}$ is uniquely determined by the image of x, which must be one of the x_i. Therefore, $X(L) = \{\xi_1, \ldots, \xi_n\}$ where $\xi_i \colon L \to F_{\mathrm{sep}}$ maps x to x_i.

If $\mathrm{char}\, F \neq 2$, an element $\gamma \in \Gamma$ induces an even permutation of $X(L)$ if and only if

$$\gamma\Big(\prod_{1 \le i < j \le n} (x_i - x_j) \Big) = \prod_{1 \le i < j \le n} (x_i - x_j),$$

since

$$\prod_{1 \le i < j \le n} (x_i - x_j) = \prod_{1 \le i < j \le n} \big(\xi_i(x) - \xi_j(x) \big)$$

and

$$\gamma\Big(\prod_{1 \le i < j \le n} (x_i - x_j) \Big) = \prod_{1 \le i < j \le n} \big({}^\gamma\xi_i(x) - {}^\gamma\xi_j(x) \big).$$

By (18.21), it follows that

$$\Delta(L) \simeq F[T]/(T^2 - d)$$

where $d = \prod_{1 \le i < j \le n} (x_i - x_j)^2 \in F$.

If $\mathrm{char}\, F = 2$, the condition that γ induces an even permutation of $X(L)$ amounts to $\gamma(s) = s$, where

$$s = \sum_{1 \le i < j \le n} \frac{x_i}{x_i + x_j},$$

hence

$$\Delta(L) \simeq F[T]/(T^2 + T + d)$$

where $d = s^2 + s = \sum_{1 \le i < j \le n} \frac{x_i x_j}{x_i^2 + x_j^2} \in F$.

The following proposition relates the discriminant $\Delta(L)$ to the determinant of the trace forms on L. Recall from (18.3) that the bilinear form T on L is nonsingular; if char $F = 2$, Proposition (18.8) shows that the quadratic form $S_{L/F}$ on L is nonsingular if $\dim_F L$ is even and the quadratic form S^0 on $L^0 = \ker T_{L/F}$ is nonsingular if $\dim_F L$ is odd.

(18.24) Proposition. *Let L be an étale F-algebra of dimension n.*
   If char $F \neq 2$,

$$\Delta(L) \simeq F[t]/(t^2 - d)$$

where $d \in F^\times$ represents the determinant of the bilinear form T.
   If char $F = 2$,

$$\Delta(L) \simeq F[t]/(t^2 + t + a)$$

where $a \in F$ represents the determinant of the quadratic form $S_{L/F}$ if n is even and $a + \frac{1}{2}(n-1)$ represents the determinant of the quadratic form S^0 if n is odd.

Proof: Let $X(L) = \{\xi_1, \dots, \xi_n\}$ and let $\Gamma_0 \subset \Gamma$ be the subgroup which acts on $X(L)$ by even permutations, so that $\Delta(L)$ is determined up to F-isomorphism by (18.21). The idea of the proof is to find an element $u \in F_{\text{sep}}$ satisfying the following conditions:

(a) if char $F \neq 2$:

$$\gamma(u) = \begin{cases} u & \text{if } \gamma \in \Gamma_0, \\ -u & \text{if } \gamma \in \Gamma \smallsetminus \Gamma_0, \end{cases}$$

   and $u^2 \in F^\times$ represents $\det T \in F^\times/F^{\times 2}$.
(b) if char $F = 2$:

$$\gamma(u) = \begin{cases} u & \text{if } \gamma \in \Gamma_0, \\ u+1 & \text{if } \gamma \in \Gamma \smallsetminus \Gamma_0, \end{cases}$$

   and $\wp(u) = u^2 + u \in F$ represents $\det S_{L/F} \in F/\wp(F)$ if n is even, $u^2 + u + \frac{1}{2}(n-1)$ represents $\det S^0 \in F/\wp(F)$ if n is odd.

The proposition readily follows, since in each case $F(u) = (F_{\text{sep}})^{\Gamma_0}$.
   Suppose first that char $F \neq 2$. Let $(e_i)_{1 \leq i \leq n}$ be a basis of L over F. Consider the matrix

$$M = \big(\xi_i(e_j)\big)_{1 \leq i, j \leq n} \in M_n(F_{\text{sep}})$$

and

$$u = \det M \in F_{\text{sep}}.$$

For $\gamma \in \Gamma$ we have $\gamma(u) = \det\big[\big({}^\gamma\xi_i(e_j)\big)_{1 \leq i,j \leq n}\big]$. Since an even permutation of the rows of a matrix does not change its determinant and an odd permutation changes its sign, it follows that $\gamma(u) = u$ if $\gamma \in \Gamma_0$ and $\gamma(u) = -u$ if $\gamma \in \Gamma \smallsetminus \Gamma_0$. Moreover, by (18.5) we have:

$$M^t \cdot M = \big(\textstyle\sum_{k=1}^n \xi_k(e_i)\xi_k(e_j)\big)_{1 \leq i,j \leq n} = \big(T(e_i, e_j)\big)_{1 \leq i,j \leq n},$$

hence u^2 represents $\det T$. This completes the proof in the case where char $F \neq 2$.

Suppose next that $\operatorname{char} F = 2$ and n is even: $n = 2m$. Let $(e_i)_{1 \leq i \leq n}$ be a symplectic basis of L for the bilinear form b_S, so that the matrix of b_S with respect to this basis is

$$A = \begin{pmatrix} J & & 0 \\ & \ddots & \\ 0 & & J \end{pmatrix} \quad \text{where} \quad J = \begin{pmatrix} 0 & 1 \\ 1 & 0 \end{pmatrix}.$$

Assume moreover that $e_1 = 1$. For $i, j = 1, \ldots, n$, define

$$b_{ij} = \begin{cases} \sum_{1 \leq k < \ell \leq n} \xi_k(e_i)\xi_\ell(e_j) & \text{if } i > j, \\ 0 & \text{if } i \leq j. \end{cases}$$

Let $B = (b_{ij})_{1 \leq i,j \leq n} \in M_n(F_{\text{sep}})$ and let

$$u = \operatorname{tr}(A^{-1}B) = \sum_{i=1}^m b_{2i,2i-1}.$$

Under the transposition permutation of $\xi_1, \ldots, \xi_n$ which exchanges ξ_r and ξ_{r+1} and fixes ξ_i for $i \neq r, r+1$, the element b_{ij} is replaced by $b_{ij} + \xi_r(e_i)\xi_{r+1}(e_j) + \xi_r(e_j)\xi_{r+1}(e_i)$, hence u becomes $u + \varepsilon$ where

$$\varepsilon = \sum_{i=1}^m \big(\xi_r(e_{2i})\xi_{r+1}(e_{2i-1}) + \xi_r(e_{2i-1})\xi_{r+1}(e_{2i})\big).$$

Claim. $\varepsilon = 1$.

Since $(e_i)_{1 \leq i \leq n}$ is a symplectic basis of L for the bilinear form b_S, it follows from (18.7) that $T_{L/F}(e_2) = 1$ and $T_{L/F}(e_i) = 0$ for $i \neq 2$. The dual basis for the bilinear form T is then $(f_i)_{1 \leq i \leq n}$ where

$$f_1 = e_1 + e_2, \ f_2 = e_1, \text{ and } f_{2i-1} = e_{2i}, \ f_{2i} = e_{2i-1} \text{ for } i = 2, \ldots, m.$$

Proposition (18.12) shows that $\sum_{i=1}^n e_i \otimes f_i \in L \otimes L$ is the separability idempotent of L. From the definition of this idempotent, it follows that

$$(\xi_r, \xi_{r+1})\Big(\sum_{i=1}^n e_i \otimes f_i\Big) = \sum_{i=1}^n \xi_r(e_i)\xi_{r+1}(f_i) = 0.$$

(See (18.9) for the notation.) On the other hand, the formulas above for $f_1, \ldots, f_n$ show:

$$\sum_{i=1}^n e_i \otimes f_i = e_1 \otimes e_1 + \sum_{i=1}^m (e_{2i} \otimes e_{2i-1} + e_{2i-1} \otimes e_{2i}).$$

Since $e_1 = 1$, the claim follows.

Since the symmetric group is generated by the transpositions $(r, r+1)$, the claim shows that u is transformed into $u + 1$ by any odd permutation of $\xi_1, \ldots, \xi_n$ and is fixed by any even permutation. Therefore,

$$\gamma(u) = \begin{cases} u & \text{if } \gamma \in \Gamma_0, \\ u+1 & \text{if } \gamma \in \Gamma \smallsetminus \Gamma_0. \end{cases}$$

We proceed to show that $u^2 + u$ represents $\det S_{L/F} \in F/\wp(F)$.

Let $C = \big(\sum_{1 \leq k < \ell \leq n} \xi_k(e_i)\xi_\ell(e_j)\big)_{1 \leq i,j \leq n} \in M_n(F_{\text{sep}})$ and let

$$D = C + B + B^t.$$

We have $D = (d_{ij})_{1 \leq i,j \leq n}$ where

$$d_{ij} = \begin{cases} 0 & \text{if } i > j, \\ \sum_{1 \leq k < \ell \leq n} \xi_k(e_i)\xi_\ell(e_i) = S_{L/F}(e_i) & \text{if } i = j, \\ \sum_{1 \leq k < \ell \leq n} \big(\xi_k(e_i)\xi_\ell(e_j) + \xi_k(e_j)\xi_\ell(e_i)\big) = b_S(e_i, e_j) & \text{if } i < j, \end{cases}$$

hence $D \in M_n(F)$ is a matrix of the quadratic form $S_{L/F}$ with respect to the basis $(e_i)_{1 \leq i \leq n}$ and $D + D^t = A$. Therefore, $s_2(A^{-1}D) \in F$ represents the determinant of $S_{L/F}$. Since $D = C + B + B^t$, Lemma (0.5) yields

$$s_2(A^{-1}D) = s_2(A^{-1}C) + \wp\big(\mathrm{tr}(A^{-1}B)\big) = s_2(A^{-1}C) + u^2 + u.$$

To complete the proof, it suffices to show $s_2(A^{-1}C) = 0$.

Let $M = \big(\xi_i(e_j)\big)_{1 \leq i,j \leq n} \in M_n(F_{\mathrm{sep}})$ and let $V = (v_{ij})_{1 \leq i,j \leq n}$ where

$$v_{ij} = \begin{cases} 0 & \text{if } i \geq j, \\ 1 & \text{if } i < j. \end{cases}$$

The matrix M is invertible, since $M^t \cdot M = \big(T(e_i, e_j)\big)_{1 \leq i,j \leq n}$ and T is nonsingular. Moreover, $C = M^t \cdot V \cdot M$, hence $A = C + C^t = M^t(V + V^t)M$ and therefore

$$s_2(A^{-1}C) = s_2\big((V + V^t)^{-1}V\big).$$

Observe that $(V + V^t + 1)^2 = 0$, hence $(V + V^t)^{-1} = V + V^t$. It follows that

$$s_2\big((V + V^t)^{-1}V\big) = s_2\big((V + V^t)V\big) = s_2\big((V + V^t + 1)V + V\big).$$

Using the relations between coefficients of characteristic polynomials (see (0.2)), we may expand the right-hand expression to obtain:

$$\begin{aligned} s_2\big((V + V^t)^{-1}V\big) = {} & s_2\big((V + V^t + 1)V\big) + s_2(V) + \\ & \mathrm{tr}\big((V + V^t + 1)V\big)\,\mathrm{tr}(V) + \mathrm{tr}\big((V + V^t + 1)V^2\big). \end{aligned}$$

Since $V + V^t + 1$ has rank 1, we have $s_2\big((V + V^t + 1)V\big) = 0$. Since V is nilpotent, we have $\mathrm{tr}(V) = s_2(V) = 0$. A computation shows that $\mathrm{tr}\big((V + V^t + 1)V^2\big) = 0$, hence $s_2\big((V + V^t)^{-1}V\big) = 0$ and the proof is complete in the case where char $F = 2$ and n is even.

Assume finally that char $F = 2$ and n is odd: $n = 2m + 1$. Let $(e_i)_{1 \leq i \leq n-1}$ be a symplectic basis of L^0 for the bilinear form b_{S^0}. We again denote by A the matrix of the bilinear form with respect to this basis and define a matrix $B = (b_{ij})_{1 \leq i,j \leq n-1} \in M_{n-1}(F_{\mathrm{sep}})$ by

$$b_{ij} = \begin{cases} \sum_{1 \leq k < \ell \leq n} \xi_k(e_i)\xi_\ell(e_j) & \text{if } i > j, \\ 0 & \text{if } i \leq j. \end{cases}$$

Let

$$u = \mathrm{tr}(A^{-1}B) = \sum_{i=1}^{m} b_{2i,2i-1}.$$

In order to extend $(e_i)_{1 \leq i \leq n-1}$ to a basis $(e_i)_{1 \leq i \leq n}$ of L, define $e_n = 1$. Since $b_{S^0}(x, y) = T(x, y)$ for all x, $y \in L^0$, by (18.7), the dual basis $(f_i)_{1 \leq i \leq n}$ for the bilinear form T is given by

$$f_{2i-1} = e_{2i}, \; f_{2i} = e_{2i-1} \text{ for } i = 1, \ldots, m, \quad f_n = e_n = 1.$$

Proposition (18.12) shows that the separability idempotent of L is

$$e = \sum_{i=1}^{m}\big(e_{2i} \otimes e_{2i-1} + e_{2i-1} \otimes e_{2i}\big) + e_n \otimes e_n$$

hence for $r = 1, \ldots, n - 2$,

$$\sum_{i=1}^{m}\big(\xi_r(e_{2i})\xi_{r+1}(e_{2i-1}) + \xi_r(e_{2i-1})\xi_{r+1}(e_{2i})\big) = \xi_r(e_n)\xi_{r+1}(e_n) = 1.$$

The same argument as in the preceding case then shows

$$\gamma(u) = \begin{cases} u & \text{if } \gamma \in \Gamma_0, \\ u+1 & \text{if } \gamma \in \Gamma \setminus \Gamma_0. \end{cases}$$

Mimicking the preceding case, we let

$$C = \left(\sum_{1 \le k < \ell \le n} \xi_k(e_i)\xi_\ell(e_j)\right)_{1 \le i,j \le n-1} \in M_{n-1}(F_{\text{sep}})$$

and

$$D = C + B + B^t = (d_{ij})_{1 \le i,j \le n-1},$$

where

$$d_{ij} = \begin{cases} 0 & \text{if } i > j, \\ S^0(e_i) & \text{if } i = j, \\ b_{S^0}(e_i, e_j) & \text{if } i < j. \end{cases}$$

The element $s_2(A^{-1}D)$ represents the determinant $\det S^0 \in F/\wp(F)$ and

$$s_2(A^{-1}D) = s_2(A^{-1}C) + \wp\big(\text{tr}(A^{-1}B)\big) = s_2(A^{-1}C) + u^2 + u,$$

hence it suffices to show $s_2(A^{-1}C) = \frac{1}{2}(n-1)$ to complete the proof.

For $j = 1, \ldots, n-1$ we have $e_j^0 \in L^0$, hence $\xi_n(e_j) = \sum_{\ell=1}^{n-1} \xi_\ell(e_j)$, and therefore

$$c_{ij} = \sum_{1 \le k < \ell \le n-1} \xi_k(e_i)\xi_\ell(e_j) + \left(\sum_{k=1}^{n-1} \xi_k(e_i)\right)\left(\sum_{\ell=1}^{n-1} \xi_\ell(e_j)\right)$$
$$= \sum_{n-1 \ge k \ge \ell \ge 1} \xi_k(e_i)\xi_\ell(e_j).$$

The matrix $M = \big(\xi_i(e_j)\big)_{1 \le i,j \le n-1} \in M_{n-1}(F_{\text{sep}})$ is invertible since

$$M^t \cdot M = \big(T(e_i, e_j)\big)_{1 \le i,j \le n-1} = \big(b_{S^0}(e_i, e_j)\big)_{1 \le i,j \le n-1}$$

and b_{S^0} is nonsingular. Moreover, $C = M^t \cdot W \cdot M$ where $W = (w_{ij})_{1 \le i,j \le n-1}$ is defined by

$$w_{ij} = \begin{cases} 1 & \text{if } i \ge j, \\ 0 & \text{if } i < j. \end{cases}$$

hence $s_2(A^{-1}C) = s_2\big((W + W^t)^{-1}W\big)$. Computations similar to those of the preceding case show that $s_2\big((W + W^t)^{-1}W\big) = \frac{1}{2}(n-1)$. $\square$

18.C. Cubic étale algebras. Cubic étale algebras, i.e., étale algebras of dimension 3, have special features with respect to the Galois S_3-closure and discriminant: if L is such an algebra, we establish below canonical isomorphisms

$$L \otimes_F L \simeq L \times \Sigma(L) \quad \text{and} \quad \Sigma(L) \simeq L \otimes \Delta(L).$$

Moreover, we show that if F is infinite of characteristic different from 3, these algebras have the form $F[X]/(X^3 - 3X + t)$ for some $t \in F$, and we set up an exact sequence relating the square class group of L to the square class group of F.

The Galois closure and the discriminant. Let L be a cubic étale F-algebra. Let $\Sigma(L)$ be the Galois S_3-closure of L and let $e \in L \otimes L$ be the separability idempotent of L.

(18.25) Proposition. *There are canonical F-algebra isomorphisms*

$$\Theta \colon (1-e) \cdot (L \otimes L) \xrightarrow{\sim} \Sigma(L) \quad and \quad \widehat{\Theta} \colon L \otimes L \xrightarrow{\sim} L \times \Sigma(L)$$

related by $\widehat{\Theta}(x \otimes y) = \big(xy, \Theta\big((1-e) \cdot (x \otimes y)\big)\big)$ *for* $x, y \in L$.

Proof: Consider the disjoint union decomposition of Γ-sets

$$X(L \otimes L) = X(L) \times X(L) = D(L) \amalg E(L)$$

where $D(L)$ is the diagonal of $X(L) \times X(L)$ and $E(L)$ is its complement. By definition, e corresponds to the characteristic function on $D(L)$ under the canonical isomorphism $L \otimes L \simeq \mathrm{Map}\big(X(L \otimes L), F_{\mathrm{sep}}\big)^\Gamma$, hence

$$E(L) = \{\, \xi \in X(L \otimes L) \mid \xi(1-e) = 1 \,\}.$$

Therefore, we may identify

$$X\big((1-e) \cdot (L \otimes L)\big) = E(L),$$

hence also $(1-e) \cdot (L \otimes L) = \mathrm{Map}\big(E(L), F_{\mathrm{sep}}\big)^\Gamma$. On the other hand, there is a canonical bijection

$$\Sigma\big(X(L)\big) \xrightarrow{\sim} E(L)$$

which maps $(\xi_i, \xi_j, \xi_k) \in \Sigma\big(X(L)\big)$ to $(\xi_i, \xi_j) \in E(L)$. Under the anti-equivalence $Et_F \equiv Sets_\Gamma$, this bijection induces an isomorphism

$$\Theta \colon (1-e) \cdot (L \otimes L) \xrightarrow{\sim} \Sigma(L).$$

By decomposing $L \otimes L = \big(e \cdot (L \otimes L)\big) \oplus \big((1-e) \cdot (L \otimes L)\big)$, and combining Θ with the isomorphism $\varepsilon^{-1} \colon e \cdot (L \otimes L) \xrightarrow{\sim} L$ of (18.10) which maps $e \cdot (x \otimes y) = e \cdot (xy \otimes 1)$ to xy, we obtain the isomorphism

$$\widehat{\Theta} \colon L \otimes L \xrightarrow{\sim} L \times \Sigma(L). \qquad \square$$

Note that there are actually three canonical bijections $\Sigma\big(X(L)\big) \xrightarrow{\sim} E(L)$, since $(\xi_i, \xi_j, \xi_k) \in \Sigma\big(X(L)\big)$ may alternately be mapped to (ξ_i, ξ_k) or (ξ_j, ξ_k) instead of (ξ_i, ξ_j); therefore, there are three canonical isomorphisms $(1-e) \cdot (L \otimes L) \xrightarrow{\sim} \Sigma(L)$ and $L \otimes L \xrightarrow{\sim} L \times \Sigma(L)$.

In view of the proposition above, there is an S_3-algebra structure on $(1-e) \cdot (L \otimes L)$ and there are three embeddings $\epsilon_1, \epsilon_2, \epsilon_3 \colon L \hookrightarrow (1-e) \cdot (L \otimes L)$, which we now describe: for $\ell \in L$, we set

$$\epsilon_1(\ell) = (1-e) \cdot (\ell \otimes 1), \quad \epsilon_2(\ell) = (1-e) \cdot (1 \otimes \ell)$$

and

(18.26) $\epsilon_3(\ell) = (1-e) \cdot \big(T_{L/F}(\ell)1 \otimes 1 - \ell \otimes 1 - 1 \otimes \ell\big).$

These isomorphisms correspond to the maps $E(L) \to X(L)$ which carry an element $(\xi_i, \xi_j) \in E(L)$ respectively to ξ_i, ξ_j and to the element ξ_k such that $X(L) = \{\xi_i, \xi_j, \xi_k\}$. Indeed, for $\ell \in L$,

$$(\xi_i, \xi_j)[(1-e) \cdot (\ell \otimes 1)] = \xi_i(\ell), \quad (\xi_i, \xi_j)[(1-e) \cdot (1 \otimes \ell)] = \xi_j(\ell)$$

and

$$(\xi_i, \xi_j)\big[(1-e) \cdot \big(T_{L/F}(\ell)1 \otimes 1 - \ell \otimes 1 - 1 \otimes \ell\big)\big] = T_{L/F}(\ell) - \xi_i(\ell) - \xi_j(\ell) = \xi_k(\ell).$$

An action of S_3 on $(1-e) \cdot (L \otimes L)$ is defined by permuting the three copies of L in $(1-e) \cdot (L \otimes L)$, so that $\sigma \circ \epsilon_i = \epsilon_{\sigma(i)}$ for $\sigma \in S_3$.

We now turn to the discriminant algebra $\Delta(L)$:

(18.27) Proposition. *Let L be a cubic étale algebra. The canonical embeddings ε_1, ε_2, $\varepsilon_3 \colon L \hookrightarrow \Sigma(L)$ define isomorphisms:*

$$\widehat{\varepsilon}_1, \widehat{\varepsilon}_2, \widehat{\varepsilon}_3 \colon \Delta(L) \otimes_F L \xrightarrow{\sim} \Sigma(L).$$

Proof: Consider the transpositions in S_3:

$$\tau_1 = (2,3) \quad \tau_2 = (1,3) \quad \tau_3 = (1,2)$$

and the subgroups of order 2:

$$H_i = \{\mathrm{Id}, \tau_i\} \subset S_3 \quad \text{for } i = 1,\, 2,\, 3.$$

For $i = 1$, 2, 3, the canonical map

$$\Sigma\big(X(L)\big) \to \big(\Sigma\big(X(L)\big)/A_3\big) \times \big(\Sigma\big(X(L)\big)/H_i\big)$$

which carries $\zeta \in \Sigma\big(X(L)\big)$ to the pair $(\zeta^{A_3}, \zeta^{H_i})$ consisting of its orbits under A_3 and under H_i is a Γ-equivariant bijection, since $\zeta^{A_3} \cap \zeta^{H_i} = \{\zeta\}$. Moreover, projection on the i-th component $\pi_i \colon \Sigma\big(X(L)\big) \to X(L)$ factors through $\Sigma\big(X(L)\big)/H_i$; we thus get three canonical Γ-equivariant bijections:

$$\widehat{\pi}_i \colon \Sigma\big(X(L)\big) \xrightarrow{\sim} \Delta\big(X(L)\big) \times X(L).$$

Under the anti-equivalence $Et_F \equiv Sets_\Gamma$, these bijections yield the required isomorphisms $\widehat{\varepsilon}_i$ for $i = 1$, 2, 3. $\qquad\square$

Combining (18.25) and (18.27), we get:

(18.28) Corollary. *For every cubic étale F-algebra L, there are canonical F-algebra isomorphisms:*

$$(\mathrm{Id} \times \widehat{\varepsilon}_i^{\,-1}) \circ \widehat{\Theta} \colon L \otimes_F L \xrightarrow{\sim} L \times \big(\Delta(L) \otimes_F L\big) \quad \text{for } i = 1,\, 2,\, 3.$$

The isomorphism $(\mathrm{Id} \times \widehat{\varepsilon}_2^{\,-1}) \circ \widehat{\Theta}$ is L-linear for the action of L on $L \otimes_F L$ and on $\Delta(L) \otimes_F L$ by multiplication on the right factor.

Proof: The first assertion follows from (18.25) and (18.27). A computation shows that $\Theta(1 \otimes \ell)$ maps $(\xi_1, \xi_2, \xi_3) \in \Sigma\big(X(L)\big)$ to $\xi_2(\ell)$, for all $\ell \in L$. Similarly, $\widehat{\varepsilon}_2(1 \otimes \ell)$ maps (ξ_1, ξ_2, ξ_3) to $\xi_2(\ell)$, hence $\widehat{\varepsilon}_2^{\,-1} \circ \Theta$ is L-linear. $\qquad\square$

We conclude with two cases where the discriminant algebra can be explicitly calculated:

(18.29) Proposition. *For every quadratic étale F-algebra K,*

$$\Delta(F \times K) \simeq K.$$

Proof: The projection on the first component $\xi \colon F \times K \to F$ is an element of $X(F \times K)$ which is invariant under the action of Γ. If $X(K) = \{\eta, \zeta\}$, then $X(F \times K) = \{\xi, \eta, \zeta\}$, and the map which carries η to $(\xi, \eta, \zeta)^{A_3}$ and ζ to $(\xi, \zeta, \eta)^{A_3}$ defines an isomorphism of Γ-sets $X(K) \xrightarrow{\sim} \Delta\big(X(F \times K)\big)$. The isomorphism $\Delta(F \times K) \simeq K$ follows from the anti-equivalence $Et_F \equiv Sets_\Gamma$. $\qquad\square$

(18.30) Proposition. *A cubic étale F-algebra L can be given an action of the alternating group A_3 which turns it into a Galois A_3-algebra over F if and only if $\Delta(L) \simeq F \times F$.*

Proof: Suppose $\Delta(L) \simeq F \times F$; then Proposition (18.27) yields an isomorphism $\Sigma(L) \simeq L \times L$. The action of A_3 on $\Sigma(L)$ preserves each term and therefore induces an action on each of them. The induced actions are not the same, but each of them defines a Galois A_3-algebra structure on L since $\Sigma(L)$ is a Galois S_3-algebra.

Conversely, suppose L has a Galois A_3-algebra structure; we have to show that Γ acts by even permutations on $X(L)$. By way of contradiction, suppose $\gamma \in \Gamma$ induces an odd permutation on $X(L)$: we may assume $X(L) = \{\xi_1, \xi_2, \xi_3\}$ and $^\gamma\xi_1 = \xi_1$, $^\gamma\xi_2 = \xi_3$, $^\gamma\xi_3 = \xi_2$. Since L is a Galois A_3-algebra, $X(L)$ is an A_3-torsor; we may therefore find $\sigma \in A_3$ such that

$$\xi_1^\sigma = \xi_2, \quad \xi_2^\sigma = \xi_3, \quad \xi_3^\sigma = \xi_1;$$

then $^\gamma(\xi_1^\sigma) = \xi_3$ whereas $(^\gamma\xi_1)^\sigma = \xi_2$, a contradiction. Therefore, Γ acts on $X(L)$ by even permutations, hence $\Delta(L) \simeq F \times F$. $\square$

Reduced equations. Let L be a cubic étale F-algebra. As a first step in finding a reduced form for L, we relate the quadratic form S^0 on the subspace $L^0 = \{\, x \in L \mid T_{L/F}(x) = 0 \,\}$ and the bilinear form T on L to the discriminant algebra $\Delta(L)$. Proposition (18.8) shows that S^0 is nonsingular if char $F \neq 3$; the bilinear form T is nonsingular in every characteristic, since L is étale.

(18.31) Lemma. (1) *The quadratic form S^0 is isometric to the quadratic form Q on $\Delta(L)$ defined by*

$$Q(x) = N_{\Delta(L)/F}(x) - T_{\Delta(L)/F}(x)^2 \quad for\ x \in \Delta(L).$$

(2) *Suppose char $F \neq 2$ and let $\delta \in F^\times$ be such that $\Delta(L) \simeq F[t]/(t^2 - \delta)$. The bilinear form T on L has a diagonalization*

$$T \simeq \langle 1, 2, 2\delta \rangle.$$

Proof: (1) By a theorem of Springer (see Scharlau [247, Corollary 2.5.4], or Baeza [28, p. 119] if char $F = 2$), it suffices to check that S^0 and Q are isometric over an odd-degree scalar extension of F. If L is a field, we may therefore extend scalars to L; then L is replaced by $L \otimes L$, which is isomorphic to $L \times (\Delta(L) \otimes L)$, by (18.28). In all cases, we may thus assume $L \simeq F \times K$, where K is a quadratic étale F-algebra. Proposition (18.29) shows that we may identify $K = \Delta(L)$.

The generic polynomial of $(\alpha, x) \in F \times K = L$ is

$$(X - \alpha)\big(X^2 - T_{K/F}(x)X + N_{K/F}(x)\big) =$$
$$X^3 - T_{L/F}(\alpha, x)X^2 + S_{L/K}(\alpha, x)X - N_{L/F}(\alpha, x),$$

hence

$$T_{L/F}(\alpha, x) = \alpha + T_{K/F}(x) \quad \text{and} \quad S_{L/F}(\alpha, x) = \alpha T_{K/F}(x) + N_{K/F}(x).$$

Therefore, the map which carries $x \in K = \Delta(L)$ to $\big(-T_{K/F}(x), x\big) \in L^0$ is an isometry $\big(\Delta(L), Q\big) \xrightarrow{\sim} (L^0, S^0)$.

(2) As in (1), we may reduce to the case where $L = F \times K$, with $K \simeq \Delta(L) \simeq F[t]/(t^2 - \delta)$. Let $\bar{t}$ be the image of t in K. A computation shows that

$$T\big((x_1, x_2 + x_3\bar{t}), (y_1, y_2 + y_3\bar{t})\big) = x_1 y_1 + 2(x_2 y_2 + \delta x_3 y_3),$$

hence T has a diagonalization $\langle 1, 2, 2\delta \rangle$ with respect to the basis $\big((1,0), (0,1), (0,\bar{t})\big)$.

$\square$

(18.32) Proposition. *Every cubic étale F-algebra L is isomorphic to an algebra of the form $F[X]/(f)$ for some polynomial f, unless $F = \mathbb{F}_2$ and $L \simeq F \times F \times F$.*

If char $F \neq 3$ and F is infinite, the polynomial f may be chosen of the form

$$f = X^3 - 3X + a \quad \text{for some } a \in F,\ a \neq \pm 2;$$

then $\Delta(L) \simeq F[t]/\big(t^2 + 3(a^2 - 4)\big)$ if char $F \neq 2$ and $\Delta(L) \simeq F[t]/(t^2 + t + 1 + a^{-2})$ if char $F = 2$. (Note that $X^3 - 3X \pm 2 = (X \mp 1)^2(X \pm 2)$, hence $F[X]/(X^3 - 3X \pm 2)$ is not étale.)

If char $F \neq 3$ (and card F is arbitrary), the polynomial f may be chosen of the form $f = X^3 - b$ for some $b \in F^\times$ if and only if $\Delta(L) \simeq F[t]/(t^2 + t + 1)$.

If char $F = 3$, let $\delta \in F^\times$ be such that $\Delta(L) \simeq F[t]/(t^2 - \delta)$; then f may be chosen of the form

$$f = X^3 - \delta X + a \quad \text{for some } a \in F.$$

Proof: If L is a field, then it contains a primitive element x (see for instance Bourbaki [49, p. V.39]); we then have $L \simeq F[X]/(f)$ where f is the minimal polynomial of x. Similarly, if $L \simeq F \times K$ for some quadratic field extension K/F, then $L \simeq F[X]/(Xg)$ where g is the minimal polynomial of any primitive element of K. If $L \simeq F \times F \times F$ and F contains at least three distinct elements a, b, c, then $L \simeq F[X]/(f)$ where $f = (X - a)(X - b)(X - c)$. This completes the proof of the first assertion. We now show that, under suitable hypotheses, the primitive element x may be chosen in such a way that its minimal polynomial takes a special form.

Suppose first that char $F \neq 3$. The lemma shows that S^0 represents -3, since

$$Q(1) = N_{\Delta(L)/F}(1) - T_{\Delta(L)/F}(1)^2 = -3.$$

Let $x \in L^0$ be such that $S^0(x) = -3$. Since $F \not\subset L^0$, the element x is a primitive element if L is a field, and its minimal polynomial coincides with its generic polynomial, which has the form $X^3 - 3X + a$ for some $a \in F$. If $L \simeq F \times F \times F$, the nonprimitive elements in L^0 have the form (x_1, x_2, x_3) where $x_1 + x_2 + x_3 = 0$ and two of the x_i are equal. The conic $S^0(X) = -3$ is nondegenerate by (18.8). Therefore it has only a finite number of intersection points with the lines $x_1 = x_2$, $x_1 = x_3$ and $x_2 = x_3$. If F is infinite we may therefore find a primitive element x such that $S^0(x) = -3$. Similarly, if $L \simeq F \times K$ where K is a field, the nonprimitive elements in L^0 have the form (x_1, x_2) where $x_1, x_2 \in F$ and $x_1 + 2x_2 = 0$. Again, the conic $S^0(X) = -3$ has only a finite number of intersection points with this line, hence we may find a primitive element x such that $S^0(x) = -3$ if F is infinite.

Example (18.23) shows how to compute $\Delta(L)$ for $L = F[X]/(f)$. If x_1, x_2, x_3 are the roots of f in an algebraic closure of F, we have

$$\Delta(L) \simeq F[t]/(t^2 - d) \quad \text{with } d = (x_1 - x_2)^2(x_1 - x_3)^2(x_2 - x_3)^2 \quad \text{if char } F \neq 2,$$

and

$$\Delta(L) \simeq F[t]/(t^2 + t + d) \quad \text{with } d = \frac{x_1 x_2}{x_1^2 + x_2^2} + \frac{x_1 x_3}{x_1^2 + x_3^2} + \frac{x_2 x_3}{x_2^2 + x_3^2} \quad \text{if char } F = 2.$$

If $f = X^3 + pX + q$, then $x_1 + x_2 + x_3 = 0$, $x_1 x_2 + x_1 x_3 + x_2 x_3 = p$ and $x_1 x_2 x_3 = -q$, and a computation shows that

$$(x_1 - x_2)^2(x_1 - x_3)^2(x_2 - x_3)^2 = -4p^3 - 27q^2.$$

Moreover, if char $F = 2$ we have $x_1^2 + x_2^2 + x_3^2 = 0$, hence

$$\frac{x_1 x_2}{x_1^2 + x_2^2} + \frac{x_1 x_3}{x_1^2 + x_3^2} + \frac{x_2 x_3}{x_2^2 + x_3^2} = \frac{x_1^3 x_2^3 + x_1^3 x_3^3 + x_2^3 x_3^3}{x_1^2 x_2^2 x_3^2} = \frac{p^3 + q^2}{q^2}.$$

Thus,

$$\Delta(L) \simeq F[t]/(t^2 + 4p^3 + 27q^2) \quad \text{if char } F \neq 2$$

and

$$\Delta(L) \simeq F[t]/(t^2 + t + 1 + p^3 q^{-2}) \quad \text{if char } F = 2.$$

In particular, we have $\Delta(L) \simeq F[t]/(t^2 + 3(a^2 - 4))$ if $f = X^3 - 3X + a$ and char $F \neq 2, 3$, and $\Delta(L) \simeq F[t]/(t^2 + t + 1)$ if $f = X^3 - b$ and char $F \neq 3$.

Conversely, if $\Delta(L) \simeq F[t]/(t^2 + t + 1)$, then the form Q on $\Delta(L)$ defined in (18.31) is isotropic: for $j = t + (t^2 + t + 1) \in \Delta(L)$ we have $T_{\Delta(L)/F}(j) = -1$ and $N_{\Delta(L)/F}(j) = 1$, hence $Q(j) = 0$. Lemma (18.31) then shows that there is a nonzero element $x \in L^0$ such that $S^0(x) = 0$. An inspection of the cases where L is not a field shows that x is primitive in all cases. Therefore, $L \simeq F[X]/(X^3 - b)$ where $b = x^3 = N_{L/F}(x)$.

Suppose finally that char $F = 3$, and let $\delta \in F^\times$ be such that $\Delta(L) \simeq F[t]/(t^2 - \delta)$. The element $d = t + (t^2 - \delta) \in \Delta(L)$ then satisfies

$$N_{\Delta(L)/F}(d) - T_{\Delta(L)/F}(d)^2 = -\delta,$$

hence (18.31) shows that there exists $x \in L^0$ such that $S^0(x) = -\delta$. This element may be chosen primitive, and its minimal polynomial then has the form

$$X^3 - \delta X + a \quad \text{for some } a \in F. \qquad \square$$

A careful inspection of the argument in the proof above shows that if char $F \neq 3$ and $L \simeq F \times F \times F$ one may find $a \in F$ such that $L \simeq F[X]/(X^3 - 3X + a)$ as soon as card $F \geq 8$. The same conclusion holds if $L \simeq F \times K$ for some quadratic field extension K when card $F \geq 4$.

The group of square classes. Let L be a cubic étale F-algebra. The inclusion $F \hookrightarrow L$ and the norm map $N_{L/F} \colon L \to F$ induce maps on the square class groups:

$$i \colon F^\times / F^{\times 2} \to L^\times / L^{\times 2}, \quad N \colon L^\times / L^{\times 2} \to F^\times / F^{\times 2}.$$

Since $N_{L/F}(x) = x^3$ for all $x \in F$, the composition $N \circ i$ is the identity on $F^\times / F^{\times 2}$:

$$N \circ i = \text{Id}_{F^\times / F^{\times 2}}.$$

In order to relate $L^\times / L^{\times 2}$ and $F^\times / F^{\times 2}$ by an exact sequence, we define a map $\# \colon L \to L$ as follows: for $\ell \in L$ we set

(18.33) $$\ell^\# = \ell^2 - T_{L/F}(\ell)\ell + S_{L/F}(\ell) \in L,$$

so that

$$\ell \ell^\# - N_{L/F}(\ell) = 0.$$

In particular, for $\ell \in L^\times$ we have $\ell^\# = N_{L/F}(\ell)\ell^{-1}$, hence $\#$ defines an endomorphism of $L^\times$. We also put $\#$ for the induced endomorphism of $L^\times / L^{\times 2}$.

(18.34) Proposition. *The following sequence is exact*:

$$1 \to F^\times/F^{\times 2} \xrightarrow{i} L^\times/L^{\times 2} \xrightarrow{\#} L^\times/L^{\times 2} \xrightarrow{N} F^\times/F^{\times 2} \to 1.$$

Moreover, for $x \in F^\times/F^{\times 2}$ and $y \in L^\times/L^{\times 2}$,

$$x = N \circ i(x) \quad and \quad (y^\#)^\# = i \circ N(y) \cdot y.$$

Proof: It was observed above that $N \circ i$ is the identity on $F^\times/F^{\times 2}$. Therefore, i is injective and N is surjective. For $\ell \in L^\times$ we have $\ell^\# = N_{L/F}(\ell)\ell^{-1}$, hence

(18.35) $$y^\# = i \circ N(y) \cdot y \qquad \text{for } y \in L^\times/L^{\times 2}.$$

Taking the image of each side under N, we obtain $N(y^\#) = 1$, since $N \circ i \circ N(y) = N(y)$. Therefore, substituting $y^\#$ for y in (18.35) we get $(y^\#)^\# = y^\# = i \circ N(y) \cdot y$. In particular, if $y^\# = 1$ we have $y = i \circ N(y)$, and if $N(y) = 1$ we have $y = y^\#$. Therefore, the kernel of $\#$ is in the image of i, and the kernel of N is in the image of $\#$. To complete the proof, observe that putting $y = i(x)$ in (18.35) yields $i(x)^\# = 1$ for $x \in F^\times/F^{\times 2}$, since $N \circ i(x) = x$.   □

§19. Central Simple Algebras of Degree Three

In this section, we turn to central simple algebras of degree 3. We first prove Wedderburn's theorem which shows that these algebras are cyclic, and we next discuss their involutions of unitary type. It turns out that involutions of unitary type on a given central simple algebra of degree 3 are classified up to conjugation by a 3-fold Pfister form. In the final subsection, we relate this invariant to cubic étale subalgebras and prove a theorem of Albert on the existence of certain cubic étale subalgebras.

19.A. Cyclic algebras. To simplify the notation, we set $C_3 = \mathbb{Z}/3\mathbb{Z}$ and $\rho = 1 + 3\mathbb{Z} \in C_3$.

Given a Galois C_3-algebra L over F and an element $a \in F^\times$, the *cyclic algebra* (L, a) is defined as follows:

$$(L, a) = L \oplus Lz \oplus Lz^2$$

where z is subject to the relations:

$$z\ell = \rho(\ell)z, \quad z^3 = a$$

for all $\ell \in L$.

(19.1) Example. Let $L = F \times F \times F$ with the C_3-structure defined by

$$\rho(x_1, x_2, x_3) = (x_3, x_1, x_2) \quad \text{for } (x_1, x_2, x_3) \in L.$$

We have $(L, a) \simeq M_3(F)$ for all $a \in F^\times$. An explicit isomorphism is given by

$$(x_1, x_2, x_3) \mapsto \begin{pmatrix} x_1 & & \\ & x_2 & \\ & & x_3 \end{pmatrix} \quad \text{and} \quad z \mapsto \begin{pmatrix} 0 & 0 & a \\ 1 & 0 & 0 \\ 0 & 1 & 0 \end{pmatrix}.$$

From this example, it readily follows that (L, a) is a central simple F-algebra for all Galois C_3-algebras L and all $a \in F^\times$, since $(L, a) \otimes_F F_{\text{sep}} \simeq (L \otimes_F F_{\text{sep}}, a)$ and $L \otimes_F F_{\text{sep}} \simeq F_{\text{sep}} \times F_{\text{sep}} \times F_{\text{sep}}$. Of course, this is also easy to prove without extending scalars: see for instance Draxl [84, p. 49].

The main result of this subsection is the following:

(19.2) Theorem (Wedderburn). *Every central simple F-algebra of degree 3 is cyclic.*

The proof below is due to Haile [113]; it is free from restrictions on the characteristic of F. However, the proof can be somewhat simplified if char $F = 3$: see Draxl [84, p. 63] or Jacobson [146, p. 80].

(19.3) Lemma. *Let A be a central simple F-algebra of degree 3 and let $x \in A^\times$. If* $\mathrm{Trd}(x) = \mathrm{Trd}(x^{-1}) = 0$, *then* $x^3 = \mathrm{Nrd}(x)$.

Proof: If the reduced characteristic polynomial of x is
$$X^3 - \mathrm{Trd}(x)X^2 + \mathrm{Srd}(x)X - \mathrm{Nrd}(x),$$
then the reduced characteristic polynomial of x^{-1} is
$$X^3 - \frac{\mathrm{Srd}(x)}{\mathrm{Nrd}(x)}X^2 + \frac{\mathrm{Trd}(x)}{\mathrm{Nrd}(x)}X - \frac{1}{\mathrm{Nrd}(x)}.$$
Therefore, $\mathrm{Trd}(x^{-1}) = \mathrm{Srd}(x)\,\mathrm{Nrd}(x^{-1})$, and it follows that the reduced characteristic polynomial of x takes the form $X^3 - \mathrm{Nrd}(x)$ if $\mathrm{Trd}(x) = \mathrm{Trd}(x^{-1}) = 0$. □

Proof of Theorem (19.2): In view of Example (19.1), it suffices to prove that central division F-algebras of degree 3 are cyclic. Let D be such a division algebra. We claim that it suffices to find elements y, $z \in D$ such that $z \notin F$, $y \notin F(z)$ and $\mathrm{Trd}(x) = \mathrm{Trd}(x^{-1}) = 0$ for $x = z$, yz, yz^2. Indeed, the lemma then shows
$$z^3 = \mathrm{Nrd}(z), \quad (yz)^3 = \mathrm{Nrd}(yz), \quad (yz^2)^3 = \mathrm{Nrd}(yz^2);$$
since $\mathrm{Nrd}(yz^2) = \mathrm{Nrd}(yz)\,\mathrm{Nrd}(z)$ it follows that
$$(yz^2)^3 = (yz)^3 z^3$$
hence, after cancellation,
$$zyz^2y = yzyz^2.$$
By dividing each side by $z^3 = \mathrm{Nrd}(z)$, we obtain
$$(zyz^{-1})y = y(zyz^{-1}),$$
which shows that $zyz^{-1} \in F(y)$. We have $zyz^{-1} \neq y$ since $y \notin F(z)$, hence we may define a Galois C_3-algebra structure on $F(y)$ by letting $\rho(y) = zyz^{-1}$; then,
$$D \simeq \big(F(y), \mathrm{Nrd}(z)\big).$$
We now proceed to construct elements y, z satisfying the required conditions.

Let $L \subset D$ be an arbitrary maximal subfield. Considering D as a bilinear space for the nonsingular bilinear form induced by the reduced trace, pick a nonzero element $u_1 \in L^\perp$. Since $\dim\big((u_1^{-1}F)^\perp \cap L\big) \geq 2$, we may find $u_2 \in (u_1^{-1}F)^\perp \cap L$ such that $u_2 \notin u_1 F$. Set $z = u_1 u_2^{-1}$. We have $\mathrm{Trd}(z) = 0$ because $u_1 \in L^\perp$ and $u_2^{-1} \in L$, and $\mathrm{Trd}(z^{-1}) = 0$ because $u_2 \in (u_1^{-1}F)^\perp$. Moreover, $z \notin F$ since $u_2 \notin u_1 F$.

Next, pick a nonzero element $v_1 \in F(z)^\perp \smallsetminus F(z)$. Since $\dim(zF + z^{-1}F)^\perp = 7$, we have
$$v_1(zF + z^{-1}F)^\perp \cap F(z) \neq \{0\};$$
we may thus find a nonzero element v_2 in this intersection. Set $y = v_2^{-1}v_1$. Since $v_1 \notin F(z)$, we have $y \notin F(z)$. On the other hand, since $v_1 \in F(z)^\perp$ and $v_2 \in F(z)$, we have
$$\mathrm{Trd}(yz) = \mathrm{Trd}(v_1 z v_2^{-1}) = 0 \quad \text{and} \quad \mathrm{Trd}(yz^2) = \mathrm{Trd}(v_1 z^2 v_2^{-1}) = 0.$$

Since $v_1^{-1}v_2 \in (zF + z^{-1}F)^\perp$ and $z^3 = \mathrm{Nrd}(z)$, we also have
$$\mathrm{Trd}(z^{-1}y^{-1}) = \mathrm{Trd}(z^{-1}v_1^{-1}v_2) = 0,$$
$$\mathrm{Trd}(z^{-2}y^{-1}) = \mathrm{Nrd}(z)\,\mathrm{Trd}(zv_1^{-1}v_2) = 0.$$

The elements y and z thus meet our requirements.   □

19.B. Classification of involutions of the second kind. Let K be a quadratic étale extension of F, and let B be a central simple[28] K-algebra of degree 3 such that $N_{K/F}(B)$ is split. By Theorem (3.1), this condition is necessary and sufficient for the existence of involutions of the second kind on B which fix the elements of F.

We aim to classify those involutions up to conjugation, by means of the associated trace form on $\mathrm{Sym}(B, \tau)$ (see §11). We therefore assume $\operatorname{char} F \neq 2$ throughout this subsection.

(19.4) Definition. Let τ be an involution of the second kind on B which is the identity on F. Recall from §11 the quadratic form Q_τ on $\mathrm{Sym}(B, \tau)$ defined by
$$Q_\tau(x) = \mathrm{Trd}_B(x^2) \quad \text{for } x \in \mathrm{Sym}(B, \tau).$$

Let $\alpha \in F$ be such that $K \simeq F[X]/(X^2 - \alpha)$. Proposition (11.22) shows that Q_τ has a diagonalization of the form:
$$Q_\tau = \langle 1, 1, 1 \rangle \perp \langle 2 \rangle \cdot \langle\!\langle \alpha \rangle\!\rangle \cdot q_\tau$$
where q_τ is a 3-dimensional quadratic form of determinant 1. The form
$$\pi(\tau) = \langle\!\langle \alpha \rangle\!\rangle \cdot q_\tau \perp \langle\!\langle \alpha \rangle\!\rangle$$
is uniquely determined by τ up to isometry and is a 3-fold Pfister form since $\det q_\tau = 1$. We call it the *Pfister form* of τ. Every F-isomorphism $(B, \tau) \xrightarrow{\sim} (B', \tau')$ of algebras with involution induces an isometry of trace forms $\big(\mathrm{Sym}(B, \tau), Q_\tau\big) \xrightarrow{\sim} \big(\mathrm{Sym}(B', \tau'), Q_{\tau'}\big)$, hence also an isometry $\pi(\tau) \xrightarrow{\sim} \pi(\tau')$. The Pfister form $\pi(\tau)$ is therefore an invariant of the conjugacy class of τ. Our main result (Theorem (19.6)) is that it determines this conjugacy class uniquely.

(19.5) Example. Let V be a 3-dimensional vector space over K with a nonsingular hermitian form h. Let $h = \langle \delta_1, \delta_2, \delta_3 \rangle_K$ be a diagonalization of this form (so that δ_1, δ_2, $\delta_3 \in F^\times$) and let $\tau = \tau_h$ be the adjoint involution on $B = \mathrm{End}_K(V)$ with respect to h.

Propositions (11.14) and (11.13) show that
$$Q_\tau \simeq \langle 1, 1, 1 \rangle \perp \langle 2 \rangle \cdot \langle\!\langle \alpha \rangle\!\rangle \cdot \langle \delta_1\delta_2, \delta_1\delta_3, \delta_2\delta_3 \rangle.$$

Therefore,
$$\pi(\tau) = \langle\!\langle \alpha \rangle\!\rangle \cdot \langle\!\langle -\delta_1\delta_2, -\delta_1\delta_3 \rangle\!\rangle.$$

(19.6) Theorem. *Let B be a central simple K-algebra of degree 3 and let τ, τ' be involutions of the second kind on B fixing the elements of F. The following conditions are equivalent:*

(1) τ and τ' are conjugate, i.e., there exists $u \in B^\times$ such that
$$\tau' = \mathrm{Int}(u) \circ \tau \circ \mathrm{Int}(u)^{-1};$$

---

[28]We allow $K \simeq F \times F$, in which case $B \simeq A \times A^{\mathrm{op}}$ for some central simple F-algebra A of degree 3.

(2) $Q_\tau \simeq Q_{\tau'}$;

(3) $\pi(\tau) \simeq \pi(\tau')$.

Proof: (1) $\Rightarrow$ (2) Conjugation by u defines an isometry from $\big(\mathrm{Sym}(B, \tau), Q_\tau\big)$ to $\big(\mathrm{Sym}(B, \tau'), Q_{\tau'}\big)$.

(2) $\Leftrightarrow$ (3) This follows by Witt cancellation, in view of the relation between Q_τ and $\pi(\tau)$.

(2) $\Rightarrow$ (1) If $K \simeq F \times F$, then all the involutions on B are conjugate to the exchange involution. We may therefore assume K is a field. If B is split, let $B = \mathrm{End}_K(V)$ for some 3-dimensional vector space V. The involutions τ and τ' are adjoint to nonsingular hermitian forms h, h' on V. Let

$$h = \langle \delta_1, \delta_2, \delta_3 \rangle_K \quad \text{and} \quad h' = \langle \delta_1', \delta_2', \delta_3' \rangle_K$$

(with $\delta_1, \ldots, \delta_3' \in F^\times$) be diagonalizations of these forms. As we observed in (19.5), we have

$$Q_\tau \simeq \langle 1, 1, 1 \rangle \perp \langle 2 \rangle \cdot \langle\!\langle \alpha \rangle\!\rangle \cdot \langle \delta_1 \delta_2, \delta_1 \delta_3, \delta_2 \delta_3 \rangle$$

and

$$Q_{\tau'} \simeq \langle 1, 1, 1 \rangle \perp \langle 2 \rangle \cdot \langle\!\langle \alpha \rangle\!\rangle \cdot \langle \delta_1' \delta_2', \delta_1' \delta_3', \delta_2' \delta_3' \rangle.$$

Therefore, condition (2) implies

$$\langle\!\langle \alpha \rangle\!\rangle \cdot \langle \delta_1 \delta_2, \delta_1 \delta_3, \delta_2 \delta_3 \rangle \simeq \langle\!\langle \alpha \rangle\!\rangle \cdot \langle \delta_1' \delta_2', \delta_1' \delta_3', \delta_2' \delta_3' \rangle.$$

It follows from a theorem of Jacobson (see Scharlau [247, Theorem 10.1.1]) that the hermitian forms $\langle \delta_1 \delta_2, \delta_1 \delta_3, \delta_2 \delta_3 \rangle_K$ and $\langle \delta_1' \delta_2', \delta_1' \delta_3', \delta_2' \delta_3' \rangle_K$ are isometric; then

$$\langle \delta_1 \delta_2 \delta_3 \rangle \cdot h \simeq \langle \delta_1 \delta_2, \delta_1 \delta_3, \delta_2 \delta_3 \rangle_K \simeq \langle \delta_1' \delta_2', \delta_1' \delta_3', \delta_2' \delta_3' \rangle_K \simeq \langle \delta_1' \delta_2' \delta_3' \rangle \cdot h',$$

hence the hermitian forms h, h' are similar. The involutions τ, τ' are therefore conjugate. This completes the proof in the case where B is split.

The general case is reduced to the split case by an odd-degree scalar extension. Suppose B is a division algebra and let $\tau' = \mathrm{Int}(v) \circ \tau$ for some $v \in B^\times$, which may be assumed symmetric under τ. By substituting $\mathrm{Nrd}(v)v$ for v, we may assume $\mathrm{Nrd}(v) = \mu^2$ for some $\mu \in F^\times$. Let L/F be a cubic field extension contained in B. (For example, we may take for L the subfield of B generated by any noncentral τ-symmetric element.) The algebra $B_L = B \otimes_F L$ is split, hence the argument given above shows that τ_L and τ_L' are conjugate:

$$\tau_L' = \mathrm{Int}(u) \circ \tau_L \circ \mathrm{Int}(u)^{-1} = \mathrm{Int}\big(u\tau_L(u)\big) \circ \tau_L \quad \text{for some } u \in B_L^\times,$$

hence

$$v = \lambda u \tau_L(u) \quad \text{for some } \lambda \in K_L = K \otimes_F L.$$

Since $\tau(v) = v$, we have in fact $\lambda \in L^\times$. Let ι be the nontrivial automorphism of K_L/L. Since $\mathrm{Nrd}(u) = \mu^2$, by taking the reduced norm on each side of the preceding equality we obtain:

$$\mu^2 = \lambda^3 \mathrm{Nrd}(u) \cdot \iota\big(\mathrm{Nrd}(u)\big),$$

hence $\lambda = \big(\mu\lambda^{-1}\mathrm{Nrd}(u)\big) \cdot \iota\big(\mu\lambda^{-1}\mathrm{Nrd}(u)\big)$ and, letting $w = \mu\lambda^{-1}\mathrm{Nrd}(u)u \in B_L^\times$,

$$v = w\tau_L(w).$$

By (6.17), there exists $w' \in B^\times$ such that $v = w'\tau(w')$. Therefore,

$$\tau' = \mathrm{Int}(w') \circ \tau \circ \mathrm{Int}(w')^{-1}. \qquad \square$$

(19.7) Remark. If (B, τ) and (B', τ') are central simple K-algebras of degree 3 with involutions of the second kind leaving F elementwise invariant, the condition $\pi(\tau) \simeq \pi(\tau')$ does not imply $(B, \tau) \simeq (B', \tau')$. For example, if $K \simeq F \times F$ all the forms $\pi(\tau)$ are hyperbolic since they contain the factor $\langle\langle \alpha \rangle\rangle$, but (B, τ) and (B', τ') are not isomorphic if $B \not\simeq B'$.

(19.8) Definition. An involution of the second kind τ on a central simple K-algebra B of degree 3 is called *distinguished* if $\pi(\tau)$ is hyperbolic. Theorem (19.6) shows that the distinguished involutions form a conjugacy class. If $K \simeq F \times F$, all involutions are distinguished (see the preceding remark).

(19.9) Example. Consider again the split case $B = \operatorname{End}_K(V)$, as in (19.5). If the hermitian form h is isotropic, then we may find a diagonalization of the form $h = \langle 1, -1, \lambda \rangle_K$ for some $\lambda \in F^{\times}$, hence the computations in (19.5) show that the adjoint involution τ_h is distinguished. Conversely, if h' is a nonsingular hermitian form whose adjoint involution $\tau_{h'}$ is distinguished, then τ_h and $\tau_{h'}$ are conjugate, hence h' is similar to h. Therefore, in the split case the distinguished involutions are those which are adjoint to isotropic hermitian forms.

We next characterize distinguished involutions by a condition on the Witt index of the restriction of Q_τ to elements of trace zero. This characterization will be used to prove the existence of distinguished involutions on arbitrary central simple K-algebras B of degree 3 such that $N_{K/F}(B)$ is split, at least when char $F \neq 3$.

Let

$$\operatorname{Sym}(B, \tau)^0 = \{\, x \in \operatorname{Sym}(B, \tau) \mid \operatorname{Trd}_B(x) = 0 \,\}$$

and let Q_τ^0 be the restriction of the bilinear form Q_τ to $\operatorname{Sym}(B, \tau)^0$. We avoid the case where char $F = 3$, since then Q_τ^0 is singular (with radical F). Assuming char $F \neq 2, 3$, let $w(Q_\tau^0)$ be the Witt index of Q_τ^0.

(19.10) Proposition. *Suppose* char $F \neq 2, 3$. *The following conditions are equivalent*:

(1) τ *is distinguished*;
(2) $w(Q_\tau^0) \geq 2$;
(3) $w(Q_\tau^0) \geq 3$.

Proof: Let $K = F(\sqrt{\alpha})$. The subspace $\operatorname{Sym}(B, \tau)^0$ is the orthogonal complement of F in $\operatorname{Sym}(B, \tau)$ for the form Q_τ; since $Q_\tau(1) = 3$, it follows that $Q_\tau = \langle 3 \rangle \perp Q_\tau^0$. Since $\langle 1, 1, 1 \rangle \simeq \langle 3, 2, 6 \rangle$, it follows from (11.22) that

$$Q_\tau^0 \simeq \langle 2 \rangle \cdot \big(\langle 1, 3 \rangle \perp \langle\langle \alpha \rangle\rangle \cdot q_\tau \big).$$

Since $\pi(\tau) = \langle\langle \alpha \rangle\rangle \cdot q_\tau \perp \langle\langle \alpha \rangle\rangle$, we have

$$Q_\tau^0 = \langle 2 \rangle \cdot \big(\langle 3, \alpha \rangle + \pi(\tau) \big) \quad \text{in } WF.$$

By comparing dimensions on each side, we obtain

$$w(Q_\tau^0) = w\big(\langle 3, \alpha \rangle \perp \pi(\tau) \big) - 1.$$

Condition (1) implies that $w\big(\langle 3, \alpha \rangle \perp \pi(\tau) \big) \geq 4$, hence the equality above yields (3). On the other hand, condition (2) shows that $w\big(\langle 3, \alpha \rangle \perp \pi(\tau) \big) \geq 3$, from which it follows that $\pi(\tau)$ is isotropic, hence hyperbolic. Therefore, (2) $\Rightarrow$ (1). Since (3) $\Rightarrow$ (2) is clear, the proof is complete. $\square$

The relation between Trd_B and Srd_B (see (0.2)) shows that for $x \in \mathrm{Sym}(B, \tau)^0$, the condition that $\mathrm{Trd}_B(x^2) = 0$ is equivalent to $\mathrm{Srd}_B(x) = 0$. Therefore, the totally isotropic subspaces of $\mathrm{Sym}(B, \tau)^0$ for the form Q_τ^0 can be described as the subspaces consisting of elements x such that $x^3 = \mathrm{Nrd}_B(x)$. We may therefore reformulate the preceding proposition as follows:

(19.11) Corollary. *Suppose* $\mathrm{char}\, F \neq 2$, 3. *The following conditions are equivalent*:

(1) τ *is distinguished;*
(2) *there exists a subspace* $U \subset \mathrm{Sym}(B, \tau)^0$ *of dimension 2 such that* $u^3 = \mathrm{Nrd}_B(u)$ *for all* $u \in U$;
(3) *there exists a subspace* $U \subset \mathrm{Sym}(B, \tau)^0$ *of dimension 3 such that* $u^3 = \mathrm{Nrd}_B(u)$ *for all* $u \in U$.

We now prove the existence of distinguished involutions:

(19.12) Proposition. *Suppose*[29] $\mathrm{char}\, F \neq 2$, 3. *Every central simple* K-*algebra* B *such that* $N_{K/F}(B)$ *is split carries a distinguished involution.*

Proof: In view of (19.9), we may assume that B is a division algebra. Let τ be an arbitrary involution of the second kind on B which is the identity on F. Let $L \subset \mathrm{Sym}(B, \tau)$ be a cubic field extension of F and let u be a nonzero element in the orthogonal complement $L^\perp$ of L for the quadratic form Q_τ. We claim that the involution $\tau' = \mathrm{Int}(u) \circ \tau$ is distinguished.

In order to prove this, consider the F-vector space

$$U = \left(L \cap (u^{-1}F)^\perp\right) \cdot u^{-1}.$$

Since Q_τ is nonsingular, we have $\dim U \geq 2$. Moreover, since $L \subset \mathrm{Sym}(B, \tau)$, we have $U \subset \mathrm{Sym}(B, \tau')$. For $x \in L \cap (u^{-1}F)^\perp$, $x \neq 0$, we have $\mathrm{Trd}(xu^{-1}) = 0$ because $x \in (u^{-1}F)^\perp$, and $\mathrm{Trd}(ux^{-1}) = 0$ because $u \in L^\perp$ and $x^{-1} \in L$. Therefore, for all nonzero $y \in U$ we have

$$\mathrm{Trd}(y) = \mathrm{Trd}(y^{-1}) = 0,$$

hence also $y^3 = \mathrm{Nrd}(y)$, by (19.3). Therefore, Corollary (19.11) shows that τ' is distinguished. $\qquad\square$

(19.13) Remark. If $\mathrm{char}\, F = 3$, the proof still shows that for every central simple K-algebra B such that $N_{K/F}(B)$ is split, there exists a unitary involution τ' on B and a 2-dimensional subspace $U \subset \mathrm{Sym}(B, \tau')$ such that $u^3 \in F$ for all $u \in U$.

19.C. Étale subalgebras. As in the preceding subsection, we consider a central simple algebra B of degree 3 over a quadratic étale extension K of F such that $N_{K/F}(B)$ is split. We continue to assume $\mathrm{char}\, F \neq 2$, 3, and let ι be the nontrivial automorphism of K/F. Our aim is to obtain information on the cubic étale F-algebras L contained in B and on the involutions of the second kind which are the identity on L.

---

[29]See (19.30) for a different proof, which works also if $\mathrm{char}\, F = 3$.

Albert's theorem. The first main result is a theorem of Albert [10] which asserts the existence of cubic étale F-subalgebras $L \subset B$ with discriminant $\Delta(L)$ isomorphic to K. For such an algebra, we have $L_K \simeq L \otimes \Delta(L) \simeq \Sigma(L)$, by (18.27), hence L_K can be endowed with a Galois S_3-algebra structure. This structure may be used to give an explicit description of (B, τ) as a cyclic algebra with involution.

(19.14) Theorem (Albert). *Suppose[30] char $F \neq 2, 3$ and let K be a quadratic étale extension of F. Every central simple K-algebra B such that $N_{K/F}(B)$ is split contains a cubic étale F-algebra L with discriminant $\Delta(L)$ isomorphic to K.*

Proof: (after Haile-Knus [114]). We first consider the easy special cases where B is not a division algebra. If $K \simeq F \times F$, then $B \simeq A \times A^{\mathrm{op}}$ for some central simple F-algebra A of degree 3. Wedderburn's theorem (19.2) shows that A contains a Galois C_3-algebra L_0 over F. By (18.30), we have $\Delta(L_0) \simeq F \times F$, hence we may set

$$L = \{ (\ell, \ell^0) \mid \ell \in L_0 \}.$$

If K is a field and B is split, then we may find in B a subalgebra L isomorphic to $F \times K$. Identifying B with $M_3(K)$, we may then choose

$$L = \left\{ \begin{pmatrix} f & 0 & 0 \\ 0 & k & 0 \\ 0 & 0 & k \end{pmatrix} \;\middle|\; f \in F,\, k \in K \right\}.$$

Proposition (18.29) shows that $\Delta(L) \simeq K$.

For the rest of the proof, we may thus assume B is a division algebra. Let τ be a distinguished involution on B. By (19.11), there exists a subspace $U \subset \mathrm{Sym}(B, \tau)$ of dimension 2 such that $u^3 = \mathrm{Nrd}(u)$ for all $u \in U$. Pick a nonzero element $u \in U$. Since $\dim U = 2$, the linear map $U \to F$ which carries $x \in U$ to $\mathrm{Trd}(u^{-1}x)$ has a nonzero kernel; we may therefore find a nonzero $v \in U$ such that $\mathrm{Trd}(u^{-1}v) = 0$.

Consider the reduced characteristic polynomial of $u^{-1}v$:

$$\mathrm{Nrd}(X - u^{-1}v) = X^3 + \mathrm{Srd}(u^{-1}v)X - \mathrm{Nrd}(u^{-1}v).$$

By substituting 1 and -1 for X and multiplying by $\mathrm{Nrd}(u)$, we obtain

$$\mathrm{Nrd}(u - v) = \mathrm{Nrd}(u) + \mathrm{Nrd}(u)\,\mathrm{Srd}(u^{-1}v) - \mathrm{Nrd}(v)$$

and

$$\mathrm{Nrd}(u + v) = \mathrm{Nrd}(u) + \mathrm{Nrd}(u)\,\mathrm{Srd}(u^{-1}v) + \mathrm{Nrd}(v),$$

hence

$$\mathrm{Nrd}(u + v) - \mathrm{Nrd}(u - v) = 2\,\mathrm{Nrd}(v).$$

On the other hand, since u, v, $u + v$, $u - v \in U$ and $x^3 = \mathrm{Nrd}(x)$ for all $x \in U$, we obtain

$$\mathrm{Nrd}(u + v) = (u + v)^3$$
$$= \mathrm{Nrd}(u) + (u^2v + uvu + vu^2) + (uv^2 + vuv + v^2u) + \mathrm{Nrd}(v)$$

---

[30]See Exercise 9 for the case where char $F = 3$.

and

$$\mathrm{Nrd}(u - v) = (u - v)^3$$
$$= \mathrm{Nrd}(u) - (u^2 v + uvu + vu^2) + (uv^2 + vuv + v^2 u) - \mathrm{Nrd}(v),$$

hence

$$\mathrm{Nrd}(u + v) - \mathrm{Nrd}(u - v) = 2(u^2 v + uvu + vu^2) + 2\,\mathrm{Nrd}(v).$$

By comparing the expressions above for $\mathrm{Nrd}(u + v) - \mathrm{Nrd}(u - v)$, it follows that

$$u^2 v + uvu + vu^2 = 0.$$

Define

$$t_1 = u^{-1} v = \mathrm{Nrd}(u)^{-1} u^2 v,$$
$$t_2 = u^{-1} t_1 u = \mathrm{Nrd}(u)^{-1} uvu,$$
$$t_3 = u^{-1} t_2 u = \mathrm{Nrd}(u)^{-1} vu^2,$$

so that

$$t_1 + t_2 + t_3 = 0$$

and conjugation by u permutes t_1, t_2, and t_3 cyclically. Moreover, since u and v are τ-symmetric, we have $\tau(t_2) = t_2$ and $\tau(t_1) = t_3$.

Let $w = t_2^{-1} t_3$. Suppose first that $w \in K$. Since $\mathrm{Nrd}(t_2) = \mathrm{Nrd}(t_3)$, we have $\mathrm{Nrd}(w) = w^3 = 1$. If $w = 1$, then $t_2 = t_3$, hence also $t_3 = t_1$, a contradiction to $t_1 + t_2 + t_3 = 0$. Therefore, w is a primitive cube root of unity. Conjugating each side of the relation $t_3 = wt_2$ by u, we find $t_2 = wt_1$; hence $K(t_1) = K(t_2) = K(t_3)$ and conjugation by u is an automorphism of order 3 of this subfield. Cubing the equations $t_2 = wt_1$ and $t_3 = wt_2$, we obtain $t_1^3 = t_2^3 = t_3^3$, hence this element is invariant under conjugation by u and therefore $t_1^3 = t_2^3 = t_3^3 \in K^\times$. Since $\tau(t_2) = t_2$, we have in fact $t_2^3 \in F^\times$, hence Proposition (18.32) shows that $F(t_2)$ is a subfield of B with discriminant $\Delta\big(F(t_2)\big) \simeq F[X]/(X^2 + X + 1)$. On the other hand, by applying τ to each side of the equation $t_2 = wt_1$, we find $t_2 = t_3\tau(w)$. Since $t_3 = wt_2$, it follows that $\tau(w) = w^{-1}$, so that $K = F(w)$ and therefore $K \simeq F[X]/(X^2 + X + 1)$. The theorem is thus proved if $w \in K$, since then $\Delta\big(F(t_2)\big) \simeq K$.

Suppose next that $w \notin K$, hence $K(w)$ is a cubic extension of K. Since $t_1 + t_2 + t_3 = 0$, we have

$$\mathrm{Int}(u^{-1})(w) = t_3^{-1} t_1 = -t_3^{-1}(t_2 + t_3) = -1 - w^{-1} \in K(w),$$

hence $\mathrm{Int}(u^{-1})$ restricts to a K-automorphism θ of $K(w)$. If $\theta = \mathrm{Id}$, then $w = -1 - w^{-1}$, hence w is a root of an equation of degree 2 with coefficients in K, a contradiction. Therefore, θ is nontrivial; it is of order 3 since $u^3 \in F^\times$.

Now consider the action of τ:

$$\tau(w) = t_1 t_2^{-1} = t_2(t_2^{-1} t_1) t_2^{-1} = -t_2(1 + w)t_2^{-1}.$$

This shows that the involution $\tau' = \mathrm{Int}(t_2^{-1}) \circ \tau$ satisfies

$$\tau'(w) = -1 - w \in K(w).$$

Therefore, τ' defines an automorphism of order 2 of $K(w)$. We claim that θ and τ' generate a group of automorphisms of $K(w)$ isomorphic to the symmetric group S_3.

Indeed,

$$\tau' \circ \theta(w) = -1 + (1 + w)^{-1} = \frac{-w}{1 + w}$$

and

$$\theta^2 \circ \tau'(w) = -(1 + w^{-1})^{-1} = \tau' \circ \theta(w).$$

Therefore, $K(w)/F$ is a Galois extension with Galois group S_3. Let $L = K(w)^{\tau'}$, the subfield of elements fixed by τ'. This subfield is a cubic étale extension of F. Since L/F is not cyclic, we have $\Delta(L) \not\simeq F \times F$, by Corollary (18.30). However, $L_K = L \otimes K \simeq K(w)$ is a cyclic extension of K, hence $\Delta(L_K) \simeq K \times K$. Therefore, $\Delta(L) \simeq K$. $\qquad\square$

Suppose $L \subset B$ is a cubic étale F-algebra with discriminant $\Delta(L)$ isomorphic to K. Let $L_K = L \otimes K \simeq L \otimes \Delta(L)$. By (18.27), we have $L \otimes \Delta(L) \simeq \Sigma(L)$, hence $L \otimes K$ can be given a Galois S_3-algebra structure over F. Under any of these S_3-algebra structures, the automorphism $\mathrm{Id} \otimes \iota$ gives the action of some transposition, and K is the algebra of invariant elements under the action of the alternating group A_3. It follows that L_K can be given a Galois C_3-algebra structure, since $A_3 \simeq C_3 = \mathbb{Z}/3\mathbb{Z}$. We fix such a structure and set $\rho = 1 + 3\mathbb{Z} \in C_3$, as in §19.A. Since conjugation by a transposition yields the nontrivial automorphism of A_3, we have

$$(\mathrm{Id} \otimes \iota)\big(\rho(x)\big) = \rho^2\big(\mathrm{Id} \otimes \iota(x)\big) \quad \text{for } x \in L_K.$$

By (4.18), there exist involutions τ of the second kind on B fixing the elements of L. We proceed to describe these involutions in terms of the cyclic algebra structure of B. It will be shown below (see (19.28)) that these involutions are all distinguished. (This property also follows from (19.11)).

(19.15) Proposition. *Suppose τ is an involution of the second kind on B such that $L \subset \mathrm{Sym}(B, \tau)$. The algebra B is a cyclic algebra:*

$$B = L_K \oplus L_K z \oplus L_K z^2$$

where z is subject to the relations: $\tau(z) = z$, $zx = \rho(x)z$ for all $x \in L_K$ and $z^3 \in F^\times$.

Proof: We first consider the case where B is split. We may then assume $B = \mathrm{End}_K(L_K)$ and identify $x \in L_K$ with the endomorphism of multiplication by x. The involution τ is the adjoint involution with respect to some nonsingular hermitian form

$$h \colon L_K \times L_K \to K.$$

Since $\mathrm{Hom}_K(L_K, K)$ is a free module of rank 1 over L_K, the linear form $x \mapsto h(1, x)$ is of the form $x \mapsto T_{L_K/K}(\ell x)$ for some $\ell \in L_K$. For $x, y \in L_K$, we then have

$$h(x, y) = h\big(1, \tau(x)y\big) = T_{L_K/K}\big(\ell\tau(x)y\big).$$

If ℓ were not invertible, then we could find $x \neq 0$ in L_K such that $\ell x = 0$. It follows that $h\big(\tau(x), y\big) = 0$ for all $y \in L_K$, a contradiction. So, $\ell \in L_K^\times$. Moreover, since $h(y, x) = \iota\big(h(x, y)\big)$ for all $x, y \in L_K$, we have

$$T_{L_K/K}\big(\ell\tau(y)x\big) = \tau\big(T_{L_K/K}\big(\ell\tau(x)y\big)\big) \quad \text{for } x, y \in L_K,$$

hence $\tau(\ell) = \ell$ since the bilinear trace form on L_K is nonsingular. Therefore, $\ell \in L^\times$.

Note that the restriction of τ to L_K is $\mathrm{Id} \otimes \iota$, hence

$$\tau \circ \rho(x) = \rho^2 \circ \tau(x) \quad \text{for } x \in L_K.$$

Consider $\beta = \ell^{-1}\rho \in \mathrm{End}_K(L_K)$. For x, $y \in L_K$ we have

$$h\big(\beta(x), y\big) = T_{L_K/K}\big(\tau \circ \rho(x)y\big) = T_{L_K/K}\big(\rho^2 \circ \tau(x)y\big)$$

and

$$h\big(x, \beta(y)\big) = T_{L_K/K}\big(\tau(x)\rho(y)\big).$$

Since $T_{L_K/K}\big(\rho^2(u)\big) = T_{L_K/K}(u)$ for all $u \in L_K$, we also have

$$h\big(x, \beta(y)\big) = T_{L_K/K}\big(\rho^2 \circ \tau(x)y\big),$$

hence $h(\beta(x), y) = h\big(x, \beta(y)\big)$ for all x, $y \in L_K$ and therefore $\tau(\beta) = \beta$. Clearly, $\beta \circ x = \rho(x) \circ \beta$ for $x \in L$, and $\beta^3 = \ell^{-1}\rho(\ell^{-1})\rho^2(\ell^{-1}) \in F^\times$. Therefore, we may choose $z = \beta$. This proves the proposition in the case where B is split.

If B is not split, consider the F-vector space:

$$S = \{\, z \in \mathrm{Sym}(B, \tau) \mid zx = \rho(x)z \text{ for } x \in L_K \,\}.$$

The invertible elements in S form a Zariski-open set. Extension of scalars to a splitting field of B shows that this open set is not empty. Since B is not split, the field F is infinite, hence the rational points in S are dense. We may therefore find an invertible element in S.

If $z \in S$, then z^3 centralizes L_K, hence $z^3 \in L_K$. Since z^3 commutes with z and $z \in \mathrm{Sym}(B, \tau)$, we have $z^3 \in F$. Therefore, every invertible element $z \in S$ satisfies the required conditions. $\square$

Étale subalgebras and the invariant $\pi(\tau)$. We now fix an involution of the second kind τ on the central simple K-algebra B of degree 3 and a cubic étale F-algebra $L \subset \mathrm{Sym}(B, \tau)$. We assume throughout that $\mathrm{char}\, F \neq 2$. We will give a special expression for the quadratic form Q_τ, hence also for the Pfister form $\pi(\tau)$, taking into account the algebra L (see Theorem (19.25)). As an application, we prove the following statements: if an involution is the identity on a cubic étale F-algebra of discriminant isomorphic to K, then it is distinguished; moreover, every cubic étale F-subalgebra in B is stabilized by some distinguished involution.

The idea to obtain the special form of Q_τ is to consider the orthogonal decomposition $\mathrm{Sym}(B, \tau) = L \perp M$ where $M = L^\perp$ is the orthogonal complement of L for the quadratic form Q_τ:

$$M = \{\, x \in \mathrm{Sym}(B, \tau) \mid \mathrm{Trd}_B(x\ell) = 0 \text{ for } \ell \in L \,\}.$$

We show that the restriction of Q_τ to M is the transfer of some hermitian form H_M on M. This hermitian form is actually defined on the whole of B, with values in $L_K \otimes_K L_K$, where $L_K = L \otimes K \subset B$.

We first make B a right $L_K \otimes L_K$-module as follows: for $b \in B$ and ℓ_1, $\ell_2 \in L_K$, we set:

$$b * (\ell_1 \otimes \ell_2) = \ell_1 b \ell_2.$$

(19.16) Lemma. *The separability idempotent $e \in L_K \otimes L_K$ satisfies the following properties relative to $*$:*

(1) $\ell * e = \ell$ *for all $\ell \in L$, and $\mathrm{Trd}(x) = \mathrm{Trd}(x * e)$ for all $x \in B$;*

(2) $(x * e)\ell = (x\ell) * e = (\ell x) * e = \ell(x * e)$ for all $x \in B$, $\ell \in L_K$;
(3) $B * e = L_K$;
(4) $x * e = 0$ for all $x \in M_K = M \otimes K$.

From (1) and (4) it follows that multiplication by e is the orthogonal projection $B \to L_K$ for the trace bilinear form.

Proof: (1) Let $e = \sum_{i=1}^{3} u_i \otimes v_i$. Proposition (18.10) shows that $\sum_i u_i v_i = 1$; since L_K is commutative, it follows that

$$\ell * e = \sum_{i=1}^{3} u_i \ell v_i = (\sum_{i=1}^{3} u_i v_i)\ell = \ell \quad \text{for } \ell \in L_K.$$

Moreover, for $x \in B$ we have

$$\text{Trd}(x * e) = \text{Trd}(\sum_{i=1}^{3} u_i x v_i) = \text{Trd}\big((\sum_{i=1}^{3} v_i u_i)x\big) = \text{Trd}(x).$$

(2) For $x \in B$ and $\ell \in L_K$,

$$(x * e)\ell = (x * e) * (1 \otimes \ell) = x * \big(e(1 \otimes \ell)\big).$$

By (18.10), we have $e(1 \otimes \ell) = (1 \otimes \ell)e = (\ell \otimes 1)e$, hence, by substituting this in the preceding equality:

$$(x * e)\ell = (x\ell) * e = (\ell x) * e.$$

Similarly, $\ell(x * e) = (x * e) * (\ell \otimes 1) = x * \big(e(\ell \otimes 1)\big)$ and $e(\ell \otimes 1) = (\ell \otimes 1)e$, hence we also have

$$\ell(x * e) = (\ell x) * e.$$

(3) Property (2) shows that $B * e$ centralizes L_K, hence $B * e = L_K$.
(4) For $x \in M_K$ and $\ell \in L_K$ we have by (1) and (2):

$$\text{Trd}\big((x * e)\ell\big) = \text{Trd}\big((x\ell) * e\big) = \text{Trd}(x\ell) = 0.$$

From (3), we obtain $(x * e) \in M_K \cap L_K = \{0\}$. $\qquad\qquad\qquad\square$

(19.17) Example. The split case. Suppose $B = \text{End}_K(V)$ for some 3-dimensional vector space V and $\tau = \tau_h$ is the adjoint involution with respect to some hermitian form h on V. Suppose also that $L \simeq F \times F \times F$ and let e_1, e_2, $e_3 \in L$ be the primitive idempotents. There is a corresponding direct sum decomposition of V into K-subspaces of dimension 1:

$$V = V_1 \oplus V_2 \oplus V_3$$

such that e_i is the projection onto V_i with kernel $V_j \oplus V_k$ for $\{i, j, k\} = \{1, 2, 3\}$. Since e_1, e_2, e_3 are τ-symmetric, we have for $x, y \in V$ and $i, j = 1, 2, 3$, $i \neq j$:

$$h\big(e_i(x), e_j(y)\big) = h\big(x, e_i \circ e_j(y)\big) = 0.$$

Therefore, the subspaces V_1, V_2, V_3 are pairwise orthogonal with respect to h. For $i = 1, 2, 3$, pick a nonzero vector $v_i \in V_i$ and let $h(v_i, v_i) = \delta_i \in F^\times$. We may use the basis (v_1, v_2, v_3) to identify B with $M_3(K)$; the involution τ is then given by

$$\tau\begin{pmatrix} x_{11} & x_{12} & x_{13} \\ x_{21} & x_{22} & x_{23} \\ x_{31} & x_{32} & x_{33} \end{pmatrix} = \begin{pmatrix} \iota(x_{11}) & \delta_1^{-1}\iota(x_{21})\delta_2 & \delta_1^{-1}\iota(x_{31})\delta_3 \\ \delta_2^{-1}\iota(x_{12})\delta_1 & \iota(x_{22}) & \delta_2^{-1}\iota(x_{32})\delta_3 \\ \delta_3^{-1}\iota(x_{13})\delta_1 & \delta_3^{-1}\iota(x_{23})\delta_2 & \iota(x_{33}) \end{pmatrix},$$

so that

$$\text{Sym}(B, \tau) = \left\{ \begin{pmatrix} x & \delta_2 a & \delta_3 b \\ \delta_1 \iota(a) & y & \delta_3 c \\ \delta_1 \iota(b) & \delta_2 \iota(c) & z \end{pmatrix} \middle| \; x, y, z \in F, \; a, b, c \in K \right\}.$$

Under this identification,

$$e_1 = \begin{pmatrix} 1 & 0 & 0 \\ 0 & 0 & 0 \\ 0 & 0 & 0 \end{pmatrix}, \quad e_2 = \begin{pmatrix} 0 & 0 & 0 \\ 0 & 1 & 0 \\ 0 & 0 & 0 \end{pmatrix}, \quad e_3 = \begin{pmatrix} 0 & 0 & 0 \\ 0 & 0 & 0 \\ 0 & 0 & 1 \end{pmatrix},$$

and L is the F-algebra of diagonal matrices in $\mathrm{Sym}(B, \tau)$. The separability idempotent is $e = e_1 \otimes e_1 + e_2 \otimes e_2 + e_3 \otimes e_3$. A computation shows that

$$M = \left\{ \begin{pmatrix} 0 & \delta_2 a & \delta_3 b \\ \delta_1 \iota(a) & 0 & \delta_3 c \\ \delta_1 \iota(b) & \delta_2 \iota(c) & 0 \end{pmatrix} \;\middle|\; a,\, b,\, c \in K \right\}.$$

The K-algebra L_K is the algebra of diagonal matrices in B and M_K is the space of matrices whose diagonal entries are all 0. Let

$$m = \begin{pmatrix} 1 & 1 & 1 \\ 1 & 1 & 1 \\ 1 & 1 & 1 \end{pmatrix} \in B.$$

For $x = (x_{ij})_{1 \le i, j \le 3} \in B$, we have

$$m * \left(\textstyle\sum_{i,j=1}^{3} x_{ij} e_i \otimes e_j \right) = x,$$

hence B is a free $L_K \otimes L_K$-module of rank 1.

We now return to the general case, and let θ be the K-automorphism of $L_K \otimes L_K$ which switches the factors:

$$\theta(\ell_1 \otimes \ell_2) = \ell_2 \otimes \ell_1 \quad \text{for } \ell_1,\, \ell_2 \in L_K.$$

As in (19.16), we call $e \in L_K \otimes L_K$ the separability idempotent of L_K. The characterization of e in (18.10) shows that e is invariant under θ.

(19.18) Proposition. *Consider B as a right $L_K \otimes L_K$-module through the $*$-multiplication. The module B is free of rank 1. Moreover, there is a unique hermitian form*

$$H : B \times B \to L_K \otimes L_K$$

with respect to θ such that for all $x \in B$,

(19.19)

$$H(x, x) = e(x' \otimes x') + (1 - e)\left((x''^2 * e) \otimes 1 + 1 \otimes (x''^2 * e) - \tfrac{1}{2} \mathrm{Trd}(x''^2) \right),$$

where $x' = x * e$ and $x'' = x * (1 - e)$. (Note that x' and $x''^2 * e$ lie in L_K, by (19.16).)

Proof: Since $L_K \otimes L_K$ is an étale F-algebra, it decomposes into a direct product of fields by (18.3). Let $L_K \otimes L_K \simeq L_1 \times \cdots \times L_n$ for some fields $L_1, \ldots, L_n$. Then $B \simeq B_1 \times \cdots \times B_n$ where B_i is a vector space over L_i for $i = 1, \ldots, n$. To see that B is a free $L_K \otimes L_K$-module of rank 1, it suffices to prove that $\dim_{L_i} B_i = 1$ for $i = 1, \ldots, n$. Since $\dim_F B = \dim_F(L_K \otimes L_K)$, it actually suffices to show that $\dim_{L_i} B_i \ne 0$ for $i = 1, \ldots, n$, which means that B is a faithful $L_K \otimes L_K$-module. This property may be checked over a scalar extension of F. Since it holds in the split case, as was observed in (19.17), it also holds in the general case. (For a slightly different proof, see Jacobson [146, p. 44].)

Now, let $b \in B$ be a basis of B (as a free $L_K \otimes L_K$-module). We define a hermitian form H on B by

$$H(b * \lambda_1, b * \lambda_2) = \theta(\lambda_1)H(b,b)\lambda_2 \quad \text{for } \lambda_1, \lambda_2 \in L_K \otimes L_K,$$

where $H(b,b)$ is given by formula (19.19). This is obviously the unique hermitian form on B for which $H(b,b)$ takes the required value. In order to show that the hermitian form thus defined satisfies formula (19.19) for all $x \in B$, we may extend scalars to a splitting field of B and L, and assume we are in the split case discussed in (19.17). With the same notation as in (19.17), define:

$$(19.20) \qquad H'(x,y) = \sum_{i,j=1}^{3} x_{ji} y_{ij} e_i \otimes e_j \in L_K \otimes L_K$$

for $x = (x_{ij})_{1 \le i,j \le 3}$, $y = (y_{ij})_{1 \le i,j \le 3} \in B$. Straightforward computations show that H' is hermitian and satisfies formula (19.19) for all $x \in B$. In particular, $H'(b,b) = H(b,b)$, hence $H' = H$. This proves the existence and uniqueness of the hermitian form H. $\square$

The hermitian form H restricts to hermitian forms on $\mathrm{Sym}(B, \tau)$ and on M_K which we discuss next.

Let ω be the K-semilinear automorphism of $L_K \otimes L_K$ defined by

$$\omega\big((\ell_1 \otimes k_1) \otimes (\ell_2 \otimes k_2)\big) = \big(\ell_2 \otimes \iota(k_2)\big) \otimes \big(\ell_1 \otimes \iota(k_1)\big)$$

for $\ell_1, \ell_2 \in L$ and $k_1, k_2 \in K$. The following property is clear from the definition:

$$\tau(x * \lambda) = \tau(x) * \omega(\lambda) \qquad \text{for } x \in B \text{ and } \lambda \in L_K \otimes L_K.$$

This shows that $\mathrm{Sym}(B, \tau)$ is a right module over the algebra $(L_K \otimes L_K)^\omega$ of ω-invariant elements in $L_K \otimes L_K$. Moreover, by extending scalars to a splitting field of B and using the explicit description of $H = H'$ in (19.20), one can check that

$$H\big(\tau(x), \tau(y)\big) = \omega\big(H(x,y)\big) \qquad \text{for } x, y \in B.$$

Therefore, the hermitian form H restricts to a hermitian form

$$H_S \colon \mathrm{Sym}(B, \tau) \times \mathrm{Sym}(B, \tau) \to (L_K \otimes L_K)^\omega.$$

Now, consider the restriction of H to M_K. By (19.16), we have $x * e = 0$ for all $x \in M_K$, hence $x * (1-e) = x$ and the $L_K \otimes L_K$-module action on B restricts to an action of $(1-e) \cdot (L_K \otimes L_K)$ on M_K. Moreover, for $x, y \in M_K$,

$$H(x,y) = H\big(x * (1-e), y\big) = (1-e)H(x,y) \in (1-e) \cdot (L_K \otimes L_K),$$

hence H restricts to a hermitian form

$$H_{M_K} \colon M_K \times M_K \to (1-e) \cdot (L_K \otimes L_K).$$

Recall from (18.26) the embedding $\epsilon_3 \colon L_K \hookrightarrow (1-e) \cdot (L_K \otimes L_K)$. By (18.25) and (18.27), ϵ_3 induces a canonical isomorphism

$$\widetilde{\epsilon_3} \colon L_K \otimes \Delta(L_K) \xrightarrow{\sim} (1-e) \cdot (L_K \otimes L_K)$$

which we use to identify $(1-e) \cdot (L_K \otimes L_K)$ with $L_K \otimes \Delta(L_K)$. Since the image of ϵ_3 is the subalgebra of $(1-e) \cdot (L_K \otimes L_K)$ of elements fixed by θ, the automorphism of $L_K \otimes \Delta(L_K)$ corresponding to θ via $\widetilde{\epsilon_3}$ is the identity on L_K and the unique nontrivial K-automorphism on $\Delta(L_K)$. We call this automorphism also θ. Thus, we may consider the restriction of H to M_K as a hermitian form with respect to θ:

$$H_{M_K} \colon M_K \times M_K \to L_K \otimes \Delta(L_K).$$

In particular, $H_{M_K}(x, x) \in L_K$ for all $x \in M_K$.

(19.21) Lemma. *For all $x \in M_K$,*

$$2T_{L_K/K}\big(H_{M_K}(x, x)\big) = \mathrm{Trd}_B(x^2)$$

and

$$N_{L_K/K}\big(H_{M_K}(x, x)\big)^2 = N_{\epsilon_3(L_K)/K}\big(H\big(x^2 * (1 - e), x^2 * (1 - e)\big)\big).$$

For all $x \in M_K$ and $\ell \in L_K$,

$$H_{M_K}(\ell x, \ell x) = \ell^{\#} H_{M_K}(x, x),$$

where $^{\#} \colon L_K \to L_K$ is the quadratic map defined in (18.33).

Proof: It suffices to verify these formulas when B and L are split. We may thus assume B and L are as in (19.17). For

$$x = \begin{pmatrix} 0 & x_{12} & x_{13} \\ x_{21} & 0 & x_{23} \\ x_{31} & x_{32} & 0 \end{pmatrix} \in M_K,$$

we have, using (18.26) and (19.20),

$$\begin{aligned}
H(x, x) &= x_{12}x_{21}(e_1 \otimes e_2 + e_2 \otimes e_1) + x_{13}x_{31}(e_1 \otimes e_3 + e_3 \otimes e_1) \\
&\quad + x_{23}x_{32}(e_2 \otimes e_3 + e_3 \otimes e_2) \\
&= \epsilon_3(x_{23}x_{32}e_1 + x_{13}x_{31}e_2 + x_{12}x_{21}e_3),
\end{aligned}$$

hence $H_{M_K}(x, x) = x_{23}x_{32}e_1 + x_{13}x_{31}e_2 + x_{12}x_{21}e_3 \in L_K$. It follows that

$$T_{L_K/K}\big(H_{M_K}(x, x)\big) = x_{23}x_{32} + x_{13}x_{31} + x_{12}x_{21} = \tfrac{1}{2}\,\mathrm{Trd}(x^2)$$

and

$$N_{L_K/K}\big(H(x, x)\big) = x_{23}x_{32}x_{13}x_{31}x_{12}x_{21}.$$

On the other hand,

$$x^2 * (1 - e) = \begin{pmatrix} 0 & x_{13}x_{32} & x_{12}x_{23} \\ x_{23}x_{31} & 0 & x_{21}x_{13} \\ x_{32}x_{21} & x_{31}x_{12} & 0 \end{pmatrix},$$

hence

$$N_{\epsilon_3(L_K)/K}\big(H\big(x^2 * (1 - e), x^2 * (1 - e)\big)\big) = (x_{23}x_{32}x_{13}x_{31}x_{12}x_{21})^2.$$

For $\ell = \ell_1 e_1 + \ell_2 e_2 + \ell_3 e_3 \in L_K$, we have

$$\ell x = \begin{pmatrix} 0 & \ell_1 x_{12} & \ell_1 x_{13} \\ \ell_2 x_{21} & 0 & \ell_2 x_{23} \\ \ell_3 x_{31} & \ell_3 x_{32} & 0 \end{pmatrix}$$

and

$$H(\ell x, \ell x) = \epsilon_3(\ell_2 \ell_3 e_1 + \ell_1 \ell_3 e_2 + \ell_1 \ell_2 e_3) H(x, x),$$

proving the last formula of the lemma, since $\ell^{\#} = \ell_2 \ell_3 e_1 + \ell_1 \ell_3 e_2 + \ell_1 \ell_2 e_3$. Alternately, the last formula follows from the fact that $\ell x = x * (\ell \otimes 1)$, hence $H(\ell x, \ell x) = (\ell \otimes \ell) \cdot H(x, x)$, together with the observation that $(\ell \otimes \ell)(1 - e) = \epsilon_3(\ell^{\#})$.

The first formula also follows from the definition of $H(x, x)$ and of ϵ_3, since (19.16) shows that $T_{L/F}(x^2 * e) = \mathrm{Trd}(x^2)$. $\qquad\square$

Finally, we combine the restrictions H_S and H_{M_K} of H to $\text{Sym}(B, \tau)$ and to M_K to describe the restriction of H to $\text{Sym}(B, \tau) \cap M_K = M$. The automorphism of $L_K \otimes \Delta(L_K) = L \otimes \Delta(L) \otimes K$ corresponding to ω under the isomorphism $\widetilde{\epsilon}_3$ is the identity on L and restricts to the unique nontrivial automorphism of $\Delta(L)$ and of K. The F-subalgebra of elements fixed by ω therefore has the form $L \otimes E$, where E is the quadratic étale F-subalgebra of ω-invariant elements in $\Delta(L) \otimes K$. If α, $\delta \in F^\times$ are such that $K \simeq F[X]/(X^2 - \alpha)$ and $\Delta(L) \simeq F[X]/(X^2 - \delta)$, then $E \simeq F[X]/(X^2 - \alpha\delta)$. The $L_K \otimes \Delta(L_K)$-module structure on M_K restricts to an $L \otimes E$-module structure on M, and the hermitian form H_{M_K} restricts to a hermitian form

$$H_M \colon M \times M \to L \otimes E$$

with respect to (the restriction of) θ. Note that θ on $L \otimes E$ is the identity on L and restricts to the nontrivial automorphism of E, hence $H_M(x, x) \in L$ for all $x \in M$.

(19.22) Proposition. *The hermitian form H_M satisfies*:

$$T_{L \otimes E/F}\big(H_M(x, x)\big) = Q_\tau(x) \quad \text{for } x \in M.$$

The $L \otimes E$-module M is free of rank 1; it contains a basis vector m such that

$$N_{L/F}\big(H_M(m, m)\big) \in F^{\times 2}.$$

Proof: The first formula readily follows from (19.21), since

$$T_{L \otimes E/F}\big(H_M(x, x)\big) = 2T_{L/F}\big(H_M(x, x)\big).$$

We claim that every element $x \in M$ such that $H_M(x, x) \in L^\times$ is a basis of the $L \otimes_F E$-module M. Indeed, if $\lambda \in L \otimes E$ satisfies $x * \lambda = 0$, then $H_M(x, x * \lambda) = H_M(x, x)\lambda = 0$, hence $\lambda = 0$. Therefore, $x * (L \otimes E)$ is a submodule of M which has the same dimension over F as M. It follows that $M = x * (L \otimes E)$, and that x is a basis of M over $L \otimes E$.

The existence of elements x such that $H_M(x, x) \in L^\times$ is clear if F is infinite, since the proof of (19.21) shows that $N_{L/F}\big(H_M(x, x)\big)$ is a nonzero polynomial function of x. It is also easy to establish when F is finite. (Note that in that case the algebra B is split).

To find a basis element $m \in M$ such that $N_{L/F}\big(H_M(m, m)\big) \in F^{\times 2}$, pick any $x \in M$ such that $H_M(x, x) \in L^\times$ and set $m = x^2 * (1 - e)$. By (19.21), we have

$$N_{L/F}\big(H_M(m, m)\big) = \big[N_{L/F}\big(H_M(x, x)\big)\big]^2 \in F^{\times 2}. \qquad \square$$

Let $m \in M$ be a basis of M over $L \otimes E$ such that $N_{L/F}\big(H_M(m, m)\big) \in F^{\times 2}$ and let $\ell = H_M(m, m) \in L^\times$. We then have a diagonalization $H_M = \langle \ell \rangle_{L \otimes E}$, and Proposition (19.22) shows that the restriction $Q_\tau|_M$ of Q_τ to M is the Scharlau transfer of the hermitian form $\langle \ell \rangle_{L \otimes E}$:

$$Q_\tau|_M = (T_{L \otimes H/F})_*(\langle \ell \rangle_{L \otimes H}).$$

We may use transitivity of the trace to represent the right-hand expression as the transfer of a 2-dimensional quadratic space over L: if $K \simeq F[X]/(X^2 - \alpha)$ and $\Delta(L) \simeq F[X]/(X^2 - \delta)$, so that $E \simeq F[X]/(X^2 - \alpha\delta)$, we have

$$(T_{L \otimes E/L})_*\big(\langle \ell \rangle_{L \otimes E}\big) \simeq \langle 2\ell, -2\alpha\delta\ell \rangle_L \simeq \langle 2 \rangle \cdot \langle\!\langle \alpha\delta \rangle\!\rangle \cdot \langle \ell \rangle_L,$$

hence

(19.23) $$Q_\tau|_M \simeq \langle 2 \rangle \cdot \langle\!\langle \alpha\delta \rangle\!\rangle \cdot (T_{L/F})_*\big(\langle \ell \rangle_L\big).$$

This formula readily yields an expression for the form Q_τ, in view of the orthogonal decomposition $\mathrm{Sym}(B, \tau) = L \perp M$. In order to get another special expression, we prove a technical result:

(19.24) Lemma. *For all $\ell \in L^\times$ such that $N_{L/F}(\ell) \in F^{\times 2}$, the quadratic form*

$$\langle\!\langle \delta \rangle\!\rangle \cdot \big((T_{L/F})_*(\langle \ell \rangle_L) \perp \langle -1 \rangle \big)$$

is hyperbolic.

Proof: By Springer's theorem on odd-degree extensions, it suffices to prove that the quadratic form above is hyperbolic after extending scalars from F to L. We may thus assume $L \simeq F \times \Delta(L)$. Let $\ell = (\ell_0, \ell_1)$ with $\ell_0 \in F$ and $\ell_1 \in \Delta(L)$; then

$$(T_{L/F})_*\big(\langle \ell \rangle_L \big) = \langle \ell_0 \rangle \perp (T_{\Delta(L)/F})_*\big(\langle \ell_1 \rangle_{\Delta(L)} \big).$$

By Scharlau [247, p. 50], the image of the transfer map from the Witt ring $W\Delta(L)$ to WF is killed by $\langle\!\langle \delta \rangle\!\rangle$, hence

$$\langle\!\langle \delta \rangle\!\rangle \cdot (T_{L/F})_*\big(\langle \ell \rangle_L \big) = \langle\!\langle \delta \rangle\!\rangle \cdot \langle \ell_0 \rangle \quad \text{in } WF.$$

On the other hand, $N_{L/F}(\ell) = \ell_0 N_{\Delta(L)/F}(\ell_1) \in F^{\times 2}$, hence ℓ_0 is a norm from $\Delta(L)$ and therefore

$$\langle\!\langle \delta \rangle\!\rangle \cdot \langle \ell_0 \rangle = \langle\!\langle \delta \rangle\!\rangle \quad \text{in } WF. \qquad \square$$

Here, finally, is the main result of this subsection:

(19.25) Theorem. *Let (B, τ) be a central simple K-algebra of degree 3 with involution of the second kind which is the identity on F and let $L \subset \mathrm{Sym}(B, \tau)$ be a cubic étale F-algebra. Let α, $\delta \in F^\times$ be such that $K \simeq F[X]/(X^2 - \alpha)$ and $\Delta(L) \simeq F[X]/(X^2 - \delta)$. Then, the quadratic form Q_τ and the invariant $\pi(\tau)$ satisfy:*

(19.26)
$$\begin{aligned}
Q_\tau &\simeq \langle 1, 2, 2\delta \rangle \perp \langle 2 \rangle \cdot \langle\!\langle \alpha\delta \rangle\!\rangle \cdot (T_{L/F})_*\big(\langle \ell \rangle_L \big) \\
&\simeq \langle 1, 1, 1 \rangle \perp \langle 2\delta \rangle \cdot \langle\!\langle \alpha \rangle\!\rangle \cdot (T_{L/F})_*\big(\langle \ell \rangle_L \big)
\end{aligned}$$

and

(19.27)
$$\pi(\tau) \simeq \langle\!\langle \alpha \rangle\!\rangle \cdot \big(\langle 1 \rangle \perp \langle \delta \rangle \cdot (T_{L/F})_*\big(\langle \ell \rangle_L \big) \big)$$

for some $\ell \in L^\times$ such that $N_{L/F}(\ell) \in F^{\times 2}$.

In particular, $\pi(\tau)$ has a factorization: $\pi(\tau) \simeq \langle\!\langle \alpha \rangle\!\rangle \cdot \varphi$ where φ is a 2-fold Pfister form such that

$$\varphi \cdot \langle\!\langle \delta \rangle\!\rangle = 0 \quad \text{in } WF.$$

Proof: Lemma (18.31) shows that the restriction of Q_τ to L has a diagonalization:

$$Q_\tau|_L \simeq \langle 1, 2, 2\delta \rangle.$$

Since $\mathrm{Sym}(B, \tau) = L \perp M$, the first formula for Q_τ follows from (19.23).

In WF, we have $\langle\!\langle \alpha\delta \rangle\!\rangle = \langle\!\langle \alpha \rangle\!\rangle \cdot \langle \delta \rangle + \langle\!\langle \delta \rangle\!\rangle$. By substituting this in the first formula for Q_τ, we obtain:

$$Q_\tau = \langle 1, 2, 2\delta \rangle + \langle 2\delta \rangle \cdot \langle\!\langle \alpha \rangle\!\rangle \cdot (T_{L/F})_*\big(\langle \ell \rangle_L \big) + \langle 2 \rangle \cdot \langle\!\langle \delta \rangle\!\rangle \cdot (T_{L/F})_*\big(\langle \ell \rangle_L \big) \quad \text{in } WF.$$

Lemma (19.24) shows that the last term on the right equals $\langle 2 \rangle \cdot \langle\!\langle \delta \rangle\!\rangle$. Since

$$\langle 2\delta \rangle + \langle 2 \rangle \cdot \langle\!\langle \delta \rangle\!\rangle = \langle 2 \rangle \quad \text{and} \quad \langle 1, 2, 2 \rangle = \langle 1, 1, 1 \rangle,$$

we find

$$Q_\tau = \langle 1, 1, 1 \rangle + \langle 2\delta \rangle \cdot \langle\!\langle \alpha \rangle\!\rangle \cdot (T_{L/F})_* \big(\langle \ell \rangle_L \big) \quad \text{in } WF.$$

Since these two quadratic forms have the same dimension, they are isometric, proving the second formula for Q_τ.

The formula for $\pi(\tau)$ readily follows, by the definition of $\pi(\tau)$ in (19.4).

According to Scharlau [247, p. 51], we have $\det(T_{L/F})_*\big(\langle \ell \rangle_L\big) = \delta N_{L/F}(\ell)$, hence the form $\varphi = \langle 1 \rangle \perp \langle \delta \rangle \cdot (T_{L/F})_*\big(\langle \ell \rangle_L\big)$ is a 2-fold Pfister form. Finally, Lemma (19.24) shows that

$$\langle\!\langle \delta \rangle\!\rangle \cdot \varphi = \langle\!\langle \delta \rangle\!\rangle + \langle \delta \rangle \cdot \langle\!\langle \delta \rangle\!\rangle = 0 \quad \text{in } WF. \qquad \square$$

(19.28) Corollary. *Every unitary involution τ such that $\mathrm{Sym}(B, \tau)$ contains a cubic étale F-algebra L with discriminant $\Delta(L)$ isomorphic to K is distinguished.*

Proof: Theorem (19.25) yields a factorization $\pi(\tau) = \langle\!\langle \alpha \rangle\!\rangle \cdot \varphi$ with $\varphi \cdot \langle\!\langle \delta \rangle\!\rangle = 0$ in WF. Therefore, $\pi(\tau) = 0$ if $\alpha = \delta$. $\qquad \square$

So far, the involution τ has been fixed, as has been the étale subalgebra $L \subset \mathrm{Sym}(B, \tau)$. In the next proposition, we compare the quadratic forms Q_τ and $Q_{\tau'}$ associated to two involutions of the second kind which are the identity on L. By (2.18), we then have $\tau' = \mathrm{Int}(u) \circ \tau$ for some $u \in L^\times$.

(19.29) Proposition. *Let $\delta \in F^\times$ be such that $\Delta(L) \simeq F[X]/(X^2 - \delta)$. Let $u \in L^\times$ and let $\tau_u = \mathrm{Int}(u) \circ \tau$. For any $\ell \in L^\times$ such that*

$$Q_\tau \simeq \langle 1, 2, 2\delta \rangle \perp \langle 2 \rangle \cdot \langle\!\langle \alpha\delta \rangle\!\rangle \cdot (T_{L/F})_*\big(\langle \ell \rangle_L\big),$$

we have

$$Q_{\tau_u} \simeq \langle 1, 2, 2\delta \rangle \perp \langle 2 \rangle \cdot \langle\!\langle \alpha\delta \rangle\!\rangle \cdot (T_{L/F})_*\big(\langle u^\# \ell \rangle_L\big).$$

Proof: Left multiplication by u gives a linear bijection $\mathrm{Sym}(B, \tau) \to \mathrm{Sym}(B, \tau_u)$ which maps L to L and the orthogonal complement M of L in $\mathrm{Sym}(B, \tau)$ for the form Q_τ to the orthogonal complement M_u of L in $\mathrm{Sym}(B, \tau_u)$ for the form Q_{τ_u}. Lemma (19.21) shows that

$$H_{M_K}(ux, ux) = u^\# H_{M_K}(x, x) \quad \text{for } x \in M_K,$$

hence multiplication by u defines a similitude $(M, H_M) \xrightarrow{\sim} (M_u, H_{M_u})$ with multiplier $u^\#$. $\qquad \square$

(19.30) Corollary. *Let L be an arbitrary cubic étale F-algebra in B with $\Delta(L) \simeq F[X]/(X^2 - \delta)$ for $\delta \in F^\times$.*

(1) *For every $\ell \in L^\times$ such that $N_{L/F}(\ell) \in F^{\times 2}$, there exists an involution τ which is the identity on L and such that Q_τ and $\pi(\tau)$ satisfy (19.26) and (19.27).*

(2) *There exists a distinguished involution which is the identity on L.*

Proof: (1) By (4.18), there is an involution of the second kind τ_0 such that $L \subset \mathrm{Sym}(B, \tau_0)$. Theorem (19.25) yields

$$Q_{\tau_0} \simeq \langle 1, 1, 1 \rangle \perp \langle 2\delta \rangle \cdot \langle\!\langle \alpha \rangle\!\rangle \cdot (T_{L/F})_*\big(\langle \ell_0 \rangle_L\big)$$

for some $\ell_0 \in L^\times$ with $N_{L/F}(\ell_0) \in F^{\times 2}$. If $\ell \in L^\times$ satisfies $N_{L/F}(\ell) \in F^{\times 2}$, then $N_{L/F}(\ell_0^{-1}\ell) \in F^{\times 2}$, hence Proposition (18.34) shows that there exists $u \in L^\times$ satisfying $u^\# \equiv \ell_0^{-1}\ell \mod L^{\times 2}$. We then have $\langle \ell \rangle_L \simeq \langle u^\# \ell_0 \rangle_L$, hence, by (19.29), the involution $\tau = \mathrm{Int}(u) \circ \tau_0$ satisfies the specified conditions.

(2) Choose $\ell_1 \in L^\times$ satisfying $T_{L/F}(\ell_1) = 0$ and let $\ell = \ell_1 N_{L/F}(\ell_1)^{-1}$; then $N_{L/F}(\ell) = N_{L/F}(\ell_1)^{-2} \in F^{\times 2}$ and $T_{L/F}(\ell) = 0$. Part (1) shows that there exists an involution τ which is the identity on L and satisfies

$$\pi(\tau) \simeq \langle\!\langle \alpha \rangle\!\rangle \cdot \big(\langle 1 \rangle \perp \langle \delta \rangle \cdot (T_{L/F})_*(\langle \ell \rangle_L)\big).$$

Since $T_{L/F}(\ell) = 0$, the form $(T_{L/F})_*(\langle \ell \rangle_L)$ is isotropic. Therefore, $\pi(\tau)$ is isotropic, hence hyperbolic since it is a Pfister form, and it follows that τ is distinguished. $\square$

As another consequence of (19.29), we obtain some information on the conjugacy classes of involutions which leave a given cubic étale F-algebra L elementwise invariant:

(19.31) Corollary. *Let $L \subset B$ be an arbitrary cubic étale F-subalgebra and let τ be an arbitrary involution which is the identity on L. For u, $v \in L^\times$, the involutions $\tau_u = \mathrm{Int}(u) \circ \tau$ and $\tau_v = \mathrm{Int}(v) \circ \tau$ are conjugate if $uv \in N_{L_K/L}(L_K^\times) \cdot F^\times$. Therefore, the map $u \in L^\times \mapsto \tau_u$ induces a surjection of pointed sets from $L^\times / N_{L_K/L}(L_K^\times) \cdot F^\times$ to the set of conjugacy classes of involutions which are the identity on L, where the distinguished involution is τ.*

Proof: We use the same notation as in (19.25) and (19.29); thus

$$Q_{\tau_u} \simeq \langle 1, 2, 2\delta \rangle \perp \langle 2 \rangle \cdot \langle\!\langle \alpha\delta \rangle\!\rangle \cdot (T_{L/F})_*(\langle u^\# \ell \rangle_L)$$
$$\simeq \langle 1, 1, 1 \rangle \perp \langle 2\delta \rangle \cdot \langle\!\langle \alpha \rangle\!\rangle \cdot (T_{L/F})_*(\langle u^\# \ell \rangle_L)$$

for some $\ell \in L^\times$ such that $N_{L/F}(\ell) \in F^{\times 2}$ and, similarly,

$$Q_{\tau_v} \simeq \langle 1, 1, 1 \rangle \perp \langle 2\delta \rangle \cdot \langle\!\langle \alpha \rangle\!\rangle \cdot (T_{L/F})_*(\langle v^\# \ell \rangle_L).$$

According to (19.6), the involutions τ_u and τ_v are conjugate if and only if $Q_{\tau_u} \simeq Q_{\tau_v}$. In view of the expressions above for Q_{τ_u} and Q_{τ_v}, this condition is equivalent to:

$$\langle\!\langle \alpha \rangle\!\rangle \cdot (T_{L/F})_*(\langle u^\# \ell, -v^\# \ell \rangle_L) = 0 \qquad \text{in } WF,$$

or, using Frobenius reciprocity, to:

$$(T_{L/F})_*(\langle\!\langle \alpha, (uv)^\# \rangle\!\rangle_L \cdot \langle u^\# \ell \rangle_L) = 0 \qquad \text{in } WF.$$

If $uv = N_{L_K/L}(\lambda)\mu$ for some $\lambda \in L_K^\times$ and some $\mu \in F^\times$, then

$$(uv)^\# = N_{L_K/L}\big(\mu N_{L_K/K}(\lambda)\lambda^{-1}\big),$$

hence $\langle\!\langle \alpha, (uv)^\# \rangle\!\rangle$ is hyperbolic. $\square$

EXERCISES

1. Let L be a finite dimensional commutative algebra over a field F. Let $\mu \colon L \otimes_F L \to L$ be the multiplication map. Suppose $L \otimes_F L$ contains an element e such that $e(x \otimes 1) = e(1 \otimes x)$ for all $x \in L$ and $\mu(e) = 1$. Show that L is étale.

   Hint: Let $(u_i)_{1 \le i \le n}$ be a basis of L and $e = \sum_{i=1}^n u_i \otimes v_i$. Show that $(v_i)_{1 \le i \le n}$ is a basis of L and that $T(u_i, v_j) = \delta_{ij}$ for all i, $j = 1, \ldots, n$, hence T is nonsingular.

   From this exercise and Proposition (18.10), it follows that L is étale if and only if $L \otimes_F L$ contains a separability idempotent of L.

2. Let $G = \{g_1, \ldots, g_n\}$ be a finite group of order n, and let L be a commutative algebra of dimension n over a field F, endowed with an action of G by F-algebra automorphisms. Show that the following conditions are equivalent:
   (a) L is a Galois G-algebra;
   (b) the map $\Psi\colon L \otimes_F L = L_L \to \operatorname{Map}(G, L)$ defined by $\Psi(\ell_1 \otimes \ell_2)(g) = g(\ell_1)\ell_2$ is an isomorphism of L-algebras;
   (c) for some basis $(e_i)_{1 \le i \le n}$ of L, the matrix $\big(g_i(e_j)\big)_{1 \le i,j \le n} \in M_n(L)$ is invertible;
   (d) for every basis $(e_i)_{1 \le i \le n}$ of L, the matrix $\big(g_i(e_j)\big)_{1 \le i,j \le n} \in M_n(L)$ is invertible.

3. Suppose L is a Galois G-algebra over a field F. Show that for all field extensions K/F, the algebra L_K is a Galois G-algebra over K.

4. Show that every étale algebra of dimension 2 is a Galois $(\mathbb{Z}/2\mathbb{Z})$-algebra.

5. (Saltman) Suppose L is an étale F-algebra of dimension n. For $i = 1, \ldots, n$, let $\pi_i \colon L \to L^{\otimes n}$ denote the map which carries $x \in L$ to $1 \otimes \cdots \otimes 1 \otimes x \otimes 1 \otimes \cdots \otimes 1$ (where x is in the i-th position). For $i < j$, let $\pi_{ij} \colon L \otimes_F L \to L^{\otimes n}$ be defined by $\pi_{ij}(x \otimes y) = \pi_i(x)\pi_j(y)$. Let $s = \prod_{1 \le i < j \le n} \pi_{ij}(1 - e)$ where e is the separability idempotent of L. Show that s is invariant under the action of the symmetric group S_n on $L^{\otimes n}$ by permutation of the factors, and that there is an isomorphism of S_n-algebras over F:
$$\Sigma(L) \simeq s \cdot L^{\otimes n}.$$

   Hint: If $L = F \times \cdots \times F$ and $(e_i)_{1 \le i \le n}$ is the canonical basis of L, show that $s = \sum_{\sigma \in S_n} e_{\sigma(1)} \otimes \cdots \otimes e_{\sigma(n)}$.

6. (Barnard [30]) Let $L = F[X]/(f)$ where
$$f = X^n - a_1 X^{n-1} + a_2 X^{n-2} - \cdots + (-1)^n a_n \in F[X]$$
is a polynomial with no repeated roots in an algebraic closure of F. For $k = 1, \ldots, n$, let $s_k \in F[X_1, \ldots, X_n]$ be the k-th symmetric polynomial: $s_k = \sum_{i_1 < \cdots < i_k} X_{i_1} \cdots X_{i_k}$. Show that the action of S_n by permutation of the indeterminates $X_1, \ldots, X_n$ induces an action of S_n on the quotient algebra
$$R = F[X_1, \ldots, X_n]/(s_1 - a_1, s_2 - a_2, \ldots, s_n - a_n).$$

   Establish an isomorphism of S_n-algebras: $\Sigma(L) \simeq R$.

7. (Bergé-Martinet [39]) Suppose L is an étale algebra of odd dimension over a field F of characteristic 2. Let $L' = L \times F$ and $S' = S_{L'/F}$. Show that
$$\Delta(L) \simeq F[t]/(t^2 + t + a)$$
where $a \in F$ is a representative of the determinant of S'.

8. Let B be a central division algebra of degree 3 over a field K of characteristic 3, and let $u \in B \smallsetminus F$ be such that $u^3 \in F^\times$. Show that there exists $x \in B^\times$ such that $ux = (x+1)u$ and $x^3 - x \in F$.

   Hint: (Jacobson [146, p. 80]) Let $\partial_u \colon B \to B$ map x to $ux - xu$. Show that $\partial_u^3 = 0$. Show that if $y \in B$ satisfies $\partial_u(y) \ne 0$ and $\partial_u^2(y) = 0$, then one may take $x = u(\partial_u y)^{-1} y$.

9. (Albert's theorem (19.14) in characteristic 3) Let B be a central division algebra of degree 3 over a field K of characteristic 3. Suppose K is a quadratic extension of some field F and $N_{K/F}(B)$ splits. Show that B contains a cubic extension of K which is Galois over F with Galois group isomorphic to S_3.

Hint: (Villa [296]) Let τ be a unitary involution on B as in (19.13). Pick $u \in \mathrm{Sym}(B, \tau)$ such that $u^3 \in F$, $u \notin F$, and use Exercise 8 to find $x \in B$ such that $ux = (x+1)u$. Show that $x + \tau(x) \in K(u)$, hence $\big(x + \tau(x)\big)^3 \in F$. Use this information to show $\mathrm{Trd}_B\big(x\tau(x) + \tau(x)x\big) = -1$, hence $\mathrm{Srd}_B\big(x - \tau(x)\big) = -1$. Conclude by proving that $K\big(x - \tau(x)\big)$ is cyclic over K and Galois over F with Galois group isomorphic to S_3.

Notes

§18. The notion of a separable algebraic field extension first occurs, under the name of *algebraic extension of the first kind*, in the fundamental paper of Steinitz [273] on the algebraic theory of fields. It was B. L. van der Waerden who proposed the term *separable* in his Moderne Algebra, Vol. I, [300]. The extension of this notion to associative (not necessarily commutative) algebras (as algebras which remain semisimple over any field extension) is already in Albert's "Structure of Algebras" [9], first edition in 1939. The cohomological interpretation (A has dimension 0 or, equivalently, A is projective as an $A \otimes A^{\mathrm{op}}$-module) is due to Hochschild [123]. A systematic study of separable algebras based on this property is given in Auslander-Goldman [26]. Commutative separable algebras over rings occur in Serre [253] as unramified coverings, and are called *étales* by Grothendieck in [109]. Étale algebras over fields were consecrated as a standard tool by Bourbaki [51].

Galois algebras are considered in Grothendieck (loc. ref.) and Serre (loc. ref.). A systematic study is given in Auslander-Goldman (loc. ref.). Further developments may be found in the Memoir of Chase, Harrison and Rosenberg [62] and in the notes of DeMeyer-Ingraham [71].

The notion of the discriminant of an étale F-algebra, and its relation to the trace form, are classical in characteristic different from 2. (In this case, the discriminant is usually defined in terms of the trace form, and the relation with permutations of the roots of the minimal polynomial of a primitive element is proved subsequently.) In characteristic 2, however, this notion is fairly recent. A formula for the discriminant of polynomials, satisfying the expected relation with the permutations of the roots (see (18.23)), was first proposed by Berlekamp [40]. For an étale F-algebra L, Revoy [230] suggested a definition based on the quadratic forms $S_{L/F}$ or S^0, and conjectured the relation, demonstrated in (18.24), between his definition and Berlekamp's. Revoy's conjecture was independently proved by Bergé-Martinet [39] and by Wadsworth [298]. Their proofs involve lifting the étale algebra to a discrete valuation ring of characteristic zero. A different approach, by descent theory, is due to Waterhouse [306]; this approach also yields a definition of discriminant for étale algebras over commutative rings. The proof of (18.24) in characteristic 2 given here is new.

Reduced equations for cubic étale algebras (see (18.32)) (as well as for some higher-dimensional algebras) can be found in Serre [255, p. 657] (in characteristic different from 2 and 3) and in Bergé-Martinet [39, §4] (in characteristic 2).

§19. The fact that central simple algebras of degree 3 are cyclic is another fundamental contribution of Wedderburn [307] to the theory of associative algebras.

Albert's difficult paper [10] seems to be the first significant contribution in the literature to the theory of algebras of degree 3 with unitary involutions. The classification of unitary involutions on such an algebra, as well as the related description of distinguished involutions, comes from Haile-Knus-Rost-Tignol [115]. See (30.21) and (31.45) for the cohomological version of this classification.

CHAPTER VI

Algebraic Groups

It turns out that most of the groups which have occurred thus far in the book are groups of points of certain algebraic group schemes. Moreover, many constructions described previously are related to algebraic groups. For instance, the Clifford algebra and the discriminant algebra are nothing but Tits algebras for certain semisimple algebraic groups; the equivalences of categories considered in Chapter IV, for example of central simple algebras of degree 6 with a quadratic pair and central simple algebras of degree 4 with a unitary involution over an étale quadratic extension (see §15.D), reflect the fact that certain semisimple groups have the same Dynkin diagram ($D_3 \simeq A_3$ in this example).

The aim of this chapter is to give the classification of semisimple algebraic groups of classical type without any field characteristic assumption, and also to study the Tits algebras of semisimple groups.

In the study of linear algebraic groups (more generally, affine group schemes) we use a functorial approach equivalent to the study of Hopf algebras. The advantage of such an approach is that nilpotents in algebras of functions are allowed (and they really do appear when considering centers of simply connected groups over fields of positive characteristic); moreover many constructions like kernels, intersections of subgroups, are very natural. A basic reference for this approach is Waterhouse [305]. The classical view of an algebraic group as a variety with a regular group structure is equivalent to what we call a smooth algebraic group scheme.

The classical theory (mostly over an algebraically closed field) can be found in Borel [45], Humphreys [125], or Springer [270]. We also refer to Springer's survey article [271]. (The new (1998) edition of [270] will contain the theory of algebraic groups over non algebraically closed fields.) We use some results in commutative algebra which can be found in Bourbaki [50], [51], [52], and in the book of Matsumura [179].

The first three sections of the chapter are devoted to the general theory of group schemes. In §23 we define the families of algebraic groups related to an algebra with involution, a quadratic form, and an algebra with a quadratic pair. After a short interlude (root systems, in §24) we come to the classification of split semisimple groups over an arbitrary field. In fact, this classification does not depend on the ground field F, and is essentially equivalent to the classification over the algebraic closure F_{alg} (see Tits [290], Borel-Tits [47]).

The central section of this chapter, §26, gives the classification of adjoint semisimple groups over arbitrary fields. It is based on the observation of Weil [310] that (in characteristic different from 2) a classical adjoint semisimple group is the connected component of the automorphism group of some algebra with involution. In arbitrary characteristic the notion of orthogonal involution has to be replaced by the notion of a quadratic pair which has its origin in the fundamental paper [291] of

Tits. Groups of type G_2 and F_4 which are related to Cayley algebras (Chapter VIII) and exceptional Jordan algebras (Chapter IX), are also briefly discussed.

In the last section we define and study Tits algebras of semisimple groups. It turns out that for the classical groups the nontrivial Tits algebras are the λ-powers of a central simple algebra, the discriminant algebra of a simple algebra with a unitary involution, and the Clifford algebra of a central simple algebra with an orthogonal pair—exactly those algebras which have been studied in the book (and nothing more!).

§20. Hopf Algebras and Group Schemes

This section is mainly expository. We refer to Waterhouse [305] for proofs and more details.

Hopf algebras. Let F be a field and let A be a commutative (unital, associative) F-algebra with multiplication $m \colon A \otimes_F A \to A$. Assume we have F-algebra homomorphisms

$$c \colon A \to A \otimes_F A \quad (comultiplication)$$
$$i \colon A \to A \quad (co\text{-}inverse)$$
$$u \colon A \to F \quad (co\text{-}unit)$$

which satisfy the following:

(a) The diagram

$$
\begin{array}{ccc}
A & \xrightarrow{\ \ c\ \ } & A \otimes_F A \\
{\scriptstyle c}\downarrow & & \downarrow{\scriptstyle c \otimes \mathrm{Id}} \\
A \otimes_F A & \xrightarrow{\ \mathrm{Id} \otimes c\ } & A \otimes_F A \otimes_F A
\end{array}
$$

commutes.

(b) The map

$$A \xrightarrow{\ c\ } A \otimes_F A \xrightarrow{\ u \otimes \mathrm{Id}\ } F \otimes_F A = A$$

equals the identity map $\mathrm{Id} \colon A \to A$.

(c) The two maps

$$A \xrightarrow{\ c\ } A \otimes_F A \xrightarrow{\ i \otimes \mathrm{Id}\ } A \otimes_F A \xrightarrow{\ m\ } A$$
$$A \xrightarrow{\ u\ } F \xrightarrow{\ \cdot 1\ } A$$

coincide.

An F-algebra A together with maps c, i, and u as above is called a (*commutative*) *Hopf algebra over F*. A *Hopf algebra homomorphism* $f \colon A \to B$ is an F-algebra homomorphism preserving c, i, and u, i.e., $(f \otimes f) \circ c_A = c_B \circ f$, $f \circ i_A = i_B \circ f$, and $u_A = u_B \circ f$. Hopf algebras and homomorphisms of Hopf algebras form a category.

If A is a Hopf algebra over F and L/F is a field extension, then A_L together with c_L, i_L, u_L is a Hopf algebra over L. If $A \to B$ and $A \to C$ are Hopf algebra homomorphisms then there is a canonical induced Hopf algebra structure on the F-algebra $B \otimes_A C$.

Let A be a Hopf algebra over F. An ideal J of A such that

$$c(J) \subset J \otimes_F A + A \otimes_F J, \quad i(J) \subset J \quad \text{and} \quad u(J) = 0$$

is called a *Hopf ideal*. If J is a Hopf ideal, the algebra A/J admits the structure of a Hopf algebra and there is a natural surjective Hopf algebra homomorphism $A \to A/J$. For example, $J = \ker(u)$ is a Hopf ideal and $A/J = F$ is the trivial Hopf F-algebra. The kernel of a Hopf algebra homomorphism $f \colon A \to B$ is a Hopf ideal in A and the image of f is a Hopf subalgebra in B.

20.A. Group schemes. Recall that Alg_F denotes the category of unital commutative (associative) F-algebras with F-algebra homomorphisms as morphisms. Let A be a Hopf algebra over F. For any unital commutative associative F-algebra R one defines a product on the set

$$G^A(R) = \mathrm{Hom}_{Alg_F}(A, R)$$

by the formula $fg = m_R \circ (f \otimes_F g) \circ c$ where $m_R \colon R \otimes_F R \to R$ is the multiplication in R. The defining properties of Hopf algebras imply that this product is associative, with a left identity given by the composition $A \xrightarrow{u} F \to R$ and left inverses given by $f^{-1} = f \circ i$; thus $G^A(R)$ is a group.

For any F-algebra homomorphism $f \colon R \to S$ there is a group homomorphism

$$G^A(f) \colon G^A(R) \to G^A(S), \quad g \mapsto f \circ g,$$

hence we obtain a functor

$$G^A \colon Alg_F \to Groups.$$

Any Hopf algebra homomorphism $A \to B$ induces a natural transformation of functors $G^B \to G^A$.

(20.1) Remark. Let A be an F-algebra with a comultiplication $c \colon A \to A \otimes_F A$. Then c yields a binary operation on the set $G^A(R)$ for any F-algebra R. If for any R the set $G^A(R)$ is a group with respect to this operation, then A is automatically a Hopf algebra, that is, the comultiplication determines uniquely the co-inverse i and the co-unit u.

An (*affine*) *group scheme* G over F is a functor $G \colon Alg_F \to Groups$ isomorphic to G^A for some Hopf algebra A over F. By Yoneda's lemma (see for example Waterhouse [305, p. 6]) the Hopf algebra A is uniquely determined by G (up to an isomorphism) and is denoted $A = F[G]$. A group scheme G is said to be *commutative* if $G(R)$ is commutative for all $R \in Alg_F$.

A *group scheme homomorphism* $\rho \colon G \to H$ is a natural transformation of functors. For any $R \in Alg_F$, let ρ_R be the corresponding group homomorphism $G(R) \to H(R)$. By Yoneda's lemma, ρ is completely determined by the unique Hopf algebra homomorphism $\rho^* \colon F[H] \to F[G]$ (called the *comorphism of ρ*) such that $\rho_R(g) = g \circ \rho^*$.

Group schemes over F and group scheme homomorphisms form a category. We denote the set of group scheme homomorphisms (over F) $\rho \colon G \to H$ by $\mathrm{Hom}_F(G, H)$ The functors

Group schemes over F		Commutative Hopf algebras over F
	$\longleftrightarrow$	

$$G \mapsto F[G]$$
$$G^A \mapsfrom A$$

define an equivalence of categories. Thus, essentially, the theory of group schemes is equivalent to the theory of (commutative) Hopf algebras.

For a group scheme G over F and for any $R \in \mathsf{Alg}_F$ the group $G(R)$ is called the *group of R-points of G*. If $f \colon R \to S$ is an injective F-algebra homomorphism, then the homomorphism $G(f) \colon G(R) \to G(S)$ is also injective. If L/E is a Galois extension of fields containing F, with Galois group $\Delta = \mathrm{Gal}(L/E)$, then Δ acts naturally on $G(L)$; Galois descent (Lemma (18.1)) applied to the algebra $L[G]$ shows that the natural homomorphism $G(E) \to G(L)$ identifies $G(E)$ with the subgroup $G(L)^\Delta$ of Galois stable elements.

(20.2) Examples. (1) The *trivial group* $\mathbf{1}(R) = 1$ is represented by the trivial Hopf algebra $A = F$.

(2) Let V be a finite dimensional vector space over F. The functor

$$\mathbf{V} \colon \mathsf{Alg}_F \to \mathsf{Groups}, \quad R \mapsto V_R = V \otimes_F R$$

(to additive groups) is represented by the symmetric algebra $F[\mathbf{V}] = S(V^*)$ of the dual space V^*. Namely one has

$$\mathrm{Hom}_{\mathsf{Alg}_F}\big(S(V^*), R\big) = \mathrm{Hom}_F(V^*, R) = V \otimes_F R.$$

for any $R \in \mathsf{Alg}_F$ The comultiplication c is given by $c(f) = f \otimes 1 + 1 \otimes f$, the co-inverse i by $i(f) = -f$, and the co-unit u by $u(f) = 0$ for $f \in V^*$.

In the particular case $V = F$ we have the *additive group*, written $\mathbf{G}_{\mathrm{a}}$. One has $\mathbf{G}_{\mathrm{a}}(R) = R$ and $F[\mathbf{G}_{\mathrm{a}}] = F[t]$.

(3) Let A be a unital associative F-algebra of dimension n. The functor

$$\mathbf{GL}_1(A) \colon \mathsf{Alg}_F \to \mathsf{Groups}, \quad R \mapsto (A_R)^\times$$

is represented by the algebra $B = S(A^*)[\frac{1}{N}]$ where $N \colon A \to F$ is the norm map considered as an element of $S^n(A^*)$. For,

$$\mathrm{Hom}_{\mathsf{Alg}_F}(B, R) = \{\, f \in \mathrm{Hom}_{\mathsf{Alg}_F}\big(S(A^*), R\big) \mid f(N) \in R^\times \,\}$$
$$= \{\, a \in A_R \mid N(a) \in R^\times \,\} = (A_R)^\times.$$

The comultiplication c is induced by the map

$$A^* \to A^* \otimes A^*$$

dual to the multiplication m. In the particular case $A = \mathrm{End}_F(V)$ we set $\mathbf{GL}(V) = \mathbf{GL}_1(A)$ (the *general linear group*), thus $\mathbf{GL}(V)(R) = \mathrm{GL}(V_R)$.

If $V = F^n$, we write $\mathbf{GL}_n(F)$ for $\mathbf{GL}(V)$. Clearly, $F[\mathbf{GL}_n(F)] = F[X_{ij}, \frac{1}{\det X}]$ where $X = (X_{ij})$.

If $A = F$ we set $\mathbf{G}_{\mathrm{m}} = \mathbf{G}_{\mathrm{m},F} = \mathbf{GL}_1(A)$ (the *multiplicative group*). Clearly, $\mathbf{G}_{\mathrm{m}}(R) = R^\times$, $F[\mathbf{G}_{\mathrm{m}}] = F[t, t^{-1}]$ with comultiplication $c(t) = t \otimes t$, co-inverse $i(t) = t^{-1}$, and co-unit $u(t) = 1$.

A group scheme G over F is said to be *algebraic* if the F-algebra $F[G]$ is finitely generated. All the examples of group schemes given above are algebraic.

Let G be a group scheme over F and let L/F be a field extension. The functor

$$G_L \colon \mathsf{Alg}_L \to \mathsf{Groups}, \quad G_L(R) = G(R)$$

is represented by $F[G]_L = F[G] \otimes_F L$, since

$$\mathrm{Hom}_{\mathsf{Alg}_L}(F[G]_L, R) = \mathrm{Hom}_{\mathsf{Alg}_F}(F[G], R) = G(R), \quad R \in \mathsf{Alg}_L.$$

The group scheme G_L is called the *restriction* of G to L. For example we have

$$\mathbf{GL}_1(A)_L = \mathbf{GL}_1(A_L).$$

Subgroups. Let G be a group scheme over F, let $A = F[G]$, and let $J \subset A$ be a Hopf ideal. Consider the group scheme H represented by A/J and the group scheme homomorphism $\rho \colon H \to G$ induced by the natural map $A \to A/J$. Clearly, for any $R \in \mathsf{Alg}_F$ the homomorphism $\rho_R \colon H(R) \to G(R)$ is injective, hence we can identify $H(R)$ with a subgroup in $G(R)$. H is called a (*closed*) *subgroup of G* and ρ a *closed embedding*. A subgroup H in G is said to be *normal* if $H(R)$ is normal in $G(R)$ for all $R \in \mathsf{Alg}_F$.

(20.3) Examples. (1) For any group scheme G, the *augmentation Hopf ideal* $I = \ker(u) \subset F[G]$ corresponds to the trivial subgroup $\mathbf{1}$ since $F[G]/I \simeq F$.

(2) Let V be an F-vector space of finite dimension. For $v \in V$, $v \neq 0$, consider the functor

$$\mathbf{S}_v(R) = \{\, \alpha \in \mathrm{GL}(V_R) \mid \alpha(v) = v \,\} \subset \mathbf{GL}(V)(R).$$

To show that $\mathbf{S}_v$ is a subgroup of $\mathbf{GL}(V)$ (called the *stabilizer of v*) consider an F-basis $(v_1, v_2, \ldots, v_n)$ of V with $v = v_1$. Then $F[\mathbf{GL}(V)] = F[X_{ij}, \frac{1}{\det X}]$ and $\mathbf{S}_v$ corresponds to the Hopf ideal in this algebra generated by $X_{11} - 1$, X_{21}, $\ldots$, X_{n1}.

(3) Let $U \subset V$ be a subspace. Consider the functor

$$\mathbf{N}_U(R) = \{\, \alpha \in \mathrm{GL}(V_R) \mid \alpha(U_R) = U_R \,\} \subset \mathbf{GL}(V)(R).$$

To show that $\mathbf{N}_U$ is a subgroup in $\mathbf{GL}(V)$ (called the *normalizer of U*) consider an F-basis $(v_1, v_2, \ldots, v_n)$ of V such that $(v_1, v_2, \ldots, v_k)$ is a basis of U. Then $\mathbf{N}_U$ corresponds to the Hopf ideal in $F[X_{ij}, \frac{1}{\det X}]$ generated by the X_{ij} for $i = k + 1$, $\ldots$, n; $j = 1, 2, \ldots, k$.

Let $f \colon G \to H$ be a homomorphism of group schemes, with comorphism $f^* \colon F[H] \to F[G]$. The ideal $J = \ker(f^*)$ is a Hopf ideal in $F[H]$. It corresponds to a subgroup in H called the *image of f* and denoted $\mathrm{im}(f)$. Clearly, f decomposes as

$$G \xrightarrow{\bar{f}} \mathrm{im}(f) \xrightarrow{h} H$$

where h is a closed embedding. A homomorphism f is said to be *surjective* if f^* is injective. Thus the $\bar{f}$ above is surjective. Note that for a surjective homomorphism, the induced homomorphism of groups of points $G(R) \to H(R)$ need not be surjective. For example, the nth power homomorphism $f \colon \mathbf{G}_m \to \mathbf{G}_m$ is surjective since its comorphism $f^* \colon F[t] \to F[t]$ given by $f^*(t) = t^n$ is injective. However for $R \in \mathsf{Alg}_F$ the nth power homomorphism $f_R \colon R^\times \to R^\times$ is not in general surjective.

A *character* of a group scheme G over F is a group scheme homomorphism $G \to \mathbf{G}_m$. Characters of G form an abelian group denoted G^*.

A character $\chi \colon G \to \mathbf{G}_m$ is uniquely determined by the element $f = \chi^*(t) \in F[G]^\times$ which satisfies $c(f) = f \otimes f$. The elements $f \in F[G]^\times$ satisfying this condition are called *group-like elements*. The group-like elements form a subgroup of $F[G]^\times$ isomorphic to G^*.

Let A be a central simple algebra over F. The reduced norm homomorphism

$$\mathrm{Nrd} \colon \mathbf{GL}_1(A) \to \mathbf{G}_m$$

is a character of $\mathbf{GL}_1(A)$.

Fiber products, inverse images, and kernels. Let $f_i \colon G_i \to H$, $i = 1, 2$, be group scheme homomorphisms. The functor

$$(G_1 \times_H G_2)(R) = G_1(R) \times_{H(R)} G_2(R)$$
$$= \{\, (x, y) \in G_1(R) \times G_2(R) \mid (f_1)_R(x) = (f_2)_R(y) \,\}$$

is called the *fiber product of G_1 and G_2 over H*. It is represented by the Hopf algebra $F[G_1] \otimes_{F[H]} F[G_2]$.

(20.4) Examples. (1) For $H = 1$, we get the *product* $G_1 \times G_2$, represented by $F[G_1] \otimes_F F[G_2]$.

(2) Let $f \colon G \to H$ be a homomorphism of group schemes and let H' be the subgroup of H given by a Hopf ideal $J \subset F[H]$. Then $G \times_H H'$ is a subgroup in G given by the Hopf ideal $f^*(J) \cdot F[G]$ in $F[G]$, called the *inverse image* of H' and denoted $f^{-1}(H')$. Clearly

$$f^{-1}(H')(R) = \{\, g \in G(R) \mid f_R(g) \in H'(R) \,\}.$$

(3) The group $f^{-1}(1)$ in the preceding example is called the *kernel of f*, $\ker(f)$,

$$\ker(f)(R) = \{\, g \in G(R) \mid f_R(g) = 1 \,\}.$$

The kernel of f is the subgroup in G corresponding to the Hopf ideal $f^*(I) \cdot F[G]$ where I is the augmentation ideal in $F[H]$.

(4) If $f_i \colon H_i \to H$ are closed embeddings, $i = 1, 2$, then $H_1 \times_H H_2 = f_1^{-1}(H_2) = f_2^{-1}(H_1)$ is a subgroup of H_1 and of H_2, called the *intersection* $H_1 \cap H_2$ of H_1 and H_2.

(5) The kernel of the nth power homomorphism $\mathbf{G}_m \to \mathbf{G}_m$ is denoted $\boldsymbol{\mu}_n = \boldsymbol{\mu}_{n,F}$ and called the *group of nth roots of unity*. Clearly,

$$\boldsymbol{\mu}_n(R) = \{\, x \in R^\times \mid x^n = 1 \,\}$$

and $F[\boldsymbol{\mu}_n] = F[t]/(t^n - 1) \cdot F[t]$.

(6) Let A be a central simple algebra over F. The kernel of the reduced norm character $\mathrm{Nrd} \colon \mathbf{GL}_1(A) \to \mathbf{G}_m$ is denoted $\mathbf{SL}_1(A)$. If $A = \mathrm{End}(V)$ we write $\mathbf{SL}(V)$ for $\mathbf{SL}_1(A)$ and call the corresponding group scheme the *special linear group*.

(7) Let $\rho \colon G \to \mathbf{GL}(V)$ be a group scheme homomorphism and let $0 \neq v \in V$. The inverse image of the stabilizer $\rho^{-1}(\mathbf{S}_v)$ is denoted $\mathbf{Aut}_G(v)$,

$$\mathbf{Aut}_G(v)(R) = \{\, g \in G(R) \mid \rho_R(g)(v) = v \,\}.$$

(8) Let A be an F-algebra of finite dimension (not necessarily unital, commutative, associative). Let $V = \mathrm{Hom}_F(A \otimes_F A, A)$ and let $v \in V$ be the multiplication map in A. Consider the group scheme homomorphism

$$\rho \colon \mathbf{GL}(A) \to \mathbf{GL}(V)$$

given by

$$\rho_R(\alpha)(f)(a \otimes a') = \alpha\big(f\big(\alpha^{-1}(a) \otimes \alpha^{-1}(a')\big)\big).$$

The group scheme $\mathbf{Aut}_{\mathbf{GL}(A)}(v)$ for this v is denoted $\mathbf{Aut}_{\mathrm{alg}}(A)$. The group of R-points $\mathbf{Aut}_{\mathrm{alg}}(A)(R)$ coincides with the group $\mathrm{Aut}_R(A)$ of R-automorphisms of the R-algebra A_R.

The corestriction. Let L/F be a finite separable field extension and let G be a group scheme over L with $A = L[G]$. Consider the functor

$$(20.5) \qquad R_{L/F}(G) \colon \mathsf{Alg}_F \to \mathsf{Groups}, \quad R \mapsto G(R \otimes_F L).$$

(20.6) Lemma. *The functor $R_{L/F}(G)$ is a group scheme.*

Proof: Let $X = X(L)$ be the set of all F-algebra homomorphisms $\tau \colon L \to F_{\mathrm{sep}}$. The Galois group $\Gamma = \mathrm{Gal}(F_{\mathrm{sep}}/F)$ acts on X by $^\gamma \tau = \gamma \circ \tau$. For any $\tau \in X$ let A_τ be the tensor product $A \otimes_L F_{\mathrm{sep}}$ where F_{sep} is made an L-algebra via τ, so that $a\ell \otimes x = a \otimes \tau(\ell)x$ for $a \in A$, $\ell \in L$ and $x \in F_{\mathrm{sep}}$. For any $\gamma \in \Gamma$ and $\tau \in X$ the map

$$\widetilde{\gamma}_\tau \colon A_\tau \to A_{\gamma\tau}, \quad a \otimes x \mapsto a \otimes \gamma(x)$$

is a ring isomorphism such that $\widetilde{\gamma}_\tau(xu) = \gamma(x) \cdot \widetilde{\gamma}_\tau(u)$ for $x \in F_{\mathrm{sep}}$, $u \in A_\tau$.

Consider the tensor product $\widetilde{B} = \otimes_{\tau \in X} A_\tau$ over F_{sep}. The group Γ acts continuously on $\widetilde{B}$ by

$$\gamma(\otimes a_\tau) = \otimes a'_\tau \quad \text{where} \quad a'_{\gamma\tau} = \widetilde{\gamma}_\tau(a_\tau).$$

The F_{sep}-algebra $\widetilde{B}$ has a natural Hopf algebra structure arising from the Hopf algebra structure on A, and the structure on $\widetilde{B}$, compatible with the action of Γ. Hence the F-algebra $B = \widetilde{B}^\Gamma$ of Γ-stable elements is a Hopf algebra and by Lemma (18.1) we get $B \otimes_F F_{\mathrm{sep}} \simeq \widetilde{B}$.

We show that the F-algebra B represents the functor $R_{L/F}(G)$. For any F-algebra R we have a canonical isomorphism

$$\mathrm{Hom}_{\mathsf{Alg}_F}(B, R) \simeq \mathrm{Hom}_{\mathsf{Alg}_{F_{\mathrm{sep}}}}(\widetilde{B}, R \otimes_F F_{\mathrm{sep}})^\Gamma.$$

A Γ-equivariant homomorphism $\widetilde{B} \to R \otimes_F F_{\mathrm{sep}}$ is determined by a collection of F_{sep}-algebra homomorphisms $\{f_\tau \colon A_\tau \to R \otimes_F F_{\mathrm{sep}}\}_{\tau \in X}$ such that, for all $\gamma \in \Gamma$ and $\tau \in X$, the diagram

$$
\begin{array}{ccc}
A_\tau & \xrightarrow{\ f_\tau\ } & R \otimes_F F_{\mathrm{sep}} \\
\widetilde{\gamma}_\tau \downarrow & & \downarrow \mathrm{Id} \otimes \gamma \\
A_{\gamma\tau} & \xrightarrow{\ f_{\gamma\tau}\ } & R \otimes_F F_{\mathrm{sep}}
\end{array}
$$

commutes. For the restrictions $g_\tau = f_\tau|_A \colon A \to R \otimes_F F_{\mathrm{sep}}$ we have

$$(\mathrm{Id} \otimes \gamma) \cdot g_\tau = g_{\gamma\tau}.$$

Hence the image of g_τ is invariant under $\mathrm{Gal}(F_{\mathrm{sep}}/\tau L) \subset \Gamma$ and $\mathrm{im}\, g_\tau \subset R \otimes_F (\tau L)$. It is clear that the map

$$h = (\mathrm{Id} \otimes \tau)^{-1} \circ g_\tau \colon A \to R \otimes_F L$$

is independent of the choice of τ and is an L-algebra homomorphism. Conversely, any L-algebra homomorphism $h \colon A \to R \otimes_F L$ defines a collection of maps f_τ by

$$f_\tau(a \otimes x) = [(\mathrm{Id} \otimes \tau)h(a)]x.$$

Thus, $\mathrm{Hom}_{\mathsf{Alg}_F}(B, R) = \mathrm{Hom}_{\mathsf{Alg}_L}(A, R \otimes_F L) = G(R \otimes_F L)$. $\qquad \square$

The group scheme $R_{L/F}(G)$ is called the *corestriction of G from L to F*.

(20.7) Proposition. *The functors restriction and corestriction are adjoint to each other, i.e., for any group schemes H over F and G over L, there is a natural bijection*

$$\mathrm{Hom}_F\big(H, R_{L/F}(G)\big) \simeq \mathrm{Hom}_L(H_L, G).$$

Furthermore we have

$$[R_{L/F}(G)]_{F_{\mathrm{sep}}} \simeq \prod_{\tau \in X} G_\tau,$$

where $G_\tau = G_{F_{\mathrm{sep}}}$, with F_{sep} made an L-algebra via τ.

Proof: Both statements follow from the proof of Lemma (20.6).   □

(20.8) Example. For a finite dimensional L-vector space V, $R_{L/F}(\mathbf{V}) = \mathbf{V}_F$ where $V_F = V$ considered as an F-vector space.

(20.9) Remark. Sometimes it is convenient to consider group schemes over arbitrary étale F-algebras (not necessarily fields) as follows. An étale F-algebra L decomposes canonically into a product of separable field extensions,

$$L = L_1 \times L_2 \times \cdots \times L_n,$$

(see Proposition (18.3)) and a *group scheme G over L* is a collection of group schemes G_i over L_i. One then defines the corestriction $R_{L/F}(G)$ to be the product of the corestrictions $R_{L_i/F}(G_i)$. For example we have

$$\mathbf{GL}_1(L) = R_{L/F}(\mathbf{G}_{\mathrm{m},L})$$

for an étale F-algebra L. Proposition (20.7) also holds in this setting.

The connected component. Let A be a finitely generated commutative F-algebra and let $B \subset A$ be an étale F-subalgebra. Since the F_{sep}-algebra $B \otimes_F F_{\mathrm{sep}}$ is spanned by its idempotents (see Proposition (18.3)), $\dim_F B$ is bounded by the (finite) number of primitive idempotents of $A \otimes_F F_{\mathrm{sep}}$. Furthermore, if B_1, $B_2 \subset A$ are étale F-subalgebras, then $B_1 B_2$ is also étale in A, being a quotient of the tensor product $B_1 \otimes_F B_2$. Hence there exists a unique largest étale F-subalgebra in A, which we denote $\pi_0(A)$.

(20.10) Proposition. (1) *The subalgebra $\pi_0(A)$ contains all idempotents of A. Hence A is connected (i.e., the affine variety $\mathrm{Spec}\,A$ is connected, resp. A has no non-trivial idempotents) if and only if $\pi_0(A)$ is a field.*
(2) *For any field extension L/F, $\pi_0(A_L) = \pi_0(A)_L$.*
(3) *$\pi_0(A \otimes_F B) = \pi_0(A) \otimes_F \pi_0(B)$.*

Reference: See Waterhouse [305, §6.5].   □

(20.11) Proposition. *Let A be a finitely generated Hopf algebra over F. Then A is connected if and only if $\pi_0(A) = F$.*

Proof: The "if" implication is part of (1) of Proposition (20.10). We show the converse: the co-unit $u\colon A \to F$ takes the field $\pi_0(A)$ to F, hence $\pi_0(A) = F$.   □

We call an algebraic group scheme G over F *connected* if $F[G]$ is connected (i.e., $F[G]$ contains no non-trivial idempotents) or, equivalently, if $\pi_0(F[G]) = F$.

Let G be an algebraic group scheme over F and let $A = F[G]$. Then $c\big(\pi_0(A)\big)$, being an étale F-subalgebra in $A \otimes_F A$, is contained in $\pi_0(A \otimes_F A) = \pi_0(A) \otimes_F \pi_0(A)$ (see (3) of Proposition (20.10)). Similarly, we have $i\big(\pi_0(A)\big) \subset \pi_0(A)$. Thus, $\pi_0(A)$

is a Hopf subalgebra of A. The group scheme represented by $\pi_0(A)$ is denoted $\pi_0(G)$. There is a natural surjection $G \to \pi_0(G)$. Clearly, G is connected if and only if $\pi_0(G) = \mathbf{1}$. Propositions (20.10) and (20.11) then imply:

(20.12) Proposition. (1) *Let L/F be a field extension and let G be an algebraic group scheme over F. Then $\pi_0(G_L) = \pi_0(G)_L$. In particular, G_L is connected if and only if G is connected.*
(2) *$\pi_0(G_1 \times G_2) = \pi_0(G_1) \times \pi_0(G_2)$. In particular, the G_i are connected if and only if $G_1 \times G_2$ is connected.*

Let G be an algebraic group scheme over F and let $A = F[G]$. The co-unit homomorphism u maps all but one primitive idempotent of A to 0, so let e be the primitive idempotent with $u(e) = 1$. Since $\pi_0(A)e$ is a field, $\pi_0(A)e = F$ and $I_0 = \pi_0(A) \cdot (1 - e)$ is the augmentation ideal in $\pi_0(A)$. Denote the kernel of $G \to \pi_0(G)$ by G^0. It is represented by the algebra $A/A \cdot I_0 = A/A(1 - e) = Ae$. Since Ae is connected, G^0 is connected; it is called the *connected component* of G. We have $(G_1 \times G_2)^0 = G_1^0 \times G_2^0$ and for any field extension L/F, $(G_L)^0 = (G^0)_L$.

(20.13) Examples. (1) $\mathbf{GL}_1(A)$ is connected.
(2) For a central simple algebra A, $G = \mathbf{SL}_1(A)$ is connected since $F[G]$ is is the quotient of a polynomial ring modulo the ideal generated by the irreducible polynomial $\mathrm{Nrd}(X) - 1$.
(3) $\boldsymbol{\mu}_n$ is an example of a non-connected group scheme.

(20.14) Lemma (Homogeneity property of Hopf algebras). *Let A be a Hopf algebra which is finitely generated over $F = F_{\mathrm{alg}}$. Then for any pair of maximal ideals M, $M' \subset A$ there exists an F-algebra automorphism $\rho\colon A \to A$ such that $\rho(M) = M'$.*

Proof: We may assume that M' is the augmentation ideal in A. Let f be the canonical projection $A \to A/M = F$, and set $\rho = (\mathrm{Id}_A \otimes f) \circ c$. One checks that the map $\big(\mathrm{Id}_A \otimes (f \circ i)\big) \circ c$ is inverse to ρ, i.e., $\rho \in \mathrm{Aut}_F(A)$. Since $(u \otimes \mathrm{Id}_A) \circ c = \mathrm{Id}_A$, it follows that $u \circ \rho = (u \otimes f) \circ c = f \circ (u \otimes \mathrm{Id}_A) \circ c = f$ and $\rho(M) = \rho(\ker f) = \ker u = M'$. $\square$

Let $\mathrm{nil}(A)$ be the set of all nilpotent elements of A; it is an ideal of A, and equals the intersection of all the prime ideals of A. The algebra $A/\mathrm{nil}(A)$ is denoted by A_{red}.

(20.15) Proposition. *Let G be an algebraic group scheme over F and let $A = F[G]$. Then the following conditions are equivalent:*
(1) *G is connected.*
(2) *A is connected.*
(3) *A_{red} is connected.*
(4) *A_{red} is a domain.*

Proof: The implications $(1) \Leftrightarrow (2) \Leftrightarrow (3) \Leftarrow (4)$ are easy.

For $(1) \Rightarrow (4)$ we may assume that $F = F_{\mathrm{alg}}$. Since G is an algebraic group the scheme A is finitely generated. Hence the intersection of all maximal ideals in A containing a given prime ideal P is P (Bourbaki [51, Ch.V, §3, no. 4, Cor. to Prop. 8 (ii)]) and there is a maximal ideal containing exactly one minimal prime ideal. By the lemma above, each maximal ideal contains exactly one minimal prime ideal. Hence any two different minimal prime ideals P and P' are coprime:

$P + P' = A$. Let P_1, P_2, ..., P_n be all minimal prime ideals. Since $\bigcap P_i = \mathrm{nil}(A)$, we have $A_{\mathrm{red}} = A/\mathrm{nil}(A) \simeq \prod A/P_i$ by the Chinese Remainder Theorem. By assumption A_{red} is connected, hence $n = 1$ and $A_{\mathrm{red}} = A/P_1$ is a domain. $\square$

Constant and étale group schemes. Let H be a finite (abstract) group. Consider the algebra

$$A = \mathrm{Map}(H, F)$$

of all functions $H \to F$. For $h \in H$, let e_h be the characteristic function of $\{h\}$; this map is an idempotent in A, and we have $A = \prod_{h \in H} F \cdot e_h$. A Hopf algebra structure on A is given by

$$c(e_h) = \sum_{xy=h} e_x \otimes e_y, \quad i(e_h) = e_{h^{-1}}, \quad u(e_h) = \begin{cases} 1 & \text{if } h = 1, \\ 0 & \text{if } h \neq 1. \end{cases}$$

The group scheme over F represented by A is denoted H_{const} and called the *constant group scheme associated to H*. For any connected F-algebra $R \in \mathsf{Alg}_F$, $H_{\mathrm{const}}(R) = H$.

A group scheme G over F is said to be *étale* if $F[G]$ is an étale F-algebra. For example, constant group schemes are étale and, for any algebraic group scheme G, the group scheme $\pi_0(G)$ is étale. If G is étale, then $G(F_{\mathrm{sep}})$ is a finite (discrete) group with a continuous action of $\Gamma = \mathrm{Gal}(F_{\mathrm{sep}}/F)$. Conversely, given a finite group H with a continuous Γ-action by group automorphisms, we have a Γ-action on the F_{sep}-algebra $A = \mathrm{Map}(H, F_{\mathrm{sep}})$. Let H_{et} be the étale group scheme represented by the (étale) Hopf algebra A^Γ. Subgroups of H_{et} are étale and correspond to Γ-stable subgroups of H.

(20.16) Proposition. *The two functors*

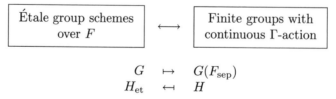

$$
\begin{array}{ccc}
G & \mapsto & G(F_{\mathrm{sep}}) \\
H_{\mathrm{et}} & \leftarrow\!\shortmid & H
\end{array}
$$

are mutually inverse equivalences of categories. In this equivalence constant group schemes correspond to finite groups with trivial Γ-action.

Proof: This follows from Theorem (18.4). $\square$

Diagonalizable group schemes and group schemes of multiplicative type. Let H be an (abstract) abelian group, written multiplicatively. We have a structure of a Hopf algebra on the group algebra $F\langle H \rangle$ over F given by $c(h) = h \otimes h$, $i(h) = h^{-1}$ and $u(h) = 1$. The group scheme represented by $F\langle H \rangle$ is said to be *diagonalizable* and is denoted H_{diag}. Clearly,

$$H_{\mathrm{diag}}(R) = \mathrm{Hom}(H, R^\times), \quad R \in \mathsf{Alg}_F.$$

The group-like elements in $F\langle H \rangle$ are of the form $h \otimes h$ for $h \in H$. Hence the character group $(H_{\mathrm{diag}})^*$ is naturally isomorphic to H. For example, we have

$$\mathbb{Z}_{\mathrm{diag}} = \mathbf{G}_{\mathrm{m}}, \quad (\mathbb{Z}/n\mathbb{Z})_{\mathrm{diag}} = \boldsymbol{\mu}_n.$$

A group scheme G over F is said to be of *multiplicative type* if $G_{\mathrm{sep}}(= G_{F_{\mathrm{sep}}})$ is diagonalizable. In particular, diagonalizable group schemes are of multiplicative type. Let G be of multiplicative type. The character group $(G_{\text{sep}})^*$ has a natural continuous action of $\Gamma = \text{Gal}(F_{\text{sep}}/F)$. To describe this action we observe that the group of characters $(G_{\text{sep}})^*$ is isomorphic to the group of group-like elements in $F_{\text{sep}}[G_{\text{sep}}]$. The action is induced from the natural action on action on $F_{\text{sep}}[G_{\text{sep}}]$. Conversely, given an abelian group H with a continuous Γ-action, the Hopf algebra of Γ-stable elements in $F_{\text{sep}}[H_{\text{diag}}] = F_{\text{sep}}\langle H \rangle$ represents a group scheme of multiplicative type which we denote H_{mult}. Clearly,

$$H_{\text{mult}}(R) = \text{Hom}_\Gamma\big(H, (R \otimes_F F_{\text{sep}})^\times\big).$$

(20.17) Proposition. *The two contravariant functors*

Group schemes of multiplicative type over F	$\longleftrightarrow$	Abelian groups with continuous Γ-action

$$\begin{array}{ccc} G & \mapsto & (G_{\text{sep}})^* \\ H_{\text{mult}} & \leftmapsto & H \end{array}$$

define an equivalence of categories. Under this equivalence diagonalizable group schemes correspond to abelian groups with trivial Γ-action. $\qquad\square$

An *algebraic torus* is a group scheme of multiplicative type H_{mult} where H is a free abelian group of finite rank. A torus T is said to be *split* if it is a diagonalizable group scheme, i.e., $T = H_{\text{diag}} \simeq (\mathbb{Z}^n)_{\text{diag}} = \mathbf{G}_{\text{m}} \times \cdots \times \mathbf{G}_{\text{m}}$ (n factors) is isomorphic to the group scheme of diagonal matrices in $\mathbf{GL}_n(F)$. Any torus T is split over F_{sep}.

Cartier Duality. Let H be a finite abelian (abstract) group with a continuous Γ-action and let $\Gamma = \text{Gal}(F_{\text{sep}}/F)$. One can associate two group schemes to H: H_{et} and H_{mult}. We discuss the relation between these group schemes. A group scheme G over F is called *finite* if $\dim_F F[G] < \infty$. The *order of* G is $\dim_F F[G]$. For example an étale group scheme G is finite. Its order is the order of $G(F_{\text{sep}})$. Let G be a finite commutative group scheme over F; then $A = F[G]$ is of finite dimension. Consider the dual F-vector space $A^* = \text{Hom}_F(A, F)$. The duals of the five structure maps on A, namely the unit map $e\colon F \to A$, the multiplication $m\colon A \otimes_F A \to A$ and the maps c, i, u defining the Hopf algebra structure on A, yield five maps which define a Hopf algebra structure on A^*. The associated group scheme is denoted G^D and is called *Cartier dual* of G. Thus, $F[G^D] = F[G]^*$ and $G^{DD} = G$.

Elements of the group $(G^D)(F)$ are represented by F-algebra homomorphisms $F[G]^* \to F$ which, as is easily seen, are given by group-like elements of $F[G]$. Hence, $G^D(F) \simeq G^*$, the character group of G.

Cartier duality is an involutory contravariant functor D on the category of finite commutative group schemes over F.

The restriction of D gives an equivalence of categories

Étale commutative group schemes over F	$\longleftrightarrow$	Finite group schemes of multiplicative type over F

More precisely, if H is a finite abelian (abstract) group with a continuous Γ-action, then

$$(H_{\text{et}})^D = H_{\text{mult}}, \quad (H_{\text{mult}})^D = H_{\text{et}}.$$

(20.18) Example.

$$(\mathbb{Z}/n\mathbb{Z})^D = \boldsymbol{\mu}_n, \quad \boldsymbol{\mu}_n^D = \mathbb{Z}/n\mathbb{Z}.$$

(We write $\mathbb{Z}/n\mathbb{Z}$ for $(\mathbb{Z}/n\mathbb{Z})_{\text{const}}$.)

§21. The Lie Algebra and Smoothness

Let M be an A-module. A *derivation* D *of* A *into* M is an F-linear map $D\colon A \to M$ such that

$$D(ab) = a \cdot D(b) + b \cdot D(a).$$

We set $\operatorname{Der}(A, M)$ for the A-module of all derivations of A into M.

21.A. The Lie algebra of a group scheme. Let G be an algebraic group scheme over F and let $A = F[G]$. A derivation $D \in \operatorname{Der}(A, A)$ is said to be *left-invariant* if $c \circ D = (id \otimes D) \circ c$. The F-vector space of left-invariant derivations is denoted $\operatorname{Lie}(G)$ and is called the *Lie algebra of* G. The Lie algebra structure on $\operatorname{Lie}(G)$ is given by $[D_1, D_2] = D_1 \circ D_2 - D_2 \circ D_1$.

Denote by $F[\varepsilon]$ the F-algebra of *dual numbers*, i.e., $F[\varepsilon] = F \cdot 1 \oplus F \cdot \varepsilon$ with multiplication given by $\varepsilon^2 = 0$. There is a unique F-algebra homomorphism $\kappa\colon F[\varepsilon] \to F$ with $\kappa(\varepsilon) = 0$. The kernel of $G(F[\varepsilon]) \xrightarrow{G(\kappa)} G(F)$ carries a natural F-vector space structure: addition is the multiplication in $G(F[\varepsilon])$ and scalar multiplication is defined by the formula $a \cdot g = G(\ell_a)(g)$ for $g \in G(F[\varepsilon])$, $a \in F$, where $\ell_a\colon F[\varepsilon] \to F[\varepsilon]$ is the F-algebra homomorphism defined by $\ell_a(\varepsilon) = a\varepsilon$.

(21.1) Proposition. *There exist natural isomorphisms between the following F-vector spaces:*

(1) $\operatorname{Lie}(G)$,
(2) $\operatorname{Der}(A, F)$ *where F is considered as an A-module via the co-unit map $u\colon A \to F$,*
(3) $(I/I^2)^*$ *where $I \subset A$ is the augmentation ideal,*
(4) $\ker\big(G(F[\varepsilon]) \xrightarrow{G(\kappa)} G(F)\big)$.

Proof: (1) $\Leftrightarrow$ (2) If $D \in \operatorname{Lie}(G)$, then $u \circ D \in \operatorname{Der}(A, F)$. Conversely, for $d \in \operatorname{Der}(A, F)$ one has $D = (\operatorname{Id} \otimes d) \circ c \in \operatorname{Lie}(G)$.

(2) $\Leftrightarrow$ (3) Any derivation $d\colon A \to F$ satisfies $d(I^2) = 0$, hence the restriction $d|_I$ induces an F-linear form on I/I^2. Conversely, if $f\colon I \to F$ is an F-linear map such that $f(I^2) = 0$, then $d\colon A = F \cdot 1 \oplus I \to F$ given by $d(\alpha + x) = f(x)$ is a derivation.

(2) $\Leftrightarrow$ (4) An element f of $\ker G(\kappa)$ is an F-algebra homomorphism $f\colon A \to F[\varepsilon]$ of the form $f(a) = u(a) + d(a) \cdot \varepsilon$ where $d \in \operatorname{Der}(A, F)$. $\qquad\square$

(21.2) Corollary. *If G is an algebraic group scheme, then $\dim_F \operatorname{Lie}(G) < \infty$.*

Proof: Since A is noetherian, I is a finitely generated ideal, hence I/I^2 is finitely generated over $A/I = F$. $\qquad\square$

(21.3) Proposition. *Let G be an algebraic group scheme over F and let $A = F[G]$. Then $\operatorname{Der}(A, A)$ is a finitely generated free A-module and*

$$\operatorname{rank}_A \operatorname{Der}(A, A) = \dim \operatorname{Lie}(G).$$

Proof: Let G be an algebraic group scheme over F. The map

$$\mathrm{Der}(A, F) \simeq (I/I^2)^* \to \mathrm{Der}(A, A), \quad d \mapsto (id \otimes d) \circ c$$

used in the proof of Proposition (21.1) extends to an isomorphism of A-modules (Waterhouse [305, 11.3.])

$$A \otimes_F (I/I^2)^* \xrightarrow{\sim} \mathrm{Der}(A, A). \qquad \square$$

The Lie algebra structure on $\mathrm{Lie}(G)$ can be recovered as follows (see Waterhouse [305, p. 94]). Consider the commutative F-algebra $R = F[\varepsilon, \varepsilon']$ with $\varepsilon^2 = 0 = \varepsilon'^2$. From $d, d' \in \mathrm{Der}(A, F)$ we build two elements $g = u + d \cdot \varepsilon$ and $g' = u + d' \cdot \varepsilon'$ in $G(R)$. A computation of the commutator of g and g' in $G(R)$ yields $gg'g^{-1}g'^{-1} = u + d'' \cdot \varepsilon\varepsilon'$ where $d'' = [d, d']$ in $\mathrm{Lie}(G)$.

Any homomorphism of group schemes $f \colon G \to H$ induces a commutative diagram

$$
\begin{array}{ccc}
G(F[\varepsilon]) & \xrightarrow{f_{F[\varepsilon]}} & H(F[\varepsilon]) \\
{\scriptstyle G(\kappa)}\big\downarrow & & \big\downarrow{\scriptstyle H(\kappa)} \\
G(F) & \xrightarrow{f_F} & H(F)
\end{array}
$$

and hence defines an F-linear map $df \colon \mathrm{Lie}(G) \to \mathrm{Lie}(H)$, which is a Lie algebra homomorphism, called the *differential* of f. If f is a closed embedding (i.e., G is a subgroup of H) then df is injective and identifies $\mathrm{Lie}(G)$ with a Lie subalgebra of $\mathrm{Lie}(H)$.

In the next proposition we collect some properties of the Lie algebra; we assume that all group schemes appearing here are algebraic.

(21.4) Proposition. (1) *For any field extension L/F, $\mathrm{Lie}(G_L) \simeq \mathrm{Lie}(G) \otimes_F L$.*
(2) *Let $f_i \colon G_i \to H$ be group scheme homomorphisms, $i = 1, 2$. Then*

$$\mathrm{Lie}(G_1 \times_H G_2) = \mathrm{Lie}(G_1) \times_{\mathrm{Lie}(H)} \mathrm{Lie}(G_2).$$

In particular:
    (a) *For a homomorphism $f \colon G \to H$ and a subgroup $H' \subset H$,*

$$\mathrm{Lie}\big(f^{-1}(H')\big) \simeq (df)^{-1}\big(\mathrm{Lie}(H')\big).$$

    (b) $\mathrm{Lie}(\ker f) = \ker(df)$.
    (c) $\mathrm{Lie}(G_1 \times G_2) = \mathrm{Lie}(G_1) \times \mathrm{Lie}(G_2)$.
(3) *For any finite separable field extension L/F and any algebraic group scheme G over L, $\mathrm{Lie}\big(R_{L/F}(G)\big) = \mathrm{Lie}(G)$ as F-algebras.*
(4) $\mathrm{Lie}(G) = \mathrm{Lie}(G^0)$.

Reference: See Waterhouse [305, Chap. 12]. $\qquad \square$

(21.5) Examples. (1) Let $G = \mathbf{V}$, V a vector space over F. The elements of $\ker G(\kappa)$ have the form $v \cdot \varepsilon$ with $v \in V$ arbitrary. Hence $\mathrm{Lie}(G) = V$ with the trivial Lie product. In particular, $\mathrm{Lie}(\mathbf{G}_a) = F$.

(2) Let $G = \mathbf{GL}_1(A)$ where A is a finite dimensional associative F-algebra. Elements of $\ker G(\kappa)$ are of the form $1 + a \cdot \varepsilon$, $a \in A$. Hence $\mathrm{Lie}\big(\mathbf{GL}_1(A)\big) = A$. One can compute the Lie algebra structure using $R = F[\varepsilon, \varepsilon']$: the commutator of

$1 + a \cdot \varepsilon$ and $1 + a' \cdot \varepsilon'$ in $G(R)$ is $1 + (aa' - a'a) \cdot \varepsilon\varepsilon'$, hence the Lie algebra structure on A is given by $[a, a'] = aa' - a'a$. In particular,

$$\mathrm{Lie}\big(\mathbf{GL}(V)\big) = \mathrm{End}(V), \quad \mathrm{Lie}(\mathbf{G}_{\mathrm{m}}) = F.$$

(3) For a central simple algebra A over F, the differential of the reduced norm homomorphism Nrd: $\mathbf{GL}_1(A) \to \mathbf{G}_{\mathrm{m}}$ is the reduced trace Trd: $A \to F$ since $\mathrm{Nrd}(1 + a \cdot \varepsilon) = 1 + \mathrm{Trd}(a) \cdot \varepsilon$. Hence,

$$\mathrm{Lie}\big(\mathbf{SL}_1(A)\big) = \{\, a \in A \mid \mathrm{Trd}(a) = 0 \,\} \subset A.$$

(4) The differential of the nth power homomorphism $\mathbf{G}_{\mathrm{m}} \to \mathbf{G}_{\mathrm{m}}$ is multiplication by $n \colon F \to F$ since $(1 + a \cdot \varepsilon)^n = 1 + na \cdot \varepsilon$. Hence

$$\mathrm{Lie}(\boldsymbol{\mu}_n) = \begin{cases} 0 & \text{if char } F \text{ does not divide } n, \\ F & \text{otherwise.} \end{cases}$$

(5) If G is an étale group scheme, then $\mathrm{Lie}(G) = 0$ since $\mathrm{Der}(A, A) = 0$ for any étale F-algebra A.

(6) Let H be an (abstract) abelian group with a continuous Γ-action and let $G = H_{\mathrm{mult}}$. An element in $\ker G(\kappa)$ has the form $1 + f \cdot \varepsilon$ where $f \in \mathrm{Hom}_\Gamma(H, F_{\mathrm{sep}})$. Hence,

$$\mathrm{Lie}(G) = \mathrm{Hom}_\Gamma(H, F_{\mathrm{sep}}) = \mathrm{Hom}_\Gamma(G^*_{\mathrm{sep}}, F_{\mathrm{sep}}).$$

(7) Let $\mathbf{S}_v \subset \mathbf{GL}(V)$ be the stabilizer of $0 \neq v \in V$. An element of $\ker \mathbf{S}_v(\kappa)$ has the form $1 + \alpha \cdot \varepsilon$ where $\alpha \in \mathrm{End}(V)$ and $(1 + \alpha \cdot \varepsilon)(v) = v$, i.e., $\alpha(v) = 0$. Thus,

$$\mathrm{Lie}(\mathbf{S}_v) = \{\, \alpha \in \mathrm{End}(V) \mid \alpha(v) = 0 \,\}.$$

(8) Let $\rho \colon G \to \mathbf{GL}(V)$ be a homomorphism, $0 \neq v \in V$. Then

$$\mathrm{Lie}\big(\mathbf{Aut}_G(v)\big) = \{\, x \in \mathrm{Lie}(G) \mid (df)(x)(v) = 0 \,\}.$$

(9) Let $G = \mathbf{Aut}_{\mathrm{alg}}(A)$ where A is a finite dimensional F-algebra and let

$$\rho \colon \mathbf{GL}(A) \to \mathbf{GL}\big(\mathrm{Hom}(A \otimes_F A, A)\big)$$

be as in Example (20.4), (8). By computing over $F[\varepsilon]$, one finds that the differential

$$d\rho \colon \mathrm{End}(A) \to \mathrm{End}\big(\mathrm{Hom}(A \otimes_F A, A)\big)$$

is given by the formula

$$(d\rho)(\alpha)(\phi)(a \otimes a') = \alpha\big(\phi(a \otimes a')\big) - \phi\big(\alpha(a) \otimes a'\big) - \phi\big(a \otimes \alpha(a')\big),$$

hence the condition $(d\rho)(\alpha)(v) = 0$, where v is the multiplication map, is equivalent to $\alpha \in \mathrm{Der}(A, A)$, i.e.,

$$\mathrm{Lie}\big(\mathbf{Aut}_{\mathrm{alg}}(A)\big) = \mathrm{Der}(A, A).$$

(10) Let $\mathbf{N}_U$ be the normalizer of a subspace $U \subset V$ (see Example (20.3.3)). Since the condition $(1 + \alpha \cdot \varepsilon)(u + u' \cdot \varepsilon) \in U + U \cdot \varepsilon$, for $\alpha \in \mathrm{End}(V)$, u, $u' \in U$, is equivalent to $\alpha(u) \in U$, we have

$$\mathrm{Lie}(\mathbf{N}_U) = \{\, \alpha \in \mathrm{End}(V) \mid \alpha(U) \subset U \,\}.$$

The dimension. Let G be an algebraic group scheme over F. If G is connected, then $F[G]_{\mathrm{red}}$ is a domain (Proposition (20.15)). The *dimension* $\dim G$ of G is the transcendence degree over F of the field of fractions of $F[G]_{\mathrm{red}}$. If G is not connected, we define $\dim G = \dim G^0$.

(21.6) Examples. (1) $\dim \mathbf{V} = \dim_F V$.
(2) $\dim \mathbf{GL}_1(A) = \dim_F A$.
(3) $\dim \mathbf{SL}_1(A) = \dim_F A - 1$.
(4) $\dim \mathbf{G}_{\mathrm{m}} = \dim \mathbf{G}_{\mathrm{a}} = 1$.
(5) $\dim \boldsymbol{\mu}_n = 0$.

The main properties of the dimension are collected in the following

(21.7) Proposition. (1) $\dim G = \dim F[G]$ *(Krull dimension)*.
(2) G *is finite if and only if* $\dim G = 0$.
(3) *For any field extension* L/F, $\dim(G_L) = \dim G$.
(4) $\dim(G_1 \times G_2) = \dim G_1 + \dim G_2$.
(5) *For any separable field extension* L/F *and any algebraic group scheme* G *over* L, $\dim R_{L/F}(G) = [L:F] \cdot \dim G$.
(6) *Let* G *be a connected algebraic group scheme with* $F[G]$ *reduced (i.e.,* $F[G]$ *has no nilpotent elements) and let* H *be a proper subgroup of* G. *Then* $\dim H < \dim G$.

Proof: (1) follows from Matsumura [179, Th. 5.6]; (3) and (6) are immediate consequences of the definition.

(2) Set $A = F[G]$. Since A is noetherian and $\dim A = 0$, A is artinian. But A is also finitely generated, hence $\dim_F A < \infty$.

(4) We may assume that the G_i are connected and $F = F_{\mathrm{alg}}$. Let L_i be the field of fractions of $F[G_i]_{\mathrm{red}}$. Since $F = F_{\mathrm{alg}}$, the ring $L_1 \otimes_F L_2$ is an integral domain and the field of fractions of $F[G_1 \times G_2]_{\mathrm{red}}$ is the field of fractions E of $L_1 \otimes_F L_2$. Thus, $\dim(G_1 \times G_2) = \mathrm{tr.deg}_F(E) = \mathrm{tr.deg}_F(L_1) + \mathrm{tr.deg}_F(L_2) = \dim G_1 + \dim G_2$.

(5) We have

$$\dim R_{L/F}(G) = \dim\big(R_{L/F}(G)_{\mathrm{sep}}\big)$$
$$= \dim(\prod_{\tau \in X} G_\tau) = \sum_{\tau \in X} \dim G_\tau = [L:F] \cdot \dim G$$

with the notation of (20.5). □

Smoothness. Let S be a commutative noetherian local ring with maximal ideal M and residue field $K = S/M$. It is known (see Matsumura [179, p. 78]) that

$$\dim_K(M/M^2) \geq \dim S.$$

The ring S is said to be *regular* if equality holds. Recall that regular local rings are integral domains (Matsumura [179, Th. 19.4]).

(21.8) Lemma. *For any algebraic group scheme* G *over* F *we have* $\dim_F \mathrm{Lie}(G) \geq \dim G$. *Equality holds if and only if the local ring* $F[G]_I$ *is regular where* I *is the augmentation ideal of* $F[G]$.

Proof: Let $A = F[G]$. The augmentation ideal $I \subset A$ is maximal with $A/I = F$. Hence, for the localization $S = A_I$ with respect to the maximal ideal $M = IS$ we have $S/M = F$ and

$$\dim_F \mathrm{Lie}(G) = \dim_F(I/I^2) = \dim_F(M/M^2) \geq \dim S = \dim A = \dim G,$$

proving the lemma. □

(21.9) Proposition. *Let G be an algebraic group scheme over F and let $A = F[G]$. Then the following conditions are equivalent:*

(1) A_L *is reduced for any field extension L/F.*
(2) $A_{F_{\mathrm{alg}}}$ *is reduced.*
(3) $\dim_F\big(\mathrm{Lie}(G)\big) = \dim G$.

If F is perfect, these conditions are also equivalent to

(4) A *is reduced.*

Proof: $(1) \Rightarrow (2)$ is trivial.

$(2) \Rightarrow (3)$ We may assume that $F = F_{\mathrm{alg}}$ and that G is connected (since $F[G^0]$ is a direct factor of A and hence is reduced). By (20.15) A is an integral domain. Let K be its field of fractions. The K-space of derivations $\mathrm{Der}(K,K)$ is isomorphic to $\mathrm{Der}(A,A) \otimes_A K$, hence by (21.3)

$$\dim_F\big(\mathrm{Lie}(G)\big) = \mathrm{rank}_A\big(\mathrm{Der}(A,A)\big) = \dim_K \mathrm{Der}(K,K).$$

But the latter is known to equal $\mathrm{tr.deg}_F(K) = \dim G$.

$(3) \Rightarrow (1)$ We may assume that $L = F = F_{\mathrm{alg}}$. By (21.8) the ring A_I is regular hence is an integral domain and is therefore reduced. By the homogeneity property (see (20.14)) A_M is reduced for every maximal ideal $M \subset A$. Hence, A is reduced.

Finally, assume F is perfect. Since the tensor product of reduced algebras over a perfect field is reduced (see Bourbaki [49, Ch.V, §15, no. 5, Théorème 3]), it follows that $A_{F_{\mathrm{alg}}}$ is reduced if A is reduced. The converse is clear. $\qquad\square$

An algebraic group scheme G is said to be *smooth* if G satisfies the equivalent conditions of Proposition (21.9). Smooth algebraic group schemes are also called *algebraic groups*.

(21.10) Proposition. (1) *Let G be an algebraic group scheme over F and let L/F be a field extension. Then G_L is smooth if and only if G is smooth.*
(2) *If G_1, G_2 are smooth then $G_1 \times G_2$ is smooth.*
(3) *If $\mathrm{char}\, F = 0$, all algebraic group schemes are smooth.*
(4) *An algebraic group scheme is smooth if and only if its connected component G_0 is smooth.*

Proof: (1) and (2) follow from the definition of smoothness and (3) (which is a result due to Cartier) is given in Waterhouse [305, §11.4]. (4) follows from the proof of (21.9). $\qquad\square$

(21.11) Examples. (1) $\mathbf{GL}_1(A)$, $\mathbf{SL}_1(A)$ are smooth for any central simple F-algebra A.
(2) Étale group schemes are smooth.
(3) H_{mult} is smooth if and only if H has no p-torsion where $p = \mathrm{char}\, F$.

Let F be a perfect field (for example $F = F_{\mathrm{alg}}$), let G be an algebraic group scheme over F and let $A = F[G]$. Since the ring $A_{\mathrm{red}} \otimes_F A_{\mathrm{red}}$ is reduced, the comultiplication c factors through

$$c_{\mathrm{red}} \colon A_{\mathrm{red}} \to A_{\mathrm{red}} \otimes_F A_{\mathrm{red}},$$

making A_{red} a Hopf algebra. The corresponding smooth algebraic group scheme G_{red} is called the *smooth algebraic group* associated to G. Clearly G_{red} is a subgroup of G and $G_{\mathrm{red}}(R) = G(R)$ for any reduced algebra $R \in \mathsf{Alg}_F$.

(21.12) Remark. The classical notion of an (affine) algebraic group over an algebraically closed field, as an affine variety Spec A endowed with a group structure corresponds to reduced finitely generated Hopf algebras A, i.e., coincides with the notion of a smooth algebraic group scheme. This is why we call such group schemes algebraic groups. Therefore, for any algebraic group scheme G, one associates a (classical) algebraic group $(G_{\mathrm{alg}})_{\mathrm{red}}$ over F_{alg}. The notions of dimension, connectedness, Lie algebra, ..., given here then coincide with the classical ones (see Borel [45], Humphreys [125]).

§22. Factor Groups

22.A. Group scheme homomorphisms.

The injectivity criterion. We will use the following

(22.1) Proposition. *Let $A \subset B$ be Hopf algebras. Then B is faithfully flat over A.*

Reference: See Waterhouse [305, §14.1].   □

A group scheme homomorphism $f \colon G \to H$ is said to be *injective* if $\ker f = \mathbf{1}$, or equivalently, if $f_R \colon G(R) \to H(R)$ is injective for all $R \in \mathsf{Alg}_F$.

(22.2) Proposition. *Let $f \colon G \to H$ be a homomorphism of algebraic group schemes. The following conditions are equivalent*:

(1) *f is injective.*
(2) *f is a closed embedding (i.e., f^* is surjective).*
(3) *$f_{\mathrm{alg}} \colon G(F_{\mathrm{alg}}) \to H(F_{\mathrm{alg}})$ is injective and df is injective.*

Proof: (1) $\Rightarrow$ (2) By replacing H by the image of f we may assume that

$$f^* \colon A = F[H] \to F[G] = B$$

is injective. The elements in $G(B \otimes_A B)$ given by the two natural maps $B \rightrightarrows B \otimes_A B$ have the same image in $H(B \otimes_A B)$, hence they are equal. But B is faithfully flat over A, hence the equalizer of $B \rightrightarrows B \otimes_A B$ is A. Thus, $A = B$.

The implication (2) $\Rightarrow$ (3) is clear.

(3) $\Rightarrow$ (1) Let $N = \ker f$. We have $\mathrm{Lie}(N) = \ker(df) = 0$, hence by Lemma (21.8) $\dim N \leq \dim \mathrm{Lie}(N) = 0$ and N is finite (Proposition (21.7)). Then it follows from Proposition (21.9) that N is smooth and hence étale, $N = H_{\mathrm{et}}$ where $H = N(F_{\mathrm{sep}})$ (see 20.16). But $N(F_{\mathrm{sep}}) \subset N(F_{\mathrm{alg}}) = \ker(f_{\mathrm{alg}}) = 1$, hence $N = 1$ and f is injective.   □

The surjectivity criterion.

(22.3) Proposition. *Let $f \colon G \to H$ be a homomorphism of algebraic group schemes. If H is smooth, the following conditions are equivalent*:

(1) *f is surjective (i.e., f^* is injective).*
(2) *$f_{\mathrm{alg}} \colon G(F_{\mathrm{alg}}) \to H(F_{\mathrm{alg}})$ is surjective.*

Proof: (1) $\Rightarrow$ (2) We may assume that $F = F_{\mathrm{alg}}$. Since $B = F[G]$ is faithfully flat over $A = F[H]$, any maximal ideal of A is the intersection with A of a maximal ideal of B (Bourbaki [50, Ch.1, §3, no. 5, Prop. 8 (iv)]), or equivalently, any F-algebra homomorphism $A \to F$ can be extended to B. (Note that we are not using the smoothness assumption here.)

(2) $\Rightarrow$ (1) Assume $F = F_{\mathrm{alg}}$. Any F-algebra homomorphism $A \to F$ factors through f^*, hence all maximal ideals in A contain $\ker f$. But the intersection of

all maximal ideals in A is zero since A is reduced, therefore f^* is injective and f is surjective.   □

(22.4) Proposition. *Let $f\colon G \to H$ be a surjective homomorphism of algebraic group schemes.*

(1) *If G is connected (resp. smooth), then H is connected (resp. smooth).*
(2) *Let H' be a subgroup of H. Then the restriction of f to $f^{-1}(H')$ is a surjective homomorphism $f^{-1}(H') \to H'$.*

Proof: (1) is clear. For (2), let $J \subset A = F[H]$ be the Hopf ideal corresponding to H'. Hence the ideal $J' = f^*(J) \cdot B \subset B = F[G]$ corresponds to $f^{-1}(H')$, and the homomorphism $F[H'] = A/J \to B/J' = F[f^{-1}(H')]$ is injective since $(f^*)^{-1}(J') = J$ (see Bourbaki [50, Ch.I, §3, no. 5, Prop. 8 (ii)]).   □

The isomorphism criterion. Propositions (22.2) and (22.3) imply that

(22.5) Proposition. *Let $f\colon G \to H$ be a homomorphism of algebraic group schemes with H smooth. Then the following conditions are equivalent*:

(1) *f is an isomorphism.*
(2) *f is injective and surjective.*
(3) *$f_{\mathrm{alg}}\colon G(F_{\mathrm{alg}}) \to H(F_{\mathrm{alg}})$ is an isomorphism and df is injective.*   □

(22.6) Example. Let $f\colon \mathbf{G}_{\mathrm{m}} \to \mathbf{G}_{\mathrm{m}}$ be the pth-power homomorphism where $p = \operatorname{char} F$. Clearly, f_{alg} is an isomorphism, but f is not since $df = 0$.

Factor group schemes.

(22.7) Proposition. *Let $f\colon G \to H$ be a surjective homomorphism of group schemes with kernel N. Then any group scheme homomorphism $f'\colon G \to H'$ vanishing on N factors uniquely through f.*

Proof: Let $A = F[H]$ and $B = F[G]$. The two natural homomorphisms $B \rightrightarrows B \otimes_A B$, being elements in $G(B \otimes_A B)$, have the same image in $H(B \otimes_A B)$ and hence they are congruent modulo $N(B \otimes_A B)$. Hence the two composite maps

$$F[H'] \xrightarrow{\ f'^*\ } B \rightrightarrows B \otimes_A B$$

coincide. By the faithful flatness of B over A the equalizer of $B \rightrightarrows B \otimes_A B$ is A, thus the image of f'^* lies in A.   □

The proposition shows that a surjective homomorphism $f\colon G \to H$ is uniquely determined (up to isomorphism) by G and the normal subgroup N. We write $H = G/N$ and call H the *factor group scheme G modulo N*.

(22.8) Proposition. *Let G be a group scheme and let N be a normal subgroup in G. Then there is a surjective homomorphism $G \to H$ with the kernel N, i.e., the factor group scheme G/N exists.*

Reference: See Waterhouse [305, §16.3].   □

Exact sequences. A sequence of homomorphisms of group schemes

$$(22.9) \qquad\qquad 1 \to N \xrightarrow{f} G \xrightarrow{g} H \to 1$$

is called *exact* if f induces an isomorphism of N with $\ker(g)$ and g is surjective or, equivalently, f is injective and $H \simeq G/\operatorname{im}(f)$. For any group scheme homomorphism $g \colon G \to H$ we have an exact sequence $1 \to \ker(g) \to G \to \operatorname{im}(g) \to 1$, i.e., $\operatorname{im}(g) \simeq G/\ker(g)$.

(22.10) Proposition. *A sequence as in* (22.9) *with H smooth is exact if and only if*

(1) $1 \to N(R) \xrightarrow{f_R} G(R) \xrightarrow{g_R} H(R)$ *is exact for every $R \in \mathsf{Alg}_F$ and*
(2) $g_{\mathrm{alg}} \colon G(F_{\mathrm{alg}}) \to H(F_{\mathrm{alg}})$ *is surjective.*

Proof: It follows from (1) that $N = \ker(g)$ and from Proposition (22.3) that g is surjective. $\qquad\qquad\qquad\qquad\qquad\qquad\qquad\qquad\qquad\qquad\qquad\qquad\Box$

(22.11) Proposition. *Suppose that* (22.9) *is exact. Then*

$$\dim G = \dim N + \dim H.$$

Proof: We may assume that $F = F_{\mathrm{alg}}$ and that G (hence also H) is connected. Put $A = F[H]$, $B = F[G]$, $C = F[N]$. We have a bijection of represented functors

$$G \times N \xrightarrow{\sim} G \times_H G, \quad (g,n) \mapsto (g, gn).$$

By Yoneda's lemma there is an F-algebra isomorphism $B \otimes_A B \simeq B \otimes_F C$. We compute the Krull dimension of both sides. Denote by $\mathrm{Quot}(S)$ the field of fractions of a domain S; let $K = \mathrm{Quot}(A_{\mathrm{red}})$ and $L = \mathrm{Quot}(B_{\mathrm{red}})$; then

$$\begin{aligned}
\dim(B \otimes_A B) &= \dim(B_{\mathrm{red}} \otimes_{A_{\mathrm{red}}} B_{\mathrm{red}}) = \mathrm{tr.deg}_F \, \mathrm{Quot}(L \otimes_K L)_{\mathrm{red}} \\
&= 2 \cdot \mathrm{tr.deg}_K(L) + \mathrm{tr.deg}_F(K) = 2 \cdot \mathrm{tr.deg}_F(L) - \mathrm{tr.deg}_F(K) \\
&= 2 \cdot \dim G - \dim H.
\end{aligned}$$

On the other hand

$$\dim(B \otimes_F C) = \dim(G \times N) = \dim G + \dim N$$

by Proposition (21.7). $\qquad\qquad\qquad\qquad\qquad\qquad\qquad\qquad\qquad\qquad\qquad\Box$

(22.12) Corollary. *Suppose that in* (22.9) *N and H are smooth. Then G is also smooth.*

Proof: By Proposition (21.4)(b), $\ker(dg) = \mathrm{Lie}(N)$. Hence

$$\begin{aligned}
\dim \mathrm{Lie}(G) &= \dim \ker(dg) + \dim \operatorname{im}(dg) \le \dim \mathrm{Lie}(N) + \dim \mathrm{Lie}(H), \\
&= \dim N + \dim H = \dim G,
\end{aligned}$$

and therefore, G is smooth. $\qquad\qquad\qquad\qquad\qquad\qquad\qquad\qquad\qquad\qquad\Box$

A surjective homomorphism $f \colon G \to H$ is said to be *separable* if the differential $df \colon \mathrm{Lie}(G) \to \mathrm{Lie}(H)$ is surjective.

(22.13) Proposition. *A surjective homomorphism $f \colon G \to H$ of algebraic groups is separable if and only if $\ker(f)$ is smooth.*

Proof: Let $N = \ker(f)$. By Propositions (21.9) and (22.11),

$$\dim \mathrm{Lie}(N) = \dim \ker(df) = \dim \mathrm{Lie}(G) - \dim \mathrm{im}(df),$$
$$= \dim G - \dim \mathrm{im}(df) = \dim N + \dim H - \dim \mathrm{im}(df),$$
$$= \dim N + \dim \mathrm{Lie}(H) - \dim \mathrm{im}(df).$$

Hence, N is smooth if and only if $\dim N = \dim \mathrm{Lie}(N)$ if and only if $\dim \mathrm{Lie}(H) - \dim \mathrm{im}(df) = 0$ if and only if df is surjective. $\square$

(22.14) Example. The natural surjection $\mathbf{GL}_1(A) \to \mathbf{GL}_1(A)/\mathbf{G}_\mathrm{m}$ is separable.

(22.15) Proposition. *Let* $1 \to N \to G \xrightarrow{f} H \to 1$ *be an exact sequence of algebraic group schemes with N smooth. Then the sequence of groups*

$$1 \to N(F_\mathrm{sep}) \to G(F_\mathrm{sep}) \to H(F_\mathrm{sep}) \to 1$$

is exact.

Proof: Since $N = \ker(f)$, it suffices to prove only exactness on the right. We may assume that $F = F_\mathrm{sep}$. Let $A = F[H]$, $B = F[G]$, (so $A \subset B$) and $C = F[N]$. Take any $h \in H(F)$ and consider the F-algebra $D = B \otimes_A F$ where F is made into an A-algebra via h. For any $R \in \mathsf{Alg}_F$ with structure homomorphism $\nu \colon F \to R$, we have

$$\mathrm{Hom}_{\mathsf{Alg}_F}(D, R) = \{\, g \in \mathrm{Hom}_{\mathsf{Alg}_F}(B, R) \mid g|_A = \nu \circ h \,\} = f_R^{-1}(\nu \circ h),$$

i.e., the F-algebra D represents the *fiber functor*

$$R \mapsto P(R) := f_R^{-1}(\nu \circ h) \subset G(R).$$

If there exists $g \in G(F)$ such that $f_F(g) = h$, i.e. $g \in P(F)$, then there is a bijection of functors $\tau \colon N \to P$, given by $\tau(R)(n) = n \cdot (\nu \circ g) \in P(R)$. By Yoneda's lemma the F-algebras C and D representing the functors N and P are then isomorphic.

We do not know yet if such an element $g \in P(F)$ exists, but it certainly exists over $E = F_\mathrm{alg}$ since $\mathrm{Hom}_{\mathsf{Alg}_E}(D_E, E) \neq \varnothing$ (a form of Hilbert Nullstellensatz). Hence the E-algebras C_E and D_E are isomorphic. In particular, D_E is reduced. Then $\mathrm{Hom}_{\mathsf{Alg}_F}(D, F) \neq \varnothing$ (see Borel [45, AG 13.3]), i.e., $P(F) \neq \varnothing$, so h belongs to the image of f_F, and the described g exists. $\square$

Isogenies. A surjective homomorphism $f \colon G \to H$ of group schemes is called an *isogeny* if $N = \ker(f)$ is finite, and is called a *central isogeny* if $N(R)$ is central in $G(R)$ for every $R \in \mathsf{Alg}_F$.

(22.16) Example. The nth-power homomorphism $\mathbf{G}_\mathrm{m} \to \mathbf{G}_\mathrm{m}$ is a central isogeny.

Representations. Let G be a group scheme over F, with $A = F[G]$. A *representation* of G is a group scheme homomorphism $\rho \colon G \to \mathbf{GL}(V)$ where V is a finite dimensional vector space over F. For any $R \in \mathsf{Alg}_F$ the group $G(R)$ then acts on $V_R = V \otimes_F R$ by R-linear automorphisms; we write

$$g \cdot v = \rho_R(g)(v), \quad g \in G(R), \quad v \in V_R.$$

By taking $R = A$, we obtain an F-linear map

$$\overline{\rho} \colon V \to V \otimes_F A, \quad \overline{\rho}(v) = \mathrm{Id}_A \cdot v$$

(where $\mathrm{Id}_A \in G(A)$ is the "generic" element), such that the following diagrams commute (see Waterhouse [305, §3.2])

(22.17)

$$
\begin{array}{ccc}
V & \xrightarrow{\overline{\rho}} & V \otimes_F A \\
\overline{\rho}\downarrow & & \downarrow{\mathrm{Id}\otimes c} \\
V \otimes_F A & \xrightarrow{\overline{\rho}\otimes\mathrm{Id}} & V \otimes_F A \otimes_F A,
\end{array}
$$

(22.18)

$$
\begin{array}{ccc}
V & \xrightarrow{\overline{\rho}} & V \otimes_F A \\
\| & & \downarrow{\mathrm{Id}\otimes u} \\
V & \xrightarrow{\sim} & V \otimes_F F.
\end{array}
$$

Conversely, a map $\overline{\rho}$ for some F-vector space V, such that the diagrams (22.17) and (22.18) commute, yields a representation $\rho\colon G \to \mathbf{GL}(V)$ as follows: given $g \in G(R)$, $\rho(g)$ is the R-linear extension of the composite map

$$
V \xrightarrow{\overline{\rho}} V \otimes_F A \xrightarrow{\mathrm{Id}\otimes g} V \otimes_F R.
$$

A finite dimensional F-vector space V together with a map $\overline{\rho}$ as above is called an *A-comodule*. There is an obvious notion of subcomodules.

A vector $v \in V$ is said to be *G-invariant* if $\overline{\rho}(v) = v \otimes 1$. Denote by V^G the F-subspace of all G-invariant elements. Clearly, $G(R)$ acts trivially on $(V^G) \otimes_F R$ for any $R \in \mathsf{Alg}_F$. For a field extension L/F one has $(V_L)^{G_L} \simeq (V^G)_L$.

A representation $\rho\colon G \to \mathbf{GL}(V)$ is called *irreducible* if the A-comodule V has no nontrivial subcomodules.

(22.19) Examples. (1) If $\dim V = 1$, then $\mathbf{GL}(V) = \mathbf{G}_{\mathrm{m}}$. Hence a 1-dimensional representation is simply a character.

(2) Let G be an algebraic group scheme over F. For any $R \in \mathsf{Alg}_F$ the group $G(R)$ acts by conjugation on

$$
\ker\big(G(R[\varepsilon]) \to G(R)\big) = \mathrm{Lie}(G) \otimes_F R.
$$

Hence we get a representation

$$
\mathrm{Ad} = \mathrm{Ad}_G\colon G \to \mathbf{GL}\big(\mathrm{Lie}(G)\big)
$$

called the *adjoint representation*. When $G = \mathbf{GL}(V)$ the adjoint representation

$$
\mathrm{Ad}\colon \mathbf{GL}(V) \to \mathbf{GL}\big(\mathrm{End}(V)\big)
$$

is given by conjugation: $\mathrm{Ad}(\alpha)(\beta) = \alpha\beta\alpha^{-1}$.

Representations of diagonalizable groups. Let $G = H_{\mathrm{diag}}$ be a diagonalizable group scheme, $A = F[G] = F\langle H \rangle$. Let $\overline{\rho}\colon V \to V \otimes_F A$ be the A-comodule structure on a finite dimensional vector space V corresponding to some representation $\rho\colon G \to \mathbf{GL}(V)$.

Write $\overline{\rho}(v) = \sum_{h \in H} f_h(v) \otimes h$ for uniquely determined F-linear maps $f_h\colon V \to V$. The commutativity of diagram (22.17) is equivalent to the conditions

$$
f_h \circ f_{h'} = \begin{cases} f_h & \text{if } h = h', \\ 0 & \text{if } h \neq h', \end{cases}
$$

and the commutativity of (22.18) gives $\sum_{h \in H} f_h(v) = v$ for all $v \in V$. Hence the maps f_h induce a decomposition

(22.20) $$V = \bigoplus_{h \in H} V_h, \quad \text{where} \quad V_h = \text{im}(f_h).$$

A character $h \in H = G^*$ is called a *weight of* ρ if $V_h \neq 0$. A representation ρ of a diagonalizable group is uniquely determined (up to isomorphism) by its weights and their *multiplicities* $m_h = \dim V_h$.

§23. Automorphism Groups of Algebras

In this section we consider various algebraic group schemes related to algebras and algebras with involution.

Let A be a *separable* associative unital F-algebra (i.e., A is a finite product of algebras which are central simple over finite separable field extensions of F, or equivalently, $A_{\widetilde{F}} = A \otimes \widetilde{F}$ is semisimple for every field extension $\widetilde{F}$ of F). Let L be the center of A (which is an étale F-algebra). The kernel of the restriction homomorphism $\mathbf{Aut}_{\text{alg}}(A) \to \mathbf{Aut}_{\text{alg}}(L)$ is denoted $\mathbf{Aut}_L(A)$. Since all L-derivations of A are inner (see for example Knus-Ojanguren [160, p. 73-74]), it follows from Example (21.5.9) that

$$\text{Lie}\big(\mathbf{Aut}_L(A)\big) = \text{Der}_L(A, A) = A/L.$$

We use the notation $\text{ad}(a)(x) = [a, x] = ax - xa$ for the inner derivation $\text{ad}(a)$ associated to $a \in A$. Consider the group scheme homomorphism

$$\text{Int}: \mathbf{GL}_1(A) \to \mathbf{Aut}_L(A), \quad a \mapsto \text{Int}(a)$$

with kernel $\mathbf{GL}_1(L) = R_{L/F}(\mathbf{G}_{\text{m},L})$. By Proposition (22.11) we have:

$$\dim \mathbf{Aut}_L(A) \geq \dim \text{im}(\text{Int}) = \dim \mathbf{GL}_1(A) - \dim \mathbf{GL}_1(L)$$
$$= \dim_F A - \dim_F L = \dim_F \text{Lie}\big(\mathbf{Aut}_L(A)\big).$$

The group scheme $\mathbf{Aut}_L(A)$ is smooth. This follows from Lemma (21.8) and Proposition (21.9). By the Skolem-Noether theorem the homomorphism Int_E is surjective for any field extension E/F, hence Int is surjective by Proposition (22.3), and $\mathbf{Aut}_L(A)$ is connected by Proposition (22.4). Thus we have an exact sequence of connected algebraic groups

(23.1) $$1 \to \mathbf{GL}_1(L) \to \mathbf{GL}_1(A) \to \mathbf{Aut}_L(A) \to 1.$$

Assume now that A is a central simple algebra over F, i.e., $L = F$. We write $\mathbf{PGL}_1(A)$ for the group $\mathbf{Aut}_{\text{alg}}(A)$, so that

$$\mathbf{PGL}_1(A) \simeq \mathbf{GL}_1(A)/\mathbf{G}_{\text{m}}, \quad \text{Lie}\big(\mathbf{PGL}_1(A)\big) = A/F,$$

and

$$\mathbf{PGL}_1(A)(R) = \text{Aut}_R(A_R), \quad R \in \mathit{Alg}_F.$$

We say that an F-algebra R *satisfies the SN-condition* if for any central simple algebra A over F all R-algebra automorphisms of A_R are inner. Fields and local rings satisfy the SN-condition (see for example Knus-Ojanguren [160, p. 107]).

If R satisfies the SN-condition then

(23.2) $$\mathbf{PGL}_1(A)(R) = (A_R)^\times / R^\times.$$

We set $\mathbf{PGL}(V) = \mathbf{PGL}_1\big(\mathrm{End}(V)\big) = \mathbf{GL}(V)/\mathbf{G}_m$ and call $\mathbf{PGL}(V)$ the *projective general linear group*; we write $\mathbf{PGL}(V) = \mathbf{PGL}_n$ if $V = F^n$.

23.A. Involutions. In this part we rediscuss most of the groups introduced in Chapter III from the point of view of group schemes. Let A be a separable F-algebra with center K and F-involution σ. We define various group schemes over F related to A. Consider the representation

$$\rho\colon \mathbf{GL}_1(A) \to \mathbf{GL}(A), \quad a \mapsto \big(x \mapsto a \cdot x \cdot \sigma(a)\big).$$

The subgroup $\mathbf{Aut}_{\mathbf{GL}_1(A)}(1)$ in $\mathbf{GL}_1(A)$ is denoted $\mathbf{Iso}(A,\sigma)$ and is called the *group scheme of isometries* of (A,σ):

$$\mathbf{Iso}(A,\sigma)(R) = \{\, a \in A_R^\times \mid a \cdot \sigma_R(a) = 1 \,\}.$$

An element $1 + a \cdot \varepsilon$, $a \in A$ lies in $\ker \mathbf{Iso}(A,\sigma)(\kappa)$ if and only if

$$(1 + a \cdot \varepsilon)\big(1 + \sigma(a) \cdot \varepsilon\big) = 1,$$

or equivalently, $a + \sigma(a) = 0$. Hence,

$$\mathrm{Lie}\big(\mathbf{Iso}(A,\sigma)\big) = \mathrm{Skew}(A,\sigma) \subset A.$$

Consider the adjoint representation

$$\rho\colon \mathbf{GL}(A) \to \mathbf{GL}\big(\mathrm{End}_F(A)\big), \quad \alpha \mapsto (\beta \mapsto \alpha\beta\alpha^{-1})$$

and denote the intersection of the subgroups $\mathbf{Aut}_{\mathrm{alg}}(A)$ and $\mathbf{Aut}_{\mathbf{GL}(A)}(\sigma)$ of $\mathbf{GL}(A)$ by $\mathbf{Aut}(A,\sigma)$:

$$\mathbf{Aut}(A,\sigma)(R) = \{\, \alpha \in \mathrm{Aut}_R(A_R) \mid \alpha \circ \sigma_R = \sigma_R \circ \alpha \,\}.$$

A derivation $x = \mathrm{ad}(a) \in \mathrm{Der}(A,A) = \mathrm{Lie}\big(\mathbf{Aut}_{\mathrm{alg}}(A)\big)$ lies in $\mathrm{Lie}\big(\mathbf{Aut}(A,\sigma)\big)$ if and only if $(1 + x \cdot \varepsilon) \circ \sigma = \sigma \circ (1 + x \cdot \varepsilon)$ if and only if $x \circ \sigma = \sigma \circ x$ if and only if $a + \sigma(a) \in K$. Hence

$$\mathrm{Lie}\big(\mathbf{Aut}(A,\sigma)\big) = \{\, a \in A \mid a + \sigma(a) \in K \,\}/K.$$

Denote the intersection of $\mathbf{Aut}(A,\sigma)$ and $\mathbf{Aut}_K(A)$ by $\mathbf{Aut}_K(A,\sigma)$. If an F-algebra R satisfies the SN-condition, then

$$\mathbf{Aut}_K(A,\sigma)(R) = \{\, a \in A_R^\times \mid a \cdot \sigma_R(a) \in K_R^\times \,\}/K_R^\times.$$

The inverse image of $\mathbf{Aut}_K(A,\sigma)$ with respect to the surjection

$$\mathrm{Int}\colon \mathbf{GL}_1(A) \to \mathbf{Aut}_K(A)$$

(see 23.2) is denoted $\mathbf{Sim}(A,\sigma)$ and called the *group scheme of similitudes of* (A,σ). Clearly,

$$\mathbf{Sim}(A,\sigma)(R) = \{\, a \in A_R^\times \mid a \cdot \sigma(a) \in K_R^\times \,\}$$

$$\mathrm{Lie}\big(\mathbf{Sim}(A,\sigma)\big) = \{\, a \in A \mid a + \sigma(a) \in K \,\}.$$

By Proposition (22.4) we have an exact sequence of group schemes

(23.3) $\qquad 1 \to \mathbf{GL}_1(K) \to \mathbf{Sim}(A,\sigma) \to \mathbf{Aut}_K(A,\sigma) \to 1.$

Let E be the F-subalgebra of K consisting of all σ-invariant elements. We have a group scheme homomorphism

$$\mu\colon \mathbf{Sim}(A,\sigma) \to \mathbf{GL}_1(E), \quad a \mapsto a \cdot \sigma(a).$$

The map μ_{alg} is clearly surjective. Hence, by Proposition (22.3), we have an exact sequence

(23.4) $$1 \to \mathbf{Iso}(A, \sigma) \to \mathbf{Sim}(A, \sigma) \xrightarrow{\mu} \mathbf{GL}_1(E) \to 1.$$

Unitary involutions. Let K/F be an étale quadratic extension, B be a central simple algebra over K of degree n with a unitary F-involution τ. We use the following notation (and definitions) for group schemes over F:

$$\mathbf{U}(B, \tau) = \mathbf{Iso}(B, \tau) \qquad \textit{Unitary group}$$
$$\mathbf{GU}(B, \tau) = \mathbf{Sim}(B, \tau) \qquad \textit{Group of unitary similitudes}$$
$$\mathbf{PGU}(B, \tau) = \mathbf{Aut}_K(B, \tau) \quad \textit{Projective unitary group}$$

Assume first that K is split, $K \simeq F \times F$. Then $B \simeq A \times A^{\mathrm{op}}$ and τ is the exchange involution. Let $b = (a_1, a_2^{\mathrm{op}}) \in B$. The condition $b \cdot \tau b = 1$ is equivalent to $a_1 a_2 = 1$. Hence we have an isomorphism

$$\mathbf{GL}_1(A) \xrightarrow{\sim} \mathbf{U}(B, \tau), \quad a \mapsto \big(a, (a^{-1})^{\mathrm{op}}\big).$$

The homomorphism

$$\mathbf{Aut}_{\mathrm{alg}}(A) \to \mathbf{PGU}(B, \tau), \quad \phi \mapsto (\phi, \phi^{\mathrm{op}})$$

is clearly an isomorphism. Hence,

$$\mathbf{PGU}(B, \tau) \simeq \mathbf{PGL}_1(A).$$

Thus the group schemes $\mathbf{U}(B, \tau)$ and $\mathbf{PGU}(B, \tau)$ are smooth and connected. This also holds when K is not split, as one sees by scalar extension. Furthermore the surjection $\mathbf{Aut}(B, \tau) \to \mathbf{Aut}_{\mathrm{alg}}(K) \simeq \mathbb{Z}/2\mathbb{Z}$ induces an isomorphism $\pi_0\big(\mathbf{Aut}(B, \tau)\big) \simeq \mathbb{Z}/2\mathbb{Z}$. Hence $\mathbf{PGU}(B, \tau)$ is the connected component of $\mathbf{Aut}(B, \tau)$ and is as a subgroup of index 2.

The kernel of the reduced norm homomorphism $\mathrm{Nrd} \colon \mathbf{U}(B, \tau) \to \mathbf{GL}_1(K)$ is denoted $\mathbf{SU}(B, \tau)$ and called the *special unitary group*. Clearly,

$$\mathbf{SU}(B, \tau)(R) = \{\, b \in (B \otimes_F R)^\times \mid b \cdot \tau_R(b) = 1, \mathrm{Nrd}_R(b) = 1 \,\},$$
$$\mathrm{Lie}\big(\mathbf{SU}(B, \tau)\big) = \{\, x \in \mathrm{Skew}(B, \tau) \mid \mathrm{Trd}(x) = 0 \,\}.$$

The group scheme $\mathbf{SU}(B, \tau)$ is smooth and connected since, when K is split, $\mathbf{SU}(B, \tau) = \mathbf{SL}_1(A)$ (as the description given above shows). The kernel N of the composition

$$f \colon \mathbf{SU}(B, \tau) \hookrightarrow \mathbf{U}(B, \tau) \to \mathbf{PGU}(B, \tau)$$

satisfies

$$N(R) = \{\, B \in (K \otimes_F R)^\times \mid b \cdot \tau_R(b) = 1, b^n = 1 \,\}.$$

In other words,

$$N = \ker\big(R_{K/F}(\boldsymbol{\mu}_{n,K}) \xrightarrow{N_{K/F}} \boldsymbol{\mu}_{n,F}\big),$$

hence N is a finite group scheme of multiplicative type and is Cartier dual to $\mathbb{Z}/n\mathbb{Z}$ where the Galois group Γ acts through $\mathrm{Gal}(K/F)$ as $x \mapsto -x$. Subgroups of N correspond to (cyclic) subgroups of $\mathbb{Z}/n\mathbb{Z}$, which are automatically Γ-invariant.

Since f_{alg} is surjective, f is surjective by Proposition (22.3). Clearly, f is a central isogeny and

$$\mathbf{PGU}(B, \tau) \simeq \mathbf{SU}(B, \tau)/N.$$

Symplectic involutions. Let A be a central simple algebra of degree $n = 2m$ over F with a symplectic involution σ. We use the following notation (and definitions):

$$\mathbf{Sp}(A, \sigma) = \mathbf{Iso}(A, \sigma) \quad \textit{Symplectic group}$$

$$\mathbf{GSp}(A, \sigma) = \mathbf{Sim}(A, \sigma) \quad \textit{Group of symplectic similitudes}$$

$$\mathbf{PGSp}(A, \sigma) = \mathbf{Aut}(A, \sigma) \quad \textit{Projective symplectic group}$$

Assume first that A is split, $A = \mathrm{End}_F(V)$, hence $\sigma = \sigma_h$ where h is a nonsingular alternating bilinear form on V. Then $\mathbf{Sp}(A, \sigma) = \mathbf{Sp}(V, h)$, the *symplectic group of* (V, h),

$$\mathbf{Sp}(V, h)(R) = \{ \, \alpha \in \mathrm{GL}(V_R) \mid h_R\big(\alpha(v), \alpha(v')\big) = h_R(v, v') \text{ for } v, \, v' \in V_R \, \}.$$

The associated classical algebraic group is connected of dimension $m(2m + 1)$ (Borel [45, 23.3]).

Coming back to the general case, we have

$$\dim \mathrm{Lie}\big(\mathbf{Sp}(A, \sigma)\big) = \dim \mathrm{Skew}(A, \sigma) = m(2m + 1) = \dim \mathbf{Sp}(A, \sigma),$$

hence $\mathbf{Sp}(A, \sigma)$ is a smooth and connected group. It follows from the exactness of

$$1 \to \mathbf{Sp}(A, \sigma) \to \mathbf{GSp}(A, \sigma) \xrightarrow{\mu} \mathbf{G}_\mathrm{m} \to 1$$

(see 23.4) and Corollary (22.12) that $\mathbf{GSp}(A, \sigma)$ is smooth.

The exactness of

$$1 \to \mathbf{G}_\mathrm{m} \to \mathbf{GSp}(A, \sigma) \to \mathbf{PGSp}(A, \sigma) \to 1$$

(see 23.3) implies that $\mathbf{PGSp}(A, \sigma)$ is smooth. Consider the composition

$$f \colon \mathbf{Sp}(A, \sigma) \hookrightarrow \mathbf{GSp}(A, \sigma) \to \mathbf{PGSp}(A, \sigma)$$

whose kernel is $\boldsymbol{\mu}_2$. Clearly, f_alg is surjective, hence f is surjective and $\mathbf{PGSp}(A, \sigma)$ is connected. Therefore, f is a central isogeny and $\mathbf{PGSp}(A, \sigma) \simeq \mathbf{Sp}(A, \sigma)/\boldsymbol{\mu}_2$.

In the split case the group $\mathbf{PGSp}(V, h) = \mathbf{PGSp}(A, \sigma)$ is called the *projective symplectic group* of (V, h).

Orthogonal involutions. Let A be a central simple algebra of degree n over F with an orthogonal involution σ. We use the following notation

$$\mathbf{O}(A, \sigma) = \mathbf{Iso}(A, \sigma)$$

$$\mathbf{GO}(A, \sigma) = \mathbf{Sim}(A, \sigma)$$

$$\mathbf{PGO}(A, \sigma) = \mathbf{Aut}(A, \sigma)$$

Consider the split case $A = \mathrm{End}_F(V)$, $\sigma = \sigma_b$, where b is a nonsingular symmetric (non-alternating, if char $F = 2$) bilinear form on V. Then

$$\mathbf{O}(A, \sigma)(R) = \{ \, \alpha \in \mathrm{GL}(V_R) \mid b(\alpha v, \alpha v') = b(v, v') \text{ for } v, \, v' \in V_R \, \}.$$

The associated classical algebraic group has dimension $\frac{n(n-1)}{2}$ (Borel [45]). On the other hand,

$$\dim \mathrm{Lie}\big(\mathbf{O}(A, \sigma)\big) = \dim \mathrm{Skew}(A, \sigma) = \begin{cases} \frac{n(n-1)}{2} & \text{if char } F \neq 2, \\ \frac{n(n+1)}{2} & \text{if char } F = 2. \end{cases}$$

Hence $\mathbf{O}(A, \sigma)$ (and the other groups) are not smooth if char $F = 2$. To get smooth groups also in characteristic 2 we use a different context, described in the next two subsections.

Orthogonal groups. Let (V, q) be a quadratic form of dimension n over F and let b_q be the polar bilinear form of q on V. We recall that the form q is *regular* if b_q is a nonsingular bilinear form except for the case n is odd and char $F = 2$. In this case b_q is symplectic and is degenerate. The *radical* of q is the space $V^\perp$ and (in case char $F = 2$ and $\dim_F V$ is odd) q is *regular* if $\dim \operatorname{rad}(b_q) = 1$, say $\operatorname{rad}(q) = F \cdot v$, $q(v) \neq 0$.

We view q as an element of $S^2(V^*)$, the space of degree 2 elements in the symmetric algebra $S^2(V^*)$. There is a natural representation

$$\rho \colon \mathbf{GL}(V) \to \mathbf{GL}\big(S^2(V^*)\big).$$

We set $\mathbf{O}(V, q)$ for the group $\mathbf{Aut}_{\mathbf{GL}(V)}(q)$ and call it the *orthogonal group of* (V, q):

$$\mathbf{O}(V, q)(R) = \{\, \alpha \in \mathrm{GL}(V_R) \mid q_R(\alpha v) = q_R(v) \text{ for } v \in V_R \,\}.$$

The associated classical algebraic group has dimension $\frac{n(n-1)}{2}$ (Borel [45, 23.6]). For $\alpha \in \operatorname{End}(V)$, we have $1 + \alpha \cdot \varepsilon \in \mathbf{O}(V, q)$ if and only if $b_q(v, \alpha v) = 0$ for all $v \in V$. Hence

$$\operatorname{Lie}\big(\mathbf{O}(V, q)\big) = \{\, \alpha \in \operatorname{End}(V) \mid b_q(v, \alpha v) = 0 \text{ for } v \in V \,\} = \mathfrak{o}(V, q).$$

The dimensions are:

$$\dim \operatorname{Lie}\big(\mathbf{O}(V, q)\big) = \begin{cases} \frac{n(n-1)}{2} & \text{if } n \text{ is even or char } F \neq 2, \\ \frac{n(n-1)}{2} + 1 & \text{if } n \text{ is odd and char } F = 2. \end{cases}$$

Hence, in the first case $\mathbf{O}(V, q)$ is a smooth group scheme. We consider now the following cases:

(a) char $F = 2$ and n is even: we define

$$\mathbf{O}^+(V, q) = \ker\big(\mathbf{O}(V, q) \xrightarrow{\Delta} \mathbb{Z}/2\mathbb{Z}\big)$$

where Δ is the *Dickson invariant*, i.e., $\Delta(\alpha) = 0$ for $\alpha \in \mathbf{O}(V, q)(R)$ if α induces the identity automorphism of the center of the Clifford algebra and $\Delta(\alpha) = 1$ if not (see (13.3)). The associated classical algebraic group is known to be connected (Borel [45, 23.6]). Hence, $\mathbf{O}^+(V, q)$ is a smooth connected group scheme.

(b) char $F \neq 2$ or n is odd: we set

$$\mathbf{O}^+(V, q) = \ker\big(\mathbf{O}(V, q) \xrightarrow{\det} \mathbf{G}_{\mathrm{m}}\big)$$

where det is the determinant map. Here also the associated classical algebraic group is known to be connected (Borel [45]).

We get in each case

$$\operatorname{Lie}\big(\mathbf{O}^+(V, q)\big) = \{\, \alpha \in \operatorname{End}(V) \mid \operatorname{tr}(\alpha) = 0,\ b_q(v, \alpha v) = 0 \text{ for } v \in V \,\}.$$

If char $F \neq 2$ this Lie algebra coincides with $\operatorname{Lie}\big(\mathbf{O}(V, q)\big)$ and $\mathbf{O}^+(V, q)$ is the connected component of $\mathbf{O}(V, q)$. If char $F = 2$, then $\operatorname{Lie}\big(\mathbf{O}^+(V, q)\big) \subsetneqq \operatorname{Lie}\big(\mathbf{O}(V, q)\big)$ hence

$$\dim \operatorname{Lie}\big(\mathbf{O}^+(V, q)\big) = \frac{n(n-1)}{2}$$

and $\mathbf{O}^+(V, q)$ is a smooth connected group scheme. Thus in every case $\mathbf{O}^+(V, q)$ is a connected algebraic group. Consider the conjugation homomorphism

$$\mathbf{GL}_1\big(C_0(V, q)\big) \to \mathbf{GL}\big(C(V, q)\big), \quad x \mapsto \operatorname{Int}(x)$$

where $C(V, q) = C_0(V, q) \oplus C_1(V, q)$ is the Clifford algebra. The inverse image of the normalizer $\mathbf{N}_V$ of the subspace $V \subset C(V, q)$ is $\mathbf{\Gamma}^+(V, q)$, the *even Clifford group* of (V, q),

$$\mathbf{\Gamma}^+(V, q)(R) = \{\, g \in C_0(V, q)_R^\times \mid q \cdot V_R \cdot g^{-1} = V_R \,\}.$$

It follows from Example (21.5.10) that

$$\mathrm{Lie}\big(\mathbf{\Gamma}^+(V, q)\big) = \{\, x \in C_0(V, q) \mid [x, V] \subset V \,\} = V \cdot V \subset C_0(V, q).$$

Let

$$\chi \colon \mathbf{\Gamma}^+(V, q) \to \mathbf{O}^+(V, q), \quad x \mapsto \mathrm{Int}(x)|_V.$$

Clearly, $\ker \chi = \mathbf{G}_{\mathrm{m}} \subset \mathbf{\Gamma}^+(V, q)$. Since χ_{alg} is surjective, we have by Proposition (22.3) an exact sequence

$$1 \to \mathbf{G}_{\mathrm{m}} \to \mathbf{\Gamma}^+(V, q) \xrightarrow{\chi} \mathbf{O}^+(V, q) \to 1.$$

Hence by Corollary (22.12) $\mathbf{\Gamma}^+(V, q)$ is smooth.

The kernel of the spinor norm homomorphism

$$\mathrm{Sn} \colon \mathbf{\Gamma}^+(V, q) \to \mathbf{G}_{\mathrm{m}}, \quad x \mapsto x \cdot \underline{\sigma}(x)$$

is the *spinor group* of (V, q) and is denoted $\mathbf{Spin}(V, q)$. Thus,

$$\mathbf{Spin}(V, q)(R) = \{\, g \in C_0(V, q)_R^\times \mid g \cdot V_R \cdot g^{-1} = V_R, \, g \cdot \underline{\sigma}(g) = 1 \,\}$$

The differential $d(\mathrm{Sn})$ is given by

$$d(\mathrm{Sn})(uv) = uv + \underline{\sigma}(uv) = uv + vu = b_q(u, v).$$

In particular, Sn is separable and

$$\mathrm{Lie}\big(\mathbf{Spin}(V, q)\big) = \{\, x \in V \cdot V \subset C_0(V, q) \mid x + \underline{\sigma}(x) = 0 \,\}.$$

Since $\mathrm{Sn}_{\mathrm{alg}}$ is surjective, we have by Proposition (22.3) an exact sequence

$$1 \to \mathbf{Spin}(V, q) \to \mathbf{\Gamma}^+(V, q) \xrightarrow{\mathrm{Sn}} \mathbf{G}_{\mathrm{m}} \to 1.$$

Hence by Proposition (22.13) $\mathbf{Spin}(V, q)$ is smooth. The classical algebraic group associated to $\mathbf{Spin}(V, q)$ is known to be connected (Borel [45, 23.3]), therefore $\mathbf{Spin}(V, q)$ is connected.

The kernel of the composition

$$f \colon \mathbf{Spin}(V, q) \hookrightarrow \mathbf{\Gamma}^+(V, q) \xrightarrow{\chi} \mathbf{O}^+(V, q)$$

is $\boldsymbol{\mu}_2$. Since f_{alg} is surjective, it follows by Proposition (22.3) that f is surjective. Hence, f is a central isogeny and

$$\mathbf{O}^+(V, q) \simeq \mathbf{Spin}(V, q)/\boldsymbol{\mu}_2.$$

(23.5) Remark. The preceding discussion focuses on orthogonal groups of *quadratic spaces*. Orthogonal groups of *symmetric bilinear spaces* may be defined in a similar fashion: every nonsingular symmetric nonalternating bilinear form b on a vector space V may be viewed as an element of $S^2(V)^*$, and letting $\mathbf{GL}(V)$ act on $S^2(V)^*$ we may set $\mathbf{O}(V, b) = \mathbf{Aut}_{\mathbf{GL}(V)}(b)$.

If char $F \neq 2$ we may identify $S^2(V)^*$ to $S^2(V^*)$ by mapping every symmetric bilinear form b to its associated quadratic form q_b defined by $q_b(x) = b(x, x)$, hence $\mathbf{O}(V, b) = \mathbf{O}(V, q_b)$. If char $F = 2$ the group $\mathbf{O}(V, b)$ is not smooth, and if F is not perfect there may be no associated smooth algebraic group, see Exercise 15.

Suppose F is perfect of characteristic 2. In that case, there is an associated smooth algebraic group $\mathbf{O}(V, b)_{\mathrm{red}}$. If $\dim V$ is odd, $\mathbf{O}(V, b)_{\mathrm{red}}$ turns out to be isomorphic to the symplectic group of an alternating space of dimension $\dim V - 1$, see Exercise 16. If $\dim V$ is even, $\mathbf{O}(V, b)_{\mathrm{red}}$ contains a nontrivial solvable connected normal subgroup, see Exercise 16; it is therefore not semisimple (see §25 for the definition of semisimple group).

23.B. Quadratic pairs. Let A be a central simple algebra of degree $n = 2m$ over F, and let (σ, f) be a quadratic pair on A. Consider the homomorphism

$$\mathbf{Aut}(A, \sigma) \to \mathbf{GL}\big(\mathrm{Sym}(A, \sigma)^*\big), \quad \alpha \mapsto (g \mapsto g \circ \alpha).$$

The inverse image of the stabilizer $\mathbf{S}_f$ of f is denoted $\mathbf{PGO}(A, \sigma, f)$ and is called the *projective orthogonal group*:

$$\mathbf{PGO}(A, \sigma, f)(R) = \{\, \alpha \in \mathrm{Aut}(A, \sigma) \mid f_R \circ \alpha = f_R \,\}.$$

If R satisfies the SN-condition, then, setting $(A, \sigma)_+ = \mathrm{Sym}(A, \sigma)$,

$$\mathbf{PGO}(A, \sigma, f)(R) =$$
$$\{\, a \in A_R^\times \mid a \cdot \sigma_R(a) \in R^\times,\ f(axa^{-1}) = f(x) \text{ for } x \in (A_R, \sigma_R)_+ \,\}/R^\times.$$

In the split case $A = \mathrm{End}(V)$, with q a quadratic form corresponding to the quadratic pair (σ, f), we write $\mathbf{PGO}(V, q)$ for this group. The inverse image of $\mathbf{PGO}(A, \sigma, f)$ under

$$\mathrm{Int}\colon \mathbf{GL}_1(A) \to \mathbf{Aut}_{\mathrm{alg}}(A)$$

is the *group of orthogonal similitudes* and is denoted $\mathbf{GO}(A, \sigma, f)$:

$$\mathbf{GO}(A, \sigma, f)(R) =$$
$$\{\, a \in A_R^\times \mid a \cdot \sigma_R(a) \in R^\times,\ f(axa^{-1}) = f(x) \text{ for } x \in (A_R, \sigma_R)_+ \,\}.$$

One sees that $1 + a \cdot \varepsilon \in \mathbf{GO}(A, \sigma, f)(F[\varepsilon])$ if and only if $a + \sigma(a) \in F$ and $f(ax - xa) = 0$ for all symmetric x. Thus

$$\mathrm{Lie}\big(\mathbf{GO}(A, \sigma, f)\big) = \{\, a \in A \mid a + \sigma(a) \in F,\ f(ax - xa) = 0 \text{ for } x \in \mathrm{Sym}(A, \sigma) \,\}.$$

An analogous computation shows that

$$\mathrm{Lie}\big(\mathbf{PGO}(A, \sigma, f)\big) = \mathrm{Lie}\big(\mathbf{GO}(A, \sigma, f)\big)/F.$$

The kernel of the homomorphism

$$\mu\colon \mathbf{GO}(A, \sigma, f) \to \mathbf{G}_{\mathrm{m}}, \quad a \mapsto a \cdot \sigma(a)$$

is denoted $\mathbf{O}(A, \sigma, f)$ and is called the *orthogonal group*,

$$\mathbf{O}(A, \sigma, f)(R) = \{\, a \in A_R^\times \mid a \cdot \sigma(a) = 1,\ f(axa^{-1}) = f(x) \text{ for } x \in \mathrm{Sym}(A, \sigma)_R \,\}.$$

Since for $a \in A$ with $a + \sigma(a) = 0$ one has $f(ax - xa) = f\big(ax + \sigma(ax)\big) = \mathrm{Trd}(ax)$, it follows that the condition $f(ax - xa) = 0$ for all $x \in \mathrm{Sym}(A, \sigma)$ is equivalent to $a \in \mathrm{Alt}(A, \sigma)$. Thus

$$\mathrm{Lie}\big(\mathbf{O}(A, \sigma, f)\big) = \mathrm{Alt}(A, \sigma)$$

(and does not depend on f!).

In the split case we have $\mathbf{O}(A, \sigma, f) = \mathbf{O}(V, q)$, hence by 23.A, $\mathbf{O}(A, \sigma, f)$ is smooth.

The sequence

$$1 \to \mathbf{O}(A, \sigma, f) \to \mathbf{GO}(A, \sigma, f) \xrightarrow{\mu} \mathbf{G}_{\mathrm{m}} \to 1$$

is exact by Proposition (22.3), since μ_{alg} is surjective. It follows from Corollary (22.12) that $\mathbf{GO}(A, \sigma, f)$ is smooth. By Proposition (22.4), the natural homomorphism $\mathbf{GO}(A, \sigma, f) \to \mathbf{PGO}(A, \sigma, f)$ is surjective, hence $\mathbf{PGO}(A, \sigma, f)$ is smooth. There is an exact sequence

$$1 \to \mathbf{G}_{\text{m}} \to \mathbf{GO}(A, \sigma, f) \to \mathbf{PGO}(A, \sigma, f) \to 1.$$

The kernel of the composition

$$g \colon \mathbf{O}(A, \sigma, f) \hookrightarrow \mathbf{GO}(A, \sigma, f) \to \mathbf{PGO}(A, \sigma, f)$$

is $\boldsymbol{\mu}_2$. Clearly, g_{alg} is surjective, hence g is surjective. Therefore, g is a central isogeny and

$$\mathbf{PGO}(A, \sigma, f) \simeq \mathbf{O}(A, \sigma, f)/\boldsymbol{\mu}_2.$$

Now comes into play the Clifford algebra $C(A, \sigma, f)$. By composing the natural homomorphism

$$\mathbf{PGO}(A, \sigma, f) \to \mathbf{Aut}_{\text{alg}}\big(C(A, \sigma, f)\big)$$

with the restriction map

$$\mathbf{Aut}_{\text{alg}}\big(C(A, \sigma, f)\big) \to \mathbf{Aut}_{\text{alg}}(Z) = \mathbb{Z}/2\mathbb{Z}$$

where Z is the center of $C(A, \sigma, f)$, we obtain a homomorphism $\mathbf{PGO}(A, \sigma, f) \to \mathbb{Z}/2\mathbb{Z}$, the kernel of which we denote $\mathbf{PGO}^+(A, \sigma, f)$. The inverse image of this group in $\mathbf{GO}(A, \sigma, f)$ is denoted $\mathbf{GO}^+(A, \sigma, f)$ and the intersection of $\mathbf{GO}^+(A, \sigma, f)$ with $\mathbf{O}(A, \sigma, f)$ by $\mathbf{O}^+(A, \sigma, f)$. In the split case $\mathbf{O}^+(A, \sigma, f) = \mathbf{O}^+(V, q)$, hence $\mathbf{O}^+(A, \sigma, f)$ is smooth and connected. In particular it is the connected component of $\mathbf{O}(A, \sigma, f)$. It follows from the exactness of

$$1 \to \boldsymbol{\mu}_2 \to \mathbf{O}^+(A, \sigma, f) \to \mathbf{PGO}^+(A, \sigma, f) \to 1$$

that $\mathbf{PGO}^+(A, \sigma, f)$ is also a connected algebraic group, namely the connected component of $\mathbf{PGO}(A, \sigma, f)$.

Let $B(A, \sigma, f)$ be the Clifford bimodule. Consider the representation

$$\mathbf{GL}_1\big(C(A, \sigma, f)\big) \to \mathbf{GL}\big(B(A, \sigma, f)\big), \quad c \mapsto \big(x \mapsto (c * x \cdot c^{-1})\big).$$

Let $b \colon A \to B(A, \sigma, f)$ be the canonical map. Let $\boldsymbol{\Gamma}(A, \sigma, f)$ be the inverse image of the normalizer $\mathbf{N}_{b(A)}$ of the subspace $b(A) \subset B(A, \sigma, f)$ and call it the *Clifford group* of (A, σ, f),

$$\boldsymbol{\Gamma}(A, \sigma, f)(R) = \{\, c \in C(A, \sigma, f)_R^\times \mid c * b(A)_R \cdot c^{-1} = b(A)_R \,\}.$$

In the split case $\boldsymbol{\Gamma}(A, \sigma, f) = \boldsymbol{\Gamma}^+(V, q)$ is a smooth group and

$$\text{Lie}\big(\boldsymbol{\Gamma}(A, \sigma, f)\big) = V \cdot V = c(A) \subset C(A, \sigma, f).$$

Hence $\boldsymbol{\Gamma}(A, \sigma, f)$ is a smooth algebraic group and

$$\text{Lie}\big(\boldsymbol{\Gamma}(A, \sigma, f)\big) = c(A).$$

For any $g \in \boldsymbol{\Gamma}(A, \sigma, f)(R)$ one has $g \cdot \underline{\sigma}(g) \in R^\times$, hence there is a *spinor norm homomorphism*

$$\text{Sn} \colon \boldsymbol{\Gamma}(A, \sigma, f) \to \mathbf{G}_{\text{m}}, \quad g \mapsto g \cdot \underline{\sigma}(g).$$

We denote the kernel of Sn by $\mathbf{Spin}(A, \sigma, f)$ and call it the *spinor group* of (A, σ, f). It follows from the split case (where $\mathbf{Spin}(A, \sigma, f) = \mathbf{Spin}(V, q)$) that $\mathbf{Spin}(A, \sigma, f)$ is a connected algebraic group.

Let $\chi \colon \mathbf{\Gamma}(A, \sigma, f) \to \mathbf{O}^+(A, \sigma, f)$ be the homomorphism defined by the formula $c^{-1} * (1)b \cdot c = \big(\chi(c)\big) \cdot b$, and let g be the composition

$$\mathbf{Spin}(A, \sigma, f) \hookrightarrow \mathbf{\Gamma}(A, \sigma, f) \xrightarrow{\chi} \mathbf{O}^+(A, \sigma, f).$$

Clearly, $\ker g = \boldsymbol{\mu}_2$ and, since g_{alg} is surjective, g is surjective, hence g is a central isogeny and

$$\mathbf{O}^+(A, \sigma, f) \simeq \mathbf{Spin}(A, \sigma, f)/\boldsymbol{\mu}_2.$$

Consider the natural homomorphism

$$C \colon \mathbf{PGO}^+(A, \sigma, f) \to \mathbf{Aut}_Z\big(C(A, \sigma, f), \underline{\sigma}\big).$$

If $n = \deg A$ with $n > 2$, then $c(A)_R$ generates the R-algebra $C(A, \sigma, f)_R$ for any $R \in \mathsf{Alg}_F$. Hence C_R is injective and C is a closed embedding by Proposition (22.2). By (23.3), there is an exact sequence

$$1 \to \mathbf{GL}_1(Z) \to \mathbf{Sim}\big(C(A, \sigma, f), \underline{\sigma}\big) \xrightarrow{\;\mathrm{Int}\;} \mathbf{Aut}_Z\big(C(A, \sigma, f), \underline{\sigma}\big) \to 1.$$

Let $\mathbf{\Omega}(A, \sigma, f)$ be the group $\mathrm{Int}^{-1}(\mathrm{im}\, C)$, which we call the *extended Clifford group*. Note that $\mathbf{\Gamma}(A, \sigma, f) \subset \mathbf{\Omega}(A, \sigma, f) \subset \mathbf{Sim}\big(C(A, \sigma, f), \underline{\sigma}\big)$. By Proposition (22.4) we have a commutative diagram with exact rows:

$$
\begin{array}{ccccccccc}
1 & \longrightarrow & \mathbf{G}_{\mathrm{m}} & \longrightarrow & \mathbf{\Gamma}(A, \sigma, f) & \longrightarrow & \mathbf{O}^+(A, \sigma, f) & \longrightarrow & 1 \\
 & & \downarrow & & \downarrow & & \downarrow & & \\
1 & \longrightarrow & \mathbf{GL}_1(Z) & \longrightarrow & \mathbf{\Omega}(A, \sigma, f) & \longrightarrow & \mathbf{PGO}^+(A, \sigma, f) & \longrightarrow & 1.
\end{array}
$$

The first two vertical maps are injective. By Corollary (22.12), the group $\mathbf{\Omega}(A, \sigma, f)$ is smooth.

(23.6) Remark. If $\mathrm{char}\, F \neq 2$, the involution σ is orthogonal and f is prescribed. We then have,

$$\mathbf{O}(A, \sigma, f) = \mathbf{O}(A, \sigma)$$
$$\mathbf{GO}(A, \sigma, f) = \mathbf{GO}(A, \sigma)$$
$$\mathbf{PGO}(A, \sigma, f) = \mathbf{PGO}(A, \sigma).$$

§24. Root Systems

In this section we recall basic results from the theory of root systems and refer to Bourbaki [53] for details. Let V be an $\mathbb{R}$-vector space of positive finite dimension. An endomorphism $s \in \mathrm{End}(V)$ is called a *reflection with respect to* $\alpha \in V$, $\alpha \neq 0$ if

(a) $s(\alpha) = -\alpha$,
(b) there is a hyperplane $W \subset V$ such that $s|_W = \mathrm{Id}$.

We will use the natural pairing

$$V^* \otimes V \to \mathbb{R}, \quad \chi \otimes v \mapsto \langle \chi, v \rangle = \chi(v).$$

A reflection s with respect to α is given by the formula $s(v) = v - \langle \chi, v \rangle \alpha$ for a uniquely determined linear form $\chi \in V^*$ with $\chi|_W = 0$ and $\langle \chi, \alpha \rangle = 2$. Note that a finite set of vectors which spans V is preserved as a set by at most one reflection with respect to any given α (see Bourbaki [53, Chapter VI, §1, Lemme 1]).

A finite subset $\Phi \subset V \neq 0$ is called a *(reduced) root system* if

(a) $0 \notin \Phi$ and Φ spans V.

(b) If $\alpha \in \Phi$ and $x\alpha \in \Phi$ for $x \in \mathbb{R}$, then $x = \pm 1$.

(c) For each $\alpha \in \Phi$ there is a reflection s_α with respect to α such that $s_\alpha(\Phi) = \Phi$.

(d) For each $\alpha, \beta \in \Phi$, $s_\alpha(\beta) - \beta$ is an integral multiple of α.

The elements of Φ are called *roots*. The reflection s_α in (c) is uniquely determined by α. For $\alpha \in \Phi$, we define $\alpha^* \in V^*$ by

$$s_\alpha(v) = v - \langle \alpha^*, v \rangle \cdot \alpha.$$

Such α^* are called *coroots*. The set $\Phi^* = \{ \alpha^* \in V \}$ forms the *dual root system* in V^*. Clearly, $\langle \alpha^*, \beta \rangle \in \mathbb{Z}$ for any $\alpha, \beta \in \Phi$ and $\langle \alpha^*, \alpha \rangle = 2$.

An *isomorphism of root systems* (V, Φ) and (V', Φ') is an isomorphism of vector spaces $f: V \to V'$ such that $f(\Phi) = \Phi'$. The automorphism group $\mathrm{Aut}(V, \Phi)$ is a finite group. The subgroup $W(\Phi)$ of $\mathrm{Aut}(V, \Phi)$ generated by all the reflections s_α, $\alpha \in \Phi$, is called the *Weyl group* of Φ.

Let Φ_i be a root system in V_i, $i = 1, 2, \ldots, n$, and $V = V_1 \oplus V_2 \oplus \cdots \oplus V_n$, $\Phi = \Phi_1 \cup \Phi_2 \cup \cdots \cup \Phi_n$. Then Φ is a root system in V, called the *sum* of the Φ_i. We write $\Phi = \Phi_1 + \Phi_2 + \cdots + \Phi_n$. A root system Φ is called *irreducible* if Φ is not isomorphic to the sum $\Phi_1 + \Phi_2$ of some root systems. Any root system decomposes uniquely into a sum of irreducible root systems.

Let Φ be a root system in V. Denote by Λ_r the (additive) subgroup of V generated by all roots $\alpha \in \Phi$; Λ_r is a lattice in V, called the *root lattice*. The lattice

$$\Lambda = \{ v \in V \mid \langle \alpha^*, v \rangle \in \mathbb{Z} \text{ for } \alpha \in \Phi \}$$

in V, dual to the root lattice generated by $\Phi^* \subset V^*$, is called the *weight lattice*. Clearly, $\Lambda_r \subset \Lambda$ and Λ/Λ_r is a finite group. The group $\mathrm{Aut}(V, \Phi)$ acts naturally on Λ, Λ_r, and Λ/Λ_r, and $W(\Phi)$ acts trivially on Λ/Λ_r.

A subset $\Pi \subset \Phi$ of the root system Φ is a *system of simple roots* or a *base of a root system* if for any $\alpha \in \Phi$,

$$\alpha = \sum_{\beta \in \Pi} n_\beta \cdot \beta$$

for some uniquely determined $n_\beta \in \mathbb{Z}$ and either $n_\beta \geq 0$ for all $\beta \in \Pi$ or $n_\beta \leq 0$ for all $\beta \in \Pi$. In particular, Π is a basis of V. For a system of simple roots $\Pi \subset \Phi$ and $w \in W(\Phi)$ the subset $w(\Pi)$ is also a system of simple roots. Every root system has a base and the Weyl group $W(\Phi)$ acts simply transitively on the set of bases of Φ.

Let Φ be a root system in V and $\Pi \subset \Phi$ be a base. We define a graph, called the *Dynkin diagram* of Φ, which has Π as its set of vertices. The vertices α and β are connected by $\langle \alpha^*, \beta \rangle \cdot \langle \beta^*, \alpha \rangle$ edges. If $\langle \alpha^*, \beta \rangle > \langle \beta^*, \alpha \rangle$, then all the edges between α and β are directed, with α the origin and β the target. This graph does not depend (up to isomorphism) on the choice of a base $\Pi \subset \Phi$, and is denoted $\mathrm{Dyn}(\Phi)$. The group of automorphisms of $\mathrm{Dyn}(\Phi)$ embeds into $\mathrm{Aut}(V, \Phi)$, and $\mathrm{Aut}(V, \Phi)$ is a semidirect product of $W(\Phi)$ (a normal subgroup) and $\mathrm{Aut}(\mathrm{Dyn}(\Phi))$. In particular, $\mathrm{Aut}(\mathrm{Dyn}(\Phi))$ acts naturally on Λ/Λ_r.

Two root systems are isomorphic if and only if their Dynkin diagrams are isomorphic. A root system is irreducible if and only if its Dynkin diagram is connected. The Dynkin diagram of a sum $\Phi_1 + \cdots + \Phi_n$ is the disjoint union of the $\mathrm{Dyn}(\Phi_i)$.

Let $\Pi \subset \Phi$ be a system of simple roots. The set

$$\Lambda_+ = \{ \chi \in \Lambda \mid \langle \alpha^*, \chi \rangle \geq 0 \text{ for } \alpha \in \Pi \}$$

is the *cone of dominant weights* in Λ (relative to Π). We introduce a partial ordering on Λ: $\chi > \chi'$ if $\chi - \chi'$ is sum of simple roots. For any $\lambda \in \Lambda/\Lambda_r$ there exists a unique *minimal dominant weight* $\chi(\lambda) \in \Lambda_+$ in the coset λ. Clearly, $\chi(0) = 0$.

24.A. Classification of irreducible root systems. There are four infinite families A_n, B_n, C_n, D_n and five exceptional irreducible root systems E_6, E_7, E_8, F_4, G_2. We refer to Bourbaki [54] for the following datas about root systems.

Type A_n, $n \geq 1$. Let $V = \mathbb{R}^{n+1}/(e_1 + e_2 + \cdots + e_{n+1})\mathbb{R}$ where $\{e_1, \ldots, e_{n+1}\}$ is the canonical basis of $\mathbb{R}^{n+1}$. We denote by $\bar{e}_i$ the class of e_i in V.

Root system   : $\Phi = \{\bar{e}_i - \bar{e}_j \mid i \neq j\}$, $n(n+1)$ roots.

Root lattice   : $\Lambda_r = \{\sum a_i \bar{e}_i \mid \sum a_i = 0\}$.

Weight lattice   : $\Lambda = \sum \bar{e}_i \mathbb{Z}$, $\Lambda/\Lambda_r \simeq \mathbb{Z}/(n+1)\mathbb{Z}$.

Simple roots   : $\Pi = \{\bar{e}_1 - \bar{e}_2, \bar{e}_2 - \bar{e}_3, \ldots, \bar{e}_n - \bar{e}_{n+1}\}$.

Dynkin diagram   :   o———o—— $\cdots$ ——o
12n

$\mathrm{Aut}(\mathrm{Dyn}(\Phi))$: $\{1\}$ if $n = 1$, $\{1, \tau\}$ if $n \geq 2$.

Dominant weights : $\Lambda_+ = \{\sum a_i \cdot \bar{e}_i \in \Lambda \mid a_1 \geq a_2 \geq \cdots \geq a_{n+1}\}$.

Minimal weights   : $\bar{e}_1 + \bar{e}_2 + \cdots + \bar{e}_i$, $i = 1, 2, \ldots, n+1$.

Type B_n, $n \geq 1$. Let $V = \mathbb{R}^n$ with canonical basis $\{e_i\}$.

Root system   : $\Phi = \{\pm e_i, \pm e_i \pm e_j \mid i > j\}$, $2n^2$ roots.

Root lattice   : $\Lambda_r = \mathbb{Z}^n$.

Weight lattice   : $\Lambda = \Lambda_r + \frac{1}{2}(e_1 + e_2 + \cdots + e_n)\mathbb{Z}$, $\Lambda/\Lambda_r \simeq \mathbb{Z}/2\mathbb{Z}$.

Simple roots   : $\Pi = \{e_1 - e_2, e_2 - e_3, \ldots, e_{n-1} - e_n, e_n\}$.

Dynkin diagram   :   o———o—— $\cdots$ ——o$\Rightarrow$o
12n-1$$n

$\mathrm{Aut}(\mathrm{Dyn}(\Phi))$: $\{1\}$.

Dominant weights : $\Lambda_+ = \{\sum a_i e_i \in \Lambda \mid a_1 \geq a_2 \geq \cdots \geq a_n \geq 0\}$.

Minimal weights   : 0, $\frac{1}{2}(e_1 + e_2 + \cdots + e_n)$.

Type C_n, $n \geq 1$. Let $V = \mathbb{R}^n$ with canonical basis $\{e_i\}$.

Root system   : $\Phi = \{\pm 2e_i, \pm e_i \pm e_j \mid i > j\}$, $2n^2$ roots.

Root lattice   : $\Lambda_r = \{\sum a_i e_i \mid a_i \in \mathbb{Z}, \sum a_i \in 2\mathbb{Z}\}$.

Weight lattice   : $\Lambda = \mathbb{Z}^n$, $\Lambda/\Lambda_r \simeq \mathbb{Z}/2\mathbb{Z}$.

Simple roots   : $\Pi = \{e_1 - e_2, e_2 - e_3, \ldots, e_{n-1} - e_n, 2e_n\}$.

Dynkin diagram   :   o———o—— $\cdots$ ——o$\Leftarrow$o
12n-1$$n

$\mathrm{Aut}(\mathrm{Dyn}(\Phi))$: $\{1\}$.

Dominant weights : $\Lambda_+ = \{\sum a_i e_i \in \Lambda \mid a_1 \geq a_2 \geq \cdots \geq a_n \geq 0\}$.

Minimal weights   : 0, e_1.

Type D_n, $n \geq 3$. (For $n = 2$ the definition works but yields $A_1 + A_1$.) Let $V = \mathbb{R}^n$ with canonical basis $\{e_i\}$.

Root system : $\Phi = \{\pm e_i \pm e_j \mid i > j\}$, $\quad 2n(n-1)$ roots.

Root lattice : $\Lambda_r = \{\sum a_i e_i \mid a_i \in \mathbb{Z}, \sum a_i \in 2\mathbb{Z}\}$.

Weight lattice : $\Lambda = \mathbb{Z}^n + \frac{1}{2}(e_1 + e_2 + \cdots + e_n)\mathbb{Z}$,

$$\Lambda/\Lambda_r \simeq \begin{cases} \mathbb{Z}/2\mathbb{Z} \oplus \mathbb{Z}/2\mathbb{Z} & \text{if } n \text{ is even,} \\ \mathbb{Z}/4\mathbb{Z} & \text{if } n \text{ is odd.} \end{cases}$$

Simple roots : $\Pi = \{e_1 - e_2, \ldots, e_{n-1} - e_n, e_{n-1} + e_n\}$.

Dynkin diagram :

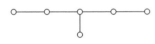

$\mathrm{Aut}\big(\mathrm{Dyn}(\Phi)\big)$: S_3 if $n = 4$, $\quad \{1, \tau\}$ if $n = 3$ or $n > 4$.

Dominant weights : $\Lambda_+ = \{\sum a_i e_i \in \Lambda \mid a_1 \geq a_2 \geq \cdots \geq a_n, a_{n-1} + a_n \geq 0\}$.

Minimal weights : $0, e_1, \frac{1}{2}(e_1 + e_2 + \cdots + e_{n-1} \pm e_n)$.

Exceptional types.

E_6: $\mathrm{Aut}\big(\mathrm{Dyn}(\Phi)\big) = \{1, \tau\}$, $\quad \Lambda/\Lambda_r \simeq \mathbb{Z}/3\mathbb{Z}$.

E_7: $\mathrm{Aut}\big(\mathrm{Dyn}(\Phi)\big) = \{1\}$, $\quad \Lambda/\Lambda_r \simeq \mathbb{Z}/2\mathbb{Z}$.

E_8: $\mathrm{Aut}\big(\mathrm{Dyn}(\Phi)\big) = \{1\}$, $\quad \Lambda/\Lambda_r = 0$.

F_4: $\mathrm{Aut}\big(\mathrm{Dyn}(\Phi)\big) = \{1\}$, $\quad \Lambda/\Lambda_r = 0$.

G_2: $\mathrm{Aut}\big(\mathrm{Dyn}(\Phi)\big) = \{1\}$, $\quad \Lambda/\Lambda_r = 0$.

§25. Split Semisimple Groups

In this section we give the classification of split semisimple groups over an arbitrary field F. The classification does not depend on the base field and corresponds to the classification over an algebraically closed field. The basic references are Borel-Tits [47] and Tits [290]. An algebraic group G over F is said to be *solvable* if the abstract group $G(F_{\mathrm{alg}})$ is solvable, and *semisimple* if $G \neq 1$, G is connected, and $G_{F_{\mathrm{alg}}}$ has no nontrivial solvable connected normal subgroups.

A subtorus $T \subset G$ is said to be *maximal* if it is not contained in a larger subtorus. Maximal subtori remain maximal over arbitrary field extensions and are conjugate over F_{alg} by an element of $G(F_{\text{alg}})$. A semisimple group is *split* if it contains a split maximal torus. Any semisimple group over a separably closed field is split.

We will classify split semisimple groups over an arbitrary field. Let G be split semisimple and let $T \subset G$ be a split maximal torus. Consider the adjoint representation (see Example (22.19.2)):

$$\text{ad} \colon G \to \mathbf{GL}\big(\text{Lie}(G)\big).$$

By the theory of representations of diagonalizable groups (see (22.20)) applied to the restriction of ad to T, we get a decomposition

$$\text{Lie}(G) = \bigoplus_\alpha V_\alpha$$

where the sum is taken over all weights $\alpha \in T^*$ of the representation $\text{ad}\,|_T$. The non-zero weights of the representation are called the *roots* of G (with respect to T). The multiplicity of a root is 1, i.e., $\dim V_\alpha = 1$ if $\alpha \neq 0$ (we use additive notation for T^*).

(25.1) Theorem. *The set $\Phi(G)$ of all roots of G is a root system in $T^* \otimes_{\mathbb{Z}} \mathbb{R}$.* $\square$

The root system $\Phi(G)$ does not depend (up to isomorphism) on the choice of a maximal split torus and is called the *root system of G*. We say that G is *of type Φ* if $\Phi \simeq \Phi(G)$. The root lattice Λ_r clearly is contained in T^*.

(25.2) Proposition. *For any $\alpha \in \Phi(G)$ and $\chi \in T^*$ one has $\langle \alpha^*, \chi \rangle \in \mathbb{Z}$. In particular $\Lambda_r \subset T^* \subset \Lambda$.* $\square$

Consider pairs (Φ, A) where Φ is a root system in some $\mathbb{R}$-vector space V and $A \subset V$ is an (additive) subgroup such that $\Lambda_r \subset A \subset \Lambda$. An *isomorphism* $(\Phi, A) \xrightarrow{\sim} (\Phi', A')$ of pairs is an $\mathbb{R}$-linear isomorphism $f \colon V \xrightarrow{\sim} V'$ such that $f(\Phi) = \Phi'$ and $f(A) = A'$. To each split semisimple group G with a split maximal torus $T \subset G$ one associates the pair $\big(\Phi(G), T^*\big)$.

(25.3) Theorem. *Let G_i be split semisimple groups with a split maximal torus T_i, $i = 1, 2$. Then G_1 and G_2 are isomorphic if and only if the pairs $\big(\Phi(G_1), T_1^*\big)$ and $\big(\Phi(G_2), T_2^*\big)$ are isomorphic.* $\square$

When are two pairs (Φ_1, A_1) and (Φ_2, A_2) isomorphic? Clearly, a necessary condition is that $\Phi_1 \simeq \Phi_2$. Assume that $\Phi_1 = \Phi_2 = \Phi$, then $\Lambda_r \subset A_i \subset \Lambda$ for $i = 1, 2$.

(25.4) Proposition. *$(\Phi, A_1) \simeq (\Phi, A_2)$ if and only if A_1/Λ_r and A_2/Λ_r are conjugate under the action of $\text{Aut}(V, \Phi)/W(\Phi) \simeq \text{Aut}\big(\text{Dyn}(\Phi)\big)$.* $\square$

Thus, to every split semisimple group G one associates two invariants: a root system $\Phi = \Phi(G)$ and a (finite) subgroup $T^*/\Lambda_r \subset \Lambda/\Lambda_r$ modulo the action of $\text{Aut}\big(\text{Dyn}(\Phi)\big)$.

(25.5) Theorem. *For any root system Φ and any additive group A with $\Lambda_r \subset A \subset \Lambda$ there exists a split semisimple group G such that $\big(\Phi(G), T^*\big) \simeq (\Phi, A)$.* $\square$

A split semisimple group G is called *adjoint* if $T^* = \Lambda_r$ and *simply connected simply connected* if $T^* = \Lambda$. These two types of groups are determined (up to isomorphism) by their root systems.

Central isogenies. Let $\pi\colon G \to G'$ be a central isogeny of semisimple groups and let $T' \subset G'$ be a split maximal torus. Then, $T = \pi^{-1}(T')$ is a split maximal torus in G and the natural homomorphism $T'^* \to T^*$ induces an isomorphism of root systems $\Phi(G') \xrightarrow{\sim} \Phi(G)$.

Let G be a split semisimple group with a split maximal torus T. The kernel $C = C(G)$ of the adjoint representation ad_G is a subgroup of T and hence is a diagonalizable group (not necessarily smooth!). The restriction map $T^* \to C^*$ induces an isomorphism $T^*/\Lambda_r \simeq C^*$. Hence, C is a Cartier dual to the constant group T^*/Λ_r. One can show that $C(G)$ is the center of G in the sense of Waterhouse [305]. The image of the adjoint representation ad_G is the adjoint group $\overline{G}$, so that $\overline{G} = G/C$.

If G is simply connected then $C^* \simeq \Lambda/\Lambda_r$ and all other split semisimple groups with root system isomorphic to $\Phi(G)$ are of the form G/N where N is an arbitrary subgroup of C, Cartier dual to a subgroup in $(\Lambda/\Lambda_r)_{\mathrm{const}}$. Thus, for any split semisimple G there are central isogenies

(25.6) $$\widetilde{G} \xrightarrow{\widetilde{\pi}} G \xrightarrow{\pi} \overline{G}$$

with $\widetilde{G}$ simply connected and $\overline{G}$ adjoint.

(25.7) Remark. The central isogenies $\widetilde{\pi}$ and π are unique in the following sense: If $\widetilde{\pi}'$ and π' is another pair of isogenies then there exist $\alpha \in \mathrm{Aut}(\widetilde{G})$ and $\beta \in \mathrm{Aut}(\overline{G})$ such that $\widetilde{\pi}' = \widetilde{\pi} \circ \alpha$ and $\pi' = \beta \circ \pi$.

25.A. Simple split groups of type A, B, C, D, F, and G. A split semisimple group G is said to be *simple* if G_{alg} has no nontrivial connected normal subgroups.

(25.8) Proposition. *A split semisimple group G is simple if and only if $\Phi(G)$ is an irreducible root system. A simply connected (resp. adjoint) split semisimple group G is the direct product of uniquely determined simple subgroups G_i and $\Phi(G) \simeq \sum \Phi(G_i)$.* $\qquad\square$

Type A_n, $n \geq 1$. Let V be an F-vector space of dimension $n + 1$ and let $G = \mathbf{SL}(V)$. A choice of a basis in V identifies G with a subgroup in $\mathbf{GL}_{n+1}(F)$. The subgroup $T \subset G$ of diagonal matrices is a split maximal torus in G. Denote by $\chi_i \in T^*$ the character

$$\chi_i\big(\mathrm{diag}(t_1, t_2, \ldots, t_{n+1})\big) = t_i, \quad i = 1, 2, \ldots, n+1.$$

The character group T^* then is identified with $\mathbb{Z}^n/(e_1 + e_2 + \cdots + e_{n+1})\mathbb{Z}$ by $\bar{e}_i \leftrightarrow \chi_i$.

The Lie algebra of G consists of the trace zero matrices. The torus T acts on $\mathrm{Lie}(G)$ by conjugation through the adjoint representation (see (22.19)). The weight subspaces in $\mathrm{Lie}(G)$ are:

(a) The space of diagonal matrices (trivial weight),
(b) $F \cdot E_{ij}$ for all $1 \leq i \neq j \leq n+1$ with weight $\chi_i \cdot \chi_j^{-1}$.

We get therefore the root system $\{\, \bar{e}_i - \bar{e}_j \mid i \neq j \,\}$ (in additive notation) in the space $T^* \otimes_{\mathbb{Z}} \mathbb{R}$, of type A_n. One can show that $\mathbf{SL}(V)$ is a simple group and since $T^* = \sum \mathbb{Z} \cdot \bar{e}_i = \Lambda$, it is simply connected. The kernel of the adjoint representation of G is $\boldsymbol{\mu}_{n+1}$. Thus:

(25.9) Theorem. *A split simply connected simple group of type A_n is isomorphic to $\mathbf{SL}(V)$ where V is an F-vector space of dimension $n + 1$. All other split semisimple groups of type A_n are isomorphic to $\mathbf{SL}(V)/\boldsymbol{\mu}_k$ where k divides $n + 1$. The group $\mathbf{SL}(V)/\boldsymbol{\mu}_{n+1} \simeq \mathbf{PGL}(V)$ is adjoint.* $\qquad\square$

Type B_n, $n \geq 1$. Let V be an F-vector space of dimension $2n+1$ with a regular quadratic form q and associated polar form b_q. Assume that b_q is of maximal Witt index. Choose a basis $(v_0, v_1, \ldots, v_{2n})$ of V such that $b_q(v_0, v_i) = 0$ for all $i \geq 1$ and

$$b_q(v_i, v_j) = \begin{cases} 1 & \text{if } i = j \pm n, \text{ with } i, j \geq 1, \\ 0 & \text{otherwise.} \end{cases}$$

Consider the group $G = \mathbf{O}^+(V, q) \subset \mathbf{GL}_{2n+1}(F)$. The subgroup T of diagonal matrices $t = \mathrm{diag}(1, t_1, \ldots, t_n, t_1^{-1}, \ldots, t_n^{-1})$ is a split maximal torus of G. Let χ_i be the character $\chi_i(t) = t_i$, $(1 \leq i \leq n)$, and identify T^* with $\mathbb{Z}^n$ via $\chi_i \leftrightarrow e_i$.

The Lie algebra of G consists of all $x \in \mathrm{End}(V) = M_{2n+1}(F)$ such that $b_q(v, xv) = 0$ for all $v \in V$ and $\mathrm{tr}(x) = 0$. The weight subspaces in $\mathrm{Lie}(G)$ with respect to $\mathrm{ad}\,|_T$ are:

(a) The space of diagonal matrices in $\mathrm{Lie}(G)$ (trivial weight),
(b) $F \cdot (E_{i,n+j} - E_{j,n+i})$ for all $1 \leq i < j \leq n$ with weight $\chi_i \cdot \chi_j$,
(c) $F \cdot (E_{i+n,j} - E_{j+n,i})$ for all $1 \leq i < j \leq n$ with weight $\chi_i^{-1} \cdot \chi_j^{-1}$,
(d) $F \cdot (E_{ij} - E_{n+j,n+i})$ for all $1 \leq i \neq j \leq n$ with weight $\chi_i \cdot \chi_j^{-1}$,
(e) $F \cdot (E_{0i} - 2aE_{n+i,0})$ where $a = q(v_0)$, for all $1 \leq i \leq n$ with weight χ_i^{-1},
(f) $F \cdot (E_{0,n+i} - 2aE_{i,0})$ for all $1 \leq i \leq n$ with weight χ_i.

We get the root system $\{\pm e_i, \pm e_i \pm e_j \mid i > j\}$ in $\mathbb{R}^n$ of type B_n. One can show that $\mathbf{O}^+(V, q)$ is a simple group, and since $T^* = \Lambda_r$, it is adjoint. The corresponding simply connected group is $\mathbf{Spin}(V, q)$. Thus:

(25.10) Theorem. *A split simple group of type B_n is isomorphic to $\mathbf{Spin}(V, q)$ (simply connected) or to $\mathbf{O}^+(V, q)$ (adjoint) where (V, q) is a regular quadratic form of dimension $2n + 1$ with polar form b_q which is hyperbolic on $V/\mathrm{rad}(b_q)$.* $\qquad\square$

Type C_n, $n \geq 1$. Let V be a F-vector space of dimension $2n$ with a nondegenerate alternating form h. Choose a basis $(v_1, v_2, \ldots, v_{2n})$ of V such that

$$h(v_i, v_j) = \begin{cases} 1 & \text{if } j = i + n, \\ -1 & \text{if } j = i - n, \\ 0 & \text{otherwise.} \end{cases}$$

Consider the group $G = \mathbf{Sp}(V, h) \subset \mathbf{GL}_{2n}(F)$. The subgroup T of diagonal matrices $t = \mathrm{diag}(t_1, \ldots, t_n, t_1^{-1}, \ldots t_n^{-1})$ is a split maximal torus in G. Let χ_i be the character $\chi_i(t) = t_i$ $(1 \leq i \leq n)$ and identify T^* with $\mathbb{Z}^n$ via $\chi_i \leftrightarrow e_i$.

The Lie algebra of G consists of all $x \in \mathrm{End}(V) = M_{2n}(F)$ such that

$$h(xv, u) + h(v, xu) = 0$$

for all $v, u \in V$. The weight subspaces in $\mathrm{Lie}(G)$ with respect to $\mathrm{ad}\,|_T$ are:

(a) The space of diagonal matrices in $\mathrm{Lie}(G)$ (trivial weight),
(b) $F \cdot (E_{i,n+j} + E_{j,n+i})$ for all $1 \leq i < j \leq n$ with weight $\chi_i \cdot \chi_j$,
(c) $F \cdot (E_{i+n,j} + E_{j+n,i})$ for all $1 \leq i < j \leq n$ with weight $\chi_i^{-1} \cdot \chi_j^{-1}$,
(d) $F \cdot (E_{ij} - E_{n+j,n+i})$ for all $1 \leq i \neq j \leq n$ with weight $\chi_i \cdot \chi_j^{-1}$,
(e) $F \cdot E_{i,n+i}$ for all $1 \leq i \leq n$ with weight χ_i^2,
(f) $F \cdot E_{n+i,i}$ for all $1 \leq i \leq n$ with weight χ_i^{-2}.

We get the root system $\{\pm 2e_i, \pm e_i \pm e_j \mid i > j\}$ in $\mathbb{R}^n$ of type C_n. One can show that $\mathbf{Sp}(V, h)$ is a simple group, and since $T^* = \Lambda$, it is simply connected. The corresponding adjoint group is $\mathbf{PGSp}(V, h)$. Thus

(25.11) Theorem. *A split simple group of type C_n is isomorphic either to $\mathbf{Sp}(V, h)$ (simply connected) or to $\mathbf{PGSp}(V, h)$ (adjoint) where (V, h) is a non-degenerate alternating form of dimension $2n$.* $\square$

Type D_n, $n \geq 2$. Let (V, q) be a hyperbolic quadratic space of dimension $2n$ over F. Choose a basis $(v_1, v_2, \ldots, v_{2n})$ in V such that

$$b_q(v_i, v_j) = \begin{cases} 1 & \text{if } i = j \pm n, \\ 0 & \text{otherwise.} \end{cases}$$

Consider the group $G = \mathbf{O}^+(V, q) \subset \mathbf{GL}_{2n}(F)$. The subgroup T of diagonal matrices $t = \operatorname{diag}(t_1, \ldots, t_n, t_1^{-1}, \ldots, t_n^{-1})$ is a split maximal torus in G. As in the preceding case we identify T^* with $\mathbb{Z}^n$.

The Lie algebra of G consists of all $x \in \operatorname{End}(V) = M_{2n}(F)$, such that $h(v, xv) = 0$ for all $v \in V$.

The weight subspaces in $\operatorname{Lie}(G)$ with respect to $\operatorname{ad}|_T$ are:

(a) The space of diagonal matrices in $\operatorname{Lie}(G)$ (trivial weight).
(b) $F \cdot (E_{i,n+j} - E_{j,n+i})$ for all $1 \leq i < j \leq n$ with the weight $\chi_i \cdot \chi_j$,
(c) $F \cdot (E_{i+n,j} - E_{j+n,i})$ for all $1 \leq i < j \leq n$ with weight $\chi_i^{-1} \cdot \chi_j^{-1}$,
(d) $F \cdot (E_{ij} - E_{j+n,i+n})$ for all $1 \leq i \neq j \leq n$ with weight $\chi_i \cdot \chi_j^{-1}$.

We get the root system $\{\pm e_i \pm e_j \mid i > j\}$ in $\mathbb{R}^n$ of type D_n. The group $\mathbf{O}^+(V, q)$ is a semisimple group (simple, if $n \geq 3$) and $\Lambda_r \subsetneq T^* \subsetneq \Lambda$. The corresponding simply connected and adjoint groups are $\mathbf{Spin}(V, q)$ and $\mathbf{PGO}^+(V, q)$, respectively. If n is odd, then Λ/Λ_r is cyclic and there are no other split groups of type D_n. If n is even, there are three proper subgroups in $\Lambda/\Lambda_r \simeq (\mathbb{Z}/2\mathbb{Z})^2$, one of which corresponds to $\mathbf{O}^+(V, q)$. The two other groups correspond to the images of the compositions

$$\mathbf{Spin}(V, q) \hookrightarrow \mathbf{GL}_1\big(C_0(V, q)\big) \to \mathbf{GL}_1\big(C^\pm(V, q)\big)$$

where $C_0(V, q) = C^+(V, q) \oplus C^-(V, q)$. We denote these groups by $\mathbf{Spin}^\pm(V, q)$. They are isomorphic under any automorphism of $C_0(V, q)$ which interchanges its two components.

(25.12) Theorem. *A split simple group of type D_n is isomorphic to one of the following groups: $\mathbf{Spin}(V, q)$ (simply connected), $\mathbf{O}^+(V, q)$, $\mathbf{PGO}^+(V, q)$ (adjoint), or (if n is even) to $\mathbf{Spin}^\pm(V, q)$ where (V, q) is a hyperbolic quadratic space of dimension $2n$.* $\square$

Type F_4 and G_2. Split simple groups of type F_4 and G_2 are related to certain types of nonassociative algebras:

(25.13) Theorem. *A split simple group of type F_4 is simply connected and adjoint and is isomorphic to $\mathbf{Aut}_{\mathrm{alg}}(J)$ where J is a split simple exceptional Jordan algebra of dimension 27.*

Reference: See Chevalley-Schafer [67], Freudenthal [101, Satz 4.11], Springer [268] or [269, 14.27, 14.28]. The proof given in [101] is over $\mathbb{R}$, the proofs in [67] and [268] assume that F is a field of characteristic different from 2 and 3. Springer's proof [269] holds for any field. $\square$

For a simple split group of type G_2 we have

(25.14) Theorem. *A split simple group of type G_2 is simply connected and adjoint and is isomorphic to* $\mathbf{Aut}_{\mathrm{alg}}(\mathfrak{C})$ *where* $\mathfrak{C}$ *is a split Cayley algebra.*

Reference: See Jacobson [128], Freudenthal [101] or Springer [268]. The proof in [101] is over $\mathbb{R}$, the one in [128] assumes that F is a field of characteristic zero and [268] gives a proof for arbitrary fields. $\square$

More on (25.13), resp. (25.14) can be found in the notes at the end of Chapter IX, resp. VIII.

25.B. Automorphisms of split semisimple groups. Let G be a split semisimple group over F, let $T \subset G$ be a split maximal torus, and Π a system of simple roots in $\Phi(G)$. For any $\varphi \in \mathrm{Aut}(G)$ there is $g \in G(F)$ such that for $\psi = \mathrm{Int}(g) \circ \varphi$, one has $\psi(T) = T$ and $\psi(\Pi) = \Pi$, hence ψ induces an automorphism of $\mathrm{Dyn}(\Phi)$. Thus, we obtain a homomorphism $\mathrm{Aut}(G) \to \mathrm{Aut}\big(\mathrm{Dyn}(\Phi)\big)$.

On the other hand, we have a homomorphism $\mathrm{Int}\colon G(F) \to \mathrm{Aut}(G)$ taking $g \in G(F)$ to the inner automorphism $\mathrm{Int}(g_R)$ of $G(R)$ for any $R \in \mathsf{Alg}_F$ where g_R is the image of g under $G(F) \to G(R)$.

(25.15) Proposition. *If G is a split semisimple adjoint group, the sequence*

$$1 \to G(F) \xrightarrow{\mathrm{Int}} \mathrm{Aut}(G) \to \mathrm{Aut}\big(\mathrm{Dyn}(\Phi)\big) \to 1$$

is split exact. $\square$

Let G be a split semisimple group (not necessarily adjoint) and let $C = C(G)$ be the kernel of ad_G. Then $\overline{G} = G/C$ is an adjoint group with $\Phi(\overline{G}) = \Phi(G)$ and we have a natural homomorphism $\mathrm{Aut}(G) \to \mathrm{Aut}(\overline{G})$. It turns out to be injective and its image contains $\mathrm{Int}\big(\overline{G}(F)\big)$.

(25.16) Theorem. *Let G be a split semisimple group. Then there is an exact sequence*

$$1 \to \overline{G}(F) \to \mathrm{Aut}(G) \to \mathrm{Aut}\big(\mathrm{Dyn}(\Phi)\big)$$

where the last map is surjective and the sequence splits provided G is simply connected or adjoint. $\square$

(25.17) Corollary. *Let G be a split simply connected semisimple group. Then the natural map* $\mathrm{Aut}(G) \to \mathrm{Aut}(\overline{G})$ *is an isomorphism.* $\square$

§26. Semisimple Groups over an Arbitrary Field

In this section we give the classification of semisimple groups over an arbitrary field which do not contain simple components of types D_4, E_6, E_7 or E_8. We recall that a category A is a groupoid if all morphisms in A are isomorphisms. A groupoid A is *connected* if all its objects are isomorphic.

Let Γ be a profinite group and let A be a groupoid. A Γ-*embedding of* A is a functor $\mathbf{i}\colon \mathsf{A} \to \widetilde{\mathsf{A}}$ where $\widetilde{\mathsf{A}}$ is a connected groupoid, such that for every X, Y in A, there is a continuous Γ-action on the set $\mathrm{Hom}_{\widetilde{\mathsf{A}}}(\mathbf{i}\,X, \mathbf{i}\,Y)$ with the discrete topology, compatible with the composition law in $\widetilde{\mathsf{A}}$, and such that the functor $\mathbf{i}$ induces a bijection

$$\mathrm{Hom}_{\mathsf{A}}(X, Y) \xrightarrow{\sim} \mathrm{Hom}_{\widetilde{\mathsf{A}}}(\mathbf{i}\,X, \mathbf{i}\,Y)^{\Gamma}.$$

It follows from the definition that a Γ-embedding is a faithful functor.

(26.1) Examples. (1) Let $^1A'_n = {}^1A'_n(F)$ be the category of all central simple algebras over F of degree $n+1$ with morphisms being isomorphisms of F-algebras. Then for the group $\Gamma = \mathrm{Gal}(F_{\mathrm{sep}}/F)$ the natural functor

$$\mathbf{j}\colon {}^1A'_n(F) \to {}^1A'_n(F_{\mathrm{sep}}), \quad A \mapsto A_{\mathrm{sep}} = A \otimes_F F_{\mathrm{sep}}$$

is a Γ-embedding.

(2) Let G be an algebraic group over a field F and let $A = A(F)$ be the groupoid of all *twisted forms* of G (objects are algebraic groups G' over F such that $G'_{\mathrm{sep}} \simeq G_{\mathrm{sep}}$ and morphisms are algebraic group isomorphisms over F). Then for $\Gamma = \mathrm{Gal}(F_{\mathrm{sep}}/F)$ the natural functor

$$\mathbf{j}\colon A(F) \to A(F_{\mathrm{sep}}), \quad G' \mapsto G'_{\mathrm{sep}}$$

is a Γ-embedding.

Let $\mathbf{i}\colon A \to \widetilde{A}$ and let $\mathbf{j}\colon B \to \widetilde{B}$ be two Γ-embeddings and let $\mathbf{S}\colon A \to B$ be a functor. A Γ-*extension of* $\mathbf{S}$ (with respect to $\mathbf{i}$ and $\mathbf{j}$) is a functor $\widetilde{\mathbf{S}}\colon \widetilde{A} \to \widetilde{B}$ such that $\mathbf{j} \circ \mathbf{S} = \widetilde{\mathbf{S}} \circ \mathbf{i}$ and for all $\gamma \in \Gamma$, X, $Y \in A$, and $f \in \mathrm{Hom}_{\widetilde{A}}(\mathbf{i}\,X, \mathbf{i}\,Y)$ one has $\gamma\widetilde{\mathbf{S}}(f) = \widetilde{\mathbf{S}}(\gamma f)$.

We call a continuous map $\gamma \in \Gamma \mapsto f_\gamma \in \mathrm{Aut}_{\widetilde{A}}(\mathbf{i}\,X)$ a 1-*cocycle* if

$$f_\gamma \circ \gamma f_\rho = f_{\gamma\rho}$$

for all γ, $\rho \in \Gamma$ and, we say that a Γ-embedding $\mathbf{i}\colon A \to \widetilde{A}$ *satisfies the descent condition* if for any object $X \in A$ and for any 1-cocycle $f_\gamma \in \mathrm{Aut}_{\widetilde{A}}(\mathbf{i}\,X)$ there exist an object $Y \in A$ and a morphism $h\colon \mathbf{i}\,Y \to \mathbf{i}\,X$ in $\widetilde{A}$ such that

$$f_\gamma = h \circ \gamma h^{-1}$$

for all $\gamma \in \Gamma$.

(26.2) Proposition. *Let* $\mathbf{i}\colon A \to \widetilde{A}$ *and* $\mathbf{j}\colon B \to \widetilde{B}$ *be two* Γ-*embeddings and let* $\mathbf{S}\colon A \to B$ *be a functor having a* Γ-*extension* $\widetilde{\mathbf{S}}\colon \widetilde{A} \to \widetilde{B}$. *Assume that the* Γ-*embedding* $\mathbf{i}$ *satisfies the descent condition. If* $\widetilde{\mathbf{S}}$ *is an equivalence of categories, then so is* $\mathbf{S}$.

Proof: Since $\mathbf{i}$, $\mathbf{j}$, and $\widetilde{\mathbf{S}}$ are faithful functors, the functor $\mathbf{S}$ is also faithful. Let $g \in \mathrm{Hom}_B(\mathbf{S}\,X, \mathbf{S}\,Y)$ be any morphism for some X and Y in A. Since $\widetilde{\mathbf{S}}$ is an equivalence of categories, we can find $f \in \mathrm{Hom}_{\widetilde{A}}(\mathbf{i}\,X, \mathbf{i}\,Y)$ such that $\widetilde{\mathbf{S}}(f) = \mathbf{j}(g)$. For any $\gamma \in \Gamma$ one has

$$\widetilde{\mathbf{S}}(\gamma f) = \gamma\widetilde{\mathbf{S}}(f) = \gamma\big(\mathbf{j}(g)\big) = \mathbf{j}(g) = \widetilde{\mathbf{S}}(f),$$

hence $\gamma f = f$. By the definition of a Γ-embedding, there exists $h \in \mathrm{Hom}_A(X, Y)$ such that $\mathbf{i}\,h = f$. The equality $\mathbf{j}\,\mathbf{S}(h) = \widetilde{\mathbf{S}}\,\mathbf{i}(h) = \widetilde{\mathbf{S}}(f) = \mathbf{j}\,g$ shows that $\mathbf{S}(h) = g$, i.e., $\mathbf{S}$ is full as a functor. In view of Maclane [176, p. 91] it remains to check that any object $Z \in B$ is isomorphic to $\mathbf{S}(Y)$ for some $Y \in A$. Take any object $X \in A$. Since $\widetilde{B}$ is a connected groupoid, the objects $\mathbf{j}\,\mathbf{S}(X)$ and $\mathbf{j}\,Z$ are isomorphic. Choose any isomorphism $g\colon \mathbf{j}\,\mathbf{S}(X) \to \mathbf{j}\,Z$ in $\widetilde{B}$ and set

$$g_\gamma = g^{-1} \circ \gamma g \in \mathrm{Aut}_{\widetilde{B}}\big(\mathbf{j}\,\mathbf{S}(X)\big)$$

for $\gamma \in \Gamma$. Clearly, g_γ is a 1-cocycle. Since $\widetilde{\mathbf{S}}$ is bijective on morphisms, there exists a 1-cocycle

$$f_\gamma \in \mathrm{Aut}_{\widetilde{A}}(\mathbf{i}\,X)$$

such that $\widetilde{\mathbf{S}}(f_\gamma) = g_\gamma$ for any $\gamma \in \Gamma$. By the descent condition for the Γ-embedding $\mathbf{i}$ one can find $Y \in A$ and a morphism $h\colon \mathbf{i}\,Y \to \mathbf{i}\,X$ in $\widetilde{A}$ such that $f_\gamma = h \circ \gamma h^{-1}$ for any $\gamma \in \Gamma$. Consider the composition

$$l = g \circ \widetilde{\mathbf{S}}(h)\colon \mathbf{j}\,\mathbf{S}(Y) = \widetilde{\mathbf{S}}\,\mathbf{i}(Y) \to \widetilde{\mathbf{S}}\,\mathbf{i}(X) = \mathbf{j}\,\mathbf{S}(X) \to \mathbf{j}\,Z$$

in $\widetilde{B}$. For any $\gamma \in \Gamma$ one has

$$\gamma l = \gamma g \circ \gamma\widetilde{\mathbf{S}}(h) = g \circ g_\gamma \circ \widetilde{\mathbf{S}}(\gamma h) = g \circ g_\gamma \circ \widetilde{\mathbf{S}}(f_\gamma^{-1} \circ h) = g \circ \widetilde{\mathbf{S}}(h) = l.$$

By the definition of a Γ-embedding, $l = \mathbf{j}(m)$ for some isomorphism $m\colon \mathbf{S}(Y) \to Z$ in B, i.e., Z is isomorphic to $\mathbf{S}(Y)$. $\square$

(26.3) Remark. Since $\widetilde{A}$ and $\widetilde{B}$ are connected groupoids, in order to check that a functor $\widetilde{\mathbf{S}}\colon \widetilde{A} \to \widetilde{B}$ is an equivalence of categories, it suffices to show that for some object $X \in \widetilde{A}$ the map

$$\mathrm{Aut}_{\widetilde{A}}(X) \to \mathrm{Aut}_{\widetilde{B}}\big(\widetilde{\mathbf{S}}(X)\big)$$

is an isomorphism, see Proposition (12.37).

We now introduce a class of Γ-embeddings satisfying the descent condition. All Γ-embeddings occurring in the sequel will be in this class.

Let F be a field and $\Gamma = \mathrm{Gal}(F_{\mathrm{sep}}/F)$. Consider a collection of F-vector spaces $V^{(1)}, V^{(2)}, \ldots, V^{(n)}$, and W (not necessarily of finite dimension). The group Γ acts on $\mathrm{GL}(V_{\mathrm{sep}}^{(i)})$ and $\mathrm{GL}(W_{\mathrm{sep}})$ in a natural way. Let

$$\rho\colon H = \mathrm{GL}(V_{\mathrm{sep}}^{(1)}) \times \cdots \times \mathrm{GL}(V_{\mathrm{sep}}^{(n)}) \to \mathrm{GL}(W_{\mathrm{sep}})$$

be a Γ-equivariant group homomorphism.

Fix an element $w \in W \subset W_{\mathrm{sep}}$ and consider the category $\widetilde{A} = \widetilde{A}(\rho, w)$ whose objects are elements $w' \in W_{\mathrm{sep}}$ such that there exists $h \in H$ with $\rho(h)(w) = w'$ (for example w is always an object of $\widetilde{A}$). The set $\mathrm{Hom}_{\widetilde{A}}(w', w'')$ consists of all those $h \in H$ such that $\rho(h)(w') = w''$. The composition law in $\widetilde{A}$ is induced by the multiplication in H. Clearly, $\widetilde{A}$ is a connected groupoid.

Let $A = A(\rho, w)$ be the subcategory of $\widetilde{A}$ consisting of all $w' \in W$ which are objects in $\widetilde{A}$. Clearly, for any w', $w'' \in A$ the set $\mathrm{Hom}_{\widetilde{A}}(w', w'')$ is Γ-invariant with respect to the natural action of Γ on H, and we set

$$\mathrm{Hom}_A(w', w'') = \mathrm{Hom}_{\widetilde{A}}(w', w'')^\Gamma \subset H^\Gamma.$$

Clearly, the embedding functor $\mathbf{i}\colon A(\rho, w) \to \widetilde{A}(\rho, w)$ is a Γ-embedding.

(26.4) Proposition. *The Γ-embedding $\mathbf{i}\colon A(\rho, w) \to \widetilde{A}(\rho, w)$ satisfies the descent condition.*

Proof: Let $w' \in W$ be an object in A and let $f_\gamma \in \mathrm{Aut}_{\widetilde{A}}(w') \subset H$ be a 1-cocycle. Let $f_\gamma^{(i)}$ be the i-th component of f_γ in $\mathrm{GL}(V_{\mathrm{sep}}^{(i)})$. We introduce a new Γ-action on each $V_{\mathrm{sep}}^{(i)}$ by the formula $\gamma * v = f_\gamma^{(i)}(\gamma v)$ where $\gamma \in \Gamma$, $v \in V_{\mathrm{sep}}^{(i)}$. Clearly, $\gamma * (xv) = \gamma x \cdot (\gamma * v)$ for any $x \in F_{\mathrm{sep}}$. Let $U^{(i)}$ be the F-subspace in $V_{\mathrm{sep}}^{(i)}$ of Γ-invariant elements with respect to the new action. In view of Lemma (18.1) the natural maps

$$\theta^{(i)}\colon F_{\mathrm{sep}} \otimes_F U^{(i)} \to V_{\mathrm{sep}}^{(i)}, \quad x \otimes u \mapsto xu$$

are F_{sep}-isomorphisms of vector spaces. For any $x \in F_{\text{sep}}$ and $u \in U^{(i)}$ one has $\gamma u = f_\gamma^{(i)^{-1}}(u)$, hence (with respect to the usual Γ-action)

$$(f_\gamma^{(i)} \circ \gamma \circ \theta^{(i)})(x \otimes u) = (f_\gamma^{(i)} \circ \gamma)(xu)$$
$$= f_\gamma^{(i)}(\gamma x \cdot \gamma u) = \gamma x \cdot f_\gamma^{(i)}\big(f_\gamma^{(i)^{-1}}(u)\big) = \gamma x \cdot u$$
$$= \theta^{(i)}(\gamma x \otimes u) = \big(\theta^{(i)} \circ \gamma\big)(x \otimes u).$$

In other words, $f_\gamma = \theta \circ \gamma\theta^{-1}$ where $\gamma\theta = \gamma \circ \theta \circ \gamma^{-1}$.

The F-vector spaces $U^{(i)}$ and $V^{(i)}$ have the same dimension and are therefore F-isomorphic. Choose any F-isomorphism $\alpha^{(i)} \colon V^{(i)} \to U^{(i)}$ and consider the composition $\beta^{(i)} = \theta^{(i)} \circ \alpha_{\text{sep}}^{(i)} \in \mathrm{GL}(V_{\text{sep}}^{(i)})$. Clearly,

$$\textbf{(26.5)} \qquad\qquad\qquad f_\gamma = \beta \circ \gamma\beta^{-1}.$$

Consider the element $w'' = \rho(\beta^{-1})(w') \in W_{\text{sep}}$. By definition, w'' is an object of $\widetilde{A}$ and β represents a morphism $w'' \to w'$ in $\widetilde{A}$. We show that $w'' \in W$, i.e., $w'' \in A$. Indeed, for any $\gamma \in \Gamma$ one has

$$\gamma(w'') = \rho(\gamma\beta^{-1})(w') = \rho(\beta^{-1} \circ f_\gamma)(w') = \rho(\beta^{-1})(w') = w''$$

since $\rho(f_\gamma)(w') = w'$. Finally, the equation (26.5) shows that the functor $\mathbf{i}$ satisfies the descent condition. $\qquad\square$

(26.6) Corollary. *The functors* $\mathbf{j}$ *in Examples* (26.1), (1) *and* (26.1), (2) *satisfy the descent condition.*

Proof: For (1) we consider the F-vector space $W = \mathrm{Hom}_F(A \otimes_F A, A)$ where $A = M_{n+1}(F)$ is the split algebra and $w = m \in W$ is the multiplication map of A. For (2), let $A = F[G]$. Consider the F-vector space

$$W = \mathrm{Hom}_F(A \otimes_F A, A) \oplus \mathrm{Hom}(A, A \otimes_F A),$$

the element $w = (m, c) \in W$ where m is the multiplication and c is the comultiplication on A. In each case we have a natural representation

$$\rho_{\text{sep}} \colon \mathrm{GL}(A_{\text{sep}}) \to \mathrm{GL}(W_{\text{sep}})$$

(see Example (20.4), (8)).

We now restrict our attention to Example (26.1),(2), since the argument for Example (26.1), (1) is similar (and even simpler). By Proposition (26.4) there is a Γ-embedding

$$\mathbf{i} \colon A(\rho_{\text{sep}}, w) \to \widetilde{A}(\rho_{\text{sep}}, w)$$

satisfying the descent condition. We have a functor

$$\mathbf{T} \colon A(\rho_{\text{sep}}, w) \to A(F)$$

taking $w' = (m', c') \in A(\rho_{\text{sep}}, w)$ to the F-vector space A with the Hopf algebra structure given by m' and comultiplication c'. Clearly A has a Hopf algebra structure (with some co-inverse map i' and co-unit u') since over F_{sep} it is isomorphic to the Hopf algebra A_{sep}. A morphism between w' and w'', being an element of $\mathrm{GL}(A)$, defines an isomorphism of the corresponding Hopf algebra structures on A. The functor $\mathbf{T}$ has an evident Γ-extension

$$\widetilde{\mathbf{T}} \colon \widetilde{A}(\rho_{\text{sep}}, w) \to A(F_{\text{sep}}),$$

which is clearly an equivalence of groupoids. Since the functor $\mathbf{i}$ satisfies the descent condition, so does the functor $\mathbf{j}$. $\square$

26.A. Basic classification results. Let G be a semisimple algebraic group over an arbitrary field F. Choose any maximal torus $T \subset G$. Then T_{sep} is a split maximal torus in G_{sep}, hence we have a root system $\Phi(G_{\mathrm{sep}})$, which we call the *root system of G* and denote $\Phi(G)$. The absolute Galois group $\Gamma = \mathrm{Gal}(F_{\mathrm{sep}}/F)$ acts naturally on $\Phi(G)$ and hence on the Dynkin diagram $\mathrm{Dyn}\big(\Phi(G)\big)$.

The group G is said to be *simply connected (resp. adjoint)* if the split group G_{sep} is so.

(26.7) Theorem. *For any semisimple group G there exists (up to an isomorphism) a unique simply connected group $\widetilde{G}$ and a unique adjoint group $\overline{G}$ such that there are central isogenies $\widetilde{G} \to G \to \overline{G}$.*

Proof: Let $C \subset G$ be the kernel of the adjoint representation ad_G. Then $\overline{G} = G/C \simeq \mathrm{im}(\mathrm{ad}_G)$ is an adjoint group with the same root system as G. Denote by $\overline{G}^d$ a split twisted form of $\overline{G}$ and by $\widetilde{G}^d$ its simply connected covering. Consider the groupoid $A(F)$ (resp. $B(F)$) of all twisted forms of $\overline{G}^d$ (resp. $\widetilde{G}^d$). The group $\overline{G}$ is an object of $A(F)$. Clearly, the natural functors

$$\mathbf{i}\colon A(F) \to A(F_{\mathrm{sep}}), \quad \mathbf{j}\colon B(F) \to B(F_{\mathrm{sep}})$$

are Γ-embeddings where $\Gamma = \mathrm{Gal}(F_{\mathrm{sep}}/F)$. The natural functor

$$\mathbf{S}(F)\colon B(F) \to A(F), \quad G' \mapsto \overline{G}' = G'/C(G')$$

has the Γ-extension $\mathbf{S}(F_{\mathrm{sep}})$. By Corollary (25.17) the functor $\mathbf{S}(F_{\mathrm{sep}})$ is an equivalence of groupoids. By Proposition (26.2) and Corollary (26.6) $\mathbf{S}(F)$ is also an equivalence of groupoids. Hence there exists a unique (up to isomorphism) simply connected group $\widetilde{G}$ such that $\widetilde{G}/C(\widetilde{G}) \simeq \overline{G}$.

Let $\widetilde{\pi}\colon \widetilde{G} \to \overline{G}$ and $\pi\colon G \to \overline{G}$ be central isogenies. Since G_{sep} is a split group there exists a central isogeny $\rho\colon \widetilde{G}_{\mathrm{sep}} \to G_{\mathrm{sep}}$ (see (25.6)). Remark (25.7) shows that after modifying ρ by an automorphism of $\widetilde{G}_{\mathrm{sep}}$ one can assume that $\pi_{\mathrm{sep}} \circ \rho = \widetilde{\pi}_{\mathrm{sep}}$. Take any $\gamma \in \Gamma$. Since $\gamma\rho\colon \widetilde{G}_{\mathrm{sep}} \to G_{\mathrm{sep}}$ is a central isogeny, by (25.7) there exists $\alpha \in \mathrm{Aut}(\widetilde{G}_{\mathrm{sep}})$ such that $\gamma\rho = \rho \circ \alpha$. Then

$$\widetilde{\pi}_{\mathrm{sep}} = \gamma\widetilde{\pi}_{\mathrm{sep}} = \gamma(\pi_{\mathrm{sep}} \circ \rho) = \pi_{\mathrm{sep}} \circ \gamma\rho = \pi_{\mathrm{sep}} \circ \rho \circ \alpha = \widetilde{\pi}_{\mathrm{sep}} \circ \alpha,$$

hence α belongs to the kernel of $\mathrm{Aut}(\widetilde{G}_{\mathrm{sep}}) \to \mathrm{Aut}(\overline{G}_{\mathrm{sep}})$, which is trivial by Corollary (25.17), i.e., $\alpha = \mathrm{Id}$ and $\gamma\rho = \rho$. Then $\rho = \delta_{\mathrm{sep}}$ for a central isogeny $\delta\colon \widetilde{G} \to G$. $\square$

The group G in Theorem (26.7) is isomorphic to $\widetilde{G}/N$ where N is a subgroup of $C = C(\widetilde{G})$. Note that the Galois group Γ acts on T_{sep}^*, leaving invariant the subset $\Phi = \Phi(G) \subset T_{\mathrm{sep}}^*$, and hence acts on the lattices Λ, Λ_r, and on the group Λ/Λ_r. Note that the Γ-action on Λ/Λ_r factors through the natural action of $\mathrm{Aut}\big(\mathrm{Dyn}(\Phi)\big)$. The group C is finite of multiplicative type, Cartier dual to $(\Lambda/\Lambda_r)_{\mathrm{et}}$ (see p. 357). Therefore, the classification problem of semisimple groups reduces to the classification of simply connected groups and Γ-submodules in Λ/Λ_r. Note that the classifications of simply connected and adjoint groups are equivalent.

A semisimple group G is called *absolutely simple* if G_{sep} is simple. For example, a split simple group is absolutely simple.

(26.8) Theorem. *A simply connected (resp. adjoint) semisimple group over F is isomorphic to the product of groups $R_{L/F}(G')$ where L/F is a finite separable field extension and G' is an absolutely simple simply connected (resp. adjoint) group over L.*

Proof: Let Δ be the set of connected components of the Dynkin diagram of G. The absolute Galois group Γ acts in a natural way on Δ making it a finite Γ-set. Since G is a simply connected or an adjoint group and G_{sep} is split, it follows from Proposition (25.8) that G_{sep} is the product of its simple components over F_{sep} indexed by the elements of Δ:

$$G_{\mathrm{sep}} = \prod_{\delta \in \Delta} G_\delta.$$

Set $A_\delta = F_{\mathrm{sep}}[G_\delta]$, then $F[G]_{\mathrm{sep}}$ is the tensor product over F_{sep} of all A_δ, $\delta \in \Delta$.

Since Γ permutes the connected components of the Dynkin diagram of G, there exist F-algebra isomorphisms

$$\widetilde{\gamma} \colon A_\delta \to A_{\gamma\delta}$$

such that $\widetilde{\gamma}(xa) = \gamma(x)\widetilde{\gamma}(a)$ for all $x \in F_{\mathrm{sep}}$ and $a \in A_\delta$, and the Γ-action on $F[G]_{\mathrm{sep}}$ is given by the formula

$$\gamma(\otimes a_\delta) = \otimes a'_\delta \quad \text{where} \quad a'_{\gamma\delta} = \widetilde{\gamma}(a_\delta).$$

Consider the étale F-algebra $L = \mathrm{Map}(\Delta, F_{\mathrm{sep}})^\Gamma$ corresponding to the finite Γ-set Δ (see Theorem (18.4)). Then Δ can be identified with the set of all F-algebra homomorphisms $L \to F_{\mathrm{sep}}$. In particular,

$$L_{\mathrm{sep}} = L \otimes_F F_{\mathrm{sep}} = \prod_{\delta \in \Delta} e_\delta L_{\mathrm{sep}}$$

where the e_δ are idempotents, and each $e_\delta L_{\mathrm{sep}} \simeq F_{\mathrm{sep}}$.

We will define a group scheme G' over L such that $G \simeq R_{L/F}(G')$. Let S be an L-algebra. The structure map $\alpha \colon L \to S$ gives a decomposition of the identity, $1 = \sum_{\delta \in \Delta} f_\delta$ where the f_δ are the orthogonal idempotents in $S_{\mathrm{sep}} = S \otimes_F F_{\mathrm{sep}}$, which are the images of the e_δ under $\alpha_{\mathrm{sep}} \colon L_{\mathrm{sep}} \to S_{\mathrm{sep}}$; they satisfy $\gamma f_\delta = f_{\gamma\delta}$ for all $\gamma \in \Gamma$. For any $\delta \in \Delta$ consider the group isomorphism

$$\overline{\gamma} \colon G_\delta(f_\delta S_{\mathrm{sep}}) \to G_{\gamma\delta}(f_{\gamma\delta} S_{\mathrm{sep}})$$

taking a homomorphism $u \in \mathrm{Hom}_{Alg_{F_{\mathrm{sep}}}}(A_\delta, f_\delta S_{\mathrm{sep}})$ to

$$\gamma \circ u \circ \widetilde{\gamma}^{-1} \in \mathrm{Hom}_{Alg_{F_{\mathrm{sep}}}}(A_{\gamma\delta}, f_{\gamma\delta} S_{\mathrm{sep}}) = G_{\gamma\delta}(f_{\gamma\delta} S_{\mathrm{sep}}).$$

The collection of $\overline{\gamma}$ defines a Γ-action on the product

$$\prod_{\delta \in \Delta} G_\delta(f_\delta S_{\mathrm{sep}}).$$

We define $G'(S)$ to be the group of Γ-invariant elements in this product. Clearly, G' is a contravariant functor $Alg_L \to Groups$.

Let $S = R \otimes_F L$ where R is an F-algebra. Then

$$S_{\mathrm{sep}} \simeq \prod_{\delta \in \Delta}(R \otimes_F e_\delta L_{\mathrm{sep}}) = \prod_{\delta \in \Delta} f_\delta S_{\mathrm{sep}}$$

where each $f_\delta = 1 \otimes e_\delta \in S \otimes_L L_{\text{sep}}$ and $f_\delta S_{\text{sep}} \simeq R_{\text{sep}}$. Hence

$$G'(R \otimes_F L) = \Big(\prod_{\delta \in \Delta} G_\delta(R_{\text{sep}}) \Big)^\Gamma = G(R_{\text{sep}})^\Gamma = G(R),$$

therefore $G = R_{L/F}(G')$.

By writing L as a product of fields, $L = \prod L_i$, we obtain

$$G \simeq \prod R_{L_i/F}(G'_i)$$

where the G'_i are components of G'. By comparing the two sides of this isomorphism over F_{sep}, we see that G'_i is a semisimple group over L_i. A count of the number of connected components of Dynkin diagrams shows that the G'_i are absolutely simple groups. $\square$

The collection of field extensions L_i/F and absolutely simple groups G'_i in Theorem (26.8) is uniquely determined by G. Thus the theorem reduces the classification problem to the classification of absolutely simple simply connected groups. In what follows we classify such groups of types A_n, B_n, C_n, D_n ($n \neq 4$), F_4, and G_2.

Classification of simple groups of type A_n. As in Chapter IV, consider the groupoid $A_n = A_n(F)$, $n > 1$, of central simple algebras of degree $n + 1$ over some étale quadratic extension of F with a unitary involution which is the identity over F, where the morphisms are the F-algebra isomorphisms which preserve the involution, consider also the groupoid $A_1 = A_1(F)$ of quaternion algebras over F where morphisms are F-algebra isomorphisms.

Let $A^n = A^n(F)$ (resp. $\overline{A}^n = \overline{A}^n(F)$) be the groupoid of simply connected (resp. adjoint) absolutely simple groups of type A_n ($n \geq 1$) over F, where morphisms are group isomorphisms. By §23.A and Theorem (25.9) we have functors

$$\mathbf{S}_n \colon A_n(F) \to A^n(F) \quad \text{and} \quad \overline{\mathbf{S}}_n \colon A_n(F) \to \overline{A}^n(F)$$

defined by $\mathbf{S}_n(B, \tau) = \mathbf{SU}(B, \tau)$, $\overline{\mathbf{S}}_n(B, \tau) = \mathbf{PGU}(B, \tau)$ if $n \geq 2$, and $\mathbf{S}_n(Q) = \mathbf{SL}_1(Q)$, $\overline{\mathbf{S}}_n(Q) = \mathbf{PGL}_1(Q)$ if $n = 1$. Observe that if $B = A \times A^{\text{op}}$ and τ is the exchange involution, then $\mathbf{SU}(B, \tau) = \mathbf{SL}_1(A)$ and

$$\mathbf{PGU}(B, \tau) = \mathbf{PGL}_1(A).$$

(26.9) Theorem. *The functors $\mathbf{S}_n \colon A_n(F) \to A^n(F)$ and $\overline{\mathbf{S}}_n \colon A_n(F) \to \overline{A}^n(F)$ are equivalences of categories.*

Proof: Since the natural functor $A^n(F) \to \overline{A}^n(F)$ is an equivalence (see the proof of Theorem (26.7)), it suffices to prove that $\overline{\mathbf{S}}_n$ is an equivalence. Let $\Gamma = \text{Gal}(F_{\text{sep}}/F)$. The field extension functor $\mathbf{j} \colon A_n(F) \to A_n(F_{\text{sep}})$ is clearly a Γ-embedding. We show that $\mathbf{j}$ satisfies the descent condition. Assume first that $n \geq 2$. Let (B, τ) be some object in $A_n(F)$ (a split object, for example). Consider the F-vector space

$$W = \text{Hom}_F(B \otimes_F B, B) \oplus \text{Hom}_F(B, B),$$

and the element $w = (m, \tau) \in W$ where m is the multiplication on B. The natural representation

$$\rho \colon \mathbf{GL}(B) \to \mathbf{GL}(W).$$

induces a Γ-equivariant homomorphism

$$\rho_{\mathrm{sep}}\colon \mathrm{GL}(B_{\mathrm{sep}}) \to \mathrm{GL}(W_{\mathrm{sep}}).$$

By Proposition (26.4) the Γ-embedding

$$\mathbf{i}\colon A(\rho_{\mathrm{sep}}, w) \to \widetilde{A}(\rho_{\mathrm{sep}}, w)$$

satisfies the descent condition. We have a functor

$$\mathbf{T} = \mathbf{T}(F)\colon A(\rho_{\mathrm{sep}}, w) \to A_n(F)$$

taking $w' \in A(\rho_{\mathrm{sep}}, w)$ to the F-vector space B with the algebra structure and involution defined by w'. A morphism from w' to w'' is an element of $\mathrm{GL}(B)$ and it defines an isomorphism of the corresponding algebra structures on B. The functor $\mathbf{T}$ has an evident Γ-extension

$$\widetilde{\mathbf{T}} = \mathbf{T}(F_{\mathrm{sep}})\colon \widetilde{A}(\rho_{\mathrm{sep}}, w) \to A_n(F_{\mathrm{sep}})$$

which is clearly an equivalence of groupoids. Since the functor $\mathbf{i}$ satisfies the descent condition, so does the functor $\mathbf{j}$.

Assume now that $n = 1$. Let Q be a quaternion algebra over F. Consider the F-vector space

$$W = \mathrm{Hom}_F(Q \otimes_F Q, Q),$$

the multiplication map $w \in W$, and the natural representation

$$\rho\colon \mathbf{GL}(Q) \to \mathbf{GL}(W).$$

By Proposition (26.4) there is a Γ-embedding $\mathbf{i}$ satisfying the descent condition and a functor $\mathbf{T}$ as above taking $w' \in A(\rho_{\mathrm{sep}}, w)$ to the F-vector space Q with the algebra structure defined by w'. The functor $\mathbf{T}$ has an evident Γ-extension which is an equivalence of groupoids. As above, we conclude that the functor $\mathbf{j}$ satisfies the descent condition.

For the rest of the proof we again treat the cases $n \geq 2$ and $n = 1$ separately. Assume that $n \geq 2$. By Remark (26.3) it suffices to show that for any $(B, \tau) \in A_n(F)$ the functor $\overline{\mathbf{S}}_n$, for F be separably closed, induces a group isomorphism

(26.10) $\mathrm{Aut}_F(B, \tau) \to \mathrm{Aut}\big(\mathbf{PGU}(B, \tau)\big).$

The restriction of this homomorphism to the subgroup $\mathrm{PGU}(B, \tau)$ of index 2, is the conjugation homomorphism. It induces an isomorphism of this group with the group of inner automorphisms $\mathrm{Int}\big(\mathbf{PGU}(B, \tau)\big)$, a subgroup of $\mathrm{Aut}\big(\mathbf{PGU}(B, \tau)\big)$, which is also of index 2 (Theorem (25.16)). We may take the split algebra $B = M_{n+1}(F) \times M_{n+1}(F)^{\mathrm{op}}$ and τ the exchange involution. Then $(x, y^{\mathrm{op}}) \mapsto (y^t, x^{t\,\mathrm{op}})$ is an outer automorphism of (B, τ). Its image in $\mathrm{Aut}\big(\mathbf{PGU}(B, \tau)\big) = \mathbf{PGL}_{n+1}$ is the class of $x \mapsto x^{-t}$, which is known to be an outer automorphism if (and only if) $n \geq 2$. Hence (26.10) is an isomorphism.

Finally, consider the case $n = 1$. As above, it suffices to show that, for a quaternion algebra Q over a separably closed field F, the natural map

$$\mathrm{PGL}_1(Q) = \mathrm{Aut}_F(Q) \to \mathrm{Aut}\big(\mathbf{PGL}_1(Q)\big)$$

is an isomorphism. But this follows from the fact that any automorphism of an adjoint simple group of type A_1 is inner (Theorem (25.16)). $\square$

(26.11) Remark. Let A be a central simple algebra of degree $n+1$ over F. Then $\mathbf{S}_n(A \times A^{\mathrm{op}}, \varepsilon) = \mathbf{SL}_1(A)$, where ε is the exchange involution. In particular, two groups $\mathbf{SL}_1(A_1)$ and $\mathbf{SL}_1(A_2)$ are isomorphic if and only if

$$(A_1 \times A_1^{\mathrm{op}}, \varepsilon_1) \simeq (A_2 \times A_2^{\mathrm{op}}, \varepsilon_2),$$

i.e. $A_1 \simeq A_2$ or $A_1 \simeq A_2^{\mathrm{op}}$.

Let B be a central simple algebra of degree $n + 1$ over an étale quadratic extension L/F. The kernel C of the universal covering

$$\mathbf{SU}(B,\tau) \to \mathbf{PGU}(B,\tau)$$

is clearly equal to

$$\ker\left(R_{L/F}(\boldsymbol{\mu}_{n+1,L}) \xrightarrow{N_{L/F}} \boldsymbol{\mu}_{n+1,F}\right).$$

It is a finite group of multiplicative type, Cartier dual to $\big(\mathbb{Z}/(n+1)\mathbb{Z}\big)_{\mathrm{et}}$. An absolutely simple group of type A_n is isomorphic to $\mathbf{SU}(B,\tau)/N_k$ where k divides $n+1$ and N_k is the unique subgroup of order k in C.

Classification of simple groups of type B_n. For $n \geq 1$, let $B_n = B_n(F)$ be the groupoid of oriented quadratic spaces of dimension $2n+1$, i.e., the groupoid of triples (V, q, ζ), where (V, q) is a regular quadratic space of trivial discriminant and $\zeta \in C(V,q)$ is an orientation (so $\zeta = 1$ if char $F = 2$). Let $B^n = B^n(F)$ (resp. $\overline{B}^n = \overline{B}^n(F)$) be the groupoid of simply connected (resp. adjoint) absolutely simple groups of type B_n ($n \geq 1$) over F. By §23.A and Theorem (25.10) we have functors

$$\mathbf{S}_n \colon B_n(F) \to B^n(F) \quad \text{and} \quad \overline{\mathbf{S}}_n \colon B_n(F) \to \overline{B}^n(F)$$

defined by $\mathbf{S}_n(V, q, \zeta) = \mathbf{Spin}(V, q)$, $\overline{\mathbf{S}}_n(V, q, \zeta) = \mathbf{O}^+(V, q)$.

(26.12) Theorem. *The functors $\mathbf{S}_n \colon B_n(F) \to B^n(F)$ and $\overline{\mathbf{S}}_n \colon B_n(F) \to \overline{B}^n(F)$ are equivalences of categories.*

Proof: Since the natural functor $B^n(F) \to \overline{B}^n(F)$ is an equivalence, it suffices to prove that $\overline{\mathbf{S}}_n$ is an equivalence. Let $\Gamma = \mathrm{Gal}(F_{\mathrm{sep}}/F)$. The field extension functor $\mathbf{j} \colon B_n(F) \to B_n(F_{\mathrm{sep}})$ is clearly a Γ-embedding. We show first that the functor $\mathbf{j}$ satisfies the descent condition. Let (V, q) be some regular quadratic space over F of trivial discriminant and dimension $n + 1$. Consider the F-vector space

$$W = S^2(V^*) \oplus F,$$

the element $w = (q, 1) \in W$, and the natural representation

$$\rho \colon \mathbf{GL}(V) \to \mathbf{GL}(W), \quad \rho(g)(x, \alpha) = \big(g(x), \det x \cdot \alpha\big)$$

where $g(x)$ is given by the natural action of $\mathbf{GL}(V)$ on $S^2(V^*)$. By Proposition (26.4) the Γ-embedding

$$\mathbf{i} \colon A(\rho_{\mathrm{sep}}, w) \to \widetilde{A}(\rho_{\mathrm{sep}}, w)$$

satisfies the descent condition. Thus, to prove that $\mathbf{j}$ satisfies the descent condition, it suffices to show that the functors $\mathbf{i}$ and $\mathbf{j}$ are equivalent. First recall that:

  (a) If $(q', \lambda) \in A(\rho_{\mathrm{sep}}, w)$, then q' has trivial discriminant.
  (b) $(q, \lambda) \simeq (q', \lambda')$ in $A(\rho_{\mathrm{sep}}, w)$ if and only if $q \simeq q'$.
  (c) $\mathrm{Aut}_{A(\rho_{\mathrm{sep}}, w)}(q, \lambda) = \mathbf{O}^+(V, q) = \mathrm{Aut}_{B_n(F)}(V, q, \zeta)$ (see (12.42)).

We construct a functor

$$\mathbf{T} = \mathbf{T}(F)\colon A(\rho_{\mathrm{sep}}, w) \to B_n(F)$$

as follows. If char $F = 2$ we put $\mathbf{T}(q', \lambda) = (V, q', 1)$. Now, assume that the characteristic of F is not 2. Choose an orthogonal basis $(v_1, v_2, \ldots, v_{2n+1})$ of V for the form q, such that the central element $\zeta = v_1 \cdot v_2 \cdot \ldots \cdot v_{2n+1} \in C(V, q)$ satisfies $\zeta^2 = 1$, i.e., ζ is an orientation. Take any $(q', \lambda) \in A(\rho_{\mathrm{sep}}, w)$ and $f \in \mathrm{GL}(V_{\mathrm{sep}})$ such that $q'_{\mathrm{sep}}\big(f(v)\big) = q_{\mathrm{sep}}(v)$ for any $v \in V_{\mathrm{sep}}$ and $\det f = \lambda$. Then the central element

$$\zeta' = f(v_1) \cdot f(v_2) \cdot \ldots \cdot f(v_{2n+1}) \in C(V_{\mathrm{sep}}, q'_{\mathrm{sep}})$$

satisfies $\zeta'^2 = 1$. In particular, $\zeta' \in C(V, q')$. It is easy to see that ζ' does not depend on the choice of f. Set $\mathbf{T}(q', \lambda) = (V, q', \zeta')$. It is immediate that $\mathbf{T}(F)$ is a well-defined equivalence of categories. Thus, the functor $\mathbf{j}$ satisfies the descent condition.

To complete the proof of the theorem, it suffices by Proposition (26.2) (and Remark (26.3)) to show that, for any $(V, q, \zeta) \in B_n(F)$, the functor $\overline{\mathbf{S}}_n$ over a separably closed field F induces a group isomorphism

$$\mathbf{O}^+(V, q) \to \mathrm{Aut}\big(\mathbf{O}^+(V, q)\big).$$

This holds since automorphisms of $\mathbf{O}^+(V, q)$ are inner (Theorem (25.16)). □

(26.13) Remark. If char $F \neq 2$, the theorem can be reformulated in terms of algebras with involution. Namely, the groupoid B_n is naturally equivalent to to the groupoid B'_n of central simple algebras over F of degree $2n + 1$ with involution of the first kind, where morphisms are isomorphisms of algebras which are compatible with the involutions (see (12.41)).

Classification of simple groups of type C_n. Consider the groupoid $C_n = C_n(F)$, $n \geq 1$, of central simple F-algebras of degree $2n$ with symplectic involution, where morphisms are F-algebra isomorphisms which are compatible with the involutions.

Let $C^n = C^n(F)$ (resp. $\overline{C}^n = \overline{C}^n(F)$) be the groupoid of simply connected (resp. adjoint) simple groups of type C_n ($n \geq 1$) over F, where morphisms are group isomorphisms. By (23.A) and Theorem (25.11) we have functors

$$\mathbf{S}_n\colon C_n(F) \to C^n(F) \quad \text{and} \quad \overline{\mathbf{S}}_n\colon C_n(F) \to \overline{C}^n(F)$$

defined by $\mathbf{S}_n(A, \sigma) = \mathbf{Sp}(A, \sigma)$, $\overline{\mathbf{S}}_n(A, \sigma) = \mathbf{PGSp}(A, \sigma)$.

(26.14) Theorem. *The functors $\mathbf{S}_n\colon C_n(F) \to C^n(F)$ and $\overline{\mathbf{S}}_n\colon C_n(F) \to \overline{C}^n(F)$ are equivalences of categories.*

Proof: Since the natural functor $C^n(F) \to \overline{C}^n(F)$ is an equivalence, it suffices to prove that $\overline{\mathbf{S}}_n$ is an equivalence. Let $\Gamma = \mathrm{Gal}(F_{\mathrm{sep}}/F)$. The field extension functor $\mathbf{j}\colon C_n(F) \to C_n(F_{\mathrm{sep}})$ is clearly a Γ-embedding. We first show that the functor $\mathbf{j}$ satisfies the descent condition. Let (A, σ) be some object in $C_n(F)$ (a split one, for example). Consider the F-vector space

$$W = \mathrm{Hom}_F(A \otimes_F A, A) \oplus \mathrm{Hom}_F(A, A),$$

the element $w = (m, \sigma) \in W$ where m is the multiplication on A, and the natural representation

$$\rho\colon \mathbf{GL}(A) \to \mathbf{GL}(W).$$

By Proposition (26.4) the Γ-embedding

$$\mathbf{i}\colon A(\rho_{\mathrm{sep}}, w) \to \widetilde{A}(\rho_{\mathrm{sep}}, w)$$

satisfies the descent condition. We have the functor

$$\mathbf{T} = \mathbf{T}(F)\colon A(\rho_{\mathrm{sep}}, w) \to C_n(F)$$

taking $w' \in A(\rho_{\mathrm{sep}}, w)$ to the F-vector space A with the algebra structure and involution defined by w'. A morphism from w' to w'' is an element of $\mathrm{GL}(A)$ and it defines an isomorphism of the corresponding algebra structures on A. The functor $\mathbf{T}$ has an evident Γ-extension

$$\widetilde{\mathbf{T}} = \mathbf{T}(F_{\mathrm{sep}})\colon \widetilde{A}(\rho_{\mathrm{sep}}, w) \to C_n(F_{\mathrm{sep}}),$$

which is clearly an equivalence of groupoids. Since the functor $\mathbf{i}$ satisfies the descent condition, so does the functor $\mathbf{j}$.

To complete the proof of the theorem, it suffices by Remark (26.3) to show that for any $(A, \sigma) \in C_n(F)$ the functor $\overline{\mathbf{S}}_n$ over a separably closed field F induces a group isomorphism

$$\mathrm{PGSp}(A, \sigma) = \mathrm{Aut}_F(A, \sigma) \to \mathrm{Aut}\big(\mathbf{PGSp}(A, \sigma)\big).$$

This follows from the fact that automorphisms of $\mathbf{PGSp}$ are inner (Theorem (25.16)). $\square$

Classification of semisimple groups of type D_n, $n \neq 4$. Consider the groupoid $D_n = D_n(F)$, $n \geq 2$, of central simple F-algebras of degree $2n$ with quadratic pair, where morphisms are F-algebra isomorphisms compatible with the quadratic pairs.

Denote by $D^n = D^n(F)$ (resp. $\overline{D}^n = \overline{D}^n(F)$) the groupoid of simply connected (resp. adjoint) semisimple (simple if $n > 2$) groups of type D_n ($n \geq 2$) over F, where morphisms are group isomorphisms. By §23.B and Theorem (25.12) we have functors

$$S_n\colon D_n(F) \to D^n(F) \quad \text{and} \quad \overline{S}_n\colon D_n(F) \to \overline{D}^n(F)$$

defined by $S_n(A, \sigma, f) = \mathbf{Spin}(A, \sigma, f)$, $\overline{S}_n(A, \sigma, f) = \mathbf{PGO}^+(A, \sigma, f)$.

(26.15) Theorem. *If $n \neq 4$, the functors $S_n\colon D_n(F) \to D^n(F)$ and $\overline{S}_n\colon D_n(F) \to \overline{D}^n(F)$ are equivalences of categories.*

Proof: Since the natural functor $D^n(F) \to \overline{D}^n(F)$ is an equivalence, it suffices to prove that $\overline{S}_n$ is an equivalence. Let $\Gamma = \mathrm{Gal}(F_{\mathrm{sep}}/F)$. The field extension functor $\mathbf{j}\colon D_n(F) \to D_n(F_{\mathrm{sep}})$ is clearly a Γ-embedding. We show first that the functor $\mathbf{j}$ satisfies the descent condition. Let (A, σ, f) be some object in $D_n(F)$ (a split one, for example). Let A_+ be the space $\mathrm{Sym}(A, \sigma)$. Consider the F-vector space

$$W = \mathrm{Hom}_F(A_+, A) \oplus \mathrm{Hom}_F(A \otimes_F A, A) \oplus \mathrm{Hom}_F(A, A) \oplus (A_+)^*,$$

which contains the element $w = (i, m, \sigma, f)$ where $i\colon A_+ \hookrightarrow A$ is the inclusion and m is the multiplication on A; we have a natural representation

$$\rho\colon \mathbf{GL}(A) \times \mathbf{GL}(A_+) \to \mathbf{GL}(W)\colon$$

$$\rho(g, h)(\lambda, x, y, p) = \big(g \circ \lambda \circ h^{-1}, g(x), g(y), p \circ h^{-1}\big)$$

where $g(x)$ and $g(y)$ are obtained by applying the natural action of $\mathbf{GL}(A)$ on the second and third summands of W. By Proposition (26.4) the Γ-embedding

$$\mathbf{i}\colon A(\rho_{\mathrm{sep}}, w) \to \widetilde{A}(\rho_{\mathrm{sep}}, w)$$

satisfies the descent condition. We have the functor

$$\mathbf{T} = \mathbf{T}(F)\colon A(\rho_{\mathrm{sep}}, w) \to D_n(F)$$

which takes $w' \in A(\rho_{\mathrm{sep}}, w)$ to the F-vector space A with the algebra structure and quadratic pair defined by w'. A morphism from w' to w'' is an element of $\mathrm{GL}(A) \times \mathrm{GL}(A_+)$ and it defines an isomorphism between the corresponding structures on A. The functor $\mathbf{T}$ has an evident Γ-extension

$$\widetilde{\mathbf{T}} = \mathbf{T}(F_{\mathrm{sep}})\colon \widetilde{A}(\rho_{\mathrm{sep}}, w) \to D_n(F_{\mathrm{sep}}),$$

which is clearly an equivalence of groupoids. Since the functor $\mathbf{i}$ satisfies the descent condition, so does the functor $\mathbf{j}$. For the proof of the theorem it suffices by Proposition (26.2) (and Remark (26.3)) to show that for any $(A, \sigma, f) \in D_n(F)$ the functor $\overline{\mathbf{S}}_n$ for a separably closed field F induces a group isomorphism

$$(26.16) \qquad \mathrm{PGO}(A, \sigma, f) = \mathrm{Aut}_F(A, \sigma, f) \to \mathrm{Aut}\big(\mathbf{PGO}^+(A, \sigma, f)\big).$$

The restriction of this homomorphism to the subgroup $\mathrm{PGO}^+(A, \sigma, f)$, which is of index 2, induces an isomorphism of this subgroup with the group of inner automorphisms $\mathrm{Int}\big(\mathbf{PGO}^+(A, \sigma, f)\big)$, which is a subgroup in $\mathrm{Aut}\big(\mathbf{PGO}^+(A, \sigma)\big)$ also of index 2 (since $n \neq 4$, see Theorem (25.16)). A straightforward computation shows that any element in $\mathrm{PGO}^-(A, \sigma, f)$ induces an outer automorphism of $\mathbf{PGO}^+(A, \sigma, f)$. Hence (26.16) is an isomorphism. $\qquad\square$

(26.17) Remark. The case of D_4 is exceptional, in the sense that the group of automorphisms of the Dynkin diagram of D_4 is S_3. Triality is needed and we refer to Theorem (44.8) below for an analogue of Theorem (26.15) for D_4.

Let C be the kernel of the adjoint representation of $\mathbf{Spin}(A, \sigma, f)$. If n is even, then C is the Cartier dual to $(\mathbb{Z}/2\mathbb{Z} \oplus \mathbb{Z}/2\mathbb{Z})_{\mathrm{et}}$, where the absolute Galois group Γ acts by the permutation of summands. This action factors through $\mathrm{Aut}\big(\mathrm{Dyn}(D_n)\big)$. On the other hand, the Γ-action on the center Z of the Clifford algebra $C(A, \sigma, f)$ given by the composition

$$\Gamma \to \mathrm{Aut}_{F_{\mathrm{sep}}}\big(C(A_{\mathrm{sep}}, \sigma_{\mathrm{sep}}, f_{\mathrm{sep}})\big) \to \mathrm{Aut}_{F_{\mathrm{sep}}}(Z_{\mathrm{sep}}) \simeq \mathbb{Z}/2\mathbb{Z}$$

also factors through $\mathrm{Aut}\big(\mathrm{Dyn}(D_n)\big)$. Hence the Cartier dual to C is isomorphic to $(\mathbb{Z}/2\mathbb{Z})[G]_{\mathrm{et}}$, where $G = \mathrm{Gal}(Z/F)$ and Γ acts by the natural homomorphism $\Gamma \to G$. By Exercise 6,

$$C = R_{Z/F}(\boldsymbol{\mu}_{2,Z}).$$

If n is odd, then C is the Cartier dual to $(\mathbb{Z}/4\mathbb{Z})_{\mathrm{et}}$ and Γ acts on $M = \mathbb{Z}/4\mathbb{Z}$ through G identified with the automorphism group of $\mathbb{Z}/4\mathbb{Z}$. We have an exact sequence

$$0 \to \mathbb{Z}/4\mathbb{Z} \to (\mathbb{Z}/4\mathbb{Z})[G] \to M \to 0,$$

where $\mathbb{Z}/4\mathbb{Z}$ is considered with the trivial Γ-action. By Cartier duality,

$$C = \ker\big(R_{Z/F}(\boldsymbol{\mu}_{4,Z}) \xrightarrow{\ N_{Z/F}\ } \boldsymbol{\mu}_{4,F}\big).$$

If n is odd, then C has only one subgroup of order 2 which corresponds to $\mathbf{O}^+(A, \sigma, f)$. If n is even, then $C_{\mathrm{sep}}^* \simeq \mathbb{Z}/2\mathbb{Z} \oplus \mathbb{Z}/2\mathbb{Z}$. If σ has nontrivial discriminant (i.e., Z is not split), then Γ acts non-trivially on C_{sep}^*, hence there is still only one proper subgroup of C corresponding to $\mathrm{GO}^+(A, \sigma, f)$. In the case where the discriminant is trivial (so Z is split), Γ acts trivially on C_{sep}^*, and there are three proper subgroups of C, one of which corresponds again to $\mathbf{O}^+(A, \sigma, f)$. The two other groups correspond to $\mathbf{Spin}^\pm(A, \sigma, f)$, which are the images of the compositions

$$\mathbf{Spin}(A, \sigma, f) \hookrightarrow \mathbf{GL}_1\big(C(A, \sigma, f)\big) \to \mathbf{GL}_1\big(C^\pm(A, \sigma, f)\big)$$

where $C(A, \sigma, f) = C^+(A, \sigma, f) \times C^-(A, \sigma, f)$.

Classification of simple groups of type F_4. Consider the groupoid $\mathsf{F}_4 = \mathsf{F}_4(F)$ of exceptional Jordan algebras of dimension 27 over F (see §37.C below if char $F \neq 2$ and §38 if char $F = 2$), where morphisms are F-algebra isomorphisms.

Denote by $\mathsf{F}^4 = \mathsf{F}^4(F)$ the groupoid of simple groups of type F_4 over F, where morphisms are group isomorphisms. By Theorem (25.13) we have a functor

$$\mathbf{S} \colon \mathsf{F}_4(F) \to \mathsf{F}^4(F), \quad \mathbf{S}(J) = \mathbf{Aut}_{\mathrm{alg}}(J).$$

(26.18) Theorem. *The functor $\mathbf{S} \colon \mathsf{F}_4(F) \to \mathsf{F}^4(F)$ is an equivalence of categories.*

Proof: Let $\Gamma = \mathrm{Gal}(F_{\mathrm{sep}}/F)$. The field extension functor $\mathbf{j} \colon \mathsf{F}_4(F) \to \mathsf{F}_4(F_{\mathrm{sep}})$ is clearly a Γ-embedding. We first show that the functor $\mathbf{j}$ satisfies the descent condition. Let J be some object in $\mathsf{F}_4(F)$ (a split one, for example). If char $F \neq 2$, consider the F-vector space

$$W = \mathrm{Hom}_F(J \otimes_F J, J),$$

the multiplication element $w \in W$ and the natural representation

$$\rho \colon \mathbf{GL}(J) \to \mathbf{GL}(W).$$

By Proposition (26.4) the Γ-embedding

$$\mathbf{i} \colon A(\rho_{\mathrm{sep}}, w) \to \widetilde{A}(\rho_{\mathrm{sep}}, w)$$

satisfies the descent condition. We have the functor

$$\mathbf{T} = \mathbf{T}(F) \colon A(\rho_{\mathrm{sep}}, w) \to \mathsf{F}_4(F)$$

taking $w' \in A(\rho_{\mathrm{sep}}, w)$ to the F-vector space J with the Jordan algebra structure defined by w'. A morphism from w' to w'' is an element of $\mathrm{GL}(J)$ and it defines an isomorphism of the corresponding Jordan algebra structures on J. The functor $\mathbf{T}$ has an evident Γ-extension

$$\widetilde{\mathbf{T}} = \mathbf{T}(F_{\mathrm{sep}}) \colon \widetilde{A}(\rho_{\mathrm{sep}}, w) \to \mathsf{F}_4(F_{\mathrm{sep}}),$$

which is clearly an equivalence of groupoids. Since the functor $\mathbf{i}$ satisfies the descent condition, so does the functor $\mathbf{j}$.

For the proof of the theorem it suffices by Proposition (26.2) to show that for any $J \in \mathsf{F}_4(F)$ the functor $\mathbf{S}$ for a separably closed field F induces a group isomorphism

$$\mathrm{Aut}_{\mathrm{alg}}(J) \to \mathrm{Aut}\big(\mathbf{Aut}_{\mathrm{alg}}(J)\big).$$

This follows from the fact that automorphisms of simple groups of type F_4 are inner (Theorem (25.16)).

If char $F = 2$, an exceptional Jordan algebra of dimension 27 is (see §38 below) a datum $(J, N, \#, T, 1)$ consisting of a space J of dimension 27, a cubic form $N\colon J \to F$, the adjoint $\#\colon J \to J$ of N, which is a quadratic map, a bilinear trace form T, and a distinguished element 1, satisfying certain properties (given in §38). In this case we consider the F-vector space

$$W = S^3(J^*) \oplus S^2(J^*) \otimes J \oplus S^2(J^*) \oplus F$$

and complete the argument as in the preceding cases. $\qquad\square$

Classification of simple groups of type G_2**.** Consider the groupoid $\mathsf{G}_2 = \mathsf{G}_2(F)$ of Cayley algebras over F, where morphisms are F-algebra isomorphisms.

Denote by $\mathsf{G}^2 = \mathsf{G}^2(F)$ the groupoid of simple groups of type G_2 over F, where morphisms are group isomorphisms. By Theorem (25.14) there is a functor

$$\mathbf{S}\colon \mathsf{G}_2(F) \to \mathsf{G}^2(F), \quad \mathbf{S}(C) = \mathbf{Aut}_{\mathrm{alg}}(C).$$

(26.19) Theorem. *The functor* $\mathbf{S}\colon \mathsf{G}_2(F) \to \mathsf{G}^2(F)$ *is an equivalence of categories.*

Proof: Let $\Gamma = \mathrm{Gal}(F_{\mathrm{sep}}/F)$. The field extension functor $\mathbf{j}\colon \mathsf{G}_2(F) \to \mathsf{G}_2(F_{\mathrm{sep}})$ is clearly a Γ-embedding. We first show that the functor $\mathbf{j}$ satisfies the descent condition. Let C be some object in $\mathsf{G}_2(F)$ (a split one, for example). Consider the F-vector space $W = \mathrm{Hom}_F(C \otimes_F C, C)$, the multiplication element $w \in W$, and the natural representation $\rho\colon \mathbf{GL}(C) \to \mathbf{GL}(W)$. By Proposition (26.4) the Γ-embedding $\mathbf{i}\colon A(\rho_{\mathrm{sep}}, w) \to \widetilde{A}(\rho_{\mathrm{sep}}, w)$ satisfies the descent condition. We have a functor

$$\mathbf{T} = \mathbf{T}(F)\colon A(\rho_{\mathrm{sep}}, w) \to \mathsf{G}_2(F)$$

which takes $w' \in A(\rho_{\mathrm{sep}}, w)$ to the F-vector space C with the Cayley algebra structure defined by w'. A morphism from w' to w'' is an element of $\mathrm{GL}(C)$ and it defines an isomorphism between the corresponding Cayley algebra structures on C. The functor $\mathbf{T}$ has an evident Γ-extension

$$\widetilde{\mathbf{T}} = \mathbf{T}(F_{\mathrm{sep}})\colon \widetilde{A}(\rho_{\mathrm{sep}}, w) \to \mathsf{G}_2(F_{\mathrm{sep}}),$$

which is clearly an equivalence of groupoids. Since the functor $\mathbf{i}$ satisfies the descent condition, so does the functor $\mathbf{j}$.

For the proof of the theorem it suffices by Proposition (26.2) to show that for any $J \in \mathsf{G}_2(F)$ the functor $\mathbf{S}$ for a separably closed field F induces a group isomorphism

$$\mathrm{Aut}_{\mathrm{alg}}(C) \to \mathrm{Aut}\big(\mathbf{Aut}_{\mathrm{alg}}(C)\big).$$

This follows from the fact that automorphisms of simple groups of type G_2 are inner (Theorem (25.16)). $\qquad\square$

26.B. Algebraic groups of small dimension. Some Dynkin diagrams of small ranks coincide:

(26.20) $\qquad\qquad\qquad\qquad A_1 = B_1 = C_1$

(26.21) $\qquad\qquad\qquad\qquad D_2 = A_1 + A_1$

(26.22) $\qquad\qquad\qquad\qquad\quad B_2 = C_2$

(26.23) $\qquad\qquad\qquad\qquad\quad A_3 = D_3$

We describe explicitly the corresponding isomorphisms for adjoint groups (analogues for algebras are in §15):

$A_1 = B_1 = C_1$. Let (V, q) be a regular quadratic form of dimension 3 over a field F. Then $C_0(V, q)$ is a quaternion algebra over F. The canonical homomorphism

$$\mathbf{O}^+(V, q) \to \mathbf{PGL}_1\big(C_0(V, q)\big) = \mathbf{PGSp}\big(C_0(V, q), \underline{\sigma}_q\big)$$

is injective (see §23.A) and hence is an isomorphism of adjoint simple groups of types B_1, A_1 and C_1 since by dimension count its image has the same dimension as the target group, and since these groups are connected they must coincide, by Propositions (21.7) and (22.2). (We will use this argument several times below.)

Let Q be a quaternion algebra over F and let $Q^0 = \{ x \in Q \mid \operatorname{Trd}_Q(x) = 0 \}$. For $x \in Q^0$, we have $x^2 \in F$, and the squaring map $s \colon Q^0 \to F$ is a quadratic form of discriminant 1 on Q^0 (see §15.A). Consider the conjugation homomorphism

$$f \colon \mathbf{GL}_1(Q) \to \mathbf{O}^+(Q^0, s).$$

Since Q^0 generates Q, $\ker(f) = \mathbf{G}_\mathrm{m}$ and the injection

$$\mathbf{PGSp}(Q, \sigma) = \mathbf{PGL}_1(Q) \to \mathbf{O}^+(Q^0, s)$$

is an isomorphism of adjoint simple groups of types C_1, A_1, and B_1.

$D_2 = A_1 + A_1$. Let A be a central simple algebra over F of degree 4 with a quadratic pair (σ, f). Then $C(A, \sigma, f)$ is a quaternion algebra over a quadratic étale extension Z of F. We have the canonical injection

$$\mathbf{PGO}^+(A, \sigma, f) \to \mathbf{Aut}_Z\big(C(A, \sigma, f)\big) = R_{Z/F}\big(\mathbf{PGL}_1\big(C(A, \sigma, f)\big)\big)$$

which is an isomorphism between adjoint groups of type D_2 and those of type $A_1 + A_1$.

Conversely, let Q be a quaternion algebra over an étale quadratic extension Z/F. The norm $A = N_{Z/F}(Q)$ is a central simple algebra of degree 4 over F with a canonical quadratic pair (σ, f) (see §15.B). We have the natural homomorphism

$$g \colon R_{Z/F}\big(\mathbf{GL}_1(Q)\big) \to \mathbf{GO}^+(A, \sigma, f), \quad x \in Q_R^\times \mapsto x \otimes x \in A_R^\times.$$

One checks that $x \otimes x \in R^\times$ if and only if $x \in Z_R^\times$, hence $g^{-1}(\mathbf{G}_\mathrm{m}) = R_{Z/F}(\mathbf{G}_{\mathrm{m}, Z})$. By factoring out these subgroups we obtain an injective homomorphism

$$R_{Z/F}\big(\mathbf{PGL}_1(Q)\big) \to \mathbf{PGO}^+(A, \sigma, f)$$

which is actually an isomorphism from an adjoint group of type $A_1 + A_1$ to one of type D_2.

$B_2 = C_2$. Let (V, q) be a regular quadratic form of dimension 5. Then $C_0(V, q)$ is a central simple algebra of degree 4 with (canonical) symplectic involution τ. There is a canonical injective homomorphism (see §23.A)

$$\mathbf{O}^+(V, q) \to \mathbf{PGSp}\big(C_0(V, q), \tau\big)$$

which is in fact an isomorphism from an adjoint simple groups of type B_2 to one of type C_2.

Conversely, for a central simple algebra A of degree 4 over F with a symplectic involution σ, the F-vector space

$$\operatorname{Symd}(A, \sigma)^0 = \{ x \in \operatorname{Symd}(A, \sigma) \mid \operatorname{Trp}_A(x) = 0 \}$$

admits the quadratic form $s_\sigma(x) = x^2 \in F$ (see §15.C). Consider the conjugation homomorphism

$$f \colon \mathbf{GSp}(A, \sigma) \to \mathbf{O}^+\big(\mathrm{Symd}(A, \sigma)^0, s_\sigma\big), \quad a \mapsto \mathrm{Int}(a).$$

Since $\mathrm{Symd}(A, \sigma)$ generates A, one has $\ker(f) = \mathbf{G}_{\mathrm{m}}$. Hence, the injection

$$\mathbf{PGSp}(A, \sigma) \to \mathbf{O}^+(V, q)$$

is an isomorphism from an adjoint simple group of type C_2 to one of type B_2.

$A_3 = D_3$. Let A be a central simple algebra of degree 6 over F with an orthogonal pair (σ, f). Then $C(A, \sigma, f)$ is a central simple algebra of degree 4 over an étale quadratic extension Z/F with a unitary involution $\underline{\sigma}$. The natural homomorphism

$$\mathbf{PGO}^+(A, \sigma, f) \to \mathbf{PGU}\big(C(A, \sigma, f), \underline{\sigma}\big)$$

is injective (see §23.B) and hence is an isomorphism from an adjoint simple group of type D_3 to one of type A_3.

Conversely, let B be a central simple algebra of degree 4 over an étale quadratic extension Z/F with a unitary involution τ. Then the discriminant algebra $D(B, \tau)$ is a central simple algebra of degree 6 over F with canonical quadratic pair $(\underline{\tau}, f)$. Consider the natural homomorphism

$$\mathbf{GU}(B, \tau) \to \mathbf{GO}^+\big(D(B, \tau), \underline{\tau}, f\big).$$

One checks (in the split case) that $g^{-1}(\mathbf{G}_{\mathrm{m}}) = \mathbf{GL}_1(Z)$. By factoring out these subgroups we obtain an injection

$$\mathbf{PGU}(B, \tau) \to \mathbf{PGO}^+\big(D(B, \tau), \underline{\tau}, f\big)$$

which is an isomorphism from an adjoint simple group of type A_3 to one of type D_3.

§27. Tits Algebras of Semisimple Groups

The Clifford algebra, the discriminant algebra, the λ-powers of a central simple algebra all arise as to be so-called Tits algebras of the appropriate semisimple groups. In this section we define Tits algebras and classify them for simple groups of the classical series.

For this we need some results on the classification of representations of split semisimple groups. Let G be a split semisimple group over F. Choose a split maximal torus $T \subset G$. Fix a system of simple roots in $\Phi(G)$, so we have the corresponding cone $\Lambda_+ \subset \Lambda$ of dominant weights.

Let $\rho \colon G \to \mathbf{GL}(V)$ be a representation. By the representation theory of diagonalizable groups (22.20) one can associate to the representation $\rho|_T$ a finite number of weights, elements of T^*. If ρ is irreducible, among the weights there is a largest (with respect to the ordering on Λ). It lies in Λ_+ and is called the *highest weight* of ρ (Humphreys [125]).

(27.1) Theorem. *The map*

$$\boxed{\begin{array}{c} \text{Isomorphism classes of} \\ \text{irreducible representations of } G \end{array}} \quad \longleftrightarrow \quad T^* \cap \Lambda_+$$

taking the class of a representation ρ to its highest weight, is a bijection.

Reference: Tits [292, Th.2.5]   □

(27.2) Remark. If G is a simply connected group (i.e., $T^* = \Lambda$), then $T^* \cap \Lambda_+ = \Lambda_+$.

(27.3) Remark. The classification of irreducible representations of a split semi-simple groups does not depend on the base field in the sense that an irreducible representation remains irreducible over an arbitrary field extension and any irreducible representation over an extension comes from the base field.

27.A. Definition of the Tits algebras.

Let G be a semisimple (not necessarily split) group over F and let $T \subset G$ be a maximal torus. Choose a system of simple roots $\Pi \subset \Phi = \Phi(G)$. The group Γ acts on T^*_{sep} leaving invariant Φ, Λ, Λ_r (but not Π).

There is another action of Γ on T^*_{sep}, called the $*$-*action*, which is defined as follows. Take any $\gamma \in \Gamma$. Since the Weyl group W acts simply transitively on the set of systems of simple roots and $\gamma\Pi$ is clearly a system of simple roots, there is a unique $w \in W$ such that $w(\gamma\Pi) = \Pi$. We set $\gamma * \alpha = w(\gamma\alpha) \in \Pi$ for any $\alpha \in \Pi$. This action, defined on Π, extends to an action on Λ which is the identity on Π, Φ, Λ_r, Λ_+. Note that since W acts trivially on Λ/Λ_r, the $*$-action on Λ/Λ_r coincides with the usual one.

Choose a finite Galois extension $F \subset L \subset F_{\mathrm{sep}}$ splitting T and hence G. The $*$-action of Γ then factors through $\mathrm{Gal}(L/F)$. Let $\rho \colon G_L \to \mathbf{GL}(V)$ be an irreducible representation over L (so V is an L-vector space) with highest weight $\lambda \in \Lambda_+ \cap T^*$ (see Theorem (27.1)). For any $\gamma \in \Gamma$ we can define the L-space $_\gamma V$ as V as an abelian group and with the L-action $x \circ v = \gamma^{-1}(x) \cdot v$, for all $x \in L$, $v \in V$. Then $v \mapsto v$ viewed as a map $V \to {}_\gamma V$ is γ-semilinear. Denote it i_γ.

Let $A = F[G]$ and let $\bar{\rho} \colon V \to V \otimes_L A_L$ be the comodule structure for ρ (see p. 343). The composite

$$_\gamma V \xrightarrow{i_\gamma^{-1}} V \xrightarrow{\bar{\rho}} V \otimes_L (L \otimes_F A) \xrightarrow{i_\gamma \otimes (\gamma \otimes \mathrm{Id})} {}_\gamma V \otimes_L (L \otimes_F A)$$

gives the comodule structure for some irreducible representation

$$_\gamma \rho \colon G_L \to \mathbf{GL}(_\gamma V).$$

(Observe that the third map is well-defined because both i_γ and $\gamma \otimes \mathrm{Id}$ are γ-semilinear.) Clearly, the weights of $_\gamma\rho$ are obtained from the weights of ρ by applying γ. Hence, the highest weight of $_\gamma\rho$ is $\gamma * \lambda$.

Assume now that $\lambda \in \Lambda_+ \cap T^*$ is invariant under the $*$-action. Consider the conjugation representation

$$\mathrm{Int}(\rho) \colon G \to \mathbf{GL}\big(\mathrm{End}_F(V)\big), \quad g \mapsto \big(\alpha \mapsto \rho(g) \circ \alpha \circ \rho(g)^{-1}\big).$$

Let $\mathrm{End}_G(V)$ be the subalgebra of G-invariant elements in $\mathrm{End}_F(V)$ under $\mathrm{Int}(\rho)$. Then

$$\textbf{(27.4)} \qquad \mathrm{End}_G(V) \otimes_F L \simeq \mathrm{End}_{G_L}(V \otimes_F L) \simeq \mathrm{End}_{G_L}\big(\textstyle\prod_{\gamma \in \mathrm{Gal}(L/F)} {}_\gamma V\big)$$

since $V \otimes_F L$ is L-isomorphic to $\prod_\gamma V$ via $v \otimes x \mapsto (\gamma^{-1}x \cdot v)_\gamma$. Since the representation $_\gamma\rho$ is of highest weight $\gamma * \lambda = \lambda$, it follows from Theorem (27.1) that $_\gamma\rho \simeq \rho$, i.e., all the G-modules $_\gamma V$ are isomorphic to V. Hence, the algebras in (27.4) are isomorphic to

$$\mathrm{End}_{G_L}(V^n) = M_n\big(\mathrm{End}_{G_L}(V)\big)$$

where $n = [L : F]$.

(27.5) Lemma. $\mathrm{End}_{G_L}(V) \simeq L$.

Proof: Since ρ is irreducible, $\mathrm{End}_{G_L}(V)$ is a division algebra over L by Schur's lemma. But ρ_{alg} remains irreducible by Remark (27.3), hence $\mathrm{End}_{G_L}(V) \otimes_L F_{\mathrm{alg}}$ is also a division algebra and therefore $\mathrm{End}_{G_L}(V) = L$. $\qquad\square$

It follows from the lemma that $\mathrm{End}_G(V) \otimes_F L \simeq M_n(L)$, hence $\mathrm{End}_G(V)$ is a central simple algebra over F of degree n. Denote its centralizer in $\mathrm{End}_F(V)$ by A_λ. This is a central simple algebra over F of degree $\dim_L V$. It is clear that A_λ is independent of the choice of L. The algebra A_λ is called the *Tits algebra* of the group G corresponding to the dominant weight λ.

Since the image of ρ commutes with $\mathrm{End}_G(V)$, it actually lies in A_λ. Thus we obtain a representation

$$\rho' \colon G \to \mathbf{GL}_1(A_\lambda).$$

By the double centralizer theorem (see (1.5)), the centralizer of $\mathrm{End}_G(V) \otimes_F L$ in $\mathrm{End}_F(V) \otimes_F L$ is $A_\lambda \otimes_F L$. On the other hand it contains $\mathrm{End}_L(V)$ (where the image of ρ lies). By dimension count we have

$$A_\lambda \otimes_F L = \mathrm{End}_L(V)$$

and hence the representation $(\rho')_L$ is isomorphic to ρ. Thus, ρ' can be considered as a descent of ρ from L to F. The restriction of ρ' to the center $C = C(G) \subset G$ is given by the restriction of λ on C, i.e., is the character of the center C given by the class of λ in $C^* = T^*/\Lambda_r \subset \Lambda/\Lambda_r$.

The following lemma shows the uniqueness of the descent ρ'.

(27.6) Lemma. *Let $\mu_i \colon G \to \mathbf{GL}_1(A_i)$, $i = 1, 2$ be two homomorphisms where the A_i are central simple algebras over F. Assume that the representations $(\mu_i)_{\mathrm{sep}}$ are isomorphic and irreducible. Then there is an F-algebra isomorphism $\alpha \colon A_1 \to A_2$ such that $\mathbf{GL}_1(\alpha) \circ \mu_1 = \mu_2$.*

Proof: Choose a finite Galois field extension L/F splitting G and the A_i, $A_i \otimes_F L \simeq \mathrm{End}_L(V_i)$. An L-isomorphism $V_1 \xrightarrow{\sim} V_2$ of G_L-representations gives rise to an algebra isomorphism

$$\alpha \colon \mathrm{End}_F(V_1) \xrightarrow{\sim} \mathrm{End}_F(V_2)$$

taking $\mathrm{End}_G(V_1)$ to $\mathrm{End}_G(V_2)$. Clearly, A_i lies in the centralizer of $\mathrm{End}_G(V_i)$ in $\mathrm{End}_F(V_i)$. By dimension count A_i coincides with the centralizer, hence $\alpha(A_1) = A_2$. $\qquad\square$

Let $\pi \colon \widetilde{G} \to G$ be a central isogeny with $\widetilde{G}$ simply connected. Then the Tits algebra built out of a representation ρ of G_L is the Tits algebra of the group $\widetilde{G}_L$ corresponding to the representation $\rho \circ \pi_L$. Hence, in order to classify Tits algebras one can restrict to simply connected groups.

Assume that G is a simply connected semisimple group. For any $\lambda \in \Lambda/\Lambda_r$ consider the corresponding (unique) minimal weight $\chi(\lambda) \in \Lambda_+$. The uniqueness shows that $\chi(\gamma\lambda) = \gamma * \chi(\lambda)$ for any $\gamma \in \Gamma$. Hence, if $\lambda \in (\Lambda/\Lambda_r)^\Gamma$, then clearly $\chi(\lambda) \in \Lambda_+^\Gamma$ (with respect to the $*$-action); the Tits algebra $A_{\chi(\lambda)}$ is called a *minimal Tits algebra* and is denoted simply by A_λ. For example, if $\lambda = 0$, then $A_\lambda = F$.

(27.7) Theorem. *The map*

$$\beta \colon (\Lambda/\Lambda_r)^\Gamma \to \mathrm{Br}\, F, \quad \lambda \mapsto [A_\lambda]$$

is a homomorphism.

Reference: Tits [292, Cor. 3.5].   □

If $\lambda \in \Lambda/\Lambda_r$ is not necessarily Γ-invariant, let

$$\Gamma_0 = \{\, \gamma \in \Gamma \mid \gamma(\lambda) = \lambda \,\} \subset \Gamma$$

and $F_\lambda = (F_{\text{sep}})^{\Gamma_0}$. Then $\lambda \in (\Lambda/\Lambda_r)^{\Gamma_0}$ and one gets a Tits algebra A_λ, which is a central simple algebra over F_λ, for the group G_{F_λ}. The field F_λ is called the *field of definition* of λ.

27.B. Simply connected classical groups. We give here the classification of the minimal Tits algebras of the absolutely simple simply connected groups of classical type.

Type A_n, $n \geq 1$. Let first $G = \mathbf{SL}_1(A)$ where A is a central simple algebra of degree $n + 1$ over F. Then $C = \boldsymbol{\mu}_{n+1}$, $C^* = \mathbb{Z}/(n+1)\mathbb{Z}$ with the trivial Γ-action. For any $i = 0, 1, \ldots, n$, consider the natural representation

$$\rho_i \colon G \to \mathbf{GL}_1(\lambda^i A).$$

In the split case ρ_i is the i-th exterior power representation with the highest weight $\bar{e}_1 + \bar{e}_2 + \cdots + \bar{e}_i$ in the notation of §24.A, which is a minimal weight. Hence, the λ-powers $\lambda^i A$, for $i = 0, 1, \ldots, n$, (see §10.A) are the minimal Tits algebras of G.

Now let $G = \mathbf{SU}(B, \tau)$ where B is a central simple algebra of degree $n+1$ with a unitary involution over a quadratic separable field extension K/F. The group Γ acts on $C^* = \mathbb{Z}/(n+1)\mathbb{Z}$ by $x \mapsto -x$ through $\text{Gal}(K/F)$. The only nontrivial element in $(C^*)^\Gamma$ is $\lambda = \frac{n+1}{2} + (n+1)\mathbb{Z}$ (n should be odd). There is a natural homomorphism

$$\rho \colon G \to \mathbf{GL}_1\big(D(B, \tau)\big)$$

which in the split case is the external $\frac{n+1}{2}$-power. Hence, the discriminant algebra (see §10.E) $D(B, \tau)$ is the minimal Tits algebra corresponding to λ for the group G.

The fields of definition F_μ of the other nontrivial characters $\mu = i + (n+1)\mathbb{Z} \in C^*$, ($i \neq \frac{(n+1)}{2}$), coincide with K. Hence, by extending the base field to K one sees that $A_\mu \simeq \lambda^i B$.

Type B_n, $n \geq 1$. Let $G = \mathbf{Spin}(V, q)$, here (V, q) is a regular quadratic form of dimension $2n + 1$. Then $C = \boldsymbol{\mu}_2$, $C^* = \mathbb{Z}/2\mathbb{Z} = \{0, \lambda\}$. The embedding

$$G \hookrightarrow \mathbf{GL}_1\big(C_0(V, q)\big)$$

in the split case is the spinor representation with highest weight $\frac{1}{2}(\bar{e}_1 + \bar{e}_2 + \cdots + \bar{e}_n)$ in the notation of §24.A, which is a minimal weight. Hence, the even Clifford algebra $C_0(V, q)$ is the minimal Tits algebra A_λ.

Type C_n, $n \geq 1$. Let $G = \mathbf{Sp}(A, \sigma)$ where A is a central simple algebra of degree $2n$ with a symplectic involution σ. Then $C = \boldsymbol{\mu}_2$, $C^* = \mathbb{Z}/2\mathbb{Z} = \{0, \lambda\}$. The embedding

$$G \hookrightarrow \mathbf{GL}_1(A)$$

in the split case is the representation with highest weight $\bar{e}_1$ in the notation of §24.A, which is a minimal weight. Hence, A is the minimal Tits algebra A_λ.

Type D_n, $n \geq 2$, $n \neq 4$. Let $G = \mathbf{Spin}(A, \sigma, f)$ where A is a central simple algebra of degree $2n$ with a quadratic pair (σ, f), $C^* = \{0, \lambda, \lambda^+, \lambda^-\}$ where λ factors through $\mathbf{O}^+(A, \sigma, f)$. The composition

$$\mathbf{Spin}(A, \sigma, f) \to \mathbf{GO}^+(A, \sigma, f) \hookrightarrow \mathbf{GL}_1(A)$$

in the split case is the representation with highest weight $\bar{e}_1$ in the notation of §24.A, which is a minimal weight. Hence, A is the minimal Tits algebra A_λ.

Assume further that the discriminant of σ is trivial (i.e., the center Z of the Clifford algebra is split). The group Γ then acts trivially on C^*. The natural compositions

$$\mathbf{Spin}(A, \sigma, f) \hookrightarrow \mathbf{GL}_1\big(C(A, \sigma, f)\big) \to \mathbf{GL}_1\big(C^\pm(A, \sigma, f)\big)$$

in the split case are the representations with highest weights $\frac{1}{2}(e_1 + \cdots + e_{n-1} \pm e_n)$ which are minimal weights. Hence, $C^\pm(A, \sigma, f)$ are the minimal Tits algebras $A_{\lambda^\pm}$.

If $\mathrm{disc}(\sigma)$ is not trivial then Γ interchanges λ^+ and λ^-, hence the field of definition of $\lambda^\pm$ is Z. By extending the base field to Z one sees that $A_{\lambda^\pm} = C(A, \sigma, f)$. Again, the case of D_4 is exceptional, because of triality, and we give on p. 563 a description of the minimal Tits algebra in this case.

27.C. Quasisplit groups. A semisimple group G is called *quasisplit* if there is a maximal torus $T \subset G$ and a system Π of simple roots in the root system Φ of G with respect to T which is Γ-invariant with respect to the natural action, or equivalently, if the $*$-action on T^*_{sep} coincides with the natural one. For example, split groups are quasisplit.

Let G be a quasisplit semisimple group. The natural action of Γ on the system Π of simple roots, which is invariant under Γ, defines an action of Γ on $\mathrm{Dyn}(G) = \mathrm{Dyn}(\Phi)$ by automorphisms of the Dynkin diagram. Simply connected and adjoint split groups are classified by their Dynkin diagrams. The following statement generalizes this result for quasisplit groups.

(27.8) Proposition. *Two quasisplit simply connected (resp. adjoint) semisimple groups G and G' are isomorphic if and only if there is a Γ-bijection between $\mathrm{Dyn}(G)$ and $\mathrm{Dyn}(G')$. For any Dynkin diagram D and any (continuous) Γ-action on D there is a quasisplit simply connected (resp. adjoint) semisimple group G and a Γ-bijection between $\mathrm{Dyn}(G)$ and D.* $\square$

The Γ-action on $\mathrm{Dyn}(G)$ is trivial if and only if Γ acts trivially on T^*_{sep}, hence T and G are split. Therefore, if $\mathrm{Aut}(\mathrm{Dyn}(G)) = 1$ (i.e., $\mathrm{Dyn}(G)$ has only irreducible components B_n, C_n, E_7, E_8, F_4, G_2) and G is quasisplit, then G is actually split.

(27.9) Example. The case A_n, $n > 1$. A non-trivial action of the Galois group Γ on the cyclic group $\mathrm{Aut}(A_n)$ of order two factors through the Galois group of a quadratic field extension L/F. The corresponding quasisplit simply connected simple group of type A_n is isomorphic to $\mathrm{SU}(V, h)$, where (V, h) is a non-degenerate hermitian form over L/F of dimension $n + 1$ and maximal Witt index.

(27.10) Example. The case D_n, $n > 1$, $n \neq 4$. As in the previous example, to give a nontrivial Γ-action on D_n is to give a quadratic Galois field extension L/F. The corresponding quasisplit simply connected simple group of type D_n is isomorphic to $\mathrm{Spin}(V, q)$, where (V, q) is a non-degenerate quadratic form of dimension $2n$ and Witt index $n - 1$ with the discriminant quadratic extension L/F.

Exercises

1. If L is an étale F-algebra, then $\mathbf{Aut}_{\mathrm{alg}}(L)$ is an étale group scheme corresponding to the finite group $\mathrm{Aut}_{F_{\mathrm{sep}}}(L_{\mathrm{sep}})$ with the natural $\mathrm{Gal}(F_{\mathrm{sep}}/F)$-action.

2. Let G be an algebraic group scheme. Prove that the following statements are equivalent:
   (a) G is étale,
   (b) $G^0 = \mathbf{1}$,
   (c) G is smooth and finite,
   (d) $\mathrm{Lie}(G) = 0$.

3. Prove that H_{diag} is algebraic if and only if H is a finitely generated abelian group.

4. Let H be a finitely generated abelian group, and let $H' \subset H$ be the subgroup of elements of order prime to char F. Prove that $(H_{\mathrm{diag}})^0 \simeq (H/H')_{\mathrm{diag}}$ and $\pi_0(H_{\mathrm{diag}}) \simeq H'_{\mathrm{diag}}$.

5. Prove that an algebraic group scheme G is finite if and only if $\dim G = 0$.

6. Let L/F be a finite Galois field extension with the Galois group G. Show that $R_{L/F}(\boldsymbol{\mu}_{n,L})$ is the Cartier dual to $(\mathbb{Z}/n\mathbb{Z})[G]_{\mathrm{et}}$, where the Γ-action is induced by the natural homomorphism $\Gamma \to G$.

7. Let $p = \mathrm{char}\, F$ and $\boldsymbol{\alpha}_p$ the kernel of the pth power homomorphism $\mathbf{G}_{\mathrm{a}} \to \mathbf{G}_{\mathrm{a}}$. Show that $(\boldsymbol{\alpha}_p)^D \simeq \boldsymbol{\alpha}_p$.

8. Let $f \colon G \to H$ be an algebraic group scheme homomorphism with G connected. Prove that if f_{alg} is surjective then H is also connected.

9. If N and G/N are connected then G is also connected.

10. Show that $F[\mathbf{PGL}_n(F)]$ is isomorphic to the subalgebra of $F[X_{ij}, \frac{1}{\det X}]$ consisting of all homogeneous rational functions of degree 0.

11. Let B be a quaternion algebra with a unitary involution τ over an étale quadratic extension of F. Prove that $\mathbf{SU}(B, \tau) \simeq \mathbf{SL}_1(A)$ for some quaternion algebra A over F.

12. Show that $\mathbf{Spin}^+(A, \sigma, f)$ and $\mathbf{Spin}^-(A, \sigma, f)$ are isomorphic if and only if $\mathrm{GO}^-(A, \sigma, f) \neq \varnothing$.

13. Show that the automorphism $x \mapsto x^{-t}$ of $\mathbf{SL}_2$ is inner.

14. $\mathrm{Nrd}(X) - 1$ is irreducible.

15. Let F be a field of characteristic 2 and $\alpha_1, \alpha_2 \in F^\times$. Let G be the algebraic group scheme of isometries of the bilinear form $\alpha_1 x_1 y_1 + \alpha_2 x_2 y_2$, so that $F[G]$ is the factor algebra of the polynomial ring $F[x_{11}, x_{12}, x_{21}, x_{22}]$ by the ideal generated by the entries of

$$\begin{pmatrix} x_{11} & x_{12} \\ x_{21} & x_{22} \end{pmatrix}^t \cdot \begin{pmatrix} \alpha_1 & 0 \\ 0 & \alpha_2 \end{pmatrix} \cdot \begin{pmatrix} x_{11} & x_{12} \\ x_{21} & x_{22} \end{pmatrix} - \begin{pmatrix} \alpha_1 & 0 \\ 0 & \alpha_2 \end{pmatrix}.$$

   (a) Show that $x_{11}x_{22} + x_{12}x_{21} + 1$ and $x_{11} + x_{22}$ are nilpotent in $F[G]$.
   (b) Assuming $\alpha_1 \alpha_2^{-1} \notin F^{\times 2}$, show that
   $$F[G]_{\mathrm{red}} = F[x_{11}, x_{21}]/(x_{11}^2 + \alpha_2 \alpha_1^{-1} x_{21}^2 + 1),$$
   and that $\left(F[G]_{\mathrm{red}}\right)_{F_{\mathrm{alg}}}$ is not reduced. Therefore, there is no smooth algebraic group associated to G.
   (c) Assuming $\alpha_1 \alpha_2^{-1} \in F^{\times 2}$, show that the additive group $\mathbf{G}_{\mathrm{a}}$ is the smooth algebraic group associated to G.

16. Let F be a perfect field of characteristic 2 and let b be a nonsingular symmetric nonalternating bilinear form on a vector space V of dimension n.

   (a) Show that there is a unique vector $e \in V$ such that $b(v,v) = b(v,e)^2$ for all $v \in V$. Let $V' = e^\perp$ be the hyperplane of all vectors which are orthogonal to e. Show that $e \in V'$ if and only if n is even, and that the restriction b' of b to V' is an alternating form.

   (b) Show that the smooth algebraic group $\mathbf{O}(V,b)_{\mathrm{red}}$ associated to the orthogonal group of the bilinear space (V,b) stabilizes e.

   (c) Suppose n is odd. Show that the alternating form b' is nonsingular and that the restriction map $\mathbf{O}(V,b)_{\mathrm{red}} \to \mathbf{Sp}(V',b')$ is an isomorphism.

   (d) Suppose n is even. Show that the radical of b' is eF. Let $V'' = V'/eF$ and let b'' be the nonsingular alternating form on V'' induced by b'. Show that the restriction map $\rho\colon \mathbf{O}(V,b)_{\mathrm{red}} \to \mathbf{Sp}(V'',b'')$ is surjective. Show that every $u \in \ker\rho$ induces on V' a linear transformation of the form $v' \mapsto v' + e\varphi'(v')$ for some linear form $\varphi' \in (V')^*$ such that $\varphi'(e) = 0$. The form φ' therefore induces a linear form $\varphi'' \in (V'')^*$; show that the map $j\colon \ker\rho \to (V'')^*$ which maps u to φ'' is a homomorphism. Show that there is an exact sequence

   $$0 \to \mathbf{G}_a \xrightarrow{i} \ker\rho \xrightarrow{j} (V'')^* \to 0$$

   where i maps $\lambda \in F$ to the endomorphism $v \mapsto v + e\lambda b(v,e)$ of V. Conclude that $\ker\rho$ is the maximal solvable connected normal subgroup of $\mathbf{O}(V,b)_{\mathrm{red}}$.

Notes

§§ 20–22. Historical comments on the theory of algebraic groups are given by Springer in his survey article [271] and we restrict to comments closely related to material given in this chapter. The functorial approach to algebraic groups was developed in the Séminaire du Bois Marie 62/64, directed by M. Demazure and A. Grothendieck [110]. The first systematic presentation of this approach is given in the treatise of Demazure-Gabriel [70]. As mentioned in the introduction to this chapter, the classical theory (mostly over an algebraically closed field) can be found for example in Borel [45] and Humphreys [125]. See also the new edition of the book of Springer [270]. Relations between algebraic structures and exceptional algebraic groups (at least from the point of view of Lie algebras) are described in the books of Jacobson [138], Seligman [251] and the survey of Faulkner and Ferrar [96].

§ 23. In his commentary (Collected Papers, Vol. II, [311, pp. 548–549]) to [310], Weil makes interesting historical remarks on the relations between classical groups and algebras with involution. In particular he attributes the idea to view classical algebraic groups as groups of automorphisms of algebras with involution to Siegel.

§ 26. Most of this comes from Weil [310] (see also the Tata notes of Kneser [155] and the book of Platonov-Rapinchuk [222]). One difference is that we use ideas from Tits [291], to give a characteristic free presentation, and that we also consider types G_2 and F_4 (see the paper [121] of Hijikata). For type D_4 (also excluded by Weil), we need the theory developed in Chapter X. The use of groupoids (categories with isomorphisms as morphisms) permits one to avoid the explicit use of non-abelian Galois cohomology, which will not be introduced until the following chapter.

§27. A discussion of the maximal possible indexes of Tits algebras can be found in Merkurjev-Panin-Wadsworth [194], [193] and Merkurjev [192].

CHAPTER VII

Galois Cohomology

In the preceding chapters, we have met groupoids $M = M(F)$ of "algebraic objects" over a base field F, for example finite dimensional F-algebras or algebraic groups of a certain type. If over a separable closure F_{sep} of F the groupoid $M(F_{\text{sep}})$ is connected, i.e., all objects over F_{sep} are isomorphic, then in many cases the objects of M are classified up to isomorphism by a cohomology set $H^1\big(\text{Gal}(F_{\text{sep}}/F), A\big)$, where A is the automorphism group of a fixed object of $M(F_{\text{sep}})$. The aim of this chapter is to develop the general theory of such cohomology sets, to reinterpret some earlier results in this setting and to give techniques (like twisting) which will be used in later parts of this book.

There are four sections. The basic techniques are explained in §28, and §29 gives an explicit description of the cohomology sets of various algebraic groups in terms of algebras with involution. In §30 we focus on the cohomology groups of $\boldsymbol{\mu}_n$, which are used in §31 to reinterpret various invariants of algebras with involution or of algebraic groups, and to define higher cohomological invariants.

§28. Cohomology of Profinite Groups

In this chapter, we let Γ denote a *profinite group*, i.e., a group which is the inverse limit of a system of finite groups. For instance, Γ may be the absolute Galois group of a field (this is the main case of interest for the applications in §§29–31), or a finite group (with the discrete topology). An action of Γ on the left on a discrete topological space is called *continuous* if the stabilizer of each point is an open subgroup of Γ; discrete topological spaces with a continuous left action of Γ are called Γ-*sets*. (Compare with §18.A, where only *finite* Γ-sets are considered.) A group A which is also a Γ-set is called a Γ-*group* if Γ acts by group homomorphisms, i.e.,

$$\sigma(a_1 \cdot a_2) = \sigma a_1 \cdot \sigma a_2 \quad \text{for } \sigma \in \Gamma,\, a_1,\, a_2 \in A.$$

A Γ-group which is commutative is called a Γ-*module*.

In this section, we review some general constructions of nonabelian cohomology: in the first subsection, we define cohomology sets $H^i(\Gamma, A)$ for $i = 0$ if A is a Γ-set, for $i = 0$, 1 if A is a Γ-group and for $i = 0$, 1, 2, ... if A is a Γ-module, and we relate these cohomology sets by exact sequences in the second subsection. The third subsection discusses the process of twisting, and the fourth subsection gives an interpretation of $H^1(\Gamma, A)$ in terms of torsors.

28.A. Cohomology sets. For any Γ-set A, we set

$$H^0(\Gamma, A) = A^\Gamma = \{\, a \in A \mid \sigma a = a \text{ for } \sigma \in \Gamma \,\}.$$

If A is a Γ-group, the subset $H^0(\Gamma, A)$ is a subgroup of A.

Let A be a Γ-group. A 1-*cocycle* of Γ with values in A is a continuous map $\alpha\colon \Gamma \to A$ such that, denoting by α_σ the image of $\sigma \in \Gamma$ in A,

$$\alpha_{\sigma\tau} = \alpha_\sigma \cdot \sigma\alpha_\tau \quad \text{for } \sigma,\, \tau \in \Gamma.$$

We denote by $Z^1(\Gamma, A)$ the set of all 1-cocycles of Γ with values in A. The constant map $\alpha_\sigma = 1$ is a distinguished element in $Z^1(\Gamma, A)$, which is called the *trivial 1-cocycle*. Two 1-cocycles $\alpha,\, \alpha' \in Z^1(\Gamma, A)$ are said to be *cohomologous* or *equivalent* if there exists $a \in A$ satisfying

$$\alpha'_\sigma = a \cdot \alpha_\sigma \cdot \sigma a^{-1} \quad \text{for all } \sigma \in \Gamma.$$

Let $H^1(\Gamma, A)$ be the set of equivalence classes of 1-cocycles. It is a pointed set whose distinguished element (or base point) is the cohomology class of the trivial 1-cocycle.

For instance, if the action of Γ on A is trivial, then $Z^1(\Gamma, A)$ is the set of all continuous group homomorphisms from Γ to A; two homomorphisms $\alpha,\, \alpha'$ are cohomologous if and only if $\alpha' = \operatorname{Int}(a) \circ \alpha$ for some $a \in A$.

If A is a Γ-module the set $Z^1(\Gamma, A)$ is an abelian group for the operation $(\alpha\beta)_\sigma = \alpha_\sigma \cdot \beta_\sigma$. This operation is compatible with the equivalence relation on 1-cocycles, hence it induces an abelian group structure on $H^1(\Gamma, A)$.

Now, let A be a Γ-module. A 2-*cocycle* of Γ with values in A is a continuous map $\alpha\colon \Gamma \times \Gamma \to A$ satisfying

$$\sigma\alpha_{\tau,\rho} \cdot \alpha_{\sigma,\tau\rho} = \alpha_{\sigma\tau,\rho}\alpha_{\sigma,\tau} \quad \text{for } \sigma,\, \tau,\, \rho \in \Gamma.$$

The set of 2-cocycles of Γ with values in A is denoted by $Z^2(\Gamma, A)$. This set is an abelian group for the operation $(\alpha\beta)_{\sigma,\tau} = \alpha_{\sigma,\tau} \cdot \beta_{\sigma,\tau}$. Two 2-cocycles $\alpha,\, \alpha'$ are said to be *cohomologous* or *equivalent* if there exists a continuous map $\varphi\colon \Gamma \to A$ such that

$$\alpha'_{\sigma,\tau} = \sigma\varphi_\tau \cdot \varphi_{\sigma\tau}^{-1} \cdot \varphi_\sigma \cdot \alpha_{\sigma,\tau} \quad \text{for all } \sigma,\, \tau \in \Gamma.$$

The equivalence classes of 2-cocycles form an abelian group denoted $H^2(\Gamma, A)$. Higher cohomology groups $H^i(\Gamma, A)$ (for $i \geq 3$) will be used less frequently in the sequel; we refer to Brown [56] for their definition.

Functorial properties. Let $f\colon A \to B$ be a homomorphism of Γ-sets, i.e., a map such that $f(\sigma a) = \sigma f(a)$ for $\sigma \in \Gamma$ and $a \in A$. If $a \in A$ is fixed by Γ, then so is $f(a) \in B$. Therefore, f restricts to a map

$$f^0 \colon H^0(\Gamma, A) \to H^0(\Gamma, B).$$

If A, B are Γ-groups and f is a group homomorphism, then f^0 is a group homomorphism. Moreover, there is an induced map

$$f^1 \colon H^1(\Gamma, A) \to H^1(\Gamma, B)$$

which carries the cohomology class of any 1-cocycle α to the cohomology class of the 1-cocycle $f^1(\alpha)$ defined by $f^1(\alpha)_\sigma = f(\alpha_\sigma)$. In particular, f^1 is a homomorphism of pointed sets, in the sense that f^1 maps the distinguished element of $H^1(\Gamma, A)$ to the distinguished element of $H^1(\Gamma, B)$.

If A, B are Γ-modules, then f^1 is a group homomorphism. Moreover, f induces homomorphisms

$$f^i \colon H^i(\Gamma, A) \to H^i(\Gamma, B)$$

for all $i \geq 0$.

Besides those functorial properties in A, the sets $H^i(\Gamma, A)$ also have functorial properties in Γ. We just consider the case of subgroups: let $\Gamma_0 \subset \Gamma$ be a closed subgroup and let A be a Γ-group. The action of Γ restricts to a continuous action of Γ_0. The obvious inclusion $A^\Gamma \subset A^{\Gamma_0}$ is called *restriction*:

$$\text{res}\colon H^0(\Gamma, A) \to H^0(\Gamma_0, A).$$

If A is a Γ-group, the restriction of a 1-cocycle $\alpha \in Z^1(\Gamma, A)$ to Γ_0 is a 1-cocycle of Γ_0 with values in A. Thus, there is a *restriction* map of pointed sets

$$\text{res}\colon H^1(\Gamma, A) \to H^1(\Gamma_0, A).$$

Similarly, if A is a Γ-module, there is for all $i \geq 2$ a restriction map

$$\text{res}\colon H^i(\Gamma, A) \to H^i(\Gamma_0, A).$$

28.B. Cohomology sequences. By definition, the *kernel* $\ker(\mu)$ of a map of pointed sets $\mu\colon N \to P$ is the subset of all $n \in N$ such that $\mu(n)$ is the base point of P. A sequence of maps of pointed sets

$$M \xrightarrow{\rho} N \xrightarrow{\mu} P$$

is said to be *exact* if $\text{im}(\rho) = \ker(\mu)$. Thus, the sequence $M \xrightarrow{\rho} N \to 1$ is exact if and only if ρ is surjective. The sequence $1 \to N \xrightarrow{\mu} P$ is exact if and only if the base point of N is the only element mapped by μ to the base point of P. Note that this condition *does not* imply that μ is injective.

The exact sequence associated to a subgroup. Let B be a Γ-group and let $A \subset B$ be a Γ-subgroup (i.e., $\sigma a \in A$ for all $\sigma \in \Gamma$, $a \in A$). Let B/A be the Γ-set of left cosets of A in B, i.e.,

$$B/A = \{\, b \cdot A \mid b \in B \,\}.$$

The natural projection of B onto B/A induces a map of pointed sets $B^\Gamma \to (B/A)^\Gamma$. Let $b \cdot A \in (B/A)^\Gamma$, i.e., $\sigma b \cdot A = b \cdot A$ for all $\sigma \in \Gamma$. The map $\alpha\colon \Gamma \to A$ defined by $\alpha_\sigma = b^{-1} \cdot \sigma b \in A$ is a 1-cocycle with values in A, whose class $[\alpha]$ in $H^1(\Gamma, A)$ is independent of the choice of b in $b \cdot A$. Hence we have a map of pointed sets

$$\delta^0\colon (B/A)^\Gamma \to H^1(\Gamma, A), \quad b \cdot A \mapsto [\alpha] \quad \text{where} \quad \alpha_\sigma = b^{-1} \cdot \sigma b.$$

(28.1) Proposition. *The sequence*

$$1 \to A^\Gamma \to B^\Gamma \to (B/A)^\Gamma \xrightarrow{\delta^0} H^1(\Gamma, A) \to H^1(\Gamma, B)$$

is exact.

Proof: For exactness at $(B/A)^\Gamma$, suppose that the 1-cocycle $\alpha_\sigma = b^{-1} \cdot \sigma b \in A$ is trivial in $H^1(\Gamma, A)$ i.e., $\alpha_\sigma = a^{-1} \cdot \sigma a$ for some $a \in A$. Then $ba^{-1} \in B^\Gamma$ and the coset $b \cdot A = ba^{-1} \cdot A$ in B/A is equal to the image of $ba^{-1} \in B^\Gamma$.

If $\alpha \in Z^1(\Gamma, A)$ satisfies $\alpha_\sigma = b^{-1} \cdot \sigma b$ for some $b \in B$, then $b \cdot A \in (B/A)^\Gamma$ and $[\alpha] = \delta^0(b \cdot A)$. $\square$

The group B^Γ acts naturally (by left multiplication) on the pointed set $(B/A)^\Gamma$.

(28.2) Corollary. *There is a natural bijection between* $\ker\big(H^1(\Gamma, A) \to H^1(\Gamma, B)\big)$ *and the orbit set of the group B^Γ in $(B/A)^\Gamma$.*

Proof: A coset $b \cdot A \in (B/A)^\Gamma$ determines the element

$$\delta^0(b \cdot A) = [b^{-1} \cdot \sigma b] \in \ker\big(H^1(\Gamma, A) \to H^1(\Gamma, B)\big).$$

One checks easily that $\delta^0(b \cdot A) = \delta^0(b' \cdot A)$ if and only if the cosets $b \cdot A$ and $b' \cdot A$ lie in the same B^Γ-orbit in $(B/A)^\Gamma$. $\square$

The exact sequence associated to a normal subgroup. Assume for the rest of this subsection that the Γ-subgroup A of B is normal in B, and set $C = B/A$. It is a Γ-group.

(28.3) Proposition. *The sequence*

$$1 \to A^\Gamma \to B^\Gamma \to C^\Gamma \xrightarrow{\delta^0} H^1(\Gamma, A) \to H^1(\Gamma, B) \to H^1(\Gamma, C)$$

is exact.

Proof: Let $\beta \in Z^1(\Gamma, B)$ where $[\beta]$ lies in the kernel of the last map. Then $\beta_\sigma \cdot A = b^{-1} \cdot \sigma b \cdot A = b^{-1} \cdot A \cdot \sigma b$ for some $b \in B$. Hence $\beta_\sigma = b^{-1} \cdot \alpha_\sigma \cdot \sigma b$ for $\alpha \in Z^1(\Gamma, A)$ and $[\beta]$ is the image of $[\alpha]$ in $H^1(\Gamma, B)$. $\square$

The group C^Γ acts on $H^1(\Gamma, A)$ as follows: for $c = b \cdot A \in C^\Gamma$ and $\alpha \in Z^1(\Gamma, A)$, set $c[\alpha] = [\beta]$ where $\beta_\sigma = b \cdot \alpha_\sigma \cdot \sigma b^{-1}$.

(28.4) Corollary. *There is a natural bijection between* $\ker\big(H^1(\Gamma, B) \to H^1(\Gamma, C)\big)$ *and the orbit set of the group* C^Γ *in* $H^1(\Gamma, A)$. $\square$

The exact sequence associated to a central subgroup. Now, assume that A lies in the center of B. Then A is an abelian group and one can define a map of pointed sets

$$\delta^1 \colon H^1(\Gamma, C) \to H^2(\Gamma, A)$$

as follows. Given any $\gamma \in Z^1(\Gamma, C)$, choose a map $\beta \colon \Gamma \to B$ such that β_σ maps to γ_σ for all $\sigma \in \Gamma$. Consider the function $\alpha \colon \Gamma \times \Gamma \to A$ given by

$$\alpha_{\sigma, \tau} = \beta_\sigma \cdot \sigma\beta_\tau \cdot \beta_{\sigma\tau}^{-1}.$$

One can check that $\alpha \in Z^2(\Gamma, A)$ and that its class in $H^2(\Gamma, A)$ does not depend on the choices of $\gamma \in [\gamma]$ and β. We define $\delta^1\big([\gamma]\big) = [\alpha]$.

(28.5) Proposition. *The sequence*

$$1 \to A^\Gamma \to B^\Gamma \to C^\Gamma \xrightarrow{\delta^0} H^1(\Gamma, A) \to H^1(\Gamma, B) \to H^1(\Gamma, C) \xrightarrow{\delta^1} H^2(\Gamma, A)$$

is exact.

Proof: Assume that for $\gamma \in Z^1(\Gamma, C)$ and β, α as above we have

$$\alpha_{\sigma, \tau} = \beta_\sigma \cdot \sigma\beta_\tau \cdot \beta_{\sigma\tau}^{-1} = a_\sigma \cdot \sigma a_\tau \cdot a_{\sigma\tau}^{-1}$$

for some $a_\sigma \in A$. Then $\beta'_\sigma = \beta_\sigma \cdot a_\sigma^{-1}$ is a 1-cocycle in $Z^1(\Gamma, B)$ and γ is the image of β'. $\square$

The group $H^1(\Gamma, A)$ acts naturally on $H^1(\Gamma, B)$ by $(\alpha \cdot \beta)_\sigma = \alpha_\sigma \cdot \beta_\sigma$.

(28.6) Corollary. *There is a natural bijection between the kernel of the connecting map* $\delta^1 \colon H^1(\Gamma, C) \to H^2(\Gamma, A)$ *and the orbit set of the group* $H^1(\Gamma, A)$ *in* $H^1(\Gamma, B)$.

Proof: Two elements of $H^1(\Gamma, B)$ have the same image in $H^1(\Gamma, C)$ if and only if they are in the same orbit under the action of $H^1(\Gamma, A)$. $\square$

(28.7) Remark. If the exact sequence of Γ-homomorphisms

$$1 \to A \to B \to C \to 1$$

is split by a Γ-map $C \to B$, then the connecting maps δ^0 and δ^1 are trivial.

28.C. Twisting. Let A be a Γ-group. We let Γ act on the group $\operatorname{Aut} A$ of automorphisms of A by

$${}^\sigma f(a) = \sigma\big(f(\sigma^{-1}a)\big) \qquad \text{for } \sigma \in \Gamma, \ a \in A \text{ and } f \in \operatorname{Aut} A.$$

(Compare with §18.A.) The subgroup $(\operatorname{Aut} A)^\Gamma$ of $\operatorname{Aut} A$ consists of all Γ-automorphisms of A.

For a fixed 1-cocycle $\alpha \in Z^1(\Gamma, \operatorname{Aut} A)$ we define a new action of Γ on A by

$$\sigma * a = \alpha_\sigma(\sigma a), \qquad \text{for } \sigma \in \Gamma \text{ and } a \in A.$$

The group A with this new Γ-action is denoted by A_α. We say that A_α is obtained by *twisting* A by the 1-cocycle α.

If 1-cocycles α, $\alpha' \in Z^1(\Gamma, \operatorname{Aut} A)$ are related by $\alpha'_\sigma = f \circ \alpha_\sigma \circ {}^\sigma f^{-1}$ for some $f \in \operatorname{Aut} A$, then f defines an isomorphism of Γ-groups $A_\alpha \xrightarrow{\sim} A_{\alpha'}$. Therefore, cohomologous cocycles define isomorphic twisted Γ-groups. However, the isomorphism $A_\alpha \xrightarrow{\sim} A_{\alpha'}$ is not canonical, hence we cannot define a twisted group $A_{[\alpha]}$ for $[\alpha] \in H^1(\Gamma, A)$.

Now, let $\alpha \in Z^1(\Gamma, A)$ and let $\overline{\alpha}$ be the image of α in $Z^1(\Gamma, \operatorname{Aut} A)$ under the map $\operatorname{Int}\colon A \to \operatorname{Aut} A$. We also write A_α for the twist $A_{\overline{\alpha}}$ of A. By definition we then have

$$\sigma * a = \alpha_\sigma \cdot \sigma a \cdot \alpha_\sigma^{-1}, \qquad \text{for } a \in A_\alpha \text{ and } \sigma \in \Gamma.$$

(28.8) Proposition. *Let A be a Γ-group and $\alpha \in Z^1(\Gamma, A)$. Then the map*

$$\theta_\alpha\colon H^1(\Gamma, A_\alpha) \to H^1(\Gamma, A) \ \text{given by } (\gamma_\sigma) \mapsto (\gamma_\sigma \cdot \alpha_\sigma)$$

is a well-defined bijection which takes the trivial cocycle of $H^1(\Gamma, A_\alpha)$ to $[\alpha]$.

Proof: Let γ be a cocycle with values in A_α. We have $\gamma_{\sigma\tau} = \gamma_\sigma \alpha_\sigma \sigma(\gamma_\tau) \alpha_\sigma^{-1}$, hence

$$\gamma_{\sigma\tau} \cdot \alpha_{\sigma\tau} = \gamma_\sigma \cdot \alpha_\sigma \cdot \sigma(\gamma_\tau \alpha_\tau)$$

and $\gamma\alpha \in Z^1(\Gamma, A)$. If $\gamma' \in Z^1(\Gamma, A_\alpha)$ is cohomologous to γ, let $a \in A$ satisfy $\gamma'_\sigma = a \cdot \gamma_\sigma \cdot (\sigma * a^{-1})$. Then $\gamma'_\sigma \alpha_\sigma = a \cdot \gamma_\sigma \alpha_\sigma \cdot \sigma a^{-1}$, hence $\gamma'\alpha$ is cohomologous to $\gamma\alpha$. This shows that θ_α is a well-defined map. To prove that θ_α is a bijection, observe that the map $\sigma \mapsto \alpha_\sigma^{-1}$ is a 1-cocycle in $Z^1(\Gamma, A_\alpha)$. The induced map $\theta_{\alpha^{-1}}\colon H^1(\Gamma, A) \to H^1(\Gamma, A_\alpha)$ is the inverse of θ_α. $\qquad\square$

(28.9) Remark. If A is abelian, we have $A = A_\alpha$ for $\alpha \in Z^1(\Gamma, A)$, and θ_α is translation by $[\alpha]$.

Functoriality. Let $f\colon A \to B$ be a Γ-homomorphism and let $\beta = f^1(\alpha) \in Z^1(\Gamma, B)$ for $\alpha \in Z^1(\Gamma, A)$. Then the map f, considered as a map $f_\alpha\colon A_\alpha \to B_\beta$, is a Γ-homomorphism, and the following diagram commutes:

$$
\begin{array}{ccc}
H^1(\Gamma, A_\alpha) & \xrightarrow{\ \theta_\alpha\ } & H^1(\Gamma, A) \\
f_\alpha^1 \downarrow & & \downarrow f^1 \\
H^1(\Gamma, B_\beta) & \xrightarrow{\ \theta_\beta\ } & H^1(\Gamma, B).
\end{array}
$$

In particular, θ_α induces a bijection between $\ker f_\alpha^1$ and the fiber $(f^1)^{-1}([\beta])$.

Let A be a Γ-subgroup of a Γ-group B, let $\alpha \in Z^1(\Gamma, A)$, and let $\beta \in Z^1(\Gamma, B)$ be the image of α. Corollary (28.2) implies:

(28.10) Proposition. *There is a natural bijection between the fiber of $H^1(\Gamma, A) \to H^1(\Gamma, B)$ over $[\beta]$ and the orbit set of the group $(B_\beta)^\Gamma$ in $(B_\beta/A_\alpha)^\Gamma$.* $\square$

Now, assume that A is a normal Γ-subgroup of B and let $C = B/A$. Let $\beta \in Z^1(\Gamma, B)$ and let $\gamma \in Z^1(\Gamma, C)$ be the image of β. The conjugation map $B \to \mathrm{Aut}\, A$ associates to β a 1-cocycle $\alpha \in Z^1(\Gamma, \mathrm{Aut}\, A)$. Corollary (28.4) implies:

(28.11) Proposition. *There is a natural bijection between the fiber of $H^1(\Gamma, B) \to H^1(\Gamma, C)$ over $[\gamma]$ and the orbit set of the group $(C_\gamma)^\Gamma$ in $H^1(\Gamma, A_\alpha)$.* $\square$

Assume further that A lies in the center of B and let $\gamma \in Z^1(\Gamma, C)$, where $C = B/A$. The conjugation map $C \to \mathrm{Aut}\, B$ induces a 1-cocycle $\beta \in Z^1(\Gamma, \mathrm{Aut}\, B)$. Let ε be the image of $[\gamma]$ under the map $\delta^1 \colon H^1(\Gamma, C) \to H^2(\Gamma, A)$.

(28.12) Proposition. *The following diagram*

$$
\begin{array}{ccc}
H^1(\Gamma, C_\gamma) & \xrightarrow{\ \theta_\gamma\ } & H^1(\Gamma, C) \\
{\scriptstyle \delta^1_\gamma}\downarrow & & \downarrow{\scriptstyle \delta^1} \\
H^2(\Gamma, A) & \xrightarrow{\ g\ } & H^2(\Gamma, A)
\end{array}
$$

commutes, where δ^1_γ is the connecting map with respect to the exact sequence

$$1 \to A \to B_\beta \to C\gamma \to 1$$

and g is multiplication by ε.

Proof: Let $\alpha \in Z^1(\Gamma, C_\gamma)$. Choose $x_\sigma \in \alpha_\sigma$ and $y_\sigma \in \gamma_\sigma$. Then

$$\varepsilon_{\sigma,\tau} = y_\sigma \cdot \sigma y_\tau \cdot y_{\sigma\tau}^{-1}$$

and

$$
\begin{aligned}
\delta^1\big(\theta_\gamma(\alpha)\big)_{\sigma,\tau} &= x_\sigma y_\sigma \cdot \sigma(x_\tau y_\tau) \cdot y_{\sigma\tau}^{-1} x_{\sigma\tau}^{-1} = x_\sigma y_\sigma \cdot \sigma x_\tau \cdot y_\sigma^{-1} \cdot \varepsilon_{\sigma,\tau} \cdot x_{\sigma\tau}^{-1} \\
&= x_\sigma \cdot (\sigma \circ x_\tau) \cdot x_{\sigma\tau}^{-1} \cdot \varepsilon_{\sigma,\tau} = \delta^1_\gamma(x)_{\sigma,\tau} \cdot \varepsilon_{\sigma,\tau},
\end{aligned}
$$

hence $\delta^1\big(\theta_\gamma(\alpha)\big) = \delta^1_\gamma(\alpha) \cdot \varepsilon = g\big(\delta^1_\gamma(\alpha)\big)$. $\square$

As in Corollary (28.6), one obtains:

(28.13) Corollary. *There is a natural bijection between the fiber over ε of the map $\delta^1 \colon H^1(\Gamma, C) \to H^2(\Gamma, A)$ and the orbit set of the group $H^1(\Gamma, A)$ in $H^1(\Gamma, B_\beta)$.* $\square$

28.D. Torsors. Let A be a Γ-group and let P be a nonempty Γ-set on which A acts on the right. Suppose that

$$\sigma(x^a) = \sigma(x)^{\sigma a} \qquad \text{for } \sigma \in \Gamma,\ x \in P \text{ and } a \in A.$$

We say that P is an *A-torsor* (or a *principal homogeneous set under A*) if the action of A on P is *simply transitive*, i.e., for any pair x, y of elements of P there exists exactly one $a \in A$ such that $y = x^a$. (Compare with (18.15), where the Γ-group A (denoted there by G) is finite and carries the trivial action of Γ.) We let $A\text{–}\mathit{Tors}_\Gamma$ denote the category of A-torsors, where the maps are the A- and Γ-equivariant functions. This category is a groupoid, since the maps are isomorphisms.

To construct examples of A-torsors, we may proceed as follows: for $\alpha \in Z^1(\Gamma, A)$, let P_α be the set A with the Γ- and A-actions

$$\sigma \star x = \alpha_\sigma \sigma x \quad \text{and} \quad x^a = xa \qquad \text{for } \sigma \in \Gamma \text{ and } x, a \in A.$$

It turns out that every A-torsor is isomorphic to some P_α:

(28.14) Proposition. *The map $\alpha \mapsto P_\alpha$ induces a bijection*

$$H^1(\Gamma, A) \xrightarrow{\sim} \mathrm{Isom}(A\text{--}\mathsf{Tors}_\Gamma).$$

Proof: If $\alpha, \alpha' \in Z^1(\Gamma, A)$ are cohomologous, let $a \in A$ satisfy $\alpha'_\sigma = a \cdot \alpha_\sigma \cdot \sigma a^{-1}$ for all $\sigma \in \Gamma$. Multiplication on the left by a is an isomorphism of torsors $P_\alpha \xrightarrow{\sim} P_{\alpha'}$. We thus have a well-defined map $H^1(\Gamma, A) \to \mathrm{Isom}(A\text{--}\mathsf{Tors}_\Gamma)$. The inverse map is given as follows: Let $P \in A\text{--}\mathsf{Tors}_\Gamma$. For a fixed $x \in P$, the map $\alpha \colon \Gamma \to A$ defined by

$$\sigma(x) = x^{\alpha_\sigma} \qquad \text{for } \sigma \in \Gamma$$

is a 1-cocycle. Replacing x with x^a changes α_σ into the cohomologous cocycle $a^{-1}\alpha_\sigma \sigma a$. $\qquad\square$

(28.15) Example. Let Γ be the absolute Galois group of a field F, and let G be a finite group which we endow with the trivial action of Γ. By combining (28.14) with (18.19), we obtain a canonical bijection

$$H^1(\Gamma, G) \xrightarrow{\sim} \mathrm{Isom}(G\text{--}\mathsf{Gal}_F),$$

hence $H^1(\Gamma, G)$ classifies the Galois G-algebras over F up to isomorphism.

Functoriality. Let $f \colon A \to B$ be a Γ-homomorphism of Γ-groups. We define a functor

$$f_* \colon A\text{--}\mathsf{Tors}_\Gamma \to B\text{--}\mathsf{Tors}_\Gamma$$

as follows: for $P \in A\text{--}\mathsf{Tors}_\Gamma$, consider the product $P \times B$ with the diagonal action of Γ. The groups A and B act on $P \times B$ by

$$(p, b)^a = \bigl(p^a, f(a^{-1})b\bigr) \quad \text{and} \quad (p, b) \diamond b' = (p, bb')$$

for $p \in P$, $a \in A$ and $b, b' \in B$, and these two actions commute. Hence there is an induced right action of B on the set of A-orbits $f_*(P) = (P \times B)/A$, making $f_*(P)$ a B-torsor.

(28.16) Proposition. *The following diagram commutes:*

$$
\begin{array}{ccc}
H^1(\Gamma, A) & \xrightarrow{\ f^1\ } & H^1(\Gamma, B) \\
\downarrow & & \downarrow \\
\mathrm{Isom}(A\text{--}\mathsf{Tors}_\Gamma) & \xrightarrow{\ f_*\ } & \mathrm{Isom}(B\text{--}\mathsf{Tors}_\Gamma),
\end{array}
$$

where the vertical maps are the natural bijections of (28.14).

Proof: Let $\alpha \in Z^1(\Gamma, A)$. Every A-orbit in $(P_\alpha \times B)/A$ can be represented by a unique element of the form $(1, b)$ with $b \in B$. The map which takes the orbit $(1, b)^A$ to $b \in P_{f^1(\alpha)}$ is an isomorphism of B-torsors $(P_\alpha \times B)/A \xrightarrow{\sim} P_{f^1(\alpha)}$. $\qquad\square$

Induced torsors. Let Γ_0 be a closed subgroup of Γ and let A_0 be a Γ_0-group. The *induced* Γ-*group* $\mathrm{Ind}_{\Gamma_0}^{\Gamma} A_0$ is defined as the group of all continuous maps $f \colon \Gamma \to A_0$ such that $f(\gamma_0 \gamma) = \gamma_0 f(\gamma)$ for all $\gamma_0 \in \Gamma_0$, $\gamma \in \Gamma$:

$$\mathrm{Ind}_{\Gamma_0}^{\Gamma} A_0 = \{\, f \in \mathrm{Map}(\Gamma, A_0) \mid f(\gamma_0 \gamma) = \gamma_0 f(\gamma) \text{ for } \gamma_0 \in \Gamma_0, \, \gamma \in \Gamma \,\}.$$

The Γ-action on $\mathrm{Ind}_{\Gamma_0}^{\Gamma} A_0$ is given by ${}^{\sigma}f(\gamma) = f(\gamma\sigma)$ for $\sigma, \gamma \in \Gamma$. We let $\pi \colon \mathrm{Ind}_{\Gamma_0}^{\Gamma} A_0 \to A_0$ be the map which takes $f \in \mathrm{Ind}_{\Gamma_0}^{\Gamma} A_0$ to $f(1)$. This map satisfies $\pi({}^{\sigma}f) = {}^{\sigma}\bigl(\pi(f)\bigr)$ for all $\sigma \in \Gamma_0$, $f \in \mathrm{Ind}_{\Gamma_0}^{\Gamma} A_0$. It is therefore a Γ_0-homomorphism. (Compare with (18.17).)

By applying this construction to A_0-torsors, we obtain $(\mathrm{Ind}_{\Gamma_0}^{\Gamma} A_0)$-torsors: for $P_0 \in A_0\text{--}\mathit{Tors}_{\Gamma_0}$, the Γ-group $\mathrm{Ind}_{\Gamma_0}^{\Gamma} P_0$ carries a right action of $\mathrm{Ind}_{\Gamma_0}^{\Gamma} A_0$ defined by

$$p^f(\gamma) = p(\gamma)^{f(\gamma)} \qquad \text{for } p \in \mathrm{Ind}_{\Gamma_0}^{\Gamma} P_0, \, f \in \mathrm{Ind}_{\Gamma_0}^{\Gamma} A_0 \text{ and } \gamma \in \Gamma.$$

This action makes $\mathrm{Ind}_{\Gamma_0}^{\Gamma} P_0$ an $(\mathrm{Ind}_{\Gamma_0}^{\Gamma} A_0)$-torsor, called the *induced torsor*. We thus have a functor

$$\mathrm{Ind}_{\Gamma_0}^{\Gamma} \colon A_0\text{--}\mathit{Tors}_{\Gamma_0} \to (\mathrm{Ind}_{\Gamma_0}^{\Gamma} A_0)\text{--}\mathit{Tors}_{\Gamma}.$$

On the other hand, the Γ_0-homomorphism $\pi \colon \mathrm{Ind}_{\Gamma_0}^{\Gamma} A_0 \to A_0$ yields a functor

$$\pi_* \colon (\mathrm{Ind}_{\Gamma_0}^{\Gamma} A_0)\text{--}\mathit{Tors}_{\Gamma} \to A_0\text{--}\mathit{Tors}_{\Gamma_0},$$

as explained above.

(28.17) Proposition. *Let Γ_0 be a closed subgroup of the profinite group Γ, let A_0 be a Γ_0-group and $A = \mathrm{Ind}_{\Gamma_0}^{\Gamma} A_0$. The functors $\mathrm{Ind}_{\Gamma_0}^{\Gamma}$ and π_* define an equivalence of categories*

$$A_0\text{--}\mathit{Tors}_{\Gamma_0} \equiv A\text{--}\mathit{Tors}_{\Gamma}.$$

Proof: Let $P_0 \in A_0\text{--}\mathit{Tors}_{\Gamma_0}$ and let $P = \mathrm{Ind}_{\Gamma_0}^{\Gamma} P_0$ be the induced A-torsor. Consider the map

$$g \colon P \times A_0 \to P_0 \quad \text{given by} \quad g(p, a_0) = p(1)^{a_0}.$$

For any $a \in A$ one has

$$g\bigl((p, a_0)^a\bigr) = g\bigl(p^a, \pi(a^{-1})a_0\bigr) = g\bigl(p^a, a(1)^{-1}a_0\bigr) = p(1)^{a_0} = g(p, a_0),$$

i.e., g is compatible with the right A-action on $P \times A_0$ and hence factors through a map on the orbit space

$$\overline{g} \colon \pi_*(P) = (P \times A_0)/A \to P_0.$$

It is straightforward to check that $\overline{g}$ is a homomorphism of A_0-torsors and hence is necessarily an isomorphism. Thus, $\pi_* \circ \mathrm{Ind}_{\Gamma_0}^{\Gamma}$ is naturally equivalent to the identity on $A_0\text{--}\mathit{Tors}_{\Gamma_0}$.

On the other hand, let $P \in A\text{--}\mathit{Tors}_{\Gamma}$. We denote the orbit in $\pi_*(P) = (P \times A_0)/A$ of a pair (p, a_0) by $(p, a_0)^A$. Consider the map

$$h \colon P \to \mathrm{Ind}_{\Gamma_0}^{\Gamma}\bigl(\pi_*(P)\bigr)$$

which carries $p \in P$ to the map h_p defined by $h_p(\sigma) = (\sigma p, 1)^A$. For any $a \in A$ one has

$$h_{p^a}(\sigma) = \bigl(\sigma(p^a), 1\bigr)^A = \bigl(\sigma(p)^{\sigma a}, 1\bigr)^A \qquad \text{for } \sigma \in \Gamma.$$

Since $\left(\sigma(p)^{\sigma a}, 1\right) = \left(\sigma(p), \sigma a(1)\right)^{\sigma a}$, the A-orbits of $\left(\sigma(p)^{\sigma a}, 1\right)$ and $\left(\sigma(p), \sigma a(1)\right)$ coincide. We have $\left(\sigma(p), \sigma a(1)\right) = \left(\sigma(p), 1\right) \diamond \sigma a(1)$, hence

$$h_{p^a}(\sigma) = h_p(\sigma) \diamond \sigma a(1) = h_p^a(\sigma).$$

Thus h is a homomorphism (hence an isomorphism) of A-torsors, showing that $\mathrm{Ind}_{\Gamma_0}^{\Gamma} \circ \pi_*$ is naturally equivalent to the identity on $A{-}\mathit{Tors}_\Gamma$. $\qquad\square$

By combining the preceding proposition with (28.14), we obtain:

(28.18) Corollary. *With the same notation as in* (28.17), *there is a natural bijection of pointed sets between* $H^1(\Gamma_0, A_0)$ *and* $H^1(\Gamma, A)$. $\qquad\square$

(28.19) Remark. If A_0 is a Γ-group (not just a Γ_0-group), there is a simpler description of the Γ-group $\mathrm{Ind}_{\Gamma_0}^{\Gamma} A_0$: let Γ/Γ_0 denote the set of left cosets of Γ_0 in Γ. On the group $\mathrm{Map}(\Gamma/\Gamma_0, A_0)$ of continuous maps $\Gamma/\Gamma_0 \to A_0$, consider the Γ-action given by $^\sigma f(x) = \sigma f(\sigma^{-1} x)$. The Γ-group $\mathrm{Map}(\Gamma/\Gamma_0, A_0)$ is naturally isomorphic to $\mathrm{Ind}_{\Gamma_0}^{\Gamma} A_0$. For, there are mutually inverse isomorphisms

$$\alpha\colon \mathrm{Ind}_{\Gamma_0}^{\Gamma} A_0 \to \mathrm{Map}(\Gamma/\Gamma_0, A_0) \quad \text{given by} \quad \alpha(a)(\sigma \cdot \Gamma_0) = \sigma a(\sigma^{-1})$$

and

$$\beta\colon \mathrm{Map}(\Gamma/\Gamma_0, A_0) \to \mathrm{Ind}_{\Gamma_0}^{\Gamma} A_0 \quad \text{given by} \quad \beta(f)(\sigma) = \sigma f(\sigma^{-1} \cdot \Gamma_0).$$

(28.20) Example. Let A be a Γ-group and n be an integer, $n \geq 1$. We let the symmetric group S_n act by permutations on the product A^n of n copies of A, and we let Γ act trivially on S_n. Any continuous homomorphism $\rho\colon \Gamma \to S_n$ is a 1-cocycle in $Z^1(\Gamma, S_n)$. It yields a 1-cocycle $\alpha\colon \Gamma \to \mathrm{Aut}\, A^n$ via the action of S_n on A^n, and we may consider the twisted group $(A^n)_\alpha$.

Assume that Γ acts transitively via ρ on the set $X = \{1, 2, \dots, n\}$. Let $\Gamma_0 \subset \Gamma$ be the stabilizer of $1 \in X$. The set X is then identified with Γ/Γ_0. It is straightforward to check that $(A^n)_\alpha$ is identified with $\mathrm{Map}(\Gamma/\Gamma_0, A) = \mathrm{Ind}_{\Gamma_0}^{\Gamma} A$.

Consider the semidirect product $A^n \rtimes S_n$ and the exact sequence

$$1 \to A^n \to A^n \rtimes S_n \to S_n \to 1.$$

By (28.11) and (28.18), there is a canonical bijection between the fiber of the map $H^1(\Gamma, A^n \rtimes S_n) \to H^1(\Gamma, S_n)$ over $[\rho]$ and the orbit set in $H^1(\Gamma_0, A)$ of the group $(S_n)_\rho^\Gamma$, which is the centralizer of the image of ρ in S_n.

§29. Galois Cohomology of Algebraic Groups

In this section, the profinite group Γ is the absolute Galois group of a field F, i.e., $\Gamma = \mathrm{Gal}(F_{\mathrm{sep}}/F)$ where F_{sep} is a separable closure of F. If A is a discrete Γ-group, we write $H^i(F, A)$ for $H^i(\Gamma, A)$.

Let G be a group scheme over F. The Galois group Γ acts continuously on the discrete group $G(F_{\mathrm{sep}})$. Hence $H^i\!\left(F, G(F_{\mathrm{sep}})\right)$ is defined for $i = 0, 1$, and it is defined for all $i \geq 2$ if G is a commutative group scheme. We use the notation

$$H^i(F, G) = H^i\!\left(F, G(F_{\mathrm{sep}})\right).$$

In particular, $H^0(F, G) = G(F)$.

Every group scheme homomorphism $f\colon G \to H$ induces a Γ-homomorphism $G(F_{\mathrm{sep}}) \to H(F_{\mathrm{sep}})$ and hence a homomorphism of groups (resp. of pointed sets)

$$f^i\colon H^i(F, G) \to H^i(F, H)$$

for $i = 0$ (resp. $i = 1$). If $1 \to N \to G \to S \to 1$ is an exact sequence of algebraic group schemes such that the induced sequence of Γ-homomorphisms

$$1 \to N(F_{\mathrm{sep}}) \to G(F_{\mathrm{sep}}) \to S(F_{\mathrm{sep}}) \to 1$$

is exact (this is always the case if N is smooth, see (22.15)), we have a connecting map $\delta^0 \colon S(F) \to H^1(F, N)$, and also, if N lies in the center of G, a connecting map $\delta^1 \colon H^1(F, S) \to H^2(F, N)$. We may thus apply the techniques developed in the preceding section.

Our main goal is to give a description of the pointed set $H^1(F, G)$ for various algebraic groups G. We first explain the main technical tool.

Let G be a group scheme over F and let $\rho \colon G \to \mathbf{GL}(W)$ be a representation with W a finite dimensional F-space. Fix an element $w \in W$, and identify W with an F-subspace of $W_{\mathrm{sep}} = W \otimes_F F_{\mathrm{sep}}$. An element $w' \in W_{\mathrm{sep}}$ is called a *twisted ρ-form of w* if $w' = \rho_{\mathrm{sep}}(g)(w)$ for some $g \in G(F_{\mathrm{sep}})$. As in §26, consider the category $\widetilde{A}(\rho, w)$ whose objects are the twisted ρ-forms of w and whose maps $w' \to w''$ are the elements $g \in G(F_{\mathrm{sep}})$ such that $\rho_{\mathrm{sep}}(g)(w') = w''$. This category is a connected groupoid. On the other hand, let $A(\rho, w)$ denote the groupoid whose objects are the twisted ρ-forms of w which lie in W, and whose maps $w' \to w''$ are the elements $g \in G(F)$ such that $\rho(g)(w') = w''$. Thus, if X denotes the Γ-set of objects of $\widetilde{A}(\rho, w)$, the set $X^\Gamma = H^0(\Gamma, X)$ is the set of objects of $A(\rho, w)$. Moreover, the set of orbits of $G(F)$ in X^Γ is the set of isomorphism classes $\mathrm{Isom}(A(\rho, w))$. It is a pointed set with the isomorphism class of w as base point.

Let $\mathbf{Aut}_G(w)$ denote the stabilizer of w; it is a subgroup of the group scheme G. Since $G(F_{\mathrm{sep}})$ acts transitively on X, the Γ-set X is identified with the set of left cosets of $G(F_{\mathrm{sep}})$ modulo $\mathbf{Aut}_G(w)(F_{\mathrm{sep}})$. Corollary (28.2) yields a natural bijection of pointed sets between the kernel of $H^1(F, \mathbf{Aut}_G(w)) \to H^1(F, G)$ and the orbit set $X^\Gamma / G(F)$. We thus obtain:

(29.1) Proposition. *If $H^1(F, G) = 1$, there is a natural bijection of pointed sets*

$$\mathrm{Isom}(A(\rho, w)) \xrightarrow{\sim} H^1(F, \mathbf{Aut}_G(w))$$

which maps the isomorphism class of w to the base point of $H^1(F, \mathbf{Aut}_G(w))$. $\square$

The bijection is given by the following rule: for $w' \in A(\rho, w)$, choose $g \in G(F_{\mathrm{sep}})$ such that $\rho_{\mathrm{sep}}(g)(w) = w'$, and let $\alpha_\sigma = g^{-1} \cdot \sigma(g)$. The map $\alpha \colon \Gamma \to \mathbf{Aut}_G(w)(F_{\mathrm{sep}})$ is a 1-cocycle corresponding to w'. On the other hand, since $H^1(F, G) = 1$, any 1-cocycle $\alpha \in Z^1(F, \mathbf{Aut}_G(w))$ is cohomologous to the base point in $Z^1(F, G)$, hence $\alpha_\sigma = g^{-1} \cdot \sigma(g)$ for some $g \in G(F_{\mathrm{sep}})$. The corresponding object in $A(\rho, w)$ is $\rho_{\mathrm{sep}}(g)(w)$.

In order to apply the proposition above, we need examples of group schemes G for which $H^1(F, G) = 1$. Hilbert's Theorem 90, which is discussed in the next subsection, provides such examples. We then apply (29.1) to give descriptions of the first cohomology set for various algebraic groups.

29.A. Hilbert's Theorem 90 and Shapiro's lemma.

(29.2) Theorem (Hilbert's Theorem 90). *For any separable and associative F-algebra A,*

$$H^1\big(F, \mathbf{GL}_1(A)\big) = 1.$$

In particular $H^1(F, \mathbf{G}_{\mathrm{m}}) = 1$.

Proof: Let $\alpha \in Z^1(\Gamma, A_{\mathrm{sep}}^{\times})$. We define a new action of Γ on A_{sep} by putting

$$\gamma * a = \alpha_{\gamma} \cdot \gamma(a) \qquad \text{for } \gamma \in \Gamma \text{ and } a \in A_{\mathrm{sep}}.$$

This action is continuous and semilinear, i.e. $\gamma * (ax) = (\gamma * a)\gamma(x)$ for $\gamma \in \Gamma$, $a \in A_{\mathrm{sep}}$ and $x \in F_{\mathrm{sep}}$. Therefore, we may apply the Galois descent Lemma (18.1): if

$$U = \{\, a \in A_{\mathrm{sep}} \mid \gamma * a = a \text{ for all } \gamma \in \Gamma \,\},$$

the map

$$f \colon U \otimes_F F_{\mathrm{sep}} \to A_{\mathrm{sep}} \quad \text{given by} \quad f(u \otimes x) = ux$$

is an isomorphism of F_{sep}-vector spaces. For $\gamma \in \Gamma$, $a \in A_{\mathrm{sep}}$ and $a_0 \in A$ we have

$$\gamma * (aa_0) = (\gamma * a)a_0$$

since $\gamma(a_0) = a_0$. Therefore, U is a right A-submodule of A_{sep}, hence $U \otimes F_{\mathrm{sep}}$ is a right A_{sep}-module, and f is an isomorphism of right A_{sep}-modules.

Since A is separable, we have $A = A_1 \times \cdots \times A_m$ for some finite dimensional simple F-algebras $A_1, \ldots, A_m$, and the A-module U decomposes as $U = U_1 \times \cdots \times U_m$ where each U_i is a right A_i-module. Since modules over simple algebras are classified by their reduced dimension (see (1.9)), and since $U \otimes F_{\mathrm{sep}} \simeq A_{\mathrm{sep}}$, we have $U_i \simeq A_i$ for $i = 1, \ldots, m$, hence the right A-modules U and A are isomorphic. Choose an A-module isomorphism $g \colon A \to U$. The composition $f \circ (g \otimes \mathrm{Id}_{F_{\mathrm{sep}}})$ is an A_{sep}-module automorphism of A_{sep} and is therefore left multiplication by the invertible element $a = g(1) \in A_{\mathrm{sep}}^{\times}$. Since $a \in U$ we have

$$a = \gamma * a = \alpha_{\gamma} \cdot \gamma(a) \qquad \text{for all } \gamma \in \Gamma,$$

hence $\alpha_{\gamma} = a \cdot \gamma(a)^{-1}$, showing that α is a trivial cocycle. $\qquad \square$

(29.3) Remark. It follows from (29.2) that $H^1\big(\mathrm{Gal}(L/F), \mathbf{GL}_n(L)\big) = 1$ for any finite Galois field extension L/F, a result due to Speiser [261] (and applied by Speiser to irreducible representations of finite groups). Suppose further that L is cyclic Galois over F, with θ a generator of the Galois group $G = \mathrm{Gal}(L/F)$. Let c be a cocycle with values in $\mathbf{G}_{\mathrm{m}}(L) = L^{\times}$. Since $c_{\theta^i} = c_{\theta} \cdot \ldots \cdot \theta^{i-1}(c_{\theta})$, the cocycle is determined by its value on θ, and $N_{L/F}(c_{\theta}) = 1$. Conversely any $\ell \in L^{\times}$ with $N_{L/F}(\ell) = 1$ defines a cocycle such that $c_{\theta} = \ell$. Thus, by (29.2), any $\ell \in L^{\times}$ such that $N_{L/F}(\ell) = 1$ is of the form $\ell = a\theta(a)^{-1}$. This is the classical Theorem 90 of Hilbert (see [122, §54]).

(29.4) Corollary. *Suppose A is a central simple F-algebra. The connecting map in the cohomology sequence associated to the exact sequence*

$$1 \to \mathbf{SL}_1(A) \to \mathbf{GL}_1(A) \xrightarrow{\mathrm{Nrd}} \mathbf{G}_{\mathrm{m}} \to 1$$

induces a canonical bijection of pointed sets

$$H^1\big(F, \mathbf{SL}_1(A)\big) \simeq F^{\times} / \mathrm{Nrd}(A^{\times}). \qquad \square$$

Let V be a finite dimensional F-vector space. It follows from (29.2) that $H^1\big(F, \mathbf{GL}(V)\big) = 1$ since $\mathbf{GL}_1(A) = \mathbf{GL}(V)$ for $A = \mathrm{End}_F(V)$. A similar result holds for flags:

(29.5) Corollary. *Let $\mathcal{F} \colon V = V_0 \supset V_1 \supset \cdots \supset V_k$ be a flag of finite dimensional F-vector spaces and let G be its group scheme of automorphisms over F. Then $H^1(F, G) = 1$.*

Proof: Let $\alpha \in Z^1(\Gamma, G(F_{\mathrm{sep}}))$. We define a new action of Γ on V_{sep} by

$$\gamma * v = \alpha_\gamma(\gamma v) \qquad \text{for } \gamma \in \Gamma \text{ and } v \in V_{\mathrm{sep}}.$$

This action is continuous and semilinear, hence we are in the situation of Galois descent. Moreover, the action preserves $(V_i)_{\mathrm{sep}}$ for $i = 0, \ldots, k$. Let

$$V_i' = \{\, v \in (V_i)_{\mathrm{sep}} \mid \gamma * v = v \text{ for all } \gamma \in \Gamma \,\}.$$

Each V_i' is an F-vector space and we may identify $(V_i')_{\mathrm{sep}} = (V_i)_{\mathrm{sep}}$ by (18.1). Clearly, $\mathcal{F}'\colon V' = V_0' \supset V_1' \supset \cdots \supset V_k'$ is a flag (see (18.2)). Let $f\colon \mathcal{F} \xrightarrow{\sim} \mathcal{F}'$ be an isomorphism of flags, i.e., an isomorphism of F-vector spaces $V \xrightarrow{\sim} V'$ such that $f(V_i) = V_i'$ for all i. Extend f by linearity to an isomorphism of F_{sep}-vector spaces $V_{\mathrm{sep}} \xrightarrow{\sim} V_{\mathrm{sep}}' = V_{\mathrm{sep}}$, and write also f for this extension. Then f is an automorphism of $\mathcal{F}_{\mathrm{sep}}$, hence $f \in G(F_{\mathrm{sep}})$. Moreover, for $v \in V$ we have $f(v) \in V'$, hence $\sigma\big(f(v)\big) = \alpha_\sigma^{-1}\big(f(v)\big)$. Therefore,

$$^\sigma f(v) = \sigma\big(f(\sigma^{-1}v)\big) = \alpha_\sigma^{-1}\big(f(v)\big) \qquad \text{for all } \sigma \in \Gamma.$$

It follows that $\alpha_\sigma = f \circ {}^\sigma f^{-1}$ for all $\sigma \in \Gamma$, hence α is cohomologous to the trivial cocycle. $\qquad\square$

Corollary (29.5) also follows from the fact that, if H is a parabolic subgroup of a connected reductive group G, then the map $H^1(F, H) \to H^1(F, G)$ is injective (see Serre [257, III, 2.1, Exercice 1]).

The next result is classical and independently due to Eckmann, Faddeev, and Shapiro. It determines the cohomology sets with coefficients in a corestriction $R_{L/F}(G)$.

Let L/F be a finite separable extension of fields and let G be a group scheme defined over L. By fixing an embedding $L \hookrightarrow F_{\mathrm{sep}}$, we consider L as a subfield of F_{sep}. Let $\Gamma_0 = \mathrm{Gal}(F_{\mathrm{sep}}/L) \subset \Gamma$ and let $A = L[G]$, so that

$$R_{L/F}(G)(F_{\mathrm{sep}}) = G(L \otimes_F F_{\mathrm{sep}}) = \mathrm{Hom}_{Alg_L}(A, L \otimes F_{\mathrm{sep}}).$$

For $h \in \mathrm{Hom}_{Alg_L}(A, L \otimes F_{\mathrm{sep}})$, define $\varphi_h\colon \Gamma \to \mathrm{Hom}_{Alg_L}(A, F_{\mathrm{sep}}) = G(F_{\mathrm{sep}})$ by $\varphi_h(\gamma) = \overline{\gamma} \circ h$, where $\overline{\gamma}(\ell \otimes x) = \ell\gamma(x)$ for $\gamma \in \Gamma$, $\ell \in L$ and $x \in F_{\mathrm{sep}}$. The map φ_h is continuous and satisfies $\varphi_h(\gamma_0 \circ \gamma) = \gamma_0 \circ \varphi_h(\gamma)$ for $\gamma_0 \in \Gamma_0$ and $\gamma \in \Gamma$, hence $\varphi_h \in \mathrm{Ind}_{\Gamma_0}^\Gamma G(F_{\mathrm{sep}})$. Since $L \otimes_F F_{\mathrm{sep}} \simeq \mathrm{Map}(\Gamma/\Gamma_0, F_{\mathrm{sep}})$ by (18.4), the map $h \mapsto \varphi_h$ defines an isomorphism of Γ-groups

$$R_{L/F}(G)(F_{\mathrm{sep}}) \xrightarrow{\sim} \mathrm{Ind}_{\Gamma_0}^\Gamma G(F_{\mathrm{sep}}).$$

The following result readily follows by (28.18):

(29.6) Lemma (Eckmann, Faddeev, Shapiro). *Let L/F be a finite separable extension of fields and let G be a group scheme defined over L. There is a natural bijection of pointed sets*

$$H^1\big(F, R_{L/F}(G)\big) \xrightarrow{\sim} H^1(L, G). \qquad\qquad \square$$

The same result clearly holds for H^0-groups, since

$$H^0\big(F, R_{L/F}(G)\big) = R_{L/F}(G)(F) = G(L) = H^0(L, G).$$

If G is a commutative group scheme, there is a group isomorphism

$$H^i\big(F, R_{L/F}(G)\big) \xrightarrow{\sim} H^i(L, G)$$

for all $i \geq 0$. (See Brown [56, Chapter 3, Proposition (6.2)].) It is the composition

$$H^i\big(F, R_{L/F}(G)\big) \xrightarrow{\text{res}} H^i\big(L, R_{L/F}(G)_L\big) \xrightarrow{f^i} H^i(L, G)$$

where $f\colon R_{L/F}(G)_L \to G$ is the group scheme homomorphism corresponding to the identity on $R_{L/F}(G)$ under the bijection

$$\operatorname{Hom}_F\big(R_{L/F}(G), R_{L/F}(G)\big) \xrightarrow{\sim} \operatorname{Hom}_L\big(R_{L/F}(G)_L, G\big)$$

of (20.7).

(29.7) Remark. If L/F is an étale algebra (not necessarily a field), one defines the pointed set $H^1(L, G)$ as the product of the $H^1(L_i, G)$ where the L_i are the field extensions of F such that $L = \prod L_i$. Lemma (29.6) remains valid in this setting (see Remark (20.9) for the definition of $R_{L/F}$).

29.B. Classification of algebras. We now apply Proposition (29.1) and Hilbert's Theorem 90 to show how étale and central simple algebras are classified by H^1-cohomology sets.

Let A be a finite dimensional algebra over F. Multiplication in A yields a linear map $w\colon A \otimes_F A \to A$. Let $W = \operatorname{Hom}_F(A \otimes A, A)$ and $G = \mathbf{GL}(A)$, the linear group of A where A is viewed as an F-vector space. Consider the representation

$$\rho\colon G \to \mathbf{GL}(W)$$

given by the formula

$$\rho(g)(\varphi)(x \otimes y) = g \circ \varphi\big(g^{-1}(x) \otimes g^{-1}(y)\big)$$

for $g \in G$, $\varphi \in W$ and x, $y \in A$. A linear map $g \in G$ is an algebra automorphism of A if and only if $\rho(g)(w) = w$, hence the group scheme $\mathbf{Aut}_G(w)$ coincides with the group scheme $\mathbf{Aut}_{\text{alg}}(A)$ of all algebra automorphisms of A. A twisted ρ-form of w is an algebra structure A' on the F-vector space A such that the F_{sep}-algebras A'_{sep} and A_{sep} are isomorphic. Thus, by Proposition (29.1) there is a bijection

(29.8)
$$\boxed{\begin{array}{c} F\text{-isomorphism classes of } F\text{-algebras } A' \\ \text{such that the } F_{\text{sep}}\text{-algebras} \\ A'_{\text{sep}} \text{ and } A_{\text{sep}} \text{ are isomorphic} \end{array}} \longleftrightarrow H^1\big(F, \mathbf{Aut}_{\text{alg}}(A)\big).$$

The bijection is given explicitly as follows: if $\beta\colon A_{\text{sep}} \xrightarrow{\sim} A'_{\text{sep}}$ is an F_{sep}-isomorphism, the corresponding cocycle is $\alpha_\gamma = \beta^{-1} \circ (\operatorname{Id} \otimes \gamma) \circ \beta \circ (\operatorname{Id} \otimes \gamma^{-1})$. Conversely, given a cocycle $\alpha \in Z^1\big(\Gamma, \mathbf{Aut}_{\text{alg}}(A)\big)$, we set

$$A' = \{\, x \in A_{\text{sep}} \mid \alpha_\gamma \circ (\operatorname{Id} \otimes \gamma)(x) = x \text{ for all } \gamma \in \Gamma \,\}.$$

We next apply this general principle to étale algebras and to central simple algebras.

Étale algebras. The F-algebra $A = F \times \cdots \times F$ (n copies) is étale of dimension n. If $\{\, e_i \mid i = 1, \ldots, n \,\}$ is the set of primitive idempotents of A, any F-algebra automorphism of A is determined by the images of the e_i. Thus $\mathbf{Aut}_{\text{alg}}(A)$ is the constant symmetric group S_n. Proposition (18.3) shows that the étale F-algebras of dimension n are exactly the twisted forms of A. Therefore, the preceding discussion with $A = F \times \cdots \times F$ yields a natural bijection

(29.9)
$$\boxed{\begin{array}{c} F\text{-isomorphism classes of} \\ \text{étale } F\text{-algebras of degree } n \end{array}} \longleftrightarrow H^1(F, S_n).$$

Since the Γ-action on S_n is trivial, the pointed set $H^1(F, S_n)$ coincides with the set of conjugacy classes of continuous maps $\Gamma \to S_n$ and hence also classifies isomorphism classes of Γ-sets X consisting of n elements (see (18.4)). The cocycle $\gamma \colon \Gamma \to S_n$ corresponds to the étale algebra $L = \mathrm{Map}(X, F_{\mathrm{sep}})^\Gamma$ where Γ acts on the set X via γ.

The sign map $\mathrm{sgn}\colon S_n \to \{\pm 1\} = S_2$ induces a map in cohomology

$$\mathrm{sgn}^1 \colon H^1(F, S_n) \to H^1(F, S_2).$$

In view of (18.21) this map sends (the isomorphism class of) an étale algebra L to (the isomorphism class of) its discriminant $\Delta(L)$.

Another interpretation of $H^1(F, S_n)$ is given in Example (28.15):

$$H^1(\Gamma, S_n) \simeq \mathrm{Isom}(S_n\text{-}\mathit{Gal}_F).$$

In fact, we may associate to any étale F-algebra L of dimension n its Galois S_n-closure $\Sigma(L)$ (see (18.20)). This construction induces a canonical bijection between the isomorphism classes of étale algebras of dimension n and isomorphism classes of Galois S_n-algebras. Note however that Σ is not a functor: an F-algebra homomorphism $L_1 \to L_2$ which is not injective does not induce any homomorphism $\Sigma(L_1) \to \Sigma(L_2)$.

Central simple algebras. Let $A = M_n(F)$, the matrix algebra of degree n. Since every central simple F-algebra is split by F_{sep}, and since every F-algebra A' such that $A'_{\mathrm{sep}} \simeq M_n(F_{\mathrm{sep}})$ is central simple (see (1.1)), the twisted forms of A are exactly the central simple F-algebras of degree n. The Skolem-Noether theorem (1.4) shows that every automorphism of A is inner, hence $\mathbf{Aut}_{\mathrm{alg}}(A) = \mathbf{PGL}_n$. Therefore, as in (29.8), there is a natural bijection

$$\boxed{\begin{array}{c} F\text{-isomorphism classes of} \\ \text{central simple } F\text{-algebras of degree } n \end{array}} \quad \longleftrightarrow \quad H^1(F, \mathbf{PGL}_n).$$

Consider the exact sequence:

(29.10) $$1 \to \mathbf{G}_{\mathrm{m}} \to \mathbf{GL}_n \to \mathbf{PGL}_n \to 1.$$

By twisting all the groups by a cocycle in $H^1(F, \mathbf{PGL}_n)$ corresponding to a central simple F-algebra B of degree n, we get the exact sequence

$$1 \to \mathbf{G}_{\mathrm{m}} \to \mathbf{GL}_1(B) \to \mathbf{PGL}_1(B) \to 1.$$

Since $H^1\big(F, \mathbf{GL}_1(B)\big) = 1$ by Hilbert's Theorem 90, it follows from Corollary (28.13) that the connecting map

$$\delta^1 \colon H^1(F, \mathbf{PGL}_n) \to H^2(F, \mathbf{G}_{\mathrm{m}})$$

with respect to (29.10) is injective. The map δ^1 is defined here as follows: if $\alpha_\gamma \in \mathrm{Aut}_{F_{\mathrm{sep}}}\big(M_n(F_{\mathrm{sep}})\big)$ is a 1-cocycle, choose $c_\gamma \in \mathbf{GL}_n(F_{\mathrm{sep}})$ such that $\alpha_\gamma = \mathrm{Int}(c_\gamma)$ (by Skolem-Noether). Then

$$c_{\gamma, \gamma'} = c_\gamma \cdot \gamma c_{\gamma'} \cdot c_{\gamma\gamma'}^{-1} \in Z^2(F, \mathbf{G}_{\mathrm{m}})$$

is the corresponding 2-cocycle. The δ^1 for different n's fit together to induce an injective homomorphism $\mathrm{Br}(F) \to H^2(F, \mathbf{G}_{\mathrm{m}})$. To prove that this homomorphism is surjective, we may reduce to the case of finite Galois extensions, since for every 2-cocycle $c_{\gamma, \gamma'}$ with values in $\mathbf{G}_{\mathrm{m}}$ there is a finite Galois extension L/F such that $c\colon \Gamma \times \Gamma \to F_{\mathrm{sep}}^\times$ factors through a 2-cocycle in $Z^2\big(\mathrm{Gal}(L/F), L^\times\big)$. Thus, the following proposition completes the proof that $\mathrm{Br}(F) \simeq H^2(F, \mathbf{G}_{\mathrm{m}})$:

(29.11) Proposition. *Let L/F be a finite Galois extension of fields of degree n, and let $G = \mathrm{Gal}(L/F)$. The map*

$$\delta^1 \colon H^1\big(G, \mathrm{PGL}_n(L)\big) \to H^2(G, L^\times)$$

is bijective.

Proof: Injectivity follows by the same argument as for the connecting map with respect to (29.10). To prove surjectivity, choose $c \in Z^2(G, L^\times)$ and let V be the n-dimensional L-vector space

$$V = \bigoplus_{\sigma \in G} e_\sigma L.$$

Numbering the elements of G, we may identify $V = L^n$, hence $\mathrm{End}_L(V) = M_n(L)$ and $\mathrm{Aut}\big(\mathrm{End}_L(V)\big) = \mathrm{PGL}_n(L)$. For $\sigma \in G$, let $a_\sigma \in \mathrm{End}_L(V)$ be defined by

$$a_\sigma(e_\tau) = e_{\sigma\tau} c_{\sigma,\tau}.$$

We have

$$a_\sigma \circ \sigma(a_\tau)(e_\nu) = e_{\sigma\tau\nu} c_{\sigma,\tau\nu} \sigma(c_{\tau,\nu}) = e_{\sigma\tau\nu} c_{\sigma\tau,\nu} c_{\sigma,\tau} = a_{\sigma\tau}(e_\nu) c_{\sigma,\tau}$$

for all σ, τ, $\nu \in G$, hence $\mathrm{Int}(a_\sigma) \in \mathrm{Aut}\big(\mathrm{End}_L(V)\big)$ is a 1-cocycle whose image under δ^1 is represented by the cocycle $c_{\sigma,\tau}$. $\square$

With the same notation as in the proof above, a central simple F-algebra A_c corresponding to the 2-cocycle $c \in Z^2(G, L^\times)$ is given by

$$A_c = \{\, f \in \mathrm{End}_L(V) \mid a_\sigma \circ \sigma(f) = f \circ a_\sigma \text{ for all } \sigma \in G \,\}.$$

This construction is closely related to the *crossed product construction*, which we briefly recall: on the L-vector space

$$C = \bigoplus_{\sigma \in G} L z_\sigma$$

with basis $(z_\sigma)_{\sigma \in G}$, define multiplication by

$$z_\sigma \ell = \sigma(\ell) z_\sigma \quad \text{and} \quad z_\sigma z_\tau = c_{\sigma,\tau} z_{\sigma\tau}$$

for σ, $\tau \in G$ and $\ell \in L$. The cocycle condition ensures that C is an associative algebra, and it can be checked that C is central simple of degree n over F (see, e.g., Pierce [219, 14.1]).

For $\sigma \in G$ and for $\ell \in L$, define y_σ, $u_\ell \in \mathrm{End}_L(V)$ by

$$y_\sigma(e_\tau) = e_{\tau\sigma} c_{\tau,\sigma} \quad \text{and} \quad u_\ell(e_\tau) = e_\tau \tau(\ell) \qquad \text{for } \tau \in G.$$

Computations show that y_σ, $u_\ell \in A_c$, and

$$u_\ell \circ y_\sigma = y_\sigma \circ u_{\sigma(\ell)}, \qquad y_\sigma \circ y_\tau = y_{\tau\sigma} \circ u_{c_{\tau,\sigma}}.$$

Therefore, the map $C \to A_c^{\mathrm{op}}$ which sends $\sum \ell_\sigma z_\sigma$ to $\sum u_{\ell_\sigma}^{\mathrm{op}} \circ y_\sigma^{\mathrm{op}}$ is an F-algebra homomorphism, hence an isomorphism since C and A_c^{op} are central simple of degree n. Thus, $C \simeq A_c^{\mathrm{op}}$, showing that the isomorphism $\mathrm{Br}(F) \simeq H^2(F, \mathbf{G}_{\mathrm{m}})$ defined by the crossed product construction is the opposite of the isomorphism induced by δ^1.

29.C. Algebras with a distinguished subalgebra. The same idea as in §29.B applies to pairs (A, L) consisting of an F-algebra A and a subalgebra $L \subset A$. An isomorphism of pairs $(A', L') \xrightarrow{\sim} (A, L)$ is an F-isomorphism $A' \xrightarrow{\sim} A$ which restricts to an isomorphism $L' \xrightarrow{\sim} L$. Let $G \subset \mathbf{GL}(B)$ be the group scheme of automorphisms of the flag of vector spaces $A \supset L$. The group G acts on the space $\mathrm{Hom}_F(A \otimes_F A, A)$ as in §29.B and the group scheme $\mathbf{Aut}_G(m)$ where $m \colon A \otimes_F A \to A$ is the multiplication map coincides with the group scheme $\mathbf{Aut}_{\mathrm{alg}}(A, L)$ of all F-algebra automorphisms of the pair (A, L). Since $H^1(F, G) = 1$ by (29.5), there is by Proposition (29.1) a bijection

$$(29.12) \quad \boxed{\begin{array}{c} \text{F-isomorphism classes of pairs} \\ \text{of F-algebras (A', L')} \\ \text{such that } (A', L')_{\mathrm{sep}} \simeq (A, L)_{\mathrm{sep}} \end{array}} \quad \longleftrightarrow \quad H^1\big(F, \mathbf{Aut}_{\mathrm{alg}}(A, L)\big).$$

The map $H^1\big(F, \mathbf{Aut}_{\mathrm{alg}}(A, L)\big) \to H^1\big(F, \mathbf{Aut}_{\mathrm{alg}}(A)\big)$ induced by the inclusion of $\mathrm{Aut}(A, L)$ in $\mathrm{Aut}(A)$ maps the isomorphism class of a pair (A', L') to the isomorphism class of A'. On the other hand, the map

$$H^1\big(F, \mathbf{Aut}(A, L)\big) \to H^1\big(F, \mathbf{Aut}_{\mathrm{alg}}(L)\big)$$

induced by the restriction map $\mathrm{Aut}(A, L) \to \mathrm{Aut}(L)$ takes the isomorphism class of (A', L') to the isomorphism class of L'.

Let $\mathbf{Aut}_L(A)$ be the kernel of the restriction map $\mathbf{Aut}(A, L) \to \mathbf{Aut}_{\mathrm{alg}}(L)$. In order to describe the set $H^1\big(F, \mathbf{Aut}_L(A)\big)$ as a set of isomorphism classes as in (29.8), let

$$W = \mathrm{Hom}_F(A \otimes_F A, A) \oplus \mathrm{Hom}_F(L, A).$$

The group $G = \mathbf{GL}(A)$ acts on W as follows:

$$\rho(g)(\psi, \varphi)(x \otimes y, z) = \big(g \circ \psi\big(g^{-1}(x) \otimes g^{-1}(y)\big), g \circ \varphi(z)\big)$$

for $g \in G$, $\psi \in \mathrm{Hom}_F(A \otimes_F A, A)$, $\varphi \in \mathrm{Hom}_F(L, A)$, $x, y \in A$ and $z \in L$. The multiplication map $m \colon A \otimes_F A \to A$ and the inclusion $i \colon L \to A$ define an element $w = (m, i) \in W$, and the group $\mathbf{Aut}_G(w)$ coincides with $\mathbf{Aut}_L(A)$. A twisted form of w is a pair (A', φ) where A' is an F-algebra isomorphic to A over F_{sep} and $\varphi \colon L \to A'$ is an F-algebra embedding of L in A'. By Proposition (29.1) there is a natural bijection

$$(29.13) \quad \boxed{\begin{array}{c} \text{F-isomorphism classes of} \\ \text{pairs } (A', \varphi) \text{ isomorphic to} \\ \text{the pair } (A, i)_{\mathrm{sep}} \text{ over } F_{\mathrm{sep}} \end{array}} \quad \longleftrightarrow \quad H^1\big(F, \mathbf{Aut}_L(A)\big).$$

The canonical map $H^1\big(F, \mathbf{Aut}_L(A)\big) \to H^1\big(F, \mathbf{Aut}(A, L)\big)$ takes the isomorphism class of a pair (A', φ) to the isomorphism class of the pair $\big(A', \varphi(L)\big)$.

The preceding discussion applies in particular to separable F-algebras. If B is a separable F-algebra with center Z, the restriction homomorphism

$$\mathbf{Aut}_{\mathrm{alg}}(B) = \mathbf{Aut}_{\mathrm{alg}}(B, Z) \to \mathbf{Aut}_{\mathrm{alg}}(Z)$$

gives rise to the map of pointed sets

$$H^1\big(F, \mathbf{Aut}_{\mathrm{alg}}(B)\big) \to H^1\big(F, \mathbf{Aut}_{\mathrm{alg}}(Z)\big).$$

which takes the class of a twisted form B' of B to the class of its center Z'. On the other hand, the natural isomorphism $\mathbf{Aut}_Z(B) \simeq R_{Z/F}(\mathbf{PGL}_1(B))$ and Lemma (29.6) give a bijection of pointed sets

$$H^1(F, \mathbf{Aut}_Z(B)) \simeq H^1(F, R_{Z/F}(\mathbf{PGL}_1(B))) \simeq H^1(Z, \mathbf{PGL}_1(B))$$

which takes the class of a pair (B', φ) to the class of the Z-algebra $B' \otimes_{Z'} Z$ (where the tensor product is taken with respect to φ).

29.D. Algebras with involution. Let (A, σ) be a central simple algebra with involution (of any kind) over a field F. In this section, we give interpretations for the cohomology sets

$$H^1(F, \mathbf{Aut}(A, \sigma)), \quad H^1(F, \mathbf{Sim}(A, \sigma)), \quad H^1(F, \mathbf{Iso}(A, \sigma)).$$

We shall discuss separately the unitary, the symplectic and the orthogonal case, but we first outline the general principles.

Let $W = \mathrm{Hom}_F(A \otimes A, A) \oplus \mathrm{End}_F(A)$ and $G = \mathbf{GL}(A)$, the linear group of A where A is viewed as an F-vector space. Consider the representation

$$\rho: G \to \mathbf{GL}(W)$$

defined by

$$\rho(g)(\varphi, \psi)(x \otimes y, z) = \left(g \circ \varphi(g^{-1}(x) \otimes g^{-1}(y)), g \circ \psi \circ g^{-1}(z)\right)$$

for $g \in G$, $\varphi \in \mathrm{Hom}_F(A \otimes A, A)$, $\psi \in \mathrm{End}_F(A)$ and x, y, $z \in A$. Let $w = (m, \sigma) \in W$, where m is the multiplication map of A. The subgroup $\mathbf{Aut}_G(w)$ of G coincides with the group scheme $\mathbf{Aut}(A, \sigma)$ of F-algebra automorphisms of A commuting with σ. A twisted form of w is the structure of an algebra with involution isomorphic over F_{sep} to $(A_{\mathrm{sep}}, \sigma_{\mathrm{sep}})$. Hence, by Proposition (29.1) there is a natural bijection

(29.14)
$$\boxed{\begin{array}{c} F\text{-isomorphism classes of} \\ F\text{-algebras with involution } (A', \sigma') \\ \text{isomorphic to } (A_{\mathrm{sep}}, \sigma_{\mathrm{sep}}) \text{ over } F_{\mathrm{sep}} \end{array}} \quad \longleftrightarrow \quad H^1(F, \mathbf{Aut}(A, \sigma)).$$

Next, let $W' = \mathrm{End}_F(A)$ and $G' = \mathbf{GL}_1(A)$, the linear group of A (i.e., the group of invertible elements in A). Consider the representation

$$\rho': G' \to \mathbf{GL}(W')$$

defined by

$$\rho'(a)(\psi) = \mathrm{Int}(a) \circ \psi \circ \mathrm{Int}(a)^{-1},$$

for $a \in G'$ and $\psi \in \mathrm{End}_F(A)$. The subgroup $\mathbf{Aut}_{G'}(\sigma)$ of G' coincides with the group scheme $\mathbf{Sim}(A, \sigma)$ of similitudes of (A, σ). A twisted ρ'-form of σ is an involution of A which, over F_{sep}, is conjugate to $\sigma_{\mathrm{sep}} = \sigma \otimes \mathrm{Id}_{F_{\mathrm{sep}}}$. By Proposition (29.1) and Hilbert's Theorem 90 (see (29.2)), we get a bijection

(29.15)
$$\boxed{\begin{array}{c} \text{conjugacy classes of involutions} \\ \text{on } A \text{ which over } F_{\mathrm{sep}} \text{ are} \\ \text{conjugate to } \sigma_{\mathrm{sep}} \end{array}} \quad \longleftrightarrow \quad H^1(F, \mathbf{Sim}(A, \sigma)).$$

The canonical homomorphism $\mathrm{Int}: \mathbf{Sim}(A, \sigma) \to \mathbf{Aut}(A, \sigma)$ induces a map

$$\mathrm{Int}^1: H^1(F, \mathbf{Sim}(A, \sigma)) \to H^1(F, \mathbf{Aut}(A, \sigma))$$

which maps the conjugacy class of an involution σ' to the isomorphism class of (A, σ').

Finally, recall from §23.A that the group scheme $\mathbf{Iso}(A, \sigma)$ is defined as the stabilizer of $1 \in A$ under the action of $\mathbf{GL}_1(A)$ on A given by

$$\rho''(a)(x) = a \cdot x \cdot \sigma(a)$$

for $a \in \mathbf{GL}_1(A)$ and $x \in A$. Twisted ρ''-forms of 1 are elements $s \in A$ for which there exists $a \in A_{\text{sep}}^{\times}$ such that $s = a \cdot \sigma(a)$. We write $\text{Sym}(A, \sigma)'$ for the set of these elements,

$$\text{Sym}(A, \sigma)' = \{\, s \in A \mid s = a \cdot \sigma(a) \text{ for some } a \in A_{\text{sep}}^{\times} \,\} \subset \text{Sym}(A, \sigma) \cap A^{\times},$$

and define an equivalence relation on $\text{Sym}(A, \sigma)'$ by

$$s \sim s' \quad \text{if and only if } s' = a \cdot s \cdot \sigma(a) \text{ for some } a \in A^{\times}.$$

The equivalence classes are exactly the ρ''-isomorphism classes of twisted forms of 1, hence Proposition (29.1) yields a canonical bijection

(29.16) $\text{Sym}(A, \sigma)'/\!\sim \quad \longleftrightarrow \quad H^1\big(F, \mathbf{Iso}(A, \sigma)\big).$

The inclusion $i\colon \mathbf{Iso}(A, \sigma) \to \mathbf{Sim}(A, \sigma)$ induces a map

$$i^1\colon H^1\big(F, \mathbf{Iso}(A, \sigma)\big) \to H^1\big(F, \mathbf{Sim}(A, \sigma)\big)$$

which maps the equivalence class of $s \in \text{Sym}(A, \sigma)'$ to the conjugacy class of the involution $\text{Int}(s) \circ \sigma$.

We now examine the various types of involutions separately.

Unitary involutions. Let (B, τ) be a central simple F-algebra with unitary involution. Let K be the center of B, which is a quadratic étale F-algebra, and let $n = \deg(B, \tau)$. (The algebra B is thus central simple of degree n if K is a field, and it is a direct product of two central simple F-algebras of degree n if $K \simeq F \times F$.)

From (29.14), we readily derive a canonical bijection

$$\boxed{\begin{array}{c} F\text{-isomorphism classes of} \\ \text{central simple } F\text{-algebras} \\ \text{with unitary involution of degree } n \end{array}} \quad \longleftrightarrow \quad H^1\big(F, \mathbf{Aut}(B, \tau)\big).$$

We have an exact sequence of group schemes

$$1 \to \mathbf{PGU}(B, \tau) \to \mathbf{Aut}(B, \tau) \xrightarrow{f} S_2 \to 1$$

where f is the restriction homomorphism to $\mathbf{Aut}_{\text{alg}}(K) = S_2$. We may view the group $\mathbf{PGU}(B, \tau)$ as the automorphism group of the pair (B, τ) over K. As in Proposition (29.1) (see also (29.13)) we obtain a natural bijection

$$\boxed{\begin{array}{c} F\text{-isomorphism classes of triples } (B', \tau', \varphi) \\ \text{consisting of a central simple } F\text{-algebra} \\ \text{with unitary involution } (B', \tau') \text{ of degree } n \\ \text{and an } F\text{-algebra isomorphism } \varphi\colon Z(B') \xrightarrow{\sim} K \end{array}} \quad \longleftrightarrow \quad H^1\big(F, \mathbf{PGU}(B, \tau)\big).$$

By Proposition (28.10) the group $\mathbf{Aut}_{\text{alg}}(K)$ acts transitively on each fiber of the map

$$H^1\big(F, \mathbf{PGU}(B, \tau)\big) \to H^1\big(F, \mathbf{Aut}(B, \tau)\big).$$

The fiber over a pair (B', τ') consists of the triples $(B', \tau', \mathrm{Id}_K)$ and (B', τ', ι), where ι is the nontrivial automorphism of K/F. These triples are isomorphic if and only if $\mathrm{PGU}(B', \tau') \subsetneqq \mathrm{Aut}_F(B', \tau')$.

After scalar extension to F_{sep}, we have $B_{\mathrm{sep}} \simeq M_n(F_{\mathrm{sep}}) \times M_n(F_{\mathrm{sep}})^{\mathrm{op}}$, and all the unitary involutions on B_{sep} are conjugate to the exchange involution ε by (2.14). Therefore, (29.15) specializes to a bijection

$$
\boxed{\begin{array}{c} \text{conjugacy classes of} \\ \text{unitary involutions on } B \\ \text{which are the identity on } F \end{array}} \quad \longleftrightarrow \quad H^1\big(F, \mathbf{GU}(B, \tau)\big).
$$

The exact sequence

$$
1 \to R_{K/F}(\mathbf{G}_{\mathrm{m},K}) \to \mathbf{GU}(B, \tau) \to \mathbf{PGU}(B, \tau) \to 1
$$

induces a connecting map in cohomology

$$
\delta^1 \colon H^1\big(F, \mathbf{PGU}(B, \tau)\big) \to H^2\big(F, R_{K/F}(\mathbf{G}_{\mathrm{m},K})\big) = H^2(K, \mathbf{G}_{\mathrm{m},K}) = \mathrm{Br}(K)
$$

where the identification $H^2\big(F, R_{K/F}(\mathbf{G}_{\mathrm{m},K})\big) = H^2(K, \mathbf{G}_{\mathrm{m},K})$ is given by Shapiro's lemma and the identification $H^2(K, \mathbf{G}_{\mathrm{m},K}) = \mathrm{Br}(K)$ by the connecting map in the cohomology sequence associated to

$$
1 \to \mathbf{G}_{\mathrm{m},K} \to \mathbf{GL}_{n,K} \to \mathbf{PGL}_{n,K} \to 1,
$$

see §29.B. Under δ^1, the class of a triple (B', τ', φ) is mapped to the Brauer class $[B' \otimes_{K'} K] \cdot [B]^{-1}$, where the tensor product is taken with respect to φ.

Our next goal is to give a description of $H^1\big(F, \mathbf{U}(B, \tau)\big)$. Every symmetric element $s \in \mathrm{Sym}\big(M_n(F_{\mathrm{sep}}) \times M_n(F_{\mathrm{sep}})^{\mathrm{op}}, \varepsilon\big)$ has the form

$$
s = (m, m^{\mathrm{op}}) = (m, 1^{\mathrm{op}}) \cdot \varepsilon(m, 1^{\mathrm{op}})
$$

for some $m \in M_n(F_{\mathrm{sep}})$. Therefore, the set $\mathrm{Sym}(B, \tau)'$ of (29.16) is the set of symmetric units,

$$
\mathrm{Sym}(B, \tau)' = \mathrm{Sym}(B, \tau)^\times \ (= \mathrm{Sym}(B, \tau) \cap B^\times),
$$

and (29.16) yields a canonical bijection

(29.17) $\mathrm{Sym}(B, \tau)^\times / \sim \quad \longleftrightarrow \quad H^1\big(F, \mathbf{U}(B, \tau)\big).$

By associating with every symmetric unit $u \in \mathrm{Sym}(B, \tau)^\times$ the hermitian form

$$
\langle u^{-1} \rangle \colon B \times B \to B
$$

defined by $\langle u^{-1} \rangle(x, y) = \tau(x) u^{-1} y$, it follows that $H^1\big(F, \mathbf{U}(B, \tau)\big)$ classifies hermitian forms on B-modules of rank 1 up to isometry.

In order to describe the set $H^1\big(F, \mathbf{SU}(B, \tau)\big)$, consider the representation

$$
\rho \colon \mathbf{GL}_1(B) \to \mathbf{GL}(B \oplus K)
$$

given by

$$
\rho(b)(x, y) = \big(b \cdot x \cdot \tau(b), \mathrm{Nrd}(b) y\big)
$$

for $b \in \mathbf{GL}_1(B)$, $x \in B$ and $y \in K$. Let $w = (1, 1) \in B \oplus K$. The group $\mathbf{Aut}_G(w)$ coincides with $\mathbf{SU}(B, \tau)$. Clearly, twisted forms of w are contained in the set[31]

$$
\mathrm{SSym}(B, \tau)^\times = \{ (s, z) \in \mathrm{Sym}(B, \tau)^\times \times K^\times \mid \mathrm{Nrd}_B(s) = N_{K/F}(z) \}.
$$

---

[31] This set plays an essential rôle in the Tits construction of exceptional simple Jordan algebras (see §39).

Over F_{sep}, we have $B_{\text{sep}} \simeq M_n(F_{\text{sep}}) \times M_n(F_{\text{sep}})^{\text{op}}$ and we may identify τ_{sep} to the exchange involution ε. Thus, for every $(s, z) \in \text{SSym}(B_{\text{sep}}, \tau_{\text{sep}})^\times$, there are $m \in M_n(F_{\text{sep}})$ and $z_1, z_2 \in F_{\text{sep}}^\times$ such that $s = (m, m^{\text{op}})$, $z = (z_1, z_2)$ and $\det m = z_1 z_2$. Let $m_1 \in \text{GL}_n(F_{\text{sep}})$ be any matrix such that $\det m_1 = z_1$, and let $m_2 = m_1^{-1} m$. Then

$$s = (m_1, m_2^{\text{op}}) \cdot \varepsilon(m_1, m_2^{\text{op}}) \quad \text{and} \quad z = \text{Nrd}(m_1, m_2^{\text{op}}),$$

hence $(s, z) = \rho_{\text{sep}}(m_1, m_2^{\text{op}})(w)$. Therefore, $\text{SSym}(B, \tau)^\times$ is the set of twisted ρ-forms of w.

Define an equivalence relation $\approx$ on $\text{SSym}(B, \tau)^\times$ by

$$(s, z) \approx (s', z') \quad \text{if and only if } s' = b \cdot s \cdot \tau(b) \text{ and } z' = \text{Nrd}_B(b)z \text{ for some } b \in B^\times$$

so that the equivalence classes under $\approx$ are exactly the ρ-isomorphism classes of twisted forms. Proposition (29.1) yields a canonical bijection

(29.18) $$\text{SSym}(B, \tau)^\times / \approx \quad \longleftrightarrow \quad H^1\big(F, \mathbf{SU}(B, \tau)\big).$$

The natural map of pointed sets

$$H^1\big(F, \mathbf{SU}(B, \tau)\big) \to H^1\big(F, \mathbf{U}(B, \tau)\big)$$

takes the class of $(s, z) \in \text{SSym}(B, \tau)^\times$ to the class of $s \in \text{Sym}(B, \tau)^\times$.

There is an exact sequence

$$1 \to \mathbf{SU}(B, \tau) \to \mathbf{U}(B, \tau) \xrightarrow{\text{Nrd}} \mathbf{G}_{\text{m},K}^1 \to 1$$

where

$$\mathbf{G}_{\text{m},K}^1 = \ker\big(R_{K/F}(\mathbf{G}_{\text{m},K}) \xrightarrow{N_{K/F}} \mathbf{G}_{\text{m},F}\big)$$

(hence $\mathbf{G}_{\text{m},K}^1(F) = K^1$ is the group of norm 1 elements in K). The connecting map

$$\mathbf{G}_{\text{m},K}^1(F) \to H^1\big(F, \mathbf{SU}(B, \tau)\big)$$

takes $x \in \mathbf{G}_{\text{m},K}^1(F) \subset K^\times$ to the class of the pair $(1, x)$.

(29.19) Example. Suppose K is a field and let (V, h) be a hermitian space over K (with respect to the nontrivial automorphism ι of K/F). We write simply $\mathbf{U}(V, h)$ for $\mathbf{U}\big(\text{End}_K(V), \sigma_h\big)$ and $\mathbf{SU}(V, h)$ for $\mathbf{SU}\big(\text{End}_K(V), \sigma_h\big)$. As in (29.17) and (29.18), we have canonical bijections

$$\text{Sym}\big(\text{End}_K(V), \sigma_h\big)^\times / \sim \quad \longleftrightarrow \quad H^1\big(F, \mathbf{U}(V, h)\big),$$
$$\text{SSym}\big(\text{End}_K(V), \sigma_h\big)^\times / \approx \quad \longleftrightarrow \quad H^1\big(F, \mathbf{SU}(V, h)\big).$$

The set $\text{Sym}\big(\text{End}_K(V), \sigma_h\big)^\times / \sim$ is also in one-to-one correspondence correspondence with the set of isometry classes of nonsingular hermitian forms on V, by mapping $s \in \text{Sym}\big(\text{End}_K(V), \sigma_h\big)^\times$ to the hermitian form $h_s \colon V \times V \to K$ defined by

$$h_s(x, y) = h\big(s^{-1}(x), y\big) = h\big(x, s^{-1}(y)\big)$$

for $x, y \in V$. Therefore, we have a canonical bijection of pointed sets

$$\boxed{\begin{array}{c} \text{isometry classes of} \\ \text{nonsingular hermitian} \\ \text{forms on } V \end{array}} \quad \longleftrightarrow \quad H^1\big(F, \mathbf{U}(V, h)\big)$$

where the base point of $H^1\big(F, \mathbf{U}(V, h)\big)$ corresponds to the isometry class of h.

To give a similar interpretation of $H^1\big(F, \mathbf{SU}(V, h)\big)$, observe that for every unitary involution τ on $\operatorname{End}_K(V)$ and every $y \in K^\times$ such that $N_{K/F}(y) = 1$ there exists $u \in \mathbf{U}\big(\operatorname{End}_K(V), \tau\big)$ such that $\det(u) = y$. Indeed, τ is the adjoint involution with respect to some hermitian form h'. If $(e_1, \ldots, e_n)$ is an orthogonal basis of V for h', we may take for u the endomorphism which leaves e_i invariant for $i = 1, \ldots, n-1$ and maps e_n to $e_n y$. From this observation, it follows that the canonical map

$$\operatorname{SSym}\big(\operatorname{End}_K(V), \sigma_h\big)^\times /\approx \quad \to \quad \operatorname{Sym}\big(\operatorname{End}_K(V), \sigma_h\big)^\times /\sim$$

given by $(s, z) \mapsto s$ is injective. For, suppose (s, z), $(s', z') \in \operatorname{SSym}\big(\operatorname{End}_K(V), \sigma_h\big)^\times$ are such that $s \sim s'$, and let $b \in \operatorname{End}_K(V)^\times$ satisfy

$$s' = b \cdot s \cdot \sigma_h(b).$$

Since $\det(s) = N_{K/F}(z)$ and $\det(s') = N_{K/F}(z')$, it follows that

$$N_{K/F}(z') = N_{K/F}\big(z \det(b)\big).$$

Choose $u \in \mathbf{U}\big(\operatorname{End}_K(V), \operatorname{Int}(s') \circ \sigma_h\big)$ such that $\det(u) = z' z^{-1} \det(b)^{-1}$. Then

$$s' = u \cdot s' \cdot \sigma_h(u) = ub \cdot s \cdot \sigma_h(ub) \quad \text{and} \quad z' = z \det(ub),$$

hence $(s', z') \approx (s, z)$.

As a consequence, the canonical map $H^1\big(F, \mathbf{SU}(V, h)\big) \to H^1\big(F, \mathbf{U}(V, h)\big)$ is injective, and we may identify $H^1\big(F, \mathbf{SU}(V, h)\big)$ to a set of isometry classes of hermitian forms on V. For $s \in \operatorname{Sym}\big(\operatorname{End}_K(V), \sigma_h\big)^\times$, we have

$$\operatorname{disc} h_s = \operatorname{disc} h \cdot \det s^{-1} \quad \text{in } F^\times / N(K/F),$$

hence there exists $z \in K^\times$ such that $\det s = N_{K/F}(z)$ if and only if $\operatorname{disc} h_s = \operatorname{disc} h$. Therefore, we have a canonical bijection of pointed sets

$$\boxed{\begin{array}{c} \text{isometry classes of nonsingular} \\ \text{hermitian forms } h' \text{ on } V \\ \text{with } \operatorname{disc} h' = \operatorname{disc} h \end{array}} \quad \longleftrightarrow \quad H^1\big(F, \mathbf{SU}(V, h)\big).$$

(29.20) Example. Consider $B = M_n(F) \times M_n(F)^{\mathrm{op}}$, with ε the exchange involution $(a, b^{\mathrm{op}}) \mapsto (b, a^{\mathrm{op}})$. We have

$$\mathbf{U}(B, \varepsilon) = \big\{ \big(u, (u^{-1})^{\mathrm{op}}\big) \mid u \in \mathbf{GL}_n(F) \big\},$$

hence $\mathbf{SU}(B, \tau) = \mathbf{SL}_n(F)$ and $\mathbf{PGU}(B, \tau) = \mathbf{PGL}_n(F)$. Therefore, by Hilbert's Theorem 90 (29.2) and (29.4),

$$H^1\big(F, \mathbf{U}(B, \varepsilon)\big) = H^1\big(F, \mathbf{SU}(B, \varepsilon)\big) = 1.$$

The map $(a, b^{\mathrm{op}}) \mapsto \big(b^t, (a^t)^{\mathrm{op}}\big)$ is an outer automorphism of order 2 of (B, ε), and we may identify

$$\mathbf{Aut}\big(M_n(F) \times M_n(F)^{\mathrm{op}}, \varepsilon\big) = \mathbf{PGL}_n \rtimes S_2$$

where the nontrivial element of S_2 acts on $\mathbf{PGL}_n$ by mapping $a \cdot F^\times$ to $(a^t)^{-1} \cdot F^\times$. The exact sequence

$$1 \to \mathbf{PGL}_n \to \mathbf{PGL}_n \rtimes S_2 \to S_2 \to 1$$

induces the following exact sequence in cohomology:

$$H^1(F, \mathbf{PGL}_n) \to H^1(F, \mathbf{PGL}_n \rtimes S_2) \to H^1(F, S_2).$$

This cohomology sequence corresponds to

central simple F-algebras of degree n	$\rightarrow$	central simple F-algebras with unitary involution of degree n	$\rightarrow$	quadratic étale F-algebras

$$A \;\mapsto\; (A \times A^{\mathrm{op}}, \varepsilon) \qquad\qquad B \;\mapsto\; Z(B)$$

where ε is the exchange involution. Observe that S_2 acts on $H^1(F, \mathbf{PGL}_n)$ by sending a central simple algebra A to the opposite algebra A^{op}, and that the algebras with involution $(A \times A^{\mathrm{op}}, \varepsilon)$ and $(A^{\mathrm{op}} \times A, \varepsilon)$ are isomorphic over F.

(29.21) Remark. Let Z be a quadratic étale F-algebra. The cohomology set $H^1\big(F, (\mathbf{PGL}_n)_{[Z]}\big)$, where the action of Γ is twisted through the cocycle defining $[Z]$, classifies triples (B', τ', ϕ) where (B', τ') is a central simple F-algebra with unitary involution of degree n and ϕ is an isomorphism $Z(B') \xrightarrow{\sim} Z$.

Symplectic involutions. Let A be a central simple F-algebra of degree $2n$ with a symplectic involution σ. The group $\mathbf{Aut}(A, \sigma)$ coincides with $\mathbf{PGSp}(A, \sigma)$. Moreover, since all the nonsingular alternating bilinear forms of dimension $2n$ are isometric, all the symplectic involutions on a split algebra of degree $2n$ are conjugate, hence (29.14) and (29.15) yield bijections of pointed sets

(29.22)

F-isomorphism classes of central simple F-algebras of degree $2n$ with symplectic involution	$\longleftrightarrow$	$H^1\big(F, \mathbf{PGSp}(A, \sigma)\big)$

(29.23)

conjugacy classes of symplectic involutions on A	$\longleftrightarrow$	$H^1\big(F, \mathbf{GSp}(A, \sigma)\big).$

The exact sequence

$$1 \to \mathbf{G}_{\mathrm{m}} \to \mathbf{GSp}(A, \sigma) \to \mathbf{PGSp}(A, \sigma) \to 1$$

yields a connecting map in cohomology

$$\delta^1 \colon H^1\big(F, \mathbf{PGSp}(A, \sigma)\big) \to H^2(F, \mathbf{G}_{\mathrm{m}}) = \mathrm{Br}(F).$$

The commutative diagram

$$
\begin{array}{ccccccccc}
1 & \longrightarrow & \mathbf{G}_{\mathrm{m}} & \longrightarrow & \mathbf{GSp}(A, \sigma) & \longrightarrow & \mathbf{PGSp}(A, \sigma) & \longrightarrow & 1 \\
& & \Big\| & & \Big\downarrow & & \Big\downarrow & & \\
1 & \longrightarrow & \mathbf{G}_{\mathrm{m}} & \longrightarrow & \mathbf{GL}_1(A) & \longrightarrow & \mathbf{PGL}_1(A) & \longrightarrow & 1
\end{array}
$$

and Proposition (28.12) show that δ^1 maps the class of (A', σ') to the Brauer class $[A'] \cdot [A]^{-1}$.

We now consider the group of isometries $\mathbf{Sp}(A, \sigma)$. Our first goal is to describe the set $\mathrm{Sym}(A, \sigma)'$. By identifying $A_{\mathrm{sep}} = M_{2n}(F_{\mathrm{sep}})$, we have $\sigma_{\mathrm{sep}} = \mathrm{Int}(u) \circ t$ for some unit $u \in \mathrm{Alt}\big(M_{2n}(F_{\mathrm{sep}}), t\big)$, where t is the transpose involution. For $x \in A_{\mathrm{sep}}$, we have

$$x + \sigma(x) = \big(xu - (xu)^t\big)u^{-1}.$$

If $x + \sigma(x)$ is invertible, then $xu - (xu)^t$ is an invertible alternating matrix. Since all the nonsingular alternating forms of dimension $2n$ are isometric, we may find $a \in \mathrm{GL}_{2n}(F_{\mathrm{sep}})$ such that $xu - (xu)^t = aua^t$. Then

$$x + \sigma(x) = a \cdot \sigma(a),$$

proving that every unit in $\mathrm{Symd}(A, \sigma)$ is in $\mathrm{Sym}(A, \sigma)'$. On the other hand, since σ is symplectic we have $1 = y + \sigma(y)$ for some $y \in A$, hence for all $a \in A_{\mathrm{sep}}$

$$a \cdot \sigma(a) = a\big(y + \sigma(y)\big)\sigma(a) = ay + \sigma(ay).$$

Therefore, $\mathrm{Sym}(A, \sigma)'$ is the set of all symmetrized units in A, i.e.,

$$\mathrm{Sym}(A, \sigma)' = \mathrm{Symd}(A, \sigma)^{\times},$$

and (29.16) yields a bijection of pointed sets

(29.24) $\mathrm{Symd}(A, \sigma)^{\times}/\!\sim \quad \longleftrightarrow \quad H^1\big(F, \mathbf{Sp}(A, \sigma)\big).$

(29.25) Example. Let a be a nonsingular alternating bilinear form on an F-vector space V. To simplify notation, write $\mathbf{GSp}(V, a)$ for $\mathbf{GSp}\big(\mathrm{End}_F(V), \sigma_a\big)$ and $\mathbf{Sp}(V, a)$ for $\mathbf{Sp}\big(\mathrm{End}_F(V), \sigma_a\big)$. Since all the nonsingular alternating bilinear forms on V are isometric to a, we have

$$H^1\big(F, \mathbf{GSp}(V, a)\big) = H^1\big(F, \mathbf{Sp}(V, a)\big) = 1.$$

Orthogonal involutions. Let A be a central simple F-algebra of degree n with an orthogonal involution σ. We have $\mathbf{Aut}(A, \sigma) = \mathbf{PGO}(A, \sigma)$. Assume that char $F \neq 2$ or that F is perfect of characteristic 2. Then F_{sep} is quadratically closed, hence all the nonsingular symmetric nonalternating bilinear forms of dimension n over F_{sep} are isometric. Therefore, all the orthogonal involutions on $A_{\mathrm{sep}} \simeq M_n(F_{\mathrm{sep}})$ are conjugate, and the following bijections of pointed sets readily follow from (29.14) and (29.15):

F-isomorphism classes of central simple F-algebras of degree n with orthogonal involution

$\longleftrightarrow \quad H^1\big(F, \mathbf{PGO}(A, \sigma)\big)$

conjugacy classes of orthogonal involutions on A

$\longleftrightarrow \quad H^1\big(F, \mathbf{GO}(A, \sigma)\big).$

The same arguments as in the case of symplectic involutions show that the connecting map

$$\delta^1 \colon H^1\big(F, \mathbf{PGO}(A, \sigma)\big) \to H^2(F, \mathbf{G}_{\mathrm{m}}) = \mathrm{Br}(F)$$

in the cohomology sequence arising from the exact sequence

$$1 \to \mathbf{G}_{\mathrm{m}} \to \mathbf{GO}(A, \sigma) \to \mathbf{PGO}(A, \sigma) \to 1$$

takes the class of (A', σ') to the Brauer class $[A'] \cdot [A]^{-1}$.

In order to give a description of $H^1\big(F, \mathbf{O}(A, \sigma)\big)$, we next determine the set $\mathrm{Sym}(A, \sigma)'$. We still assume that char $F \neq 2$ or that F is perfect. By identifying $A_{\mathrm{sep}} = M_n(F_{\mathrm{sep}})$, we have $\sigma_{\mathrm{sep}} = \mathrm{Int}(u) \circ t$ for some symmetric nonalternating matrix $u \in \mathrm{GL}_n(F_{\mathrm{sep}})$. For $s \in \mathrm{Sym}(A, \sigma)$, we have $su \in \mathrm{Sym}\big(M_n(F_{\mathrm{sep}}), t\big)$. If $su = x - x^t$ for some $x \in M_n(F_{\mathrm{sep}})$, then $s = xu^{-1} - \sigma(xu^{-1})$. Therefore, su is not alternating if $s \notin \mathrm{Alt}(A, \sigma)$. Since all the nonsingular symmetric nonalternating bilinear forms of dimension n over F_{sep} are isometric, we then have $su = vuv^t$ for some $v \in \mathrm{GL}_n(F_{\text{sep}})$, hence $s = v\sigma(v)$. This proves

$$\mathrm{Sym}(A, \sigma)' \subset \mathrm{Sym}(A, \sigma)^\times \smallsetminus \mathrm{Alt}(A, \sigma).$$

To prove the reverse inclusion, observe that if $a\sigma(a) = x - \sigma(x)$ for some $a \in \mathrm{GL}_n(F_{\text{sep}})$, then

$$1 = a^{-1}x\sigma(a)^{-1} - \sigma\big(a^{-1}x\sigma(a)^{-1}\big) \in \mathrm{Alt}(A_{\text{sep}}, \sigma).$$

This is impossible since σ is orthogonal (see (2.6)).

By (29.16), we have a bijection of pointed sets

$$\big(\mathrm{Sym}(A, \sigma)^\times \smallsetminus \mathrm{Alt}(A, \sigma)\big)/\!\sim \quad \longleftrightarrow \quad H^1\big(F, \mathbf{O}(A, \sigma)\big)$$

where the base point in the left set is the equivalence class of 1. Of course, if char $F \neq 2$, then $\mathrm{Sym}(A, \sigma) \cap \mathrm{Alt}(A, \sigma) = \{0\}$ hence the bijection above takes the form

(29.26) $$\mathrm{Sym}(A, \sigma)^\times/\!\sim \quad \longleftrightarrow \quad H^1\big(F, \mathbf{O}(A, \sigma)\big).$$

Assuming char $F \neq 2$, let

$$\mathrm{SSym}(A, \sigma)^\times = \big\{\, (s, z) \in \mathrm{Sym}(A, \sigma)^\times \times F^\times \mid \mathrm{Nrd}_A(s) = z^2 \,\big\}$$

and define an equivalence relation $\approx$ on this set by

$$(s, z) \approx (s', z') \quad \text{if and only if} \quad s' = a \cdot s \cdot \sigma(a) \text{ and } z' = \mathrm{Nrd}_A(a)z \text{ for some } a \in A^\times.$$

The same arguments as in the proof of (29.18) yield a canonical bijection of pointed sets

(29.27) $$\mathrm{SSym}(A, \sigma)^\times/\!\approx \quad \longleftrightarrow \quad H^1\big(F, \mathbf{O}^+(A, \sigma)\big).$$

29.E. Quadratic spaces. Let (V, q) be a nonsingular quadratic space of dimension n over an arbitrary field F. Let $W = S^2(V^*)$, the second symmetric power of the dual space of V. Consider the representation

$$\rho \colon G = \mathbf{GL}(V) \to \mathbf{GL}(W)$$

defined by

$$\rho(\alpha)(f)(x) = f\big(\alpha^{-1}(x)\big)$$

for $\alpha \in G$, $f \in W$ and $x \in V$ (viewing $S^2(V^*)$ as a space of polynomial maps on V). The group scheme $\mathbf{Aut}_G(q)$ is the orthogonal group $\mathbf{O}(V, q)$.

We postpone until the end of this subsection the discussion of the case where n is odd and char $F = 2$. Assume thus that n is even or that char $F \neq 2$. Then, all the nonsingular quadratic spaces of dimension n over F_{sep} are isometric, hence Proposition (29.1) yields a canonical bijection

(29.28)
$$\boxed{\begin{array}{c} \text{isometry classes of} \\ n\text{-dimensional nonsingular} \\ \text{quadratic spaces over } F \end{array}} \quad \longleftrightarrow \quad H^1\big(F, \mathbf{O}(V, q)\big).$$

To describe the pointed set $H^1\big(F, \mathbf{O}^+(V, q)\big)$, we first give another description of $H^1\big(F, \mathbf{O}(V, q)\big)$. Consider the representation

$$\rho \colon G = \mathbf{GL}(V) \times \mathbf{GL}\big(C(V, q)\big) \to \mathbf{GL}(W)$$

where

$$W = S^2(V^*) \oplus \operatorname{Hom}_F\big(V, C(V,q)\big) \oplus \operatorname{Hom}_F\big(C(V,q) \otimes C(V,q), C(V,q)\big)$$

and

$$\rho(\alpha, \beta)(f, g, h) = \big(f\alpha^{-1}, \beta g\alpha^{-1}, \alpha h(\alpha^{-1} \otimes \alpha^{-1})\big).$$

Set $w = (q, i, m)$, where $i \colon V \to C(V,q)$ is the canonical map and m is the multiplication of $C(V,q)$. Then, we obviously also have $\mathbf{Aut}_G(w) = \mathbf{O}(V,q)$ since automorphisms of (C,q) which map V to V are in $\mathbf{O}(V,q)$. If n is even, let Z be the center of the even Clifford algebra $C_0(V,q)$; if n is odd, let Z be the center of the full Clifford algebra $C(V,q)$. The group G also acts on

$$W^+ = W \oplus \operatorname{End}_F(Z)$$

where the action of G on $\operatorname{End}_F(Z)$ is given by $\rho(\alpha, \beta)(j) = j\beta|_Z^{-1}$. If we set $w^+ = (w, \operatorname{Id}_Z)$, with w as above, we obtain $\mathbf{Aut}_G(w^+) = \mathbf{O}^+(V,q)$. By Proposition (29.1) the set $H^1\big(F, \mathbf{O}^+(V,q)\big)$ classifies triples (V', q', φ) where $\varphi \colon Z' \xrightarrow{\sim} Z$ is an isomorphism from the center of $C(V', q')$ to Z. We claim that in fact we have a bijection

(29.29)
$$\boxed{\begin{array}{c} \text{isometry classes of} \\ n\text{-dimensional nonsingular} \\ \text{quadratic spaces } (V', q') \text{ over } F \\ \text{such that } \operatorname{disc} q' = \operatorname{disc} q \end{array}} \quad \longleftrightarrow \quad H^1\big(F, \mathbf{O}^+(V,q)\big).$$

Since the F-algebra Z is determined up to isomorphism by $\operatorname{disc} q$ (see (8.2) when n is even), the set on the left corresponds to the image of $H^1\big(F, \mathbf{O}^+(V,q)\big)$ in $H^1\big(F, \mathbf{O}(V,q)\big)$. Hence we have to show that the canonical map

$$H^1\big(F, \mathbf{O}^+(V,q)\big) \to H^1\big(F, \mathbf{O}(V,q)\big)$$

is injective. If $\operatorname{char} F = 2$ (and n is even), consider the exact sequence

(29.30)
$$1 \to \mathbf{O}^+(V,q) \to \mathbf{O}(V,q) \xrightarrow{\Delta} \mathbb{Z}/2\mathbb{Z} \to 0$$

where Δ is the Dickson invariant, and the induced cohomology sequence

$$\mathrm{O}(V,q) \to \mathbb{Z}/2\mathbb{Z} \to H^1\big(F, \mathbf{O}^+(V,q)\big) \to H^1\big(F, \mathbf{O}(V,q)\big).$$

Since $\Delta \colon \mathrm{O}(V,q) \to \mathbb{Z}/2\mathbb{Z}$ is surjective, we have the needed injectivity at the base point. To get injectivity at a class $[x]$, we twist the sequence (29.30) by a cocycle x representing $[x] = [(V', q')]$; then $[x]$ is the new base point and the claim follows from $\mathrm{O}(V,q)_x = \mathrm{O}(V', q')$.

If $\operatorname{char} F \neq 2$ (regardless of the parity of n), the arguments are the same, substituting for (29.30) the exact sequence

$$1 \to \mathbf{O}^+(V,q) \to \mathbf{O}(V,q) \xrightarrow{\det} \boldsymbol{\mu}_2 \to 1.$$

We now turn to the case where $\operatorname{char} F = 2$ and n is odd, which was put aside for the preceding discussion. Nonsingular quadratic spaces of dimension n become isometric over F_{sep} if and only if they have the same discriminant, hence

Proposition (29.1) yields a bijection

$$\boxed{\begin{array}{c} \text{isometry classes of} \\ n\text{-dimensional nonsingular} \\ \text{quadratic spaces } (V', q') \text{ over } F \\ \text{such that disc } q' = \text{disc } q \end{array}} \quad \longleftrightarrow \quad H^1\big(F, \mathbf{O}(V, q)\big).$$

The description of $H^1\big(F, \mathbf{O}^+(V, q)\big)$ by triples (V', q', φ) where $\varphi \colon Z' \to Z$ is an isomorphism of the centers of the full Clifford algebras $C(V', q')$, $C(V, q)$ still holds, but in this case $Z' = F(\sqrt{\text{disc } q'})$, $Z = F(\sqrt{\text{disc } q})$ are purely inseparable quadratic F-algebras, hence the isomorphism $\varphi \colon Z' \to Z$ is unique when it exists, i.e., when disc $q' = $ disc q. Therefore, we have

$$H^1\big(F, \mathbf{O}^+(V, q)\big) = H^1\big(F, \mathbf{O}(V, q)\big).$$

This equality also follows from the fact that $\mathbf{O}^+(V, q)$ is the smooth algebraic group associated to $\mathbf{O}(V, q)$, hence the groups of points of $\mathbf{O}^+(V, q)$ and of $\mathbf{O}(V, q)$ over F_{sep} (as over any reduced F-algebra) coincide.

29.F. Quadratic pairs. Let A be a central simple F-algebra of degree $2n$ with a quadratic pair (σ, f). Consider the representation already used in the proof of (26.15):

$$\rho \colon G = \mathbf{GL}(A) \times \mathbf{GL}\big(\text{Sym}(A, \sigma)\big) \to \mathbf{GL}(W)$$

where

$$W = \text{Hom}_F\big(\text{Sym}(A, \sigma), A\big) \oplus \text{Hom}_F(A \otimes_F A, A) \oplus \text{End}_F(A) \oplus \text{Sym}(A, \sigma)^*,$$

with ρ given by

$$\rho(g, h)(\lambda, \psi, \varphi, p) = \big(g \circ \lambda \circ h^{-1}, g(\psi), g \circ \varphi \circ g^{-1}, p \circ h\big)$$

where $g(\psi)$ arises from the natural action of $\text{GL}(A)$ on $\text{Hom}_F(A \otimes A, A)$. Consider also the element $w = (i, m, \sigma, f) \in W$ where $i \colon \text{Sym}(A, \sigma) \to A$ is the inclusion. The group $\mathbf{Aut}_G(w)$ coincides with $\mathbf{PGO}(A, \sigma, f)$ (see §23.B).

Every twisted ρ-form $(\lambda, \psi, \varphi, p)$ of w defines a central simple F-algebra with quadratic pair (A', σ', f') as follows: on the set $A' = \{ x' \mid x \in A \}$, we define the multiplication by $x'y' = \psi(x \otimes y)'$ and the involution by $\sigma'(x') = \varphi(x)'$. Then $\text{Sym}(A', \sigma') = \{ \lambda(s)' \mid s \in \text{Sym}(A, \sigma) \}$, and we define f' by the condition $f'\big(\lambda(s)'\big) = p(s)$ for $s \in \text{Sym}(A, \sigma)$.

Conversely, to every central simple F-algebra with quadratic pair (A', σ', f') of degree $2n$, we associate an element $(\lambda, \psi, \varphi, p) \in W$ as follows: we choose arbitrary bijective F-linear maps $\nu \colon A \xrightarrow{\sim} A'$ and $\lambda \colon \text{Sym}(A, \sigma) \xrightarrow{\sim} \nu^{-1}\big(\text{Sym}(A', \sigma')\big)$, and define ψ, φ, p by

$$\psi(x \otimes y) = \nu^{-1}\big(\nu(x)\nu(y)\big) \qquad \text{for } x, y \in A,$$

$$\varphi = \nu^{-1} \circ \sigma' \circ \nu \qquad \text{and} \qquad p = f' \circ \nu \circ \lambda.$$

Over F_{sep}, all the algebras with quadratic pairs of degree $2n$ become isomorphic to the split algebra with the quadratic pair associated to the hyperbolic quadratic form. If

$$\theta \colon (A_{\text{sep}}, \sigma_{\text{sep}}, f_{\text{sep}}) \xrightarrow{\sim} (A'_{\text{sep}}, \sigma'_{\text{sep}}, f'_{\text{sep}})$$

is an isomorphism, then we let $g = \nu^{-1} \circ \theta \in \mathrm{GL}(A_{\mathrm{sep}})$ and define

$$h \colon \mathrm{Sym}(A_{\mathrm{sep}}, \sigma_{\mathrm{sep}}) \to A_{\mathrm{sep}} \qquad \text{by} \qquad \nu \circ \lambda \circ h = \theta \circ i.$$

Then $\rho(g, h)(i, m, \sigma, f) = (\lambda, \psi, \varphi, p)$, proving that $(\lambda, \psi, \varphi, p)$ is a twisted ρ-form of w. Thus, twisted ρ-forms of w are in one-to-one correspondence with isomorphism classes of central simple F-algebras with quadratic pair of degree $2n$. By (29.1) there is a canonical bijection

$$\boxed{\begin{array}{c} \text{F-isomorphism classes of} \\ \text{central simple F-algebras with} \\ \text{quadratic pair of degree $2n$} \end{array}} \quad \longleftrightarrow \quad H^1\big(F, \mathbf{PGO}(A, \sigma, f)\big).$$

The center Z of the Clifford algebra $C(A, \sigma, f)$ is a quadratic étale F-algebra which we call the *discriminant quadratic extension*. The class $[Z]$ of Z in $H^1(F, S_2)$ is the *discriminant class of* (σ, f). We have an exact sequence of group schemes

$$1 \to \mathbf{PGO}^+(A, \sigma, f) \to \mathbf{PGO}(A, \sigma, f) \xrightarrow{d} S_2 \to 1$$

where d is the natural homomorphism

$$\mathbf{PGO}(A, \sigma, f) \to \mathbf{Aut}_{\mathrm{alg}}(Z) \simeq S_2.$$

Thus the map

$$H^1\big(F, \mathbf{PGO}(A, \sigma, f)\big) \to H^1(F, S_2)$$

takes (A', σ', f') to $[Z'] - [Z]$ where Z' is the center of $C(A', \sigma', f')$. As in (29.29), we obtain a natural bijection

$$\boxed{\begin{array}{c} \text{F-isomorphism classes of 4-tuples $(A', \sigma', f', \varphi)$} \\ \text{with a central simple F-algebra A'} \\ \text{of degree $2n$, a quadratic pair (σ', f')} \\ \text{and an F-algebra isomorphism $\varphi \colon Z' \to Z$ of} \\ \text{the centers of the Clifford algebras} \end{array}} \quad \longleftrightarrow \quad H^1\big(F, \mathbf{PGO}^+(A, \sigma, f)\big).$$

(29.31) Remark. In particular, if $Z \simeq F \times F$, the choice of φ amounts to a designation of the two components $C_+(A, \sigma, f)$, $C_-(A, \sigma, f)$ of $C(A, \sigma, f)$.

In order to obtain similar descriptions for the cohomology sets of $\mathbf{GO}(A, \sigma, f)$ and $\mathbf{GO}^+(A, \sigma, f)$, it suffices to let $\mathbf{GL}_1(A) \times \mathbf{GL}\big(\mathrm{Sym}(A, \sigma)\big)$ act on W via ρ and the map $\mathbf{GL}_1(A) \to \mathbf{GL}(A)$ which takes $x \in A^\times$ to $\mathrm{Int}(x)$. As in (29.15), we obtain bijections

$$\boxed{\begin{array}{c} \text{conjugacy classes of} \\ \text{quadratic pairs on A} \end{array}} \quad \longleftrightarrow \quad H^1\big(F, \mathbf{GO}(A, \sigma, f)\big),$$

$$\boxed{\begin{array}{c} \text{conjugacy classes of triples (σ', f', φ)} \\ \text{where (σ', f') is a quadratic pair on A} \\ \text{and $\varphi \colon Z' \to Z$ is an isomorphism of} \\ \text{the centers of the Clifford algebras} \end{array}} \quad \longleftrightarrow \quad H^1\big(F, \mathbf{GO}^+(A, \sigma, f)\big).$$

If char $F \neq 2$, the quadratic pair (σ, f) is completely determined by the orthogonal involution σ, hence $\mathbf{O}(A, \sigma, f) = \mathbf{O}(A, \sigma)$ and we refer to §29.D for a description of $H^1\big(F, \mathbf{O}(A, \sigma)\big)$ and $H^1\big(F, \mathbf{O}^+(A, \sigma)\big)$. For the rest of this subsection, we assume char $F = 2$. As a preparation for the description of $H^1\big(F, \mathbf{O}(A, \sigma, f)\big)$, we make an observation on involutions on algebras over separably closed fields.

(29.32) Lemma. *Suppose that* char $F = 2$. *Let* σ *be an involution of the first kind on* $A = M_{2n}(F_{\mathrm{sep}})$. *For all* a, $b \in A$ *such that* $a + \sigma(a)$ *and* $b + \sigma(b)$ *are invertible, there exists* $g \in A^\times$ *and* $x \in A$ *such that*

$$b = ga\sigma(g) + x + \sigma(x).$$

Moreover, if $\mathrm{Srd}\big((a + \sigma(a))^{-1}a\big) = \mathrm{Srd}\big((b + \sigma(b))^{-1}b\big)$, *then we may assume* $\mathrm{Trd}\big((b + \sigma(b))^{-1}x\big) = 0$.

Proof: Let $u \in A^\times$ satisfy $\sigma = \mathrm{Int}(u) \circ t$, where t is the transpose involution. Then $u = u^t$ and $a + \sigma(a) = u\big(u^{-1}a + (u^{-1}a)^t\big)$, hence $u^{-1}a + (u^{-1}a)^t$ is invertible. Therefore, the quadratic form $q(X) = Xu^{-1}aX^t$, where $X = (x_1, \ldots, x_{2n})$, is nonsingular. Similarly, the quadratic form $Xu^{-1}bX^t$ is nonsingular. Since all the nonsingular quadratic forms of dimension $2n$ over a separably closed field are isometric, we may find $g_0 \in A^\times$ such that

$$u^{-1}b \equiv g_0 u^{-1}a g_0^t \mod \mathrm{Alt}(A, \sigma)$$

hence

$$b \equiv g_1 a \sigma(g_1) \mod \mathrm{Alt}(A, \sigma)$$

for $g_1 = ug_0u^{-1}$. This proves the first part.

To prove the second part, choose $g_1 \in A^\times$ as above and $x_1 \in A$ such that

$$b = g_1 a \sigma(g_1) + x_1 + \sigma(x_1).$$

Let $g_0 = u^{-1}g_1 u$ and $x_0 = u^{-1}x_1$, so that

(29.33) $$u^{-1}b = g_0 u^{-1}a g_0^t + x_0 + x_0^t.$$

Let also $v = u^{-1}b + (u^{-1}b)^t$. We have $b + \sigma(b) = uv$ and, by the preceding equation,

$$v = g_0\big(u^{-1}a + (u^{-1}a)^t\big)g_0^t.$$

From (0.5), we derive

(29.34) $$s_2(v^{-1}u^{-1}b) = s_2(v^{-1}g_0 u^{-1}a g_0^t) + \wp\big(\mathrm{tr}(v^{-1}x_0)\big).$$

The left side is $\mathrm{Srd}\big((b + \sigma(b))^{-1}b\big)$. On the other hand, $v^{-1}g_0 u^{-1}a g_0^t$ is conjugate to

$$\big(u^{-1}a + (u^{-1}a)^t\big)u^{-1}a = \big(a + \sigma(a)\big)^{-1}a.$$

Therefore, if $\mathrm{Srd}\big((b + \sigma(b))^{-1}b\big) = \mathrm{Srd}\big((a + \sigma(a))^{-1}a\big)$, equation (29.34) yields $\wp\big(\mathrm{tr}(v^{-1}x_0)\big) = 0$, hence $\mathrm{tr}(v^{-1}x_0) = 0$ or 1. In the former case, we are finished since $v^{-1}x_0 = \big(b + \sigma(b)\big)^{-1}x_1$. In the latter case, let g_2 be an improper isometry of the quadratic form $Xg_0 u^{-1}a g_0^t X^t$ (for instance a hyperplane reflection, see (12.13)). We have

$$g_0 u^{-1}a g_0^t = g_2 g_0 u^{-1}a g_0^t g_2^t + x_2 + x_2^t$$

for some $x_2 \in A$ such that $\mathrm{tr}(v^{-1}x_2) = 1$, by definition of the Dickson invariant in (12.12), and by substituting in (29.33),

$$u^{-1}b = g_2 g_0 u^{-1}a g_0^t g_2^t + (x_0 + x_2) + (x_0 + x_2)^t.$$

Now, $\mathrm{tr}\big(v^{-1}(x_0 + x_2)\big) = 0$, hence we may set $g = ug_2 g_0 u^{-1}$ and $x = u(x_0 + x_2)$. $\square$

Now, let (A, σ, f) be a central simple algebra with quadratic pair over a field F of characteristic 2. Let $G = \mathbf{GL}_1(A)$ act on the vector space $W = A/\operatorname{Alt}(A, \sigma)$ by

$$\rho(g)\big(a + \operatorname{Alt}(A, \sigma)\big) = ga\sigma(g) + \operatorname{Alt}(A, \sigma)$$

for $g \in G$ and $a \in A$. Let $\ell \in A$ satisfy $f(s) = \operatorname{Trd}_A(\ell s)$ for all $s \in \operatorname{Sym}(A, \sigma)$ (see (5.7)). We next determine the stabilizer $\mathbf{Aut}_G\big(\ell + \operatorname{Alt}(A, \sigma)\big)$. For every rational point g of this stabilizer we have

$$g\ell\sigma(g) = \ell + x + \sigma(x) \qquad \text{for some } x \in A.$$

Applying σ and using $\ell + \sigma(\ell) = 1$ (see (5.7)), it follows that $g\sigma(g) = 1$. Moreover, since $\operatorname{Alt}(A, \sigma)$ is orthogonal to $\operatorname{Sym}(A, \sigma)$ for the bilinear form T_A (see (2.3)) we have

$$\operatorname{Trd}_A\big(g\ell\sigma(g)s\big) = \operatorname{Trd}_A(\ell s) \qquad \text{for all } s \in \operatorname{Sym}(A, \sigma)$$

hence

$$f\big(\sigma(g)sg\big) = f(s) \qquad \text{for all } s \in \operatorname{Sym}(A, \sigma).$$

Therefore, $g \in \mathbf{O}(A, \sigma, f)$. Conversely, if $g \in \mathbf{O}(A, \sigma, f)$ then $f\big(\sigma(g)sg\big) = f(s)$ for all $s \in \operatorname{Sym}(A, \sigma)$, hence $\operatorname{Trd}_A\big(g\ell\sigma(g)s\big) = \operatorname{Trd}_A(\ell s)$ for all $s \in \operatorname{Sym}(A, \sigma)$, and it follows that $g\ell\sigma(g) \equiv \ell \mod \operatorname{Alt}(A, \sigma)$. Therefore,

(29.35) $$\mathbf{Aut}_G\big(\ell + \operatorname{Alt}(A, \sigma)\big) = \mathbf{O}(A, \sigma, f).$$

On the other hand, Lemma (29.32) shows that the twisted ρ-forms of $\ell + \operatorname{Alt}(A, \sigma)$ are the elements $a + \operatorname{Alt}(A, \sigma)$ such that $a + \sigma(a) \in A^\times$. Let

$$Q(A, \sigma) = \{\, a + \operatorname{Alt}(A, \sigma) \mid a + \sigma(a) \in A^\times \,\} \subset A/\operatorname{Alt}(A, \sigma)$$

and define an equivalence relation $\sim$ on $Q(A, \sigma)$ by $a + \operatorname{Alt}(A, \sigma) \sim a' + \operatorname{Alt}(A, \sigma)$ if and only if $a' \equiv ga\sigma(g) \mod \operatorname{Alt}(A, \sigma)$ for some $g \in A^\times$. Proposition (29.1) and Hilbert's Theorem 90 yield a bijection

(29.36) $$Q(A, \sigma)/\!\sim \quad \longleftrightarrow \quad H^1\big(F, \mathbf{O}(A, \sigma, f)\big)$$

which maps the base point of $H^1\big(F, \mathbf{O}(A, \sigma, f)\big)$ to the equivalence class of $\ell + \operatorname{Alt}(A, \sigma)$. (Compare with (29.26) and (29.28).) Note that if $A = \operatorname{End}_F(V)$ the set $Q(A, \sigma)$ is in one-to-one correspondence with the set of nonsingular quadratic forms on V, see §5.B.

In order to give a similar description of the set $H^1\big(F, \mathbf{O}^+(A, \sigma, f)\big)$ (still assuming $\operatorname{char} F = 2$), we consider the set

$$Q_0(A, \sigma, \ell) = \{\, a \in A \mid a + \sigma(a) \in A^\times \text{ and } \operatorname{Srd}_A\big((a + \sigma(a))^{-1}a\big) = \operatorname{Srd}_A(\ell) \,\}$$

and the set of equivalence classes

$$Q^+(A, \sigma, \ell) = \{\, [a] \mid a \in Q_0(A, \sigma, \ell) \,\}$$

where $[a] = [a']$ if and only if $a' = a + x + \sigma(x)$ for some $x \in A$ such that

$$\operatorname{Trd}_A\big((a + \sigma(a))^{-1}x\big) = 0.$$

We thus have a natural map $Q^+(A, \sigma, \ell) \to Q(A, \sigma)$ which maps $[a]$ to $a + \operatorname{Alt}(A, \sigma)$.

For simplicity of notation, we set

$$Q^+(A, \sigma, \ell)_{\mathrm{sep}} = Q^+(A_{\mathrm{sep}}, \sigma_{\mathrm{sep}}, \ell) \quad \text{and} \quad \mathbf{O}(A, \sigma, f)_{\mathrm{sep}} = \mathbf{O}(A_{\mathrm{sep}}, \sigma_{\mathrm{sep}}, f_{\mathrm{sep}}).$$

Since $Q^+(A, \sigma, \ell)$ is not contained in a vector space, we cannot apply the general principle (29.1). Nevertheless, we may let $A_{\mathrm{sep}}^\times$ act on $Q^+(A, \sigma, \ell)_{\mathrm{sep}}$ by

$$g[a] = [ga\sigma(g)] \qquad \text{for } g \in A_{\mathrm{sep}}^\times \text{ and } a \in Q_0(A, \sigma, \ell)_{\mathrm{sep}}.$$

As observed in (29.35), for $g \in A_{\mathrm{sep}}^\times$ we have $g\ell\sigma(g) = \ell + x + \sigma(x)$ for some $x \in A_{\mathrm{sep}}$ if and only if $g \in O(A, \sigma, f)_{\mathrm{sep}}$. Moreover the definition of the Dickson invariant in (12.32) yields $\Delta(g) = \mathrm{Trd}(x)$, hence we have $g[\ell] = [\ell]$ if and only if $g \in O^+(A, \sigma, f)_{\mathrm{sep}}$. On the other hand, Lemma (29.32) shows that the $A_{\mathrm{sep}}^\times$-orbit of $[\ell]$ is $Q^+(A, \sigma, \ell)_{\mathrm{sep}}$, hence the action on ℓ yields a bijection

(29.37) $$A_{\mathrm{sep}}^\times / O^+(A, \sigma, f)_{\mathrm{sep}} \quad \longleftrightarrow \quad Q^+(A, \sigma, \ell)_{\mathrm{sep}}.$$

Therefore, by (28.2) and Hilbert's Theorem 90 we obtain a bijection between the pointed set $H^1\big(F, \mathbf{O}^+(A, \sigma, f)\big)$ and the orbit set of $A^\times$ in $Q^+(A, \sigma, \ell)_{\mathrm{sep}}^\Gamma$, with the orbit of $[\ell]$ as base point.

Claim. $Q^+(A, \sigma, \ell)_{\mathrm{sep}}^\Gamma = Q^+(A, \sigma, \ell)$.

Let $a \in A$ satisfy $\gamma[a] = [a]$ for all $\gamma \in \Gamma$. This means that for all $\gamma \in \Gamma$ there exists $x_\gamma \in A_{\mathrm{sep}}$ such that

$$\gamma(a) = a + x_\gamma + \sigma(x_\gamma) \quad \text{and} \quad \mathrm{Trd}\big((a + \sigma(a))^{-1} x_\gamma\big) = 0.$$

The map $\gamma \mapsto \gamma(a) - a$ is a 1-cocycle of Γ in $\mathrm{Alt}(A, \sigma)_{\mathrm{sep}}$. For any finite Galois extension L/F, the normal basis theorem (see Bourbaki [49, §10]) shows that $\mathrm{Alt}(A, \sigma) \otimes_F L$ is an induced Γ-module, hence $H^1\big(\Gamma, \mathrm{Alt}(A, \sigma)_{\mathrm{sep}}\big) = 0$. Therefore, there exists $y \in \mathrm{Alt}(A, \sigma)_{\mathrm{sep}}$ such that

$$\gamma(a) - a = x_\gamma + \sigma(x_\gamma) = y - \gamma(y) \qquad \text{for all } \gamma \in \Gamma.$$

Choose $z_0 \in A_{\mathrm{sep}}$ such that $y = z_0 + \sigma(z_0)$. Then $a + z_0 + \sigma(z_0)$ is invariant under Γ, hence $a + z_0 + \sigma(z_0) \in A$. Moreover, $x_\gamma + z_0 + \gamma(z_0) \in \mathrm{Sym}(A, \sigma)_{\mathrm{sep}}$, hence the condition $\mathrm{Trd}\big((a + \sigma(a))^{-1} x_\gamma\big) = 0$ implies

$$\gamma\big(\mathrm{Trd}\big((a + \sigma(a))^{-1} z_0\big)\big) = \mathrm{Trd}\big((a + \sigma(a))^{-1} z_0\big),$$

i.e., $\mathrm{Trd}\big((a + \sigma(a))^{-1} z_0\big) \in F$. Let $z_1 \in A$ satisfy

$$\mathrm{Trd}\big((a + \sigma(a))^{-1} z_0\big) = \mathrm{Trd}\big((a + \sigma(a))^{-1} z_1\big).$$

Then $\big(a + z_0 + \sigma(z_0)\big) + z_1 + \sigma(z_1) \in A$ and

$$[a] = \big[a + (z_0 + z_1) + \sigma(z_0 + z_1)\big] \in Q^+(A, \sigma, \ell),$$

proving the claim.

In conclusion, we obtain from (29.37) via (28.2) and Hilbert's Theorem 90 a canonical bijection

(29.38) $$Q^+(A, \sigma, \ell)/\!\sim \quad \longleftrightarrow \quad H^1\big(F, \mathbf{O}^+(A, \sigma, f)\big)$$

where the equivalence relation $\sim$ is defined by the action of $A^\times$, i.e.,

$$[a] \sim [a'] \quad \text{if and only if} \quad [a'] = [ga\sigma(g)] \text{ for some } g \in A^\times.$$

§30. Galois Cohomology of Roots of Unity

Let F be an arbitrary field. As in the preceding section, let $\Gamma = \mathrm{Gal}(F_{\mathrm{sep}}/F)$ be the absolute Galois group of F. Let n be an integer which is not divisible by $\mathrm{char}\, F$. The *Kummer sequence* is the exact sequence of group schemes

$$\textbf{(30.1)} \qquad\qquad 1 \to \boldsymbol{\mu}_n \to \mathbf{G}_{\mathrm{m}} \xrightarrow{\;(\ )^n\;} \mathbf{G}_{\mathrm{m}} \to 1.$$

Since $H^1(F, \mathbf{G}_{\mathrm{m}}) = 1$ by Hilbert's Theorem 90, the induced long exact sequence in cohomology yields isomorphisms

$$H^1(F, \boldsymbol{\mu}_n) \simeq F^\times/F^{\times n} \quad \text{and} \quad H^2(F, \boldsymbol{\mu}_n) \simeq {}_n H^2(F, \mathbf{G}_{\mathrm{m}}),$$

where, for any abelian group H, ${}_n H$ denotes the n-torsion subgroup of H. Since $H^2(F, \mathbf{G}_{\mathrm{m}}) \simeq \mathrm{Br}(F)$ (see (29.11)), we also have

$$H^2(F, \boldsymbol{\mu}_n) \simeq {}_n\mathrm{Br}(F).$$

This isomorphism suggests deep relations between central simple algebras and the cohomology of $\boldsymbol{\mu}_n$, which are formalized through the cyclic algebra construction in §30.A.

If $\mathrm{char}\, F \neq 2$, we may identify $\mu_2 \otimes \mu_2$ with μ_2 through the map $(-1)^a \otimes (-1)^b \mapsto (-1)^{ab}$ and define a cup product

$$\cup \colon H^i(F, \boldsymbol{\mu}_2) \times H^j(F, \boldsymbol{\mu}_2) \to H^{i+j}(F, \boldsymbol{\mu}_2).$$

For $\alpha \in F^\times$, we set $(\alpha) \in H^1(F, \boldsymbol{\mu}_2)$ for the image of $\alpha \cdot F^{\times 2}$ under the isomorphism $H^1(F, \boldsymbol{\mu}_2) \simeq F^\times/F^{\times 2}$.

The following theorem shows that the Galois cohomology of $\boldsymbol{\mu}_2$ also has a far-reaching relationship with quadratic forms:

(30.2) Theorem. *Let F be a field of characteristic different from 2. For $\alpha_1, \ldots, \alpha_n \in F^\times$, the cup product $(\alpha_1) \cup \cdots \cup (\alpha_n) \in H^n(F, \boldsymbol{\mu}_2)$ depends only on the isometry class of the Pfister form $\langle\!\langle \alpha_1, \ldots, \alpha_n \rangle\!\rangle$. We may therefore define a map e_n on the set of isometry classes of n-fold Pfister forms by setting*

$$e_n\big(\langle\!\langle \alpha_1, \ldots, \alpha_n \rangle\!\rangle\big) = (\alpha_1) \cup \cdots \cup (\alpha_n).$$

Moreover, the map e_n is injective: n-fold Pfister forms π, π' are isometric if and only if $e_n(\pi) = e_n(\pi')$.

Reference: The first assertion appears in Elman-Lam [91, (3.2)], the second in Arason-Elman-Jacob [23, Theorem 1] for $n \leq 4$ (see also Lam-Leep-Tignol [170, Theorem A5] for $n = 3$). The second assertion was proved by Rost (unpublished) for $n \leq 6$, and a proof for all n was announced by Voevodsky in 1996. (In this book, the statement above is not used for $n > 3$.) $\qquad\square$

By combining the interpretations of Galois cohomology in terms of algebras and in terms of quadratic forms, we translate the results of §19.B to obtain in §30.C a complete set of cohomological invariants for central simple F-algebras with unitary involution of degree 3. We also give a cohomological classification of cubic étale F-algebras. The cohomological invariants discussed in §30.C use cohomology groups with twisted coefficients which are introduced in §30.B. Cohomological invariants will be discussed in greater generality in §31.B.

Before carrying out this programme, we observe that there is an analogue of the Kummer sequence in characteristic p. If char $F = p$, the *Artin-Schreier sequence* is the exact sequence of group schemes

$$0 \to \mathbb{Z}/p\mathbb{Z} \to \mathbf{G}_a \xrightarrow{\wp} \mathbf{G}_a \to 0$$

where $\wp(x) = x^p - x$. The normal basis theorem (Bourbaki [49, §10]) shows that the additive group of any finite Galois extension L/F is an induced $\mathrm{Gal}(L/F)$-module, hence $H^\ell\big(\mathrm{Gal}(L/F), L\big) = 0$ for all $\ell > 0$. Therefore,

$$H^\ell(F, \mathbf{G}_a) = 0 \qquad \text{for all } \ell > 0$$

and the cohomology sequence induced by the Artin-Schreier exact sequence yields

(30.3) $H^1(F, \mathbb{Z}/p\mathbb{Z}) \simeq F/\wp(F)$ and $H^\ell(F, \mathbb{Z}/p\mathbb{Z}) = 0$ for $\ell \geq 2$.

30.A. Cyclic algebras. The construction of cyclic algebras, already introduced in §19.A in the particular case of degree 3, has a close relation with Galois cohomology which is described next.

Let n be an arbitrary integer and let L be a Galois $(\mathbb{Z}/n\mathbb{Z})$-algebra over F. We set $\rho = 1 + n\mathbb{Z} \in \mathbb{Z}/n\mathbb{Z}$. For $a \in F^\times$, the *cyclic algebra* (L, a) is

$$(L, a) = L \oplus Lz \oplus \cdots \oplus Lz^{n-1}$$

where $z^n = a$ and $z\ell = \rho(\ell)z$ for $\ell \in L$. Every cyclic algebra (L, a) is central simple of degree n over F. Moreover, it is easy to check, using the Skolem-Noether theorem, that every central simple F-algebra of degree n which contains L has the form (L, a) for some $a \in F^\times$ (see Albert [9, Chapter 7, §1]).

We now give a cohomological interpretation of this construction. Let $[L] \in H^1(F, \mathbb{Z}/n\mathbb{Z})$ be the cohomology class corresponding to L by (28.15). Since the Γ-action on $\mathbb{Z}/n\mathbb{Z}$ is trivial, we have

$$H^1(F, \mathbb{Z}/n\mathbb{Z}) = Z^1(F, \mathbb{Z}/n\mathbb{Z}) = \mathrm{Hom}(\Gamma, \mathbb{Z}/n\mathbb{Z}),$$

so $[L]$ is a continuous homomorphism $\Gamma \to \mathbb{Z}/n\mathbb{Z}$. If L is a field (viewed as a subfield of F_{sep}), this homomorphism is surjective and its kernel is the absolute Galois group of L. For $\sigma \in \Gamma$, define $f(\sigma) \in \{0, 1, \ldots, n-1\}$ by the condition

$$[L](\sigma) = f(\sigma) + n\mathbb{Z} \in \mathbb{Z}/n\mathbb{Z}.$$

Since $[L]$ is a homomorphism, we have $f(\sigma\tau) \equiv f(\sigma) + f(\tau) \mod n$.

Now, assume that n is not divisible by char F. We may then use the Kummer sequence to identify $F^\times/F^{\times n} = H^1(F, \boldsymbol{\mu}_n)$ and $_n\mathrm{Br}(F) = H^2(F, \boldsymbol{\mu}_n)$. For this last identification, we actually have two canonical (and opposite) choices (see §29.B); we choose the identification afforded by the crossed product construction. Thus, the image in $H^2(F, \mathbf{G}_m)$ of the class $\big[(L, a)\big] \in H^2(F, \boldsymbol{\mu}_n)$ corresponding to (L, a) is represented by the cocycle $h \colon \Gamma \times \Gamma \to F_{\mathrm{sep}}^\times$ defined as follows:

$$h(\sigma, \tau) = z^{f(\sigma)} \cdot z^{f(\tau)} \cdot z^{-f(\sigma\tau)} = a^{\big(f(\sigma) + f(\tau) - f(\sigma\tau)\big)/n}.$$

(See Pierce [219, p. 277] for the case where L is a field.)

The bilinear pairing $(\mathbb{Z}/n\mathbb{Z}) \times \boldsymbol{\mu}_n(F_{\mathrm{sep}}) \to \boldsymbol{\mu}_n(F_{\mathrm{sep}})$ which maps $(i + n\mathbb{Z}, \zeta)$ to ζ^i induces a cup product

$$\cup \colon H^1(F, \mathbb{Z}/n\mathbb{Z}) \times H^1(F, \boldsymbol{\mu}_n) \to H^2(F, \boldsymbol{\mu}_n).$$

(30.4) Proposition. *The homomorphism* $[L] \in H^1(F, \mathbb{Z}/n\mathbb{Z})$, *the class* $(a) \in H^1(F, \boldsymbol{\mu}_n)$ *corresponding to* $a \cdot F^{\times n}$ *under the identification* $H^1(F, \boldsymbol{\mu}_n) = F^\times/F^{\times n}$ *and the class* $\big[(L, a)\big] \in H^2(F, \boldsymbol{\mu}_n)$ *corresponding to* (L, a) *by the crossed product construction are related by*

$$\big[(L, a)\big] = [L] \cup (a).$$

Proof: Since the canonical map $H^2(F, \boldsymbol{\mu}_n) \to H^2(F, \mathbf{G}_\mathrm{m})$ is injective, it suffices to compare the images of $[L] \cup (a)$ and of $\big[(L, a)\big]$ in $H^2(F, \mathbf{G}_\mathrm{m})$. Let $\xi \in F_\mathrm{sep}$ satisfy $\xi^n = a$. The class (a) is then represented by the cocycle $\sigma \mapsto \sigma(\xi)\xi^{-1}$, and the cup product $[L] \cup (a)$ by the cocycle $g \colon \Gamma \times \Gamma \to F_\mathrm{sep}^\times$ defined by

$$g(\sigma, \tau) = \sigma\big(\tau(\xi)\xi^{-1}\big)^{[L](\sigma)} = \sigma\big(\tau(\xi)\xi^{-1}\big)^{f(\sigma)} \quad \text{for } \sigma, \tau \in \Gamma.$$

Consider the function $c \colon \Gamma \to F_\mathrm{sep}^\times$ given by $c_\sigma = \sigma(\xi)^{f(\sigma)}$. We have

$$g(\sigma, \tau)c_\sigma \sigma(c_\tau)c(\sigma\tau)^{-1} = \sigma\tau(\xi)^{f(\sigma)+f(\tau)-f(\sigma\tau)} = a^{\big(f(\sigma)+f(\tau)-f(\sigma\tau)\big)/n}.$$

Therefore, the cocyles g and h are cohomologous in $H^2(F, \mathbf{G}_\mathrm{m})$. $\qquad\square$

(30.5) Corollary. *For* a, $b \in F^\times$,

$$\big[(L, a) \otimes (L, b)\big] = \big[(L, ab)\big] \qquad in \ H^2(F, \boldsymbol{\mu}_n). \qquad\square$$

In order to determine when two cyclic algebras (L, a), (L, b) are isomorphic, we first give a criterion for a cyclic algebra to be split:

(30.6) Proposition. *Let* L *be a Galois* $(\mathbb{Z}/n\mathbb{Z})$-*algebra over* F *and* $a \in F^\times$. *The cyclic algebra* (L, a) *is split if and only if* $a \in N_{L/F}(L^\times)$.

Proof: A direct proof (without using cohomology) can be found in Albert [9, Theorem 7.6] or (when L is a field) in Pierce [219, p. 278]. We next sketch a cohomological proof. Let $A = \mathrm{End}_F L$. We embed L into A by identifying $\ell \in L$ with the map $x \mapsto \ell x$. Let $\overline{\rho} \in A$ be given by the action of $\rho = 1 + n\mathbb{Z} \in \mathbb{Z}/n\mathbb{Z}$ on L. From Dedekind's lemma on the independence of automorphisms, we have

$$A = L \oplus L\overline{\rho} \oplus \cdots \oplus L\overline{\rho}^{n-1}$$

(so that A is a cyclic algebra $A = (L, 1)$). Let $L^1 = \{ u \in L \mid N_{L/F}(u) = 1 \}$. For $u \in L^1$, define $\psi(u) \in \mathrm{Aut}(A)$ by

$$\psi(u)(\textstyle\sum_i \ell_i \overline{\rho}^i) = \sum_i \ell_i u^i \overline{\rho}^i \qquad \text{for } \ell_0, \ldots \ell_{n-1} \in L.$$

Clearly, $\psi(u)$ is the identity on L, hence $\psi(u) \in \mathrm{Aut}(A, L)$. In fact, every automorphism of A which preserves L has the form $\psi(u)$ for some $u \in L^1$, and the restriction map $\mathrm{Aut}(A, L) \to \mathrm{Aut}(L)$ is surjective by the Skolem-Noether theorem, hence there is an exact sequence

$$1 \to L^1 \xrightarrow{\psi} \mathrm{Aut}(A, L) \to \mathrm{Aut}(L) \to 1.$$

More generally, there is an exact sequence of group schemes

$$1 \to \mathbf{G}_{\mathrm{m}, L}^1 \xrightarrow{\psi} \mathbf{Aut}(A, L) \to \mathbf{Aut}(L) \to 1,$$

where $\mathbf{G}_{\mathrm{m}, L}^1$ is the kernel of the norm map $N_{L/F} \colon R_{L/F}(\mathbf{G}_{\mathrm{m}, L}) \to \mathbf{G}_{\mathrm{m}, L}$. Since the restriction map $\mathrm{Aut}(A, L) \to \mathrm{Aut}(L)$ is surjective, the induced cohomology sequence shows that the map $\psi^1 \colon H^1(F, \mathbf{G}_{\mathrm{m}, L}^1) \to H^1\big(F, \mathbf{Aut}(A, L)\big)$ has trivial kernel. On the other hand, by Shapiro's lemma and Hilbert's Theorem 90, the cohomology sequence induced by the exact sequence

$$1 \to \mathbf{G}_{\mathrm{m},L}^1 \to R_{L/F}(\mathbf{G}_{\mathrm{m},L}) \xrightarrow{N_{L/F}} \mathbf{G}_{\mathrm{m}} \to 1$$

yields an isomorphism $H^1(F, \mathbf{G}_{\mathrm{m},L}^1) \simeq F^\times / N_{L/F}(L^\times)$.

Recall from (29.12) that $H^1\big(F, \mathbf{Aut}(A, L)\big)$ is in one-to-one correspondence with the set of isomorphism classes of pairs (A', L') where A' is a central simple F-algebra of degree n and L' is an étale F-subalgebra of A' of dimension n. The map ψ^1 associates to $a \cdot N_{L/F}(L^\times) \in F^\times / N_{L/F}(L^\times)$ the isomorphism class of the pair $\big((L, a), L\big)$. The algebra (L, a) is split if and only if this isomorphism class is the base point in $H^1\big(F, \mathbf{Aut}(A, L)\big)$. Since ψ^1 has trivial kernel, the proposition follows. $\square$

(30.7) Corollary. *Two cyclic algebras (L, a) and (L, b) are isomorphic if and only if $a/b \in N_{L/F}(L^\times)$.*

Proof: This readily follows from (30.5) and (30.6). $\square$

30.B. Twisted coefficients. The automorphism group of $\mathbb{Z}$ is the group S_2 of two elements, generated by the automorphism $x \mapsto -x$. We may use cocycles in $Z^1(F, S_2) = H^1(F, S_2)$ to twist the (trivial) action of $\Gamma = \mathrm{Gal}(F_{\mathrm{sep}}/F)$ on $\mathbb{Z}$, hence also on every Γ-module. Since $H^1(F, S_2)$ is in one-to-one correspondence with the isomorphism classes of quadratic étale F-algebras by (29.9), we write $\mathbb{Z}_{[K]}$ for the module $\mathbb{Z}$ with the action of Γ twisted by the cocycle corresponding to a quadratic étale F-algebra K. Thus, $\mathbb{Z}_{[F \times F]} = \mathbb{Z}$ and, if K is a field with absolute Galois group $\Gamma_0 \subset \Gamma$, the Γ-action on $\mathbb{Z}_{[K]}$ is given by

$$\sigma x = \begin{cases} x & \text{if } \sigma \in \Gamma_0 \\ -x & \text{if } \sigma \in \Gamma \smallsetminus \Gamma_0. \end{cases}$$

We may similarly twist the Γ-action of any Γ-module M. We write $M_{[K]}$ for the twisted module; thus

$$M_{[K]} = \mathbb{Z}_{[K]} \otimes_{\mathbb{Z}} M.$$

Clearly, $M_{[K]} = M$ if $2M = 0$.

Since S_2 is commutative, there is a group structure on the set $H^1(F, S_2)$, which can be transported to the set of isomorphism classes of quadratic étale F-algebras. For K, K' quadratic étale F-algebras, the sum $[K] + [K']$ is the class of the quadratic étale F-algebra

(30.8) $K * K' = \{\, x \in K \otimes K' \mid (\iota_K \otimes \iota_{K'})(x) = x \,\}$

where ι_K, $\iota_{K'}$ are the nontrivial automorphisms of K and K' respectively. We say that $K * K'$ is the *product algebra* of K and K'. If $\mathrm{char}\, F \neq 2$, we have

$$H^1(F, S_2) = H^1(F, \boldsymbol{\mu}_2) \simeq F^\times / F^{\times 2}.$$

To any $\alpha \in F^\times$, the corresponding quadratic étale algebra is

$$F(\sqrt{\alpha}) = F[X]/(X^2 - \alpha).$$

In this case

$$F(\sqrt{\alpha}) * F(\sqrt{\alpha'}) = F(\sqrt{\alpha \alpha'}).$$

If char $F = 2$, we have

$$H^1(F, S_2) = H^1(F, \mathbb{Z}/2\mathbb{Z}) \simeq F/\wp(F).$$

To any $\alpha \in F$, the corresponding quadratic étale algebra is

$$F\big(\wp^{-1}(\alpha)\big) = F[X]/(X^2 + X + \alpha).$$

In this case

$$F\big(\wp^{-1}(\alpha)\big) * F\big(\wp^{-1}(\alpha')\big) = F\big(\wp^{-1}(\alpha + \alpha')\big).$$

A direct computation shows:

(30.9) Proposition. *Let K, K' be quadratic étale F-algebras. For any Γ-module M,*

$$M_{[K][K']} = M_{[K*K']}.$$

In particular, $M_{[K][K]} = M$. □

Now, let K be a quadratic étale F-algebra which is a field and let $\Gamma_0 \subset \Gamma$ be the absolute Galois group of K. Let M be a Γ-module. Recall from §28.D the induced Γ-module $\mathrm{Ind}_{\Gamma_0}^{\Gamma} M$, which in this case can be defined as

$$\mathrm{Ind}_{\Gamma_0}^{\Gamma} M = \mathrm{Map}(\Gamma/\Gamma_0, M)$$

(see (28.19)). The map $\varepsilon \colon \mathrm{Ind}_{\Gamma_0}^{\Gamma} M \to M$ which takes f to $\sum_{x \in \Gamma/\Gamma_0} f(x)$ is a Γ-module homomorphism. Its kernel can be identified to $M_{[K]}$ by mapping $m \in M_{[K]}$ to the map $i(m)$ which carries the trivial coset to m and the nontrivial coset to $-m$. Thus, we have an exact sequence of Γ-modules

$$0 \to M_{[K]} \xrightarrow{i} \mathrm{Ind}_{\Gamma_0}^{\Gamma} M \xrightarrow{\varepsilon} M \to 0.$$

For $\ell \geq 0$, Shapiro's lemma yields a canonical isomorphism $H^\ell(F, \mathrm{Ind}_{\Gamma_0}^{\Gamma} M) = H^\ell(K, M)$. Under this isomorphism, the map induced by i (resp. ε) is the restriction (resp. corestriction) homomorphism (see Brown [56, p. 81]). The cohomology sequence associated to the sequence above therefore takes the form

$$0 \to H^0(F, M_{[K]}) \xrightarrow{\mathrm{res}} H^0(K, M) \xrightarrow{\mathrm{cor}} H^0(F, M) \xrightarrow{\delta^0} \ldots$$

(30.10)

$$\ldots$$

$$\ldots \xrightarrow{\delta^{\ell-1}} H^\ell(F, M_{[K]}) \xrightarrow{\mathrm{res}} H^\ell(K, M) \xrightarrow{\mathrm{cor}} H^\ell(F, M) \xrightarrow{\delta^\ell} \ldots.$$

By substituting $M_{[K]}$ for M in this sequence, we obtain the exact sequence

$$0 \to H^0(F, M) \xrightarrow{\mathrm{res}} H^0(K, M) \xrightarrow{\mathrm{cor}} H^0(F, M_{[K]}) \xrightarrow{\delta^0} \ldots$$

$$\ldots$$

$$\ldots \xrightarrow{\delta^{\ell-1}} H^\ell(F, M) \xrightarrow{\mathrm{res}} H^\ell(K, M) \xrightarrow{\mathrm{cor}} H^\ell(F, M_{[K]}) \xrightarrow{\delta^\ell} \ldots$$

since $M_{[K][K]} = M$ by (30.9) and since $M = M_{[K]}$ as Γ_0-module.

(30.11) Proposition. *Assume K is a field. Then*

$$H^1(F, \mathbb{Z}_{[K]}) \simeq \mathbb{Z}/2\mathbb{Z}.$$

Moreover, for all $\ell \geq 0$, the connecting map $\delta^\ell \colon H^\ell(F, M) \to H^{\ell+1}(F, M_{[K]})$ in (30.10) is the cup product with the nontrivial element ζ_K of $H^1(F, \mathbb{Z}_{[K]})$, i.e.,

$$\delta^\ell(\xi) = \zeta_K \cup \xi \qquad \text{for } \xi \in H^\ell(F, M).$$

Proof: The first part follows from the long exact sequence (30.10) with $M = \mathbb{Z}$. The second part is verified by an explicit cochain calculation. $\qquad\square$

A cocycle representing the nontrivial element $\zeta_K \in H^1(F, \mathbb{Z}_{[K]})$ is given by the map

$$\sigma \mapsto \begin{cases} 0 & \text{if } \sigma \in \Gamma_0, \\ 1 & \text{if } \sigma \in \Gamma \smallsetminus \Gamma_0. \end{cases}$$

Therefore, the map $H^1(F, \mathbb{Z}_{[K]}) \to H^1(F, \mathbb{Z}/2\mathbb{Z}) = H^1(F, S_2)$ induced by reduction modulo 2 carries ζ_K to the cocycle associated to K.

(30.12) Corollary. *Assume K is a field. Write $[K]$ for the cocycle in $H^1(F, S_2)$ associated to K.*

(1) *For any Γ-module M such that $2M = 0$, there is a long exact sequence*

$$0 \to H^0(F, M) \xrightarrow{\text{res}} H^0(K, M) \xrightarrow{\text{cor}} H^0(F, M) \xrightarrow{[K]\cup} \dots$$
$$\dots$$
$$\dots \xrightarrow{[K]\cup} H^\ell(F, M) \xrightarrow{\text{res}} H^\ell(K, M) \xrightarrow{\text{cor}} H^\ell(F, M) \xrightarrow{[K]\cup} \dots.$$

(2) *Suppose M is a Γ-module for which multiplication by 2 is an isomorphism. For all $\ell \geq 0$, there is a split exact sequence*

$$0 \to H^\ell(F, M_{[K]}) \xrightarrow{\text{res}} H^\ell(K, M) \xrightarrow{\text{cor}} H^\ell(F, M) \to 0.$$

The splitting maps are $\frac{1}{2}\text{cor}: H^\ell(K, M) \to H^\ell(F, M_{[K]})$ and $\frac{1}{2}\text{res}: H^\ell(F, M) \to H^\ell(K, M)$.

Proof: (1) follows from (30.11) and the description of ζ_K above.

(2) follows from $\text{cor} \circ \text{res} = [K : F] = 2$ (see Brown [56, Chapter 3, Proposition (9.5)]), since multiplication by 2 is an isomorphism. $\qquad\square$

For the sequel, the case where $M = \boldsymbol{\mu}_n(F_{\text{sep}})$ is particularly relevant. The Γ-module $\boldsymbol{\mu}_n(F_{\text{sep}})_{[K]}$ can be viewed as the module of F_{sep}-points of the group scheme

$$\boldsymbol{\mu}_{n[K]} = \ker\big(R_{K/F}(\boldsymbol{\mu}_{n,K}) \xrightarrow{N_{K/F}} \boldsymbol{\mu}_n\big).$$

We next give an explicit description of the group $H^1(F, \boldsymbol{\mu}_{n[K]})$.

(30.13) Proposition. *Let F be an arbitrary field and let n be an integer which is not divisible by $\operatorname{char} F$. For any étale quadratic F-algebra K, there is a canonical isomorphism*

$$H^1(F, \boldsymbol{\mu}_{n[K]}) \simeq \frac{\{(x, y) \in F^\times \times K^\times \mid x^n = N_{K/F}(y)\}}{\{(N_{K/F}(z), z^n) \mid z \in K^\times\}}.$$

Proof: Assume first $K = F \times F$. Then $H^1(F, \boldsymbol{\mu}_{n[K]}) = H^1(F, \boldsymbol{\mu}_n) \simeq F^\times/F^{\times n}$. On the other hand, the map $(x, (y_1, y_2)) \mapsto y_2$ induces an isomorphism from the factor group on the right side to $F^\times/F^{\times n}$, and the proof is complete.

Assume next that K is a field. Define a group scheme $\mathbf{T}$ over F as the kernel of the map $\mathbf{G}_{\text{m}} \times R_{K/F}(\mathbf{G}_{\text{m},K}) \to \mathbf{G}_{\text{m}}$ given by $(x, y) \mapsto x^n N_{K/F}(y)^{-1}$ and define

$$\theta: R_{K/F}(\mathbf{G}_{\text{m},K}) \to \mathbf{T}$$

by $\theta(z) = \big(N_{K/F}(z), z^n\big)$. The kernel of θ is $\boldsymbol{\mu}_{n[K]}$, and we have an exact sequence

$$1 \to \boldsymbol{\mu}_{n[K]} \to R_{K/F}(\mathbf{G}_{\mathrm{m},K}) \xrightarrow{\theta} \mathbf{T} \to 1.$$

By Hilbert's Theorem 90 and Shapiro's lemma (29.2) and (29.6), we have

$$H^1\big(F, R_{K/F}(\mathbf{G}_{\mathrm{m},K})\big) = 1,$$

hence the induced cohomology sequence yields an exact sequence

$$K^\times \xrightarrow{\theta} \mathbf{T}(F) \to H^1(F, \boldsymbol{\mu}_{n[K]}) \to 1. \qquad\qquad \square$$

If K is a field with absolute Galois group $\Gamma_0 \subset \Gamma$, an explicit description of the isomorphism

$$\frac{\{\, (x,y) \in F^\times \times K^\times \mid x^n = N_{K/F}(y) \,\}}{\{\, \big(N_{K/F}(z), z^n\big) \mid z \in K^\times \,\}} \xrightarrow{\sim} H^1(F, \boldsymbol{\mu}_{n[K]})$$

is given as follows: for $(x,y) \in F^\times \times K^\times$ such that $x^n = N_{K/F}(y)$, choose $\xi \in F^\times_{\mathrm{sep}}$ such that $\xi^n = y$. A cocycle representing the image of (x,y) in $H^1(F, \boldsymbol{\mu}_{n[K]})$ is given by

$$\sigma \mapsto \begin{cases} \sigma(\xi)\xi^{-1} & \text{if } \sigma \in \Gamma_0, \\ x\big(\sigma(\xi)\xi\big)^{-1} & \text{if } \sigma \in \Gamma \smallsetminus \Gamma_0. \end{cases}$$

Similarly, if $K = F \times F$, the isomorphism

$$F^\times / F^{\times n} \xrightarrow{\sim} H^1(F, \boldsymbol{\mu}_n)$$

associates to $x \in F^\times$ the cohomology class of the cocycle $\sigma \mapsto \sigma(\xi)\xi^{-1}$, where $\xi \in F^\times_{\mathrm{sep}}$ is such that $\xi^n = x$.

(30.14) Corollary. *Suppose K is a quadratic separable field extension of F. Let $K^1 = \{\, x \in K^\times \mid N_{K/F}(x) = 1 \,\}$. For every odd integer n which is not divisible by* char F, *there is a canonical isomorphism*

$$H^1(F, \boldsymbol{\mu}_{n[K]}) \simeq K^1/(K^1)^n.$$

Proof: For $(x,y) \in F^\times \times K^\times$ such that $x^n = N_{K/F}(y)$, let

$$\psi(x,y) = y \cdot (x^{\frac{n-1}{2}} y^{-1})^n \in K^1.$$

A computation shows that ψ induces an isomorphism

$$\frac{\{\, (x,y) \in F^\times \times K^\times \mid x^n = N_{K/F}(y) \,\}}{\{\, \big(N_{K/F}(z), z^n\big) \mid z \in K^\times \,\}} \xrightarrow{\sim} K^1/(K^1)^n.$$

The corollary follows from (30.13). (An alternate proof can be derived from (30.12).)
$$\square$$

Finally, we use Corollary (30.12) to relate $H^2(F, \boldsymbol{\mu}_{n[K]})$ to central simple F-algebras with unitary involution with center K.

(30.15) Proposition. *Suppose K is a quadratic separable field extension of F. Let n be an odd integer which is not divisible by* char F. *There is a natural bijection between the group $H^2(F, \boldsymbol{\mu}_{n[K]})$ and the set of Brauer classes of central simple K-algebras of exponent dividing n which can be endowed with a unitary involution whose restriction to F is the identity.*

Proof: The norm map $N_{K/F} \colon {}_n \mathrm{Br}(K) \to {}_n \mathrm{Br}(F)$ corresponds to the corestriction map cor: $H^2(K, \boldsymbol{\mu}_n) \to H^2(F, \boldsymbol{\mu}_n)$ under any of the canonical (opposite) identifications ${}_n \mathrm{Br}(K) = H^2(K, \boldsymbol{\mu}_n)$ (see Riehm [231]). Therefore, by Theorem (3.1), Brauer classes of central simple K-algebras which can be endowed with a unitary involution whose restriction to F is the identity are in one-to-one correspondence with the kernel of the corestriction map. Since n is odd, Corollary (30.12) shows that this kernel can be identified with $H^2(F, \boldsymbol{\mu}_{n[K]})$.   □

(30.16) Definition. Let K be a quadratic separable field extension of F and let n be an odd integer which is not divisible by char F. For any central simple F-algebra with unitary involution (B, τ) of degree n with center K, we denote by $g_2(B, \tau)$ the cohomology class in $H^2(F, \boldsymbol{\mu}_{n[K]})$ corresponding to the Brauer class of B under the bijection of the proposition above, identifying ${}_n \mathrm{Br}(K)$ to $H^2(K, \boldsymbol{\mu}_n)$ by the crossed product construction. This cohomology class is given by

$$g_2(B, \tau) = \tfrac{1}{2} \operatorname{cor}[B] \in H^2(F, \boldsymbol{\mu}_{n[K]}),$$

where cor: $H^2(K, \boldsymbol{\mu}_n) \to H^2(F, \boldsymbol{\mu}_{n[K]})$ is the corestriction[32] map; it is uniquely determined by the condition

$$\operatorname{res}\bigl(g_2(B, \tau)\bigr) = [B] \in H^2(K, \boldsymbol{\mu}_n),$$

where res: $H^2(F, \boldsymbol{\mu}_{n[K]}) \to H^2(K, \boldsymbol{\mu}_n)$ is the restriction map.

From the definition, it is clear that for (B, τ), (B', τ') central simple F-algebras with unitary involution of degree n with the same center K, we have

$$g_2(B, \tau) = g_2(B', \tau') \text{ if and only if } B \simeq B' \text{ as } K\text{-algebras.}$$

Thus, $g_2(B, \tau)$ does not yield any information on the involution τ.

Note that $g_2(B, \tau)$ is the opposite of the Tits class $t(B, \tau)$ defined in a more general situation in (31.8), because we are using here the identification $H^2(K, \boldsymbol{\mu}_n) = {}_n \mathrm{Br}(K)$ afforded by the crossed product construction instead of the identification given by the connecting map of (29.10).

If the center K of B is not a field, then $(B, \tau) \simeq (E \times E^{\mathrm{op}}, \varepsilon)$ for some central simple F-algebra E of degree n, where ε is the exchange involution. In this case, we define a class $g_2(B, \tau) \in H^2(F, \boldsymbol{\mu}_{n[K]}) = H^2(F, \boldsymbol{\mu}_n)$ by

$$g_2(B, \tau) = [E],$$

the cohomology class corresponding to the Brauer class of E by the crossed product construction.

30.C. Cohomological invariants of algebras of degree three. As a first illustration of Galois cohomology techniques, we combine the preceding results with those of Chapter V to obtain cohomological invariants for cubic étale algebras and for central simple algebras with unitary involution of degree 3. Cohomological invariants will be discussed from a more general viewpoint in §31.B.

---

[32]By contrast, observe that $[B]$ is in the kernel of the corestriction map cor: $H^2(K, \boldsymbol{\mu}_n) \to H^2(F, \boldsymbol{\mu}_n)$, by Theorem (3.1).

Étale algebras of degree 3. Cubic étale algebras, i.e., étale algebras of dimension 3, are classified by $H^1(F, S_3)$ (see (29.9)). Let L be a cubic étale F-algebra and let $\phi \colon \Gamma \to S_3$ be a cocycle defining L. Since the Γ-action on S_3 is trivial, the map ϕ is a homomorphism which is uniquely determined by L up to conjugation. We say that L is of type ${}^iS_3$ (for $i = 1, 2, 3$ or 6) if $\mathrm{im}\,\phi \subset S_3$ is a subgroup of order i. Thus,

L is of type ${}^1S_3$ if and only if $L \simeq F \times F \times F$,

L is of type ${}^2S_3$ if and only if $L \simeq F \times K$ for some quadratic separable
field extension K of F,

L is of type ${}^3S_3$ if and only if L is a cyclic field extension of F,

L is of type ${}^6S_3$ if and only if L is a field extension of F which is not Galois.

Let A_3 be the alternating group on 3 elements. The group S_3 is the semidirect product $A_3 \rtimes S_2$, so the exact sequence

$$(30.17) \qquad\qquad 1 \to A_3 \xrightarrow{i} S_3 \xrightarrow{\mathrm{sgn}} S_2 \to 1$$

is split. In the induced sequence in cohomology

$$1 \to H^1(F, A_3) \xrightarrow{i^1} H^1(F, S_3) \xrightarrow{\mathrm{sgn}^1} H^1(F, S_2) \to 1$$

the map sgn^1 associates to an algebra L its discriminant algebra $\Delta(L)$. The (unique) section of sgn^1 is given by $[K] \mapsto [F \times K]$, for any quadratic étale algebra K. The set $H^1(F, A_3)$ classifies Galois A_3-algebras (see (28.15)); it follows from the sequence above (by an argument similar to the one used for $H^1(F, \mathbf{Aut}_Z(B))$ in (29.13)) that they can as well be viewed as pairs (L, ψ) where L is cubic étale over F and ψ is an isomorphism $\Delta(L) \xrightarrow{\sim} F \times F$. The group S_2 acts on $H^1(F, A_3)$ by $(L, \phi) \mapsto (L, \iota \circ \phi)$ where ι is the exchange map. Let K be a quadratic étale F-algebra. We use the associated cocycle $\Gamma \to S_2$ to twist the action of Γ on the sequence (30.17). In the corresponding sequence in cohomology:

$$1 \to H^1(F, A_{3[K]}) \xrightarrow{i^1} H^1(F, S_{3[K]}) \xrightarrow{\mathrm{sgn}^1} H^1(F, S_2) \to 1$$

the distinguished element of $H^1(F, S_{3[K]})$ is the class of $F \times K$ and the pointed set $H^1(F, A_{3[K]})$ classifies pairs (L, ψ) with ψ an isomorphism $\Delta(L) \xrightarrow{\sim} K$. Note that, again, S_2 operates on $H^1(F, A_{3[K]})$. We now define two cohomological invariants for cubic étale F-algebras: $f_1(L) \in H^1(F, S_2)$ is the class of the discriminant algebra $\Delta(L)$ of L, and $g_1(L)$ is the class of L in the orbit space $H^1\big(F, A_{3[\Delta(L)]}\big)/S_2$.

(30.18) Proposition. *Cubic étale algebras are classified by the cohomological invariants $f_1(L)$ and $g_1(L)$. In particular L is of type ${}^1S_3$ if $g_1(L) = 0$ and $f_1(L) = 0$, of type ${}^2S_3$ if $g_1(L) = 0$ and $f_1(L) \neq 0$, of type ${}^3S_3$ if $g_1(L) \neq 0$ and $f_1(L) = 0$, and of type ${}^6S_3$ if $g_1(L) \neq 0$ and $f_1(L) \neq 0$.*

Proof: The fact that cubic étale algebras are classified by the cohomological invariants $f_1(L)$ and $g_1(L)$ follows from the exact sequence above. In particular L is a field if and only if $g_1(L) \neq 0$ and is a cyclic algebra if and only if $f_1(L) = 0$. $\qquad\square$

Let F be a field of characteristic not 3 and let $F(\omega) = F[X]/(X^2 + X + 1)$, where ω is the image of X in the factor ring. (Thus, $F(\omega) \simeq F(\sqrt{-3})$ if $\mathrm{char}\,F \neq 2$.) We have $\boldsymbol{\mu}_3 = A_{3[F(\omega)]}$ so that $H^1(F, \boldsymbol{\mu}_3)$ classifies pairs (L, ψ) where L is a cubic étale F-algebra and ψ is an isomorphism $\Delta(L) \xrightarrow{\sim} F(\omega)$. The action of S_2 interchanges the pairs (L, ψ) and $(L, \iota_{F(\omega)} \circ \psi)$. In particular $H^1(F, \boldsymbol{\mu}_3)$ modulo the action of S_2 classifies cubic étale F-algebras with discriminant algebra $F(\omega)$.

(30.19) Proposition. *Let K be a quadratic étale F-algebra and let $x \in K^1$ be an element of norm 1 in K. Let $K(\sqrt[3]{x}) = K[t]/(t^3 - x)$ and let $\xi = \sqrt[3]{x}$ be the image of t in $K(\sqrt[3]{x})$. Extend the nontrivial automorphism ι_K to an automorphism ι of $K(\sqrt[3]{x})$ by setting $\iota(\xi) = \xi^{-1}$. Then, the F-algebra $L = \{ u \in K(\sqrt[3]{x}) \mid \iota(u) = u \}$ is a cubic étale F-algebra with discriminant algebra $K * F(\omega)$. Conversely, suppose L is a cubic étale F-algebra with discriminant $\Delta(L)$, and let $K = F(\omega) * \Delta(L)$. Then, there exists $x \in K^1$ such that*

$$L \simeq \{ u \in K(\sqrt[3]{x}) \mid \iota(u) = u \}.$$

Proof: If $x = \pm 1$, then $K(\sqrt[3]{x}) \simeq K \times K(\omega)$ and $L \simeq F \times (K * F(\omega))$, hence the first assertion is clear. Suppose $x \neq \pm 1$. Since every element in $K(\sqrt[3]{x})$ has a unique expression of the form $a + b\xi + c\xi^{-1}$ with a, b, $c \in K$, it is easily seen that $L = F(\eta)$ with $\eta = \xi + \xi^{-1}$. We have

$$\eta^3 - 3\eta = x + x^{-1}$$

with $x + x^{-1} \neq \pm 2$, hence Proposition (18.32) shows that $\Delta(L) \simeq K * F(\omega)$. This completes the proof of the first assertion.

To prove the second assertion, we use the fact that cubic étale F-algebras with discriminant $\Delta(L)$ are in one-to-one correspondence with the orbit space $H^1(F, A_{3[\Delta(L)]})/S_2$. For $K = F(\omega) * \Delta(L)$, we have $A_{3[\Delta(L)]} = \boldsymbol{\mu}_{3[K]}$, hence Corollary (30.14) yields a bijection $H^1(F, A_{3[\Delta(L)]}) \simeq K^1/(K^1)^3$. If $x \in K^1$ corresponds to the isomorphism class of L, then $L \simeq \{ u \in K(\sqrt[3]{x}) \mid \iota(u) = u \}$. $\qquad\square$

Central simple algebras with unitary involution. To every central simple F-algebra with unitary involution (B, τ) we may associate the cocycle $[K]$ of its center K. We let

$$f_1(B, \tau) = [K] \in H^1(F, S_2).$$

Now, assume $\operatorname{char} F \neq 2, 3$ and let (B, τ) be a central simple F-algebra with unitary involution of degree 3. Let K be the center of B. A secondary invariant $g_2(B, \tau) \in H^2(F, \boldsymbol{\mu}_{3[K]})$ is defined in (30.16). The results in §19 show that $g_2(B, \tau)$ has a special form:

(30.20) Proposition. *For any central simple F-algebra with unitary involution (B, τ) of degree 3 with center K, there exist $\alpha \in H^1(F, \mathbb{Z}/3\mathbb{Z}_{[K]})$ and $\beta \in H^1(F, \boldsymbol{\mu}_3)$ such that*

$$g_2(B, \tau) = \alpha \cup \beta.$$

Proof: Suppose first that $K \simeq F \times F$. Then $(B, \tau) \simeq (E \times E^{\mathrm{op}}, \varepsilon)$ for some central simple F-algebra E of degree 3, where ε is the exchange involution, and $g_2(B, \tau) = [E]$. Wedderburn's theorem on central simple algebras of degree 3 (see (19.2)) shows that E is cyclic, hence Proposition (30.4) yields the required decomposition of $g_2(B, \tau)$.

Suppose next that K is a field. Albert's theorem on central simple algebras of degree 3 with unitary involution (see (19.14)) shows that B contains a cubic étale F-algebra L with discriminant isomorphic to K. By (30.18), we may find a corresponding cocycle $\alpha \in H^1(F, \mathbb{Z}/3\mathbb{Z}_{[K]}) = H^1(F, A_{3[K]})$ (whose orbit under S_2 is $g_1(L)$). The K-algebra $L_K = L \otimes_F K$ is cyclic and splits B, hence by (30.4)

$$[B] = \mathrm{res}(\alpha) \cup (b) \in H^2(K, \boldsymbol{\mu}_3)$$

for some $(b) \in H^1(K, \boldsymbol{\mu}_3)$, where $\mathrm{res}\colon H^1(F, \mathbb{Z}/3\mathbb{Z}_{[K]}) \to H^1(K, \mathbb{Z}/3\mathbb{Z})$ is the restriction map. By taking the image of each side under the corestriction map $\mathrm{cor}\colon H^2(K, \boldsymbol{\mu}_3) \to H^2(F, \boldsymbol{\mu}_{3[K]})$, we obtain by the projection (or transfer) formula (see Brown [56, (3.8), p. 112])

$$\mathrm{cor}[B] = \alpha \cup \mathrm{cor}(b)$$

hence $g_2(B, \tau) = \alpha \cup \beta$ with $\beta = \frac{1}{2}\mathrm{cor}(b)$. (Here, $\mathrm{cor}(b) = \big(N_{K/F}(b)\big)$ is the image of (b) under the corestriction map $\mathrm{cor}\colon H^1(K, \boldsymbol{\mu}_3) \to H^1(F, \boldsymbol{\mu}_3)$.) $\square$

The 3-fold Pfister form $\pi(\tau)$ of (19.4) yields a third cohomological invariant of (B, τ) via the map e_3 of (30.2). We set

$$f_3(B, \tau) = e_3\big(\pi(\tau)\big) \in H^3(F, \boldsymbol{\mu}_2).$$

This is a Rost invariant in the sense of §31.B, see (31.45).

Since the form $\pi(\tau)$ classifies the unitary involutions on B up to isomorphism by (19.6), we have a complete set of cohomological invariants for central simple F-algebras with unitary involution of degree 3:

(30.21) Theorem. *Let F be a field of characteristic different from* 2, 3. *Triples* (B, K, τ), *where K is a quadratic separable field extension of F and (B, τ) is a central simple F-algebra with unitary involution of degree 3 with center K, are classified over F by the three cohomological invariants $f_1(B, \tau)$, $g_2(B, \tau)$, and $f_3(B, \tau)$.* $\square$

§31. Cohomological Invariants

In this section, we show how cohomology can be used to define various canonical maps and to attach invariants to algebraic groups. In §31.A, we use cohomology sequences to relate multipliers and spinor norms to connecting homomorphisms. We also define the Tits class of a simply connected semisimple group; it is an invariant of the group in the second cohomology group of its center. In §31.B, we take a systematic approach to the definition of cohomological invariants of algebraic groups and define invariants of dimension 3.

Unless explicitly mentioned, the base field F is arbitrary throughout this section. However, when using the cohomology of $\boldsymbol{\mu}_n$, we will need to assume that char F does not divide n.

31.A. Connecting homomorphisms. Let G be a simply connected semisimple group with center C and let $\overline{G} = G/C$. The exact sequence

(31.1) $1 \to C \to G \to \overline{G} \to 1$

yields connecting maps in cohomology

$$\delta^0\colon H^0(F, \overline{G}) \to H^1(F, C) \qquad \text{and} \qquad \delta^1\colon H^1(F, \overline{G}) \to H^2(F, C).$$

We will give an explicit description of δ^0 for each type of classical group and use δ^1 to define the *Tits class* of G.

Unitary groups. Let (B, τ) be a central simple F-algebra with unitary involution of degree n, and let K be the center of B. As observed in §23.A, we have an exact sequence of group schemes

$$(31.2) \qquad 1 \to \boldsymbol{\mu}_{n[K]} \to \mathbf{SU}(B, \tau) \to \mathbf{PGU}(B, \tau) \to 1$$

since the kernel N of the norm map $R_{K/F}(\boldsymbol{\mu}_{n,K}) \to \boldsymbol{\mu}_{n,F}$ is $\boldsymbol{\mu}_{n[K]}$.

Suppose $\mathrm{char}\, F$ does not divide n, so that $\boldsymbol{\mu}_{n[K]}$ is smooth. By Proposition (22.15), we derive from (31.2) an exact sequence of Galois modules. The connecting map

$$\delta^0 \colon \mathrm{PGU}(B, \tau) \to H^1(F, \boldsymbol{\mu}_{n[K]})$$

can be described as follows: for $g \in \mathrm{GU}(B, \tau)$,

$$\delta^0(g \cdot K^\times) = \big[\mu(g), \mathrm{Nrd}_B(g)\big]$$

where $\mu(g) = \tau(g)g \in F^\times$ is the multiplier of g, and $[x, y]$ is the image of $(x, y) \in F^\times \times K^\times$ in the factor group

$$\frac{\{\, (x, y) \in F^\times \times K^\times \mid x^n = N_{K/F}(y)\,\}}{\{\, \big(N_{K/F}(z), z^n\big) \mid z \in K^\times \,\}} \simeq H^1(F, \boldsymbol{\mu}_{n[K]})$$

(see (30.13)).

If $K \simeq F \times F$, then $(B, \tau) \simeq (A \times A^{\mathrm{op}}, \varepsilon)$ for some central simple F-algebra A of degree n, where ε is the exchange involution. We have $\mathbf{PGU}(B, \tau) \simeq \mathbf{PGL}_1(A)$ and $\mathbf{SU}(B, \tau) \simeq \mathbf{SL}_1(A)$, and the exact sequence (31.2) takes the form

$$1 \to \boldsymbol{\mu}_n \to \mathbf{SL}_1(A) \to \mathbf{PGL}_1(A) \to 1.$$

The connecting map $\delta^0 \colon \mathrm{PGL}_1(A) \to H^1(F, \boldsymbol{\mu}_n) = F^\times/F^{\times n}$ is given by the reduced norm map.

Orthogonal groups. For groups of type B_n, the exact sequence (31.1) takes the form

$$1 \to \boldsymbol{\mu}_2 \to \mathbf{Spin}(V, q) \to \mathbf{O}^+(V, q) \to 1$$

where (V, q) is a quadratic space of dimension $2n + 1$ (see §23.A). If $\mathrm{char}\, F \neq 2$, the connecting map $\delta^0 \colon \mathrm{O}^+(V, q) \to H^1(F, \boldsymbol{\mu}_2) = F^\times/F^{\times 2}$ associated to this exact sequence is the spinor norm. (The same result holds if $\dim V$ is even.)

We next consider groups of type D_n. Let (A, σ, f) be a central simple F-algebra of even degree $2n$ with quadratic pair. The center C of the spin group $\mathbf{Spin}(A, \sigma, f)$ is determined in §26.A: if Z is the center of the Clifford algebra $C(A, \sigma, f)$, we have

$$C = \begin{cases} R_{Z/F}(\boldsymbol{\mu}_2) & \text{if } n \text{ is even,} \\ \boldsymbol{\mu}_{4[Z]} & \text{if } n \text{ is odd.} \end{cases}$$

Therefore, the exact sequence (31.1) takes the form

$$1 \to R_{Z/F}(\boldsymbol{\mu}_2) \to \mathbf{Spin}(A, \sigma, f) \to \mathbf{PGO}^+(A, \sigma, f) \to 1 \quad \text{if } n \text{ is even,}$$

$$1 \to \boldsymbol{\mu}_{4[Z]} \to \mathbf{Spin}(A, \sigma, f) \to \mathbf{PGO}^+(A, \sigma, f) \to 1 \quad \text{if } n \text{ is odd.}$$

Suppose $\mathrm{char}\, F \neq 2$. The connecting maps δ^0 in the associated cohomology sequences are determined in §13.B. If n is even, the map

$$\delta^0 \colon \mathrm{PGO}^+(A, \sigma, f) \to H^1\big(F, R_{Z/F}(\boldsymbol{\mu}_2)\big) = H^1(Z, \boldsymbol{\mu}_2) = Z^\times/Z^{\times 2}$$

coincides with the map S of (13.32), see Proposition (13.34). If n is odd, the map

$$\delta^0 \colon \mathrm{PGO}^+(A, \sigma, f) \to H^1(F, \boldsymbol{\mu}_{4[Z]})$$

is defined in (13.35), see Proposition (13.37). Note that the discussion in §13.B does *not* use the hypothesis that char $F \neq 2$. This hypothesis is needed here because we apply (22.15) to derive exact sequences of Galois modules from the exact sequences of group schemes above.

Symplectic groups. Let (A, σ) be a central simple F-algebra with symplectic involution. We have an exact sequence of group schemes

$$1 \to \boldsymbol{\mu}_2 \to \mathbf{Sp}(A, \sigma) \to \mathbf{PGSp}(A, \sigma) \to 1$$

(see §23.A). If char $F \neq 2$, the connecting homomorphism

$$\delta^0 \colon \mathrm{PGSp}(A, \sigma) \to H^1(F, \boldsymbol{\mu}_2) = F^\times / F^{\times 2}$$

is induced by the multiplier map: it takes $g \cdot F^\times \in \mathrm{PGSp}(A, \sigma)$ to $\mu(g) \cdot F^{\times 2}$.

The Tits class. Let G be a split simply connected or adjoint semisimple group over F, let $T \subset G$ be a split maximal torus, Π a system of simple roots in the root system of G with respect to T. By (25.15) and (25.16), the homomorphism

(31.3) $\mathrm{Aut}(G) \to \mathrm{Aut}\big(\mathrm{Dyn}(G)\big)$

is a split surjection. A splitting $i\colon \mathrm{Aut}\big(\mathrm{Dyn}(G)\big) \to \mathrm{Aut}(G)$ can be chosen in such a way that any automorphism in the image of i leaves the torus T invariant. Assume that the Galois group Γ acts on the Dynkin diagram $\mathrm{Dyn}(G)$, or equivalently, consider a continuous homomorphism

$$\varphi \in \mathrm{Hom}\big(\Gamma, \mathrm{Aut}(\mathrm{Dyn}(G))\big) = H^1\big(F, \mathrm{Aut}(\mathrm{Dyn}(G))\big).$$

Denote by γ a cocycle representing the image of φ in $H^1\big(F, \mathrm{Aut}(G_{\mathrm{sep}})\big)$ under the map induced by the splitting i. Since γ normalizes T, the twisted group G_γ contains the maximal torus T_γ. Moreover, the natural action of Γ on T_γ leaves Π invariant, hence G_γ is a quasisplit group. In fact, up to isomorphism G_γ is the unique simply connected quasisplit group with Dynkin diagram $\mathrm{Dyn}(G)$ and with the given action of Γ on $\mathrm{Dyn}(G)$ (see (27.8)). Twisting G in (31.3) by γ, we obtain:

(31.4) Proposition. *Let G be a quasisplit simply connected group. Then the natural homomorphism $\mathrm{Aut}(G) \to \mathrm{Aut}\big(\mathrm{Dyn}(G)\big)$ is surjective.* $\qquad\square$

Let G be semisimple group over F. By §29 a twisted form G' of G corresponds to an element $\xi \in H^1\big(F, \mathrm{Aut}(G_{\mathrm{sep}})\big)$. We say that G' is an *inner form of G* if ξ belongs to the image of the map

$$\alpha_G \colon H^1(F, \overline{G}) \longrightarrow H^1\big(F, \mathrm{Aut}(G_{\mathrm{sep}})\big)$$

induced by the homomorphism $\mathrm{Int}\colon \overline{G} = G/C \to \mathrm{Aut}(G)$, where C is the center of G. Since $\overline{G}$ acts trivially on C, the centers of G and G' are isomorphic (as group schemes of multiplicative type).

(31.5) Proposition. *Any semisimple group is an inner twisted form of a unique quasisplit group up to isomorphism.*

Proof: Since the centers of inner twisted forms are isomorphic and all the groups which are isogenous to a simply connected group correspond to subgroups in its center, we may assume that the given group G is simply connected. Denote by G^d the split twisted form of G, so that G corresponds to some element $\rho \in$

$H^1\big(F, \mathrm{Aut}(G_{\mathrm{sep}}^d)\big)$. Denote by $\gamma \in H^1\big(F, \mathrm{Aut}(G_{\mathrm{sep}}^d)\big)$ the image of ρ under the composition induced by

$$\mathrm{Aut}(G^d) \to \mathrm{Aut}\big(\mathrm{Dyn}(G^d)\big) \xrightarrow{i} \mathrm{Aut}(G^d),$$

where i is the splitting considered above. We prove that the quasisplit group G_γ^d is an inner twisted form of $G = G_\rho^d$. By (28.8), there is a bijection

$$\theta_\rho \colon H^1\big(F, \mathrm{Aut}(G_{\mathrm{sep}})\big) \xrightarrow{\sim} H^1\big(F, \mathrm{Aut}(G_{\mathrm{sep}}^d)\big)$$

taking the trivial cocycle to ρ. Denote by γ_0 the element in $H^1\big(F, \mathrm{Aut}(G_{\mathrm{sep}})\big)$ such that $\theta(\gamma_0) = \gamma$. Since ρ and γ have the same image in $H^1\big(F, \mathrm{Aut}\big(\mathrm{Dyn}(G_{\mathrm{sep}}^d)\big)\big)$, the trivial cocycle and γ_0 have the same images in $H^1\big(F, \mathrm{Aut}\big(\mathrm{Dyn}(G_{\mathrm{sep}})\big)\big)$, hence Theorem (25.16) shows that γ_0 belongs to the image of $H^1(F, \overline{G}) \to H^1\big(F, \mathrm{Aut}(G_{\mathrm{sep}})\big)$, i.e., the group $G_{\gamma_0} \simeq G_\gamma^d$ is a quasisplit inner twisted form of G. $\qquad\square$

Until the end of the subsection we shall assume that G is a simply connected semisimple group. Denote by ξ_G the element in $H^1\big(F, \mathrm{Aut}(G_{\mathrm{sep}})\big)$ corresponding to the (unique) quasisplit inner twisted form of G. In general, the map α_G is not injective. Nevertheless, we have

(31.6) Proposition. *There is only one element $\nu_G \in H^1(F, \overline{G})$ such that*

$$\alpha_G(\nu_G) = \xi_G.$$

Proof: Denote G^q the quasisplit inner twisted form of G. By (28.10), there is a bijection between $\alpha_G^{-1}(\xi_G)$ and the factor group of $\mathrm{Aut}\big(\mathrm{Dyn}(G^q)\big)$ by the image of $\mathrm{Aut}(G^q) \to \mathrm{Aut}\big(\mathrm{Dyn}(G^q)\big)$. But the latter map is surjective (see (31.4)). $\qquad\square$

Let C be the center of G. The exact sequence $1 \to C \to G \to \overline{G} \to 1$ induces the connecting map

$$\delta^1 \colon H^1(F, \overline{G}) \longrightarrow H^2(F, C).$$

The *Tits class* of G is the element $t_G = -\delta^1(\nu_G) \in H^2(F, C)$.

(31.7) Proposition. *Let $\chi \in C^*$ be a character. Denote by F_χ the field of definition of χ and by A_χ its minimal Tits algebra. The image of the Tits class t_G under the composite map*

$$H^2(F, C) \xrightarrow{\mathrm{res}} H^2(F_\chi, C) \xrightarrow{\chi_*} H^2(F_\chi, \mathbf{G}_\mathrm{m}) = \mathrm{Br}(F_\chi)$$

is $[A_\chi]$. (We use the canonical identification $H^2(F_\chi, \mathbf{G}_\mathrm{m}) = \mathrm{Br}(F_\chi)$ given by the connecting map of (29.10), which is the opposite of the identification given by the crossed product construction.)

Proof: There is a commutative diagram (see §27)

$$
\begin{array}{ccccccc}
1 & \longrightarrow & C_{F_\chi} & \longrightarrow & G_{F_\chi} & \longrightarrow & \overline{G}_{F_\chi} & \longrightarrow & 1 \\
 & & \chi\downarrow & & \downarrow & & \downarrow & & \\
1 & \longrightarrow & \mathbf{G}_\mathrm{m} & \longrightarrow & \mathbf{GL}_1(A) & \longrightarrow & \mathbf{PGL}_1(A) & \longrightarrow & 1
\end{array}
$$

where $A = A_\chi$. Therefore, it suffices to prove that the image of $\mathrm{res}(\nu_G)$ under the composite map

$$H^1(F_\chi, \overline{G}_{F_\chi}) \to H^1\big(F_\chi, \mathbf{PGL}_1(A)\big) \to H^2(F_\chi, \mathbf{G}_\mathrm{m}) = \mathrm{Br}(F_\chi)$$

is $-[A]$. The twist of the algebra A by a cocycle representing $\mathrm{res}(\nu_G)$ is the Tits algebra of the quasisplit group $(G_{\nu_G})_{F_\chi}$, hence it is trivial. Therefore the image γ of $\mathrm{res}(\nu_G)$ in $H^1\big(F_\chi, \mathbf{PGL}_1(A)\big)$ corresponds to the split form $\mathbf{PGL}_n(F_\chi)$ of $\mathbf{PGL}_1(A)$, where $n = \deg(A)$. By (28.12), there is a commutative square

$$
\begin{array}{ccc}
H^1\big(F_\chi, \mathbf{PGL}_n(F_\chi)\big) & \longrightarrow & \mathrm{Br}(F_\chi) \\
\theta_\gamma \downarrow & & \downarrow g \\
H^1\big(F_\chi, \mathbf{PGL}_1(A)\big) & \longrightarrow & \mathrm{Br}(F_\chi)
\end{array}
$$

where $g(v) = v + u$ and u is the image of γ in $\mathrm{Br}(F_\chi)$. Pick $\varepsilon \in Z^1\big(F_\chi, \mathbf{PGL}_n(F_\chi)\big)$ such that the twisting of $\mathbf{PGL}_n(F_\chi)$ by ε equals $\mathbf{PGL}_1(A)$. In other words, $\theta_\gamma(\varepsilon) = 1$ and the image of ε in $\mathrm{Br}(F_\chi)$ equals $[A]$. The commutativity of the diagram then implies that $u = -[A]$. $\qquad\square$

(31.8) Example. Let $G = \mathbf{SU}(B, \tau)$ where (B, τ) is a central simple F-algebra with unitary involution of degree $n > 2$. Assume that char F does not divide n and let K be the center of B. Since the center C of $\mathbf{SU}(B, \tau)$ is $\boldsymbol{\mu}_{n[K]}$, the Tits class t_G belongs to $H^2(F, \boldsymbol{\mu}_{n[K]})$. Abusing terminology, we call it the *Tits class* of (B, τ) and denote it by $t(B, \tau)$, i.e.,

$$
t(B, \tau) = t_{\mathbf{SU}(B,\tau)} \in H^2(F, \boldsymbol{\mu}_{n[K]}).
$$

Suppose K is a field. For $\chi = 1 + n\mathbb{Z} \in \mathbb{Z}/n\mathbb{Z} = C^*$, the field of definition of χ is K and the minimal Tits algebra is B, see §27.B. Therefore, Proposition (31.7) yields

$$
\mathrm{res}_{K/F}\big(t(B, \tau)\big) = [B] \in \mathrm{Br}(K).
$$

If n is odd, it follows from Corollary (30.12) that

$$
t(B, \tau) = \tfrac{1}{2} \mathrm{cor}[B].
$$

On the other hand, if n is even we may consider the character $\lambda = \frac{n}{2} + n\mathbb{Z} \in C^*$ of raising to the power $\frac{n}{2}$. The corresponding minimal Tits algebra is the discriminant algebra $D(B, \tau)$, see §27.B. By (31.7) we obtain

$$
\lambda_*\big(t(B, \tau)\big) = \big[D(B, \tau)\big] \in \mathrm{Br}(F).
$$

If $K \simeq F \times F$ we have $(B, \tau) = (A \times A^{\mathrm{op}}, \varepsilon)$ for some central simple F-algebra A of degree n, where ε is the exchange involution. The commutative diagram with exact rows

$$
\begin{array}{ccccccccc}
1 & \longrightarrow & \boldsymbol{\mu}_n & \longrightarrow & \mathbf{SL}_n & \longrightarrow & \mathbf{PGL}_n & \longrightarrow & 1 \\
& & \downarrow & & \downarrow & & \| & & \\
1 & \longrightarrow & \mathbf{G}_{\mathrm{m}} & \longrightarrow & \mathbf{GL}_n & \longrightarrow & \mathbf{PGL}_n & \longrightarrow & 1
\end{array}
$$

shows that $t(B, \tau) = [A] \in H^2(F, \boldsymbol{\mu}_n) = {}_n\mathrm{Br}(F)$.

(31.9) Example. Let $G = \mathbf{Spin}(V, q)$, where (V, q) is a quadratic space of odd dimension. Suppose char $F \neq 2$. The center of G is $\boldsymbol{\mu}_2$ and the Tits class $t_G \in H^2(F, \boldsymbol{\mu}_2) = {}_2\mathrm{Br}(F)$ is the Brauer class of the even Clifford algebra $C_0(V, q)$, since the minimal Tits algebra for the nontrivial character is $C_0(V, q)$, see §27.B.

(31.10) Example. Let $G = \mathbf{Sp}(A, \sigma)$, where (A, σ) is a central simple F-algebra with symplectic involution. Suppose char $F \neq 2$. The center of G is $\boldsymbol{\mu}_2$ and the Tits class $t_G \in H^2(F, \boldsymbol{\mu}_2) = {}_2\mathrm{Br}(F)$ is the Brauer class of the algebra A, see §27.B.

(31.11) Example. Let $G = \mathbf{Spin}(A, \sigma, f)$, where (A, σ, f) is a central simple F-algebra of even degree $2n$ with quadratic pair. Let Z be the center of the Clifford algebra $C(A, \sigma, f)$ and assume that char $F \neq 2$.

Suppose first that n is even. Then the center of G is $R_{Z/F}(\boldsymbol{\mu}_2)$, hence

$$t_G \in H^2\big(F, R_{Z/F}(\boldsymbol{\mu}_2)\big) = H^2(Z, \boldsymbol{\mu}_2) = {}_2\mathrm{Br}(Z).$$

The minimal Tits algebra corresponding to the norm character

$$\lambda \colon R_{Z/F}(\boldsymbol{\mu}_2) \to \boldsymbol{\mu}_2 \hookrightarrow \mathbf{G}_m$$

is A, hence

(31.12) $\mathrm{cor}_{Z/F}(t_G) = [A] \in H^2(F, \boldsymbol{\mu}_2) = {}_2\mathrm{Br}(F).$

On the other hand, the minimal Tits algebras for the two other nontrivial characters $\lambda^\pm$ are $C(A, \sigma, f)$ (see §27.B), hence

(31.13) $t_G = \big[C(A, \sigma, f)\big] \in H^2(Z, \boldsymbol{\mu}_2) = {}_2\mathrm{Br}(Z).$

Now, assume that n is odd. Then the center of G is $\boldsymbol{\mu}_{4[Z]}$, hence

$$t_G \in H^2(F, \boldsymbol{\mu}_{4[Z]}).$$

By applying Proposition (31.7), we can compute the image of t_G under the squaring map

$$\lambda_* \colon H^2(F, \boldsymbol{\mu}_{4[Z]}) \to H^2(F, \boldsymbol{\mu}_2) = {}_2\mathrm{Br}(F)$$

and under the restriction map

$$\mathrm{res} \colon H^2(F, \boldsymbol{\mu}_{4[Z]}) \to H^2(Z, \boldsymbol{\mu}_4) = {}_4\mathrm{Br}(Z)$$

(or, equivalently, under the map $H^2(F, \boldsymbol{\mu}_{4[Z]}) \to H^2\big(F, R_{Z/F}(\mathbf{G}_{m,Z})\big) = \mathrm{Br}(Z)$ induced by the inclusion $\boldsymbol{\mu}_{4[Z]} \hookrightarrow R_{Z/F}(\mathbf{G}_{m,Z})$). We obtain

$$\lambda_*(t_G) = [A] \qquad \text{and} \qquad \mathrm{res}(t_G) = \big[C(A, \sigma, f)\big].$$

Note that the fundamental relations (9.12) of Clifford algebras readily follow from the computations above (under the hypothesis that char $F \neq 2$). If n is even, (31.13) shows that $\big[C(A, \sigma, f)\big]^2 = 1$ in $\mathrm{Br}(Z)$ and (31.12) (together with (31.13)) implies

$$\big[N_{Z/F}\big(C(A, \sigma, f)\big)\big] = [A].$$

If n is odd we have

$$[A_Z] = \mathrm{res} \circ \lambda_*(t_G) = \mathrm{res}(t_G)^2 = \big[C(A, \sigma, f)\big]^2$$

and

$$\big[N_{Z/F}\big(C(A, \sigma, f)\big)\big] = \mathrm{cor} \circ \mathrm{res}(t_G) = 0,$$

by (30.12). (See Exercise 15 for a cohomological proof of the fundamental relations without restriction on char F.)

31.B. Cohomological invariants of algebraic groups. The final part of this chapter has the character of survey. Many statements appear here without proof.

Let G be an algebraic group over a field F. For any field extension E of F we consider the pointed set

$$H^1(E, G) = H^1(E, G_E)$$

of G_E-torsors over E. A homomorphism $E \to L$ of fields over F induces a map of pointed sets

$$H^1(E, G) \to H^1(L, G).$$

Thus, $H^1(?, G)$ is a functor from the category of field extensions of F (with morphisms being field homomorphisms over F) to the category of pointed sets.

Let M be a torsion discrete Galois module over F, i.e., a discrete module over the absolute Galois group $\Gamma = \mathrm{Gal}(F_{\mathrm{sep}}/F)$. For a field extension E of F, M can be endowed with a structure of a Galois module over E and hence the ordinary cohomology groups $H^d(E, M)$ are defined. Thus, for any $d \geq 0$, we obtain a functor $H^d(?, M)$ from the category of field extensions of F to the category of pointed sets (actually, the category of abelian groups). A *cohomological invariant of the group G of dimension d with coefficients in M* is a natural transformation of functors

$$a \colon H^1(?, G) \to H^d(?, M).$$

In other words, the cohomological invariant a assigns to any field extension E of F a map of pointed sets

$$a_E \colon H^1(E, G) \to H^d(E, M),$$

such that for any field F-homomorphism $E \to L$ the following diagram commutes

$$
\begin{array}{ccc}
H^1(E, G) & \xrightarrow{\ a_E\ } & H^d(E, M) \\
\downarrow & & \downarrow \\
H^1(L, G) & \xrightarrow{\ a_L\ } & H^d(L, M).
\end{array}
$$

The set $\mathrm{Inv}^d(G, M)$ of all cohomological invariants of the group G of dimension d with coefficients in M forms an abelian group in a natural way.

(31.14) Example. Let $G = \mathbf{GL}(V)$ or $\mathbf{SL}(V)$. By Hilbert's Theorem 90 (see (29.2) and (29.4)) we have $H^1(E, G) = 1$ for any field extension L of F. Hence $\mathrm{Inv}^d(G, M) = 0$ for any d and any Galois module M.

A group homomorphism $\alpha \colon G \to G'$ over F induces a natural homomorphism

$$\alpha^* \colon \mathrm{Inv}^d(G', M) \to \mathrm{Inv}^d(G, M).$$

A homomorphism of Galois modules $g \colon M \to M'$ yields a group homomorphism

$$g_* \colon \mathrm{Inv}^d(G, M) \to \mathrm{Inv}^d(G, M').$$

For a field extension L of F there is a natural *restriction homomorphism*

$$\mathrm{res} \colon \mathrm{Inv}^d(G, M) \to \mathrm{Inv}^d(G_L, M).$$

Let L be a finite separable extension of F. If G is an algebraic group over L and M is a Galois module over F, then the corestriction homomorphism for cohomology groups and Shapiro's lemma yield the *corestriction homomorphism*

$$\text{cor} \colon \text{Inv}^d(G, M) \to \text{Inv}^d\big(R_{L/F}(G), M\big).$$

Invariants of dimension 1. Let G be an algebraic group over a field F. As in §20, write $\pi_0(G)$ for the factor group of G modulo the connected component of G. It is an étale group scheme over F.

Let M be a discrete Galois module over F and let

$$g \colon \pi_0(G_{\text{sep}}) \to M$$

be a Γ-homomorphism. For any field extension E of F we then have the following composition

$$a_E^g \colon H^1(E, G) \to H^1\big(E, \pi_0(G)\big) \to H^1(E, M)$$

where the first map is induced by the canonical surjection $G \to \pi_0(G)$ and the second one by g. We can view a^g as an invariant of dimension 1 of the group G with coefficients in M.

(31.15) Proposition. *The map*

$$\text{Hom}_\Gamma\big(\pi_0(G_{\text{sep}}), M\big) \to \text{Inv}^1(G, M) \quad \text{given by } g \mapsto a^g$$

is an isomorphism. $\qquad\qquad\qquad\qquad\qquad\qquad\qquad\qquad\qquad\qquad\qquad\qquad\square$

In particular, a connected group has no nonzero invariants of dimension 1.

(31.16) Example. Let (A, σ, f) be a central simple F-algebra with quadratic pair of degree $2n$ and let $G = \mathbf{PGO}(A, \sigma, f)$ be the corresponding projective orthogonal group. The set $H^1(F, G)$ classifies triples (A', σ', f') with a central simple F-algebra A' of degree $2n$ with an quadratic pair (σ', f') (see §29.F). We have $\pi_0(G) \simeq \mathbb{Z}/2\mathbb{Z}$ and the group $\text{Inv}^1(G, \mathbb{Z}/2\mathbb{Z})$ is isomorphic to $\mathbb{Z}/2\mathbb{Z}$. The nontrivial invariant

$$H^1(F, G) \to H^1(F, \mathbb{Z}/2\mathbb{Z})$$

associates to any triple (A', σ', f') the sum $[Z'] + [Z]$ of corresponding classes of the discriminant quadratic extensions.

(31.17) Example. Let K be a quadratic étale F-algebra, let (B, τ) be a central simple K-algebra of degree n with a unitary involution and set $G = \mathbf{Aut}(B, \tau)$. The set $H^1(F, G)$ classifies algebras of degree n with a unitary involution (see (29.D)). Then $\pi_0(G) \simeq \mathbb{Z}/2\mathbb{Z}$ and, as in the previous example, the group $\text{Inv}^1(G, \mathbb{Z}/2\mathbb{Z})$ is isomorphic to $\mathbb{Z}/2\mathbb{Z}$. The nontrivial invariant

$$H^1(F, G) \to H^1(F, \mathbb{Z}/2\mathbb{Z})$$

associates to any central simple F-algebra with unitary involution (B', τ') with center K' the class $[K'] + [K]$.

Invariants of dimension 2. For any natural numbers i and n let $\mu_n^{\otimes i}(F)$ be the i-th tensor power of the group $\mu_n(F)$. If n divides m, there is a natural injection

$$\mu_n^{\otimes i}(F) \to \mu_m^{\otimes i}(F).$$

The groups $\mu_n^{\otimes i}(F)$ form an injective system with respect to the family of injections defined above. We denote the direct limit of this system, for all n prime to the characteristic of F, by $\mathbb{Q}/\mathbb{Z}(i)(F)$. For example, $\mathbb{Q}/\mathbb{Z}(1)(F)$ is the group of all roots of unity in F.

The group $\mathbb{Q}/\mathbb{Z}(i)(F_{\mathrm{sep}})$ is endowed in a natural way with a structure of a Galois module. We set

$$H^d\big(F, \mathbb{Q}/\mathbb{Z}(i)\big) = H^d\big(F, \mathbb{Q}/\mathbb{Z}(i)(F_{\mathrm{sep}})\big).$$

In the case where $\operatorname{char} F = p > 0$, this group can be modified by adding an appropriate p-component. In particular, the group $H^2\big(F, \mathbb{Q}/\mathbb{Z}(1)\big)$ is canonically isomorphic to $\operatorname{Br}(F)$, while before the modification it equals $\varinjlim H^2\big(F, \mu_n(F_{\mathrm{sep}})\big)$ which is the subgroup of elements in $\operatorname{Br}(F)$ of exponent prime to p.

Let G be a connected algebraic group over a field F. Assume that we are given an exact sequence of algebraic groups

(31.18) $1 \to \mathbf{G}_{\mathrm{m},F} \to G' \to G \to 1.$

For any field extension E of F, this sequence induces a connecting map $H^1(E, G) \to H^2(E, \mathbf{G}_{\mathrm{m},F})$ which, when composed with the identifications

$$H^2(E, \mathbf{G}_{\mathrm{m},F}) = \operatorname{Br}(E) = H^2\big(E, \mathbb{Q}/\mathbb{Z}(1)\big),$$

provides an invariant a_E of dimension 2 of the group G. On the other hand, the sequence (31.18) defines an element of the Picard group $\operatorname{Pic}(G)$. It turns out that the invariant a depends only on the element of the Picard group and we have a well defined group homomorphism

$$\beta \colon \operatorname{Pic}(G) \to \operatorname{Inv}^2\big(G, \mathbb{Q}/\mathbb{Z}(1)\big).$$

(31.19) Proposition. *The map β is an isomorphism.* $\qquad\square$

Since the n-torsion part of $\mathbb{Q}/\mathbb{Z}(1)$ equals μ_n, we have

(31.20) Corollary. *If n is not divisible by* $\operatorname{char} F$, *then*

$$\operatorname{Inv}^2\big(G, \mu_n(F_{\mathrm{sep}})\big) \simeq {}_n\operatorname{Pic}(G). \qquad\square$$

(31.21) Example. Let G be a semisimple algebraic group, let $\pi \colon \widetilde{G} \to G$ be a universal covering and set $Z = \ker(\pi)$. There is a natural isomorphism

$$Z^* \xrightarrow{\sim} \operatorname{Pic}(G) \xrightarrow{\sim} \operatorname{Inv}^2\big(G, \mathbb{Q}/\mathbb{Z}(1)\big).$$

Hence attached to each character $\chi \in Z^*$ is an invariant which we denote by a^χ. The construction is as follows. Consider the group $G' = (\widetilde{G} \times \mathbf{G}_{\mathrm{m},F})/Z$ where Z is embedded into the product canonically on the first factor and by the character χ on the second. There is an exact sequence

$$1 \to \mathbf{G}_{\mathrm{m},F} \to G' \to G \to 1.$$

We define a^χ to be the invariant associated to this exact sequence as above.

The conjugation homomorphism $G \to \operatorname{Aut}(G)$ induces the map

$$H^1(F, G) \to H^1\big(F, \operatorname{Aut}(G_{\mathrm{sep}})\big).$$

Hence, associated to each $\gamma \in H^1(F,G)$ is a twisted form G_γ of G (called an inner form of G). If we choose γ such that G_γ is quasisplit (i.e., G_γ contains a Borel subgroup defined over F), then

$$a_F^\chi(\gamma) = [A_\chi] \in \mathrm{Br}(F)$$

where A_χ is the Tits algebra associated to the character χ (see §27).

(31.22) Example. Let T be an algebraic torus over F. Then

$$\mathrm{Pic}(T) \xrightarrow{\sim} H^1(F, T^*(F_{\mathrm{sep}}))$$

and all the cohomological invariants of dimension 2 of T with coefficients in $\mathbb{Q}/\mathbb{Z}(1)$ are given by the cup product

$$H^1(F,T) \otimes H^1(F, T^*(F_{\mathrm{sep}})) \to H^2(F, F_{\mathrm{sep}}^\times) = H^2(F, \mathbb{Q}/\mathbb{Z}(1))$$

associated to the natural pairing

$$T(F_{\mathrm{sep}}) \otimes T^*(F_{\mathrm{sep}}) \to F_{\mathrm{sep}}^\times.$$

Invariants of dimension 3. Let G be an algebraic group over a field F. Assume first that F is separably closed. A *loop* in G is a group homomorphism $\mathbf{G}_{\mathrm{m},F} \to G$ over F. Write G_* for the set of all loops in G. In general there is no group structure on G_*, but if f and h are two loops with commuting images, then the pointwise product fh is also a loop. In particular, for any integer n and any loop f the nth power f^n is defined. For any $g \in G(F)$ and any loop f, write ${}^g f$ for the loop

$$\mathbf{G}_{\mathrm{m},F} \xrightarrow{f} G \xrightarrow{\mathrm{Int}(g)} G.$$

Consider the set $Q(G)$ of all functions $q \colon G_* \to \mathbb{Z}$, such that

(a) $q({}^g f) = q(f)$ for all $g \in G(F)$ and $f \in G_*$,
(b) for any two loops f and h with commuting images, the function

$$\mathbb{Z} \times \mathbb{Z} \to \mathbb{Z}, \quad (k,m) \mapsto q(f^k h^m)$$

    is a quadratic form.

There is a natural abelian group structure on $Q(G)$.

Assume now that F is an arbitrary field. There is a natural action of the absolute Galois group Γ on the set of loops in G_{sep} and hence on $Q(G_{\mathrm{sep}})$. We set

$$Q(G) = Q(G_{\mathrm{sep}})^\Gamma.$$

(31.23) Example. Let T be an algebraic torus. Then $T_*(F_{\mathrm{sep}})$ is the group of cocharacters of T and

$$Q(T) = S^2(T^*(F_{\mathrm{sep}}))^\Gamma$$

is the group of Galois invariant integral quadratic forms on $T_*(F_{\mathrm{sep}})$.

(31.24) Example. Let $G = \mathbf{GL}(V)$ and $f \in G_*(F_{\mathrm{sep}})$ be a loop. We can view f as a representation of $\mathbf{G}_{\mathrm{m},F}$. By the theory of representations of diagonalizable groups (see §22), f is uniquely determined by its weights χ^{a_i}, $i = 1, 2, \ldots, n = \dim(V)$ where χ is the canonical character of $\mathbf{G}_{\mathrm{m},F}$. Let

$$q_V \colon G_*(F_{\mathrm{sep}}) \to \mathbb{Z}, \qquad q_V(f) = \sum_{i=1}^n a_i^2.$$

The function q_V clearly belongs to $Q(G)$. It is *positive definite* in the sense that $q_v(f) > 0$ if $f \neq 1$.

A group homomorphism $G \to G'$ over F induces a map of loop sets $G_*(F_{\mathrm{sep}}) \to G'_*(F_{\mathrm{sep}})$ and hence a group homomorphism

$$Q(G') \to Q(G),$$

making Q a contravariant functor from the category of algebraic groups over F to the category of abelian groups.

Let G and G' be two algebraic groups over F. The natural embeddings of G and G' in $G \times G'$ and both projections from the product $G \times G'$ to its factors induce a natural isomorphism of $Q(G) \oplus Q(G')$ with a direct summand of $Q(G \times G')$.

(31.25) Lemma. *If $G_*(F_{\mathrm{sep}}) \neq 1$, then $Q(G) \neq 0$.*

Proof: Choose an embedding $G \hookrightarrow \mathbf{GL}(V)$. Since $G_*(F_{\mathrm{sep}}) \neq 1$, the restriction of the positive definite function q_V (see Example (31.24)) on this set is nonzero. □

Assume now that G is a semisimple algebraic group over F.

(31.26) Lemma. *$Q(G)$ is a free abelian group of rank at most the number of simple factors of G_{sep}.*

Proof: We may assume that F is separably closed. Let T be a maximal torus in G defined over F. Since any loop in G is conjugate to a loop with values in T, the restriction homomorphism $Q(G) \to Q(T)$ is injective. By Example (31.23), the group $Q(T)$ is free abelian of finite rank, hence so is $Q(G)$.

The Weyl group W acts naturally on $Q(T)$ and the image of the restriction homomorphism belongs to $Q(T)^W$. Hence any element in $Q(G)$ defines a W-invariant quadratic form on T_* and hence on the $\mathbb{Q}$-vector space $T_* \otimes_{\mathbb{Z}} \mathbb{Q}$. This space decomposes as a direct sum of subspaces according to the decomposition of G into the product of simple factors and such a quadratic form (with values in $\mathbb{Q}$) is known to be unique (up to a scalar) on each component. Hence, the rank of $Q(G)$ is at most the number of simple components. □

From Lemma (31.26) and the proof of Lemma (31.25) we obtain

(31.27) Corollary. *If G is an absolutely simple algebraic group, then $Q(G)$ is an infinite cyclic group with a canonical generator which is a positive definite function.*
□

(31.28) Corollary. *Under the hypotheses of the previous corollary the homomorphism $Q(G) \to Q(G_L)$ is an isomorphism for any field extension L of F.*

Proof: It suffices to consider the case $L = F_{\mathrm{sep}}$. Since the group $Q(G)$ is nontrivial, the Galois action on the infinite cyclic group $Q(G_L)$ must be trivial, and hence $Q(G) = Q(G_L)^\Gamma = Q(G_L)$. □

(31.29) Example. Let $G = \mathbf{SL}(V)$. As in Example (31.24), for $i = 1, \ldots, n = \dim V$, one associates integers a_i to a loop f. In our case the sum of all the a_i is even (in fact, zero), hence the sum of the squares of the a_i is even. Therefore

$$q'_V = \tfrac{1}{2} q_V \in Q(G).$$

It is easy to show that q'_V is the canonical generator of $Q(G)$.

(31.30) Corollary. *If F is separably closed, then the rank of $Q(G)$ equals the number of simple factors of G.*

Proof: Let $G = G_1 \times \cdots \times G_m$ where the G_i are simple groups. By Lemma (31.26), $\operatorname{rank}(Q(G)) \leq m$. On the other hand, the group $Q(G)$ contains the direct sum of the $Q(G_i)$ which is a free group of rank m by Corollary (31.27). $\square$

(31.31) Proposition. *Let G and G' be semisimple algebraic groups. Then*
$$Q(G \times G') = Q(G) \oplus Q(G'). \qquad \square$$

Proof: Clearly, we may assume that F is separably closed. The group $Q(G) \oplus Q(G')$ is a direct summand of the free group $Q(G \times G')$ and has the same rank by Corollary (31.30), hence the claim. $\square$

Let L be a finite separable field extension of F and let G be a semisimple group over L. Then the transfer $R_{L/F}(G)$ is a semisimple group over F.

(31.32) Proposition. *There is a natural isomorphism $Q(G) \xrightarrow{\sim} Q\big(R_{L/F}(G)\big)$.*

Proof: Choose an embedding $\rho \colon L \to F_{\mathrm{sep}}$ and set
$$\Gamma' = \operatorname{Gal}\big(F_{\mathrm{sep}}/\rho(L)\big) \subset \Gamma.$$

The group $R_{L/F}(G)_{\mathrm{sep}}$ is isomorphic to the direct product of groups G_τ as τ varies over the set X of all F-embeddings of L into F_{sep} (see Proposition (20.7)). Hence, by Proposition (31.31),
$$Q\big(R_{L/F}(G)_{\mathrm{sep}}\big) = \bigoplus_{\tau \in X} Q(G_\tau).$$

The Galois group Γ acts naturally on the direct sum, transitively permuting components. Hence it is the induced Γ-module from the Γ'-module $Q(G_\rho)$. The proposition then follows from the fact that for any Γ'-module M there is a natural isomorphism between the group of Γ'-invariant elements in M and the group of Γ-invariant elements in the induced module $\operatorname{Map}_{\Gamma'}(\Gamma, M)$. $\square$

By Theorem (26.8), a simply connected semisimple group over F is isomorphic to a product of groups of the form $R_{L/F}(G')$ where L is a finite separable field extension of F and G' is an absolutely simple simply connected group over L. Hence, Corollary (31.27) and Propositions (31.31), (31.32) yield the computation of $Q(G)$ for any simply connected semisimple group G.

A relation between $Q(G)$ and cohomological invariants of dimension 3 of simply connected semisimple groups is given by the following

(31.33) Theorem. *Let G be a simply connected semisimple algebraic group over a field F. Then there is a natural surjective homomorphism*
$$\gamma(G) \colon Q(G) \to \operatorname{Inv}^3\big(G, \mathbb{Q}/\mathbb{Z}(2)\big). \qquad \square$$

The naturality of γ in the theorem means, first of all, that for any group homomorphism $\alpha \colon G \to G'$ the following diagram commutes:

(31.34)
$$
\begin{array}{ccc}
Q(G') & \xrightarrow{\;\gamma(G')\;} & \operatorname{Inv}^3\big(G', \mathbb{Q}/\mathbb{Z}(2)\big) \\
\Big\downarrow & & \Big\downarrow{\scriptstyle \alpha^*} \\
Q(G) & \xrightarrow{\;\gamma(G)\;} & \operatorname{Inv}^3\big(G, \mathbb{Q}/\mathbb{Z}(2)\big).
\end{array}
$$

For any field extension L of F the following diagram also commutes:

(31.35)
$$\begin{array}{ccc} Q(G) & \xrightarrow{\gamma(G)} & \operatorname{Inv}^3\big(G, \mathbb{Q}/\mathbb{Z}(2)\big) \\ \downarrow & & \downarrow{\scriptstyle\text{res}} \\ Q(G_L) & \xrightarrow{\gamma(G_L)} & \operatorname{Inv}^3\big(G_L, \mathbb{Q}/\mathbb{Z}(2)\big). \end{array}$$

In addition, for a finite separable extension L of F and an algebraic group G over L the following diagram is also commutative:

(31.36)
$$\begin{array}{ccc} Q(G) & \xrightarrow{\gamma(G)} & \operatorname{Inv}^3\big(G, \mathbb{Q}/\mathbb{Z}(2)\big) \\ \wr\downarrow & & \downarrow{\scriptstyle\text{cor}} \\ Q\big(R_{L/F}(G)\big) & \xrightarrow{\gamma(R_{L/F}(G))} & \operatorname{Inv}^3\big(R_{L/F}(G), \mathbb{Q}/\mathbb{Z}(2)\big). \end{array}$$

Let G be an absolutely simple simply connected group over F. By Corollary (31.27) and Theorem (31.33), the group $\operatorname{Inv}^3\big(G, \mathbb{Q}/\mathbb{Z}(2)\big)$ is cyclic, generated by a canonical element which we denote $i(G)$ and call the *Rost invariant of the group G*. The commutativity of diagram (31.35) and Corollary (31.28) show that for any field extension L of F,

$$\operatorname{res}_{L/F}\big(i(G)\big) = i(G_L).$$

Let L be a finite separable field extension of F and let G be an absolutely simple simply connected group over L. It follows from the commutativity of (31.36) and Proposition (31.32) that the group $\operatorname{Inv}^3\big(R_{L/F}(G), \mathbb{Q}/\mathbb{Z}(2)\big)$ is cyclic and generated by $\operatorname{cor}_{L/F}\big(i(G)\big)$.

Let G be a simply connected semisimple group over F and let $\rho\colon G \to \mathbf{SL}(V)$ be a representation. The triviality of the right-hand group in the top row of the following commutative diagram (see Example (31.14)):

$$\begin{array}{ccc} Q\big(\mathbf{SL}(V)\big) & \longrightarrow & \operatorname{Inv}^3\big(\mathbf{SL}(V), \mathbb{Q}/\mathbb{Z}(2)\big) \\ \downarrow & & \downarrow \\ Q(G) & \longrightarrow & \operatorname{Inv}^3\big(G, \mathbb{Q}/\mathbb{Z}(2)\big) \end{array}$$

shows that the image of $Q\big(\mathbf{SL}(V)\big) \to Q(G)$ belongs to the kernel of

$$\gamma\colon Q(G) \to \operatorname{Inv}^3\big(G, \mathbb{Q}/\mathbb{Z}(2)\big).$$

One can prove that all the elements in the kernel are obtained in this way.

(31.37) Theorem. *The kernel of $\gamma\colon Q(G) \to \operatorname{Inv}^3\big(G, \mathbb{Q}/\mathbb{Z}(2)\big)$ is generated by the images of $Q\big(\mathbf{SL}(V)\big) \to Q(G)$ for all representations of G.* $\square$

(31.38) Corollary. *Let G and G' be simply connected semisimple groups over F. Then*
$$\operatorname{Inv}^3\big(G \times G', \mathbb{Q}/\mathbb{Z}(2)\big) = \operatorname{Inv}^3\big(G, \mathbb{Q}/\mathbb{Z}(2)\big) \oplus \operatorname{Inv}^3\big(G', \mathbb{Q}/\mathbb{Z}(2)\big). \qquad \square$$

(31.39) Corollary. *Let L/F be a finite separable field extension, G be a simply connected semisimple group over L. Then the corestriction map*

$$\operatorname{cor}\colon \operatorname{Inv}^3\big(G, \mathbb{Q}/\mathbb{Z}(2)\big) \to \operatorname{Inv}^3\big(R_{L/F}(G), \mathbb{Q}/\mathbb{Z}(2)\big)$$

is an isomorphism. $\square$

These two corollaries reduce the study of the group $\mathrm{Inv}^3\big(G, \mathbb{Q}/\mathbb{Z}(2)\big)$ to the case of an absolutely simple simply connected group G.

Let $\alpha\colon G \to G'$ be a homomorphism of absolutely simple simply connected groups over F. There is a unique integer n_α such that the following diagram commutes:

$$
\begin{array}{ccc}
\mathbb{Z} & \overset{=}{\longrightarrow} & Q(G') \\
{\scriptstyle n_\alpha}\big\downarrow & & \big\downarrow{\scriptstyle Q(\alpha)} \\
\mathbb{Z} & \overset{=}{\longrightarrow} & Q(G).
\end{array}
$$

If we have another homomorphism $\beta\colon G' \to G''$, then clearly $n_{\beta\alpha} = n_\beta n_\alpha$. Assume that $G'' = \mathbf{GL}(V)$ and $\beta \neq 1$. It follows from the proof of Lemma (31.25) that $n_\beta > 0$ and $n_{\beta\alpha} \geq 0$, hence n_α is a natural number for any group homomorphism α.

Let $\rho\colon G \to \mathbf{SL}(V)$ be a representation. As we observed above, $n_\rho \cdot i_G = 0$. Denote n_G the greatest common divisor of n_ρ for all representations ρ of the group G. Clearly, $n_G \cdot i(G) = 0$. Theorem (31.37) then implies

(31.40) Proposition. *Let G be an absolutely simple simply connected group. Then* $\mathrm{Inv}^3\big(G, \mathbb{Q}/\mathbb{Z}(2)\big)$ *is a finite cyclic group of order n_G.* $\qquad\square$

Let n be any natural number prime to char F. The exact sequence

$$1 \to \mu_n^{\otimes 2} \to \mathbb{Q}/\mathbb{Z}(2) \overset{n}{\to} \mathbb{Q}/\mathbb{Z}(2) \to 1$$

yields the following exact sequence of cohomology groups

$$
\begin{aligned}
H^2\big(F, \mathbb{Q}/\mathbb{Z}(2)\big) &\overset{n}{\to} H^2\big(F, \mathbb{Q}/\mathbb{Z}(2)\big) \to H^3(F, \mu_n^{\otimes 2}) \to \\
&\to H^3\big(F, \mathbb{Q}/\mathbb{Z}(2)\big) \overset{n}{\to} H^3\big(F, \mathbb{Q}/\mathbb{Z}(2)\big).
\end{aligned}
$$

Since the group $H^2\big(F, \mathbb{Q}/\mathbb{Z}(2)\big)$ is n-divisible (see Merkurjev-Suslin [195]), the group $H^3(F, \mu_n^{\otimes 2})$ is identified with the subgroup of elements of exponent n in $H^3\big(F, \mathbb{Q}/\mathbb{Z}(2)\big)$.

Now let G be an absolutely simple simply connected group over F. By Proposition (31.40), the values of the invariant $i(G)$ lie in $H^3(F, \mu_{n_G}^{\otimes 2})$, so that

$$\mathrm{Inv}^3\big(G, \mathbb{Q}/\mathbb{Z}(2)\big) = \mathrm{Inv}^3(G, \mu_{n_G}^{\otimes 2}).$$

In the following sections we give the numbers n_G for all absolutely simple simply connected groups. In some cases we construct the Rost invariant directly.

Spin groups of quadratic forms. Let F be a field of characteristic different from 2. Let WF be the Witt ring of F and let IF be the fundamental ideal of even-dimensional forms. The nth power $I^n F$ of this ideal is generated by the classes of n-fold Pfister forms.

To any 3-fold Pfister form $\langle\langle a, b, c \rangle\rangle$ the *Arason invariant* associates the class

$$(a) \cup (b) \cup (c) \in H^3(F, \mathbb{Z}/2\mathbb{Z}) = H^3(F, \mu_2^{\otimes 2}),$$

see (30.2). The Arason invariant extends to a group homomorphism

$$e_3\colon I^3 F \to H^3(F, \mathbb{Z}/2\mathbb{Z})$$

(see Arason [21]). Note that $I^3 F$ consists precisely of the classes $[q]$ of quadratic forms q having even dimension, trivial discriminant, and trivial Hasse-Witt invariant (see Merkurjev [189]).

Let q be a non-degenerate quadratic form over F. The group $G = \mathbf{Spin}(q)$ is a simply connected semisimple group if $\dim q \geq 3$ and is absolutely simple if $\dim q \neq 4$. It is a group of type B_n if $\dim q = 2n + 1$ and of type D_n if $\dim q = 2n$. Conversely, any absolutely simple simply connected group of type B_n is isomorphic to $\mathbf{Spin}(q)$ for some q. (The same property does *not* hold for D_n.)

The exact sequence

(31.41) $1 \to \boldsymbol{\mu}_2 \to \mathbf{Spin}(q) \xrightarrow{\pi} \mathbf{O}^+(q) \to 1$

gives the following exact sequence of pointed sets

$$H^1\big(F, \mathbf{Spin}(q)\big) \xrightarrow{\pi_*} H^1\big(F, \mathbf{O}^+(q)\big) \xrightarrow{\delta^1} H^2(F, \boldsymbol{\mu}_2) = {}_2\mathrm{Br}(F).$$

The set $H^1\big(F, \mathbf{O}^+(q)\big)$ classifies quadratic forms of the same dimension and discriminant as q, see (29.29). The connecting map δ^1 takes such a form q' to the Hasse-Witt invariant $e_2\big([q'] - [q]\big) \in \mathrm{Br}(F)$. Thus, the image of π_* consists of classes of forms having the same dimension, discriminant, and Hasse-Witt invariant as q. Therefore, $\pi_*(u) - [q] \in I^3 F$ for any $u \in H^1\big(F, \mathbf{Spin}(q)\big)$.

The map

(31.42) $i\big(\mathbf{Spin}(q)\big) \colon H^1\big(F, \mathbf{Spin}(q)\big) \to H^3(F, \mu_2^{\otimes 2})$

defined by $u \mapsto e_3\big(\pi_*(u) - [q]\big)$ gives rise to an invariant of $\mathbf{Spin}(q)$. It turns out that this is the Rost invariant if $\dim q \geq 5$. If $\dim q = 5$ or 6 and q is of maximal Witt index, the anisotropic form representing $\pi_*(u) - [q]$ is of dimension less than 8 and hence is trivial by the Arason-Pfister Hauptsatz. In these cases the invariant is trivial and $n_G = 1$. Otherwise the invariant is not trivial and $n_G = 2$.

In the case where $\dim q = 4$ the group G is a product of two groups of type A_1 if $\mathrm{disc}\, q$ is trivial and otherwise is isomorphic to $R_{L/F}\big(\mathbf{SL}_1(C_0)\big)$ where L/F is the discriminant field extension and $C_0 = C_0(q)$ (see (15.10) and §26.B). In the latter case the group $\mathrm{Inv}^3\big(G, \mathbb{Q}/\mathbb{Z}(2)\big)$ is cyclic and generated by the invariant described above. This invariant is trivial if and only if the even Clifford algebra C_0 is split.

If $\dim q = 3$, then the described invariant is trivial since it is twice the Rost invariant. In this case $G = \mathbf{SL}_1(C_0)$ is a group of type A_1 and the Rost invariant is described below.

Type A_n.

Inner forms. Let G be an absolutely simple simply connected group of inner type A_n over F, so that $G = \mathbf{SL}_1(A)$ for a central simple F-algebra of degree $n + 1$. It turns out that $n_G = e = \exp(A)$, and the Rost invariant

$$i(G) \colon H^1(F, G) = F^\times / \mathrm{Nrd}(A^\times) \to H^3(F, \mu_e^{\otimes 2})$$

is given by the formula

$$i(G)\big(a \cdot \mathrm{Nrd}(A^\times)\big) = (a) \cup [A]$$

where $(a) \in H^1(F, \mu_e) = F^\times / F^{\times e}$ is the class of $a \in F^\times$ and $[A] \in H^2(F, \mu_e) = {}_e\mathrm{Br}(F)$ is the class of the algebra A.

Outer forms. Let G be an absolutely simple simply connected group of outer type A_n over F with $n \geq 2$, so that $G = \mathbf{SU}(B, \tau)$ where (B, τ) is a central simple F-algebra with unitary involution of degree $n + 1$ and the center K of B is a quadratic separable field extension of F. If n is odd, let $D = D(B, \tau)$ be the discriminant algebra (see §10.E).

(31.43) Proposition. *The number n_G equals either $\exp(B)$ or $2\exp(B)$. The first case occurs if and only if $(n+1)$ is a 2-power and either*

(1) $\exp(B) = n+1$ *or*

(2) $\exp(B) = \frac{n+1}{2}$ *and D is split.* $\qquad\qquad\qquad\qquad\qquad$ $\square$

An element of the set $H^1\big(F, \mathbf{SU}(B,\tau)\big)$ is represented by a pair (s,z) where $s \in \mathrm{Sym}(B,\tau)^\times$ and $z \in K^\times$ satisfy $\mathrm{Nrd}(s) = N_{K/F}(z)$ (see (29.18)). Since $\mathbf{SU}(B,\tau)_K \simeq \mathbf{SL}_1(B)$, it follows from the description of the Rost invariant in the inner case that

$$i(G)\big((s,z)/\approx\big)_K = (z) \cup [B] \in H^3\big(K, \mathbb{Q}/\mathbb{Z}(2)\big).$$

(31.44) Example. Assume that char $F \neq 2$ and B is split, i.e., $B = \mathrm{End}_K(V)$ for some vector space V of dimension $n+1$ over K. The involution τ is adjoint to some hermitian form h on V over K (Theorem (4.2)). Considering V as a vector space over F we have a quadratic form q on V given by $q(v) = h(v,v) \in F$ for $v \in V$. Any isometry of h is also an isometry of q, hence we have the embedding

$$\mathbf{U}(B,\tau) \hookrightarrow \mathbf{O}^+(V,q).$$

Since $\mathbf{SU}(B,\tau)$ is simply connected, the restriction of this embedding lifts to a group homomorphism

$$\alpha\colon \mathbf{SU}(B,\tau) \to \mathbf{Spin}(V,q)$$

(see Borel-Tits [48, Proposition 2.24(i), p. 262]). One can show that $n_\alpha = 1$, so that $i(G)$ is the composition

$$H^1\big(F, \mathbf{SU}(B,\tau)\big) \to H^1\big(F, \mathbf{Spin}(V,q)\big) \to H^3\big(F, \mathbb{Q}/\mathbb{Z}(2)\big)$$

where the latter map is the Rost invariant of $\mathbf{Spin}(V,q)$ which was described in (31.42). An element of the first set in the composition is represented by a pair $(s,z) \in \mathrm{Sym}(B,\tau)^\times \times K^\times$ such that $\mathrm{Nrd}(s) = N_{K/F}(z)$. The symmetric element s defines another hermitian form h_s on V by

$$h_s(u,v) = h\big(s^{-1}(u), v\big)$$

which in turn defines, as described above, a quadratic form q_s on V considered as a vector space over F. The condition on the reduced norm of s shows that the discriminants of h_s and h are equal, see (29.19), hence $[q_s] - [q] \in I^3 F$. It follows from the description of the Rost invariant for the group $\mathbf{Spin}(V,q)$ (see (31.42)) that the invariant of the group G is given by the formula

$$i(G)\big((s,z)/\approx\big) = e_3\big([q_s] - [q]\big).$$

If $\dim V$ is odd (i.e., n is even), the canonical map

$$H^1\big(F, \mathbf{SU}(B,\tau)\big) \to H^1\big(F, \mathbf{GU}(B,\tau)\big)$$

is surjective, since every unitary involution $\tau' = \mathrm{Int}(u) \circ \tau$ on B may be written as

$$\tau' = \mathrm{Int}\big(u\,\mathrm{Nrd}_B(u)\big) \circ \tau,$$

showing that the conjugacy class of τ' is the image of $\big(u\,\mathrm{Nrd}_B(u), \mathrm{Nrd}_B(u)^{(n/2)+1}\big)$. The invariant $i(G)$ induces an invariant

$$i\big(\mathbf{GU}(B,\tau)\big)\colon H^1\big(F, \mathbf{GU}(B,\tau)\big) \to H^3(F, \mu_2^{\otimes 2})$$

which can be explicitly described as follows: given a unitary involution τ' on B, represent τ' as the adjoint involution with respect to some hermitian form h' with disc $h' = \text{disc } h$, and set

$$i\big(\mathbf{GU}(B,\tau)\big)(\tau') = e_3\big([q'] - [q]\big)$$

where q' is the quadratic form on V defined by $q'(v) = h'(v,v)$. Alternatively, consider the quadratic trace form $Q_{\tau'}(x) = \text{Trd}_B(x^2)$ on $\text{Sym}(B,\tau')$. If h' has a diagonalization $\langle \delta_1', \ldots, \delta_{n+1}' \rangle$ and $K \simeq F[X]/(X^2 - \alpha)$, Propositions (11.13) and (11.14) show that

$$Q_{\tau'} = (n+1)\langle 1 \rangle \perp \langle 2 \rangle \cdot \langle\!\langle \alpha \rangle\!\rangle \cdot \big(\perp_{1 \le i < j \le n+1} \langle \delta_i' \delta_j' \rangle\big).$$

On the other hand,

$$q' = \langle\!\langle \alpha \rangle\!\rangle \cdot \langle \delta_1', \ldots, \delta_{n+1}' \rangle.$$

Since disc $h = $ disc h', we may find a diagonalization $h = \langle \delta_1, \ldots, \delta_{n+1} \rangle$ such that $\delta_1 \ldots \delta_{n+1} = \delta_1' \ldots \delta_{n+1}'$. Using the formulas for the Hasse-Witt invariant of a sum in Lam [169, p. 121], we may show that

$$e_2\big(\big[\perp_{1 \le i < j \le n+1} \langle \delta_i' \delta_j' \rangle\big] - \big[\perp_{1 \le i < j \le n+1} \langle \delta_i \delta_j \rangle\big]\big) = $$
$$e_2\big(\big[\langle \delta_1', \ldots, \delta_{n+1}' \rangle\big] - \big[\langle \delta_1, \ldots, \delta_{n+1} \rangle\big]\big),$$

hence

$$i\big(\mathbf{GU}(B,\tau)\big)(\tau') = e_3\big([q'] - [q]\big) = e_3\big([Q_{\tau'}] - [Q_\tau]\big).$$

(31.45) Example. Assume that $(n+1)$ is odd and B has exponent e. Assume also that char F does not divide $2e$. For $G = \mathbf{SU}(B,\tau)$ we have $n_G = 2e$. Since e is odd we have $\mu_{2e}^{\otimes 2} = \mu_2^{\otimes 2} \times \mu_e^{\otimes 2}$, hence the Rost invariant $i(G)$ may be viewed as a pair of invariants

$$\big(i_1(G), i_2(G)\big) \colon H^1\big(F, \mathbf{SU}(B,\tau)\big) \to H^3(F, \mu_2^{\otimes 2}) \times H^3(F, \mu_e^{\otimes 2}).$$

Since B is split by a scalar extension of odd degree, we may use (31.44) to determine $i_1(G)$:

$$i_1(G)\big((s,z)/\approx\big) = e_3\big([Q_{\text{Int}(s)\circ\tau}] - [Q_\tau]\big) \in H^3(F, \mu_2^{\otimes 2}).$$

(By (11.22), it is easily seen that $[Q_{\text{Int}(s)\circ\tau}] - [Q_\tau] \in I^3 F$.)

On the other hand, we have $\mathbf{SU}(B,\tau)_K \simeq \mathbf{SL}_1(B)$ hence we may use the invariant of $\mathbf{SL}_1$ and Corollary (30.12) to determine $i_2(G)$:

$$i_2(G)\big((s,z)/\approx\big) = \tfrac{1}{2}\,\text{cor}_{K/F}\big((s) \cup [B]\big) \in H^3(F, \mu_e^{\otimes 2}).$$

Note that the canonical map $H^1\big(F, \mathbf{SU}(B,\tau)\big) \to H^1\big(F, \mathbf{GU}(B,\tau)\big)$ is surjective, as in the split case (Example (31.44)), and the invariant $i_1(G)$ induces an invariant

$$i\big(\mathbf{GU}(B,\tau)\big) \colon H^1\big(F, \mathbf{GU}(B,\tau)\big) \to H^3(F, \mu_2^{\otimes 2})$$

which maps the conjugacy class of any unitary involution τ' to $e_3\big([Q_{\tau'}] - [Q_\tau]\big)$.

In the particular case where $\deg(B,\tau) = 3$, we also have a Pfister form $\pi(\tau)$ defined in (19.4) and a cohomological invariant $f_3(B,\tau) = e_3\big(\pi(\tau)\big)$, see (30.21). From the relation between $\pi(\tau)$ and Q_τ, it follows that

$$[Q_{\tau'}] - [Q_\tau] = [\langle 2 \rangle] \cdot \big([\pi(\tau')] - [\pi(\tau)]\big),$$

hence

$$i\big(\mathbf{GU}(B,\tau)\big)(\tau') = e_3\big([\pi(\tau')] - [\pi(\tau)]\big) = f_3(B,\tau') - f_3(B,\tau).$$

Type C_n. Let G be an absolutely simple simply connected group of type C_n over F, so that $G = \mathbf{Sp}(A,\sigma)$ where A is a central simple algebra of degree $2n$ over F with a symplectic involution σ.

Assume first that the algebra A is split, i.e., $G = \mathbf{Sp}_{2n}$. Since all the nonsingular alternating forms are pairwise isomorphic, the set $H^1(E,G)$ is trivial for any field extension E of F. Hence $n_G = 1$ and the invariant $i(G)$ is trivial.

Assume now that A is nonsplit, so that $\exp(A) = 2$. Consider the natural embedding $\alpha\colon G \hookrightarrow \mathbf{SL}_1(A)$. One can check that $n_\alpha = 1$, hence the Rost invariant of G is given by the composition of α and the invariant of $\mathbf{SL}_1(A)$, so that $n_G = 2$. By (29.24), we have $H^1(F,G) = \mathrm{Symd}(A,\sigma)^\times/\sim$, and the following diagram commutes:

$$
\begin{array}{ccc}
H^1(F,G) & \xrightarrow{\ \alpha^1\ } & H^1\big(F,\mathbf{SL}_1(A)\big) \\
\big\| & & \big\| \\
\mathrm{Symd}(A,\sigma)^\times/\!\sim & \xrightarrow{\ \mathrm{Nrp}_\sigma\ } & F^\times/\mathrm{Nrd}(A^\times),
\end{array}
$$

(where Nrp_σ is the pfaffian norm). Hence the invariant

$$i(G)\colon H^1(F,G) \to H^3(F,\mu_2^{\otimes 2})$$

is given by the formula

$$i(G)(u/\!\sim) = \big(\mathrm{Nrp}_\sigma(u)\big) \cup [A].$$

The exact sequence

$$1 \to \mathbf{Sp}(A,\sigma) \to \mathbf{GSp}(A,\sigma) \xrightarrow{\ \mu\ } \mathbf{G}_{\mathrm{m}} \to 1,$$

where μ is the multiplier map, induces the following exact sequence in cohomology:

$$H^1\big(F,\mathbf{Sp}(A,\sigma)\big) \to H^1\big(F,\mathbf{GSp}(A,\sigma)\big) \to 1$$

since $H^1(F,\mathbf{G}_{\mathrm{m}}) = 1$ by Hilbert's Theorem 90. If $\deg A$ is divisible by 4, it turns out that the invariant $i(G)$ induces an invariant

$$i\big(\mathbf{GSp}(A,\sigma)\big)\colon H^1\big(F,\mathbf{GSp}(A,\sigma)\big) \to H^3(F,\mu_2^{\otimes 2}).$$

Indeed, viewing $H^1\big(F,\mathbf{GSp}(A,\sigma)\big)$ as the set of conjugacy classes of symplectic involutions on A (see (29.23)), the canonical map

$$\mathrm{Symd}(A,\sigma)^\times/\!\sim\, = H^1\big(F,\mathbf{Sp}(A,\sigma)\big) \to H^1\big(F,\mathbf{GSp}(A,\sigma)\big)$$

takes $u/\!\sim$ to the conjugacy class of $\mathrm{Int}(u) \circ \sigma$. For $z \in F^\times$ and $u \in \mathrm{Symd}(A,\sigma)^\times$ we have $\mathrm{Nrp}_\sigma(zu) = z^{\deg A/2}\,\mathrm{Nrp}_\sigma(u)$, hence $\big(\mathrm{Nrp}_\sigma(zu)\big) = \big(\mathrm{Nrp}_\sigma(u)\big)$ in $H^1(F,\mu_2)$ if $\deg A$ is divisible by 4. Therefore, in this case we may set

$$i\big(\mathbf{GSp}(A,\sigma)\big)\big(\mathrm{Int}(u) \circ \sigma\big) = i\big(\mathbf{Sp}(A,\sigma)\big)(u/\!\sim) = \big(\mathrm{Nrp}_\sigma(u)\big) \cup [A].$$

(31.46) Example. Consider the particular case where $\deg A = 4$. Since the quadratic form Nrp_σ is an Albert form of A by (16.8), its Hasse-Witt invariant is $[A]$. Therefore,

$$\big(\mathrm{Nrp}_\sigma(u)\big) \cup [A] = e_3\big(\langle\!\langle\mathrm{Nrp}_\sigma(u)\rangle\!\rangle \cdot \mathrm{Nrp}_\sigma\big)$$

and it follows by (16.18) that

$$i\big(\mathbf{GSp}(A,\sigma)\big)(\tau) = e_3\big(j_\sigma(\tau)\big)$$

for every symplectic involution τ on A.

(31.47) Example. Let $A = \mathrm{End}_Q(V)$ where V is a vector space of even dimension over a quaternion division algebra Q, and let σ be a hyperbolic involution on A. For every nonsingular hermitian form h on V (with respect to the conjugation involution on Q), the invariant $i\big(\mathbf{GSp}(A,\sigma)\big)(\sigma_h)$ of the adjoint involution σ_h is the cohomological version of the Jacobson discriminant of h, see the notes to Chapter II. Indeed, if h has a diagonalization $\langle \alpha_1, \ldots, \alpha_n \rangle$, then we may assume $\sigma_h = \mathrm{Int}(u) \circ \sigma$ where u is the diagonal matrix

$$u = \mathrm{diag}(\alpha_1, -\alpha_2, \ldots, \alpha_{n-1}, -\alpha_n).$$

Then $\mathrm{Nrp}_\sigma(u) = (-1)^{n/2}\alpha_1 \ldots \alpha_n$, hence

$$i\big(\mathbf{GSp}(A,\sigma)\big)(\sigma_h) = \big((-1)^{n/2}\alpha_1 \ldots \alpha_n\big) \cup [Q].$$

Type D_n. Assume that $\mathrm{char}\, F \neq 2$. Let G be an absolutely simple simply connected group of type D_n ($n \geq 5$) over F, so that $G = \mathbf{Spin}(A,\sigma)$ where A is a central simple algebra of degree $2n$ over F with an orthogonal involution σ. The case where A is split, i.e., $G = \mathbf{Spin}(q)$ for some quadratic form q, has been considered in (31.42).

Assume that the algebra A is not split. In this case $n_G = 4$. The exact sequence similar to (31.41) yields a map

$$i^1 \colon F^\times / F^{\times 2} = H^1(F, \boldsymbol{\mu}_2) \to H^1\big(F, \mathbf{Spin}(A,\sigma)\big).$$

The image $i^1(a \cdot F^{\times 2})$ for $a \in F^\times$ corresponds to the torsor X_a given in the Clifford group $\Gamma(A,\sigma)$ by the equation $\underline{\sigma}(x)x = a$. The Rost invariant $i(G)$ on X_a is given by the formula

$$i(G)(X_a) = (a) \cup [A]$$

and therefore it is in general nontrivial. Hence the invariant does not factor through the image of

$$H^1\big(F, \mathbf{Spin}(A,\sigma)\big) \to H^1\big(F, \mathbf{O}^+(A,\sigma)\big)$$

as is the case when A is split.

Exceptional types.

G_2. Let G be an absolutely simple simply connected group of type G_2 over F, so that $G = \mathbf{Aut}(\mathfrak{C})$ where $\mathfrak{C}$ is a Cayley algebra over F. The set $H^1(F, G)$ classifies Cayley algebras over F. One has $n_G = 2$ and the Rost invariant

$$i(G) \colon H^1(F, G) \to H^3(F, \mu_2^{\otimes 2})$$

is given by the formula

$$i(G)(\mathfrak{C}') = e_3(n_{\mathfrak{C}'}) + e_3(n_{\mathfrak{C}})$$

where $n_{\mathfrak{C}}$ is the norm form of the Cayley algebra $\mathfrak{C}$ (which is a 3-fold Pfister form) and e_3 is the Arason invariant.

D₄. An absolutely simple simply connected algebraic group of type D_4 over F is isomorphic to $\mathbf{Spin}(T)$ where $T = (E, L, \sigma, \alpha)$ is a trialitarian algebra (see §44.A). Here E is a central simple algebra with an orthogonal involution σ over a cubic étale extension L of F.

Assume first that L splits completely, i.e., $L = F \times F \times F$. Then $E = A_1 \times A_2 \times A_3$ where the A_i are central simple algebras of degree 8 over F. In this case $n_G = 2$ or 4. The first case occurs if and only if at least one of the algebras A_i is split.

Assume now that L is not a field but does not split completely, i.e., $L = F \times K$ where K is a quadratic field extension of F, hence $E = A \times C$ where A and C are central simple algebras of degree 8 over F and K respectively (see §43.C). In this case also $n_G = 2$ or 4 and the first case takes place if and only if A is split.

Finally assume that L is a field (this is the trialitarian case). In this case $n_G = 6$ or 12. The first case occurs if and only if E is split.

F₄. $n_G = 6$. The set $H^1(F, G)$ classifies absolutely simple groups of type F_4 and also exceptional Jordan algebras. The cohomological invariant is discussed in Chapter IX.

E₆. $n_G = 6$ when G is an inner form and $n_G = 12$ otherwise.

E₇. $n_G = 12$.

E₈. $n_G = 60$.

EXERCISES

1. Let G be a profinite group and let A be a (continuous) G-group. Show that there is a natural bijection between the pointed set $H^1(G, A)$ and the direct limit of $H^1(G/U, A^U)$ where U ranges over all open normal subgroups in G.

2. Let $\hat{\mathbb{Z}}$ be the inverse limit of $\mathbb{Z}/n\mathbb{Z}$, $n \in \mathbb{N}$, and A be a $\hat{\mathbb{Z}}$-group such that any element of A has a finite order. Show that there is a natural bijection between the pointed set $H^1(\hat{\mathbb{Z}}, A)$ and the set of equivalence classes of A where the equivalence relation is given by $a \sim a'$ if there is $b \in A$ such that $a' = b^{-1} \cdot a \cdot \sigma(b)$ (σ is the canonical topological generator of $\hat{\mathbb{Z}}$).

3. Show that $\mathrm{Aut}(\mathbf{GL}_2) = \mathrm{Aut}(\mathbf{SL}_2) \times \mathbb{Z}/2\mathbb{Z}$. Describe the twisted forms of $\mathbf{GL}_2$.

4. Let S_n act on $(\mathbb{Z}/2\mathbb{Z})^n$ through permutations and let $G = (\mathbb{Z}/2\mathbb{Z})^n \rtimes S_n$. Let F be an arbitrary field. Show that $H^1(F, G)$ classifies towers $F \subset L \subset E$ with L/F étale of dimension n and E/L quadratic étale.

5. Let $G = \mathbf{GL}_n / \boldsymbol{\mu}_2$. Show that there is a natural bijection between $H^1(F, G)$ and the set of isomorphism classes of triples (A, V, ρ) where A is a central simple F-algebra of degree n, V is an F-vector space of dimension n^2 and $\rho \colon A \otimes_F A \to \mathrm{End}_F(V)$ is an isomorphism of F-algebras.

   Hint: For an n-dimensional F-vector space U there is an associated triple (A_U, V_U, ρ_U) where $A_U = \mathrm{End}_F(U)$, $V_U = U^{\otimes 2}$ and where

   $$\rho \colon \mathrm{End}_F(U) \otimes_F \mathrm{End}_F(U) \to \mathrm{End}_F(U^{\otimes 2})$$

   is the natural map. If F is separably closed, then any triple (A, V, ρ) is isomorphic to (A_U, V_U, ρ_U). Moreover the homomorphism

   $$\mathrm{GL}(U) \to \{\, (\alpha, \beta) \in \mathrm{Aut}_F(A_U) \times \mathrm{GL}(V_U) \mid \rho \circ (\alpha \otimes \alpha) = \mathrm{Ad}(\beta) \circ \mu \,\}$$

   given by $\gamma \mapsto \big(\mathrm{Ad}(\gamma), \gamma^{\otimes 2}\big)$ is surjective with kernel μ_2.

6. Let G be as in Exercise 5. Show that the sequence

$$H^1(F, G) \xrightarrow{\lambda} H^1(F, \mathbf{PGL}_n) \xrightarrow{2\delta^1} H^2(F, \mathbf{G}_m)$$

is exact. Here λ is induced from the natural map $\mathbf{GL}_n \to \mathbf{PGL}_n$ and δ^1 is the connecting homomorphism for (29.10).

   Using this result one may restate Albert's theorem on the existence of involutions of the first kind (Theorem (3.1)) by saying that the natural inclusion $\mathbf{PGO}_n \to G$ induces a surjection

$$H^1(F, \mathbf{PGO}_n) \to H^1(F, G).$$

The construction of Exercise 10 in Chapter I can be interpreted in terms of Galois cohomology via the natural homomorphism $GL(U) \to PGO\big(\mathbb{H}(U)\big)$ where U is an n-dimensional vector space and $\mathbb{H}(U)$ is the hyperbolic quadratic space defined in §6.B.

7. Let K/F be separable quadratic extension of fields. Taking transfers, the exact sequence (29.10) induces an exact sequence

$$1 \to R_{K/F}(\mathbf{G}_m) \to R_{K/F}(\mathbf{GL}_n) \to R_{K/F}(\mathbf{PGL}_n) \to 1.$$

Let $N\colon R_{K/F}(\mathbf{G}_m) \to \mathbf{G}_m$ be the transfer map and set

$$G = R_{K/F}(\mathbf{GL}_n)/\ker N.$$

Show that there is a natural bijection between $H^1(F, G)$ and the set of isomorphism classes of triples (A, V, ρ) where A is a central simple K-algebra of degree n, V is an F-vector space of dimension n^2 and

$$\rho\colon N_{K/F}(A) \to \operatorname{End}_F(V)$$

is an isomorphism of F-algebras. Moreover show that the sequence

$$H^1(F, G) \xrightarrow{\lambda} H^1(K, \mathbf{PGL}_n) \xrightarrow{\operatorname{cor}_{K/F} \circ \delta^1} H^2(F, \mathbf{G}_m)$$

is exact. Here δ^1 is the connecting homomorphism for the sequence (29.10) and λ is given by

$$H^1(F, G) \to H^1\big(F, R_{K/F}(\mathbf{PGL}_n)\big) = H^1(K, \mathbf{PGL}_n).$$

Using this result one may restate the theorem on the existence of involutions of the second kind (Theorem (3.1)) by saying that the natural inclusion $\mathbf{PGU}_n = \mathbf{SU}_n/\ker N \to G$ induces a surjection

$$H^1(F, \mathbf{PGU}_n) \to H^1(F, G).$$

The construction of Exercise 11 in Chapter I can be interpreted in terms of Galois cohomology via the natural homomorphism $GL(U_K) \to PGU\big(\mathbb{H}_1(U_K)\big)$ where U is an n-dimensional F-vector space and $\mathbb{H}_1(U_K)$ is the hyperbolic hermitian space defined in §6.B.

8. Let (A, σ, f) be a central simple F-algebra with quadratic pair. Let $\mathbf{GL}_1(A)$ act on the vector space $\operatorname{Symd}(A, \sigma) \oplus \operatorname{Sym}(A, \sigma)^*$ by

$$\rho(a)(x, g) = \big(ax\sigma(a), {}^a g\big),$$

where ${}^a g(y) = g\big(\sigma(a)ya\big)$ for $y \in \operatorname{Sym}(A, \sigma)$. Show that the stabilizer of $(1, f)$ is $\mathbf{O}(A, \sigma, f)$ and that the twisted ρ-forms of $(1, f)$ are the pairs (x, g) such that $x \in A^\times$ and $g\big(y + \sigma(y)\big) = \operatorname{Trd}_A(y)$ for all $y \in A$. Use these results to give an alternate description of $H^1(F, \mathbf{O}(A, \sigma, f))$, and describe the canonical map induced by the inclusion $\mathbf{O}(A, \sigma, f) \hookrightarrow \mathbf{GO}(A, \sigma, f)$.

9. Let L be a Galois $\mathbb{Z}/n\mathbb{Z}$-algebra over a field F of arbitrary characteristic. Using the exact sequence $0 \to \mathbb{Z} \to \mathbb{Q} \to \mathbb{Q}/\mathbb{Z} \to 0$, associate to L a cohomology class $[L]$ in $H^2(F, \mathbb{Z})$ and show that the class $[(L, a)] \in H^2(F, \mathbf{G}_\mathrm{m})$ corresponding to the cyclic algebra (L, a) under the crossed product construction is the cup product $[L] \cup a$, for $a \in F^\times = H^0(F, \mathbf{G}_\mathrm{m})$.

10. Let K/F be a separable quadratic extension of fields with nontrivial automorphism ι, and let n be an integer which is not divisible by $\mathrm{char}\, F$. Use Proposition (30.13) to identify $H^1(F, \boldsymbol{\mu}_{n[K]})$ to the factor group

$$\frac{\{\, (x, y) \in F^\times \times K^\times \mid x^n = N_{K/F}(y) \,\}}{\{\, (N_{K/F}(z), z^n) \mid z \in K^\times \,\}}.$$

For $(x, y) \in F^\times \times K^\times$ such that $x^n = N_{K/F}(y)$, let $[x, y] \in H^1(F, \boldsymbol{\mu}_{[K]})$ be the corresponding cohomology class.

   (a) Suppose $n = 2$. Since $\boldsymbol{\mu}_{2[K]} = \boldsymbol{\mu}_2$, there is a canonical isomorphism $H^1(F, \boldsymbol{\mu}_{2[K]}) \simeq F^\times/F^{\times 2}$. Show that this isomorphism takes $[x, y]$ to $N_{K/F}(z) \cdot F^{\times 2}$, where $z \in K^\times$ is such that $x^{-1}y = z\iota(z)^{-1}$.

   (b) Suppose $n = rs$ for some integers r, s. Consider the exact sequence

   $$1 \to \boldsymbol{\mu}_{r[K]} \xrightarrow{i} \boldsymbol{\mu}_{n[K]} \xrightarrow{j} \boldsymbol{\mu}_{s[K]} \to 1.$$

   Show that the induced maps

   $$H^1(F, \boldsymbol{\mu}_{r[K]}) \xrightarrow{i^1} H^1(F, \boldsymbol{\mu}_{n[K]}) \xrightarrow{j^1} H^1(F, \boldsymbol{\mu}_{s[K]})$$

   can be described as follows:

   $$i^1[x, y] = [x, y^s] \qquad \text{and} \qquad j^1[x, y] = [x^r, y].$$

   (Compare with (13.36).)

   (c) Show that the restriction map

   $$\mathrm{res}\colon H^1(F, \boldsymbol{\mu}_{n[K]}) \to H^1(K, \boldsymbol{\mu}_n) = K^\times/K^{\times n}$$

   takes $[x, y]$ to $y \cdot K^{\times n}$ and the corestriction map

   $$\mathrm{cor}\colon H^1(K, \boldsymbol{\mu}_n) \to H^1(F, \boldsymbol{\mu}_{n[K]})$$

   takes $z \cdot K^{\times n}$ to $[1, z\iota(z)^{-1}]$.

11. Show that for n dividing 24, $\boldsymbol{\mu}_n \otimes \boldsymbol{\mu}_n$ and $\mathbb{Z}/n\mathbb{Z}$ are isomorphic as Galois modules.

12. Let (A, σ) be a central simple algebra over F with a symplectic involution σ. Show that the map

   $$\mathrm{Symd}(A, \sigma)^\times/\!\sim\; = H^1(F, \mathbf{Sp}(A, \sigma)) \to H^1(F, \mathbf{SL}_1(A)) = F^\times/\mathrm{Nrd}(A^\times)$$

   induced by the inclusion $\mathbf{Sp}(A, \sigma) \hookrightarrow \mathbf{SL}_1(A)$ takes $a \in \mathrm{Sym}(A, \sigma)^\times$ to its pfaffian norm $\mathrm{Nrp}_A(a)$ modulo $\mathrm{Nrd}(A^\times)$.

13. Let A be a central simple algebra over F. For any $c \in F^\times$ write X_c for the set of all $x \in A_\mathrm{sep}^\times$ such that $\mathrm{Nrd}(x) = c$. Prove that
   (a) X_c is a $\mathrm{SL}_1(A_\mathrm{sep})$-torsor.
   (b) Any $\mathrm{SL}_1(A_\mathrm{sep})$-torsor is isomorphic to X_c for some c.
   (c) $X_c \simeq X_d$ if and only if $cd^{-1} \in \mathrm{Nrd}(A^\times)$.

14. Describe $H^1(F, \mathbf{Spin}(V, q))$ in terms of twisted forms of tensors.

15. Let (A, σ, f) be a central simple F-algebra with quadratic pair of even degree $2n$ over an arbitrary field F. Let Z be the center of the Clifford algebra $C(A, \sigma, f)$ and let $\mathbf{\Omega}(A, \sigma, f)$ be the extended Clifford group.

   (a) Show that the connecting map

   $$\delta^1 \colon H^1\big(F, \mathbf{PGO}^+(A, \sigma, f)\big) \to H^2\big(F, R_{Z/F}(\mathbf{G}_{\mathrm{m},Z})\big) = \mathrm{Br}(Z)$$

   in the cohomology sequence associated to

   $$1 \to R_{Z/F}(\mathbf{G}_{\mathrm{m},Z}) \to \mathbf{\Omega}(A, \sigma, f) \xrightarrow{\chi'} \mathbf{PGO}^+(A, \sigma, f) \to 1$$

   maps the 4-tuple $(A', \sigma', f', \varphi)$ to $\big[C(A', \sigma', f') \otimes_{Z'} Z\big]\big[C(A, \sigma, f)\big]^{-1}$, where the tensor product is taken with respect to φ.

   (b) Show that the multiplication homomorphism

   $$\mathbf{Spin}(A, \sigma, f) \times R_{Z/F}(\mathbf{G}_{\mathrm{m},Z}) \to \mathbf{\Omega}(A, \sigma, f)$$

   induces an isomorphism

   $$\mathbf{\Omega}(A, \sigma, f) \simeq \big(\mathbf{Spin}(A, \sigma, f) \times R_{Z/F}(\mathbf{G}_{\mathrm{m},Z})\big)/C$$

   where C is isomorphic to the center of $\mathbf{Spin}(A, \sigma, f)$. Similarly, show that

   $$\mathbf{GO}^+(A, \sigma, f) \simeq \big(\mathbf{O}^+(A, \sigma, f) \times \mathbf{G}_{\mathrm{m}}\big)/\boldsymbol{\mu}_2$$

   where $\boldsymbol{\mu}_2$ is embedded diagonally in the product.

   (c) Assume that n is even. Let $\alpha \colon \mathbf{\Omega}(A, \sigma, f) \to \mathbf{GO}^+(A, \sigma, f)$ be the homomorphism which, under the isomorphism in (b), is the vector representation χ on $\mathbf{Spin}(A, \sigma, f)$ and the norm map on $R_{Z/F}(\mathbf{G}_{\mathrm{m},Z})$. By relating via α the exact sequence in (a) to a similar exact sequence for $\mathbf{GO}^+(A, \sigma, f)$, show that for all 4-tuple $(A', \sigma', f', \varphi)$ representing an element of $H^1\big(F, \mathbf{PGO}^+(A, \sigma, f)\big)$,

   $$N_{Z/F}\big(\big[C(A', \sigma', f') \otimes_{Z'} Z\big]\big[C(A, \sigma, f)\big]^{-1}\big) = [A'][A]^{-1} \quad \text{in } \mathrm{Br}(F).$$

   In particular, $N_{Z/F}\big(\big[C(A, \sigma, f)\big]\big) = [A]$.
   Similarly, using the homomorphism $\mathbf{\Omega}(A, \sigma, f) \to R_{Z/F}(\mathbf{G}_{\mathrm{m},Z})$ which is trivial on $\mathbf{Spin}(A, \sigma, f)$ and the squaring map on $R_{Z/F}(\mathbf{G}_{\mathrm{m},Z})$, show that

   $$\big(\big[C(A', \sigma', f') \otimes_{Z'} Z\big]\big[C(A, \sigma, f)\big]^{-1}\big)^2 = 1.$$

   In particular, $\big[C(A, \sigma, f)\big]^2 = 1$. (Compare with (9.12).)

   (d) Assume that n is odd. Let $G = \big(\mathbf{O}^+(A, \sigma, f) \times R_{Z/F}(\mathbf{G}_{\mathrm{m},Z})\big)/\boldsymbol{\mu}_2$. Using the homomorphism $\alpha \colon \mathbf{\Omega}(A, \sigma, f) \to G$ which is the vector representation χ on $\mathbf{Spin}(A, \sigma, f)$ and the squaring map on $R_{Z/F}(\mathbf{G}_{\mathrm{m},Z})$, show that for all 4-tuple $(A', \sigma', f', \varphi)$ representing an element of $H^1\big(F, \mathbf{PGO}^+(A, \sigma, f)\big)$,

   $$\big(\big[C(A', \sigma', f') \otimes_{Z'} Z\big]\big[C(A, \sigma, f)\big]^{-1}\big)^2 = [A'_Z][A_Z]^{-1} \quad \text{in } \mathrm{Br}(Z).$$

   In particular, $\big[C(A, \sigma, f)\big]^2 = [A_Z]$.
   Using the character of $\mathbf{\Omega}(A, \sigma, f)$ which is trivial on $\mathbf{Spin}(A, \sigma, f)$ and is the norm on $R_{Z/F}(\mathbf{G}_{\mathrm{m},Z})$, show that

   $$N_{Z/F}\big(\big[C(A', \sigma', f') \otimes_{Z'} Z\big]\big[C(A, \sigma, f)\big]^{-1}\big) = 1.$$

   In particular, $N_{Z/F}\big[C(A, \sigma, f)\big] = 1$. (Compare with (9.12).)

16. (Quéguiner [227]) Let (B, τ) be a central simple F-algebra with unitary involution of degree n. Let K be the center of B and let $\tau' = \mathrm{Int}(u) \circ \tau$ for some unit $u \in \mathrm{Sym}(B, \tau)$. Assume that char F does not divide n. Show that the Tits classes $t(B, \tau)$ and $t(B, \tau')$ in $H^2(F, \boldsymbol{\mu}_{n[K]})$ are related by $t(B, \tau') = t(B, \tau) + \zeta_K \cup \big(\mathrm{Nrd}_B(u)\big)$ where ζ_K is the nontrivial element of $H^1(F, \mathbb{Z}_{[K]})$ and $\big(\mathrm{Nrd}_B(u)\big) = \mathrm{Nrd}_B(u) \cdot F^{\times n} \in F^\times / F^{\times n} = H^1(F, \boldsymbol{\mu}_n)$. (Compare with (10.36).)

NOTES

§28. The concept of a nonabelian cohomology set $H^1(\Gamma, A)$ has its origin in the theory of principal homogeneous spaces (or torsors) due to Grothendieck [108], see also Frenkel [99] and Serre [254]. The first steps in the theory of principal homogeneous spaces attached to an algebraic group (in fact a commutative group variety) are found in Weil [308].

Galois descent was implicitly used by Châtelet [64], in the case where A is an elliptic curve (see also [63]). An explicit formulation (and proof) of Galois descent in algebraic geometry was first given by Weil [309]. The idea of twisting the action of the Galois group using automorphisms appears also in this paper, see Weil's commentaries in [311, pp. 543–544].

No Galois cohomology appears in the paper [308] on principal homogeneous spaces mentioned above. The fact that Weil's group of classes of principal homogeneous spaces for a commutative group variety A over a field F stands in bijection with the Galois 1-cohomology set $H^1(F, A)$ was noticed by Serre; details are given in Lang and Tate [171], see also Tate's Bourbaki talk [282].

The first systematic treatment of Galois descent, including nonabelian cases (linear groups, in particular $\mathbf{PGL}_n$ with application to the Brauer group), appeared in Serre's book "Corps locaux" [254], which was based on a course at the Collège de France in 1958/59. Twisted forms of algebraic structures viewed as tensors are mentioned as examples. Applications to quadratic forms are given in Springer [262]. Another early application is the realization by Weil [310], following an observation of "un amateur de cocycles très connu"[33], of Siegel's idea that classical groups can be described as automorphism groups of algebras with involution (Weil [311, pp. 548–549]).

Since then this simple but very useful formalism found many applications. See the latest revised and completed edition of the Lecture Notes of Serre [257] and his Bourbaki talk [258] for more information and numerous references. A far-reaching generalization of nonabelian Galois cohomology, which goes beyond Galois extensions and applies in the setting of schemes, was given by Grothendieck [109].

Our presentation in this section owes much to Serre's Lecture Notes [257] and to the paper [46] of Borel and Serre. The technique of changing base points by twisting coefficients in cohomology, which we use systematically, was first developed there. Note that the term "co-induced module" is used by Serre [254] and by Brown [56] for the modules which we call "induced," following Serre [257].

---

[33]also referred to as "Mr. P. (the famous winner of many cocycle races)"

§29. Lemma (29.6), the so-called "Shapiro lemma," was independently proved by Eckmann [87, Theorem 4], D. K. Faddeev [94], and Arnold Shapiro. Shapiro's proof appears in Hochschild-Nakayama [124, Lemma 1.1].

Besides algebras and quadratic forms, Severi-Brauer varieties also have a nice interpretation in terms of Galois cohomology: the group scheme $\mathbf{PGL}_n$ occurs not only as the automorphism group of a split central simple algebra of degree n, but also as the automorphism group of the projective space $\mathbb{P}^{n-1}$. The Severi-Brauer variety $\mathrm{SB}(A)$ attached to a central simple algebra A is a twisted form of the projective space, given by the cocycle of A (see Artin [25]).

For any quadratic space (V, q) of even dimension $2n$, the Clifford functor defines a homomorphism $C \colon \mathbf{PGO}(V, q) \to \mathbf{Aut}_{\mathrm{alg}}\big(C_0(V, q)\big)$ (see (13.1)). The induced map in cohomology $C^1 \colon H^1\big(F, \mathbf{PGO}(V, q)\big) \to H^1\big(F, \mathbf{Aut}_{\mathrm{alg}}\big(C_0(V, q)\big)\big)$ associates to every central simple F-algebra with quadratic pair of degree $2n$ a separable F-algebra of dimension 2^{2n-1}; this is the definition of the Clifford algebra of a central simple algebra with quadratic pair by Galois descent.

§30. Although the cyclic algebra construction is classical, the case considered here, where L is an arbitrary Galois $\mathbb{Z}/n\mathbb{Z}$-algebra, is not so common in the literature. It can be found however in Albert [9, Chapter VII]. Note that if L is a field, its Galois $\mathbb{Z}/n\mathbb{Z}$-algebra structure designates a generator of the Galois group $\mathrm{Gal}(L/F)$.

The exact sequence (30.10) was observed by Arason-Elman [22, Appendix] and by Serre [257, Fifth edition, Chapter I, §2, Exercise 2]. The special case where $M = \boldsymbol{\mu}_2(F_{\mathrm{sep}})$ (Corollary (30.12)) plays a crucial rôle in Arason [21].

The cohomological invariants f_1, g_2, f_3 for central simple F-algebras with unitary involution of degree 3 are discussed in Haile-Knus-Rost-Tignol [115, Corollary 32]. It is also shown in [115] that these invariants are not independent and that the invariant $g_2(B, \tau)$ gives information on the étale F-subalgebras of B. To state precise results, recall from (30.18) that cubic étale F-algebras with discriminant Δ are classified by the orbit set $H^1(F, A_{3[\Delta]})/S_2$. Suppose $\mathrm{char}\, F \neq 2$, 3 and let $F(\omega) = F[X]/(X^2 + X + 1)$, so that $\boldsymbol{\mu}_3 = A_{3[F(\omega)]}$. Let (B, τ) be a central simple F-algebra with unitary involution of degree 3 and let L be a cubic étale F-algebra with discriminant Δ. Let K be the center of B and let $c_L \in H^1(F, A_{3[\Delta]})$ be a cohomology class representing L. The algebra B contains a subalgebra isomorphic to L if and only if $g_2(B, \tau) = c_L \cup d$ for some $d \in H^1(F, A_{3[K*F(\omega)*\Delta]})$. (Compare with Proposition (30.20).) If this condition holds, then B also contains an étale subalgebra L' with associated cohomology class d (hence with discriminant $K*F(\omega)*\Delta$). Moreover, there exists an involution τ' such that $\mathrm{Sym}(B, \tau')$ contains L and L'. See [115, Proposition 31].

§31. Let (A, σ) be a central simple algebra with orthogonal involution of even degree over a field F of characteristic different from 2. The connecting homomorphism

$$\delta^1 \colon H^1\big(F, \mathbf{O}^+(A, \sigma)\big) \to H^2(F, \boldsymbol{\mu}_2) = {}_2\mathrm{Br}(F)$$

in the cohomology sequence associated to the exact sequence

$$1 \to \boldsymbol{\mu}_2 \to \mathbf{Spin}(A, \sigma) \to \mathbf{O}^+(A, \sigma) \to 1$$

is described in Garibaldi-Tignol-Wadsworth [105]. Recall from (29.27) the bijection

$$H^1\big(F, \mathbf{O}^+(A, \sigma)\big) \simeq \mathrm{SSym}(A, \sigma)^\times/\approx.$$

For $(s, z) \in \mathrm{SSym}(A, \sigma)^\times$, consider the algebra $A' = M_2(A) \simeq \mathrm{End}_A(A^2)$ with the involution σ' adjoint to the hermitian form $\langle 1, -s^{-1} \rangle$, i.e.,

$$\sigma' \begin{pmatrix} a & b \\ c & d \end{pmatrix} = \begin{pmatrix} \sigma(a) & -\sigma(c)s^{-1} \\ -s\sigma(b) & s\sigma(d)s^{-1} \end{pmatrix} \qquad \text{for } a, b, c, d \in A.$$

Let $s' = \begin{pmatrix} 0 & 1 \\ s & 0 \end{pmatrix} \in A'$. We have $s' \in \mathrm{Skew}(A', \sigma')$ and $\mathrm{Nrd}_{A'}(s') = \mathrm{Nrd}_A(s) = z^2$. Therefore, letting Z be the center of the Clifford algebra $C(A', \sigma')$ and

$$\pi \colon \mathrm{Skew}(A', \sigma') \to Z$$

the generalized pfaffian of (A', σ') (see (8.24)), we have $\pi(s')^2 = z^2$. It follows that $\frac{1}{2}(1 + z^{-1}\pi(s'))$ is a nonzero central idempotent of $C(A', \sigma')$. Set

$$E(s, z) = (1 + z^{-1}\pi(s')) \cdot C(A', \sigma'),$$

a central simple F-algebra with involution of the first kind of degree $2^{\deg A - 1}$. We have

$$C(A', \sigma') = E(s, z) \times E(s, -z)$$

and it is shown in Garibaldi-Tignol-Wadsworth [105, Proposition 4.6] that

$$\delta^1((s, z)/\approx) = [E(s, z)] \in {}_2\mathrm{Br}(F).$$

In particular, the images under δ^1 of (s, z) and $(s, -z)$ are the two components of $C(A', \sigma')$. By (9.12), it follows that $[E(s, z)][E(s, -z)] = [A]$, hence the Brauer class $[E(s, z)]$ is uniquely determined by $s \in \mathrm{Sym}(A, \sigma)^\times$ up to a factor $[A]$. This is the invariant of hermitian forms defined by Bartels [31]. Explicitly, let D be a division F-algebra with involution of the first kind and let h be a nonsingular hermitian or skew-hermitian form on a D-vector space V such that the adjoint involution $\sigma = \sigma_h$ on $A = \mathrm{End}_D(V)$ is orthogonal. Let $S = \{1, [D]\} \subset \mathrm{Br}(F)$. To every nonsingular form h' on V of the same type and discriminant as h, Bartels attaches an invariant $c(h, h')$ in the factor group $\mathrm{Br}(F)/S$ as follows: since h and h' are nonsingular and of the same type, there exists $s \in \mathrm{Sym}(A, \sigma)^\times$ such that

$$h'(x, y) = h(s^{-1}(x), y) \qquad \text{for all } x, y \in V.$$

We have $\mathrm{Nrd}_A(s) \in F^{\times 2}$ since h and h' have the same discriminant. We may then set

$$c(h, h') = [E(s, z)]S = [E(s, -z)]S \in \mathrm{Br}(F)/S$$

where $z \in F^\times$ is such that $z^2 = \mathrm{Nrd}_A(s)$.

The Tits class $t(B, \tau) \in H^2(F, \boldsymbol{\mu}_{n[K]})$ for (B, τ) a central simple F-algebra with unitary involution of degree n with center K was defined by Quéguiner [226, §3.5.2], [227, §2.2], who called it the *determinant class*. (Actually, Quéguiner's determinant class differs from the Tits class by a factor which depends only on n.)

The discovery of the $H^3(\mathbb{Q}/\mathbb{Z}(2))$-invariant for simply connected algebraic groups G was initiated by Serre, who suggested to define an $H^3(\mathbb{Z}/3)$-invariant for F_4 and an $H^3(\mu_5^{\otimes 2})$-invariant for E_8. The mod 3-invariant for F_4 was established in Rost [234] by an ad hoc method, see also Petersson-Racine [215, 216]. The existence of the mod 5-invariant for E_8 appeared later as a special case of the $H^3(\mathbb{Q}/\mathbb{Z}(2))$-invariant for arbitrary G. There are various definitions of this invariant. In one method (due to Rost, unpublished) one computes for a G-torsor P the kernel of the restriction map $H^3(F, \mathbb{Q}/\mathbb{Z}(2)) \to H^3(F(P), \mathbb{Q}/\mathbb{Z}(2))$. In the case $G = \mathbf{SL}_1(A)$ this is due to Suslin [276]. In the general case one uses here the computation of the K-cohomology group $H^1(G, K_2) = \mathbb{Z}$ in case G is split and simple (this computation was first done by Deligne). At the same time (end of 1992) Serre suggested in a letter to Rost a definition of an $H^3(\mu_p^{\otimes 2})$-invariant based on the computation of $H^3_{\text{ét}}(G, \mu_p^{\otimes 2})$. This definition turned out later to produce the desired invariants for F_4 and E_8. It is very simple, but seems not to provide an easy direct way of computation. A definition (due to Rost) of the $H^3(\mathbb{Q}/\mathbb{Z}(2))$-invariant which is relatively simple with respect to the involved tools and with respect to computation is described in Serre's Bourbaki talk [258]. From the heuristic point of view, the best definition is based on motivic cohomology; here the $H^3(\mathbb{Q}/\mathbb{Z}(2))$-invariant turns out to be just the local torsion version of the standard topological $H^4(\mathbb{Z})$-characteristic class. This interpretation (also due to Rost) is used in Esnault-Kahn-Levine-Viehweg [93]. The basic tool to compute the order of the invariant (Theorem (31.37)) is due to Merkurjev. For a description of the kernel of the invariant $H^1(F, G) \to H^3(F, \mu_5^{\otimes 2})$ for G split of type E_8 see Chernousov [65]. For more information about cohomological invariants for F_4 see Chapter IX and Serre [257, Fith Edition, Annexe, §3].

Finally, we note that getting information for special fields F on the set $H^1(F, G)$, for G an algebraic group, gives rise to many important questions which are not addressed here. Suppose that G is semisimple and simply connected. If F is a p-adic field, then $H^1(F, G)$ is trivial, as was shown by Kneser [155]. If F is a number field, the "Hasse principle" due to Kneser, Springer, Harder and Chernousov shows that the natural map $H^1(F, G) \to \prod_v H^1(F_v, G)$ is injective, where v runs over the real places of F and F_v is the completion of F at v. We refer to Platonov-Rapinchuk [222, Chap. 6] for a general survey. If F is a perfect field of cohomological dimension at most 2 and G is of classical type, Bayer-Fluckiger and Parimala [34] have shown that $H^1(F, G)$ is trivial, proving Serre's "Conjecture II" [257, Chap. III, §3] for classical groups. Analogues of the Hasse principle for fields of virtual cohomological dimension 1 or 2 were obtained by Ducros [86], Scheiderer [249] and Bayer-Fluckiger-Parimala [35].

CHAPTER VIII

Composition and Triality

The main topic of this chapter is composition algebras. Of special interest from the algebraic group point of view are *symmetric* compositions. In dimension 8 there are two such types: *Okubo algebras*, related to algebras of degree 3 with unitary involutions (type A_2), and *para-Cayley algebras* related to Cayley algebras (type G_2). The existence of these two types is due to the existence of inequivalent outer actions of the group $\mathbb{Z}/3\mathbb{Z}$ on split simply connected simple groups of type D_4 ("triality" for Spin_8), for which the fixed elements define groups of type A_2, resp. G_2. Triality is defined here through an explicit computation of the Clifford algebra of the norm of an 8-dimensional symmetric composition. As a step towards exceptional simple Jordan algebras, we introduce in the last section twisted compositions, generalizing a construction of Springer. The corresponding group of automorphisms is the semidirect product $\mathrm{Spin}_8 \rtimes S_3$.

§32. Nonassociative Algebras

In this and the following chapter, by an *F-algebra A* we mean (unless further specified) a finite dimensional vector space over F equipped with an F-bilinear multiplication $m\colon A \times A \to A$. We shall use different notations for the multiplication: $m(x, y) = xy = x \diamond y = x \star y$. We do not assume in general that the multiplication has an identity. An algebra with identity 1 is *unital*. An *ideal* of A is a subspace M such that $ma \in M$ and $am \in M$ for all $m \in M$, $a \in A$. The algebra A is *simple* if the multiplication on A is not trivial (i.e., there are elements a, b of A such that $ab \neq 0$) and 0, A are the only ideals of A. The *multiplication algebra* $M(A)$ is the subalgebra of $\mathrm{End}_F(A)$ generated by left and right multiplications with elements of A. The *centroid* $Z(A)$ is the centralizer of $M(A)$ in $\mathrm{End}_F(A)$:

$$Z(A) = \{\, f \in \mathrm{End}_F(A) \mid f(ab) = f(a)b = af(b) \text{ for } a,\, b \in A \,\}$$

and A is *central* if $F \cdot 1 = Z(A)$. If $Z(A)$ is a field, the algebra A is central over $Z(A)$. Observe that a commutative algebra may be central if it is not associative.

The algebra A is *strictly power-associative* if, for every $R \in \mathsf{Alg}_F$, the R-subalgebra of A_R generated by one element is associative. We then write a^n for n^{th}-power of $a \in A$, independently of the notation used for the multiplication of A. Examples are associative algebras, Lie algebras (trivially), *alternative algebras*, i.e., such that

$$x(xy) = (xx)y \quad \text{and} \quad (yx)x = y(xx)$$

for all x, $y \in A$, and Jordan algebras in characteristic different from 2 (see Chapter IX). Let A be strictly power-associative and unital. Fixing a basis $(u_i)_{1 \leq i \leq r}$ of A and taking indeterminates $\{x_1, \ldots, x_r\}$ we have a *generic element*

$$x = \sum x_i u_i \in A \otimes F(x_1, \ldots, x_r)$$

and there is a unique monic polynomial

$$P_{A,x}(X) = X^m - s_1(x)X^{m-1} + \cdots + (-1)^m s_m(x) \cdot 1$$

of least degree which has x as a root. This is the *generic minimal polynomial* of A. The coefficients s_i are homogeneous polynomials in the x_i's, $s_1 = T_A$ is the *generic trace*, $s_m = N_A$ the *generic norm* and m is the *degree* of A. In view of McCrimmon [184, Theorem 4, p. 535] we have

$$N_A(X \cdot 1 - x) = P_{A,x}(X)$$

for a strictly power-associative algebra A. For any element $a \in A$ we can special-ize the generic minimal polynomial $P_{A,x}(X)$ to a polynomial $P_{A,a}(X) \in F[X]$ by writing $a = \sum_i a_i u_i$ and substituting a_i for x_i. Let $\alpha \colon A \xrightarrow{\sim} A'$ be an isomor-phism of unital algebras. Uniqueness of the generic minimal polynomial implies that $P_{A',\alpha(x)} = P_{A,x}$, in particular $T_{A'}\big(\alpha(x)\big) = T_A(x)$ and $N_{A'}\big(\alpha(x)\big) = N_A(x)$.

(32.1) Examples. (1) We have $P_{A \times B,(x,y)} = P_{A,x} \cdot P_{B,y}$ for a product algebra $A \times B$.

(2) For a central simple associative algebra A the generic minimal polynomial is the reduced characteristic polynomial and for a commutative associative algebra it is the characteristic polynomial.

(3) For a central simple algebra with involution we have a generic minimal poly-nomial on the Jordan algebra of symmetric elements depending on the type of involution:

A_n: If $J = \mathcal{H}(B,\tau)$, where (B,τ) is central simple of degree $n+1$ with a unitary involution over a quadratic étale F-algebra K, $P_{J,a}(X)$ is the restriction of the reduced characteristic polynomial of B to $\mathcal{H}(B,\tau)$. The coefficients of $P_{J,a}(X)$, a priori in K, actually lie in F since they are invariant under ι. The degree of J is the degree of B.

B_n and D_n: For $J = \mathcal{H}(A,\sigma)$, A central simple over F with an orthogonal involution of degree $2n+1$, or $2n$, $P_{J,a}(X)$ is the reduced characteristic polynomial, so that the degree of J is the degree of A.

C_n: For $J = \mathcal{H}(A,\sigma)$, A central simple of degree $2n$ over F with a symplec-tic involution, $P_{J,a}(X)$ is the polynomial $\text{Prp}_{\sigma,a}$ of (2.10). Here the degree of J is $\frac{1}{2}\deg(A)$.

We now describe an invariance property of the coefficients $s_i(x)$. Let $s \in S(A^*)$ be a polynomial function on A, let $d \colon A \to A$ be an F-linear transformation, and let $F[\varepsilon]$ be the F-algebra of dual numbers. We say that s is *Lie invariant* under d if

$$s\big(a + \varepsilon d(a)\big) = s(a)$$

holds in $A[\varepsilon] = A \otimes F[\varepsilon]$ for all $a \in A$. The following result is due to Tits [288]:

(32.2) Proposition. *The coefficients $s_i(x)$ of the generic minimal polynomial of a strictly power-associative F-algebra A are Lie invariant under all derivations d of A.*

Proof: Let F' be an arbitrary field extension of F. The extensions of the forms s_i and d to $A_{F'}$ will be denoted by the same symbols s_i and d. We define forms $\{a,b\}_i$ and $\mu_i(a,b)$ by

$$(a + \varepsilon b)^i = a^i + \varepsilon\{a,b\}_i \quad \text{and} \quad s_i(a + \varepsilon b) = s_i(a) + \varepsilon\mu_i(a,b).$$

It is easy to see (for example by induction) that $d(a^i) = \{a, d(a)\}_i$ for any derivation d. We obtain

$$0 = P_{A[\varepsilon],a+\varepsilon b}(a + \varepsilon b)$$

$$= a^n + \varepsilon\{a, b\}_n + \sum_{i=1}^{n}(-1)^i\big(s_i(a) + \varepsilon\mu_i(a, b)\big)\big(a^{n-i} + \varepsilon\{a, b\}_{n-i}\big),$$

where n is the degree of the generic minimal polynomial, so that

(1) $$\{a, b\}_n + \sum_{i=1}^{n}(-1)^i s_i(a)\{a, b\}_{n-i} + \sum_{i=1}^{n}(-1)^i \mu_i(a, b)a^{n-i} = 0.$$

On the other hand we have

(2) $$d\big(P_{A,a}(a)\big) = \{a, d(a)\}_n + \sum_{i=1}^{n}(-1)^i s_i(a)\{a, d(a)\}_{n-i} = 0.$$

Setting $b = d(a)$ in (1) and subtracting (2) gives

$$\sum_{i=1}^{n}(-1)^i \mu_i\big(a, d(a)\big)a^{n-i} = 0.$$

If a is generic over F, it does not satisfy any polynomial identity of degree $n - 1$. Thus $\mu_i\big(a, d(a)\big) = 0$. This is the Lie invariance of the s_i under the derivation d. $\square$

(32.3) Corollary. *The identity $s_1(a\cdot b) = s_1(b\cdot a)$ holds for any associative algebra and the identity $s_1\big(a \cdot (b\cdot c)\big) = s_1\big((a\cdot b) \cdot c\big)$ holds for any Jordan algebra over a field of characteristic not 2.*

Proof: The maps $d_a(b) = a \cdot b - b \cdot a$, resp. $d_{b,c}(a) = a \cdot (b \cdot c) - (a \cdot b) \cdot c$ are derivations of the corresponding algebras (see for example Schafer [244, p. 92] for the last claim). $\square$

An F-algebra A is *separable* if $A \otimes \widetilde{F}$ is a direct sum of simple ideals for every field extension $\widetilde{F}$ of F. The following criterion (32.4) for separability is quite useful; it applies to associative algebras and Jordan algebras in view of Corollary (32.3) and to alternative algebras (see McCrimmon [182, Theorem 2.8]). For alternative algebras of degree 2 and 3, which are the cases we shall consider, the lemma also follows from (33.14) and Proposition (34.15). We first give a definition: a symmetric bilinear form T on an algebra A is called *associative* or *invariant* if

$$T(xy, z) = T(x, yz) \quad \text{for } x, y, z \in A.$$

(32.4) Lemma (Dieudonné). *Let A be a strictly power-associative algebra with generic trace T_A. If the bilinear form $T\colon (x, y) \mapsto T_A(xy)$ is symmetric, nonsingular and associative, then A is separable.*

Proof: This is a special case of a theorem attributed to Dieudonné, see for example Schafer [244, p. 24]. Let I be an ideal. The orthogonal complement $I^\perp$ of I (with respect to the bilinear form T) is an ideal since T is associative. For $x, y \in J = I \cap I^\perp$ and $z \in A$, we have $T(xy, z) = T(x, yz) = 0$, hence $J^2 = 0$ and elements of J are nilpotent. Nilpotent elements have generic trace 0 (see Jacobson [136, p. 226, Cor. 1(2)]); thus $T(x, z) = T_A(xz) = 0$ for all $z \in A$ and $x \in J$. This implies $J = 0$ and $A = I \oplus I^\perp$. It then follows that A (and $A \otimes \widetilde{F}$ for all field extensions $\widetilde{F}/F$) is a direct sum of simple ideals, hence separable. $\square$

A converse of Lemma (32.4) also holds for associative algebras, alternative algebras and Jordan algebras; a proof can be obtained by using Theorems (32.5) and (37.2).

Alternative algebras. The structure of finite dimensional separable alternative algebras is similar to that of finite dimensional separable associative algebras:

(32.5) Theorem. (1) *Any separable alternative F-algebra is the product of simple alternative algebras whose centers are separable field extensions of F.*
(2) *A central simple separable alternative algebra is either associative central simple or is a Cayley algebra.*

Reference: A reference for (1) is Schafer [244, p. 58]; (2) is a result due to Zorn, see for example Schafer [244, p. 56]. We shall only use Theorem (32.5) for algebras of degree 3. A description of Cayley algebras is given in the next section.   □

For nonassociative algebras the *associator*

$$(x, y, z) = (xy)z - x(yz)$$

is a useful notion. Alternative algebras are defined by the identities

$$(x, x, y) = 0 = (x, y, y).$$

Linearizing we obtain

(32.6) $$(x, y, z) + (y, x, z) = 0 = (x, y, z) + (x, z, y),$$

i.e., in an alternative algebra the associator is an alternating function of the three variables. The following result is essential for the study of alternative algebras:

(32.7) Theorem (E. Artin). *Any subalgebra of an alternative algebra A generated by two elements is associative.*

Reference: See for example Schafer [244, p. 29] or Zorn [320].   □

Thus we have $N_A(xy) = N_A(x)N_A(y)$ and $T_A(xy) = T_A(yx)$ for x, $y \in A$, A a alternative algebra, since both are true for an associative algebra (see Jacobson [136, Theorem 3, p. 235]). The symmetric bilinear form $T(x, y) = T_A(xy)$ is the *bilinear trace form* of A.

In the next two sections separable alternative F-algebras of degree 2 and 3 are studied in detail. We set $\mathsf{Sepalt}_n(m)$ for the groupoid of separable alternative F-algebras of dimension n and degree m with isomorphisms as morphisms.

§33. Composition Algebras

33.A. Multiplicative quadratic forms. Let C be an F-algebra with multiplication $(x, y) \mapsto x \diamond y$ (but not necessarily with identity). We say that a quadratic form q on C is *multiplicative* if

(33.1) $$q(x \diamond y) = q(x)q(y)$$

for all x, $y \in C$. Let $b_q(x, y) = q(x + y) - q(x) - q(y)$ be the polar of q and let

$$C^\perp = \{\, z \in C \mid b_q(z, C) = 0 \,\}.$$

(33.2) Proposition. *The space $C^\perp$ is an ideal in C.*

Proof: This is clear if $q = 0$. So let $x \in C$ be such that $q(x) \neq 0$. Linearizing (33.1) we have

$$b_q(x \diamond y, x \diamond z) = q(x)b_q(y, z).$$

Thus $x \diamond y \in C^\perp$ implies $y \in C^\perp$. It follows that the kernel of the composed map (of F-spaces)

$$\phi_x \colon C \xrightarrow{\ell_x} C \xrightarrow{p} C/C^\perp,$$

where $\ell_x(y) = x \diamond y$ and p is the projection, is contained in $C^\perp$. By dimension count it must be equal to $C^\perp$, so $x \diamond C^\perp \subset C^\perp$ and similarly $C^\perp \diamond x \subset C^\perp$. Since $C^\perp \otimes L = (C \otimes L)^\perp$ for any field extension L/F, the claim now follows from the next lemma. $\qquad\square$

(33.3) Lemma. *Let $q \colon V \to F$ be a nontrivial quadratic form. There exists a field extension L/F such that $V \otimes L$ is generated as an L-linear space by anisotropic vectors.*

Proof: Let $n = \dim_F V$ and let $L = F(t_1, \ldots, t_n)$. Taking n generic vectors in $V \otimes L$ gives a set of anisotropic generators of $V \otimes L$. $\qquad\square$

Let

$$R(C) = \{\, z \in C^\perp \mid q(z) = 0 \,\}.$$

(33.4) Proposition. *If (C, q) is a multiplicative quadratic form, then either $C^\perp = R(C)$ or char $F = 2$ and $C = C^\perp$.*

Proof: We show that $q|_{C^\perp} \neq 0$ implies that char $F = 2$ and $C = C^\perp$. If char $F \neq 2$, then $q(x) = \frac{1}{2}b_q(x, x) = 0$ for $x \in C^\perp$, hence $q|_{C^\perp} \neq 0$ already implies char $F = 2$. To show that $C = C^\perp$ we may assume that F is algebraically closed. Since char $F = 2$ the set $R(C)$ is a linear subspace of $C^\perp$; by replacing C by $C/R(C)$ we may assume that $R(C) = 0$. Then $q \colon C^\perp \to F$ is injective and semilinear with respect to the isomorphism $F \xrightarrow{\sim} F$, $x \mapsto x^2$. It follows that $\dim_F C^\perp = 1$; let $u \in C^\perp$ be a generator, so that $q(u) \neq 0$. For $x \in C$ we have $x \diamond u \in C^\perp$ by Proposition (33.2) and we define a linear form $f \colon C \to F$ by $x \diamond u = f(x)u$. Since

$$q(x)q(u) = q(x \diamond u) = q\big(f(x)u\big) = f(x)^2 q(u),$$

we have $q(x) = f(x)^2$ and the polar $b_q(x, y)$ is identically zero. This implies $C = C^\perp$, hence the claim. $\qquad\square$

(33.5) Example. Let (C, q) be multiplicative and regular of odd rank (defined on p. xvii) over a field of characteristic 2. Since $\dim_F C^\perp = 1$ and $R(C) = 0$, Proposition (33.4) implies that $C = C^\perp$ and C is of dimension 1.

(33.6) Corollary. *The set $R(C)$ is always an ideal of C and q induces a multiplicative form $\overline{q}$ on $\overline{C} = C/R(C)$ such that either*

(1) *$(\overline{C}, \overline{q})$ is regular, or*
(2) *char $F = 2$ and $\overline{C}$ is a purely inseparable field extension of F of exponent 1 of dimension 2^n for some n and $\overline{q}(x) = x^2$.*

Proof: If $R(C) = C^\perp$, $R(C)$ is an ideal in C by Proposition (33.2) and the polar of $\overline{q}$ is nonsingular. Then (1) follows from Corollary (33.19) except when $\dim_F C = 1$ in characteristic 2. If $R(C) \neq C^\perp$, then char $F = 2$ and $C = C^\perp$ by Proposition (33.4). It follows that the polar $b_q(x, y)$ is identically zero, $q \colon C \to F$ is a homomorphism and $R(C)$ is again an ideal. For the description of the induced form $\overline{q}\colon \overline{C} \to F$ we follow Kaplansky [149, p. 95]: the map $\overline{q}\colon \overline{C} \to F$ is an injective homomorphism, thus $\overline{C}$ is a commutative associative integral domain. Moreover, for x such that $\overline{q}(x) \neq 0$, $x^2/\overline{q}(x)$ is an identity element 1 with $\overline{q}(1) = 1$ and $\overline{C}$ is even a field. Since $\overline{q}(\lambda \cdot 1) = \lambda^2 \cdot 1$ for all $\lambda \in F$, we have

(33.7) $$\overline{q}(x^2) = \overline{q}\big(\overline{q}(x) \cdot 1\big)$$

for all $x \in \overline{C}$. Let $C_0 = \overline{q}(\overline{C})$, let $x_0 = \overline{q}(x)$, and let $\diamond_0$ be the induced multiplication. It follows from (33.7) that

$$x_0 \diamond_0 x_0 = \overline{q}(x) \cdot 1.$$

If $\dim_F \overline{C} = 1$ we have the part of assertion (1) in characteristic 2 which was left over. If $\dim_F \overline{C} > 1$, then $\overline{C}$ is a purely inseparable field extension of F of exponent 1, as claimed in (2). $\qquad\square$

(33.8) Remark. In case (1) of (33.6) $\overline{C}$ has dimension 1, 2, 4 or 8 in view of the later Corollary (33.28).

33.B. Unital composition algebras. Let C be an F-algebra with identity and multiplication $(x, y) \mapsto x \diamond y$ and let n be a regular multiplicative quadratic form on C. We call the triple $(C, \diamond, n)$ a *composition algebra*. If char $\neq 2$ the form $\langle 1 \rangle$ is the unique regular multiplicative quadratic form of odd dimension (see Example (33.5)). Thus it suffices to consider composition algebras with nonsingular bilinear forms. We have the following equivalent properties:

(33.9) Proposition. *Let $(C, \diamond)$ be a unital F-algebra with $\dim_F C \geq 2$. The following properties are equivalent:*

(1) *There exists a nonsingular multiplicative quadratic form n on C.*
(2) *C is alternative separable of degree 2.*
(3) *C is alternative and has an involution $\pi\colon x \mapsto \overline{x}$ such that*

$$x + \overline{x} \in F \cdot 1, \quad n(x) = x \diamond \overline{x} \in F \cdot 1,$$

and n is a nonsingular quadratic form on C.

Moreover, the quadratic form n in (1) and the involution π in (3) are uniquely determined by $(C, \diamond)$.

Proof: $(1) \Rightarrow (2)$ Let $(C, \diamond, n)$ be a composition algebra. To show that C is alternative we reproduce the proof of van der Blij and Springer [42], which is valid for any characteristic. Let

$$b_n(x, y) = n(x + y) - n(x) - n(y)$$

be the polar of n. The following formulas are deduced from $n(x \diamond y) = n(x)n(y)$ by linearization:

$$b_n(x \diamond y, x \diamond z) = n(x)b_n(y, z)$$
$$b_n(x \diamond y, u \diamond y) = n(y)b_n(x, u)$$

and

(33.10) $$b_n(x \diamond z, u \diamond y) + b_n(x \diamond y, u \diamond z) = b_n(x, u)b_n(y, z).$$

We have $n(1) = 1$. By putting $z = x$ and $y = 1$ in (33.10), we obtain

$$b_n\big(x^2 - b_n(1, x)x + n(x) \cdot 1, u\big) = 0$$

for all $u \in C$. Since n is nonsingular any $x \in C$ satisfy the quadratic equation

(33.11) $$x^2 - b_n(1,x)x + n(x) \cdot 1 = 0.$$

Hence C is of degree 2 and C is strictly power-associative. Furthermore $b_n(1,x)$ is the trace $T_C(x)$ and n is the norm N_C of C (as an algebra of degree 2). Let $\overline{x} = T_C(x) \cdot 1 - x$. It follows from (33.11) that

$$x \diamond \overline{x} = \overline{x} \diamond x = n(x) \cdot 1$$

and it is straightforward to check that

$$\overline{\overline{x}} = x \quad \text{and} \quad \overline{1} = 1.$$

Hence $x \mapsto \overline{x}$ is bijective. We claim that

(33.12) $$b_n(x \diamond y, z) = b_n(y, \overline{x} \diamond z) = b_n(x, z \diamond \overline{y}).$$

The first formula follows from

$$b_n(x \diamond y, z) + b_n(y, x \diamond z) = b_n(x, 1)b_n(z, y) = T_C(x)b_n(z, y),$$

which is a special case of (33.10), and the proof of the second is similar. We further need the formulas

$$x \diamond (\overline{x} \diamond y) = n(x)y \quad \text{and} \quad (y \diamond x) \diamond \overline{x} = n(x)y.$$

For the proof of the first one, we have

$$b_n\big(x \diamond (\overline{x} \diamond y), z\big) = b_n(\overline{x} \diamond y, \overline{x} \diamond z) = b_n\big(n(x)y, z\big) \quad \text{for} \quad z \in C.$$

The proof of the other one is similar. It follows that $x \diamond (\overline{x} \diamond y) = (x \diamond \overline{x}) \diamond y$. Therefore

$$x \diamond (x \diamond y) = x \diamond \big(T_C(x)y - \overline{x} \diamond y\big) = \big(T_C(x)x - x \diamond \overline{x}\big) \diamond y = (x \diamond x) \diamond y$$

and similarly $(y \diamond x) \diamond x = y \diamond (x \diamond x)$. This shows that C is an alternative algebra. To check that the bilinear trace form $T(x,y) = T_C(x \diamond y)$ is nonsingular, we first verify that π satisfies $\pi(x \diamond y) = \pi(y) \diamond \pi(x)$, so that π is an involution of C. By linearizing the generic polynomial (33.11) we obtain

(33.13) $$x \diamond y + y \diamond x - \big(T_C(y)x + T_C(x)y\big) + b_n(x,y)1 = 0.$$

On the other hand, putting $u = z = 1$ in (33.10) we obtain

$$b_n(x,y) = T_C(x)T_C(y) - T_C(x \diamond y)$$

(which shows that $T(x,y) = T_C(x \diamond y)$ is a symmetric bilinear form). By substituting this in (33.13), we find that

$$\big(T_C(x) - x\big) \diamond \big(T_C(y) - y\big) = T_C(y \diamond x) - y \diamond x,$$

thus $\pi(x \diamond y) = \pi(y) \diamond \pi(x)$. It now follows that

$$T_C(x \diamond y) = \overline{x \diamond y} + x \diamond y = \overline{y} \diamond \overline{x} + x \diamond y = x \diamond \overline{\overline{y}} + \overline{y} \diamond \overline{x} = b_n(x, \overline{y}),$$

hence the bilinear form T is nonsingular if n is nonsingular. Furthermore $T_C(x \diamond y) = b_n(x, \overline{y})$ and (33.12) imply that

(33.14) $$T(x \diamond y, z) = T(x, y \diamond z) \quad \text{for } x, y, z \in C,$$

hence, by Lemma (32.4), C is separable.

(2) $\Rightarrow$ (3) Let

$$X^2 - T_C(x)X + N_C(x) \cdot 1$$

be the generic minimal polynomial of C. We define $\pi(x) = T_C(x) - x$ and we put $n = N_C$; then

$$x \diamond \pi(x) = \pi(x) \diamond x = n(x) \cdot 1 \in F \cdot 1$$

follows from $x^2 - \big(x + \pi(x)\big) \diamond x + n(x) \cdot 1 = 0$. Since $b_n(x, y) = T(x, \overline{y})$ and C is separable, n is nonsingular. The fact that π is an involution follows as in the proof of $(1) \Rightarrow (2)$.

$(3) \Rightarrow (1)$ The existence of an involution with the properties given in (3) implies that C admits a generic minimal polynomial as given in (33.11). Since C is alternative we have

$$x \diamond (\overline{x} \diamond y) = n(x)y = (y \diamond \overline{x}) \diamond x$$

Using that the associator (x, y, z) is an alternating function we obtain

$$
\begin{aligned}
n(x \diamond y) &= (x \diamond y) \diamond (\overline{x \diamond y}) = (x \diamond y) \diamond (\overline{y} \diamond \overline{x}) \\
&= \big((x \diamond y) \diamond \overline{y}\big) \diamond \overline{x} - (x \diamond y, \overline{y}, \overline{x}) = n(x)n(y) - (\overline{x}, x \diamond y, \overline{y}) \\
&= n(x)n(y) - \big(\overline{x} \diamond (x \diamond y)\big) \diamond \overline{y} + \overline{x} \diamond \big((x \diamond y) \diamond \overline{y}\big) \\
&= n(x)n(y) - n(x)n(y) + n(x)n(y) = n(x)n(y)
\end{aligned}
$$

so that n is multiplicative.

The fact that n and π are uniquely determined by $(C, \diamond)$ follows from the uniqueness of the generic minimal polynomial. $\qquad\square$

Let Comp_m be the groupoid of composition algebras of dimension m with isomorphisms as morphisms and let Comp_m^+ be the groupoid of unital composition algebras with isomorphisms as morphisms.

(33.15) Corollary. *The identity map $C \mapsto C$ induces an isomorphism of groupoids $\mathsf{Comp}_m^+ \equiv \mathsf{Sepalt}_m(2)$ for $m \geq 2$.* $\qquad\square$

33.C. Hurwitz algebras. Let (B, π) be a unital F-algebra of dimension m with an involution π such that

$$x + \pi(x) \in F \cdot 1 \quad \text{and} \quad x \diamond \pi(x) = \pi(x) \diamond x \in F \cdot 1$$

for all $x \in B$. Assume further that the quadratic form $n(x) = x \diamond \pi(x)$ is nonsingular. Let $\lambda \in F^\times$. The *Cayley-Dickson algebra* $CD(B, \lambda)$ associated to (B, π) and λ is the vector space

$$CD(B, \lambda) = B \oplus vB$$

where v is a new symbol, endowed with the multiplication

$$(a + vb) \diamond (a' + vb') = a \diamond a' + \lambda b' \diamond \pi(b) + v\big(\pi(a) \diamond b' + a' \diamond b\big),$$

for a, a', b and $b' \in A$. In particular $CD(B, \lambda)$ contains B as a subalgebra and $v^2 = \lambda$.

Further we set

$$n(a + vb) = n(a) - \lambda n(b) \quad \text{and} \quad \pi(a + vb) = \pi(a) - vb.$$

(33.16) Lemma. *The algebra $C = CD(B, \lambda)$ is an algebra with identity $1 + v0$ and π is an involution such that*

$$T_C(x) = x + \pi(x) \in F \cdot 1, \quad N_C(x) = n(x) = x \diamond \pi(x) = \pi(x) \diamond x \in F \cdot 1.$$

The algebra B is contained in $CD(B, \lambda)$ as a subalgebra and

(1) C is alternative if and only if B associative,
(2) C is associative if and only if B is commutative,
(3) C is commutative if and only if $B = F$.

Proof: The fact that $C = CD(B, \lambda)$ is an algebra follows immediately from the definition of C. Identifying v with $v1$ we have $vB = v \diamond B$ and we view v as an element of C. We leave the "if" directions as an exercise. The assertions about T_C and N_C are easy to check, so that elements of C satisfy

$$x^2 - T_C(x)x + N_C(x)1 = 0$$

and C is of degree 2. Thus, if C is alternative, $n = N_C$ is multiplicative by Proposition (33.9). We have

$$n\big((a + v \diamond b) \diamond (c + v \diamond d)\big) = n\big(a \diamond c + \lambda d \diamond \overline{b} + v \diamond (c \diamond b + \overline{a} \diamond d)\big)$$
$$= n(a \diamond c + \lambda d \diamond \overline{b}) - \lambda n(c \diamond b + \overline{a} \diamond d),$$

on the other hand,

$$n\big((a + v \diamond b) \diamond (c + v \diamond d)\big) = n(a + v \diamond b)n(c + v \diamond d)$$
$$= \big(n(a) - \lambda n(b)\big)\big(n(c) - \lambda n(d)\big).$$

Comparing both expressions and using once more that n is multiplicative, we obtain

$$b_n(a \diamond c, \lambda d \diamond \overline{b}) + n(v)b_n(c \diamond b, \overline{a} \diamond d) = 0$$

or, since $n(v) = -\lambda$,

$$b_n(a \diamond c, d \diamond \overline{b}) = b_n(\overline{a} \diamond d, c \diamond b)$$

so that

$$b_n\big((a \diamond c) \diamond b, d\big) = b_n\big(a \diamond (c \diamond b), d\big)$$

for all a, b, c, $d \in B$ by (33.12). Thus we obtain $(a \diamond c) \diamond b = a \diamond (c \diamond b)$ and B is associative. If C is associative, we have $(v \diamond a) \diamond b = v \diamond (a \diamond b) = v(b \diamond a)$ and $b \diamond a = a \diamond b$. Therefore B is commutative. Claim (3) is evident. □

The passage from B to $CD(B, \lambda)$ is sometimes called a *Cayley-Dickson process*. A quadratic étale algebra K satisfies the conditions of Lemma (33.16) and the corresponding Cayley-Dickson algebra $Q = CD(K, \lambda)$ is a quaternion algebra over F for any $\lambda \in F^\times$. Repeating the process leads to an alternative algebra $CD(Q, \mu)$. A *Cayley algebra* is a unital F-algebra isomorphic to an algebra of the type $CD(Q, \mu)$ for some quaternion algebra Q over F and some $\mu \in F^\times$.

In view of Lemma (33.16) and Proposition (33.9), the Cayley-Dickson process applied to a Cayley algebra does not yield a composition algebra. We now come to the well-known classification of unital composition algebras:

(33.17) Theorem. *Composition algebras with identity element over F are either F, quadratic étale F-algebras, quaternion algebras over F, or Cayley algebras over F.*

Proof: As already observed, all algebras in the list are unital composition algebras. Conversely, let C be a composition algebra with identity element over F. If $C \neq F$, let $c \in C$ be such that $\{1, c\}$ generates a nonsingular quadratic subspace of (C, n): choose $c \in 1^\perp$ if char $F \neq 2$ and c such that $b_n(1, c) = 1$ if char $F = 2$. Then $B = F \cdot 1 \oplus F \cdot c$ is a quadratic étale subalgebra of C. Thus we may assume that C contains a unital composition algebra with nonsingular norm and it suffices to show that if $B \neq C$, then C contains a Cayley-Dickson process $B + vB$. If $B \neq C$, we have $C = B \oplus B^\perp$, $B^\perp$ is nonsingular and there exists $v \in B^\perp$ such that $n(v) = -\lambda \neq 0$. We claim that $B \oplus v \diamond B$ is a subalgebra of C obtained by a Cayley-Dickson process, i.e., that

$$(v \diamond a) \diamond b = v \diamond (b \diamond a), \quad a \diamond (v \diamond b) = v \diamond (\overline{a} \diamond b)$$

and

$$(v \diamond a) \diamond (v \diamond b) = \lambda b \diamond \overline{a}$$

for a, $b \in B$. We only check the first formula. The proofs of the others are similar. We have $\overline{v} = -v$, since $b_n(v, 1) = 0$, and $0 = b_n(v, a) \cdot 1 = \overline{v} \diamond a + \overline{a} \diamond v = -v \diamond a + \overline{a} \diamond v$, thus $v \diamond a = \overline{a} \diamond v = -\overline{a} \diamond \overline{v}$ for $a \in B$. Further

$$b_n\big((v \diamond b) \diamond a, z\big) = b_n(v \diamond b, z \diamond \overline{a}) = b_n(\overline{b} \diamond v, z \diamond \overline{a}) = -b_n(\overline{b} \diamond \overline{a}, z \diamond v).$$

The last equality follows from formula (33.10), putting $x = \overline{b}$, $u = z$, $y = \overline{a}$, $z = v$, and using that $b_n(v, \overline{a}) = 0$ for $a \in B$. On the other hand we have

$$-b_n(\overline{b} \diamond \overline{a}, z \diamond v) = -b_n\big((\overline{b} \diamond \overline{a}) \diamond \overline{v}, z\big) = b_n\big(v \diamond (a \diamond b), z\big)$$

where the last equality follows from the fact that $v \diamond a = -\overline{a} \diamond \overline{v}$ for all $a \in B$. This holds for all $z \in C$, hence $(v \diamond b) \diamond a = v \diamond (a \diamond b)$ as claimed. The formulas for the norm and the involution are easy and we do not check them. $\qquad \square$

The classification of composition algebras with identity is known as the Theorem of Hurwitz and the algebras occurring in Theorem (33.17) are called *Hurwitz algebras*.

From now on we set $\mathsf{Comp}_m^+ = \mathsf{Hurw}_m$ for $m = 1$, 2, 4, and 8. If S_m, A_1, resp. G_2, are the groupoids of étale algebras of dimension m, quaternion algebras, resp. Cayley algebras over F, then $\mathsf{Hurw}_2 = \mathsf{S}_2$, $\mathsf{Hurw}_4 = \mathsf{A}_1$, and $\mathsf{Hurw}_8 = \mathsf{G}_2$.

Hurwitz algebras are related to Pfister forms. Let PQ_m be the groupoid of Pfister quadratic forms of dimension m with isometries as morphisms.

(33.18) Proposition. (1) *Norms of Hurwitz algebras are 0-, 1-, 2-, or 3-fold Pfister quadratic forms and conversely, all 0-, 1-, 2- or 3-fold Pfister quadratic forms occur as norms of Hurwitz algebras.*
(2) *For any Hurwitz algebra (C, N_C) the space*

$$(C, N_C)^0 = \{\, x \in C \mid T_C(x) = 0 \,\},$$

where T_C is the trace, is regular.

Proof: (1) This is clear for quadratic étale algebras. The higher cases follow from the Cayley-Dickson construction.

Similarly, (2) is true for quadratic étale algebras, hence for Hurwitz algebras of higher dimension by the Cayley-Dickson construction. $\qquad \square$

(33.19) Theorem. *Let C, C' be Hurwitz algebras. The following claims are equivalent*:

(1) *The algebras C and C' are isomorphic.*
(2) *The norms N_C and $N_{C'}$ are isometric.*
(3) *The norms N_C and $N_{C'}$ are similar.*

Proof: (1) ⇒ (2) follows from the uniqueness of the generic minimal polynomial and (2) ⇒ (3) is obvious. Let now $\alpha \colon (C, N_C) \xrightarrow{\sim} (C', N_{C'})$ be a similitude with factor λ. Since $N_{C'}\big(\alpha(1_C)\big) = \lambda N(1_C) = \lambda$, λ is represented by $N_{C'}$. Since $N_{C'}$ is a Pfister quadratic form, $\lambda N_{C'}$ is isometric to $N_{C'}$ (Baeza [28, p. 95, Theorem 2.4]). Thus we may assume that α is an isometry. Let $\dim_F C \geq 2$ and let B_1 be a quadratic étale subalgebra of C such that its norm is of the form $[1, b] = X^2 + XY + bY^2$ with respect to the basis $(1, u)$ for some $u \in B_1$. Let $\alpha(1) = e$, $\alpha(u) = w$, and let $\ell_{\bar{e}}$ be the left multiplication with $\bar{e}$. Then $u' = \ell_{\bar{e}}(w)$ generates a quadratic étale subalgebra B_1' of C' and $\beta = \ell_{\bar{e}} \circ \alpha$ is an isometry $N_C \xrightarrow{\sim} N_{C'}$ which restricts to an isomorphism $B_1 \xrightarrow{\sim} B_1'$. Thus we may assume that the isometry α restricts to an isomorphism on a pair of quadratic étale algebras B_1 and B_1'. Then α is an isometry $N_{B_1} \xrightarrow{\sim} N_{B_1'}$, hence induces an isometry $B_1^\perp \xrightarrow{\sim} B_1'^\perp$. If $B_1 \neq C$, choose $v \in B_1^\perp$ such that $N(v) \neq 0$ and put $v' = \alpha(v)$. By the Cayley-Dickson construction (33.16) we may define an isomorphism

$$\alpha_0 \colon B_2 = B_1 \oplus v \diamond B_1 \xrightarrow{\sim} B_2' = B_1' \oplus v' \diamond B_1'$$

by putting $\alpha_0(a + v \diamond b) = \alpha(a) + v' \diamond \alpha(b)$ (which is not necessarily equal to $\alpha(a + v \diamond b)$!). Assume that $B_2 \neq C$. Since α_0 is an isometry, it can be extended by Witt's Theorem to an isometry $C \xrightarrow{\sim} C'$. We now conclude by repeating the last step. □

(33.20) Corollary. *There is a natural bijection between the isomorphism classes of* Hurw$_m$ *and the isomorphism classes of* PQ$_m$ *for $m = 1, 2, 4$, and 8.*

Proof: By (33.18) and (33.19). □

The following "Skolem-Noether" type of result is an immediate consequence of the proof of the implication (3) ⇒ (1) of (33.19):

(33.21) Corollary. *Let C_1, C_2 be separable subalgebras of a Hurwitz algebra C. Any isomorphism $\phi \colon C_1 \xrightarrow{\sim} C_2$ extends to an isomorphism or an anti-isomorphism of C.* □

(33.22) Remark. It follows from the proof of Theorem (33.19) that an isometry of a quadratic or quaternion algebra which maps 1 to 1 is an isomorphism or an anti-isomorphism ("$A_1 \equiv B_1$"). This is not true for Cayley algebras ("$G_2 \not\equiv B_3$").

(33.23) Proposition. *If the norm of a Hurwitz algebra is isotropic, it is hyperbolic.*

Proof: This is true in general for Pfister quadratic forms (Baeza [28, Corollary 3.2, p. 105]), but we still give a proof, since it is an easy consequence of the Cayley-Dickson process. We may assume that $\dim_F C \geq 2$. If the norm of a Hurwitz algebra C is isotropic, it contains a hyperbolic plane and we may assume that 1_C lies in this plane. Hence C contains the split separable F-algebra $B = F \times F$. But then any $B \oplus vB$ obtained by the Cayley-Dickson process is a quaternion algebra with zero divisors, hence a matrix algebra, and its norm is hyperbolic. Applying once more the Cayley-Dickson process if necessary shows that the norm must be hyperbolic if $\dim_F C = 8$. □

It follows from Theorem (33.19) and Proposition (33.23) that in each possible dimension there is only one isomorphism class of Hurwitz algebras with isotropic norms. For Cayley algebras a model is the Cayley algebra

$$\mathfrak{C}_s = CD\big(M_2(F), -1\big).$$

We call it the *split Cayley algebra*. Its norm is the hyperbolic space of dimension 8. The group of F-automorphisms of the split Cayley algebra $\mathfrak{C}_s$ over F is an exceptional simple split group G of type G_2 (see Theorem (25.14)).

(33.24) Proposition. *Let G be a simple split algebraic group of type G_2. Cayley algebras over a field F are classified by the pointed set $H^1(F, G)$.*

Proof: Since all Cayley algebras over a separable closure F_s of F are split, any Cayley algebra over F is a form of the split algebra $\mathfrak{C}_s$. Thus we are in the situation of (29.8), hence the claim.   □

If the characteristic of F is different from 2, norms of Hurwitz algebras correspond to n-fold (bilinear) Pfister forms for $n = 0, 1, 2$, and 3. We recall that for any n-fold Pfister form $q_n = \langle\!\langle a_1, \ldots, a_n \rangle\!\rangle$ the element $f_n(q_n) = (\alpha_1) \cup \cdots \cup (\alpha_n) \in H^n(F, \boldsymbol{\mu}_2)$ is an invariant of the isometry class of q_n and classifies the form (see Theorem (30.2)). Thus in characteristic not 2, the cohomological invariant $f_i(N_C)$ of the norm N_C of a Hurwitz algebra C of dimension $2^i \geq 2$ is an invariant of the algebra. We denote it by $f_i(C) \in H^i(F)$.

(33.25) Corollary. *Let C, C' be Hurwitz algebras of dimension 2^i, $i \geq 1$. The following conditions are equivalent:*

(1) *The algebras C and C' are isomorphic.*
(2) *$f_i(C) = f_i(C')$.*

Proof: By Theorem (33.19) and Theorem (30.2).   □

(33.26) Remark. There is also a cohomological invariant for Pfister quadratic forms in characteristic 2 (see for example Serre [258]). For this invariant, Theorem (30.2) holds, hence, accordingly, Corollary (33.25) also.

33.D. Composition algebras without identity. We recall here some general facts about composition algebras without identity, as well as consequences of previous results for such algebras.

The norm n of a composition algebra is determined by the multiplication even if C does not have an identity:

(33.27) Proposition. (1) *Let $(C, \star, n)$ be a composition algebra with multiplication $(x, y) \mapsto x \star y$, not necessarily with identity. Then there exists a multiplication $(x, y) \mapsto x \diamond y$ on C such that $(C, \diamond, n)$ is a unital composition algebra with respect to the new multiplication.*
(2) *Let $(C, \star, n)$, $(C', \star', n')$ be composition algebras (not necessarily with identity). Any isomorphism of algebras $\alpha \colon (C, \star) \xrightarrow{\sim} (C', \star')$ is an isometry $(C, n) \xrightarrow{\sim} (C', n')$.*

Proof: (1) (Kaplansky [149]) Let $a \in C$ be such that $n(a) \neq 0$ and let $u = n(a)^{-1}a^2$, so that $n(u) = 1$. The linear maps $\ell_u \colon x \to u \star x$ and $r_u \colon x \to x \star u$ are isometries, hence bijective. We claim that $v = u^2$ is an identity for the multiplication

$$x \diamond y = (r_u^{-1}x) \star (\ell_u^{-1}y).$$

We have $r_u^{-1}v = \ell_u^{-1}v = u$, hence $x \diamond v = (r_u^{-1}x) \star (\ell_u^{-1}v) = r_u(r_u^{-1}x) = x$ and similarly $v \diamond x = x$. Furthermore,

$$n(x \diamond y) = n\big((r_u^{-1}x) \star (\ell_u^{-1}y)\big) = n(r_u^{-1}x)n(\ell_u^{-1}y) = n(x)n(y).$$

(2) (Petersson [208]) The claim follows from the uniqueness of the degree two generic minimal polynomial if α is an isomorphism of unital algebras. In particular n is uniquely determined by $\star$ if there is a multiplication with identity. Assume now that C and C' do not have identity elements. We use α to transport n' to C, so that we have one multiplication $\star$ on C which admits two multiplicative norms n, n'. If there exists some $a \in C$ with $n(a) = n'(a) = 1$, we modify the multiplication as in (1) to obtain a multiplication $\diamond$ with 1 which admits n and n', so $n = n'$. To find a, let $u \in C$ be such that $n(u) = 1$ (such an element exists by (1)). The map $\ell_u \colon x \to u \star x$ is then an isometry of (C, n), and in particular it is bijective. This implies that $n'(u) \neq 0$. The element a such $u \star a = u$ has the desired property $n(a) = n'(a) = 1$.   □

(33.28) Corollary. *The possible dimensions for composition algebras (not necessarily unital) are 1, 2, 4 or 8.*

Proof: The claim follows from Theorem (33.17) for unital algebras and hence from Proposition (33.27) in general.   □

(33.29) Corollary. *Associating a unital composition algebra $(C, \diamond, n)$ to a composition algebra $(C, \star, n)$ defines a natural map to the isomorphism classes of Hurw_m from the isomorphism classes of Comp_m.*

Proof: By Proposition (33.27) we have a unital multiplication on (C, n) which, by Theorem (33.19), is determined up to isomorphism.   □

(33.30) Remark. As observed in Remark (33.22) an isometry of unital composition algebras which maps 1 to 1 is not necessarily an isomorphism, however isometric unital composition algebras are isomorphic. This is not necessarily the case for algebras without identity (see Remark (34.25)).

§34. Symmetric Compositions

In this section we discuss a special class of composition algebras without identity, independently considered by Petersson [207], Okubo [199], Faulkner [95] and recently by Elduque-Myung [89], Elduque-Pérez [90]. Let (S, n) be a finite dimensional F-algebra with a quadratic form $n \colon S \to F$. Let $b_n(x, y) = n(x + y) - n(x) - n(y)$ for x, $y \in S$ and let $(x, y) \mapsto x \star y$ be the multiplication of S. We recall that the quadratic form n is called associative or invariant with respect to the multiplication $\star$ if

$$b_n(x \star y, z) = b_n(x, y \star z)$$

holds for all x, y, $z \in S$.

(34.1) Lemma. *Assume that n is nonsingular. The following conditions are equivalent:*

(1) *n is multiplicative and associative.*
(2) *n satisfies the relations $x \star (y \star x) = n(x)y = (x \star y) \star x$ for x, $y \in S$.*

Proof: (Okubo-Osborn [202, Lemma II.2.3]) Assume (1). Linearizing $n(x \star y) = n(x)n(y)$, we obtain

$$b_n(x \star y, x \star z) = b_n(y \star x, z \star x) = n(x)b_n(y, z).$$

Since n is associative, this yields

$$0 = b_n\big((x \star y) \star x - n(x)y, z\big) = b_n\big(y, x \star (z \star x) - n(x)z\big),$$

hence (2), n being nonsingular.

Conversely, if (2) holds, linearizing gives

(34.2) $\qquad x \star (y \star z) + z \star (y \star x) = b_n(x, z)y = (x \star y) \star z + (z \star y) \star x.$

By substituting $x \star y$ for x in the first equation and $y \star z$ for z in the second equation, we obtain

$$(x \star y) \star (y \star z) = b_n(x \star y, z)y - z \star (y \star x \star y) = b_n(x, y \star z)y - (y \star z \star y) \star x$$
$$= b_n(x \star y, z)y - n(y)z \star x = b_n(x, y \star z)y - n(y)z \star x,$$

hence $b_n(x \star y, z)y = b_n(x, y \star z)y$ and $b_n(x, y)$ is associative. Finally, we have

$$n(x \star y)y = (x \star y) \star [y \star (x \star y)] = (x \star y) \star \big(n(y)x\big) = n(y)n(x)y$$

and the form n is multiplicative. If $2 \neq 0$, we can also argue as follows:

$$n(x \star y) = \tfrac{1}{2}b_n(x \star y, x \star y) = \tfrac{1}{2}b_n\big(x, y \star (x \star y)\big) = \tfrac{1}{2}b_n(x, x)n(y) = n(x)n(y). \qquad \square$$

We call a composition algebra with an associative norm a *symmetric composition algebra* and denote the full subcategory of Comp_m consisting of symmetric composition algebras by Scomp_m. In a symmetric composition algebra we have

$$x \star (x \star x) = (x \star x) \star x = n(x)x,$$
$$x \star \big(x \star (x \star x)\big) = n(x)x \star x.$$

This and (34.2) yield

(34.3) $\qquad (x \star x) \star (x \star x) = b_n(x, x \star x)x - n(x)x \star x.$

These identities show that the subalgebra generated by an element x is as a vector space spanned by x and $x * x$.

A symmetric composition algebra is not power-associative in general. A complete list of power-associative symmetric composition algebras is given in Exercise 8 of this chapter.

The field F is a symmetric composition algebra with identity. However it can be shown that a symmetric composition algebra of dimension ≥ 2 never admits an identity.

34.A. Para-Hurwitz algebras. Let $(C, \diamond, n)$ be a Hurwitz algebra. The multiplication

$$(x, y) \mapsto x \star y = \overline{x} \diamond \overline{y}$$

also permits composition and it follows from $b_n(x \diamond y, z) = b_n(x, z \diamond \overline{y})$ (see (33.12)) that the norm n is associative with respect to $\star$ (but not with respect to $\diamond$ if $C \neq F$). Thus $(C, \star, n)$ is a symmetric composition algebra. We say that $(C, \star, n)$ is the *para-Hurwitz algebra* associated with $(C, \diamond, n)$ (resp. the *para-quadratic algebra*, the *para-quaternion algebra* or the *para-Cayley algebra*). We denote the corresponding full subcategories of Scomp by $\overline{\mathsf{Hurw}}$, resp. $\overline{S_2}$, $\overline{A_1}$, and $\overline{G_2}$.

Observe that the unital composition algebra associated with $(C, \star)$ by the construction given in the proof of Proposition (33.27) is the Hurwitz algebra $(C, \diamond)$ if we set $a = 1$.

(34.4) Proposition. *Let $(C_1, \diamond, n_1)$ and $(C_2, \diamond, n_2)$ be Hurwitz algebras and let*

$$\alpha \colon C_1 \xrightarrow{\sim} C_2$$

be an isomorphism of vector spaces such that $\alpha(1_{C_1}) = 1_{C_2}$. Then α is an isomorphism $(C_1, \diamond) \xrightarrow{\sim} (C_2, \diamond)$ of algebras if and only if it is an isomorphism $(C_1, \star) \xrightarrow{\sim} (C_2, \star)$ of para-Hurwitz algebras. Moreover

(1) *Any isomorphism of algebras $(C_1, \diamond) \xrightarrow{\sim} (C_2, \diamond)$ is an isomorphism of the corresponding para-Hurwitz algebras.*
(2) *If $\dim C_1 \geq 4$, then an isomorphism $(C_1, \star) \xrightarrow{\sim} (C_2, \star)$ of para-Hurwitz algebras is an isomorphism of the corresponding Hurwitz algebras.*

Proof: Let $\alpha \colon C_1 \to C_2$ be an isomorphism of algebras. By uniqueness of the quadratic generic polynomial we have $\alpha(\overline{x}) = \overline{\alpha(x)}$ and α is an isomorphism of para-Hurwitz algebras. Conversely, an isomorphism of para-Hurwitz algebras is an isometry by Proposition (33.27) (or since $x \star (y \star x) = n(x)y$), and we have $T_{C_2}(\alpha(x)) = T_{C_1}(x)$, since $T_{C_1}(x) = b_{c_1}(1, x)$ and $\alpha(1_{C_1}) = 1_{C_2}$. As above it follows that $\alpha(\overline{x}) = \overline{\alpha(x)}$ and α is an isomorphism of Hurwitz algebras.

Claim (1) obviously follows from the first part and claim (2) will also follow from the first part if we show that $\alpha(1_{C_1}) = 1_{C_2}$. We use Okubo-Osborn [202, p. 1238]: we have $1 \star x = -x$ for $x \in 1^\perp$ and the claim follows if we show that there exists exactly one element $u \in C_1$ such that $u \star x = -x$ for $x \in u^\perp$. Let u be such an element. Since by Corollary (33.18.2), $1^\perp$ is nondegenerate, the maximal dimension of a subspace of $1^\perp$ on which the form is trivial is $\frac{1}{2}(\dim_F C_1 - 2)$. If $\dim_F C_1 \geq 4$, there exists some $x \in 1^\perp \cap u^\perp$ with $n_1(x) \neq 0$. For this element x we have

$$n_1(x)1 = x \star (1 \star x) = x \star (-x) = x \star (u \star x) = n_1(x)u,$$

so that, as claimed, $1 = u$. $\qquad\square$

For quadratic algebras the following nice result holds:

(34.5) Proposition. *Let C_1, C_2 be quadratic algebras and assume that there exists an isomorphism of para-quadratic algebras $\alpha \colon (C_1, \star) \xrightarrow{\sim} (C_2, \star)$, which is not an isomorphism of algebras. Then $u = \alpha(1) \notin F \cdot 1$ is such that $u^3 = 1$ and $\beta(x) = \alpha(x)u^2$ is an algebra isomorphism $C_1 \xrightarrow{\sim} C_2$. Conversely, if $\beta \colon C_1 \xrightarrow{\sim} C_2$ is an isomorphism of algebras, then, for any $u \in C_2$ such that $u \notin F \cdot 1$ and $u^3 = 1$, the map α defined by $\alpha(x) = \beta(x)u$ is an isomorphism $(C_1, \star) \xrightarrow{\sim} (C_2, \star)$ of para-quadratic algebras. In particular an isomorphism $C_1 \xrightarrow{\sim} C_2$ of para-Hurwitz algebras which is not an isomorphism of algebras can only occur if $C_1 \simeq F[X]/(X^2 + X + 1)$.*

Proof: The proof of (34.4) shows that $u = \alpha(1) \notin F$. We show that $u^3 = 1$. We have $u \star u = \overline{u}^2 = u$. Thus multiplying by $\overline{u}$ and conjugating gives $u^3 = u\overline{u} = n_2(u) = n_2(\alpha(1)) = n_1(1) = 1$ by Proposition (33.27). It then follows that $C_2 \simeq F(u)$. The condition $\alpha(x) \star \alpha(y) = \alpha(x \star y)$ with $y = 1$ gives $\overline{\alpha(x)u} = \alpha(\overline{x})$ and, replacing x by xy,

$$\overline{\alpha(xy)u} = \alpha(\overline{xy}) = \overline{\alpha(x)\alpha(y)}.$$

By conjugating and multiplying both sides with $u^4 = u$ we obtain

$$[\alpha(x)u^2][\alpha(y)u^2] = \alpha(xy)u^2,$$

so that the map $\beta \colon C_1 \to C_2$ defined by $\beta(x) = \alpha(x)u^2$ is an isomorphism of algebras. Conversely, if $C_1 \simeq C_2 = F(u)$ with $u^3 = 1$ and if $\beta \colon C_1 \xrightarrow{\sim} C_2$ is an isomorphism, then $\alpha \colon C_1 \to C_2$ defined by $\alpha(x) = \beta(x)u$ is an isomorphism $(C_1, \star) \xrightarrow{\sim} (C_2, \star)$. $\qquad\square$

Observe that $r_u \colon x \mapsto x \star u$ is an automorphism of $\big(F(u), \star\big)$ of order 3. In fact we have $\mathrm{Aut}_F\big(F(u), \star\big) = S_3$, generated by the conjugation and r_u. This is in contrast with the quadratic algebra $\big(F(u), \cdot\big)$ for which $\mathrm{Aut}_F\big(F(u)\big) = \mathbb{Z}/2\mathbb{Z}$.

(34.6) Corollary. *The map* $\mathbf{P} \colon (C, \diamond) \mapsto (C, \star)$ *is an equivalence* $\mathsf{Hurw}_m \equiv \overline{\mathsf{Hurw}}_m$ *of groupoids if* $m = 4$, 8, *and* $\mathbf{P}$ *is bijective on isomorphism classes if* $m = 2$. $\qquad\square$

In view of Corollary (34.6) we call a n-dimensional para-Hurwitz composition algebra of type A_1 if $n = 4$ and of type G_2 if $n = 8$.

(34.7) Remark. It follows from Corollary (34.6) that

$$\mathrm{Aut}_F(C, \diamond) = \mathrm{Aut}_F(C, \star)$$

for any Hurwitz algebra C of dimension ≥ 4. Thus the classification of twisted forms of para-Hurwitz algebras is equivalent to the classification of Hurwitz algebras in dimensions ≥ 4. In particular any twisted form of a para-Hurwitz algebra of dimension ≥ 4 is again a para-Hurwitz algebra. The situation is different in dimension 2: There exist forms of para-quadratic algebras which are not para-quadratic algebras (see Theorem (34.37)).

The identity 1 of a Hurwitz algebra C plays a special role also for the associated para-Hurwitz algebra: it is an idempotent and satisfies $1 \star x = x \star 1 = -x$ for all $x \in C$ such that $b_{n_C}(x, 1) = 0$. Let $(S, \star, n)$ be a symmetric composition algebra. An *idempotent* e of S (i.e., an element such that $e \star e = e$) is called a *para-unit* if

$$e \star x = x \star e = -x \quad \text{for } x \in S,\ b_n(e, x) = 0.$$

(34.8) Lemma. *A symmetric composition algebra is para-Hurwitz if and only if it admits a para-unit.*

Proof: If $(S, \star)$ is para-Hurwitz, then $1 \in S$ is a para-unit. Conversely, for any para-unit e in a symmetric composition algebra $(S, \star, n)$, we have $n(e) = 1$ and

$$x \diamond y = (e \star x) \star (y \star e)$$

defines a multiplication with identity element e on S. We have $x \star y = \overline{x} \diamond \overline{y}$ where $\overline{x} = b_n(e, x)e - x$. $\qquad\square$

34.B. Petersson algebras. Let $(C, \diamond, n)$ be a Hurwitz algebra and let φ be an F-automorphism of C such that $\varphi^3 = 1$. Following Petersson [207] we define a new multiplication on C by

$$x \star y = \varphi(\overline{x}) \diamond \varphi^2(\overline{y}).$$

This algebra, denoted C_φ, is a composition algebra for the same norm n and we call it a *Petersson algebra*. It is straightforward to check that

$$(x \star y) \star x = n(x)y = x \star (y \star x)$$

so that Petersson algebras are symmetric composition algebras. Observe that φ is automatically an automorphism of $(C, \star)$. For $\varphi = 1$, $(C, \star)$ is para-Hurwitz.

Conversely, symmetric composition algebras with nontrivial idempotents are Petersson algebras (Petersson [207, Satz 2.1], or Elduque-Pérez [90, Theorem 2.5]):

(34.9) Proposition. *Let $(S, \star, n)$ be a symmetric composition algebra and let $e \in S$ be a nontrivial idempotent.*
(1) *The product $x \diamond y = (e \star x) \star (y \star e)$ gives S the structure of a Hurwitz algebra with identity e, norm n, and conjugation $x \mapsto \overline{x} = b_n(x, e)e - x$.*
(2) *The map*
$$\varphi(x) = e \star (e \star x) = b_n(e, x)e - x \star e = \overline{x} \star e$$
is an automorphism of $(S, \diamond)$ (and $(S, \star)$) of order ≤ 3 and $(S, \star) = S_\varphi$ is a Petersson algebra with respect to φ.

Proof: (1) is easy and left as an exercise.
   (2) Replacing x by $e \star x$ and z by e in the identity (34.2):
$$b_n(x, z)y = x \star (y \star z) + z \star (y \star x)$$

gives

$$x \diamond y = b_n(e, x)y - e \star \big(y \star (e \star x)\big)$$

hence

$$(x \diamond y) \star e = y \star \big(b_n(x, e)e - e \star x\big) = \big(e \star (y \star e)\big) \star \big((x \star e) \star e\big) = (y \star e) \diamond (x \star e).$$

Thus φ is an automorphism of $(S, \diamond)$, $\varphi^3(x) = \overline{\overline{x}} = x$, $x \star y = \varphi(\overline{x}) \diamond \varphi^2(\overline{y})$ and $(S, \star) = S_\varphi$ as claimed.   □

   In general a symmetric composition may not contain an idempotent. However:

(34.10) Lemma. *Let $(S, \star, n)$ be a symmetric composition algebra.*
(1) *If the cubic form $b_n(x \star x, x)$ is isotropic on S, then $(S, \star)$ contains an idempotent. In particular there always exists a field extension L/F of degree 3 such that $(S, \star)_L$ contains an idempotent e.*
(2) *For any nontrivial idempotent $e \in S$ we have $n(e) = 1$.*

Proof: (1) It suffices to find $f \neq 0$ with $f \star f = \lambda f$, $\lambda \in F^\times$ so that $e = f\lambda^{-1}$ then is an idempotent. Let $x \neq 0$ be such that $b_n(x \star x, x) = 0$. We have
$$(x \star x) \star (x \star x) = -n(x)(x \star x)$$
by (34.3), so we take $f = x \star x$ if $n(x) \neq 0$. If $n(x) = 0$, we may also assume that $x \star x = 0$: if $x \star x \neq 0$ we replace x by $x \star x$ and use again (34.3). Since x is isotropic and n is nonsingular, there exists some $y \in S$ such that $n(y) = 0$ and $b_n(x, y) = -1$. A straightforward computation using (34.2) shows that
$$(x \star y + y \star x) \star (x \star y + y \star x) = (x \star y + y \star x) + 3b_n(y, y \star x)x,$$
and
$$e = x \star y + y \star x + b_n(y, y \star x)x$$
is an idempotent and is nonzero since
$$e \star x = (y \star x) \star x = b_n(x, y)x = -x.$$
   (2) Since $e = (e \star e) \star e = n(e)e$, we have $n(e) = 1$.   □

(34.11) Remark. Lemma (34.10.1) is in fact a special case of Theorem (36.24) and its proof is copied from the proof of implication (2) $\Rightarrow$ (3) of (36.24).

Assume that char $F \neq 3$ and that F contains a primitive cube root of unity ω. The existence of an automorphism of order 3 on a Hurwitz algebra C is equivalent with the existence of a $\mathbb{Z}/3\mathbb{Z}$-grading:

(34.12) Lemma. *Suppose that F contains a primitive cube root of unity ω.*

(1) *If φ is an automorphism of C of order 3, then C (or C_φ) admits a decomposition*

$$C = C_\varphi = S_0 \oplus S_1 \oplus S_2,$$

with

$$S_i = \{\, x \in C \mid \varphi(x) = \omega^i x \,\}$$

and such that

    (a) *$S_i \diamond S_j \subset S_{i+j}$ (resp. $S_i \star S_j \subset S_{i+j}$), with subscripts taken modulo 3,*

    (b) *$b_n(S_i, S_j) = 0$ unless $i + j \equiv 0 \mod 3$.*

In particular $(S_0, \star, n) \subset C_\varphi$ is a para-Hurwitz algebra of even dimension and S_1 (resp. S_2) is a maximal isotropic subspace of $S_1 \oplus S_2$.

(2) *Conversely, any $\mathbb{Z}/3\mathbb{Z}$-grading of C defines an automorphism φ of order 3 of C, hence a Petersson algebra C_φ.*

Proof: Claim (1) follows easily from the fact that ω^i, $i = 0$, 1, 2, are the eigenvalues of the automorphism φ. For (2) we take the identity on degree 0 elements, multiplication by ω on degree 1 elements and multiplication by ω^2 on degree 2 elements. $\square$

If $\varphi \neq 1$, S_0 in Lemma (34.12) must have dimension 2 or 4 (being a para-Hurwitz algebra). We show next that C_φ is para-Hurwitz if $\dim S_0 = 2$. The case $\dim S_0 = 4$ and $\dim C_\varphi = 8$ corresponds to a different type of symmetric composition, discussed in the next subsection.

(34.13) Proposition (Elduque-Pérez). *Let F be a field of characteristic not 3, let C be a Hurwitz algebra over F, let φ be an F-automorphism of C of order 3. Then*

$$S_0 = \{\, x \in C \mid \varphi(x) = x \,\}$$

is a para-Hurwitz algebra of dimension 2 or 4. The Petersson algebra C_φ is isomorphic to a para-Hurwitz algebra if and only if $\dim S_0 = 2$.

Proof: The first claim is clear. For the second claim we use an argument in Elduque-Pérez [90, proof of Proposition 3.4]. If $\dim C = 2$ there is nothing to prove. Thus by Remark (34.7) we may assume that F contains a primitive cube root of unity. To simplify notations we denote the multiplication in C by $(x, y) \mapsto xy$ and we put $n = N_C$ for the norm of C. Let $x_i \in S_i$, $i = 1$, 2; we have $x_i^2 = -n(x_i) = 0$ by Lemma (34.12), so that

$$b_n(x_1 x_2, x_2 x_1) = b_n(x_1, x_2)^2$$

by (33.10). Furthermore $(x_1 x_2)(x_2 x_1) = x_1(x_2 x_2)x_1 = 0$ (by Artin's theorem, see (32.7)) and

$$(x_1 x_2)^2 - b_n(x_1 x_2, 1)x_1 x_2 + n(x_1 x_2) \cdot 1 = 0$$

implies that $(x_1x_2)^2 = -b_n(x_1, x_2)x_1x_2$. Choosing x_1, x_2 such that $b_n(x_1, x_2) = -1$, we see that $e_1 = x_1x_2$ is an idempotent of C and it is easily seen that $e_2 = 1 - e_1 = x_2x_1$. We claim that if $\dim S_0 = 2$, then $e_1 = y_1y_2$ for any pair $(y_1, y_2) \in S_1 \times S_2$ such that $b_n(y_1, y_2) = -1$. We have $S_0 = F \cdot e_1 \oplus F \cdot e_2$ if $\dim S_0 = 2$, so that the claim will follow if we can show that $b_n(e_1, y_1y_2) = 0$. Let $y_1 = \lambda x_1 + x_1'$ with $b_n(x_1', x_2) = 0$. By using (33.10) and the fact that $n(S_i) = 0$ for $i = 1$, 2, we deduce

$$b_n(e_1, y_1y_2) = b_n\big(x_1x_2, (\lambda x_1 + x_1')y_2\big)$$
$$= n(x_1)b_n(x_2, \lambda y_2) + b_n(x_1x_2, x_1'y_2)$$
$$= -b_n(x_1y_2, x_1'x_2).$$

However $x_1'x_2$ satisfies

$$(x_1'x_2)^2 - b_n(x_1'x_2, 1)x_1'x_2 + n(x_1'x_2) = 0$$

hence $(x_1'x_2)^2 = 0$, since $b_n(x_1'x_2, 1) = -b_n(x_1', x_2) = 0$. Since the algebra S_0 is étale, we must have $x_1'x_2 = 0$ and, as claimed, $b_n(e_1, y_1y_2) = 0$. Similarly we have $e_2 = y_2y_1$ for $(y_1, y_2) \in S_1 \times S_2$ such that $b_n(y_1, y_2) = -1$. It follows that

$$e_1y_1 = (1 - e_2)y_1 = (1 - y_2y_1)y_1 = y_1,$$
$$y_1e_1 = y_1(y_1y_2) = 0 = e_2y_1,$$
$$y_1e_2 = y_1, \quad e_2y_2 = y_2 = y_2e_1 \quad \text{and} \quad e_1y_2 = 0 = y_2e_2,$$

so that

$$S_1 = \{\, x \in C \mid e_1x = x = xe_2 \,\}$$

and

$$S_2 = \{\, x \in C \mid e_2x = x = xe_1 \,\}.$$

The element

$$e = \omega^2 e_1 + \omega e_2$$

is a para-unit of C_φ, since

$$e \star x = (\overline{\omega^2 e_1 + \omega e_2})(\omega^{2i}\overline{x}) = -(\omega e_1 + \omega^2 e_2)\omega^{2i}x = -x$$

for $x \in S_i$ and since

$$e \star (\omega e_1 - \omega^2 e_2) = (\omega^2 e_1 + \omega e_2)(-\omega^2 e_1 + \omega^2 e_2) = (-\omega e_1 + \omega^2 e_2)$$

for $\omega e_1 - \omega^2 e_2 \in e^\perp \subset S_0$. The claim then follows from Lemma (34.8). $\qquad \square$

34.C. Cubic separable alternative algebras. Following Faulkner [95] we now give another approach to symmetric composition algebras over fields of characteristic not 3. We first recall some useful identities holding in cubic alternative algebras. Let A be a finite dimensional unital separable alternative F-algebra of degree 3 and let

$$P_{A,a}(X) = X^3 - T_A(a)X^2 + S_A(a)X - N_A(a)1$$

be its generic minimal polynomial. The trace T_A is linear, the form S_A is quadratic and the norm N_A is cubic. As was observed in the introduction to this chapter we have

$$N_A(X - a \cdot 1) = P_{A,a}(X), \quad N_A(xy) = N_A(x)N_A(y), \quad \text{and} \quad T_A(xy) = T_A(yx).$$

Let

$$b_{S_A}(x, y) = S_A(x + y) - S_A(x) - S_A(y),$$
$$x^{\#} = x^2 - T_A(x)x + S_A(x) \cdot 1$$

and

$$x \times y = (x + y)^{\#} - x^{\#} - y^{\#}.$$

Note that

$$N_A(x) = xx^{\#} = x^{\#}x.$$

Observe that the #-operation and the ×-product are defined for any cubic algebra. They will be systematically used in Chapter IX for cubic Jordan algebras.

(34.14) Lemma. (1) $N_A(1) = 1$, $S_A(1) = T_A(1) = 3$, $1^{\#} = 1$, $1 \times 1 = 2$,
(2) $(xy)^{\#} = y^{\#}x^{\#}$,
(3) $S_A(x) = T_A(x^{\#})$, $S_A(x^{\#}) = T_A(x)N_A(x)$, $N_A(x^{\#}) = N_A(x)^2$,
(4) $b_{S_A}(x, y) = T_A(x \times y)$.
(5) $N_A(x + \lambda y) = \lambda^3 N_A(y) + \lambda^2 T_A(x \cdot y^{\#}) + \lambda T_A(x^{\#} \cdot y) + N_A(x)$ for x, $y \in A$, and $\lambda \in F$.
(6) *The coefficient of $\alpha\beta\gamma$ in $N_A(\alpha x + \beta y + \gamma z)$ is $T_A\big(x(y \times z)\big)$ and $T_A\big(x(y \times z)\big)$ is symmetric in x, y, and z.*
(7) $b_{S_A}(x, 1) = 2T_A(x)$,
(8) $x \times 1 = T_A(x) \cdot 1 - x$,
(9) $T_A(xy) = T_A(x)T_A(y) - b_{S_A}(x, y)$,
(10) $T_A\big((xz)y\big) = T_A\big(x(zy)\big)$.

Proof: We may assume that F is infinite and identify polynomials through their coefficients. $N_A(1) = 1$ follows from the multiplicativity of N_A, so that $1^{\#} = 1$ and $1 \times 1 = 2$. Putting $a = 1$ in $N_A(X - a \cdot 1) = P_{A,a}(X)$ gives $P_{A,1}(X) = (X - 1)^3 N_A(1) = (X - 1)^3$, hence $S_A(1) = T_A(1) = 3$.

By density it suffices to prove (2) for x, y such that $N_A(x) \neq 0 \neq N_A(y)$. Then $(xy)^{\#} = (xy)^{-1}N_A(xy) = y^{-1}N_A(y)x^{-1}N_A(x) = y^{\#}x^{\#}$.

Again by density, it suffices to prove (3) for x such that $N_A(x) \neq 0$. We then have $N_A(x - \lambda) = N_A(1 - \lambda x^{-1})N_A(x)$. Comparing the coefficients of λ gives (3), and (4) follows by linearizing (3).

(5) follows by computing $N_A(x + \lambda y) = N_A(xy^{-1} + \lambda)N_A(y)$.

The first claim of (6) follows by computing the coefficient of $\alpha\beta\gamma$ in $N_A\big(\alpha x + (\beta y + \gamma z)\big)$ (and using (5)). The last claim of (6) then is clear by symmetry.

(7) follows from (6), since

$$b_{S_A}(x, 1) = T_A(x \times 1) = T_A\big(x(1 \times 1)\big) = 2T_A(x)$$

and (7) implies (8).

For (9) we have

$$b_{S_A}(x, y) = T_A\big((x \times y)1\big) = T_A\big((x \times 1)y\big)$$
$$= T_A\big((T_A(x) \cdot 1 - x)y\big) = T_A(x)T_A(y) - T_A(xy).$$

Finally, by linearizing

$$T_A\big(x(yx)\big) = T_A\big((xy)x\big) = T_A(yx^2) = T_A\big(y(x^{\#} + xT_A(x) - S_A(x)1)\big),$$

we obtain

$$T_A\big(x(yz)\big) + T_A\big(z(yx)\big) = T_A\big((xy)z\big) + T_A\big((zy)x\big)$$
$$= T_A\big(y(x \times z)\big) + T_A(yx)T_A(z) + T_A(yz)T_A(x)$$
$$- T_A(y)b_{S_A}(x, z)$$
$$= T_A\big(y(x \times z)\big) + T_A(yx)T_A(z) + T_A(yz)T_A(x)$$
$$+ T_A(xz)T_A(y) - T_A(x)T_A(y)T_A(z),$$

so that by (6) $T_A\big(x(yz)\big) + T_A\big(z(yx)\big) = T_A\big((xy)z\big) + T_A\big((zy)x\big)$ is symmetric in x, y and z. It follows that

$$T_A\big(x(yz)\big) + T_A\big(z(yx)\big) = T_A\big(y(xz)\big) + T_A\big(z(xy)\big)$$

and $T_A\big((xz)y\big) = T_A\big(x(zy)\big)$, as claimed. $\square$

(34.15) Proposition. *A cubic alternative algebra is separable if and only if the bilinear trace form $T(x, y) = T_A(xy)$ is nonsingular.*

Proof: By (34.14) T is associative, hence the claim follows from Dieudonné's Theorem (32.4). $\square$

We recall:

(34.16) Proposition. *For any isomorphism $\alpha\colon A \xrightarrow{\sim} A'$ of cubic unital alternative algebras we have*

$$T_{A'}\big(\alpha(x)\big) = T_A(x), \quad S_{A'}\big(\alpha(x)\big) = S_A(x), \quad N_{A'}\big(\alpha(x)\big) = N_A(x).$$

Proof: The polynomial $p_{A', \alpha(a)}(X)$ is a minimal generic polynomial for A, hence the claim by uniqueness. $\square$

(34.17) Theorem. *Let A be a cubic separable unital alternative algebra over F of dimension > 1. Then either:*

(1) *$A \simeq L$, for some unique (up to isomorphism) cubic étale algebra L over F,*
(2) *$A \simeq F \times Q$ where Q is a unique (up to isomorphism) quaternion algebra over F,*
(3) *$A \simeq F \times \mathfrak{C}$ where $\mathfrak{C}$ is a unique (up to isomorphism) Cayley algebra over F,*
(4) *A is isomorphic to a unique (up to isomorphism) central simple associative algebra of degree 3.*

In particular such an algebra has dimension 3, 5 or 9. In case (1) the generic minimal polynomial is the characteristic polynomial, in case (2) and (3) the product of the generic minimal polynomial $p_{F,a}(X) = X - a$ of F with the generic minimal polynomial $p_{C,c}(X) = X^2 - T_C(c)X + N_C(c) \cdot 1$ of $C = Q$ or $C = \mathfrak{C}$ and in case (4) the reduced characteristic polynomial.

Proof: The claim is a special case of Theorem (32.5). $\square$

Let $^1A'_n$ denote the category of central simple algebras of degree $n+1$ over F. Let $\mathbf{I}\colon \mathsf{Sepalt}_{m-1}(2) \to \mathsf{Sepalt}_m(3)$ be the functor $C \mapsto F \times C$. Theorem (34.17) gives equivalences of groupoids $\mathsf{Sepalt}_3(3) \equiv S_3$, $\mathsf{Sepalt}_5(3) \equiv \mathbf{I}(^1A'_1)$, and $\mathsf{Sepalt}_9(3) \simeq \mathbf{I}(G_2) \sqcup {}^1A'_2$.

We assume from now on (and till the end of the section) that F is a field of characteristic different from 3. Let A be cubic alternative separable over F and let

$$A^0 = \{\, x \in A \mid T_A(x) = 0 \,\}.$$

Since $x = \frac{1}{3}T_A(x) \cdot 1 + x - \frac{1}{3}T_A(x) \cdot 1$ and $T_A\big(x - \frac{1}{3}T_A(x)\big) = 0$ we have $A = F \cdot 1 \oplus A^0$ and the bilinear trace form $T \colon (x, y) \mapsto T_A(xy)$ is nonsingular on A^0. By Lemma (34.14) the polar of the quadratic form S_A on A^0 is $-T$. Thus the restriction of S_A to A^0 is a nonsingular quadratic form.

We further assume that F contains a primitive cube root of unity ω and set $\mu = \frac{1-\omega}{3}$. We define a multiplication $\star$ on A^0 by

(34.18) $$x \star y = \mu xy + (1 - \mu)yx - \tfrac{1}{3}T_A(yx)1.$$

This type of multiplication was first considered by Okubo [199] for matrix algebras and by Faulkner [95] for cubic alternative algebras.

(34.19) Proposition. *The algebra $(A^0, \star)$ is a symmetric composition algebra with norm $n(x) = -\frac{1}{3}S_A(x)$.*

Proof: The form n is nonsingular, since S_A is nonsingular. We check that
$$(x \star y) \star x = x \star (y \star x) = -\tfrac{1}{3}S_A(x)y = n(x)y.$$

Lemma (34.1) will then imply that $(A^0, \star)$ is a symmetric composition algebra. We have $3\mu(1 - \mu) = 1$. It follows that

(34.20)
$$
\begin{aligned}
(x \star y) \star x = x \star (y \star x) &= \mu^2 xyx + (1 - \mu)^2 xyx + \mu(1 - \mu)(yx^2 + x^2 y) \\
&\quad - \tfrac{1}{3}T_A(xy)x - \tfrac{1}{3}\mu T_A(xyx)1 - \tfrac{1}{3}(1 - \mu)T_A(xyx)1 \\
&= [1 - 2\mu(1 - \mu)]xyx + \mu(1 - \mu)(yx^2 + x^2 y) \\
&\quad - \tfrac{1}{3}T_A(xy)x - \tfrac{1}{3}T_A(xyx)1 \\
&= \tfrac{1}{3}(xyx + yx^2 + x^2 y) - \tfrac{1}{3}T_A(xy)x - \tfrac{1}{3}T_A(xyx)1.
\end{aligned}
$$

By evaluating T_A on the generic polynomial, we obtain $3N_A(x) = T_A(x^3)$ for elements in A^0. Thus

(34.21) $$x^3 + S_A(x)x - \tfrac{1}{3}T_A(x^3)1 = 0$$

holds for all $x \in A^0$. Since it suffices to prove (34.19) over a field extension, we may assume that F is infinite. Replacing x by $x + \lambda y$ in (34.21), the coefficient of λ must then be zero. Hence we are lead to the identity
$$xyx + yx^2 + x^2 y - T_A(xy)x + S_A(x)y - T_A(xyx)1 = 0$$

for all $x, y \in A^0$, taking into account that $b_{S_A}(x, y) = -T_A(xy)$ on A^0. Combining this with equation (34.20) shows that
$$(x \star y) \star x = x \star (y \star x) = -\tfrac{1}{3}S_A(x)y = n(x)y$$

as claimed. $\qquad\qquad\qquad\qquad\qquad\qquad\qquad\qquad\qquad\qquad\qquad\square$

Hence we have a functor ${}^1\mathbf{C} \colon \mathit{Sepalt}_{m+1}(3) \to \mathit{Scomp}_m$ for $m = 2$, 4 and 8 given by $A \mapsto (A^0, n)$. We now construct a functor ${}^1\mathbf{A}$ in opposite direction; a straightforward computation shows that (34.18) is equivalent to

(34.22) $$xy = (1 + \omega)x \star y - \omega y \star x + b_n(x, y) \cdot 1$$

for the multiplication in $A^0 \subset A$. Thus, given a symmetric composition $(S, \star)$, it is natural to define a multiplication $(x, y) \mapsto x \cdot y = xy$ on $A = F \oplus S$ by (34.22) for $x, y \in S$, and by $1 \cdot x = x = x \cdot 1$. Let ${}^1\mathbf{A}$ be the functor $(S, \star) \mapsto (F \oplus S, \cdot)$.

(34.23) Theorem (Elduque-Myung). *The functors* $^1\mathbf{C}$ *and* $^1\mathbf{A}$ *define an equivalence of groupoids*

$$\mathsf{Sepalt}_{m+1}(3) \equiv \mathsf{Scomp}_m$$

for $m = 2$, 4 *and* 8.

Proof: We first show that $A = {}^1\mathbf{A}(S) = F \oplus S$ is a separable alternative algebra of degree 3: Let $x = \alpha 1 + a$, $\alpha \in F$ and $a \in S$. We have

$$x^2 = \big(\alpha^2 + b_n(a,a)\big)1 + 2\alpha a + a \star a$$

and

$$xx^2 = x^2 x = [\alpha^3 + 3\alpha b_n(a,a) + b_n(a \star a, a)]1 + [3\alpha^2 + 3n(a)]a + 3\alpha(a \star a).$$

It follows that

$$x^3 - 3\alpha x^2 + \big(3\alpha^2 - 3n(a)\big)x = [\alpha^3 - 3n(a)\alpha + b_n(a \star a, a)]1$$

so that elements of A satisfy a polynomial condition of degree 3

$$p_{A,x}(X) = X^3 - T_A(x)X^2 + S_A(x)X - N_A(x)1 = 0$$

with

$$T_A(x) = 3\alpha, \quad S_A(x) = 3\alpha^2 - 3n(a)$$

and

$$N_A(x) = \alpha^3 - 3\alpha n(a) + b_n(a \star a, a)$$

for $x = \alpha 1 + a$. To show that A is of degree 3 we may assume that the ground field F is infinite and we need an element $x \in A$ such that the set $\{1, x, x^2\}$ is linearly independent. Because $x^2 = x \star x + (x,x)1$ for $x \in S$, it suffices to have $x \in S$ such that $\{1, x, x \star x\}$ is linearly independent. Since $T_A(1) = 3$, while $T_A(x) = 0 = T_A(x \star x)$, the only possible linear dependence is between x and $x \star x$. If $\{x, x \star x\}$ is linearly dependent for all $x \in S$, there is a map $f \colon S \to F$ such that $x \star x = f(x)x$ for $x \in S$. By the following Lemma (34.24) f is linear. Since $x \star x \star x = n(x)x$ we get $n(x) = f(x)^2$. This is only possible if $\dim_F S = 1$. We next check that A is alternative. It suffices to verify that

$$a^2 b = a(ab) \quad \text{and} \quad ba^2 = (ba)a \quad \text{for } a,\, b \in S.$$

We have

$$a^2 b = [a \star a + (a,a)]b = (1 + \omega)(a \star a) \star b - \omega b \star (a \star a) + b_n(a \star a, b)$$

and

$$\begin{aligned} a(ab) &= a[(1 + \omega)a \star b - \omega b \star a + b_n(a,b)] \\ &= (1 + \omega)[a \star (a \star b) - \omega(a \star b) \star a + b_n(a, a \star b)] \\ &\quad - \omega[(1 + \omega)a \star (b \star a) - \omega(b \star a) \star a + b_n(a, b \star a)] + b_n(a,b)a. \end{aligned}$$

By (34.2) we have

$$(a \star a) \star b + (b \star a) \star a = b_n(a,b)a = b \star (a \star a) + a \star (a \star b).$$

This, together with the identities

$$b_n(a, a \star b) = b_n(a \star a, b) = b_n(b, a \star a) = b_n(b \star a, a) = b_n(a, b \star a)$$

which follow from the associativity of n, implies that $a^2 b = a(ab)$. The proof of $ba^2 = (ba)a$ is similar. Thus A is alternative of degree 3. We next check that A is separable. We have for $x = \alpha + a$, $y = \beta + b$,

$$xy = \alpha\beta + \beta a + \alpha b + (1 + \omega)a \star b - \omega b \star a + b_n(a, b)$$

so that

$$T(x, y) = T_A(xy) = 3\alpha\beta + 3b_n(a, b)$$

is a nonsingular bilinear form. Since the trace form of a cubic alternative algebra is associative (Lemma (34.14)), A is separable by Dieudonné's Theorem (32.4). We finally have an equivalence of groupoids since

$$(F \oplus S)^0 = S \quad \text{and} \quad F \oplus A^0 = A$$

and since formulas (34.22) and (34.18) are equivalent.   □

(34.24) Lemma. *Let F be an infinite field and let $(S, \star)$ be an F-algebra. If there exists a map $f \colon S \to F$ such that $x \star x = f(x)x$ for all $x \in S$, then f is linear.*

Proof: (Elduque) If S is 1-dimensional, the claim is clear. So let $(e_1, \ldots, e_n)$ be a basis of S. For $x = \sum x_i e_i$ and $e_i \star e_j = \sum_k a_{ij}^k e_k$, we have $\sum_{i,j} a_{ij}^k x_i x_j = f(x)x_k$ for any k. Thus $f(x) = g_k(x)x_k^{-1}$ for some quadratic homogeneous polynomial $g_k(x)$ and $k = 1, \ldots, n$ in the Zariski open set

$$D(x_k) = \{\, x \in S \mid x_k \neq 0 \,\}.$$

For any pair i, j we have $g_i(x)x_j = g_j(x)x_i$ in $D(x_i) \cap D(x_j)$, so by density $g_i(x)x_j = g_j(x)x_i$ holds for any $x \in S$. Unique factorization over the polynomial ring $F[x_1, \ldots x_n]$ shows that there exists a linear map $\phi \colon S \to F$ such that $g_i(x) = x_i \phi(x)$. It is clear that $f = \phi$.   □

(34.25) Remark. Let A be central simple of degree 3 over F and assume that F has characteristic different from 3 and that F contains a primitive cube root of unity. The form n from (34.19) is then hyperbolic on A^0: by Springer's Theorem (see [28, p. 119]) we may assume that A is split, and in that case the claim is easy to check directly. Hence, if A and A' are of degree 3 and are not isomorphic, the compositions $(A^0, \star)$ and $(A'^0, \star)$ are nonisomorphic (by (34.23)) but have isometric norms. This is in contrast with Cayley (or para-Cayley) composition algebras.

(34.26) Remark. The *polar* of a cubic form N is

$$N(x, y, z) = N(x + y + z) - N(x + y) - N(x + z) - N(y + z)$$
$$+ N(x) + N(y) + N(z)$$

and N is *nonsingular* if its polar is nonsingular, i.e., if $N(x, y, z) = 0$ for all x, y implies that $z = 0$. Let A be an F-algebra. If char $F \neq 2, 3$ a necessary and sufficient condition for A to admit a nonsingular cubic form N which admits composition (i.e., such that $N(xy) = N(x)N(y)$) is that A is cubic separable alternative and $N \simeq N_A$ (see Schafer [243, Theorem 3]). Thus, putting $x = \alpha \cdot 1 + a \in F \cdot 1 \oplus A^0$, the multiplicativity of $N_A(x) = \alpha^3 - 3\alpha n(a) + b_n(a \star a, a)$ for the multiplication $(x, y) \mapsto xy$ of A is equivalent by Proposition (34.23) to the multiplicativity of $n = -\frac{1}{3}S_A$ for the multiplication $(a, b) \mapsto a \star b$ of A^0. It would be nice to have a direct proof!

A symmetric composition algebra isomorphic to a composition $(A^0, \star)$ for A central simple of degree 3 is called an *Okubo composition algebra* or a *composition algebra of type* $^1A_2$ since its automorphism group is a simple adjoint algebraic group of type $^1A_2$. Twisted forms of Okubo algebras are again Okubo algebras. The groupoid of Okubo composition algebras over a field F containing a primitive cube root of unity is denoted ^1Oku. We have an equivalence of groupoids $^1Oku \equiv {}^1A'_2$ (if F contains a primitive cube root of unity).

For para-Hurwitz compositions of dimension 4 or 8 we have the following situation:

(34.27) Proposition. *Let* $\mathbf{I}\colon Hurw_m \to Sepalt_{m+1}(3)$ *be the functor* $C \mapsto F \times C$, $\mathbf{P}\colon Hurw_m \to \overline{Hurw}_m$ *the para-Hurwitz functor and* $\mathbf{J}\colon \overline{Hurw}_m \to Scomp_m$ *the inclusion. Then the map*

$$\eta_C\colon C \to (F \times C)^0 \quad \text{given by} \quad z \mapsto \left(T_C(z), \omega z + \omega^2 \bar{z}\right)$$

is a natural transformation between the functors $^1\mathbf{C} \circ \mathbf{I}$ *and* $\mathbf{J} \circ \mathbf{P}$, *i.e., the diagram*

$$
\begin{array}{ccc}
Sepalt_{m+1}(3) & \xrightarrow{\;{}^1\mathbf{C}\;} & Scomp_m \\[2pt]
{\scriptstyle \mathbf{I}}\big\uparrow & & \big\uparrow{\scriptstyle \mathbf{J}} \\[2pt]
Hurw_m & \xrightarrow{\;\mathbf{P}\;} & \overline{Hurw}_m
\end{array}
$$

commutes up to η_C.

Proof: It suffices to check that η_C is an isomorphism of the para-Hurwitz algebra $(C, \star)$ with the symmetric composition algebra $\left((F \times C)^0, \star\right)$. We shall use that $T_A(x) = \xi + T_C(c)$, $S_A(x) = N_C(c) + \xi T_C(c)$ and $N_A(x) = \xi N_C(c)$ for $A = F \times C$, $\xi \in F$, $c \in C$ and T_C the trace and N_C the norm of C. If char $F \neq 2$, we decompose $C = F \cdot 1 \oplus C^0$ and set $u = (2, -1) \in A^0$. We then have

$$\eta_C(\beta e + x) = \beta u + (1 + 2\omega)x.$$

The element u satisfies $u \star u = u$ and $(0, x) \star u = u \star (0, x) = (0, -x)$ for $x \in C^0$. Thus it suffices to check the multiplicativity of η_C on products of elements in C^0, in which case the claim follows by a tedious but straightforward computation. If char $F = 2$, we choose $v \in C$ with $T_C(v) = 1$, to have $C = F \cdot v \oplus C^0$. We then have

$$\eta_C(\beta v + x) = (\beta, \beta v + x + \omega^2 \beta)$$

and the proof is reduced to checking the assertion for the products $v \star v$, $v \star x$, $x \star v$ and $x \star x$, where it is easy. $\qquad\square$

For symmetric compositions of dimension 2, we do not have to assume in Proposition (34.23) that F contains a contains a primitive cube root of unity since a separable alternative algebra A of degree 3 and dimension 3 is étale commutative and the multiplication $\star$ reduces to $a \star b = ab - \frac{1}{3}T_A(ab)$. Thus (34.23) and (34.27) imply:

(34.28) Proposition. *Let F be a field of characteristic different from 3 and let $(S, \star, n)$ be a symmetric composition algebra over F of dimension 2. Then $S \simeq (L^0, \star)$ for a unique cubic étale F-algebra L. The algebra S is para-quadratic if and only if L is not a field.* $\qquad\square$

To obtain a complete description of symmetric compositions in dimension 4 and 8 we may take the Structure Theorem (34.17) for cubic alternative algebras into account:

(34.29) Theorem (Elduque-Myung). *Let F be field with* char $\neq 3$ *which contains a primitive cube root of unity. There exist equivalences of groupoids*

$$Scomp_4 \equiv \overline{Hurw}_4 \equiv Hurw_4 \equiv A'_1$$

and

$$Scomp_8 \equiv \overline{Hurw}_8 \sqcup {}^1Oku \equiv G_2 \sqcup {}^1A'_2. \qquad \square$$

However we did not prove (34.17) and we shall give an alternate proof of these equivalences. Observe that this will yield, in turn, a proof of Theorem (34.17)!

We postpone the proof of (34.29) and begin with an example:

(34.30) Example. Let V be a 2-dimensional vector space over F. We view elements of $\mathrm{End}_F(V \oplus F)$ of trace zero as block matrices

$$\begin{pmatrix} \phi & v \\ f & -\mathrm{tr}(\phi) \end{pmatrix} \in \begin{pmatrix} \mathrm{End}(V) & V \\ V^* & F \end{pmatrix}$$

The product of such two blocks is given by

$$\begin{pmatrix} \phi & v \\ f & -\mathrm{tr}(\phi) \end{pmatrix} \cdot \begin{pmatrix} \phi' & v' \\ f' & -\mathrm{tr}(\phi') \end{pmatrix} = \begin{pmatrix} \phi \circ \phi' + v \circ f' & \phi(v') - \mathrm{tr}(\phi')v \\ f \circ \phi' - \mathrm{tr}(\phi)f' & f(v') + \mathrm{tr}(\phi)\,\mathrm{tr}(\phi') \end{pmatrix}$$

where $(v \circ f')(x) = vf'(x)$. With the multiplication $(x, y) \mapsto x \star y$ defined in (34.18) and the quadratic form $n = -\frac{1}{3}S_{\mathrm{End}_F(V \oplus F)}$,

$$S_V = {}^1\mathbf{C}\big(\mathrm{End}_F(V \oplus F)\big) = \big(\mathrm{End}_F(V \oplus F)^0, \star\big)$$

is a symmetric composition algebra and

$$e = \begin{pmatrix} -1_V & 0 \\ 0 & 2 \end{pmatrix} \in \begin{pmatrix} \mathrm{End}(V) & V \\ V^* & F \end{pmatrix}$$

is a nontrivial idempotent. By Proposition (34.9), (2), the map

$$\varphi(x) = e \star (e \star x) = b_n(e, x)e - x \star e$$

is an automorphism of $(S_V, \star)$ of order 3, such that $(S_V, \star)$ reduces to the Petersson algebra $S_{V\varphi}$. The corresponding $\mathbb{Z}/3\mathbb{Z}$-grading is

$$S_0 = \mathrm{End}_F(V), \quad S_1 = V, \quad S_2 = V^*.$$

In particular we have $\dim_F S_0 = 4$. The converse also holds (Elduque-Pérez [90, Theorem 3.5]) in view of the following:

(34.31) Proposition. *Let F be a field of characteristic not 3 which contains a primitive cube root of unity ω. Let $(S, \star, n)$ be a symmetric composition algebra of dimension 8 with a nontrivial idempotent e. Let $\varphi(x) = e \star (e \star x)$ and let $S = S_0 \oplus S_1 \oplus S_2$ be the $\mathbb{Z}/3\mathbb{Z}$-grading of $(S, \star, n)$ defined by φ (see (34.9)). If $\dim_F S_0 = 4$, there exists a 2-dimensional vector space V such that $(S, \star) \simeq S_{V\varphi} = {}^1\mathbf{C}\big(\mathrm{End}(V \oplus F)\big)$.*

Proof: Since $\dim_F S_0 = 4$ we must have $\dim_F S_1 = \dim_F S_2 = 2$, and S_1, S_2 are maximal isotropic direct summands of S by Lemma (34.12). Let (x_1, x_2) be a basis of S_1 and let (f_1, f_2) be a basis of S_2 such that $b_n(x_i, f_j) = \delta_{ij}$. Since

$$(x_i + f_i) \star \big(u \star (x_i + f_i)\big) = \big((x_i + f_i) \star u\big) \star (x_i + f_i) = u$$

holds for all u we have

(1) $$x_i \star (u \star x_i) = 0 = (x_i \star u) \star x_i$$

(2) $$f_j \star (u \star f_j) = 0 = (f_j \star u) \star f_j$$

and

(3) $$x_i \star (u \star f_i) + f_i \star (u \star x_i) = u$$

for all $u \in S_0$. Thus, by choosing $u = e$ and using that

$$x = -\omega e \star x = -\omega^2 x \star e, \quad f = -\omega f \star e = -\omega^2 e \star f$$

for $x \in S_1$, resp. $f \in S_2$, we see that $x_i \star x_i = 0 = f_j \star f_j$. Moreover $x_1 \star x_2 = 0 = x_2 \star x_1$ and $f_1 \star f_2 = 0 = f_2 \star f_1$, so $S_1 \star S_1 = 0 = S_2 \star S_2$. Since $u \star x_i \in S_i$, $u \star f_j \in S_2$ for $u \in S_0$, (3) implies that

$$(f_1 \star x_1, f_1 \star x_2, x_1 \star f_1, x_1 \star f_2)$$

generates S_0, hence is a basis of S_0 and $S_1 \star S_2 = S_0 = S_2 \star S_1$. We now define an F-linear map $\psi \colon S \to M_3(F)$ on basis elements by

$$x_1 \mapsto \begin{pmatrix} 0 & 0 & 1 \\ 0 & 0 & 0 \\ 0 & 0 & 0 \end{pmatrix}, \quad x_2 \mapsto \begin{pmatrix} 0 & 0 & 0 \\ 0 & 0 & 1 \\ 0 & 0 & 0 \end{pmatrix}, \quad f_1 \mapsto \begin{pmatrix} 0 & 0 & 0 \\ 0 & 0 & 0 \\ 3 & 0 & 0 \end{pmatrix}, \quad f_2 \mapsto \begin{pmatrix} 0 & 0 & 0 \\ 0 & 0 & 0 \\ 0 & 3 & 0 \end{pmatrix}$$

and

$$x_i \star f_j \mapsto \mu x_i f_j + (1 - \mu) f_j x_i - \tfrac{1}{3} \operatorname{tr}(x_i f_j),$$
$$f_j \star x_i \mapsto \mu f_j x_i + (1 - \mu) x_i f_j - \tfrac{1}{3} \operatorname{tr}(x_i f_j)$$

where multiplication on the right is in $M_3(F)$ and $\mu = \frac{1-\omega}{3}$. We leave it as a (lengthy) exercise to check that ψ is an isomorphism of composition algebras $S \xrightarrow{\sim} {}^1\mathbf{C}\big(M_3(F)\big) = S_V$ with $V = F^2$. $\square$

Proof of Theorem (34.29): Let $(S, \star, n)$ be a symmetric composition algebra. Since twisted forms of Okubo algebras are Okubo and twisted forms of para-Hurwitz algebras of dimension ≥ 4 are para-Hurwitz, we may assume by Lemma (34.10) that S contains a nontrivial idempotent. Let $S = S_0 \oplus S_1 \oplus S_2$ be the grading given by Lemma (34.12). Assume first that $\dim S = 4$. If $\dim S_0 = 4$, then $S = S_0$ is para-Hurwitz; and if $\dim S_0 = 2$, S is para-Hurwitz by Proposition (34.13). If $\dim S = 8$ and $\dim S_0 = 2$, S is para-Hurwitz by Proposition (34.13); if $\dim S = 8$ and $\dim S_0 = 4$, S is Okubo by Proposition (34.31). $\square$

34.D. Alternative algebras with unitary involutions. To overcome the condition on the existence of a primitive cube root of unity in F, one considers separable cubic alternative algebras B over the quadratic extension $K = F(\omega) = F[X]/(X^2 + X + 1)$ which admit a unitary involution τ.

(34.32) Proposition. *Let K be a quadratic separable field extension of F with conjugation ι. Cubic separable unital alternative K-algebras with a unitary involution τ are of the following types:*

(1) $L = K \otimes L_0$ *for L_0 cubic étale over F and $\tau = \iota \otimes 1$.*

(2) *Central simple associative algebras of degree 3 over K with a unitary involution.*

(3) *Products $K \times (K \otimes C)$ where C is a Hurwitz algebra of dimension 4 or 8 over F and $\tau = (\iota, \iota \otimes \pi)$ where π is the conjugation of the Hurwitz algebra C.*

Proof: A cubic separable unital K-algebra is of the types described in Theorem (34.17). (1) then follows by Galois descent and (3) follows from Proposition (2.22) (the case of a Cayley algebra being proved as the quaternion case). □

Assume that F has characteristic different from 3 and that $K = F(\omega)$ is a field. Let ${}^2\mathit{Sepalt}_m(3)$ be the groupoid of alternative F-algebras (B, τ) which are separable cubic of dimension m over K and have unitary involutions. The generic polynomial of degree 3 on B with coefficients in K restricts to a polynomial function on $\mathrm{Sym}(B, \tau)$ with coefficients in F. Let

$$\mathrm{Sym}(B, \tau)^0 = \{\, x \in \mathrm{Sym}(B, \tau) \mid T_B(x) = 0 \,\}.$$

We define a multiplication $\star$ on $\mathrm{Sym}(B, \tau)^0$ as in (34.18):

(34.33) $x \star y = \mu x y + (1 - \mu) y x - \tfrac{1}{3} T_B(yx) 1.$

The element $x \star y$ lies in $\mathrm{Sym}(B, \tau)^0$ since $\iota(\mu) = 1 - \mu$. Let $n(x) = -\tfrac{1}{3} T_B(x^2)$.

A description of $\mathrm{Sym}(B, \tau)$ (and of $\star$) in cases (1) and (2) is obvious, however less obvious in case (3).

(34.34) Lemma. *Let C be a Hurwitz algebra over F, let K be a quadratic étale F-algebra, and let B be the alternative K-algebra $B = K \times (K \otimes C)$ with unitary involution $\tau = (\iota, \iota \otimes \pi)$. Then:*
(1)

$$\mathrm{Sym}(B, \tau)^0 = \{\, \big(T_C(z), \xi \otimes z + \iota(\xi) \otimes \pi(z)\big) \mid z \in C \,\}$$

where ξ is a (fixed) generator of K such that $T_K(\xi) = -1$.
(2) If $K = F(\omega)$ where ω is a primitive cube root of unity, the map $z \mapsto \omega \otimes z + \iota(\omega) \otimes \pi(z)$ is an isomorphism $(C, \star) \xrightarrow{\sim} \big(\mathrm{Sym}(B, \tau)^0, \star\big)$.

Proof: (1) We obviously have

$$\mathrm{Sym}(B, \tau)^0 \supset \{\, \big(T_C(z), \xi \otimes z + \iota(\xi) \otimes \pi(z)\big) \mid z \in C \,\},$$

hence the claim by dimension count, since the map

$$\mathrm{Sym}(B, \tau)^0 \to \{\, \big(T_C(z), \xi \otimes z + \iota(\xi) \otimes \pi(z)\big) \mid z \in C \,\}$$

given by $z \mapsto \xi \otimes z + \iota(\xi) \otimes \pi(z)$ is an isomorphism of vector spaces.
(2) follows from (1) and Proposition (34.27). □

(34.35) Theorem. *Let F be a field of characteristic not 3 which does not contain a primitive cube root of unity ω and let $K = F(\omega)$. For any cubic separable alternative K-algebra with a unitary involution τ, the F-vector space $\big(\mathrm{Sym}(B, \tau)^0, \star\big)$ is a symmetric composition algebra. Conversely, for any symmetric composition algebra $(S, \star)$, the unital alternative K-algebra $B = K \cdot 1 \oplus (K \otimes S)$ with the multiplication*

$$xy = (1 + \omega) x \star y - \omega y \star x + b_n(x, y) \cdot 1, \quad x \cdot 1 = 1 \cdot x = x$$

for x, $y \in B^0$, admits the unitary involution $(\iota, \iota \otimes 1_S)$ and the functors

$$\mathbf{{}^2C} \colon (B, \tau) \mapsto \big(\mathrm{Sym}(B, \tau)^0, \star\big),$$
$$\mathbf{{}^2A} \colon (S, \star) \mapsto \big(B = K \cdot 1 \oplus (K \otimes S), (\iota, \iota \otimes 1)\big)$$

define an equivalence of groupoids

$${}^2\mathit{Sepalt}_{m+1}(3) \equiv \mathit{Scomp}_m$$

for $m = 2, 4$ and 8.

Proof: To check that $\left(\mathrm{Sym}(B,\tau)^0, \star\right)$ is a symmetric composition algebra over F, we may assume that $K = F \times F$. Then B is of the form (A, A^{op}) and $\tau(a, b^{\mathrm{op}}) = (b, a^{\mathrm{op}})$ for A separable alternative of degree 3 as in the associative case (see (2.14)). By projecting on the first factor, we obtain an isomorphism

$$\mathrm{Sym}(B, \tau)^0 \simeq A^0 = \{\, x \in A \mid T_A(x) = 0 \,\},$$

and the product $\star$ on A^0 is as in (34.18). Thus the composition $\left(\mathrm{Sym}(B, \tau)^0, \star\right)$ is isomorphic to $(A^0, \star)$ and hence is a symmetric composition.

Conversely, if $(S, \star)$ is a symmetric composition algebra over F, then $(S, \star) \otimes K$ is a symmetric composition algebra over K and $(K \cdot 1 \oplus K \otimes S, \cdot)$ is a K-alternative algebra by Theorem (34.23). The fact that $(\iota, \iota \otimes 1)$ is a unitary involution on B follows from the definition of the multiplication of B. That $^2\mathbf{C}$ and $^2\mathbf{A}$ define an equivalence of groupoids then follows as in (34.23). $\qquad\square$

(34.36) Corollary. *Let $(S, \star)$ be a symmetric composition algebra with norm n. The following conditions are equivalent*:

(1) *S contains a nontrivial idempotent.*
(2) *the cubic form $N(x) = b_n(x \star x, x)$ is isotropic.*
(3) *the alternative algebra $^1\mathbf{A}(S)$ (if F contains a primitive cube root of unity, otherwise $^2\mathbf{A}(S)$) has zero divisors.*

Proof: (1) $\Rightarrow$ (2) By Proposition (34.9) the map $\varphi(x) = e \star (e \star x)$ is an automorphism of $(S, \star)$ of order ≤ 3 and the corresponding subalgebra S_0 (see Lemma (34.12)) has dimension at least 2. For every nonzero $x \in S_0$ with $b_n(x, e) = 0$ we have $x \star e = e \star x = -x$, so that $b_n(x \star x, x) = -b_n(x \star x, e \star x) = n(x)b_n(x, e) = 0$.

The implication (2) $\Rightarrow$ (1) is Lemma (34.10).

We check that (3) $\Leftrightarrow$ (2): If S is Okubo, then $^2\mathbf{A}(S)$ is central simple with zero divisors and $\mathrm{Nrd}_B(x) = b_n(x \star x, x)$ for $x \in \mathrm{Sym}(B, \tau)^0$, hence the claim in this case. Since $K \times K \otimes S$ always has zero divisors we are left with showing that N is always isotropic on a para-Hurwitz algebra. Take $x \neq 0$ with $b_n(x, 1) = 0$; then $b_n(x \star x, x) = b_n(-n(x)1, x) = 0$. $\qquad\square$

In the following classification of Elduque-Myung [89, p. 2487] "unique" always means up to isomorphism:

(34.37) Theorem (Classification of symmetric compositions). *Let F be a field of characteristic $\neq 3$ and let $(S, \star)$ be a symmetric composition algebra over F.*

(1) *If $\dim_F S = 2$, then $S \simeq (L^0, \star)$ for a unique cubic étale F-algebra L. The algebra S is para-quadratic if and only if L is not a field.*

(2) *If $\dim_F S = 4$, then S is isomorphic to a para-quaternion algebra $(Q, \star)$ for a unique quaternion algebra Q.*

(3) *If $\dim_F S = 8$, then S is either isomorphic to*
   (a) *a para-Cayley algebra $(\mathfrak{C}, \star)$ for a unique Cayley algebra $\mathfrak{C}$,*
   (b) *an algebra of the form $(A^0, \star)$ for a unique central simple F-algebra A of degree 3 if F contains a primitive cube root of unity, or*
   (c) *S is of the form $\left(\mathrm{Sym}(B, \tau)^0, \star\right)$ for a unique central simple $F(\omega)$-algebra B of degree 3 with an involution of the second kind if F does not contain a primitive cube root of unity ω.*

Proof: (1) was already proved in Proposition (34.28). Let $K = F(\omega)$. By Theorem (34.35) we have $(S, \star) \simeq \left(\mathrm{Sym}(B, \tau)^0, \star\right)$ for some alternative K-algebra B

with a unitary involution τ. By Proposition (34.32) cases (3.b) and (3.c) occur if B is central simple over K, and cases (2) and (3.a) occur when $B \simeq K \oplus K \otimes C'$ for some Hurwitz algebra C' of dimension 4 or 8. In view of Lemma (34.34), we must have in the two last cases $(S, \star) \simeq (C', \star)$ so that $(S, \star)$ is a para-Hurwitz algebra, as asserted. $\square$

We call symmetric composition algebras as in (3.b) or (3.c) *Okubo algebras* (the case where ω lies in F is not new!), we denote the corresponding groupoids by ^1Oku (when $K = F \times F$), ^2Oku (when $\omega \notin F$) and we set $Oku = {}^1Oku \sqcup {}^2Oku$. Assume that F does not contain a primitive cube root of unity ω and let $^2A_2$ be the full subgroupoid of A_2 whose objects are algebras of degree 3 over $F(\omega)$ with involution of the second kind. Since we have equivalences of groupoids $^1Oku \equiv {}^1A_2'$ and $^2Oku \equiv {}^2A_2'$, we also call symmetric composition in ^iOku symmetric compositions of type $^iA_2'$.

(34.38) Corollary. *Let F be field of characteristic different from 3. There exist equivalences of groupoids*

$$Scomp_4 \equiv \overline{Hurw}_4 \equiv Hurw_4 \equiv A_1'$$

and

$$Scomp_8 \equiv \overline{Hurw}_8 \sqcup Oku \equiv G_2 \sqcup {}^1A_2' \sqcup {}^2A_2'.$$ $\square$

34.E. Cohomological invariants of symmetric compositions. Three cohomological invariants classify central simple algebras (B, τ) of degree 3 with unitary involutions τ (see Theorem (30.21)): $f_1 \in H^1(F, \mu_2)$ (which determines the center K), $g_2 \in H^2\big(F, (\mu_{3[K]})\big)$ (which determines the K-algebra B) and $f_3 \in H^3(F, \mu_2)$ (which determines the involution). We have a corresponding classification for symmetric compositions:

(34.39) Proposition. *Let F be a field of characteristic not 2, 3. Dimension 8 symmetric compositions of*

(1) *type G_2 are classified by one cohomological invariant $f_3 \in H^3(F, \mu_2)$.*
(2) *type $^1A_2'$ (if $\omega \in F$) by one invariant $g_2 \in H^2(F, \mu_3)$.*
(3) *type $^2A_2'$ by two cohomological invariants $g_2 \in H^2(F, \mathbb{Z}/3\mathbb{Z})$ and $f_3 \in H^3(F, \mu_2)$.*

Proof: The claims follow from Theorem (34.37) and the corresponding classifications of Cayley algebras, resp. central simple algebras. For (3), one also has to observe that if $K = F(\omega)$, then the action on μ_3 twisted by a cocycle γ defining K is the usual action of the Galois group. Thus $(\mu_3)_\gamma = \mathbb{Z}/3\mathbb{Z}$. $\square$

Observe that a symmetric composition algebra S of type $^1A_2$ with $g_2 \neq 0$ comes from a division algebra A over F. However, its norm is always hyperbolic (see Example (34.25)), hence $f_3(S) = 0$. On the other hand, in view of the following example, there exist composition algebras of type $^2A_2$ with invariants $g_2(S) = 0$ and $f_3(S) \neq 0$.

(34.40) Example. Over $\mathbb{R}$ there are no compositions of type $^1A_2$ but there exist compositions of type $^2A_2$ with $K = \mathbb{C}$ and $B = M_3(\mathbb{C})$. There are two classes of involutions of the second type on $M_3(\mathbb{C})$, the standard involution $x \mapsto \tau(x)$ where $\tau(x)$ is the hermitian conjugate, and the involution $\text{Int}(d) \circ \tau \colon x \mapsto d\tau(x)d^{-1}$, with $d = \text{diag}(-1, 1, 1)$, which is distinguished. Since $\text{tr}(x^2) > 0$ for any nonzero hermitian 3×3 matrix x, the norm on $\text{Sym}(B, \tau)^0$ is anisotropic. The restriction of the norm to $\mathrm{Sym}\big(B, \mathrm{Int}(d) \circ \tau\big)^0$ is hyperbolic. Observe that $f_3\big(\mathrm{Sym}(B, \tau)^0\big) \neq 0$ and $f_3\big(\mathrm{Sym}(B, \mathrm{Int}(d) \circ \tau)^0\big) = 0$.

§35. Clifford Algebras and Triality

35.A. The Clifford algebra. Let $(S, \star)$ be a symmetric composition algebra of dimension 8 over F with norm n. Let $C(n)$ be the Clifford algebra and $C_0(n)$ the even Clifford algebra of (S, n). Let τ be the involution of $C(n)$ which is the identity on S. Let $r_x(y) = y \star x$ and $\ell_x(y) = x \star y$.

(35.1) Proposition. *For any* $\lambda \in F^\times$, *the map* $S \to \mathrm{End}_F(S \oplus S)$ *given by*

$$ x \mapsto \begin{pmatrix} 0 & \lambda\ell_x \\ r_x & 0 \end{pmatrix} $$

induces isomorphisms

$$ \alpha_S \colon \big(C(\lambda n), \tau\big) \xrightarrow{\sim} \big(\mathrm{End}_F(S \oplus S), \sigma_{n \perp \lambda n}\big) $$

and

$$ \alpha_S \colon \big(C_0(\lambda n), \tau\big) \xrightarrow{\sim} \big(\mathrm{End}_F(S), \sigma_n\big) \times \big(\mathrm{End}_F(S), \sigma_n\big), $$

of algebras with involution.

Proof: We have $r_x \circ \lambda\ell_x(y) = \lambda\ell_x \circ r_x(y) = \lambda n(x) \cdot y$ by Lemma (34.1). Thus the existence of the map α_S follows from the universal property of the Clifford algebra. The fact that α_S is compatible with involutions is equivalent to

$$ b_n\big(x \star (z \star y), u\big) = b_n\big(z, y \star (u \star x)\big) $$

for all x, y, z, u in S. This formula follows from the associativity of n, since

$$ b_n\big(x \star (z \star y), u\big) = b_n(u \star x, z \star y) = b_n\big(z, y \star (u \star x)\big). $$

The map α_S is an isomorphism by dimension count, since $C(n)$ is central simple. $\square$

Let (V, q) be a quadratic space of even dimension. We call the class of the Clifford algebra $[C(V, q)] \in \mathrm{Br}(F)$ the *Clifford invariant* of (V, q). It follows from Proposition (35.1) that, for any symmetric composition $(S, \star, n)$ of dimension 8, the discriminant and the Clifford invariant of $(S, \lambda n)$ are trivial. Conversely, a quadratic form of dimension 8 with trivial discriminant and trivial Clifford invariant is, by the following Proposition, similar to the norm $\mathfrak{n}$ of a Cayley (or para-Cayley) algebra $\mathfrak{C}$:

(35.2) Proposition. *Let* (V, q) *be a quadratic space of dimension* 8. *The following condition are equivalent*:

(1) (V, q) *has trivial discriminant and trivial Clifford invariant,*

(2) (V, q) *is similar to the norm* $\mathfrak{n}$ *of a Cayley algebra* $\mathfrak{C}$.

Proof: This is a classical result of the theory of quadratic forms, due to A. Pfister (see for example Scharlau [247, p. 90]). As we already pointed out, the implication (2) $\Rightarrow$ (1) follows from Proposition (35.1). We include a proof of the converse which is much in the spirit of this chapter, however we assume that char $F \neq 2$. The idea is to construct a Cayley algebra structure on V such that the corresponding norm is a multiple of q. This construction is similar to the construction given in Chevalley [66, Chap. IV] for forms of maximal index. Let

$$ \alpha \colon \big(C(q), \tau\big) \xrightarrow{\sim} \big(\mathrm{End}_F(U), \sigma_k\big) $$

be an isomorphism of algebras with involution where τ is the involution of $C(V,q)$ which is the identity on V. Let ϵ be a nontrivial idempotent generating the center of $C_0(q)$. By putting $U_1 = \alpha(\epsilon)U$ and $U_2 = \alpha(1-\epsilon)U$, we obtain a decomposition

$$(U,k) = (U_1, q_1) \perp (U_2, q_2)$$

such that α is an isomorphism of graded algebras where $\mathrm{End}_F(U_1 \oplus U_2)$ is "checkerboard" graded. For any $x \in V$, let

$$\alpha(x) = \begin{pmatrix} 0 & \rho_x \\ \lambda_x & 0 \end{pmatrix} \in \mathrm{End}_F(U_1 \oplus U_2)$$

so that $\lambda_x \in \mathrm{Hom}_F(U_2, U_1)$ and $\rho_x \in \mathrm{Hom}_F(U_1, U_2)$ are such that $\lambda_x \circ \rho_x = q(x) \cdot 1_{U_1}$ and $\rho_x \circ \lambda_x = q(x) \cdot 1_{U_2}$. Let $\hat{b}_i \colon U_i \xrightarrow{\sim} U_i^*$ be the adjoints of q_i, i.e., the isomorphisms induced by $b_i = b_{q_i}$. We have

$$\begin{pmatrix} \hat{b}_1 & 0 \\ 0 & \hat{b}_2 \end{pmatrix}^{-1} \begin{pmatrix} 0 & \rho_x \\ \lambda_x & 0 \end{pmatrix}^t \begin{pmatrix} \hat{b}_1 & 0 \\ 0 & \hat{b}_2 \end{pmatrix} = \begin{pmatrix} 0 & \rho_x \\ \lambda_x & 0 \end{pmatrix}$$

hence

$$\hat{b}_1 \circ \rho_x = \lambda_x^t \circ \hat{b}_2 \quad \text{and} \quad \hat{b}_2 \circ \lambda_x = \rho_x^t \circ \hat{b}_1$$

or, putting $\rho_x(u_2) = \rho(x, u_2)$ and $\lambda_x(u_1) = \lambda(x, u_1)$, we obtain maps

$$\lambda \colon V \times U_1 \to U_2 \quad \text{and} \quad \rho \colon V \times U_2 \to U_1$$

such that

$$b_1\big(\rho(x, u_2), u_1\big) = b_2\big(u_2, \lambda(x, u_1)\big).$$

If we set $u_2 = \lambda(x, u_1)$ we then have

$$b_2\big(\lambda(x, u_1), \lambda(x, u_1)\big) = b_1\big(b_q(x)u_1, u_1\big)$$

so that, since we are assuming that char $F \neq 2$,

$$q_2\big(\lambda(x, u_1)\big) = q(x)q_1(u_1)$$

for $x \in V$ and $u_1 \in U_1$. Similarly the equation $q_1\big(\rho(x, u_2)\big) = q(x)q_2(u_2)$ holds for $x \in V$ and $u_2 \in U_2$. By linearizing the first formula, we obtain

$$b_2\big(\lambda(x, u_1), \lambda(x, v_1)\big) = q(x)b_1(u_1, v_1)$$

and

$$b_2\big(\lambda(x, u), \lambda(y, u)\big) = b_q(x, y)q_1(u)$$

for x, $y \in V$ and u_1, $v_1 \in U_1$. By replacing q by a multiple, we may assume that q represents 1, say $q(e) = 1$. We may do the same for q_1, say $q_1(e_1) = 1$. We then have $q_2(e_2) = 1$ for $e_2 = \lambda(e, e_1)$. We claim that $\rho(e, e_2) = e_1$. For any $u_1 \in U_1$, we have

$$\begin{aligned}
b_1\big(\rho(e, e_2), u_1\big) &= b_2\big(e_2, \mu(e, u_1)\big) \\
&= b_2\big(\lambda(e, e_1), \lambda(e, u_1)\big) \\
&= b_1(e_1, u_1)q(e) = b_1(e_1, u_1).
\end{aligned}$$

Since q_1 is nonsingular, $\rho(e, e_2)$ equals e_1 as claimed. The maps $s_1 \colon V \to U_1$ and $s_2 \colon V \to U_2$ given by $s_1(x) = \rho(x, e_2)$ and $s_2(y) = \lambda(y, e_1)$ are clearly isometries. Let

$$x \diamond y = s_2^{-1}\big(\lambda\big(x, s_1(y)\big)\big) \quad \text{for } x, y \in V.$$

We have

$$q(x \diamond y) = q_2\big(\lambda(x, s_1(y))\big) = q(x)q_1\big(s_1(y)\big) = q(x)q(y),$$
$$x \diamond e = s_2^{-1}\big(\lambda(x, s_1(e))\big) = s_2^{-1}\big(\lambda(x, e_1)\big) = x$$

and

$$b_q(v, e \diamond y) = b_2\big(\lambda(v, e_1), \mu(x, e_1)\big) = b_q(v, e)$$

for all $v \in V$, so that $e \diamond y = y$. Thus V is a composition algebra with identity element e. By Theorem (33.17) $(V, \diamond)$ is a Cayley algebra. $\square$

As an application we give another proof of a classical result of the theory of quadratic forms (see for example Scharlau [247, p. 89]).

(35.3) Corollary. *Let (V, q) be an 8-dimensional quadratic space with trivial discriminant and trivial Clifford invariant. Then (V, q) is hyperbolic if and only if it is isotropic.*

Proof: By Proposition (35.2), q is similar to the norm of a Cayley algebra, so that the claim follows from Proposition (33.23). $\square$

35.B. Similitudes and triality. Let $(S, \star, n)$ be a symmetric composition algebra of dimension 8 over F. In view of Proposition (13.1) any similitude t of n induces an automorphism $C_0(t)$ of $C_0(n)$ and t is proper, resp. improper if $C_0(t)$ restricts to the identity of the center Z of $C_0(n)$, resp. the nontrivial F-automorphism ι of Z. Let $\mathbf{GO}^+(n)$ be the group scheme of proper similitudes of n and $\mathrm{GO}^-(n)$ the set of improper similitudes. The "triality principle" for similitudes of (S, n) is the following result:

(35.4) Proposition. *Let t be a proper similitude of (S, n) with multiplier $\mu(t)$. There exist proper similitudes t^+, t^- of (S, n) such that*

$$(1) \qquad\qquad \mu(t^+)^{-1}t^+(x \star y) = t(x) \star t^-(y),$$
$$(2) \qquad\qquad \mu(t)^{-1}t(x \star y) = t^-(x) \star t^+(y)$$

and

$$(3) \qquad\qquad \mu(t^-)^{-1}t^-(x \star y) = t^+(x) \star t(y).$$

Let t be an improper similitude with multiplier $\mu(t)$. There exist improper similitudes t^+, t^- such that

$$(4) \qquad\qquad \mu(t^+)^{-1}t^+(x \star y) = t(y) \star t^-(x),$$
$$(5) \qquad\qquad \mu(t)^{-1}t(x \star y) = t^-(y) \star t^+(x)$$

and

$$(6) \qquad\qquad \mu(t^-)^{-1}t^-(x \star y) = t^+(y) \star t(x).$$

The pair (t^+, t^-) is determined by t up to a factor (μ, μ^{-1}), $\mu \in F^\times$, and we have

$$\mu(t^+)\mu(t)\mu(t^-) = 1.$$

Furthermore, any of the formulas (1) to (3) (resp. (4) to (6)) implies the others. If t is in $\mathrm{O}^+(n)$, the spinor norm $\mathrm{Sn}(t)$ of t is the class in $F^\times/F^{\times 2}$ of the multiplier of t^+ (or t^-).

Proof: Let t be a proper similitude with multiplier $\mu(t)$. The map $S \to \operatorname{End}_F(S \oplus S)$ given by

$$\varphi(t)\colon x \mapsto \begin{pmatrix} 0 & \ell_{t(x)} \\ \mu(t)^{-1} r_{t(x)} & 0 \end{pmatrix} = \begin{pmatrix} 1 & 0 \\ 0 & \mu(t)^{-1} \end{pmatrix} \alpha_S\big(t(x)\big)$$

is such that $(\varphi(t)(x))^2 = \mu(t)^{-1} n\big(t(x)\big) = n(x)$, and so it induces a homomorphism

$$\widetilde{\varphi}(t)\colon C(n) \xrightarrow{\sim} \operatorname{End}_F(S \oplus S).$$

By dimension count $\widetilde{\varphi}(t)$ is an isomorphism. By the Skolem-Noether Theorem, the automorphism $\widetilde{\varphi}(t) \circ \alpha_S^{-1}$ of $\operatorname{End}_F(S \oplus S)$ is inner. Let $\widetilde{\varphi}(t) \circ \alpha_S^{-1} = \operatorname{Int}\big(\begin{smallmatrix} s_0 & s_1 \\ s_3 & s_2 \end{smallmatrix}\big)$. Computing $\alpha_S^{-1} \circ \widetilde{\varphi}(t)$ on a product xy for x, $y \in S$ shows that $\alpha_S^{-1} \circ \widetilde{\varphi}(t)|_{C_0} = C_0(t)$. Since t is proper, $C_0(t)$ is Z-linear. Again by Skolem-Noether we may write $\alpha_S \circ C_0(t) \circ \alpha_S^{-1} = \operatorname{Int}\big(\begin{smallmatrix} s_0' & 0 \\ 0 & s_2' \end{smallmatrix}\big)$. This implies $s_1 = s_3 = 0$ and we may choose $s_0' = s_0$, $s_2' = s_2$. We deduce from $\varphi(t)(x) = \operatorname{Int}\big(\begin{smallmatrix} s_0 & 0 \\ 0 & s_2 \end{smallmatrix}\big) \circ \big(\alpha_S(x)\big)$ that

$$\ell_{t(x)} = s_0 \ell_x s_2^{-1} \quad \text{and} \quad \mu(t)^{-1} r_{t(x)} = s_2 r_x s_0^{-1}$$

or

$$s_0(x \star y) = t(x) \star s_2(y) \quad \text{and} \quad s_2(y \star x) = \mu(t)^{-1} s_0(y) \star t(x), \quad x, y \in S.$$

The fact that $C_0(t)$ commutes with the involution τ of $C_0(n)$ implies that s_0, s_2 are similitudes and we have $\mu(s_0) = \mu(t)\mu(s_2)$. Putting $t^+ = \mu(s_0)^{-1} s_0$ and $t^- = s_2$ gives (1) and (3).

To obtain (2), we replace x by $y \star x$ in (1). We have

$$\mu(t^+)^{-1} n(y) t^+(x) = t(y \star x) \star t^-(y).$$

Multiplying with $t^-(y)$ on the left gives

$$\mu(t^+)^{-1} n(y) t^-(y) \star t^+(x) = t(y \star x) \mu(t^-) n(y).$$

By viewing y as "generic" (apply (33.3)), we may divide both sides by $n(y)$. This gives (2).

If t is improper, then $C_0(t)$ switches the two factors of $Z = F \times F$ and, given t, we get t^+, t^- such that $\mu(t^+) t^+(x \star y) = t(y) \star t^-(x)$.

Formulas (5) and (6) follow similarly.

Conversely, if t satisfies (4), then $C_0(t)$ switches the two factors of $Z = F \times F$, hence is not proper. This remark and the above formulas then show that t^+, t^- are proper if t is proper. To show uniqueness of t^+, t^- up to a unit, we first observe that t^+, t^- are unique up to a pair (r^+, r^-) of scalars, since

$$\alpha_S C_0(t) \alpha_S^{-1} = \operatorname{Int}\big(\begin{smallmatrix} t^+ & 0 \\ 0 & t^- \end{smallmatrix}\big).$$

Replacing (t^+, t^-) by $(r^+ t^+, r^- t^-)$ gives

$$\mu(t^+)(r^+)^{-1} t^+(x \star y) = r^- t^-(x) \star t(y) = \mu(t^+)^{-1} r^- t^+(x \star y).$$

This implies ${r^+}^{-1} = r^-$. We finally check that $\operatorname{Sn}(t)$ is the multiplier of t^+ (or t^-) for $t \in \operatorname{O}^+(n)$. The transpose of a linear map t is denoted by t^*. Putting $\alpha_S(c) = (t^+, t^-)$ and writing $\hat{b}\colon S \to S^*$ for the isomorphism induced by b_n, we have

$$\alpha_S\big(c\tau(c)\big) = (t^+ \hat{b}^{-1} t^{+*} \hat{b}, t^- \hat{b}^{-1} t^{-*} \hat{b}) = (\hat{b}^{-1} t^{+*} \hat{b} t^+, \hat{b}^{-1} t^{-*} \hat{b} t^-)$$

(since $c\tau(c) \in F$). Then the claim follows from $t^{+*} \hat{b} t^+ = \mu(t^+) \hat{b}$ and $t^{-*} \hat{b} t^- = \mu(t^-) \hat{b}$, since $\operatorname{Sn}(t) = \operatorname{Sn}(t^{-1}) = c\tau(c) F^{\times 2} \in F^\times / F^{\times 2}$. The other claims can be checked by similar computations. $\qquad\square$

(35.5) Corollary. *For any pair λ, $\lambda^+ \in D(n)$, the set of values represented by n, there exists a triple of similitudes t, t^+, t^- such that Proposition (35.4) holds and such that λ, λ^+ are the multipliers of t, resp. t^+.*

Proof: Given $\lambda \in D(n)$, let t be a similitude with multiplier λ, for example $t(x) = u \star x$ with $n(u) = \lambda$, and let t^+, t^- be given by triality. If t is replaced by ts with $s \in O^+(n)$, the multiplier of t will not be changed and the multiplier of t^+ will be multiplied by the multiplier $\mu(s^+)$ of s^+. By Proposition (35.4) we have $\mu(s^+)F^{\times 2} = \text{Sn}(s)$. Since n is multiplicative, $\text{Sn}(O^+(n)) \equiv D(n) \mod F^{\times 2}$ and we can choose s (as a product of reflections) such that $\text{Sn}(s) = \left(\lambda^+ \mu(t^+)^{-1}\right)F^{\times 2}$, hence the claim. $\qquad\square$

Using triality we define an action of $A_3 \simeq \mathbb{Z}/3$ on $\mathbf{PGO}^+(n)(F) = \text{PGO}^+(n)$: Let $[t]$ be the class of $t \in \mathbf{GO}^+(n)(F)$ modulo the center. We put $\theta^+([t]) = [t^+]$ and $\theta^-([t]) = [t^-]$ where t^+, t^- are as in Proposition (35.4).

(35.6) Proposition. *The maps θ^+ and θ^- are outer automorphisms of the group $\text{PGO}^+(n)$. They satisfy $(\theta^+)^3 = 1$ and $\theta^+\theta^- = 1$, and they generate a subgroup of $\text{Aut}\left(\text{PGO}^+(n)\right)$ isomorphic to A_3.*

Proof: It follows from the multiplicativity of the formulas in Proposition (35.4) that the maps θ^+ and θ^- are group homomorphisms. The relations among them also follow from (35.4). Hence they are automorphisms and generate a homomorphic image of A_3. The fact that θ^+ is not inner follows from Proposition (35.14). $\qquad\square$

We shall see that the action of A_3 is, in fact, defined on the group scheme $\mathbf{PGO}^+(n) = \mathbf{GO}^+(n)/\mathbf{G}_\mathrm{m}$. For this we need triality for $\mathbf{Spin}(n)$.

35.C. The group Spin and triality. The group scheme $\mathbf{Spin}(S, n)$ is defined as

$$\mathbf{Spin}(S, n)(R) = \{\, c \in C_0(n)_R^\times \mid cS_Rc^{-1} \subset S_R \quad \text{and} \quad c\tau(c) = 1 \,\}$$

for all $R \in \mathsf{Alg}_F$. The isomorphism α_S of (35.1) can be used to give a nice description of $\text{Spin}(S, n) = \mathbf{Spin}(S, n)(F)$.

(35.7) Proposition. *Assume that* $\text{char } F \neq 2$. *There is an isomorphism*

$$\text{Spin}(S, n) \simeq \{\, (t, t^+, t^-) \mid t, t^+, t^- \in O^+(S, n), \, t(x \star y) = t^-(x) \star t^+(y) \,\}$$

such that the vector representation $\chi\colon \text{Spin}(S, n) \to O^+(S, n)$ corresponds to the map $(t, t^+, t^-) \mapsto t$. The other projections $(t, t^+, t^-) \mapsto t^+$ and $(t, t^+, t^-) \mapsto t^-$ correspond to the half-spin representations χ^{pm} of $\text{Spin}(S, n)$.

Proof: Let $c \in \text{Spin}(n)$ and let $\alpha_S(c) = \left(\begin{smallmatrix} t^+ & 0 \\ 0 & t^- \end{smallmatrix}\right)$. The condition $cxc^{-1} = \chi_c(x) \in S$ for all $x \in S$ is equivalent to the condition $t^+(x \star y) = \chi_c(x) \star t^-(y)$ or, by Proposition (35.4), to $\chi_c(x \star y) = t^-(x) \star t^+(y)$ for all $x, y \in S$. Since by Proposition (35.1) we have

$$\alpha_S\tau(c) = \begin{pmatrix} \hat{b}^{-1}t^{+*}\hat{b} & 0 \\ 0 & \hat{b}^{-1}t^{-*}\hat{b} \end{pmatrix}$$

where $\hat{b}\colon S \xrightarrow{\sim} S^*$ is the adjoint of b_n, the condition $c\tau(c) = 1$ is equivalent to $t^{+*}\hat{b}t^+ = \hat{b}$ and $t^{-*}\hat{b}t^- = \hat{b}$, i.e., the $t^\pm$ are isometries of $b = b_n$, hence of n since $\text{char } F \neq 2$. They are proper by Proposition (35.4). Thus, putting

$$\mathcal{T}(S, n) = \{\, (t, t^+, t^-) \mid t^+, t, t^- \in O^+(S, n), \, t(x \star y) = t^-(x) \star t^+(y) \,\},$$

$c \mapsto (\chi_c, t^+, t^-)$ defines an injective group homomorphism $\phi \colon \mathrm{Spin}(S,n) \to \mathcal{T}(S,n)$. It is also surjective, since, given $(t, t^+, t^-) \in \mathcal{T}(S,n)$, we have $(t, t^+, t^-) = \phi(c)$ for $\alpha_S(c) = \left(\begin{smallmatrix} t^+ & 0 \\ 0 & t^- \end{smallmatrix}\right)$. $\qquad\square$

From now on we assume that char $F \neq 2$. The isomorphism (35.7) can be defined on the level of group schemes: let G be the group scheme

$$G(R) = \left\{ (t, t^+, t^-) \mid t, t^+, t^- \in \mathbf{O}^+(S,n)(R), \ t(x \star y) = t^-(x) \star t^+(y) \right\}.$$

(35.8) Proposition. *There exists an isomorphism $\beta \colon G \xrightarrow{\sim} \mathbf{Spin}(S,n)$ of group schemes.*

Proof: By definition we have

$$(\alpha_S \otimes 1_R)^{-1} \left(\begin{smallmatrix} t^+ & 0 \\ 0 & t^- \end{smallmatrix}\right) = c \in \mathbf{Spin}(S,n)(R),$$

so that α_S induces a morphism

$$\beta \colon G \to \mathbf{Spin}(S,n).$$

Proposition (35.7) implies that $\beta(F_{\mathrm{alg}})$ is an isomorphism. Thus, in view of Proposition (22.5), it suffices to check that $d\beta$ is injective. It is easy to check that

$$\mathrm{Lie}(G) = \left\{ (\lambda, \lambda^+, \lambda^-) \in \mathfrak{o}(n) \times \mathfrak{o}(n) \times \mathfrak{o}(n) \mid \lambda(x \star y) = \lambda^-(x) \star y + x \star \lambda^+(y) \right\}.$$

On the other hand we have (see §23.A)

$$\mathrm{Lie}\big(\mathrm{Spin}(S,n)\big) = \left\{ x \in S \cdot S \subset C_0(S,n) \mid x + \underline{\sigma}(x) = 0 \right\} = [S,S]$$

where multiplication is in $C_0(S,n)$ (recall that we are assuming that char $F \neq 2$ here) and the proof of Proposition (45.5) shows that that $d\beta$ is an isomorphism. $\quad\square$

Identifying $\mathbf{Spin}(S,n)$ with G through β we may define an action of A_3 on $\mathbf{Spin}(n)$:

(35.9) Proposition. *The transformations θ^+, resp. θ^- induced by*

$$(t, t^+, t^-) \mapsto (t^+, t^-, t), \quad \text{resp.} \quad (t, t^+, t^-) \mapsto (t^-, t, t^+)$$

are outer automorphisms of $\mathbf{Spin}(S,n)$ and satisfy the relations $\theta^{+3} = 1$ and $\theta^{+-1} = \theta^-$. They generate a subgroup of $\mathrm{Aut}\big(\mathbf{Spin}(S,n)\big)$ isomorphic to A_3. Furthermore $\mathbf{Spin}(S,n)^{A_3}$ is isomorphic to $\mathbf{Aut}_{\mathrm{alg}}(\mathfrak{C})$, if S is a para-Cayley algebra $\mathfrak{C}$, and isomorphic to $\mathbf{Aut}_{\mathrm{alg}}(A)$, resp. to $\mathbf{Aut}_{\mathrm{alg}}(B, \tau)$, for a central simple algebra A of degree 3, resp. a central simple algebra (B, τ) of degree 3 with an involution of second kind over $K = F[X]/(X^2 + X + 1)$, if $(S, \star)$ is of type ${}^1A_2$, resp. of type ${}^2A_2$.

Proof: Let R be an F-algebra. It follows from the multiplicativity of the formulas of Proposition (35.4) that the maps θ_R^+ and θ_R^- are automorphisms of $\mathbf{Spin}(S,n)(R)$. The relations among them also follow from (35.4). They are outer automorphisms since they permute the vector and the two half-spin representations of the group $\mathbf{Spin}(S,n)(R)$ (since char $F \neq 2$, this also follows from the fact that they act nontrivially on the center C, see Lemma (35.11)). Now let $(t, t^+, t^-) \in \mathbf{Spin}(S,n)^{A_3}(R)$. We have $t = t^+ = t^-$ and t is an automorphism of S_R. Conversely, any automorphism of $(S,n)_R$ is an isometry and, since $\alpha(x \diamond y) = \alpha(x) \diamond \alpha(y)$, α is proper by (35.4). $\qquad\square$

Let $\mathbf{Spin}_8 = \mathbf{Spin}(V, q)$ where V is 8-dimensional and q is hyperbolic.

(35.10) Corollary. (1) *There exists an action of A_3 on $\mathbf{Spin}_8$ such that $\mathbf{Spin}_8^{A_3}$ is split of type G_2.*

(2) *There exists an action of A_3 on $\mathbf{Spin}_8$ such that $\mathbf{Spin}_8^{A_3} = \mathbf{PGU}_3(K)$ where $K = F[X]/(X^2+X+1)$. In particular, if F contains a primitive cube root of unity, there exists an action of A_3 on $\mathbf{Spin}_8$ such that $\mathbf{Spin}_8^{A_3} = \mathbf{PGL}_3$.*

Proof: Take $A = F \times \mathfrak{C}_s$, resp. $A = M_3(K)$ in Proposition (35.9). $\qquad\square$

As we shall see in Proposition (36.14), the actions described in (1) and (2) of Corollary (35.10) are (up to isomorphism) the only possible ones over F_{sep}.

Let again (S, n) be a symmetric composition of dimension 8 and norm n. In view of (13.31) (and (22.10)) the group scheme $\mathbf{Spin}(S, n)$ fits into the exact sequence

$$1 \to \boldsymbol{\mu}_2 \to \mathbf{Spin}(S, n) \xrightarrow{\chi} \mathbf{O}^+(n) \to 1$$

where χ is the *vector representation*, i.e.,

$$(\chi_c)_R(x) = cxc^{-1} \quad \text{for } x \in S_R, \, c \in \mathbf{Spin}(S, n)(R).$$

Let $\chi' \colon \mathbf{Spin}(S, n) \to \mathbf{PGO}^+(n)$ be the composition of the vector representation χ with the canonical map $\mathbf{O}^+(n) \to \mathbf{PGO}^+(n)$. The kernel C of χ' is the center of $\mathbf{Spin}(n)$ and is isomorphic to $\boldsymbol{\mu}_2 \times \boldsymbol{\mu}_2$. The action of A_3 on $\mathbf{Spin}(S, n)$ restricts to an action of A_3 on $\boldsymbol{\mu}_2 \times \boldsymbol{\mu}_2$. We recall that we are still assuming char $F \neq 2$.

(35.11) Lemma. *The action of A_3 on $C \simeq \boldsymbol{\mu}_2 \times \boldsymbol{\mu}_2$ is described by the exact sequence*

$$1 \to C \to \boldsymbol{\mu}_2 \times \boldsymbol{\mu}_2 \times \boldsymbol{\mu}_2 \to \boldsymbol{\mu}_2 \to 1$$

where A_3 acts on $\boldsymbol{\mu}_2 \times \boldsymbol{\mu}_2 \times \boldsymbol{\mu}_2$ through permutations and the map $\boldsymbol{\mu}_2 \times \boldsymbol{\mu}_2 \times \boldsymbol{\mu}_2 \to \boldsymbol{\mu}_2$ is the multiplication map.

Proof: In fact we have

(35.12)
$$C_R = \{(1,1,1), \epsilon_0 = (1,-1,-1), \epsilon_1 = (-1,1,-1), \epsilon_2 = \epsilon_0\epsilon_1 = (-1,-1,1)\},$$

hence the description of C through the exact sequence. For the claim on the action of A_3, note that $\theta^+ = (\theta^-)^{-1}$ maps ϵ_i to ϵ_{i+1} with subscripts taken modulo 3. Observe that the full group S_3 acts on C and that $\mathrm{Aut}(C) = S_3$. $\qquad\square$

In view of (13.34) (and (22.10)) we have an exact sequence

(35.13) $$1 \to C \to \mathbf{Spin}(S, n) \xrightarrow{\chi'} \mathbf{PGO}^+(n) \to 1.$$

(35.14) Proposition. *There is an outer action of A_3 on $\mathbf{PGO}^+(n)$ such that the maps in the exact sequence (35.13) above are A_3-equivariant.*

Proof: The existence of the A_3-action follows from Proposition (35.9), Lemma (35.11) and the universal property (22.8) of factor group schemes. The action is outer, since it is outer on $\mathbf{Spin}(S, n)$. Observe that for F-valued points the action is the one defined in Proposition (35.6). $\qquad\square$

Let $(\mathfrak{C}, \star, \mathfrak{n})$ be a para-Cayley algebra over F. The conjugation map $\pi \colon x \mapsto \overline{x}$ can be used to extend the action of A_3 to an action of $S_3 = A_3 \rtimes \mathbb{Z}/2\mathbb{Z}$ on $\mathbf{Spin}(\mathfrak{n})$ and $\mathbf{PGO}^+(\mathfrak{n})$: Let $\alpha_{\mathfrak{C}} \colon C(\mathfrak{C}, \mathfrak{n}) \xrightarrow{\sim} \left(\mathrm{End}_F(\mathfrak{C} \oplus \mathfrak{C}), \sigma_{\mathfrak{n} \perp \mathfrak{n}}\right)$ be the isomorphism of Proposition (35.1). The conjugation map $x \mapsto \overline{x}$ is an isometry and since

$\alpha_{\mathfrak{C}} C(\pi) \alpha_{\mathfrak{C}}^{-1} = \operatorname{Int}\left(\begin{smallmatrix} 0 & \pi \\ \pi & 0 \end{smallmatrix}\right)$, π is improper. For any similitude t with multiplier $\mu(t)$, $\widehat{t} = \mu(t)^{-1}\pi t\pi$ is a similitude with multiplier $\mu(t)^{-1}$ and is proper if and only if t is proper. Proposition (35.4) implies that

(1) $$\mu(\widehat{t})^{-1}\widehat{t}(x \star y) = \widehat{t}^+(x) \star \widehat{t}^-(y)$$

(2) $$\mu(\widehat{t}^-)^{-1}\widehat{t}^-(x \star y) = \widehat{t}(x) \star \widehat{t}^+(y)$$

(3) $$\mu(\widehat{t}^+)^{-1}\widehat{t}^+(x \star y) = \widehat{t}^-(x) \star \widehat{t}(y)$$

hold in $(\mathfrak{C}, \star)$ if t is proper. Let θ^+ and θ^- be as defined in (35.6) and (35.9). We further define for R an F-algebra, $\theta([t]) = [\widehat{t}]$ for $[t] \in \mathbf{PGO}^+(\mathfrak{n})(R)$, $\theta(t, t^+, t^-) = (\widehat{t}, \widehat{t}^-, \widehat{t}^+)$ for $(t, t^+, t^-) \in G(R) \simeq \mathbf{Spin}(\mathfrak{n})(R)$ and $\theta(\epsilon_0) = \epsilon_0$, $\theta(\epsilon_1) = \epsilon_2$, $\theta(\epsilon_2) = \epsilon_1$ for ϵ_i as in (35.12).

(35.15) Proposition. (1) *Let G be $\mathbf{Spin}(\mathfrak{n})$ or $\mathbf{PGO}^+(\mathfrak{n})$. The maps θ, θ^+ and θ^- are outer automorphisms of G. They satisfy the relations*

$$\theta^{+3} = 1, \quad \theta^{+-1} = \theta^- \quad \text{and} \quad \theta^+\theta = \theta\theta^-,$$

and they generate a subgroup of $\operatorname{Aut}(G)$ isomorphic to S_3. In particular $\operatorname{Out}(G) \simeq S_3$ and $\operatorname{Aut}(G) \simeq \mathbf{PGO}^+(\mathfrak{n}) \rtimes S_3$.
(2) *The exact sequence*

$$1 \to C \to \mathbf{Spin}(\mathfrak{C}, \mathfrak{n}) \xrightarrow{\chi'} \mathbf{PGO}^+(\mathfrak{C}, \mathfrak{n}) \to 1$$

is S_3-equivariant.

Proof: The proof is similar to the proof of Proposition (35.14) using the above formulas (1) to (3). In (2) the action of S_3 on C is as defined in (35.11). The fact that S_3 is the full group $\operatorname{Out}(G)$ follows from the fact that the group of automorphisms of the Dynkin diagram of $\mathbf{Spin}(\mathfrak{C}, \mathfrak{n})$, which is of type D_4, is S_3 (see §24.A). $\square$

Observe that for the given action of S_3 on $\mathbf{Spin}(\mathfrak{C}, \mathfrak{n})$ we have

(35.16) $$\mathbf{Spin}(\mathfrak{C}, \mathfrak{n})^{S_3} = \mathbf{Spin}(\mathfrak{C}, \mathfrak{n})^{A_3} = \mathbf{Aut}_{\mathrm{alg}}(\mathfrak{C}).$$

The action of S_3 on $\mathbf{Spin}_8$ induces an action on $H^1(F, \mathbf{Spin}_8)$. We now describe the objects classified by $H^1(F, \mathbf{Spin}_8)$ and the action of S_3 on $H^1(F, \mathbf{Spin}_8)$. A triple of quadratic spaces (V_i, q_i), $i = 1, 2, 3$, together with a bilinear map $\beta\colon V_0 \times V_1 \to V_2$ such that $q_2(\beta(v_0, v_1)) = q_0(v_0)q_1(v_1)$ for $v_i \in V_i$ is a *composition of quadratic spaces*. Examples are given by $V_i = \mathfrak{C}$, $\mathfrak{C}$ a Cayley algebra, $\beta(x, y) = x \diamond y$, and by $V_i = \mathfrak{C}$, $\beta(x, y) = x \star y$. An *isometry* $\psi\colon (V_0, V_1, V_2) \xrightarrow{\sim} (V_0', V_1', V_2')$ is a triple of isometries $(\psi_i\colon V_i \xrightarrow{\sim} V_i')$ such that $\beta' \circ (\psi_0, \psi_1) = \psi_2 \circ \beta$. The triple $(\pi, \pi, 1_{\mathfrak{C}})$ is an isometry

(35.17) $$\bigl(V_i, \beta(x, y) = x \diamond y\bigr) \xrightarrow{\sim} \bigl(V_i, \beta(x, y) = x \star y\bigr).$$

A *similitude* $(\psi_i\colon V_i \xrightarrow{\sim} V_i')$, with *multiplier* $(\lambda_0, \lambda_1, \lambda_2)$ is defined in a similar way. Observe that the equation $\lambda_2 = \lambda_0\lambda_1$ holds for a similitude. The main steps in the proof of Proposition (35.2) were to associate to a quadratic space (V, q) of rank 8, with trivial discriminant and trivial Clifford invariant, a composition of quadratic spaces $V \times U_1 \to U_2$ similar to a composition of type $\mathfrak{C} \times \mathfrak{C} \to \mathfrak{C}$.

(35.18) Proposition. (1) *Compositions of quadratic spaces of dimension 8 are classified by $H^1(F, \mathbf{Spin}_8)$.*

(2) *Let* $\beta_{012}\colon V_0 \times V_1 \to V_2$ *be a fixed composition of quadratic spaces of dimension* 8. *The action of* S_3 *on the set* $H^1(F, \mathbf{Spin}_8)$ *is given by* $\beta_{ijk} \mapsto \beta_{s(i)s(j)s(j)}$ *where the* β_{ijk} *are defined by*

$$b_{q_0}\big(v_0, \beta_{120}(v_1, v_2)\big) = b_{q_2}\big(\beta_{012}(v_0, v_1), v_2\big) = b_{q_1}\big(\beta_{201}(v_2, v_0), v_2\big)$$

and

$$\beta_{102}(v_1, v_0) = \beta_{012}(v_0, v_1), \quad \beta_{021}(v_0, v_2) = \beta_{201}(v_2, v_0).$$

Proof: By the proof of Proposition (35.2), any composition of 8-dimensional quadratic spaces is isometric over a separable closure of F to the composition

$$\big(\mathfrak{C}_s, \mathfrak{C}_s, \mathfrak{C}_s, \beta(x, y) = x \diamond y\big)$$

where $\mathfrak{C}_s$ is the split Cayley algebra. By Proposition (35.7), Spin_8 is the group of automorphisms of the composition

$$\big(\mathfrak{C}_s, \mathfrak{C}_s, \mathfrak{C}_s, \beta(x, y) = x \diamond y\big) \simeq \big(\mathfrak{C}_s, \mathfrak{C}_s, \mathfrak{C}_s, \beta(x, y) = x \star y\big).$$

Consider the representation

$$\rho\colon G = \mathbf{GL}(\mathfrak{C}_s) \times \mathbf{GL}(\mathfrak{C}_s) \times \mathbf{GL}(\mathfrak{C}_s) \to \mathbf{GL}(W)$$

where

$$W = S^2(\mathfrak{C}_s^*) \oplus S^2(\mathfrak{C}_s^*) \oplus S^2(\mathfrak{C}_s^*) \oplus \mathrm{Hom}_F(\mathfrak{C}_s \otimes_F \mathfrak{C}_s, \mathfrak{C}_s),$$

given by the formula

$$\rho(\alpha_0, \alpha_1, \alpha_2)(f, g, h, \phi) = \big(f \circ \alpha_0^{-1}, g \circ \alpha_1^{-1}, h \circ \alpha_2^{-1}, \alpha_2 \circ \phi \circ (\alpha_0^{-1} \otimes \alpha_1^{-1})\big).$$

Let $w = (\mathfrak{n}, \mathfrak{n}, \mathfrak{n}, m)$ where $m\colon \mathfrak{C}_s \otimes \mathfrak{C}_s \to \mathfrak{C}_s$ is the product $m(x, y) = x \star y$ in the para-Cayley algebra $\mathfrak{C}_s$. The group scheme $\mathbf{Aut}_G(w)$ coincides with the group scheme $\mathbf{Spin}(S, n)$ in view of Proposition (35.8). (Observe that by (35.4) elements $t, t^+, t^- \in \mathrm{O}(\mathfrak{C}_s, \mathfrak{n}_s)$ such that $t(x \star y) = t^-(x) \star t^+(y)$ are already in $\mathrm{O}^+(\mathfrak{C}_s, \mathfrak{n}_s)$.) Thus (1) follows from (29.8). Claim (2) follows from Proposition (35.9). $\qquad \square$

§36. Twisted Compositions

Let L be a cubic étale F-algebra with norm N_L, trace T_L, and let ${}^\#\colon L \to L$ be the quadratic map such that $\ell\ell^\# = N_L(\ell)$ for $\ell \in L$. Let (V, Q) be a nonsingular quadratic space over L, let $b_Q(x, y) = Q(x + y) - Q(x) - Q(y)$, and let $\beta\colon V \to V$ be a quadratic map such that

(1) $$\beta(\ell v) = \ell^\# \beta(v)$$

(2) $$Q\big(\beta(v)\big) = Q(v)^\#$$

for all $v \in V$, $\ell \in L$. We define $N(v) = b_Q\big(v, \beta(v)\big)$ and further require that

(3) $$N(v) \in F$$

for all $v \in V$. We call the datum $\Gamma = (V, L, Q, \beta)$ a *twisted composition over* L with *quadratic norm* Q, *quadratic map* β, and *cubic norm* N. A morphism

$$\phi\colon (V, L, Q, \beta) \to (V', L', Q', \beta')$$

is a pair (t, ϕ), $t\colon V \xrightarrow{\sim} V'$, $\phi\colon L \xrightarrow{\sim} L'$, with ϕ an F-algebra isomorphism, t ϕ-semilinear, and such that $t\beta = \beta't$ and $\phi Q = Q't$.

(36.1) Lemma. *Let* $\Gamma = (V, L, Q, \beta)$ *be a twisted composition. Then for any* $\lambda \in L^\times$,

$$\Gamma_\lambda = (V, L, \lambda^\# Q, \lambda\beta)$$

is again a twisted composition and, conversely, if $(V, L, \mu Q, \lambda\beta)$ *is a twisted composition, then* $\mu = \lambda^\#$.

Proof: The first claim is straightforward. If $(V, L, \mu Q, \lambda\beta)$ is a twisted composition, the equality $\lambda^2 \mu = \mu^\#$ follows from (1), (2). Since $(\mu^\#)^\# = \mu^\# \mu^2$, it follows that $(\lambda^\#)^2 = \mu^2$, thus $\mu = \pm\lambda^\#$. Since $\lambda^2 \mu = \mu^\#$, multiplying both sides by μ and using that $(\lambda^\#)^2 = \mu^2$ shows that $N_{L/F}(\lambda^\#) = N_{L/F}(\mu)$. Since L is cubic over F, $\mu = \lambda^\#$. $\qquad\square$

Observe that the map $v \mapsto \lambda v$ is an isomorphism $\Gamma_\lambda \xrightarrow{\sim} \Gamma$ if $\lambda \in F^\times$. A morphism

$$\varphi = (t, \phi) \colon (V, L, \lambda^\# Q, \lambda\beta) \xrightarrow{\sim} (V', L', Q', \beta')$$

is a *similitude* with *multiplier* $\lambda^\#$ of (V, L, Q, β) with (V', L', Q', β'). For any field extension K of F and any twisted composition Γ over F, we have a canonical twisted composition $\Gamma \otimes K$ over K.

Examples of twisted compositions arise from symmetric composition algebras:

(36.2) Examples. (1) Let $(S, \star, n)$ be a symmetric composition algebra. Let $L = F \times F \times F$, let $V = S \otimes L = S \times S \times S$ and let $Q = (n, n, n)$. We have $(x_0, x_1, x_2)^\# = (x_1 x_2, x_0 x_2, x_0 x_1)$ and putting

$$\beta_S(v_0, v_1, v_2) = (v_1 \star v_2, v_2 \star v_0, v_0 \star v_1)$$

gives a twisted composition $\widetilde{S} = (S \otimes L, L, Q, \beta_S)$. Condition (3) of the definition of a twisted composition is equivalent with the associativity of the norm. In particular

$$\beta_{\mathfrak{C}}(v_0, v_1, v_2) = (\overline{v}_1 \diamond \overline{v}_2, \overline{v}_2 \diamond \overline{v}_0, \overline{v}_0 \diamond \overline{v}_1)$$

defines a twisted composition associated with the Cayley algebra $(\mathfrak{C}, \diamond)$. We call $\widetilde{\mathfrak{C}}$ a *twisted Cayley composition*. A twisted Cayley composition with L and $\mathfrak{C}$ split is a *split twisted composition* of rank 8.

(2) Let $\big((V_0, V_1, V_2), \beta_{012}\big)$ be a composition of quadratic spaces as in Proposition (35.18). If we view $V_0 \oplus V_1 \oplus V_2$ as a module V over $L = F \times F \times F$ and group the β_{ijk} to a quadratic map $\beta \colon V \to V$ we obtain a twisted composition over $F \times F \times F$; any twisted composition over $F \times F \times F$ is of this form:

(36.3) Proposition. *A twisted composition over* $F \times F \times F$ *is of the form* $V_0 \oplus V_1 \oplus V_2$ *as in* (36.2.2) *and is similar to a twisted composition* $\widetilde{C}$ *for some Hurwitz algebra* (C, n).

Proof: Let $V_0 = (1, 0, 0)V$, $V_1 = (0, 1, 0)V$, $V_2 = (0, 0, 1)V$, so that $V = V_0 \oplus V_1 \oplus V_2$. We first construct a multiplication on V_0. We use the notations $x = (x_0, x_1, x_2)$ for $x \in L$ and $v = (v_0, v_1, v_2)$ for $v \in V$. We have $x^\# = (x_1 x_2, x_2 x_0, x_0 x_1)$. Let

$$\beta(v) = \big(\beta_0(v), \beta_1(v), \beta_2(v)\big).$$

It follows from

$$\beta\big((0, 1, 1)(v_0, v_1, v_2)\big) = (1, 0, 0)\beta(v_0, v_1, v_2)$$

that

$$\beta_0(v_0, v_1, v_2) = \beta(0, v_1, v_2)$$
$$\beta_1(v_0, v_1, v_2) = \beta(v_0, 0, v_2)$$
$$\beta_2(v_0, v_1, v_2) = \beta(v_0, v_1, 0).$$

Furthermore, the F-bilinearity of

$$\beta(x, y) = \beta(x + y) - \beta(x) - \beta(y)$$

implies that $\beta(0, v_1, v_2) = \beta((0, v_1, 0), (0, 0, v_2))$ is F-bilinear in the variables v_1 and v_2. Thus there are three F-linear maps $\beta_i \colon V_{i+1} \otimes V_{i+2} \to V_i$ where i, $i+1$, $i+2$ are taken mod 3, such that

$$\beta_0(v_1 \otimes v_2) = \beta(0, v_1, v_2)$$
$$\beta_1(v_2 \otimes v_0) = \beta(v_0, 0, v_2)$$
$$\beta_2(v_0 \otimes v_1) = \beta(v_0, v_1, 0)$$

and such that

$$\beta(v) = \big(\beta_0(v_1 \otimes v_2), \beta_1(v_2 \otimes v_0), \beta_2(v_0 \otimes v_1)\big).$$

Since Q is $F \times F \times F$-linear, we may write $Q(v) = \big(q_0(v_0), q_1(v_1), q_2(v_2)\big)$. Condition (2) of the definition of a twisted composition reads

$$q_0\big(\beta_0(v_1 \otimes v_2)\big) = q_1(v_1)q_2(v_2), \quad q_1\big(\beta_1(v_2 \otimes v_0)\big) = q_2(v_2)q_0(v_0)$$

and $q_2\big(\beta_2(v_0 \otimes v_1)\big) = q_0(v_0)q_1(v_1)$. This is the first claim. Linearizing gives

$$b_{q_0}\big(\beta_0(v_1 \otimes v_2), \beta_0(w_1 \otimes v_2)\big) = q_2(v_2)b_{q_1}(v_1, w_1)$$
$$b_{q_1}\big(\beta_1(v_2 \otimes v_0), \beta_1(w_2 \otimes v_0)\big) = q_0(v_0)b_{q_2}(v_2, w_2)$$
$$b_{q_2}\big(\beta_2(v_0 \otimes v_1), \beta_2(w_0 \otimes v_1)\big) = q_1(v_1)b_{q_0}(v_0, w_0).$$

Similarly, Condition (3) reduces to

$$b_{q_0}\big(v_0, \beta_0(v_1 \otimes v_2)\big) = b_{q_1}\big(v_1, \beta_1(v_2 \otimes v_0)\big) = b_{q_2}\big(v_2, \beta_2(v_0 \otimes v_1)\big).$$

If ν_1, $\nu_2 \in F^\times$ and $e_1 \in V_1$, $e_2 \in V_2$ are such that $\nu_1 q_1(e_1) = 1$ and $\nu_2 q_2(e_2) = 1$, then $\nu_1 \nu_2 q_0(e_0) = 1$ for $e_0 = \beta_0(e_1 \otimes e_2)$. Replacing β by $(1, \nu_2, \nu_1)\beta$ and Q by $(\nu_1 \nu_2, \nu_1, \nu_2)Q$, we may assume that there exists an element $e = (e_0, e_1, e_2)$ such that $Q(e) = 1$. We claim that

$$\beta_1(e_2 \otimes e_0) = e_1 \quad \text{and} \quad \beta_2(e_0 \otimes e_1) = e_2.$$

We have for all $(v_0, v_1, v_2) \in V$

$$\begin{aligned} b_{q_1}\big(v_1, \beta_1(e_2 \otimes e_0)\big) &= b_{q_0}\big(e_0, \beta_0(v_1 \otimes e_2)\big) \\ &= b_{q_0}\big(\beta_0(e_1 \otimes e_2), \beta_0(v_1 \otimes e_2)\big) \\ &= b_{q_1}(v_1, e_1)q_2(e_2) = b_{q_1}(v_1, e_1) \end{aligned}$$

Since $b_{q_1}(x, y)$ is nonsingular, we must have $\beta_1(e_2 \otimes e_0) = e_1$. A similar computation shows that $\beta_2(e_0 \otimes e_1) = e_2$. We define isometries $\alpha_1 \colon (V_0, q_0) \xrightarrow{\sim} (V_1, q_1)$ and $\alpha_2 \colon (V_0, q_0) \xrightarrow{\sim} (V_2, q_2)$ by $\alpha_1(v_0) = \beta_1(e_2 \otimes v_0)$ and $\alpha_2(v_0) = \beta_2(v_0 \otimes e_1)$ and define a multiplication $\diamond$ on V_0 by

$$x \diamond y = \beta_0\big(\alpha_1(x) \otimes \alpha_2(y)\big).$$

We have $q_0(x \diamond y) = q_1\big(\alpha_1(x)\big)q_2\big(\alpha_2(y)\big) = q_0(x)q_0(y)$ and, for all $y \in V_0$,

$$
\begin{aligned}
b_{q_0}(y, x \diamond e_0) &= b_{q_0}\big(y, \beta_0\big(\beta_1(e_2 \otimes x) \otimes e_2\big)\big) \\
&= b_{q_1}\big(\beta_1(e_2 \otimes x), \beta_1(e_2 \otimes y)\big) \\
&= q_2(e_2)b_{q_0}(x, y) = b_{q_0}(x, y)
\end{aligned}
$$

Thus $x \diamond e_0 = x$, e_0 is an identity for $\diamond$ and V_0 is, by Theorem (33.17), a Hurwitz algebra with multiplication $\diamond$, identity e_0 and norm q_0. We call it (C, n). Let $\widetilde{C} = C \times C \times C$ be the associated twisted composition over $F \times F \times F$. We check finally that $\widetilde{C} \simeq V$ via the map $(I_C, \alpha_1\pi, \alpha_2\pi)$ where π is the conjugation in C. It suffices to verify that

$$
\begin{aligned}
\beta_0(\alpha_1\pi x \otimes \alpha_2\pi y) &= \overline{x} \diamond \overline{y} \\
\beta_1(\alpha_2\pi x \otimes y) &= \alpha_1\pi(\overline{x} \diamond \overline{y}) \\
\beta_2(x \otimes \alpha_1\pi y) &= \alpha_2\pi(\overline{x} \diamond \overline{y}).
\end{aligned}
$$

The first formula follows from the definition of $\diamond$. For the second we have

$$
\begin{aligned}
b_{q_1}\big(\beta_1(\alpha_2\pi x \otimes y), \alpha_1\pi z\big) &= b_{q_0}\big(y, \beta_0(\alpha_1\pi z \otimes \alpha_2\pi x)\big) \\
&= b_{q_0}(y, \overline{z} \diamond \overline{x}) \\
&= b_{q_0}(y \diamond x, \overline{z}) \\
&= b_{q_1}\big(\alpha_1\pi(\overline{x} \diamond \overline{y}), \alpha_1\pi z\big)
\end{aligned}
$$

hence the claim. The proof of the third one is similar. $\qquad\square$

(36.4) Corollary. *For any twisted composition* (V, L, Q, β) *we have* $\dim_L V = 1$, *2, 4, or 8.* $\qquad\square$

We observe that the construction of the multiplication in the proof of Proposition (36.3) is similar to the construction in Proposition (35.2) or to the construction given by Chevalley [66, Chap. IV] for forms of dimension 8 of maximal index.

(36.5) Proposition. *Let* $\mathfrak{C}$ *be a Cayley algebra over* F *with norm* $\mathfrak{n}$. *The group scheme* $\mathbf{Aut}(\widetilde{\mathfrak{C}})$ *of* F-*automorphisms of the twisted composition* $\widetilde{\mathfrak{C}}$ *is isomorphic to the semidirect product* $\mathbf{Spin}(\mathfrak{C}, \mathfrak{n}) \rtimes S_3$. *In particular the group scheme of automorphisms of a split twisted composition of rank 8 is isomorphic to* $\mathbf{Spin}_8 \rtimes S_3$.

Proof: Let R be an F-algebra such that $L \otimes R = R \times R \times R$ (the other cases are let as exercises). Let

$$
p \colon \mathbf{Aut}(\widetilde{\mathfrak{C}})(R) \to \mathbf{Aut}_{\mathrm{alg}}(L)(R)
$$

be the restriction map. Since $\mathbf{Aut}_{\mathrm{alg}}(L)(R) = S_3$, we have to check that p has a section and that $\ker p = \mathbf{Spin}(\mathfrak{n})$. The permutations $\rho \colon (x_0, x_1, x_2) \mapsto (x_1, x_2, x_0)$ and $\epsilon \colon (x_0, x_1, x_2) \mapsto (x_0, x_2, x_1)$ generate S_3. A section is given by $\rho \mapsto \widetilde{\rho}$, $\widetilde{\rho}(v_0, v_1, v_2) = (v_1, v_2, v_0)$, and $\epsilon \mapsto \widetilde{\epsilon}$, $\widetilde{\epsilon}(v_0, v_1, v_2) = (\overline{v}_2, \overline{v}_1, \overline{v}_0)$. Now let $t \in \mathbf{Aut}(\widetilde{\mathfrak{C}})(R)$ be such that $p(t) = 1$, i.e., t is L-linear. Putting $t = (t_0, t_1, t_2)$, the t_i are isometries of $\mathfrak{n}$ and the condition $t\beta_{\mathfrak{C}} = \beta_{\mathfrak{C}}t$ is equivalent to

$$
\begin{aligned}
t_0(v_1 v_2) &= \hat{t}_1(v_1)\hat{t}_2(v_2) \\
t_1(v_2 v_0) &= \hat{t}_2(v_2)\hat{t}_0(v_0) \\
t_2(v_0 v_1) &= \hat{t}_0(v_0)\hat{t}_1(v_1).
\end{aligned}
$$

In fact any of these three conditions is equivalent to the two others (see Proposition (35.4)). By (35.4) the t_i are proper, so that $t_i \in \mathbf{O}^+(\mathfrak{n})(R)$ and by Proposition (35.7) we have $t = (t_0, t_1, t_2) \in \mathbf{Spin}(\mathfrak{n})(R)$. $\qquad\square$

The split exact sequence

$$(36.6) \qquad\qquad 1 \to \mathbf{Spin}_8 \to \mathbf{Spin}_8 \rtimes S_3 \xrightarrow{p} S_3 \to 1$$

induces a sequence in cohomology

$$\to H^1(F, \mathbf{Spin}_8) \to H^1(F, \mathbf{Spin}_8 \rtimes S_3) \xrightarrow{p^1} H^1(F, S_3).$$

(36.7) Proposition. (1) *Twisted compositions of dimension* 8 *over* F *are classified by the pointed set* $H^1(F, \mathbf{Spin}_8 \rtimes S_3)$.
(2) *The map* $H^1(F, \mathbf{Spin}_8) \to H^1(F, \mathbf{Spin}_8 \rtimes S_3)$ *is induced by* $\beta_{012} \mapsto (V, \beta)$ *with* $V = V_0 \oplus V_1 \oplus V_2$ *and* $\beta = (\beta_{120}, \beta_{201}, \beta_{012})$ *(with the notations of* (35.18)).
(3) *For any class* $\gamma = [L] \in H^1(F, S_3)$ *the fiber* $(p^1)^{-1}([L])$ *is in bijection with the orbits of* $(S_3)_\gamma^{\mathrm{Gal}(F_{\mathrm{sep}}/F)}$ *in* $H^1\big(F, (\mathbf{Spin}_8)_\gamma\big)$.

Proof: In view of Proposition (36.3) any twisted composition over F_{sep} is a twisted composition $\widetilde{\mathfrak{C}}_s$ for the split Cayley algebra $\mathfrak{C}_s$. Thus (1) will follow from Proposition (36.5) and Proposition (29.1) if we may identify $\mathbf{Aut}(\widetilde{\mathfrak{C}})$ with $\mathbf{Aut}_G(w)$ for some tensor w and some representation $\rho\colon G \to \mathbf{GL}(W)$ where $H^1(F, G) = 0$. Let $\widetilde{\mathfrak{C}}_s = (V, L, Q, \beta)$, let W be the F-space

$$W = S_L^2(V^*) \oplus S_L^2(V^*) \otimes V \oplus \mathrm{Hom}_F(L \otimes L, L)$$

and let $G = R_{L/F}\big(\mathbf{GL}_L(V)\big) \times \mathbf{GL}(L)$. Then G acts on W as

$$\rho(\alpha, \phi)(f, g, h) = \big(\phi \circ f \circ \alpha^{-1}, \alpha \circ g \circ \alpha^{-1}, \phi \circ h \circ (\alpha^{-1} \otimes \alpha^{-1})\big).$$

We have $H^1\big(F, R_{L/F}(\mathbf{GL}_L(V))\big) = H^1\big(L, \mathbf{GL}_L(V)\big)$ by (29.6), hence $H^1(F, G) = 0$ by Hilbert 90. We now choose $w = (Q, \beta, m)$, where m is the multiplication of L, getting $\mathbf{Aut}(\widetilde{\mathfrak{C}}) = \mathbf{Aut}_G(w)$ and (1) is proved.

(2) follows from the description of $H^1(F, \mathbf{Spin}_8)$ given in Proposition (35.18) and (3) is an example of twisting in cohomology (see Proposition (28.11)). $\qquad\square$

36.A. Multipliers of similitudes of twisted compositions. We assume in this section that $\mathrm{char}\, F \neq 2, 3$. Let $\Gamma = (V, L, Q, \beta)$ be a twisted composition over F and let

$$G = \mathbf{Aut}(V, L, Q, \beta)^0 = \mathbf{Aut}_L(V, L, Q, \beta)$$

be the connected component of $\mathbf{Aut}(V, L, Q, \beta)$. If L is split and V is of split Cayley type, the proof of Proposition (36.5) shows that $G = \mathbf{Spin}_8$, hence G is always a twisted form of $\mathbf{Spin}_8$ and we write $G = \mathbf{Spin}(V, L, Q, \beta)$. If $L = F \times K$ for K quadratic étale, and accordingly $(V, Q) = (V_1, Q_1) \oplus (V_2, Q_2)$ for (V_1, Q_1) a quadratic space over F and (V_2, Q_2) a quadratic space over K, the projection $(V, Q) \to (V_1, Q_1)$ induces an isomorphism $\mathbf{Spin}(V, L, Q, \beta) \xrightarrow{\sim} \mathbf{Spin}(V_1, Q_1)$ by Propositions (21.7) and (22.2).

The pointed set $H^1(F, G)$ classifies twisted compositions (V', L, Q', β') with fixed L. Let $\gamma\colon \mathrm{Gal}(F_{\mathrm{sep}}/F) \to S_3$ be a cocycle defining L. By twisting the exact sequence (36.6) we see that $G = (\mathbf{Spin}_8)_\gamma$. Let $\widetilde{C}$ be the center of G; the center C of $\mathbf{Spin}_8$ fits into the exact sequence

$$1 \to C \to \boldsymbol{\mu}_2 \times \boldsymbol{\mu}_2 \times \boldsymbol{\mu}_2 \xrightarrow{m} \boldsymbol{\mu}_2 \to 1$$

(see Lemma (35.11)) hence, twisting with γ gives an exact sequence

$$1 \to \widetilde{C} \to R_{L/F}(\boldsymbol{\mu}_{2,L}) \xrightarrow{N_{L/F}} \boldsymbol{\mu}_{2,F} \to 1.$$

Thus

$$H^1(F, \widetilde{C}) = \ker\left(L^\times/L^{\times 2} \xrightarrow{N_{L/F}} F^\times/F^{\times 2}\right)$$

and the exact sequence (18.34)

$$1 \to F^\times/F^{\times 2} \to L^\times/L^{\times 2} \xrightarrow{\#} L^\times/L^{\times 2} \xrightarrow{N_{L/F}} F^\times/F^{\times 2} \to 1$$

gives the identification

$$H^1(F, \widetilde{C}) = L^\times/F^\times \cdot L^{\times 2}.$$

The group $H^1(F, \widetilde{C})$ acts on $H^1(F, G)$ through the rule

(36.8) $$\lambda \cdot (V, L, Q, \beta) = (V, L, \lambda^\# Q, \lambda\beta), \quad \lambda \in L^\times,$$

hence by Proposition (28.11), $(V, L, \lambda^\# Q, \lambda\beta) \simeq (V, L, Q, \beta)$ if and only if the class $\lambda \cdot F^\times \cdot L^{\times 2} \in H^1(F, C)$ belongs to the image of the connecting homomorphism

$$\delta_F = \delta \colon \overline{G}(F) \to H^1(F, C)$$

associated to the exact sequence (recall that char $F \neq 2$)

$$1 \to \widetilde{C} \to G \to \overline{G} \to 1.$$

We would like to compute the image of δ. For this we first consider the case $L = F \times K$; then $H^1(F, \widetilde{C}) = K^\times/K^{\times 2}$ and, since $G = \mathbf{Spin}(V_1, Q_1)$, $\overline{G} = \mathbf{PGO}^+(Q_1)$ and δ is the map

$$S \colon \mathbf{PGO}^+(Q_1)(F) \to K^\times/K^{\times 2}$$

defined in (13.32). The composition of S with the norm map $N_{K/F} \colon K^\times/K^{\times 2} \to F^\times/F^{\times 2}$ gives the multiplier map (Proposition (13.33)). If L is a field, let Δ be the discriminant of L, so that $L \otimes L = L \times L \otimes \Delta$ (see Corollary (18.28)). We would like to compare the images of δ_F and δ_L. Since $[L : F] = 3$, we have

$$\operatorname{im} \delta_F = N_{L/F} \circ \operatorname{res}_{F/L}(\operatorname{im} \delta_F) \subset N_{L/F}(\operatorname{im} \delta_L).$$

By Gille's norm principle (see the following Remark) we have

$$N_{L/F}(\operatorname{im} \delta_L) \subset \operatorname{im} \delta_F,$$

hence

$$\operatorname{im} \delta_F = N_{L/F}(\operatorname{im} \delta_L).$$

The group scheme $\overline{G}_L$ is isomorphic to $\mathbf{PGO}^+(Q)$, so that

$$\operatorname{im} \delta_L = \operatorname{im} S \subset (L \otimes \Delta)^\times/(L \otimes \Delta)^{\times 2} = H^1(L, \widetilde{C})$$

(since $L \otimes L = L \times L \otimes \Delta$). One can check that the diagram

$$
\begin{array}{ccc}
H^1(L, \widetilde{C}) & \xrightarrow{N^1_{L/F}} & H^1(F, \widetilde{C}) \\
\| & & \| \\
(L \otimes \Delta)^\times/(L \otimes \Delta)^{\times 2} & \xrightarrow{(N_{L/F})_*} & L^\times/F^\times \cdot L^{\times 2}
\end{array}
$$

commutes. It follows that

$$\operatorname{im}\delta_F = N_{L/F}(\operatorname{im}\delta_L) = N_{L\otimes K/L}(\operatorname{im}S)$$

is the image of the group of multipliers $G(Q)$ in $L^\times/F^\times \cdot L^{\times 2}$. We have proved:

(36.9) Theorem. *There exists an L-isomorphism $(V, L, \lambda^\# Q, \lambda\beta) \simeq (V, L, Q, \beta)$ if and only if $\lambda \in F^\times \cdot G(Q)$.* $\square$

(36.10) Remark. The condition $\lambda \in F^\times \cdot G(Q)$ does not depend on β. One can show that the condition is also equivalent to the fact that the quadratic forms $\lambda^\# \cdot Q$ and Q are isomorphic over L. The theorem says that this isomorphism can be chosen in such a way that it takes $\lambda \cdot \beta$ to β. This is the hardest part of the theorem, where we use Gille's result [106]: it says that the norm $N_{L/F}$ takes R-trivial elements in $\operatorname{im}\delta_L \subset H^1(L, \widetilde{C})$ to R-trivial elements of $\operatorname{im}\delta_F \subset H^1(F, \widetilde{C})$. So it suffices to prove that the image of S in $(L \otimes \Delta)^\times/(L \otimes \Delta)^{\times 2}$ consists of R-trivial elements. Let $\alpha \in \mathrm{PGO}^+(Q)(L)$, let $x = \delta_L(\alpha) \in (L \otimes \Delta)^\times/(L \otimes \Delta)^{\times 2}$ and let $y = N_{L\otimes\Delta/L}(x) \in L^\times/L^{\times 2}$. The square class y is a similarity factor of Q. Note that Q is "almost" a Pfister form, i.e., there exists a 3-fold Pfister form q over L and $a \in L^\times$ such that $Q = q + \langle a, -ad \rangle \in W(L)$ where $d \in F^\times$ is such that $\Delta = F(\sqrt{d})$. Hence y is a norm from $L \otimes \Delta/L$ and a similarity factor of the Pfister form q, in particular a norm from a quadratic extension which splits q. Thus y is a norm from a biquadratic extension containing $L \otimes \Delta$ which splits Q. It follows that y is the norm of an R-trivial element in the image of $N_{L\otimes\Delta\otimes\Delta/L\otimes\Delta} \circ \delta_{L\otimes\Delta}$, so that (by Gille) y is an R-trivial element in the image of $N_{L\otimes\Delta/L} \circ \delta_L$. Therefore we may assume $y = 1$. But then x is the image of a spinor norm $x' \in L^\times/L^{\times 2}$ of Q. The claim follows, since spinor norms are R-trivial.

36.B. Cyclic compositions. Twisted compositions of dimension 8 over cyclic cubic extensions were introduced in Springer [268]. We first recall Springer's definition. Let $(L/F, \rho)$ be a cyclic F-algebra of degree 3 with ρ a generator of the group $\mathrm{Gal}(L/F) = A_3$. A *cyclic composition* is a nonsingular quadratic space (V, Q) over L, together with an F-bilinear multiplication $(x, y) \mapsto x * y$, which is ρ-semilinear in x and ρ^2-semilinear in y, and such that

(1) $$Q(x * y) = \rho\big(Q(x)\big) \cdot \rho^2\big(Q(y)\big),$$

(2) $$b_Q(x * y, z) = \rho\big(b_Q(y * z, x)\big) = \rho^2\big(b_Q(z * x, y)\big)$$

where $b_Q(x, y) = Q(x + y) - Q(x) - Q(y)$. Observe that the choice of a generator ρ of the group $\mathrm{Gal}(L/F)$ is part of the datum defining a cyclic composition. Morphisms of cyclic compositions are defined accordingly.

Linearizing (1) gives

(3) $$b_Q(x * z, x * y) = \rho\big(Q(x)\big)\rho^2\big(b_Q(z, y)\big),$$

(4) $$b_Q(x * z, y * z) = \rho\big(b_Q(x, y)\big)\rho^2\big(Q(x)\big).$$

It then follows that

$$\begin{aligned}
b_Q\big((x * y) * x, z\big) &= \rho\big(b_Q(x * z, x * y)\big) \\
&= \rho^2\big(Q(x)\big)b_Q(y, z) \\
&= b_Q\big(\rho^2\big(Q(x)\big)y, z\big)
\end{aligned}$$

so that

(5) $$(x * y) * x = \rho^2\big(Q(x)\big)y$$

and similarly

(6) $$x * (y * x) = \rho\big(Q(x)\big)y.$$

Linearizing conditions (5) and (6) gives

(7) $$(x * y) * z + (z * y) * x = \rho^2\big(b_Q(x,z)\big)y,$$

(8) $$x * (y * z) + z * (y * x) = \rho\big(b_Q(x,z)\big)y.$$

By (2) we have $b_Q(x * x, x) \in F$, thus (V, L, β, Q) with $\beta(x) = x * x$ is a twisted composition. Conversely, we shall see in Proposition (36.12) that any twisted composition over a cyclic cubic extension comes from a cyclic composition.

(36.11) Example. Let $(S, \star, n)$ be a symmetric composition algebra over F and let L be a cyclic cubic algebra. Let ρ be a generator of $\mathrm{Gal}(L/F)$. It is easy to check that $V = S_L = S \otimes L$, with the product

$$x * y = (1 \otimes \rho)(x) \star (1 \otimes \rho^2)(y)$$

and the norm $Q(x) = (n \otimes 1)(x)$, is cyclic. Thus, by putting $\beta(x) = x * x$, we may associate to any symmetric composition algebra $(S, \star, n)$ and any cubic cyclic extension L a twisted composition $\Gamma(S, \star, L)$. If, furthermore, $L = F \times F \times F$ and $\rho_1(\xi_0, \xi_1, \xi_2) = (\xi_1, \xi_2, \xi_0)$, we have a product

$$\beta_1(x, y) = x *_1 y = (x_1 \star y_2, x_2 \star y_0, x_0 \star y_1)$$

for $x = (x_0, x_1, x_2)$ and $y = (y_0, y_1, y_2)$ in $S_{F \times F \times F} = S \times S \times S$ such that

$$\beta_1(x, x) = \beta(x_0, x_1, x_2) = (x_1 \star x_2, x_2 \star x_0, x_0 \star x_1)$$

and we obtain a twisted composition as defined in Example (36.2). By taking $\rho_2 = \rho_1^2$ as generator of $\mathrm{Gal}(F \times F \times F/F)$, we have another product

$$\beta_2(x, y) = x *_2 y = (y_1 \star x_2, y_2 \star x_0, y_0 \star x_1)$$

such that $\beta_2(x, x) = \beta(x)$. Observe that $\beta_1(x, y) = \beta_2(y, x)$.

(36.12) Proposition. *Let L/F be a cyclic cubic algebra, let ρ_1, ρ_2 be different generators of the group $\mathrm{Gal}(L/F) = A_3$ and let (V, L, Q, β) be a twisted composition over L. There is a unique pair of cyclic compositions $\beta_i(x, y) = x *_i y$ over V, $i = 1, 2$, with β_i ρ_i-semilinear with respect to the first variable, such that $\beta_i(x, x) = \beta(x)$. Furthermore we have $\beta_1(x, y) = \beta_2(y, x)$.*

Proof: There exists at most one F-bilinear map $\beta_1 \colon V \otimes V \to V$ which is ρ_1-semilinear with respect to the first variable and ρ_2-semilinear with respect to the second variable and such that $\beta_1(x, x) = \beta(x)$ for all $x \in V$: the difference $\widehat{\beta}$ of two such maps would be such that $\widehat{\beta}(x, x) = 0$, hence $\widehat{\beta}(x, y) = -\widehat{\beta}(y, x)$. This is incompatible with the semilinearity properties of $\widehat{\beta}$. Over a separable closure F_{sep} of F such a map exists, since $L \otimes F_{\mathrm{sep}}$ is split and, by Proposition (36.3), $V_{F_{\mathrm{sep}}}$ is of Cayley type. Thus, by descent, such a β_1 resp. β_2 exists. Furthermore $x *_i y = \beta_i(x \otimes y)$ satisfies the identities of a cyclic composition, since it does over F_{sep}. The last claim follows from Example (36.11). $\qquad\square$

It follows from Proposition (36.12) that the twisted composition $(S \otimes L, L, Q, \beta)$ is independent of the choice of a generator of $\mathrm{Gal}(L/F)$.

(36.13) Proposition. *Cyclic compositions over F are classified by the pointed set $H^1(F, \mathbf{Spin}_8 \rtimes \mathbb{Z}/3\mathbb{Z})$.*

Proof: In view of Propositions (36.3) and (36.12) any cyclic composition over a separable closure F_{sep} of F is isomorphic to a Cayley composition with multiplication $*$ as described in Example (36.11) and ρ either given by $(x_0, x_1, x_2) \mapsto (x_1, x_2, x_0)$ or $(x_0, x_1, x_2) \mapsto (x_2, x_0, x_1)$. It then follows as in the proof of Proposition (36.5) that the group of automorphisms of the cyclic composition

$$\left(\mathfrak{C} \otimes (F \times F \times F), \mathfrak{n} \perp \mathfrak{n} \perp \mathfrak{n}, F \times F \times F, \rho, * \right)_R$$

is isomorphic to $\mathbf{Spin}(\mathfrak{n}_s)(R) \rtimes \mathbb{Z}/3\mathbb{Z}$. The claim then follows by constructing a representation $G \to \mathbf{GL}(W)$ such that $\mathbf{Spin}(\mathfrak{n}_s) \rtimes \mathbb{Z}/3\mathbb{Z} = \mathbf{Aut}_G(w)$ as in the proof of (36.7), (2). We let this construction as an exercise. $\qquad\square$

Let $(V, L, Q, \rho, *)$ be a cyclic composition. We have a homomorphism

$$p \colon \mathbf{Aut}(V, L, \rho, *) \to \mathbf{Aut}(L) = A_3$$

induced by restriction. Assume that p is surjective and has a section s. Then s defines an action of A_3 on $(V, L, Q, \rho, *)$.

(36.14) Proposition. *Let F be a field of characteristic not equal to 3. For any faithful ρ-semilinear action of A_3 on $(V, L, Q, \rho, *)$, $V_0 = V^{A_3}$ carries the structure of a symmetric composition algebra and $V = V_0 \otimes L$. In particular, if $\dim_L V = 8$, then p has only two possible sections (up to isomorphism) over F_{sep}.*

Proof: The first claim follows by Galois descent and the second from the fact that over F_{sep} we have only two types of symmetric composition algebras (Theorem (34.37)). $\qquad\square$

(36.15) Corollary. *Datas $\big((V, L, Q, \rho, *), p, s\big)$ where $(V, L, Q, \rho, *)$ is a cyclic composition, are classified either by $H^1(F, G \times A_3)$ where G is of type G_2, if the section s defines a para-Hurwitz composition, or by $H^1\big(F, \mathbf{PGU}_3(K) \times A_3\big)$ where $K = F[X]/(X^2 + X + 1)$.*

Proof: This corresponds to the two possible structures of symmetric composition algebras over F_{sep}. $\qquad\square$

Let $(V, L, \rho, *)$ be a cyclic composition of dimension 8. We write $^\rho V$ for the L-space V with the action of L twisted through ρ and put $^\rho Q(x) = \rho\big(Q(x)\big)$. Let $\ell_x(y) = x * y$ and $r_x(y) = y * x$.

(36.16) Proposition. *The map*

$$x \mapsto \begin{pmatrix} 0 & \ell_x \\ r_x & 0 \end{pmatrix} \in \mathrm{End}_L(^\rho V \oplus {}^{\rho^2} V), \quad x \in V$$

extends to an isomorphism of algebras with involution

$$\alpha_V \colon \big(C(V, Q), \tau\big) \xrightarrow{\sim} \big(\mathrm{End}_L(^\rho V \oplus {}^{\rho^2} V), \sigma_{\widetilde{Q}}\big)$$

where $\sigma_{\widetilde{Q}}$ is the involution associated with the quadratic form $\widetilde{Q} = {}^\rho Q \perp {}^{\rho^2} Q$. In particular α_V restricts to an isomorphism

$$\big(C_0(V, Q), \tau\big) \xrightarrow{\sim} \big(\mathrm{End}_L(^\rho V), \sigma_{\rho Q}\big) \times \big(\mathrm{End}_L({}^{\rho^2} V), \sigma_{\rho^2 Q}\big).$$

Proof: The existence of α_V follows from the universal property of the Clifford algebra, taking the identities $(x * y) * x = {}^\rho Q(x)y$ and $x * (y * x) = {}^{\rho^2} Q(x)y$ into account. It is an isomorphism because $C(V, Q)$ is central simple over F. The fact that α is compatible with involutions is a consequence of the formula (2) above. □

Proposition (36.16) is a twisted version of Proposition (35.1) and can be used to deduce analogues of Propositions (35.4) and (35.7). The proofs of the following two results will only be sketched.

(36.17) Proposition. *Let t be a proper similitude of (V, Q), with multiplier $\mu(t)$. There exist proper similitudes u, v of (V, Q) such that*

$$\mu(v)^{-1}v(x * y) = u(x) * t(y),$$
$$\mu(u)^{-1}u(x * y) = t(x) * v(y),$$
$$\mu(t)^{-1}t(x * y) = v(x) * u(y)$$

for all x, $y \in V$. If $t \in R_{L/F}\big(\mathbf{O}^+(V, Q)\big)(F)$ is such that $\mathrm{Sn}(t) = 1$, then u, v can be chosen in $R_{L/F}(\mathbf{O}^+(V, Q))(F)$.

Proof: We identify $C(V, Q)$ with $\big(\mathrm{End}_L({}^\rho V \oplus {}^{\rho^2}V), \sigma_{\widetilde{Q}}\big)$ through α_V. The map

$$x \mapsto \begin{pmatrix} 0 & \ell_{tx} \\ \mu(t)^{-1}r_{tx} & 0 \end{pmatrix} \in \mathrm{End}_L({}^\rho V \oplus {}^{\rho^2}V)$$

extends to an automorphism $\widetilde{t}$ of $C(V, Q)$ whose restriction to $C_0(V, Q)$ is $C_0(t)$. Thus we may write $\widetilde{t} = \mathrm{Int}\big(\begin{smallmatrix} u' & 0 \\ 0 & v' \end{smallmatrix}\big)$ where u', v' are similitudes of Q and

$$\begin{pmatrix} u' & 0 \\ 0 & v' \end{pmatrix} \begin{pmatrix} 0 & \ell_x \\ r_x & 0 \end{pmatrix} \begin{pmatrix} u' & 0 \\ 0 & v' \end{pmatrix}^{-1} = \begin{pmatrix} 0 & \ell_{tx} \\ \mu(t)^{-1}r_{tx} & 0 \end{pmatrix}$$

or

$$u'(x * y) = t(x) * v'(y) \quad \text{and} \quad v'(y * x) = u'(y) * t(x)\mu(t)^{-1}.$$

Putting $u = u'\mu(u')^{-1}$ and $v = v'$ and using that

$$\mu(u)^{-1} = \mu(u') = {}^\rho\mu(t)\, {}^{\rho^2}\mu(v)$$

gives the first two formulas. The third formula follows by replacing x by $y * x$ in the first one (compare with the proof of Proposition (35.4)). □

(36.18) Proposition. *Assume that* $\mathrm{char}\, F \neq 2$. *For $R \in \mathsf{Alg}_F$ we have*

$$R_{L/F}\big(\mathbf{Spin}(V, Q)\big)(R) \simeq$$
$$\{ (v, u, t) \in \big(R_{L/F}(\mathbf{O}^+(V))(R)\big)^3 \mid v(x * y) = u(x) * t(y) \},$$

the group A_3 operates on $\mathbf{Spin}(V, Q)$, and $\mathbf{Spin}(V, Q)^{A_3}$ is a group scheme over F such that $\big(\mathbf{Spin}(V, Q)^{A_3}\big)_L = \mathbf{Spin}(V, Q)$.

Proof: Proposition (36.18) follows from Proposition (36.17) as Proposition (35.9) follows from Proposition (35.7). □

The computation given above of the Clifford algebra of a cyclic composition can be used to compute the even Clifford algebra of an arbitrary twisted composition:

(36.19) Proposition. *Let* $\Gamma = (V, L, Q, \beta)$ *be a twisted composition. There exists an isomorphism*

$$\alpha_V : C_0(V, Q) = C\big(\mathrm{End}_L(V), \sigma_Q\big) \to {}^\rho\big(\mathrm{End}_L(V) \otimes \Delta\big)$$

where $\Delta(L)$ *is the discriminant algebra of* L *and* ρ *is a generator of the group* $\mathrm{Gal}\big(L \otimes \Delta(L)/L\big) \simeq A_3$.

Proof: By Proposition (36.12) there exists exactly one cyclic composition (with respect to ρ) on $(V, L, Q, \beta) \otimes \Delta$ and by Proposition (36.16) we then have an isomorphism

$$\alpha_{V \otimes \Delta} : C_0(V, Q) \otimes \Delta \xrightarrow{\sim} \mathrm{End}_{L \otimes \Delta}\big({}^\rho(V \otimes \Delta)\big) \times \mathrm{End}_{L \otimes \Delta}\big({}^{\rho^2}(V \otimes \Delta)\big)$$

of $L \otimes \Delta$-algebras with involution. By composing with the canonical map

$$C_0(V, Q) \to C_0(V, Q) \otimes \Delta$$

and the projection ${}^\rho\big(\mathrm{End}_L(V) \otimes \Delta\big) \times {}^{\rho^2}\big(\mathrm{End}_L(V) \otimes \Delta\big) \to {}^\rho\big(\mathrm{End}_L(V) \otimes \Delta\big)$ we obtain a homomorphism of central simple algebras

$$\alpha_V : C_0(V, Q) = C\big(\mathrm{End}_L(V), \sigma_Q\big) \to {}^\rho\big(\mathrm{End}_L(V) \otimes \Delta\big)$$

which is an isomorphism by dimension count.   □

36.C. Twisted Hurwitz compositions. In this section we first extend the construction of a twisted composition $C \otimes L$ given in Example (36.11) for L cyclic to an arbitrary cubic étale algebra L and a para-Hurwitz algebra $(C, \star, n)$. We refer to 18.C for details on cubic étale algebras.

Let Δ be the discriminant (as a quadratic étale algebra) of L, let ι be the generator of $\mathrm{Gal}(\Delta/F)$ and let C be a Hurwitz algebra over F. Since $L \otimes \Delta$ is cyclic over Δ, by Proposition (36.12) there exists a cyclic composition $(x, y) \mapsto x * y$ on $C \otimes L \otimes \Delta$ for each choice of a generator ρ of $\mathrm{Gal}(L \otimes \Delta/\Delta)$. The automorphism $\tilde{\iota} = \pi \otimes 1 \otimes \iota$ of $C \otimes L \otimes \Delta$ is ι-semilinear and satisfies $\tilde{\iota}^2 = 1$ and $\tilde{\iota}(x * y) = \tilde{\iota}(y) * \tilde{\iota}(x)$ for $x, y \in C \otimes L \otimes \Delta$. Let

$$V = \{\, x \in C \otimes L \otimes \Delta \mid \tilde{\iota}(x) = x \,\}$$

be the corresponding descent (from $L \otimes \Delta$ to L). Since $\tilde{\iota}(x * y) = \tilde{\iota}(y) * \tilde{\iota}(x)$, the map $\beta(x) = x * x$ restricts to a quadratic map β on V. The restriction Q of $n \otimes 1 \otimes 1$ to V takes values in L and is nonsingular. Thus (V, L, Q, β) is a twisted composition. We write it $\Gamma(C, L)$. A twisted composition similar to a composition $\Gamma(C, L)$ is called a *twisted Hurwitz composition* over L. If C is a Cayley algebra we also say that $\Gamma(C, L)$ is a *twisted composition of type* G_2. The underlying quadratic space (V, Q) of $\Gamma(C, L)$ is extended from the quadratic space (V_0, Q_0) over F with

$$V_0 = \{\, x \in C \otimes \Delta \mid \pi \otimes \iota(x) = x \,\}$$

and Q_0 is the restriction of Q to V_0. The space (V_0, Q_0) is called the Δ-*associate* of (C, n) by Petersson-Racine [214] (see also Loos [175]) and is denoted by $(C, N)_\Delta$. In [175] a K-associate $(U, q)_K$ is attached to any pointed quadratic space (U, q, e) and any étale quadratic algebra K:

$$(U, q)_K = \{\, x \in U \otimes K \mid \pi \otimes \iota_K(x) = x \,\}$$

where π is the reflection with respect to the point e. For any pair of quadratic étale algebras K_1, K_2 with norms n_1, n_2, $(K_1, n_1)_{K_2} = (K_2, n_2)_{K_1}$ is the étale quadratic algebra

$$K_1 * K_2 = \{ x \in K_1 \otimes K_2 \mid (\iota_1 \otimes \iota_2)(x) = x \}.$$

Recall that $K * K \simeq F \times F$.

(36.20) Lemma. *Let (U, p) be a pointed quadratic space. There are canonical isomorphisms*

$$\big((U, p)_{K_1}\big)_{K_2} \simeq (U, p)_{K_1 * K_2} \quad and \quad (U, p)_{F \times F} \simeq (U, p).$$

Reference: See Loos [175]. □

In what follows we use the notation $[a]$ for the 1-dimensional regular quadratic form $q(x) = ax^2$ (and, as usual, $\langle a \rangle$ for the bilinear form $b(x, y) = axy$, $a \in F^\times$).

(36.21) Lemma. *Let (C, n) be a Hurwitz algebra (with unit 1) and let $(C^0, n^0) = 1^\perp$ be the subspace of (C, n) of elements of trace zero. We have, with the above notations, $(V_0, Q_0) = (C, n)_\Delta$ and:*
(1) *If $\operatorname{char} F \neq 2$, $V_0 = F \cdot 1 \oplus C^0$, $Q_0 = [1] \perp \langle \delta \rangle \otimes (C^0, n^0)$.*
(2) *If $\operatorname{char} F = 2$, let $u \in C$, $\xi \in \Delta$ with $T_C(u) = 1 = T_\Delta(\xi)$ and let $w = u \otimes 1 + 1 \otimes \xi$. Then $V_0 = F \cdot w \oplus C^0$, $Q_0(w) = n(u) + \xi^2 + \xi$, $b_{Q_0}(w, x) = b_n(x, u)$ and $Q_0(x) = n(x)$ for $x \in C^0$.*

Proof: Claims (1) and (2) can be checked directly. The first claim then follows easily. □

(36.22) Proposition. *Two twisted Hurwitz compositions $\Gamma(C_1, L_1)$ and $\Gamma(C_2, L_2)$ are isomorphic if and only if $L_1 \simeq L_2$ and $C_1 \simeq C_2$.*

Proof: The if direction is clear. We show the converse. If the compositions $\Gamma(C_1, L_1)$ and $\Gamma(C_2, L_2)$ are isomorphic, we have by definition an isomorphism $L_1 \xrightarrow{\sim} L_2$ and replacing the L_2-action by the L_1-action through this isomorphism, we may assume that $L_1 = L_2 = L$. In particular the quadratic forms Q_1 and Q_2 are then isometric as quadratic spaces over L. By Springer's theorem there exists an isometry $(C_1, n_1)_\Delta \simeq (C_2, n_2)_\Delta$. Since $\Delta * \Delta \simeq F \times F$, $(C_1, n_1) \simeq (C_2, n_2)$ follows from Lemma (36.20), hence $C_1 \simeq C_2$ by Theorem (33.19). □

Let G be the automorphism group of the split Cayley algebra. Since $G = \mathbf{Spin}_8^{S_3}$ (see 35.16), we have a homomorphism

$$G \times S_3 \to \mathbf{Spin}_8 \rtimes S_3.$$

(36.23) Proposition. *Twisted compositions of type G_2 are classified by the image of $H^1(F, G \times S_3)$ in $H^1(F, \mathbf{Spin}_8 \rtimes S_3)$.*

Proof: A pair (ϕ, ψ) where ϕ is an automorphism of $\mathfrak{C}$ and ψ is an automorphism of L defines an automorphism $\Gamma(\phi, \psi)$ of $\Gamma(\mathfrak{C}, L)$. The map

$$H^1(F, G \times S_3) = H^1(F, G) \times H^1(F, S_3) \to H^1(F, \mathbf{Spin}_8 \rtimes S_3)$$

corresponds to $[\mathfrak{C}] \times [L] \to [\Gamma(\mathfrak{C}, L)]$. □

The following result is due to Springer [268] for L cyclic and V of dimension 8.

(36.24) Theorem. *Let (V, L, Q, β) be a twisted composition and let N be the cubic form $N(x) = Q\big(x, \beta(x)\big)$, $x \in V$. The following conditions are equivalent:*
(1) *The twisted composition (V, L, Q, β) is similar to a twisted Hurwitz composition.*

(2) *The cubic form $N(x)$ is isotropic.*
(3) *There exists $e \in V$ with $\beta(e) = \lambda e$, $\lambda \neq 0$, and $Q(e) = \lambda^{\#}$.*

Proof of (1) $\Rightarrow$ (2): We may assume that (V, L, Q, β) is a Hurwitz composition. The element $x = c \otimes 1 \otimes \xi \in C \otimes L \otimes \Delta$ lies in V if $t_C(c) = 0 = t_\Delta(\xi)$. For such an element we have $\beta(x) = \bar{c}^2 \otimes 1 \otimes \xi^2 \in L$. Since $t_C(c) = 0$, $\bar{c}^2 = n_C(c)$ and so $N(x) = 0$ if char $F \neq 2$. If char $F = 2$, $x = 1 \otimes 1 \otimes 1$ lies in V, $\beta(x) = x$, and $N(x) = b_Q(x, x) = 0$. $\square$

For the proof of (2) $\Rightarrow$ (3) we need the following lemma:

(36.25) Lemma. *Any element x of a twisted composition (V, L, Q, β) satisfies the identity*
$$\beta^2(x) + Q(x)\beta(x) = N(x)x.$$

Proof: Since over the separable closure F_{sep} any composition is cyclic, it suffices to prove the formula for a cyclic composition and for $\beta(x) = x * x$. In view of Formulas (7) and (6) (p. 496), we have
$$(x * x) * (x * x) + \big((x * x) * x\big) * x = \rho\big(b_Q(x * x, x)\big)x$$
hence the assertion, since $(x*x)*x = \big(\rho^2(Q(x))\big)x$ (and the product $*$ is ρ-semilinear with respect to the first variable). $\square$

Proof of (2) $\Rightarrow$ (3): First let $x \neq 0$ be such that $\beta(x) \neq 0$ and $N(x) = 0$. If $Q(x) \neq 0$ we take $e = \beta(x)$ and apply Lemma (36.25). If $Q(x) = 0$, we replace x by $\beta(x)$ and apply (36.25) to see that our new x satisfies $Q(x) = 0$ and $\beta(x) = 0$. Thus we may assume that $Q(x) = 0$ and $\beta(x) = 0$. Let $y \in V$ be such that $Q(y) = 0$ and $b_Q(x, y) = -1$. Let
$$\beta(u, v) = \beta(u + v) - \beta(u) - \beta(v).$$

We claim that

(36.26) $\beta\big(\beta(x, y)\big) = \beta(x, y) + f \cdot x$

for some $f \in F$. We extend scalars to $L \otimes \Delta$ and so assume that $\beta(x) = x * x$ is cyclic. It suffices to check that $f \in \Delta \cap L$. Since $\beta(x, y) = x * y + y * x$, we have
$$\beta\big(\beta(x, y)\big) = (x * y + y * x) * (x * y + y * x).$$
Applying formulas (7) and (8) (p. 496), we first obtain
$$(y * x) * x + (x * x) * y = (y * x) * x = -x$$
$$x * (x * y) + y * (x * x) = x * (x * y) = -x$$
since $\beta(x) = x * x = 0$ and $b_Q(x, y) = -1$. Applying again formulas (7) and (8) then give
$$(x * y) * (x * y) = \rho\big(b_Q(x * y, y)\big)x + y * x$$
$$(x * y) * (y * x) = 0$$
$$(y * x) * (x * y) = \rho\big(b_Q(y * x, y)\big)x$$
$$(y * x) * (y * x) = \rho^2\big(b_Q(y, y * x)\big)x + x * y$$

and
$$\beta\big(\beta(x, y)\big) = \beta(x, y) + f \cdot x$$

with

$$f = \rho\big(b_Q(x * y, y)\big) + \rho\big(b_Q(y * x, y)\big) + \rho^2\big(b_Q(y, y * x)\big)$$
$$= \rho^2\big(b_Q(\beta(y), x)\big) + b_Q\big(\beta(y), x\big) + \rho\big(b_Q(\beta(y), x)\big) = T_{L \otimes \Delta/\Delta}\big(b_Q(\beta(y), x)\big)$$

hence $f \in \Delta \cap L$. Let $\mu \in L$ be such that $T_L(\mu) = f$; we claim that $e = \mu x + \beta(x, y)$ satisfies $\beta(e) = e$:

$$\beta(e) = \beta(\mu x) + \beta\big(\beta(x, y)\big) + \beta\big(\mu x, \beta(x, y)\big)$$
$$= \beta(x, y) + f \cdot x + \beta\big(\mu x, \beta(x, y)\big).$$

To compute $\beta\big(\mu x, \beta(x, y)\big)$ we assume that $\beta(x) = x * x$ is cyclic. We have:

$$\beta\big(\mu x, \beta(x, y)\big) = (\mu x) * (x * y + y * x) + (x * y + y * x) * (\mu x)$$
$$= \rho(\mu)[x * (x * y + y * x)] + \rho^2(\mu)[(x * y + y * x) * x]$$
$$= -\rho(\mu)x - \rho^2(\mu)x = (\mu - f)x.$$

This implies that $\beta(e) = \beta(x, y) + \mu x = e$. The relation $\lambda^\# = Q(e)$ follows from $\beta(e) = \lambda e$ and Lemma (36.25), since replacing $\beta(e)$ by λe we see that $\lambda^\# \lambda e + Q(e)\lambda e = 2Q(e)\lambda e$. $\square$

Proof of (3) $\Rightarrow$ (1): We first construct the Hurwitz algebra C. Replacing β by the similar composition $\lambda^{-1}\beta$, we may assume that $\beta(e) = e$. This implies that $Q(e) = 1$. Let Δ be the discriminant algebra of L. We also call β the extension $\beta \otimes 1$ of β to $L \otimes \Delta$. Let ρ_1, ρ_2 be the generators of $\mathrm{Gal}(L \otimes \Delta/\Delta)$ and let ι be the generator of $\mathrm{Gal}(\Delta/F)$, so that (putting $\iota = 1 \otimes \iota$ on $L \otimes \Delta$ and $V \otimes \Delta$), we have $\rho_i^2 = \rho_{i+1}$ and $\iota\rho_i = \rho_{i+1}\iota$. Let $\beta_i(x, y) = x *_i y$ be the two unique extensions of β to $L \otimes \Delta$ as cyclic compositions with respect to ρ_i. By uniqueness we have (see Examples (36.11) and (36.2))

$$\beta_1(x, y) = \beta_2(y, x) \quad \text{and} \quad \iota\beta_1(x, y) = \beta_2(\iota x, \iota y).$$

Let $\pi\colon V \to V$ be the hyperplane reflection

$$\pi(x) = \overline{x} = -x + b_Q(e, x)e$$

as well as its extension to $V \otimes \Delta$ and let $\varphi_i\colon V \otimes \Delta \xrightarrow{\sim} V \otimes \Delta$ be the ρ_i-semilinear map given by $\varphi_i(x) = \beta_i(\overline{x}, e)$. $\square$

(36.27) Lemma. *The following identities hold for the maps φ_i:*

(1) $\varphi_i\pi = \pi\varphi_i$

(2) $\varphi_i^2 = \varphi_{i+1}$

(3) $\iota\varphi_i\iota = \varphi_{i+1}$

Proof: (1) We have $\beta_1(\overline{x}, e) = -\beta_1(x, e) + \rho_1\big(b_Q(e, x)\big)e$ so that

$$\overline{\beta_1(\overline{x}, e)} = \beta_1(x, e) - b_Q\big(e, \beta_1(x, e)\big)e$$
$$- \rho_1\big(b_Q(e, x)\big)e + b_Q\big(e, \rho_1(b_Q(e, x))e\big)e = \beta_1(x, e)$$

(2) By (1)

$$\beta_1\big(\overline{\beta_1(\overline{x}, e)}, e\big) = \beta_1\big(\beta_1(x, e), e\big) = \rho_2\big(b_Q(e, x)\big)e - \beta_1(e, x) = \beta_1(e, \overline{x})$$

using that $\beta_1\big(\beta_1(x, y), z\big) + \beta_1\big(\beta_1(z, y), x\big) = \rho_2\big(b_Q(x, z)\big)y$, a formula which follows from (7) (p. 496).

(3) follows from $\iota\beta_1(x, y) = \beta_2(\iota x, \iota y)$. □

We next define a multiplication γ_1 on $V \otimes \Delta$ by

$$\gamma_1(x, y) = \beta_1\big(\varphi_2(\overline{x}), \varphi_1(\overline{y})\big).$$

(36.28) Lemma. *The multiplication γ_1 satisfies the following properties*:

(1) $$Q\big(\gamma_1(x, y)\big) = Q(x)Q(y)$$

(2) $$\gamma_1(x, e) = \gamma_1(e, x) = x$$

(3) $$\overline{\gamma_1(x, y)} = \gamma_1(\overline{y}, \overline{x})$$

(4) $$\iota\gamma_1(x, y) = \gamma_1(\iota y, \iota x)$$

(5) $$\gamma_1\big(\varphi_1(x), \varphi_1(y)\big) = \varphi_1\big(\gamma_1(x, y)\big)$$

Proof: (1) We have

$$Q\big(\gamma_1(x, y)\big) = Q\big(\beta_1\big(\varphi_2(\overline{x}), \varphi_1(\overline{y})\big)\big) = \rho_1\big(Q\big(\varphi_2(\overline{x})\big)\big)\rho_2\big(Q\big(\varphi_1(\overline{y})\big)\big)$$
$$= \rho_1\rho_2\big(Q(\overline{\overline{x}})\big)\rho_2\rho_1\big(Q(\overline{\overline{y}})\big) = Q(x)Q(y)$$

(2) We have $\gamma_1(x, e) = \beta_1\big(\beta_2(x, e), e\big) = x$ and similarly $\gamma_1(e, x) = x$.

(3) In view of (1) and (2) γ_1 is a Hurwitz multiplication with conjugation $x \mapsto \overline{x}$ and (3) holds for any multiplication of a Hurwitz algebra, see (33.9).

(4) We have

$$\iota\gamma_1(x, y) = \iota\beta_1\big(\varphi_2(\overline{x}), \varphi_1(\overline{y})\big) = \beta_2\big(\iota\varphi_2(\overline{x}), \iota\varphi_1(\overline{y})\big)$$
$$= \beta_2\big(\varphi_1\iota(\overline{x}), \varphi_2\iota(\overline{y})\big) = \beta_1\big(\varphi_2\iota(\overline{y}), \varphi_1\iota(\overline{x})\big) = \gamma_1(\iota y, \iota x)$$

(5) Using (3) we have

$$\varphi_1\big(\gamma_1(x, y)\big) = \beta_1\big(\gamma_1(\overline{y}, \overline{x}), e\big) = \beta_1\big(\beta_1\big(\varphi_2(y), \varphi_1(x)\big), e\big)$$
$$= \rho_2\big(b_Q\big(\varphi_2(y), e\big)\big)\varphi_1(x) - \beta_1\big(\beta_1\big(e, \varphi_1(x)\big), \varphi_2(y)\big)$$
$$= \beta_1\big(\beta_1\big(e, \varphi_1(x)\big), b_Q\big(\varphi_2(y), e\big)e\big) - \beta_1\big(\beta_1\big(e, \varphi_1(x)\big), \varphi_2(y)\big)$$
$$= \beta_1\big(\beta_1\big(e, \varphi_1(x)\big), \overline{\varphi_2(y)}\big) = \beta_1\big(\varphi_2\overline{\varphi_1(x)}, \varphi_2(\overline{y})\big)$$
$$= \gamma_1\big(\varphi_1(x), \varphi_1(y)\big)$$

□

We go back to the proof of Theorem (36.24). It follows from Lemma (36.28) that

$$\gamma_1\big(\iota\pi(x), \iota\pi(y)\big) = \iota\pi\big(\gamma_1(x, y)\big)$$

Since $\iota\pi\varphi_1 = \varphi_1^2\iota\pi$, the automorphisms $\{\varphi_1, \iota\pi\}$ of $V \otimes \Delta$ define a Galois action of the group $\mathrm{Gal}(L \otimes \Delta/F) = S_3$ on the Hurwitz algebra $V \otimes \Delta$. Let C be the descended Hurwitz algebra over F, so that

$$V \otimes \Delta = C \otimes L \otimes \Delta$$

Observe that V on the left is the subspace of elements of $V \otimes \Delta$ fixed by $\mathrm{Id}_V \otimes \iota$ and that $C \otimes L$ on the right is the subspace of elements of $C \otimes L \otimes \Delta$ fixed by $\iota\pi$. We only consider the case where char $F \neq 2$ and leave the case char $F = 2$ as an exercise. Let $V = Le \perp V^0$, $V^0 = e^\perp$, and let $d \in L \otimes \Delta$ be a generator of the discriminant algebra Δ such that $(1 \otimes \iota)(d) = -d$. The map

$$\phi : V \otimes \Delta \to C \otimes L \otimes \Delta$$

given by $\phi\big((\ell e + v') \otimes s\big) = 1 \otimes \ell \otimes s + v' \otimes ds$ is such that $\phi \circ \iota = (\pi \otimes \iota) \circ \phi$. Thus the image of V in $C \otimes L \otimes \Delta$ can be identified with $L \otimes 1 \perp C^0 \otimes L \otimes 1$ and β is the restriction of the twisted Hurwitz composition on $C \otimes L \otimes \Delta$. This shows that (3) implies (1).   □

(36.29) Corollary. *If L is not a field, any twisted composition over L is similar to a twisted Hurwitz composition $\Gamma(C, L)$.*

Proof: The claim for compositions over $F \times F \times F$ follows from Proposition (36.3) (a result used in the proof of Theorem (36.24)). Let (V, L, Q, β) be a twisted composition over $L = F \times \Delta$, Δ a quadratic field extension of F. We have to check that (V, L, Q, β) is similar to a composition with an element e such that $\beta(e) = e$. By decomposing $V = (V_0, V_1)$, $Q = (Q_0, Q_1)$ and $\beta = (\beta_0, \beta_1)$ according to the decomposition $F \times \Delta$, we see (compare with the proof of (36.3)) that β_0 is a quadratic map $V_1 \to V_0$ such that $Q_0 \beta_0(v_1) = n_{K/F}\big(Q_1(v_1)\big)$ and β_1 is a Δ-linear map ${}^\iota V_1 \otimes V_0 \to V_1$ such that $Q_1 \beta_1(v_1 \otimes v_0) = \iota\big(Q_1(v_1)\big) Q_0(v_0)$. Let $b_{Q_1}(x, y)$ be the Δ-bilinear polar of Q_1 and let $b_{Q_0}(x, y)$ be F-bilinear polar of Q_0 as well as its extension to $V_0 \otimes \Delta$ as a Δ-bilinear form. By an argument similar to the argument in the proof of proposition (36.12), there exists a unique extension $\widehat{\beta}_0 \colon V_1 \otimes_\Delta {}^\iota V_1 \to V_0 \otimes \Delta$ of β_0 as a Δ-hermitian map. Property (3) of the definition of a twisted composition implies that

$$(36.30) \qquad b_{Q_0}\big(v_0, \beta_0(v_1)\big) = b_{Q_1}\big(v_1, \beta_1(v_1 \otimes v_0)\big) \quad \text{for} \quad v_0 \in V_0, v_1 \in V_1.$$

For $v = (v_0, 0)$, $v_0 \neq 0$ in V_0 we have $b_Q\big(v, \beta(v)\big) = 0$, so that the claim follows from Theorem (36.24).   □

(36.31) Remark. In the next section we give examples of twisted compositions of dimension 8 over a field L which are not induced by Cayley algebras.

(36.32) Remark. Let $\Gamma = (V, L, Q, \beta)$ be a twisted composition. Since $L \otimes L \simeq L \times L \otimes \Delta$ (where the first projection is multiplication), $\Gamma \otimes L$ is similar to a Hurwitz twisted composition $\Gamma(C, L \otimes L)$ for some unique Hurwitz algebra $C(\Gamma)$ over L. The algebra $C(\Gamma)$ over L is in fact extended from a Hurwitz algebra C over F. In dimension 8, the algebra C is determined by a cohomological invariant f_3 (see Proposition (40.16)).

36.D. Twisted compositions of type A_2'. We finish this chapter by showing how to associate twisted compositions to symmetric compositions arising from central simple algebras of degree 3 or cubic étale algebras. We assume that char $F \neq 3$ and first suppose that F contains a primitive cube root of unity ω. Let $\lambda \in F^\times$ and let $L = F[X]/(X^3 - \lambda) = F(\sqrt[3]{\lambda})$ where $\sqrt[3]{\lambda}$ is the class of X modulo $(X^3 - \lambda)$. Since char $F \neq 3$ and $\mu_3 \subset F^\times$, $F(\sqrt[3]{\lambda})$ is a cyclic cubic F-algebra and $u \mapsto \omega u$ defines a generator of $\mathrm{Gal}(L/F)$. Let A be a central simple F-algebra of degree three or a cubic étale algebra and let $(A^0, n, \star)$ be the corresponding composition algebra as defined in Proposition (34.19). As in Example (36.11) we define a cyclic composition on $A^0 \otimes L$ by

$$x \ast y = [(1 \otimes \rho)(x)] \star [(1 \otimes \rho^2)(y)]$$

and $Q(x) = (n \otimes 1)(x)$. Let $v = \sqrt[3]{\lambda}$. Taking $(1, v, v^{-1})$ as a basis of L over F, we can write any element of $A^0 \otimes L$ as a sum $x = a + bv + cv^{-1}$ with a, b, $c \in A^0$, and we have, (using that $\mu\omega^2 + \omega(1-\mu) = 0$ and $\mu\omega + \omega^2(1-\mu) = -1$),

$$\beta(x) = x * x = (a + b\omega^2 v + c\omega v^{-1}) \star \otimes 1(a + b\omega v + c\omega^2 v^{-1})$$

$$= a^2 - \tfrac{1}{3}T_A(a^2) - bc + \tfrac{1}{3}T_A(bc)$$

$$+ v[\lambda^{-1}\big(c^2 - \tfrac{1}{3}T_A(c^2)\big) - ab + \tfrac{1}{3}T_A(ab)]$$

$$+ v^{-1}[\lambda\big(b^2 - \tfrac{1}{3}T_A(b^2)\big) - ca + \tfrac{1}{3}T_A(ca)]$$

$$= a \star a - (bc)^0 + v\lambda^{-1}[c \star c - (ba)^0] + v^{-1}\lambda[b \star b - (ca)^0]$$

where $x^0 = x - \tfrac{1}{3}T_A(x)$ for $x \in A$. The form Q is given by

$$Q(a + bv + cv^{-1}) = -\tfrac{1}{3}S_{A \otimes L}(a + bv + cv^{-1})$$

$$= -\tfrac{1}{3}[S_A(a) + S_A(c)\lambda^{-1}v + S_A(b^2)\lambda v^{-1}]$$

$$+ \tfrac{1}{3}[T_A(bc) + T_A(ab)v + T_A(ca)v^{-1}]$$

and the norm N by

$$N(a + bv + cv^{-1}) = b_Q\big(a + bv + cv^{-1}, \beta(a + bv + cv^{-1})\big)$$

$$= b_n(a, a \star a) + \lambda b_n(b, b \star b) + \lambda^{-1}b_n(c, c \star c)$$

$$+ \tfrac{1}{3}[b_{S_A}\big(a, (bc)^0\big) + b_{S_A}\big(b, (ca)^0\big) + b_{S_A}\big(c, (ab)^0\big)]$$

$$= N_A(a) + \lambda N_A(b) + \lambda^{-1}N_A(c) - T_A(abc)$$

since $b_n(a, a \star a) = N_A(a)$ and $B_{S_A}\big(a, (bc)^0\big) = -T\big(a, (bc)^0\big) = -T_A(abc)$.

Assume now that F does not necessarily contain a primitive cube root of unity ω. Replacing F by $F(\omega) = F[X]/(X^2 + X + 1)$, we may define $*$ on $A^0 \otimes F(\omega) \otimes L$. However, since ω does not explicitly appear in the above formulas for β and Q restricted to $A^0 \otimes L$, we obtain for any algebra A of degree 3 over F of characteristic 3 and for any $\lambda \in F^\times$ a twisted composition $\Gamma(A, \lambda) = (A^0 \otimes L, L, Q, \beta)$ over $L = F(\sqrt[3]{\lambda})$. A twisted composition Γ similar to a composition $\Gamma(A, \lambda)$ for A associative central simple and $\lambda \in F^\times$ is said to be a *composition of type* $^1A_2'$.

Any pair $(\phi, \psi) \in \mathrm{Aut}_F(A) \times \mathrm{Aut}_F(L)$ induces an automorphism of $\Gamma(A, L)$. Thus we have a morphism of group schemes $\mathbf{PGL}_3 \times \boldsymbol{\mu}_3 \to \mathbf{Spin}_8 \rtimes S_3$ and:

(36.33) Proposition. *Twisted compositions* $\Gamma(A, \lambda)$ *of type* $^1A_2'$ *are classified by the image of* $H^1(F, \mathbf{PGL}_3 \times \boldsymbol{\mu}_3)$ *in* $H^1(F, \mathbf{Spin}_8 \rtimes S_3)$. $\square$

(36.34) Remark. If F contains a primitive cube root of unity, $\boldsymbol{\mu}_3 = A_3$ and the image under the morphism $\mathbf{PGL}_3 \times \boldsymbol{\mu}_3 \to \mathbf{Spin}_8 \rtimes S_3$ of the group $\mathbf{PGL}_3 = \mathbf{Spin}_8^{A_3}$ is contained in $\mathbf{Spin}_8$.

Let now (B, τ) be central simple of degree 3 over a quadratic étale F-algebra K, with a unitary involution τ. For $\nu \in K^\times$ we have a twisted K-composition $\Gamma(B, \nu)$ over $K(\sqrt[3]{\nu})$ which we would like (under certain conditions) to descent to a twisted F-composition.

(36.35) Proposition. *If* $N_K(\nu) = 1$, *then:*

(1) *There is an* ι-*semilinear automorphism* ι' *of* $K(\sqrt[3]{\nu})$ *of order 2 which maps* ν *to* ν^{-1}; *its set of fixed elements is a cubic étale* F-*algebra* L *with* $\mathrm{disc}(L) \simeq K * F(\omega)$ *(where* ω *is a primitive cube root of unity).*

(2) *There is an* ι-*semilinear automorphism of order 2 of the twisted* K-*composition* $\Gamma(B, \tau)$ *such that its set of fixed elements is a twisted* F-*composition* $\Gamma(B, \tau, \nu)$

over L with $\Gamma(B, \tau, \nu) \simeq \operatorname{Sym}(B, \tau)^0 \oplus B^0$; under this isomorphism we have, for $z = (x, y) \in \operatorname{Sym}(B, \tau)^0 \oplus B^0$ and $v = \sqrt[3]{\nu} \in K(\sqrt[3]{\nu}) = L \otimes K$,

$$Q(z) = \tfrac{1}{3}\big[S_B(x) + T_{L \otimes K/L}\big(S_B(y)\lambda v^{-1} + T_B(xy)v\big) + T_B\big(y\tau(y)\big)\big],$$

$$\beta(z) = x^2 - \tfrac{1}{3}T_B(x^2) - y\tau(y) + \tfrac{1}{3}T_B\big(y\tau(y)\big)$$
$$+ \nu[\tau(y)^2 - \tfrac{1}{3}T_B\big(\tau(y)^2\big)] - xy + \tfrac{1}{3}T_B(xy),$$

$$N(z) = N_B(x) + \nu N_B(y) + \nu^{-1} N_B\big(\tau(y)\big) - T_B\big(xy\tau(y)\big).$$

Proof: (1) This is Proposition (30.19).

(2) We have $\Gamma(B, \nu) = B^0 \otimes_K K(\sqrt[3]{\nu})$ and take as our τ-semilinear automorphism the map $\overline{\tau} = \tau \otimes \tau'$. We write $B^0 \otimes_K K(\sqrt[3]{\nu}) = B^0 \oplus B^0 v \oplus B^0 v^{-1}$. The isomorphism $\operatorname{Sym}(B, \tau)^0 \oplus B^0 \simeq \Gamma(B, \tau, \nu)$ is then given by $(x, y) \mapsto x + yv + \tau(y)v^{-1}$ and it is easy to check that its image lies in the descended object $\Gamma(B, \tau, \nu)$. The formulas for Q, β, and N follow from the corresponding formulas for type ${}^1A_2'$. $\qquad\square$

(36.36) Example. If $K = F(\omega)$, ω a primitive cubic root of 1, then the L given by Proposition (36.35) is cyclic and the twisted composition $\Gamma(B, \tau, \nu)$ is the twisted composition associated to the cyclic composition $\operatorname{Sym}(B, \tau)^0 \otimes L$.

A twisted composition isomorphic to a composition $\Gamma(B, \tau, \nu)$ is said to be of *type* ${}^2A_2$. We have a homomorphism

$$\mathbf{GU}_3(K) \times (\boldsymbol{\mu}_3)_\gamma \to \mathbf{Spin}_8 \rtimes S_3$$

where γ is a cocycle defining K and the analogue of Proposition (36.33) is:

(36.37) Proposition. *Twisted compositions $\Gamma(B, \tau, \nu)$ of type ${}^2A_2$ are classified by the image of $H^1\big(F, \mathbf{PGU}_3(K) \times (\boldsymbol{\mu}_3)_\gamma\big)$ in $H^1(F, \mathbf{Spin}_8 \rtimes S_3)$.* $\qquad\square$

36.E. The dimension 2 case. If (V, L, Q, β) is a twisted composition with $\operatorname{rank}_L V = 2$, then V admits, in fact, more structure:

(36.38) Proposition. *Let (V, L, Q, β) be a twisted composition with $\dim_L V = 2$. There exists a quadratic étale F-algebra K which operates on V and a nonsingular $L \otimes K$-hermitian form $h \colon V \times V \to L \otimes K$ of rank 1 such that $Q(x) = h(x, x)$, $x \in V$. Hence $Q \simeq N_K \otimes \langle\lambda\rangle$ where λ can be chosen such that $N_{L/F}(\lambda) \in F^{\times 2}$. Furthermore the algebra K is split if Q is isotropic.*

Proof: For generic $v \in V$, we may assume that $Q(v) = \lambda \in L^\times$, $b_Q\big(v, \beta(v)\big) = a \in F^\times$, and that $\{v, \beta(v)\}$ are linearly independent over L (see also the following Remark). Then $v_1 = v$, $v_2 = \lambda\beta(v)$ is an L-basis of V, and

$$Q(x_1 v_1 + x_2 v_2) = \big(x_1^2 + a x_1 x_2 + n_{L/F}(\lambda)x_2^2\big) \cdot \lambda$$

Thus $4N_{L/F}(\lambda) - a^2 = \det_L Q$ is nonzero and the quadratic F-algebra

$$K = F[x]/\big(x^2 + ax + N_{L/F}(\lambda)\big)$$

is étale. Let ι_K be the conjugation map of K. Let $\xi = x + \big(x^2 + ax + N_{L/F}(\lambda)\big) \in K$. We define a K-module structure on V by putting

$$\xi v_1 = v_2 \quad \text{and} \quad \xi v_2 = -a v_1 - N_{L/F}(\lambda)v_2.$$

Thus $v = v_1$ is a basis element for the $L \otimes K$-module V. We then define

$$h(\eta_1 v, \eta_2 v) = \eta_1 \lambda \bar{\eta}_2$$

for η_1, $\eta_2 \in L \otimes K$, and $\bar{\eta}_1 = (1 \otimes \iota_K)(\eta_1)$. In particular we have $\lambda = h(v,v)$ for the chosen element v. The fact that $Q(x) = h(x,x)$, $x \in V$, follows from the formula $Q(x_1 v_1 + x_2 v_2) = \left(x_1^2 + a x_1 x_2 + N_{L/F}(\lambda) x_2^2\right) \cdot \lambda$. The last claim, i.e., that $N_{L/F}(\lambda) \in F^{\times 2}$, follows by choosing v of the form $v = \beta(u)$. If Q is isotropic, Q is hyperbolic by Proposition (36.16) and Corollary (35.3), hence $Q \simeq N_K \otimes \langle 1 \rangle$ and the claim follows from Springer's theorem. $\qquad \square$

(36.39) Remark. If $b_Q\big(v, \beta(v)\big) = 0$ for $v \neq 0$ or if $\{v, \beta(v)\}$ is linearly dependent over L, the twisted composition is induced from a Hurwitz algebra (see Theorem (36.24)).

EXERCISES

1. Let C be a separable alternative algebra of degree 2 over F. Show that $\pi : x \mapsto \bar{x}$ is the unique F-linear automorphism of C such that $x + \pi(x) \in F \cdot 1$ for all $x \in C$.

2. Let (C, N), (C', N') be Hurwitz algebras of dimension ≤ 4. Show that an isometry $N \xrightarrow{\sim} N'$ which maps 1 to 1 is either an isomorphism or an anti-isomorphism. Give an example where this is not the case for Cayley algebras.

3. A symmetric composition algebra with identity is 1-dimensional.

4. (Petersson [208]) Let K be quadratic étale with norm $N = N_K$ and conjugation $x \mapsto \bar{x}$. Composition algebras $(K, \star)$ are either K (as a Hurwitz algebra) or—up to isomorphism—of the form
   (a) $x \star y = x\bar{y}$,
   (b) $x \star y = \bar{x}y$, or
   (c) $x \star y = u\overline{xy}$ for some $u \in K$ such that $N(u) = 1$.
   Compositions of type (c) are symmetric.

5. The split Cayley algebra over F can be regarded as the set of all matrices (*Zorn matrices*) $\left(\begin{smallmatrix} \alpha & a \\ b & \beta \end{smallmatrix} \right)$ with α, $\beta \in F$ and a, $b \in F^3$, with multiplication
$$
\begin{pmatrix} \alpha & a \\ b & \beta \end{pmatrix} \begin{pmatrix} \gamma & c \\ d & \delta \end{pmatrix} = \begin{pmatrix} \alpha\gamma + a \cdot d & \alpha c + \delta a - (b \wedge d) \\ \gamma b + \beta d + (a \wedge c) & \beta\delta + b \cdot c \end{pmatrix}
$$
   where $a \cdot d$ is the standard scalar product in F^3 and $b \wedge d$ the standard vector product (cross product). The conjugation is given by
$$
\pi \begin{pmatrix} \alpha & a \\ b & \beta \end{pmatrix} = \begin{pmatrix} \beta & -a \\ -b & \alpha \end{pmatrix}
$$
   and the norm by
$$
\mathfrak{n} \begin{pmatrix} \alpha & a \\ b & \beta \end{pmatrix} = \alpha\beta - a \cdot b.
$$

6. Let K be a quadratic étale F-algebra and let (V, h) be a ternary hermitian space over K with trivial (hermitian) discriminant, i.e., there exists an isomorphism $\phi : \wedge^3 (V, h) \xrightarrow{\sim} \langle 1 \rangle$. For any v, $w \in V$, let $v \times w \in V$ be determined by the condition $h(u, v \times w) = \phi(u \wedge v \wedge w)$.
   (a) Show that the vector space $\mathfrak{C}(K, V) = K \oplus V$ is a Cayley algebra under the multiplication
$$
(a, v) \diamond (b, w) = \big(ab - h(v, w),\, aw + \bar{b}v + v \times w\big)
$$

and the norm $\mathfrak{n}\big((a,v)\big) = N_{K/F}(a) + h(v,v)$.

   (b) Conversely, if $\mathfrak{C}$ is a Cayley algebra and K is a quadratic étale subalgebra, then $V = K^{\perp}$ admits the structure of a hermitian space over K and $\mathfrak{C} \simeq \mathfrak{C}(K, V)$.

   (c) $\operatorname{Aut}_F(\mathfrak{C}, K) = \mathbf{SU}_3(K)$.

   (d) There exists a monomorphism $\mathbf{SL}_3 \rtimes \mathbb{Z}/2\mathbb{Z} \to G$ where G is split simple of type G_2 (i.e., "$A_2 \subset G_2$") such that $H^1(F, \mathbf{SL}_3 \rtimes \mathbb{Z}/2\mathbb{Z}) \to H^1(F, G)$ is surjective.

7. (a) Let Q be a quaternion algebra and let $\mathfrak{C} = \mathfrak{C}(Q, a)$ be the Cayley algebra $Q \oplus vQ$ with $v^2 = a$. Let $\operatorname{Aut}_F(\mathfrak{C}, Q)$ be the subgroup of automorphisms of $\operatorname{Aut}_F(\mathfrak{C})$ which map Q to Q. Show that there is an exact sequence

$$1 \to \mathrm{SL}_1(Q) \xrightarrow{\phi} \operatorname{Aut}_F(\mathfrak{C}, Q) \to \operatorname{Aut}_F(Q) \to 1$$

where $\phi(y)(a + vb) = a + (vy)b$ for $y \in \mathrm{SL}_1(Q)$.

   (b) The map $\mathrm{SL}_1(Q) \times \mathrm{SL}_1(Q) \to \operatorname{Aut}_F(\mathfrak{C})$ induced by

$$(u, x) \mapsto \big[(a + vb) \mapsto \big(ua\overline{u} + (vx)(ub\overline{u})\big)\big]$$

is a group homomorphism (i.e., "$A_2 \times A_2 \subset G_2$").

8. (Elduque) Let $S = (\mathbb{F}_4, \star)$ be the unique para-quadratic $\mathbb{F}_2$-algebra. Show that 1-dimensional algebras and S are the only examples of power-associative symmetric composition algebras.

9. Let F be a field of characteristic not 3. Let A be a central simple F-algebra of degree 3. Compute the quadratic forms $T_A(x^2)$ and $S_A(x)$ on A and on A^0, and determine their discriminants and their Clifford invariants.

10. Let $\lambda \in F^{\times}$ and let (Q, n) be a quaternion algebra. Construct an isomorphism $\big(C(\lambda Q, n), \underline{\sigma}\big) \simeq \big(M_2(Q), \sigma_{n \perp n}\big)$. *Hint*: Argue as in the proof of (35.1).

11. Let $(\mathfrak{C}, \diamond, \mathfrak{n})$ be a Cayley algebra and let $(\mathfrak{C}, \star)$ be the associated para-Cayley algebra, with multiplication $x \star y = \overline{x} \diamond \overline{y}$. Show that

$$(x \star a) \star (a \star y) = \overline{a} \star \big(a \star (x \star y)\big).$$

(By using the Theorem of Cartan-Chevalley this gives another approach to triality.)

12. (Elduque) Let $\mathfrak{C}$ be a Cayley algebra, let $(\mathfrak{C}, \star)$ be the associated para-Cayley algebra, and let $(\mathfrak{C}_{\varphi}, \overline{\star})$ be a Petersson algebra. Let t be a proper similitude of $(\mathfrak{C}, \mathfrak{n})$, with multiplier $\mu(t)$.

   (a) If t^+, t^- are such that $\mu(t)^{-1} t(x \star y) = t^-(x) \star t^+(y)$, show that

$$\mu(t)^{-1} t(x \overline{\star} y) = \varphi^{-1} t^- \varphi(x) \overline{\star} \varphi t^+ \varphi^{-1}(y).$$

   (b) If θ^+ is the automorphism of $\operatorname{Spin}(\mathfrak{C}, \mathfrak{n})$ as defined in Proposition (35.9) and if $\overline{\theta}^+$ is the corresponding automorphism with respect to $\overline{\star}$, show that $\overline{\theta}^+ = C(\varphi)\theta^+ = \theta^+ C(\varphi)$.

13. Compute $\operatorname{Spin}(C, n)$ for (C, n) a symmetric composition algebra of dimension 2, resp. 4.

14. Let $\widetilde{C}$ be a twisted Hurwitz composition over $F \times F \times F$.

   (a) If C is a quaternion algebra, show that

$$\operatorname{Aut}_F(\widetilde{C}) = \big((C^{\times} \times C^{\times} \times C^{\times})^{\operatorname{Det}}/F^{\times}\big) \rtimes S_3$$

where

$$(C^{\times} \times C^{\times} \times C^{\times})^{\operatorname{Det}} = \{\, (a, b, c) \in C^{\times} \mid N_C(a) = N_C(b) = N_C(c) \,\}$$

and S_3 acts by permuting the factors.

(b) If C is quadratic,

$$\mathrm{Aut}_F(\widetilde{C}) = \big(\mathrm{SU}_1(C) \times \mathrm{SU}_1(C)\big) \rtimes (\mathbb{Z}/2\mathbb{Z} \times S_3)$$

where $\mathbb{Z}/2\mathbb{Z}$ operates on $\mathrm{SU}_1(C) \times \mathrm{SU}_1(C)$ through $(a, b) \mapsto (\overline{a}, \overline{b})$ and S_3 operates on $\mathrm{SU}_1(C) \times \mathrm{SU}_1(C)$ as in Lemma (35.11).

15. Describe the action of S_3 (triality) on the Weyl group $(\mathbb{Z}/2\mathbb{Z})^3 \rtimes S_4$ of a split simple group of type D_4.

NOTES

§ 32. The notion of a generic polynomial, which is classical for associative algebras, was extended to strictly power-associative algebras by Jacobson. A systematic treatment is given in Chap. IV of [136], see also McCrimmon [184].

§ 33. Octonions (or the algebra of *octaves*) were discovered by Graves in 1843 and described in letters to Hamilton (see Hamilton [116, Vol. 3, Editor's Appendix 3, p. 648]); however Graves did not publish his result and octonions were rediscovered by Cayley in 1845 [61, I, p. 127, XI, p. 368–371]. Their description as pairs of quaternions (the "Cayley-Dickson process") can be found in Dickson [76, p. 15]. Dickson was also the first to notice that octonions with positive definite norm function form a division algebra [75, p. 72].

The observation that $\overline{x}(xa) = (\overline{x}x)a = (ax)\overline{x}$ holds in an octonion algebra dates back to Kirmse [154, p. 76]. The fact that Cayley algebras satisfy the alternative law was conjectured by E. Artin and proved by Artin's student Max Zorn in [320]. Artin's theorem (that a subalgebra of an alternative algebra generated by two elements is associative) and the structure theorem (32.5) first appeared in [320]. The description of split octonions as "vector matrices," as well as the abstract Cayley-Dickson process, are given in a later paper [321] of Zorn. The fact that the Lie algebra of derivations of a Cayley algebra is of type G_2 and the fact that the group of automorphisms of the Lie algebra of derivations of a Cayley algebra is isomorphic to the group of automorphisms of the Cayley algebra if F is a field of characteristic zero, is given in Jacobson [128]. In this connection we observe that the Lie algebra of derivations of the split Cayley algebra over a field of characteristic 3 has an ideal of dimension 7, hence is not simple. The fact that the group of automorphisms of a Cayley algebra is of type G_2 is already mentioned without proof by E. Cartan [57, p. 298] [60, p. 433]. Other proofs are found in Freudenthal [101], done by computing the root system, or in Springer [268], done by computing the dimension of the group and applying the classification of simple algebraic groups. In [268] no assumption on the characteristic of the base field is made.

Interesting historical information on octonions can be found in the papers of van der Blij [41] and Veldkamp [295], see also the book of van der Waerden [302, Chap. 10]. The problem of determining all composition algebras has been treated by many authors (see Jacobson [131] for references). Hurwitz [126] showed that the equation

$$(x_1^2 + \cdots + x_n^2)(y_1^2 + \cdots + y_n^2) = z_1^2 + \cdots + z_n^2$$

has a solution given by bilinear forms $z_1, \ldots, z_n$ in the variables $x = (x_1, \ldots, x_n)$, $y = (y_1, \ldots, y_n)$ exactly for $n = 1, 2, 4$, and 8. The determination of all composition algebras with identity over a field of characteristic not 2 is due to Jacobson [131]. We used here the proof of van der Blij-Springer [42], which is also valid in characteristic 2. A complete classification of composition algebras (even those without an identity) is known in dimensions 2 (Petersson [208]) and 4 (Stampfli-Rollier [272]).

The restrictions $n = 1, 2, 4, 8$ for the possible dimensions of a composition algebra C can also be explained in a tensor categorial manner. In char $\neq 2$ the space $V = \langle 1 \rangle^\perp$ carries the structure of a vector product algebra. If one considers $d = \dim V$ as an abstract tensor (the contraction of the bilinear form by itself), one can derive the identity

$$d(d-1)(d-3)(d-7) = 0$$

(in the ground field) in an "abstract" way, i.e., just using the axioms of a vector product algebra, but not refering to a particular vector space, see [235, 44, 180].

§34. Compositions algebras with associative norms were considered independently by Petersson [207], Okubo [199], and Faulkner [95]. We suggest calling them *symmetric composition algebras* in view of their very nice (and symmetric) properties. Applications of these algebras in physics can be found in a recent book [200] by S. Okubo.

Petersson showed that over an algebraically closed field symmetric compositions are either para-Hurwitz or, as we call them, Petersson compositions. Okubo described para-Cayley Algebras and "split Okubo algebras" as examples of symmetric composition algebras. In the paper [202] of Okubo-Osborn it is shown that over an algebraically closed field these two types are the only examples of symmetric composition algebras.

The fact that the trace zero elements in a cubic separable alternative algebra carry the structure of a symmetric algebra was noticed by Faulkner [95]. The classification of symmetric compositions, as given in Theorem (34.37), is due to Elduque-Myung [89]. However they applied the Zorn Structure Theorem for separable alternative algebras, instead of invoking (as we do) the eigenspace decomposition of the operator ℓ_e for e an idempotent. The idea to consider such eigenspaces goes back to Petersson [207]. A similar decomposition for the operator ad_e is used by Elduque-Myung in [88]. Connections between the different constructions of symmetric algebras are clearly described in Elduque-Pérez [90]. We take the opportunity to thank A. Elduque, who detected an error in our first draft and who communicated [90] to us before its publication.

Let $(A^0, \star)$ be a composition of type ${}^1A_2$. It follows from Theorem (34.23) that $\mathrm{Aut}_F(A^0, \star) \simeq \mathrm{Aut}_F(A)$. This can also be viewed in terms of Lie algebras: Since $x \star y - y \star x = \mu(xy - yx)$, any isomorphism of compositions $\alpha \colon (A^0, \star) \xrightarrow{\sim} (A'^0, \star)$ also induces a Lie algebra isomorphism $\mathfrak{L}(A^0) \xrightarrow{\sim} \mathfrak{L}(A'^0)$. Conversely, (and assuming that F has characteristic 0) any isomorphism of Lie algebras $\mathfrak{L}(A^0) \xrightarrow{\sim} \mathfrak{L}(A'^0)$ extends to an algebra isomorphism $A \xrightarrow{\sim} A'$ or the negative of an anti-isomorphism of algebras $A \xrightarrow{\sim} A'$ (Jacobson [140, Chap. X, Theorem 10]). However the negative of an anti-isomorphism of algebras cannot restrict to an isomorphism of composition algebras. In particular we see that $\mathrm{Aut}_F(A^0, \star)$ is isomorphic to the connected component $\mathrm{Aut}_F\big(\mathfrak{L}(A^0)\big)^0$ of $\mathrm{Aut}_F\big(\mathfrak{L}(A^0)\big)$.

§35. We introduce triality using symmetric composition algebras of dimension 8 and their Clifford algebras. Most of the results for compositions of type G_2 can already be found in van der Blij-Springer [43], Springer [268], Wonenburger [317], or Jacobson [132, p. 78], [135]. However the presentation through Clifford algebras given here, which goes back to [165], is different. The use of symmetric compositions also has the advantage of giving very symmetric formulas for triality. The isomorphism of algebras $C(S, n) \xrightarrow{\sim} \mathrm{End}_F(S \oplus S)$ for symmetric compositions of dimension 8 can already be found in the paper [201] of Okubo and Myung. A different approach to triality can be found in the book of Chevalley [66].

Triality in relation to Lie groups is discussed briefly by E. Cartan [59, Vol. II, §139] as an operation permuting the vector and the $\frac{1}{2}$-spinor representations of D_4. The first systematic treatment is given in Freudenthal [101], where local triality (for Lie algebras) and global triality is discussed.

There is also an (older) geometric notion of triality between points and spaces of two kinds on a (complex) 6-dimensional quadric in $\mathbb{P}^7$. These spaces correspond to maximal isotropic spaces of the quadric given by the norm of octonions. Geometric triality goes back to Study [274] and E. Cartan [58, pp. 369-370], see also [60, I, pp. 563–565]; A systematic study of geometric triality is given in Vaney [294], Weiss [312], see also Kuiper [167]. Geometric applications can be found in the book on "Punktreihengeometrie" of Weiss [313].

The connection between triality and octonions, already noticed by Cartan, is used systematically by Vaney and Weiss. The existence of triality is, in fact, "responsible" for the existence of Cayley algebras (see Tits [286]). A systematic description of triality in projective geometry in relation to the theory of groups is given in Tits [287].

The paper of van der Blij-Springer [43] gives a very nice introduction to triality in algebra and geometry. There is also another survey article, by Adams [1].

§36. The notion of a twisted composition was suggested by the construction of cyclic compositions, due to Springer [268]. Many results of this section, for example Theorem (36.24), were inspired by the notes [268].

CHAPTER IX

Cubic Jordan Algebras

The set of symmetric elements in an associative algebra with involution admits the structure of a Jordan algebra. One aim of this chapter is to give some insight into the relationship between involutions on central simple algebras and Jordan algebras. After a short survey on central simple Jordan algebras in §37, we specialize to Jordan algebras of degree 3 in §38; in particular, we discuss extensively "Freudenthal algebras," a class of Jordan algebras connected with Hurwitz algebras and we describe the Springer construction, which ties twisted compositions with cubic Jordan algebras. On the other hand, cubic Jordan algebras are also related to cubic associative algebras through the Tits constructions (§39). Of special interest, and the main object of study of this chapter, are the exceptional simple Jordan algebras of dimension 27, whose automorphism groups are of type F_4. The different constructions mentioned above are related to interesting subgroups of F_4. For example, the automorphism group of a split twisted composition is a subgroup of F_4 and outer actions on Spin_8 (triality) become inner over F_4. Tits constructions are related to the action of the cyclic group $\mathbb{Z}/3\mathbb{Z}$ on Spin_8 which yields invariant subgroups of classical type A_2, and Freudenthal algebras are related to the action of the group S_3 on Spin_8 which yields invariant subgroups of exceptional type G_2.

Cohomological invariants of exceptional simple Jordan algebras are discussed in the last section.

§37. Jordan Algebras

We assume in this section that F is a field of characteristic different from 2. A *Jordan algebra* J is a commutative finite dimensional unitary F-algebra such that the multiplication $(a, b) \mapsto a \cdot b$ satisfies

(37.1) $$((a \cdot a) \cdot b) \cdot a = (a \cdot a)(b \cdot a)$$

for all a, $b \in J$. For any associative algebra A, the product

$$a \cdot b = \tfrac{1}{2}(ab + ba)$$

gives A the structure of a Jordan algebra, which we write A^+. If B is an associative algebra with involution τ, the set $\mathrm{Sym}(B, \tau)$ of symmetric elements is a Jordan subalgebra of B^+ which we denote $\mathcal{H}(B, \tau)$.

Observe that $A^+ \simeq \mathcal{H}(B, \tau)$ if $B = A \times A^{\mathrm{op}}$ and τ is the exchange involution.

A Jordan algebra A is *special* if there exists an injective homomorphism $A \to D^+$ for some associative algebra D and is *exceptional* otherwise.

A Jordan algebra is strictly power-associative and we write a^n for the nth power of an element a. Hence it admits a generic minimal polynomial

$$P_{J,x}(X) = X^m - s_1(x)X^{m-1} + \cdots + (-1)^m s_m(x)1,$$

where $T_J = s_1$ is the generic trace and $N_J = s_m$ the generic norm. The bilinear trace form $T(x, y) = T_J(xy)$ is associative (see Corollary (32.3)). By Dieudonné's theorem (32.4), a Jordan algebra is separable if T is nonsingular. The converse is a consequence of the following structure theorem:

(37.2) Theorem. (1) *Any separable Jordan F-algebra is the product of simple Jordan algebras whose centers are separable field extensions of F.*
(2) *A central simple Jordan algebra is either*
    (a) *the Jordan algebra of a nondegenerate quadratic space of dimension ≥ 2,*
    (b) *a Jordan algebra $\mathcal{H}(B, \tau)$ where B is associative and K-central simple as an algebra with involution τ, and where K is either quadratic étale and τ is unitary with respect to K or $K = F$ and τ is F-linear, or*
    (c) *an exceptional Jordan algebra of dimension 27.*

Reference: (1) is [136, Theorem 4, p. 239], and (2) (which goes back to Albert [6]) follows from [136, Corollary 2, p. 204] and [136, Theorem 11, p. 210]. We define and discuss the different types occurring in (37.2) in the following sections.   □

Let $\mathsf{Sepjord}_n(m)$ be the groupoid of separable Jordan F-algebras of dimension n and degree m with isomorphisms as morphisms.

37.A. Jordan algebras of quadratic forms. Let (V, q) be a nonsingular finite dimensional quadratic space with polar $b_q(x, y) = q(x + y) - q(x) - q(y)$. We define a multiplication on $J(V, q) = F \oplus V$ by setting

$$(\lambda, v) \cdot (\mu, w) = \left(\lambda\mu + \tfrac{1}{2}b_q(v, w), \lambda w + \mu v\right)$$

for v, $w \in V$ and λ, $\mu \in F$. The element $(1, 0)$ is an identity and the canonical embedding of $J(V, q) = F \cdot 1 \oplus V$ into the Clifford algebra $C(V, q)$ shows that $J(V, q)$ is a Jordan algebra (and is special). The generic minimal polynomial of $J(V, q)$ is

$$P_{J,a}(X) = X^2 - 2\xi X + \left(\xi^2 - q(v)\right)1$$

where $a = (\xi, v) \in F \cdot 1 \oplus V$, hence $J(V, q)$ has degree 2, the trace is given by $T_J(\xi, v) = 2\xi$ and the norm by $N_J(\xi, v) = \xi^2 - q(v)$. Thus N_J is a nonsingular quadratic form. The bilinear trace form $T \colon (x, y) \mapsto T_J(x \cdot y)$ is isomorphic to $\langle 2 \rangle \perp b_q$, furthermore T is associative, hence by (32.4) J is separable if and only if q is nonsingular. We set $\mathbf{J} \colon Q_n \to \mathsf{Sepjord}_{n+1}(2)$ for the functor $(V, q) \mapsto J(V, q)$. Let J be a separable Jordan algebra of degree 2, with generic minimal polynomial

$$P_{J,a}(X) = X^2 - T_J(a)X + N_J(a)1.$$

Linearizing and taking traces shows that

$$2T_J(x \cdot y) - 2T_J(x)T_J(y) + 2b_{N_J}(x, y) = 0,$$

with b_{N_J} the polar of N_J; hence

(37.3) $$b_{N_J}(x, y) = T_J(x)T_J(y) - T_J(x \cdot y).$$

For $J^0 = \{\, x \in J \mid T_J(x) = 0 \,\}$, we have an orthogonal decomposition

$$J = F \cdot 1 \perp J^0$$

with respect to the bilinear trace form T as well as with respect to N_J and, in view of (37.3), N_J is nonsingular on J^0 if and only if T is nonsingular on J^0 if and only if J is separable. Let

$$\mathbf{Q} \colon \mathsf{Sepjord}_{n+1}(2) \to Q_n$$

be the functor given by $J \mapsto (J^0, -N_J)$.

(37.4) Proposition. *The functors* **J** *and* **Q** *define an equivalence of groupoids*

$$Q_n \equiv \mathsf{Sepjord}_{n+1}(2).$$

In particular we have $\mathbf{Aut}_{\mathrm{alg}}\big(J(V,q)\big) = \mathbf{O}(V,q)$, *so that Jordan algebras of type* $\mathsf{Sepjord}_{n+1}(2)$ *are classified by* $H^1(F, \mathbf{O}_n)$.

Proof: The claim follows easily from the explicit definitions of **J** and **Q**. □

(37.5) Remark. If $\dim V \geq 2$, $J(V,q)$ is a simple Jordan algebra. If $\dim V = 1$, $J(V,q)$ is isomorphic to the quadratic algebra $F(\sqrt{\lambda}) = F(X)/(X^2 - \lambda)$ where $q \simeq \langle \lambda \rangle$.

We next consider Jordan algebras of degree ≥ 3 and begin with Jordan algebras associated to central simple algebras with involution.

37.B. Jordan algebras of classical type. Let K be an étale quadratic algebra over F with conjugation ι or let $K = F$ and $\iota = 1$. Let (B, τ) be a K-central simple algebra with τ an ι-linear involution. As in Chapter VI we denote the groupoids corresponding to different types of involutions by A_n, B_n, C_n, and D_n. We set A_n^+, B_n^+, C_n^+, resp. D_n^+ for the groupoids of Jordan algebras whose objects are sets of symmetric elements $\mathcal{H}(B, \tau)$ for $(B, \tau) \in A$, B, C, resp. D. For each of these categories A, B, C, D, we have functors $\mathbf{S} \colon A \to A^+$, ..., $D \to D^+$ induced by $(B, \tau) \mapsto \mathcal{H}(B, \tau)$.

(37.6) Proposition. *Let* B, B' *be* K-*central simple with involutions* τ, τ', *of degree* ≥ 3. *Any isomorphism* $\mathcal{H}(B, \tau) \xrightarrow{\sim} \mathcal{H}(B', \tau')$ *of Jordan algebras extends to a unique isomorphism* $(B, \tau) \xrightarrow{\sim} (B', \tau')$ *of* K-*algebras with involution. In particular* $\mathcal{H}(B, \tau)$ *and* $\mathcal{H}(B', \tau')$ *are isomorphic Jordan algebras if and only if* (B, τ) *and* (B', τ') *are isomorphic as* K-*algebras with involution and the functor* **S** *induces an isomorphism of corresponding groupoids.*

Reference: See Jacobson [136, Chap. V, Theorem 11, p. 210]. □

Thus, in view of Theorem (37.2), the classification of special central simple Jordan algebras of degree ≥ 3 is equivalent to the classification of central simple associative algebras with involution of degree ≥ 3.

If (B, τ) is a central simple algebra with a unitary involution over K, we have an exact sequence of group schemes

$$1 \to \mathbf{Aut}_K(B, \tau) \to \mathbf{Aut}(B, \tau) \to \mathbf{Aut}_{\mathrm{alg}}(K)(\simeq \mathbb{Z}/2\mathbb{Z}) \to 1$$

Thus there is a sequence

$$1 \to \mathbf{Aut}_K(B, \tau) \to \mathbf{Aut}\big(\mathcal{H}(B, \tau)\big) \to \mathbb{Z}/2\mathbb{Z} \to 1.$$

If $B = A \times A^{\mathrm{op}}$ and τ is the exchange involution, we obtain

(37.7) $$1 \to \mathbf{Aut}(A) \to \mathbf{Aut}(A^+) \to \mathbb{Z}/2\mathbb{Z} \to 1$$

and the sequence splits if A admits an anti-automorphism. The group scheme $\mathbf{Aut}\big(\mathcal{H}(B, \tau)\big)$ is smooth in view of Proposition (21.10), (4), since its connected component $\mathbf{PGU}(B, \tau) = \mathbf{Aut}_K(B, \tau)$ is smooth. Thus $\mathbf{Aut}(A^+)$ is smooth too.

37.C. Freudenthal algebras. Let C be a Hurwitz algebra with norm N_C and trace T_C over a field F of characteristic not 2 and let

$$M_n(C) = M_n(F) \otimes C.$$

For $X = (c_{ij}) \in M_n(C)$, let $\overline{X} = (\bar{c}_{ij})$ where $c \mapsto \bar{c}$, $c \in C$, is conjugation. Let $\alpha = \mathrm{diag}(\alpha_1, \alpha_2, \ldots, \alpha_n) \in \mathrm{GL}_n(F)$. Let

$$\mathcal{H}_n(C, \alpha) = \{\, X \in M_n(C) \mid \alpha^{-1}\overline{X}^t\alpha = X \,\}.$$

Let $n \geq 3$. If C is associative, $\mathcal{H}_n(C, \alpha)$ and twisted forms of $\mathcal{H}_n(C, \alpha)$ are Jordan algebras of classical type for the product $X \cdot Y = \frac{1}{2}(XY + YX)$ where XY is the usual matrix product. In particular they are special. If $n = 3$ and $C = \mathfrak{C}$ is a Cayley algebra, $\mathcal{H}_3(\mathfrak{C}, \alpha)$ (and twisted forms of $\mathcal{H}_3(\mathfrak{C}, \alpha)$) are Jordan algebras for the same multiplication (see for example Jacobson [136, Chap. III, Theorem 1, p. 127]). For $n = 2$ we still get Jordan algebras since $\mathcal{H}_2(\mathfrak{C}, \alpha)$ can be viewed as a subalgebra of $\mathcal{H}_3(\mathfrak{C}, \alpha)$ with respect to a *Peirce decomposition* ([136, Chap. III, Sect. 1]) relative to the idempotent $\mathrm{diag}(1, 0, 0)$. In fact, $\mathcal{H}_2(C, \alpha)$ is, for any Hurwitz algebra C, separable of degree 2 hence special (see Exercise 3). However the algebra $\mathcal{H}_3(\mathfrak{C}, \alpha)$ and twisted forms of $\mathcal{H}_3(\mathfrak{C}, \alpha)$ are exceptional Jordan algebras (Albert [4]). In fact they are not even homomorphic images of special Jordan algebras (Albert-Paige [15] or Jacobson [136, Chap. I, Sect. 11, Theorem 11]). Conversely, any central simple exceptional Jordan algebra is a twisted form of $\mathcal{H}_3(\mathfrak{C}, \alpha)$ for some Cayley algebra $\mathfrak{C}$ (Albert [6, Theorem 17]).

The elements of $J = \mathcal{H}_3(C, \alpha)$ can be represented as matrices

(37.8) $\qquad a = \begin{pmatrix} \xi_1 & c_3 & \alpha_1^{-1}\alpha_3\bar{c}_2 \\ \alpha_2^{-1}\alpha_1\bar{c}_3 & \xi_2 & c_1 \\ c_2 & \alpha_3^{-1}\alpha_2\bar{c}_1 & \xi_3 \end{pmatrix}, \quad c_i \in C,\ \xi_i \in F$

and the generic minimal polynomial is (Jacobson [136, p. 233]):

$$P_{J,a}(X) = X^3 - T_J(a)X^2 + S_J(a)X - N_J(a)1$$

where

$T_J(a) = \xi_1 + \xi_2 + \xi_3,$

$S_J(a) = \xi_1\xi_2 + \xi_2\xi_3 + \xi_1\xi_3 - \alpha_3^{-1}\alpha_2 N_C(c_1) - \alpha_1^{-1}\alpha_3 N_C(c_2) - \alpha_2^{-1}\alpha_1 N_C(c_3),$

$N_J(a) = \xi_1\xi_2\xi_3 - \alpha_3^{-1}\alpha_2\xi_1 N_C(c_1) - \alpha_1^{-1}\alpha_3\xi_2 N_C(c_2) - \alpha_2^{-1}\alpha_1\xi_3 N_C(c_3)$
$\qquad\qquad + T_C(c_3 c_1 c_2).$

Let

$$b = \begin{pmatrix} \eta_1 & d_3 & \alpha_1^{-1}\alpha_3\bar{d}_2 \\ \alpha_2^{-1}\alpha_1\bar{d}_3 & \eta_2 & d_1 \\ d_2 & \alpha_3^{-1}\alpha_2\bar{d}_1 & \eta_3 \end{pmatrix}, \quad d_i \in C,\ \eta_i \in F.$$

Let b_C be the polar of N_C. The bilinear trace form $T \colon (a, b) \mapsto T_J(a \cdot b)$ is given by

$T(a, b) =$

$\qquad \xi_1\eta_1 + \xi_2\eta_2 + \xi_3\eta_3 + \alpha_3^{-1}\alpha_2 b_C(c_1, d_1) + \alpha_1^{-1}\alpha_3 b_C(c_2, d_2) + \alpha_2^{-1}\alpha_1 b_C(c_3, d_3)$

or

(37.9) $\qquad\qquad T = \langle 1, 1, 1 \rangle \perp b_C \otimes \langle \alpha_3^{-1}\alpha_2, \alpha_1^{-1}\alpha_3, \alpha_2^{-1}\alpha_1 \rangle,$

Thus T is nonsingular. The quadratic form S_J is the *quadratic trace* which is a regular quadratic form. Furthermore one can check that

(37.10) $b_{S_J}(a, b) = T_J(a)T_J(b) - T_J(a \cdot b).$

We have $T(1, 1) = 3$; hence there exists an orthogonal decomposition

$$\mathcal{H}_3(C, \alpha) = F \cdot 1 \perp \mathcal{H}_3(C, \alpha)^0, \quad \mathcal{H}_3(C, \alpha)^0 = \{ x \in \mathcal{H}_3(C, \alpha) \mid T_J(x) = 0 \}$$

if char $F \neq 3$.

We call Jordan algebras isomorphic to algebras $\mathcal{H}_3(C, \alpha)$, for some Hurwitz algebra C, *reduced Freudenthal algebras* and we call twisted forms of $\mathcal{H}_3(C, \alpha)$ *Freudenthal algebras*. If we allow C to be 0 in $\mathcal{H}_3(C, \alpha)$, the split cubic étale algebra $F \times F \times F$ can also be viewed as a special case of a reduced Freudenthal algebra. Hence cubic étale algebras are Freudenthal algebras of dimension 3. Furthermore, if char $F \neq 3$, it is convenient to view F as a Freudenthal algebra with norm $N_F(x) = x^3$. Freudenthal algebras $\mathcal{H}_3(C, s)$, with $C = 0$ or C a split Hurwitz algebra and $s = \mathrm{diag}(1, -1, 1)$ are called *split*. A Freudenthal algebra can have dimension 1, 3, 6, 9, 15, or 27. In dimension 3 Freudenthal algebras are commutative cubic étale F-algebras and in dimension greater than 3 central simple Jordan algebras of degree 3 over F. Freudenthal algebras of dimension 27 are also called *Albert algebras*. The group scheme G of F-automorphisms of the split Freudenthal algebra of dimension 27 is simple exceptional split of type F_4 (see Theorem (25.13)). By (29.8) we have:

(37.11) Proposition. *Let G be a simple split group of type F_4. Albert algebras (= simple exceptional Jordan algebras of dimension 27) are classified by $H^1(F, G)$.* $\square$

It is convenient to distinguish between Freudenthal algebras with zero divisors and Freudenthal algebras without zero divisors ("division algebras").

(37.12) Theorem. *Let J be a Freudenthal algebra.*

(1) *If J has zero divisors, then $J \simeq F \times K$, K a quadratic étale F-algebra, if $\dim_F J = 3$, and $J \simeq \mathcal{H}_3(C, \alpha)$ for some Hurwitz algebra C, i.e., J is reduced if $\dim_F J > 3$. Moreover, a Freudenthal algebra J of dimension > 3 is reduced if and only if J contains a split étale algebra $L = F \times F \times F$. More precisely, if e_i, $i = 1, 2, 3$, are primitive idempotents generating L, then there exist a Hurwitz algebra C, a diagonal matrix $\alpha = \mathrm{diag}(\alpha_1, \alpha_2, \alpha_3) \in \mathrm{GL}_3(F)$ and an isomorphism $\phi \colon J \xrightarrow{\sim} \mathcal{H}_3(C, \alpha)$ such that $\phi(e_i) = E_{ii}$.*

(2) *If J does not have zero divisors, then either $J = F^+$ (if char $F \neq 3$), $J = L^+$ for a cubic (separable) field extension L of F, $J = D^+$ for a central division algebra D, $J = \mathcal{H}(B, \tau)$ for a central division algebra B of degree 3 over a quadratic field extension K of F and τ a unitary involution or J is an exceptional Jordan division algebra of dimension 27 over F.*

Reference: The first part of (1) and the last claim of (2) follow from the classification theorem (37.2) and the fact, due to Schafer [242], that Albert algebras with zero divisors are of the form $\mathcal{H}_3(C, \alpha)$. The last claim in (1) is a special case of the coordinatization theorem of Jacobson [141, Theorem 5.4.2]. $\square$

In view of a deep result of Springer [265, Theorem 1, p. 421], the bilinear trace form is an important invariant for reduced Freudenthal algebras. Springer's result and the theory of Pfister forms yield the following generalization (see also Serre [258, §9.2, §9.4]):

(37.13) Theorem. *Let F be a field of characteristic not 2. Let J, J' be reduced Freudenthal algebras. Let T, resp. T', be the corresponding bilinear trace forms. The following conditions are equivalent:*

(1) *J and J' are isomorphic.*
(2) *T and T' are isometric.*

Furthermore, if (1) (or (2)) holds, $J \simeq \mathcal{H}_3(C, \alpha)$ and $J' \simeq \mathcal{H}_3(C', \alpha')$, then C and C' are isomorphic.

Proof: We may assume that $J = \mathcal{H}_3(C, \alpha)$ and $J' = \mathcal{H}_3(C', \alpha')$ with C, $C' \neq 0$. (1) implies (2) by uniqueness of the generic minimal polynomial. Assume now that T and T' are isometric. The bilinear trace of $\mathcal{H}_3(C, \alpha)$ is of the form

$$T = \langle 1, 1, 1 \rangle \perp b_C \otimes \langle \alpha_3^{-1}\alpha_2, \alpha_1^{-1}\alpha_3, \alpha_2^{-1}\alpha_1 \rangle$$

and a similar formula holds for T'. Thus

$$(37.14) \qquad b_C \otimes \langle \alpha_3^{-1}\alpha_2, \alpha_1^{-1}\alpha_3, \alpha_2^{-1}\alpha_1 \rangle \simeq b_{C'} \otimes \langle {\alpha'_3}^{-1}\alpha'_2, {\alpha'_1}^{-1}\alpha'_3, {\alpha'_2}^{-1}\alpha'_1 \rangle.$$

We show in the following Lemma (37.15) that (37.14) implies $N_C \simeq N_{C'}$, hence $C \simeq C'$ holds by Proposition (33.19), and we may identify C and C'. Assume next that C is associative. By Jacobson [129], (37.14) implies that the C-hermitian forms $\langle \alpha_3^{-1}\alpha_2, \alpha_1^{-1}\alpha_3, \alpha_2^{-1}\alpha_1 \rangle_C$ and $\langle {\alpha'_3}^{-1}\alpha'_2, {\alpha'_1}^{-1}\alpha'_3, {\alpha'_2}^{-1}\alpha'_1 \rangle_C$ are isometric. They are similar to $\langle \alpha_1, \alpha_2, \alpha_3 \rangle_C$, resp. $\langle \alpha'_1, \alpha'_2, \alpha'_3 \rangle_C$. Thus α, α' define isomorphic unitary involutions on $M_3(C)$ and the Jordan algebras $\mathcal{H}_3(C, \alpha)$ and $\mathcal{H}_3(C, \alpha')$ are isomorphic. If C is a Cayley algebra, the claim is much deeper and we need Springer's result, which says that $\mathcal{H}_3(C, \alpha)$ and $\mathcal{H}_3(C, \alpha')$ are isomorphic if their trace forms are isometric (Springer [265, Theorem 1, p. 421]), to finish the proof. $\square$

For any Pfister form φ, let $\varphi' = \langle 1 \rangle^\perp$.

(37.15) Lemma. *Let ϕ_n, ψ_n be n-Pfister bilinear forms and χ_p, φ_p p-Pfister bilinear forms for $p \geq 2$. If $\phi_n \otimes \chi'_p \simeq \psi_n \otimes \varphi'_p$, then $\phi_n \simeq \psi_n$ and $\phi_n \otimes \chi_p \simeq \psi_n \otimes \varphi_p$.*

Proof: We make computations in the Witt ring WF and use the same notation for a quadratic form and its class in WF. Let $q = \phi_n \otimes \chi'_p = \psi_n \otimes \varphi'_p$. Adding ϕ_n, resp. ψ_n on both sides , we get that $q + \phi_n$ and $q + \psi_n$ lie in $I^{n+p}F$, so that $\psi_n - \phi_n \in I^{n+p}F$. Since $\psi_n - \phi_n$ can be represented by a form of rank $2^{n+1} - 2$, it follows from the Arason-Pfister Hauptsatz (Lam [169, Theorem 3.1, p. 289]), that $\psi_n - \phi_n = 0$. $\square$

(37.16) Corollary. *Let $T = \langle 1, 1, 1 \rangle \perp b_{N_C} \otimes \langle -b, -c, bc \rangle$ be the trace form of $J = \mathcal{H}_3(C, \alpha)$ and let q be the bilinear Pfister form $b_{N_C} \otimes \langle\!\langle b, c \rangle\!\rangle$. The isometry class of T determines the isometry classes of N_C and q. Conversely, the classes of N_C and q determine the class of T.*

Proof: The claim is a special case of Lemma (37.15). $\square$

(37.17) Remark. Theorem (37.13) holds more generally for separable Jordan algebras of degree 3: In view of the structure theorem (37.2) the only cases left are algebras of the type $F \times J(V, q)$, where the claim follows from Proposition (37.4), and étale algebras of dimension 3 with zero divisors. Here the claim follows from the fact that quadratic étale algebras are isomorphic if and only if their norms are isomorphic (see Proposition (33.19)).

An immediate consequence of Theorem (37.13) is:

(37.18) Corollary. $\mathcal{H}_3(\mathfrak{C}, \alpha)$ *is split for any* α *if* $\mathfrak{C}$ *is split.*   □

(37.19) Remark. Conditions on α, α' so that $\mathcal{H}_3(\mathfrak{C}, \alpha)$ and $\mathcal{H}_3(\mathfrak{C}, \alpha')$ are isomorphic for a Cayley division algebra $\mathfrak{C}$ are given in Albert-Jacobson [14, Theorem 5].

We conclude this section with a useful "Skolem-Noether" theorem for Albert algebras:

(37.20) Proposition. *Let* I, I' *be reduced simple Freudenthal subalgebras of degree* 3 *of a reduced Albert algebra* J. *Any isomorphism* $\phi\colon I \xrightarrow{\sim} I'$ *can be extended to an automorphism of* J.

Reference: See Jacobson [136, Theorem 3, p. 370].   □

However, for example, split cubic étale subalgebras of a reduced Albert algebra J are not necessarily conjugate by an automorphism of J. Necessary and sufficient conditions are given in Albert-Jacobson [14, Theorem 9]. It would be interesting to have a corresponding result for a pair of arbitrary isomorphic cubic étale subalgebras.

Another Skolem-Noether type of theorem for Albert algebras is given in (40.15).

§38. Cubic Jordan Algebras

A separable Jordan algebra of degree 3 is either a Freudenthal algebra or is of the form $F^+ \times J(V, q)$ where $J(V, q)$ is the Jordan algebra of a quadratic space of dimension ≥ 2 (see the structure theorem (37.2)); if J is a Freudenthal algebra, then J is of the form F^+ (assuming char $F \neq 3$), L^+ for L cubic étale, classical of type A_2, B_1, C_3 or exceptional of dimension 27. Let

$$P_{J,a}(X) = X^3 - T_J(a)X^2 + S_J(a)X - N_J(a)1$$

be the generic minimal polynomial of a separable Jordan algebra J of degree 3. The element

$$x^{\#} = x^2 - T_J(x)x + S_J(x)1 \in J$$

obviously satisfies $x \cdot x^{\#} = N_J(x)1$. It is the (*Freudenthal*) *adjoint* of x and the linearization of the quadratic map $x \mapsto x^{\#}$

$$x \times y = (x + y)^{\#} - x^{\#} - y^{\#}$$
$$= 2x \cdot y - T_J(x)y - T_J(y)x + b_{S_J}(x, y)1$$

is the *Freudenthal* "$\times$"-*product*.[34] Let $T(x, y) = T_J(x \cdot y)$ be the bilinear trace form. The datum $(J, N_J, \#, T, 1)$ has the following properties (see McCrimmon [185, Section 1]):

(a) the form $N_J\colon J \to F$ is cubic, the adjoint $\#\colon J \to J$, $x \mapsto x^{\#}$, is a quadratic map such that $x^{\#\#} = N(x)x$ and $1 \in J$ is a base point such that $1^{\#} = 1$;

(b) the nonsingular bilinear trace form T is such that

$$N_J(x + \lambda y) = \lambda^3 N_J(y) + \lambda^2 T(x^{\#}, y) + \lambda T(x, y^{\#}) + N_J(x)$$

and $T(x, 1)1 = 1 \times x + x$ for x, $y \in J$ and $\lambda \in F$.

---

[34]The $\times$-product is sometimes defined as $\frac{1}{2}[(x + y)^{\#} - x^{\#} - y^{\#}]$.

These properties are characteristic-free. Following McCrimmon [185] and Petersson-Racine [212] (see also Jacobson [141, 2.4]), we define a *cubic norm structure* over any field F (even if char $F = 2$) as a datum $(J, N, \#, T, 1)$ with properties (a) and (b). An isomorphism

$$\phi: (J, N, \#, T, 1) \xrightarrow{\sim} (J', N', \#, T', 1')$$

is an F-isomorphism $J \xrightarrow{\sim} J'$ of vector spaces which is an isometry $(J, N, T) \xrightarrow{\sim} (J', N', T')$, such that $\phi(1) = 1'$ and $\phi(x^\#) = \phi(x)^\#$ for all $x \in J$. We write *Cubjord* for the groupoid of cubic norm structures with isomorphisms as morphisms.

(38.1) Examples. Forgetting the Jordan multiplication and just considering generic minimal polynomials, we get cubic norm structures on $J = \mathcal{H}_3(C, \alpha)$ and on twisted forms of these. If $J = L$ is cubic étale over F, $N_L = N_{L/F}$, $T_L = T_{L/F}$, and T is the trace form. If J is of classical type A, B, or D, then N_J is the reduced norm and T_J is the reduced trace. If J is of classical type C, then N_J is the reduced pfaffian and T_J is the reduced pfaffian trace. We also have cubic structures associated to quadratic forms, as in the case of the Jordan algebra $J = F^+ \times J(V, q)$. More generally, let $J' = (V', q', 1')$ be a *pointed quadratic space*, i.e., $1' \in V'$ is such that $q'(1') = 1$, and let b' be the polar of q'. On $J = F \oplus J'$ we define $N_J(x, v) = xq'(v)$, $1 = (1, 1')$, $T_J(x, v) = x + b'(1', v)$, $T\big((x, v), (y, w)\big) = xy + b'(v, \overline{w})$ where $\overline{w} = b'(1', w)1' - w$ and $(x, v)^\# = \big(q'(v), x\overline{v}\big)$. Conversely, any cubic norm structure is of one of the types described above, see for example Petersson-Racine [212, Theorem 1.1]. We refer to cubic norm structures associated with Freudenthal algebras as *Freudenthal algebras* (even if they do not necessarily admit a multiplication!). Cubic norm structures of the form $\mathcal{H}_3(C, \alpha)$ for arbitrary C and α are called *reduced Freudenthal algebras*.

(38.2) Lemma. *Let $(J, N, \#, T, 1)$ be a cubic norm structure and set*

$$x^2 = T(x, 1)x - x^\# \times 1 \quad and \quad x^3 = T(x, x)x - x^\# \times x.$$

(1) *Any element $x \in J$ satisfies the cubic equation*

$$P(x) = x^3 - T_J(x)x^2 + S_J(x)x - N_J(x)1 = 0$$

where $T_J(x) = T(x, 1)$ and $S_J(x) = T_J(x^\#)$. Furthermore we have

$$x^\# = x^2 - T_J(x)x + S_J(x)1.$$

In particular, for any element $x \in J$ the vector space $F[x] \subset J$ spanned by 1, x, x^2 is a commutative associative cubic unital subalgebra.

(2) *There is a Zariski-open, non-empty subset U of J such that $F[x]$ is étale for $x \in U$.*

(3) *The identities*
  (a) $S_J(1) = T_J(1) = 3$, $N_J(1) = 1$, $1^\# = 1$,
  (b) $S_J(x) = T_J(x^\#)$, $b_{S_J}(x, y) = T_J(x \times y)$,
  (c) $b_{S_J}(x, 1) = 2T_J(x)$,
  (d) $2S_J(x) = T_J(x)^2 - T_J(x^2)$,
  (e) $T_J(x \times y) = T_J(x)T_J(y) - T(x, y)$,
  (f) $x^{\#\#} = N_J(x)x$,
  (g) $T(x \times y, z) = T(x, y \times z)$
hold in J.

Proof: (1) can be directly checked. For (2) we observe that $F[x]$ is étale if and only if the generic minimal polynomial $P_{J,x}$ of x has pairwise distinct roots (in an algebraic closure), i.e., the discriminant of $P_{J,x}$ (as a function of x) is not zero. This defines the open set U. It can be explicitly shown that the set U is non-empty if J is reduced, i.e., is not a division algebra. Thus we may assume that J is a division algebra. Then, by the following lemma (38.3), F is infinite. Again by (38.3) J is reduced over an algebraic closure F_{alg} of F. The set U being non-empty over F_{alg} and F being infinite, it follows that U is non-empty. We refer to [186] or [228] for (3). A proof for cubic alternative algebras is in (34.14). $\qquad\square$

An element $x \in J$ is *invertible* if it is invertible in the algebra $F[x] \subset J$. We say that a cubic norm structure is a *division cubic norm structure* if every nonzero element has an inverse. Such structures are (non-reduced) Freudenthal algebras and can only exist in dimensions 1, 3, 9, and 27. In dimension 3 we get separable field extensions and in dimension 9 central associative division algebras of degree 3 or symmetric elements in central associative division algebras of degree 3 over quadratic separable field extensions, with unitary involutions. Corresponding examples in dimension 27 will be given later using Tits constructions.

(38.3) Lemma. *An element $x \in J$ is invertible if and only if $N_J(x) \neq 0$ in F. In that case we have $x^{-1} = N_J(x)^{-1}x^{\#}$. Thus a cubic norm structure J is a division cubic norm structure if and only if $N_J(x) \neq 0$ for $x \neq 0$ in J, i.e., N_J is anisotropic. In particular a cubic norm structure J of dimension > 3 is reduced (i.e., is not a division algebra) if F is finite or algebraically closed.*

Proof: If $N_J(x) = 0$ for $x \neq 0$, we have by Lemma (38.2) $x^{\#\#} = N_J(x)x = 0$ hence either $u = x^{\#}$ or $u = x$ satisfies $u^{\#} = 0$ and $u \neq 0$. We then have $S_J(u) = T_J(u^{\#}) = 0$ so that u satisfies

$$0 = u^{\#} = u^2 - T_J(u)u = 0.$$

If $T_J(u) = 0$ we have $u^2 = 0$; if $T_J(u) \neq 0$ we may assume that $T_J(u) = 1$ and $u^2 = u$, however $u \neq 1$. Thus in both cases u is not invertible (see also Exercise 6 of this chapter). The claim for F finite or algebraically closed follows from the fact that such a field is C_i, $i \leq 1$ (see for example the book of Greenberg [107, Chap. 2] or Scharlau [247, §2.15]). Thus N_J, which is a form of degree 3 in 9 or 27 variables cannot be anisotropic over a finite field or an algebraically closed field. $\qquad\square$

(38.4) Proposition. *If $\operatorname{char} F \neq 2$, the categories* Cubjord *and* Sepjord(3) *are isomorphic.*

Proof: Any separable cubic Jordan algebra determines a cubic norm structure and an isomorphism of separable cubic Jordan algebras is an isomorphism of the corresponding structures. Conversely,

$$x \cdot y = \tfrac{1}{2}[(x+y)^2 - x^2 - y^2]$$

defines on the underlying vector space J of a cubic norm structure J a Jordan multiplication and an isomorphism of cubic norm structures is an isomorphism for this multiplication. $\qquad\square$

38.A. The Springer decomposition. Let $L = F[x]$ be a cubic étale subalgebra of a Freudenthal algebra J (in the new sense). Since L is étale, the bilinear trace form $T_J|_L$ is nonsingular and there is an orthogonal decomposition

$$J = L \perp V \quad \text{with} \quad V = L^\perp \subset J^0 = \{\, x \in J \mid T_J(x) = 0 \,\}.$$

We have $T_J(\ell) = T(\ell, 1) = T_{L/F}(\ell)$ and $N_J(\ell) = N_{L/F}(\ell)$ for $\ell \in L$. It follows from $T(\ell_1 \times \ell_2, v) = T(\ell_1, \ell_2 \times v) = 0$ for $v \in V$ that $\ell \times v \in V$ for $\ell \in L$ and $v \in V$. We define

$$\ell \circ v = -\ell \times v,$$

so that $\ell \circ v \in V$ for $\ell \in L$ and $v \in V$. Further, let $Q\colon V \to L$ and $\beta\colon V \to V$ be the quadratic maps defined by setting

$$v^\# = \bigl(-Q(v), \beta(v)\bigr) \in L \oplus V,$$

so that

$$(\ell, v)^\# = \bigl(\ell^\# - Q(v), \beta(v) - \ell \circ v\bigr).$$

We have

$$S_J(v) = T_J(v^\#) = -T_{L/F}\bigl(Q(v)\bigr)$$

since $T\bigl(\beta(v)\bigr) = 0$. Furthermore, putting $\beta(v, w) = \beta(v + w) - \beta(v) - \beta(w)$, we get

$$v \times w = \bigl(-b_Q(v, w), \beta(v, w)\bigr).$$

(38.5) Example. Let $J = \mathcal{H}_3(C, 1)$ be a reduced Freudenthal algebra and let $L = F \times F \times F \subset J$ be the set of diagonal elements. Then V is the space of matrices

$$v = \begin{pmatrix} 0 & \bar{c}_3 & c_2 \\ c_3 & 0 & \bar{c}_1 \\ \bar{c}_2 & c_1 & 0 \end{pmatrix}, \quad c_i \in C$$

and the "$\circ$"-action of L on V is given by

$$(\lambda_1, \lambda_2, \lambda_3) \circ \begin{pmatrix} 0 & \bar{c}_3 & c_2 \\ c_3 & 0 & \bar{c}_1 \\ \bar{c}_2 & c_1 & 0 \end{pmatrix} = \begin{pmatrix} 0 & \lambda_3\bar{c}_3 & \lambda_2 c_2 \\ \lambda_3 c_3 & 0 & \lambda_1\bar{c}_1 \\ \lambda_2\bar{c}_2 & \lambda_1 c_1 & 0 \end{pmatrix}$$

Identifying $C \oplus C \oplus C$ with V through the map

$$v = (c_1, c_2, c_3) \mapsto \begin{pmatrix} 0 & \bar{c}_3 & c_2 \\ c_3 & 0 & \bar{c}_1 \\ \bar{c}_2 & c_1 & 0 \end{pmatrix}$$

the action of L on V is diagonal, hence V is an L-module. We have

$$Q(v) = (c_1\bar{c}_1, c_2\bar{c}_2, c_3\bar{c}_3)$$

for $v = (c_1, c_2, c_3)$, so that (V, Q) is a quadratic space over L. Furthermore we get

$$\beta(v) = (\bar{c}_2\bar{c}_3, \bar{c}_3\bar{c}_1, \bar{c}_1\bar{c}_2),$$

hence $\beta(\ell \circ v) = \ell^\# \circ \beta(v)$, $Q\bigl(\beta(v)\bigr) = Q(v)^\#$, and $b_Q\bigl(v, \beta(v)\bigr) = N_J(v) \in F$. Thus (V, L, Q, β) is a twisted composition.

The properties of the "$\circ$"-action described in Example (38.5) hold in general:

(38.6) Theorem. (1) *Let* $(J, N, \#, T, 1)$ *be a cubic Freudenthal algebra, let* L *be a cubic étale subalgebra of dimension 3 and let* $V = L^{\perp}$ *for the bilinear trace form* T. *The operation* $L \times V \to V$ *given by* $(\ell, v) \mapsto \ell \circ v$ *defines the structure of an* L-*module on* V *such that* (L, Q) *is a quadratic space and* (V, L, Q, β) *is a twisted composition.* (2) *For any twisted composition* (V, L, Q, β), *the cubic structure* $(J, N, \#, T, 1)$ *on the vector space* $J(L, V) = L \oplus V$ *given by*

$$N(\ell, v) = N_{L/F}(\ell) + b_Q\big(v, \beta(v)\big) - T_L\big(\ell Q(v)\big),$$

$$(\ell, v)^{\#} = \big(\ell^{\#} - Q(v), \beta(v) - \ell \circ v\big) \in L \oplus V$$

$$T\big((\ell_1, v_1), (\ell_2, v_2)\big) = T_{L/F}(\ell_1 \ell_2) + T_{L/F}\big(b_Q(v_1, v_2)\big)$$

is a Freudenthal algebra. Furthermore we have

$$S_J(\ell, v) = T_J\big((\ell, v)^{\#}\big) = T_J(\ell^{\#}) - T_J\big(Q(v)\big).$$

Proof: (1) It suffices to check that V is an L-module over a separable closure, and there we may assume by (37.12) (which also holds for Freudenthal algebras in the new sense) that $L = F \times F \times F$ is diagonal in some $\mathcal{H}_3(C, \alpha)$. The claim then follows from Example (38.5). Claim (1) can also be checked rationally, see Petersson-Racine [210, Proposition 2.1] (or Springer [268], if char $F \neq 2$).

(2) If L is split and $V = C \oplus C \oplus C$, we may identify $J(V, L)$ with $\mathcal{H}_3(C, 1)$ as in Example (38.5). The general case then follows by descent.   □

We say that the Freudenthal algebra $J(V, L)$ is the *Springer construction* associated with the twisted composition (V, L, Q, β). This construction was introduced by Springer for cyclic compositions of dimension 8, in relation to exceptional Jordan algebras. Conversely, given $L \subset J$ étale of dimension 3, we get a *Springer decomposition* $J = L \oplus V$. Springer decompositions for arbitrary cubic structures were first considered by Petersson-Racine [210, Section 2]. Any Freudenthal algebra is (in many ways) a Springer construction.

Let $\Gamma_s = (V, L, Q, \beta)$ be a split twisted composition of dimension 8. Its associated algebraic group of automorphisms is $\mathrm{Spin}_8 \rtimes S_3$ (see (36.5)). The corresponding Freudenthal algebra $J_s = J(V, L)$ is split; we recall that by Theorem (25.13) its automorphism group defines a simple split algebraic group G of type F_4.

(38.7) Corollary. *The map* $\Gamma_s \mapsto J_s$ *induces an injective group homomorphism* $\mathrm{Spin}_8 \rtimes S_3 \to G$. *The corresponding map in cohomology* $H^1(F, \mathrm{Spin}_8 \rtimes S_3) \to H^1(F, G)$, *which associates the class of* $J(V, L)$ *to a twisted composition* (V, L, Q, β), *is surjective.*

Proof: Let (V, L, Q, β) be a twisted split composition of dimension 8. Clearly any automorphism of (V, L, Q, β) extends to an automorphism of $J(L, V)$ and conversely any automorphism of $J(L, V)$ which maps L to L restricts to an automorphism of (V, L, Q, β). This shows the injectivity of $\mathrm{Spin}_8 \rtimes S_3 \to G$. The second claim follows from the facts that $H^1(F, \mathrm{Spin}_8 \rtimes S_3)$ classifies twisted compositions of dimension 8 (Proposition (36.7)), that $H^1(F, G)$ classifies Albert algebras (Proposition (37.11)) and that any Albert algebra admits a Springer decomposition.   □

(38.8) Theorem. *Let* $J(V, L)$ *be the Springer construction associated with a twisted composition* (V, L, Q, β). *Then* $J(V, L)$ *has zero divisors if and only if the twisted composition* (V, L, Q, β) *is similar to a Hurwitz composition* $\Gamma(C, L)$ *for some Hurwitz algebra* C. *Furthermore, we have* $J(V, L) \simeq \mathcal{H}_3(C, \alpha)$ *for some* α *(and the same Hurwitz algebra* C*).*

Proof: By Theorem (36.24) the composition (V, L, Q, β) is similar to a Hurwitz composition $\Gamma(C, L)$ if and only if there exists $v \in V$ such that $\beta(v) = \lambda v$ and $N(v) = \lambda^{\#}$ for some $\lambda \neq 0 \in L$; we then have $(v, \lambda)^{\#} = 0$ in $J(V, L)$. By Exercise 6 of this chapter, this is equivalent with the existence of zero divisors in $J(V, L)$ (see also the proof of (38.3)). Hence Theorem (37.12) implies that $J(V, L)$ is a reduced Freudenthal algebra $\mathcal{H}_3(C', \alpha)$ for some Hurwitz algebra C'. It remains to be shown that $C \simeq C'$. We consider the case where F has characteristic different from 2 (and leave the other case as an exercise). For any bilinear form $(x, y) \mapsto b(x, y)$ over L, let $(T_{L/F})_*(b)$ be its *transfer* to F, i.e.,

$$(T_{L/F})_*(b)(x, y) = T_{L/F}\big(b(x, y)\big).$$

The bilinear trace form of $J(V, L)$ is the bilinear form:

$$T = (T_{L/F})_*(\langle 1 \rangle_L) \perp (T_{L/F})_*(b_Q)$$

and b_Q is extended from the bilinear form $b_{Q_0} = \langle 2 \rangle_L \perp \delta b_{N_C^0}$ over F (see Lemma (36.21)). Let $b_{N_C} = b_C$ and $b_{N_C^0} = b_C^0$. By Frobenius reciprocity (see Scharlau [247, Theorem 5.6, p. 48]) we get

$$T \simeq (T_{L/F})_*(\langle 1 \rangle_L) \perp (T_{L/F})_*(\langle 1 \rangle_L) \otimes (\langle 2 \rangle_L \perp \delta b_C^0).$$

Since $(T_{L/F})_*(\langle 1 \rangle_L) = \langle 1, 2, 2\delta \rangle$ (see (18.31.2)), it follows that

$$\begin{aligned}
T &\simeq \langle 1, 2, 2\delta \rangle \perp \delta b_C^0 \otimes \langle 1, 2, 2\delta \rangle \perp \langle 2, 1, \delta \rangle \\
&\simeq \langle 1, 2, 2 \rangle \perp b_C \otimes \langle 2, \delta, 2\delta \rangle \\
&\simeq \langle 1, 1, 1 \rangle \perp b_C \otimes \langle 2, \delta, 2\delta \rangle.
\end{aligned}$$

Thus

$$b_C \otimes \langle 2, \delta, 2\delta \rangle \simeq b_{C'} \otimes \langle \alpha_1, \alpha_2, \alpha_3 \rangle,$$

since an isomorphism $J(V, L) \simeq \mathcal{H}_3(C', \alpha)$ implies that the corresponding trace forms are isomorphic. The last claim then follows from Lemma (37.15) and Theorem (33.19). □

§39. The Tits Construction

Let K be a quadratic étale algebra with conjugation ι and let B be an associative separable algebra of degree 3 over K with a unitary involution τ (according to an earlier convention, we also view K as a cubic separable K-algebra if char $F \neq 3$). The generic norm N_B of B defines a cubic structure on B (as a K-algebra) and restricts to a cubic structure on $\mathcal{H}(B, \tau)$. Let $(u, \nu) \in \mathcal{H}(B, \tau) \times K^{\times}$ be such that

$$N_B(u) = \nu\tau(\nu).$$

One can take for example $(u, \nu) = (1, 1)$. On the set

$$J(B, \tau, u, \nu) = \mathcal{H}(B, \tau) \oplus B,$$

let $1 = (1, 0)$ and

$$N(a, b) = N_B(a) + T_{K/F}\big(\nu N_B(b)\big) - T_B\big(abu\tau(b)\big)$$

$$(a, b)^{\#} = \big(a^{\#} - bu\tau(b), \tau(\nu)\tau(b)^{\#} u^{-1} - ab\big)$$

for $(a, b) \in \mathcal{H}(B, \tau) \oplus B$. Further let

$$T\big((a_1, b_1), (a_2, b_2)\big) = T_B(a_1 a_2) + T_{K/F}\big(T_B\big(b_1 u\tau(b_2)\big)\big).$$

(39.1) Theorem. *The space $J(B, \tau, u, \nu)$ admits a Freudenthal cubic structure with 1 as unit, N as norm, $\#$ as Freudenthal adjoint and T as bilinear trace form. Furthermore we have $S_J\big((a,b)\big) = S_B(a) - T_B\big(bu\tau(b)\big)$ for the quadratic trace S_J.*

Reference: A characteristic-free proof is in Petersson-Racine [212, Theorem 3.4], see also McCrimmon [185, Theorem 7]. The claim is also a consequence of Proposition (39.7) (see Corollary (39.8)) if char $F \neq 3$. The last claim follows from $S_J(x) = T_J(x^\#)$ and $T_J(x) = T(x, 1)$.   □

If char $F \neq 2$, the Jordan product of $J(B, \tau, u, \nu)$ is given by (see p. 544):

	a_1	b_1
a_2	$\frac{1}{2}(a_1 a_2 + a_2 a_1)$	$(\overline{a_2}b_1)_*$
b_2	$(\overline{a_1}b_2)_*$	$\big(\overline{b_1 \nu \tau(b_2) + b_2 \nu \tau(b_1)}\big) + \frac{1}{2}\big(\tau(u)\big(\tau(b_1) \times \tau(b_2)\big)\nu^{-1}\big)_*$

where $\overline{x} = \frac{1}{2}\big(\mathrm{Trd}_B(x) - x\big)$ and x_* denotes x as an element of the second component B. The cubic structure $J(B, \tau, u, \nu)$ described in Theorem (39.1) is a *Tits construction* or a *Tits process* and the pair (u, ν) is called an *admissible pair* for (B, τ). The following lemma describe some useful allowed changes for admissible pairs.

(39.2) Lemma. (1) *Let (u, ν) be an admissible pair for (B, τ). For any $w \in B^\times$, $\big(wu\tau(w), \nu N_B(w)\big)$ is an admissible pair for (B, τ) and $(a, b) \mapsto (a, bw)$ is an isomorphism*

$$J(B, \tau, u, \nu) \xrightarrow{\sim} J\big(B, \tau, wu\tau(w), \nu N_B(w)\big).$$

(2) *For any Tits construction $J(B, \tau, u, \nu)$, there is an isomorphic Tits construction $J(B, \tau, u', \nu')$ with $N_B(u') = 1 = \nu'\tau(\nu')$.*

Proof: The first claim reduces to a tedious computation, which we leave as an exercise (see also Theorem (39.18)). For the second, we take $w = \nu^{-1}u$ in (1).   □

An exceptional Jordan algebra of dimension 27 of the form $J(B, \tau, u, \nu)$ where B is a central simple algebra over a quadratic field extension K of F, is classically called a *second Tits construction*. The case where K is not a field also has to be considered. Let $J(B, \tau, u, \nu)$ be a Tits process with $K = F \times F$, $B = A \times A^{\mathrm{op}}$ where A is either central simple or cubic étale over F and τ is the exchange involution. By Lemma (39.2) we may assume that the admissible pair (u, ν) is of the form $\big(1, (\lambda, \lambda^{-1})\big)$, $\lambda \in F^\times$. Projecting B onto the first factor A induces an isomorphism of vector spaces

$$J(B, \tau, u, \nu) \simeq A \oplus A \oplus A,$$

the norm is given by $N_J(a, b, c) = N_A(a) + \lambda N_A(b) + \lambda^{-1} N_A(c) - T_A(abc)$ and the Freudenthal adjoint on $A \oplus A \oplus A$ reduces to

$$(a, b, c)^\# = (a^\# - bc, \lambda^{-1}c^\# - ab, \lambda b^\# - ca)$$

where $a \mapsto a^\#$ is the Freudenthal adjoint of A^+ (which is a cubic algebra!); thus we have $S_J(a, b, c) = S_A(a) - T_A(bc)$ and the bilinear trace form is given by

$$T_J\big((a_1, b_1, c_1), (a_2, b_2, c_2)\big) = T_A(a_1 b_1) + T_A(a_2 b_3) + T_A(a_3 b_2).$$

If char $F \neq 2$, the Jordan product is

$$(a_1, b_1, c_1) \cdot (a_2, b_2, c_2) =$$
$$\left(a_1 \cdot a_2 + \overline{b_1 c_2} + \overline{b_2 c_1}, \bar{a}_1 b_2 + \bar{a}_2 b_1 + (2\lambda)^{-1}(c_1 \times c_2), c_2 \bar{a}_1 + c_1 \bar{a}_2 + \tfrac{1}{2}\lambda(b_1 \times b_2)\right),$$

where

$$\bar{a} = \tfrac{1}{2} a \times 1 = \tfrac{1}{2} T_A(a) \cdot 1 - \tfrac{1}{2} a.$$

Conversely we can associate to a pair (A, λ), A central simple of degree 3 or cubic étale over F and $\lambda \in F^{\times}$, a Freudenthal algebra $J(A, \lambda) = A \oplus A \oplus A$, with norm, Freudenthal product and trace as given above. The algebra $J(A, \lambda)$ is (classically) a *first Tits construction* if A is central simple. Any first Tits construction $J(A, \lambda)$ extends to a Tits process $J\left(A \times A^{\mathrm{op}}, \sigma, 1, (\lambda, \lambda^{-1})\right)$ over $F \times F$. According to the classical definitions, we shall say that $J(B, \tau, u, \nu)$ is a *second Tits process* if K is a field and that $J(A, \lambda)$ is a *first Tits process*.

(39.3) Remark. (See [186, p. 308].) Let $(A, \lambda) = A \oplus A \oplus A$ be a first Tits process. To distinguish the three copies of A in $J(A, \lambda)$, we write

$$J(A, \lambda) = A^+ \oplus A_1 \oplus A_2$$

and denote $a \in A$ as a, a_1, resp. a_2 if we consider it as an element of A^+, A_1, or A_2. The first copy admits the structure of an associative algebra, A_1 (resp. A_2) can be characterized by the fact that it is a subspace of $(A^+)^{\perp}$ (for the bilinear trace form) such that $a \cdot a_1 = -a \times a_1$ (resp. $a_2 \cdot a = -a \times a_2$) defines the structure of a left A-module on A_1 (resp. right A-module on A_2).

(39.4) Proposition. *For any second Tits process $J(B, \tau, u, \nu)$ over F, B a K-algebra, $J(B, \tau, u, \nu) \otimes K$ is isomorphic to the first Tits process $J(B, \nu)$ over K. Conversely, any second Tits process $J(B, \tau, u, \nu)$ over F is the Galois descent of the first Tits process (B, ν) over K under the ι-semilinear automorphism*

$$(a, b, c) \mapsto \left(\tau(a), \tau(c)u^{-1}, u\tau(b)\right).$$

Proof: An isomorphism

$$J(B, \tau, u, \nu) \otimes K \xrightarrow{\sim} J(B, \nu)$$

is induced by $(a, b) \mapsto \left(a, b, u\tau(b)\right)$. The last claim follows by straightforward computations. □

(39.5) Examples. Assume that char $F \neq 3$.

(1) Any cubic étale F-algebra L can be viewed as a Tits construction over F; if $L = F(\sqrt[3]{\lambda})$, then L is isomorphic to the first Tits construction (F, λ). In general there exist a quadratic étale F-algebra K and some element $\nu \in K$ with $N_K(\nu) = 1$ such that $L \otimes K \simeq K(\sqrt[3]{\nu})$ and L is the second Tits construction $(K, \iota, 1, \nu)$ (see Proposition (30.19)).

(2) Let A be central simple of degree 3 over F. We write A as a crossed product $A = L \oplus Lz \oplus Lz^2$ with L cyclic and $z^3 = \lambda \in F^{\times}$, $z\ell = \rho(\ell)z$ and ρ a generator of $\mathrm{Gal}(L/F)$; the map $A \to L \oplus L \oplus L$ given by $a + bz + cz^2 \mapsto \left(a, \rho(b), \lambda\rho^2(c)\right)$ is an isomorphism of A^+ with the first Tits construction (L, λ).

(3) Let (B, τ) be central simple of degree 3 with a distinguished unitary involution τ over K. In view of Proposition (19.15) and Corollary (19.28), there exists a cubic étale F-algebra L with discriminant $\Delta(L) \simeq K$ such that

$$B = L \otimes K \oplus (L \otimes K)z \oplus (L \otimes K)z^2 \quad \text{with} \quad z^3 = \lambda \in F^\times, \quad \tau(z) = z.$$

The K-algebra $L \otimes K$ is cyclic over K; let $\rho \in \text{Gal}(L \otimes K/K)$ be such that $z\xi z^{-1} = \rho(\xi)$ for $\xi \in L \otimes K$. We have

$$L_1 = \{\, \xi \in L \otimes K \mid \rho \circ (1 \otimes \iota)(\xi) = \xi \,\} \simeq L,$$
$$L_2 = \{\, \xi \in L \otimes K \mid \rho^2 \circ (1 \otimes \iota)(\xi) = \xi \,\} \simeq L$$

and $(1 \otimes \iota)(L_1) = L_2$, so that

$$\mathcal{H}(B, \tau) = L \oplus L_1 \oplus L_2 \simeq L \oplus L \oplus L$$

and a check shows this is an isomorphism of $\mathcal{H}(B, \tau)$ with the first Tits construction (L, λ). Since the exchange involution on $A \times A^{\text{op}}$ is distinguished, we see that $\mathcal{H}(B, \tau)$ is a first Tits construction if and only if τ is distinguished, if and only if $S_B|_{\mathcal{H}(B,\tau)^0}$ has Witt index at least 3 (see Proposition (19.10) for the last equivalence).

(4) Let (B, τ) be central simple with a unitary involution over K and assume that $\mathcal{H}(B, \tau)$ contains a cyclic étale algebra L over F. By Albert [10, Theorem 1] we may write B as a crossed product

(39.6) $$B = L \otimes K \oplus (L \otimes K)z \oplus (L \otimes K)z^2$$

with $z^3 = \nu \in K^\times$ such that $N_K(\nu) = 1$; furthermore the involution τ is determined by $\tau(z) = uz^{-1}$ with $u \in L$ such that $N_B(u) = 1$. In this case $\mathcal{H}(B, \tau)$ is isomorphic to the second Tits process $(L \otimes K, 1 \otimes \iota_K, u, \nu)$.

(5) A Tits construction $J = J(L \otimes K, 1 \otimes \iota_K, u, \nu)$ with L cubic étale is of dimension 9, hence by Theorem (37.12) it is of the form $\mathcal{H}(B, \tau)$ for a central simple algebra B of degree 3 over an étale quadratic F-algebra K_1 and a unitary involution τ. We may describe (B, τ) more explicitly: if L is cyclic, $K_1 = K$, and B is as in (39.6). If L is not cyclic, we replace L by $L_2 = L \otimes \Delta(L)$, where $\Delta(L)$ is the discriminant of L, and obtain (B_2, τ_2) over $\Delta(L)$ from (39.6). Let ϕ be the descent on B_2 given by $\phi = 1 \otimes \iota_{\Delta(L)} \otimes \iota_K$ on $L \otimes \Delta(L) \otimes K$ and $\phi(z) = z^{-1}$. Then $B = B_2^\phi$ and $K_1 = \Delta(L) * K$. In particular we have $(B, \tau) \simeq (A \times A^{\text{op}}, \text{exchange})$ if and only if $\Delta(L) \simeq K$.

(6) Let J be a Freudenthal algebra of dimension 9 over F and let L be a cubic étale subalgebra of J. We may describe J as a Tits construction $J(L \otimes K, 1 \otimes \iota_K, u, \nu)$ as follows. Let $J = L \oplus V$ be the Springer decomposition induced by L. Then V is a twisted composition (V, L, Q, β) and V is of dimension 2 over L; by Proposition (36.38) V admits the structure of a hermitian $L \otimes K$-space for some quadratic étale F-algebra K. Let $V = (L \otimes K)v$; let $u = Q(v) \in L^\times$ and let $\beta(v) = xv$, $x \in L \otimes K$. It follows from $b_Q(v, \beta(v)) \in F$ that $(x + \bar{x})u \in F$, where $x \mapsto \bar{x}$ is the extension of the conjugation ι_K of K to $L \otimes K$. Similarly $Q(\beta(v)) = u^\#$ implies that $x\bar{x}u^2 = N_{L/F}(u) \in F^\times$. Both imply that xu (or $\bar{x}u$) lies in K and $N_{K/F}(\bar{x}u) = \mathfrak{n}_{L/F}(u)$. Let J' be the Tits construction $J(L \otimes K, 1 \otimes \iota_K, u, \bar{x}u)$. The map $J' \to J$ given by $(a, b) \mapsto (a, bv)$ is an isomorphism of Jordan algebras.

(7) A first Tits construction $J(A, 1)$ with A cubic étale or central simple of degree 3 is always a split Freudenthal algebra: this is clear for cubic étale algebras by Example (2). So let A be central simple. Taking $a \in A$ such that $a^3 \in F^\times$, we see that $a^\# = a^2$, so that $(a, a, a)^\# = 0$ and, by Exercise 6 of this chapter $J(A, 1)$ is reduced. Theorem (37.12) then implies that $J(A, 1) \simeq \mathcal{H}_3(\mathfrak{C}, \alpha)$. Let L_1 be a cubic extension which splits A. By Theorem (37.13) $\mathfrak{C} \otimes L_1$ is split, hence by Springer's theorem for quadratic forms, $\mathfrak{C}$ is split. The claim then follows from Corollary (37.18).

39.A. Symmetric compositions and Tits constructions. In this section, we assume that char $F \neq 3$. The aim is to show that Tits constructions with admissible pairs $(1, \nu)$ are also Springer constructions. We start with a first Tits construction (A, λ); let $L = F[X]/(X^3 - \lambda)$ be the cubic Kummer extension associated with $\lambda \in F^\times$, set, as usual, $A^0 = \{ x \in A \mid T_A(x) = 0 \}$ and let $\Gamma(A, \lambda) = (A_0 \otimes L, L, N, \beta)$ be the twisted composition of type $^1A_2$ induced by A and λ (see (36.33)). Let

$$J\big(\Gamma(A, \lambda)\big) = L \oplus \Gamma(A, \lambda) = L \oplus A_0 \otimes L = L \otimes A$$

be the Freudenthal algebra obtained from $\Gamma(A, \lambda)$ by the Springer construction. If v is the class of X modulo $(X^3 - \lambda)$ in L, $(1, v, v^{-1})$ is a basis of L as vector space over F and we write elements of $L \otimes A$ as linear combinations $a + v \otimes b + v^{-1} \otimes c$, with a, b, $c \in A$.

(39.7) Proposition. *The isomorphism* $\phi \colon J\big(\Gamma(A, \lambda)\big) \xrightarrow{\sim} J(A, \lambda) = A \oplus A \oplus A$ *given by*

$$a + v \otimes b + v^{-1} \otimes c \mapsto (a, b, c) \quad \text{for } a, b, c \in A$$

is an isomorphism of Freudenthal algebras.

Proof: We use the map ϕ to identify L as an étale subalgebra of $J(A, \lambda)$ and get a corresponding Springer decomposition

$$J(A, \lambda) = L \oplus L \otimes A^0.$$

It follows from Theorem (38.6) and from the description of a twisted composition $\Gamma(A, \lambda)$ of type $^1A_2$ given in §36.D that ϕ restricts to an isomorphism of twisted compositions $\Gamma(A, \lambda) \xrightarrow{\sim} L \otimes A^0$, hence the claim. $\square$

(39.8) Corollary. *Tits constructions are Freudenthal algebras.*

Proof: We assume char $F \neq 3$. By descent we are reduced to first Tits constructions, hence the claim follows from Proposition (39.7). $\square$

(39.9) Corollary. *Let G be a split simple group scheme of type F_4. Jordan algebras which are first Tits constructions are classified by the image of the pointed set $H^1(F, \mathrm{PGL}_3 \times \boldsymbol{\mu}_3)$ in $H^1(F, G)$ under the map $\mathrm{PGL}_3 \times \boldsymbol{\mu}_3 \to G$ induced by $(A, \lambda) \mapsto J(A, \lambda)$.* $\square$

Not all exceptional Jordan algebras are first Tits construction (see Petersson-Racine [213] or Proposition (40.5)). Thus the cohomology map in (39.9) is in general not surjective (see also Proposition (39.23)).

We now show that the Springer construction associated with a twisted composition of type $^2A_2$ is always a second Tits construction. Let (B, τ) be a central simple algebra with a unitary involution over a quadratic étale F-algebra K. Let $\nu \in K$ be such that $N_K(\nu) = 1$; let L be as in Proposition (36.35.1), and let $\Gamma(B, \tau, \nu)$ be the corresponding twisted composition, as given in Proposition (36.35.2).

(39.10) Proposition. *There exists an isomorphism $L \oplus \Gamma(B, \tau, \nu) \xrightarrow{\sim} J(B, \tau, 1, \nu)$.*

Proof: The twisted composition $\Gamma(B, \tau, \nu)$ over F is defined by descent from the twisted composition $\Gamma(B, \nu)$ over K (see Proposition (36.35)); similarly $J(B, \tau, 1, \nu)$ is defined by descent from $J(B, \nu)$ (see Proposition (39.4)). The descents are compatible with the isomorphism given in Proposition (39.7), hence the claim. $\square$

39.B. Automorphisms of Tits constructions. If J is a vector space over F with some algebraic structure and A is a substructure of J, we write $\mathrm{Aut}_F(J, A)$ for the group of F-automorphisms of J which maps A to A and by $\mathrm{Aut}_F(J/A)$ the group of automorphisms of J which restrict to the identity on A. The corresponding group schemes are denoted $\mathbf{Aut}(J, A)$ and $\mathbf{Aut}(J/A)$.

(39.11) Proposition (Ferrar-Petersson, [97]). *Let A be central simple of degree 3 and let $J_0 = J(A, \lambda_0)$ be a first Tits construction. The sequence of group schemes*

$$1 \to \mathbf{SL}_1(A) \xrightarrow{\gamma} \mathbf{Aut}(J_0, A^+) \xrightarrow{\rho} \mathbf{Aut}(A^+) \to 1$$

where $\gamma(u)(a, a_1, a_2) = (a, a_1 u^{-1}, u a_2)$ for $u \in \mathbf{SL}_1(A)(R)$, $R \in \mathsf{Alg}_F$, and ρ is the restriction map, is exact.

Proof: Let $R \in \mathsf{Alg}_F$; exactness on the left (over R) and $\rho_R \circ \gamma_R = 1$ is obvious. Let $J_0 = A^+ \oplus A_1 \oplus A_2$ and let η be an automorphism of J_{0R} which restricts to the identity on A_R^+. It follows from Remark (39.3) that η stabilizes A_{1R} and A_{2R}, so there exist linear bijections $\eta_i \colon A_{iR} \to A_{iR}$ such that $\eta(a, a_1, a_2) = \big(a, \eta_1(a_1), \eta_2(a_2)\big)$. Expanding $\eta\big(a \times (0, a_1, a_2)\big)$ in two different ways shows that

$$\eta_1(a a_1) = a \eta_1(a_1) \quad \text{and} \quad \eta_2(a_2 a) = \eta_2(a_2) a.$$

Hence there are $u, v \in A_R^\times$ such that $\eta_1(a_1) = a_1 v$ and $\eta_2(a_2) = u a_2$. Comparing the first components of $\eta\big((0, 1, 1)^{\#}\big) = \big(\eta(0, 1, 1)\big)^{\#}$ yields $v = u^{-1}$. Since η preserves the norm we have $u \in \mathbf{SL}_1(A)(R)$. To conclude, since $\mathbf{Aut}(A^+)$ is smooth (see the comments after the exact sequence (37.7)), it suffices to check by (22.10) that ρ_{alg} is surjective. In fact ρ is already surjective: let $\phi \in \mathbf{Aut}(A^+)$, hence, by the exact sequence (37.7), ϕ is either an automorphism or an anti-automorphism of A. In the first case, $\widetilde{\phi}(a, a_1, a_2) = \big(\phi(a), \phi(a_1), \phi(a_2)\big)$ extends ϕ to an element of $\mathbf{Aut}(J_0, A^+)$. In the second case, A is split, so some $u \in A^\times$ has $N_A(u) \in F^{\times 2}$ and $\widetilde{\phi}(a, a_1, a_2) = \big(\phi(a), \phi(a_2) u^{-1}, u \phi(a_1)\big)$ extends ϕ. $\square$

Now, let $L = F(\sqrt[3]{\lambda})$. We embed $L = F(v)$, $v = \sqrt[3]{\lambda}$, in $J(A, \lambda) = A \oplus A \oplus A$ through $v \mapsto (0, 1, 0)$ and $v^{-1} \mapsto (0, 0, 1)$. Furthermore we set

$$(A^\times \times A^\times)^{\mathrm{Det}} = \{ (f, g) \in A^\times \times A^\times \mid \mathrm{Nrd}_A(f) = \mathrm{Nrd}_A(g) \}.$$

(39.12) Corollary. (1) *We have*

$$\mathrm{Aut}_F\big(J(A, \lambda), A^+\big) = (A^\times \times A^\times)^{\mathrm{Det}} / F^\times,$$

where $F^\times$ operates diagonally, if A is a division algebra and

$$\mathrm{Aut}_F\big(J(A, \lambda), A^+\big) = \big(\mathrm{GL}_3(F) \times \mathrm{GL}_3(F)\big)^{\mathrm{Det}} / F^\times \rtimes \mathbb{Z}/2\mathbb{Z}$$

if $A = M_3(F)$. The action of $\mathbb{Z}/2\mathbb{Z}$ on a pair (f, g) is given by

$$(f, g) \mapsto \big((f^{-1})^t, (g^{-1})^t\big).$$

The action of (f, g) on $J(A, \lambda)$ is given by

$$(f, g)(a, b, c) = (f a f^{-1}, f b g^{-1}, g c f^{-1})$$

and the action of $\mathbb{Z}/2\mathbb{Z}$ by $\tau(a,b,c) = (a^t, c^t, b^t)$.

(2) We have

$$\text{Aut}_F\big(J(A,\lambda), A^+, L\big) \simeq \text{Aut}_F(A)/F^\times \times \mu_3$$

if A is a division algebra and

$$\text{Aut}_F\big(J(A,\lambda), A^+, L\big) \simeq \big(\mathbf{PGL}_3(F) \times \mu_3\big) \rtimes \mathbb{Z}/2\mathbb{Z}$$

where the action of $\mathbb{Z}/2\mathbb{Z}$ is given by $\tau(f,\mu) = ([f^t]^{-1}, \mu^{-1})$ if $A = M_3(F)$.

Proof: (1) If $\phi \in \text{Aut}_F\big(J(A,\lambda), A^+\big)$ restricts to an automorphism ϕ' of A, we write $\phi' = \text{Int}(f)$ and (1) follows from Proposition (39.11). If ϕ restricts to an anti-automorphism ϕ' of A, we replace ϕ by $\phi \circ \tau$ and apply the preceding case.

(2) We assume that $A = M_3(F)$. By (1) we can write any element ϕ of $\text{Aut}_F\big(J(A,\lambda), A^+, L\big)$ as $[f,g]$ with f, $g \in \text{GL}_3(F)$. Since ϕ restricts to an automorphism of L, we must have $\phi(u) = \rho u^{\pm 1}$, $\rho \in F^\times$. Since $\tau(u) = u^{-1}$, we may assume that $\phi(u) = \rho u$ (replace ϕ by $\phi\tau$). It follows that $\rho^3 = 1$ and $\rho \in \boldsymbol{\mu}_3(F)$. Since $\phi\big((0,1,0)\big) = (0, fg^{-1}, 0) = (0, \rho^{-1}, 0)$ we get $g = \rho f$ with $\rho \in \boldsymbol{\mu}_3(F)$. The map $(f,\rho) \mapsto (f, \rho f)$ then induces the desired isomorphism. $\square$

(39.13) Remark. If F contains a primitive cubic root of unity, we may identify $\boldsymbol{\mu}_3$ with A_3 (as Galois-modules) and $\big(\mathbf{PGL}_3(F) \times \mu_3\big) \rtimes \mathbb{Z}/2\mathbb{Z}$ with $\mathbf{PGL}_3(F) \rtimes S_3$ where S_3 operates through its projection on $\mathbb{Z}/2\mathbb{Z}$. In particular we get for the split Jordan algebra J_s

$$\text{Aut}_F\big(J_s, M_3(F)^+, F \times F \times F\big) = \mathbf{PGL}_3(F) \rtimes S_3.$$

On the other hand we have

$$\text{Aut}_F(J_s, F \times F \times F) = \mathbf{Spin}_8(F) \rtimes S_3$$

(see Corollary (38.7)), so that

$$\mathbf{PGL}_3(F) \rtimes S_3 = \big(\mathbf{GL}_3(F) \times \mathbf{GL}_3(F)\big)^{\text{Det}}/F^\times \rtimes \mathbb{Z}/2\mathbb{Z} \cap \mathbf{Spin}_8(F) \rtimes S_3 \subset G(F)$$

where $G = \text{Aut}(J_s)$ is a simple split group scheme of type F_4.

(39.14) Theorem. (1) *(Ferrar-Petersson) Let $J_0 = J(A, \lambda_0)$ be a first Tits construction with A a central simple associative algebra of degree 3. The cohomology set $H^1\big(F, \mathbf{Aut}(J_0, A^+)\big)$ classifies pairs (J', I') with J' an Albert algebra over F and I' is a central simple Jordan algebra of dimension 9 over F. The cohomology set $H^1\big(F, \mathbf{Aut}(A^+)\big)$ classifies central simple Jordan algebra of dimension 9 over F. The sequence of pointed sets*

$$1 \to F^\times/N_A(A^\times) \xrightarrow{\psi} H^1\big(F, \mathbf{Aut}(J_0, A^+)\big) \xrightarrow{\rho^1} H^1\big(F, \mathbf{Aut}(A^+)\big)$$

is exact and $\psi([\lambda]) = [J(A, \lambda\lambda_0), A^+]$, $\rho^1([J', A']) = [A']$.

(2) *Let J be an Albert algebra containing a subalgebra A^+ for A central simple of degree 3. There exist $\lambda \in F^\times$ and an isomorphism $\phi\colon J \xrightarrow{\sim} J(A, \lambda)$ which restricts to the identity on A^+.*

(3) *$J(A, \lambda)$ is a division algebra if and only if λ is not the reduced norm of an element from A.*

Proof: We follow Ferrar-Petersson [97]. (1) We assume for simplicity that F is a field of characteristic not 2, so that J_0 is an F-algebra with a multiplication m. Let $\mathcal{F}$ be the flag $A^+ \subset J_0$ and let $W = \text{Hom}_F(J_0 \otimes J_0, J_0)$. We let $G = \text{Aut}_F(\mathcal{F})$ act on $\mathcal{F} \oplus W$ through the natural action. Let $w = (0, m)$. Since $\mathbf{Aut}_G(w) =$

$\mathbf{Aut}(J_0, A^+)$ and since $(J_0, A^+)_{F_{\mathrm{sep}}} \simeq (J', I')_{F_{\mathrm{sep}}} (\simeq (J_s, M_3^+)_{F_{\mathrm{sep}}})$, the first claim follows from (29.12) and from Corollary (39.12). The fact that $H^1(F, \mathbf{Aut}(A^+))$ classifies central simple Jordan algebra of dimension 9 over F then is clear.

The exact sequence is the cohomology exact sequence associated with the sequence (39.11), where the identification (29.4) of $F^\times / N_A(A^\times)$ with $H^1(F, \mathbf{SL}_1(A))$ is as follows: let $\lambda \in F^\times$ and let $v \in A_{\mathrm{sep}}^\times$ be such that $N_{A_{\mathrm{sep}}}(v) = \lambda$. Then $\alpha \colon \mathrm{Gal}(F_{\mathrm{sep}}/F) \to \mathbf{SL}_1(A)(F_{\mathrm{sep}})$ such that $\alpha(g) = v^{-1}g(v)$ is the cocycle induced by λ. The image of the class of $\lambda \in F^\times$ in $H^1(F, \mathbf{Aut}(J_0, A^+))$ is the class of the cocycle β given by $\beta(g)(a, a_1, a_2) = (a, a_1 g(v^{-1})v, v^{-1}g(v)a_2)$. Let $\gamma \in \mathrm{GL}((J_0)_{\mathrm{sep}})$ be given by

$$\gamma(a, a_1, a_2) = (a, a_1 v^{-1}, v a_2),$$

then $\beta(g) = g(\gamma^{-1})\gamma$, and, setting $J = J(A, \lambda\lambda_0)$, one can check that

$$\gamma \colon (J_0, A^+)_{F_{\mathrm{sep}}} \xrightarrow{\sim} (J, A^+)_{F_{\mathrm{sep}}}$$

is an isomorphism, hence (J, A^+) is the F-form of (J_0, A^+) given by the image of λ.

(2) We set $\lambda_0 = 1$ in (1). Let J' be a reduced Freudenthal algebra. By Theorem (37.20), we have $(J, A^+)_{F_{\mathrm{sep}}} \simeq (J_0, A^+)_{F_{\mathrm{sep}}}$. Therefore (J, A^+) is a form of (J_0, A^+) and its class in $H^1(F, \mathbf{Aut}(J_0, A^+))$ belongs to the kernel of ρ^1, hence in the image of ψ. Thus by (1) there exists $\lambda \in F^\times$ such that $(J(A, \lambda), A^+) \simeq (J, A^+)$, as claimed.

(3) follows similarly from (1), since $J(A, 1)$ is split (see Example (39.5.7)). $\quad\square$

(39.15) Remark. By (39.14.3), $J(A, \lambda)$ is a division algebra if and only if A is a division algebra and λ is not a reduced norm of A. Examples can be given over a purely transcendental extension of degree 1: Let F_0 be a field which admits a division algebra A_0 of degree 3 and let $A = A_0 \otimes F_0(t)$. Then the Albert algebra $J(A, t)$ is a division algebra (see Jacobson, [136, p. 417]).

The analogue of Proposition (39.11) for second Tits constructions is

(39.16) Proposition. Let $J_0 = J(B, \tau, u_0, \nu_0)$ be a second Tits construction and let $\tau' = \mathrm{Int}(u_0) \circ \tau$. The sequence

$$1 \to \mathbf{SU}(B, \tau') \xrightarrow{\gamma} \mathbf{Aut}(J_0, \mathcal{H}(B, \tau)) \xrightarrow{\rho} \mathbf{Aut}(\mathcal{H}(B, \tau)) \to 1,$$

where $\gamma_R(u)(a, b)_R = (a, bu^{-1})_R$ and ρ_R is restriction, is exact.

Proof: (39.16) follows from (39.11) by descent, using Proposition (39.4). $\quad\square$

To get a result corresponding to Theorem (39.14) for second Tits constructions, we recall that the pointed set $H^1(F, \mathbf{SU}(B, \tau))$ classifies pairs $(u, \nu) \in \mathrm{Sym}(B, \tau)^\times \times K^\times$ with $N_B(u) = N_K(\nu)$ under the equivalence $\approx$, where

(39.17) $\quad (u, \nu) \approx (u', \nu') \iff u' = bu\tau(b), \nu' = \nu \cdot N_B(b)$ for some $b \in B^\times$

(see (29.18)). As in (29.18) we set

$$\mathrm{SSym}(B, \tau)^\times = \{ (u, \nu) \in \mathcal{H}(B, \tau) \times K^\times \mid N_B(u) = N_K(\nu) \}.$$

(39.18) Theorem. Let $J_0 = J(B, \tau, u_0, \nu_0)$.

(1) *The sequence of pointed sets*

$$1 \to \mathrm{SSym}(B,\tau)^\times/\approx \xrightarrow{\psi}$$

$$H^1\big(F, \mathbf{Aut}\big(J_0, \mathcal{H}(B,\tau)\big)\big) \xrightarrow{\rho^1} H^1\big(F, \mathbf{Aut}\big(\mathcal{H}(B,\tau)\big)\big),$$

where $\psi([u,\nu]) = [J(B,\tau,uu_0,\nu\nu_0), \mathcal{H}(B,\tau)]$ *and* $\rho^1([J',A']) = [A']$, *is exact.*

(2) *Let J be a Freudenthal algebra of dimension 27 containing a subalgebra $\mathcal{H}(B,\tau)$ for (B,τ) central simple of degree 3 with a unitary involution. There exist an admissible pair $(u,\nu) \in \mathrm{SSym}(B,\tau)^\times$ and an isomorphism $\phi\colon J \xrightarrow{\sim} J(B,\tau,u,\nu)$ which restricts to the identity on $\mathcal{H}(B,\tau)$.*

(3) *$J(B,\tau,u,\nu)$ is a division algebra if and only if u is not the reduced norm of an element from $B^\times$.*

Proof: The proof of (39.18) is similar to the one of Theorem (39.14) and we skip it.   □

Any Hurwitz algebra can be obtained by successive applications of the Cayley-Dickson process, starting with F. The next result, which is a special case of a theorem of Petersson-Racine [211, Theorem 3.1], shows that a similar result holds for Freudenthal algebras of dimension 3, 9, and 27 if Cayley-Dickson processes are replaced by Tits processes:

(39.19) Theorem (Petersson-Racine). *Assume that* char $F \neq 3$. *Any Freudenthal algebra of dimension 3, 9, or 27 can be obtained by successive applications of the Tits process. In particular any exceptional Jordan algebra of dimension 27 is of the form $\mathcal{H}(B,\tau) \oplus B$ where B is a central simple of degree 3 over a quadratic étale F-algebra K with a unitary involution.*

Proof: A Freudenthal algebra of dimension 3 is a cubic étale algebra, hence the claim follows from Example (39.5.1), if $\dim J = 3$. The case $\dim J = 9$ is covered by Example (39.5.6). If J has dimension 27 and J contains a Freudenthal subalgebra of dimension 9 of the type $\mathcal{H}(B,\tau)$, then by Theorem (39.18), there exists a pair (u,ν) such that $J \simeq (B,\tau,u,\nu)$. Thus we are reduced to showing that J contains some $\mathcal{H}(B,\tau)$. If J is reduced this is clear, hence we may assume that J is not reduced. Then (see the proof of (39.20)) F is an infinite field. Let L be a cubic étale F-subalgebra of J and let $J = L \oplus V$, $V = (V, L, Q, \beta)$, be the corresponding Springer decomposition. For some $v \in V$, the set $\{v, \beta(v)\}$ is linearly independent over L since $Q\big(v, \beta(v)\big)$ is anisotropic and by a density argument (F is infinite) we may also assume that Q restricted to $U = Lv \oplus L\beta(v)$ is L-nonsingular. Thus J contains a Springer construction $J_1 = J(L, U)$ of dimension 9. In view of the 9-dimensional case J_1 is a Tits construction and by Example (39.5.5), $J_1 \simeq \mathcal{H}(B,\tau)$ for some central simple algebra (B,τ) of degree 3 with unitary involution, hence the claim.   □

Jordan algebras of the form L^+ (L cubic étale of dimension 3), A^+ (A central simple of degree 3), or $\mathcal{H}(B,\tau)$ (B central simple of degree 3 with an involution of the second type) are "generic subalgebras" of Albert algebras in the following sense:

(39.20) Proposition. *Let J be an Albert algebra.*

(1) *There is a Zariski-open subset U of J such that the subalgebra generated by x is étale for all $x \in U$.*

(2) *There is a Zariski-open subset U' of J such that the subalgebra generated by $x \in U$ and $y \in U'$ is of the form A^+, for A central simple of degree 3 over F, or of the form $\mathcal{H}(B, \tau)$ for B central simple over a quadratic separable field extension K and τ a unitary involution.*

Proof: The first claim is already in (38.2.2). The second follows from the proof of (39.19). $\square$

(39.21) Remark. The element v in the proof of (39.19) is such that v and $\beta(v)$ are linearly independent over L (see Proposition (36.38)). If v is such that $\beta(v) = \lambda v$ for λ in L, then J is reduced by Theorem (38.8) and (L, v) generates a 6-dimensional subalgebra of J of the form $\mathcal{H}_3(F, \alpha)$ (Soda [260, Theorem 2]). Such an algebra is not "generic."

(39.22) Remark. If char $F = 3$, the only Freudenthal algebras which cannot be obtained by iterated Tits constructions are separable field extensions of degree 3 (see [211, Theorem 3.1]). We note that Petersson and Racine consider the more general case of simple cubic Jordan structures (not just Freudenthal algebras) in [211, Theorem 3.1].

The Albert algebra $J_s = J\big(M_3(F), 1\big)$ is split, thus $G = \mathbf{Aut}\big(J\big(M_3(F), 1\big)\big)$ is a simple split group scheme of type F_4.

(39.23) Proposition. *The pointed set $H^1\big(F, (\mathbf{GL}_3 \times \mathbf{GL}_3)^{\mathrm{Det}} / \mathbf{G}_{\mathrm{m}} \rtimes \mathbb{Z}/2\mathbb{Z}\big)$ classifies pairs $\big(J, \mathcal{H}(B, \tau)\big)$ where J is an Albert algebra, $B \subset J$ is central simple with unitary involution τ over a quadratic étale algebra K. The map*

$$H^1\big(F, (\mathbf{GL}_3 \times \mathbf{GL}_3)^{\mathrm{Det}} / \mathbf{G}_{\mathrm{m}} \rtimes \mathbb{Z}/2\mathbb{Z}\big) \to H^1(F, G),$$

induced by $\mathrm{Aut}_F\big(J\big(M_3(F), 1\big), M_3(F)^+\big) \to \mathrm{Aut}_F\big(J\big(M_3(F), 1\big)\big)$ and which associates the class of J to the class of $\big(J, \mathcal{H}(B, \tau)\big)$ is surjective.

Proof: The first claim follows from Corollary (39.12) and Theorem (39.14), the second then is a consequence of Theorem (39.19). $\square$

(39.24) Remark. Let J be an Albert algebra. We know that $J \simeq J(B, \tau, u, \beta)$ for some datum (B, τ, u, ν). The datum can be reconstructed cohomologically as follows. Let $[\alpha] \in H^1\big(F, (\mathbf{GL}_3 \times \mathbf{GL}_3)^{\mathrm{Det}} / \mathbf{G}_{\mathrm{m}} \rtimes \mathbb{Z}/2\mathbb{Z}\big)$ be a class mapping to $[J]$. The image $[\gamma] \in H^1(F, \mathbb{Z}/2\mathbb{Z})$ of $[\alpha]$ under the map in cohomology induced by the projection $(\mathbf{GL}_3 \times \mathbf{GL}_3)^{\mathrm{Det}} / \mathbf{G}_{\mathrm{m}} \rtimes \mathbb{Z}/2\mathbb{Z} \to \mathbb{Z}/2\mathbb{Z}$ defines the quadratic extension K. Pairs $\big(J, \mathcal{H}(B, \tau)\big)$ with fixed K are classified by

$$H^1\big(F, \big((\mathbf{GL}_3 \times \mathbf{GL}_3)^{\mathrm{Det}} / \mathbf{G}_{\mathrm{m}}\big)_\gamma\big)$$

and the projection on the first factor gives an element of $H^1\big(F, (\mathbf{PGL}_3)_\gamma\big)$, hence by Remark (29.21) a central simple K-algebra B with unitary involution τ. We finally get (u, ν) from the exact sequence (39.18.1).

§40. Cohomological Invariants

In this section we assume that F is a field of characteristic not 2. Let $J = \mathcal{H}_3(C, \alpha)$ be a reduced Freudenthal algebra of dimension > 3. Its bilinear trace form is given by

$$T = \langle 1, 1, 1 \rangle \perp b_C \otimes \langle -b, -c, bc \rangle$$

where b_C is the polar of N_C. As known from Corollary (37.16) and Theorem (37.13), the Pfister forms b_C and $b_C \otimes \langle\langle b, c \rangle\rangle$ determine the isomorphism class of J. Let $\dim_F C = 2^i$, let $f_i(J) \in H^i(F, \mathbb{Z}/2\mathbb{Z})$ be the cohomological invariant of the Pfister form b_C and let $f_{i+2}(J) \in H^{i+2}(F, \mathbb{Z}/2\mathbb{Z})$ be the cohomological invariant of $b_C \otimes \langle\langle b, c \rangle\rangle$. These two invariants determine J up to isomorphism. Observe that Freudenthal algebras of dimension 3 with zero divisors are also classified by a cohomological invariant: Such an algebra is of the form $F^+ \times K^+$ and is classified by the class of K $f_1(K) \in H^1(F, S_2)$. We now define the invariants $f_3(J)$ and $f_5(J)$ for division algebras J of dimension 27 (and refer to Proposition (30.18), resp. Theorem (30.21) for the corresponding invariants of algebras of dimension 3, resp. 9). We first compute the bilinear trace form of J.

(40.1) Lemma. (1) *Let (B, τ) be a central simple algebra of degree 3 over a quadratic étale F-algebra K with a unitary involution and let T_τ be the bilinear trace form of the Jordan algebra $\mathcal{H}(B, \tau)$. Then*

$$T_\tau \simeq \langle 1, 1, 1 \rangle \perp b_{K/F} \otimes \langle -b, -c, bc \rangle \quad \text{for } b, \, c \in F^\times$$

where $b_{K/F}$ stands for the polar of the norm of K.
(2) *Let J be a Freudenthal algebra of dimension 27 and let T be the bilinear trace form of J. There exist $a, b, c, e, f \in F^\times$ such that*

$$T \simeq \langle 1, 1, 1 \rangle \perp \langle 2 \rangle \otimes \langle\langle a, e, f \rangle\rangle \otimes \langle -b, -c, bc \rangle.$$

Proof: (1) follows from Proposition (11.22).
(2) By Theorem (39.19) we may assume that J is a second Tits construction $J(B, \tau, u, \mu)$, so that

$$T\big((x, y), (x', y')\big) = T_\tau(x, x') + T_{K/F}\big(\mathrm{Trd}_B(yu\tau(y'))\big)$$

for $x, \, x' \in \mathcal{H}(B, \tau)$ and $y, \, y' \in B$. By Lemma (39.2.2), we may assume that $\mathrm{Nrd}_B(u) = 1$. Let $\tau' = \mathrm{Int}(u^{-1}) \circ \tau$. By (1) the trace form of $\mathcal{H}(B, \tau')$ is of the form

$$T_{\tau'} = \langle 1, 1, 1 \rangle \perp b_{K/F} \otimes \langle -e, -f, ef \rangle$$

for $e, f \in F^\times$. Let $T_{\tau, \tau'}(x, y) = T_{K/F}\big(\mathrm{Trd}_B(xu\tau(y))\big)$ for $x, y \in B$. We claim that

$$T_{\tau, \tau'} \simeq b_{K/F} \otimes \langle -b, -c, bc \rangle \otimes \langle -e, -f, ef \rangle.$$

The involution τ is an isometry of the bilinear form $T_{\tau, \tau'}$ with the bilinear form $(T_{K/F})_*(T_{B, \tau, u})$ where $T_{B, \tau, u}(x, y) = \mathrm{Trd}_B\big(\tau(x)uy\big)$. Thus it suffices to have an isomorphism of hermitian forms

$$T_{B, \tau, u} \simeq \langle -b, -c, bc \rangle_K \otimes_K \langle -e, -f, ef \rangle_K$$

since

$$(T_{K/F})_*\big(\langle \alpha_1, \dots \alpha_n \rangle_K\big) = b_{K/F} \otimes \langle \alpha_1, \dots \alpha_n \rangle.$$

In view of Proposition (11.1) the unitary involution $\tau' \otimes \tau$ on $B \otimes_K {}^\iota B$ corresponds to the adjoint involution on $\mathrm{End}_K(B)$ of the hermitian form $T_{(B, \tau, u)}$ under the isomorphism $\tau_* : B \otimes_K {}^\iota B \to \mathrm{End}_K(B)$. By the Bayer-Lenstra extension (6.18) of Springer's theorem, we may now assume that $B = M_3(K)$ is split, so that by Example (19.5) τ is the adjoint involution of $\langle -b, -c, bc \rangle_K$ and τ' is the adjoint involution of $\langle -e, -f, ef \rangle_K$. This shows that $T_{B, \tau, u}$ and $\langle -b, -c, bc \rangle_K \otimes_K \langle -e, -f, ef \rangle_K$

are similar hermitian forms and it suffices to show that they have the same determinant. By Corollary (11.16) the form $T_{(B,\tau,u)}$ has determinant the class of $\mathrm{Nrd}(u)$, which, by the choice of u, is 1. Thus we get

$$T \simeq T_\tau \perp T_{\tau,\tau'} \simeq \langle 1, 1, 1 \rangle \perp \langle 2 \rangle \otimes \langle\!\langle a, e, f \rangle\!\rangle \otimes \langle -b, -c, bc \rangle$$

where $K = F(\sqrt{a})$. $\qquad\qquad\qquad\square$

(40.2) Theorem. (1) *Let F be a field of characteristic not 2. For any Freudenthal algebra J of dimension $3 + 3 \cdot 2^i$, $1 \le i \le 3$, there exist cohomological invariants $f_i(J) \in H^i(F, \mathbb{Z}/2\mathbb{Z})$ and $f_{i+2}(J) \in H^{i+2}(F, \mathbb{Z}/2\mathbb{Z})$ which coincide with the invariants defined above if J is reduced.*
(2) *If $J = J(B, \tau, u, \nu)$ is a second Tits construction of dimension 27, then $f_3(J)$ is the f_3-invariant of the involution $\tau' = \mathrm{Int}(u) \circ \tau$ of B.*

Proof: (1) Let $K = F(\sqrt{a})$. With the notations of Lemma (40.1), the invariants are given by the cohomological invariants of the Pfister forms $\langle\!\langle a \rangle\!\rangle$, resp. $\langle\!\langle a, b, c \rangle\!\rangle$ if J has dimension 9 and the cohomological invariants of $\langle\!\langle a, e, f \rangle\!\rangle$, resp. $\langle\!\langle a, e, f \rangle\!\rangle \otimes \langle\!\langle b, c \rangle\!\rangle$ if J has dimension 27. The fact that these are f_i-, resp. f_{i+2}-invariants of J follows as in Corollary (37.16).

Claim (2) follows from the computation of $T_{\tau'}$. $\qquad\qquad\qquad\square$

(40.3) Corollary. *If two second Tits constructions $J(B, \tau, u_1, \nu_1)$, $J(B, \tau, u_2, \nu_2)$ of dimension 27 corresponding to different admissible pairs (u_1, ν_1), (u_2, ν_2) are isomorphic, then there exist $w \in B^\times$ and $\lambda \in F^\times$ such that $\lambda u_2 = w u_1 \tau(w)$. If furthermore $\mathrm{Nrd}(u_1) = \mathrm{Nrd}(u_2)$, then w can be chosen such that $u_2 = w u_1 \tau(w)$.*

Proof: Let $\tau_i = \mathrm{Int}(u_i) \circ \tau$, $i = 1, 2$. In view of Theorem (40.2.2), and Theorem (30.21), the involutions τ_1 and τ_2 of B are isomorphic, hence the first claim. Taking reduced norms on both sides of $\lambda u_2 = w u_1 \tau(w)$, we get

$$\lambda^3 = \mathrm{Nrd}_B(w) \mathrm{Nrd}_B\big(\tau(w)\big)$$

and λ is of the form $\lambda' \overline{\lambda'}$. Replacing w by $w \lambda'^{-1}$, we get the second claim. $\qquad\square$

(40.4) Remark. Corollary (40.3) is due to Parimala, Sridharan and Thakur [206]. As we shall see in Theorem (40.12) (which is due to the same authors) w can in fact be chosen such that $u_2 = w u_1 \tau(w)$ and $\nu_2 = \nu_1 \mathrm{Nrd}_B(w)$ so that the converse of Lemma (39.2) holds.

For an Albert algebra J with invariants $f_3(J)$ and $f_5(J)$, the condition $f_3(J) = 0$ obviously implies $f_5(J) = 0$. More interesting are the following two propositions:

(40.5) Proposition. *Let J be an Albert algebra. The following conditions are equivalent:*
(1) *J is a first Tits construction, $J = J(A, \lambda)$.*
(2) *There exists a cubic extension L/F such that J_L splits over L.*
(3) *The Witt index $w(T)$ of the bilinear trace form T of J is at least 12.*
(4) *$f_3(J) = 0$.*
(5) *For any Springer decomposition $J = J(V, L)$ with corresponding twisted composition $\Gamma = (V, L, Q, \beta)$, we have $w_L(Q) \ge 3$.*

Proof: (1) $\Rightarrow$ (2) Choose L which splits A.

(2) $\Rightarrow$ (3) By Springer's Theorem we may assume that J is split. The claim then follows from the explicit computation of the bilinear trace form given in (37.9).

(3) $\Rightarrow$ (4) We have

$$T \simeq \langle 1, 1, 1 \rangle \perp \langle\langle a, e, f \rangle\rangle \otimes \langle -b, -c, bc \rangle.$$

Thus, if $w(T) \geq 12$, the anisotropic part b_{an} of $\langle\langle a, e, f \rangle\rangle \otimes \langle -b, -c, bc \rangle$ has at most dimension 6; since $b_{\mathrm{an}} \in I^3 F$, the theorem of Arason-Pfister (see Lam [169, p. 289]) shows that $b_{\mathrm{an}} = 0$ in WF. Lemma (37.15) then implies that $\langle\langle a, e, f \rangle\rangle$ is hyperbolic, hence $f_3(J) = 0$.

(4) $\Rightarrow$ (5) Let $J = J(V, L)$ be a Springer decomposition of J for a twisted composition $\Gamma = (V, L, Q, \beta)$. To check (5), we may assume that L is not a field: if L is a field we may replace L by $L \otimes L$. Then (V, Q) is similar to a Cayley composition (V_0, Q_0) with

$$Q_0 = \langle 1 \rangle \perp \langle \delta \rangle \otimes (\mathfrak{C}, \mathfrak{n})^0$$

(see Theorem (36.24) and Lemma (36.21)). Since $f_3(J)$ is the cohomological invariant of the norm of $\mathfrak{C}$, we get (5).

(5) $\Rightarrow$ (1) We may assume that J is a division algebra. (5) also implies (4) and a reduced algebra with $f_3 = 0$ is split. Let $x \in V$ with $Q(x) = 0$. We have $Q(\beta(x)) = 0$ and $Q(x, \beta(x)) \neq 0$ (by Proposition (36.24) and Theorem (38.8), since J is a division algebra). Thus $U = Lx \oplus L\beta(x)$ is a 2-dimensional twisted composition. By Proposition (36.38), $Q|_U$ is the trace of a hermitian 1-form over $L \otimes K$ for some quadratic étale F-algebra K. Furthermore K is split if $Q|_U$ is isotropic. Now the Springer construction $J_1 = L \oplus U$ is a 9-dimensional Freudenthal algebra and the Witt index of the bilinear trace form of J_1 is at least 2. As shown in Example (39.5.6), J_1 is a second Tits construction $J_1 = J(L \otimes K, 1 \otimes \iota_K, u_1, \nu_1)$ and by Example (39.5.5), $J_1 \simeq \mathcal{H}(B_1, \tau_1)$ where B_1 is central simple of degree 3 with a unitary involution τ_1. Moreover the center K_1 of B_1 is the discriminant algebra $\Delta(L)$ (since K as above is split). Since the trace on $\mathcal{H}(B_1, \tau_1)$ is of Witt index ≥ 2, τ_1 is distinguished (Proposition (19.10)). Furthermore, by Theorem (39.18), J is a second Tits construction $J = J(B_1, \tau_1, u, \nu)$ for the given (B_1, τ_1). We have $f_3(\mathrm{Int}(u) \circ \tau_1) = f_3(J)$ and since (4) implies (3), $\tau' = \mathrm{Int}(u) \circ \tau_1$ is also distinguished. By Theorem (19.6) $\tau' \simeq \tau$ and there exist $\lambda \in F^\times$ and $w \in B^\times$ such that $u = \lambda w \tau(w)$. By Lemma (39.2) we may assume that $u = \lambda \in F^\times$. Then the Tits construction $J_2 = J(L \otimes K_1, 1 \otimes \iota_{K_1}, \lambda, \nu)$ is a subalgebra of $J = J(B_1, \tau_1, u, \nu)$. By Example (39.5.5), $J_2 \simeq \mathcal{H}(B_2, \tau_2)$ and the center of B_2 is $K_1 * \Delta(L)$. Since $K_1 \simeq \Delta(L)$, $J_2 \simeq (A \times A^{\mathrm{op}}, \text{exchange})$ and we conclude using Theorem (39.14). $\square$

(40.6) Remark. The equivalence of (1) and (2) in (40.5) is due to Petersson-Racine [210, Theorem 4.7] if F contains a primitive cube root of unity. The trace form then has maximal Witt index.

(40.7) Proposition. *Let J be an Albert algebra. The following conditions are equivalent*:

(1) *$J = J(B, \tau, u, \nu)$ is a second Tits construction with τ a distinguished unitary involution of B.*

(2) *The Witt index $w(T)$ of the bilinear trace form T of J is at least 8.*

(3) *$f_5(J) = 0$.*

Proof: We use the notations of the proof of Lemma (40.1).

(1) $\Rightarrow$ (2) The bilinear form $b_{K/F} \otimes \langle\langle b, c \rangle\rangle$ is hyperbolic if τ is distinguished. Thus $T_{\tau, \tau'}$ has Witt index at least 6. By Proposition (19.10), T_τ has Witt index at least 2, hence the claim.

(2) $\Rightarrow$ (3) If $w(T) \geq 8$, $\langle\langle a, e, f \rangle\rangle$ is isotropic, hence $f_5(J) = 0$.

The proof of (3) $\Rightarrow$ (2) goes along the same lines.

(3) $\Rightarrow$ (1) We assume that J is a division algebra. Let $J = L \oplus V$ be a Springer decomposition of J; since (3) $\Rightarrow$ (2) holds, we get that $T|_V$ is isotropic. We may choose $x \in V$ such that $T(x, x) = 0$ and such that $U = Lx \oplus L\beta(x)$ is a 2-dimensional twisted composition. Then $J_1 = L \oplus U$ is a Freudenthal subalgebra of J of dimension 9, hence of the form $H(B, \tau)$. Since $w(T|_{J_1}) \geq 2$, τ is distinguished. The claim then follows from Theorem (39.18). $\qquad\square$

We now indicate how one can associate a third cohomological invariant $g_3(J)$ to an Albert algebra J (see also §30.C and §31.B). We refer to Rost [234] for more information (see the paper [215] of Petersson and Racine for an elementary approach). By Theorem (39.19), we may assume that $J = J(B, \tau, u, \nu)$ is a second Tits process and by Lemma (39.2) that $\mathrm{Nrd}_B(u) = \nu\iota(\nu) = 1$. Let L_ν be the descent of $K(\sqrt[3]{\nu})$ under the action given by ι_K on K and $\xi \mapsto \xi^{-1}$ for $\xi = \sqrt[3]{\nu}$. Then L_ν defines a class in $H^1(F, \boldsymbol{\mu}_{3[K]})$ by Proposition (30.19). On the other hand, the algebra with involution (B, τ) determines a class $g_2(B, \tau) \in H^2(F, \boldsymbol{\mu}_{3[K]})$ by Proposition (30.15). Since there exists a canonical isomorphism of Galois modules $\boldsymbol{\mu}_{3[K]} \otimes \boldsymbol{\mu}_{3[K]} = \mathbb{Z}/3\mathbb{Z}$ (with trivial Galois action on $\mathbb{Z}/3\mathbb{Z}$), the cup product $g_2(B, \tau) \cup [\nu]$ defines a cohomology class g_3 in $H^3(F, \mathbb{Z}/3\mathbb{Z})$. If $K = F \times F$, $B = A \times A^{\mathrm{op}}$ and $\nu = (\lambda, \lambda)^{-1}$, then $[A] \in H^2(F, \boldsymbol{\mu}_3)$, $[\lambda] \in H^1(F, \boldsymbol{\mu}_3)$ and we have $g_3 = [A] \cup [\lambda] \in H^3(F, \mathbb{Z}/3\mathbb{Z})$. For the following result see [234].

(40.8) Theorem. (1) *The cohomology class $g_3 \in H^3(F, \mathbb{Z}/3\mathbb{Z})$ is an invariant of the Albert algebra $J = J(B, \tau, u, \nu)$, denoted $g_3(J)$.*
(2) *We have $g_3(J) = 0$ if and only if J has zero divisors.* $\qquad\square$

(40.9) Remark. By definition we have $g_3 = g_2(B, \tau) \cup [\nu]$ and by Proposition (30.20) we know that $g_2(B, \tau) = \alpha \cup \beta$ with $\alpha \in H^1(F, \mathbb{Z}/3\mathbb{Z}_{[K]})$ and $\beta \in H^1(F, \boldsymbol{\mu}_3)$; thus

$$g_3 \in H^1(F, \mathbb{Z}/3\mathbb{Z}_{[K]}) \cup H^1(F, \boldsymbol{\mu}_3) \cup H^1(F, \boldsymbol{\mu}_{3[K]})$$

is a decomposable class.

It is conjectured that the three invariants $f_3(J)$, $f_5(J)$ and $g_3(J)$ classify Freudenthal algebras of dimension 27. This is the case if $g_3 = 0$; then J is reduced, $J \simeq \mathcal{H}_3(\mathfrak{C}, \alpha)$, $f_3(J) = f_3(\mathfrak{C})$ determines $\mathfrak{C}$, $f_3(J)$, $f_5(J)$ determine the trace and the claim follows from Theorem (37.13).

Theorem (40.8) allows to prove another part of the converse to Lemma (39.2) which is due to Petersson-Racine [213, p. 204]:

(40.10) Proposition. *If $J(B, \tau, u, \nu) \simeq J(B, \tau, u', \nu')$, then $\nu' = \nu \mathrm{Nrd}(w)$ for some $w \in B^\times$.*

Proof: The claim is clear if B is not a division algebra, since then the reduced norm map is surjective. Assume now that $J = J(B, \tau, u, \nu) \simeq J' = J(B, \tau, u', \nu')$. By (39.2.2), we may assume that $N_{K/F}(\nu) = 1 = N_{K/F}(\nu')$. Let L, resp. L', be the cubic extensions of F determined by ν, resp. ν', as in Proposition (30.19). We have

$$[B] \cup [L] = g_3(J) = g_3(J') = [B] \cup [L'],$$

hence $[B] \cup ([L'][L]^{-1}) = 0$ in $H^3(F, \mathbb{Z}/3\mathbb{Z})$. The class $[L'][L]^{-1}$ comes from $\nu'\nu^{-1}$. Since $(u, \nu'\nu^{-1})$ is obviously admissible we have a Tits construction $J'' =$

$J(B, \tau, u, \nu'\nu^{-1})$ whose invariant $g_3(J'')$ is zero. By Theorem (40.8.2) J'' has zero divisors and by Theorem (39.18) $\nu'\nu^{-1}$ is a norm of B.   □

Let $J(A, \lambda)$ be a first Tits construction. Since an admissible pair for this construction can be assumed to be of the form $(1, (\lambda, \lambda^{-1}))$ we get

(40.11) Corollary. *Let A be a central division algebra of degree 3 and let λ, $\lambda' \in F^\times$. The Albert algebras $J(A, \lambda)$ and $J(A, \lambda')$ are isomorphic if and only if $\lambda'\lambda^{-1} \in \mathrm{Nrd}_A(A^\times)$.*

We now prove the result of Parimala, Sridharan and Thakur [206] quoted in Remark (40.4).

(40.12) Theorem. *Let (B, τ) be a degree 3 central simple K-algebra with a unitary involution. Then $J(B, \tau, u_1, \nu_1) \simeq J(B, \tau, u_2, \nu_2)$ if and only if there exists some $w \in B^\times$ such that $u_2 = wu_1\tau(w)$ and $\nu_2 = \nu_1\, \mathrm{Nrd}_B(w)$.*

Proof: Let (u_1, ν_1), (u_2, ν_2) be admissible pairs. Recall from (39.17) the equivalence relation $\approx$ on admissible pairs. Assume that $J(B, \tau, u_1, \nu_1) \simeq J(B, \tau, u_2, \nu_2)$. By (40.10), we have some u_3 such that $(u_1, \nu_1) \approx (u_3, \nu_2)$ and by (40.3) $(u_3, \nu_2) \approx (u_2, \mathrm{Nrd}_B(w)^{-1}\nu_2)$ for some $w \in B^\times$ such that $u_2 = wu_3\tau(w)$. One has $\mathrm{Nrd}_B(u_3) = \nu_2\bar{\nu}_2 = \mathrm{Nrd}_B(u_3)$ since (u_3, ν_2) and (u_2, ν_2) are admissible pairs, thus $\lambda\bar{\lambda} = 1$ for $\lambda = \mathrm{Nrd}_B(w)$. Let $\tau_2 = \mathrm{Int}(u_2) \circ \tau$. By the next lemma applied to τ_2, there exists $w' \in B^\times$ such that $w'\tau_2(w') = 1$ and $\lambda = \mathrm{Nrd}_B(w')$. It follows from $w'\tau_2(w') = 1$ that $w'u_2\tau(w') = u_2$, hence

$$\left(u_2, \mathrm{Nrd}_B(w)^{-1}\nu_2\right) \approx \left(u_2, \mathrm{Nrd}_B(w')\, \mathrm{Nrd}_B(w)^{-1}\nu_2\right) = (u_2, \nu_2)$$

and $(u_1, \nu_1) \approx (u_2, \nu_2)$ as claimed. The converse is (39.2.1).   □

(40.13) Lemma. *Let (B, τ) be a degree 3 central simple K-algebra with a unitary involution. Let $w \in B^\times$ be such that $\lambda = \mathrm{Nrd}_B(w) \in K$ satisfies $\lambda\bar{\lambda} = 1$. Then there exists $w' \in B^\times$ such that $w'\tau(w') = 1$ and $\lambda = \mathrm{Nrd}_B(w')$.*

Proof: Assume that an element w' as desired exists and assume that $M = K[w'] \subset B$ is a field. We have $\tau(M) = M$, so let $H = M^\tau$ be the subfield of fixed elements under τ. The extension M/H is of degree 2 and by Hilbert's Theorem 90 (29.2) we may write $w' = u\tau(u)^{-1}$. Since $M = H \otimes K$ and $K = F(\sqrt{a})$ for some $a \in F^\times$, we may choose u of the form $u = h + \sqrt{a}$ with $h \in H$. Then

$$\lambda = \mathrm{Nrd}(w') = \mathrm{Nrd}_B(h + \sqrt{a})\, \mathrm{Nrd}_B(h - \sqrt{a})^{-1}.$$

On the other hand $\lambda = y\tau(y)^{-1}$ by Hilbert's Theorem 90 (29.2), so that $h \in \mathcal{H}(B, \tau)$ is a zero of

$$\varphi(h) = \left(y\, \mathrm{Nrd}_B(h - \sqrt{a}) - \bar{y}\, \mathrm{Nrd}_B(h + \sqrt{a})\right)\sqrt{a}^{-1}.$$

(the factor $\sqrt{a}^{-1}$ is to get an F-valued function on $\mathcal{H}(B, \tau)$). Reversing the argument, if φ is isotropic on $\mathcal{H}(B, \tau)$, then $w' = (h + \sqrt{a})(h - \sqrt{a})^{-1}$ is as desired. The function φ is polynomial of degree 3 and it is easily seen that φ is isotropic over K. It follows that φ is isotropic over F (see Exercise 1 of this chapter).   □

(40.14) Remark. Suresh has extended Lemma (40.13) to algebras of odd degree with unitary involution (see [206], see also Exercise 12, (b) in Chapter III).

Theorem (40.4) has a nice Skolem-Noether type application, which is also due to Parimala, Sridharan, and Thakur [206]:

(40.15) Corollary. *Let* (B_1, τ_1), (B_2, τ_2) *be degree* 3 *central simple algebras over* K *with unitary involutions. Suppose that* $\mathcal{H}(B_1, \tau_1)$ *and* $\mathcal{H}(B_2, \tau_2)$ *are subalgebras of an Albert algebra* J *and that* $\alpha\colon \mathcal{H}(B_1, \tau_1) \simeq \mathcal{H}(B_2, \tau_2)$ *is an isomorphism of Jordan algebras. Then* α *extends to an automorphism of* J.

Proof: In view of Theorem (39.18.2), we have isomorphisms

$$\psi_1\colon J(B_1, \tau_1, u_1, \nu_1) \xrightarrow{\sim} J, \quad \psi_2\colon J(B_2, \tau_2, u_2, \nu_2) \xrightarrow{\sim} J.$$

By Proposition (37.6) α extends to an isomorphism $\widetilde{\alpha}\colon (B_1, \tau_1) \xrightarrow{\sim} (B_2, \tau_2)$, thus we get an isomorphism of Jordan algebras

$$J(\alpha)\colon J(B_1, \tau_1, u_1, \nu_1) \xrightarrow{\sim} J\big(B_2, \tau_2, \alpha(u_1), \nu_1\big).$$

But $J\big(B_2, \tau_2, \alpha(u_1), \nu_1\big) \simeq J(B_2, \tau_2, u_2, \nu_2)$, since both are isomorphic to J. By Theorem (40.12), there exists $w \in B_2^\times$ such that $u_2 = w\alpha(u_1)\tau_2(w)$ and $\nu_2 = \mathrm{Nrd}_B(w)\nu_1$. Let

$$\phi\colon J\big(B_2, \tau_2, \alpha(u_1), \nu_1\big) \xrightarrow{\sim} J(B_2, \tau_2, u_2, \nu_2)$$

be given by $(a, b) \mapsto (a, bw)$. Then ϕ restricts to the identity on $\mathcal{H}(B_2, \tau_2)$ and

$$\psi = \psi_2 \circ \phi \circ J(\alpha) \circ \psi_1^{-1}$$

is an automorphism of J extending α. $\square$

40.A. Invariants of twisted compositions. Let F be a field of characteristic not 2. To a twisted composition (V, L, Q, β) we may associate the following cohomological invariants:

(a) a class $f_1 = [\delta] \in H^1(F, \mu_2)$, which determines the discriminant Δ of L;
(b) a class $g_1 \in H^1\big(F, (\mathbb{Z}/3\mathbb{Z})_\delta\big)$ which determines L (with the fixed discriminant Δ given by the cocycle δ);
(c) invariants $f_3 \in H^3(F, \mu_2)$, $f_5 \in H^5(F, \mu_2)$, and $g_3 \in H^3(F, \mathbb{Z}/3\mathbb{Z})$ which are the cohomological invariants associated with the Freudenthal algebra $J(L, V)$ (see Theorem (38.6)).

As for Freudenthal algebras, it is unknown if these invariants classify twisted compositions, however:

(40.16) Proposition. *The invariant* g_3 *of a twisted composition* (V, L, Q, β) *is trivial if and only if* (V, L, Q, β) *is similar to a composition* $\Gamma(\mathfrak{C}, L)$ *of type* G_2, *in which case* (V, L, Q, β) *is classified up to similarity by* f_1 *and* g_1 *(which determine* L*) and by* f_3 *(which determines* $\mathfrak{C}$*).*

Proof: By Theorem (38.8) $J(V, L)$ has zero divisors if and only if (V, L, Q, β) is similar to a composition of type G_2, hence the claim by Theorem (40.8). $\square$

§41. Exceptional Simple Lie Algebras

There exists a very explicit construction, due to Tits [289], of models for all exceptional simple Lie algebras. This construction is based on alternative algebras of degree 2 or 1 and Jordan algebras of degree 3 or 1. We sketch it and refer to [289], to the book of Schafer [244] or to the notes of Jacobson [138] for more details. We assume throughout that the characteristic of F is different from 2 and 3. Let A, B be Hurwitz algebras over F and let $J = \mathcal{H}_3(B, \alpha)$ be the Freudenthal algebra associated to B and $\alpha = \operatorname{diag}(\alpha_1, \alpha_2, \alpha_3)$. As usual we write A^0, resp. J^0 for the trace zero elements in A, resp. J. We define a bilinear product $*$ in A^0 by

$$a * b = ab - \tfrac{1}{2}T(a, b)$$

where $T(a, b) = T_A(ab)$, $a, b \in A$, is the bilinear trace form of A. Let ℓ_a, resp. $r_a \in \operatorname{End}_F(A)$ be the left multiplication map $\ell_a(x) = ax$, resp. the right multiplication map $r_a(x) = xa$. For $f, g \in \operatorname{End}_F(A)$ we put $[f, g] = f \circ g - g \circ f$ for the Lie product in $\operatorname{End}_F(A)$. It can be checked that in any alternative algebra A

$$D_{a,b} = [\ell_a, \ell_b] + [\ell_a, r_b] + [r_a, r_b]$$

is a derivation. Similarly we may define a product on J^0:

$$x * y = xy - \tfrac{1}{3}T(x, y)$$

where $T(x, y) = T_J(xy)$. We now define a bilinear and skew-symmetric product $[\,,\,]$ on the direct sum of F-vector spaces

$$\mathfrak{L}(A, J) = \operatorname{Der}(A, A) \oplus A^0 \otimes J^0 \oplus \operatorname{Der}(J, J)$$

as follows:

(1) $[\,,\,]$ is the usual Lie product in $\operatorname{Der}(A, A)$ and $\operatorname{Der}(J, J)$ and satisfies $[D, D'] = 0$ for $D \in \operatorname{Der}(A, A)$ and $D' \in \operatorname{Der}(J, J)$,

(2) $[a \otimes x, D + D'] = D(a) \otimes x + a \otimes D'(x)$ for $a \in A^0$, $x \in J^0$, $D \in \operatorname{Der}(A, A)$ and $D' \in \operatorname{Der}(J, J)$,

(3) $[a \otimes x, b \otimes y] = \tfrac{1}{12}T(x, y)D_{a,b} + (a * b) \otimes (x * y) + \tfrac{1}{2}T(a, b)[r_x, r_y]$ for $a, b \in B^0$ and $x, y \in J^0$.

With this product $\mathfrak{L}(A, J)$ is a Lie algebra. As A and B vary over the possible composition algebras the types of $\mathfrak{L}(A, J)$ can be displayed in a table, whose last four columns are known as Freudenthal's "magic square":

dim A	F	$F \times F \times F$	$\mathcal{H}_3(F, \alpha)$	$\mathcal{H}_3(K, \alpha)$	$\mathcal{H}_3(Q, \alpha)$	$\mathcal{H}_3(\mathfrak{C}, \alpha)$
1	0	0	A_1	A_2	C_3	F_4
2	0	$\mathfrak{U}$	A_2	$A_2 \oplus A_2$	A_5	E_6
4	A_1	$A_1 \oplus A_1 \oplus A_1$	C_3	A_5	D_6	E_7
8	G_2	D_4	F_4	E_6	E_7	E_8

Here K stands for a quadratic étale algebra, Q for a quaternion algebra and $\mathfrak{C}$ for a Cayley algebra; $\mathfrak{U}$ is a 2-dimensional abelian Lie algebra. The fact that D_4 appears in the last row is one more argument for considering D_4 as exceptional.

EXERCISES

1. (Springer [268, p. 63]) A cubic form over a field is isotropic if and only if it is isotropic over a quadratic extension.

2. For any alternative algebra A over a field of characteristic not 2, A^+ is a special Jordan algebra.

3. Show that in all cases considered in §37.A, §37.B, and §37.C the generic norm $N_J(\sum_i x_i u_i)$ is irreducible in $F[x_1, \ldots, x_n]$.

4. Let C be a Hurwitz algebra. Show that $\mathcal{H}_2(\mathfrak{C}, \alpha)$ is the Jordan algebra of a quadratic form.

5. Show that a Jordan division algebra of degree 2 is the Jordan algebra $J(V, q)$ of a quadratic form (V, q) such that $b_q(x, x) \neq 1$ for all $x \in V$.

6. Let J be a cubic Jordan structure. The following conditions are equivalent:
   (a) J contains an idempotent (i.e., an element e with $e^2 = e$) such that $S_J(e) = 1$.
   (b) J contains a nontrivial idempotent e.
   (c) J contains nontrivial zero divisors.
   (d) There is some nonzero $a \in J$ such that $N_J(a) = 0$.
   (e) There is some nonzero $a \in J$ such that $a^{\#} = 0$.

7. (a) Let A be a cubic separable alternative algebra and let $\lambda \in F^{\times}$. Check that the norm N_A induces a cubic structure $J(A, \lambda)$ on $A \oplus A \oplus A$.
   (b) Show that $J(A, \lambda) \simeq \mathcal{H}_3(C, \alpha)$ for some α if $A = F \times C$, with C a Hurwitz algebra over F.

8. Let A be central simple of degree 3, $\lambda \in F^{\times}$, and $J = J(A, \lambda)$ the corresponding first Tits construction. Put:
$$V(J) = \{ [\xi] \in \mathbb{P}(J) \mid \xi^{\#} = 0 \}$$
$$\mathrm{PGL}_1(A) = \{ [x] \in \mathbb{P}(A) \mid \mathrm{Nrd}_A(x) \neq 0 \}$$
$$\mathrm{SL}_1(A)^{\lambda} = \{ x \in A \mid \mathrm{Nrd}_A(x) = \lambda \}.$$
Show that
   (a) $V(J)$ is the projective variety with coordinates $[a, b, c] \in \mathbb{P}(A \oplus A \oplus A)$ and equations
$$a^{\#} = bc, \quad b^{\#} = \lambda^{-1}ca, \quad c^{\#} = \lambda ab$$
   and $V(J)$ is smooth.
   (b) The open subvariety U of $V(J)$ given by
$$\mathrm{Nrd}_A(a) \, \mathrm{Nrd}_A(b) \, \mathrm{Nrd}_A(c) \neq 0$$
   is parametrized by coordinates
$$[a, b] \in \mathbb{P}(A \oplus A)$$
   with $\mathrm{Nrd}_A(a) = \lambda \, \mathrm{Nrd}_A(b)$ and $\mathrm{Nrd}_A(a) \, \mathrm{Nrd}_A(b) \neq 0$.
   (c) $\mathrm{SL}_1(A)^{\lambda} \times \mathrm{PGL}_1(A)$ is an open subset of $V(J)$.
   (d) $\mathrm{SL}_1(A) \times \mathrm{PGL}_1(A)$ and $\mathrm{SL}_3(F) \times \mathrm{PGL}_3(F)$ are birationally equivalent (and rational). *Hint*: Use that $J(A, 1) \simeq J(M_3(F), 1)$.

9. Show that a special Jordan central division algebra over $\mathbb{R}$ is either isomorphic to $\mathbb{R}$ or to to the Jordan algebra of a negative definite quadratic form of dimension ≥ 2 over $\mathbb{R}$.

10. Let J be an Albert algebra over F. Show that:
    (a) J is split if F is finite or p-adic.
    (b) J is reduced if $F = \mathbb{R}$ or if F is a field of algebraic numbers (Albert-Jacobson [14]).

11. Let $\mathfrak{C}_a$ be the nonsplit Cayley algebra over $\mathbb{R}$. Show that the Albert algebras $\mathcal{H}_3(\mathfrak{C}_a, 1)$, $\mathcal{H}_3(\mathfrak{C}_a, \mathrm{diag}(1, -1, 1))$, and $\mathcal{H}_3(\mathfrak{C}_s, \mathrm{diag}(1, -1, 1))$ are up to isomorphism all Albert algebras over $\mathbb{R}$.

12. Let F be a field of characteristic not 2 and $J = \mathcal{H}_3(\mathfrak{C}, 1)$, $J_1 = \mathcal{H}_3(Q, 1)$, $J_2 = \mathcal{H}_3(K, 1)$ and $J_3 = \mathcal{H}_3(F, 1)$ for $\mathfrak{C}$ a Cayley algebra, Q a quaternion algebra, and $K = F(i)$, $i^2 = a$, a quadratic étale algebra. Show that
    (a) $\mathrm{Aut}_F(J/J_1) \simeq \mathrm{SL}_1(Q)$.

(b) $\text{Aut}_F(J/J_2) \simeq \text{SU}(M,h)$ where $M = K^\perp \subset \mathfrak{C}$ (with respect to the norm) and

$$h(x,y) = N_{\mathfrak{C}}(x,y) + a^{-1}iN_{\mathfrak{C}}(ix,y).$$

In particular $\text{Aut}_F(J/J_2) \simeq \text{SL}_3(F)$ if $K = F \times F$.

(c) $\text{Aut}_F(J/J_3) \simeq \text{Aut}_F(\mathfrak{C})$.

(d) $\text{Aut}_F(J_1) \times \text{SL}_1(Q) \hookrightarrow \text{Aut}_F(J)$ ("$C_3 \times A_1 \subset F_4$").

(e) $\text{Aut}_F(J_2) \times \text{SU}(M,h) \hookrightarrow \text{Aut}_F(J)$ ("$A_2 \times A_2 \subset F_4$").

(f) $\text{Aut}_F(J_3) \times \text{SL}_2(F) \hookrightarrow \text{Aut}_F(J)$ ("$G_2 \times A_1 \subset F_4$").

(g) Let

$$V = \left\{ \begin{pmatrix} 0 & 0 & 0 \\ 0 & x & c \\ 0 & \bar{c} & -x \end{pmatrix} \;\middle|\; x \in F, c \in \mathfrak{C} \right\} \subset J.$$

Show that $\text{Aut}_F(J/F \cdot E_{11}) \simeq \text{Spin}_9(V, T|_V)$. ("$B_4 \subset F_4$").

(h) $\text{Aut}_F(J/F \cdot E_{11} \oplus F \cdot E_{22} \oplus F \cdot E_{33}) \simeq \text{Spin}(\mathfrak{C}, \mathfrak{n})$.

Observe that (d), (e), and (g) give the possible types of maximal subgroups of maximal rank for F_4.

13. (Parimala, Sridharan, Thakur) Let $J = J(B, \tau, u\nu)$ be a second Tits construction with $B = M_3(K)$ and $u \in B$ such that $\text{Nrd}_B(u) = 1$. Let $\langle u \rangle_K$ be the hermitian form of rank 3 over K determined by u and let $\mathfrak{C} = C(\langle u \rangle_K, K)$ be the corresponding Cayley algebra, as given in Exercise 6 of Chapter VIII. Show that the class of $\mathfrak{C}$ is the f_3-invariant of J.

NOTES

§37. The article of Paige [203] provides a nice introduction to the theory of Jordan algebras. Jacobson's treatise [136] gives a systematic presentation of the theory over fields of characteristic not 2. Another important source is the book of Braun-Koecher [55] and a forthcoming source is a book by McCrimmon [181]. If 2 is not invertible the Jordan identity (37.1) is unsuitable and a completely new characteristic-free approach was initiated by McCrimmon [183]. The idea is to axiomatize the quadratic product aba instead of the Jordan product $a \cdot b = \frac{1}{2}(ab+ba)$. McCrimmon's theory is described for example in Jacobson's lecture notes [137] and [141]. Another approach to Jordan algebras based on an axiomatization of the notion of inverse is provided in the book of Springer [269]. The treatment in degree 3 is similar to that given by McCrimmon for exceptional Jordan algebras (see [269, §5]). A short history of Jordan algebras can be found in Jacobson's obituary of Albert [139], and a survey for non-experts is given in the paper by McCrimmon [187]. For more recent developments by the Russian School, especially on infinite dimensional Jordan algebras, see McCrimmon [188].

A complete classification of finite dimensional simple formally real Jordan algebras over $\mathbb{R}$ appears already in Jordan, von Neumann, and Wigner [147][35]. They conjectured that $\mathcal{H}_3(\mathfrak{C}, 1)$ is exceptional and proposed it as a problem to Albert. Albert's solution appeared as a sequel [4] to their paper. Much later, Albert again took up the theory of Jordan algebras; in [6] he described the structure of simple

---

[35] A Jordan algebra is said to be *formally real* if $\sum a_i^2 = 0$ implies every $a_i = 0$.

Jordan algebras over algebraically closed fields of characteristic not 2, assuming that the algebras admit an idempotent different from the identity. (The existence of an identity in a simple finite dimensional Jordan algebra was showed by Albert in [7].) The gap was filled by Jacobson in [130]. In [242] Schafer showed that reduced exceptional Jordan algebras of dimension 27 are all of the form $\mathcal{H}_3(\mathfrak{C}, \alpha)$. A systematic study of algebras $\mathcal{H}_3(\mathfrak{C}, \alpha)$ is given in Freudenthal's long paper [101], for example the fact that they are of degree 3. In the same paper Freudenthal showed that the automorphism group of a reduced simple exceptional Jordan algebra over $\mathbb{R}$ is of type F_4 by computing the root system explicitly. In a different way, Springer [267, Theorem 3] or [268], showed that the automorphism group is simply connected of dimension 52, assuming that F is a field of characteristic different from 2 and 3, and deduced its type using the classification of simple algebraic groups. The fact that the derivation algebra of an exceptional Jordan algebra is a Lie algebra of type F_4 can already be found in Chevalley-Schafer [67]. Here also it was assumed that the base field has characteristic different from 2 and 3. Observe that this Lie algebra is not simple in characteristic 3. Split simple groups of type E_6 also occur in connection with simple split exceptional Jordan algebras, namely as automorphism groups of the cubic form N, see for example Chevalley-Schafer [67], Freudenthal [101] and Jacobson [133].

The structure of algebras $\mathcal{H}_3(\mathfrak{C}, \alpha)$ over fields of characteristic not 2, 3 was systematically studied by Springer in a series of papers ([263], [264], [265], and [266]). Some of the main results are the fact that the bilinear trace form and $\mathfrak{C}$ determine $\mathcal{H}_3(\mathfrak{C}, \alpha)$ (Theorem (37.13), see [265, Theorem 1, p. 421]) and the fact that the cubic norm determines $\mathfrak{C}$ (see [264, Theorem 1]). Thus the cubic norm and the trace form determine the algebra. The fact that the trace form alone determines the algebra was noticed in Serre [258, §9.2, §9.4]. The fact that the isomorphism class of $\mathfrak{C}$ is determined by the isomorphism class of $\mathcal{H}_3(\mathfrak{C}, \alpha)$ is a result due to Albert-Jacobson [14]. For this reason $\mathfrak{C}$ is usually called the *coordinate algebra* of $\mathcal{H}_3(\mathfrak{C}, \alpha)$. A recent survey of the theory of Albert algebras has been given by Petersson and Racine [213].

It is unknown if a division Albert algebra J always contains a cyclic cubic field extension (as does an associative central simple algebra of degree 3). However this is true (Petersson-Racine [209, Theorem 4]) if char $F \neq 3$ and F contains a primitive cube root of unity: it suffices to show that J contains a Kummer extension $F[X]/(X^3 - \lambda)$, hence that S_J restricted to $J^0 = \{\, x \in J \mid T_J(x) = 0 \,\}$ is isotropic. In view of Springer's theorem, we may replace J by $J \otimes L$ where L is a cubic étale subalgebra of J. But then J is reduced and then $S_J|J^0$ is isotropic.

§38. There are a number of characterizations of cubic Jordan algebras. One is due to Springer [263], assuming that char $F \neq 2, 3$: Let J be a finite dimensional F-algebra with 1, equipped with a quadratic form Q such that

(a) $Q(x)^2 = Q(x^2)$ if $b_Q(x, 1) = 0$,
(b) $b_Q(xy, z) = b_Q(x, yz)$,
(c) $Q(1) = \frac{3}{2}$.

Then J is a cubic Jordan algebra and $Q(x) = \frac{1}{2}T_J(x^2)$.

The characterization we use in §38 was first suggested by Freudenthal [100] and was established by Springer [266] for fields of characteristic not 2 and 3. We follow the description of McCrimmon [185], which is used by Petersson-Racine in their study of cubic Jordan algebras (see for example [212] and [211]). The Springer decomposition is given in the Göttinger notes of Springer [268]. Applications were given by Walde [303] to the construction of exceptional Lie algebras. The construction was formalized and applied to cubic forms by Petersson and Racine (see for example [210]).

§39. Tits constructions for fields of characteristic not 2 first appeared in print in Jacobson's book [136], as did the fact, also due to Tits, that any Albert algebra is a first or second Tits construction. These results were announced by Tits in a talk at the Oberwolfach meeting "Jordan-Algebren und nicht-assoziativen Algebren, 17–26.8.1967." With the kind permission of J. Tits and the Research Institute in Oberwolfach, we reproduce Tits' Résumé:

Exceptional simple Jordan Algebras

(I) Denote by k a field of characteristic not 2, by A a central simple algebra of degree 3 over k, by $n \colon A \to A$, $tr \colon A \to A$ the reduced norm and reduced trace, and by $\times \colon A \times A \to A$ the symmetric bilinear product defined by $(x \times x)x = n(x)$. For $x \in A$, set $\bar{x} = \frac{1}{2}(tr(x) - x)$. Let $c \in k^*$. In the sum $A_0 + A_1 + A_2$ of three copies of A, introduce the following product:

	x_0	y_1	z_2
x_0'	$\frac{1}{2}(xx' + x'x)_0$	$(\overline{x'y})_1$	$(z\overline{x'})_2$
y_1'	$(\overline{x}y')_1$	$c(y \times y')_2$	$(\overline{y'z})_0$
z_2'	$(z'\overline{x})_2$	$(\overline{yz'})_0$	$\frac{1}{c}(z \times z')_1$

(II) Denote by ℓ a quadratic extension of k, by B a central simple algebra of degree 3 over ℓ, and by $\sigma \colon B \to B$ an involution of the second kind kind such that $k = \{x \in \ell \mid x^\sigma = x\}$. Set $B^{\mathrm{Sym}} = \{x \in B \mid x^\sigma = x\}$. Let $b \in B^{\mathrm{Sym}}$ and $c \in l^*$ be such that $n(b) = c^\sigma c$. In the sum $B^{\mathrm{Sym}} + B_*$ of B^{Sym} and a copy B_* of B, define a product by

	x	y
x'	$\frac{1}{2}(xx' + x'x)$	$(\overline{x'y})_*$
y'	$(\overline{x}y')_*$	$(\overline{yby'^\sigma + y'by^\sigma}) + \left(c^\sigma(y^\sigma \times y'^\sigma)b^{-1}\right)_*$

Theorem 1. *The 27-dimensional algebras described under* (I) *and* (II) *are exceptional simple Jordan algebras over* k. *Every such algebra is thus obtained.*

Theorem 2. *The algebra* (I) *is split if* $c \in n(A)$ *and division otherwise. The algebra* (II) *is reduced if* $c \in n(B)$ *and division otherwise.*

Theorem 3. *There exists an algebra of type* (II) *which does not split over any cyclic extension of degree 2 or 3 of* k. *(Notice that such an algebra is necessarily division and is not of type* (I)*).*

(For more details, cf. N. Jacobson. Jordan algebras, a forthcoming book).

J. Tits

Observe that the ×-product used by Tits is our ×-product divided by 2. The extension of Tits constructions to cubic structures was carried out by McCrimmon [186]. Tits constructions were systematically used by Petersson and Racine, see for example [212] and [211]. Petersson and Racine showed in particular that (with a few exceptions) simple cubic Jordan structures can be constructed by iteration of the Tits process ([211], Theorem 3.1). The result can be viewed as a cubic analog to the theorem of Hurwitz, proved by iterating the Cayley-Dickson process.

Tits constructions can be used to give simple examples of exceptional division Jordan algebras of dimension 27. The first examples of such division algebras were constructed by Albert [8]. They were significantly more complicated than those through Tits constructions. Assertions (2) and (3) of Theorem (39.14) and (39.18) are due to Tits. The nice cohomological proof given here is due to Ferrar and Petersson [97] (for first Tits constructions).

§40. The existence of the invariants f_3 and f_5 was first noticed by Serre (see [258] and [257, Fith Edition, Annexe, §3]). The direct computation of the trace form given here, as well as Propositions (40.5) and (40.7) are due to Rost. Serre suggested the existence of the invariant g_3. Its definition is due to Rost [234]. An elementary approach to that invariant can be found in Petersson-Racine [215] and a description in characteristic 3 can be found in Petersson-Racine [216].

CHAPTER X

Trialitarian Central Simple Algebras

We assume in this chapter that F is a field of characteristic not 2. Triality for $\mathbf{PGO}_8^+$, i.e., the outer action of S_3 on $\mathbf{PGO}_8^+$ and its consequences, is the subject of this chapter. In the first section we describe the induced action on $H^1(F, \mathbf{PGO}_8^+)$. This cohomology set classifies ordered triples (A, B, C) of central simple algebras of degree 8 with involutions of orthogonal type such that $\big(C(A, \sigma_A), \underline{\sigma}\big) \xrightarrow{\sim} (B, \sigma_B) \times (C, \sigma_C)$. Triality implies that this property is symmetric in A, B and C, and the induced action of S_3 on $H^1(F, \mathbf{PGO}_8^+)$ permutes A, B, and C. As an application we give a criterion for an orthogonal involution on an algebra of degree 8 to decompose as a tensor product of three involutions.

We may view a triple (A, B, C) as above as an algebra over the split étale algebra $F \times F \times F$ with orthogonal involution $(\sigma_A, \sigma_B, \sigma_C)$ and some additional structure (the fact that $\big(C(A, \sigma_A), \underline{\sigma}\big) \xrightarrow{\sim} (B, \sigma_B) \times (C, \sigma_C)$). Forms of such "algebras," called trialitarian algebras, are classified by $H^1(F, \mathbf{PGO}_8^+ \rtimes S_3)$. Trialitarian algebras are central simple algebras with orthogonal involution of degree 8 over cubic étale F-algebras with a condition relating the central simple algebra and its Clifford algebra. Connected components of automorphism groups of such trialitarian algebras give the outer forms of simple adjoint groups of type D_4.

Trialitarian algebras also occur in the construction of Lie algebras of type D_4. Some indications in this direction are given in the last section.

§42. Algebras of Degree 8

42.A. Trialitarian triples. The pointed set $H^1(F, \mathbf{PGO}_8)$ classifies central simple algebras of degree 8 over F with an involution of orthogonal type and the image of the pointed set $H^1(F, \mathbf{PGO}_8^+)$ in $H^1(F, \mathbf{PGO}_8)$ classifies central simple algebras of degree 8 over F with an involution of orthogonal type having trivial discriminant (see Remark (29.31)). More precisely, each cocycle x in $H^1(F, \mathbf{PGO}_8^+)$ determines a central simple F-algebra $A(x)$ of degree 8 with an orthogonal involution $\sigma_{A(x)}$ having trivial discriminant, together with a designation of the two components $C^+\big(A(x), \sigma_{A(x)}\big)$ and $C^-\big(A(x), \sigma_{A(x)}\big)$ of the Clifford algebra $C\big(A(x), \sigma_{A(x)}\big)$. Thus, putting $\big(B(x), \sigma_{B(x)}\big) = \big(C^+\big(A(x), \sigma_{A(x)}\big), \underline{\sigma}\big)$ and $\big(C(x), \sigma_{C(x)}\big) = \big(C^-\big(A(x), \sigma_{A(x)}\big), \underline{\sigma}\big)$, x determines an ordered triple

$$\big[\big(A(x), \sigma_{A(x)}\big), \big(B(x), \sigma_{B(x)}\big), \big(C(x), \sigma_{C(x)}\big)\big]$$

of central simple F-algebras of degree 8 with orthogonal involution. The two components of the Clifford algebra $C\big(A(x), \sigma_{A(x)}\big)$ are determined by a nontrivial central idempotent e, say $B(x) = C\big(A(x), \sigma_{A(x)}\big)e$ and $C(x) = C\big(A(x), \sigma_{A(x)}\big)(1 - e)$. Thus two triples $[(A, \sigma_A), (B, \sigma_B), (C, \sigma_C)]$ and $[(A', \sigma_{A'}), (B', \sigma_{B'}), (C', \sigma_{C'})]$, where $B = C(A, \sigma_A)e$ and $C = C(A, \sigma_A)(1 - e)$, resp. $B' = C(A', \sigma_A)e'$ and

$C = C(A', \sigma_{A'})(1 - e')$, determine the same class in $H^1(F, \mathbf{PGO}_8^+)$ if there exists an isomorphism $\phi \colon (A, \sigma_A) \to (A', \sigma_{A'})$ such that $C(\phi)(e) = e'$. Now let (A, B, C) be an ordered triple of central simple algebras of degree 8 with orthogonal involution such that there exists an isomorphism

$$\alpha_A \colon \big(C(A, \sigma_A), \underline{\sigma}\big) \xrightarrow{\sim} (B, \sigma_B) \times (C, \sigma_C).$$

We call such a triple (A, B, C) a *trialitarian triple*. The element $e = \alpha_A^{-1}\big((1, 0)\big)$ is a central idempotent of (A, σ_A), hence determines a designation of the two components of $C(A, \sigma_A)$. Moreover this designation is independent of the particular choice of α_A, since it depends only on the ordering of the triple. Two trialitarian triples (A, B, C) and (A', B', C') are called isomorphic if there exist isomorphisms of algebras with involution

$$(\phi_1 \colon A \xrightarrow{\sim} A', \ \phi_2 \colon B \xrightarrow{\sim} B', \ \phi_3 \colon C \xrightarrow{\sim} C')$$

and isomorphisms α_A, resp. $\alpha_{A'}$ as above, such that

$$\alpha_{A'} \circ C(\phi_1) = (\phi_2, \phi_3) \circ \alpha_A.$$

We have:

(42.1) Lemma. *The set $H^1(F, \mathbf{PGO}_8^+)$ classifies isomorphism classes of trialitarian triples.* □

Observe that the ordered triples (A, B, C) and (A, C, B) determine in general different objects in $H^1(F, \mathbf{PGO}_8^+)$ since they correspond to different designations of the components of $C(A, \sigma_A)$. In fact the action of S_2 on $H^1(F, \mathbf{PGO}_8^+)$ induced by the exact sequence of group schemes

$$1 \to \mathbf{PGO}_8^+ \to \mathbf{PGO}_8 \xrightarrow{d} S_2 \to 1$$

permutes the classes of (A, B, C) and (A, C, B).

(42.2) Example. Let $A_1 = \mathrm{End}_F(\mathfrak{C})$ and $\sigma_1 = \sigma_{\mathfrak{n}}$ where $\mathfrak{C}$ is a split Cayley algebra with norm $\mathfrak{n}$. In view of proposition (35.1) we have a canonical isomorphism

$$\alpha_{\mathfrak{C}} \colon \big(C(A_1, \sigma_1), \underline{\sigma}\big) \to (A_2, \sigma_2) \times (A_3, \sigma_3)$$

where (A_2, σ_2) and (A_3, σ_3) are copies of the split algebra (A_1, σ_1). Thus the ordered triple (A_1, A_2, A_3) determines a class in $H^1(F, \mathbf{PGO}_8^+)$. Since $\mathfrak{n}$ is hyperbolic, it corresponds to the trivial class.

The group S_3 acts as outer automorphisms on the group scheme $\mathbf{PGO}_8^+$ (see Proposition (35.15)). It follows that S_3 acts on $H^1(F, \mathbf{PGO}_8^+)$.

(42.3) Proposition. *The action of S_3 on $H^1(F, \mathbf{PGO}_8^+)$ induced by the action of S_3 on $\mathbf{PGO}_8^+$ is by permutations on the triples (A, B, C). More precisely, the choice of an isomorphism*

$$\alpha_A \colon \big(C(A, \sigma_A), \underline{\sigma}\big) \xrightarrow{\sim} (B, \sigma_B) \times (C, \sigma_C)$$

determines isomorphisms

$$\alpha_B \colon \big(C(B, \sigma_B), \underline{\sigma}\big) \xrightarrow{\sim} (C, \sigma_C) \times (A, \sigma_A),$$
$$\alpha_C \colon \big(C(C, \sigma_C), \underline{\sigma}\big) \xrightarrow{\sim} (A, \sigma_A) \times (B, \sigma_B).$$

Moreover any one of the three α_A, α_B or α_C determines the two others.

Proof: We have $\mathbf{PGO}_8^+(F_{\text{sep}}) = \mathbf{PGO}^+(\mathfrak{C}_s, \mathfrak{n}_s)(F_{\text{sep}})$ and we can use the description of the action of S_3 on $\mathbf{PGO}^+(\mathfrak{C}_s, \mathfrak{n}_s)(F_{\text{sep}})$ given in Proposition (35.6). Let θ and $\theta^{\pm}$ be the automorphisms of $\mathbf{PGO}^+(\mathfrak{C}_s, \mathfrak{n}_s)$ as defined in (35.6). Let $x = (\gamma_g)_{g \in \text{Gal}(F_{\text{sep}}/F)}$ with $\gamma_g = [t_g]$, $t_g \in \mathbf{O}_8^+(F_{\text{sep}})$, be a cocycle in $H^1(F, \mathbf{PGO}_8^+)$ which defines (A, σ_A). By definition of (θ^+, θ^-) the map $(\theta^+, \theta^-)\colon \mathbf{PGO}^+ \to \mathbf{PGO}^+ \times \mathbf{PGO}^+$ factors through $\mathbf{Aut}_{\text{alg}}\big(C_0^+(\mathfrak{n}), \underline{\sigma}\big) \times \mathbf{Aut}_{\text{alg}}\big(C_0^-(\mathfrak{n}), \underline{\sigma}\big)$. Hence the cocycle $\theta^+ x = \theta^+([t_g])$ defines the triple (B, σ_B, α_B) and $\theta^- x = \theta^-([t_g])$ defines (C, σ_C, α_C). The last assertion follows by triality. $\qquad\square$

(42.4) Example. In the situation of Example (42.2), where $A_1 = A_2 = A_3 = \text{End}_F(\mathfrak{C})$ and $\alpha_{A_1} = \alpha_{\mathfrak{C}}$ we obtain $\alpha_{A_2} = \alpha_{A_3} = \alpha_{\mathfrak{C}}$ since the trivial cocycle represents the triple (A_1, A_2, A_3).

For any $\phi \in S_3$, we write α_A^ϕ for the map α induced from α_A by the action of S_3. For example we have $\alpha_A^{\theta^+} = \alpha_B$.

(42.5) Proposition. *Let (A, B, C) be a trialitarian triple. Triality induces isomorphisms*

$$\mathbf{Spin}(A, \sigma_A) \simeq \mathbf{Spin}(B, \sigma_B) \simeq \mathbf{Spin}(C, \sigma_C),$$
$$\mathbf{PGO}^+(A, \sigma_A) \simeq \mathbf{PGO}^+(B, \sigma_B) \simeq \mathbf{PGO}^+(C, \sigma_C).$$

Proof: Let $\gamma = \gamma_g = [t_g]$, $t_g \in \mathbf{GO}_8^+(F_{\text{sep}})$, be a 1-cocycle defining (A, σ_A, α_A) so that $\gamma^+ = \theta^+ \gamma$ defines (B, σ_B, α_B). Since $\text{Int}\big(\mathbf{PGO}_8^+(F_{\text{sep}})\big) = \mathbf{PGO}_8^+(F_{\text{sep}})$ we may use γ to twist the Galois action on $\mathbf{PGO}_8^+$. The isomorphism

$$\theta^+\colon \mathbf{PGO}_8^+ \to (\mathbf{PGO}_8^+)_\gamma$$

then is a Galois equivariant map, which descends to an isomorphism

$$\mathbf{PGO}^+(A, \sigma_A) \xrightarrow{\sim} \mathbf{PGO}^+(B, \sigma_B).$$

The existence of an isomorphism between corresponding simply connected groups then follows from Theorem (26.7). $\qquad\square$

(42.6) Remark. If (A, σ) is central simple of degree $2n$ with an orthogonal involution, the space

$$\mathfrak{L}(A, \sigma) = \{\, x \in A \mid \sigma(x) = -x \,\}$$

of skew-symmetric elements is a Lie algebra of type D_n under the product $[x, y] = xy - yx$ (since it is true over a separable closure of F, see [140, Theorem 9, p. 302]). In fact $\mathfrak{L}(A, \sigma)$ is the Lie algebra of the groups $\mathbf{Spin}(A, \sigma)$ or $\mathbf{PGO}^+(A, \sigma)$ (see 23.B), so that Proposition (42.5) implies that

$$\mathfrak{L}(A, \sigma_A) \simeq \mathfrak{L}(B, \sigma_B) \simeq \mathfrak{L}(C, \sigma_C)$$

if A is of degree 8 and $C(A, \sigma_A) \xrightarrow{\sim} (B, \sigma_B) \times (C, \sigma_C)$. An explicit example where $(A, \sigma_A) \not\simeq (B, \sigma_B)$, but $\mathfrak{L}(A, \sigma_A) \simeq \mathfrak{L}(B, \sigma_B)$ is in Jacobson [140, Exercise 3, p. 316].

(42.7) Proposition. *Let (A, B, C) be a trialitarian triple. We have*
(1) $[A][B][C] = 1$ *in* $\text{Br}(F)$.
(2) $A \simeq \text{End}_F(V)$ *if and only if* $B \simeq C$.
(3) $A \simeq B \simeq C$ *if and only if* $(A, \sigma_A) \simeq \big(\text{End}_F(\mathfrak{C}), \sigma_{\mathfrak{n}}\big)$ *for some Cayley algebra* $\mathfrak{C}$ *with norm* $\mathfrak{n}$.

Proof: (1) is a special case of Theorem (9.12), see also Example (31.11), and (2) is an immediate consequence of (1).

The "if" direction of (3) is a special case of Proposition (35.1). For the converse, it follows from $[A] = [B] = [C] = 1$ in $\mathrm{Br}(F)$ that $(A, \sigma_A) = (\mathrm{End}_F(V), \sigma_q)$ and that (V, q) has trivial discriminant and trivial Clifford invariant. In view of Proposition (35.2) (V, q) is similar to the norm $\mathfrak{n}$ of a Cayley algebra $\mathfrak{C}$ over F. This implies $(A, \sigma_A) \simeq (\mathrm{End}_F(\mathfrak{C}), \sigma_\mathfrak{n})$. $\square$

(42.8) Remark. As observed by A. Wadsworth [297], there exist examples of trialitarian triples (A, B, C) such that all algebras A, B, C are division algebras: Since there exist trialitarian triples $(\mathrm{End}_F(V), B, B)$ with B a division algebra (see Dherte [73], Tao [281], or Yanchevskiĭ [319]), taking B to be generic with an involution of orthogonal type and trivial discriminant (see Saltman [241]) will give such triples.

42.B. Decomposable involutions. We consider central simple F-algebras of degree 8 with involutions of orthogonal type which decompose as a tensor product of three involutions. In view of Proposition (7.3) such involutions have trivial discriminant.

(42.9) Proposition. *Let A be a central simple F-algebra of degree 8 with an involution σ of orthogonal type. Then $(A, \sigma) \simeq (A_1, \sigma_1) \otimes (A_2, \sigma_2) \otimes (A_3, \sigma_3)$ with A_i, $i = 1, 2, 3$, quaternion algebras and σ_i an involution of the first kind on A_i, if and only if (A, σ) is isomorphic to $(C(q_0), \tau')$ where $C(q_0)$ is the Clifford algebra of a quadratic space (V_0, q_0) of rank 6 and τ' is the involution which is $-\mathrm{Id}$ on V_0.*

Proof: We first check that the Clifford algebra $C(q_0)$ admits such a decomposition. Let $q_0 = q_4 \perp q_2$ be an orthogonal decomposition of q_0 with q_4 of rank 4 and q_2 of rank 2. Accordingly, we have a decomposition $C(q_0) = C(q_4) \widehat{\otimes} C(q_2)$ where $\widehat{\otimes}$ is the $\mathbb{Z}/2\mathbb{Z}$-graded tensor product (see for example Scharlau [247, p. 328]). Let z be a generator of the center of $C_0(q_4)$ such that $z^2 = \delta_4 \in F^\times$. The map

$$\phi(x \widehat{\otimes} 1 + 1 \widehat{\otimes} y) = x \otimes 1 + z \otimes y$$

induces an isomorphism

$$C(q_0) = C(q_4) \widehat{\otimes} C(q_2) \xrightarrow{\sim} C(\delta_4 q_4) \otimes C(q_2)$$

by the universal property of the Clifford algebra. The canonical involution of $C(q_0)$ is transported by ϕ to the tensor product of the two canonical involutions, since z is invariant by the canonical involution of $C(q_4)$. Similarly, we may decompose q_4 as $q_4 = q' \perp q''$ and write

$$C(q_4) = C(q') \widehat{\otimes} C(q'') \xrightarrow{\sim} C(q') \otimes C(\delta' q'')$$

where δ' is the discriminant of q'. In this case the canonical involution of $C(q')$ maps a generator z' of the center of $C_0(q')$ such that $z'^2 = \delta' \in F^\times$ to $-z'$. We then have to replace the canonical involution of $C(q'')$ (which is of orthogonal type) by the "second" involution of $C(q'')$, i.e., the involution τ' such that $\tau'(x) = -x$ on V''. This involution is of symplectic type. Conversely, let

$$(A, \sigma) \simeq (A_1, \sigma_1) \otimes (A_2, \sigma_2) \otimes (A_3, \sigma_3).$$

Renumbering the algebras if necessary, we may assume that σ_1 is of orthogonal type and that there exists a quadratic space (V_1, q_1) such that $(A_1, \sigma_1) \simeq (C(q_1), \tau_1)$ with τ_1 the canonical involution of $C(q_1)$. We may next assume that σ_2 and σ_3 are of symplectic type: if σ_2 and σ_3 are both of orthogonal type, $\sigma_2 \otimes \sigma_3$ is of orthogonal type and has trivial discriminant by Proposition (7.3). Corollary (15.12) implies that

$$(A_2, \sigma_2) \otimes (A_3, \sigma_3) \simeq (B, \sigma_B) \otimes (C, \sigma_C)$$

where σ_B, σ_C are the canonical involutions of the quaternion algebras B, C, and we replace (A_2, σ_2) by (B, σ_B), (A_3, σ_3) by (C, σ_C). Then there exist quadratic forms q_2, q_3 such that $(A_2, \sigma_2) \simeq (C(q_2), \tau_2)$ and $(A_3, \sigma_3) \simeq (C(q_3), \tau_3)$, with τ_2, τ_3 "second involutions," as described above. Let δ_i be the discriminant of q_i and let $q_0 = q_3 \perp \delta_3 q_2 \perp \delta_3 \delta_2 q_1$, then $(A, \sigma) \simeq (C(q_0), \tau')$. $\qquad\square$

Algebras $(C(q_0), \tau')$ occur as factors in special trialitarian triples:

(42.10) Proposition. *A triple $(\operatorname{End}_F(V), B, B)$ is trialitarian if and only if*

$$(B, \sigma_B) \simeq (C(V_0, q_0), \tau'),$$

where τ' is the involution of $C(q_0)$ which is $-\operatorname{Id}$ on V_0, for some quadratic space (V_0, q_0) of dimension 6.

Proof: Let $(A, \sigma) = (\operatorname{End}_F(V), \sigma_q)$ be split of degree 8, so that $C(A, \sigma) = C_0(q)$, and assume that q has trivial discriminant. Replacing q by λq for some $\lambda \in F^\times$, if necessary, we may assume that q represents 1. Putting $q = \langle 1 \rangle \perp q_1$, we define an isomorphism $\rho \colon C(-q_1) \xrightarrow{\sim} C_0(q)$ by $\rho(x) = xv_1$ where v_1 is a generator of $\langle 1 \rangle$. Since the center Z of $C_0(q)$ splits and since $C(-q_1) \simeq Z \otimes C_0(-q_1)$, we may view ρ^{-1} as an isomorphism $C_0(q) \xrightarrow{\sim} C_0(-q_1) \times C_0(-q_1)$. The center of $C_0(q)$ is fixed under the canonical involution of $C_0(q)$ since $8 \equiv 0 \mod 4$. Thus, with the canonical involution on all three algebras, ρ^{-1} is an isomorphism of algebras with involution and the triple

$$(\operatorname{End}_F(V), C_0(-q_1), C_0(-q_1))$$

is a trialitarian triple. Let $-q_1 = \langle a \rangle \perp q_2$ with q_2 of rank 6 and let $q_0 = -aq_2$, then $C_0(-q_1) \simeq C(q_0)$ as algebras with involution where the involution on $C(q_0)$ is the "second involution." Thus the triple $(\operatorname{End}_F(V), C(q_0), C(q_0))$ is trialitarian. $\qquad\square$

We now characterize fully decomposable involutions on algebras of degree 8:

(42.11) Theorem. *Let A be a central simple F-algebra of degree 8 and σ an involution of orthogonal type on A. The following conditions are equivalent:*
(1) *$(A, \sigma) \simeq (A_1, \sigma_1) \otimes (A_2, \sigma_2) \otimes (A_3, \sigma_3)$ for some quaternion algebras with involution (A_i, σ_i), $i = 1$, 2, 3.*
(2) *The involution σ has trivial discriminant and there exists a trialitarian triple $(\operatorname{End}_F(V), A, A)$.*
(3) *The involution σ has trivial discriminant and one of the factors of $C(A, \sigma)$ splits.*

Proof: The algebra (A, σ) decomposes if and only if $(A, \sigma) \simeq C(q_0)$ by Lemma (42.9). Thus the equivalence of (1) and (2) follows from (42.10).

The equivalence of (2) and (3) follows from the fact that S_3 operates through permutations on trialitarian triples. $\qquad\square$

(42.12) Remark. (Parimala) It follows from Theorem (42.11) that the condition

$$[A][B][C] = 1 \in \operatorname{Br}(F)$$

for a trialitarian triple is necessary but not sufficient. In fact, there exist a field F and a central division algebra B of degree 8 with an orthogonal involution over F which is not a tensor product of three quaternion algebras (see Amitsur-Rowen-Tignol [20]). That such an algebra always admits an orthogonal involution with trivial discriminant follows from Parimala-Sridharan-Suresh [205]. Thus, by Theorem (42.11) there are no orthogonal involutions on B such that $(M_8(F), B, B)$ is a trialitarian triple.

(42.13) Remark. If (A, σ) is central simple with an orthogonal involution which is hyperbolic, then $\mathrm{disc}(\sigma)$ is trivial and one of the factors of the Clifford algebra $C(A, \sigma)$ splits (see (8.31)). These conditions are also sufficient for A to have an orthogonal hyperbolic involution if A has degree 4 (see Proposition (15.14.4)) but they are not sufficient if A has degree 8 by Theorem (42.11).

§43. Trialitarian Algebras

43.A. A definition and some properties. Let L be a cubic étale F-algebra. We call an L-algebra D a *central simple L-algebra* if $D \otimes F' \simeq A' \times B' \times C'$ with A', B', C' central simple over F' for every field extension F'/F which splits L. For example any trialitarian triple (A, B, C) is a central simple L-algebra with an involution of orthogonal type over the split cubic algebra $L = F \times F \times F$. Conversely, let L be a cubic étale F-algebra and let E be a central simple algebra of degree 8 with an involution of orthogonal type over L. We want to give conditions on E/L such that E defines a trialitarian triple over any extension which splits L. Such a structure will be called a *trialitarian algebra*. In view of the decomposition $L \otimes L \simeq L \times L \otimes \Delta$ where Δ is the discriminant algebra of L (see (18.28)), we obtain a decomposition

$$(E, \sigma) \otimes L \simeq (E, \sigma) \times (E_2, \sigma_2)$$

and (E_2, σ_2) is an $(L \otimes \Delta)$-central simple algebra with involution of degree 8 over $L \otimes \Delta$, in particular is an L-algebra through the canonical map $L \to L \otimes \Delta$, $\ell \mapsto \ell \otimes 1$. As a first condition we require the existence of an isomorphism of L-algebras with involution

$$\alpha_E \colon \big(C(E, \sigma), \underline{\sigma}\big) \xrightarrow{\sim} (E_2, \sigma_2).$$

Fixing a generator $\rho \in \mathrm{Gal}(L \otimes \Delta/\Delta)$, this is equivalent by Corollary (18.28) to giving an isomorphism of L-algebras with involution

$$\alpha_E \colon \big(C(E, \sigma), \underline{\sigma}\big) \xrightarrow{\sim} {}^\rho(E \otimes \Delta, \sigma \otimes 1)$$

where ${}^\rho(E \otimes \Delta, \sigma \otimes 1)$ denotes $(E \otimes \Delta, \sigma \otimes 1)$ with the action of $L \otimes \Delta$ twisted through ρ. An isomorphism

$$\Phi \colon T = (E, L, \sigma, \alpha_E) \xrightarrow{\sim} T' = (E', L', \sigma', \alpha_{E'})$$

of such "data" is a pair (ϕ, ψ) where $\psi \colon L \xrightarrow{\sim} L'$ is an isomorphism of F-algebras, $\Delta(\psi) \colon \Delta(L) \xrightarrow{\sim} \Delta(L')$ is the induced map of discriminant algebras and $\phi \colon E \xrightarrow{\sim} E'$ is ψ-semilinear, such that

$$\phi \circ \sigma' = \sigma \circ \phi \quad \text{and} \quad \big(\phi \otimes \Delta(\psi)\big) \circ \alpha_E = \alpha_{E'} \circ C(\phi).$$

(43.1) Remark. The definition of α_E depends on the choice of a generator ρ of the group $\mathrm{Gal}(L \otimes \Delta/\Delta)$ and such a choice is in fact part of the structure of T. Since $1 \otimes \iota$ is an isomorphism

$$^\rho(E \otimes \Delta) \xrightarrow{\sim} {}^{\rho^2}(E \otimes \Delta),$$

there is a canonical way to change generators.

If $L = F \times F \times F$ is split, then

$$(E, \sigma) = (A, \sigma_A) \times (B, \sigma_B) \times (C, \sigma_C)$$

with (A, σ_A), (B, σ_B), (C, σ_C) algebras over F of degree 8 with orthogonal involutions and

$$^\rho\bigl(E \otimes \Delta(L)\bigr) \xrightarrow{\sim} \begin{cases} (B \times C, C \times A, A \times B) \text{ or} \\ (C \times B, A \times C, B \times A), \end{cases}$$

respectively, according to the choice of ρ. Thus an isomorphism α_E is a triple of isomorphisms

$$(\alpha_A, \alpha_B, \alpha_C) \colon \bigl(C(A, \sigma_A), C(B, \sigma_B), C(C, \sigma_C)\bigr) \xrightarrow{\sim} \begin{cases} (B \times C, C \times A, A \times B) \text{ or} \\ (C \times B, A \times C, B \times A), \end{cases}$$

respectively. Given one of the isomorphisms α_A, α_B, or α_C, there is by Proposition (42.3) a "canonical" way to obtain the two others, hence to extend it to an isomorphism α_E. We write such an induced isomorphism as $\alpha_{(A,B,C)}$ and we say that a datum

$$T' = (A' \times B' \times C', F \times F \times F, \sigma', \alpha')$$

isomorphic to

$$T = \bigl(A \times B \times C, F \times F \times F, (\sigma_A, \sigma_B, \sigma_C), \alpha_{(A,B,C)}\bigr)$$

is a *trialitarian F-algebra* over $F \times F \times F$ or that $\alpha = \alpha_{(A,B,C)}$ is a *trialitarian isomorphism*. If L is not necessarily split, $T = (E, L, \sigma, \alpha)$ is a *trialitarian algebra over L* if over any field extension $\widetilde{F}/F$ which splits L, i.e., $L \otimes \widetilde{F} \simeq \widetilde{F} \times \widetilde{F} \times \widetilde{F}$, $T \otimes \widetilde{F}$ is isomorphic to a trialitarian algebra over $\widetilde{F} \times \widetilde{F} \times \widetilde{F}$.

(43.2) Example. Let $(\mathfrak{C}, \mathfrak{n})$ be a Cayley algebra over F and let $A = \mathrm{End}_F(\mathfrak{C})$. By Proposition (35.1), we have an isomorphism

$$\alpha_{\mathfrak{C}} \colon C_0(\mathfrak{C}, \mathfrak{n}) = C(A, \sigma_{\mathfrak{n}}) \xrightarrow{\sim} (A, \sigma_{\mathfrak{n}}) \times (A, \sigma_{\mathfrak{n}}),$$

which, by Proposition (42.3), extends to define a trialitarian structure

$$T = (A \times A \times A, F \times F \times, \sigma_{\mathfrak{n}} \times \sigma_{\mathfrak{n}} \times \sigma_{\mathfrak{n}}, \alpha_{\widetilde{\mathfrak{C}}})$$

on the product $A \times A \times A$. More precisely, if $\alpha_{\mathfrak{C}}(x) = (x_+, x_-) \in A \times A$, we may take

(43.3) $$\alpha_{\widetilde{\mathfrak{C}}}(x, y, z) = \bigl((y_+, z_-), (z_+, x_-), (x_+, y_-)\bigr)$$

as a trialitarian isomorphism, in view of Example (42.4). It corresponds to the action ρ on $(F \times F)^3$ given by $(x_i, y_i) \mapsto (x_{i+1}, y_{i+2})$, $i = 1, 2, 3 \pmod 3$. We say that such a trialitarian algebra T is of *type G_2* and write it $\mathrm{End}(\widetilde{\mathfrak{C}})$. If $\mathfrak{C} = \mathfrak{C}_s$ is split, $T = T_s$ is the *split trialitarian algebra*. Triality induces an action of S_3 on T_s.

Assume that L/F is cyclic with generator ρ of the Galois group. The isomorphism

$$L \otimes L \xrightarrow{\sim} L \times L \times L, \quad x \otimes y \mapsto \big(xy, x\rho(y), x\rho^2(y)\big)$$

induces an isomorphism $L \otimes \Delta \xrightarrow{\sim} L \times L$ and any $\alpha_E : C(E, \sigma) \xrightarrow{\sim} {}^\rho(E \otimes \Delta)$ can be viewed as an isomorphism

$$\alpha_E : C(E, \sigma) \xrightarrow{\sim} {}^\rho E \times {}^{\rho^2} E.$$

Thus $(E, {}^\rho E, {}^{\rho^2} E)$ is a trialitarian triple over $L \times L \times L$ and by Proposition (42.3) α_E determines an isomorphism

$$\alpha_{\rho E} : C({}^\rho E, \sigma) \xrightarrow{\sim} {}^{\rho^2} E \times E.$$

The isomorphism $\alpha_{\rho E}$ is (tautologically) also an isomorphism

$$C(E, \sigma) \xrightarrow{\sim} {}^\rho E \times {}^{\rho^2} E.$$

We denote it by ${}^{\rho^{-1}}\alpha_{\rho E}$.

(43.4) Proposition. *The isomorphism α_E is trialitarian if and only if ${}^{\rho^{-1}}\alpha_{\rho E} = \alpha_E$.*

Proof: It suffices to check the claim for a trialitarian triple (A, B, C), where it is straightforward. $\qquad \square$

For trialitarian algebras over arbitrary cubic étale algebras L we have:

(43.5) Corollary. *An isomorphism $\alpha_E : C(E, \sigma) \xrightarrow{\sim} {}^\rho(E \otimes \Delta)$ extends to an isomorphism*

$$\alpha_{E \otimes \Delta} : C(E \otimes \Delta, \sigma \otimes 1) \xrightarrow{\sim} {}^\rho(E \otimes \Delta) \times {}^{\rho^2}(E \otimes \Delta)$$

and α_E is trialitarian if and only if ${}^{\rho^{-1}}\alpha_{\rho(E \otimes \Delta)} = \alpha_{E \otimes \Delta}$. $\qquad \square$

The norm map

$$N_{L/F} : \operatorname{Br}(L) \to \operatorname{Br}(F),$$

which is defined for finite separable field extensions L/F can be extended to étale F-algebras L: if $L = L_1 \times \cdots \times L_r$ where L_i/F, $i = 1, \ldots, r$, are separable field extensions and if $A = A_1 \times \ldots A_r$ is L-central simple (i.e., A_i is central simple over L_i), then, for $[A] \in \operatorname{Br}(L) = \operatorname{Br}(L_1) \times \cdots \times \operatorname{Br}(L_r)$, we define

$$N_{L/F}([A]) = [A_1] \cdot \ldots \cdot [A_r] \in \operatorname{Br}(F).$$

(43.6) Proposition. *For any trialitarian algebra $T = (E, L, \sigma, \alpha_E)$ the central simple L-algebra E satisfies $N_{L/F}([E]) = 1 \in \operatorname{Br}(F)$.*

Proof: The algebra $C(E, \sigma)$ is $L \otimes \Delta$-central simple and $L \otimes \Delta$ is étale. We compute the class of $N_{L \otimes \Delta/F}\big(C(E, \sigma)\big)$ in the Brauer group $\operatorname{Br}(F)$ in two different ways: on one hand, by using that $[N_{L \otimes \Delta/L}\big(C(E, \sigma)\big)] = [E]$ in $\operatorname{Br}(L)$ (see Theorem (9.12) or Example (31.11)), we see that

$$[N_{L \otimes \Delta/F}\big(C(E, \sigma)\big)] = [N_{L/F} \circ N_{L \otimes \Delta/L}\big(C(E, \sigma)\big)]$$
$$= [N_{L/F}(E)]$$

and on the other hand we have

$$[N_{L\otimes\Delta/F}(C(E,\sigma))] = [N_{\Delta/F}(N_{L\otimes\Delta/\Delta}(C(E,\sigma)))]$$
$$= [N_{\Delta/F}(N_{L\otimes\Delta/\Delta}(^\rho(E\otimes\Delta)))]$$
$$= [N_{\Delta/F}(N_{L/F}(E)\otimes\Delta)]$$
$$= [N_{L/F}(E)]^2,$$

so that, as claimed $[N_{L/F}(E)] = 1$. $\square$

(43.7) Example. A trialitarian algebra can be associated to any twisted composition $\Gamma = (V, L, Q, \beta)$: Let ρ be a fixed generator of the cyclic algebra $L\otimes\Delta/\Delta$, Δ the discriminant of L. By Proposition (36.12) there exists exactly one cyclic composition (with respect to ρ) on $(V, L, Q, \beta)\otimes\Delta$. By Proposition (36.19) we then have an isomorphism

$$\alpha_V: C_0(V, Q) = C(\mathrm{End}_L(V), \sigma_Q) \xrightarrow{\sim} {}^\rho(\mathrm{End}_L(V)\otimes\Delta).$$

We claim that the datum $(\mathrm{End}_L(V), L, \sigma_Q, \alpha_V)$ is a trialitarian algebra. By descent it suffices to consider the case where $\Gamma = \widetilde{\mathfrak{C}}$ is of type G_2. Then the claim follows from Example (43.2).

We set $\mathrm{End}(\Gamma)$ for the trialitarian algebra associated to the twisted composition Γ.

43.B. Quaternionic trialitarian algebras. The proof of Proposition (43.6) shows that the sole existence of a map α_E implies that $N_{L/F}([E]) = 1$. In fact, the condition $N_{L/F}([E]) = 1$ is necessary for E to admit a trialitarian structure, but not sufficient, even if L is split, see Remark (42.12). We now give examples where the condition $N_{L/F}([E]) = 1$ is sufficient for the existence of a trialitarian structure on E.

(43.8) Theorem. *Let Q be a quaternion algebra over a cubic étale algebra L. Then $M_4(Q)$ admits a trialitarian structure $T(Q)$ if and only if $N_{L/F}([Q]) = 1$ in $\mathrm{Br}(F)$.*

Before proving Theorem (43.8) we observe that over number fields any central simple algebra which admits an involution of the first kind is of the form $M_n(Q)$ for some quaternion algebra Q (Albert, [9, Theorem 20, p. 161]). Thus, for such fields, the condition $N_{L/F}([E]) = 1$ is necessary and sufficient for E to admit a trialitarian structure (see Allison [19] and the notes at the end of the chapter).

The first step in the proof of Theorem (43.8) is the following reduction:

(43.9) Proposition. *Let L/F be a cubic étale algebra and let Q be a quaternion algebra over L. The following conditions are equivalent:*

(1) $N_{L/F}([Q]) = 1$.
(2) $Q \simeq (a, b)_L$ *with* $b \in F^\times$ *and* $N_L(a) = 1$.

Proof: (2) $\Rightarrow$ (1) follows from the projection (or transfer) formula (see for example Brown [56, V, (3.8)]). For the proof of (1) $\Rightarrow$ (2) it suffices to show that $Q \simeq (a, b)_L$ with $b \in F^\times$: The condition $N_{L/F}([Q]) = 1$ then implies $N_L(a) = N_{F(\sqrt{b})}(z)$ for some $z \in F(\sqrt{b})$, again by the projection formula. Replacing a by $a^3 N_{F(\sqrt{b})}(z)^{-1}$ gives a as wanted. We first consider the case $L = F \times K$, K quadratic étale. Let $Q_1 \times Q_2$ be the corresponding decomposition of Q. The condition $N_{L/F}([Q]) = 1$

is equivalent with $N_{K/F}([Q_2]) = [Q_1]$ or $N_{K/F}(Q_2) \simeq M_2(Q_1)$. In this case the claim follows from Corollary (16.28). Let now $Q = (\alpha, \beta)_L$, for L a field. We have to check that the L-quadratic form $q = \langle \alpha, \beta, -\alpha\beta \rangle$ represents a nonzero element of F. Let $L = F(\theta)$ and $q(x) = q_1(x) + q_2(x)\theta + q_3(x)\theta^2$ with q_i quadratic forms over F. In view of the case $L = F \times K$, q_2 and q_3 have a nontrivial common zero over L, hence the claim by Springer's theorem for pairs of quadratic forms (see Pfister [218, Corollary 1.1, Chap. 9]). $\qquad\square$

(43.10) Remark. Proposition (43.9) in the split case $L = F \times F \times F$ reduces to the classical result of Albert that the condition $[Q_1][Q_2][Q_3] = 1$ for quaternions algebras Q_i over F is equivalent to the existence of a, b, c such that $[Q_1] = (a, b)_F$, $[Q_2] = (a, c)_F$, $[Q_3] = (a, bc)_F$. In particular, the algebras Q_i have a common quadratic subalgebra (see Corollary (16.29)). Thus (43.9) can be viewed as a "twisted" version of Albert's result.

Theorem (43.8) now is a consequence of the following:

(43.11) Proposition. *Let K/F be quadratic étale, let L/F be cubic étale and let $a \in L^\times$ be such that $N_L(a) = 1$. Let Q be the quaternion algebra $(K \otimes L/L, a)_L$ and let $E(a) = M_4(Q)$. There exists a trialitarian structure $T = \big(E(a), L, \sigma, \alpha\big)$ on $E(a)$.*

The main step in the construction of T is a result of Allen and Ferrar [18]. To describe it we need some notations. Let $(\mathfrak{C}, \mathfrak{n})$ be the split Cayley algebra with norm $\mathfrak{n}$. The vector space $\mathfrak{C}$ has a basis $(u_1, \dots, u_8)$ (use Exercise 5 of Chapter VIII) such that

(a) the multiplication table of $\mathfrak{C}$ is

	u_1	u_2	u_3	u_4	u_5	u_6	u_7	u_8
u_1	0	u_7	$-u_6$	u_1	$-u_8$	0	0	0
u_2	$-u_7$	0	u_5	u_2	0	$-u_8$	0	0
u_3	u_6	$-u_5$	0	u_3	0	0	$-u_8$	0
u_4	0	0	0	u_4	u_5	u_6	u_7	0
u_5	$-u_4$	0	0	0	0	u_3	$-u_2$	u_5
u_6	0	$-u_4$	0	0	$-u_3$	0	u_1	u_6
u_7	0	0	$-u_4$	0	u_2	$-u_1$	0	u_7
u_8	u_1	u_2	u_3	0	0	0	0	u_8

(b) $1 = u_4 + u_8$ and the conjugation map π is given by $\pi(u_i) = -u_i$ for $i \neq 4, 8$ and $\pi(u_4) = u_8$.

(c) $b_{\mathfrak{n}}(u_i, u_j) = \delta_{i+4, j}$, $i + 4$ being taken mod 8, in particular $\{u_1, \dots, u_4\}$ and $\{u_5, \dots, u_8\}$ span complementary totally isotropic subspaces of $\mathfrak{C}$.

(43.12) Lemma. *Let a_1, a_2, $a_3 \in F^\times$ be such that $a_1 a_2 a_3 = 1$, let*

$$A_i = \mathrm{diag}(a_i, a_i, a_i, a_{i+2}^{-1}), \quad B_i = \mathrm{diag}(1, 1, 1, a_{i+1}^{-1})$$

in $M_4(F)$ and let $t_i = \big(\begin{smallmatrix} 0 & A_i \\ B_i & 0 \end{smallmatrix}\big) \in M_8(F)$, $i = 1, 2, 3$. Also, write t_i for the F-vector space automorphism of $\mathfrak{C}$ induced by t_i with respect to the basis $(u_1, \dots, u_8)$. Then t_i is a similitude of $(\mathfrak{C}, \mathfrak{n})$ with multiplier a_i such that

(1) $a_1 t_1(x \star y) = t_2(x) \star t_3(y)$ *where $\star$ is the multiplication in the para-Cayley algebra $\mathfrak{C}$.*

(2) $t_i \in \mathrm{Sym}\big(\mathrm{End}(\mathfrak{C}), \sigma_{\mathfrak{n}}\big)$, *in particular $t_i^2 = a_i \cdot 1$, $i = 1, 2, 3$.*

Proof: A lengthy computation! See Allen-Ferrar [18, p. 480-481]. □

Proof of (43.11): Let $\Delta = \Delta(L)$ be the discriminant algebra of L. The F-algebra $P = L \otimes \Delta \otimes K$ is a G-Galois algebra where $G = S_3 \times \mathbb{Z}/2\mathbb{Z}$, S_3 acts on the Galois S_3-closure $L \otimes \Delta$ and $\mathbb{Z}/2\mathbb{Z}$ acts on K. We have $L \otimes P \simeq P \times P \times P$ and we may view $L \otimes P$ as a Galois G-algebra over L. The group S_3 acts through permutations of the factors. Let ι_K be a generator of $\mathrm{Gal}(K/F)$ and ρ be a generator of $\mathrm{Gal}(L \otimes \Delta/\Delta)$. Let $\sigma = \rho \otimes \iota_K$, so that σ generates a cyclic subgroup of $S_3 \times \mathbb{Z}/2\mathbb{Z}$ of order 6 and G is generated by σ and $1 \otimes \iota_\Delta \otimes 1$. Let $(\mathfrak{C}_s, \mathfrak{n}_s)$ be the split Cayley algebra over F and $(\mathfrak{C}, \mathfrak{n}) = (\mathfrak{C}_s, \mathfrak{n}_s) \otimes P$. As in §34.A, let $(x, y) \mapsto x \star y = \overline{x} \diamond \overline{y}$ be the symmetric composition on $\mathfrak{C}$. The trialitarian structure on $E(a)$ over L is constructed by Galois descent from the split trialitarian structure $\mathrm{End}(\mathfrak{C}) \times \mathrm{End}(\mathfrak{C}) \times \mathrm{End}(\mathfrak{C})$ over $P \times P \times P$:

(43.13) Lemma. *Let* $a \in L^\times$ *be such that* $N_{L/F}(a) = 1$. *There exist similitudes* t, t^+, t^- *of* $(\mathfrak{C}, \mathfrak{n})$ *with multipliers* a, $\sigma(a)$, $\sigma^2(a)$, *respectively, such that:*

(1) $at(x \star y) = t^+(x) \star t^-(y)$.
(2) t, t^+, $t^- \in \mathrm{Sym}\big(\mathrm{End}_P(\mathfrak{C}), \sigma_\mathfrak{n}\big)$, *and* $(t, t^+, t^-)^2 = \big(a, \sigma(a), \sigma^2(a)\big)$.
(3) $\sigma t = t^+ \sigma$, $\sigma t^+ = t^- \sigma$, $\sigma t^- = t \sigma$.
(4) *One has*

$$(\pi \otimes \iota_\Delta \otimes 1)t = t(\pi \otimes \iota_\Delta \otimes 1),$$
$$(\pi \otimes \iota_\Delta \otimes 1)t^+ = t^-(\pi \otimes \iota_\Delta \otimes 1),$$
$$(\pi \otimes \iota_\Delta \otimes 1)t^- = t^+(\pi \otimes \iota_\Delta \otimes 1),$$
$$(\pi \otimes \iota_\Delta \otimes 1)(1 \otimes \sigma) = (1 \otimes \sigma^2)(\pi \otimes \iota_\Delta \otimes 1).$$

Proof: Lemma (43.12) applied over P to $a_1 = a$, $a_2 = \sigma(a)$, $a_3 = \sigma^2(a)$ gives (1) and (2).

(3) and (4) can easily be verified using the explicit form of t, t^+, and t^- given in Lemma (43.12). □

We now describe the descent defining $E(a)$. Let $\widetilde{t}$, $\widetilde{\sigma}$ and $\widetilde{\pi}$ be the automorphisms of $\mathfrak{C} \times \mathfrak{C} \times \mathfrak{C}$ given by $\widetilde{t} = (t, t^-, t^+)$, $\widetilde{\sigma}(x, y, z) = (\sigma y, \sigma z, \sigma x)$, and

$$\widetilde{\pi}(x, y, z) = \big(\pi \otimes \iota_\Delta \otimes 1(x), \pi \otimes \iota_\Delta \otimes 1(z), \pi \otimes \iota_\Delta \otimes 1(y)\big).$$

It follows from the description of (t, t^-, t^+) that $\widetilde{t}\widetilde{\sigma} = \widetilde{\sigma}\widetilde{t}$, $\widetilde{t}\widetilde{\pi} = \widetilde{\pi}\widetilde{t}$, and $\widetilde{\sigma}\widetilde{\pi} = \widetilde{\pi}\widetilde{\sigma}^2$. Further $\widetilde{t}\widetilde{\sigma}$ is σ-linear, $\widetilde{t}\widetilde{\pi}$ is ι-linear and, by (1) of Lemma (43.13), $\mathrm{Int}(\widetilde{t})$ is an automorphism of the trialitarian algebra $\mathrm{End}(\mathfrak{C}) \times \mathrm{End}(\mathfrak{C}) \times \mathrm{End}(\mathfrak{C})$ over $P \times P \times P$. Thus $\{\mathrm{Int}(\widetilde{t}\widetilde{\sigma}), \mathrm{Int}(\widetilde{t}\widetilde{\pi})\}$ gives a G-Galois action on $\mathrm{End}(\mathfrak{C}) \times \mathrm{End}(\mathfrak{C}) \times \mathrm{End}(\mathfrak{C})$. By Galois descent we obtain a trialitarian algebra $E(a) = (E, L, \sigma, \alpha)$ over L. We claim that $E \simeq M_4\big((K \otimes L/L, a)\big)$. Since L/F is cubic, it suffices to check that

$$E \otimes L \simeq M_4\big((K \otimes L/L, a)\big) \times E_2$$

for some $L \otimes \Delta$-algebra E_2. Let $E \otimes L = E_1 \times E_2$. The $(L \otimes L)$-algebra $E \otimes L$ is the descent of $\mathrm{End}(\widetilde{\mathfrak{C}})$ under $\{\mathrm{Int}(\widetilde{t}\widetilde{\sigma})^3, \mathrm{Int}(\widetilde{t}\widetilde{\pi})\}$. Since $[\widetilde{t}\widetilde{\sigma}^3]^2 = a \in L$, we have

$$[E \otimes L] = \big([(K \otimes L/L, a)], [E_2]\big) \in \mathrm{Br}(L \times L \otimes \Delta),$$

hence the claim. □

For fixed extensions K and L over F, the trialitarian algebras $E(a)$ are classified by $L^\times / N_{K \otimes L/L}\big((K \otimes L)^\times\big)$:

(43.14) Proposition. *The following conditions are equivalent*:

(1) $E(a_1) \simeq E(a_2)$ *as trialitarian algebras.*
(2) $E(a_1) \simeq E(a_2)$ *as L-algebras (without involutions).*
(3) $a_1 a_2^{-1} \in N_{K \otimes L/L}\big((K \otimes L)^\times\big)$.

Proof: (1) implies (2) and the equivalence of (2) and (3) is classical for cyclic algebras, see for example Corollary (30.7).

We show that (3) implies (1) following [18]. Assume that $a_1 = a_2 \cdot \lambda \iota_K(\lambda)$ for $\lambda \in L \otimes K$. We have $\lambda \iota_K(\lambda) = \lambda \sigma^3(\lambda) \in P^\times$. It follows from $N_{L/F}(a_1) = 1 = N_{L/F}(a_2)$ that $N_{P/F}(\lambda) = 1$, so that, by choosing $\mu = a_2\big(\sigma^4(\lambda)\sigma^5(\lambda)\big)^{-1}$, we deduce $\mu \sigma^2(\mu)\sigma^4(\mu) = 1$. Now let $t = (t_1, t_2, t_3)$ be given by Lemma (43.12) for $a_1 = \mu$, $a_2 = \sigma^2(\mu)$, and $a_3 = \sigma^4(\mu)$. Let $c(a_i)$ be the map $\widetilde{t\sigma}$ as used in the descent defining $E(a)$ for $a = a_i$, $i = 1, 2$. A straightforward computation shows that

$$T^{-1}c(a_1)T = c(a_2)\big(\sigma(\lambda)\sigma^2(\lambda), \sigma^3(\lambda)\sigma^4(\lambda), \sigma^5(\lambda)\lambda\big).$$

This implies (by descent) that $E(a_1) \simeq E(a_2)$. □

43.C. Trialitarian algebras of type $^2D_4$. We say that a trialitarian algebra $T = (E, L, \Delta, \sigma, \alpha)$ of *type* $^1D_4$ if L is split, $^2D_4$ if $L = F \times K$ for K a quadratic separable field extension over F isomorphic to Δ, $^3D_4$ if L is a cyclic field extension of F and $^6D_4$ if $L \otimes \Delta$ is a Galois field extension with group S_3 over F.

We now describe trialitarian algebras over an algebra $L = F \times \Delta$ where Δ is quadratic (and is the discriminant algebra of L), i.e., is of type $^1D_4$ or $^2D_4$. The results of this section were obtained in collaboration with R. Parimala and R. Sridharan.

(43.15) Proposition. *Let (A, σ) be a central simple F-algebra of degree 8 with an orthogonal involution and let Z be the center of $C(A, \sigma)$.*

(1) *The central simple algebra with involution $(A, \sigma) \times \big(C(A, \sigma), \underline{\sigma}\big)$ over $F \times Z$ admits the structure of a trialitarian algebra $T(A, \sigma)$ and is functorial in (A, σ).*
(2) *If $T = (A \times B, F \times \Delta, \sigma_A \times \sigma_B, \alpha)$ is a trialitarian algebra over $L = F \times \Delta$ for Δ a quadratic étale F-algebra, then there exists, after fixing a generator ρ of $\mathrm{Gal}(L \otimes \Delta/\Delta)$, a unique isomorphism $\phi \colon T \xrightarrow{\sim} T(A, \sigma)$ of trialitarian algebras such that $\phi|_A = 1|_A$.*

Proof: (1) Let ι be the conjugation on Z. The isomorphism $Z \otimes Z \xrightarrow{\sim} Z \times Z$ given by $x \otimes y \mapsto \big(xy, x\iota(y)\big)$ induces an isomorphism

$$\alpha_1 \colon C(A \otimes Z) \xrightarrow{\sim} C(A, \sigma) \times {}^\iota C(A, \sigma).$$

Thus $\big(A \otimes Z, C(A, \sigma), {}^\iota C(A, \sigma)\big)$ is a trialitarian triple over Z. By triality α_1 induces a Z-isomorphism

$$\alpha_2 = {}^{\theta^+}\alpha_1 \colon C\big(C(A, \sigma), \underline{\sigma}\big) \xrightarrow{\sim} {}^\iota C(A, \sigma) \times A \otimes Z,$$

so that $\alpha = (1, \alpha_2)$ is an $(F \times Z)$-isomorphism

$$\alpha \colon C\big(A \times C(A, \sigma)\big) \xrightarrow{\sim} C(A, \sigma) \times {}^\iota C(A, \sigma) \times A \otimes Z.$$

On the other hand we have

$$C(A,\sigma) \times {}^{\iota}C(A,\sigma) \times A \otimes Z \xrightarrow{\sim} {}^{\rho}\big(A \otimes Z \times C(A,\sigma) \times {}^{\iota}C(A,\sigma)\big)$$
$$\xrightarrow{\sim} {}^{\rho}\big((A \times C(A,\sigma)) \otimes Z\big)$$

for $\rho \in \mathrm{Aut}_Z(Z \times Z \times Z) = \mathrm{Gal}\big((F \times Z) \otimes Z/Z\big)$ given by $\rho(z_0, z_1, z_2) = (z_1, z_2, z_0)$. Thus α can be viewed as an isomorphism

$$\alpha \colon C\big(A \times C(A,\sigma)\big) \xrightarrow{\sim} {}^{\rho}\big((A \times C(A,\sigma)) \otimes Z\big).$$

It is easy to check that α is trialitarian by splitting Z.

(2) Let

$$\beta \colon C(A \times B) \xrightarrow{\sim} {}^{\rho}\big((A \times B) \otimes Z\big)$$

be a trialitarian structure for $(A \times B, \sigma_A \times \sigma_B)$. Then β is an $L \otimes Z$ isomorphism

$$\beta \colon C(A) \times C(B) \xrightarrow{\sim} B \times {}^{\iota}B \times A \otimes Z$$

and splits as (β_1, β_2) where $\beta_1 \colon C(A) \xrightarrow{\sim} B$ and β_2 is determined by β_1 through triality. Then

$$\widetilde{\beta} = (1, \beta_1) \colon A \times C(A) \xrightarrow{\sim} A \times B$$

is an isomorphism of $T(A,\sigma)$ with $(A \times B, \sigma_A \times \sigma_B, \beta)$. This follows from the fact that a trialitarian algebra over a product $F \times Z$ is determined by the first component. $\qquad\square$

(43.16) Corollary. *Let (E, σ) be such that there exists an isomorphism*

$$\alpha \colon C(E, \sigma) \xrightarrow{\sim} {}^{\rho}(E \otimes \Delta)$$

(not necessarily trialitarian). If L is not a field, then there exists a trialitarian isomorphism $\alpha_E \colon C(E, \sigma) \xrightarrow{\sim} {}^{\rho}(E \otimes \Delta)$.

Proof: Let $E = A \times B$ and write $L = F \times K = Z(A) \times Z(B)$. Then $\alpha = (\alpha_1, \alpha_2)$ with $\alpha_1 \colon C(A, \sigma_A) \xrightarrow{\sim} (B, \sigma_B)$ and $\alpha_2 \colon C(B, \sigma_B) \xrightarrow{\sim} {}^{\iota}B \times A \otimes K$. On the other hand

$$\alpha_1 \otimes 1_K \colon C(A \otimes K, \sigma \otimes 1) \xrightarrow{\sim} (B, \sigma_B) \otimes K = (B, \sigma_B) \times {}^{\iota}(B, \sigma_B)$$

induces by triality an isomorphism

$$\widetilde{\alpha}_2 \colon C(B, \sigma_B) \xrightarrow{\sim} {}^{\iota}B \times A \otimes K.$$

The pair $\alpha_E = (\alpha_1, \widetilde{\alpha}_2)$ is trialitarian. $\qquad\square$

(43.17) Corollary. *Let A, A' be central simple F-algebras of degree 8 with orthogonal involutions σ, σ' and let Z, Z' be the centers of $C(A, \sigma)$, resp. $C(A', \sigma')$. Then the F-algebras $\big(C(A, \sigma), \underline{\sigma}\big)$ and $\big(C(A', \sigma'), \underline{\sigma'}\big)$ are isomorphic (as algebras with involution) if and only if $(A, \sigma) \otimes Z$ and $(A', \sigma') \otimes Z'$ are isomorphic (as F-algebras).*

Proof: Any isomorphism $\phi \colon C(A, \sigma) \xrightarrow{\sim} C(A', \sigma')$ induces an isomorphism

$${}^{\iota}C(A, \sigma) \times A \otimes Z \xrightarrow{\sim} C\big(C(A, \sigma), \underline{\sigma}\big) \xrightarrow{\sim} C\big(C(A', \sigma'), \underline{\sigma}\big) \xrightarrow{\sim} {}^{\iota}C(A', \sigma') \times A' \otimes Z'.$$

Looking at all possible components of $C(\phi)$ and taking in account that by assumption $C(A, \sigma) \xrightarrow{\sim} C(A', \sigma')$ gives an isomorphism $(A, \sigma) \otimes Z \xrightarrow{\sim} (A', \sigma') \otimes Z'$. Conversely, any isomorphism $(A, \sigma) \otimes Z \xrightarrow{\sim} (A', \sigma') \otimes Z'$ induces an isomorphism $C(A, \sigma) \otimes Z \xrightarrow{\sim} C(A', \sigma') \otimes Z'$. Since $C(A, \sigma) \otimes Z \xrightarrow{\sim} C(A, \sigma) \times^{\iota} C(A, \sigma)$, composing with the inclusion $C(A, \sigma) \to C(A, \sigma) \otimes Z$ and the projection $C(A', \sigma') \otimes Z' \to$

$C(A', \sigma')$ gives a homomorphism $C(A, \sigma) \to C(A', \sigma')$ of algebras with involution. This must be an isomorphism since $C(A, \sigma)$ is central simple over Z. $\quad\square$

(43.18) Corollary ([165]). *Let (V, q) and (V', q') be quadratic spaces of rank 8 and let Z, Z' be the centers of $C(V, q)$, resp. $C(V', q')$. Then $C_0(V, q)$ and $C_0(V', q')$ are isomorphic (as algebras over F with involution) if and only if $(V, q) \otimes Z$ and $(V', q') \otimes Z'$ are similar.*

Proof: Since any isomorphism $(\operatorname{End}_F(V), \sigma_q) \xrightarrow{\sim} (\operatorname{End}_F(V'), \sigma_{q'})$ is induced by a similitude $(V, q) \xrightarrow{\sim} (V', q')$ and vice versa the result follows from Corollary (43.17). $\quad\square$

§44. Classification of Algebras and Groups of Type D_4

Let $(\mathfrak{C}, \mathfrak{n})$ be a Cayley algebra with norm $\mathfrak{n}$ over F, let $\widetilde{\mathfrak{C}} = \mathfrak{C} \otimes (F \times F \times F)$ be the induced twisted composition and let $\operatorname{End}(\widetilde{\mathfrak{C}})$ be the induced trialitarian algebra (see Example (43.2)). Since S_3 acts by triality on $\mathbf{PGO}(\mathfrak{C}, \mathfrak{n})$, we have a split exact sequence

(44.1) $$1 \to \mathbf{PGO}(\mathfrak{C}, \mathfrak{n}) \to \mathbf{PGO}(\mathfrak{C}, \mathfrak{n}) \rtimes S_3 \xrightarrow{p} S_3 \to 1$$

where p is the projection.

(44.2) Proposition. *We have*

$$\operatorname{Aut}_F\big(\operatorname{End}(\widetilde{\mathfrak{C}})\big) \simeq \mathbf{PGO}_+(\mathfrak{C}, \mathfrak{n})(F) \rtimes S_3.$$

Proof: One shows as in the proof of Proposition (36.5) that the restriction map

$$\rho \colon \operatorname{Aut}_F\big(\operatorname{End}(\widetilde{\mathfrak{C}})\big) \to \operatorname{Aut}_F(F \times F \times F) = S_3$$

has a section. Thus it suffices to check that $\ker \rho = \mathbf{PGO}_+(\mathfrak{C}, \mathfrak{n})(F)$. Any β in $\ker \rho$ is of the form $\operatorname{Int}(t)$ where $t = (t_0, t_1, t_2)$ is a $(F \times F \times F)$-similitude of $\mathfrak{C} \times \mathfrak{C} \times \mathfrak{C}$ with multiplier $\lambda = (\lambda_0, \lambda_1, \lambda_2)$, such that

$$\alpha_{\widetilde{\mathfrak{C}}} \circ C_0(t) = \big(\operatorname{Int}(t) \otimes 1\big) \circ \alpha_{\widetilde{\mathfrak{C}}}.$$

It follows from the explicit description of $\alpha_{\widetilde{\mathfrak{C}}}$ given in (43.3) that

(44.3) $$\lambda^{-1} t(x) * \big(z * t(y)\big) = t\big(x * (t^{-1}(z) * y)\big)$$

for all x, y, $z \in \mathfrak{C}$, where $x * y = (\bar{x}_1 \bar{y}_2, \bar{x}_2 \bar{y}_0, \bar{x}_0 \bar{y}_1)$ for $x = (x_0, x_1, x_2)$, $y = (y_0, y_1, y_2)$, multiplication is in the Cayley algebra and $x \mapsto \bar{x}$ is conjugation. Condition (44.3) gives three relations for (t_0, t_1, t_2):

(44.4) $$t_i\big(\bar{x}_{i+1}(y_{i+1} z_i)\big) = \lambda_i^{-1} \overline{t_{i+1}(x_{i+1})}\big(t_{i+1}(y_{i+1}) t_i(z_i)\big), \quad i = 0, 1, 2.$$

We claim that the group homomorphism

$$\operatorname{Int}(t) \in \ker \rho \mapsto [t_0] \in \mathbf{PGO}_+(\mathfrak{C}, \mathfrak{n})(F)$$

is an isomorphism. It is surjective since, by triality, there exist $t_1 = (t_0)^-$, $t_2 = (t_0)^+$ such that $t = (t_0, t_1, t_2)$ (see Proposition (35.4)). We check that it is injective: let $[t_0] = 1$, so that $t_0 = \mu_0 \cdot 1_{\mathfrak{C}}$ for some $\mu_0 \in F^\times$. It follows from Equation (44.4) (for $i = 2$) that

$$t_2\big(x(yz)\big) = \lambda_2^{-1} \mu_0^2 x\big(y t_2(z)\big)$$

holds for all x, y, $z \in \mathfrak{C}$. By putting $y = z = 1$ we obtain $t_2(x) = \lambda_2^{-1} \mu_0^2 x t_2(1)$. This implies, with $a = t_2(1)$, that $\big(x(yz)\big) a = x\big(x(za)\big)$. Hence $a = t_2(1)$ is central in $\mathfrak{C}$ and the class of t_2 in $\mathbf{PGO}_+(\mathfrak{C},\mathfrak{n})(F)$ is trivial. One shows similarly that the class of t_1 is trivial and, as claimed, $\ker \rho \simeq \mathbf{PGO}_+(\mathfrak{C},\mathfrak{n})(F)$. $\qquad\square$

(44.5) Corollary. *The pointed set $H^1(F, \mathbf{PGO}_8^+ \rtimes S_3)$ classifies trialitarian F-algebras up to isomorphism. In the exact sequence*

$$H^1(F, \mathbf{PGO}_8^+) \to H^1(F, \mathbf{PGO}_8^+ \rtimes S_3) \to H^1(F, S_3)$$

induced by the exact sequence (44.1), the first map associates the trialitarian algebra

$$T = \big(A \times B \times C, F \times F \times F, (\sigma_A, \sigma_B, \sigma_C), \alpha_{(A,B,C)}\big)$$

where $\alpha_{(A,B,C)}$ is determined as in Proposition (42.3), to the triple (A, B, C). The second map associates the class of L to the trialitarian algebra $T = (E, L, \sigma, \alpha)$.

Proof: Over a separable closure of F, L and E split, hence the trialitarian algebra is isomorphic to a split trialitarian algebra T_s. We let it as an exercise to identify $\mathbf{Aut}(T_s)$ with $\mathbf{Aut}_G(w)$ for some tensor $w \in W$ and some representation $G \to \mathbf{GL}(W)$ such that $H^1(F, G) = 0$ (see the proof of Theorem (44.8)). Then (44.5) follows from Proposition (44.2). $\qquad\square$

44.A. Groups of trialitarian type D_4. Let $T = (E, L, \sigma, \alpha)$ be a trialitarian F-algebra. The group scheme $\mathbf{Aut}_L(T)$ of automorphisms of T which are the identity on L is the connected component of the identity of $\mathbf{Aut}_F(T)$. We have, for $R \in Alg_F$,

$$\mathbf{Aut}_L(T)(R) = \{\, \phi \in R_{L/F}\big(\mathbf{PGO}^+(E,\sigma)\big)(R) \mid \alpha_E \circ C(\phi) = (\phi \otimes 1) \circ \alpha_E \,\}$$

and we set $\mathbf{PGO}^+(T) = \mathbf{Aut}_L(T)$. Similarly we set

$$\mathbf{GO}^+(T)(R) =$$
$$\{\, x \in R_{L/F}\big(\mathbf{GO}^+(E,\sigma)\big)(R) \mid \alpha_{E_R} \circ C\big(\mathrm{Int}(x)\big) = \big(\mathrm{Int}(x) \otimes 1\big) \circ \alpha_{E_R} \,\},$$

so that $\mathbf{PGO}^+(T) = \mathbf{GO}^+(T)/\mathbf{G}_m$, and

$$\mathbf{Spin}(T)(R) = \{\, x \in R_{L/F}\big(\mathbf{Spin}(E,\sigma)\big)(R) \mid \alpha_{E_R}(x) = \chi(x) \otimes 1 \,\}.$$

(44.6) Lemma. *For the split trialitarian algebra T_s we have*

$$\mathbf{Spin}(T_s) \simeq \mathbf{Spin}(\mathfrak{C}_s, \mathfrak{n}_s) \quad and \quad \mathbf{PGO}^+(T_s) \simeq \mathbf{PGO}^+(\mathfrak{C}_s, \mathfrak{n}_s).$$

Proof: Let $T_s = (E, F \times F \times F, \sigma, \alpha_E)$ with $E = A \times A \times A$, $A = \mathrm{End}_F(\mathfrak{C})$ the split trialitarian algebra. For $x \in \mathbf{Spin}(E,\sigma)(R)$, we have $\alpha_{E_R}(x) = \chi(x) \otimes 1$ if and only if $x = (t, t_1, t_2)$ and $t_1(x \star y) = t(x) \star t_2(y)$, hence the claim for $\mathbf{Spin}(T_s)$. The claim for $\mathbf{PGO}^+(T_s)$ follows along similar lines. $\qquad\square$

Since $\mathbf{Spin}(T_s) \simeq \mathbf{Spin}(\mathfrak{C}_s, \mathfrak{n}_s)$, $\mathbf{Spin}(T)$ is simply connected of type D_4 and the vector representation induces a homomorphism

$$\chi' \colon \mathbf{Spin}(T) \to \mathbf{PGO}^+(T)$$

which is a surjection of algebraic group schemes. Thus $\mathbf{Spin}(T)$ is the simply connected cover of $\mathbf{PGO}^+(T)$. Let γ be a cocycle in $H^1(F, \mathbf{PGO}_8^+ \rtimes S_3)$ defining the trialitarian algebra T. Since

$$\mathbf{PGO}_8^+ \rtimes S_3 = \mathrm{Aut}(\mathbf{Spin}_8) = \mathrm{Aut}(\mathbf{PGO}_8^+),$$

we may use γ to twist the Galois action on $\mathbf{Spin}_8$ or $\mathbf{PGO}_8^+$ and we have

$$(\mathbf{Spin}_8)_\gamma(F) \simeq \mathbf{Spin}(T) \quad and \quad (\mathbf{PGO}_8^+)_\gamma(F) \simeq \mathbf{PGO}^+(T).$$

(44.7) Remark. If G is of type $^1D_4$ or $^2D_4$, i.e., if $L = F \times Z$, then $E = A \times C(A, \sigma)$ and $\mathbf{PGO}^+(T) \simeq \mathbf{PGO}^+(A)$, $\mathbf{Spin}(T) \simeq \mathbf{Spin}(A)$.

Classification of simple groups of type D_4. Consider the groupoid $D_4 = D_4(F)$, of trialitarian F-algebras. Denote by $D^4 = D^4(F)$ (resp. $\overline{D}^4 = \overline{D}^4(F)$) the groupoid of simply connected (resp. adjoint) simple groups of type D_4 over F where morphisms are group isomorphisms. We have functors

$$S_4 \colon D_4(F) \to D^4(F) \quad \text{and} \quad \overline{S}_4 \colon D_4(F) \to \overline{D}^4(F)$$

defined by $S_4(T) = \mathbf{Spin}(T)$, $\overline{S}_4(T) = \mathbf{PGO}^+(T)$.

(44.8) Theorem. *The functors $S_4 \colon D_4(F) \to D^4(F)$ and $\overline{S}_4 \colon D_4(F) \to \overline{D}^4(F)$ are equivalences of categories.*

Proof: Since the natural functor $D^4(F) \to \overline{D}^4(F)$ is an equivalence by Theorem (26.7), it suffices to prove that $\overline{S}_4$ is an equivalence. Let $\Gamma = \mathrm{Gal}(F_{\mathrm{sep}}/F)$. The field extension functor $\mathbf{j} \colon D_4(F) \to D_4(F_{\mathrm{sep}})$ is clearly a Γ-embedding. We show first that the functor $\mathbf{j}$ satisfies the descent condition. Let $T = (E, L, \sigma, \alpha)$ be some object in $D_4(F)$ (split, for example). Consider the F-space

$$W = \mathrm{Hom}_F(E \otimes_F E, E) \oplus \mathrm{Hom}_F(E, E) \oplus \mathrm{Hom}_F\big(C(E, \sigma), E \otimes_F \Delta(L)\big),$$

the element $w = (m, \sigma, \alpha) \in W$ where m is the multiplication in E, and the representation

$$\rho \colon \mathbf{GL}(E) \times \mathbf{GL}\big(C(E, \sigma)\big) \to \mathbf{GL}(W)$$

given by

$$\rho(g, h)(x, y, p) = \big(g(x), g(y), h \circ p \circ (g \otimes 1)^{-1}\big)$$

where $g(x)$ and $g(y)$ is the result of the natural action of $\mathbf{GL}(E)$ on the first and second summands. By Proposition (26.4) the Γ-embedding

$$\mathbf{i} \colon A(\rho_{\mathrm{sep}}, w) \to \widetilde{A}(\rho_{\mathrm{sep}}, w)$$

satisfies the descent condition. We have a functor

$$\mathbf{T} = \mathbf{T}(F) \colon A(\rho_{\mathrm{sep}}, w) \to D_4(F)$$

taking $w' \in A(\rho_{\mathrm{sep}}, w)$ to the F-space E with the trialitarian structure defined by w'. A morphism between w' and w'' defines an isomorphism of the corresponding structures on D. The functor $\mathbf{T}$ has an evident Γ-extension

$$\widetilde{\mathbf{T}} = \mathbf{T}(F_{\mathrm{sep}}) \colon \widetilde{A}(\rho_{\mathrm{sep}}, w) \to D_4(F_{\mathrm{sep}}),$$

which is clearly an equivalence of groupoids. Since the functor $\mathbf{i}$ satisfies the descent condition, so does the functor $\mathbf{j}$.

For the proof of the theorem it suffices by Proposition (26.2) (and the following Remark (26.3)) to show that for some $T \in D_4(F)$ the functor $\mathbf{T}(F)$ for a separably closed field F induces a group isomorphism

(44.9) $$\mathrm{PGO}(T) = \mathrm{Aut}_{D_4(F)}(T) \to \mathrm{Aut}\big(\mathbf{PGO}^+(T)\big).$$

The restriction of this homomorphism to the subgroup $\mathrm{PGO}^+(T)$, which is of index 6, induces an isomorphism of this subgroup with the group of inner automorphisms $\mathrm{Int}\big(\mathbf{PGO}^+(A, \sigma, f)\big)$, which is a subgroup in $\mathrm{Aut}\big(\mathbf{PGO}^+(A, \sigma)\big)$ also of index 6 (see Theorem (25.16)). A straightforward computation shows that the elements θ, θ^+ in $\mathrm{PGO}^-(A, \sigma, f)$ induce outer automorphisms of $\mathbf{PGO}^+(A, \sigma, f)$ and (44.9) is an isomorphism. $\square$

Tits algebras. If (A, σ) is a degree 8 algebra with an orthogonal involution, the description of the Tits algebra of $G = \mathbf{Spin}(A, \sigma)$ is given in (27.B). Now let $T = (E, L, \sigma, \alpha)$ be a trialitarian algebra with L is a cubic field extension and let $G = \mathbf{Spin}(T)$. The Galois group Γ acts on C^* through $\mathrm{Gal}(L \otimes Z/F)$. There exists some $\chi \in C^*$ such that $F_\chi = L$ is the field of definition of χ. Since $G_L \simeq \mathbf{Spin}(E, \sigma)$ by Remark (44.7), we have $A_\chi = E$ for the corresponding Tits algebra.

44.B. The Clifford invariant. The exact sequence (35.11) of group schemes

$$1 \to C \to \mathbf{Spin}_8 \xrightarrow{\chi'} \mathbf{PGO}_8^+ \to 1$$

where C is the center of $\mathbf{Spin}_8$, induces an exact sequence

$$1 \to C \to \mathbf{Spin}_8 \rtimes S_3 \xrightarrow{\chi' \rtimes 1} \mathbf{PGO}_8^+ \rtimes S_3 \to 1$$

which leads to an exact sequence in cohomology

(44.10) $\qquad \to H^1(F, C) \to H^1(F, \mathbf{Spin}_8 \rtimes S_3) \xrightarrow{(\chi' \rtimes 1)^1} H^1(F, \mathbf{PGO}_8^+ \rtimes S_3).$

Since C is not central in $\mathbf{Spin}_8 \rtimes S_3$, there is no connecting homomorphism from the pointed set $H^1(F, \mathbf{PGO}_8^+ \rtimes S_3)$ to $H^2(F, C)$. However we can obtain a connecting homomorphism over a *fixed* cubic extension L_0 by "twisting" the action of $\mathrm{Gal}(F_{\mathrm{sep}}/F)$ on each term of the exact sequence (35.11) through the cocycle $\delta \colon \mathrm{Gal}(F_{\mathrm{sep}}/F) \to S_3$ defining L_0. We have a sequence of Galois modules

(44.11) $\qquad 1 \to (C)_\delta \to (\mathbf{Spin}_8)_\delta \xrightarrow{\chi'_\delta} (\mathbf{PGO}_8^+)_\delta \to 1.$

In turn (44.11) leads to a sequence in cohomology

(44.12) $\quad H^1(F, C_\delta) \to H^1\big(F, (\mathbf{Spin}_8)_\delta\big) \xrightarrow{\chi'^1_\delta} H^1\big(F, (\mathbf{PGO}_8^+)_\delta\big) \xrightarrow{\mathrm{Sn}^1} H^2(F, C_\delta).$

The set $H^1\big(F, (\mathbf{PGO}_8^+)_\delta\big)$ classifies pairs $(T, \phi \colon L \xrightarrow{\sim} L_0)$ where $T = (E, L, \sigma, \alpha)$ is a trialitarian algebra. Moreover the group $\mathrm{Aut}_F(L_0)$ acts on the pointed set $H^1\big(F, (\mathbf{PGO}_8^+)_\delta\big)$ and $H^1\big(F, (\mathbf{PGO}_8^+)_\delta\big)/\mathrm{Aut}_F(L_0)$ classifies trialitarian algebras (E, L, σ, α) with $L \simeq L_0$.

The map $H^1\big(F, (\mathbf{PGO}_8^+)_\delta\big) \to H^1(F, \mathbf{PGO}_8^+ \rtimes S_3)$, $[T, \phi \colon L \xrightarrow{\sim} L_0] \mapsto [T]$ has $(p^1)^{-1}([L_0])$ as image where

$$p^1 \colon H^1(F, \mathbf{PGO}_8^+ \rtimes S_3) \to H^1(F, S_3)$$

maps the class of a trialitarian algebra (E, L, σ, α_E) to the class of the cubic extension L. Corresponding results hold for $(\mathbf{Spin}_8)_\delta$; in particular $H^1\big(F, (\mathbf{Spin}_8)_\delta\big)$ classifies pairs

$$\big(\Gamma = (V, L, Q, \beta), \phi \colon L \xrightarrow{\sim} L_0\big)$$

where $\Gamma = (V, L, Q, \beta)$ is a twisted composition. The map

$$H^1\big(F, (\mathbf{Spin}_8)_\delta\big) \to H^1\big(F, (\mathbf{PGO}_8^+)_\delta\big)$$

associates to (Γ, ϕ) the pair $\big(\mathrm{End}(\Gamma), \phi\big)$. (See Example (43.7) for the definition of $\mathrm{End}(\Gamma)$.)

We call the class $\mathrm{Sn}^1([T, \phi]) \in H^2(F, C_\delta)$ the *Clifford invariant* of T and denote it by $c(T)$. Observe that it depends on the choice of a fixed L_0-structure on E.

(44.13) Proposition. *If* $L = F \times Z$ *and* $T = T(A, \sigma) = \big((A, \sigma) \times C(A, \sigma), \underline{\sigma}\big)$, *then* $c(T) = [C(A, \sigma)] \in \mathrm{Br}(Z)$.

Proof: The image of the homomorphism $\delta\colon \mathrm{Gal}(F_{\mathrm{sep}}/F) \to S_3$ is a subgroup of order 2 and

$$C(F_{\mathrm{sep}}) = \mu_2 \times \mu_2$$

(see Proposition (13.34)). Therefore we have $H^2(F, C_\delta) = H^2(Z, \mu_2)$ and it follows from the long exact sequence (44.12) that $c(T) = [C(A, \sigma)]$ in $\mathrm{Br}(Z)$. $\qquad\square$

The exact sequence (35.11)

$$1 \to C \to \mu_2 \times \mu_2 \times \mu_2 \xrightarrow{m} \mu_2 \to 1$$

was used to define the action of S_3 on C. As above, if L_0 is a fixed cubic étale F-algebra and $\delta\colon \mathrm{Gal}(F_{\mathrm{sep}}/F) \to S_3$ is a cocycle which defines L_0, we may use δ to twist the action of $\mathrm{Gal}(F_{\mathrm{sep}}/F)$ on the above sequence and consider the induced sequence in cohomology:

(44.14) Lemma. *For $i \geq 1$, there exists a commutative diagram*

$$
\begin{array}{ccccc}
H^i(F, C_\delta) & \longrightarrow & H^i\big(F, (\mu_2 \times \mu_2 \times \mu_2)_\delta\big) & \longrightarrow & H^i(F, \mu_2) \\
\downarrow & & \simeq\downarrow & & \| \\
H^i\big(L \otimes \Delta(L), \mu_2\big) & \xrightarrow{N_{L\otimes\Delta/L}} & H^i(L, \mu_2) & \xrightarrow{N_{L/F}^i} & H^i(F, \mu_2)
\end{array}
$$

where the first row is exact, the first vertical map is injective and the second is an isomorphism. In particular we have

$$H^i(F, C_\delta) \simeq \ker[N_{L/F}^i\colon H^i(L, \mu_2) \to H^i(F, \mu_2)].$$

Proof: The first vertical map is the composition of the restriction homomorphism

$$H^i(F, C_\delta) \to H^i(L, C_\delta),$$

which is injective since $[L:F] = 3$, with the isomorphism

$$\varphi_i\colon H^i(L, C_\delta) = H^i\big(L, R_{L\otimes\Delta/L}(\mu_2)\big) \xrightarrow{\sim} H^i(L \otimes \Delta, \mu_2)$$

(see Lemma (29.6) and Remark (29.7)). The map $\mu_2(L) \to \mu_2(L \otimes F_{\mathrm{sep}})$ yields an isomorphism

$$R_{L/F}(\mu_2) \xrightarrow{\sim} (\mu_2 \times \mu_2 \times \mu_2)_\delta$$

so that, by Lemma (29.6), we have an isomorphism

$$H^i\big(F, (\mu_2 \times \mu_2 \times \mu_2)_\delta\big) \xrightarrow{\sim} H^i(L, \mu_2).$$

Commutativity follows from the definition of the corestriction. $\qquad\square$

By Lemma (44.14) we have maps

$$\nu_1\colon H^2(F, C_\delta) \hookrightarrow H^2(L, C_\delta) \xrightarrow{\sim} H^2\big(L \otimes \Delta(L), \mu_2\big),$$
$$\nu_2\colon H^2(F, C_\delta) \to H^2\big(F, (\mu_2 \times \mu_2 \times \mu_2)_\delta\big) \xrightarrow{\sim} H^2(L, \mu_2),$$

(44.15) Proposition. *The image of the Clifford invariant $c(T)$ under ν_1 is the class $[C(E, \sigma)] \in \mathrm{Br}(L \otimes \Delta)$ and its image under ν_2 is the class $[E] \in \mathrm{Br}(L)$.*

Proof: The claim follows from Proposition (44.13) if L is not a field and the general case follows by tensoring with L. $\qquad\square$

Twisted compositions and trialitarian algebras. We conclude this section with a characterization of trialitarian algebras $T = (E, L, \sigma, \alpha_E)$ such that $[E] = 1 \in \mathrm{Br}(L)$.

(44.16) Proposition. (1) *If $T = (E, L, \sigma, \alpha_E)$ is a trialitarian algebra such that $[E] = 1 \in \mathrm{Br}(L)$, then there exists a twisted composition $\Gamma = (V, L, N, \beta)$ such that $T = \mathrm{End}(\Gamma)$.*
(2) *Γ, Γ' are twisted compositions such that $\mathrm{End}(\Gamma) \simeq \mathrm{End}(\Gamma')$ if and only if there exists $\lambda \in L^\times$ such that $\Gamma' \simeq \Gamma_\lambda$.*

Proof: (1) The trialitarian algebra (E, L, σ, α) is of the form $\mathrm{End}(\Gamma)$ if and only if its class is in the image of the map $(\chi' \rtimes 1)^1$ of sequence (44.12). Thus, in view of (44.12), the assertion will follow if we can show that the condition $[E] = 1$ in $\mathrm{Br}(L)$ implies $\mathrm{Sn}^1([x]) = 0$ for $[x] = [T, \phi] = [(E, L, \sigma, \alpha), \phi] \in H^1\big(F, (\mathbf{PGO}_8^+)_\delta\big)$. We first consider the case where $L = F \times \Delta$, so that $E = \big(A, C(A)\big)$ (see Proposition (43.15)). The homomorphism δ factors through S_2 and the action on $C = \mu_2 \times \mu_2$ in the sequence (44.11) is the twist. Thus $C = \mu_2 \times \mu_2$ is a permutation module. By Lemma (29.6) and Remark (29.7) , we have

$$H^2(F, C_\delta) \simeq H^2(\Delta, \mu_2)$$

and $\mathrm{Sn}^1([x]) = [C(A, \sigma)]$ (see Proposition (44.13)). Thus $[E] = 1$ implies $\mathrm{Sn}^1([x]) = 1$ as wanted. If L is a field, we extend scalars from F to L. Since L is a cubic extension, the restriction map $H^2(F, C_\delta) \to H^2(L, C_\delta)$ is injective and, since $L \otimes L \simeq L \times L \otimes \Delta$, we are reduced to the case $L = F \times \Delta$.
  (2) The group $H^1(F, C_\delta)$ operates transitively on the fibers of $(\chi' \rtimes 1)^1$; recall that by (44.14)

$$H^1(F, C_\delta) = \ker[N_{L/F}^1 \colon L^\times / L^{\times 2} \to F^\times / F^{\times 2}].$$

On the other hand we have an exact sequence

$$1 \to F^\times / F^{\times 2} \to L^\times / L^{\times 2} \xrightarrow{\#} L^\times / L^{\times 2} \xrightarrow{N_{L/F}^1} F^\times / F^{\times 2}$$

by Proposition (18.34), hence $H^1(F, C_\delta) \simeq \mathrm{im}(\#) \subset L^\times / L^{\times 2}$. One can then check that, for $[\lambda^\#] \in H^1(F, C_\delta)$, $[\lambda^\#]$ acts on $[\Gamma, \phi]$ as $[\lambda^\#] \cdot [\Gamma, \phi] = [\Gamma_\lambda, \phi]$.
  Now let Γ, Γ' be such that $\mathrm{End}(\Gamma) \simeq \mathrm{End}(\Gamma')$. We may assume that Γ, Γ' are defined over the same étale algebra L. Furthermore, since the action of $\mathrm{Aut}_F(L)$ is equivariant with respect to the map $(\chi')_\delta^1$ of sequence (44.11), we may assume that we have pairs (Γ, ϕ), (Γ', ϕ') such that $\big(\mathrm{End}(\Gamma), \phi\big) \simeq \big(\mathrm{End}(\Gamma'), \phi'\big)$. Then (Γ, ϕ), (Γ', ϕ') are in the same fiber and the claim follows from the definition of the action of $H^1(F, C_\delta)$ on this fiber. $\qquad\square$

§45. Lie Algebras and Triality

In this section we describe how trialitarian algebras are related to Lie algebras of type D_4. Most of the proofs will only be sketched. We still assume that char $F \neq 2$. We write $\mathfrak{o}_8$ for the Lie algebra of the orthogonal group $O(V, q)$ where q is a hyperbolic quadratic form of rank 8. As for the groups $\mathbf{Spin}_8$ and $\mathbf{PGO}_8^+$, there exists an S_3-action on the Lie algebra $\mathfrak{o}_8$, which is known as "local triality." Its description will again use Clifford algebras. For any quadratic space (V, q) we have

$$\mathfrak{o}(V, q) = \{\, f \in \mathrm{End}_F(V) \mid b_q(fx, y) + b_q(x, fy) = 0 \text{ for all } x, y \in V \,\}.$$

It turns out that this Lie algebra can be identified with a (Lie) subalgebra of the Clifford algebra $C(V, q)$, as we now show. (Compare Jacobson [140, pp. 231–232].)

(45.1) Lemma. *For x, y, $z \in V$ we have in $C(V, q)$:*

$$[[x, y], z] = 2\big(xb_q(y, z) - yb_q(x, z)\big) \in V.$$

Proof: This is a direct computation based on the fact that for v, $w \in V$, $b_q(v, w) = vw + wv$ in $C(V, q)$: one finds

$$
\begin{aligned}
[[x, y], z] &= (xyz + xzy + yzx + zyx) \\
&\quad - (yxz + yzx + xzy + zxy) \\
&= 2\big(xb_q(y, z) - yb_q(x, z)\big) \in V.
\end{aligned}
$$

for x, y, $z \in V$. $\qquad\qquad\qquad\qquad\qquad\qquad\qquad\qquad\qquad\qquad\qquad\qquad\qquad\square$

Let $[V, V] \subset C(V, q)$ be the subspace spanned by the brackets $[x, y] = xy - yx$ for x, $y \in V$. In view of (45.1) we may define a linear map

$$\mathrm{ad}\colon [V, V] \to \mathrm{End}_F(V)$$

by: $\mathrm{ad}_\xi(z) = [\xi, z]$ for $\xi \in [V, V]$ and $z \in V$. Lemma (45.1) yields:

(45.2) $\qquad\qquad \mathrm{ad}_{[x,y]} = 2\big(x \otimes \hat{b}_q(y) - y \otimes \hat{b}_q(x)\big)$ for x, $y \in V$.

(45.3) Lemma. (1) *The following diagram is commutative:*

$$
\begin{array}{ccc}
[V, V] & \longrightarrow & C_0(V, q) \\
{\scriptstyle \mathrm{ad}}\downarrow & & \downarrow{\scriptstyle \eta_q} \\
\mathrm{End}_F(V) & \xrightarrow{\ \frac{1}{2}c\ } & C(\mathrm{End}_F(V), \sigma_q)
\end{array}
$$

where c is the canonical map and η_q is the canonical identification of Proposition (8.8).

(2) *The subspace $[V, V]$ is a Lie subalgebra of $\mathfrak{L}\big(C_0(V, q)\big)$, and ad induces an isomorphism of Lie algebras:*

$$\mathrm{ad}\colon [V, V] \xrightarrow{\sim} \mathfrak{o}(V, q).$$

(3) *The restriction of the canonical map c to $\mathfrak{o}(V, q)$ yields an injective Lie algebra homomorphism:*

$$\tfrac{1}{2}c\colon \mathfrak{o}(V, q) \hookrightarrow \mathfrak{L}\big(C(\mathrm{End}(V), \sigma_q)\big).$$

Proof: (1) follows from (45.2) and from the definitions of c and η_q.

(2) Jacobi's identity yields for x, y, u, $v \in V$:

$$[[u, v], [x, y]] = [[[x, y], v], u] - [[[x, y], u], v].$$

Since Lemma (45.1) shows that $[[x, y], z] \in V$ for all x, y, $z \in V$, it follows that

$$[[u, v], [x, y]] \in [V, V].$$

Therefore, $[V, V]$ is a Lie subalgebra of $\mathfrak{L}\big(C_0(V, q)\big)$. Jacobi's identity also yields:

$$\mathrm{ad}_{[\xi,\zeta]} = [\mathrm{ad}_\xi, \mathrm{ad}_\zeta] \text{ for } \xi, \zeta \in [V, V],$$

hence ad is a Lie algebra homomorphism. From (45.2) it follows for x, y, u, $v \in V$ that:

$$b_q\big(\mathrm{ad}_{[x,y]}(u), v\big) = 2(b_q(x,v)b_q(y,u) - b_q(y,v)b_q(x,u))$$
$$= -b_q\big(u, \mathrm{ad}_{[x,y]}(v)\big),$$

hence $\mathrm{ad}_{[x,y]} \in \mathfrak{o}(V, q)$. Therefore, we may consider ad as a map:

$$\mathrm{ad}\colon [V, V] \to \mathfrak{o}(V, q).$$

It only remains to prove that this map is bijective. Let $n = \dim V$. Using an orthogonal basis of V, it is easily verified that $\dim[V,V] = n(n-1)/2 = \dim \mathfrak{o}(V, q)$. On the other hand, since η_q is an isomorphism, (1) shows that ad is injective; it is therefore also surjective.

(3) Using η_q to identify $[V, V]$ with a Lie subalgebra of $C(\mathrm{End}(V), \sigma_q)$, we derive from (1) and (2) that the restriction of $\frac{1}{2}c$ to $\mathfrak{o}(V, q)$ is the inverse of ad. Therefore, $\frac{1}{2}c$ is injective on $\mathfrak{o}(V, q)$ and is a Lie algebra homomorphism. $\square$

We have more in dimension 8:

(45.4) Lemma. *Let Z be the center of the even Clifford algebra $C_0(q)$. If V has dimension 8, the embedding $[V,V] \subset \mathfrak{L}\big(C_0(q), \tau\big)$ induces a canonical isomorphism of Lie Z-algebras $[V, V] \otimes Z \xrightarrow{\sim} \mathfrak{L}\big(C_0(q), \tau\big)$. Thus the adjoint representation induces an isomorphism* ad: $\mathfrak{L}\big(C_0(q), \tau\big) \xrightarrow{\sim} \mathfrak{o}(q) \otimes Z$.

Proof: Fixing an orthogonal basis of V, it is easy to check that $[V, V]$ and Z are linearly disjoint over F in $C_0(q)$, so that the canonical map is injective. It is surjective by dimension count. $\square$

45.A. Local triality. Let $(S, \star)$ be a symmetric composition algebra with norm n. The following proposition is known as the "triality principle" for the Lie algebra $\mathfrak{o}(n)$ or as "local triality."

(45.5) Proposition. *For any $\lambda \in \mathfrak{o}(n)$, there exist unique elements λ^+, $\lambda^- \in \mathfrak{o}(n)$ such that*

(1) $$\lambda^+(x \star y) = \lambda(x) \star y + x \star \lambda^-(y),$$

(2) $$\lambda^-(x \star y) = \lambda^+(x) \star y + x \star \lambda(y),$$

(3) $$\lambda(x \star y) = \lambda^-(x) \star y + x \star \lambda^+(y)$$

for all x, $y \in \mathfrak{o}(n)$.

Proof: Let $\lambda = \mathrm{ad}_\xi|_S$ for $\xi \in [S, S]$, so that ad_ξ extends to an inner derivation of $C_0(n)$, also written ad_ξ. Let $\alpha_S\colon C_0(n) \xrightarrow{\sim} \mathrm{End}_F(S) \times \mathrm{End}_F(S)$ be as in Proposition (35.1). The derivation $\alpha_S \circ \mathrm{ad}_\xi \circ \alpha_S^{-1}$ is equal to $\mathrm{ad}_{\alpha(\xi)}$; we write $\alpha(\xi)$ as (λ^+, λ^-) and, since ad_ξ commutes with τ, we see that λ^+, $\lambda^- \in \mathfrak{o}(n)$. For any $x \in S$ we have

$$\begin{pmatrix} 0 & \ell_{\lambda x} \\ r_{\lambda x} & 0 \end{pmatrix} = \begin{pmatrix} \lambda^+ & 0 \\ 0 & \lambda^- \end{pmatrix}\begin{pmatrix} 0 & \ell_x \\ r_x & 0 \end{pmatrix} - \begin{pmatrix} 0 & \ell_x \\ r_x & 0 \end{pmatrix}\begin{pmatrix} \lambda^+ & 0 \\ 0 & \lambda^- \end{pmatrix}$$

by definition of α_S, or

$$\lambda^+(x \star y) - x \star \lambda^-(y) = \lambda(x) \star y$$
$$\lambda^-(y \star x) - \lambda^+(y) \star x = y \star \lambda(x).$$

This gives formulas (1) and (2).

From (1) we obtain

$$b_n\big(\lambda^+(x \star y), z\big) = b_n\big(\lambda(x) \star y, z\big) + b_n\big(x \star \lambda^-(y), z\big).$$

Since $b_n(x \star y, z) = b_n(x, y \star z)$ and since λ^-, λ and λ^+ are in $\mathfrak{o}(n)$, this implies

$$-b_n\big(x, y \star \lambda^+(z)\big) = -b_n\big(x, \lambda(y \star z)\big) + b_n\big(x, \lambda^-(y) \star z\big)$$

for all x, y, and z in $\mathfrak{o}(n)$, hence (3). We leave uniqueness as an exercise. $\qquad\square$

Proposition (45.5) is a Lie analogue of Proposition (35.4). We have obvious Lie analogues of (35.7) and (35.9). Let $\theta^+(\lambda) = \lambda^+$, $\theta^-(\lambda) = \lambda^-$.

(45.6) Corollary. *For all x, $y \in \mathfrak{o}(n)$ we have*

$$\mathfrak{o}(n) \simeq \big\{ (\lambda, \lambda^+, \lambda^-) \in \mathfrak{o}(n) \times \mathfrak{o}(n) \times \mathfrak{o}(n) \mid \lambda(x \star y) = \lambda^-(x) \star y + x \star \lambda^+(y) \big\}$$

and the projections ρ, ρ^+, ρ^-: $(\lambda, \lambda^+, \lambda^-) \mapsto \lambda$, λ^+, λ^- give the three irreducible representations of $\mathfrak{o}(n)$ of degree 8. The maps θ^+, θ^- permute the representations ρ^+, ρ, ρ^-, hence are outer automorphisms of $\mathfrak{o}(n)$. They generate a group isomorphic to A_3 and $\mathfrak{o}(n)^{A_3}$ is the Lie algebra of derivations of the composition algebra S.

Proof: The projection ρ is the natural representation of $\mathfrak{o}(n)$ and $\rho^\pm$ correspond to the half-spin representations. These are the three non-equivalent irreducible representations of $\mathfrak{o}(n)$ of degree 8 (see Jacobson [140]). Since θ^+, θ^- permute these representations, they are outer automorphisms. $\qquad\square$

(45.7) Remark. The Lie algebra of derivations of a symmetric composition S is a simple Lie algebra of type A_2 if S is of type A_2 or is of type G_2 if S is of type G_2.

If the composition algebra $(S, \star, n)$ is a para-Cayley algebra $(\mathfrak{C}, \star, \mathfrak{n})$ with conjugation $\pi \colon x \mapsto \bar{x}$, we have, as in the case of $\mathbf{Spin}(\mathfrak{C}, \mathfrak{n})$, not only an action of A_3, but of S_3. For any $\lambda \in \mathfrak{o}(\mathfrak{n})$ the element $\theta(\lambda) = \pi\lambda\pi$ belongs to $\mathfrak{o}(\mathfrak{n})$. The automorphisms θ, θ^+ and θ^- of $\mathfrak{o}(\mathfrak{n})$ generate a group isomorphic to S_3.

(45.8) Theorem ([138, Theorem 5, p. 26]). *The group of F-automorphisms of the Lie algebra $\mathfrak{o}(\mathfrak{n})$ is isomorphic to the semidirect product $\mathbf{PGO}^+(\mathfrak{n}) \rtimes S_3$ where $\mathbf{PGO}^+$ operates through inner automorphisms and S_3 operates through θ^+, θ^- and θ.*

Proof: Let φ be an automorphism of $\mathfrak{o}(\mathfrak{n})$ and let ρ_i, $i = 1$, 2, 3, be the three irreducible representations of degree 8. Then $\rho_i \circ \varphi$ is again an irreducible representation of degree 8. By Jacobson [140, Chap. 9], there exist $\psi \in \mathrm{GL}(\mathfrak{C})$ and $\pi \in S_3$ such that

$$\rho_i \circ \varphi = \mathrm{Int}(\psi) \circ \rho_{\pi(i)}.$$

By Corollary (45.6) there exists some $\widetilde{\pi} \in \mathrm{Aut}\big(\mathfrak{o}(\mathfrak{n})\big)$ such that $\rho_{\pi(i)} = \rho_i \circ \widetilde{\pi}$. Hence we obtain

$$\rho_i \circ \varphi = \mathrm{Int}(\psi) \circ \rho_i \circ \widetilde{\pi}.$$

It follows in particular for the natural representation $\mathfrak{o}(\mathfrak{n}) \hookrightarrow \mathrm{End}_F(\mathfrak{C})$ that $\varphi = \mathrm{Int}(\psi) \circ \widetilde{\pi}$. It remains to show that $\mathrm{Int}(\psi) \in \mathbf{PGO}^+(\mathfrak{n})$ or that $\psi \in \mathrm{GO}_+(\mathfrak{n})$. For any $x \in \mathfrak{o}(\mathfrak{n})$, we have $\mathrm{Int}(\psi)(x) \in \mathfrak{o}(\mathfrak{n})$, hence

$$\hat{b}_\mathfrak{n}^{-1}(\psi x \psi^{-1})^* \hat{b}_\mathfrak{n} = \hat{b}_\mathfrak{n}^{-1} \psi^{*-1} x^* \psi^* \hat{b}_\mathfrak{n} = -\psi x \psi^{-1} = -\hat{b}_\mathfrak{n}^{-1} \psi^{*-1} \hat{b}_\mathfrak{n} x \hat{b}_\mathfrak{n}^{-1} \psi^* \hat{b}_\mathfrak{n},$$

so that $\psi^*\hat{b}_\mathfrak{n}\psi\hat{b}_\mathfrak{n}^{-1}$ is central in $\mathrm{End}_F(\mathfrak{C})$. Thus there exists some $\lambda \in F^\times$ such that $\psi^*\hat{b}_\mathfrak{n}\psi = \lambda\hat{b}_\mathfrak{n}$ and ψ is a similitude. The fact that ψ is proper follows from the fact that $\mathrm{Int}(\psi)$ does not switch the two half-spin representations. $\qquad\square$

A Lie algebra $\mathfrak{L}$ is of *type* D_4 if $\mathfrak{L} \otimes F_{\mathrm{sep}} \simeq \mathfrak{o}_8$. In particular $\mathfrak{o}(\mathfrak{n})$ is of type D_4.

(45.9) Corollary. *The pointed set $H^1(F, \mathbf{PGO}_8^+ \rtimes S_3)$ classifies Lie algebras of type D_4 over F.*

Proof: If F is separably closed, we have $\mathbf{PGO}^+(\mathfrak{n}) = \mathbf{PGO}_8^+$, so that Corollary (45.9) follows from Theorem (45.8) and (29.8). $\qquad\square$

45.B. Derivations of twisted compositions. Let $\Gamma = (V, L, Q, \beta)$ be a twisted composition and let $\beta(x, y) = \beta(x + y) - \beta(x) - \beta(y)$ for x, $y \in V$. An L-linear map $d\colon V \to V$ such that $d \in \mathfrak{o}(Q)$ and

(45.10) $$d\big(\beta(x, y)\big) = \beta(dx, y) + \beta(x, dy)$$

is a *derivation* of Γ. The set $\mathrm{Der}(\Gamma) = \mathrm{Der}(V, L, Q, \beta)$ of all derivations of Γ is a Lie algebra under the operation $[x, y] = x \circ y - y \circ x$. In fact we have

$$\mathrm{Der}(\Gamma) = \mathrm{Lie}\big(\mathbf{Spin}(V, L, Q, \beta)\big)$$

where $\mathbf{Spin}(V, L, Q, \beta)$ is as in §36.A.

If L/F is cyclic with ρ a generator of $\mathrm{Gal}(L/F)$ and $\beta(x) = x * x$, comparing the ρ-semilinear parts on both sides of (45.10) shows that (45.10) is equivalent with $d(x * y) = x * dy + dx * y$. If $\Gamma = \widetilde{\mathfrak{C}}$ for $\mathfrak{C}$ a Cayley algebra, the formula $d(x * y) = x * dy + dx * y$ and Corollary (45.6) implies that $\mathrm{Der}(\widetilde{\mathfrak{C}}) \simeq \mathfrak{o}(\mathfrak{n})$. Hence, by descent, $\mathrm{Der}(\Gamma)$ is always a Lie algebra of type D_4.

Let J be an Albert algebra over a field F of characteristic $\neq 2, 3$. The F-vector space $\mathrm{Der}(J)$ is a Lie algebra of type F_4 (see Chevalley-Schafer [67] or Schafer [244, Theorem 4.9, p. 112]). Let L be a cubic étale subalgebra of J and let $J = J(V, L) = L \oplus V$ be the corresponding Springer decomposition. Let $\mathrm{Der}(J/L)$ be the F-subspace of $\mathrm{Der}(J)$ of derivations which are zero on L. We have an obvious isomorphism

$$\mathrm{Der}(\Gamma) \simeq \mathrm{Der}(J/L)$$

obtained by extending any derivation of Γ to a derivation of J by mapping L to zero. Thus $\mathrm{Der}(J/L)$ is a Lie algebra of type D_4. Such a Lie algebra is said of *Jordan type*. We have thus shown the following:

(45.11) Proposition. *Every Lie algebra of Jordan type is isomorphic to $\mathrm{Der}(\Gamma)$ for some twisted composition Γ.*

45.C. Lie algebras and trialitarian algebras. We may also associate a Lie algebra $\mathfrak{L}(T)$ to a trialitarian algebra $T = (E, L, \sigma, \alpha)$:

$$\mathfrak{L}(T) = \{\, x \in \mathfrak{L}(E, \sigma) \mid \alpha(x) = x \otimes 1 \,\}$$

where $\mathfrak{L}(E, \sigma)$ is the Lie algebra of skew-symmetric elements in (E, σ) and can be identified with a Lie subalgebra of $C(E, \sigma)$ in view of Lemma (45.3). For $T = \mathrm{End}(\widetilde{\mathfrak{C}})$ we obtain

$$\mathfrak{L}(T) \simeq \mathfrak{L}\big(\mathrm{End}_F(\mathfrak{C}), \sigma_\mathfrak{n}\big) \simeq \mathfrak{o}(\mathfrak{n})$$

by (43.3), hence $\mathfrak{L}(T)$ is of type D_4. We shall see that any simple Lie algebra of type D_4 is of the form $\mathfrak{L}(T)$ for some trialitarian algebra T.

(45.12) Proposition. *The restriction map induces an isomorphism of algebraic group schemes*

$$\mathbf{Aut}_{\mathrm{alg}}\big(\mathrm{End}(\widetilde{\mathfrak{C}})\big) \xrightarrow{\sim} \mathbf{Aut}_{\mathrm{alg}}\big(\mathfrak{o}(\mathfrak{n})\big).$$

Proof: The restriction map induces a group homomorphism

$$\mathrm{Aut}_F\big(\mathrm{End}(\widetilde{\mathfrak{C}})\big) \to \mathrm{Aut}_F\big(\mathfrak{o}(\mathfrak{n})\big).$$

Since $\mathfrak{o}(\mathfrak{n})$ generates $C_0(\mathfrak{n})$ over F it generates $C_0(\mathfrak{n})_{(F \times F \times F)}$ over $F \times F \times F$ and the map is injective. To prove surjectivity, we show that any automorphism of $\mathfrak{o}(\mathfrak{n})$ extends to an automorphism of $\mathrm{End}(\widetilde{\mathfrak{C}})$. The group $\mathrm{Aut}_F\big(\mathfrak{o}(\mathfrak{n})\big)$ is the semidirect product of the group of inner automorphisms with the group S_3 where S_3 acts as in Corollary (45.6). An inner automorphism is of the form $\mathrm{Int}(f)$ where f is a direct similitude of $(\mathfrak{C}, \mathfrak{n})$ with multiplier λ. By Equation (45.2) we see that in $C(\mathfrak{C}, \mathfrak{n})$

$$\lambda^{-1} \mathrm{ad}_{[f(x), f(y)]}\, z = 2\lambda^{-1}\big(f(x)b_{\mathfrak{n}}\big(f(y), z\big) - f(y)b_{\mathfrak{n}}\big(f(x), z\big)\big)$$
$$= f\big(\mathrm{ad}_{[x,y]}\big(f^{-1}(z)\big)\big).$$

Thus

$$\mathrm{ad} \circ C_0(f) = \mathrm{Int}(f) \circ \mathrm{ad}$$

holds in the Lie algebra $[\mathfrak{C}, \mathfrak{C}] \subset C_0(\mathfrak{C}, \mathfrak{n})$. Since $[\mathfrak{C}, \mathfrak{C}]$ generates $C_0(\mathfrak{C}, \mathfrak{n})$, the automorphism $\big(C_0(f), \mathrm{Int}(f)\big)$ of $\mathrm{End}(\widetilde{\mathfrak{C}})$ extends $\mathrm{Int}(f)$. We now extend the automorphisms $\theta^\pm$ of $\mathfrak{o}(\mathfrak{n})$ to automorphisms of $\mathrm{End}(\widetilde{\mathfrak{C}})$. Let $\nu \colon \mathfrak{o}(\mathfrak{n}) \to C_0(\mathfrak{n})_{(F \times F \times F)}$, $\xi \mapsto \big(\xi, \rho_1(\xi), \rho_2(\xi)\big)$ be the canonical embedding. Since $\rho_1 \nu = \nu\theta^+$ and $\rho_2 \nu = \nu\theta^-$, the extension of θ^+ is (ρ_1, ρ_1) and the extension of θ^- is (ρ_2, ρ_2). Let $\rho_\epsilon(x_0, x_1, x_2) = (x_0, x_2, x_1)$. The fact that $\epsilon \in \mathrm{Aut}_F\big(\mathfrak{o}(\mathfrak{n})\big)$ extends follows from $\nu\epsilon = \mathrm{Int}(\pi)\rho_\epsilon\nu$. $\qquad\square$

(45.13) Corollary. *Any Lie algebra $\mathfrak{L}$ of type D_4 over F is of the form $\mathfrak{L}(T)$ for some trialitarian algebra T which is uniquely determined up to isomorphism by $\mathfrak{L}$.*

Proof: By (45.12) trialitarian algebras and Lie algebras of type D_4 are classified by the same pointed set $H^1(F, \mathbf{PGO}_8^+ \rtimes S_3)$ and, in view of (45.12), the same descent datum associated to a cohomology class gives the trialitarian algebra T and its Lie subalgebra $\mathfrak{L}(T)$. $\qquad\square$

(45.14) Remark. We denote the trialitarian algebra $T = (E, L, \sigma, \alpha)$ corresponding to the Lie algebra $\mathfrak{L}$ by $T(\mathfrak{L}) = \big(E(\mathfrak{L}), L(\mathfrak{L}), \sigma, \alpha\big)$. The semisimple F-algebra $E(\mathfrak{L})$ (and its center $L(\mathfrak{L})$) was already defined by Jacobson [135] and Allen [16] through Galois descent for any Lie algebra $\mathfrak{L}$ of type D_4. More precisely, if $\mathfrak{L}$ is a Lie algebra of type D_4, then $\mathfrak{L}_s = \mathfrak{L} \otimes F_{\mathrm{sep}}$ can be identified with

$$\mathfrak{o}(\mathfrak{n}_s) \simeq \mathfrak{S}(\mathfrak{n}_s) \subset \mathrm{End}_{F_s}(\mathfrak{C}_s) \times \mathrm{End}_{F_s}(\mathfrak{C}_s) \times \mathrm{End}_{F_s}(\mathfrak{C}_s)$$

where

$$\mathfrak{S}(\mathfrak{n}_s) = \big\{\, (\lambda, \lambda^+, \lambda^-) \in \mathfrak{o}(\mathfrak{n}_s) \times \mathfrak{o}(\mathfrak{n}_s) \times \mathfrak{o}(\mathfrak{n}_s) \mid \lambda(x \star y) = \lambda^-(x) \star y + x \star \lambda^+(y) \,\big\}$$

(see (45.6)) and $E(\mathfrak{L})$ is the associative F-subalgebra of $\mathrm{End}_{F_s}(\mathfrak{C}_s) \times \mathrm{End}_{F_s}(\mathfrak{C}_s) \times \mathrm{End}_{F_s}(\mathfrak{C}_s)$ generated by the image of $\mathfrak{L}$. The algebra $E(\mathfrak{L})$ is called the *Allen invariant* of $\mathfrak{L}$ in Allison [19].

In particular:

(45.15) Proposition (Jacobson [135, §4]). *For (A, σ) a central simple algebra of degree 8 over F with orthogonal involution,*

$$\mathfrak{L}\big(T(A, \sigma)\big) \simeq \mathfrak{L}(A, \sigma)$$

where $T(A, \sigma)$ is as in (43.15). *In particular any Lie algebra $\mathfrak{L}$ of type $^1D_4$ or $^2D_4$ is of the form $\mathfrak{L}(A, \sigma)$. The algebra $\mathfrak{L}$ is of type $^1D_4$ if and only if the discriminant of the involution σ is trivial.* ☐

We conclude with a result of Allen [16, Theorem I, p. 258]:

(45.16) Proposition (Allen). *The Allen invariant of a Lie algebra $\mathfrak{L}$ of type D_4 is a full matrix ring over its center if and only if the algebra is a Lie algebra of Jordan type.*

Proof: Let $\mathfrak{L}$ be of type D_4. If $[E(\mathfrak{L})] = 1$ in $\mathrm{Br}(L)$ then by Proposition (44.16) $T(\mathfrak{L}) \simeq \mathrm{End}(\Gamma)$ for some twisted composition Γ. Then $\mathfrak{L} \simeq \mathfrak{L}\big(\mathrm{End}(\Gamma)\big)$, which is isomorphic to $\mathrm{Der}(\Gamma)$, and the assertion follows by Proposition (45.11) Conversely, if $\mathfrak{L}$ is of Jordan type, we have $\mathfrak{L} \simeq \mathfrak{L}(T)$ for $T \simeq \mathrm{End}(\Gamma)$, Γ a twisted composition, hence the claim. ☐

EXERCISE

1. Describe all real and $\mathfrak{p}$-adic trialitarian algebras.

NOTES

The notion of a trialitarian algebra defined here seems to be new, and our definition may be not the final one. The main reason for assuming characteristic different from 2, is that in characteristic 2 we need to work with quadratic pairs. The involution $\underline{\sigma}$ of $C(A, \sigma, f)$ is part of a quadratic pair if A has degree 8 (see the notes of Chapter II). Thus, if $C(A, \sigma, f) \xrightarrow{\sim} (B, \sigma_B) \times (C, \sigma_C)$, the involutions σ_B and σ_C will also be parts of quadratic pairs (as it should be by triality!). However we did not succeed in giving a rational definition of the quadratic pair on $C(A, \sigma, f)$.

It may be still useful to explain how we came to the concept of trialitarian algebras, out of three different situations:

(I) Having the notion of a twisted composition $\Gamma = (V, L, Q, \beta)$, which is in particular a quadratic space (V, Q) over a cubic étale algebra L, it is tempting to consider the algebra with involution $\big(\mathrm{End}_L(V), \sigma_L\big)$ and to try to describe the structure induced from the existence of β.

(II) In the study of outer forms of Lie algebras of type D_4 Jacobson [135] introduced the semisimple algebra $E(\mathfrak{L})$, as defined in Remark (45.14), and studied the cases $^1D_4$ and $^2D_4$; in particular he proved Proposition (45.15). The techniques of Jacobson were then applied by Allen [16] to arbitrary outer forms. Allen proved in particular that $N_{L/F}\big(E(\mathfrak{L})\big) = 1$ (see Proposition (43.6)) and associated a cohomological invariant in $H^2(L, \mathbf{G}_{\mathrm{m}})$ to the Lie algebra $\mathfrak{L}$. In fact this invariant is just the image in $H^2(L, \mathbf{G}_{\mathrm{m}})$ of our Clifford invariant. It is used by Allen in his proof of Proposition (45.16). As an application, Allen obtained the classification of Lie algebras of type D_4 over finite and p-adic fields. In [19] Allison used the algebra $E(\mathfrak{L})$ (which he called the Allen algebra) to construct all Lie algebras of type D_4 over a number field. One step in his proof is Proposition (43.8) in the special case of number fields (see [19, Proposition 6.1]).

(III) For any central simple algebra (A, σ) of degree 8 with an orthogonal involution having trivial discriminant, we have $C(A, \sigma) \simeq B \times C$, with B, C of degree 8 with an orthogonal involution having trivial discriminant. At this stage one can easily suspect that triality permutes A, B and C. In connection with (I) and (II), the next step is to view the triple A, B, C as an algebra over $F \times F \times F$, and this explains how the Clifford algebra comes into the picture.

Quaternionic trialitarian algebras (see §43.B) were recently used by Garibaldi [103] to construct all isotropic algebraic groups of type $^3D_4$ and $^6D_4$ over a field of characteristic not 2.

Bibliography

Page references at the end of entries indicate the places in the text where the entry is quoted.

1. J. F. Adams, *Spin(8), triality, F_4 and all that*, Superspace and Supergravity (S. W. Hawking and M. Roček, eds.), Cambridge University Press, Cambridge, 1981, Proceedings of the Nuffield Workshop, Cambridge, June 16 - July 12, 1980, pp. 435–445. *511.*

2. A. A. Albert, *On the Wedderburn norm condition for cyclic algebras*, Bull. Amer. Math. Soc. **37** (1931), 301–312, (also: [12, pp. 185–196]). *205, 236, 275.*

3. _____, *Normal division algebras of degree four over an algebraic field*, Trans. Amer. Math. Soc. **34** (1932), 363–372, (also: [12, pp. 279–288]). *233.*

4. _____, *On a certain algebra of quantum mechanics*, Ann. of Math. (2) **35** (1934), 65–73, (also: [13, pp. 85–93]). *516, 542.*

5. _____, *Involutorial simple algebras and real Riemann matrices*, Ann. of Math. (2) **36** (1935), 886–964, (also: [12, pp. 475–553]). *68.*

6. _____, *A structure theory for Jordan algebras*, Ann. of Math. (2) **48** (1947), 546–567, (also: [13, pp. 401–422]). *514, 516, 542.*

7. _____, *A theory of power-associative commutative algebras*, Trans. Amer. Math. Soc. **69** (1950), 503–527, (also: [13, pp. 515–539]). *543.*

8. _____, *A construction of exceptional Jordan division algebras*, Ann. of Math. (2) **67** (1958), 1–28, (also: [13, pp. 701–728]). *545.*

9. _____, *Structure of algebras*, American Mathematical Society, Providence, R.I., 1961, Revised printing of the 1939 edition. American Mathematical Society Colloquium Publications, Vol. XXIV. *67, 68, 191, 270, 321, 414, 415, 447, 555.*

10. _____, *On involutorial associative division algebras*, Scripta Math. **26** (1963), 309–316, (also: [12, pp. 679–686]). *308, 322, 527.*

11. _____, *Tensor products of quaternion algebras*, Proc. Amer. Math. Soc. **35** (1972), 65–66, (also: [12, pp. 739–740]). *251, 275.*

12. _____, *Collected mathematical papers. Part 1*, American Mathematical Society, Providence, RI, 1993, Associative algebras and Riemann matrices, Edited by R. E. Block, N. Jacobson, J. Marshall Osborn, D. J. Saltman and D. Zelinsky.

13. _____, *Collected mathematical papers. Part 2*, American Mathematical Society, Providence, RI, 1993, Nonassociative algebras and miscellany, Edited by R. E. Block, N. Jacobson, J. Marshall Osborn, D. J. Saltman and D. Zelinsky.

14. A. A. Albert and N. Jacobson, *On reduced exceptional simple Jordan algebras*, Ann. of Math. (2) **66** (1957), 400–417, (also: [13, pp. 677–694], [144, pp. 310–327]). *519, 541, 543.*

15. A. A. Albert and L. J. Paige, *On a homomorphism property of certain Jordan algebras*, Trans. Amer. Math. Soc. **93** (1959), 20–29, (also: [13, pp. 755–764]). *516.*

16. H. P. Allen, *Jordan algebras and Lie algebras of type D_4*, J. Algebra **5** (1967), 250–265. *570, 571.*

17. _____, *Hermitian forms. II*, J. Algebra **10** (1968), 503–515. *149.*

18. H. P. Allen and J. C. Ferrar, *New simple Lie algebras of type D_4*, Bull. Amer. Math. Soc. **74** (1968), 478–483. *556–558.*

19. B. N. Allison, *Lie algebras of type D_4 over number fields*, Pacific J. Math. **156** (1992), no. 2, 209–250. *555, 570–572.*

20. S. A. Amitsur, L. H. Rowen, and J.-P. Tignol, *Division algebras of degree 4 and 8 with involution*, Israel J. Math. **33** (1979), no. 2, 133–148. *27, 552.*

21. J. K. Arason, *Cohomologische Invarianten quadratischer Formen*, J. Algebra **36** (1975), no. 3, 448–491. *436, 447.*

22. J. K. Arason and R. Elman, *Nilpotence in the Witt ring*, Amer. J. Math. **113** (1991), no. 5, 861–875. *447*.

23. J. K. Arason, R. Elman, and B. Jacob, *Fields of cohomological 2-dimension three*, Math. Ann. **274** (1986), no. 4, 649–657. *276, 413*.

24. E. Artin, *Geometric algebra*, Wiley Classics Library, John Wiley & Sons Inc., New York, 1988, Reprint of the 1957 original, A Wiley-Interscience Publication. *19, 154*.

25. M. Artin, *Brauer-Severi varieties* (Notes by A. Verschoren), Brauer groups in ring theory and algebraic geometry (Wilrijk, 1981) (F. M. J. van Oystaeyen and A. H. M. J. Verschoren, eds.), Lecture Notes in Math., vol. 917, Springer, Berlin, 1982, pp. 194–210. *67, 447*.

26. M. Auslander and O. Goldman, *The Brauer group of a commutative ring*, Trans. Amer. Math. Soc. **97** (1960), 367–409. *321*.

27. R. Baer, *Linear algebra and projective geometry*, Academic Press Inc., New York, N. Y., 1952. *8*.

28. R. Baeza, *Quadratic forms over semilocal rings*, Springer-Verlag, Berlin, 1978, Lecture Notes in Mathematics, Vol. 655. *xix, 183, 299, 461, 474*.

29. A. Bak, *K-theory of forms*, Annals of Mathematics Studies, vol. 98, Princeton University Press, Princeton, N.J., 1981. *69*.

30. A. D. Barnard, *Commutative rings with operators (Galois theory and ramification)*, Proc. London Math. Soc. (3) **28** (1974), 274–290. *320*.

31. H.-J. Bartels, *Invarianten hermitescher Formen über Schiefkörpern*, Math. Ann. **215** (1975), 269–288. *448*.

32. E. Bayer-Fluckiger, *Multiplicateurs de similitudes*, C. R. Acad. Sci. Paris Sér. I Math. **319** (1994), no. 11, 1151–1153. *204*.

33. E. Bayer-Fluckiger and H. W. Lenstra, Jr., *Forms in odd degree extensions and self-dual normal bases*, Amer. J. Math. **112** (1990), no. 3, 359–373. *79, 80*.

34. E. Bayer-Fluckiger and R. Parimala, *Galois cohomology of the classical groups over fields of cohomological dimension ≤ 2*, Invent. Math. **122** (1995), no. 2, 195–229. *202, 449*.

35. _____, *Classical groups and Hasse principle*, preprint, 1998. *449*.

36. E. Bayer-Fluckiger, D. B. Shapiro, and J.-P. Tignol, *Hyperbolic involutions*, Math. Z. **214** (1993), no. 3, 461–476. *145, 149*.

37. K. I. Beĭdar, W. S. Martindale III, and A. V. Mikhalëv, *Lie isomorphisms in prime rings with involution*, J. Algebra **169** (1994), no. 1, 304–327. *68*.

38. A. Berele and D. J. Saltman, *The centers of generic division algebras with involution*, Israel J. Math. **63** (1988), no. 1, 98–118. *31*.

39. A.-M. Bergé and J. Martinet, *Formes quadratiques et extensions en caractéristique 2*, Ann. Inst. Fourier (Grenoble) **35** (1985), no. 2, 57–77. *320, 321*.

40. E. R. Berlekamp, *An analog to the discriminant over fields of characteristic two*, J. Algebra **38** (1976), no. 2, 315–317. *321*.

41. F. van der Blij, *History of the octaves*, Simon Stevin **34** (1960/1961), 106–125. *509*.

42. F. van der Blij and T. A. Springer, *The arithmetics of octaves and of the group G_2*, Nederl. Akad. Wetensch. Proc. Ser. A 62 = Indag. Math. **21** (1959), 406–418. *456, 510*.

43. _____, *Octaves and triality*, Nieuw Arch. Wisk. (3) **8** (1960), 158–169. *510, 511*.

44. D. Boos, *Ein tensorkategorieller Zugang zum Satz von Hurwitz*, Diplomarbeit, ETH Zürich, 1998.

45. A. Borel, *Linear algebraic groups*, second ed., Graduate Texts in Mathematics, vol. 126, Springer-Verlag, New York, 1991. *323, 339, 342, 347–349, 381*.

46. A. Borel and J.-P. Serre, *Théorèmes de finitude en cohomologie galoisienne*, Comment. Math. Helv. **39** (1964), 111–164. *446*.

47. A. Borel and J. Tits, *Groupes réductifs*, Inst. Hautes Études Sci. Publ. Math. (1965), no. 27, 55–150. *149, 323, 355*.

48. _____, *Compléments à l'article: "Groupes réductifs"*, Inst. Hautes Études Sci. Publ. Math. (1972), no. 41, 253–276. *438*.

49. N. Bourbaki, *Éléments de mathématique*, Masson, Paris, 1981, Algèbre. Chapitres 4 à 7. *281, 300, 338, 412, 414*.

50. _____, *Éléments de mathématique*, Masson, Paris, 1985, Algèbre commutative. Chapitres 1 à 4. *323, 339, 340*.

51. _____, *Éléments de mathématique*, Masson, Paris, 1985, Algèbre commutative. Chapitres 5 à 7. *321, 323, 331*.

52. ———, *Éléments de mathématique*, Masson, Paris, 1983, Algèbre commutative. Chapitres 8 et 9. *323*.

53. ———, *Éléments de mathématique*, Masson, Paris, 1981, Groupes et algèbres de Lie. Chapitres 4, 5 et 6. *352*.

54. ———, *Éléments de mathématique*, Hermann, Paris, 1975, Groupes et algèbres de Lie. Chapitre 7 et 8. *354*.

55. H. Braun and M. Koecher, *Jordan-Algebren*, Grundlehren der Mathematischen Wissenschaften, vol. 128, Springer-Verlag, Berlin, 1966. *542*.

56. K. S. Brown, *Cohomology of groups*, Graduate Texts in Mathematics, vol. 87, Springer-Verlag, New York, 1982. *384, 395, 417, 418, 423, 446, 555*.

57. É. Cartan, *Les groupes réels simples finis et continus*, Ann. Sci. École Norm. Sup. **31** (1914), 263–355, (also: [60, pp. 399–491]). *509*.

58. ———, *Le principe de dualité et la théorie des groupes simples et semi-simples*, Bull. Sci. Math. **49** (1925), 361–374, (also: [60, pp. 555–568]). *511*.

59. ———, *Leçons sur la théorie des spineurs*, Hermann, Paris, 1938, english transl.: The theory of spinors, Dover Publications Inc., New York, 1981. *511*.

60. ———, *Œuvres complètes. Partie I. Groupes de Lie*, Gauthier-Villars, Paris, 1952. *509, 511*.

61. A. Cayley, *Collected mathematical papers. Vol. I-XIII*, Johnson Reprint Corporation, New York and London, 1961. *509*.

62. S. U. Chase, D. K. Harrison, and A. Rosenberg, *Galois theory and Galois cohomology of commutative rings*, Mem. Amer. Math. Soc. No. **52** (1965), 15–33. *321*.

63. F. Châtelet, *Variations sur un thème de H. Poincaré*, Ann. Sci. École Norm. Sup. (3) **61** (1944), 249–300. *10, 446*.

64. ———, *Méthode galoisienne et courbes de genre un*, Ann. Univ. Lyon. Sect. A. (3) **9** (1946), 40–49. *446*.

65. V. I. Chernousov, *A remark on the* (mod 5)-*invariant of Serre for groups of type* E_8, Mat. Zametki **56** (1994), no. 1, 116–121, 157 (Russian), english transl.: Mathematical Notes (Institute of Mathematics, Belorussian Academy of Science, Minsk), Vol. **56** (1994), Nos. 1–2, 730–733. *449*.

66. C. C. Chevalley, *The algebraic theory of spinors*, Columbia University Press, New York, 1954, also: Collected Works, vol. 2, Springer-Verlag, Berlin, 1996. *481, 492, 511*.

67. C. C. Chevalley and R. D. Schafer, *The exceptional simple Lie algebras* F_4 *and* E_6, Proc. Nat. Acad. Sci. U.S.A. **36** (1950), 137–141. *359, 543, 569*.

68. I. Dejaiffe, *Somme orthogonale d'algèbres à involution et algèbre de Clifford*, Preprint, 1996. *149*.

69. P. Deligne and D. Sullivan, *Division algebras and the Hausdorff-Banach-Tarski paradox*, Enseign. Math. (2) **29** (1983), no. 1-2, 145–150. *31*.

70. M. Demazure and P. Gabriel, *Groupes algébriques. Tome I: Géométrie algébrique, généralités, groupes commutatifs*, Masson & Cie, Éditeur, Paris, 1970, Avec un appendice *Corps de classes local* par Michiel Hazewinkel. *381*.

71. F. DeMeyer and E. Ingraham, *Separable algebras over commutative rings*, Springer-Verlag, Berlin, 1971, Lecture Notes in Mathematics, Vol. 181. *321*.

72. M. Deuring, *Algebren*, Springer-Verlag, Berlin, 1968, Zweite, korrigierte Auflage. Ergebnisse der Mathematik und ihrer Grenzgebiete, Band 41. *67*.

73. H. Dherte, *Decomposition of Mal'cev-Neumann division algebras with involution*, Math. Z. **216** (1994), no. 4, 629–644. *550*.

74. L. E. Dickson, *Linear groups*, Teubner, Leipzig, 1901. *203*.

75. ———, *Linear algebras*, Trans. Amer. Math. Soc. **13** (1912), 59–73. *509*.

76. ———, *Linear algebras*, Hafner Publishing Co., New York, 1960, Reprint of Cambridge Tracts in Mathematics and Mathematical Physics, No. 16, 1914. *509*.

77. J. Dieudonné, *Les extensions quadratiques des corps non commutatifs et leurs applications*, Acta Math. **87** (1952), 175–242, (also: [82, pp. 296–363]). *274, 275*.

78. ———, *On the structure of unitary groups*, Trans. Amer. Math. Soc. **72** (1952), 367–385, (also: [82, pp. 277–295]). *266*.

79. ———, *Sur les multiplicateurs des similitudes*, Rend. Circ. Mat. Palermo (2) **3** (1954), 398–408 (1955), (also: [82, pp. 408–418]). *203*.

80. _____, *Pseudo-discriminant and Dickson invariant*, Pacific J. Math. **5** (1955), 907–910, (also: [82, pp. 419–422]). *203*.

81. _____, *Sur les groupes classiques*, Hermann, Paris, 1967, Publications de l'Institut de Mathématique de l'Université de Strasbourg, VI, Actualités Scientifiques et Industrielles, No. 1040. *178, 180*.

82. _____, *Choix d'œuvres mathématiques. Tome II*, Hermann, Paris, 1981.

83. P. K. Draxl, *Über gemeinsame separabel-quadratische Zerfällungskörper von Quaternionenalgebren*, Nachr. Akad. Wiss. Göttingen Math.-Phys. Kl. II **16** (1975), 251–259. *275*.

84. _____, *Skew fields*, London Mathematical Society Lecture Note Series, vol. 81, Cambridge University Press, Cambridge, 1983. *4, 5, 18, 25, 27, 36, 38, 39, 50, 163, 302, 303*.

85. P. K. Draxl and M. Kneser (eds.), SK_1 *von Schiefkörpern*, Lecture Notes in Mathematics, vol. 778, Berlin, Springer, 1980, Seminar held at Bielefeld and Göttingen, 1976. *253*.

86. A. Ducros, *Principe de Hasse pour les espaces principaux homogènes sous les groupes classiques sur un corps de dimension cohomologique virtuelle au plus 1*, Manuscripta Math. **89** (1996), no. 3, 335–354. *449*.

87. B. Eckmann, *Cohomology of groups and transfer*, Ann. of Math. (2) **58** (1953), 481–493. *447*.

88. A. Elduque and H. C. Myung, *Flexible composition algebras and Okubo algebras*, Comm. Algebra **19** (1991), no. 4, 1197–1227. *510*.

89. _____, *On flexible composition algebras*, Comm. Algebra **21** (1993), no. 7, 2481–2505. *463, 479, 510*.

90. A. Elduque and J. M. Pérez, *Composition algebras with associative bilinear form*, Comm. Algebra **24** (1996), no. 3, 1091–1116. *463, 467, 468, 476, 510*.

91. R. Elman and T. Y. Lam, *Pfister forms and K-theory of fields*, J. Algebra **23** (1972), 181–213. *276, 413*.

92. _____, *Classification theorems for quadratic forms over fields*, Comment. Math. Helv. **49** (1974), 373–381. *203*.

93. H. Esnault, B. Kahn, M. Levine, and E. Viehweg, *The Arason invariant and mod 2 algebraic cycles*, J. Amer. Math. Soc. **11** (1998), no. 1, 73–118. *449*.

94. D. K. Faddeev, *On the theory of homology in groups*, Izv. Akad. Nauk SSSR Ser. Mat. **16** (1952), 17–22. *447*.

95. J. R. Faulkner, *Finding octonion algebras in associative algebras*, Proc. Amer. Math. Soc. **104** (1988), no. 4, 1027–1030. *463, 469, 472, 510*.

96. J. R. Faulkner and J. C. Ferrar, *Exceptional Lie algebras and related algebraic and geometric structures*, Bull. London Math. Soc. **9** (1977), no. 1, 1–35. *381*.

97. J. C. Ferrar and H. P. Petersson, *Exceptional simple Jordan algebras and Galois cohomology*, Arch. Math. (Basel) **61** (1993), no. 6, 517–520. *529, 530, 545*.

98. R. W. Fitzgerald, *Witt kernels of function field extensions*, Pacific J. Math. **109** (1983), no. 1, 89–106. *276*.

99. J. Frenkel, *Cohomologie non abélienne et espaces fibrés*, Bull. Soc. Math. France **85** (1957), 135–220. *446*.

100. H. Freudenthal, *Beziehungen der E_7 und E_8 zur Oktavenebene. I*, Nederl. Akad. Wetensch. Proc. Ser. A. 57 = Indag. Math. **16** (1954), 218–230. *543*.

101. _____, *Oktaven, Ausnahmegruppen und Oktavengeometrie*, Geom. Dedicata **19** (1985), no. 1, 7–63, Reprint of Utrecht Lecture Notes, 1951. *359, 360, 509, 511, 543*.

102. A. Fröhlich and A. M. McEvett, *Forms over rings with involution*, J. Algebra **12** (1969), 79–104. *68*.

103. R. Garibaldi, *Isotropic trialitarian algebraic groups*, Preprint, 1997. *572*.

104. _____, *Clifford algebras of hyperbolic involutions*, Preprint, 1998. *150*.

105. R. Garibaldi, J.-P. Tignol, and A. R. Wadsworth, *Galois cohomology of special orthogonal groups*, Manuscripta Math. **93** (1997), no. 2, 247–266. *447, 448*.

106. P. Gille, *R-équivalence et principe de norme en cohomologie galoisienne*, C. R. Acad. Sci. Paris Sér. I Math. **316** (1993), no. 4, 315–320. *495*.

107. M. J. Greenberg, *Lectures on forms in many variables*, W. A. Benjamin, Inc., New York-Amsterdam, 1969. *521*.

108. A. Grothendieck, *A general theory of fiber spaces with structure sheaf*, Tech. report, University of Kansas, 1955. *446*.

109. ———, *Technique de descente et théorèmes d'existence en géométrie algébrique I.*, 1959/1960, Séminaire Bourbaki n°190. *321, 446.*

110. ———, *Schémas en groupes. I, II, III*, Springer-Verlag, Berlin, 1962/1964, Séminaire de Géométrie Algébrique du Bois Marie 1962/64 (SGA 3). Dirigé par M. Demazure et A. Grothendieck. Lecture Notes in Mathematics, Vol. 151, 152, 153. *381.*

111. A. J. Hahn and O. T. O'Meara, *The classical groups and K-theory*, Grundlehren der Mathematischen Wissenschaften, vol. 291, Springer-Verlag, Berlin, 1989, With a foreword by J. Dieudonné. *69.*

112. D. E. Haile, *On central simple algebras of given exponent*, J. Algebra **57** (1979), no. 2, 449–465. *150.*

113. ———, *A useful proposition for division algebras of small degree*, Proc. Amer. Math. Soc. **106** (1989), no. 2, 317–319. *303.*

114. D. E. Haile and M.-A. Knus, *On division algebras of degree 3 with involution*, J. Algebra **184** (1996), no. 3, 1073–1081. *308.*

115. D. E. Haile, M.-A. Knus, M. Rost, and J.-P. Tignol, *Algebras of odd degree with involution, trace forms and dihedral extensions*, Israel J. Math. **96** (1996), part B, 299–340. *322, 447.*

116. W. R. Hamilton, *Mathematical papers*, vol. 3, Cambridge University Press, 1963. *509.*

117. J. Harris, *Algebraic geometry. A first course*, Graduate Texts in Mathematics, vol. 133, Springer-Verlag, New York, 1992. *9, 11, 12.*

118. A. Held and W. Scharlau, *On the existence of involutions on simple algebras IV*, Preprint, 1995. *68.*

119. I. N. Herstein, *Topics in ring theory*, The University of Chicago Press, Chicago, Ill.-London, 1969. *68.*

120. ———, *Rings with involution*, The University of Chicago Press, Chicago, Ill.-London, 1976, Chicago Lectures in Mathematics. *68.*

121. H. Hijikata, *A remark on the groups of type G_2 and F_4*, J. Math. Soc. Japan **15** (1963), 159–164. *381.*

122. D. Hilbert, *Die Theorie der algebraischen Zahlkörper*, Jahresbericht der Deutschen Mathematikervereinigung **4** (1897), 175–546, (pp. 63–363 in: Gesammelte Abhandlungen. Bd. I., Springer-Verlag, Berlin, 1970). *393.*

123. G. Hochschild, *On the cohomology groups of an associative algebra*, Ann. of Math. (2) **46** (1945), 58–67. *321.*

124. G. Hochschild and T. Nakayama, *Cohomology in class field theory*, Ann. of Math. (2) **55** (1952), 348–366. *447.*

125. J. E. Humphreys, *Linear algebraic groups*, Springer-Verlag, New York, 1975, Graduate Texts in Mathematics, No. 21. *323, 339, 375, 381.*

126. A. Hurwitz, *Über die Komposition der quadratischen Formen von beliebigen vielen Variablen*, Nachrichten Ges. der Wiss. Göttingen, Math.-Phys. Klasse (1898), 309–316, (pp. 565–571 in: Mathematische Werke. Bd. II, Birkhäuser Verlag, Basel, 1963). *509.*

127. B. Jacob and A. Rosenberg (eds.), *K-theory and algebraic geometry: connections with quadratic forms and division algebras*, Proceedings of Symposia in Pure Mathematics, vol. 58, Part 2, Providence, RI, American Mathematical Society, 1995.

128. N. Jacobson, *Cayley numbers and normal simple Lie algebras of type G*, Duke Math. J. **5** (1939), 775–783, (also: [143, pp. 191–199]). *360, 509.*

129. ———, *A note on hermitian forms*, Bull. Amer. Math. Soc. **46** (1940), 264–268, (also: [143, pp. 208–212]). *151, 239, 518.*

130. ———, *A theorem on the structure of Jordan algebras*, Proc. Nat. Acad. Sci. U.S.A. **42** (1956), 140–147, (also: [144, pp. 290–297]). *543.*

131. ———, *Composition algebras and their automorphisms*, Rend. Circ. Mat. Palermo (2) **7** (1958), 55–80, (also: [144, pp. 341–366]). *509, 510.*

132. ———, *Some groups of transformations defined by Jordan algebras. II. Groups of type F_4*, J. reine angew. Math. **204** (1960), 74–98, (also: [144, pp. 406–430]). *510.*

133. ———, *Some groups of transformations defined by Jordan algebras. III. Groups of type E_6*, J. reine angew. Math. **207** (1961), 61–85, (also: [144, pp. 431–455]). *543.*

134. ———, *Clifford algebras for algebras with involution of type D*, J. Algebra **1** (1964), 288–300, (also: [144, pp. 516–528]). *87, 149, 150, 203.*

135. ———, *Triality and Lie algebras of type D_4*, Rend. Circ. Mat. Palermo (2) **13** (1964), 129–153, (also: [144, pp. 529–553]). *510, 570, 571.*

136. _____, *Structure and representations of Jordan algebras*, American Mathematical Society, Providence, R.I., 1968, American Mathematical Society Colloquium Publications, Vol. XXXIX. *21, 29, 275, 453, 454, 509, 514–516, 519, 531, 542, 544.*

137. _____, *Lectures on quadratic Jordan algebras*, Tata Institute of Fundamental Research, Bombay, 1969, Tata Institute of Fundamental Research Lectures on Mathematics, No. 45. *542.*

138. _____, *Exceptional Lie algebras*, Marcel Dekker Inc., New York, 1971, Lecture Notes in Pure and Applied Mathematics, 1. *381, 539, 568.*

139. _____, *Abraham Adrian Albert (1905–1972)*, Bull. Amer. Math. Soc. **80** (1974), 1075–1100, (also: [145, pp. 449–474]). *542.*

140. _____, *Lie algebras*, Dover Publications Inc., New York, 1979, Republication of the 1962 original. *29, 203, 274, 510, 549, 566, 568.*

141. _____, *Structure theory of Jordan algebras*, University of Arkansas Lecture Notes in Mathematics, vol. 5, University of Arkansas, Fayetteville, Ark., 1981. *517, 520, 542.*

142. _____, *Some applications of Jordan norms to involutorial simple associative algebras*, Adv. in Math. **48** (1983), no. 2, 149–165, (also: [145, pp. 251–267]). *275.*

143. _____, *Collected mathematical papers. Vol. 1*, Contemporary Mathematicians, Birkhäuser Boston Inc., Boston, MA, 1989, (1934–1946).

144. _____, *Collected mathematical papers. Vol. 2*, Contemporary Mathematicians, Birkhäuser Boston Inc., Boston, MA, 1989, (1947–1965).

145. _____, *Collected mathematical papers. Vol. 3*, Contemporary Mathematicians, Birkhäuser Boston Inc., Boston, MA, 1989, (1965–1988). *275.*

146. _____, *Finite-dimensional division algebras over fields*, Springer-Verlag, Berlin, 1996. *31, 67, 191, 303, 313, 320.*

147. P. Jordan, J. von Neumann, and E. Wigner, *On an algebraic generalization of the quantum mechanical formalism*, Ann. of Math. (2) **35** (1934), 29–64. *542.*

148. B. Kahn, M. Rost, and R. Sujatha, *Unramified cohomology of quadrics I*, 1997, to appear in Amer. J. Math. *276.*

149. I. Kaplansky, *Infinite-dimensional quadratic forms admitting composition*, Proc. Amer. Math. Soc. **4** (1953), 956–960. *456, 462.*

150. _____, *Linear algebra and geometry. A second course*, Allyn and Bacon Inc., Boston, Mass., 1969. *30, 52.*

151. N. Karpenko and A. Quéguiner, *A criterion of decomposability for degree 4 algebras with unitary involution*, Preprint, 1997. *270.*

152. I. Kersten, *On involutions of the first kind*, Diskrete Strukturen in der Mathematik, vol. 343, 1996, Preprintreihe des SFB, pp. 96–107. *149.*

153. I. Kersten and U. Rehmann, *Generic splitting of reductive groups*, Tôhoku Math. J. (2) **46** (1994), no. 1, 35–70. *149.*

154. J. Kirmse, *Über die Darstellbarkeit natürlicher ganzer Zahlen als Summen von acht Quadraten und über ein mit diesem Problem zusammenhängendes nichtkommutatives und nichtassoziatives Zahlensystem*, Berichte über die Verh. der Sächs. Akad. der Wiss. Leipzig, math.-phys. Klasse **76** (1924), 63–82. *509.*

155. M. Kneser, *Lectures on Galois cohomology of classical groups*, Tata Institute of Fundamental Research, Bombay, 1969, With an appendix by T. A. Springer, Notes by P. Jothilingam, Tata Institute of Fundamental Research Lectures on Mathematics, No. 47. *50, 68, 204, 381, 449.*

156. M.-A. Knus, *Pfaffians and quadratic forms*, Adv. in Math. **71** (1988), no. 1, 1–20. *274, 275.*

157. _____, *Quadratic and Hermitian forms over rings*, Grundlehren der Mathematischen Wissenschaften, vol. 294, Springer-Verlag, Berlin, 1991, With a foreword by I. Bertuccioni. *xix, 45, 68, 69, 87–89, 96, 177, 274, 275.*

158. _____, *Sur la forme d'Albert et le produit tensoriel de deux algèbres de quaternions*, Bull. Soc. Math. Belg. Sér. B **45** (1993), no. 3, 333–337. *275.*

159. M.-A. Knus, T. Y. Lam, D. B. Shapiro, and J.-P. Tignol, *Discriminants of involutions on biquaternion algebras*, In Jacob and Rosenberg [127], pp. 279–303. *275.*

160. M.-A. Knus and M. Ojanguren, *Théorie de la descente et algèbres d'Azumaya*, Springer-Verlag, Berlin, 1974, Lecture Notes in Mathematics, Vol. 389. *32, 344.*

161. M.-A. Knus, R. Parimala, and R. Sridharan, *A classification of rank 6 quadratic spaces via pfaffians*, J. reine angew. Math. **398** (1989), 187–218. *275.*

162. _____, *Involutions on rank* 16 *central simple algebras*, J. Indian Math. Soc. (N.S.) **57** (1991), no. 1-4, 143–151. *27, 274.*

163. _____, *On the discriminant of an involution*, Bull. Soc. Math. Belg. Sér. A **43** (1991), no. 1-2, 89–98, Contact Franco-Belge en Algèbre (Antwerp, 1990). *149, 275.*

164. _____, *Pfaffians, central simple algebras and similitudes*, Math. Z. **206** (1991), no. 4, 589–604. *149, 271, 274, 275.*

165. _____, *On compositions and triality*, J. reine angew. Math. **457** (1994), 45–70. *511, 560.*

166. M.-A. Knus, R. Parimala, and V. Srinivas, *Azumaya algebras with involutions*, J. Algebra **130** (1990), no. 1, 65–82. *64.*

167. N. H. Kuiper, *On linear families of involutions*, Amer. J. Math. **72** (1950), 425–441. *511.*

168. V. V. Kursov, *Commutators of the multiplicative group of a finite-dimensional central division algebra*, Dokl. Akad. Nauk BSSR **26** (1982), no. 2, 101–103, 187. *277.*

169. T. Y. Lam, *The algebraic theory of quadratic forms*, W. A. Benjamin, Inc., Reading, Mass., 1973, Mathematics Lecture Note Series. *88, 141, 142, 235, 243, 439, 518, 536.*

170. T. Y. Lam, D. B. Leep, and J.-P. Tignol, *Biquaternion algebras and quartic extensions*, Inst. Hautes Études Sci. Publ. Math. (1993), no. 77, 63–102. *68, 276, 413.*

171. S. Lang and J. Tate, *Principal homogeneous spaces over abelian varieties*, Amer. J. Math. **80** (1958), 659–684. *446.*

172. D. W. Lewis, *A note on trace forms and involutions of the first kind*, Exposition. Math. **15** (1997), no. 3, 265–272. *151.*

173. D. W. Lewis and J. F. Morales, *The Hasse invariant of the trace form of a central simple algebra*, Théorie des nombres, Années 1992/93–1993/94, Publ. Math. Fac. Sci. Besançon, Univ. Franche-Comté, Besançon, 1994, pp. 1–6. *142.*

174. D. W. Lewis and J.-P. Tignol, *On the signature of an involution*, Arch. Math. (Basel) **60** (1993), no. 2, 128–135. *151.*

175. O. Loos, *Tensor products and discriminants of unital quadratic forms over commutative rings*, Monatsh. Math. **122** (1996), no. 1, 45–98. *499, 500.*

176. S. MacLane, *Categories for the working mathematician*, Springer-Verlag, New York, 1971, Graduate Texts in Mathematics, Vol. 5. *170, 361.*

177. P. Mammone and D. B. Shapiro, *The Albert quadratic form for an algebra of degree four*, Proc. Amer. Math. Soc. **105** (1989), no. 3, 525–530. *275.*

178. W. S. Martindale III, *Lie and Jordan mappings in associative rings*, Ring theory (Proc. Conf., Ohio Univ., Athens, Ohio, 1976), Dekker, New York, 1977, pp. 71–84. Lecture Notes in Pure and Appl. Math., Vol. 25. *68.*

179. H. Matsumura, *Commutative algebra*, second ed., Mathematics Lecture Note Series, vol. 56, Benjamin/Cummings Publishing Co., Inc., Reading, Mass., 1980. *323, 337.*

180. S. Maurer, *Vektorproduktalgebren*, Diplomarbeit, Universität Regensburg, 1998.

181. K. McCrimmon, *Jordan book*, http://www.math.virginia.edu/Faculty/KMcCrimmon/. *542.*

182. _____, *Norms and noncommutative Jordan algebras*, Pacific J. Math. **15** (1965), 925–956. *453.*

183. _____, *A general theory of Jordan rings*, Proc. Nat. Acad. Sci. U.S.A. **56** (1966), 1072–1079. *542.*

184. _____, *Generically algebraic algebras*, Trans. Amer. Math. Soc. **127** (1967), 527–551. *452, 509.*

185. _____, *The Freudenthal-Springer-Tits constructions of exceptional Jordan algebras*, Trans. Amer. Math. Soc. **139** (1969), 495–510. *519, 520, 525, 543.*

186. _____, *The Freudenthal-Springer-Tits constructions revisited*, Trans. Amer. Math. Soc. **148** (1970), 293–314. *521, 526, 544.*

187. _____, *Jordan algebras and their applications*, Bull. Amer. Math. Soc. **84** (1978), no. 4, 612–627. *542.*

188. _____, *The Russian revolution in Jordan algebras*, Algebras Groups Geom. **1** (1984), 1–61. *542.*

189. A. S. Merkurjev, *On the norm residue symbol of degree* 2, Dokl. Akad. Nauk SSSR **261** (1981), no. 3, 542–547, english transl.: Soviet Math. Dokl. **24** (1981), 546–551. *27, 45, 436.*

190. _____, *K-theory of simple algebras*, In Jacob and Rosenberg [127], pp. 65–83. *276, 277.*

191. _____, *The norm principle for algebraic groups*, Algebra i Analiz **7** (1995), no. 2, 77–105, english transl.: St. Petersburg Math. J. **7** (1996), no. 2, 243–264. *202.*

192. _____, *Maximal indexes of Tits algebras*, Doc. Math. **1** (1996), No. 12, 229–243 (electronic). *382*.

193. A. S. Merkurjev, I. A. Panin, and A. R. Wadsworth, *Index reduction formulas for twisted flag varieties. II*, 1997, to appear. *149, 382*.

194. _____, *Index reduction formulas for twisted flag varieties. I*, K-Theory **10** (1996), no. 6, 517–596. *149, 382*.

195. A. S. Merkurjev and A. A. Suslin, *K-cohomology of Severi-Brauer varieties and the norm residue homomorphism*, Izv. Akad. Nauk SSSR Ser. Mat. **46** (1982), no. 5, 1011–1046, 1135–1136, english transl.: Math. USSR Izv. **21** (1983), 307–340. *436*.

196. A. S. Merkurjev and J.-P. Tignol, *The multipliers of similitudes and the Brauer group of homogeneous varieties*, J. reine angew. Math. **461** (1995), 13–47. *132, 141, 147, 149, 150, 203, 204, 274*.

197. A. P. Monastyrnyĭ and V. I. Yanchevskiĭ, *Whitehead groups of spinor groups*, Izv. Akad. Nauk SSSR Ser. Mat. **54** (1990), no. 1, 60–96, 221, english transl.: Math. USSR Izv. **36** (1991), 61–100. *277*.

198. P. J. Morandi and B. A. Sethuraman, *Kummer subfields of tame division algebras*, J. Algebra **172** (1995), no. 2, 554–583. *273*.

199. S. Okubo, *Pseudo-quaternion and pseudo-octonion algebras*, Hadronic J. **1** (1978), no. 4, 1250–1278. *463, 472, 510*.

200. _____, *Introduction to octonion and other non-associative algebras in physics*, Montroll Memorial Lecture Series in Mathematical Physics, vol. 2, Cambridge University Press, Cambridge, 1995. *510*.

201. S. Okubo and H. C. Myung, *Some new classes of division algebras*, J. Algebra **67** (1980), no. 2, 479–490. *511*.

202. S. Okubo and J. M. Osborn, *Algebras with nondegenerate associative symmetric bilinear forms permitting composition*, Comm. Algebra **9** (1981), no. 12, 1233–1261. *464, 465, 510*.

203. L. J. Paige, *Jordan algebras*, Studies in Modern Algebra (A. A. Albert, ed.), Mathematical Association of America, Prentice-Hall, N. J., 1963, pp. 144–186. *542*.

204. G. Papy, *Sur l'arithmétique dans les algèbres de Grassman*, Acad. Roy. Belgique. Cl. Sci. Mém. Coll. 8° **26** (1952), no. 8, 108. *151*.

205. R. Parimala, R. Sridharan, and V. Suresh, *A question on the discriminants of involutions of central division algebras*, Math. Ann. **297** (1993), no. 4, 575–580. *149, 275, 552*.

206. R. Parimala, R. Sridharan, and M. L. Thakur, *A classification theorem for Albert algebras*, Trans. Amer. Math. Soc. **350** (1998), no. 3, 1277–1284. *202, 535, 538*.

207. H. P. Petersson, *Eine Identität fünften Grades, der gewisse Isotope von Kompositions-Algebren genügen*, Math. Z. **109** (1969), 217–238. *463, 466, 467, 510*.

208. _____, *Quasi composition algebras*, Abh. Math. Sem. Univ. Hamburg **35** (1971), 215–222. *463, 507, 510*.

209. H. P. Petersson and M. L. Racine, *Cubic subfields of exceptional simple Jordan algebras*, Proc. Amer. Math. Soc. **91** (1984), no. 1, 31–36. *543*.

210. _____, *Springer forms and the first Tits construction of exceptional Jordan division algebras*, Manuscripta Math. **45** (1984), no. 3, 249–272. *523, 536, 544*.

211. _____, *Classification of algebras arising from the Tits process*, J. Algebra **98** (1986), no. 1, 244–279. *532, 533, 543–545*.

212. _____, *Jordan algebras of degree 3 and the Tits process*, J. Algebra **98** (1986), no. 1, 211–243. *520, 525, 543, 544*.

213. _____, *Albert algebras*, Jordan algebras (Oberwolfach, 1992) (W. Kaup and K. McCrimmon, eds.), de Gruyter, Berlin, 1994, pp. 197–207. *528, 537, 543*.

214. _____, *On the invariants mod 2 of Albert algebras*, J. Algebra **174** (1995), no. 3, 1049–1072. *499*.

215. _____, *An elementary approach to the Serre-Rost invariant of Albert algebras*, Indag. Mathem., N.S. **7** (1996), no. 3, 343–365. *537, 545*.

216. _____, *The Serre-Rost invariant of Albert algebras in characteristic 3*, Indag. Mathem., N.S. **8** (1997), no. 4, 543–548. *545*.

217. A. Pfister, *Quadratische Formen in beliebigen Körpern*, Invent. Math. **1** (1966), 116–132. *275*.

218. _____, *Quadratic forms with applications to algebraic geometry and topology*, London Mathematical Society Lecture Note Series, vol. 217, Cambridge University Press, Cambridge, 1995. *243, 556.*

219. R. S. Pierce, *Associative algebras*, Graduate Texts in Mathematics, vol. 88, Springer-Verlag, New York, 1982, Studies in the History of Modern Science, 9. *5, 253, 267, 397, 414, 415.*

220. V. P. Platonov, *The Dieudonné conjecture, and the nonsurjectivity of coverings of algebraic groups at k-points*, Dokl. Akad. Nauk SSSR **216** (1974), 986–989, english transl.: Soviet Math. Dokl. **15** (1974), 927–931. *68.*

221. _____, *The Tannaka-Artin problem, and reduced K-theory*, Izv. Akad. Nauk SSSR Ser. Mat. **40** (1976), no. 2, 227–261, 469, english transl.: Math. USSR Izv. **10** (1976), no. 2, 211–243. *277.*

222. V. P. Platonov and A. Rapinchuk, *Algebraic groups and number theory*, Pure and Applied Mathematics, vol. 139, Academic Press Inc., Boston, MA, 1994, Translated from the 1991 Russian original by Rachel Rowen. *381, 449.*

223. V. P. Platonov and V. I. Yanchevskiĭ, *Finite-dimensional division algebras*, Algebra IX, Finite Groups of Lie Type, Finite-Dimensional Division Algebras (Berlin) (A. I. Kostrikin and I. R. Shafarevich, eds.), Encyclopaedia of Mathematical Sciences, vol. 77, Springer-Verlag, 1995, pp. 121–233. *266.*

224. A. Quéguiner, *Invariants d'algèbres à involution*, Thèse, Besançon, 1996. *151.*

225. _____, *Signature des involutions de deuxième espèce*, Arch. Math. (Basel) **65** (1995), no. 5, 408–412. *151.*

226. _____, *Cohomological invariants of algebras with involution*, J. Algebra **194** (1997), no. 1, 299–330. *151, 448.*

227. _____, *On the determinant class of an algebra with unitary involution*, Prépublications Mathématiques de l'Université Paris 13, 1998. *151, 446, 448.*

228. M. L. Racine, *A note on quadratic Jordan algebras of degree 3*, Trans. Amer. Math. Soc. **164** (1972), 93–103. *234, 521.*

229. I. Reiner, *Maximal orders*, Academic Press, London-New York, 1975, London Mathematical Society Monographs, No. 5. *5, 22, 67.*

230. P. Revoy, *Remarques sur la forme trace*, Linear and Multilinear Algebra **10** (1981), no. 3, 223–233. *321.*

231. C. Riehm, *The corestriction of algebraic structures*, Invent. Math. **11** (1970), 73–98. *68, 420.*

232. C. Rosati, *Sulle matrici di Riemann*, Rend. Circ. Mat. Palermo **53** (1929), 79–134. *68.*

233. M. P. Rosen, *Isomorphisms of a certain class of prime Lie rings*, J. Algebra **89** (1984), no. 2, 291–317. *68.*

234. M. Rost, *A (mod 3) invariant for exceptional Jordan algebras*, C. R. Acad. Sci. Paris Sér. I Math. **313** (1991), no. 12, 823–827. *448, 537, 545.*

235. _____, *On the dimension of a composition algebra*, Doc. Math. **1** (1996), No. 10, 209–214 (electronic).

236. L. H. Rowen, *Central simple algebras*, Israel J. Math. **29** (1978), no. 2-3, 285–301. *68.*

237. _____, *Ring theory. Vol. II*, Pure and Applied Mathematics, vol. 128, Academic Press Inc., Boston, MA, 1988. *32, 36, 39.*

238. L. H. Rowen and D. J. Saltman, *The discriminant of an involution*, Preprint, 1990. *63.*

239. C. H. Sah, *Symmetric bilinear forms and quadratic forms*, J. Algebra **20** (1972), 144–160. *275.*

240. D. J. Saltman, *The Brauer group is torsion*, Proc. Amer. Math. Soc. **81** (1981), no. 3, 385–387. *150.*

241. _____, *A note on generic division algebras*, Abelian groups and noncommutative rings (L. Fuchs, K. R. Goodearl, J. T. Stafford, and C. Vinsonhaler, eds.), Contemp. Math., vol. 130, Amer. Math. Soc., Providence, RI, 1992, A collection of papers in memory of Robert B. Warfield, Jr, pp. 385–394. *65, 550.*

242. R. D. Schafer, *The exceptional simple Jordan algebras*, Amer. J. Math. **70** (1948), 82–94. *517, 543.*

243. _____, *Forms permitting composition*, Advances in Math. **4** (1970), 127–148 (1970). *474.*

244. _____, *An introduction to nonassociative algebras*, Dover Publications Inc., New York, 1995, Corrected reprint of the 1966 original. *453, 454, 539, 569.*

245. W. Scharlau, *Induction theorems and the structure of the Witt group*, Invent. Math. **11** (1970), 37–44. *79.*

246. _____, *Zur Existenz von Involutionen auf einfachen Algebren*, Math. Z. **145** (1975), no. 1, 29–32. *68*.

247. _____, *Quadratic and Hermitian forms*, Grundlehren der mathematischen Wissenschaften, vol. 270, Springer-Verlag, Berlin, 1985. *4, 5, 19, 25, 36, 38, 50, 68, 69, 77, 79, 88, 136, 142, 144, 162, 163, 177, 178, 187, 201, 234, 243, 261, 276, 299, 305, 317, 318, 481, 483, 521, 524, 550*.

248. C. Scheiderer, *Real and étale cohomology*, Lecture Notes in Mathematics, vol. 1588, Springer-Verlag, Berlin, 1994. *132*.

249. _____, *Hasse principles and approximation theorems for homogeneous spaces over fields of virtual cohomological dimension one*, Invent. Math. **125** (1996), no. 2, 307–365. *449*.

250. E. A. M. Seip-Hornix, *Clifford algebras of quadratic quaternion forms. I, II*, Nederl. Akad. Wetensch. Proc. Ser. A 68 = Indag. Math. **27** (1965), 326–363. *69, 149*.

251. G. B. Seligman, *Modular Lie algebras*, Springer-Verlag New York, Inc., New York, 1967, Ergebnisse der Mathematik und ihrer Grenzgebiete, Band 40. *381*.

252. _____, *Rational methods in Lie algebras*, Marcel Dekker Inc., New York, 1976, Lecture Notes in Pure and Applied Mathematics, Vol. 17. *275*.

253. J.-P. Serre, *Espaces fibrés algébriques*, 1958, Séminaire Chevalley, No. 2, Exposé 1. *321*.

254. _____, *Corps locaux*, Hermann, Paris, 1968, Deuxième édition, Publications de l'Université de Nancago, No. VIII, english transl.: Local fields, Graduate Texts in Mathematics 67, Springer-Verlag, New York, 1979. *446*.

255. _____, *L'invariant de Witt de la forme* $\mathrm{Tr}(x^2)$, Comment. Math. Helv. **59** (1984), no. 4, 651–676, (also: [256, pp. 675–700]). *321*.

256. _____, *Œuvres. Vol. III*, Springer-Verlag, Berlin, 1986, 1972–1984.

257. _____, *Cohomologie galoisienne, cinquième édition, révisée et complétée*, fifth ed., Lecture Notes in Mathematics, vol. 5, Springer-Verlag, Berlin, 1994, english transl.: Galois cohomology, Springer-Verlag, Berlin, 1997. *142, 394, 446, 447, 449, 545*.

258. _____, *Cohomologie galoisienne: progrès et problèmes*, Astérisque (1995), no. 227, Exp. No. 783, 4, 229–257, Séminaire Bourbaki, Vol. 1993/94. *446, 449, 462, 517, 543, 545*.

259. D. B. Shapiro, *Symmetric and skew-symmetric linear transformations*, in preparation. *69*.

260. D. Soda, *Some groups of type* D_4 *defined by Jordan algebras*, J. reine angew. Math. **223** (1966), 150–163. *533*.

261. A. Speiser, *Zahlentheoretische Sätze aus der Gruppentheorie*, Math. Z. **5** (1919), 1–6. *393*.

262. T. A. Springer, *On the equivalence of quadratic forms*, Nederl. Akad. Wetensch. Proc. Ser. A 62 = Indag. Math. **21** (1959), 241–253. *446*.

263. _____, *On a class of Jordan algebras*, Nederl. Akad. Wetensch. Proc. Ser. A 62 = Indag. Math. **21** (1959), 254–264. *543*.

264. _____, *The projective octave plane. I, II*, Nederl. Akad. Wetensch. Proc. Ser. A 63 = Indag. Math. **22** (1960), 74–101. *543*.

265. _____, *The classification of reduced exceptional simple Jordan algebras*, Nederl. Akad. Wetensch. Proc. Ser. A 63 = Indag. Math. **22** (1960), 414–422. *517, 518, 543*.

266. _____, *Characterization of a class of cubic forms*, Nederl. Akad. Wetensch. Proc. Ser. A 65 = Indag. Math. **24** (1962), 259–265. *543*.

267. _____, *On the geometric algebra of the octave planes*, Nederl. Akad. Wetensch. Proc. Ser. A 65 = Indag. Math. **24** (1962), 451–468. *543*.

268. _____, *Oktaven, Jordan-Algebren und Ausnahmegruppen*, Mathematisches Institut der Universität Göttingen, 1963, english transl. in preparation. *359, 360, 495, 500, 509–511, 523, 540, 543, 544*.

269. _____, *Jordan algebras and algebraic groups*, Springer-Verlag, New York, 1973, Ergebnisse der Mathematik und ihrer Grenzgebiete, Band 75, new edition to appear. *359, 542*.

270. _____, *Linear algebraic groups*, Progress in Mathematics, vol. 9, Birkhäuser Boston, Mass., 1981, new edition to appear. *323, 381*.

271. _____, *Linear algebraic groups*, Algebraic Geometry IV, Linear Algebraic Groups, Invariant Theory (Berlin) (A. N. Parshin and I. R. Shafarevich, eds.), Encyclopaedia of Mathematical Sciences, vol. 55, Springer-Verlag, 1994, pp. 1–121. *323, 381*.

272. C. Stampfli-Rollier, *4-dimensionale Quasikompositionsalgebren*, Arch. Math. (Basel) **40** (1983), no. 6, 516–525. *510*.

273. E. Steinitz, *Algebraische theorie der Körper*, J. reine angew. Math. **137** (1910), 167–309, (New editions: de Gruyter, Berlin, 1930, Chelsea, New York, 1950, H. Hasse, R. Baer, eds.). *321.*

274. E. Study, *Grundlagen und Ziele der analytischen Kinematik*, Sitz. Ber. Berliner Math. Gesellschaft **12** (1913), 36–60. *511.*

275. A. A. Suslin (ed.), *Algebraic K-theory*, Advances in Soviet Mathematics, vol. 4, Providence, RI, American Mathematical Society, 1991, Papers from the seminar held at Leningrad State University, Leningrad, Translation from the Russian edited by A. B. Sossinsky.

276. _____, *K-theory and K-cohomology of certain group varieties*, In Algebraic K-theory [275], pp. 53–74. *151, 449.*

277. _____, *SK₁ of division algebras and Galois cohomology*, In Algebraic K-theory [275], pp. 75–99. *277.*

278. T. Tamagawa, *On Clifford algebra*, Unpublished notes, ∼1971. *150.*

279. _____, *On quadratic forms and pfaffians*, J. Fac. Sci. Univ. Tokyo Sect. IA Math. **24** (1977), no. 1, 213–219. *275.*

280. D. Tao, *A variety associated to an algebra with involution*, J. Algebra **168** (1994), no. 2, 479–520. *149.*

281. _____, *The generalized even Clifford algebra*, J. Algebra **172** (1995), no. 1, 184–204. *93, 150, 550.*

282. J. Tate, *WC-groups over p-adic fields*, Secrétariat mathématique, Paris, 1958, Séminaire Bourbaki; 10e année: 1957/1958. Textes des conférences; Exposés 152 à 168; 2e éd. corrigée, Exposé 156, 13 pp. *446.*

283. R. C. Thompson, *Commutators in the special and general linear groups*, Trans. Amer. Math. Soc. **101** (1961), 16–33. *277.*

284. J.-P. Tignol, *On the corestriction of central simple algebras*, Math. Z. **194** (1987), no. 2, 267–274. *39.*

285. _____, *La norme des espaces quadratiques et la forme trace des algèbres simples centrales*, Théorie des nombres, Années 1992/93–1993/94, Publ. Math. Fac. Sci. Besançon, Univ. Franche-Comté, Besançon, 1994, 19 pp. *142.*

286. J. Tits, *Sur la trialité et les algèbres d'octaves*, Acad. Roy. Belg. Bull. Cl. Sci. (5) **44** (1958), 332–350. *511.*

287. _____, *Sur la trialité et certains groupes qui s'en déduisent*, Inst. Hautes Études Sci. Publ. Math. (1959), no. 2, 13–60. *511.*

288. _____, *A theorem on generic norms of strictly power associative algebras*, Proc. Amer. Math. Soc. **15** (1964), 35–36. *452.*

289. _____, *Algèbres alternatives, algèbres de Jordan et algèbres de Lie exceptionnelles. I. Construction*, Nederl. Akad. Wetensch. Proc. Ser. A 69 = Indag. Math. **28** (1966), 223–237. *539.*

290. _____, *Classification of algebraic semisimple groups*, Algebraic Groups and Discontinuous Subgroups (Proc. Sympos. Pure Math., Boulder, Colo., 1965) (A. Borel and G. D. Mostow, eds.), vol. 9, Amer. Math. Soc., Providence, R.I., 1966, 1966, pp. 33–62. *149, 323, 355.*

291. _____, *Formes quadratiques, groupes orthogonaux et algèbres de Clifford*, Invent. Math. **5** (1968), 19–41. *xii, 68, 69, 87, 146, 149, 150, 203, 323, 381.*

292. _____, *Représentations linéaires irréductibles d'un groupe réductif sur un corps quelconque*, J. reine angew. Math. **247** (1971), 196–220. *150, 151, 375, 378.*

293. _____, *Sur les produits tensoriels de deux algèbres de quaternions*, Bull. Soc. Math. Belg. Sér. B **45** (1993), no. 3, 329–331. *275.*

294. F. Vaney, *Le parallélisme absolu dans les espaces elliptiques réels à 3 et à 7 dimensions et le principe de trialité*, Thèse, Paris, 1929. *511.*

295. F. D. Veldkamp, *Freudenthal and the octonions*, Nieuw Arch. Wisk. (4) **9** (1991), no. 2, 145–162. *509.*

296. O. Villa, *On division algebras of degree 3 over a field of characteristic 3*, Master's thesis, ETH, Zurich, March 1997. *321.*

297. A. R. Wadsworth, Oral communication, July 1992. *550.*

298. _____, *Discriminants in characteristic two*, Linear and Multilinear Algebra **17** (1985), no. 3-4, 235–263. *321.*

299. _____, *16-dimensional algebras with involution*, http://math.ucsd.edu/~wadswrth/, Unpublished notes, 1990. *246, 271.*

300. B. L. van der Waerden, *Algebra I*, Grundlehren der Mathematischen Wissenschaften, vol. 33, Springer-Verlag, Berlin, 1930. *321*.

301. _____, *Gruppen von linearen Transformationen*, Springer-Verlag, Berlin, 1935, Ergebnisse der Mathematik und ihrer Grenzgebiete, Band 4. *274, 275*.

302. _____, *A history of algebra, From al-Khwārizmī to Emmy Noether*, Springer-Verlag, Berlin, 1985. *509*.

303. R. Walde, *Jordan algebras and exceptional Lie algebras*, Preprint, 1967. *544*.

304. S. Wang, *On the commutator subgroup of a simple algebra*, Amer. J. Math. **72** (1950), 323–334. *253, 267*.

305. W. C. Waterhouse, *Introduction to affine group schemes*, Graduate Texts in Mathematics, vol. 66, Springer-Verlag, New York, 1979. *281, 323–325, 330, 335, 338–340, 343, 357*.

306. _____, *Discriminants of étale algebras and related structures*, J. reine angew. Math. **379** (1987), 209–220. *321*.

307. J. H. M. Wedderburn, *On division algebras*, Trans. Amer. Math. Soc. **22** (1921), 129–135. *321*.

308. A. Weil, *On algebraic groups and homogeneous spaces*, Amer. J. Math. **77** (1955), 493–512, (also: [311, pp. 235–254]). *446*.

309. _____, *The field of definition of a variety*, Amer. J. Math. **78** (1956), 509–524, (also: [311, pp. 291–306]). *446*.

310. _____, *Algebras with involutions and the classical groups*, J. Indian Math. Soc. (N.S.) **24** (1960), 589–623 (1961), (also: [311, pp. 413–447]). *xii, 151, 323, 381, 446*.

311. _____, *Scientific works. Collected papers. Vol. II (1951–1964)*, Springer-Verlag, New York, 1979. *381, 446*.

312. E. A. Weiss, *Oktaven, Engelscher Komplex, Trialitätsprinzip*, Math. Z. **44** (1938), 580–611. *511*.

313. _____, *Punktreihengeometrie*, Teubner, Leipzig, Berlin, 1939. *511*.

314. H. Weyl, *On generalized Riemann matrices*, Ann. of Math. (2) **35** (1934), 714–729, (also: [315, Vol. III, pp. 400–415]). *67, 68*.

315. _____, *Gesammelte Abhandlungen. Bände I, II, III, IV*, Springer-Verlag, Berlin, 1968, Herausgegeben von K. Chandrasekharan.

316. M. J. Wonenburger, *The Clifford algebra and the group of similitudes*, Canad. J. Math. **14** (1962), 45–59. *175, 200, 203*.

317. _____, *Triality principle for semisimilarities*, J. Algebra **1** (1964), 335–341. *510*.

318. V. I. Yanchevskiĭ, *Simple algebras with involutions, and unitary groups*, Mat. Sb. (N.S.) **93** **(135)** (1974), 368–380, 487, english transl.: Math. USSR Sb. **22** (1974), 372–385. *277*.

319. _____, *Symmetric and skew-symmetric elements of involutions, associated groups and the problem of decomposability of involutions*, In Jacob and Rosenberg [127], pp. 431–444. *148, 266, 268, 277, 550*.

320. M. Zorn, *Theorie der alternativen Ringe*, Abh. Math. Semin. Hamburg. Univ. **8** (1930), 123–147. *454, 509*.

321. _____, *Alternativkörper und quadratische Systeme*, Abh. Math. Semin. Hamburg. Univ. **9** (1933), 395–402. *509*.

Index

Notation

Spin(A, σ, f), 187
Spin(A, σ, f), 351
Spin(V, q), 187
Spin(V, q), 349
Spin$_8$, 486
Spin$^{\pm}(V, q)$, 359
Srd$_A(a)$, 5
SSym$(A, \sigma)^{\times}$, 406
SSym$(B, \tau)^{\times}$, 531
SSym$(B, \tau)^{\times}$, 401
SU(B, τ), 194
SU(B, τ), 346
SU(V, h), 155
S$_v$, 327
Sym(A, σ) (symmetric elements in A), 14
Sym$(A, \sigma)'$, 400
Sym$(B, \tau)^0$, 306
Symd(A, σ) (symmetrized elements in A), 14

$T(\ell)$ (trace of ℓ), xvi
$T(A, \sigma)$ (trialitarian algebra), 558
$t(B, \tau)$ (Tits class), 427
$T(\mathfrak{L})$ (trialitarian algebra of a Lie algebra), 570
$T(x, y)$ (bilinear trace form), 281
t^+, 483
t^-, 483
T_A (generic trace), 452
τ', 550
$T_C(x)$ (trace of x), 457
θ^+, 485
θ^-, 485
θ_α, 387
$T_J(a)$, 516, 519
$T_{L/F}(\ell)$ (trace of ℓ), xvi
$(T_{L/F})_*(b)$, 524
tr(m) (trace of m), xv
Trd$_A(a)$ (reduced trace of a), 5
T_τ, 534
$T_{\tau, \tau'}(x, y)$, 534

U(B, τ), 346
UK$_1(B)$ (unitary Whitehead group), 267
U(A, σ), 159
U(B, τ), 193
U(V, h), 155

$V^{\perp}$, xvii
V^* (dual space of V), xvi
V^G (G-invariant subspace), 343

$W(\Phi)$ (Weyl group), 353
WF (Witt ring of F), 141

$x^{\#}$, 470, 519
$X(f)$, 281
$X(L)$, 280
$x \star y$, 472
X_c, 444

x^f (left module homomorphism), xv
$\chi(\lambda)$, 354
χ_i, 357
$x *_\sigma (a \otimes b)$ (right module structure), 32

$Z(A)$ (centroid), 451
$Z(A, \sigma, f)$, 187
$Z^1(\Gamma, A)$ (1-cocycles), 384
$Z^2(\Gamma, A)$ (2-cocycles), 384